D0706941

Modern Plastics Handbook

Other McGraw-Hill Books of Interest

MARTIN • *Electronic Failure Analysis Handbook: Techniques and Applications for Electronic and Electrical Packages, Components, and Assemblies*

HARPER, SAMPSON • *Electronic Materials and Processes Handbook*

PETRIE • *Handbook of Adhesives and Sealants*

HARPER • *Handbook of Plastics, Elastomers, and Composites*

Modern Plastics Handbook

Modern Plastics
and
Charles A. Harper Editor in Chief

Technology Seminars, Inc.
Lutherville, Maryland

McGraw-Hill

New York San Francisco Washington, D.C. Auckland Bogotá
Caracas Lisbon London Madrid Mexico City Milan
Montreal New Delhi San Juan Singapore
Sydney Tokyo Toronto

Library of Congress Cataloging-in-Publication Data

Modern plastics handbook / Modern Plastics, Charles A. Harper (editor in chief).
p. cm.
ISBN 0-07-026714-6
1. Plastics. I. Modern Plastics. II. Harper, Charles A.
TA455.P5 M62 1999
668.4—dc21 99-056522
CIP

McGraw-Hill

A Division of The ***McGraw-Hill*** *Companies*

2 3 4 5 6 7 8 9 0 DOC/DOC 0 6 5 4 3 2 1

ISBN 0-07-026714-6

The sponsoring editor of this book was Robert Esposito. The editing supervisor was David E. Fogarty, and the production supervisor was Sherri Souffrance. It was set in New Century Schoolbook per the MHT design by Paul Scozzari and Deirdre Sheean of McGraw-Hill's Professional Book Group, in Hightstown, N.J.

Printed and bound by R. R. Donnelley & Sons Company..

This book was printed on recycled, acid-free paper containing a minimum of 50% recycled, de-inked fiber.

Contents

Contributors

Anne-Marie Baker *University of Massachusetts, Lowell, Mass.* (CHAP. 1)

Carol M. F. Barry *University of Massachusetts, Lowell, Mass.* (CHAP. 5)

Allison A. Cacciatore *TownsendTarnell, Inc., Mt. Olive, N.J.* (CHAP. 4)

Fred Gastrock *TownsendTarnell, Inc., Mt. Olive, N.J.* (CHAP. 4)

John L. Hull *Hall/Finmac, Inc., Warminster, Pa.* (CHAP. 6)

Carl P. Izzo *Consultant, Murrysville, Pa.* (CHAP. 10)

Louis N. Kattas *TownsendTarnell, Inc., Mt. Olive, N.J.* (CHAP. 4)

Peter Kennedy *Moldflow Corporation, Lexington, Mass.* (CHAP. 7, SEC. 3)

Inessa R. Levin *TownsendTarnell, Inc., Mt. Olive, N.J.* (CHAP. 4)

William R. Lukaszyk *Universal Dynamics, Inc., North Plainfield, N.J.* (CHAP. 7, SEC. 1)

Joey Meade *University of Massachusetts, Lowell, Mass.* (CHAP. 1)

James Margolis *Montreal, Quebec, Canada* (CHAP. 3)

Stephen A. Orroth *University of Massachusetts, Lowell, Mass.* (CHAP. 5)

Edward M. Petrie *ABB Transmission Technology Institute, Raleigh, N.C.* (CHAP. 9)

Jordon I. Rotheiser *Rotheiser Design, Inc., Highland Park, Ill.* (CHAP. 8)

Susan E. Selke *Michigan State University, School of Packaging, East Lansing, Mich.* (CHAP. 12)

Ranganath Shastri *Dow Chemical Company, Midland, Mich.* (CHAP. 11)

Peter Stoughton *Conair, Pittsburgh, Pa.* (CHAP. 7, SEC. 2)

Ralph E. Wright *Consultant, Yarmouth, Maine* (CHAP. 2)

Preface

The Modern Plastics Handbook has been prepared as a third member of the well-known and highly respected team of publications which includes *Modern Plastics* magazine and *Modern Plastics World Encyclopedia*. The *Modern Plastics Handbook* offers a thorough and comprehensive technical coverage of all aspects of plastics materials and processes, in all of their forms, along with coverage of additives, auxiliary equipment, plastic product design, testing, specifications and standards, and the increasingly critical subject of plastics recycling and biodegradability. Thus, this Handbook will serve a wide range of interests. Likewise, with presentations ranging from terms and definitions and fundamentals, to clearly explained technical discussions, to extensive data and guideline information, this Handbook will be useful for all levels of interest and backgrounds. These broad objectives could only have been achieved by an outstanding and uniquely diverse group of authors with a combination of academic, professional, and business backgrounds. It has been my good fortune to have obtained such an elite group of authors, and it has been a distinct pleasure to have worked with this group in the creation of this Handbook. I would like to pay my highest respects and offer my deep appreciation to all of them.

The Handbook has been organized and is presented as a thorough sourcebook of technical explanations, data, information, and guidelines for all ranges of interests. It offers an extensive array of property and performance data as a function of the most important product and process variables. The chapter organization and coverage is well suited to reader convenience for the wide range of product and equipment categories. The first three chapters cover the important groups of plastic materials, namely, thermoplastics, thermosets, and elastomers. Then comes a chapter on the all important and broad based group of additives, which are so critical for tailoring plastic properties. Following this are three chapters covering processing technologies and equipment for all types of plastics, and the all important subject of auxiliary equipment and components for optimized plastics processing. Next is a most thorough and comprehensive chapter on design of plastic products, rarely treated in such a practical manner. After this, two chapters are devoted to the highly important plastic materials and process topics of coatings and adhesives, including surface finishing and fabricating of plastic parts. Finally, one chapter is devoted to the fundamentally important areas of testing and standards, and one chapter to the increasingly critical area of plastic recycling and biodegradability.

Needless to say, a book of this caliber could not have been achieved without the guidance and support of many people. While it is not possible to name all of the advisors and constant supporters, I feel that I must highlight a few. First, I would like to thank the Modern Plastics team, namely, Robert D. Leaversuch, Executive Editor of *Modern Plastics* magazine, Stephanie Finn, Modern Plastics Events Manager, Steven J. Schultz, Managing Director, *Modern Plastics World Encyclopedia*, and William A. Kaplan, Managing Editor of *Modern Plastics World Encyclopedia*. Their advice and help was constant. Next, I would like to express my very great appreciation to the team from Society of Plastics Engineers, who both helped me get off the ground and supported me readily all through this project. They are Michael R. Cappelletti, Executive Director, David R. Harper, Past President, John L. Hull, Honored Service Member, and Glenn L. Beall, Distinguished Member. In addition, I would like to acknowledge, with deep appreciation, the advice and assistance of Dr. Robert Nunn and Dr. Robert Malloy of University of Massachusetts, Lowell for their guidance and support, especially in selection of chapter authors. Last, but not least, I am indebted to Robert Esposito, Executive Editor of the McGraw-Hill Professional Book Group, for both his support and patience in my editorial responsibilities for this *Modern Plastics Handbook*.

It is my hope, and expectation, that this book will serve its reader well. Any comments or suggestions will be welcomed.

Charles A. Harper

Chapter

1

Thermoplastics

A.-M. M. Baker
Joey Mead

Plastics Engineering Department
University of Massachusetts, Lowell

1.1 Introduction

Plastics are an important part of everyday life; products made from plastics range from sophisticated products, such as prosthetic hip and knee joints, to disposable food utensils. One of the reasons for the great popularity of plastics in a wide variety of industrial applications is due to the tremendous range of properties exhibited by plastics and their ease of processing. Plastic properties can be tailored to meet specific needs by varying the atomic makeup of the repeat structure; by varying molecular weight and molecular weight distribution; by varying flexibility as governed by presence of side chain branching, as well as the lengths and polarities of the side chains; and by tailoring the degree of crystallinity, the amount of orientation imparted to the plastic during processing and through copolymerization, blending with other plastics, and through modification with an enormous range of additives (fillers, fibers, plasticizers, stabilizers). Given all of the avenues available to pursue tailoring any given polymer, it is not surprising that such a variety of choices available to us today exist.

Polymeric materials have been used since early times, even though their exact nature was unknown. In the 1400s Christopher Columbus found natives of Haiti playing with balls made from material obtained from a tree. This was natural rubber, which became an important product after Charles Goodyear discovered that the addition of sulfur dramatically improved the properties. However, the use of polymeric materials was still limited to natural-based materials. The first true synthetic polymers were prepared in the early 1900s using phenol and formaldehyde to form resins—Baekeland's Bakelite. Even with the development of synthetic polymers, scientists were still unaware of the true nature of the materials they had prepared. For many years scientists believed they were *colloids*—aggregates of molecules with a particle size of 10- to 1000-nm diameter. It was not until the 1920s that Herman Staudinger showed that polymers were giant molecules or macromolecules. In 1928 Carothers developed linear polyesters and then polyamides, now known as nylon. In the 1950s Ziegler and Natta's work on anionic coordination catalysts led to the development of polypropylene, high-density linear polyethylene, and other stereospecific polymers.

Polymers come in many forms including plastics, rubber, and fibers. Plastics are stiffer than rubber, yet have reduced low-temperature properties. Generally, a plastic differs from a rubbery material due to the location of its glass transition temperature (T_g). A plastic has a T_g above room temperature, while a rubber will have a T_g below room temperature. T_g is most clearly defined by evaluating the classic relationship of elastic modulus to temperature for polymers as presented in Fig. 1.1. At low temperatures, the material can best be described as a glassy solid. It has a high modulus and behavior in this state is characterized ideally as a purely elastic solid. In this temperature regime, materials most closely obey Hooke's law:

$$\sigma = E\varepsilon$$

where σ is the stress being applied and ε is the strain. Young's modulus, E, is the proportionality constant relating stress and strain.

In the leathery region, the modulus is reduced by up to three orders of magnitude for amorphous polymers. The temperature at which the polymer behavior changes from glassy to leathery is known as the *glass transition temperature*, T_g. The rubbery plateau has a relatively stable modulus until as the temperature is further increased, a rubbery flow begins. Motion at this point does not involve entire molecules, but in this region deformations begin to become nonrecoverable as permanent set takes place. As temperature is further increased, eventually the onset of liquid flow takes place. There is little elastic recovery in this region, and the flow involves entire molecules slipping past each other. Ideally, this region is modeled as representing viscous materials which obey Newton's law :

$$\sigma = \eta\dot{\varepsilon}$$

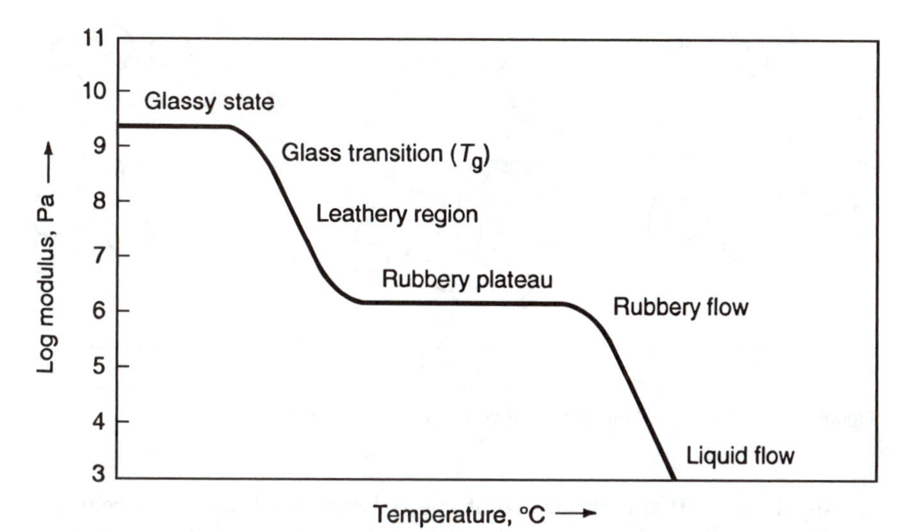

Figure 1.1 Relationship between elastic modulus and temperature.

Plastics can also be separated into thermoplastics and thermosets. A thermoplastic material is a high molecular weight polymer that is not cross-linked. A thermoplastic material can exist in a linear or branched structure. Upon heating a thermoplastic, a highly viscous liquid is formed that can be shaped using plastics processing equipment. A thermoset has all of the chains tied together with covalent bonds in a network (cross-linked). A thermoset cannot be reprocessed once cross-linked, but a thermoplastic material can be reprocessed by heating to the appropriate temperature. The different types of structures are shown in Fig. 1.2.

A polymer is prepared by stringing together a series of low molecular weight species (such as ethylene) into an extremely long chain (polyethylene) much as one would string together a series of beads to make a necklace. The chemical characteristics of the starting low molecular weight species will determine the properties of the final polymer. When two different low molecular weight species are polymerized, the resulting polymer is termed a *copolymer* such as ethylene vinylacetate.

The properties of different polymers can vary widely, for example, the modulus can vary from 1 MN/m^2 to 50 GN/m^2. Properties can be varied for each individual plastic material as well, simply by varying the microstructure of the material.

In its solid form a polymer can take up different structures depending on the structure of the polymer chain as well as the processing conditions. The polymer may exist in a random unordered structure termed an *amorphous* polymer. An example of an amorphous polymer

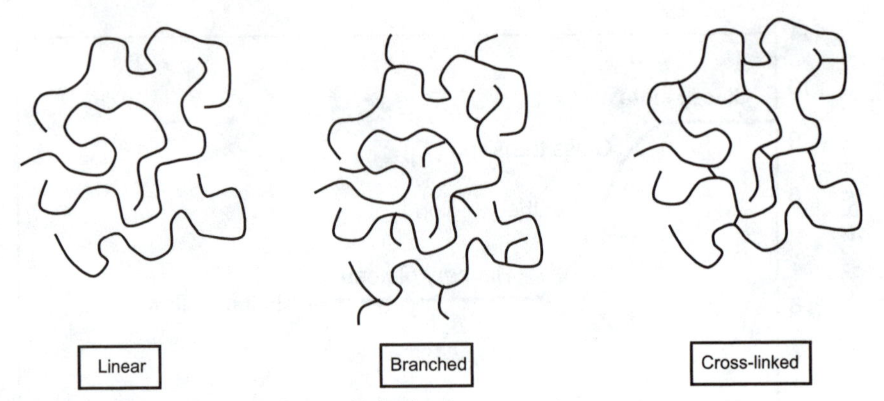

Figure 1.2 Linear, branched, cross-linked polymer structures.

is polystyrene. If the structure of the polymer backbone is a regular, ordered structure, then the polymer can tightly pack into an ordered crystalline structure, although the material will generally be only semicrystalline. Examples are polyethylene and polypropylene. The exact makeup and details of the polymer backbone will determine whether or not the polymer is capable of crystallizing. This microstructure can be controlled by different synthetic methods. As mentioned previously, the Ziegler-Natta catalysts are capable of controlling the microstructure to produce stereospecific polymers. The types of microstructure that can be obtained for a vinyl polymer are shown in Fig. 1.3. The isotactic and syndiotactic structures are capable of crystallizing because of their highly regular backbone. The atactic form would produce an amorphous material.

1.2 Polymer Categories

1.2.1 Acetal (POM)

Acetal polymers are formed from the polymerization of formaldehyde. They are also known by the name polyoxymethylenes (POM). Polymers prepared from formaldehyde were studied by Staudinger in the 1920s, but thermally stable materials were not introduced until the 1950s when DuPont developed Delrin.[1] Homopolymers are prepared from very pure formaldehyde by anionic polymerization, as shown in Fig. 1.4. Amines and the soluble salts of alkali metals catalyze the reaction.[2] The polymer formed is insoluble and is removed as the reaction proceeds. Thermal degradation of the acetal resin occurs by unzipping with the release of formaldhyde. The thermal stability of the polymer is increased by esterification of the hydroxyl ends with acetic anhydride. An alternative method to improve the thermal stability is copoly-

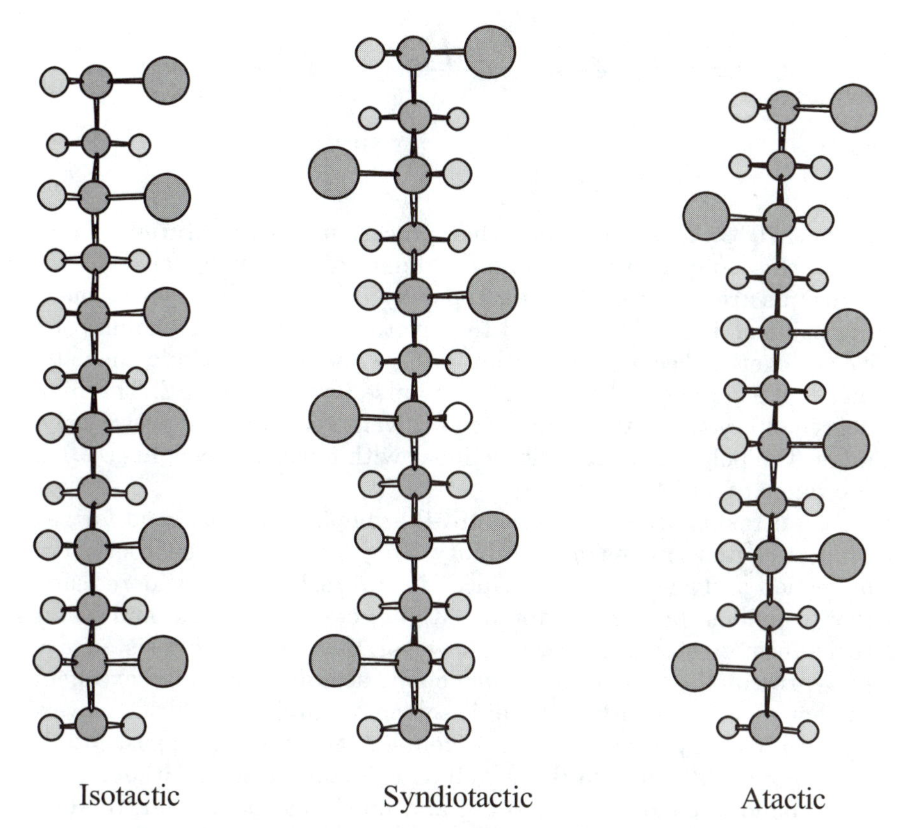

Figure 1.3 Isotactic, syndiotactic, and atactic polymer chains.

merization with a second monomer such as ethylene oxide. The copolymer is prepared by cationic methods.[3] This was developed by Celanese and marketed under the tradename Celcon. Hostaform is another copolymer marketed by Hoescht. The presence of the second monomer reduces the tendency for the polymer to degrade by unzipping.[4]

There are four processes for the thermal degradation of acetal resins. The first is thermal or base-catalyzed depolymerization from the chain, resulting in the release of formaldehyde. End capping the polymer chain will reduce this tendency. The second is oxidative attack at random positions, again leading to depolymerization. The use of antioxidants will reduce this degradation mechanism. Copolymerization is also helpful. The third mechanism is cleavage of the acetal linkage by acids. It is, therefore, important not to process acetals in equipment used for polyvinyl chloride (PVC), unless it has been cleaned, due to the possible presence of traces of HCl. The fourth degradation mechanism is thermal depolymerization at temperatures

$$n \; H_2C{=}O \longrightarrow \left(CH_2{-}O \right)_n$$

Figure 1.4 Polymerization of formaldehyde to polyoxymethylene.

above 270°C. It is important that processing temperatures remain below this temperature to avoid degradation of the polymer.[5]

Acetals are highly crystalline, typically 75% crystalline, with a melting point of 180°C.[6] Compared to polyethylene (PE), the chains pack closer together because of the shorter C—O bond. As a result, the polymer has a higher melting point. It is also harder than PE. The high degree of crystallinity imparts good solvent resistance to acetal polymers. The polymer is essentially linear with molecular weights (M_n) in the range of 20,000 to 110,000.[7]

Acetal resins are strong and stiff thermoplastics with good fatigue properties and dimensional stability. They also have a low coefficient of friction and good heat resistance.[8] Acetal resins are considered similar to nylons, but are better in fatigue, creep, stiffness, and water resistance.[9] Acetal resins do not, however, have the creep resistance of polycarbonate. As mentioned previously, acetal resins have excellent solvent resistance with no organic solvents found below 70°C, however, swelling may occur in some solvents. Acetal resins are susceptible to strong acids and alkalis, as well as oxidizing agents. Although the C—O bond is polar, it is balanced and much less polar than the carbonyl group present in nylon. As a result, acetal resins have relatively low water absorption. The small amount of moisture absorbed may cause swelling and dimensional changes, but will not degrade the polymer by hydrolysis.[10] The effects of moisture are considerably less dramatic than for nylon polymers. Ultraviolet light may cause degradation, which can be reduced by the addition of carbon black. The copolymers generally have similar properties, but the homopolymer may have slightly better mechanical properties, and higher melting point, but poorer thermal stability and poorer alkali resistance.[11] Along with both homopolymers and copolymers, there are also filled materials (glass, fluoropolymer, aramid fiber, and other fillers), toughened grades, and ultraviolet (UV) stabilized grades.[12] Blends of acetal with polyurethane elastomers show improved toughness and are available commercially.

Acetal resins are available for injection molding, blow molding, and extrusion. During processing it is important to avoid overheating or the production of formaldehyde may cause serious pressure buildup. The polymer should be purged from the machine before shutdown to avoid excessive heating during startup.[13] Acetal resins should be stored in a dry place. The apparent viscosity of acetal resins is less dependent on shear stress and temperature than polyolefins, but the melt has low elasticity and melt strength. The low melt strength is a problem for blow molding applications. For blow molding applications, copolymers with branched structures are available. Crystallization occurs rapidly with postmold shrinkage complete within 48 h of molding. Because of the rapid crystallization it is difficult to obtain clear films.[14]

The market demand for acetal resins in the United States and Canada was 368 million pounds in 1997.[15] Applications for acetal resins include gears, rollers, plumbing components, pump parts, fan blades, blow-molded aerosol containers, and molded sprockets and chains. They are often used as direct replacements for metal. Most of the acetal resins are processed by injection molding, with the remainder used in extruded sheet and rod. Their low coefficient of friction make acetal resins good for bearings.[16]

1.2.2 Biodegradable polymers

Disposal of solid waste is a challenging problem. The United States consumes over 53 billion pounds of polymers a year for a variety of applications.[17] When the life cycle of these polymeric parts is completed they may end up in a landfill. Plastics are often selected for applications based on their stability to degradation, however, this means degradation will be very slow, adding to the solid waste problem. Methods to reduce the amount of solid waste include either recycling or biodegradation.[18] Considerable work has been done to recycle plastics, both in the manufacturing and consumer area. Biodegradable materials offer another way to reduce the solid waste problem. Most waste is disposed of by burial in a landfill. Under these conditions oxygen is depleted and biodegradation must proceed without the presence of oxygen.[19] An alternative is aerobic composting. In selecting a polymer that will undergo biodegradation it is important to ascertain the method of disposal. Will the polymer be degraded in the presence of oxygen and water, and what will be the pH level? Biodegradation can be separated into two types—chemical and microbial degradation. Chemical degradation includes degradation by oxidation, photodegradation, thermal degradation, and hydrolysis. Microbial degradation can include both fungi and bacteria. The susceptibility of a polymer to biodegradation depends on the structure of the backbone.[20] For example, polymers with hydrolyzable backbones can be attacked by acids or bases, breaking down the molecular weight. They are, therefore, more likely to be degraded. Polymers that fit into this category include most natural-based polymers, such as polysaccharides, and synthetic materials, such as polyurethanes, polyamides, polyesters, and polyethers.

Polymers that contain only carbon groups in the backbone are more resistant to biodegradation.

Photodegradation can be accomplished by using polymers that are unstable to light sources or by the use of additives that undergo photodegradation. Copolymers of divinyl ketone with styrene, ethylene, or polypropylene (Eco Atlantic) are examples of materials that are susceptible to photodegradation.[21] The addition of a UV-absorbing material will also act to enhance photodegradation of a polymer. An example is the addition of iron dithiocarbamate.[22] The degradation must be controlled to ensure that the polymer does not degrade prematurely.

Many polymers described elsewhere in this book can be considered for biodegradable applications. Polyvinyl alcohol has been considered in applications requiring biodegradation because of its water solubility. However, the actual degradation of the polymer chain may be slow.[23] Polyvinyl alcohol is a semicrystalline polymer synthesized from polyvinyl acetate. The properties are governed by the molecular weight and by the amount of hydrolysis. Water soluble polyvinyl alcohol has a degree of hydrolysis 87 to 89%. Water insoluble polymers are formed if the degree of hydrolysis is greater than 89%.[24]

Cellulose-based polymers are some of the more widely available, naturally based polymers. They can, therefore, be used in applications requiring biodegradation. For example, regenerated cellulose is used in packaging applications.[25] A biodegradable grade of cellulose acetate is available from Rhone-Poulenc (Bioceta and Biocellat), where an additive acts to enhance the biodegradation.[26] This material finds application in blister packaging, transparent window envelopes, and other packaging applications.

Starch-based products are also available for applications requiring biodegradability. The starch is often blended with polymers for better properties. For example, polyethylene films containing between 5 to 10% cornstarch have been used in biodegradable applications. Blends of starch with vinyl alcohol are produced by Fertec (Italy) and used in both film and solid product applications.[27] The content of starch in these blends can range up to 50% by weight and the materials can be processed on conventional processing equipment. A product developed by Warner-Lambert, called Novon, is also a blend of polymer and starch, but the starch contents in Novon are higher than in the material by Fertec. In some cases the content can be over 80% starch.[28]

Polylactides (PLA) and copolymers are also of interest in biodegradable applications. This material is a thermoplastic polyester synthesized from the ring opening of lactides. Lactides are cyclic diesters of lactic acid.[29] A similar material to polylactide is polyglycolide (PGA).

PGA is also a thermoplastic polyester, but one that is formed from glycolic acids. Both PLA and PGA are highly crystalline materials. These materials find application in surgical sutures, resorbable plates and screws for fractures, and new applications in food packaging are also being investigated.

Polycaprolactones are also considered in biodegradable applications such as films and slow-release matrices for pharmaceuticals and fertilizers.[30] Polycaprolactone is produced through ring opening polymerization of lactone rings with a typical molecular weight in the range of 15,000 to 40,000.[31] It is a linear, semicrystalline polymer with a melting point near 62°C and a glass transition temperature about −60°C.[32]

A more recent biodegradable polymer is polyhydroxybutyrate-valerate copolymer (PHBV). These copolymers differ from many of the typical plastic materials in that they are produced through biochemical means. It is produced commercially by ICI using the bacteria *Alcaligenes eutrophus,* which is fed a carbohydrate. The bacteria produce polyesters, which are harvested at the end of the process.[33] When the bacteria are fed glucose, the pure polyhydroxybutyrate polymer is formed, while a mixed feed of glucose and propionic acid will produce the copolymers.[34] Different grades are commercially available that vary in the amount of hydroxyvalerate units and the presence of plasticizers. The pure hydroxybutyrate polymer has a melting point between 173 and 180°C and a T_g near 5°C.[35] Copolymers with hydroxyvalerate have reduced melting points, greater flexibility and impact strength, but lower modulus and tensile strength. The level of hydroxyvalerate is 5 to 12%. These copolymers are fully degradable in many microbial environments. Processing of PHBV copolymers requires careful control of the process temperatures. The material will degrade above 195°C, so processing temperatures should be kept below 180°C and the processing time kept to a minimum. It is more difficult to process unplasticized copolymers with lower hydroxyvalerate content because of the higher processing temperatures required. Applications for PHBV copolymers include shampoo bottles, cosmetic packaging, and as a laminating coating for paper products.[36]

Other biodegradable polymers include Konjac, a water-soluble natural polysaccharide produced by FMC, Chitin, another polysaccharide that is insoluble in water, and Chitosan, which is soluble in water.[37] Chitin is found in insect exoskeletons and in shellfish. Chitosan can be formed from chitin and is also found in fungal cell walls.[38] Chitin is used in many biomedical applications, including dialysis membranes, bacteriostatic agents, and wound dressings. Other applications include cosmetics, water treatment, adhesives, and fungicides.[39]

1.2.3 Cellulosics

Cellulosic polymers are the most abundant organic polymers in the world, making up the principal polysaccharide in the walls of almost all of the cells of green plants and many fungi species.[40] Plants produce cellulose through photosynthesis. Pure cellulose decomposes before it melts, and must be chemically modified to yield a thermoplastic. The chemical structure of cellulose is a heterochain linkage of different anhydroglucose units into high molecular weight polymer, regardless of plant source. The plant source, however, does affect molecular weight, molecular weight distribution, degrees of orientation, and morphological structure. Material described commonly as "cellulose" can actually contain hemicelluloses and lignin.[41] Wood is the largest source of cellulose and is processed as fibers to supply the paper industry and is widely used in housing and industrial buildings. Cotton-derived cellulose is the largest source of textile and industrial fibers, with the combined result being that cellulose is the primary polymer serving the housing and clothing industries. Crystalline modifications result in celluloses of differing mechanical properties, and Table 1.1 compares the tensile strengths and ultimate elongations of some common celluloses.[42]

Cellulose, whose repeat structure features three hydroxyl groups, reacts with organic acids, anhydrides, and acid chlorides to form esters. Plastics from these cellulose esters are extruded into film and sheet, and are injection-molded to form a wide variety of parts. Cellulose esters can also be compression-molded and cast from solution to form a coating. The three most industrially important cellulose ester plastics are cellulose acetate (CA), cellulose acetate butyrate (CAB), and cellulose acetate propionate (CAP), with structures as shown below in Fig. 1.5.

These cellulose acetates are noted for their toughness, gloss, and transparency. CA is well suited for applications requiring hardness and stiffness, as long as the temperature and humidity conditions don't cause the CA to be too dimensionally unstable. CAB has the best environmental stress cracking resistance, low-temperature impact

TABLE 1.1 Selected Mechanical Properties of Common Celluloses

Form	Tensile strength, MPa		Ultimate elongation, %	
	Dry	Wet	Dry	Wet
Ramie	900	1060	2.3	2.4
Cotton	200–800	200–800	12–16	6–13
Flax	824	863	1.8	2.2
Viscose rayon	200–400	100–200	8–26	13–43
Cellulose acetate	150–200	100–120	21–30	29–30

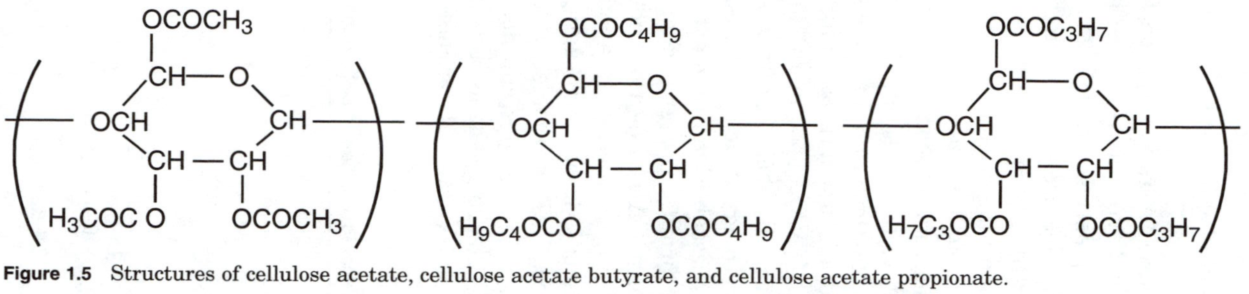

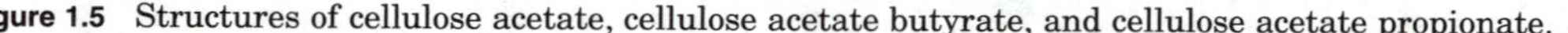

Figure 1.5 Structures of cellulose acetate, cellulose acetate butyrate, and cellulose acetate propionate.

strength, and dimensional stability. CAP has the highest tensile strength and hardness. Comparison of typical compositions and properties for a range of formulations are given in Table 1.2.[43] Properties can be tailored by formulating with different types and loadings of plasticizers.

Formulation of cellulose esters is required to reduce charring and thermal discoloration, and typically includes the addition of heat stabilizers, antioxidants, plasticizers, UV stabilizers, and coloring agents.[44] Cellulose molecules are rigid due to the strong intermolecular hydrogen bonding which occurs. Cellulose itself is insoluble and reaches its decomposition temperature prior to melting. The acetylation of the hydroxyl groups reduces intermolecular bonding, and increases free volume depending upon the level and chemical nature of the alkylation.[45] Cellulose acetates are thus soluble in specific solvents, but still require plasticization for rheological properties appropriate to molding and extrusion processing conditions. Blends of ethylene vinyl acetate (EVA) copolymers and CAB are available. Cellulose acetates have also been graft-copolymerized with alkyl esters of acrylic and methacrylic acid and then blended with EVA to form a clear, readily processable, thermoplastic.

CA is cast into sheet form for blister packaging, window envelopes, and file tab applications. CA is injection-molded into tool handles, tooth brushes, ophthalmic frames, and appliance housings and is extruded into pens, pencils, knobs, packaging films, and industrial pressure-sensitive tapes. CAB is molded into steering wheels, tool handles, camera parts, safety goggles, and football nose guards. CAP is injection-molded into steering wheels, telephones, appliance housings, flashlight cases, screw and bolt anchors, and is extruded into

TABLE 1.2 Selected Mechanical Properties of Cellulose Esters

Composition, %	Cellulose acetate	Cellulose acetate butyrate	Cellulose acetate propionate
Acetyl	38–40	13–15	1.5–3.5
Butyrl	—	36–38	—
Propionyl	—	—	43–47
Hydroxyl	3.5–4.5	1–2	2–3
Tensile strength at fracture, 23°C, MPa	13.1–58.6	13.8–51.7	13.8–51.7
Ultimate elongation, %	6–50	38–74	35–60
Izod impact strength, J/m			
notched, 23°C	6.6–132.7	9.9–149.3	13.3–182.5
notched, −40°C	1.9–14.3	6.6–23.8	1.9–19.0
Rockwell hardness, R scale	39–120	29–117	20–120
Percent moisture absorption at 24 h	2.0–6.5	1.0–4.0	1.0–3.0

pens, pencils, tooth brushes, packaging fim, and pipe.[46] Cellulose acetates are well suited for applications which require machining and then solvent vapor polishing, such as in the case of tool handles, where the consumer market values the clarity, toughness, and smooth finish. CA and CAP are likewise suitable for ophthalmic sheeting and injection-molding applications which require many postfinishing steps.[47]

Cellulose acetates are also commercially important in the coatings arena. In this synthetic modification, cellulose is reacted with an albrecht halide, primarily methylchloride to yield methylcellulose or sodium chloroacetate to yield sodium cellulose methylcellulose (CMC). The structure of CMC is shown in Fig. 1.6. CMC gums are water soluble and are used in food contact and packaging applications. Its outstanding film-forming properties are used in paper sizings and textiles and its thickening properties are used in starch adhesive formulations, paper coatings, toothpaste, and shampoo. Other cellulose esters, including cellulosehydroxyethyl, hydroxypropylcellulose, and ethylcellulose, are used in film and coating applications, adhesives, and inks.

1.2.4 Fluoropolymers

Fluoropolymers are noted for their heat-resistance properties. This is due to the strength and stability of the carbon-fluorine bond.[48] The first patent was awarded in 1934 to IG Farben for a fluorine-containing polymer, polychlorotrifluoroethylene (PCTFE). This polymer had limited application and fluoropolymers did not have wide application until the discovery of polytetrafluorethylene (PTFE) in 1938.[49] In addition to their high-temperature properties, fluoropolymers are known for their chemical resistance, very low coefficient of friction, and good dielectric properties. Their mechanical properties are not high unless reinforcing fillers, such as glass fibers, are added.[50] The compressive properties of fluoropolymers are generally superior to their tensile properties. In addition to their high-

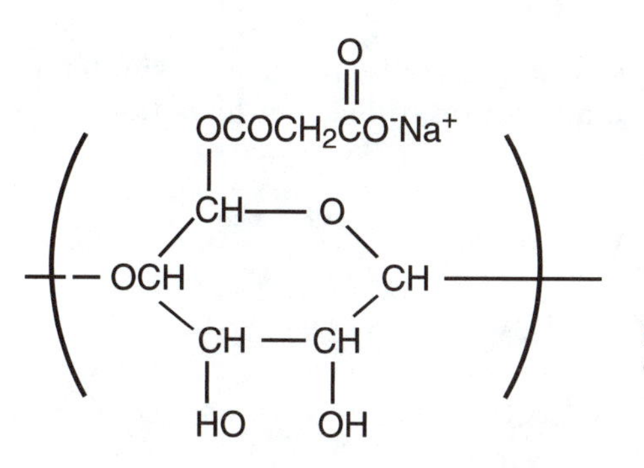

Figure 1.6 Sodium cellulose methylcellulose structure.

temperature resistance, these materials have very good toughness and flexibility at low temperatures.[51] A wide variety of fluoropolymers are available, PTFE, PCTFE, fluorinated ethylene propylene (FEP), ethylene chlorotrifluoroethylene (ECTFE), ethylene tetrafluoroethylene (ETFE), polyvinylindene fluoride (PVDF), and polyvinyl fluoride (PVF).

Copolymers. FEP is a copolymer of tetrafluoroethylene and hexafluoropropylene. It has properties similar to PTFE, but with a melt viscosity suitable for molding with conventional thermoplastic processing techniques.[52] The improved processability is obtained by replacing one of the fluorine groups on PTFE with a trifluoromethyl group as shown in Fig. 1.7.[53]

FEP polymers were developed by DuPont, but other commercial sources are available, such as Neoflon (Daikin Kogyo) and Teflex (Niitechem, formerly USSR).[54] FEP is a crystalline polymer with a melting point of 290°C, which can be used for long periods at 200°C with good retention of properties.[55] FEP has good chemical resistance, a low dielectric constant, low friction properties, and low gas permeability. Its impact strength is better than PTFE, but the other mechanical properties are similar to PTFE.[56] FEP may be processed by injection, compression, or blow molding. FEP may be extruded into sheets, films, rods, or other shapes. Typical processing temperatures for injection molding and extrusion are in the range of 300 to 380°C.[57] Extrusion should be done at low shear rates because of the polymer's high melt viscosity and melt fracture at low shear rates. Applications for FEP include chemical process pipe linings, wire and cable, and solar collector glazing.[58] A material similar to FEP, Hostaflon TFB (Hoechst), is a terpolymer of tetrafluoroethylene, hexafluoropropene, and vinylidene fluoride.

ECTFE is an alternating copolymer of chlorotrifluoroethylene and ethylene. It has better wear properties than PTFE along with good flame resistance. Applications include wire and cable jackets, tank linings, chemical process valve and pump components, and corrosion-resistant coatings.[59]

ETFE is a copolymer of ethylene and tetrafluoroethylene similar to ECTFE, but with a higher use temperature. It does not have the flame

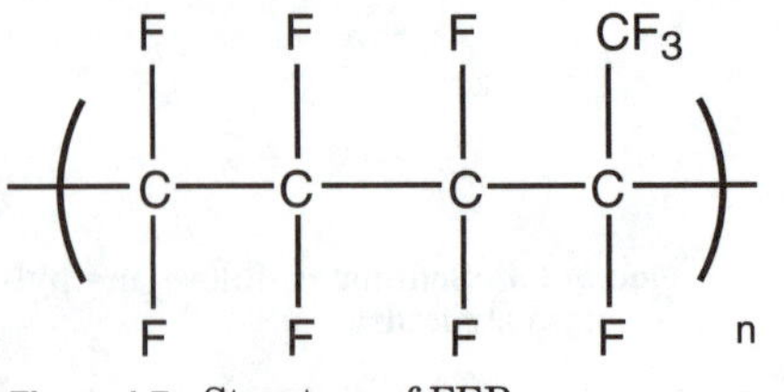

Figure 1.7 Structure of FEP.

resistance of ECTFE, however, and will decompose and melt when exposed to a flame.[60] The polymer has good abrasion resistance for a fluorine-containing polymer, along with good impact strength. The polymer is used for wire and cable insulation where its high-temperature properties are important. ETFE finds application in electrical systems for computers, aircraft, and heating systems.[61]

Polychlorotrifluoroethylene. Polychlorotrifluoroethylene (PCTFE) is made by the polymerization of chlorotrifluoroethylene, which is prepared by the dechlorination of trichlorotrifluoroethane. The polymerization is initiated with redox initiators.[62] The replacement of one fluorine atom with a chlorine atom, as shown in Fig. 1.8, breaks up the symmetry of the PTFE molecule, resulting in a lower melting point and allowing PCTFE to be processed more easily than PTFE. The crystalline melting point of PCTFE at 218°C is lower than PTFE. Clear sheets of PCTFE with no crystallinity may also be prepared.

PCTFE is resistant to temperatures up to 200°C and has excellent solvent resistance with the exception of halogenated solvents or oxygen containing materials, which may swell the polymer.[63] The electrical properties of PCTFE are inferior to PTFE, but PCTFE is harder and has higher tensile strength. The melt viscosity of PCTFE is low enough that it may be processing using most thermoplastic processing techniques.[64] Typical processing temperatures are in the range of 230 to 290°C.[65] PCTFE is higher in cost than PTFE, somewhat limiting its use. Applications include gaskets, tubing, and wire and cable insulation. Very low vapor transmission films and sheets may also be prepared.[66]

Polytetrafluoroethylene. Polytetrafluoroethylene (PTFE) is polymerized from tetrafluoroethylene by free radical methods.[67] The reaction is shown in Fig. 1.9. Commercially, there are two major processes for the polymerization of PTFE, one yielding a finer particle size dispersion polymer with lower molecular weight than the second method, which yields a "granular" polymer. The weight average molecular weights of commercial materials range from 400,000 to 9,000,000.[68] PTFE is a linear crystalline polymer with a melting point of 327°C.[69]

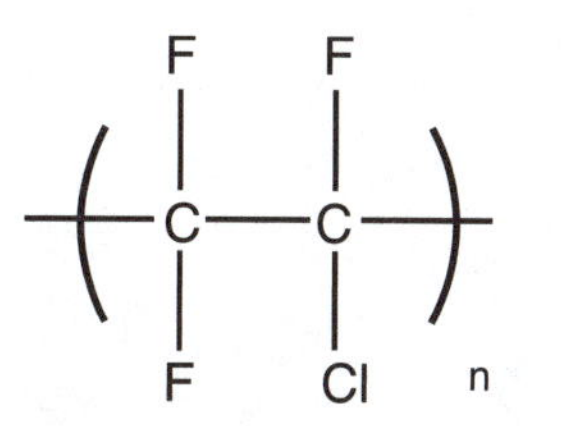

Figure 1.8 Structure of PCTFE.

Because of the larger fluorine atoms, PTFE assumes a twisted zigzag in the crystalline state, while polyethylene assumes the planar zigzag form.[70] There are several crystal forms for PTFE, and some of the transitions from one crystal form to another occur near room temperature. As a result of these transitions, volumetric changes of about 1.3% may occur.

PTFE has excellent chemical resistance, but may go into solution near its crystalline melting point. PTFE is resistant to most chemicals. Only alkali metals (molten) may attack the polymer.[71] The polymer does not absorb significant quantities of water and has low permeability to gases and moisture vapor.[72] PTFE is a tough polymer with good insulating properties. It is also known for its low coefficient of friction, with values in the range of 0.02 to 0.10.[73] PTFE, like other fluoropolymers, has excellent heat resistance and can withstand temperatures up to 260°C. Because of the high thermal stability, the mechanical and electrical properties of PTFE remain stable for long times at temperatures up to 250°C. However, PTFE can be degraded by high energy radiation.

One disadvantage of PTFE is that it is extremely difficult to process by either molding or extrusion. PFTE is processed in powder form by either sintering or compression molding. It is also available as a dispersion for coating or impregnating porous materials.[74] PTFE has a very high viscosity, prohibiting the use of many conventional processing techniques. For this reason techniques developed for the processing of ceramics are often used. These techniques involve preforming the powder, followed by sintering above the melting point of the polymer. For granular polymers, the preforming is carried out with the powder compressed into a mold. Pressures should be controlled as too low a pressure may cause voids, while too high a pressure may result in cleavage planes. After sintering, thick parts should be cooled in an oven at a controlled cooling rate, often under pressure. Thin parts may be cooled at room temperature. Simple shapes may be made by this technique, but more detailed parts should be machined.[75]

Extrusion methods may be used on the granular polymer at very low rates. In this case the polymer is fed into a sintering die that is heated. A typical sintering die has a length about 90 times the internal

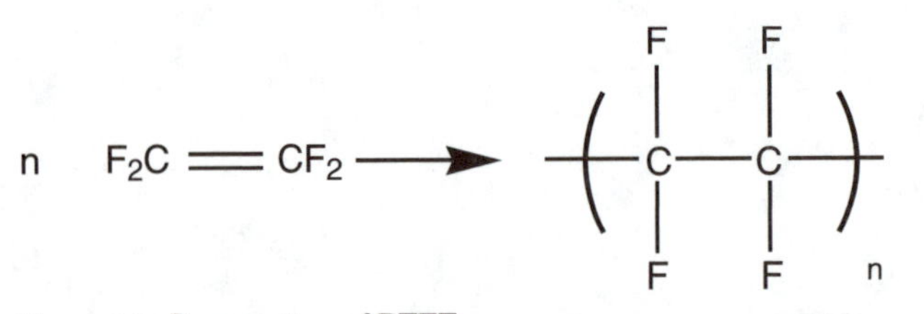

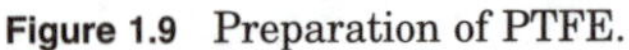
Figure 1.9 Preparation of PTFE.

diameter. Dispersion polymers are more difficult to process by the techniques previously mentioned. The addition of a lubricant (15 to 25%) allows the manufacture of preforms by extrusion. The lubricant is then removed and the part sintered. Thick parts are not made by this process because the lubricant must be removed. PTFE tapes are made by this process, however, the polymer is not sintered and a non-volatile oil is used.[76] Dispersions of PTFE are used to impregnate glass fabrics and to coat metal surfaces. Laminates of the impregnated glass cloth may be prepared by stacking the layers of fabric, followed by pressing at high temperatures.

Processing of PTFE requires adequate ventilation for the toxic gases that may be produced. In addition, PTFE should be processed under high cleanliness standards because the presence of any organic matter during the sintering process will result in poor properties as a result of the thermal decomposition of the organic matter. This includes both poor visual qualities and poor electrical properties.[77] The final properties of PTFE are dependent on the processing methods and the type of polymer. Both particle size and molecular weight should be considered. The particle size will affect the amount of voids and the processing ease, while crystallinity will be influenced by the molecular weight.

Additives for PTFE must be able to undergo the high processing temperatures required, which limits the range of additives available. Glass fiber is added to improve some mechanical properties. Graphite or molybdenum disulfide may be added to retain the low coefficient of friction while improving the dimensional stability. Only a few pigments are available that can withstand the processing conditions. These are mainly inorganic pigments such as iron oxides and cadmium compounds.[78]

Because of the excellent electrical properties, PTFE is used in a variety of electrical applications such as wire and cable insulation and insulation for motors, capacitors, coils, and transformers. PTFE is also used for chemical equipment, such as valve parts and gaskets. The low friction characteristics make PTFE suitable for use in bearings, mold release devices, and antistick cookware. Low molecular weight polymers may be used in aerosols for dry lubrication.[79]

Polyvinylindene fluoride. Polyvinylindene fluoride (PVDF) is crystalline with a melting point near 170°C.[80] The structure of PVDF is shown in Fig. 1.10. PVDF has good chemical and weather resistance, along with good resistance to distortion and creep at low and high temperatures. Although the chemical resistance is good, the polymer can be affected by very polar solvents, primary amines, and concentrated acids. PVDF has limited use as an insulator because the dielectric properties are frequency dependent. The polymer is important because of its relatively

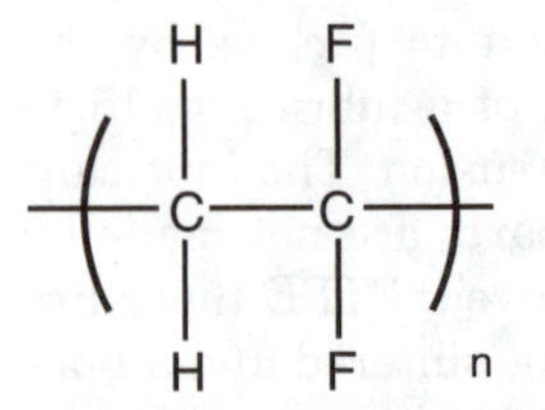

Figure 1.10 Structure of PVDF.

low cost compared to other fluorinated polymers.[81] PVDF is unique in that the material has piezoelectric properties, meaning that it will generate electric current when compressed.[82] This unique feature has been utilized for the generation of ultrasonic waves.

PVDF can be melt processed by most conventional processing techniques. The polymer has a wide range between the decomposition temperature and the melting point. Melt temperatures are usually 240 to 260°C.[83] Processing equipment should be extremely clean as any contaminants may affect the thermal stability. As with other fluorinated polymers, the generation of HF is a concern. PVDF is used for applications in gaskets, coatings, wire and cable jackets, and chemical process piping and seals.[84]

Polyvinyl fluoride. Polyvinyl fluoride (PVF) is a crystalline polymer available in film form and used as a lamination on plywood and other panels.[85] The film is impermeable to many gases. PVF is structurally similar to polyvinyl chloride (PVC) except for the replacement of a chlorine atom with a fluorine atom. PVF exhibits low moisture absorption, good weatherability, and good thermal stability. Similar to PVC, PVF may give off hydrogen halides in the form of HF at elevated temperatures. However, PVF has a greater tendency to crystallize and better heat resistance than PVC.[86]

1.2.5 Nylons

Nylons were one of the early polymers developed by Carothers.[87] Today, nylons are an important thermoplastic with consumption in the United States of about 1.2 billion pounds in 1997.[88] Nylons, also known as polyamides, are synthesized by condensation polymerization methods, often reacting an aliphatic diamine and a diacid. Nylon is a crystalline polymer with high modulus, strength, impact properties, low coefficient of friction, and resistance to abrasion.[89] Although the materials possess a wide range of properties, they all contain the amide (—CONH—) linkage in their backbone. Their general structure is shown in Fig. 1.11.

There are five primary methods to polymerize nylon. They are reaction of a diamine with a dicarboxylic acid, condensation of the appropriate amino acid, ring opening of a lactam, reaction of a diamine with a diacid chloride, and reaction of a diisocyanate with a dicarboxylic acid.[90]

The type of nylon (nylon 6, nylon 10, etc.) is indicative of the number of carbon atoms. There are many different types of nylons that can be prepared, depending on the starting monomers used. The type of nylon is determined by the number of carbon atoms in the monomers used in the polymerization. The number of carbon atoms between the amide linkages also controls the properties of the polymer. When only one monomer is used (lactam or amino acid), the nylon is identified with only one number (nylon 6, nylon 12). When two monomers are used in the preparation, the nylon will be identified using two numbers (nylon 6/6, nylon 6/12).[91] This is shown in Fig. 1.12. The first number refers to the number of carbon atoms in the diamine used (a) and the second number refers to the number of carbon atoms in the diacid monomer ($b + 2$), due to the two carbons in the carbonyl group.[92]

The amide groups are polar groups and significantly affect the polymer properties. The presence of these groups allows for hydrogen bonding between chains, improving the interchain attraction. This gives nylon polymers good mechanical properties. The polar nature of nylons also improves the bondability of the materials, while the flexible aliphatic carbon groups give nylons low melt viscosity for easy processing.[93] This structure also gives polymers that are tough above their glass transition temperature.[94]

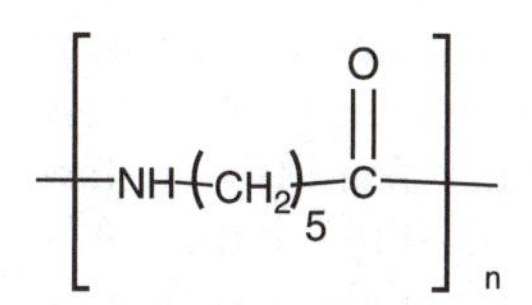

Figure 1.11 General structure of nylons.

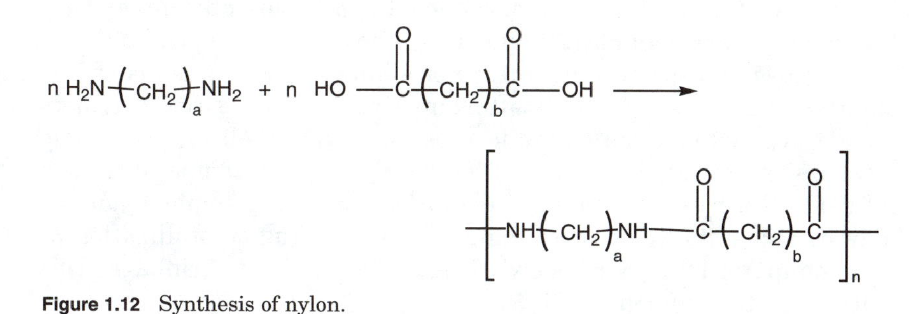

Figure 1.12 Synthesis of nylon.

Nylons are relatively insensitive to nonpolar solvents, however, because of the presence of the polar groups, nylons can be affected by polar solvents, particularly water.[95] The presence of moisture must be considered in any nylon application. Moisture can cause changes in part dimensions and reduce the properties, particularly at elevated temperatures.[96] As a result, the material should be dried before any processing operations. In the absence of moisture nylons are fairly good insulators, but as the level of moisture or the temperature increases, nylons are less insulating.[97]

The strength and stiffness will be increased as the number of carbon atoms between amide linkages is decreased because there are more polar groups per unit length along the polymer backbone.[98] The degree of moisture absorption is also strongly influenced by the number of polar groups along the backbone of the chain. Nylon grades with fewer carbon atoms between the amide linkages will absorb more moisture than grades with more carbon atoms between the amide linkages (nylon 6 will absorb more moisture than nylon 12). Furthermore, nylon types with an even number of carbon atoms between the amide groups have higher melting points than those with an odd number of carbon atoms. For example, the melting point of nylon 6/6 is greater than either nylon 5/6 or nylon 7/6.[99] Ring-opened nylons behave similarly. This is due to the ability of the nylons with the even number of carbon atoms to pack better in the crystalline state.[100]

Nylon properties are affected by the amount of crystallinity. This can be controlled, to a great extent, in nylon polymers by the processing conditions. A slowly cooled part will have significantly greater crystallinity (50 to 60%) than a rapidly cooled, thin part (perhaps as low as 10%).[101] Not only can the degree of crystallinity be controlled, but also the size of the crystallites. In a slowly cooled material the crystal size will be larger than for a rapidly cooled material. In injection-molded parts where the surface is rapidly cooled the crystal size may vary from the surface to internal sections.[102] Nucleating agents can be utilized to create smaller spherulites in some applications. This creates materials with higher tensile yield strength and hardness, but lower elongation and impact.[103] The degree of crystallinity will also affect the moisture absorption, with less crystalline polyamides being more prone to moisture pickup.[104]

The glass transition temperature of aliphatic polyamides is of secondary importance to the crystalline melting behavior. Dried polymers have T_g values near 50°C, while those with absorbed moisture may have T_gs in the range of 0°C.[105] The glass transition temperature can influence the crystallization behavior of nylons; for example, nylon 6/6 may be above its T_g at room temperature, causing crystallization at room temperature to occur slowly leading to postmold shrinkage. This is less significant for nylon 6.[106]

Nylons are processed by extrusion, injection molding, blow molding, and rotational molding among other methods. Nylon has a very sharp melting point and low melt viscosity, which is advantageous in injection molding, but causes difficulty in extrusion and blow molding. In extrusion applications a wide molecular weight distribution (MWD) is preferred, along with a reduced temperature at the exit to increase melt viscosity.[107]

When used in injection-molding applications, nylons have a tendency to drool due to their low melt viscosity. Special nozzles have been designed for use with nylons to reduce this problem.[108] Nylons show high mold shrinkage as a result of their crystallinity. Average values are about 0.018 cm/cm for nylon 6/6. Water absorption should also be considered for parts with tight dimensional tolerances. Water will act to plasticize the nylon, relieving some of the molding stresses and causing dimensional changes. In extrusion a screw with a short compression zone is used, with cooling initiated as soon as the extrudate exits the die.[109]

A variety of commercial nylons are available including nylon 6, nylon 11, nylon 12, nylon 6/6, nylon 6/10, and nylon 6/12. The most widely used nylons are nylon 6/6 and nylon 6.[110] Specialty grades with improved impact resistance, improved wear, or other properties are also available. Polyamides are used most often in the form of fibers, primarily nylon 6,6 and nylon 6, although engineering applications are also of importance.[111]

Nylon 6/6 is prepared from the polymerization of adipic acid and hexamethylenediamine. The need to control a 1:1 stoichiometric balance between the two monomers can be improved by the fact that adipic acid and hexamethylenediamine form a 1:1 salt that can be isolated. Nylon 6/6 is known for high strength, toughness, and abrasion resistance. It has a melting point of 265°C and can maintain properties up to 150°C.[112] Nylon 6/6 is used extensively in nylon fibers that are used in carpets, hose and belt reinforcements, and tire cord. Nylon 6/6 is used as an engineering resin in a variety of molding applications, such as gears, bearings, rollers, and door latches, because of its good abrasion resistance and self-lubricating tendencies.[113]

Nylon 6 is prepared from caprolactam. It has properties similar to those of nylon 6/6, but with a lower melting point (255°C). One of the major applications is in tire cord. Nylon 6/10 has a melting point of 215°C and lower moisture absorption than nylon 6/6.[114] Nylon 11 and nylon 12 have lower moisture absorption and also lower melting points than nylon 6/6. Nylon 11 has found applications in packaging films. Nylon 4/6 is used in a variety of automotive applications due to its ability to withstand high mechanical and thermal stresses. It is used in gears, gearboxes, and clutch areas.[115] Other applications for nylons include brush bristles, fishing line, and packaging films.

Additives, such as glass or carbon fibers, can be incorporated to improve the strength and stiffness of nylon. Mineral fillers are also used. A variety of stabilizers can be added to nylon to improve the heat and hydrolysis resistance. Light stabilizers are often added as well. Some common heat stabilizers include copper salts, phosphoric acid esters, and phenyl-β-naphthylamine. In bearing applications self-lubricating grades are available which may incorporate graphite fillers. Although nylons are generally impact resistant, rubber is sometimes incorporated to improve the failure properties.[116] Nylon fibers do have a tendency to pick up a static charge, so antistatic agents are often added for carpeting and other applications.[117]

Aromatic polyamides. A related polyamide is prepared when aromatic groups are present along the backbone. This imparts a great deal of stiffness to the polymer chain. One difficulty encountered in this class of materials is their tendency to decompose before melting.[118] However, certain aromatic polyamides have gained commercial importance. The aromatic polyamides can be classified into three groups: amorphous copolymers with a high T_g, crystalline polymers that can be used as a thermoplastic, and crystalline polymers used as fibers.

The copolymers are noncrystalline and clear. The rigid aromatic chain structure gives the materials a high T_g. One of the oldest types is poly(trimethylhexamethylene terephthalatamide) (Trogamid T). This material has an irregular chain structure, restricting the material from crystallizing, but with a T_g near 150°C.[119] Other glass-clear polyamides include Hostamid with a T_g also near 150°C, but with better tensile strength than Trogamid T. Grilamid TR55 is a third polyamide copolymer with a T_g about 160°C and the lowest water absorption and density of the three.[120] The aromatic polyamides are tough materials and compete with polycarbonate, poly(methyl methacrylate), and polysulfone. These materials are used in applications requiring transparency. They have been used for solvent containers, flowmeter parts, and clear housings for electrical equipment.[121]

An example of a crystallizable aromatic polyamide is poly-*m*-xylylene adipamide. It has a T_g near 85 to 100°C and a T_m of 235 to 240°C.[122] To obtain high heat deflection temperature the filled grades are normally sold. Applications include gears, electrical plugs, and mowing machine components.[123] Crystalline aromatic polyamides are also used in fiber applications. An example of this type of material is Kevlar, a high-strength fiber used in bulletproof vests and in composite structures. A similar material, which can be processed more easily, is Nomex, which can be used to give flame retardance to cloth when used as a coating.[124]

1.2.6 Polyacrylonitrile

Polyacrylonitrile is prepared by the polymerization of acrylonitrile monomer using either free-radical or anionic initiators. Bulk, emulsion, suspension, solution, or slurry methods may be used for the polymerization. The reaction is shown in Fig. 1.13.

Polyacrylonitrile will decompose before reaching its melting point, making the materials difficult to form. The decomposition temperature is near 300°C.[125] Suitable solvents, such as dimethylformamide and tetramethylenesulphone, have been found for polyacrylonitrile, allowing the polymer to be formed into fibers by dry and wet spinning techniques.[126]

Polyacrylonitrile is a polar material, giving the polymer good resistance to solvents, high rigidity, and low gas permeability.[127] Although the polymer degrades before melting, special techniques allowed a melting point of 317°C to be measured. The pure polymer is difficult to dissolve, but the copolymers can be dissolved in solvents such as methyl ethyl ketone, dioxane, acetone, dimethyl formamide, and tetrahydrofuran. Polyacrylonitrile exhibits exceptional barrier properties to oxygen and carbon dioxide.[128]

Copolymers of acrylonitrile with other monomers are widely used. Copolymers of vinylidene chloride and acrylonitrile find application in low gas permeability films. Styrene-acrylonitrile (SAN polymers) copolymers have also been used in packaging applications. Although the gas permeability of the copolymers is higher than for pure polyacrylonitrile, the acrylonitrile copolymers have lower gas permeability than many other packaging films. A number of acrylonitrile copolymers were developed for beverage containers, but the requirement for very low levels of residual acrylonitrile monomer in this application led to many products being removed from the market.[129] One copolymer currently available is Barex (BP Chemicals). The copolymer has better barrier properties than both polypropylene and polyethylene terephthalate.[130] Acrylonitrile is also used with butadiene and styrene to form ABS polymers. Unlike the homopolymer, copolymers of acrylonitrile can be processed by many methods including extrusion, blow molding, and injection molding.[131]

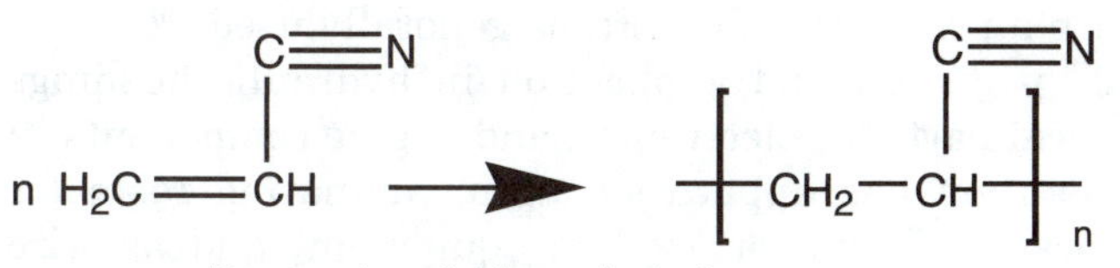

Figure 1.13 Preparation of polyacrylonitrile.

Acrylonitrile is often copolymerized with other monomers to form fibers. Copolymerization with monomers such as vinyl acetate, vinyl pyrrolidone, and vinyl esters gives the fibers the ability to be dyed using normal textile dyes. The copolymer generally contains at least 85% acrylonitrile.[132] Acrylic fibers have good abrasion resistance, flex life, toughness, and high strength. They have good resistance to stains and moisture. Modacrylic fibers contain between 35 and 85% acrylonitrile.[133]

Most of the acrylonitrile consumed goes into the production of fibers. Copolymers also consume large amounts of acrylonitrile. In addition to their use as fibers, polyacrylonitrile polymers can be used as precursors to carbon fibers.

1.2.7 Polyamide-imide

Polyamide-imide (PAI) is a high-temperature amorphous thermoplastic that has been available since the 1970s under the trade name of Torlon.[134] PAI can be produced from the reaction of trimellitic trichloride with methylenedianiline, as shown in Fig. 1.14.

Polyamide-imides can be used from cryogenic temperatures to nearly 260°C. They have the temperature resistance of the polyimides, but with better mechanical properties, including good stiffness and creep resistance. PAI polymers are inherently flame retardant with little smoke produced when they are burned. The polymer has good chemical resistance, but at high temperatures it can be affected by strong acids, bases, and steam.[135] PAI has a heat-deflection temperature of 280°C, along with good wear and friction properties.[136] Polyamide-imides also have good radiation resistance and are more stable than standard nylons under different humidity conditions. The polymer has one of the highest glass transition temperatures, in the range of 270 to 285°C.[137]

Polyamide-imide can be processed by injection molding, but special screws are needed due to the reactivity of the polymer under molding conditions. Low compression ratio screws are recommended.[138] The parts should be annealed after molding at gradually increased temperatures.[139] For injection molding the melt temperature should be near 355°C, with mold temperatures of 230°C. PAI can also be processed by compression molding or used in solution form. For compression molding, preheating at 280°C, followed by molding between 330 and 340°C with a pressure of 30 MPa, is generally used.[140]

Polyamide-imide polymers find application in hydraulic bushings and seals, mechanical parts for electronics, and engine components.[141] The polymer in solution has application as a laminating resin for spacecraft, a decorative finish for kitchen equipment, and as wire enamel.[142] Low coefficient of friction materials may be prepared by blending PAI with polytetrafluoroethylene and graphite.[143]

1.2.8 Polyarylate

Polyarylates are amorphous, aromatic polyesters. Polyarylates are polyesters prepared from dicarboxylic acids and bisphenols.[144] Bisphenol A is commonly used along with aromatic dicarboxylic acids, such as mixtures of isophthalic acid and terephthalic acid. The use of two different acids results in an amorphous polymer; however, the presence of the aromatic rings gives the polymer a high T_g and good temperature resistance. The temperature resistance of polyarylates lies between polysulfone and polycarbonate. The polymer is flame retardant and shows good toughness and UV resistance.[145] Polyarylates are transparent and have good electrical properties. The abrasion resistance of polyarylates is superior to polycarbonate. In addition, the polymers show very high recovery from deformation.

Polarylates are processed by most of the conventional methods. Injection molding should be performed with a melt temperature of 260 to 382°C with mold temperatures of 65 to 150°C. Extrusion and blow molding grades are also available. Polyarylates can react with water at processing temperatures and they should be dried prior to use.[146]

Polyarylates are used in automotive applications such as door handles, brackets, and headlamp and mirror housings. Polyarylates are also used in electrical applications for connectors and fuses. The polymer can be used in circuit board applications because its high-temperature resistance allows the part to survive exposure to the temperatures generated during soldering.[147] The excellent UV resistance of these polymers allows them to be used as a coating for other thermoplastics for improved UV resistance of the part. The good heat resistance of polyarylates allows them to be used in applications such as fire helmets and shields.[148]

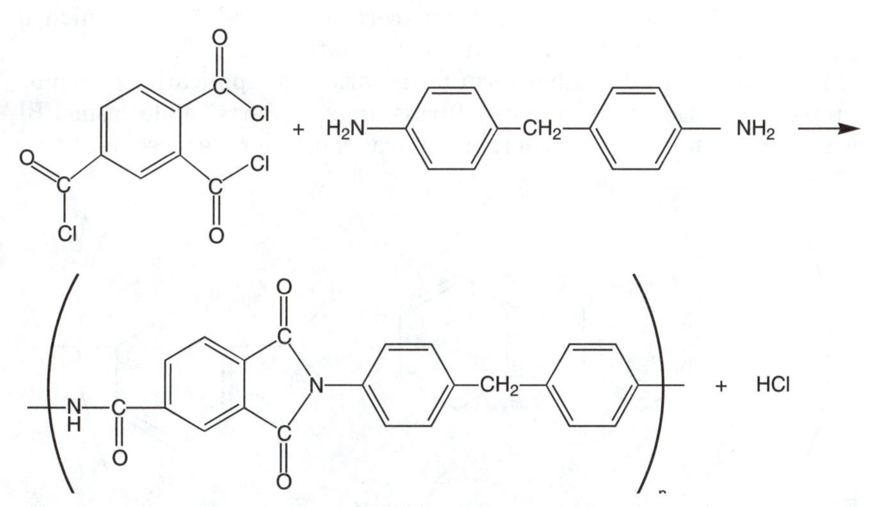

Figure 1.14 Preparation of polyamide-imide.

1.2.9 Polybenzimidazole

Polybenzimidazoles (PBI) are high-temperature resistant polymers. They are prepared from aromatic tetramines (for example, tetra amino-biphenol) and aromatic dicarboxylic acids (diphenylisophthalate).[149] The reactants are heated to form a soluble prepolymer that is converted to the insoluble polymer by heating at temperatures above 300°C.[150] The general structure of PBI is shown in Fig. 1.15.

The resulting polymer has high-temperature stability, good chemical resistance, and nonflammability. The polymer releases very little toxic gas and does not melt when exposed to pyrolysis conditions. The polymer can be formed into fibers by dry-spinning processes. Polybenzimidazole is usually amorphous with a T_g near 430°C.[151] Under certain conditions crystallinity may be obtained. The lack of many single bonds and the high glass transition temperature give this polymer its superior high-temperature resistance. In addition to the high-temperature resistance, the polymer exhibits good low-temperature toughness. PBI polymers show good wear and frictional properties along with excellent compressive strength and high surface hardness.[152] The properties of PBI at elevated temperatures are among the highest of the thermoplastics. In hot, aqueous solutions the polymer may absorb water with a resulting loss in mechanical properties. Removal of moisture will restore the mechanical properties. The heat-deflection temperature of PBI is higher than most thermoplastics and this is coupled with a low coefficient of thermal expansion. PBI can withstand temperatures up to 760°C for short durations and exposure to 425°C for longer durations.

The polymer is not available as a resin and is generally not processed by conventional thermoplastic processing techniques, but rather by a high-temperature and pressure sintering process.[153] The polymer is available in fiber form, certain shaped forms, finished parts, and solutions for composite impregnation.

PBI is often used in fiber form for a variety of applications such as protective clothing and aircraft furnishings.[154] Parts made from PBI are used as thermal insulators, electrical connectors, and seals.[155]

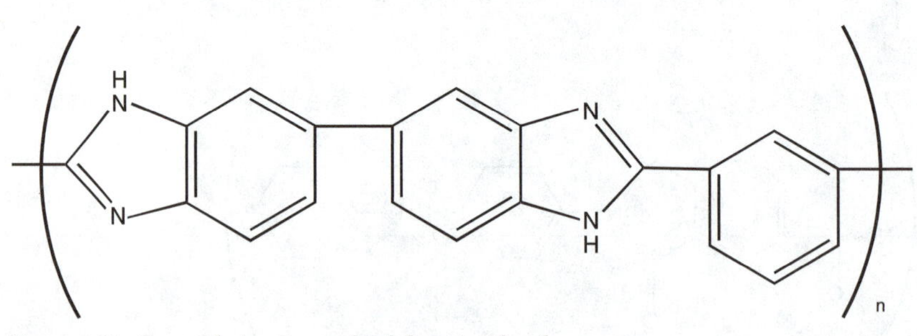

Figure 1.15 General structure of polybenzimidazoles.

1.2.10 Polybutylene (PB)

Polybutylene polymers are prepared by the polymerization of 1-butene using Ziegler-Natta catalysts The molecular weights range from 770,000 to 3,000,000.[156] Copolymers with ethylene are often prepared as well. The chain structure is mainly isotactic and is shown in Fig. 1.16.[157]

The glass transition temperature for this polymer ranges from −17 to −25°C. Polybutylene resins are linear polymers exhibiting good resistance to creep at elevated temperatures and good resistance to environmental stress cracking.[158] They also show high impact strength, tear resistance, and puncture resistance. As with other polyolefins, polybutylene shows good resistance to chemicals, good moisture barrier properties, and good electrical insulation properties. Pipes prepared from polybutylene can be solvent welded, yet the polymer still exhibits good environmental stress cracking resistance.[159] The chemical resistance is quite good below 90°C, but at elevated temperatures the polymer may dissolve in solvents such as toluene, decalin, chloroform, and strong oxidizing acids.[160]

Polybutylene is a crystalline polymer with three crystalline forms. The first crystalline form is obtained when the polymer is cooled from the melt. The first crystalline form is unstable and will change to a second crystalline form after standing over a period of 3 to 10 days. The third crystalline form is obtained when polybutylene is crystallized from solution. The melting point and density of the first crystalline form are 124°C and 0.89 g/cm^3, respectively.[161] On transformation to the second crystalline form, the melting point increases to 135°C and the density is increased to 0.95 g/cm^3. The transformation to the second crystalline form increases the polymer's hardness, stiffness, and yield strength.

Polybutylene can be processed on equipment similar to that used for low-density polyethylene. Polybutylene can be extruded and injection-molded. Film samples can be blown or cast. The slow transformation from one crystalline form to another allows polybutylene to undergo postforming techniques such as cold forming of molded parts or sheeting.[162] A range of 160 to 240°C is typically used to process polybutylene.[163] The die swell and shrinkage are generally greater for polybutylene than

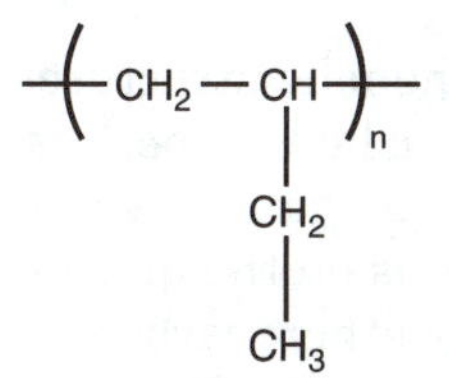

Figure 1.16 General structure for polybutylene.

for polyethylene. Because of the crystalline transformation, initially molded samples should be handled with care.

An important application for polybutylene is plumbing pipe for both commercial and residential use. The excellent creep resistance of polybutylene allows for the manufacture of thinner wall pipes compared to pipes made from polyethylene or polypropylene. Polybutylene pipe can also be used for the transport of abrasive fluids. Other applications for polybutylene include hot melt adhesives and additives for other plastics. The addition of polybutylene improves the environmental stress cracking resistance of polyethylene and the impact and weld line strength of polypropylene.[164] Polybutylene is also used in packaging applications.[165]

1.2.11 Polycarbonate

Polycarbonate (PC) is often viewed as the quintessential engineering thermoplastic, due to its combination of toughness, high strength, high heat-deflection temperatures, and transparency. The worldwide growth rate, predicted in 1999 to be between 8 and 10%, is hampered only by the resin cost and is paced by applications where PC can replace ferrous or glass products. Global consumption is anticipated to be more than 1.4 billion kg (3 billion lb) by the year 2000.[166] The polymer was discovered in 1898 and by the year 1958 both Bayer in Germany and General Electric in the United States had commenced production. Two current synthesis processes are commercialized, with the economically most successful one said to be the “interface” process, which involves the dissolution of bisphenol A in aqueous caustic soda and the introduction of phosgene in the presence of an inert solvent such as pyridine. The bisphenol A monomer is dissolved in the aqueous caustic soda, then stirred with the solvent for phosgene. The water and solvent remain in separate phases. Upon phosgene introduction, the reaction occurs at the interface with the ionic ends of the growing molecule being soluble in the catalytic caustic soda solution and the remainder of the molecule soluble in the organic solvent.[167] An alternative method involves transesterification of bisphenol A with diphenyl carbonate at elevated temperatures.[168] Both reactions are shown in Fig. 1.17. Molecular weights of between 30,000 and 50,000 g/mol can be obtained by the second route, while the phosgenation route results in higher molecular weight product.

The structure of PC with its carbonate and bisphenolic structures has many characteristics which promote its distinguished properties. The *para* substitution on the phenyl rings results in a symmetry and lack of stereospecificity. The phenyl and methyl groups on the quartenary carbon promote a stiff structure. The ester-ether carbonate groups —OCOO— are polar, but their degree of intermolecular polar

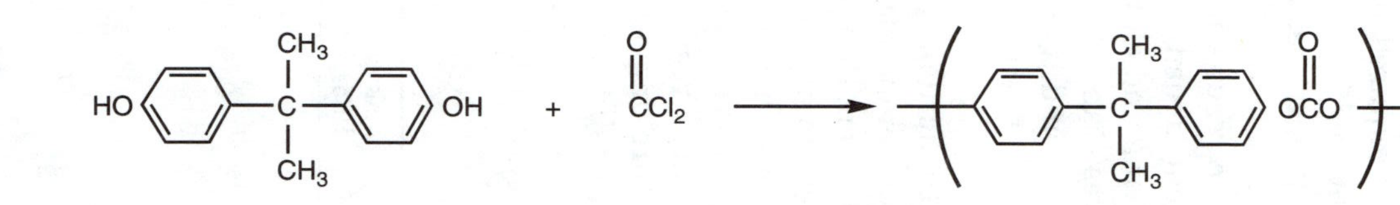

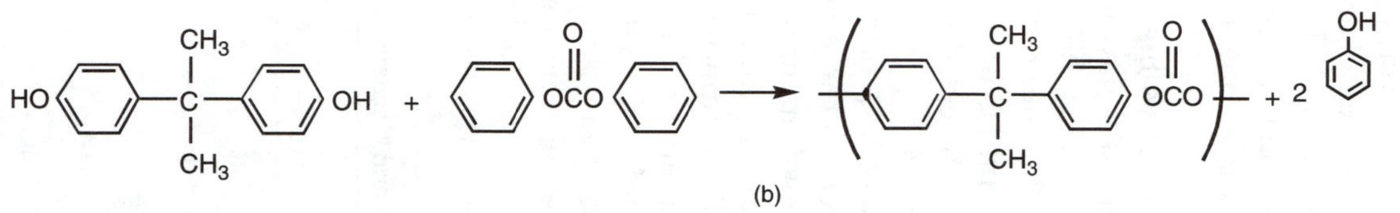

Figure 1.17 Synthesis routes for PC: (*a*) interface process and (*b*) transesterification reaction.

bond formation is minimized due to the steric hindrance posed by the benzene rings. The high level of aromaticity on the backbone, and the large size of the repeat structure, yield a molecule of very limited mobility. The ether linkage on the backbone permits some rotation and flexibility, producing high impact strength. Its amorphous nature with long, entangled chains, contributes to the unusually high toughness. Upon crystallization, however, PC is brittle. PC is so reluctant to crystallize that films must be held at 180°C for several days in order to impart the flexibility and thermal mobility required to conform to a structured three-dimensional crystalline lattice.[169] The rigidity of the molecule accounts for strong mechanical properties, elevated heat-deflection temperatures, and high dimensional stability at elevated temperatures. The relative high free volume results in a low-density polymer, with unfilled PC having a 1.22-g/cm^3 density.

A disadvantage includes the need for drying and elevated temperature processing. PC has limited chemical resistance to numerous aromatic solvents, including benzene, toluene, and xylene and has a weakness to notches. Selected mechanical and thermal properties are given in Table 1.3.[170]

Applications where PC is blended with acrilonitrile butiadiene styrene (ABS) increases the heat distortion temperature of the ABS and improves the low-temperature impact strength of PC. The favorable ease of processing and improved economics makes PC/ABS blends well suited for thin-walled electronic housing applications such as laptop computers. Blends with polybutylene terephthalate (PBT) are useful for improving the chemical resistance of PC to petroleum products and its low-temperature impact strength. PC alone is widely used as vacuum cleaner housings, household appliance housings, and power tools. These are arenas where PC's high impact strength, heat resistance, durability, and high-quality finish

TABLE 1.3 PC Thermal and Mechanical Properties

	Polycarbonate	30% glass-filled polycarbonate	Makroblend PR51 Bayer	Xenoy, CL101 GE
Heat-deflection temperature, °C, method A	138	280	90	95
Heat-deflection temperature, °C, method B	142	287	105	105
Ultimate tensile strength, N/mm^2	>65	70	56	>100
Ultimate elongation, %	110	3.5	120	>100
Tensile modulus, N/mm^2	2300	5500	2200	1900

justify its expense. It is also used in safety helmets, riot shields, aircraft canopies, traffic light lens housings, and automotive battery cases. Design engineers take care not to design with tight radii where PC's tendency to stress crack could be a hindrance. PC cannot withstand constant exposure to hot water and can absorb 0.2% of its weight of water at 33°C and 65% relative humidity. This does not impair its mechanical properties but at levels greater than 0.01% processing results in streaks and blistering.

1.2.12 Polyester thermoplastic

The broad class of organic chemicals, called polyesters, is characterized by the fact that they contain an ester linkage,

$$—(\overset{\overset{\displaystyle O}{\|}}{C} — O)—$$

and may have either aliphatic or aromatic hydrocarbon units. As an introduction, Table 1.4 offers some selected thermal and mechanical properties as a means of comparing polybutylene terephthalate (PBT), polycyclohexylenedimethylene terephthalate (PCT), and poly(ethylene terephthalate) (PET).

Liquid Crystal Polymers (LCP). Liquid crystal polyesters, known as liquid crystal polymers, are aromatic copolyesters. The presence of phenyl rings in the backbone of the polymer gives the chain rigidity, forming a rodlike chain structure. Generally, the phenyl rings are arranged in *para* linkages to yield rodlike structures.[171] This chain structure orients itself in an ordered fashion both in the melt and in the solid state, as shown in Figure 1.18. The materials are self-reinforcing with high mechanical properties, but as a result of the oriented liquid crystal behavior, the properties will be anisotropic. The designer must be aware of this in order to properly design the part and gate the molds.[172] The phenyl ring also helps increase the heat distortion temperature.[173]

The basic building blocks for liquid crystal polyesters are *p*-hydroxybenzoic acid, terephthalic acid, and hydroquinone. Unfortunately, the use of these monomers alone gives materials that are difficult to process with very high melting points. The polymers often degraded before melting.[174] Various techniques have been developed to give materials with lower melting points and better processing behavior. Some methods include the incorporation of flexible units in the chain (copolymerizing with ethylene glycol), the addition of nonlinear rigid structures, and the addition of aromatic groups to the side of the chain.[175]

Liquid crystal polymers based on these techniques include Victrex (ICI), Vectra (Hoescht Celanese), and Xydar (Amoco). Xydar is based

TABLE 1.4 Comparison between Thermal and Mechanical Properties of PBT, PCT, PCTA, PET, PETG, and PCTG

	PBT unfilled	30% glass-filled PBT	30% glass-filled PCT	30% glass-filled PCTA	PET unfilled	30% glass-filled PET	PETG unfilled	PCTG unfilled
T_m, °C	220–267	220–267	—	285	212–265	245–265	—	—
Tensile modulus, MPa	1,930–3,000	8,960–10,000	—	—	2,760–4,140	8,960–9,930	—	—
Ultimate tensile strength, MPa	56–60	96–134	124–134	97	48–72	138–165	28	52
Ultimate elongation, %	50–300	2–4	1.9–2.3	3.1	30–300	2–7	110	330
Specific gravity	1.30–1.38	1.48–1.54	1.45	1.41	1.29–1.40	1.55–1.70	1.27	1.23
HDT, °C								
264 lb/in²	50–85	196–225	260	221	21–65	210–227	64	65
66 lb/in²	115–190	216–260	>260	268	75	243–249	70	74

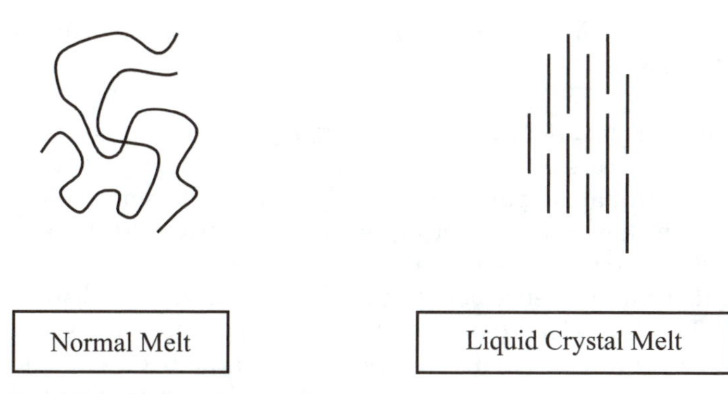

Figure 1.18 Melt configurations.

on terephthalic acid, *p*-hydroxybenzoic acid, and *p,p′*-dihydroxybiphenyl, while Vectra is based on *p*-hydroxybenzoic acid and hydroxynaphthoic acid.[176] These materials are known for their high-temperature resistance, particularly heat-distortion temperature. The heat-distortion temperature can vary from 170 to 350°C. They also have excellent mechanical properties, especially in the flow direction. For example, the tensile strength varies from 165 to 230 MPa, the flexural strength varies from 169 to 256 MPa, and the flexural modulus varies from 9 to 12.5 GPa.[177] Filled materials exhibit even higher values. LCPs are also known for good solvent resistance and low water absorption compared to other heat-resistant polymers. They have good electrical insulation properties, low flammability with a limiting oxygen index in the range of 35 to 40, but a high specific gravity (about 1.40).[178] LCPs show little dimensional change when exposed to high temperatures and a low coefficient of thermal expansion.[179]

These materials can be high priced and often exhibit poor abrasion resistance, due to the oriented nature of the polymer chains.[180] Surface fibrillation may occur quite easily.[181] The materials are processable on a variety of conventional equipment. Process temperatures are normally below 350°C, although some materials may need to be processed higher. They generally have low melt viscosity as a result of their ordered melt and should be dried before use to avoid degradation.[182] LCPs can be injection-molded on conventional equipment and regrind may be used. Mold release is generally not required.[183] Part design for LCPs requires careful consideration of the anisotropic nature of the polymer. Weld lines can be very weak if the melt meets in a “butt” type of weld line. Other types of weld lines show better strength.[184]

Liquid crystal polymers are used in automotive, electrical, chemical processing, and household applications. One application is for oven and microwave cookware.[185] Because of their higher costs the material will be used in applications only where their superior performance justifies the additional expense.

Polybutylene terephthalate (PBT). With the expiry of the original PET patents, manufacturers pursued the polymerization of other polyalkene terephthalates, particularly polybutylene terephthalate. The polymer is synthesized by reacting terephthalic acid with butane 1,4-diol to yield the structure shown in Fig. 1.19.

The only structural difference between PBT and PET is the substitution in PBT of four methylene repeat units rather than the two present in PET. This feature imparts additional flexibility to the backbone and reduces the polarity of the molecule resulting in similar mechanical properties to PET (high strength, stiffness, and hardness). PBT growth is at least 10% annually, in large part due to automotive exterior and under-the-hood applications such as electronic stability control and housings which are made out of a PBT/ASA (acrylonitrile/styrene/acrylic ester) blend. PBT/ASA blends are sold by BASF and GE Plastics Europe. Another development involving the use of PBT is coextrusion of PBT and a copolyester thermoplastic elastomer. This can then be blow-molded into under-the-hood applications which minimize noise vibration. Highly filled PBTs are also making inroads into the kitchen and bathroom tile industries.[186] As with PET, PBT is also often glass fiber filled in order to increase its flexural modulus, creep resistance, and impact strength. PBT is suitable for applications requiring dimensional stability, particularly in water, and resistance to hydrocarbon oils without stress-cracking.[187] Hence PBT is used in pump housings, distributors, impellers, bearing bushings, and gear wheels.

To improve PBT's poor notched impact strength, copolymerization with 5% ethylene and vinyl acetate onto the polyester backbone improves its toughness. PBT is also blended with PMMA, PET, PC, and polybutadiene in order to provide enhanced properties tailored to specific applications. Table 1.5 shows a breakdown of the U.S. market use for PBT.[188]

Polycyclohexylenedimethylene terephthalate (PCT). Another polyalkylene terephthalate polyester of significant commercial importance is PCT; a condensation product of the reaction between dimethyl terephthalate and 1,4-cyclohexylene glycol, as shown in Fig. 1.20.

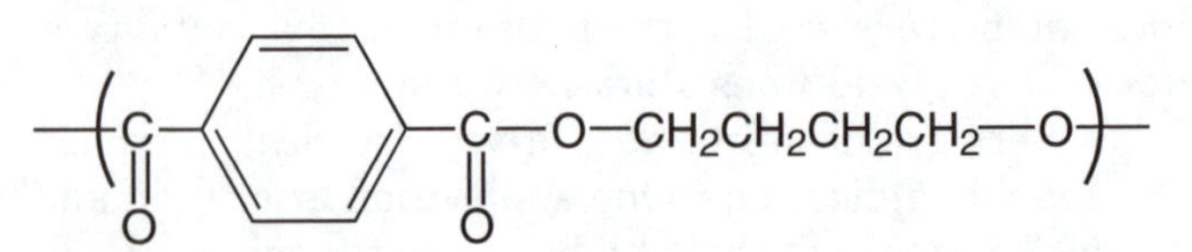

Figure 1.19 Repeat structure of PBT.

TABLE 1.5 U.S. Markets for PBT Use in 1997 and 1998

Market	Millions of pounds, 1997	Millions of pounds, 1998
Appliances	29	31
Consumer/ recreational	13	14
Electrical/electronic	58	65
Industrial	36	40
Transportion, including PC/PBT blends	136	154
Other	13	14
Export	28	28
Total	313	346

This material is biaxially oriented into films and while it is mechanically weaker than PET, it offers superior water resistance and weather resistance.[189] As seen in the introductory Table 1.4, PCT differentiates itself from PET and PBT by its high heat-distortion temperature. As with PET and PBT, PCT has low moisture absorption and its good chemical resistance to engine fluids and organic solvents lend it to under-the-hood applications such as alternator armatures and pressure sensors.[190]

Copolymers of PCT include PCTA, an acid-modified polyester, and PCTG, a glycol-modified polyester. PCTA is used primarily for extruded film and sheet for packaging applications. PCTA has high clarity, tear strength, chemical resistance, and when PCTA is filled it is used for dual ovenable cookware. PCTG is primarily injection-molded and PCTG parts have notched Izod impact strengths similar to polycarbonate, against which it often competes. It also competes with ABS, another clear polymer. It finds use in medical and optical applications.[191]

Poly(ethylene terephthalate) (PET). There are tremendous commercial applications for PET; as an injection-molding grade material, for blow-molded bottles, and for oriented films. In 1998, the U.S. consumption of PET was 4330 million pounds, while domestic consumption of PBT was 346 million pounds.[192] PET, also known as poly(oxyethylene oxyterephthaloyl), can be synthesized from dimethyl terephthalate and ethylene glycol by a two-step ester interchange process, as shown in Fig. 1.21.[193] The first stage involves a solution polymerization of 1 mole of dimethyl terephthalate with 2.1 to 2.2 moles of ethylene glycol.[194] The excess ethylene glycol increases the rate of formation of bis(2-hydroxyethyl) terephthalate. Small amounts of trimer, tetramer, and other oligomers are formed. A metal alkanoate, such as manganese acetate, is often added as a catalyst; this is later deactivated by

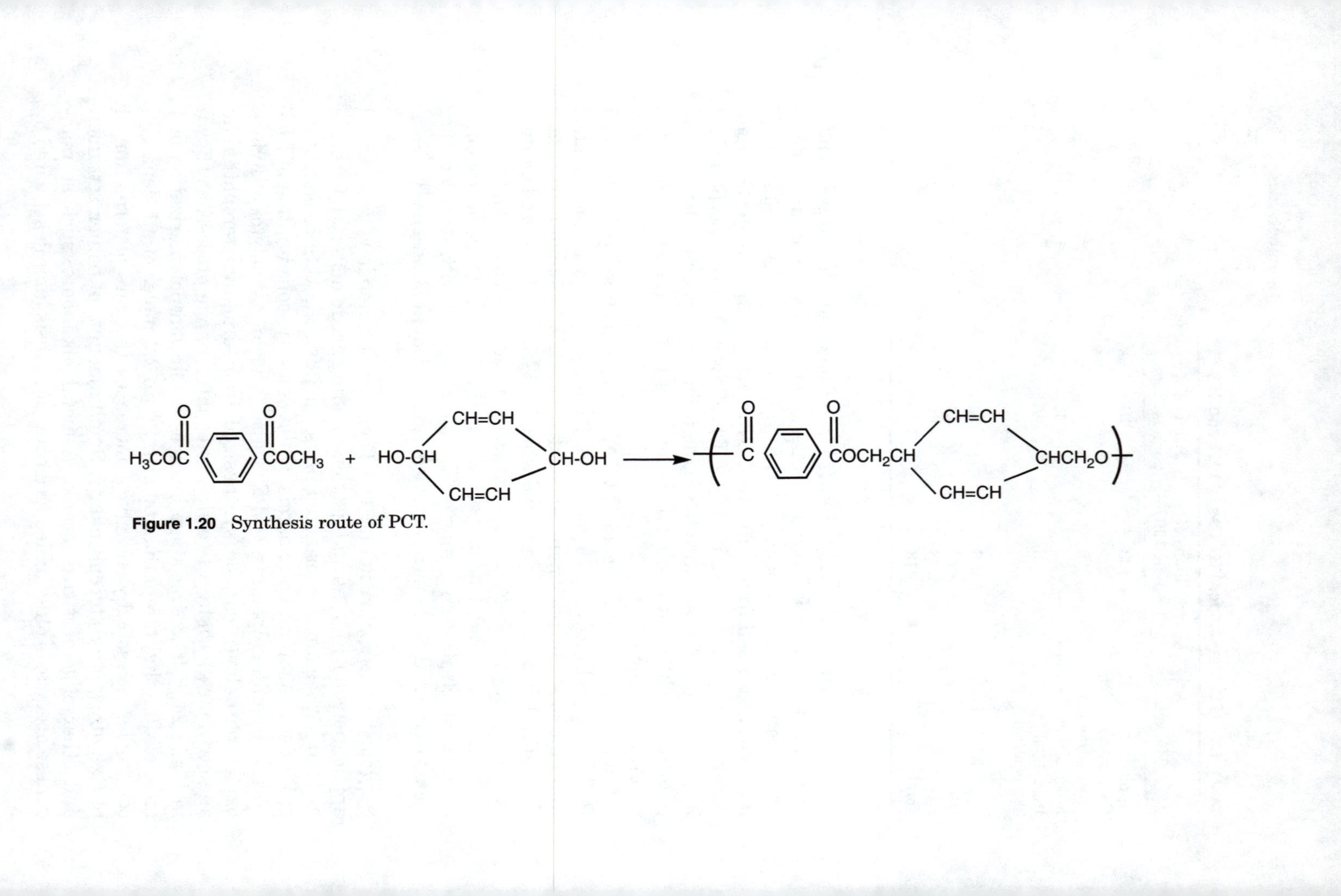

Figure 1.20 Synthesis route of PCT.

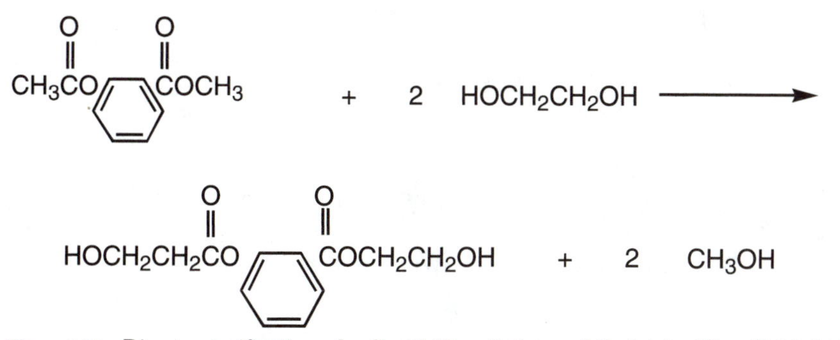

Figure 1.21 Direct esterification of a diacid (dimethyl terephthalate) with a diol (ethylene glycol) in the first stages of PET polymerization.

the addition of a phosphorous compound such as phosphoric acid. The antioxidant phosphate improves the thermal and color stability of the polymer during the higher-temperature second-stage process.[195] The first stage of the reaction is run at 150 to 200°C with continuous methanol distillation and removal.[196]

The second step of the polymerization, shown in Fig. 1.22, is a melt polymerization as the reaction temperature is raised to 260 to 290°C. This second stage is carried out under either partial vacuum (0.13 kPa)[197] to facilitate the removal of ethylene glycol or with an inert gas being forced through the reaction mixture. Antimony trioxide is often used as a polymerization catalyst for this stage.[198] It is critical that excess ethylene glycol be completely removed during this alcoholysis stage of the reaction in order to proceed to high molecular weight products; otherwise equilibrium is established at an extent of reaction of less than 0.7. This second stage of the reaction proceeds until a number-average molecular weight, M_n, of about 20,000 g/mol is obtained. The very high temperatures at the end of this reaction cause thermal decomposition of the end groups to yield acetaldehyde. Thermal ester scission also occurs, which competes with the polymer step-growth reactions. It is this competition which limits the ultimate M_n which can be achieved through this melt condensation reaction.[199] Weight-average molecular weights of oriented films are around 35,000 g/mol.

Other commercial manufacturing methods have evolved to a direct esterification of acid and glycol in place of the ester-exchange process. In direct esterification, terephthalic acid and ethylene glycol are reacted, rather than esterifying terephthalic acid with methanol to produce the dimethyl terephthalate intermediate. The ester is easier to purify than the acid, which sublimes at 300°C and is insoluble, however, better catalysts and purer terephthalic acid offer the elimination of the intermediate use of methanol.[200] Generally, PET resins made by direct esterification of terephthalic acid contain more diethylene glycol,

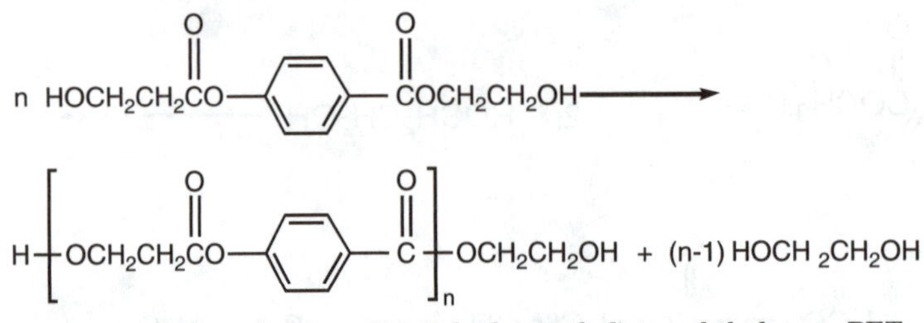

Figure 1.22 Polymerization of bis(2-hydroxyethyl) terephthalate to PET.

which is generated by an intermolecular ether-forming reaction between β-hydroxyethyl ester end groups. Oriented films produced from these resins have reduced mechanical strength and melting points as well as decreased thermo-oxidative resistance and poorer UV stability.[201]

The degree of crystallization and direction of the crystallite axis governs all of the resin's physical properties. The percentage of a structure existing in crystalline domains is primarily determined through density measurements or by thermal means using a differential scanning calorimeter (DSC). The density of amorphous PET is 1.333 g/cm^3, while the density of a PET crystal is 1.455 g/cm^3.[202] Once the density is known, the fraction of crystalline material can be determined.

An alternate means of measuring crystallinity involves comparing the ratio of the heat of cold crystallization, ΔH_{cc}, of amorphous polymer to the heat of fusion, ΔH_f, of crystalline polymer. This ratio is 0.61 for an amorphous PET and a fully crystalline PET sample should yield a value close to zero.[203] After the sample with its initial morphology has been run once in the DSC, the heat of fusion determined in the next run can be considered as ΔH_{cc}. The lower the $\Delta H_{cc}/\Delta H_f$ ratio, the more crystalline the original sample was.

In the absence of nucleating agents and plasticizers, PET crystallizes slowly which is a hindrance in injection-molding applications as either hot molds or costly extended cooling times are required. In the case of films, however, where crystallinity can be mechanically induced, PET resins combine rheological properties which lend themselves to melt extrusion with a well-defined melting point, making them ideally suited for biaxially oriented film applications. The attachment of the ester linkage directly to the aromatic component of the backbone means that these linear, regular PET chains have enough flexibility to form stress-induced crystals and achieve enough molecular orientation to form strong, thermally stable films.[204]

Methods for producing oriented PET films have been well documented and will only be briefly discussed here. The process as described in the *Encyclopedia of Polymer Science and Engineering* usually involves a sequence of five steps which include: melt extrusion and slot casting, quenching, drawing in the longitudinal machine direction (MD), drawing in the transverse direction (TD), and annealing.[205] Dried, highly viscous polymer melt is extruded through a slot die with an adjustable gap width onto a highly polished quenching drum. If very high output rates are required, a cascade system of extruders can be set up to first melt and homogenize the PET granules, then to use the next in-line extruder to meter the melt to the die. Molten resin is passed through filter packs with average pore sizes of 5 to 30 μm. Quenching to nearly 100% amorphous morphology is critical to avoid embrittlement; films which have been allowed to form spherulites are brittle, translucent, and are unable to be further processed.

The sheet is then heated to about 95°C (above the glass transition point of approximately 70°C), where thermal mobility allows the material to be stretched to three or four times its original dimension in the MD. This uniaxially oriented film has stress-induced crystals whose main axes are aligned in the machine direction. The benzene rings, however, are aligned parallel to the surface of the film in the <1,0,0> crystal plane. The film is then again heated, generally to above 100°C, and stretched to three to four times its initial dimension in the TD. This induces further crystallization bringing the degree of crystallinity to 25 to 40% and creates a film which has isotropic tensile strength and elongation properties in the machine and transverse directions. The film at this point is thermally unstable above 100°C, and must be annealed in the tenter frame in order to partially relieve the stress.

The annealing involves heating to 180 to 220°C for several seconds to allow amorphous chain relaxation, partial melting, recrystallization, and crystal growth to occur.[206] The resultant film is approximately 50% crystalline and possesses good mechanical strength, a smooth surface which readily accepts a wide variety of coatings, and has good winding and handling characteristics. PET films are produced from 1.5-μm thick as capacitor films to 350-μm thick for use as electrical insulation in motors and generators.[207]

Due to the chemically inert nature of PET, films which are used in coatings applications are often treated with a variety of surface modifiers. Organic and inorganic fillers are often incorporated in relatively thick films in order to improve handling characteristics by roughening the surface slightly. For thin films, however, many applications require transparency which would be marred by the incorporation of fillers. Therefore an in-line coating step of either aqueous or solvent-based coatings is set up between the MD and TD drawing stations. The drawing of the film after the coating has been applied helps to achieve very thin coatings.

1.2.13 Polyetherimide

Polyetherimides (PEI) are a newer class of amorphous thermoplastics with high-temperature resistance, impact strength, creep resistance, and rigidity. They are transparent with an amber color.[208] The polymer is sold under the trade name of Ultem (General Electric) and has the structure shown in Fig. 1.23. It is prepared from the condensation polymerization of diamines and dianhydrides.[209]

The material can be melt processed because of the ether linkages present in the backbone of the polymer, but still maintains properties similar to the polyimides.[210] The high-temperature resistance of the polymer allows it to compete with the polyketones, polysulfones, and poly(phenylene sulfides). The glass transition temperature of PEI is 215°C. The polymer has very high tensile strength, a UL temperature index of 170°C, flame resistance, and low smoke emission.[211] The polymer is resistant to alcohols, acids, and hydrocarbon solvents, but will dissolve in partially halogenated solvents.[212] Both glass and carbon fiber–reinforced grades are available.[213]

The polymer should be dried before processing and typical melt temperatures are 340 to 425°C.[214] Polyetherimides can be processed by injection molding and extrusion. In addition, the high melt strength of the polymer allows it to be thermoformed and blow molded. Annealing of the parts is not required.

Polyetherimide is used in a variety of applications. Electrical applications include printed circuit substrates and burn-in sockets. In the automotive industry PEI is used for under-the-hood temperature sensors and lamp sockets. PEI sheet has also been used to form an aircraft cargo vent.[215] The dimensional stability of this polymer allows its use for large flat parts such as in hard disks for computers.

1.2.14 Polyethylene

Polyethylene (PE) is the highest-volume polymer in the world. Its high toughness, ductility, excellent chemical resistance, low water vapor permeability, and very low water absorption, combined with the ease with which it can be processed, make PE of all different density grades an attractive choice for a variety of goods. PE is limited by its relatively low modulus, yield stress, and melting point. PE is used to make containers, bottles, film, and pipes, among other things. It is an incredibly versatile polymer with almost limitless variety due to copolymerization potential, a wide density range, a MW which ranges from very low (waxes have a MW of a few hundred) to very high (6×10^6), and the ability to vary MWD.

Its repeat structure is $(\text{—}CH_2CH_2\text{—})_x$, which is written as polyethylene rather than polymethylene $(\text{—}CH_2)_x$, in deference to the various

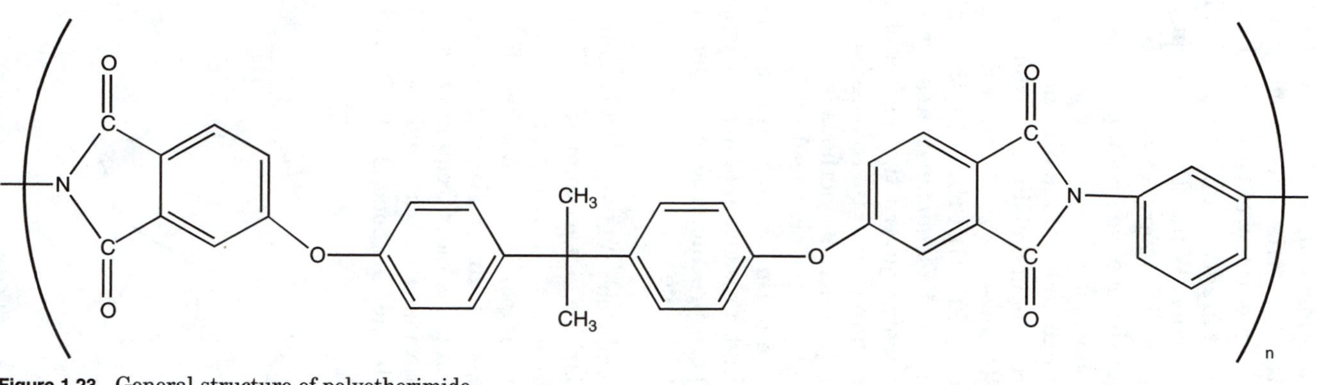

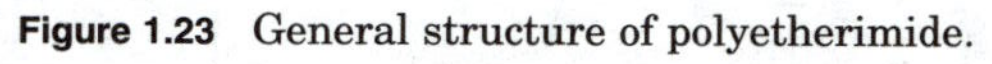

Figure 1.23 General structure of polyetherimide.

ethylene polymerization mechanisms. PE has a deceptive simplicity. PE homopolymers are made up exclusively of carbon and hydrogen atoms, and just as the properties of diamond and graphite (which are also materials made up entirely of carbon and hydrogen atoms) vary tremendously, different grades of PE have markedly different thermal and mechanical properties. Whilst PE is generally a whitish, translucent polymer, it is available in grades of density that range from 0.91 to 0.97 g/cm^3. The density of a particular grade is governed by the morphology of the backbone; long, linear chains with very few side branches can assume a much more three-dimensionally compact, regular, crystalline structure. Commercially available grades are: very-low-density PE (VLDPE), low-density PE (LDPE), linear low-density PE (LLDPE), high-density PE (HDPE) and ultra-high molecular weight PE (UHMWPE). Figure 1.24 demonstrates figurative differences in chain configuration which govern the degree of crystallinity, which, along with MW, determines final thermomechanical properties.

Four established production methods are: a gas phase method known as the Unipol process practiced by Union Carbide, a solution method used by Dow and DuPont, a slurry emulsion method practiced by Phillips, and a high-pressure method.[216] Generally, yield strength and the melt temperature increase with density, while elongation decreases with increased density.

Very-low-density polyethylene (VLDPE). This material was introduced in 1985 by Union Carbide, is very similar to LLDPE, and is principally used in film applications. VLDPE grades vary in density from 0.880 to 0.912 g/cm^3.[217] Its properties are marked by high elongation, good environmental stress cracking resistance, excellent low-temperature properties, and it competes most frequently as an alternative to plasticised polyvinyl chloride (PVC) or ethylene-vinyl acetate (EVA). The inherent flexibility in the backbone of VLDPE circumvents plasticizer

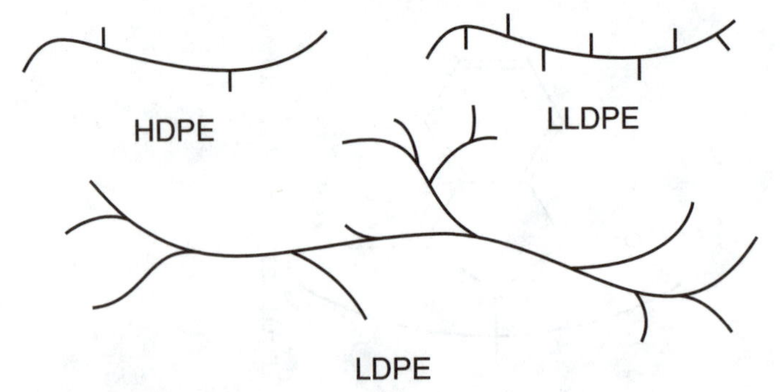

Figure 1.24 Chain configurations of polyethylene.

stability problems which can plague PVC, and it avoids odor and stability problems which are often associated with molding EVAs.[218]

Low-density polyethylene (LDPE). LDPE combines high impact strength, toughness, and ductility to make it the material of choice for packaging films, which is one of its largest applications. Films range from shrink film, thin film for automatic packaging, heavy sacking, and multilayer films (both laminated and coextruded) where LDPE acts as a seal layer or a water vapor barrier.[219] It has found stiff competition from LLDPE in these film applications due to LLDPE's higher melt strength. LDPE is still very widely used, however, and is formed via free-radical polymerization, with alkyl branch groups [given by the structure $—(CH_2)_xCH_3$] of two to eight carbon atom lengths. The most common branch length is four carbons long. High reaction pressures encourage crystalline regions. The reaction to form LDPE is shown in Fig. 1.25, where *n* approximately varies in commercial grades between 400 and 50,000.[220]

Medium-density PE is produced via the previous reaction, carried out at lower polymerization temperatures.[221] The reduced temperatures are postulated to reduce the randomizing Brownian motion of the molecules and this reduced thermal energy allows crystalline formation more readily at these lowered temperatures.

Linear low-density polyethylene (LLDPE). This product revolutionized the plastics industry with its enhanced tensile strength for the same density compared to LDPE. Table 1.6 compares mechanical properties of LLDPE to LDPE. As is the case with LDPE, film accounts for approximately three-quarters of the consumption of LLDPE. As the name implies, it is a long linear chain without long side chains or branches. The short chains which are present disrupt the polymer chain uniformity enough to prevent crystalline formation and hence prevent the polymer from achieving high densities. Developments of the past decade have enabled production economies compared to LDPE due to lower polymerization pressures and temperatures. A typical LDPE process requires 35,000 lb/in² which is reduced to 300 lb/in² in the case of LLDPE and reaction temperatures as low as 100°C rather than 200 or 300°C are used. LLDPE is actually a copolymer

$$n\ CH_2{=}CH_2 \xrightarrow[\text{small amounts of } O_2 \text{ or organic peroxide present}]{200°C,\ 20{,}000-35{,}000\ \text{psi}} —(CH_2CH_2)_n—$$

Figure 1.25 Polymerization of PE.

TABLE 1.6 Comparison of Blown Film Properties of LLDPE and LDPE[412]

	LLDPE	LDPE
Density, g/cm^3	0.918	0.918
Melt index, g/10 min	2.0	2.0
Dart impact, g	110	110
Puncture energy, J/mm	60	25
Machine direction tensile strength, MPa	33	20
Cross-direction tensile strength, MPa	25	18
Machine direction tensile elongation, %	690	300
Cross-direction tensile elongation, %	740	500
Machine direction modulus, MPa	210	145
Cross-direction modulus, MPa	250	175

containing side branches of 1-butene most commonly, with 1-hexene or 1-octene also present. Density ranges of 0.915 to 0.940 g/cm^3 are polymerized with Ziegler catalysts which orient the polymer chain and govern the tacticity of the pendant side groups.[222]

High-density polyethylene (HDPE). HDPE is one of the highest-volume commodity chemicals produced in the world, in 1998 the worldwide demand was 1.8×10^{10} kg.[223] The most common method of processing HDPE is blow molding, where resin is turned into bottles (especially for milk and juice), housewares, toys, pails, drums, and automotive gas tanks. It is also commonly injection-molded into housewares, toys, food containers, garbage pails, milk crates, and cases. HDPE films are commonly found as bags in supermarkets, department stores, and as garbage bags.[224] Two commercial polymerization methods are most commonly practiced; one involves Phillips catalysts (chromium oxide) and the other involves the Ziegler-Natta catalyst systems (supported heterogeneous catalysts such as titanium halides, titanium esters, and aluminum alkyls on a chemically inert support such as PE or PP). Molecular weight is governed primarily through temperature control, with elevated temperatures resulting in reduced molecular weights. The catalyst support and chemistry also play an important factor in controlling molecular weight and molecular weight distribution.

Ultra-high molecular weight polyethylene (UHMWPE). UHMWPE is identical to HDPE, but rather than having a MW of 50,000 g/mol, it typically has a MW of between 3×10^6 and 6×10^6. The high MW imparts outstanding abrasion resistance, high toughness, even at cryogenic temperatures, and excellent stress cracking resistance, but does not generally allow the material to be processed conventionally. The polymer chains are so entangled due to their considerable length that the conventionally considered melt point doesn't exist practically, as it is too close to the degradation temperature, although an injection-molding grade is marketed by Hoechst. Hence, UHMWPE is often processed as a fine powder that can be ram extruded or compression molded. Its properties are taken advantage of in uses which include liners for chemical processing equipment, lubrication coatings in railcar applications to protect metal surfaces, recreational equipment such as ski bases, and medical devices.[225] A recent product has been developed by Allied Chemical which involves gel spinning UHMWPE into light weight, very strong fibers which compete with Kevlar in applications for protective clothing.

1.2.15 Polyethylene copolymers

Ethylene is copolymerized with many nonolefinic monomers, particularly acrylic acid variants and vinyl acetate, with EVA polymers being the most commercially significant. All of the copolymers discussed in this section necessarily involve disruption of the regular, crystallizable PE homopolymer and as such feature reduced yield stresses and moduli, with improved low-temperature flexibility.

Ethylene-acrylic acid (EAA) copolymers. EAA copolymers enjoy a renewed interest since they were first identified in the 1950s when in 1974 Dow introduced new grades characterized by outstanding adhesion to metallic and nonmetallic substrates.[226] The presence of the carboxyl and hydroxyl functionalities promotes hydrogen bonding and these strong intermolecular interactions are taken advantage of to bond aluminum foil to polyethylene in multilayer extrusion-laminated toothpaste tubes and as tough coatings for aluminum foil pouches.

Ethylene-ethyl acrylate (EEA) copolymers. EEA copolymers typically contain 15 to 30% by weight of ethyl acrylate (EA) and are flexible polymers of relatively high molecular weight suitable for extrusion, injection molding, and blow molding. Products made of EEA have high environmental stress cracking resistance, excellent resistance to flexural fatigue, and low-temperature properties down to as low as −65°C. Applications include molded rubberlike parts, flexible film for disposable gloves and hospital sheeting, extruded hoses, gaskets, and bumpers.[227] Typical applications include polymer modifications where EEA is blended with olefin polymers (since it is compatible with VLDPE, LLDPE, LDPE, HDPE, and PP[228]) to yield a blend with a specific modulus, yet with the advantages inherent in EEA's polarity. The EA presence promotes toughness, flexibility, and greater adhesive properties. EEA blending can provide a cost-effective way to improve the impact resistance of polyamides and polyesters.[229]

The similarity of ethyl-acrylate monomer to vinyl acetate predicates that these copolymers have very similar properties, although EEA is considered to have higher abrasion and heat resistance while EVA tends to be tougher and of greater clarity.[230] EEA copolymers are FDA approved up to 8% EA content in food contact applications.[231]

Ethylene-methyl acrylate (EMA) copolymers. EMA copolymers are often blown into film with very rubbery mechanical properties and outstanding dart-drop impact strength. The latex-rubberlike properties of EMA film lend to its use in disposable gloves and medical devices without the associated hazards to people with allergies to latex rubber. Due to their adhesive properties, EMA copolymers like their EAA and EEA counterparts are used in extrusion coating, coextrusions, and laminating applications as heat-seal layers. EMA is one of the most thermally stable of this group, and as such it is commonly used to form heat and RF seals, as well in multiextrusion tie-layer applications. This copolymer is also widely used as a blending compound with olefin homopolymers (VLDPE, LLDPE, LDPE, and PP) as well as with polyamides, polyesters, and polycarbonate to improve impact strength and toughness and to increase either heat-seal response or to promote adhesion.[232] EMA is also used in soft blow-molded articles such as squeeze toys, tubing, disposable medical gloves, and foamed sheet. EMA copolymers and EEA copolymers containing up to 8% ethyl acrylate are approved by the FDA for food packaging.[233]

Ethylene-*n*-butyl acrylate (EBA) copolymers. EBA copolymers are also widely blended with olefin homopolymers to improve impact strength, toughness, heat sealability, and to promote adhesion. The polymerization process and resultant repeat unit of EBA are shown in Fig. 1.26.

Ethylene-vinyl acetate (EVA) copolymers. EVA copolymers are given by the structure shown in Fig. 1.27 and find commercial importance in the coating, laminating, and film industries. EVA copolymers typically contain between 10 and 15 mol % vinyl acetate, which provides a bulky, polar pendant group to the ethylene and provides an opportunity to tailor the end properties by optimizing the vinyl acetate content. Very low vinyl-acetate content (approximately 3 mol %) results in a copolymer which is essentially a modified low-density polyethylene,[234] with an even further reduced regular structure. The resultant copolymer is used as a film due to its flexibility and surface gloss. Vinyl acetate is a

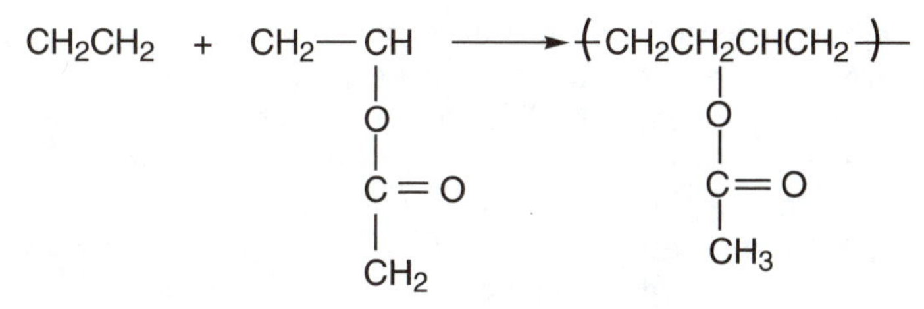

Figure 1.26 Polymerization and structure of EBA.

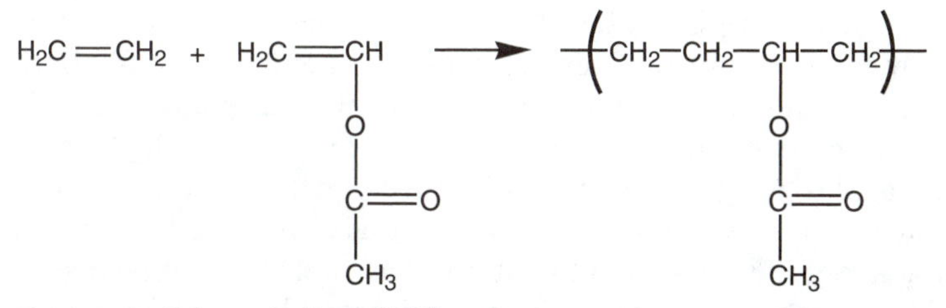

Figure 1.27 Polymerization of EVA.

low-cost comonomer which is nontoxic and allows for this copolymer to be used in many food packaging applications. These films are soft and tacky and therefore appropriate for cling-wrap applications (they are more thermally stable than the PVDC films often used as cling wrap) as well as interlayers in coextruded and laminated films.

EVA copolymers with approximately 11 mol % vinyl acetate are widely used in the hot-melt coatings and adhesives arena where the additional intermolecular bonding promoted by the polarity of the vinyl-acetate ether and carbonyl linkages enhances melt strength, while still enabling low melt processing temperatures. At 15 mol % vinyl acetate, a copolymer with very similar mechanical properties to plasticized PVC is formed. There are many advantages to an inherently flexible polymer for which there is no risk of plasticizer migration, and PVC alternatives is the area of largest growth opportunity. These copolymers have higher moduli than standard elastomers and are preferable in that they are more easily processed without concern for the need to vulcanize.

Ethylene-vinyl alcohol (EVOH) copolymers. Poly(vinyl alcohol) is prepared through alcoholysis of poly(vinyl acetate). PVOH is an atactic polymer, but since the crystal lattice structure is not disrupted by hydroxyl groups, the presence of residual acetate groups greatly diminishes the crystal formation and the degree of hydrogen bonding. Polymers which are highly hydrolyzed (have low residual acetate content) have a high tendency to crystallize and for hydrogen bonding to occur. As the degree of hydrolysis increases, the molecules will very readily crystallize and hydrogen bonds will keep them associated if they are not fully dispersed prior to dissolution. At degrees of hydrolysis above 98%, manufacturers recommend a minimum temperature of 96°C in order to ensure that the highest molecular weight components have enough thermal energy to go into solution. Polymers with low degrees of residual acetate have high humidity resistance.

1.2.16 Modified polyethylenes

The properties of PE can be tailored to meet the needs of a particular application by a variety of different methods. Chemical modification, copolymerization, and compounding can all dramatically alter specific properties. The homopolymer itself has a range of properties depending upon the molecular weight, the number and length of side branches, the degree of crystallinity, and the presence of additives such as fillers or reinforcing agents. Further modification is possible by chemical substitution of hydrogen atoms; this occurs preferentially at the tertiary carbons of a branching point and primarily involves chlorination, sulphonation, phosphorylination, and intermediate combinations.

Chlorinated polyethylene (CPE). The first patent on the chlorination of PE was awarded to ICI in 1938.[235] CPE is polymerized by substituting select hydrogen atoms on the backbone of either HDPE or LDPE with chlorine. Chlorination can occur in the gaseous phase, in solution, or as an emulsion. In the solution phase, chlorination is random, while the emulsion process can result in uneven chlorination due to the crystalline regions. The chlorination process generally occurs by a free-radical mechanism, shown in Fig. 1.28, where the chlorine free radical is catalyzed by ultraviolet light or initiators.

Interestingly, the properties of CPE can be adjusted to almost any intermediary position between PE and PVC by varying the properties of the parent PE and the degree and tacticity of chlorine substitution. Since the introduction of chlorine reduces the regularity of the PE, crystallinity is disrupted, and at up to a 20% chlorine level the modified material is rubbery (if the chlorine was randomly substituted). When the level of chlorine reaches 45% (approaching PVC), the material is stiff at room temperature. Typically HDPE is chlorinated to a chlorine content of 23 to 48%.[236] Once the chlorine substitution reaches 50%, the polymer is identical to PVC, although the polymerization route differs. The largest use of CPE is as a blending agent with PVC to promote flexibility and thermal stability for increased ease of processing. Blending CPE with PVC essentially plasticizes the PVC without adding double-bond unsaturation prevalent with rubber-modified

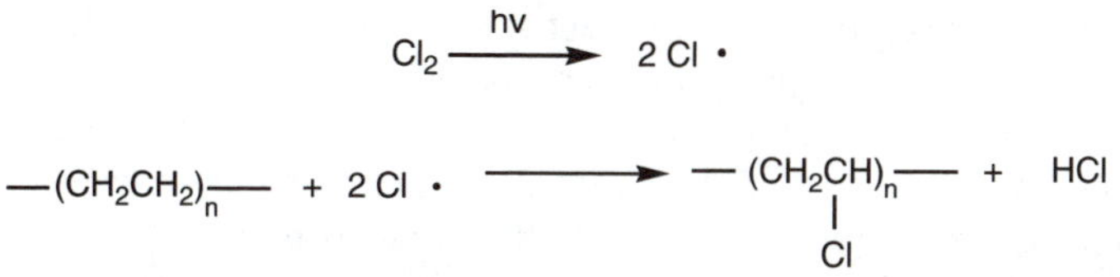

$$Cl_2 \xrightarrow{h\nu} 2\,Cl\cdot$$

$$-(CH_2CH_2)_n- + 2\,Cl\cdot \longrightarrow -(CH_2\underset{\displaystyle Cl}{\underset{|}{C}}H)_n- + HCl$$

Figure 1.28 Chlorination process of CPE.

PVCs and results in a more UV-stable, weather-resistant polymer. While rigid PVC is too brittle to be machined, the addition of as little as three to six parts per hundred CPE in PVC allows extruded profiles such as sheets, films, and tubes to be sawed, bored, and nailed.[237] Higher CPE content blends result in improved impact strength of PVC and are made into flexible films which don't have plasticizer migration problems. These films find applications in roofing, water and sewage-treatment pond covers, and sealing films in building construction.

CPE is used in highly filled applications, often using $CaCO_3$ as the filler, and finds use as a homopolymer in industrial sheeting, wire and cable insulations, and solution applications. When PE is reacted with chlorine in the presence of sulfur dioxide, a chlorosulfonyl substitution takes place, yielding an elastomer.

Chlorosulfonated polyethylenes (CSPE). Chlorosulfonation introduces the polar, cross-linkable SO_2 group onto the polymer chain, with the unavoidable introduction of chlorine atoms as well. The most common method involves exposing LDPE, which has been solubilized in a chlorinated hydrocarbon, to SO_2 and Cl in the presence of UV or high-energy radiation.[238] Both linear and branched PEs are used, and CSPEs contain 29 to 43% chlorine and 1 to 1.5% sulfur.[239] As in the case of CPEs, the introduction of Cl and SO_2 functionalities reduces the regularity of the PE structure hence reducing the degree of crystallinity, and the resultant polymer is more elastomeric than the unmodified homopolymer. CSPE is manufactured by DuPont under the tradename Hypalon and is used in protective coating applications such as the lining for chemical processing equipment, as the liners and covers for waste-containment ponds, as cable jacketing and wire insulation, spark plug boots, power steering pressure hoses, and in the manufacture of elastomers.

Phosphorylated polyethylenes. Phosphorylated PEs have higher ozone and heat resistance than ethylene propylene copolymers due to the fire-retardant nature provided by phosphor.[240]

Ionomers. Acrylic acid can be copolymerized with polyethylene to form an ethylene-acrylic acid copolymer (EAA) through addition or chain growth polymerization. It is structurally similar to ethylene-vinyl acetate, but with acid groups off the backbone. The concentration of acrylic acid groups is generally in the range of 3 to 20%.[241] The acid groups are then reacted with a metal-containing base, such as sodium methoxide or magnesium acetate, to form the metal salt as depicted in Fig. 1.29.[242] The ionic groups can associate with each other forming a cross-link between chains. The resulting materials are called ionomers in reference to the ionic bonds formed between chains. They were originally developed by DuPont under the trade name of Surlyn.

The association of the ionic groups forms a thermally reversible cross-link that can be broken when exposed to heat and shear. This allows ionomers to be processed on conventional thermoplastic processing equipment, while still maintaining some of the behavior of a thermoset at room temperature.[243] The association of ionic groups is generally believed to take two forms: multiplets and clusters.[244] Multiplets are considered to be a small number of ionic groups dispersed in the matrix, while clusters are phase-separated regions containing many ion pairs and also hydrocarbon backbone.

A wide range of properties can be obtained by varying the ethylene/methacrylic acid ratios, molecular weight, and the amount and type of metal cation used. Most commercial grades use either zinc or sodium for the cation. Materials using sodium as the cation generally have better optical properties and oil resistance, while those using zinc usually have better adhesive properties, lower water absorption, and better impact strength.[245]

The presence of the comonomer breaks up the crystallinity of the polyethylene, so that ionomer films have lower crystallinity and better clarity compared to polyethylene.[246] Ionomers are known for their toughness and abrasion resistance, and the polar nature of the polymer improves both its bondability and paintability. Ionomers have good low-temperature flexibility and resistance to oils and organic solvents. Ionomers show a yield point with considerable cold drawing. In

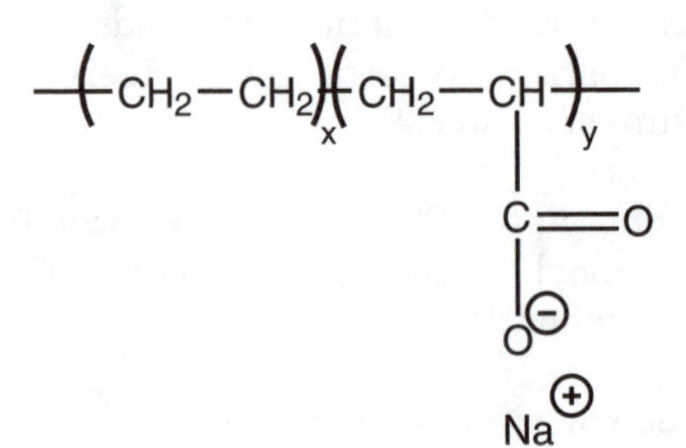

Figure 1.29 Structure of an ionomer.

contrast to PE, the stress increases with strain during cold drawing, giving a very high energy to break.[247]

Ionomers can be processed by most conventional extrusion and molding techniques using conditions similar to other olefin polymers. For injection molding the melt temperatures are in the range 210 to 260°C.[248] The melts are highly elastic due to the presence of the metal ions. Increasing temperatures rapidly decrease the melt viscosity, with the sodium- and zinc-based ionomers showing similar rheological behavior. Typical commercial ionomers have melt index values between 0.5 and 15.[249] Both unmodified and glass-filled grades are available.

Ionomers are used in applications, such as golf ball covers and bowling pin coatings, where their good abrasion resistance is important.[250] The puncture resistance of films allows these materials to be widely used in packaging applications. One of the early applications was the packaging of fish hooks.[251] They are often used in composite products as an outer heat-seal layer. Their ability to bond to aluminum foil is also utilized in packaging applications.[252] Ionomers also find application in footwear for shoe heels.[253]

1.2.17 Polyimide (PI)

Thermoplastic polyimides are linear polymers noted for their high-temperature properties. Polyimides are prepared by condensation polymerization of pyromellitic anhydrides and primary diamines. A polyimide contains the structure —CO—NR—CO as a part of a ring structure along the backbone. The presence of ring structures along the backbone, as depicted in Fig. 1.30, gives the polymer good high-temperature properties.[254] Polyimides are used in high-performance applications as replacements for metal and glass. The use of aromatic diamines gives the polymer exceptional thermal stability. An example of this is the use of di-(4-amino-phenyl) ether, which is used in the manufacture of Kapton (DuPont).

Although called thermoplastics, some polyimides must be processed in precursor form because they will degrade before their softening point.[255] Fully imidized injection-molding grades are available along with powder forms for compression molding and cold forming. However, injection molding of polyimides requires experience on the part of the molder.[256] Polyimides are also available as films and preformed stock shapes. The polymer may also be used as a soluble prepolymer, where heat and pressure are used to convert the polymer into the final, fully imidized form. Films can be formed by casting soluble polymers or precursors. It is generally difficult to form good films by melt extrusion. Laminates of polyimides can also be formed by impregnating fibers such as glass or graphite.

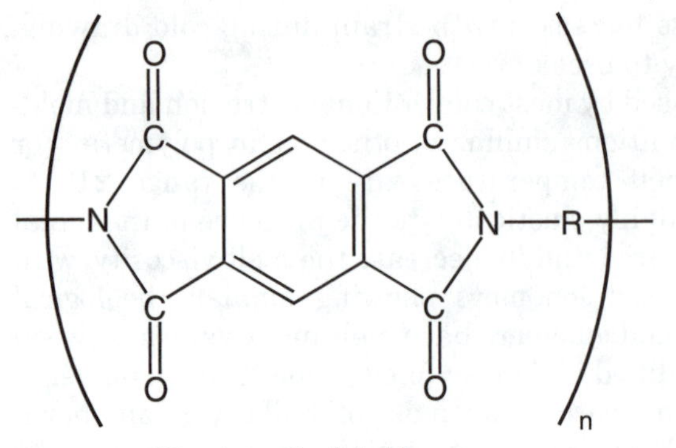

Figure 1.30 Structure of polyimide.

Polyimides have excellent physical properties and are used in applications where parts are exposed to harsh environments. They have outstanding high-temperature properties and their oxidative stability allows them to withstand continuous service in air at temperatures of 260°C.[257] Polyimides will burn, but they have self-extinguishing properties.[258] They are resistant to weak acids and organic solvents but are attacked by bases. The polymer also has good electrical properties and resistance to ionizing radiation.[259] A disadvantage of polyimides is their hydrolysis resistance. Exposure to water or steam above 100°C may cause parts to crack.[260]

The first application of polyimides was for wire enamel.[261] Applications for polyimides include bearings for appliances and aircraft, seals, and gaskets. Film versions are used in flexible wiring and electric motor insulation. Printed circuit boards are also fabricated with polyimides.[262]

1.2.18 Polyketones

The family of aromatic polyether ketones includes structures which vary in the location and number of ketonic and ether linkages on their repeat unit and, therefore, include polyether ketone (PEK), polyether ether ketone (PEEK), polyether ether ketone ketone (PEEKK), as well as other combinations. Their structures are as shown in Fig. 1.31. All have very high thermal properties due to the aromaticity of their backbones and are readily processed via injection molding and extrusion, although their melt temperatures are very high—370°C for unfilled PEEK and 390°C for filled PEEK, and both unfilled and filled PEK. Mold temperatures as high as 165°C are also used.[263] Their toughness, surprisingly high for such high heat-resistant materials, high dynamic cycles and fatigue resistance capabilities, low moisture absorption, and good hydrolytic stability lend these materials to applications such as parts found in nuclear plants, oil wells, high-pressure steam valves, chemical plants, and airplane and automobile engines.

One of the two ether linkages in PEEK is not present in PEK and the ensuing loss of some molecular flexibility results in PEK having an even higher T_m and heat-distortion temperature than PEEK. A relatively higher ketonic concentration in the repeat unit results in high ultimate tensile properties as well. A comparison of different aromatic polyether ketones is given in Table 1.7.[264,265] As these properties are from different sources, strict comparison between the data is not advisable due to likely differing testing techniques.

Glass and carbon fiber reinforcements are the most important fillers for all of the PEK family. While elastic extensibility is sacrificed, the additional heat resistance and moduli improvements allow glass or carbon fiber formulations entry into many applications. PEK is polymerized either through self-condensation of the structure in Fig. 1.32*a*, or via the reaction of intermediates as in Fig. 1.32*b*. Since these polymers can crystallize and tend, therefore, to precipitate from the reactant mixture, they must be reacted in high boiling solvents close to the 320°C melt temperature.[266]

1.2.19 Poly(methyl methacrylate) (PMMA)

Poly(methyl methacrylate) is a transparent thermoplastic material of moderate mechanical strength and outstanding outdoor weather resistance. It is available as sheet, tubes, or rods which can be machined, bonded, and formed into a variety of different parts and in bead form which can be conventionally processed via extrusion or

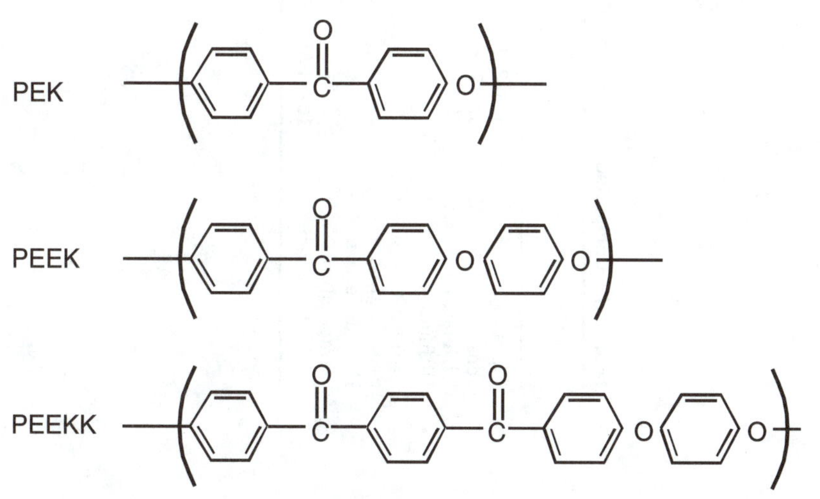

Figure 1.31 Structures of PEK, PEEK, and PEEKK.

TABLE 1.7 Comparison of Selected PEK, PEEK, and PEEKK Properties

	PEK unfilled	30% glass-filled PEK	PEEK unfilled	30% glass-filled PEEK	PEEKK unfilled	30% glass-filled PEEKK
T_m, °C	323–381	329–381	334	334	365	—
Tensile modulus, MPa	3,585–4,000	9,722–12,090	—	8,620–11,030	4,000	13,500
Ultimate elongation, %	50	2.2–3.4	30–150	2–3	—	—
Ultimate tensile strength, MPa	103	—	91	—	86	168
Specific gravity	1.3	1.47–1.53	1.30–1.32	1.49–1.54	1.3	1.55
Heat-deflection temperature, °C at 264 lb/in^2	162–170	326–350	160	288–315	160	>320

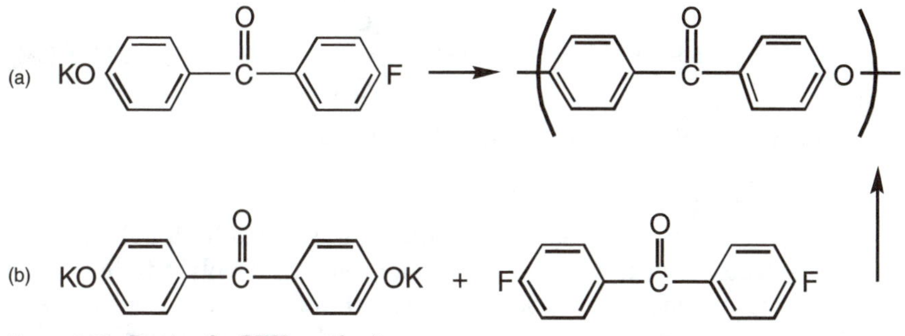

Figure 1.32 Routes for PEK synthesis.

injection molding. The sheet form material is polymerized in situ by casting a monomer which has been partly prepolymerized by removing any inhibitor, heating, and adding an agent to initiate the free-radical polymerization. This agent is typically a peroxide. This mixture of polymer and monomer is then poured into the sheet mold and the plates are brought together and reinforced to prevent bowing to ensure the final product will be of uniform thickness and flatness. This bulk polymerization process generates such high molecular weight material that the sheet or rod will decompose prior to melting. As such, this technique is not suitable for producing injection-molding grade resin, but it does aid in producing material which has a large rubbery plateau and has a high enough elevated temperature strength to allow for band-sawing, drilling, and other common machinery practices as long as the localized heating doesn't reach the polymer's decomposition temperature.

Suspension polymerization provides a final polymer with a low enough molecular weight to allow for typical melt processing. In this process, methyl methacrylate monomer is suspended in water to which the peroxide is added along with emulsifying/suspension agents, protective colloids, lubricants, and chain transfer agents to aid in molecular weight control. The resultant bead can then be dried and is ready for injection molding, or it can be further compounded with any desired colorants, plasticizers, rubber-modifier, as required.[267] Number-average molecular weights from the suspension process are approximately 60,000 g/mol, while the bulk polymerization process can result in number-average molecular weights of approximately 1 million g/mol.[268]

Typically, applications for PMMA optimize use of its clarity, with an up to 92% light transmission, depending upon the thickness of the sample. Again, because it has such strong weathering behavior it is well suited for applications such as automobile rear-light housings, lenses, aircraft cockpits, helicopter canopies, dentures, steering wheel bosses, and windshields. Cast PMMA is used extensively as bathtub materials, in showers, and in whirlpools.[269]

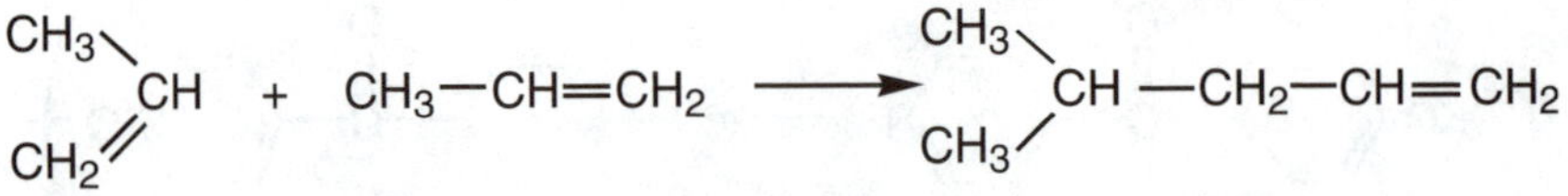

$$\begin{matrix} CH_3 \diagdown \\ \quad CH \\ CH_2 \diagup\!\!\diagup \end{matrix} + CH_3{-}CH{=}CH_2 \longrightarrow \begin{matrix} CH_3 \diagdown \\ \quad CH \\ CH_3 \diagup \end{matrix} {-}CH_2{-}CH{=}CH_2$$

Figure 1.33 Polymerization route for polymethylpentene.

Since the homopolymer is fairly brittle, PMMA can be toughened via copolymerization with another monomer (such as polybutadiene), or blended with an elastomer in the same way that high-impact polystyrene is to enable better stress distribution via the elastomeric domain.

1.2.20 Polymethylpentene (PMP)

Polymethylpentene was introduced in the mid-1960s by ICI, and is now marketed under the same tradename, TPX, by Mitsui Petrochemical Industries. The most significant commercial polymerization method involves the dimerization of propylene, as shown in Fig. 1.33.

As a polyolefin, this material offers chemical resistance to mineral acids, alkaline solutions, alcohols, and boiling water. It is not resistant to ketones, or aromatic and chlorinated hydrocarbons and like polyethylene and polypropylene it is susceptible to environmental stress cracking[270] and requires formulation with antioxidants. Its use is primarily in injection-molding and thermoforming applications, where the additional cost incurred compared to other polyolefins is justified by its high melt point (245°C), transparency, low density, and good dielectric properties. The high degree of transparency of polymethylpentene is attributed both to the similarities of the refractive indices of the amorphous and crystalline regions, as well as to the large coil size of the polymer due to the bulky branched four-carbon side chain. The free volume regions are large enough to allow light of visible region wavelengths to pass unimpeded. This degree of free volume is also responsible for the 0.83-g/cm^3 low density. As typically cooled, the polymer achieves about 40% crystallinity, although with annealing can reach 65% crystallinity.[271] The structure of the polymer repeat unit is shown in Fig. 1.34.

Voids are frequently formed at the crystalline/amorphous region interfaces during injection molding, rendering an often undesirable lack of transparency. In order to counter this, polymethylpentene is often copolymerized with hex-1-ene, oct-1-ene, dec-1-ene, and octadec-1-ene which reduces the voids and concomitantly reduces the melting point and degree of crystallinity.[272] Typical products made from polymethylpentene include transparent pipes and other chemical plant applications, sterilizable medical equipment, light fittings, and transparent housings.

1.2.21 Polyphenylene oxide

The term polyphenylene oxide (PPO) is a misnomer for a polymer which is more accurately named poly-(2,6-dimethyl-*p*-phenylene ether) and which in Europe is more commonly known as a polymer covered by the more generic term polyphenyleneether (PPE). This engineering polymer has high-temperature properties due to the large degree of aromaticity on the backbone, with dimethyl-substituted benzene rings joined by an ether linkage, as shown in Fig. 1.35.

The stiffness of this repeat unit results in a heat-resistant polymer with a T_g of 208°C and a T_m of 257°C. The fact that these two thermal transitions occur within such a short temperature span of each other means that PPO does not have time to crystallize while it cools before reaching a glassy state and as such is typically amorphous after processing.[273] Commercially available as PPO from General Electric, the polymer is sold in molecular weight ranges of 25,000 to 60,000 g/mol.[274] Properties which distinguish PPO from other engineering polymers are its high degree of hydrolytic and dimensional stabilities, which enable it to be molded with precision, albeit high processing temperatures are required. It finds application as television tuner strips, microwave insulation components, and transformer housings, which take advantage of its strong dielectric properties over wide temperature ranges. It is also used in applications which benefit from its hydrolytic stability including pumps, water meters, sprinkler systems, and hot-water tanks.[275] Its greater use is limited by the often-prohibitive cost, and General Electric responded by commercializing a PPO/PS blend marketed under the tradename Noryl. GE sells many

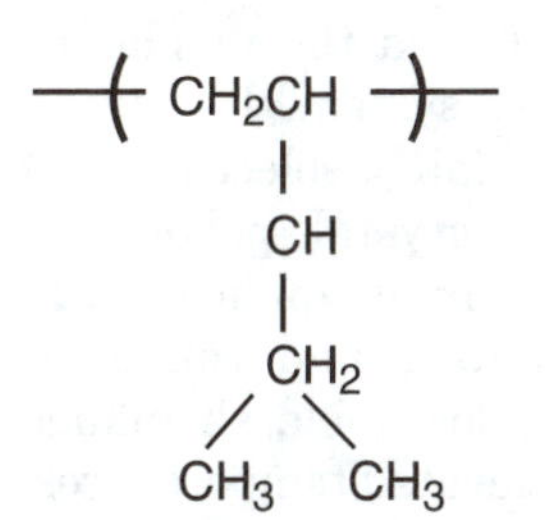

Figure 1.34 Repeat structure of polymethylpentene.

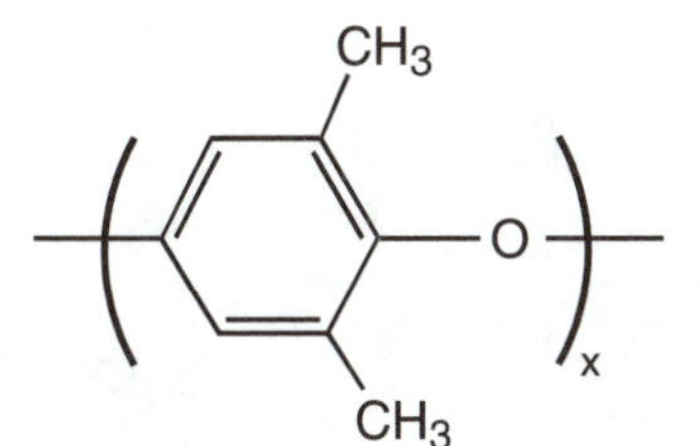

Figure 1.35 Repeat structure of PPO.

grades of Noryl based on different blend ratios and specialty formulations. The styrenic nature of PPO leads one to surmise very close compatibility (similar solubility parameters) with PS, although strict thermodynamic compatibility is questioned due to the presence of two distinct T_g peaks when measured by mechanical rather than calorimetric means.[276] The blends present the same high degree of dimensional stability, low water absorption, excellent resistance to hydrolysis, and good dielectric properties offered by PS, yet with the elevated heat-distortion temperatures which result from PPO's contribution. These polymers are more cost competitive than PPO and are used in moldings for dishwashers, washing machines, hair dryers, cameras, instrument housings, and as television accessories.[277]

1.2.22 Polyphenylene sulfide (PPS)

The structure of PPS, shown in Fig. 1.36, clearly indicates high temperature, high strength, and high chemical resistance due to the presence of the aromatic benzene ring on the backbone linked with the electronegative sulfur atom. In fact, the melt point of PPS is 288°C, and the tensile strength is 70 MPa at room temperature. The brittleness of PPS, due to the highly crystalline nature of the polymer, is often overcome by compounding with glass fiber reinforcements. Typical properties of PPS and a commercially available 40% glass-filled polymer blend are shown in Table 1.8.[278] The mechanical properties of PPS are similar to other engineering thermoplastics, such as polycarbonate and polysulphones, except that, as mentioned, the PPS suffers from the brittleness arising from its crystallinity, but does, however, offer improved resistance to environmental stress cracking.[279]

PPS is of most significant commercial interest as a thermoplastic, although it can be cross-linked into a thermoset system. Its strong inherent flame retardance puts this polymer in a fairly select class of polymers, including polyethersulphones, liquid crystal polyesters, polyketones, and polyetherimides.[280] As such, PPS finds application in electrical components, printed circuits, and contact and connector encapsulation. Other uses take advantage of the low mold shrinkage values and strong mechanical properties even at elevated temperatures. These include pump housings, impellers, bushings, and ball valves.[281]

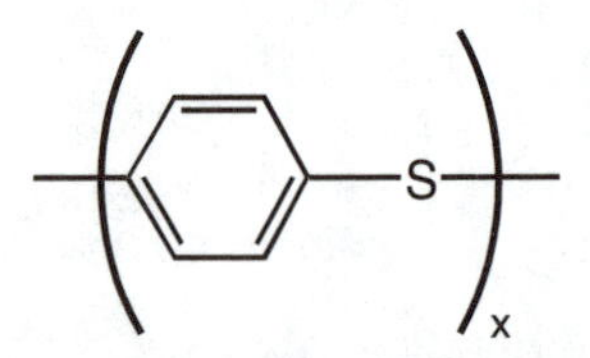

Figure 1.36 Repeat structure of polyphenylene sulfide.

TABLE 1.8 Selected Properties of PPS and GF PPS

Property, units	PPS	40% glass-filled PPS
T_g, °C	85	—
Heat distortion temperature, method A, °C	135	265
Tensile strength		
21°C, MPa	64–77	150
204°C, MPa	33	33
Elongation at break, %	3	2
Flexural modulus, MPa	3,900	10,500
Limiting oxygen index, %	44	47

1.2.23 Polyphthalamide (PPA)

Polyphthalamides were originally developed for use as fibers and later found application in other areas. They are semi-aromatic polyamides based on the polymerization of terephthalic acid or isophthalic acid and an amine.[282] Both amorphous and crystalline grades are available. Polyphthalamides are polar materials with a melting point near 310°C and a glass transition temperature of 127°C.[283] The material has good strength and stiffness along with good chemical resistance. Polyphthalamides can be attacked by strong acids or oxidizing agents and are soluble in cresol and phenol.[284] Polyphthalamides are stronger, less moisture sensitive, and possess better thermal properties when compared to the aliphatic polyamides such as nylon 6,6. However, polyphthalamide is less ductile than nylon 6,6, although impact grades are available.[285] Polyphthalamides will absorb moisture, decreasing the glass transition temperature and causing dimensional changes. The material can be reinforced with glass and has extremely good high-temperature performance. Reinforced grades of polyphthalamides are able to withstand continuous use at 180°C.[286]

The crystalline grades are generally used in injection molding, while the amorphous grades are often used as barrier materials.[287] The recommended mold temperatures are 135 to 165°C, with recommended melt temperatures of 320 to 340°C.[288] The material should have a moisture content of 0.15% or less for processing.[289] Because mold temperature is important to surface finish, higher mold temperatures may be required for some applications. Both crystalline and amorphous grades are available under the trade names Amodel (Amoco), and amorphous grades are available under the names Zytel (DuPont) and Trogamid (Dynamit Nobel). Crystalline grades are available under the trade name Arlen (Mitsui).[290]

Polyphthalamides are used in automotive applications where their chemical resistance and temperature stability are important.[291]

Examples include sensor housings, fuel line components, head lamp reflectors, electrical components, and structural components. Electrical components attached by infrared and vapor phase soldering are applications utilizing PPA's high-temperature stability. Switching devices, connectors, and motor brackets are often made from PPA. Mineral-filled grades are used in applications which require plating such as decorative hardware and plumbing. Impact modified grades of unreinforced PPA are used in sporting goods, oil field parts, and military applications.

1.2.24 Polypropylene (PP)

Polypropylene is a versatile polymer used in applications from films to fibers with a worldwide demand of over 21 million pounds.[292] It is similar to polyethylene in structure, except for the substitution of one hydrogen group with a methyl group on every other carbon. On the surface this change would appear trivial, but this one replacement now changes the symmetry of the polymer chain. This allows for the preparation of different stereoisomers, namely, syndiotactic, isotactic, and atactic chains. These configurations are shown in Figure 1.3 in the Introduction. Polypropylene (PP) is synthesized by the polymerization of propylene, a monomer derived from petroleum products through the reaction shown in Fig. 1.37.

It was not until Ziegler-Natta catalysts became available that polypropylene could be polymerized into a commercially viable product. These catalysts allowed the control of stereochemistry during polymerization to form polypropylene in the isotactic and syndiotactic forms, both capable of crystallizing into a more rigid, useful polymeric material.[293] The first commercial method for the production of polypropylene was a suspension process. Current methods of production include a gas phase process and a liquid slurry process.[294] New grades of polypropylene are now being polymerized using metallocene catalysts.[295] The range of molecular weights for PP is M_n = 38,000 to 60,000 and M_w = 220,000 to 700,000. The molecular weight distribution (M_n/M_w) can range from 2 to about 11.[296]

Different behavior can be found for each of the three stereoisomers. Isotactic and syndiotactic polypropylene can pack into a regular crystalline array giving a polymer with more rigidity. Both materials are crystalline, however, syndiotactic polypropylene has a lower T_m than the isotactic polymer.[297] The isotactic polymer is the most commercially used

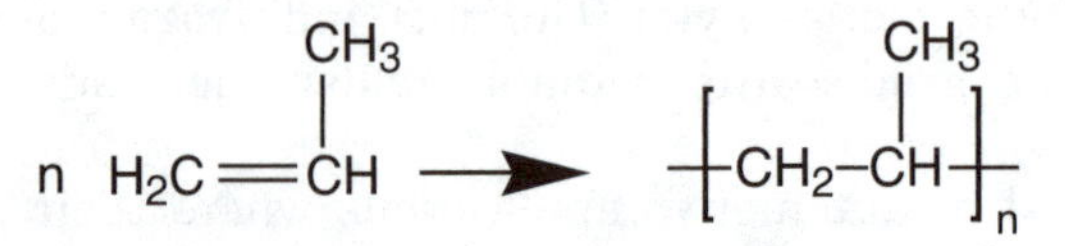

Figure 1.37 The reaction to prepare polypropylene.

form with a melting point of 165°C. Atactic polypropylene has a very small amount of crystallinity (5 to 10%) because its irregular structure prevents crystallization, thus, it behaves as a soft flexible material.[298] It is used in applications such as sealing strips, paper laminating, and adhesives.

Unlike polyethylene, which crystallizes in the planar zigzag form, isotactic polypropylene crystallizes in a helical form because of the presence of the methyl groups on the chain.[299] Commercial polymers are about 90 to 95% isotactic. The amount of isotacticity present in the chain will influence the properties. As the amount of isotactic material (often quantified by an isotactic index) increases, the amount of crystallinity will also increase, resulting in increased modulus, softening point, and hardness.

Although, in many respects, polypropylene is similar to polyethylene, since both are saturated hydrocarbon polymers, they differ in some significant properties. Isotactic polypropylene is harder and has a higher softening point than polyethylene, so it is used where higher stiffness materials are required. Polypropylene is less resistant to degradation, particularly high-temperature oxidation, than polyethylene, but has better environmental stress cracking resistance.[300] The decreased degradation resistance of PP is due to the presence of a tertiary carbon in PP, allowing for easier hydrogen abstraction compared to PE.[301] As a result, antioxidants are added to polypropylene to improve the oxidation resistance. The degradation mechanisms of the two polymers are also different. PE cross-links on oxidation, while PP undergoes chain scission. This is also true of the polymers when exposed to high energy radiation, a method commonly used to cross-link PE.

Polypropylene is one of the lightest plastics with a density of 0.905.[302] The nonpolar nature of the polymer gives PP low water absorption. Polypropylene has good chemical resistance, but liquids, such as chlorinated solvents, gasoline, and xylene, can affect the material. Polypropylene has a low dielectric constant and is a good insulator. Difficulty in bonding to polypropylene can be overcome by the use of surface treatments to improve the adhesion characteristics.

With the exception of UHMWPE, polypropylene has a higher T_g and melting point than polyethylene. Service temperature is increased, but PP needs to be processed at higher temperatures. Because of the higher softening, PP can withstand boiling water and can be used in applications requiring steam sterilization.[303] Polypropylene is also more resistant to cracking in bending than PE and is preferred in applications that require tolerance to bending. This includes applications such as ropes, tapes, carpet fibers, and parts requiring a living hinge. Living hinges are integral parts of a molded piece that are thinner and allow for bending.[304] One weakness of polypropylene is its low-temperature brittleness behavior, with the polymer becoming brittle near 0°C.[305] This can be improved through copolymerization with other polymers such as ethylene.

Comparing the processing behavior of PP to PE, it is found that polypropylene is more non-Newtonian than PE and that the specific heat of PP is lower than polyethylene.[306] The melt viscosity of PE is less temperature sensitive than PP.[307] Mold shrinkage is generally less than for PE, but is dependent on the actual processing conditions. Unlike many other polymers, an increase in molecular weight of polypropylene does not always translate into improved properties. The melt viscosity and impact strength will increase with molecular weight, but often with a decrease in hardness and softening point. A decrease in the ability of the polymer to crystallize as molecular weight increases is often offered as an explanation for this behavior.[308]

The molecular weight distribution (MWD) has important implications for processing. A PP grade with a broad MWD is more shear sensitive than a grade with a narrow MWD. Broad MWD materials will generally process better in injection-molding applications. In contrast, a narrow MWD may be preferred for fiber formation.[309] Various grades of polypropylene are available tailored to a particular application. These grades can be classified by flow rate, which depends on both average molecular weight and MWD. Lower flow rate materials are used in extrusion applications. In injection-molding applications, low flow rate materials are used for thick parts and high flow rate materials are used for thin-wall molding.

Polypropylene can be processed by methods similar to those used for PE. The melt temperatures are generally in the range of 210 to 250°C.[310] Heating times should be minimized to reduce the possibility of oxidation. Blow molding of PP requires the use of higher melt temperatures and shear, but these conditions tend to accelerate the degradation of PP. Because of this, blow molding of PP is more difficult than for PE. The screw metering zone should not be too shallow in order to avoid excessive shear. For a 60-mm screw the flights depths are typically about 2.25 and 3.0 mm for a 90-mm screw.[311]

In film applications, film clarity requires careful control of the crystallization process to ensure that small crystallites are formed. This is accomplished in blown film by extruding downwards into two converging boards. In the Shell TQ process the boards are covered with a film of flowing, cooling water. Oriented films of PP are manufactured by passing the PP film into a heated area and stretching the film both transversely and longitudinally. To reduce shrinkage the film may be annealed at 100°C while under tension.[312] Highly oriented films may show low transverse strength and a tendency to fibrillate. Other manufacturing methods for polypropylene include extruded sheet for thermoforming applications and extruded profiles.

If higher stiffness is required short glass reinforcement can be added. The use of a coupling agent can dramatically improve the properties of glass-filled PP.[313] Other fillers for polypropylene include calcium carbonate and talc, which can also improve the stiffness of PP.

Other additives such as pigments, antioxidants, and nucleating agents can be blended into polypropylene to give the desired properties. Carbon black is often added to polypropylene to impart UV resistance in outdoor applications. Antiblocking and slip agents may be added for film applications to decrease friction and prevent sticking. In packaging applications antistatic agents may be incorporated.

The addition of rubber to polypropylene can lead to improvements in impact resistance. One of the most commonly added elastomers is ethylene-propylene rubber. The elastomer is blended with polypropylene, forming a separate elastomer phase. Rubber can be added in excess of 50% to give elastomeric compositions. Compounds with less than 50% added rubber are of considerable interest as modified thermoplastics. Impact grades of PP can be formed into films with good puncture resistance.

Copolymers of polypropylene with other monomers are also available, the most common monomer being ethylene. Copolymers usually contain between 1 and 7 wt % of ethylene randomly placed in the polypropylene backbone. This disrupts the ability of the polymer chain to crystallize, giving more flexible products. This also improves the impact resistance of the polymer, decreases the melting point, and increases flexibility. The degree of flexibility increases with ethylene content, eventually turning the polymer into an elastomer (ethylene propylene rubber). The copolymers also exhibit increased clarity and are used in blow molding, injection molding, and extrusion.

Polypropylene has many applications. Injection-molding applications cover a broad range from automotive uses such as dome lights, kick panels, and car battery cases to luggage and washing machine parts. Filled PP can be used in automotive applications such as mounts and engine covers. Elastomer-modified PP is used in the automotive area for bumpers, fascia panels, and radiator grills. Ski boots are another application for these materials.[314] Structural foams, prepared with glass-filled PP, are used in the outer tank of washing machines. New grades of high-flow PPs are allowing manufacturers to mold high-performance housewares.[315] Polypropylene films are used in a variety of packaging applications. Both oriented and nonoriented films are used. Film tapes are used for carpet backing and sacks. Foamed sheet is used in a variety of applications including thermoformed packaging. Fibers are another important application for polypropylene, particularly in carpeting because of its low cost and wear resistance. Fibers prepared from polypropylene are used in both woven and nonwoven fabrics.

1.2.25 Polyurethanes (PUR)

Polyurethanes are very versatile polymers. They are used as flexible and rigid foams, elastomers, and coatings. Polyurethanes are available as both thermosets and thermoplastics; in addition, their hardnesses span the range from rigid material to elastomer. Thermoplastic polyurethanes will be the focus of this section.

The term *polyurethane* is used to cover materials formed from the reaction of isocyanates and polyols.[316] The general reaction for a polyurethane produced through the reaction of a diisocyanate with a diol is shown in Fig. 1.38.

Polyurethanes are phase-separated block copolymers, as depicted in Fig. 1.39, where the *A* and *B* portions represent different polymer segments. One segment, called the hard segment, is rigid, while the other, the soft segment, is elastomeric. In polyurethanes the soft segment is prepared from an elastomeric long-chain polyol, generally a polyester or polyether, but other rubbery polymers end-capped with a hydroxyl group could be used. The hard segment is composed of the diisocyanate and a short chain diol called a *chain extender.* The hard segments have high interchain attraction due to hydrogen bonding between the urethane groups; in addition, they may be capable of crystallizing.[317] The soft elastomeric segments are held together by the hard phases, which are rigid at room temperature and act as physical cross-links. The hard segments hold the material together at room temperature, but at processing temperatures the hard segments can flow and be processed.

The properties of polyurethanes can be varied by changing the type or amount of the three basic building blocks of the polyurethane—diisocyanate, short-chain diol, or long-chain diol. Given the same starting materials the polymer can be varied simply by changing the ratio of the hard and soft segments. This allows the manufacturer a great deal of flexibility in compound development for specific applications. The materials are typically manufactured by reacting a linear polyol with an excess of diisocyanate. The polyol is end-capped with isocyanate groups. The end-capped polyol and free isocyanate are then reacted with a chain extender, usually a short-chain diol to form the polyurethane.[318]

There are a variety of starting materials available for use in the preparation of polyurethanes, some of which are listed here:

Diisocyanates	Chain extenders	Polyols
4,4′-diphenylmethane diisocyanate (MDI)	1,4 butanediol	Polyesters
Hexamethylene diisocyanate (HDI)	Ethylene glycol	Polyethers
Hydrogenated 4,4′-diphenylmethane diisocyanate (HMDI)	1,6 hexanediol	

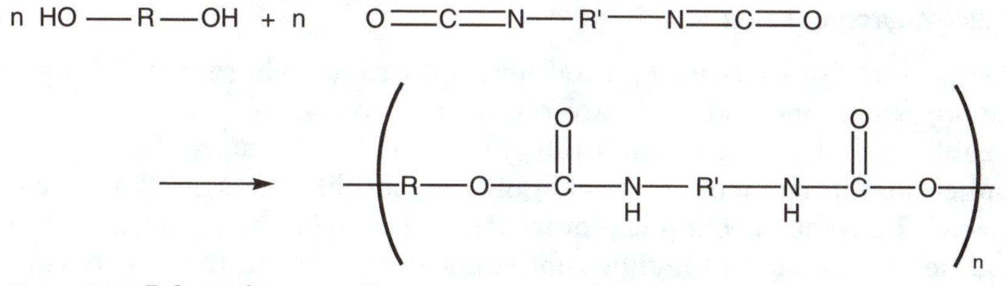

Figure 1.38 Polyurethane reaction.

[—A—B—A—B—A—B—]$_n$

Figure 1.39 Block structure of polyurethanes.

Polyurethanes are generally classified by the type of polyol used, for example, polyester polyurethane or polyether polyurethane. The type of polyol can affect certain properties. For example, polyether polyurethanes are more resistant to hydrolysis than polyester-based urethanes, while the polyester polyurethanes have better fuel and oil resistance.[319] Low-temperature flexibility can be controlled by proper selection of the long-chain polyol. Polyether polyurethanes generally have lower glass transition temperatures than polyester polyurethanes. The heat resistance of the polyurethane is governed by the hard segments. Polyurethanes are noted for their abrasion resistance, toughness, low-temperature impact strength, cut resistance, weather resistance, and fungus resistance.[320] Specialty polyurethanes include glass-reinforced products, fire-retardant grades, and UV-stabilized grades.

Polyurethanes find application in many areas. They can be used as impact modifiers for other plastics. Other applications include rollers or wheels, exterior body parts, drive belts, and hydraulic seals.[321] Polyurethanes can be used in film applications such as textile laminates for clothing and protective coatings for hospital beds. They are also used in tubing and hose in both unreinforced and reinforced forms because of their low-temperature properties and toughness. Their abrasion resistance allows them to be used in applications such as athletic shoe soles and ski boots. Polyurethanes are also used as coatings for wire and cable.[322]

Polyurethanes can be processed by a variety of methods including: extrusion, blow molding, and injection molding. They tend to pick up moisture and must be thoroughly dried prior to use. The processing conditions vary with the type of polyurethane; higher hardness grades usually require higher processing temperatures. Polyurethanes tend to exhibit shear sensitivity at lower melt temperatures. Postmold heating in an oven, shortly after processing, can often improve the properties of the finished product. A cure cycle of 16 to 24 h at 100°C is typical.[323]

1.2.26 Styrenic resins

The styrene family is well suited for applications where rigid, dimensionally stable molded parts are required. Polystyrene (PS) is a transparent, brittle, high modulus material with a multitude of applications, primarily in packaging, disposable cups, and medical ware. When the mechanical properties of the PS homopolymer are modified to produce a tougher, more ductile blend as in the case of rubber-modified high-impact grades of PS (HIPS), a far wider range of applications becomes available. HIPS is preferred for durable molded items including radio, television, and stereo cabinets as well as compact disk jewel cases. Copolymerization is also used to produce engineering grade plastics of higher performance as well as higher price, with acrylonitrile butadiene styrene (ABS) and styrene acrylonitrile (SAN) plastics being of greatest industrial importance.

Acrylonitrile butadiene styrene (ABS) terpolymer. As with any copolymers, there is tremendous flexibility in tailoring the properties of ABS by varying the ratios of the three monomers: acrylonitrile, butadiene, and styrene. The acrylonitrile component contributes heat resistance, strength, and chemical resistance. The elastomeric contribution of butadiene imparts higher-impact strength, toughness, low-temperature property retention and flexibility, while the styrene contributes rigidity, glossy finish, and ease of processability. As such, worldwide usage of ABS is surpassed only by that of the "big four" commodity thermoplastics (polyethylene, polypropylene, polystyrene, and polyvinyl chloride). Primary drawbacks to ABS include opacity, poor weather resistance, and poor flame resistance. Flame retardance can be improved by the addition of fire-retardant additives, or by blending ABS with PVC, with some reduction in ease of processability.[324] As its use is widely prevalent as equipment housings (such as telephones, televisions, and computers), these disadvantages are tolerated. Figure 1.40 shows the repeat structure of ABS.

Most common methods of manufacturing ABS include graft polymerization of styrene and acrylonitrile onto a polybutadiene latex, blending with a styrene acrylonitrile latex, and then coagulating and drying the resultant blend. Alternatively, the graft polymer of styrene, acrylonitrile, and polybutadiene can be manufactured separately from the styrene acrylonitrile latex and the two grafts blended and granulated after drying.[325]

Its ease of processing by a variety of common methods (including injection molding, extrusion, thermoforming, compression molding, and blow molding), combined with a good economic value for the mechanical properties achieved, results in widespread use of ABS. It is commonly found in under-the-hood automotive applications and in refrigerator linings, radios, computer housings, telephones, business machine housings, and television housings.

Acrylonitrile-chlorinated polyethylene-styrene (ACS) terpolymer. While ABS itself can be readily tailored by modifying the ratios of the three monomers and by modifying the lengths of each grafted segment, several companies are pursuing the addition of a fourth monomer, such as alpha-methylstyrene for enhanced heat resistance and methylmethacrylate to produce a transparent ABS. One such modification involves using chlorinated polyethylene in place of the butadiene segments. This terpolymer, ACS, has very similar properties to the engineering terpolymer ABS, but the addition of chlorinated polyethylene imparts improved flame retardance, weatherability, and resistance to electrostatic deposition of dust, without the addition of antistatic agents. The addition of the chlorinated olefin requires more care when injection molding to ensure that the chlorine does not dehydrohalogenate. Mold temperatures are recommended to be kept at between 190 and 210°C and not to exceed 220°C, and as with other chlorinated polymers such as polyvinyl chloride, that residence times be kept relatively short in the molding machine.[326] Applications for ACS include housings and parts for office machines such as desk-top calculators, copying machines, electronic cash registers, as well as housings for television sets, and video cassette recorders.[327]

Acrylic styrene acrylonitrile (ASA) terpolymer. Like ACS, ASA is a specialty product with similar mechanical properties to ABS but which offers improved outdoor weathering properties. This is due to the grafting of an acrylic ester elastomer onto the styrene acrylonitrile backbone. Sunlight usually combines with atmospheric oxygen to result in embrittlement and yellowing of thermoplastics and this process takes a much longer time in the case of ASA and, therefore, ASA finds applications in gutters, drain pipe fittings, signs, mail boxes, shutters, window trims, and outdoor furniture.[328]

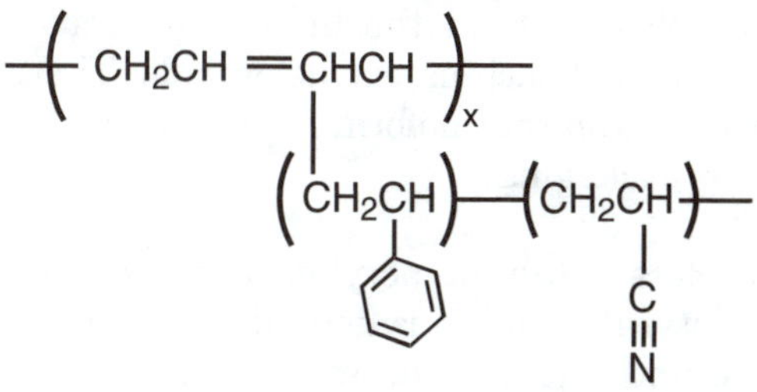

Figure 1.40 Repeat structure of ABS.

General purpose polystyrene (PS). PS is one of the four plastics whose combined usage accounts for 75% of the worldwide usage of plastics.[329] These four commodity thermoplastics are PE, PP, PVC, and PS. Although it can be polymerized via free-radical, anionic, cationic, and Ziegler mechanisms, commercially available PS is produced via free-radical addition polymerization. PS's popularity is due to its transparency, low density, relatively high modulus, excellent electrical properties, low cost, and ease of processing. The steric hindrance caused by the presence of the bulky benzene side groups results in brittle mechanical properties, with ultimate elongations only around 2 to 3%, depending upon molecular weight and additive levels. Most commercially available PS grades are atatic and, in combination with the large benzene groups, results in an amorphous polymer. The amorphous morphology provides not only transparency, but also the lack of crystalline regions means that there is no clearly defined temperature at which the plastic melts. PS is a glassy solid until its T_g of ~100°C is reached whereupon further heating softens the plastic gradually from a glass to a liquid. Advantage is taken of this gradual transition by molders who can eject parts which have cooled to beneath the relatively high Vicat temperature. Also, the lack of a heat of crystallization means that high heating and cooling rates can be achieved, which reduces cycle time and also promotes an economical process. Lastly, upon cooling PS does not crystallize the way PE and PP do. This gives PS low shrinkage values (0.004 to 0.005 mm/mm) and high dimensional stability during molding and forming operations.

Commercial PS is segmented into easy flow, medium flow, and high heat-resistance grades. Comparison of these three grades is made in Table 1.9. The easy flow grades have the lowest molecular weight to which 3 to 4% mineral oil have been added. The mineral oil reduces melt viscosity, which is well suited for increased injection speeds while molding inexpensive thin-walled parts such as disposable dinnerware, toys, and packaging. The reduction in processing time comes at the cost of a reduced softening temperature and a more brittle polymer. The medium flow grades have a slightly higher molecular weight and contain only 1 to 2% mineral oil. Applications include injection-molded tumblers, medical ware, toys, injection-blow–molded bottles, and extruded food packaging. The high heat-resistance plastics have the highest molecular weight and the lowest level of additives such as extrusion aids. These products are used in sheet extrusion and thermoforming, and extruded film applications for oriented food packaging.[330]

Styrene acrylonitrile (SAN) copolymers. Styrene acrylonitrile polymers are copolymers prepared from styrene and acrylonitrile monomers. The polymerization can be done under emulsion, bulk, or suspension conditions.[331] The polymers generally contain between 20 and 30% acrylonitrile.[332] The acrylonitrile content of the polymer influences the final properties with tensile strength, elongation, and heat distortion temperature increasing as the amount of acrylonitrile in the copolymer increases.

SAN copolymers are linear, amorphous materials with improved heat resistance over pure polystyrene.[333] The polymer is transparent, but may have a yellow color as the acrylonitrile content increases. The addition of a polar monomer, acrylonitrile, to the backbone gives these polymers better resistance to oils, greases, and hydrocarbons when compared to polystyrene.[334] Glass-reinforced grades of SAN are available for applications requiring higher modulus combined with lower mold shrinkage and lower coefficient of thermal expansion.[335]

As the polymer is polar, it should be dried before processing. It can be processed by injection molding into a variety of parts. SAN can also be processed by blow molding, extrusion, casting, and thermoforming.[336]

SAN competes with polystyrene, cellulose acetate, and polymethyl methacrylate. Applications for SAN include injection-molded parts for medical devices, PVC tubing connectors, dishwasher-safe products, and refrigerator shelving.[337] Other applications include packaging for the pharmaceutical and cosmetics markets, automotive equipment, and industrial uses.

Olefin-modified SAN. SAN can be modified with olefins, resulting in a polymer that can be extruded and injection molded. The polymer has good weatherability and is often used as a capstock to provide weatherability to less expensive parts such as swimming pools, spas, and boats.[338]

Styrene butadiene copolymers. Styrene butadiene polymers are block copolymers prepared from styrene and butadiene monomers. The

TABLE 1.9 Properties of Commercial Grades of General Purpose PS[413]

Property	Easy flow PS	Medium flow PS	High heat-resistance PS
M_w	218,000	225,000	300,000
M_n	74,000	92,000	130,000
Melt flow index, g/10 min	16	7.5	1.6
Vicat softening temperature, °C	88	102	108
Tensile modulus, MPa	3,100	2,450	3,340
Ultimate tensile strength, MPa	1.6	2.0	2.4

polymerization is performed using sequential anionic polymerization.[339] The copolymers are better known as thermoplastic elastomers, but copolymers with high styrene contents can be treated as thermoplastics. The polymers can be prepared as either a star block form or as a linear, multiblock polymer. The butadiene exists as a separate dispersed phase in a continuous matrix of polystyrene.[340] The size of the butadiene phase is controlled to be less than the wavelength of light resulting in clear materials. The resulting amorphous polymer is tough with good flex life, and low mold shrinkage. The copolymer can be ultrasonically welded, solvent welded, or vibration welded. The copolymers are available in injection-molding grades and thermoforming grades. The injection-molding grades generally contain a higher styrene content in the block copolymer. Thermoforming grades are usually mixed with pure polystyrene.

Styrene butadiene copolymers can be processed by injection molding, extrusion, thermoforming, and blow molding. The polymer does not need to be dried prior to use.[341] Styrene butadiene copolymers are used in toys, housewares, and medical applications.[342] Thermoformed products include disposable food packaging such as cups, bowls, "clam shells," deli containers, and lids. Blister packs and other display packaging also use styrene butadiene copolymers. Other packaging applications include shrink wrap and vegetable wrap.[343]

1.2.27 Sulfone-based resins

Sulfone resins refer to polymers containing -SO_2 groups along the backbone as depicted in Fig. 1.41. The *R* groups are generally aromatic. The polymers are usually yellowish, transparent, amorphous materials and are known for their high stiffness, strength, and thermal stability.[344] The polymers have low creep over a large temperature range. Sulfones can compete against some thermoset materials in performance, while their ability to be injection-molded offers an advantage.

The first commercial polysulfone was Udel (Union Carbide, now Amoco), followed by Astrel 360 (3M Company), which is termed a polyarylsulfone, and finally Victrex (ICI), a polyethersulfone.[345] Current manufacturers also include Amoco, Carborundum, and BASF, among others. The different polysulfones vary by the spacing between the aromatic groups, which in turn affects their T_g values and their heat-distortion temperatures. Commercial polysulfones are linear with high T_g values in the range of 180 to 250°C, allowing for continuous use from 150 to 200°C.[346] As a result, the processing temperatures of polysulfones are above 300°C.[347] Although the polymer is polar, it still has good electrical insulating properties. Polysulfones are resistant to high thermal and ionizing radiation. They are also resistant to most aqueous acids and alkalis, but may be attacked by concentrated sulfuric acid. The polymers have good hydrolytic stability and can withstand hot water and steam.[348] Polysulfones are tough materials, but they do exhibit notch sensitivity. The presence of the aromatic rings causes the polymer chain to be rigid. Polysulfones generally do not require the addition of flame retardants and usually emit low smoke.

The properties of the main polysulfones are generally similar, although polyethersulfones have better creep resistance at high temperatures, higher heat-distortion temperature, but more water absorption and higher density than the Udel type materials.[349] Glass fiber–filled grades of polysulfone are available as are blends of polysulfone with ABS.

Polysulfones may absorb water, leading to potential processing problems such as streaks or bubbling.[350] The processing temperatures are quite high and the melt is very viscous. Polysulfones show little change in melt viscosity with shear. Injection-molding melt temperatures are in the range of 335 to 400°C and mold temperatures in the range of 100 to 160°C. The high viscosity necessitates the use of large cross-sectional runners and gates. Purging should be done periodically as a layer of black, degraded polymer may build up on the cylinder wall, yielding parts with black marks. Residual stresses may be reduced by higher mold temperatures or by annealing. Extrusion and blow-molding grades of polysulfones have a higher molecular weight with blow-molding melt temperatures in the range of 300 to 360°C and mold temperatures between 70 and 95°C.

The good heat resistance and electrical properties of polysulfones allows them to be used in applications such as circuit boards and TV components.[351] Chemical and heat resistance are important properties for automotive applications. Hair dryer components can also be made from polysulfones. Polysulfones find application in ignition components and structural foams.[352] Another important market for polysulfones is microwave cookware.[353]

Polyaryl sulfone (PAS). This polymer differs from the other polysulfones in the lack of any aliphatic groups in the chain. The lack of aliphatic groups gives this polymer excellent oxidative stability as the aliphatic groups are more susceptible to oxidative degradation.[354]

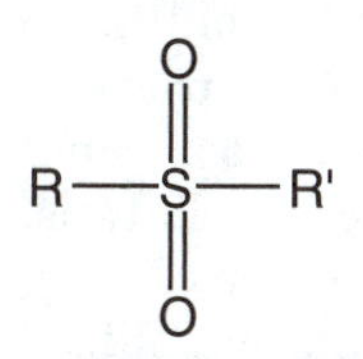

Figure 1.41 General structure of a polysulfone.

Polyaryl sulfones are stiff, strong, and tough polymers with very good chemical resistance. Most fuels, lubricants, cleaning agents, and hydraulic fluids will not affect the polymer.[355] However, methylene chloride, dimethyl acetamide, and dimethyl formamide will dissolve the polymer.[356] The glass transition temperature of these polymers is about 210°C with a heat-deflection temperature of 205°C at 1.82 MPa.[357] PAS also has good hydrolytic stability. Polyarylsulfone is available in filled and reinforced grades as well as both opaque and transparent versions.[358] This polymer finds application in electrical applications for motor parts, connectors, and lamp housings.[359]

The polymer can be injection-molded, provided the cylinder and nozzle are capable of reaching 425°C.[360] It may also be extruded. The polymer should be dried prior to processing. Injection-molding barrel temperatures should be 270 to 360°C at the rear, 295 to 390°C in the middle, and 300 to 395°C at the front.[361]

Polyether sulfone (PES). Polyether sulfone is a transparent polymer with high-temperature resistance and self-extinguishing properties.[362] It gives off little smoke when burned. Polyether sulfone has the basic structure shown in Fig. 1.42.

Polyether sulfone has a T_g near 225°C and is dimensionally stable over a wide range of temperatures.[363] It can withstand long-term use up to 200°C and can carry loads for long times up to 180°C.[364] Glass fiber–reinforced grades are available for increased properties. It is resistant to most chemicals with the exception of polar, aromatic hydrocarbons.[365]

Polyether sulfone can be processed by injection molding, extrusion, blow molding, or thermoforming.[366] It exhibits low mold shrinkage. For injection molding, barrel temperatures of 340 to 380°C with melt temperatures of 360°C are recommended.[367] Mold temperatures should be in the range of 140 to 180°C. For thin-walled molding higher temperatures may be required. Unfilled PES can be extruded into sheets, rods, films, and profiles.

PES finds application in aircraft interior parts due to its low smoke emission.[368] Electrical applications include switches, integrated circuit carriers, and battery parts.[369] The high-temperature oil and gas resistance allow polyether sulfone to be used in the automotive markets for water pumps, fuse housings, and car heater fans. The ability of PES to endure repeated sterilization allows PES to be used in a variety of medical applications, such as parts for centrifuges and root canal drills. Other applications include membranes for kidney dialysis, chemical separation, and desalination. Consumer uses include cooking equipment and lighting fittings. PES can also be vacuum metallized for a high-gloss mirror finish.

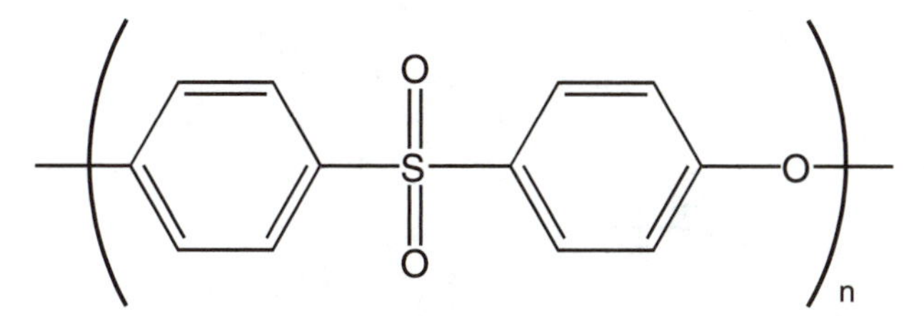

Figure 1.42 Structure of polyether sulfone.

Polysulfone (PSU). Polysulfone is a transparent thermoplastic prepared from bisphenol A and 4,4′-dichlorodiphenylsulfone.[370] The structure is shown in Fig. 1.43. It is self-extinguishing and has a high heat-distortion temperature. The polymer has a glass transition temperature of 185°C.[371] Polysulfones have impact resistance and ductility below 0°C. Polysulfone also has good electrical properties. The electrical and mechanical properties are maintained to temperatures near 175°C. Polysulfone shows good chemical resistance to alkali, salt, and acid solutions.[372] It has resistance to oils, detergents, and alcohols, but polar organic solvents and chlorinated aliphatic solvents may attack the polymer. Glass- and mineral-filled grades are available.[373]

Properties, such as physical aging and solvent crazing, can be improved by annealing the parts.[374] This also reduces molded-in stresses. Molded-in stresses can also be reduced by using hot molds during injection molding. As mentioned previously, runners and gates should be as large as possible due to the high melt viscosity. The polymer should hit a wall or pin shortly after entering the cavity of the mold as polysulfone has a tendency toward jetting. For thin-walled or long parts, multiple gates are recommended.

For injection-molding barrel temperatures should be in the range of 310 to 400°C, with mold temperatures of 100 to 170°C.[375] In blow molding the screw type should have a low compression ratio, 2.0:1 to 2.5:1. Higher compression ratios will generate excessive frictional heat. Mold temperatures of 70 to 95°C with blow air pressures of 0.3 to 0.5 MPa are generally used. Polysulfone can be extruded into films, pipe, or wire coatings. Extrusion melt temperatures should be from 315 to 375°C. High compression ratio screws should not be used for extrusion. Polysulfone shows high melt strength, allowing for good drawdown and the manufacture of thin films. Sheets of polysulfone can be thermoformed, with surface temperatures of 230 to 260°C recommended. Sheets may be bonded by heat sealing, adhesive bonding, solvent fusion, or ultrasonic welding.

Polysulfone is used in applications requiring good high-temperature resistance such as coffee carafes, piping, sterilizing equipment, and microwave oven cookware.[376] The good hydrolytic stability of polysulfone

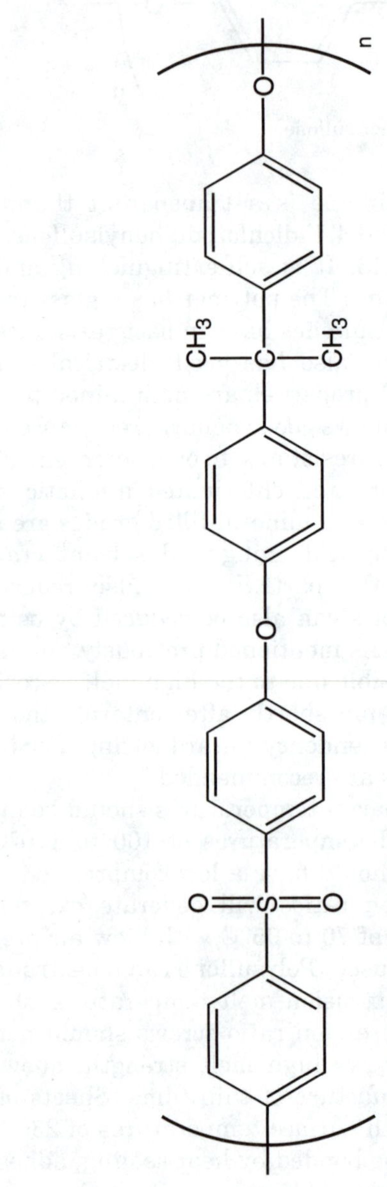

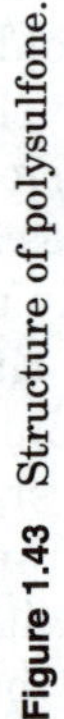

Figure 1.43 Structure of polysulfone.

is important in these applications. Polysulfone is also used in electrical applications for connectors, switches, and circuit boards and in reverse osmosis applications as a membrane support.[377]

1.2.28 Vinyl-based resins

Polyvinyl chloride. Polyvinyl chloride polymers (PVC), generally referred to as *vinyl resins,* are prepared by the polymerization of vinyl chloride in a free-radical addition polymerization reaction. The vinyl chloride monomer is prepared by reacting ethylene with chlorine to form 1,2-dichloroethane.[378] The 1,2 dichloroethane is then cracked to give vinyl chloride. The polymerization reaction is depicted in Fig. 1.44.

The polymer can be made by suspension, emulsion, solution, or bulk polymerization methods. Most of the PVC used in calendering, extrusion, and molding is prepared by suspension polymerization. Emulsion-polymerized vinyl resins are used in plastisols and organisols.[379] Only a small amount of commercial PVC is prepared by solution polymerization. The microstructure of PVC is mostly atactic, but a sufficient quantity of syndiotactic portions of the chain allows for a low fraction of crystallinity (about 5%). The polymers are essentially linear, but a low number of short-chain branches may exist.[380] The monomers are predominantly arranged head to tail along the backbone of the chain. Due to the presence of the chlorine group PVC polymers are more polar than polyethylene. The molecular weights of commercial polymers are M_w = 100,000 to 200,000 and M_n = 45,000 to 64,000.[381] Thus, $M_w/M_n = 2$ for these polymers. The polymeric PVC is insoluble in the monomer, therefore, bulk polymerization of PVC is a heterogeneous process.[382] Suspension PVC is synthesized by suspension polymerization. These are suspended droplets, approximately 10 to 100 nm in diameter, of vinyl chloride monomer in water. Suspension polymerizations allow control of particle size, shape, and size distribution by varying the dispersing agents and stirring rate. Emulsion polymerization results in much smaller particle sizes than suspension polymerized PVC, but soaps used in the emulsion polymerization process can affect the electrical and optical properties.

The glass transition temperature of PVC varies with the polymerization method, but falls within the range of 60 to 80°C.[383] PVC is a self-extinguishing polymer and, therefore, has application in the field of wire and cable. PVC's good flame resistance results from removal of HCl from the chain, releasing HCl gas.[384] Air is restricted from reaching the flame because HCl gas is denser than air. Because PVC is

$$n\ CH_2{=}CHCl \rightarrow -(CH_2\text{-}CHCl)_n-$$

Figure 1.44 Synthesis of polyvinyl chloride.

thermally sensitive, the thermal history of the polymer must be carefully controlled to avoid decomposition. At temperatures above 70°C degradation of PVC by loss of HCl can occur, resulting in the generation of unsaturation in the backbone of the chain. This is indicated by a change in the color of the polymer. As degradation proceeds, the polymer changes color from yellow to brown to black, visually indicating that degradation has occurred. The loss of HCl accelerates the further degradation and is called *autocatalytic decomposition.* The degradation can be significant at processing temperatures if the material has not been heat stabilized so thermal stabilizers are often added at additional cost to PVC to reduce this tendency. UV stabilizers are also added to protect the material from ultraviolet light, which may also cause the loss of HCl.

There are two basic forms of PVC—rigid and plasticized. Rigid PVC, as its name suggests, is an unmodified polymer and exhibits high rigidity.[385] Unmodified PVC is stronger and stiffer than PE and PP. Plasticized PVC is modified by the addition of a low molecular weight species (plasticizer) to flexibilize the polymer.[386] Plasticized PVC can be formulated to give products with rubbery behavior.

PVC is often compounded with additives to improve the properties. A wide variety of applications for PVC exist because one can tailor the properties by proper selection of additives. As mentioned previously, one of the principal additives is stabilizers. Lead compounds are often added for this purpose, reacting with the HCl released during degradation.[387] Among the lead compounds commonly used are basic lead carbonate or white lead and tribasic lead sulfate. Other stabilizers include metal stearates, ricinoleates, palmitates, and octoates. Of particular importance are the cadmium-barium systems with synergistic behavior. Organo-tin compounds are also used as stabilizers to give clear compounds. In addition to stabilizers, other additives such as fillers, lubricants, pigments, and plasticizers are used. Fillers are often added to reduce cost and include talc, calcium carbonate, and clay.[388] These fillers may also impart additional stiffness to the compound.

The addition of plasticizers lowers the T_g of rigid PVC, making it more flexible. A wide range of products can be manufactured by using different amounts of plasticizer. As the plasticizer content increases, there is usually an increase in toughness and a decrease in the modulus and tensile strength.[389] Many different compounds can be used to plasticize PVC, but the solvent must be miscible with the polymer. A compatible plasticizer is considered a nonvolatile solvent for the polymer. The absorption of solvent may occur automatically at room temperature or may require the addition of slight heat and mixing. PVC plasticizers are divided into three groups depending on their compatibility with the polymer: primary plasticizers, secondary plasticizers, and extenders. *Primary plasticizers* are compatible (have similar solubility parameters) with the polymer and should not exude. If the plasticizer and polymer have differences in their solubility parameters, they tend to be incompatible or have limited compatibility and are called *secondary plasticizers.* Secondary plasticizers are added along with the primary plasticizer to meet a secondary performance requirement (cost, low-temperature properties, permanence). The plasticizer can still be used in mixtures with a primary plasticizer provided the mixture has a solubility parameter within the desired range. *Extenders* are used to lower the cost and are generally not compatible when used alone. Common plasticizers for PVC include dioctyl phthalate, and di-iso-octyl phthalate, and dibutyl phthalate among others.[390]

The plasticizer is normally added to the PVC before processing. Since the plasticizers are considered solvents for PVC, they will normally be absorbed by the polymer with only a slight rise in temperature.[391] This reduces the time the PVC is exposed to high temperatures and potential degradation. In addition, the plasticizer reduces the T_g and T_m, therefore, lowering the processing temperatures and thermal exposure. Plasticized PVC can be processed by methods, such as extrusion and calendering, into a variety of products.

Rigid PVC can be processed using most conventional processing equipment. Because HCl can be given off in small amounts during processing, corrosion of metal parts is a concern. Metal molds, tooling, and screws should be inspected regularly. Corrosion-resistant metals and coatings are available but add to the cost of manufacturing. Rigid PVC products include house siding, extruded pipe, thermoformed, and injection-molded parts. Rigid PVC is calendered into credit cards. Plasticized PVC is used in applications such as flexible tubing, floor mats, garden hose, shrink wrap, and bottles. PVC joints can be solvent welded, rather than heated in order to fuse the two part together. This can be an advantage when heating the part is not feasible.

Chlorinated PVC (CPVC). Postchlorination of PVC was practiced during World War II.[392] CPVC can be prepared by passing chlorine through a solution of PVC. The chlorine adds to the carbon that does not already have a chlorine atom present. Commercial materials have chlorine contents around 66 to 67%. The materials have a higher softening point and higher viscosity than PVC, and are known for good chemical resistance. Compared to PVC, chlorinated PVC has a higher modulus and tensile strength. Compounding processes are similar to those for PVC but are more difficult.

Chlorinated PVC can be extruded, calendered, or injection-molded.[393] The extrusion screw should be chrome-plated or stainless steel. Dies should be streamlined. Injection molds should be chrome or nickel plated or stainless steel. CPVC is used for water distribution piping, industrial chemical liquid piping, outdoor skylight frames, automotive interior parts, and a variety of other applications.

Copolymers. Vinyl chloride can be copolymerized with vinyl acetate giving a polymer with a lower softening point and better stability than pure PVC.[394] The compositions can vary from 5 to 40% vinyl acetate content. This material has application in areas where PVC is too rigid and the use of plasticized PVC is unacceptable. Flooring is one application for these copolymers. Copolymers with about 10% vinylidene chloride and copolymers with 10 to 20% diethyl fumarate or diethyl maleate are also available.

Dispersion PVC. If a sufficient quantity of solvent is added to PVC, it can become suspended in the solvent, giving a fluid that can be used in coating applications.[395] This form of PVC is called a *plastisol* or *ogan-isol.* PVC in the fluid form can be processed by methods such as spread coating, rotational casting, dipping, and spraying. The parts are then dried with heat to remove any solvent and fuse the polymer. Parts, such as handles for tools and vinyl gloves, are produced by this method.

The plastisols or organisols are prepared from PVC produced through emulsion polymerization.[396] The latex is then spray dried to form particles from 0.1 to 1 μm. These particles are then mixed with plasticizers to make plastisols or with plasticizers and other volatile organic liquids to make organisols. Less plasticizer is required with the organisols so that harder coatings can be produced. The polymer particles are not dissolved in the liquid, but remain dispersed until the material is heated and fused. Other additives, such as stabilizers and fillers, may be compounded into the dispersion.

As plasticizer is added, the mixture goes through different stages as the voids between the polymer particles are filled.[397] Once all the voids between particles have been filled, the material is considered a paste. In these materials the size of the particle is an important variable. If the particles are too large, they may settle out so small particles are preferred. Very small particles have the disadvantage that the particles will absorb the plasticizer with time, giving a continuous increase in viscosity of the mixture. Paste polymers have particle sizes in the range of 0.2 to 1.5 μm. Particle size distribution will also affect the paste. It is usually better to have a wide particle size distribution so that particles can pack efficiently. This reduces the void space that must be filled by the plasticizer, and any additional plasticizer will act as a lubricant. For a fixed particle/plasticizer ratio a wide distribution will generally have a lower viscosity than for a constant particle size. In some cases very large particles are added to the paste as they will take up volume, again

$$n\ CH_2{=}CCl_2 \rightarrow (\text{-}CH_2\text{-}CCl_2\text{-})_n$$

Figure 1.45 Preparation of vinylindene chloride polymers.

reducing the amount of plasticizer required. These particles are made by suspension polymerization. With the mixture of particle sizes these larger particles will not settle out as they would if used alone. Plastisols and organisols require the addition of heat to fuse. Temperatures in the range of 300 to 410°F are used to form the polymer.

Vinylidene chloride. Polyvinylidene chloride (PVDC) is similar to PVC except that two chlorine atoms are present on one of the carbon groups.[398] Like PVC, PVDC is also polymerized by addition polymerization methods. Both emulsion and suspension polymerization methods are used. The reaction is shown in Fig. 1.45. The emulsion polymers are either used directly as a latex or dried for use in coatings or melt processing.

This material has excellent barrier properties and is frequently used in food packaging applications. Films made from PVDC have good cling properties, which is an advantage for food wraps. Commercial polymers are all copolymers of vinylidene chloride with vinyl chloride, acrylates, or nitriles. Copolymerization of vinylidene chloride with other monomers reduces the melting point to allow easier processing. Corrosion-resistant materials should be considered for use when processing PVDC.

1.3 Comparative Properties of Thermoplastics

Representative properties of selected thermoplastics are shown in Table 1.10. In cases where a range of values were given, the average value was listed.

1.4 Additives

There is a broad range of additives for thermoplastics. Some of the more important additives include plasticizers, lubricants, anti-aging additives, colorants, flame retardants, blowing agents, cross-linking agents, and UV protectants. Fillers are also considered additives but are covered separately later.

Plasticizers are considered nonvolatile solvents.[399] They act to soften a material by separating the polymer chains allowing them to be more flexible. As a result, the plasticized polymer is softer with greater extensibility. Plasticizers reduce the melt viscosity and glass transition temperature of the polymer. In order for the plasticizer to be a "solvent" for the polymer, it is necessary for the solubility parameter of

TABLE 1.10 Comparative Properties of Thermoplastics[414,415]

Material	Heat deflection temperature @1.82 MPa, °C	Tensile strength, MPa	Tensile modulus, GPa	Impact strength, J/m	Density, g/cm³	Dielectric strength, MV/m	Dielectric constant @ 60 Hz
ABS	99	41	2.3	347	1.18	15.7	3.0
CA	68	37.6	1.26	210	1.30	16.7	5.5
CAB	69	34	0.88	346	1.19	12.8	4.8
PTFE		17.1	0.36	173	2.2	17.7	2.1
PCTFE		50.9	1.3	187	2.12	22.2	2.6
PVDF	90	49.2	2.5	202	1.77	10.2	10.0
PB	102	25.9	0.18	NB*	0.91		2.25
LDPE	43	11.6	0.17	NB	0.92	18.9	2.3
HDPE	74	38.2		373	0.95	18.9	2.3
PMP		23.6	1.10	128	0.83	27.6	
PI		42.7	3.7	320	1.43	12.2	4.1
PP	102	35.8	1.6	43	0.90	25.6	2.2
PUR	68	59.4	1.24	346	1.18	18.1	6.5
PS	93	45.1	3.1	59	1.05	19.7	2.5
PVC–rigid	68	44.4	2.75	181	1.4	34.0	3.4
PVC–flexible		9.6		293	1.4	25.6	5.5
POM	136	69	3.2	133	1.42	19.7	3.7
PMMA	92	72.4	3	21	1.19	19.7	3.7
Polyarylate	155	68	2.1	288	1.19	15.2	3.1
LCP	311	110	11	101	1.70	20.1	4.6
Nylon 6	65	81.4	2.76	59	1.13	16.5	3.8
Nylon 6/6	90	82.7	2.83	53	1.14	23.6	4.0
PBT	54	52	2.3	53	1.31	15.7	3.3
PC	129	69	2.3	694	1.20	15	3.2
PEEK	160	93.8	3.5	59	1.32		
PEI	210	105	3	53	1.27	28	3.2
PES	203	84.1	2.6	75	1.37	16.1	3.5
PET	224	159	8.96	101	1.56	21.3	3.6
PPO (modified)	100	54	2.5	267	1.09	15.7	3.9
PPS	260	138	11.7	69	1.67	17.7	3.1
PSU	174	73.8	2.5	64	1.24	16.7	3.5

*No break.

the plasticizer to be similar to the polymer. As a result, the plasticizer must be selected carefully so it is compatible with the polymer. One of the primary applications of plasticizers is for the modification of PVC. In this case the plasticizers are divided into three classes, namely, primary and secondary plasticizers and extenders.[400] Primary plasticizers are compatible, can be used alone, and will not exude from the polymer. They should have a solubility parameter similar to the polymer. Secondary plasticizers have limited compatibility and are generally used with a primary plasticizer. Extenders have limited compatibility and will exude from the polymer if used alone. They are usually used along with the primary plasticizer. Plasticizers are usually in the form of high-viscosity liquids. The plasticizer should be capable of withstanding the high processing temperatures without degradation and discoloration which would adversely affect the end product. The plasticizer should be capable of withstanding any environmental conditions that the final product will see. This might include UV exposure, fungal attack, or water. In addition, it is important that the plasticizer show low volatility and migration so that the properties of the plasticized polymer will remain relatively stable over time. There is a wide range of plasticizer types. Some typical classes include phthalic esters, phosphoric esters, fatty acid esters, fatty acid esters, polyesters, hydrocarbons, aromatic oils, and alcohols.

Lubricants are added to thermoplastics to aid in processing. High molecular weight thermoplastics have high viscosity. The addition of lubricants acts to reduce the melt viscosity to minimize machine wear and energy consumption.[401] Lubricants may also be added to prevent friction between molded products. Examples of these types of lubricants include graphite and molybdenum disulfide.[402] Lubricants that function by exuding from the polymer to the interface between the polymer and machine surface are termed *external lubricants.* Their presence at the interface between the polymer and metal walls acts to ease the processing. They have low compatibility with the polymer and may contain polar groups so that they have an attraction to metal. Lubricants must be selected based on the thermoplastic used. Lubricants may cause problems with clarity, ability to heat seal, and printing on the material. Examples of these lubricants include stearic acid or other carboxylic acids, paraffin oils, and certain alcohols and ketones for PVC. Low molecular weight materials that do not affect the solid properties, but act to enhance flow in the melt state, are termed *internal lubricants.* Internal lubricants for PVC include amine waxes, montan wax ester derivatives, and long-chain esters. Polymeric flow promoters are also examples of internal lubricants. They have solubility parameters similar to the thermoplastic, but lower viscosity at processing temperatures. They have little effect on the mechanical properties of the solid polymer. An example is the use of ethylene-vinyl acetate copolymers with PVC.

Anti-aging additives are incorporated to improve the resistance of the formulation. Examples of aging include attack by oxygen, ozone, dehydrochlorination, and UV degradation. Aging often results in changes in the structure of the polymer chain such as cross-linking, chain scission, addition of polar groups, or the addition of groups that cause discoloration. Additives are used to help prevent these reactions. Antioxidants are added to the polymer to stop the free-radical reactions that occur during oxidation. Antioxidants include compounds such as phenols and amines. Phenols are often used because they have less of a tendency to stain.[403] Peroxide decomposers are also added to improve the aging properties of thermoplastics. These include mecaptans, sulfonic acids, and zinc dialkylthiophosphate. The presence of metal ions can act to increase the oxidation rate, even in the presence of antioxidants. Metal deactivators are often added to prevent this from taking place. Chelating agents are added to complex with the metal ion.

The absorption of ultraviolet light by a polymer may lead to the production of free radicals. These radicals react with oxygen resulting in what is termed *photodegradation.* This leads to the production of chemical groups that tend to absorb ultraviolet light, increasing the amount of photodegradation. To reduce this effect UV stabilizers are added. One way to accomplish UV stabilization is by the addition of UV absorbers such as benzophenones, salicylates, and carbon black.[404] They act to dissipate the energy in a harmless fashion. Quenching agents react with the activated polymer molecule. Nickel chelates and hindered amines can be used as quenching agents. Peroxide decomposers may be used to aid in UV stability.

In certain applications flame resistance can be important. In this case flame retarders may be added.[405] They act by one of four possible mechanisms. They may act to chemically interfere with the propagation of flame, react or decompose to absorb heat, form a fire-resistant coating on the polymer, or produce gases which reduce the supply of air. Phosphates are an important class of flame retarders. Tritolyl phosphate and trixylyl phosphate are often used in PVC. Halogenated compounds, such as chlorinated paraffins, may also be used. Antimony oxide is often used in conjunction to obtain better results. Other flame retarders include titanium dioxide, zinc oxide, zinc borate, and red phosphorus. As with other additives the proper selection of a flame retarder will depend on the particular thermoplastic.

Colorants are added to produce color in the polymeric part. They are separated into pigments and dyes. Pigments are insoluble in the polymer, while dyes are soluble in the polymer. The particular color desired and the type of polymer will affect the selection of the colorants.

Blowing agents are added to the polymer to produce a foam or cellular structure.[406] They may be chemical blowing agents which decompose at certain temperatures and release a gas or they may be low-boiling liquids which become volatile at the processing temperatures. Gases may be introduced into the polymer under pressure and expand when the polymer is depressurized. Mechanical whipping and the incorporation of hollow glass spheres can also be used to produce cellular materials.

Peroxides are often added to produce cross-linking in a system. Peroxides can be selected to decompose at a particular temperature for the application. Peroxides can be used to cross-link saturated polymers.

1.5 Fillers

The term *fillers* refers to solid additives, which are incorporated into the plastic matrix.[407] They are generally inorganic materials, and can be classified according to their effect on the mechanical properties of the resulting mixture. Inert or extender fillers are added mainly to reduce the cost of the compound, while reinforcing fillers are added in order to improve certain mechanical properties such as modulus or tensile strength.

Although termed inert, inert fillers can nonetheless affect other properties of the compound besides cost. In particular, they may increase the density of the compound, lower the shrinkage, increase the hardness, and increase the heat-deflection temperature. Reinforcing fillers typically will increase the tensile, compressive, and shear strength; increase the heat-deflection temperature; lower shrinkage; increase the modulus; and improve the creep behavior. Reinforcing fillers improve the properties via several mechanisms. In some cases a chemical bond is formed between the filler and the polymer, while in other cases the volume occupied by the filler affects the properties of the thermoplastic. As a result, the surface properties and interaction between the filler and the thermoplastic are of great importance. Certain properties of the fillers are of particular importance. These include the particle shape, the particle size and distribution of sizes, and the surface chemistry of the particle. In general, the smaller the particle, the higher the mechanical property of interest (such as tensile strength).[408] Larger particles may give reduced properties compared to the pure thermoplastic. Particle shape can also influence the properties. For example, platelike particles or fibrous particles may be oriented during processing. This may result in properties that are anisotropic. The surface chemistry of the particle is important to promote interaction with the polymer and allow for good interfacial adhesion. It is important that the polymer wet the particle surface and have good interfacial bonding in order to obtain the best property enhancement.

Examples of inert or extender fillers include china clay (kaolin), talc, and calcium carbonate. Calcium carbonate is an important filler with a particle size of about 1 μm.[409] It is a natural product from sedimentary rocks and is separated into chalk, limestone, and marble. In some cases the calcium carbonate may be treated to improve the bonding with the thermoplastic. Glass spheres are also used as thermoplastic fillers. They may be either solid or hollow, depending on the particular application. Talc is an important filler with a lamellar particle shape.[410] It is a natural, hydrated magnesium silicate with good slip properties. Kaolin and mica are also natural materials with lamellar structure. Other fillers include wollastonite, silica, barium sulfate, and metal powders. Carbon black is used as a filler primarily in the rubber industry, but it also finds application in thermoplastics for conductivity, UV protection, and as a pigment. Fillers in fiber form are often used in thermoplastics. Types of fibers include cotton, wood flour, fiberglass, and carbon. Table 1.11 shows the fillers and their forms.

1.6 Polymer Blends

There is considerable interest in polymer blends. This is driven by consideration of the difficulty in developing new polymeric materials from

TABLE 1.11 Forms of Various Fillers

Spherical	Lamellar	Fibrous
Sand/quartz powder	Mica	Glass fibers
Silica	Talc	Asbestos
Glass spheres	Graphite	Wollastonite
Calcium carbonate	Kaolin	Carbon fibers
Carbon black		Whiskers
Metallic oxides		Cellulose
		Synthetic fibers

monomers. In many cases it can be more cost effective to tailor the properties of a material through the blending of existing materials. One of the most basic questions in blends is whether or not the two polymers are miscible or exist as a single phase. In many cases the polymers will exist as two separate phases. In this case the morphology of the phases is of great importance. In the case of a miscible single-phase blend there is a single T_g, which is dependent on the composition of the blend.[411] Where two phases exist, the blend will exhibit two separate T_g values, one for each of the phases present. In the case where the polymers can crystallize, the crystalline portions will exhibit a melting point (T_m), even in the case where the two polymers are a miscible blend.

Although miscible blends of polymers exist, most blends of high molecular weight polymers exist as two-phase materials. Control of the morphology of these two-phase systems is critical to achieve the desired properties. A variety of morphologies exist such as dispersed spheres of one polymer in another, lamellar structures, and co-continuous phases. As a result, the properties depend in a complex manner on the types of polymers in the blend, the morphology of the blend, and the effects of processing, which may orient the phases by shear.

Miscible blends of commercial importance include PPO-PS, PVC-nitrile rubber, and PBT-PET. Miscible blends show a single T_g that is dependent on the ratios of the two components in the blend and their respective T_g values. In immiscible blends the major component has a great effect on the final properties of the blend. Immiscible blends include toughened polymers in which an elastomer is added, existing as a second phase. The addition of the elastomer phase dramatically improves the toughness of the resulting blend as a result of the crazing and shear yielding caused by the rubber phase. Examples of toughed polymers include high impact polystyrene (HIPS), modified polypropylene, ABS, PVC, nylon, and others. In addition to toughened polymers, a variety of other two phase blends are commercially available. Examples include PC-PBT, PVC-ABS, PC-PE, PP-EPDM, and PC-ABS.

References

1. Carraher, C. E., *Polymer Chemistry, An Introduction,* 4th ed., Marcel Dekker, Inc., New York, 1996, p. 238.
2. Brydson, J. A., *Plastics Materials,* 6th ed., Butterworth-Heinemann, Oxford, 1995, p. 516.
3. Kroschwitz, J. I., *Concise Encyclopedia of Polymer Science and Engineering,* John Wiley and Sons, New York, 1990, p. 4.
4. Brydson, *Plastics Materials,* 6th ed., p. 517.
5. Ibid., p. 518.
6. Billmeyer, F. W., Jr., *Textbook of Polymer Science,* 2d ed., John Wiley and Sons, Inc., New York, 1962, p. 439.
7. Brydson, *Plastics Materials,* 5th ed., p. 519.
8. Berins, M. L., *Plastics Engineering Handbook of the Society of the Plastics Industry,* 5th ed., Chapman and Hall, New York, 1991, p. 61.
9. Brydson, *Plastics Materials,* 6th ed., p. 521.
10. Ibid., p. 523.
11. Ibid., p. 524.
12. Berins, *Plastics Engineering Handbook,* p. 62.
13. Strong, A. B., *Plastics: Materials and Processing,* Prentice-Hall, Englewood Cliffs, N.J., 1996, p. 193.
14. Brydson, *Plastics Materials,* 6th ed., p. 525.
15. "Resins '98," *Modern Plastics,* vol. 75, no. 1, January 1998, p. 76.
16. Brydson, *Plastics Materials,* 6th ed., p. 527.
17. Carraher, *Polymer Chemistry,* p. 524.
18. Ibid., p. 524.
19. McCarthy, S. P., "Biodegradable Polymers for Packaging," in *Biotechnological Polymers,* C. G. Gebelein, ed., Technomic Publishing Co., Lancaster, Pa., 1993, p. 215.
20. Carraher, *Polymer Chemistry,* p. 525.
21. Brydson, *Plastics Materials,* 6th ed., p. 858.
22. Ibid., p. 858.
23. Ibid., p. 859.
24. McCarthy, "Biodegradable Polymers," p. 220.
25. Ibid., p. 217.
26. Brydson, *Plastics Materials,* 6th ed., p. 608.
27. Byrom, D., "Miscellaneous Biomaterials," in *Biomaterials,* D. Byrom, ed., Stockton Press, New York, 1991, p. 341.
28. Ibid., p. 341.
29. Ibid., p. 343.
30. Brydson, *Plastics Materials,* 6th ed., p. 859.
31. Ibid., p. 718.
32. McCarthy, "Biodegradable Polymers," p. 220.
33. Brydson, *Plastics Materials,* 6th ed., p. 860.
34. Byrom, D., "Miscellaneous Biomaterials," p. 338.
35. Brydson, *Plastics Materials,* 6th ed., p. 860.
36. Ibid., p. 862.
37. McCarthy, "Biodegradable Polymers," pp. 218–219.
38. Byrom, "Miscellaneous Biomaterials," p. 351.
39. Ibid., p. 353.
40. *Encyclopedia of Polymer Science and Engineering,* Mark, Bilkales, Overberger, Menges, Kroschwitz, eds., 2d ed., Vol. 3, Wiley Interscience, 1986, p. 60.
41. Ibid., p. 68.
42. Ibid., p. 92.
43. Ibid., p. 182.
44. Ibid., p. 182.
45. Brydson, J. A., *Plastics Materials,* 5th ed., Butterworths, Oxford, 1989, p. 583.
46. Ibid., p. 187.
47. Williams, R. W., "Cellulosics," in *Modern Plastics Encyclopedia Handbook,* McGraw-Hill, New York, 1994, p. 8.

48. Brydson, *Plastics Materials,* 6th ed., p. 349.
49. Ibid., p. 349.
50. Berins, *Plastics Engineering Handbook,* p. 62.
51. Billmeyer, *Polymer Science,* p. 423.
52. Berins, *Plastics Engineering Handbook,* p. 63.
53. Carraher, *Polymer Chemistry,* p. 319.
54. Brydson, *Plastics Materials,* 6th ed., p. 359.
55. Billmeyer, *Polymer Science,* p. 426.
56. Brydson, *Plastics Materials,* 6th ed., p. 359.
57. Brydson, *Plastics Materials,* 6th ed., p. 359.
58. Berins, *Plastics Engineering Handbook,* p. 63.
59. Ibid., p. 63.
60. Ibid., p. 63.
61. Brydson, *Plastics Materials,* 6th ed., p. 360.
62. Billmeyer, *Polymer Science,* p. 427.
63. Berins, *Plastics Engineering Handbook,* p. 62.
64. Billmeyer, *Polymer Science,* p. 428.
65. Brydson, *Plastics Materials,* 6th ed., p. 361.
66. Berins, *Plastics Engineering Handbook,* p. 62.
67. Billmeyer, *Polymer Science,* p. 423.
68. Brydson, *Plastics Materials,* 6th ed., p. 352.
69. Billmeyer, *Polymer Science,* p. 424.
70. Brydson, *Plastics Materials,* 6th ed., p. 351.
71. Billmeyer, *Polymer Science,* p. 425.
72. Brydson, *Plastics Materials,* 6th ed., p. 355.
73. Ibid., p. 353.
74. Berins, *Plastics Engineering Handbook,* p. 62.
75. Brydson, *Plastics Materials,* 6th ed., p. 356.
76. Ibid., p. 357.
77. Ibid., p. 357.
78. Ibid., p. 357.
79. Billmeyer, *Polymer Science,* p. 426.
80. Ibid., p. 428.
81. Brydson, *Plastics Materials,* 6th ed., p. 362.
82. Carraher, *Polymer Chemistry,* p. 319.
83. Brydson, *Plastics Materials,* 6th ed., p. 363.
84. Berins, *Plastics Engineering Handbook,* p. 63.
85. Ibid., p. 63.
86. Brydson, *Plastics Materials,* 6th ed., p. 362.
87. Billmeyer, *Polymer Science,* p. 434.
88. *Modern Plastics,* January 1998, p. 76.
89. Berins, *Plastics Engineering Handbook,* p. 64.
90. Brydson, *Plastics Materials,* 6th ed., p. 462.
91. Berins, *Plastics Engineering Handbook,* p. 64.
92. Billmeyer, *Polymer Science,* p. 433.
93. Deanin, R. D., *Polymer Structure, Properties and Applications,* Cahners Publishing Company, Inc., York, Pa., 1972, p. 455.
94. Brydson, *Plastics Materials,* 6th ed., p. 470.
95. Strong, *Plastics,* p. 190.
96. Berins, *Plastics Engineering Handbook,* p. 64.
97. Brydson, *Plastics Materials,* 6th ed., p. 477.
98. Strong, *Plastics,* p. 191.
99. Brydson, *Plastics Materials,* 6th ed., p. 471.
100. Carraher, *Polymer Chemistry,* p. 233.
101. Brydson, *Plastics Materials,* 6th ed., p. 472.
102. Ibid., p. 472.
103. Galanty, P. G., and G. A. Bujtas, "Nylon," in *Modern Plastics Encyclopedia Handbook,* McGraw-Hill, New York, 1994, p. 12.
104. Brydson, *Plastics Materials,* 6th ed., p. 473.

105. Ibid., p. 472.
106. Ibid., p. 473.
107. Strong, *Plastics,* p. 190.
108. Brydson, *Plastics Materials,* 6th ed., p. 484.
109. Ibid., p. 484.
110. Berins, *Plastics Engineering Handbook,* p. 64.
111. Brydson, *Plastics Materials,* 6th ed., p. 461.
112. Billmeyer, *Polymer Science,* p. 435.
113. Ibid., p. 436.
114. Ibid., p. 437.
115. Brydson, *Plastics Materials,* 6th ed., p. 486.
116. Ibid., p. 480.
117. Strong, *Plastics,* p. 191.
118. Brydson, *Plastics Materials,* 6th ed., p. 492.
119. Ibid., p. 492.
120. Ibid., pp. 494–495.
121. Ibid., p. 493.
122. Ibid., p. 496.
123. Ibid., p. 497.
124. Strong, *Plastics,* p. 192.
125. Brydson, *Plastics Materials,* 6th ed., p. 400.
126. Billmeyer, *Polymer Science,* p. 414.
127. Kroschwitz, *Polymer Science and Engineering,* p. 28.
128. Ibid., p. 29.
129. Brydson, *Plastics Materials,* 6th ed., p. 401.
130. Ibid., p. 402.
131. Kroschwitz, *Polymer Science and Engineering,* p. 29.
132. Billmeyer, *Polymer Science,* p. 413.
133. Kroschwitz, *Polymer Science and Engineering,* p. 23.
134. Brydson, *Plastics Materials,* 6th ed., p. 507.
135. Berins, *Plastics Engineering Handbook,* p. 65.
136. Carraher, *Polymer Chemistry,* p. 533.
137. Johson, S.H., "Polyamide-imide," in *Modern Plastics Encyclopedia Handbook,* McGraw-Hill, New York, 1994, p. 14.
138. Ibid., p. 14.
139. Berins, *Plastics Engineering Handbook,* p. 65.
140. Brydson, *Plastics Materials,* 6th ed., p. 507.
141. Berins, *Plastics Engineering Handbook,* p. 65.
142. Brydson, *Plastics Materials,* 6th ed., p. 507.
143. Ibid., p. 507
144. Ibid., p. 708.
145. Berins, *Plastics Engineering Handbook,* p. 66.
146. Dunkle, S. R. and B. D. Dean, "Polyarylate," in *Modern Plastics Encyclopedia Handbook,* McGraw-Hill, New York, 1994, p. 15.
147. Berins, *Plastics Engineering Handbook,* p. 66.
148. Dunkle, "Polyarylate," p. 16.
149. DiSano, L., "Polybenzimidazole," in *Modern Plastics Encyclopedia Handbook,* McGraw-Hill, New York, 1994, p. 16.
150. Carraher, *Polymer Chemistry,* p. 236.
151. Kroschwitz, *Polymer Science and Engineering,* p. 772.
152. DiSano, "Polybenzimidazole," p. 16.
153. Ibid., p. 16.
154. Kroschwitz, *Polymer Science and Engineering,* p. 773.
155. DiSano, "Polybenzimidazole," p. 17.
156. Brydson, *Plastics Materials,* 6th ed., p. 259.
157. Kroschwitz, *Polymer Science and Engineering,* p. 100.
158. Berins, *Plastics Engineering Handbook,* p. 55.
159. Brydson, *Plastics Materials,* 6th ed., p. 259.
160. Kroschwitz, *Polymer Science and Engineering,* p. 100.

161. Brydson, *Plastics Materials,* 6th ed., p. 259.
162. Berins, *Plastics Engineering Handbook,* p. 55.
163. Brydson, *Plastics Materials,* 6th ed., p. 260.
164. Berins, *Plastics Engineering Handbook,* p. 55.
165. Kroschwitz, *Polymer Science and Engineering,* p. 101.
166. "Resin supply at the crossroads," *Modern Plastics,* vol. 79, no. 1, January 1999, p. 64.
167. Brydson, *Plastics Materials,* 5th ed., p. 525.
168. Domininghaus, H., *Plastics for Engineers, Materials, Properties, Applications,* Hanser Publishers, New York, 1988, p. 423.
169. Ibid., p. 424.
170. Ibid., p. 426.
171. Brydson, *Plastics Materials,* 6th ed., p. 711.
172. Berins, *Plastics Engineering Handbook,* p. 67.
173. Brydson, *Plastics Materials,* 6th ed., p. 707.
174. Dominghaus, *Plastics for Engineers,* p. 477.
175. Brydson, *Plastics Materials,* 6th ed., p. 712.
176. Ibid., p. 713.
177. Ibid., p. 714.
178. Ibid., p. 712.
179. McChesney, C. E, *Engineering Plastics, Engineering Materials Handbook,* vol. 2, ASM International, Metals Park, Ohio, 1988, p. 181.
180. Brydson, *Plastics Materials,* 6th ed., p. 712.
181. McChesney, *Engineering Plastics,* p. 181.
182. Brydson, *Plastics Materials,* 6th ed., p. 713.
183. *Modern Plastics Encyclopedia Handbook,* McGraw-Hill, New York, 1994. p. 20.
184. McChesney, *Engineering Plastics,* p. 181.
185. Berins, *Plastics Engineering Handbook,* p. 67.
186. *Modern Plastics,* January 1999, p. 65.
187. Brydson, *Plastics Materials,* 5th ed., p. 681.
188. *Modern Plastics,* January 1999, p. 75.
189. Brydson, *Plastics Materials,* 5th ed., p. 677.
190. *Modern Plastics Encyclopedia,* p. 23.
191. Ibid., p. 23.
192. *Modern Plastics,* January 1999, pp. 74 and 75.
193. Odian, G., *Principles of Polymerization,* 2d ed., John Wiley and Sons, Inc., New York, 1981, p. 103.
194. Brydson, *Plastic Materials,* 5th ed., p. 675.
195. *Polymer Science & Engineering,* vol. 12, p. 223.
196. Odian, p. 103.
197. *Polymer Science & Engineering,* vol. 12, p. 223.
198. Odian, p. 105.
199. *Polymer Science & Engineering,* vol. 12, p. 223.
200. Ibid., p. 222.
201. Ibid., p. 223.
202. Ibid., p. 195.
203. Ibid., p. 228.
204. Ibid., p. 194.
205. Ibid., pp. 195, 204–209.
206. Ibid., p. 197.
207. Ibid., p. 213.
208. Berins, *Plastics Engineering Handbook,* p. 67.
209. Kroschwitz, *Polymer Science and Engineering,* p. 327.
210. Brydson, *Plastics Materials,* 6th ed., p. 508.
211. Ibid., p. 508.
212. Berins, *Plastics Engineering Handbook,* p. 68.
213. Ibid., p. 68.
214. Brydson, *Plastics Materials,* 6th ed., p. 508.
215. Berins, *Plastics Engineering Handbook,* p. 68.
216. Domininghaus, *Plastics for Engineers,* p. 24.

217. *Modern Plastics Encyclopedia,* vol. 74, no. 13, McGraw-Hill, Inc., New York, 1998, p. B-4.
218. Brydson, *Plastics Materials,* 5th ed., p. 217.
219. Domininghaus, *Plastics for Engineers,* p. 55.
220. Mark, *Polymer Science and Engineering,* p. 383.
221. *McGraw-Hill Encyclopedia of Science & Technology,* 5th ed., vol. 10, 1982, p. 647.
222. Mark, *Polymer Science and Engineering,* p. 385.
223. *Modern Plastics Encyclopedia,* 1998, p. A-15.
224. Mark, *Polymer Science and Engineering,* p. 486.
225. Ibid., p. 493.
226. Brydson, *Plastics Materials,* 5th ed., p. 262.
227. Mark, *Polymer Science and Engineering,* p. 422.
228. Kung, D. M., "Ethylene-ethyl acrylate," in *Modern Plastics Encyclopedia Handbook,* McGraw-Hill, New York, 1994, p. 38.
229. Ibid., p. 38.
230. Brydson, *Plastics Materials,* 5th ed., p. 262.
231. Kung, "Ethylene-ethyl acrylate," p. 38.
232. Baker, G., "Ethylene-methyl acrylate," in *Modern Plastics Encyclopedia Handbook,* McGraw-Hill, New York, 1994, p. 38.
233. Mark, *Polymer Science and Engineering,* p. 422.
234. Brydson, *Plastic Materials,* 5th ed., p. 261.
235. Ibid., p. 229.
236. Kroschwitz, *Polymer Science and Engineering,* p. 357.
237. Domininghaus, *Plastics for Engineers,* p. 65.
238. Ibid., p. 67.
239. Brydson, *Plastic Materials,* 5th ed., p. 284.
240. Domininghaus, *Plastics for Engineers,* p. 68.
241. Strong, *Plastics,* p. 165.
242. Brydson, *Plastics Materials,* 6th ed., p. 268.
243. Ibid., p. 268.
244. MacKnight, W. J. and R. D. Lundberg, "Research and ionomeric systems," in *Thermoplastic Elastomers,* 2d ed., G. Holden et. al., eds., Hanser Publishers, New York, 1996, p. 279.
245. Kroschwitz, *Polymer Science and Engineering,* p. 126.
246. Strong, *Plastics,* p. 165.
247. Rees, R. W., "Ionomers," in *Engineering Plastics,* vol 2, *Engineering Materials Handbook,* ASM International, Metals Park, Ohio, 1988, p 120.
248. Ibid., p. 122.
249. Ibid., p. 123.
250. Strong, *Plastics,* p. 165.
251. Rees, *Thermoplastic Elastomers,* p. 263.
252. Brydson, *Plastics Materials,* 6th ed., p. 268.
253. Ibid., p. 269.
254. Kroschwitz, *Polymer Science and Engineering,* p. 827.
255. Berins, *Plastics Engineering Handbook,* p. 69.
256. Albermarle, "Polyimide, Thermoplastic," in *Modern Plastics Encyclopedia Handbook,* McGraw-Hill, New York, 1994, p. 43.
257. Berins, *Plastics Engineering Handbook,* p. 69.
258. Kroschwitz, *Polymer Science and Engineering,* p. 827.
259. Berins, *Plastics Engineering Handbook,* p. 69.
260. Brydson, *Plastics Materials,* 6th ed., p. 504.
261. Ibid., p. 501.
262. Berins, *Plastics Engineering Handbook,* p. 69.
263. Brydson, *Plastics Materials,* 5th ed., p. 565.
264. *Modern Plastics Encyclopedia,* mid-November 1997 issue/vol. 74, no. 13, McGraw-Hill, New York, 1998, pp. B-162, B-163.
265. Brydson, *Plastics Materials,* 6th ed., p. 586.
266. Ibid., p. 564.
267. Ibid., p. 389.

268. Ibid., p. 391.
269. Domininghaus, *Plastics for Engineers,* p. 280.
270. Ibid., p. 122.
271. Brydson, *Plastics Materials,* 6th ed., p. 261.
272. Ibid., p. 263.
273. Ibid., p. 567.
274. Ibid., p. 568.
275. Ibid., p. 570.
276. Ibid., p. 570.
277. Domininghaus, *Plastics for Engineers,* p. 490.
278. Brydson, *Plastics Materials,* 6th ed., p. 575.
279. Ibid., p. 576.
280. Ibid., p. 575.
281. Domininghaus, *Plastics for Engineers,* p. 529.
282. Harris, J. H. and J. A. Reksc, "Polyphthalamide," in *Modern Plastics Encyclopedia Handbook,* McGraw-Hill, New York, 1994, p. 47.
283. Brydson, *Plastics Materials,* 6th ed., p. 499.
284. Harris, "Polyphthalamide," p. 47.
285. Ibid., p. 47.
286. Brydson, *Plastics Materials,* 6th ed., p. 499.
287. Harris, "Polyphthalamide," p. 47.
288. Brydson, *Plastics Materials,* 6th ed., p. 499.
289. Harris, "Polyphthalamide," p. 47.
290. Ibid., p. 48.
291. Ibid., p. 48.
292. *Modern Plastics,* January 1998, p. 58.
293. Brydson, *Plastics Materials,* 6th ed., p. 244.
294. Cradic, G. W., "PP Homopolymer," in *Modern Plastics Encyclopedia Handbook,* McGraw-Hill, New York, 1994, p. 49.
295. Colvin, R., *Modern Plastics,* May 1997, p. 62.
296. Brydson, *Plastics Materials,* 6th ed., p. 245.
297. Odian, *Principles of Polymerization,* 2d ed., John Wiley and Sons, Inc., New York, 1981, p. 581.
298. Brydson, *Plastics Materials,* 6th ed., p. 258.
299. Ibid., p. 244.
300. Strong, *Plastics,* p. 168.
301. Billmeyer, *Polymer Science,* p. 388.
302. Ibid., p. 387.
303. Brydson, *Plastics Materials,* 6th ed., p. 256.
304. Strong, *Plastics,* p. 169.
305. Brydson, *Plastics Materials,* 6th ed., p. 245.
306. Ibid., p. 246.
307. Ibid., p. 248.
308. Ibid., p. 245.
309. Cradic, G.W., "PP Homopolymer," in *Modern Plastics Encyclopedia Handbook,* McGraw-Hill, Inc., New York, 1994, p. 49.
310. Brydson, *Plastics Materials,* 6th ed., p. 253.
311. Ibid., p. 254.
312. Ibid., p. 255.
313. Ibid., p. 251.
314. Ibid., p. 257.
315. Leaversuch, R. D., *Modern Plastics,* December 1996, p. 52.
316. Brydson, *Plastics Materials,* 6th ed., p. 756.
317. Ibid., p. 767.
318. Ibid., p. 768.
319. Sardanopoli, A. A., "Thermoplastic Polyurethanes," in *Engineering Plastics,* vol. 2, *Engineering Materials Handbook,* ASM International, Metals Park, Ohio, 1988, p 203.
320. Ibid., p. 206.

321. Ibid., p. 205.
322. Ibid., p. 205.
323. Ibid., p. 207.
324. Brydson, *Plastics Materials,* 6th ed., p. 427.
325. Domininghaus, *Plastics for Engineers,* p. 226.
326. Akane, J., "ACS," in *Modern Plastics Encyclopedia Handbook,* McGraw-Hill, Inc., New York, 1994, p. 54.
327. Ibid., p. 54.
328. Ostrowski, S., "Acrylic-styrene-acrylonitrile," in *Modern Plastics Encyclopedia Handbook,* McGraw-Hill, New York, 1994, p. 54.
329. McCrum, N. G., C. P. Buckley, and C. B. Bucknall, *Principles of Polymer Engineering,* 2nd ed., Oxford Science Publications, New York, 1997, p. 372.
330. Mark, *Polymer Science and Engineering,* 2d ed., p. 65.
331. Kroschwitz, *Polymer Science and Engineering,* p. 30.
332. Brydson, *Plastics Materials,* 6th ed., p. 426.
333. Berins, *Plastics Engineering Handbook,* p. 57.
334. Brydson, *Plastics Materials,* 6th ed., p. 426.
335. Ibid., p. 426.
336. Kroschwitz, *Polymer Science and Engineering,* p. 30.
337. Berins, *Plastics Engineering Handbook,* p. 57.
338. Ibid., p. 57.
339. Brydson, *Plastics Materials,* 6th ed., p. 435.
340. Salay, J. E., and D. J. Dougherty, "Styrene-butadiene Copolymers," in *Modern Plastics Encyclopedia Handbook,* McGraw-Hill, Inc., New York, 1994, p. 60.
341. Ibid., p. 60.
342. Brydson, *Plastics Materials,* 6th ed., p. 435.
343. Salay, "Styrene-butadiene Copolymers," p. 60.
344. Strong, *Plastics,* p. 205.
345. Brydson, *Plastics Materials,* 6th ed., p. 577.
346. Kroschwitz, *Polymer Science and Engineering,* p. 886.
347. Brydson, *Plastics Materials,* 6th ed., p. 580.
348. Kroschwitz, *Polymer Science and Engineering,* p. 886.
349. Brydson, *Plastics Materials,* 6th ed., p. 582.
350. Ibid., p. 582.
351. Ibid., p. 583.
352. Carraher, *Polymer Chemistry,* p. 240.
353. Kroschwitz, *Polymer Science and Engineering,* p. 888.
354. Berins, *Plastics Engineering Handbook,* p. 71.
355. Ibid., p. 71.
356. Sauers, M. E., "Polyaryl Sulfones," in *Engineering Plastics,* vol. 2, *Engineered Materials Handbook,* ASM International, Metals Park, Ohio, 1988, p. 146.
357. Ibid., p. 145.
358. Berins, *Plastics Engineering Handbook,* p. 72.
359. Ibid., p. 72.
360. Berins, *Plastics Engineering Handbook,* p. 71.
361. Sauers, "Polyaryl Sulfones," p. 146.
362. Berins, *Plastics Engineering Handbook,* p. 72.
363. Watterson, E. C., "Polyether Sulfones," in *Engineering Plastics,* vol. 2, *Engineered Materials Handbook,* ASM International, Metals Park, Ohio, 1988, p. 161.
364. Ibid., p. 160.
365. Berins, *Plastics Engineering Handbook,* p. 72.
366. Ibid., p. 72.
367. Watterson, "Polyether Sulfones," p. 161.
368. Berins, *Plastics Engineering Handbook,* p. 72.
369. Watterson, "Polyether Sulfones," p. 159.
370. Dunkle, "Polysulfones," p. 200.
371. Ibid., p. 200.
372. Berins, *Plastics Engineering Handbook,* p. 71.
373. Ibid., p. 71.

374. Dunkle, "Polysulfones," p. 200.
375. Ibid., p. 201.
376. Berins, *Plastics Engineering Handbook,* p. 71.
377. Dunkle, "Polysulfones," p. 200.
378. Brydson, *Plastics Materials,* 6th ed., p. 301.
379. Billmeyer, *Polymer Science,* p. 420.
380. Brydson, *Plastics Materials,* 6th ed., p. 304.
381. Ibid., p. 307.
382. Ibid., pp. 302–304.
383. Strong, *Plastics,* p. 171.
384. Ibid., p. 170.
385. Billmeyer, *Polymer Science,* p. 420.
386. Strong, *Plastics,* p. 172.
387. Brydson, *Plastics Materials,* 6th ed., pp. 314–316.
388. Strong, *Plastics,* p. 171.
389. Ibid., p. 172.
390. Brydson, *Plastics Materials,* 6th ed., pp. 317–319.
391. Strong, *Plastics,* p. 173.
392. Brydson, *Plastics Materials,* 6th ed., p. 346.
393. Martello, G. A., "Chlorinated PVC," in *Modern Plastics Encyclopedia Handbook,* McGraw-Hill, Inc., New York, 1994, p. 71.
394. Brydson, *Plastics Materials,* 6th ed., p. 341.
395. Strong, *Plastics,* p. 173.
396. Hurter, D., "Dispersion PVC," in *Modern Plastics Encyclopedia Handbook,* McGraw-Hill, Inc., New York, 1994, p. 72.
397. Brydson, *Plastics Materials,* 6th ed., p. 309.
398. Ibid., p. 450.
399. Ibid., p. 127.
400. Sommer, I. W., "Plasticizers," in *Plastics Additives,,* 2d ed., R. Gachter and H. Muller, eds., Hanser Publishers, New York, 1987, pp. 253–255.
401. Brotz, W., "Lubricants and Related Auxiliaries for Thermoplastic Materials," in *Plastics Additives,* 2d ed., R. Gachter and H. Muller, eds., Hanser Publishers, New York, 1987, p. 297.
402. Brydson, *Plastics Materials,* 6th ed., p. 129.
403. Ibid., p. 136.
404. Ibid., pp. 130–141.
405. Ibid., pp. 141–145.
406. Ibid., pp. 146–149.
407. Bosshard, A. W., and H. P. Schlumpf, "Fillers and Reinforcements," in *Plastics Additives,* 2d ed., R. Gachter and H. Muller, eds., Hanser Publishers, New York, 1987, p. 397.
408. Brydson, *Plastics Materials,* 6th ed., p. 122.
409. Bosshard, "Fillers and Reinforcements," p. 407.
410. Ibid., p. 420.
411. Kroschwitz, *Polymer Science and Engineering,* pp. 830–835.
412. Mark, *Polymer Science and Engineering,* p. 433.
413. Ibid., p. 65.
414. Maccani, R. R., "Characteristics Crucial to the Application of Engineering Plastics," in *Engineering Plastics,* vol. 2, *Engineering Materials Handbook,* ASM International, Metals Park, Ohio, 1988, p. 69.
415. Berins, *Plastics Engineering Handbook,* pp. 48–49.
416. Mark, *Polymer Science and Engineering,* p. 433.
417. Maccani, "Application of Engineering Plastics," p. 69.
418. Berins, *Plastics Engineering Handbook,* pp. 48–49.

Chapter

2

Thermosets, Reinforced Plastics, and Composites

Ralph E. Wright

R. E. Wright Associates
Yarmouth, Maine

2.1 Resins

One definition of *resin* is "any class of solid, semi-solid, or liquid organic material, generally the product of natural or synthetic origin with a high molecular weight and with no melting point." The 10 basic thermosetting resins all possess a commonality in that they will, upon exposure to elevated temperature from ambient to upwards of 450°F, undergo an irreversible chemical reaction often referred to as *polymerization* or cure. Each family member has its own set of individual chemical characteristics based upon their molecular makeup and their ability to either homopolymerize, copolymerize, or both.

This transformation process represents the line of demarcation separating the thermosets from the thermoplastic polymers. Crystalline thermoplastic polymers are capable of a degree of crystalline cross-linking but there is little, if any, of the chemical cross-linking that occurs during the thermosetting reaction. The important beneficial factor here lies in the inherent enhancement of thermoset resins in their physical, electrical, thermal, and chemical properties due to that chemical cross-linking polymerization reaction which, in turn, also contributes to their ability to maintain and retain these enhanced properties when exposed to severe environmental conditions.

2.2 Thermosetting Resin Family

2.2.1 Allyls: Diallyl ortho phthalate (DAP) and diallyl iso phthalate (DAIP)

Chemical characteristics.[1] The most broadly used allyl resins are prepared from the prepolymers of either DAP or DAIP which have been condensed from dibasic acids. The diallyl phthalate monomer is an ester produced by the esterification process involving a reaction between a dibasic acid (phthalic anhydride) and an alcohol (allyl alcohol) which yields the DAP ortho monomer, as shown in Fig. 2.1. Similar reactions with dibasic acids will yield the DAIP (iso) prepolymer. Both prepolymers are white, free-flowing powders and are relatively stable whether catalyzed or not, with the DAP being more stable and showing negligible change after storage of several years in temperatures up to 90°C.

The monomer is capable of cross-linking and will polymerize in the presence of certain peroxide catalysts such as

- Dicumyl peroxide (DICUP)
- t-Butyl perbenzoate (TBP)
- t-Butylperoxyisopropyl carbonate (TBIC)

2.2.2 Aminos: Urea and melamine

Chemical characteristics.[2] Both resins will react with formaldehyde to initiate and form monomeric addition products. Six molecules of formaldehyde added to one single molecule of melamine will form hexamethylol melamine, whereas a single molecule of urea will combine with two molecules of formaldehyde to form dimethylolurea. If carried on further, these condensation reactions produce an infusible polymer network. The urea/formaldehyde and melamine/formaldehyde reactions are illustrated in Fig. 2.2.

2.2.3 Bismaleimides (BMIs)

Chemical characteristics.[3] The bismaleimides (BMIs) are generally prepared by the condensation reaction of a diamine with maleic anhydride. A typical BMI based on methylene dianiline (MDA) is illustrated in Fig. 2.3.

2.2.4 Epoxies

Chemical characteristics.[4] *Epoxy resins* are a group of cross-linking polymers and are sometimes known as the oxirane group which is reactive toward a broad range of curing agents. The curing reactions

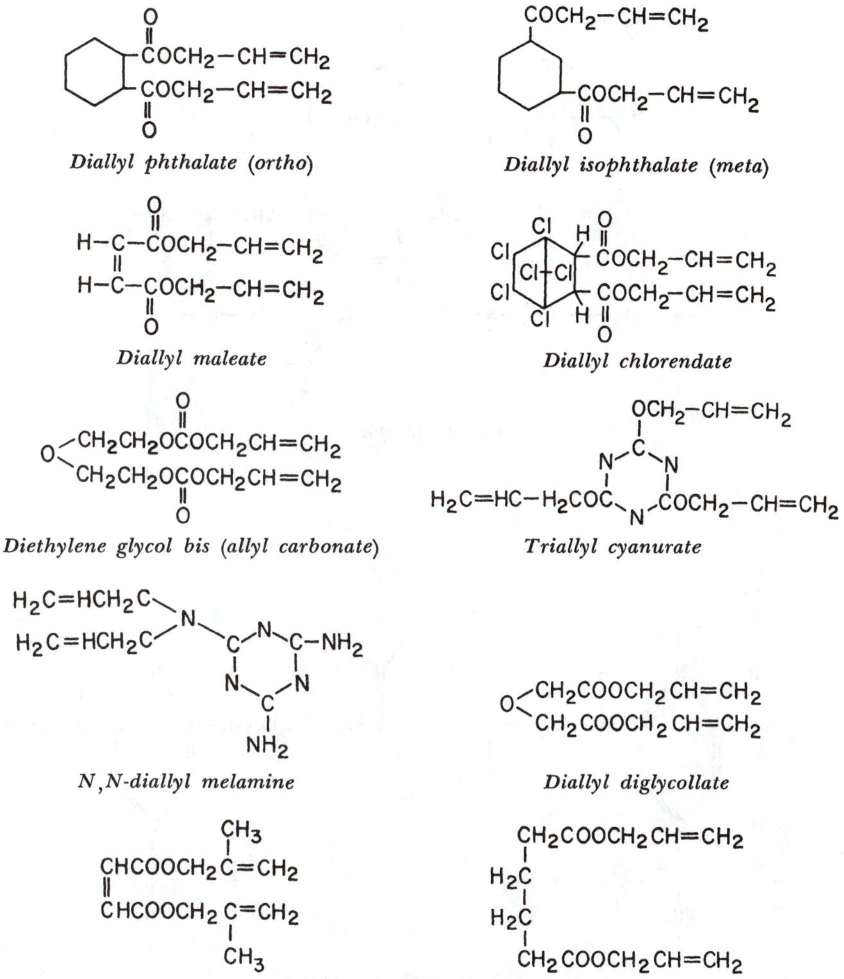

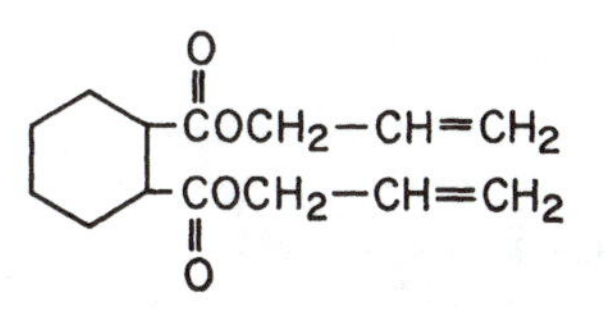

Diallyl phthalate (ortho)

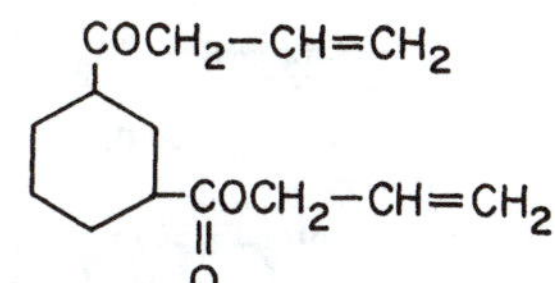

Diallyl isophthalate (meta)

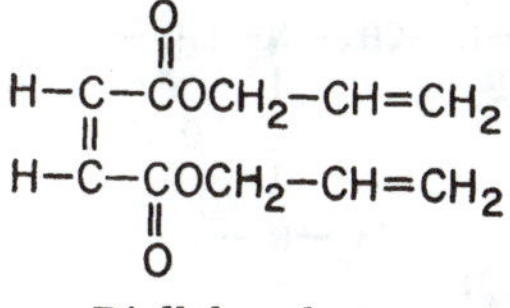

Diallyl maleate

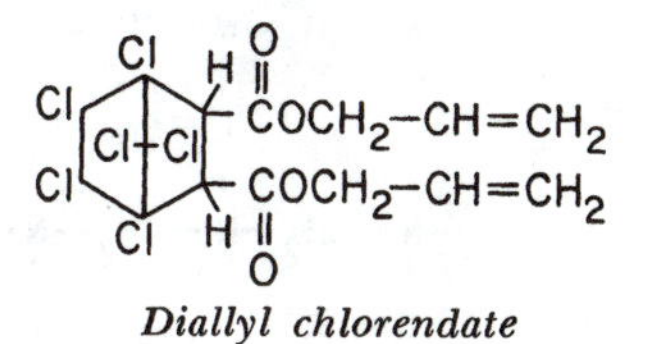

Diallyl chlorendate

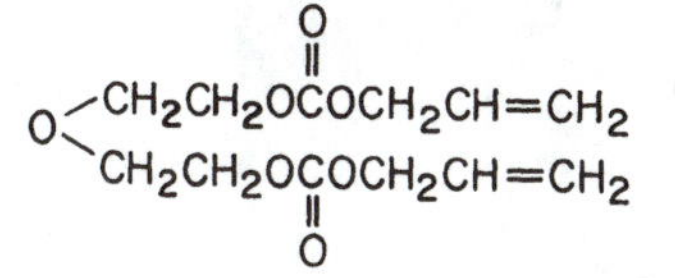

Diethylene glycol bis (allyl carbonate)

Triallyl cyanurate

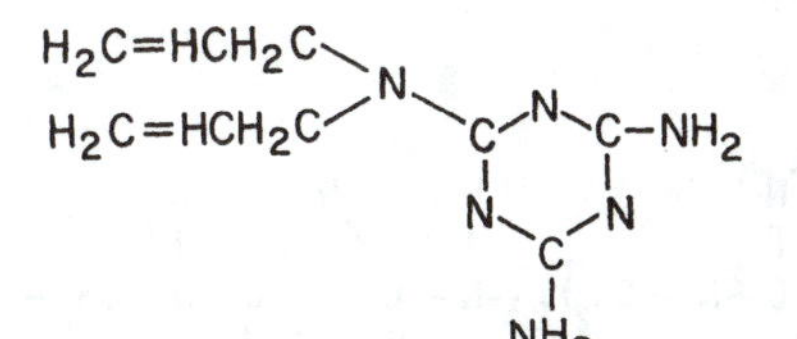

N,N-diallyl melamine

Diallyl diglycollate

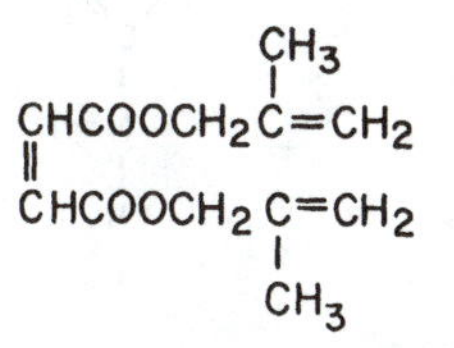

Figure 2.1 Structural formula for allylic resins.

convert the low molecular weight resins into three-dimensional thermoset structures exhibiting valuable properties. The standard epoxy resins used in molding compounds meeting Mil-M-24325 (ships) are based on bisphenol A and epichlorohydrin as raw materials with anhydride catalysts, as illustrated in Fig. 2.4. The low-pressure encapsulation compounds consist of an epoxy-novolac resin system using amine catalysts.

2.2.5 Phenolics: Resoles and novolacs

Chemical characteristics.[5] Phenol and formaldehyde when reacted together will produce condensation products when there are free

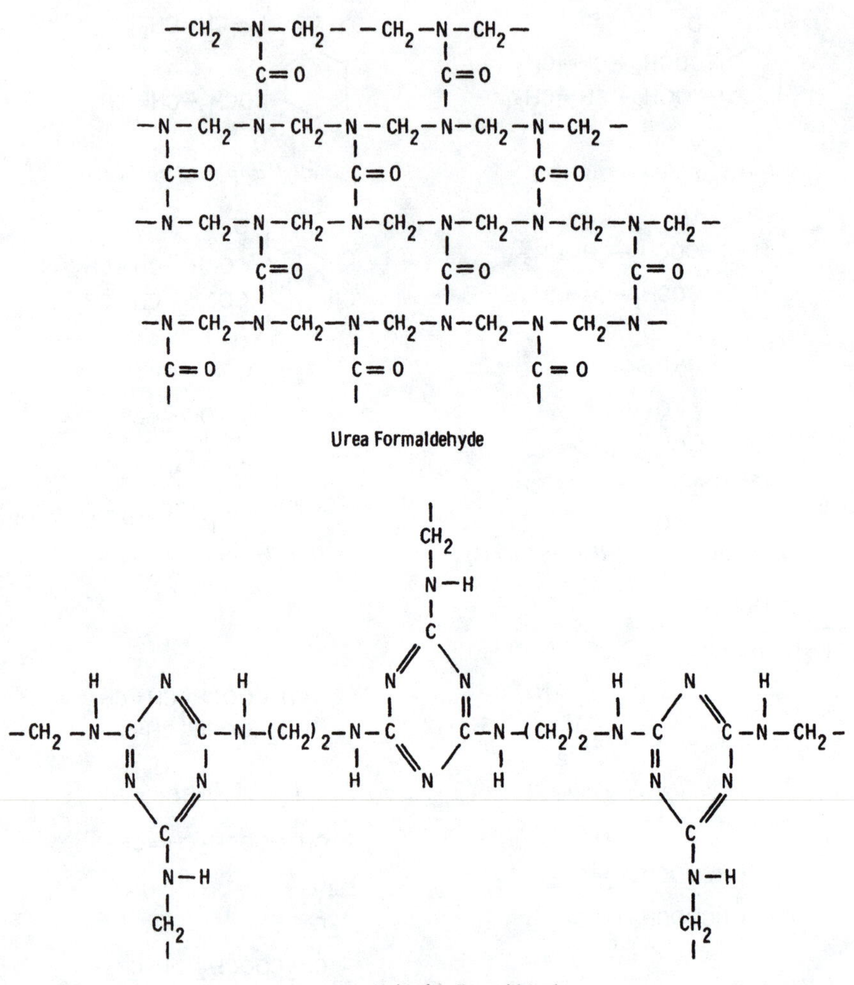

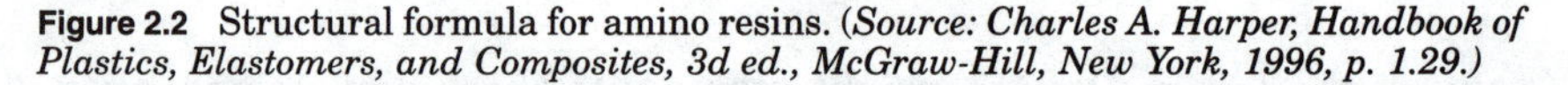

Figure 2.2 Structural formula for amino resins. (*Source: Charles A. Harper, Handbook of Plastics, Elastomers, and Composites, 3d ed., McGraw-Hill, New York, 1996, p. 1.29.*)

positions on the benzene ring—*ortho* and *para* to the hydroxyl group. Formaldehyde is by far the most reactive and is used almost exclusively in commercial applications. The product is greatly dependent upon the type of catalyst and the mole ratio of the reactants. Although there are four major reactions in the phenolic resin chemistry, the resole (single stage) and the novolac (two stage) are the two primarily used in the manufacture of phenolic molding compounds.

Novolacs (two-stage). In the presence of acid catalysts, and with a mole ratio of formaldehyde to phenol of less than 1, the methylol derivatives

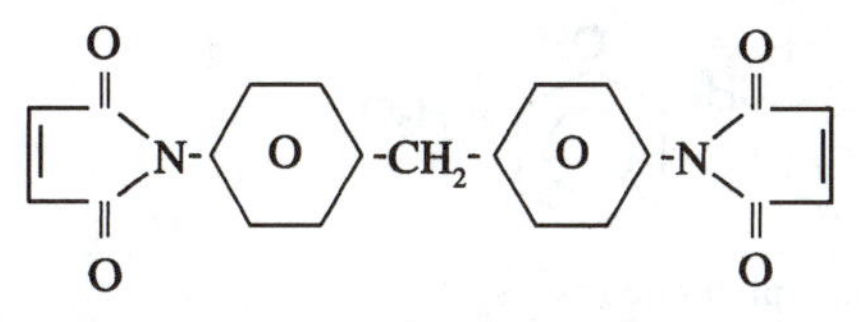

Figure 2.3 Structural formulas for bismaleimide resins. (*From Plastics Handbook, Modern Plastics Magazine, McGraw-Hill, New York, 1994, p. 76.*)

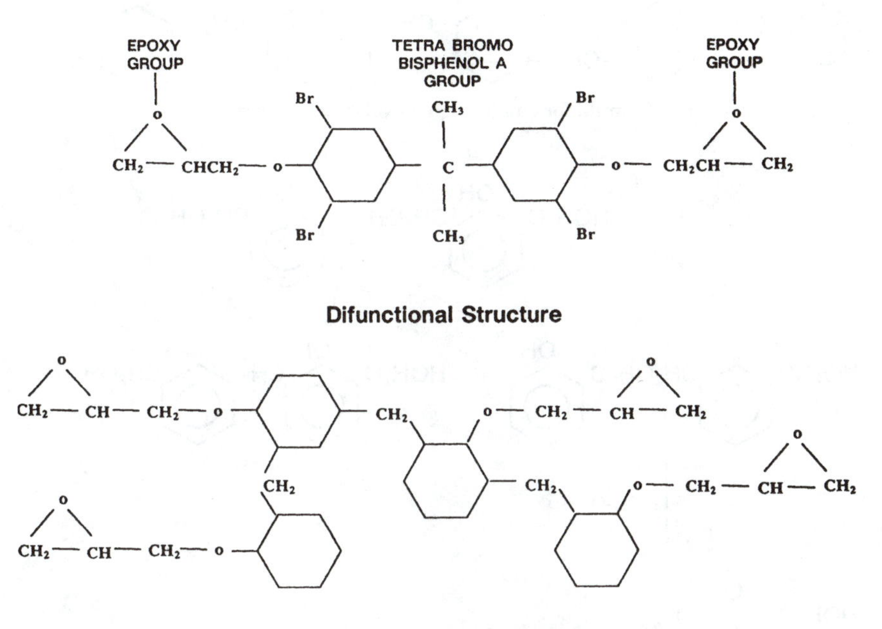

Figure 2.4 Structural formulas for epoxy resins.

condense with phenol to form, first, dihydroxydiphenyl methane, as shown in Fig. 2.5. On further condensation and with methyl bridge formation, fusible and soluble linear low polymers called *novolacs* are formed with the structure where the *ortho* and *para* links occur at random, as shown in Fig. 2.6.

The novolac (two-stage) resins are made with an acid catalyst, and only part of the necessary formaldehyde is added to the reaction kettle, producing a mole ratio of about 0.81. The rest is added later: a hexamethylenetetramine (hexa), which decomposes in the final curing step, with heat and moisture present, to yield formaldehyde and ammonia, which act as the catalyst for curing.

Resole (single-stage). In the presence of alkaline catalysts and with more formaldehyde, the methylol phenols can condense either through

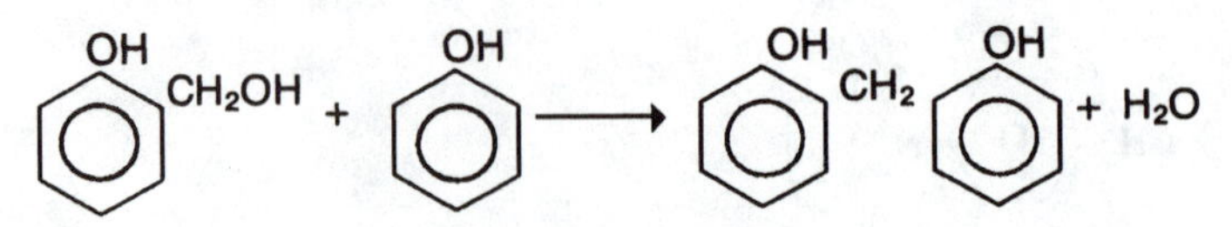

Figure 2.5 Structural formulas for single-stage phenolic resins.

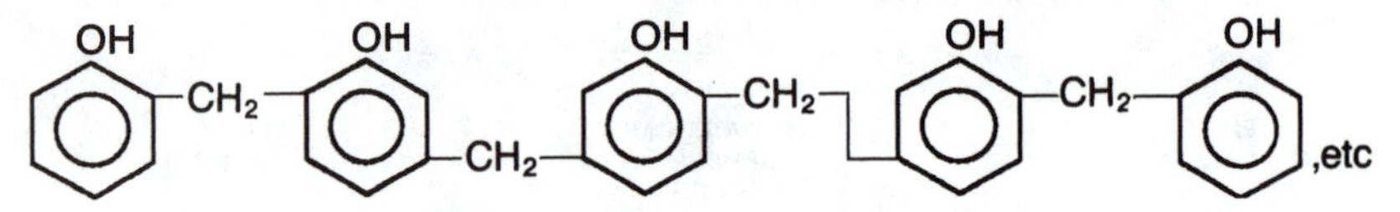

Figure 2.6 Structural formula for single-stage novolac phenolic resins.

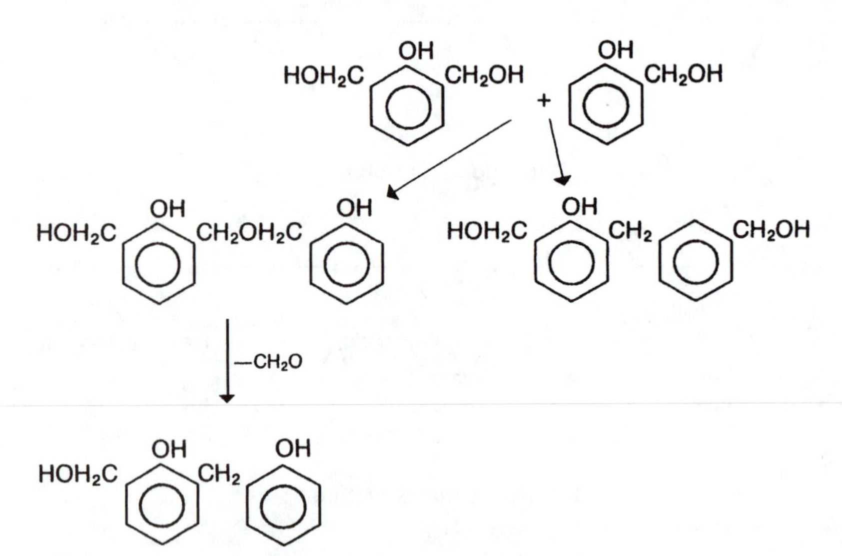

Figure 2.7 Structural formula for two-stage resole phenolic resin.

methylene linkages or through ether linkages. In the latter case, subsequent loss of formaldehyde may occur with methylene bridge formation. Products of this type, soluble and fusible but containing alcohol groups, are called *resoles* and are shown in Fig. 2.7. If the reactions leading to their formation are carried further, large numbers of phenol nuclei can condense to give a network formation.

In the production of a single-stage phenolic resin, all the necessary reactants for the final polymer (phenol, formaldehyde, and catalyst) are charged into the resin kettle and reacted together. The ratio of formaldehyde to phenol is about 1.25:1, and an alkaline catalyst is used. Resole resin-based molding compounds will not outgas ammonia during the molding process as do the novolac compounds, with the result that the resole compounds are used for applications where their lack of outgassing is most beneficial and, in same cases, essential. The chief drawback lies in their sensitivity to temperatures in excess of 45°F (7.2°C). This makes it imperative to produce these compounds in air-conditioned rooms and to store the finished goods in similar conditions in sealed containers.

The novolac (two-stage) resin based compounds are the most widely used despite the ammonia outgassing, which can be controlled through proper mold venting. They have an outstanding property profile, can be readily molded in all thermosetting molding techniques, and are one of the least costly compounds available. Their chief drawback has been their limited color range but that has been improved with the advent of phenolic-melamine alloying.

2.2.6 Polyesters (thermosetting)

Chemical characteristics.[6] Thermosetting polyesters can be produced from phthalic or maleic anhydrides and polyfunctional alcohols with catalyzation achieved by the use of free radical-producing peroxides. The chemical linkages shown in Fig. 2.8 will form a rigid, cross-linked, thermosetting molecular structure. Thermosetting polyesters are derived from the condensation effects of combining an unsaturated dibasic acid (maleic anhydride) with a glycol (propylene glycol, ethylene glycol, and dipropylene glycol). When utilized as a molding compound, the unsaturated polyester resin is dissolved in a cross-linking monomer (styrene) with the addition of an inhibitor (hydroquinone) to prevent the cross-linking until the compound is ready for use in the molding process. The compounds are further enhanced with additives, such as chlorendic anhydride for flame retardance, isophthalic acid for chemical resistance, and neopentyl glycol for weathering resistance.

The compounder uses the free radical addition to polymerize the resin. The catalyst (organic peroxide) becomes the source for the free radicals and with elevated temperature the heat decomposes the peroxide, producing free radicals. Peroxyesters and benzoyl peroxide are the organic peroxides primarily used at elevated temperatures.

2.2.7 Polyimides (thermosetting)

Chemical characteristics.[7] *Polyimides* are heterocyclic polymers with a noncarbon atom of nitrogen in one of the rings in the molecular chains, as shown in Fig. 2.9. The fused rings provide chain stiffness essential to high-temperature strength retention. The low concentration of hydrogen provides oxidation resistance by preventing thermal degradative fracture of the chain.

Reaction A

One quantity of unsaturated acid reacts with two quantities of glycol to yield linear polyester (alkyd) polymer of n polymer units

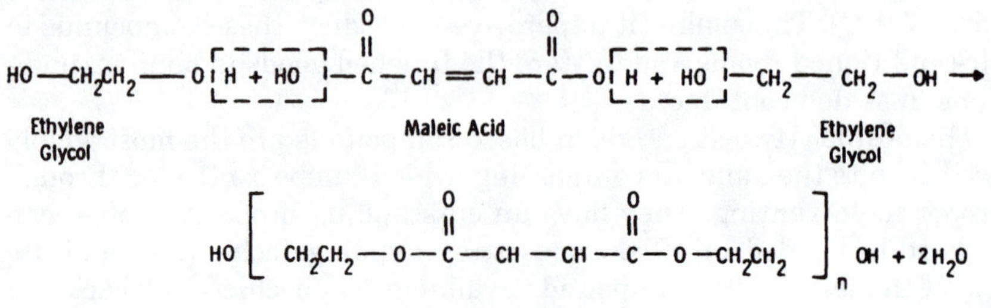

Ethylene Glycol Maleate Polyester

Reaction B

Polyester polymer units react (copolymerize) with styrene monomer in presence of catalyst and/or heat to yield styrene-polyester copolymer resin or, more simply, a cured polyester. (Asterisk indicates points capable of further cross-linking.)

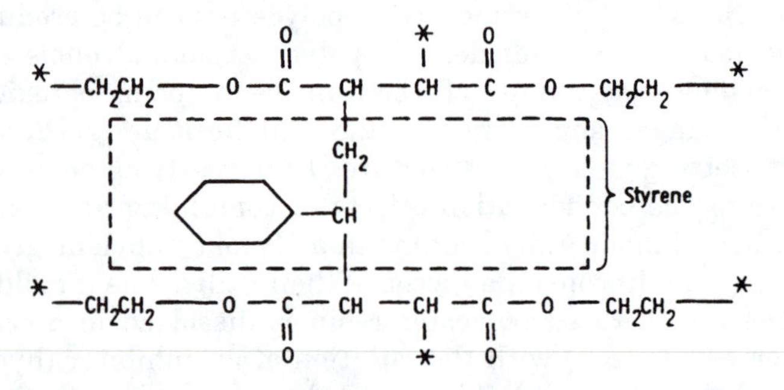

Styrene-Polyester Copolymer

Figure 2.8 Simplified diagrams showing how cross-linking reactions produce polyester resin (styrene-polyester copolymer) from basic chemicals. *(From Charles A. Harper, Handbook of Plastics, Elastomers, and Composites, 3d ed., McGraw-Hill, New York, 1996, p. 1.3.)*

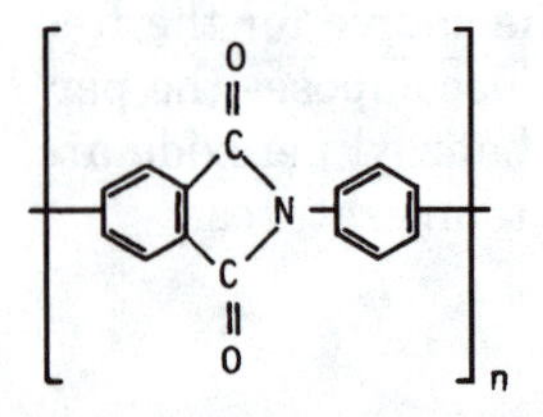

Figure 2.9 Structural formula for polyimide resin. (*From Charles A. Harper, Handbook of Plastics, Elastomers, and Composites, 3d ed., McGraw-Hill, New York, 1996, p. 1.63.*)

Poly(amide-imide) resins contain aromatic rings and the characteristic nitrogen linkages, as shown in Fig. 2.10. There are two basic types of polyimides:

1. *Condensation resin* is based upon a reaction of an aromatic diamine with an aromatic dianhydride, producing a fusible polyamic acid intermediate which is converted by heat to an insoluble and infusible polyimide, with water being given off during the cure.
2. *Addition resins* are based on short, preimidized polymer chain segments similar to those comprising condensation polyimides. The prepolymer chains, which have unsaturated aliphatic end groups, are capped by termini which polymerize thermally without the loss of volatiles.

While condensation polyimides are available as both thermosetting and thermoplastic, the addition polyimides are available only as thermosets.

2.2.8 Polyurethanes

Chemical characteristics.[8] The true foundation of the polyurethane industry is the *isocyanate,* an organic functional group that is capable of an enormously diverse range of chemical reactions. The reactions of isocyanates fall into two broad categories, with the most important category being the reactions involving active hydrogens. This category of reactions requires at least one co-reagent containing one or more hydrogens which are potentially exchangeable under conditions of reaction. The familiar reaction of isocyanates with polyols, to form polycarbamates, is of the "active hydrogen" type. The active groups are, in this case, the hydroxyl groups on the polyol. *Nonactive hydrogen reactions* constitute the second broad category, including cycloaddition reactions and linear polymerizations (which may or may not involve co-reagents.

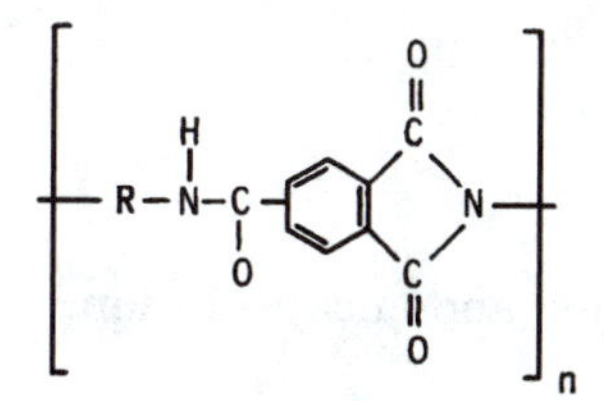

Figure 2.10 Structural formula for poly(amide-imide) resin. (*From Charles A. Harper, Handbook of Plastics, Elastomers, and Composites, 3d ed., McGraw-Hill, New York, 1996, p. 1.64.*)

The diversity of isocyanate chemistry, combined with the availability of selective catalysts, has made it possible to "select" reactions which "fit" desired modes and rates of processing. Table 2.1 lists some of the known polymer-forming reactions of the isocyanates.

2.2.9 Silicones

Chemical characteristics.[9] The silicone family consists of three forms, as shown in:

Fluids. The *fluids* are linear chains of dimethyl siloxane whose molecular weights determine their viscosity. They are supplied as both neat fluids and as water emulsions, and as nonreactive and reactive fluids.

Resins. The *resins* are highly branched polymers that cure to solids. They resemble glass but are somewhat softer and usually soluble in solvent until cured. Their degree of hardness when cured depends on the extent of cross-linking.

Elastomers. The *elastomers* are prepared from linear silicone oils or gums and reinforced with a filler and then vulcanized (cured or cross-linked). The base resin for silicone molding compounds are the result of reacting silicone monomer with methylchloride to produce methyl chlorosilane:

$$\text{Si} + \text{RCl (catalyst/heat)} \rightarrow \text{RSiCl}$$

2.3 Resin Characteristics

1. *A stage.* Resin is still soluble in certain liquids and still fusible.
2. *B stage.* An intermediate stage in the reaction process when the material softens with heat and swells in contact with certain liquids but does not entirely fuse or dissolve.
3. *C stage.* Final stage of the reaction when the resin becomes completely and solidly cured and is relatively insoluble and infusible.

2.4 Resin Forms

1. *Liquid.* Naked for coatings.
2. *Liquid.* Catalyzed for castings, foundry resins, and encapsulation.
3. *Liquid.* In molding compounds.
4. *Solid.* Flakes, granules, powders in molding compounds, and/or foundry resins.

TABLE 2.1 Examples of Polymer-Forming Reactions of Isocyanates

Co-reactant	Product(s)	Catalysts	Category
Alcohols	Carbamates	3° amines Tin soaps Alkalai soaps	Active-H
Carbamates	Allophanates	3° amines Tin soaps Alkalai soaps	Active-H
Primary amines	Ureas	Carboxylic acids	Active-H
Secondary amines	Ureas	Carboxylic acids	Active-H
Ureas	Biurates	Carboxylic acids	Active-H
Imines	Ureas Amides Triazines	Carboxylic acids	Either category
Enamines	Amides Heterocycles	Carboxylic acids	Either category
Carboxylic acids	Amides	Phospholene oxides	Active-H
Amides	Acyl ureas	Acids/bases	Active-H
H_2O	Ureas CO_2	3° amines Alkalai soaps	Active-H
Anhydrides (cyclic)	Imides	Phospholene oxides	Nonactive hydrogen
Ketones	CO_2		
	Imines	Alkalai soaps	Either category
	Heterocycles	Alkalai soaps	Either category
	CO_2		
Aldehydes	Imines	Alkalai soaps	Either category
	Heterocycles	Alkalai soaps	Either category
	CO_2		
Active methylene (active methine) compounds	Amides Heterocycles	Bases	Either category
Isocyanates	Carbodiimides	Phosphorous-oxygen compounds	Nonactive Nonactive Nonactive
	Dimers	Pyridines	Nonactive
	Trimers	Alkalai soaps	Nonactive
	Polymers	Strong bases	Nonactive
Carbodiimides	Uretonimines	—	Nonactive
Epoxides	Oxazolidones	Organoantimony iodides	Either category
2 or B hydroxy acids (esters)	Heterocycles ROH or H_2O	Acids/bases	Active-H
2 or B amino acids (esters)	Heterocycles ROH or H_2O	Acids/bases	Active-H
Cyanohydrins	Heterocycles	Acids/bases	Active-H
2-Cyano amines	Heterocycles	Acids/bases	Active-H

TABLE 2.1 Examples of Polymer-Forming Reactions of Isocyanates (*Continued*)

Co-reactant	Product(s)	Catalysts	Category
2-Amidno esters	Heterocycles ROH	Acids/bases	Active-H
Orthoformates	Heterocycles ROH	Acids/bases	Active-H
Oxazolines; imidazolines	Various heterocycles	Acids/bases	Either category
Cyclic carbonates	Heterocycles CO2	Bases	Either category
Acetylenes	Various heterocycles	Various	Either category
Pyrroles	Amides	Acids/bases	Active-H
Carbamic acid	Ureas	—	Active-H
Amine salts	CO_2		

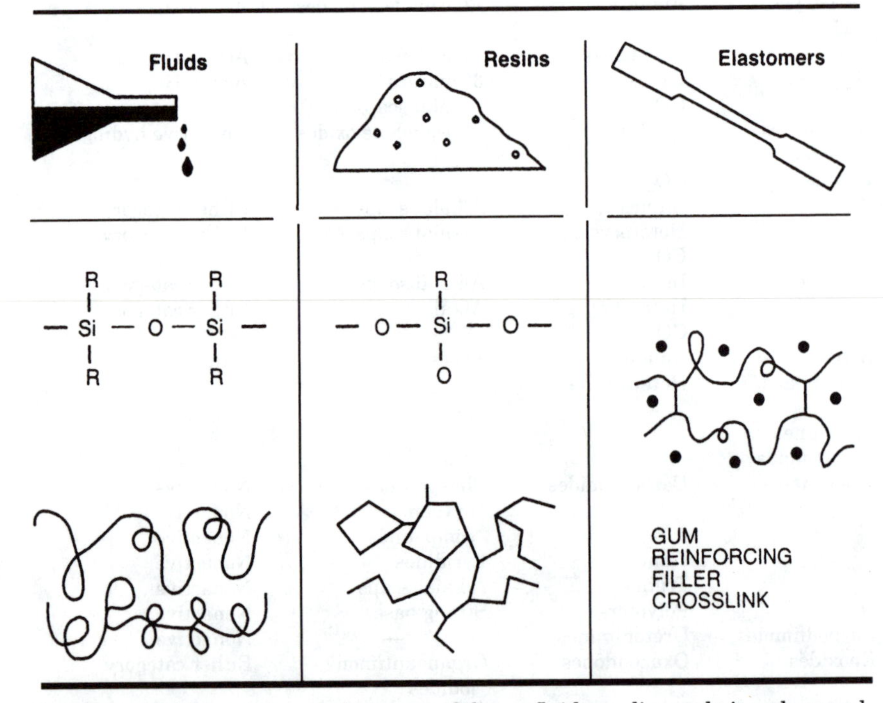

Figure 2.11 Fluids, resins, and elastomers. *Silicone fluids* are linear chains where molecular weight determines viscosity, *silicone resins* are branched polymers and thus glass-like solids; and *silicone elastomers* are composed of long, linear polysiloxane chains reinforced with an inorganic filler and cross-linked. (*From Plastics Handbook, Modern Plastics Magazine, McGraw-Hill, New York, 1994, p. 86.*)

2.5 Liquid Resin Processes[10]

2.5.1 Potting process

The processing of liquid potting compounds commences with dispersing curing agents and other ingredients with a simple propeller mixer, producing potting compounds capable of a variety of properties achieved by changes in formulations to gain stiffness, color, strength, and electrical properties meeting specific needs and applications. The liquid potting compound is poured into a receptacle or jacket where the compound cures in place—either with added heat or at ambient temperature. The receptacle or jacket remains as the permanent outer skin of the final product. Proper compounding can produce articles with negligible shrinkage and optical clarity along with a minimum of internal stress. Part and mold design, coupled with the proper placement of the reinforcement, are calculated with sophisticated computer software.

2.5.2 Potting resin selection

Tables 2.2 to 2.5 provide helpful information for guidance in the selection process for potting resins.

2.5.3 Casting process

The process for casting fluid monomeric resin compounds involves pouring the premixed compound into a stationary mold (metal or glass) and then allowing the compound to cure at ambient or elevated temperatures. The mold may contain objects that will have been prepositioned in the mold and they become embedded by the resin as it cures. This technique can be done with minimal equipment and delicate inserts can be embedded or they can be introduced into the compound prior to the curing stage. The chief disadvantages are that high-viscosity resins are difficult to handle and the occurrence of voids or bubbles can present problems.

Cast epoxies. *Cast epoxy resins* have proven to be very popular in a wide variety of applications because of their versatility, excellent adhesion, low cure shrinkage, good electrical properties, compatibility with many other materials, resistance to weathering and chemicals, dependability, and ability to cure under adverse conditions. Some of their application fields are adhesives, coatings, castings, pottings, building construction, chemical resistant equipment, and marine applications. The most widely used epoxy resins in the casting field are the epi-bis and cycloaliphatic epoxies. Table 2.6 lists the properties of typical cured epi-bis resins with a variety of curing agents, and Table 2.7 provides information on the properties of blends of cycloaliphatic epoxy resins.

TABLE 2.2 Comparison of Properties of Liquid Resins

Material	Cure shrinkage	Adhesion	Thermal shock	Electrical properties	Mechanical properties	Handling properties	Cost
Epoxy:							
Room temperature cure	Low	Good	Fair*	Fair	Fair	Good	Moderate
High temperature	Low	Good	Fair*	Good	Good	Fair to good	Moderate
Flexible	Low	Excellent	Good	Fair	Fair	Good	Moderate
Polyesters:							
Rigid	High	Fair	Fair*	Fair to good	Good	Fair	Low to moderate
Flexible	Moderate	Fair	Fair	Fair	Fair	Fair	Low to moderate
Silicones:							
Rubbers	Very low	Poor	Excellent	Good	Poor	Good	High
Rigid	High	Poor	Poor	Excellent	Poor	Fair	High
Polyurethanes:							
Solid	Low	Excellent	Good	Fair	Good	Fair	Moderate
Foams	Variable	Good	Good	Fair	Good for density	Fair	Moderate
Butyl LM	Low	Good	Excellent	Good	Poor	Fair	Low to moderate
Butadienes	Moderate	Fair	Poor*	Excellent	Fair	Fair	Moderate

*Depends on filler.
SOURCE: From Ref. 10, p. 4.3.

TABLE 2.3 Properties of Various Liquid Resins

Property	ASTM	Rigids (silica-filled)			Flexibilized epoxies			Liquid elastomers				Rigid polyurethane
		Styrene polyester	Epoxy	Silicone	Epoxy-polysulfide (50–50)	Epoxy-polyamide (50–50)	Epoxy-polyurethane (50–50, diamine cure)	Poly-sulfide	Silicone	Poly-urethane-(diamine cure)	Plas-tisol	Prepolymer
Electrical Properties												
Dielectric strength, V/mil*	D 149	425	425	350	350	430	640	340	400	350	200	
Dielectric constant:												
60 Hz	D 150	3.7	3.8	3.7	5.6	3.2		7.2	3	8.2	5.6	1.05
10^3 Hz	D 150	3.7	3.6	3.6	5.4	3.2	4.9	7.2	3	7.3	4.9	1.06
10^6 Hz	D 150	3.6	3.4	3.6	4.8	3.1		7.2	3		3.6	1.04
Dissipation factor:												
60 Hz	D 150	0.01	0.02	0.008	0.02	0.01		0.01	0.005	0.08	0.12	0.004
10^3 Hz	D 150	0.02	0.02	0.004	0.02	0.01	0.04	0.01	0.004	0.09	0.1	0.003
10^6 Hz	D 150	0.02	0.03	0.01	0.06	0.02		0.02	0.003		0.12	0.003
Surface resistivity, Ω/square:												
Dry	D 257	10^{15}	10^{15}	10^{15}	10^{12}	10^{14}		10^{11}	10^{13}		10^{10}	$>10^{12}$
After 96 h at 95°F, 90% RH	D 257	10^{11}	10^{13}	10^{15}	10^{10}			10^{10}	10^{12}		10^{10}	$>10^{12}$
Volume resistivity (dry), Ω-cm	D 257	$>10^{15}$	$>10^{15}$	10^{15}	10^{12}	10^{14}	10^{14}	10^{11}	10^{14}	10^{12}	10^{10}	$>10^{14}$
Mechanical Properties												
Tensile strength, lb/in²	D 638, D 412	10,000	9,000	4,000	1,800	4,600	6,000	800	275	4,000	2,400	110
Elongation, %	D 412				30		10	450	200	450	300	
Compression strength, lb/in²	D 695, D 575	25,000	16,000	13,000		7,000						85
Flexural modulus of rupture, lb/in²	D 790	10,000	12,000	8,000		8,300						
Izod impact strength (notched), ft-lb	D 256	0.3	0.4	0.3								
Shore hardness	D 676, D 1484				40D		80D	45A	35A	90A	80A	
Penetration at 77°F, mils	D 5											
Shrinkage on cooling or curing, % by vol		6	3.5	8	3	3	6	1	2	4.5		

TABLE 2.3 Properties of Various Liquid Resins *(Continued)*

Property	ASTM	Rigids (silica-filled)			Flexibilized epoxies			Liquid elastomers				Rigid polyurethane
		Styrene polyester	Epoxy	Silicone	Epoxy-polysulfide (50–50)	Epoxy-polyamide (50–50)	Epoxy-polyurethane (50–50, diamine cure)	Poly-sulfide	Silicone	Poly-urethane-(diamine cure)	Plas-tisol	Prepolymer
Physical Properties												
Specific gravity	D 71, D 792	1.6	1.6	1.8	1.2	1.0	1.2	1.2	1.1	1.1	1.2	0.1†
Minimum cold flow, °F												
Softening or drip point, °F	D 36											
Heat-distortion temp, °F	D 648	230	165	300			100					
Maximum continuous service temp, °F		250	250	480	225	175	250	200	350	210	150	165
Coefficient of thermal expansion per °F×10^{-6}	D 696	26	22	44	44	44	110		128	110		19
Thermal conductivity, Btu/(ft^2)(h)(°F/ft)		0.193	0.29	0.23			0.77		0.13		0.09	0.02
Cost, $/lb		0.25	0.50	2.75	0.90	0.80	1.10	0.85	4.25	1.30	0.30	1.75

*Short-time test on $^1/_8$-in specimen.
†6 lb/ft^3.
SOURCE: C. V. Lundberg, "A Guide to Potting and Encapsulation Materials," *Material Engineering,* May 1960.

TABLE 2.4 Characteristics Influencing Choice of Resins in Electrical Applications

Resin	Cure and handling characteristics	Final part properties
Epoxies	Low shrinkage, compatible with a wide variety of modifiers, very long storage stability, moderate viscosity, cure under adverse conditions	Excellent adhesion, high strength, available clear, resistant to solvents and strong bases, sacrifice of properties for high flexibility
Polyesters	Moderate to high shrinkage, cure cycle variable over wide range, very low viscosity possible, limited compatibility, low cost, long pot life, easily modified, limited shelf life, strong odor with styrene	Fair adhesion, good electrical properties, water-white, range of flexibilities
Polyurethanes	Free isocyanate is toxic, must be kept water-free, low cure shrinkage, solid curing agent for best properties	Wide range of hardness, excellent wear, tear, and chemical resistance, fair electricals, excellent adhesion, reverts in humidity
Silicones (flexible)	Some are badly cure-inhibited, some have uncertain cure times, low cure shrinkage, room temperature cure, long shelf life, adjustable cure times, expensive	Properties constant with temperature, excellent electrical properties, available from soft gels to strong elastomers, good release properties
Silicones (rigid)	High cure shrinkage, expensive	Brittle, high-temperature stability, electrical properties excellent, low tensile and impact strength
Polybutadienes	High viscosity, reacted with isocyanates, epoxies, or vinyl monomers, moderate cure shrinkage	Excellent electrical properties and low water absorption, lower strength than other materials
Polysulfides	Disagreeable odor, no cure exotherm, high viscosity	Good flexibility and adhesion, excellent resistance to solvents and oxidation, poor physical properties
Depolymerized rubber	High viscosity, low cure shrinkage, low cost, variable cure times and temperature, one-part material available	Low strength, flexible, low vapor transmission, good electrical properties
Allylic resins	High viscosity for low-cure-shrinkage materials, high cost	Excellent electrical properties, resistant to water and chemicals

SOURCE: From Ref. 10, p. 4.6.

TABLE 2.5 Characteristics of Liquid Resins for Nonelectrical Applications

Resin	Handling characteristics	Final part properties	Typical applications
Cast acrylics	Low to high viscosity, long cure times, bubbles a problem, special equipment necessary for large parts	Optical clarity, excellent weathering, resistance to chemicals and solvents	In glazing, furniture, embedments, impregnation
Cast nylon	Complex casting procedure, very large parts possible	Strong, abrasion-resistant, wear-resistant, resistance to chemicals and solvents, good lubricity	Gears, bushings, wear plates, stock shapes, bearings
Phenolics	Acid catalyst used, water given off in cure	High density unfilled, brittle, high temperature, brilliance	Billiard balls, beads
Vinyl plastisol	Inexpensive, range of viscosities, fast set, cure in place, one-part system, good shelf life	Same as molded vinyls, flame-resistant, range of hardness	Sealing gaskets, hollow toys, foamed carpet backing
Epoxies	Cures under adverse conditions, over a wide temperature range, compatible with many modifiers, higher filler loadings	High strength, good wear, chemical and abrasion resistance, excellent adhesion	Tooling, fixtures, road and bridge repairs, chemical-resistant coating, laminates, and adhesives
Polyesters	Inexpensive, low viscosity, good pot life, fast cures, high exotherm, high cure shrinkage, some cure inhibition possible, wets fibers easily	Moderate strength, range of flexibilities, water-white available, good chemical resistance, easily made fire retardant	Art objects, laminates for boats, chemical piping, tanks, aircraft, and building panels

Polyurethanes	Free isocyanate is toxic, must be kept water-free, low cure shrinkage, cast hot	Wide range of hardnesses, excellent wear, tear and, chemical resistance, very strong	Press pads, truck wheels, impellers, shoe heels and soles
Flexible soft silicone	High-strength materials, easily inhibited, low cure shrinkage, adjustable cure time, no exotherm	Good release properties, flexible and useful over wide temperature range, resistant to many chemicals, good tear resistance	Casting molds for plastics and metals, high-temperature seals
Allylic resins	Long pot life, low vapor pressure monomers, high cure temperature, low viscosity for monomers	Excellent clarity, abrasion resistant,- color stability, resistant to solvents and acids	Safety lenses, face shields, casting impregnation, as monomer in polyester
Depolymerized rubber	High viscosity, low cost, adjustable cure time	Low strength, low vapor transmission, resistant to reversion in high humidity	Roofing coating, sealant, in reservoir liners
Polysulfide	High viscosity, characteristic odor, low cure exotherm	Good flexibility and adhesion, excellent resistance to solvents and oxidation, poor physical properties	Sealants, leather impregnation
Epoxy vinyl esters	Low viscosity, high cure shrinkage, wets fibers easily, fast cures	Excellent corrosion resistance, high impact resistance, excellent electrical insulation properties	Absorption towers, process vessels, storage tanks, piping, hood scrubbers, ducts and exhaust stacks
Cyanate esters	Low viscosity at room temperature or heated, rapid fiber wetting, high temperature cure	High operating temperature, good water resistance, excellent adhesive properties and low dielectric constant	Structural fiber-reinforced products, high temperature film adhesive, pultrusion and filament winding

SOURCE: From Ref. 10, p. 4.7.

TABLE 2.6 Properties of Epi-Bis Resin with Various Hardeners

Hardener	phr*	Tensile strength, lb/in^2 at 25°C	Tensile modulus, lb/in^2 $\times 10^{-6}$	Tensile elonga-tion,%	Dielectric constant			Dielectric strength S/T, V/mil	Dissipation factor, 60 Hz at				Volume resistivity, MΩ-cm	Arc resistance, s (ASTM D 495)
					25°C	60 Hz at 50°C	100°C		25°C	50°C	100°C	150°C		
Aliphatic amines:														
Diethyl amino propylamine	7	8,500												58
Aminoethyl piperazine	20	10,000		5–6										62
Aromatic amines:														
Mixture of aromatic amines	23	12,600		6–5										80
p,p′-Methylene dianiline	27	9,500	0.4	4–5	4.4		4.6	420	0.007		0.002		$>1 \times 10^{10}$	83
m-Phenylene diamine	14.5	13,000	0.5	4	4.4		4.7	483	0.007		0.003		$>1 \times 10^{10}$	78
Diaminodiphenyl sulfone	30	7,000	0.47		4.22		4.71	410	0.004		0.068		$>2 \times 10^{8}$	65
Anhydrides:														
Methyl tetrahydrophthalic anhydride	84	12,200	0.44	6–4	3.0	3–0	3.0	377	0.006	0.005	0.005	0.09	$>2 \times 10^{9}$	110
Phthalic anhydride	78	8,500	0.5		3.5		3.8	390	0.004		0.006		$>5 \times 10^{9}$	95
"Nadic" methyl anhydride	80	11,400	0.40	5–6	3.3	3–4	3.4	443	0.003	0.002	0.004		$>4.7 \times 10^{10}$	110
BF_3MEA	3	5,900	0.4	1–8	3.6		4.1	480	0.004		0.049		13×10^{9}	120

*Parts hardener per hundred parts resin.
SOURCE: "Bahelite Liquid Epoxy Resins and Hardeners," Technical Bulletin, Union Carbide Corporation.

TABLE 2.7 Cast-Resin Data on Blends of Cycloaliphatic Epoxy Resins

Resin ERL-4221,* parts	100	75	50	25	0
Resin ERR-4090,* parts	0	25	50	75	100
Hardener, hexa-hydrophthalic (HHPA), phr†	100	83	65	50	34
Catalyst (BDMA), phr†	1	1	1	1	1
Cure, h/°C	2/120	2/120	2/120	2/120	2/120
Postcure, h/°C	4/160	4/160	4/160	4/160	4/160
Pot life, h/°C	>8/25	>8/25	>8/25	>8/25	>8/25
HDT (ASTM D 648), °C	190	155	100	30	−25
Flexural strength (D 790), lb/in^2	14,000	17,000	13,500	5,000	Too soft
Compressive strength (D 695) lb/in^2	20,000	23,000	20,900	Too soft	Too soft
Compressive yield (D 695), lb/in^2	18,800	17,000	12,300	Too soft	Too soft
Tensile strength (D 638), lb/in^2	8,000–10,000	10,500	8,000	4,000	500
Tensile elongation (D 638), %	2	6	27	70	115
Dielectric constant (D 150), 60 Hz:					
25°C	2.8	2.7	2.9	3.7	5.6
50°C	3.0	3.1	3.4	4.5	6.0
100°C	2.7	2.8	3.2	4.9	Too high
150°C	2.4	2.6	3.3	4.6	Too high
Dissipation factor:					
25°C	0.008	0.009	0.010	0.020	0.090
100°C	0.007	0.008	0.030	0.30	Too high
150°C	0.003	0.010	0.080	0.80	
Volume resistivity (D 257), Ω-cm	1×10^{13}	1×10^{12}	1×10^{11}	1×10^{8}	1×10^{6}
Arc resistance (D 495), s	>150‡	>150	>150	>150	>150

*Union Carbide Corp.
†Parts per 100 resin.
‡Systems started to burn at 120 s. All tests stopped at 150 s.
SOURCE: "Bakelite Cycloaliphatic Epoxides," Technical Bulletin, Union Carbide Corporation.

Novolac epoxy resins, phenolic or cresol novolacs, are reacted with epichlorohydrin to produce these novolac epoxy resins which cure more rapidly than the epi-bis epoxies and have higher exotherms. These cured novolacs have higher heat-deflection temperatures than the epi-bis resins as shown in Table 2.8. The novolacs also have excellent resistance to solvents and chemicals when compared with that of an epi-bis resin as seen in Table 2.9.

Cast polyesters. General purpose polyester, when blended with a monomer such as polystyrene and then cured, will produce rigid, rapidly curing transparent castings exhibiting the properties shown in

Table 2.10. Other monomers, in conjunction with polystyrene, such as alpha methyl styrene, methyl methacrylate, vinyl toluene, diallyl phthalate, triallyl cyanurate, divinyl benzene, and chlorostyrene, can be blended to achieve specific property enhancements. The reactivity of the polyester used, as well as the configuration of the product, affect the choice of systems.

Flexible polyester resins are available which are tougher and slower curing and produce lower exotherms and less cure shrinkage. They absorb more water and are more easily scratched but show more abrasion resistance than the rigid type. Their property profile is shown in

TABLE 2.8 Heat-Deflection Temperatures* of Blends of Novolac Epoxy and Epi-Bis Resins

Hardener	D. E. N. 438†	75/25	50/50	25/75	D. E. R. 332‡
TETA	§	§	133	126	127
MPDA	202	192	180		165
MDA	205	193	190	186	168
5% BF_3MEA	235			204	160
HET	225	213	205	203	196

*Heat-distortion temperature, °C (stoichiometric amount of curing agent—except BF_3 MEA—cured 15 h at 180°C).

†Novolac resin, Dow Chemical Co.

‡Epi-bis resin, Dow Chemical Co.

§The mixture reacts too quickly to permit proper mixing by hand.

SOURCE: "Dow Epoxy Novolac Resins," Technical Bulletin, Dow Chemical Company.

TABLE 2.9 Comparison of Chemical Resistance for Cured Epoxy Novolac and Epi-Bis Resin

Chemical	Weight gain,* %	
	D.E.N 438†	D.E.R. 331‡
Acetone	1.9	12.4
Ethyl alcohol	1.0	1.5
Ethylene dichloride	2.6	6.5
Distilled water	1.6	1.5
Glacial acetic acid	0.3	1.0
30% sulfuric acid	1.9	2.1
3% sulfuric acid	1.6	1.2
10% sodium hydroxide	1.4	1.2
1% sodium hydroxide	1.6	1.3
10% ammonium hydroxide	1.1	1.3

One-year immersion at 25°C; cured with methylene dianiline; gelled 16 h at 25°C, postcured $4^1/_2$ h at 166°C; D.E.N. 438 cured additional $3^1/_2$ h at 204°C.

*Sample size $^1/_2$ by $^1/_2$ by 1 in.

†Novolac epoxy resin, Dow Chemical Co.

‡Epi-bis resin, Dow Chemical Co.

SOURCE: "Dow Epoxy Novolac Resins," Technical Bulletin, Dow Chemical Company.

TABLE 2.10 Properties of Typical Polyester Producing Rigid Castings

Product specifications at 25°C	
Flash point, Seta closed cup, °F	89
Shelf life, minimum, months	3
Specific gravity	1.10–1.20
Weight per gallon, lb	9.15–10.0
% styrene monomer	31–35
Viscosity, Brookfield model LVF, #3 spindle at 60 r/min, cP	650–850
Gel time	
150–190°F, min	4–7
190°F to peak exotherm, min	1–3
Peak exotherm, °F	385–425
Color	Amber clear
Typical physical properties (clear casting)	
Barcol hardness (ASTM D 2583)	47
Heat-deflection temperature, °C (°F) (ASTM D 648)	87 (189)
Tensile strength, lb/in^2 (ASTM D 638)	8000
Tensile modulus, 10^5 lb/in^2 (ASTM D 638)	5.12
Flexural strength, lb/in^2 (ASTM D 790)	13,500
Flexural modulus, 10^5 lb/in^2 (ASTM D 790)	6.0
Compressive strength (ASTM D 695)	22,000
Tensile elongation, % at break (ASTM D 638)	1.5
Dielectric constant (ASTM D 150)	
At 60 Hz	2.97
At 1 MHz	2.87
Power factor	
At 1 kHz	0.005
At 1 MHz	0.017
Loss tangent at 1 MHz	0.017

SOURCE: From Ref. 10, p. 4.24.

Table 2.11. The two types, rigid and flexible, can be blended to produce intermediate properties, as shown in Table 2.12.

Cast polyurethanes. *Polyurethanes* are reaction products of an isocyanate, a polyol, and a curing agent. Because of the hazards involved in handling free isocyanate, prepolymers of the isocyanate and the polyol are generally used.

The choice of curing agent influences the curing characteristics and final properties. Diamines are the best general-purpose curing agent, as shown by Table 2.13. The highest physical properties are produced using MOCA 4,4-methyl-bis (2chloroaniline). The other major class of curing agents, the polyols, are more convenient to use but the final products have lower physical properties. By providing good abrasion resistance and a low coefficient of friction, polyurethanes find application in roller coatings and press pads as well as gaskets, casting molds, timing belts, wear strips, liners, and heels and soles.

TABLE 2.11 Properties of Typical Polyester Producing Flexible Material

Product specifications at 25°C	
Flash point, Seta closed cup, °F	89
Shelf life, minimum, months	3
Specific gravity	1.13–1.25
Weight per gallon, lb	9.4–10.4
% styrene monomer	18–22
Viscosity, Brookfield model	
LVF, #3 spindle at 60 r/min, cP	1100–1400
Gel time	
SPI 150–190 °F, min	6–8
190°F to peak exotherm, min	4.5–6.0
Peak temperature, °C (°F)	102–116 (215–240)
Color	Amber clear
Typical physical properties (clear casting)	
Tensile strength, lb/in^2 (ASTM D 638)	50
Tensile elongation, % (ASTM D 638)	10
Flexural strength, lb/in^2 (ASTM D 790)	Yields
Hardness, Shore D (ASTM D 2240)	15

SOURCE: From Ref. 10, p. 4.25.

TABLE 2.12 Properties of Blend of Rigid and Flexible Polyesters

Flexible polyester*	30%	20%	10%	5%	—
Rigid polyester†	70%	80%	90%	95%	100%
Tensile strength, lb/in^2 (ASTM D 638)	5200	8100	7300	6800	6500
Tensile elongation, % (ASTM D 638)	10.0	4.8	1.7	1.3	1.0
Flexural strength, lb/in^2 (ASTM D 790)	8100	13,200	15,600	14,600	13,500
Flexural modulus, 10^5 lb/in^2 (ASTM D 790)	2.40	3.75	5.60	5.80	6.00
Barcol hardness (ASTM D 2583)	0–5	20–25	30–35	35–40	40–45
Heat-deflection temperature, °C (°F) (ASTM D 648)	51 (124)	57.5 (136)	63 (145)	85 (185)	88 (190)

*Polylite 31-820, Reichhold Chemicals, Inc.
†Polylite 31-000, Reichhold Chemicals, Inc.
SOURCE: From Ref. 10, p. 4.25.

Cast phenolics. *Phenolic casting resins* are available as syrupy liquids produced in huge kettles by the condensation of formaldehyde and phenol at high temperature in the presence of a catalyst and the removal of excess moisture by vacuum distillation. These resins, when blended with a chemically active hardener, can be cast and cured solid in molds constructed from various materials and of a variety of mold designs. They will exhibit a broad-based property profile as described in Table 2.14.

TABLE 2.13 Properties of Polyurethane Resins Cured with Diamines and Polyols

Property	Diamine cures	Polyol cures
Resilience	Medium	Medium to high
Reactivity	Medium to high	Low to medium
Modulus	Medium to high	Low
Tensile strength	Medium to high	Low to medium
Ultimate elongation	Medium to high	High
Tear strength	High	Low to medium
Abrasion resistance:		
Sliding	High to very high	Low to medium
Impact	Medium to high	High to very high
Compression set	Medium	Low
Hardness	High	Low to medium

SOURCE: "Adiprene," Technical Bulletin, Elastomers Chemicals Dept., du Pont de Nemours & Company.

The mold designs available are draw molds, split molds, cored molds, flexible molds, and plaster molds.

Cast allylics.[11] The *allylic ester resins* possess excellent clarity, hardness, and color stability and thus are used to cast them into optical parts. These castings can be either homo or co-polymers. The free radical addition polymerization of the allylic ester presents some casting difficulties such as exotherm control, monomer shrinkage during curing and the interaction between the exotherm, the free radical source, and the environmental heat required to decompose the peroxide and initiate the reaction.

A simple casting formulation is as follows:

Prepolymer:	60 parts/wt
Monomer:	40 parts/wt
Tert-butyl perbenzoate:	2 parts/wt
Tert-butylcatechol:	0.1 parts/wt

The function of the catechol is to retard the polymerization and the exothermic heat over a longer time period, allowing the heat to dissipate and minimize cracking. Monomer catalyzed with benzoyl peroxide or tert-butyl perbenzoate may be stored at room temperature from 2 weeks to 1 year, but at 120°F (49°C) the catalyzed resin will gel in a few hours.

Molds for casting allylics may be ground and polished metal or glass, with glass being the preferred type, since it is scratch resistant and able to take a high polish.

Cast allylics are noted for their hardness, heat resistance, electrical properties, and chemical resistance as shown in Tables 2.15 and 2.16. They do lack strength and so their usage is confined to optical parts and some small electrical insulators.

TABLE 2.14 Properties of Cured Cast-Phenolic Resins

Property	Test method	Test values
Inherent properties:		
Specific gravity	D 792	1.30–1.32
Specific volume, in^3/lb	D 792	21.3–20.9
Refractive index, n_D	D 542	1.58–1.66
Cured properties:		
Tensile strength, lb/in^2	D 638–D 651	6,000–9,000
Elongation, %	D 638	1.5–2.0
Modulus of elasticity in tension, 10^5 lb/in^2	D 638	4–5
Compressive strength, lb/in^2	D 695	12,000–15,000
Flexural strength, lb/in^2	D 790	11,000–17,000
Impact strength, ft-lb/in notch ($^1/_2 \times {}^1/_2$-in notched bar, Izod test)	D 256	0.25–0.40
Hardness, Rockwell	D 785	M93–M120
Thermal conductivity, 10^4 cal/(s)(cm^2)(°C)(cm)	C 177	3–5
Specific heat, cal/(°C)(g)		0.3–0.4
Thermal expansion, 10^{-5} per °C	D 696	6–8
Resistance to heat (continuous), °F		160
Heat-distortion temperature, °F	D 648	165–175
Volume resistivity (50% RH and 23°C), Ω-cm	D 257	10^{12}–10^{13}
Dielectric strength ($^1/_8$-in thickness), V/mil:		
Short time	D 149	350–400
Step-by-step	D 149	250–300
Dielectric constant:		
60 Hz	D 150	5.6–7.5
10^5 Hz	D 150	5.5–6.0
10^6 Hz	D 150	4.0–5.5
Dissipation (power) factor:		
60 Hz	D 150	0.10–0.15
10^3 Hz	D 150	0.01–0.05
10^6 Hz	D 150	0.04–0.05
Arc resistance, s	D 495	200–250
Water absorption (24 h, $^1/_8$-in thickness), %	D 570	0.3–0.4
Burning rate	D 635	Very low
Effect of sunlight		Colors may fade
Effect of weak acids	D 543	None to slight, depending on acid
Effect of strong acids	D 543	Decomposed by oxidizing acids; reducing and organic acids, none to slight effect
Effect of weak alkalies	D 543	Slight to marked, depending on alkalinity
Effect of strong alkalies	D 543	Decomposes
Effect of organic solvents	D 543	Attacked by some
Machining qualities		Excellent
Clarity		Transparent, translucent, opaque

SOURCE: "Plastics Property Chart," *Modern Plastics,* MPE Supplement, 1970–1971.

TABLE 2.15 Properties of Allyl Ester Monomers

Property	Diallyl phthalate	Diallyl iso-phthalate	Diallyl maleate	Diallyl chloren-date	Diallyl adipate	Diallyl diglycollate	Triallyl cyanurate	Diethylene glycol bis-(allyl carbonate)
Density, at 20°C	1.12	1.12	1.076	1.47	1.025	1.113	1.113	1.143
Molecular wt	246.35	246.35	196	462.76	226.14	214.11	249.26	274.3
Boiling pt, °C, 4 mm Hg	160	181	111		137	135	162	160
Freezing pt, °C	−70	−3	−47	29.5	−33		27	−4
Flash pt, °C	166	340	122	210	150	146	>80	177
Viscosity at 20°C, cP	12	16.9	4.5	4.0	4.12	7.80	12	9
Vapor pressure at 20°C, mm Hg	27							
Surface tension at 20°C, dynes/cm	34.4	35.4	33		32.10	34.37		35
Solubility in gasoline, %	24	100	23		100	4.8	80	
Thermal expansion, in/(in)(°C)	0.00076							

SOURCE: From Ref. 11.

TABLE 2.16 Properties of Two Cured Allyl Esters

Property by ASTM procedures	Diallyl phthalate	Diallyl isophthalate
Dielectric constant:		
25°C and 60 Hz	3.6	3.5
25°C and 10^6 Hz	3.4	3.2
Dissipation factor:		
25°C and 60 Hz	0.010	0.008
25°C and 10^6 Hz	0.011	0.009
Volume resistivity, Ω-cm at 25°C	1.8×10^{16}	3.9×10^{17}
Volume resistivity, Ω-cm at 25°C (wet)*	1.0×10^{14}	
Surface resistivity, Ω at 25°C	9.7×10^{15}	8.4×10^{12}
Surface resistivity, Ω at 25°C (wet)*	4.0×10^{13}	
Dielectric strength, V/mil at 25°C†	450	422
Arc resistance, s	118	123–128
Moisture absorption, %, 24 h at 25°C	0.09	0.1
Tensile strength, lb/in²	3,000–4,000	4,000–4,500
Specific gravity	1.270	1.264
Heat-distortion temperature, °C at 264 lb/in²	155 (310°F)	238‡ (460°F)
Heat-distortion temperature, °C at 546 lb/in²	125 (257°F)	184–211 (364–412°F)
Chemical resistance, % gain in wt		
After 1 month immersion at 25°C in:		
Water	0.9	0.8
Acetone	1.3	−0.03
1% NaOH	0.7	0.7
10% NaOH	0.5	0.6
3% H_2SO_4	0.8	0.7
30% H_2SO_4	0.4	0.4

*Tested in humidity chamber after 30 days at 70°C (158°F) and 100% relative humidity.

†Step by step.

‡No deflection.

SOURCE: Directory/Encyclopedia Issue, *Insulation/Circuits,* June/July 1972.

Allylic monomers are sometimes used with alkyds to produce polyesters, with the orthophthalate resin being the most widely used because of its lower cost and very low water vapor pressures. Alkyd-diallylphthalate copolymers have significantly lower exotherm than an alkyd-styrene copolymer. The electrical properties of allylic resins are excellent and the variations of dissipation factor, dielectric constant, and dielectric strength with temperature and frequency are given in Figs. 2.12 and 2.13. The surface and volume resistivities remain

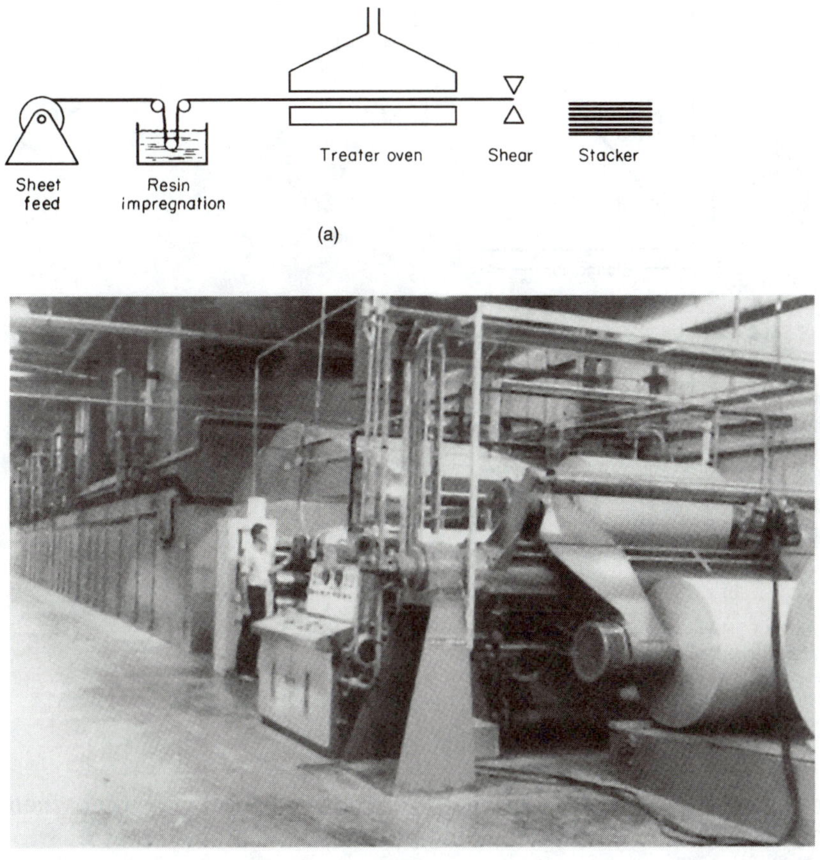

Figure 2.12 (*a*) Horizontal treater; (*b*) decorative laminate treater. (*From Charles A. Harper, Handbook of Plastics, Elastomers, and Composites, 3d ed., McGraw-Hill, New York, 1996, p. 2.2.*)

high after prolonged exposure to high humidity. Resistance to solvents and acids is excellent along with good resistance to alkalies. Tables 2.17 and 2.18 compare the chemical resistance of some plastics with the chemical resistance of the DAIP formulations.

2.6 Laminates[12]

2.6.1 Laminates

Laminates can be defined as combinations of liquid thermosetting resins with reinforcing materials that are bonded together by the application of heat and pressure, forming an infusible matrix. Plywood is a good example of a thermosetting laminate with the phenolic resin

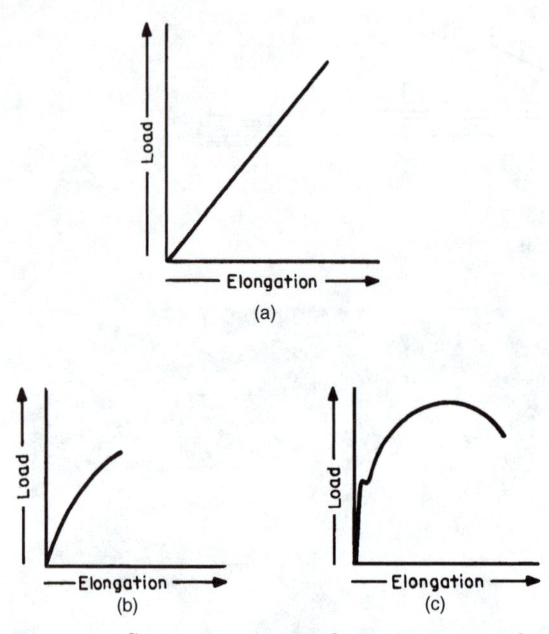

Figure 2.13 Stress-strain curves for various materials: (*a*) Reinforced plastics; (*b*) wood and most metals; and (*c*) steel. (*From Charles A. Harper, Handbook of Plastics, Elastomers, and Composites, 3d ed., McGraw-Hill, New York, 1996, p. 2.3.*)

serving as the binder to bond the layers of wood sheets together when compressed with heat in a molding press.

2.6.2 Resins

The resin systems primarily used for laminates are bismaleimides, epoxies, melamines, polyesters, polyimides, silicones, and phenolics and cyanate esters. These resins are described in Sec. 2.7.8, and the reinforcements are described in Sec. 2.6.3.

2.6.3 Reinforcements

Glass fibers are the most commonly used laminate reinforcement and are available in some six formulations with the E glass providing excellent moisture resistance, which, in turn, results in superior electrical properties along with other valuable properties. Properties of reinforced plastics using a variety of reinforcing fibers are shown in Table 2.19. Compositions of the major types of glass fibers are shown in Table 2.20. The glass fibers are available in filaments, chopped strands, mats, and fabrics in a wide variety of diameters, as described in Table 2.21.

TABLE 2.17 Chemical Resistance of Some Plastics (Change in wt on exposure, % after 24 hr)

	Melamine cellulose-filled	Poly-styrene high impact	Styrene polyester unfilled medium unsaturation	Epoxy resin unfilled	Diallyl ortho-phthalate
Acetone at 25°C	−0.75	*	+2.5	*	+0.05
Benzene at 25°C	+0.05	*	−0.02	+0.15	+0.01
Carbon tetrachloride at 25°C	−0.11	*	−0.01	−0.09	+0.01
95% Ethanol at 25°C	+0.013	+0.32	+0.22	−0.03	+0.06
Heptane at 25°C	+0.48	+70.0*(5)	−0.02	+0.21	+0.03
n-Butyl acetate at 70°C	−3.3	*	+2.82	+10.9	−3.3
1% Sodium hydroxide at 70°C	+3.10	+1.00	+1.49	−2.67	+1.07
20% Sodium hydroxide at 70°C	+2.10	+0.16	+0.89	−2.90	+0.43
1% Sulfuric acid at 70°C	+3.50	+1.00	+1.63	+0.10	+0.99
30% Sulfuric acid at 70°C	+4.6	+0.17	+1.12	+0.05	+0.61
1% Nitric acid at 70°C	+3.40	+1.0	+1.85	+0.66	+1.10
10% Nitric acid at 70°C	+8.52	+0.98	+1.73	−1.9	+1.0
3% Citric acid at 70°C	+3.1	+0.9	+1.53	+0.49	+1.1
Chromic acid at 70°C	+2.8	+1.1	+1.52	+0.65	+1.1
Mazola oil at 70°C	−1.89	+0.91	−0.25	−0.08	+0.05

*Disintegrated.
SOURCE: From Ref. 11.

TABLE 2.18 Chemical Resistance of Cast Diallyl Metaphthalate

Material	Time at 25°C	% wt Gain
Water	24 h	0.2
	7 days	0.4
	1 month	0.8
	Reconditioned*	0.6
Acetone	24 h	−0.03
	7 days	−0.03
	1 month	−0.03
	Reconditioned*	−0.07
1% NaOH	24 h	0.0
	7 days	0.3
	1 month	0.7
	Reconditioned*	0.6
10% NaOH	24 h	0.1
	7 days	0.3
	1 month	0.6
	Reconditioned*	0.5
3% H_2SO_4	24 h	0.1
	7 days	0.3
	1 month	0.7
	Reconditioned*	0.5
30% H_2SO_4	24 h	0.1
	7 days	0.2
	1 month	0.4
	Reconditioned*	0.4

*Followed after the immersion period by 2 h drying at 70°C.
SOURCE: From Ref. 11.

Glass fabrics: Many different fabrics are made for reinforced plastics, with E glass being the most common. Filament laminates using glass types of D, G, H, and K are also common with the filaments combined into strands, and the strands plied into yarns. These yarns can be woven into fabrics on looms.

2.6.4 Processes

The liquid resins are poured onto the reinforcement material and the combined resin-reinforcement sheet is then placed in a "horizontal treater," as shown in Fig. 2.12, where the resin is dissolved in a solvent to achieve optimum wetting or saturation of the resin into the reinforcement. The sheets come out of the treater and are sheared to size and stacked and stored in temperature- and humidity-controlled rooms. The preimpregnated, stacked sheets are further stacked into packs with each pack containing 10 laminates. The stacked units are then placed in between two polished steel plates and the press is closed. This process involves molding temperatures ranging from 250 to 400°F (122 to 204°C), with molding pressures in the 200- to 3000-lb/in² range.

TABLE 2.19 Properties of Reinforced Plastic and of Reinforcing Fiber

	Laminate properties									
Reinforcing fiber	Mechanical strength	Electrical properties	Impact resistance	Chemical resistance	Machining and punching	Heat resistance	Moisture resistance	Abrasion resistance	Low cost	Stiffness
Glass strands	X		X	X		X	X		X	X
Glass fabric	X	X	X	X		X	X			X
Glass mat			X	X		X	X		X	X
Asbestos		X	X			X				
Paper		X			X				X	
Cotton/linen	X	X	X		X				X	
Nylon		X	X	X				X		
Short inorganic fibers	X		X						X	
Organic fibers	X	X	X	X	X			X		
Ribbons		X					X			
Metals	X		X				X			X
Polyethylene	X		X		X					X
Aramid	X	X	X			X		X		X
Boron	X		X			X				X
Carbon/graphite	X		X			X				X
Ceramic	X		X						X	

SOURCE: From Ref. 12, p. 2.5.

TABLE 2.20 Composition of Glass Fibers

	Glass type*					
Ingredient	A	C	D	E	R	S and S2
SiO_2	72	65	74	52–56	60	65
CaO	10	14	0.5	16–25	9	—
Al_2O_3	0.6	4	0.3	12–16	25	25
MgO	2.5	3	—	0–5	6	10
B_2O_3	—	6	22	5–10	—	—
TiO_2	—	—	—	0–1.5	—	—
Na_2O	14.2	8	1.0	0–2	—	—
K_2O	—	—	1.5	—	—	—
Fe_2O_3	—	0.2	—	0–0.8	—	—
So_3	0.7	0.1	—	—	—	—
F_2	—	—	—	0–1.0	—	—

*Values represent weight percent of total.
SOURCE: From Ref. 12, p. 2.6.

The press may have 24 platens making a press load of some 240 laminates each cycle. After molding, the laminates are trimmed to final size and sometimes postcured by heating them in ovens.

2.6.5 Properties

Physical and mechanical. The use of reinforcements in combination with thermosetting resins will produce laminates exhibiting higher tensile, compressive, and flexural strengths due to the polymerization of the resin with its property enhancement contribution to the matrix. The impact strength sometimes improves by a factor of 10 and the laminates have no clear yield point, as shown in Fig. 2.13 (stress and strain).

Reinforced thermosetting laminates are anisotropic, with their properties differing depending upon the direction of measurement. The properties of a fabric laminate are controlled by the weave of the fabric and the number and density of the threads in the warp and woof directions, with these values differing from the values in the *z* of thickness direction. Both thermal expansion and conductivity properties are also anisotropic, as displayed in Table 2.22.

Electrical. The dielectric strength of laminates will decrease with increasing thickness and is highly dependent upon the direction of the electric field stress. This property will show higher when tested across the sample's thickness whereas end-to-end testing will show lower values. Laminates with higher resin content will show better electrical properties but poorer physical properties than laminates with lower resin content.

TABLE 2.21 Glass Fiber Reinforcement Forms

Form of fiberglass reinforcement	Definition and description	Range of grades available	General types of sizing applied	General usage in RP/C (and secondary uses if any)
Twisted yarns	Single-end fiberglass strands twisted on standard textile tube-drive machinery.	B to K fiber, S or Z twist, 0.25–10.0 twists/in, many fiber and yardage variations.	Starch.	Into single and plied yarns for weaving; many other industrial uses (and decorative uses).
Plied yarns	Twisted yarns plied with reverse twist on standard textile ply frames.	B to K fiber, up to $^4/_{18}$ ply—many fiber and yardage variations.	Starch.	Weaving industrial fabrics and tapes in many different cloth styles; also heavy cordage.
Fabrics	Yarns woven into a multiplicity of cloth styles with various thicknesses and strength orientations.	D to K fibers, 2.5–40 oz/yd^2 in weight.	Starch size removed and compatible finish; applied after weaving	Wet lay-up for open molding, prepreg, high pressure lamination, and also some press molding.
Chopped strands	Filament bundles (strands) bonded by sizings, subsequently cured and cut or chopped into short lengths. Also the reverse, i.e., chopped and cured.	G to M fiber, $^1/_8$- to $^1/_2$-in or longer lengths, various yardages.	LSB and RTP.	Compounding, compression, transfer and injection press molding.
Roving	Gathered bundle of one or more continuous strands wound in parallel and in an untwisted manner into a cylindrical package.	G or T fiber used. Roving yields 1800 to 28 yd/lb and packages 15- to 450-lb weight (size of up to 24 × 24 in).	HSB, LSB, and RTP.	Used in all phases of RP/C. Some (HSB) are used in continuous form, e.g., filament winding, and others (LSB) are chopped, as in sheet molding compound for compression molding.
Woven roving	Coarse fabric, bidirectional reinforcement, mostly plain weave, but some twill. Uni- and multidirectional nonwoven rovings are also produced.	K to T fiber, fabric weights 10–48 oz/yd^2.	HSB.	Mostly wet lay-up, but some press molding.

TABLE 2.21 Glass Fiber Reinforcement Forms *(Continued)*

Form of fiberglass reinforcement	Definition and description	Range of grades available	General types of sizing applied	General usage in RP/C (and secondary uses if any)
Chopped strand mats	Strands from forming packages chopped and collected in a random pattern with additional binder applied and cured; some "needled" mat produced with no extra binder required.	G to K fiber, weights 0.75–6.0 oz/ft^2.	HSB and LSB.	Both wet lay-up and press molding.
Mats, continuous strand (swirl)	Strands converted directly into mat form without cutting with additional binder applied and cured, or needled.	Nominally M to R fiber diameter, weights 0.75–4.5 oz/ft^2.	LSB.	Compression and press molding, resin transfer molding, and pultrusion.
Mat-woven roving combinations	Chopped strand mat and woven roving combined into a drapable reinforcement by addition of binder or by stitching.	30–62 oz/yd^2.	HSB.	Wet lay-up to save time in handling.
Three- dimensional reinforcements	Woven, knitted, stitched, or braided strands or yarns in bulky, continuous shapes.	—	HSB.	Molding, pultrusion.

Milled fibers	Fibers reduced by mechanical attrition to short lengths in powder or nodule form.	Screened from $^1/_{32}$–$^1/_4$ in. Actual lengths range 0.001–$^1/_4$ in. Several grades.	None, HSB, and RTP.	Casting, potting, injection molding, reinforced reaction injection molding (RRIM).
Related forms: Glass beads	Small solid or hollow spheres of glass.	Range 1–53 μm in diameter, bulk densities; hollow = 0.15–0.38 g/cm^3, solid = 1.55 g/cm^3.	Usually treated with cross-linking additives.	Used as filler, flow aid, or weight reduction medium in casting, lamination, and press molding.
Glass flake	Thin glass platelets of controlled thickness and size.	0.0001-in and up thick.	None, or treated with doupling agent.	Used as a barrier in, or to enhance abrasion resistance of, linear resins; coatings used for corrosion-resistance applications. Also used in RRIM for increased dimensional stability.

SOURCE: C. V. Lundberg, "A Guide to Potting and Encapsulation Materials," *Material Engineering,* May 1960.

TABLE 2.22 Thermal Coefficient of Expansion of Epoxy Glass Fabric Laminate and of Kevlar Fiber

Material	Direction	Coefficient of thermal expansion, 10^{-6}/°C
Epoxy	x	16.0
	y	12.7
	z	200
Kevlar fiber	x	−2
	z	60

SOURCE: From Ref. 12, p. 2.4.

Dimensional stability. Laminates using thermosetting resins are superior in stability, thermal resistance, and electrical properties, with dimensional stability the most important due to stresses incurred within the laminate during the molding and curing process. Laminate thickness is limited by the thermal conductivity of the polymerizing resin and the thermal conductivity of the cured laminate. The laminate derives its heat from the press platens to begin the curing action, and the resin, as it polymerizes, will produce substantial reactive heat, making it difficult to produce very thick laminates. The normal thickness range for laminates will range from 0.002 to about 2.000 in with special grades produced at 4 to 10 in.

2.7 Molding Compounds

2.7.1 Resin systems

Thermosetting resin systems are the backbone of a large, versatile, and important family of molding compounds which provides the industrial, military, and commercial markets with plastic molding materials exhibiting exceptional electrical, mechanical, thermal, and chemical properties. These important property values enable the product designer, manufacturing engineer, and research and design engineer to select from a wide choice of products, and enable them to choose the most suitable molding compound to meet their specific needs and requirements. These versatile materials are covered by military, industrial, and commercial specifications that are designed to ensure the quality of the molded articles utilizing thermosetting molding compounds.

The following sections will take each of the family members, in alphabetical order, and outline their history; specifications; and reinforcements, fillers, and data sheet value applications.

2.7.2 Allyls: Diallyl-ortho-phthalate (DAP) and diallyl-iso-phthalate (DAIP)

History.[13] The most sophisticated work on saturated polyesters is usually traced to W. H. Carothers, who, from 1928 to 1935 with DuPont, studied polyhydroxy condensates of carbolic acids. Unable to achieve suitable heat and chemical resistance from these esters, he turned to polyamide-carboxylic acid reaction products, which later became *nylon*. Carother's concepts on saturated polyesters yielded several other polyester products such as Terylene which became patented and introduced in the United States as *dacron* and *mylar*..

In 1937 Carlton Ellis found that unsaturated polyesters, which are condensation products of unsaturated dicarboxylic acids and dihydroxy alcohols, would freely co-polymerize with monomers which contained double bond unsaturation, yielding rigid thermosetting resins. The allylic resins became commercial resins suitable for compounding into a broad range of products with exceptional electrical, mechanical, thermal, and chemical properties.

Physical forms. Molding compounds using either the DAP or the DAIP resin systems are available in free-flowing granular form and also in high-bulk factor flake form. The resins are in a white powder form, which makes it possible to provide a broad opaque color range. Both compounds in their granular form are readily molded, preformed, or preplasticated automatically, whereas the high-bulk factor compounds generally require auxiliary equipment for such operations.

Reinforcements. Compounds of either DAP or DAIP resin systems utilize a variety of reinforcement materials ranging from the granular compounds with mineral, glass, and synthetic fibers to the high-strength, high-bulk factor types employing long fibers of cotton flock, glass, acrylic, and polyethylene terephthalate.

Specifications. DAP and DAIP molding compounds meet the requirements of several commercial and military specifications as well as those of certain industrial and electronic firms who will often use the Mil spec in conjunction with their own special requirements.

Military. Mil-M-14 DAP (ortho) compounds:

MDG: Mineral filler, general purpose

SDG: Short glass filler, general purpose

SDG-F: Short glass filler, flame resistant

SDI-5: Orlon filler, impact value of 0.5 ft-lb

SDI-30: Dacron filler, impact value of 3.0 ft-lb

GDI-30: Long glass filler, impact value of 3.0 ft-lb

GDI-30F: Long glass filler, impact value of 3.0 ft-lb, flame resistant

GDI-300: Long glass filler, impact value of 30 ft-lb

GDI-300F: Long glass filler, impact value of 30 ft-lb, flame resistant

DAIP (iso) compounds. Mil-M-14 H lists designations for the (iso) compounds and refers to them as heat-resistant compounds:

MIG: Mineral filler, heat resistant

MIG-F: Mineral filler, heat and flame resistant

SIG: Short glass filler, heat resistant

SIG-F: Short glass filler, heat and flame resistant

GII-30: Long glass filler, heat resistant, impact value of 3.0 ft-lb

GII-30-F: Long glass filler, heat and flame resistant, impact value of 30 ft-lb

Note: All of these DAIP compounds will perform without substantial property loss during exposure to elevated temperatures of 490°F (200°C).

ASTM. Under D-1636-75 A, both DAP and DAIP are listed by type, class and grade as shown here:

Type	Class	Grade
I	A Ortho (DAP)	1. Long glass
	B Ortho FR (DAP)	2. Medium glass
	C Iso (DAIP)	3. Short glass
II	A Ortho (DAP)	1. Mineral
	B Ortho FR (DAP)	2. Mineral and organic fiber
	C Iso (DAIP)	1. Mineral
		2. Mineral and organic fiber
III	A Ortho (DAP)	1. Acrylic fiber
		2. Polyester fiber, long
		3. Polyester fiber, milled

Underwriters Laboratory (UL). UL rates DAP and DAIP thermally at 266°F (130°C), with a flame resistance rating of 94 VO in a $^{1}/_{16}$-in section. Neither the DAP nor DAIP compounds have been submitted to UL's long-term thermal testing. However, MIL-M-14 H does recognize their excellent long-term heat resistance and ability to retain their initial property profile under adverse thermal conditions over lengthy time spans.

Data sheet values. Table 2.23 includes a typical property value sheet for a wide range of allyl molding compounds, and Table 2.24 summarizes the allyl molding compound properties.

Applications. The DAP and DAIP molding compounds are used primarily in applications requiring superior electrical and electronic insulating properties with high reliability when under the most severe environmental conditions. A few such applications are military

TABLE 2.23 Allyl Molding Compound Data Sheet Values *(Continued)*

Property*	ASTM test method	Grade and Mil-M-14 type 73-70-70 C SDG-F	73-70-70 R SDG-F	FS-10 VO SDG-F	52-20-30 GDI-30	FS-4 GDI-30	52-40-40 GDI-30F	FS-80 GDI-30F
Reinforcement	—	Short glass	Short glass	Short glass	Long glass	Long glass	Long glass	Long glass
Resin-isomer (DAP)	—	Ortho	Ortho	Iso	Ortho	Iso	Ortho	Iso
Form	—	Granular	Granular	Granular	Flake	Flake	Flake	Flake
Bulk factor	—	2.4	2.3	2.3	6	6	6	6
Specific gravity, g/cm^3	D-792A	1.91	1.85	1.91	1.72	1.64	1.79	1.74
Shrinkage (comp), in/in	D-955	0.002–0.004	0.002–0.004	0.002–0.004	0.0025	0.0025	0.0025	0.0025
Izod impact, ft/lb/in	D-256A	0.6	0.6	0.6	4.0	3.6	3.9	3.4
Flexural strength, lb/in^2	D-790	13,000	14,000	13,000	15,000	15,000	15,000	15,000
Flexural modulus, $lb/in^2 \times 10^6$	D-790	2.0	1.9	2.0	1.3	1.4	1.3	1.4
Tensile strength, lb/in^2	D-638	7,000	6,500	8,000	—	—	—	—
Compressive strength, lb/in^2	D-695	—	—	—	—	—	—	—
Water absorption, %, 24 h at 100°C and 48 h at 50°C in H_2O	D-570	0.25	0.30	0.35	0.35	0.35	0.32	0.35
CTE, in/in °C $\times 10^{-5}$	D-696	—	—	—	—	—	—	—
Deflection temperature, °F	D-648A	500	450	500	500	500	500	500
Flammability (ign./burn), s	Fed. 2023	90A 90	90A 90	90A 90	—	—	90A 90	90A 90
UL flammability rating								
$^1/_8$ in	UL 94	94V-O	94V-O	94V-O	—	—	—	—
$^1/_{16}$ in	UL 94	94V-O	94V-O	94V-O	—	—	—	—
Comparative tracking index, s	—	600+	600+	600+	600+	600+	600+	600+
Dielectric strength 60 Hz ST/SS wet, V/mil	D-149	350	350	350	350	350	350	340
Dielectric constant, 1 kHz/1 MHz wet	D-150	4.4	4.2	4.6	4.2	4.0	4.3	4.1
Dissipation factor, 1 kHz/1 MHz wet	D-150	0.014	0.015	0.016	0.016	0.014	0.016	0.015
Arc resistance, s	D-495	130	125	175	130	138	133	145

TABLE 2.23 Allyl Molding Compound Data Sheet Values *(Continued)*

		Grade and Mil-M-14 type						
Property*	ASTM test method	73-70-70 C SDG-F	73-70-70 R SDG-F	FS-10 VO SDG-F	52-20-30 GDI-30	FS-4 GDI-30	52-40-40 GDI-30F	FS-80 GDI-30F
Reinforcement	—	Short glass	Short glass	Short glass	Long glass	Long glass	Long glass	Long glass
Resin-isomer (DAP)	—	Ortho	Ortho	Iso	Ortho	Iso	Ortho	Iso
Form	—	Granular	Granular	Granular	Flake	Flake	Flake	Flake
Bulk factor	—	2.4	2.3	2.3	6	6	6	6
Specific gravity, g/cm^3	D-792A	1.91	1.85	1.91	1.72	1.64	1.79	1.74
Shrinkage (comp), in/in	D-955	0.002–0.004	0.002–0.004	0.002–0.004	0.0025	0.0025	0.0025	0.0025
Izod impact, ft/lb/in	D-256A	0.6	0.6	0.6	4.0	3.6	3.9	3.4
Flexural strength, lb/in^2	D-790	13,000	14,000	13,000	15,000	15,000	15,000	15,000
Flexural modulus, lb/in^2 × 10^6	D-790	2.0	1.9	2.0	1.3	1.4	1.3	1.4
Tensile strength, lb/in^2	D-638	7,000	6,500	8,000	—	—	—	—
Compressive strength, lb/in^2	D-695	—	—	—	—	—	—	—
Water absorption, %, 24 h at 100°C and 48 h at 50°C in H_2O	D-570	0.25	0.30	0.35	0.35	0.35	0.32	0.35
CTE, in/in °C × 10^{-5}	D-696	—	—	—	—	—	—	—
Deflection temperature, °F	D-648A	500	450	500	500	500	500	500
Flammability (ign./burn), s	Fed. 2023	90A 90	90A 90	90A 90	—	—	90A 90	90A 90
UL flammability rating								
1/8 in	UL 94	94V-O	94V-O	94V-O	—	—	—	—
1/16 in	UL 94	94V-O	94V-O	94V-O	—	—	—	—
Comparative tracking index, s	—	600+	600+	600+	600+	600+	600+	600+
Dielectric strength 60 Hz ST/SS wet, V/mil	D-149	350	350	350	350	350	350	340
Dielectric constant, 1 kHz/1 MHz wet	D-150	4.4	4.2	4.6	4.2	4.0	4.3	4.1
Dissipation factor, 1 kHz/1 MHz wet	D-150	0.014	0.015	0.016	0.016	0.014	0.016	0.015
Arc resistance, s	D-495	130	125	175	130	138	133	145

Property*	ASTM test method	Grade and Mil-M-14 type							
		RX 2-520	RX 1-520 RX 1310	RX 3-2-520F	RX 3-1-525 RX 1366FR	RX-1-530	RX-2-530	RX 3-1-530	RX-2-530
		SDG	SDG	SDG-F	SDG-F	GDI-30	GDI-30	GDI-30F	GDI-30F
Reinforcement	—	Short glass	Short glass	Short glass	Short glass	Long glass	Long glass	Long glass	Long glass
Resin-isomer (DAP)	—	Ortho	Ortho	Iso	Ortho	Ortho	Iso	Ortho	—
Form	—	Granular	Granular	Granular	Granular	Flake	Flake	Flake	Flake
Bulk factor	—	2.5	2.4	2.4	2.3	3.5	3.5	3.5	—
Specific gravity, g/cm^3	D-792A	1.75	1.83	1.90	1.87	1.74	1.73	1.76	1.00
Shrinkage (comp), in/in	D-955	0.001–0.003	0.001–0.003	0.001–0.003	0.001–0.003	0.001–0.004	0.001–0.004	0.001–0.004	0.000–0.000
Izod impact, ft/lb/in	D-256A	0.80	0.70	0.80	0.80	3.0–7.0	3.0–7.0	3.0–7.0	3.0–0.0
Flexural strength, lb/in^2	D-790	22,000	22,000	23,000	23,000	23,000	23,000	23,000	23,000
Flexural modulus, lb/in^2 $\times 10^6$	D-790	2.6	2.6	2.6	2.6	1.1	2.0	2.0	0.0
Tensile strength, lb/in^2	D-638	12,000	11,000	12,000	12,000	10,000	10,000	10,000	10,000
Compressive strength, lb/in^2	D-695	25,000	25,000	25,000	25,000	27,000	27,000	27,000	27,000
Water absorption, %, 24 h at 100°C and 48 h at 50°C in H_2O	D-570	0.25	0.25	0.25	0.25	0.35	0.35	0.35	0.00
CTE, in in °C $\times 10^{-5}$	D-696	1.7	1.5	1.2	1.5	1.6	1.5	2.2	0.0
Deflection temperature, °F	D-648A	525	400	525	400	500	550	500	550
Flammability (ign./burn), s	Fed. 2023	—	—	1120 40	110 40	—	—	110 40	110 00
UL flammability rating									
$^1/_8$ in	UL 94	94 HB	—	94 VO	94 VO	—	—	94 VO	
$^1/_{16}$ in	UL 94	94 HB	—	94 VO	94 VO	—	—	94 VO	
Comparative tracking index, s	—	—	—	600+	600+	—	—	600+	
Dielectric strength 60 Hz ST/SS wet, V/mil	D-149	450/400	450/400	450/400	450/400	400/375	400/375	400/375	400/375
Dielectric constant, 1 kHz/1 MHz wet	D-150	3.9/3.7	4.0/3.5	4.1/3.9	4.2/3.5	0.0/0.0	3.8/3.6	4.1/3.5	4.1/0.0
Dissipation factor, 1 kHz/1 MHz wet	D-150	0.011–0.015	0.009–0.016	0.010–0.013	0.010–0.016	0.011–0.019	0.010–0.019	0.010–0.018	0.010–0.019
Arc resistance, s	D-495	150	130	150	130	135	135	130	135

*Certification to Mil-M-14 requires batch testing. All testing in accordance with Mil-M-14. Values are typical and not statistical minimums.
SOURCE: FROM REF. 11.

TABLE 2.24 Summary of Properties of Allyls

Summary of properties of allyls
Physical:
Excellent long-term dimensional stability
Virtually no postmold shrinkage
Chemically inert
Mechanical:
Excellent flexural strength
High impact resistance
Highly resistant to sudden, extreme jolts and severe stresses
Electrical:
Retains high insulation resistance at elevated temperatures
Retains performance characteristics at high ambient humidity
Thermal:
Successful performance in vapor-phase soldering environments—419° and 487°F (215° and 253°C)
FS-10VO:200°C UL temperature index
Isophthalate materials are recommended for very high temperature applications
Chemical:
Highly resistant to solvents, acids, alkalies, fuels, hydraulic fluids, plating chemicals, and sterilizing solutions

SOURCE: From Ref. 11.

and commercial connectors, potentiometer housings, insulators, switches, circuit boards and breakers, x-ray tube holders, and TV components.

Molded articles of either the DAP or the DAIP compounds have a nil lifetime shrinkage after their cooling-off shrinkage. This feature has been one of the main reasons for the use of DAP and DAIP compounds in the military connector field. Parts molded today will fit parts molded years ago.

Suppliers

Rogers Corporation

Cosmic Plastics, Inc.

(See App. C for supplier addresses.)

2.7.3 Aminos (Urea Melamine)

History[14]

Urea. Urea formaldehyde resins came into being through work done by Fritz Pollack and Kurt Ritter along with Carleton Ellis in the 1920s. This was followed up by the introduction of a white urea compound, by Dr. A. M. Howald, named *Plaskon* and was sold by Toledo Synthetic Products Company which, in turn, became a part of Allied Chemical Company.

Melamines. Melamine was isolated in 1834 and it wasn't until 1933 that Palmer Griffith produced dicyanamide and found that it contained melamine. The addition of formaldehyde produced a resin which could be compounded into a desirable molding compound. This new compound had a number of desirable qualities superior to phenolics and ureas of that time. The colorability and surface hardness led to its use in molded dinnerware along with some very important military and electrical applications.

Physical forms. The urea molding compounds are available as free-flowing granular products that are readily preformable and can be preheated and preplasticated prior to molding. They mold very easily in all thermosetting molding methods. The melamine compounds are available as both free-flowing granular products and as high-strength, high-bulk factor materials. The bulky materials require auxiliary equipment for preforming or preplasticating.

Reinforcements. The reinforcements are available as purified cellulose fibers, minerals, chopped cotton flock, wood flour, and glass fibers (short and long).

Specifications

Military (Mil-M-14). Urea compounds are not listed.

Melamine

CMG: Cellulose filler, general purpose

CMI-5: Cellulose filler, impact value of 0.5 ft-lb

CMI-10: Cellulose filler, impact value of 1.0 ft-lb

MMD: Mineral filler, general purpose

MMI-5: Glass filler, impact value of 0.5 ft-lb

MMI-30: Glass filler, impact value of 3.0 ft-lb

ASTM

Urea D-705

Melamine D-704

UL

	Thermal index	Flame resistance
Urea	221°F (105°C) maximum	94VO in $^1/_{16}$-in section
Melamine	220 to 292°F (130 to 150°C)	94 VO in $^1/_{16}$-in section

Data sheet values. Table 2.25 provides a list of property values for a range of urea compounds and Table 2.26 shows a similar list for melamine compounds.

TABLE 2.25 Amino Molding Compound Data Sheet Values—Urea

	Filler				
	α Cellulose	Mineral	Chopped cotton fabric	Cellulose	Glass
Specific gravity	1.49	1.78	1.5	1.5	1.97
Flex strength, lb/in^2	10,000	7,600	12,000	7,000	10,500
Impact strength, ft-lb	0.25	0.3	0.6	0.25	3.0–4.0
Water absorption, % gain (24 h at 25°C)	0.4	0.15	0.5	0.65	0.2
Dielectric strength, V/mil					
Short time	300	350	250	350	170
Step × step	250	250	100	200	170
Arc resistance, s	120	120	125	70	180

SOURCE: From Ref. 11.

TABLE 2.26 Amino Molding Compound Data Sheet Values—Melamine

	Filler	
	α Cellulose	Cellulose
Specific gravity	1.49	1.49
Flex strength, lb/in^2	8000	7500
Impact strength, ft-lb	0.2–0.3	0.2–0.3
Water absorption, % gain (24 h at 25°C)	0.6	0.6
Dielectric strength, V/mil		
Short time	350	350
Step × step	275	275
Arc resistance, s	120	120

SOURCE: From Ref. 17.

Applications

Urea. Molded urea articles will have a high-gloss finish that is scratch resistant and used in the following applications: closures, control housings, wiring devices, control buttons, electric shaver housings, and knobs.

Melamines. Melamine molded products are used in commercial, industrial, and military applications, taking advantage of the variety of reinforcements available, scratch resistance, and very wide color range. Typical military uses are for connector bodies and circuit breaker housings. Typical commercial uses are for dinnerware, shavers, knobs, buttons, ashtrays, and connector bodies.

Note: Combinations of phenolic and melamine resins in molding compounds have produced materials with excellent color stability, ease of moldability, and good heat resistance. Their usage is in appliance components such as pot or pan handles.

Suppliers. Amino compounds:.

American Cyanamide Company, Polymer Products Division

ICI/Fiberite

Plenco Engineering Company

(See App. C for supplier addresses.)

2.7.4 Epoxies

History. The Shell Chemical Corporation introduced the *epoxy resin systems* into the United States in 1941, and their good property profile has been utilized in a wide range of applications. The molding compounds are available in extreme soft flows and long gelation times, which make them very adaptable for encapsulation molding techniques in the encapsulation of electronic components such as integrated circuits, resistors, diode capacitors, relays, and bobbins.

The compounds also found a market in the commercial and military industrial areas for connector bodies, potting shells, printed circuit boards, coils, and bobbins with the higher-strength and higher filler and pressure compounds that compete with the phenolics and DAPs.

Physical forms. The molding compounds are available in a free-flowing granular form suitable for automatic preforming or preplasticating, and are readily moldable in all thermosetting molding techniques. The higher-impact, bulkier compounds mold readily but require special auxiliary equipment for either preforming or preplasticating.

Reinforcements. The standard compounds (Mil-M-24325) are available with mineral fillers and also with short and long glass fibers. The low-pressure encapsulation compounds have fused or chrystalline silica reinforcements and are in a free-flowing granulate state.

Specifications

Military (Mil-M-24325)

MEC: Mineral filler, encapsulation grade, low pressure

MEE: Mineral filler, electrical grade, low pressure

MEG: Mineral filler, general purpose

MEH: Mineral filler, heat resistant

GEI: Glass filler, impact value of 0.5 ft-lb

GEI-20: Glass filler, impact value of 2.0 ft-lb

GEI-100: Glass filler, impact value of 10.0 ft-lb

ASTM. These specifications are listed under D-3013-77 and D-1763.

UL. These specifications have a thermal index rating of 266°F (130°C) and a flammability rating of 94 VO.

Data sheet values. Table 2.27 contains a list of property values for a wide range of epoxy molding compounds—both for the low-pressure encapsulation types and the high-pressure types.

Applications. Due to retention of their excellent electrical, mechanical, and chemical properties at elevated temperature and very high moisture resistance, the high-pressure Mil spec. molding compounds have found their market niche in high-performance military and commercial applications such as connectors, potting shells, relays, printed circuit boards, switches, coils, and bobbins.

The low-pressure encapsulation compounds, which exhibit the same high-performance characteristics as the high-pressure compounds, have become the primary insulating medium for the encapsulation of components such as integrated circuits, resistors, coils, diodes, capacitors, relays, and bobbins.

Suppliers

Cytec Fiberite

Dexter Electronics, Materials Division

Morton Chemical

(See App. C for supplier addresses.)

2.7.5 Phenolics

History. *Phenolic resins* came into being when Dr. Leo Baekeland, in the early 1900s, discovered that a successful reaction between phenol and formaldehyde in a heated pressure kettle produced an amber-colored liquid thermosetting resin. This resin became the foundation of the entire thermosetting molding compound industry, and an entire family of thermosetting resins and compounds were developed over the next several decades. Phenolic molding compounds became the primary insulating material for a wide and diversified range of applications for industrial, commercial, and military applications.

Physical forms. The novolac-based molding compounds are available as free-flowing granular powders or pellets in their general purpose grades and in a variety of large pellets and flakes in the high-strength bulky grades. The general purpose grades are easily preformed and

TABLE 2.27 Epoxy Molding Compound Data Sheet Values

Property*	ASTM test method	Class					
		Hardware grade					
		1904B	1906	2004B	1907	1908 1908B	1914
Reinforcement		Mineral glass	Mineral glass	Mineral glass	Short-glass	Short-glass	Short-glass
Bulk factor	—	2.4	2.3	2.3	2.2	2.3	2.1
Form	—	Granular	Granular	Granular	Granular	Granular	Granular
Specific gravity, g/cm^3	D-792	2.05	1.90	1.95	1.95	1.85	1.94
Mold shrinkage, in/in	D-955	0.001–0.003	0.001–0.003	0.002–0.004	0.002–0.004	0.002–0.004	0.001–0.003
Impact strength, ft/lb/in	D-256	0.61	0.40	0.60	0.58	0.55	0.61
Flexural strength, RT, lb/in^2	D-790	17,000	15,000	16,000	15,000	17,000	17,000
Flexural modulus, $lb/in^2 \times 10^6$	D-790	1.9	2.1	1.8	1.9	1.9	2.2
Tensile strength, lb/in^2	D-651	10,000	10,000	10,000	10,000	9,000	10,500
Compressive strength, lb/in^2	D-695	32,000	32,000	32,000	32,000	33,000	31,000
Water absorption, %, 48 h at 122°F (50°C)	D-570	0.2	0.2	0.2	0.3	0.3	0.3
Barcol hardness	D-2583	72	70	72	73	72	70
Deflection temperature, °F	D-648	500	340	500	450	500	500
Coefficient of linear thermal expansion, in/in/°C $\times 10^{-6}$ (−30°C±30°C) (−30°C–+30°C)	D-696	28	37	33	47	37	41
Coefficient of thermal conductivity, $cal/s/cm^3$/°C/cm$\times 10^{-4}$	—	17.5	24	14.5	14	17.5	14

TABLE 2.27 Epoxy Molding Compound Data Sheet Values *(Continued)*

Property*	ASTM test method	Class: Electrical encapulation grade 1960B	1961B	2060B	2061B
Reinforcement		Mineral glass	Mineral glass	Mineral glass	Mineral glass
Bulk factor	—	2.5	2.3	2.5	2.3
Form	—	Granular	Granular	Granular	Granular
Specific gravity, g/cm^3	D-792	1.75	1.90	1.80	1.95
Mold shrinkage, in/in	D-955	0.004–0.006	0.002–0.001	0.004–0.006	0.002–0.001
Impact strength, ft/lb/in	D-256	0.56	0.63	0.60	0.61
Flexural strength, RT, lb/in^2	D-790	15,000	14,000	15,000	14,000
Flexural modulus, lb/in^2 $\times 10^6$	D-790	1.7	2.0	1.7	2.0
Tensile strength, lb/in^2	D-651	9,000	10,500	9,000	11,000
Compressive strength, lb/in^2	D-695	30,000	31,500	30,500	31,000
Water absorption, %, 48 h at 122°F (50°C)	D-570	0.3	0.2	0.3	0.3
Barcol hardness	D-2583	65	65	65	70
Deflection temperature, °F	D-648	350	425	375	450
Coefficient of linear thermal expansion, in/in/°C $\times 10^{-6}$ (−30°C±30°C) (−30°C−+30°C)	D-696	38	38	35	42
Coefficient of thermal conductivity, cal/s/cm^3/°C/cm $\times 10^{-4}$	—	14	16	17	13
Dielectric constant (1 MHz/24 h at 23°C H_2O)	D-150	4.2	4.5	4.0	4.2
Dissipation factor (1 MHz/24 h at 23°C H_2O)	D-150	0.02	0.02	0.02	0.02
Dielectric strength (S/S) (VPM/48 h at 50°C H_2O)	D-149	325	325	320	320
Arc resistance, s	D-495	185	186	170	170
Insulation resistance (MΩ/30 days at 150°F, 95% RH)	D-257	10	10	10	10
Oxygen index	D-2863	50	3208	—	45.0
UL flammability rating					
1/4 in	UL 94	—	—	—	—
1/16 in	UL 94	—	—	—	94 VO
Mil-M-14, type	—	—	—	—	

*Certification to Mil-M-14 requires batch testing. All testing in accordance with Mil-M-14. Values are typical and not statistical minimums.

SOURCE: From Ref. 17.

preplasticated while the bulky products often require auxiliary equipment for such operations.

The resole-based compounds are generally only available as granular powder or small pellets. Both resin system compounds are easily molded in all thermosetting molding procedures.

Reinforcements. Both the resole and novolac compounds use a broad array of reinforcements to meet the demands of the market place: wood flour, cotton flock, minerals, chopped fabric, Teflon, glass fibers (long and short), nylon, rubber, and kevlar. Asbestos, which had been a widely used filler in many thermosetting compounds, has been replaced over the past 20 years with glass fiber–reinforced phenolic compounds in many applications.

Specifications

Military (Mil-M-14)

CFG: Cellulose filler, general purpose

CFI-5: Cellulose filler, impact value of 0.5 ft-lb

CFI-10: Cellulose filler, impact value of 1.0 ft-lb

CFI-20: Cellulose filler, impact value of 2.0 ft-lb

CFI-30: Cellulose filler, impact value of 3.0 ft-lb

CFI-40: Cellulose filler, impact value of 4.0 ft-lb

MFE: Mineral filler, best electrical grade

MFH: Mineral filler, heat resistant

GPG: Glass filler, general purpose

GPI-10: Glass filler, impact value of 1.0 ft-lb

GPI-20: Glass filler, impact value of 2.0 ft-lb

GPI-30: Glass filler, impact value of 3.0 ft-lb

GPI-40: Glass filler, impact value of 4.0 ft-lb

ASTM. The ASTM specifications can be found in D-700-65 and D-4617.

UL. The UL specification has a thermal index rating of 155°C (378°F) for most compounds but the glass-reinforced grades and some phenolic alloys carry a 185°C (417°F) rating. The flame retardant rating is at 94 VO in a $^1/_{16}$-in section.

Data sheet values. Table 2.28 lists the property values for a broad range of phenolic molding compounds, including both the resole (single-stage) and the novolac (two-stage) compounds.

Applications. The property profile, which includes a broad range of reinforcements, provides design engineers with great flexibility in

TABLE 2.28 Phenolic Molding Compound Data Sheet Values

Grade	Special self-lube RX 342	RX 525	RX 448	RX 431	RX 475	RX 466	RX 468
Reinforcement type	Cellulose	Cellulose	Cellulose	Cellulose	Cellulose	Asbestos	Asbestos
Phenolic resin type	Two-step	Two-step	Two-step	Two-step	Two-step	Two-step	Two-step
Form	Nodular	Nodular	Nodular	Nodular	Nodular	Nodular	Nodular
Bulk factor	2.8	2.7	2.6	3.5	3.3	2.1	2.2
Specific gravity, g/cm^3	1.44	1.39	1.45	1.41	1.40	1.69	1.72
Shrinkage, in/in	0.006	0.006	0.006	0.005	0.005	0.002	0.002
Izod impact, ft-lb/in notch	0.60	0.50	0.70	1.2	1.8	0.70	0.75
Flexural strength, lb/in^2	10,000	10,000	11,000	10,000	10,500	11,000	10,500
Flexural modulus, lb/in^2	1.3×10^6	1.3×10^6	1.3×10^6	1.3×10^6	1.3×10^6	1.8×10^6	1.8×10^6
Tensile strength, lb/in^2	5,500	6,500	6,000	5,700	6,000	6,500	6,000
Compressive strength, lb/in^2	24,000	32,000	28,000	26,000	27,000	24,000	26,000
Water absorption, %	0.4	0.4	0.3	1.0	1.0	0.5	0.35
Deflection temperature @ 264 lb/in^2	350°F	325°F	325°F	335°F	330°F	525°F	500+°F
Continuous use temperature	300°F	300°F	300°F	300°F	310°F	—	—
UL flammability rating @ $^1/_8$ in	—	94 VO	94 HB	94 HB	—	94 VO	94 VO
Dielectric strength 60Hz ST/SS, V/mil	—	325/250	270/175	250/190	250/190	150/100	150/100
Arc resistance, s	—	90	130	125	75	165	175

TABLE 2.28 Phenolic Molding Compound Data Sheet Values *(Continued)*

Grade	RX 655	RX 630	RX 611	RX 660	RX 862	RX 865	RX 867	Ammonia Free RX 640
Reinforcement type	Glass	Glass	Glass	Glass	Glass	Glass	Glass	Glass
Phenolic resin type	Two-step	Two-step	Two-step	Two-step	Two-step	Two-step	Two-step	One-step
Form	Granular	Granular	Granular	Granular	Coarse granular	Coarse granular	Nodular	Granular
Bulk factor	2.1	2.7	2.3	2.2	2.5	2.4	2.7	2.2
Specific gravity, g/cm^3	2.08	1.75	1.75	1.79	1.88	1.88	1.78	1.73
Shrinkage, in/in	0.0015	0.001	0.0015	0.0015	0.0015	0.001	0.0015	0.0015
Izod impact, ft-lb/in notch	0.45	1.20	0.90	0.75	0.90	1.3	0.70	0.90
Flexural strength, lb/in^2	12,000	23,000	18,000	18,000	12,000	15,000	11,500	16,000
Flexural modulus, lb/in^2	2.7×10^6	2.2×10^6	2.0×10^6	2.0×10^6	2.3×10^6	2.5×10^6	2.0×10^6	1.8×10^6
Tensile strength, lb/in^2	6,400	12,000	10,000	8,500	6,500	7,500	6,000	9,000
Compressive strength, lb/in^2	30,000	40,000+	40,000	40,000	28,000	33,000	28,000	35,000
Water absorption, %	0.04	0.07	0.07	0.08	0.07	0.05	0.17	0.15
Deflection temperature @ 264 lb/in^2	550+°F	450°F	440°F	410°F	500°F	550°F	500°F	500°F
Continuous use temperature	385°F	390°F	380°F	375°F	380°F	390°F	350°F	380°F
UL flammability rating @ $^1/_8$ in	—	94 VO	94 VO	94 VO	94 VO	94 VO	94 VO	—
Dielectric strength 60Hz ST/SS, V/mil	—	500/440	450/425	450/425	300/250	300/250	300/230	475/425
Arc resistance, s	—	180	181	180	183	185	184	180

SOURCE: From Ref. 19.

their compound selection process. Phenolic molding compounds have found worldwide acceptance in such diverse market areas as.

- Automotive (as seen in Figs. 2.14*a*, *b*, and *c*)
- General transportation
- Electronics
- Aeronautics
- Arospace
- Electrical
- Appliances
- Business equipment

Suppliers

Cytec Fiberite

Occidental Chemical

Plastics Engineering Company

Plaslok Corporation

Rogers Corporation

(a)

Figure 2.14 (*a*) Transmission stator and reactor.

(b)

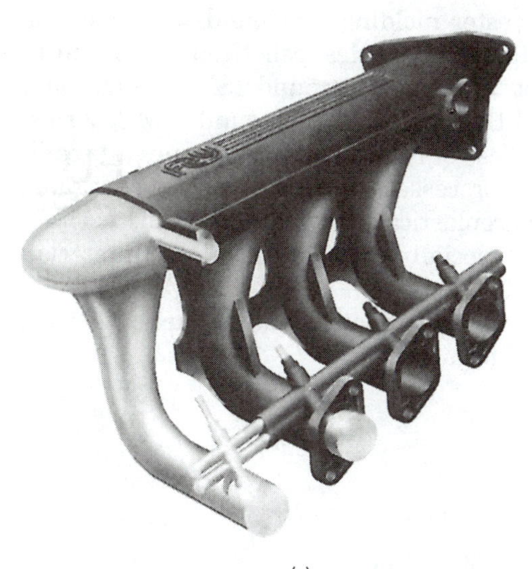

(c)

Figure 2.14 *(Continued)* (*b*) Water pump housing; (*c*) manifold housing.

Resinoid Engineering Corporation

Valentine Sugars Inc., Valite Division

(See App. C for supplier addresses.)

2.7.6 Thermoset polyesters

History. Polyester-based thermosetting molding compounds have been an important component of the thermosetting molding industry for many years but the past decade has seen a marked increase in their use in many market areas. This increase has come about because of their low cost, wide range of colors, high strength/weight ratio, and, more importantly, because of the introduction of molding equipment capable of injection molding these bulky, dough-like compounds by using stuffing mechanisms that augment the passage of the compound from the hopper into the barrel for delivery into the mold. Lower molding costs and reduced finishing costs are among the benefits derived with the use of these versatile molding compounds with the injection molding process.

Physical forms. Thermoset polyester molding compounds are available in several physical forms: free-flowing granules, pelletized (PMG), putty or rope-type extrudates, sheet molding compound (SMC), high-bulk molding compound (BMC), and thick molding compound (TMC) forms. The molding compounds, regardless of reinforcement type, are all readily moldable in all thermosetting processes. When compression or transfer molded, the preforming and preplasticating operation will necessitate the use of auxiliary equipment, especially the bulky dough-like material.

Reinforcements. The types of reinforcements available are minerals, long and short glass fibers, and organic fibers.

Specifications

Military (Mil-M-14)

MAG: Mineral filler, general purpose

MAI-30: Glass filler, impact value of 3.0 ft-lb

MAI 60: Glass filler, impact value of 6.0 ft-lb

MAT 30: Glass filler, impact value of 3.0 ft-lb

ASTM. The specifications are covered under D-1201-80 and D-1201-62 9 (reapproved 1975).

UL. The thermoset polyester compounds generally carry a 365°F (180°C) thermal index rating, with a flammability rating of 94 VO in $\frac{1}{16}$-in sections.

Data sheet values. Table 2.29 includes a list of the property values for a range of thermoset polyester compounds.

Applications. The applications are automotive components, circuit breaker housings, brush holders, commercial connectors, battery racks, business machine housings, marine structures, and household articles.

Suppliers

American Cyanamid Company

BMC Inc.

(See App. C for supplier addresses.)

2.7.7 Silicones

History. Silicone fluids, resins, and elastomers have been in use for over 50 years, originating with the discovery by E Rochow (GE Company) in 1940 in what was designated as the "direst process" in which elemental silicon was obtained by the reduction of silicon dioxide in an electric furnace. The resultant silicon was then pulverized and reacted with gaseous methyl chloride in the presence of a copper catalyst.

Physical forms. The molding compound will consist of 20 to 25% resin (phenyl and methyl siloxanes), 75% filler (glass fiber and fused silica mix), a lead-based catalyst pigment, and lubricants. The compounds are free-flowing granular in form and are available in opaque colors (mostly red). They are readily moldable in compression, transfer, and injection molding processes.

Reinforcements. The reinforcements available are quartz, "E" type glass fibers, and fused silica.

Specifications

Military (Mil-M-14)

MSG: Mineral filler, heat resistant

MSI-30: Glass filler, impact value of 3.0 ft-lb, heat resistant

UL. The UL ratings are: thermal index rating of 464°F (240°C) and flammability rating of 94 VO in $^{1}/_{16}$-in sections.

Data sheet values. Table 2.30 provides a property value list for the silicone family of molding compounds.

TABLE 2.29 Thermoset Polyester Molding Compound Data Sheet Values

Sheet molding compounds (typical ASTM data)	ASTM test method	UL recognized electrical		
		4000-( )-( )	42000-( )-( )	47000-( )-( )
Water absorption, %	D-570	0.2	0.18	0.15
Specific gravity	—	1.85	1.75	1.85
UL 94 flame classification—UL File E-27875	—	94 VO	94 VO	94 VO, 945-V
Tensile strength, lb/in^2	D-638	14,000	10,000	15,000
Flexural strength, lb/in^2	D-790	30,000	25,000	30,000
Compressive strength, flatwise, lb/in^2	D-695	30,000	28,000	30,000
Impact strength, izod, edgewise, ft/lb/in notch	D-256	13.5	18.5	20.0
Arc resistance, s	D-495	180	180	180
Track resistance, min	D-2303	450	>600	500
Dielectric strength, perpendicular, short time, V/mil in oil	D-149	420	425	420
Dielectric constant at 60 Hz	D-150	4.7	4.4	4.4
Dissipation factor at 60 Hz	D-150	0.96	0.96	0.95
Glass content, % for data shown	—	22	22	33
Range of glass content available, %	—	15–35	15–35	33
Mold shrinkage, in/in	—	0.002	0.001	0.002

Sheet molding compounds (typical ASTM data)	ASTM test method	UL recognized electrical		
		13600-( )-( )	14100-( )-( )	14100-( )-( )
Water absorption, %	D-570	0.35	0.1	0.10
Specific gravity	—	1.94	1.89	1.9
UL 94 flame classification—UL File E-27875	—	94 VO	94 VO	94 VO, 945-V
Tensile strength, lb/in^2	D-638	6,000	6,000	7,000
Flexural strength, lb/in^2	D-790	15,000	12,500	15,000
Compressive strength, flatwise, lb/in^2	D-695	15,000	13,000	20,000
Impact strength, izod, edgewise, ft/lb/in notch	D-256	3.5	3.5	5.0
Arc resistance, s	D-495	190	192	180
Track resistance, min	D-2303	>600	>600	600
Dielectric strength, perpendicular, short time, V/mil in oil	D-149	375	400	400
Dielectric constant at 60 Hz	D-150	5.8	5.0	5.0
Dissipation factor at 60 Hz	D-150	295	3.25	3.0
Glass content, % for data shown	—	15	15	20
Range of glass content available, %	—	5–35	5–35	20
Mold shrinkage, in/in	—	0.002	0.0001	0.002

SOURCE: From Ref. 10.

TABLE 2.30 Silicone Molding Compound Data Sheet Values

	Dow Corning silicone molding compound					Test method	
Properties[a]	306	307	308	480	1-5021	CTM[b]	ASTM
Physical							
Specific gravity	1.88	1.85	1.86	1.92	1.92	0540	D-792
Flexural strength (lb/in^2 × 10^3)	8.8±0.7	7.5±0.8	9.2±1.4	9.1±1.1	8.9±1.1	0491A	D-790
Flexural modulus (lb/in^2 × 10^6)	1.2	0.7	1.2	1.5	1.5	0491A	D-790
Compressive strength, lb/in^2	14,000	10,000	14,000	18,000	18,000	0533	D-695
Tensile strength, lb/in^2	5,000	4,500	5,200	5,000	5,000	—	D-638
Izod impact strength, ft/lb/in notch[c]	0.34	0.34	0.26	0.24	0.24	0498	D-256
Water absorption, %	0.1	0.1	0.1	0.1	0.1	0248	D-570
Flammability, UL 94, 0.0625 in[d]	V-0	V-0	V-0	V-0	V-0	—	—
Thermal conductivity, cal/s-°C-cm × 10^{-4}	15	14	16	18	18	Colora	—
Thermal expansion,[e]							
Measured parallel to flow							
50–150°C, in/in°C × 10^{-6}	22	22	24	25	25 0563	—	
150–250°C, in/in°C × 10^{-6}	27	29	28	26	28 0563	—	
Measured perpendicular to flow							
50–150°C, in/in°C × 10^{-6}	40	43	38	34	35 0563	—	
150–250°C, in/in°C × 10^{-6}	46	52	39	35	38 0563	—	
Stability to 70°C/93% RH after 10,000 h weight change, %	+1.4	+1.5	—	—	—	—	—
Dimension change, %	+0.14	+0.16	—	—	—	—	—
Electrical							
Arc resistance, s	290	270	280	300	—	0171	D-495
Dielectric strength, in oil, V/mil[f]	320	310	290	300	370	0114	D-149
Volume resistivity							
Condition A[g] Ω-cm × 10^{15}	2.8	2.7	2.8	2.8	2.6	0249	D-257
Condition D[h] Ω-cm × 10^{15}	2.7	2.7	2.8	2.8	2.5	0249	D-257
Dielectric constant at 10^6 Hz							
Condition A[g]	3.55	3.61	3.68	3.60	3.49	0543	D-150
Condition D[h]	3.60	3.64	3.72	3.65	3.53	0543	D-150
Dissipation factor at 10^6 Hz							
Condition A[g]	0.0016	0.0018	0.0018	0.0014	0.0012	0543	D-150
Condition D[h]	0.0019	0.0027	0.0019	0.0019	0.0015	0543	D-150

[a] All values shown are determined periodically and must be considered typical. Should a property be judged to be pertinent in a given application, verification must be made by the user prior to use.
[b] Dow Corning Test Method. Similar to ASTM method shown. Copies available upon request.
[c] Not corrected for energy to toss.
[d] UL yellow card listed.
[e] Test specimens 1.5 in × 0.25 in × 0.10 in. Transfer molded bars postcured 2 h at 200°C.
[f] Tested under ¼-in electrodes, 500-V/s rise, ⅛-in-thick specimen.
[g] Condition A=as received.
[h] Condition D=after 24-h immersion in distilled water at 23°C (75°F).
SOURCE: From Ref. 10.

Applications. The silicones are nonconductors of either heat or electricity; have good resistance to oxidation, ozone, and ultra-violet radiation (weatherability); and are generally inert. They have a constant property profile of tensile, modulus, and viscosity values over a broad temperature range 60 to 390°F (13 to 166°C). They also have a low glass transition temperature (T_g) of −185°F. Encapsulation of semiconductor devices such as microcircuits, capacitors, and resistors, electrical connectors seals, gaskets, O-rings, and terminal and plug covers all take advantage of these excellent properties.

Suppliers

Dow Corning Corporation

Cytec Fiberite

General Electric Company Silicone Products Division

(See App. C for supplier addresses.)

2.7.8 Composites

History.[15] The introduction of fiberglass-reinforced structural applications in 1949 brought a new plastics application field which began with the consumption of 10 lb and burgeoned into the annual usage of over 1 to 2 billion lb over the next several decades. This usage has been, and still is, taking place in application areas that take advantage of the extraordinary low-weight–high-strength ratio inherent in these composite materials.

A *thermosetting matrix* is defined as a composite matrix capable of curing at some temperature from ambient to several hundred degrees of elevated temperature and cannot be reshaped by subsequent reheating. In general, thermosetting polymers contain two or more ingredients—a resinous matrix with a curing agent which causes the matrix to polymerize (cure) at room temperature or a resinous matrix and curing agent that, when subjected to elevated temperatures, will commence to polymerize and cure.

Resins (matrices). The available resins are polyester and vinyl esters, polyureas, epoxy, bismaleimides, polyimides, cyanate ester, and phenyl triazine.

Polyester and vinyl esters. Polyester matrices have had the longest period of use, with wide application in many large structural applications. (See Table 2.31.) They will cure at room temperature with a catalyst (peroxide) which produces an exothermic reaction. The resultant polymer is nonpolar and very water resistant, making it an excellent choice in the marine construction field. The isopolyester resins, regarded as the most water-resistant polymer in the polymer group, has been chosen as the prime matrix material for use on a fleet of U.S. Navy mine hunters.

Epoxy. The most widely used matrices for advanced composites are the epoxy resins even though they are more costly and do not have the high-temperature capability of the bismaleimides or polyimide; the advantages listed in Table 2.32 show why they are widely used.

TABLE 2.31 Neat Resin Casting Properties of Polyester-Related Matrices

Material	Barcol hardness	Tensile strength		Tensile modulus		Elongation, %	Flexural strength		Flexural modulus		Compressive strength		Heat deflection temperature	
		MPa	$kips/in^2$	10^{-2} Pa	10^{-5} $kips/in^2$		MPa	$kips/in^2$	10^{-2} Pa	10^{-5} $kips/in^2$	MPa	$kips/in^2$	°C	°F
Orthophthalic	—	55	8	34.5	5.0	2.1	80	12	34.5	5.0	—	—	80	175
Isophthalic	40	75	11	33.8	4.9	3.3	130	19	35.9	5.2	120	17	90	195
BPA fumarate	34	40	6	28.3	4.1	1.4	110	16	33.8	4.9	100	15	130	265
Chlorendic	40	20	3	33.8	4.9	—	120	17	39.3	5.7	100	15	140	285
Vinylester	35	80	12	35.9	5.2	4.0	140	20	37.2	5.4	—	—	100	212

SOURCE: Charles D. Dudgeon in *Engineering Materials Handbook*, vol. 1, Theodore Reinhart, Tech. Chairman, ASM International, 1987, p. 91.

TABLE 2.32 Epoxy Resin Selection Factors

Advantages
- Adhesion to fibers and to resin
- No by-products formed during cure
- Low shrinkage during cure
- Solvent and chemical resistance
- High or low strength and flexibility
- Resistance to creep and fatigue
- Good electrical properties
- Solid or liquid resins in uncured state
- Wide range of curative options

Disadvantages
- Resins and curatives somewhat toxic in uncured form
- Absorb moisture
  Heat distortion point lowered by moisture absorption
  Change in dimensions and physical properties due to moisture absorption
- Limited to about 200°C upper temperature use (dry)
- Difficult to combine toughness and high temperature resistance
- High thermal coefficient of expansion
- High degree of smoke liberation in a fire
- May be sensitive to ultraviolet light degradation
- Slow curing

SOURCE: From Ref. 15, p. 3.15.

Bismaleimides (BMI). The *bismaleimide resins* have found their niche in the high-temperature aircraft design applications where temperature requirements are in the 177°C (350°F) range. BMI is the primary product and is based upon the reaction product from methylene dianiline (MDA) and maleic anhydride: bis (4 maleimidophynyl) methane (MDA BMI). Variations of this polymer with compounded additives to improve impregnation are now on the market and can be used to impregnate suitable reinforcements to result in high-temperature mechanical properties (Table 2.33).

Polyimides. *Polyimides* are the highest-temperature polymer in the general advanced composite with a long-term upper temperature limit of 232 to 316°C (450 to 600°F). Table 2.34 is a list of commercial polyimides being used in structural composites.

Polyureas. Polyureas involve the combination of novel MDI polymers and either amine or imino-functional polyether polyols. The resin systems can be reinforced with milled glass fibers, flaked glass, Wollastanite, or treated mica, depending upon the compound requirements as to processability or final product.

Cyanate ester and phenolic triazine (PT). The cyanate ester resins have shown superior dielectric properties and much lower moisture absorption than any other structural resin for composites. The physical properties of cyanate ester resins are compared to those of a representative

TABLE 2.33 Approximation of Mechanical Properties of BMI Composites

Property	Unreinforced homopolymer	Glass-reinforced homopolymer	Carbon-reinforced homopolymer	Carbon-reinforced homopolymer
Reinforcement, vol %	0	60	60	70
Service temperature, °C (°F)	260 (500)	177–232 (350–450)	177–232 (350–450)	149–204 (300–400)
Flexural strength, MPA (kips/in^2)				
At room temperature	210 (30)	480 (70)	2000 (290)	725 (105)
At 230°C (450°F)	105 (15)	290 (42)	1340 (194)	—
Flexural modulus, GPa (10^6 lb/in^2)				
At room temperature	4.8 (0.7)	17.2 (2.5)	126 (18.4)	71 (10.3)
At 230°C (450°F)	3.4 (0.5)	15.1 (2.2)	57 (8.2)	—
Interlaminar shear strength, MPa (kips/in^2)				
At room temperature	—	—	117 (17)	—
At 230°C (450°F)	—	—	59 (8.6)	—
Tensile strength, MPa (kips/in^2)				
At room temperature	97 (14)	—	1725 (250)	570 (83)
At 230°C (450°F)	76 (11)	—	—	—
Tensile modulus, GPa (10^6 lb/in^2)				
At room temperature	4.1 (0.6)	—	148 (21.5)	15.9 (2.3)
At 230°C (450°F)	2.8 (0.4)	—	—	—

SOURCE: James A. Harvey in *Engineering Materials Handbook,* vol. 1, Theodore Reinhart, Tech. Chairman, ASM International, 1987, p. 256

BMI resin in Table 2.33. The PT resins also possess superior elevated temperature properties, along with excellent properties at cryogenic temperatures. They are available in several viscosities, ranging from a viscous liquid to powder, which facilitates their use in applications that use liquid resins such as filament winding and transfer molding.

Reinforcements

Fiberglass. Fiberglass possesses high tensile strength and strain to failure but the real benefits of its use relates to its heat and fire resistance, chemical resistance, moisture resistance, and very good thermal and electrical values. Some important properties of glass fibers are shown in Table 2.35.

Graphite. Graphite fibers have the widest variety of strength and moduli and also the greatest number of suppliers. These fibers start out as organic fiber, rayon, polyacrylonitrile, or pitch called the *precursor.* The precursor is stretched, oxidized, carbonized, and graphitized. The relative amount of exposure to temperatures from 2500 to 3000°C will then determine the graphitization level of the fiber. A higher degree of graphitization will usually result in a stiffer (higher

TABLE 2.34 Commercial Polyimides Used for Structural Composites

	Upper temperature capability	
	°C	°F
Condensation		
Monsanto Skybond 700, 703	316	600
DuPont NR-150B2 (Avimid N)	316	600
LARC TPI	300	572
Avimid K-III	225	432
Ultem	200	400
Addition		
PMR-15 (Reverse Diels-Adler nadic end-capped)	316	600
LARC 160 (Reverse Diels-Adler nadic end-capped)	316	600
Thermid 600 (Acetylene end-capped)	288	550
BMIs (Bismaleimides, maleimide end-capped)	232	450

SOURCE: D. A. Scola in *Engineering Materials Handbook,* J. N. Epel et al. (eds.), vol. 2, ASM International, 1988, p. 241.

modulus) fiber with greater electrical and thermal conductivities. Some important properties of carbon and graphite fibers are shown in Table 2.36.

Aramid. The organic fiber kevlar 49, an aramid, essentially revolutionized pressure vessel technology because of its great tensile strength and consistency coupled with low density, resulting in much more weight-effective designs for rocket motors.

Boron. Boron fibers, the first fibers to be used in production aircraft, are produced as individual monofilaments upon a tungsten or carbon substrate by pyrolytic reduction of boron trichloride (BCL) in a sealed glass chamber. Some important properties of boron fibers are shown in Table 2.37.

2.7.9 Molding compound production[16]

Introduction. The selection of production equipment and processes for thermosetting molding compounds commences with the compound designer's formulation which designates the type and quantity of the various ingredients that make up the compound. These molding compounds are a physical mixture of resin, reinforcement or filler, catalyst, lubricant, and color. The resin, by far, is the *key* component in any thermosetting molding compound since it is the only component that actually goes through the chemical reaction known as *polymerization* or *cure* during the molding process. Also, because of this curing quality,

TABLE 2.35 Glass Fibers in Order of Ascending Modulus Normalized to 100% Fiber Volume (Vendor Data)

Type	Nominal tensile modulus, GPa (kips/in^2 × 10^6) strand	Nominal tensile strength, MPa (kips/in^2 × 10^6) strand	Ultimate strain, %	Fiber density, kg/m^3 (lb/in^3)	Typical suppliers
E	72.5 (10.5)	3447 (500)	4.8	2600 (0.093)	Pittsburgh Plate Glass, Manville Co., Owens Corning Fiberglass
R	86.2 (12.5)	4400 (638)	5.1	2530 (0.089)	Vetrotex St. Gobain, Certainteed
Te	84.3 (12.2)	4660 (675)	5.5	2530 (0.089)	Nittobo
S-2, S	88 (12.6)	4600 (665)	5.2	2490 (0.090)	Owens Corning

SOURCE: S. T. Peters, W. D. Humphrey, and R. F. Foral, *Filament Winding: Composite Structure Fabrications,* SAMPE Publishers, Covina, Calif., 1991.

TABLE 2.36 Carbon and Graphite Fibers in Order of Ascending Modulus Normalized to 100 Percent Fiber Volume (Vendor Data)

Class of fiber	Nominal tensile modulus GPa (lb/in^2 × 10^6) strand	Nominal tensile strength MPa (lb/in^2 × 10^3) strand	Ultimate strain (%)	Fiber density kg/m^3 (lb/in^3)	Suppliers/ typical products
High tensile strength	227 (33)	3996 (580)	1.60	1750 (0.063)	Amoco, T-300; Hercules, AS-4
High strain	234 (34)	4100 (594)	1.95	1790 (0.064)	Courtaulds Grafil, 33-600
Intermediate modulus	275 (40)	5133 (745)	1.75	1740 (0.062)	Hercules, IM-6; Amoco, T-40; Courtaulds Grafil, 42-500
Very high strength	289 (42)	7027 (1020)	1.82	1820 (0.066)	Toray, T-1000
High modulus	358	2482 (360)	0.70	1810 (0.065)	Amoco, T-50; Celanese, G-50
High modulus (pitch)	379 (55)	2068 (300)	0.50	2000 (0.072)	Amoco, P-55
Ultrahigh modulus (pan)	517 (75)	1816 (270)	0.36	1960 (0.070)	Celanese, GY-70
Ultrahigh modulus (pitch)	517 (75)	2068 (300)	0.40	2000 (0.072)	Amoco, P-75
Extremely high modulus (pitch)	689 (100)	2240 (325)	0.31	2150 (0.077)	Amoco, P-100

SOURCE: S. T. Peters, W. D. Humphrey, and R. F. Foral, *Filament Winding: Composite Structure Fabrications,* SAMPE Publishers, Covina, Calif., 1991.

the resin, production process, and equipment is governed by the need to understand this chemical reaction with its effects upon the production process and/or equipment. Also, the resin is the primary flow promoter and chief provider of the desired electrical insulating properties of the final molded product.

The next most important component is the reinforcement or filler because the type and quantity of either will determine the manufacturing process and equipment as will be seen in the following sections which describe the various processes and their equipment. The compounds utilizing “fillers” as opposed to “reinforcements” can be processed with either the “dry” or the “wet” (solvent) process since the compound formulations include free-flowing granular fillers that are not as susceptible to degradation when exposed to the hot roll mill phase of the operation. The catalyst in each compound serves as the *reaction controller,* with the type and

TABLE 2.37 Boron and Ceramic Fibers, Normalized to 100 Percent Fiber Volume (Vendor Data)

Type	Nominal tensile modulus, GPa ($lb/in^2 \times 10^6$)	Nominal tensile strength, MPa ($lb/in^2 \times 10^3$)	Fiber density kg/m^3 (lb/in^3)	Suppliers
Boron	400 (58)	3520 (510)	2.55–3.30 (0.093)	Textron, Huber, Nippon, Tokai
Silicon carbide	425 (62)	611	3.56 (0.125)	Textron, Dow Corning, Nippon Carbon
Silicon nitride	300	2500	2.5	Tonen
Silica	66		1.80–2.50	Enka, Huber
Alumina	345 (50)	1380 (200)	3.71 (0.134)	DuPont, FP
Alumina boria silica	(27)	300	2.71 (0.098)	3MS, Nexel 312

quantity of catalyst acting to either accelerate or inhibit the curing rate in both the production and the molding phases.

Lubricants which provide a measure of flow promotion, mold release, and barrel life during molding are generally internally supplied but are occasionally provided as an external addition. All thermosetting molding compound colors are opaque, with the pigments or dyes heat stable within the molding process temperature range of 200 to 400°. Coloring does have a large effect upon the manufacturing process when the product line includes a wide variety of colors such as are common to the DAP, melamine, urea, and thermoset polyester compounds. Choice of production equipment has to be designed to meet the need for quick and easy color changes.

Production processes

Dry process (batch and blend). The *dry* or *nonsolvent process,* illustrated in Fig. 2.15, employs low-strength, low-cost, free-flowing granular fillers, and involves the use of ribbon or conical mixers to homogenize the dry ingredients prior to feeding the mix onto the heated roll mills where the mix is compounded (worked) for a specific time and temperature. Once the mix or batch has been worked to the proper consistency and temperature, it is then fed onto a three-roll calendering mill where it is shaped to a specific width and thickness to allow the sheet to pass into a grinder and then onto screens to obtain the desired granulation and for dust removal. The thickness and temperature of the calendered sheet is controlled for ease of granulation. If the sheet is too warm, it will not cut cleanly; if the sheet is too cool, it will be too fragile to produce clean, even size particles.

An individual batch is generally 200 lb, which eventually is blended with other batches into 2500- to 5000-lb blends that are ready for shipment to the customer. The batch and blend process is employed for

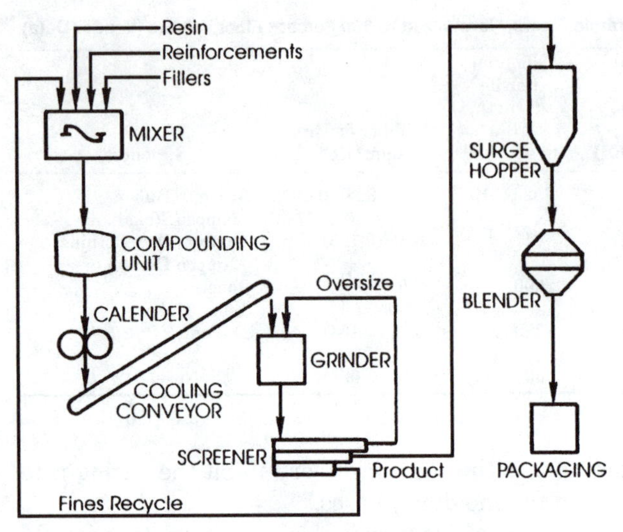

Figure 2.15 Batch and blend dry production method.

short runs, especially where the production schedules call for a variety of colors. The equipment must be such that it allows for relatively quick color changeovers.

Wet process (pelletized). The *wet process,* shown in Fig. 2.16, when low-strength, low-cost, free-flowing granular fillers are used, can produce free-flowing pelletized material with water as the solvent and the ingredients thoroughly mixed in a kneader and then auger fed to a heated extruder. The extruder screw densifies the wet mass, forcing it out of the extruder head which contains many small through orifices that determine the diameter of the pellets, and the fly-cutters working across the face of the extruder head determine the pellet length. The extruded pellets then require a drying operation prior to final blending for shipment.

Wet process (high strength). As can be seen in Fig. 2.17, this process involves the use of mixers, mobile carts, air-drying rooms, prebreakers, hammer mills, lenders, and extruders. The basic purpose of the entire process is to provide minimum reinforcement degradation so as to maintain sufficient fiber integrity to meet the various mechanical strength requirements which are generally set by military or commercial specifications.

The process begins with the mixing of all the ingredients, except the reinforcement, into a suitable solvent. Once the mix has been properly dispersed, the reinforcement is added, keeping the mixing time to a minimum to preserve the fiber integrity. Solvent recovery is possible during this phase of the operation as well as later on in the drying phase. The wet mass is removed from the mixer and spread onto wire

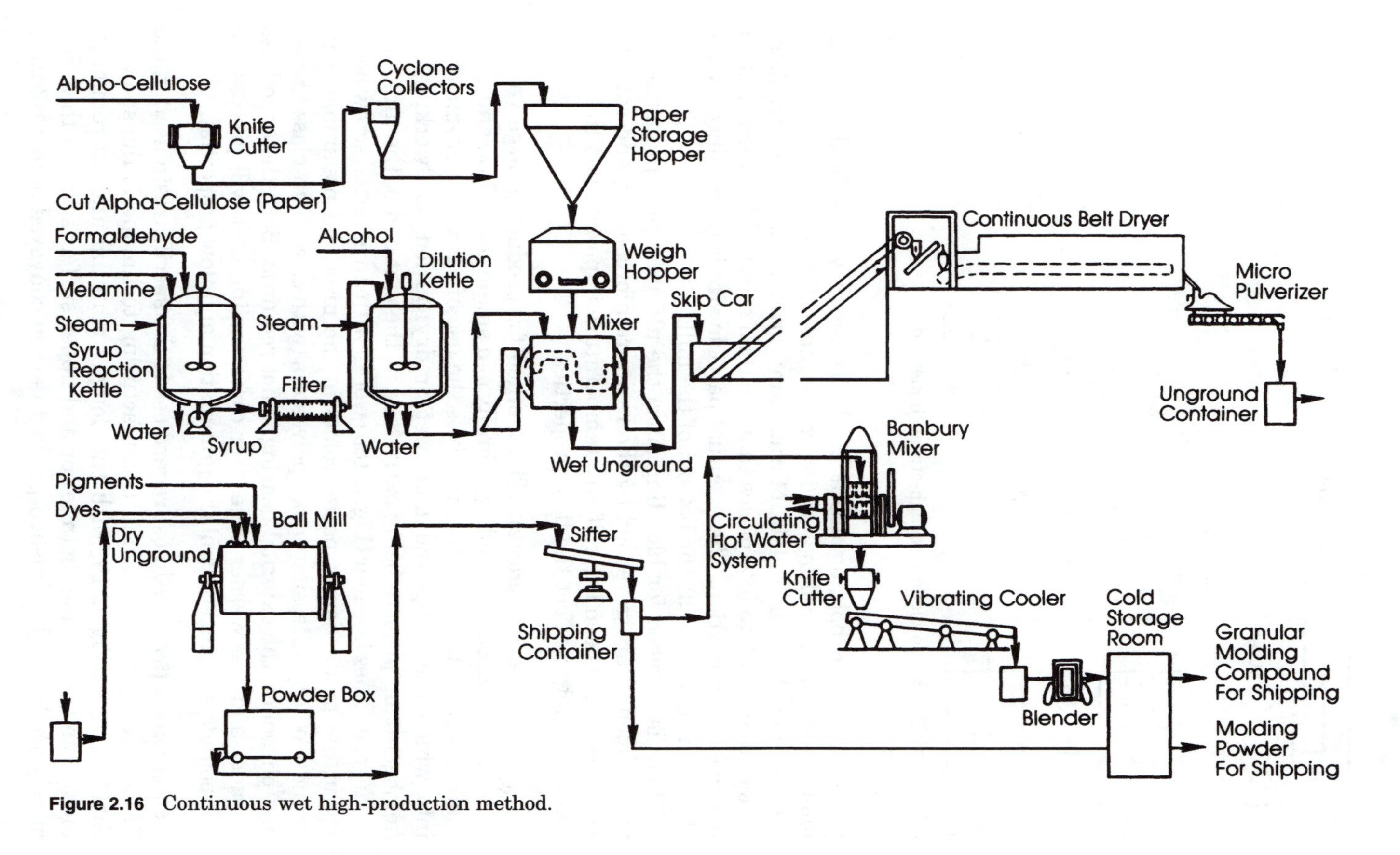

Figure 2.16 Continuous wet high-production method.

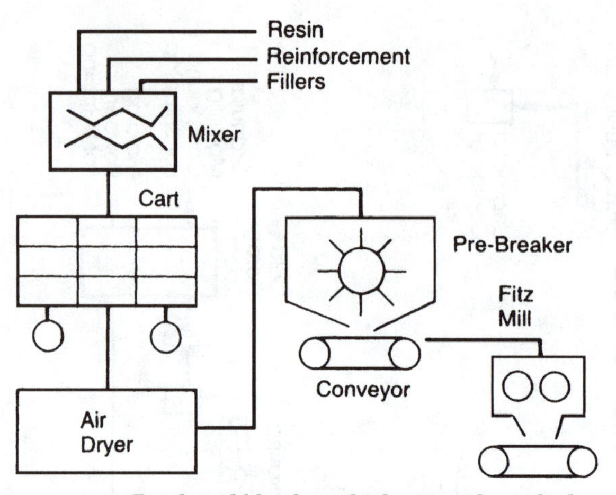

Figure 2.17 Batch and blend wet high-strength method.

trays capable of holding about 25 lb. These trays are loaded into a mobile cart and placed in a drying room, generally overnight. After drying, the now-hardened slabs of compound are fed into a "prebreaker," which tears the slab into pieces that are then sent to a hammer mill for particle size reduction. The eventual particle size is determined by the size of the openings at the bottom of the mill.

There is no need for blending since the ultimate flow properties of the compound are governed by the resin, reinforcement, and catalyst mix. There has been little, if any, temperature imposed upon the materials that might affect the flow properties.

High volume (general purpose). The method for producing large volume runs of granular compounds, generally of a single color, flow, and granulation, shown in Fig. 2.18*a*, involves the use of extruders or kneaders into which the compound mix (wet or dry) is fed for "working" or homogenizing prior to being extruded out the exit end of the unit.

The open buss kneader (Fig. 2.18*b*) works with an external screw bearing in the production of pastel colors in the urea and melamine compounds. When processing epoxy, polyesters (with or without glass fibers), and phenolics, the external bearing is not required. Both the open buss kneader and the Werner Pfliederer compounding unit will process the compounds and also control the granulation and flow properties.

Sheet process (SMC). Almost regardless of the specific resin used in the SMC process, the compound manufacturing technique is the same as that shown in Fig. 2.19, with the doughy material and its reinforcement being covered on both lower and upper surfaces with a thin film of polyethylene. The finished product is then conveyed onto a rotating

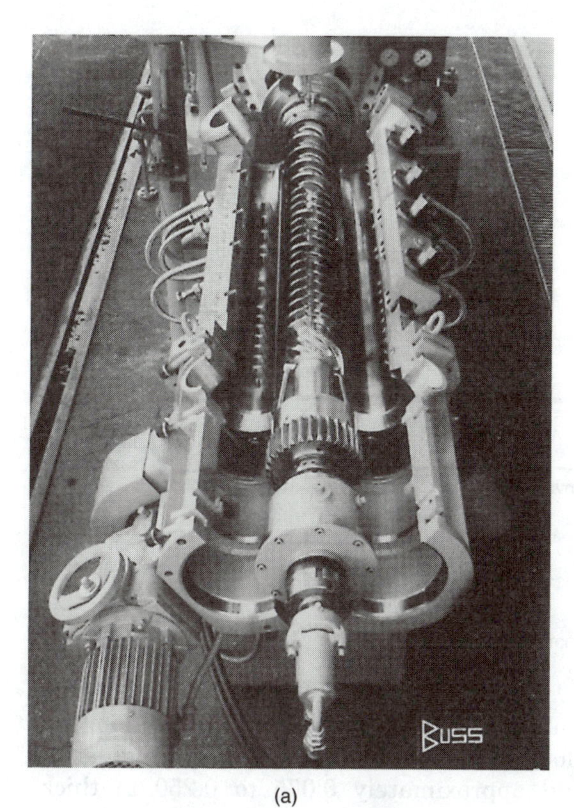

(a)

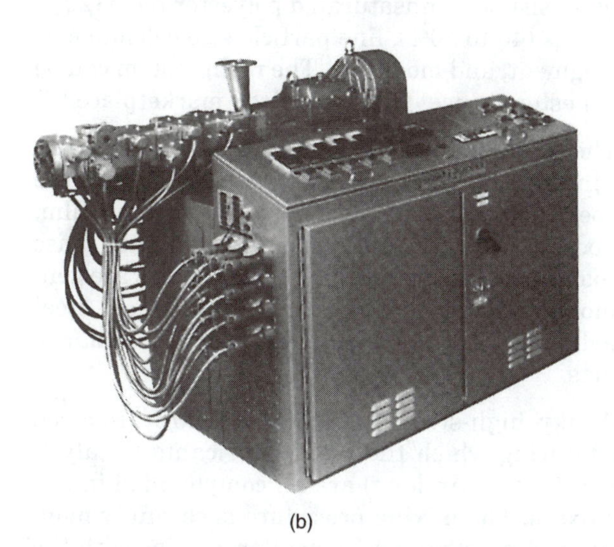

(b)

Figure 2.18 (*a*) Continuous dry high-volume method; (*b*) open bus kneader.

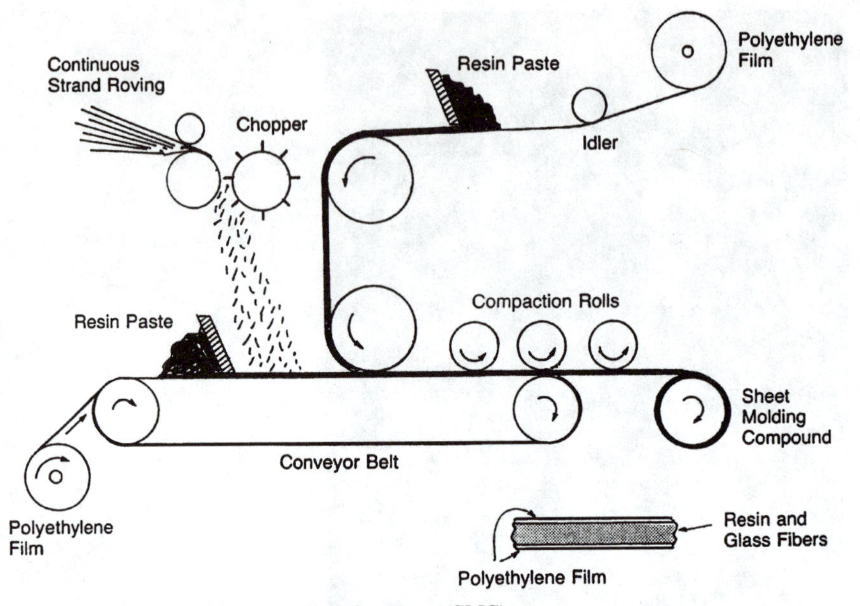

Figure 2.19 Sheet molding compound process (SMC).

mandrel and wound up until it reaches a preset weight and then it is cut off. The sheets ready for use or for shipment generally weigh about 50 lb, are 4 ft wide, and approximately 0.075 to 0.250 in thick. Formulations generally consist of an unsaturated polyester resin (20 to 30%), chopped glass rovings (40 to 50%), fine particle size calcium carbonate, filler, catalyst, pigment, and modifiers. The resin system can be epoxy, polyester, or vinyl ester to meet the need of the marketplace.

Sheet process (TMC). The production of TMC, shown in Fig. 2.20, differs from that of SMC in that the glass fibers are wetted between the impregnating rollers before being deposited onto the moving film. TMC sheets can be produced in thicknesses of up to 2 in with glass lengths of 1 in/min at loading levels of 20 to 30%. These sheets are generally compression molded using matched metal-hardened steel molds. Packaging and shipment of the unmolded product is similar to that of the SMC products.

High strength (BMC). Bulky high-strength compounds are produced with the batch method during which the resin, lubricants, catalyst, and chopped glass fibers ($^1/_8$ to $^1/_2$ in long) are all compounded in relatively low-intensity mixers. The mixing procedure is carefully monitored to achieve the highest possible mechanical properties with the least amount of fiber degradation. The finished product is shipped in bulk form using vapor barrier cartons with the compound in a sealed

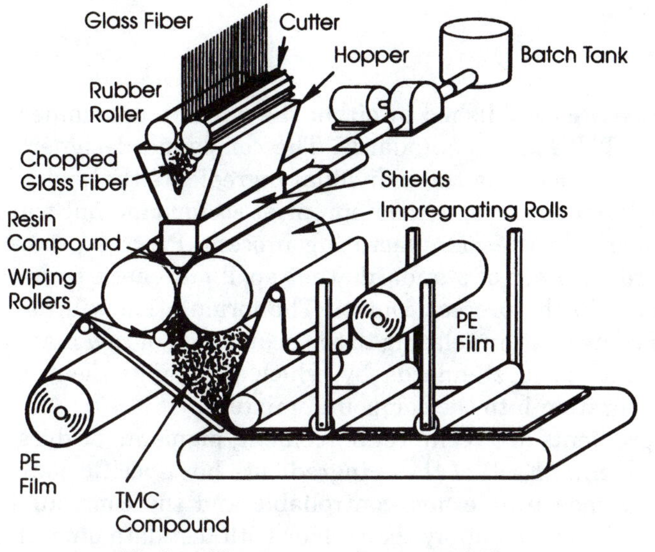

Figure 2.20 Thick molding compound process (TMC).

polyethylene bag. These compounds are also available in a "rope" form in any length or diameter specified by the customer. The shape is attained by feeding the doughy material through an extruder, with the extruder nozzle providing the desired shape and length.

Quality assurance. Thermosetting compound production uses comprehensive quality assurance programs to ensure that the customer receives products that meet their specific specifications, regardless of the process or equipment employed. Formulation and processing specifications require that the incoming raw materials meet specific quality standards and that the manufacturing processes be carefully monitored to make certain that the final compound meets a designated property profile.

The formulation requirements are based upon meeting certain standards relating to the following property standards:

- Flow properties
- Electrical (insulation)
- Mechanical (strength)
- Chemical resistance
- Weather resistance
- Thermal resistance
- Flame resistance

- Surface finish
- Color

These specifications are established by either the military or commercial users or by ASTM and UL standards. The compound designer's *primary* task is one of selecting the individual ingredients that, within a cost range, will not only create the appropriate compound but will also function properly in the manufacturing process. Process procedures are governed by a set of standards that spell out check points along the line to aid in the process control. The formulation will furnish the necessary information relating to the acquisition of raw materials along with the basic standards by which each ingredient is accepted for incorporation into the compound mixture.

The specific ingredients are resin, reinforcement, pigments or dyes, lubricants, and solvents. Each of these ingredients has specific characteristics that are measurable and controllable and the compound producer can, and often will, supply the molder with test data on each production blend the customer has received. The special characteristics of each of these ingredients are as follows:

Resin. Viscosity, gel time, cure rate, and solubility

Reinforcements and fillers. Aspect ratio, moisture content, fiber size and length, purity, and color

Pigments and dyes. Solubility, coatability, and thermal stability

Catalysts. Solubility, reaction temperature, and purity

Lubricants. Solubility, melting point, and purity

Solvents. Purity and toxicity

Compounds manufactured to meet Mil specifications will be subjected to a certification process involving the documentation of actual property values derived from the testing of the compound both during, and upon completion of the manufacturing procedure. The values called for are:

- Specific gravity
- Shrinkage
- Arc resistance
- Dielectric constant
- Dielectric strength
- Impact strength
- Flexural strength
- Volume and surface resistivity

Certification will cover a blend of 2500 lb or more, with the actual values recorded and furnished to the customer upon request. The alternate certification process involves the compounder furnishing a letter of compliance which confirms that the actual blend involved does meet all the requirements of the specification.

Testing equipment and procedures. Most compounders employ a variety of testing tools to control and monitor their manufacturing processes, as well as their R&D programs, and for trouble-shooting problems encountered in the field. A list of such equipment follows.

Unitron metallograph. A sophisticated metal detection device used to check production as well as trouble-shoot areas.

Scanning electron microscope (SEM). An *SEM* is employed to examine surfaces from low magnification up to a 100,000 × enlargement. It is an excellent research and problem-solving tool.

Dynamic mechanical spectrophotometer (DMS). The *DMS* measures the viscoelastic properties of polymers, thus determining a compound's viscosity and elastic modulus following the change of these properties over time and changes in temperature.

Infrared spectrophotometer. The *infrared spectrophotometer* can provide information concerning the composition of a compound such as degradation, replacement for fillers and reinforcing agents, and to evaluate the purity of resins, fillers, catalysts, reinforcements, solvents, and lubricants.

Capillary rheometer. The Monsanto *capillary rheometer* measures the viscosity properties of polymers, and provides a direct measure of viscosity and the change in viscosity with time and flow rate at plastication temperatures. The capillary orifice simulates the gate and runner system of actual molding conditions, thus providing valuable flow information for molding compounds.

Brabender plasticorder. The *plasticorder* is a small mixer capable of measuring the viscosity and the gel time of thermosetting molding compounds with results that can be correlated to the performance of a compound during molding conditions.

High-pressure liquid chromatograph. This device is available in two forms:

1. *Gel permeation.* Measures the molecular weight distribution and average molecular weight of the molecules in a sample of a compound.
2. *Liquid.* Detects and measures the amount of chemical constituents present in a compound.

Thermal analysis (TA). DuPont's TA equipment is available in four modules that provide information regarding the effect of temperature on a compound's physical properties.

1. *Differential scanning calorimeter (DSC).* Measures heat uptake or heat release of a compound as the temperature is raised and also the heat effects associated with material transitions such as melting.
2. *Thermal analyzer (TMA).* Measures the variation in the length of a sample as temperature is increased. It is good for comparing this property with a sample of another compound. TMA also measures thermal transition points by predicting the point and rate at which a compound will melt as well as determining the temperature at which blistering will occur if a molded part has not been properly postbaked.
3. *Dynamic mechanical analyzer (DMA).* Measures a compound's modulus (stiffness) as its temperature is raised. This instrument has provided interesting insights into the properties of phenolics as well as those of DAPs, thermoset polyesters, silicones, and epoxies, by indicating the ability of thermosets to retain their modulus at elevated temperatures.
4. *Thermogravimetric analyzer (TGA).* TGA measures weight changes in a sample as the temperature is varied, providing a useful means to determine degradative processes and heat resistance in polymeric compounds.

Particle size analyzer (PSA). A *particle size analyzer* is an accurate and automatic development tool that allows for a very rapid measurement of particle size distribution in powder or slurry compounds.

Humidity chamber. A *humidity chamber* is employed to measure the effects of temperature and humidity cycles on molded parts.

Instron testers. *Instron testers* will measure flexural, tensile, and compressive strength as well as stress-strain curves at ambient temperatures, and, when fitted with an environmental chamber, the flexural and tensile tests can run at elevated temperatures.

Rheology (flow testing). Easily, the foremost characteristic of a thermosetting molding compound is its ability to flow under pressure within the confines of the heated mold. This property value is of utmost importance in the eyes of the molder and will vary according to the molding method, mold design, molding equipment, and certainly the configuration of the molded part.

Since the molding compound is subjected to elevated temperature and pressure in the molding process, its ability to flow is greatly affected by the chemical reaction taking place as a result of these conditions. As a general rule, the speed of this chemical reaction will double with every 10° increase in temperature. In every thermosetting molding cycle, regardless of the type of compound used, the type of mold used, or the type of molding press used, the molding compound will go through the typical thermosetting reaction curve shown in Fig. 2.21. At the top left side (A), the compound is at room temperature and 0 pressure. With pressure applied and the compound increasing in temperature, its viscosity decreases as shown on the slope at B. This decrease in viscosity continues along the B slope as the compound temperature increases until the compound reaches its peak of flow at C, just prior to a rapid acceleration of the reaction as the curve turns upward to D, thus completing the cure.

Every individual molding compound possesses its own flow characteristics which are affected by the resin, catalyst, and reinforcement ratio as well as the type and content of all the ingredients that make up the complete formulation. The *desired* rheology or flow property requirements of a specific compound will be determined by

Molding method. Compression, transfer, injection, mold designs, part configuration, number of cavities and their location in the mold, and size and location of the runner and gate system.

Flow specifications. Generally identified as stiff, medium, or soft or by a designated cure rate or flow time. All thermosetting molding compounds possess flow characteristics that are both measurable and controllable with the important characteristics being the rate of curing, speed of the flow, distance of the flow, and finally the amount of compound used during the flow time.

Flow testing procedures. The five most widely used flow testing procedures have one main purpose—to provide specific and detailed information based upon the compound's intended use. Compounds that are designated for use in compression molds will have decidedly different flow requirements than if the intended use is in either a transfer or an injection molding process.

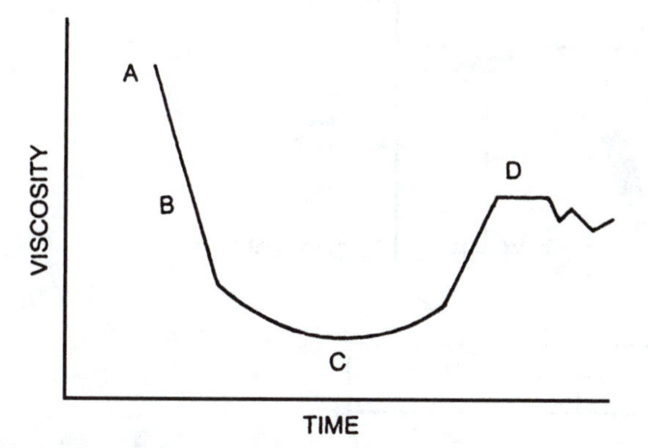

Figure 2.21 Typical thermosetting curve.

In all of the following flow tests, with the exception of the Brabender, three elements are always kept constant during the testing procedure:

- *Amount of compound.* Charge weight
- *Mold temperature.* Usually 300°F
- *Molding pressure.* Usually 1000 lb/in^2

Cup closing (test and mold) (Fig. 2.22). With the mold set at 300°F (148°C) and a molding pressure of 1000 lb/in,^{2} a room temperature charge of compound is placed in the lower half of the mold and the mold is closed. The time required for the mold to completely close is recorded in seconds. The longer the time, the stiffer the compound; the shorter the time, the shorter the flow. Generally speaking stiff flow compounds will be 15 s or more, whereas the medium flow compounds will be in the 8- to 14-s range and the soft flow compounds closing in less than 8 s.

Disk flow I (test and mold) (Fig. 2.23). With the mold in the open position, a measured amount of room temperature compound is placed on the lower half of the mold and the mold is closed and then reopened as soon as the compound is cured. The molded disk is then measured for diameter and thickness. The thinner the disk, the softer the flow; the thicker the disk, the stiffer the flow.

Disk flow II (test and mold) (Fig. 2.24). With the diameter of the disk being used as the gauge, the molded disk is placed on a target of concentric

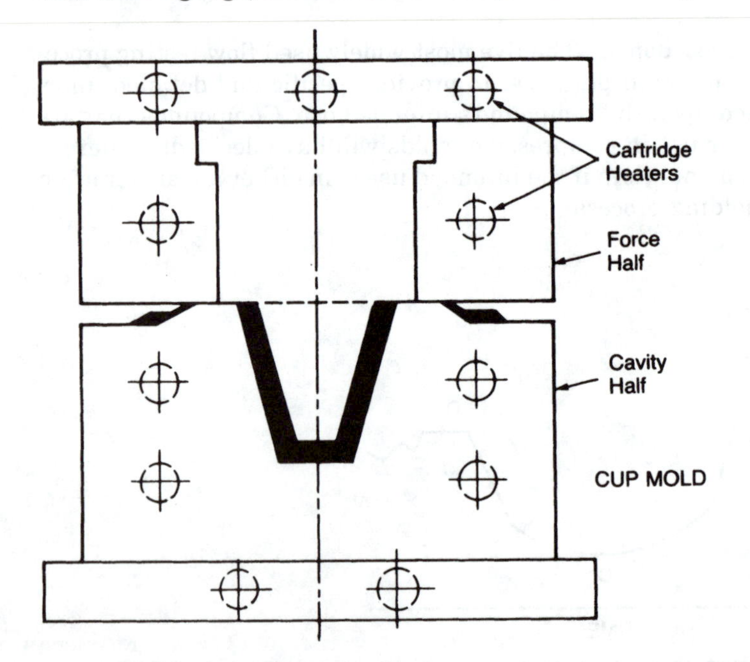

Figure 2.22 Cup closing test and mold.

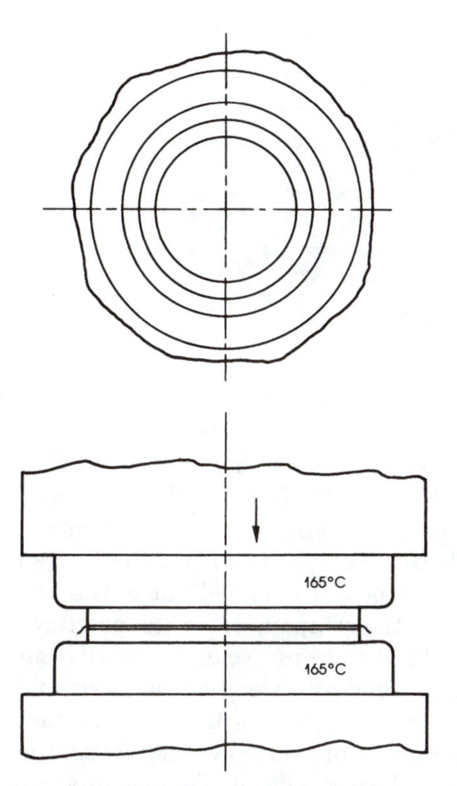

Figure 2.23 Disk flow test and mold (I).

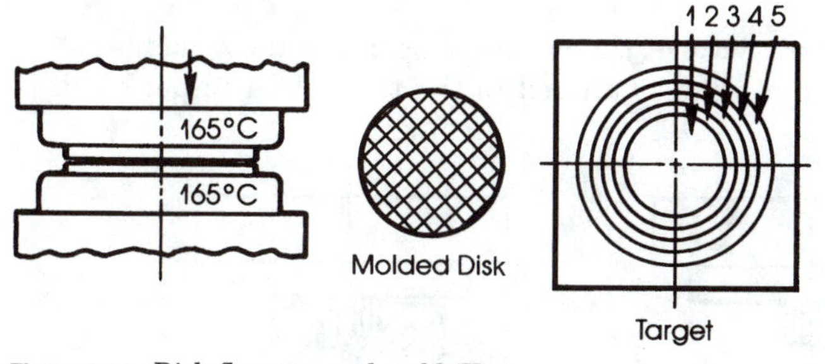

Figure 2.24 Disk flow test and mold (II).

circles numbered 1 to 5. A disk matching the #1 circle on the target will be designated as 1S flow, whereas a disk matching the #5 circle will carry a 5S flow. The higher the number, the softer the flow.

Orifice flow I (test and mold) (Fig. 2.25). This flow test involves the use of a mold with a lower plate containing a cavity into which a measured quantity of room temperature compound is placed. The upper plate has a plunger with two small orifices cut into the outer circumference,

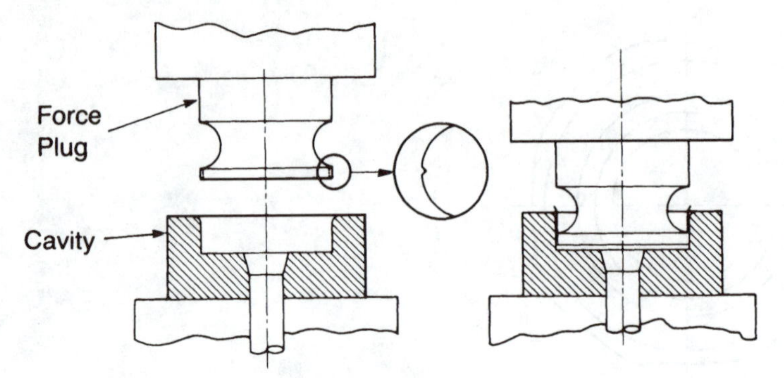

Figure 2.25 Orifice flow test and mold (I).

as shown in Fig. 2.25. The test generally uses a charge of 12 to 15 g and a mold temperature of 300°F (148°C). The molding pressure employed can be 600, 900, 1800, or 2700 lb/in², depending upon the molding process to be used. With the charged mold on, the heated platens of the press close and the compound is forced out of the two orifices. The mold is kept closed until the compound has stopped flowing and is cured. Upon completion of the molding cycle, the cured compound remaining in the mold is extracted and weighed to determine the percentage of the flow. For example, if 30% of the original shot weight is left in the mold and the molding pressure was 900 lb/in², the compound would be designated as 30% @ 900 lb/in². The use of varying pressures reflects the need to consider the flow types normally used for either compression, transfer, or injection molding.

Orifice flow II (test and mold) (Fig. 2.26). As shown, the mold has a pot in the lower half and a plunger affixed to the top half. The pot block is

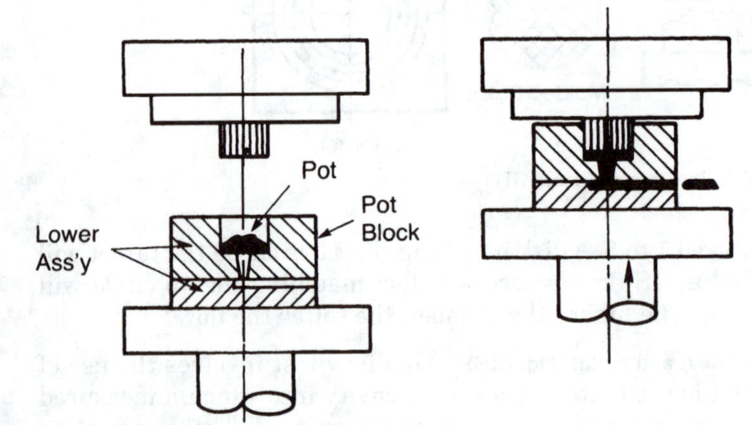

Figure 2.26 Orifice flow test and mold (II).

designed with a sprue hole in the bottom of the cavity which feeds the molding compound into a runner that comes out one side of the mold. The pot is charged with 90 g of compound and the mold is closed under 1000 lb/in^2 with a mold temperature of 300°F (148°C). The compound exiting from the runner can be weighed once the flow has ceased and if the time of flow from the start has been timed from start to finish, the rate of flow can be calculated as x g/s. This is a very useful tool for use in transfer or injection molding of thermosets.

If desired, the extrudate from the mold can be cut off every 20 s, and each segment can then be weighed. The result will provide a time-weight ratio that depicts the decrease in viscosity and increase in weight as the compound reacts to the increased mold temperature.

Spiral flow (test and mold). There are two types of spiral flow molds—one for the very soft flow encapsulation compound generally associated with the encapsulation grades of the epoxy family of compounds and a spiral flow mold, which is used when testing the high-pressure phenolic, DAP, melamine, urea, epoxy, and thermoset polyester compounds.

Figure 2.27 illustrates the *Emmi mold* which is used for testing the very soft flow encapsulation compounds. It contains a runner system cut in an Archimedes spiral starting from the center of the mold. The runner configuration is $^1/_8$-in half-round 100 in in length. The mold is heated to 300°F and run @ 300 lb/in^2. The runner is marked in linear inches and the test results are recorded as a flow of x in.

Figure 2.28 depicts the *Mesa flow mold* which has a different runner configuration. The runner is cut 0.250 in wide by 0.033 in deep and is

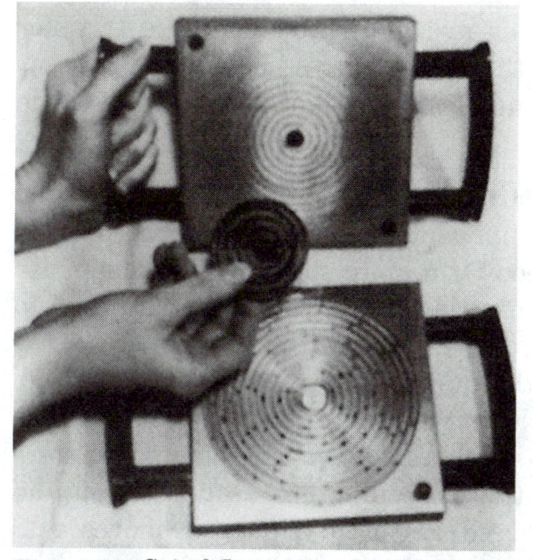

Figure 2.27 Spiral flow test and mold (Emmi).

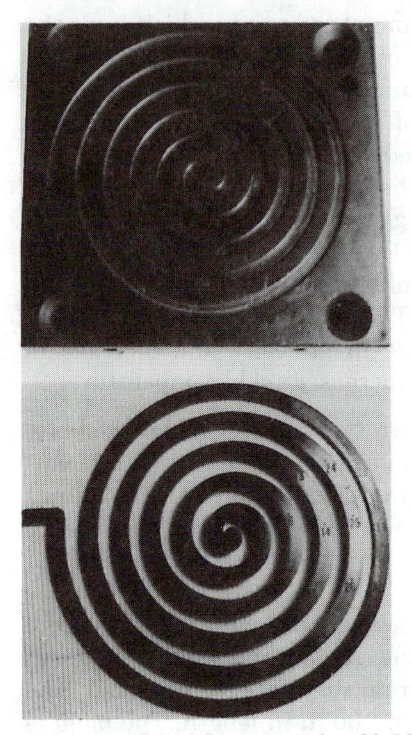

Figure 2.28 Spiral flow test and mold (Mesa).

50 lin in long. The molding cycle is run with the molding pressure set at 1000 lb/in^2 and the mold temperature is 300°F.

The flow measurement results from either the Emmi or Mesa molds can be enhanced by using a timer in conjunction with the transfer stroke during the molding cycle, thus gathering additional data relating to the quantity of the compound used as well as the distance of the compound flow.

Brabender plasticorder (Fig. 2.29 and Fig. 2.30). The introduction of the extruder screw as an integral component of the so-called *closed mold method (injection)* brought the need for a more sophisticated means for measuring the rheology of thermosetting compounds when exposed to the different molding conditions encountered with this molding technique. There is a need to know the duration of a compound's flow life when exposed to both the initial barrel temperatures as the compound is prepared for its movement into the mold and when the compound enters the much hotter mold. Generally, the compound temperature within the barrel or "reservoir" will not exceed 250°F and once it moves into the mold, it will be met with temperatures in the 325 to 400°F range. The

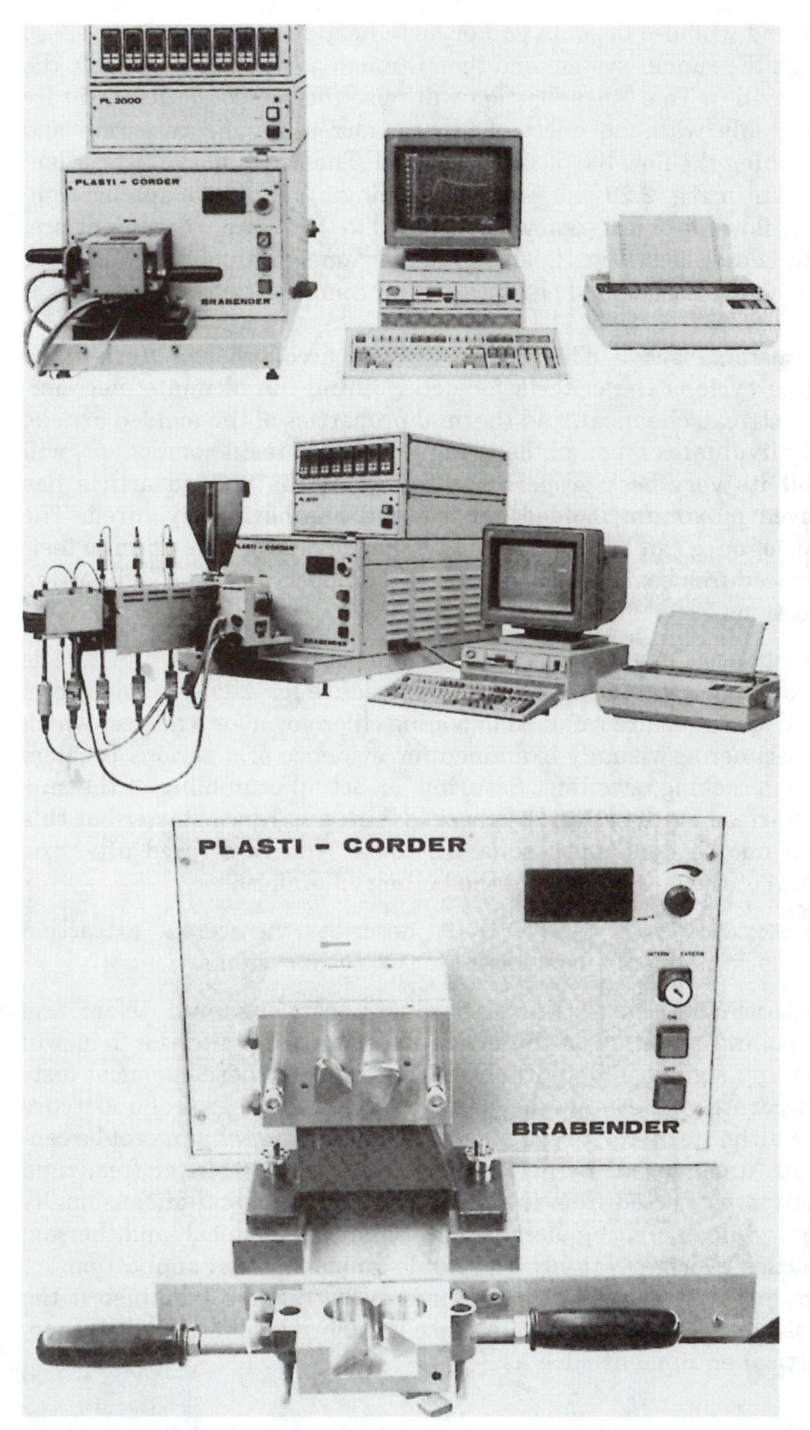

Figure 2.29 Brabender Plasticorder.

compound will also begin to gather more heat as it flows at great speed through the runner system and then through a small gate area into the mold cavities. This "shear" action will raise the compound temperature considerably with the effect of greatly increasing the cure rate and diminishing the flow life of the compound. The Brabender Plasticorder* as shown in Fig. 2.29 will provide meaningful data for a specific compound's flow life or duration when exposed to the thermal conditions previously described (Fig. 2.30). The instrument supplier, Brabender Instruments, Inc., should be contacted for complete technical details.

Cure characteristics. The degree of cure accomplished during the molding cycle is the criteria for determining the ultimate mechanical, electrical, chemical, and thermal properties of the molded article. The individual compound, depending upon its resin component, will exhibit its very best property profile when the molded article has achieved maximum molded density and has been fully cured. The degree of cure can be ascertained by one or more of the specific testing procedures based on the resin used as described in the following subsections.

Allyls (DAP and DAIP). The reflux apparatus described in Fig. 2.31 is the generally accepted cure testing procedure for Allyls. Sections of a molded specimen are refluxed in boiling chloroform for 3 h, after which the specimen is visually examined for evidence of a serious problem such as cracking, swelling, fissuring, or actual crumbling of the surface. Surface hardness can be checked with a hardness tester but this should not be done until some 12 to 14 h have elapsed after the removal of the specimen from the boiling chloroform.

Phenolics and epoxies. ASTM D-494 describes the acetone extraction test that is commonly used for these two resin systems.

Thermoset polyesters. A Barcol hardness test, employed before and after boiling a section of the molded specimen in water for 1 h, will detect any sign of undercure. Note that none of these chemical tests will accurately determine the actual effects of the degree of undercure on the ultimate integrity of a molded article. Often, what might be considered "undercured" by one of these tests will be blister free, rigid enough to be ejected from the mold, warpage free, and dimensionally correct while exhibiting electrical, mechanical, chemical, and thermal properties which are highly acceptable for commercial applications.

The integrity of a molded article may be further examined if the end-use application requires it, through more sophisticated thermoanalytical equipment such as

*Tradename of Brabender Instruments, Inc., P.O. Box 2127, South Hackensack, NJ 07606.

BRABENDER

Data-Processing Plasti-Corder PL2000 and Mixer Measuring Head
Flow-Curing Behav. of Crosslink. Polymers acc. to DIN suggestion

Test-Conditions

Order	: BRABENDER	Mixer-Temp.	:	140	°C
Operator	: EICKMEIER	Speed	:	30	1/min
Check-Date	: 16. MAY '88	Meas. Range	:	50	Nm
PL-Type	: 2000-3	Zero-Suppr.	:	0	%
Mixer-Type	: MB 30	Damping	:	3	
Load. Chute	: MANUAL + 5 KG	Test-Time	:	5.0	min
Sample	: THERMOSETTING	Sample-Weight	:	24.00	g
Additive	:	Codenumber	:	1	
		Start-Temp.	:	137	°C

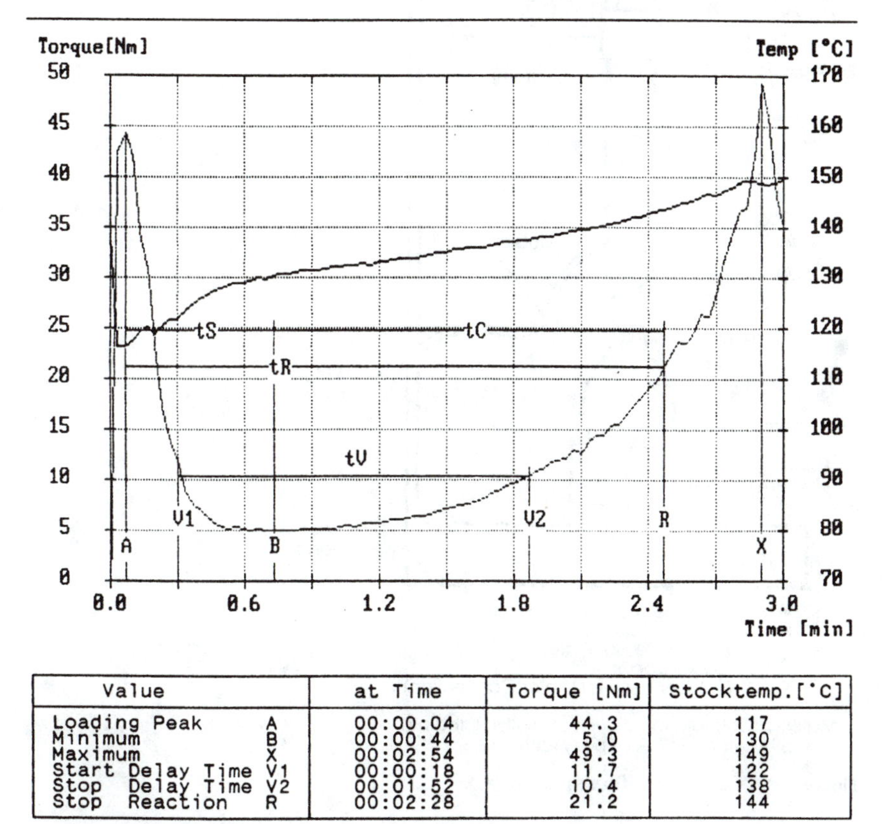

Value		at Time	Torque [Nm]	Stocktemp. [°C]
Loading Peak	A	00:00:04	44.3	117
Minimum	B	00:00:44	5.0	130
Maximum	X	00:02:54	49.3	149
Start Delay Time	V1	00:00:18	11.7	122
Stop Delay Time	V2	00:01:52	10.4	138
Stop Reaction	R	00:02:28	21.2	144

Integration / Energy

- Load.Peak to Minimum	A - B :	W1 =	1.6	[kNm]
- Minimum to Reaction	B - R :	W2 =	3.1	[kNm]
- Delay Time	tV :	W3 =	1.9	[kNm]
- Load.Peak to Reaction	A - R :	W4 =	4.7	[kNm]
- Specific Energy (W4/Sample-Weight)	:	W5 =	0.2	[kNm/g]

Results

- Delay Time	f(B+ 5.0 Nm)	V1- V2:	tV =	00:01:34
- Melt Time		A - B :	tS =	00:00:40
- Curing Time		B - R :	tC =	00:01:44
- Reaction Time	f(B+ 15.0 Nm)	A - R :	tR =	00:02:24

Figure 2.30 Brabender Plasticorder data output.

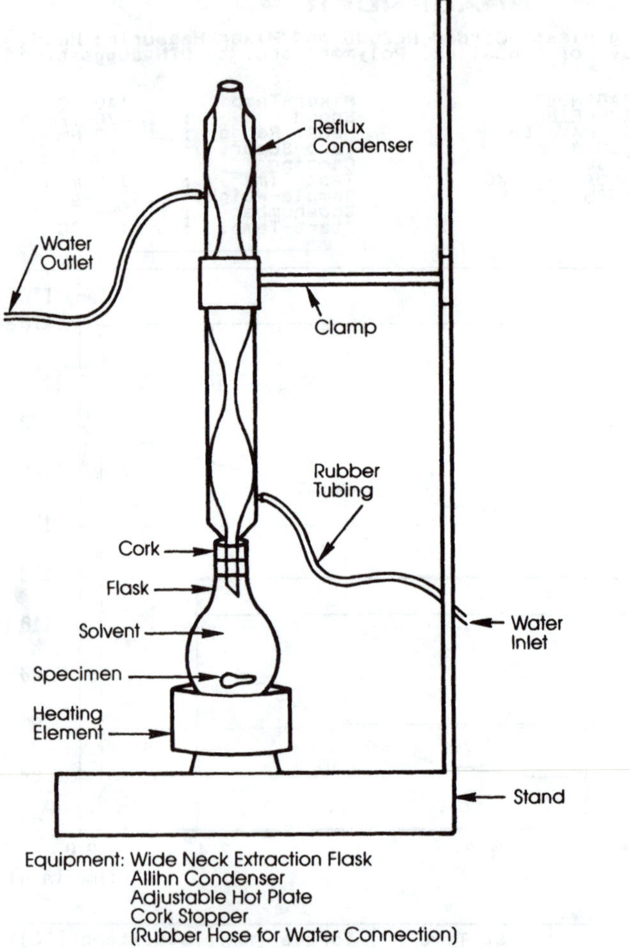

Figure 2.31 Cure test reflux apparatus.

- Differential scanning calorimetry (DSC)
- Thermo-mechanical analysis (TMA)
- Thermo-gravimetric analysis (TGA)

These techniques involve the analysis of the changes in the chemical and physical properties of a molded specimen as a function of temperature. They are not generally used for production or quality control but are quite useful in resolving problems that could be the result of insufficient cure that are not readily detected by the more common visual or chemical tests. There is a "rule-of-thumb" test procedure that is a fairly reliable cure test, and it is based upon a visual inspection of the molded article for obvious defects such as

- Lack of rigidity when ejected from the mold
- Cracks, surface porosity, and/or blisters and swelling
- Lack of flatness and/or signs of warpage that affect the ultimate dimensional stability

Postbaking. Thermosetting molded articles are frequently postcured by after-baking at recommended times and temperatures for the purpose of enhancing mechanical and thermal properties, particularly when the end-use application requires optimum performance. The usual end results of postbaking are

- Improved creep resistance
- Reduction of molded-in volatiles and/or stresses
- Improved dimensional stability

In a recent study on postcuring the following observations were made:

1. The starting temperature should be well below the glass transition (T_g) of the compound involved. A temperature below the actual molding temperature will suffice.
2. Parts of uneven cross-sectional thickness will exhibit uneven shrinkage.
3. Postbaking should be done with a multistage temperature cycle as shown here.
4. The reinforcement system of the compound will, to a large extent, distort the temperature and time cycles. Parts molded with organic reinforcements must be postcured at lower temperatures than those molded with glass and mineral reinforcement systems.

Postcure cycles1

Phenolics

One-step resole compounds: At $^1/_8$ in or less thick—2 h @ 280°F (138°C), 4 h @ 330°F (166°C), and 4 h @ 375°F (191°C)

Two-step novolac compounds: At $^1/_8$ in or less thick—2 h @ 300°F (149°C), 4 h @ 350°F (177°C), and 4 h @ 375°F (191°C)

For parts exceeding the $^1/_8$-in thickness, it is recommended that the time be doubled for each $^1/_{16}$ in of added thickness for best results. Longer times are more effective than increased temperature.

Allyls (DAP and DAIP). The recommendations are 8 h at a range starting at 275°F (135°C) and raising the temperature 20°F/h up to 415°F (213°C).

Epoxies. The recommendations are 8 h at a range starting at 275°F (135°C) and raising the temperature 20°F/h up to 415°F (213°C).

Thermoset polyesters. The recommendations are 8 h at 250°F (121°C) and raising the temperature 20°F up to a maximum of 350°F (177°C).

Note that much of the information on postbaking procedures have come from a study conducted by Bruce Fitts and David Daniels of the Rogers Corporation.[18]

References

1. R. E. Wright, Chap. 1 in *Molded Thermosets,* Hanser Verlag, Munich, 1991.
2. B. J. Shupp, *Plastics Handbook, Modern Plastics Magazine,* McGraw-Hill, New York, 1994.
3. M. A. Chaudhari, *Plastics Handbook, Modern Plastics Magazine,* McGraw-Hill, New York, 1994.
4. J. Gannon, *Plastics Handbook, Modern Plastics Magazine,* McGraw-Hill, New York, 1994.
5. W. Ayles, *Plastics Handbook, Modern Plastics Magazine,* McGraw-Hill, New York, 1994.
6. W. McNeil, *Plastics Handbook, Modern Plastics Magazine,* McGraw-Hill, New York, 1994.
7. R. Ray Patrylak, *Plastics Handbook, Modern Plastics Magazine,* McGraw-Hill, New York, 1994.
8. H. R. Gillis, *Plastics Handbook, Modern Plastics Magazine,* McGraw-Hill, New York, 1994.
9. R. Bruce Frye, *Plastics Handbook, Modern Plastics Magazine,* McGraw-Hill, New York, 1994.
10. Charles A. Harper, Chap. 4 in *Handbook of Plastics, Elastomers, and Composites,* McGraw-Hill, New York, 1996.
11. Harry Raech, Jr., Chap. 1 in *Allylic Resins and Monomers,* Reinhold Publishing Corporation, New York, 1965.
12. Charles A. Harper, Chap. 2 in *Handbook of Plastics, Elastomers, and Composites,* McGraw-Hill, New York, 1996.
13. Harry Raech, Jr., Chap. 1 in *Allylic Resins and Monomers,* Reinhold Publishing Corporation, New York, 1965.
14. Harry Dubois, *Plastics History USA,* Cahners Publishers, Boston, 1972.
15. Charles A. Harper, Chap. 3 in *Handbook of Plastics, Elastomers, and Composites,* McGraw-Hill, New York, 1996.
16. R. E. Wright, Chap. 1 in *Molded Thermosets,* Hanser Verlag, Munich, 1991.
17. Charles A. Harper, Chap. 1 in *Handbook of Plastics, Elastomers, and Composites,* McGraw-Hill, New York, 1996.
18. Bruce Fitts and David Daniels, Applications Report, Rogers Corporation, Rogers, Conn., 1991.
19. Rogers Corporation, *Molding Compound Product Bulletin,* Rogers, Conn.

Chapter

3

Elastomeric Materials and Processes*

James M. Margolis
Montreal, Quebec, Canada

3.1 Introduction

Another important group of polymers is that group which is elastic or rubberlike, known as elastomers. This chapter discusses this group of materials, including TPEs, MPRs, TPVs, synthetic rubbers, and natural rubber.

3.2 Thermoplastic Elastomers (TPEs)

Worldwide consumption of TPE for the year 2000 is estimated to be about 2.5 billion pounds, primarily due to new polymer and processing technologies, with an annual average growth rate of about 6% between 1996 and 2000.[4] About 40% of this total is consumed in North America.[4]

TPE grades are often characterized by their hardness, resistance to abrasion, cutting, scratching, local strain (deformation), and wear. A

*The chapter author, editors, publisher, and companies referred to are not responsible for the use or accuracy of information in this chapter, such as property data, processing parameters, aplications.

conventional measure of hardness is Shore A and Shore D shown in Fig. 3.1. Shore A is a softer and Shore D is a harder TPE, with ranges from as soft as Shore A 40 to as hard as Shore D 82. Durometer hardness (ASTM D 2240) is an industry standard test method for rubbery materials, covering two types of durometers, A and D. The *durometer* is the hardness measuring apparatus; and the term *durometer hardness* is often used with Shore hardness values. There are other hardness test methods such as Rockwell hardness for plastics and electrical insulating materials (ASTM D 785 and ISO 2039), and Barcol hardness (ASTM D 2583) for rigid plastics. While hardness is often a quantifying distinction between grades, it does not indicate comparisons between physical/mechanical, chemical, and electrical properties.

Drying times depend on moisture absorption of a given resin. TPE producers suggest typical drying times and processing parameters. Actual processing temperature and pressure settings are determined by resin melt temperatures and rheological properties, mold cavity design, and equipment design such as screw configuration.

Performance property tables provided by suppliers usually refer to compounded grades containing property enhancers (additives) such as stabilizers, modifiers, and flame retardants. Sometimes the suppliers' property tables refer to a polymer, rather than a formulated compound.

3.2.1 Styrenics

Styrene block copolymers are the most widely used TPEs, accounting for close to 45% of total TPE consumption worldwide at the close of the twentieth century.[1] They are characterized by their molecular architecture which has a "hard" thermoplastic segment (block) and a "soft" elastomeric segment (block) (see Fig. 3.2). Styrenic TPEs are usually styrene butadiene styrene (SBS), styrene ethylene/butylene styrene (SEBS), and styrene isoprene styrene (SIS). Styrenic TPEs usually have about 30 to 40% (wt) bound styrene; certain grades have a higher bound styrene content. The polystyrene endblocks create a network of reversible physical

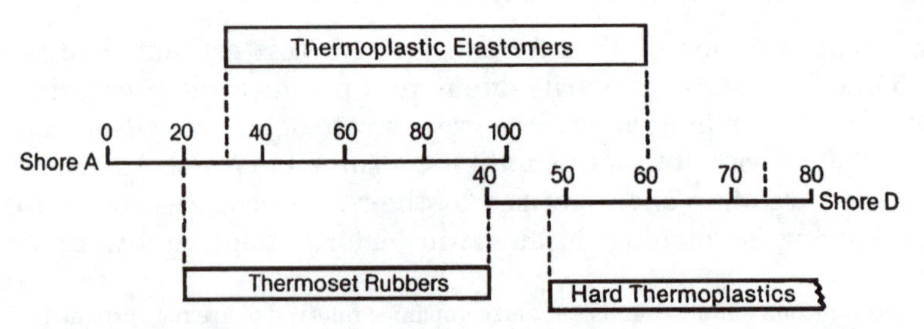

Figure 3.1 TPEs bridge the hardness ranges of rubbers and plastics. (*Source: Ref. 10, p. 5.2.*)

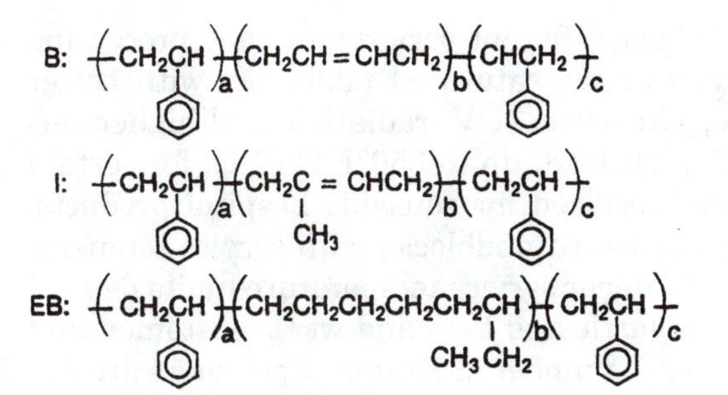

Figure 3.2 Structures of three common styrenic block copolymer TPEs: a and c = 50 to 80; b = 20 to 100. (*Source: Ref. 10, p. 5.12.*)

cross-links which allow thermoplasticity for melt processing or solvation. With cooling or solvent evaporation, the polystyrene domains reform and harden, and the rubber network is fixed in position.[2]

Principal styrenic TPE markets are: molded shoe soles and other footwear; extruded film/sheet and wire/cable covering; and pressure-sensitive adhesives (PSA) and hot-melt adhesives, viscosity index (VI) improver additives in lube oils, resin modifiers, and asphalt modifiers. They are also popular as grips (bike handles), kitchen utensils, clear medical products, and personal care products.[1,4] Adhesives and sealants are the largest single market.[1] Styrenic TPEs are useful in adhesive compositions in web coatings.[1]

Styrenic block copolymer (SBC) thermoplastic elastomers are produced by Shell Chemical (Kraton®*), Firestone Synthetic Rubber and Latex, Division of Bridgestone/Firestone (Stereon®†), Dexco Polymers (Vector®‡), EniChem Elastomers (Europrene®§), and other companies. SBC properties and processes are described for these four producers' TPEs.

Kraton TPEs are usually SBS, SEBS, and SIS, as are SEP (styrene ethylene/propylene) and SEB (styrene ethylene/butylene).[2] The polymers can be precisely controlled during polymerization to meet property requirements for a given application.[2]

Two Kraton types are chemically distinguished: Kraton G and Kraton D. A third type, Kraton Liquid®,¶ poly(ethylene/butylene), is described

---

*Kraton is a registered trademark of Shell Chemical Company.

†Stereon is a registered trademark of Firestone Synthetic Rubber and Latex Company, Division of Bridgestone/Firestone.

‡Vector is a registered trademark of Dexco, A Dow/Exxon Partnership.

§Europrene is a registered trademark of EniChem Elastomers.

¶Kraton Liquid is a registered trademark of Shell Chemical Company.

on page 3.5. Kraton G and D have different performance and processing properties. Kraton G polymers have saturated midblocks with better resistance to oxygen, ozone, ultraviolet (UV) radiation, and higher service temperatures, depending on load, up to 350°F (177°C) for certain grades.[2] They can be steam sterilized for reusable hospital products. Kraton D polymers have unsaturated midblocks with service temperatures up to 150°F (66°C).[2] SBC upper service temperature limits depend on the type and wt % thermoplastic and type and wt % elastomer, and the addition of heat stabilizers. A number of Kraton G polymers are linear SEBS, while several Kraton D polymers are linear SIS.[2] Kraton G polymer compounds' melt process is similar to polypropylene; Kraton D polymer compounds' process is comparable to polystyrene (PS).[2]

Styrenic TPEs have strength properties equal to vulcanized rubber, but they do not require vulcanization.[2] Properties are determined by polymer type and formulation. There is a wide latitude in compounding to meet a wide variety of application properties.[2] According to application-driven formulations, Kratons are compounded with a hardness range from Shore A 28 to 95 (Shore A 95 is approximately equal to Shore D 40), sp gr from 0.90 to 1.18, tensile strengths from 150 to 5000 lb/in^2 (1.03 to 34.4 MPa), and flexibility down to −112°F (−80°C) (see Table 3.1).[2]

Kratons are resistant to acids, alkalis, and water, but long soaking in hydrocarbon solvents and oils deteriorate the polymers.[2]

Automotive applications range from window seals and gasketing to enhanced noise/vibration attenuation.[1] The polymers are candidates for automotive seating, interior padded trim and insulation, hospital padding, and topper pads.[1] SEBS is extruded/blown into 1-mil films for disposable gloves for surgical/hospital/dental, food/pharmaceutical, and household markets.[1]

TABLE 3.1 Typical Properties of a Kraton D and Kraton G Polymer for Use as Formulation Ingredients and as Additives (U.S. FDA Compliance)

Property [74°F (23°C)]	Kraton D D1101 (linear SBS)	Kraton G G1650 (linear SEBS)
Specific gravity, g/cm^3	0.94	0.91
Hardness, Shore A	71	75
Tensile strength, lb/in^2 (MPa)	4600 (32)	5000 (34)
300% modulus, lb/in^2 (MPa)	400 (2.7)	800 (5.5)
Elongation, %	880	550
Set @ break, %	10	—
Melt flow index, g/10 min	<1	—
Brookfield viscosity, Hz @ 77°F (25°C), toluene solution	4000	8000
Styrene/rubber ratio	31/69	29/71

Kratons are used in PSAs, hot-melt adhesives, sealants, solution-applied coatings, flexible oil gels, modifiers in asphalt, thermoplastics, and thermosetting resins.[2] When Kratons are used as an impact modifier in nylon 66, notched Izod impact strength can be increased from 0.8 ft-lb/in for unmodified nylon 66 to 19 ft-lb/in. Flexural modulus may decrease from 44,000 lb/in^2 (302 MPa) for unmodified nylon 66 to about 27,000 lb/in^2 (186 MPa) for impact-modified nylon 66.

SBCs are injection molded, extruded, blow molded, and compression molded.[2]

Kraton Liquid polymers are polymeric diols with an aliphatic, primary OH^- group on each terminal end of the poly(ethylene/butylene) elastomer. They are used in formulations for adhesives, sealants, coatings, inks, foams, fibers, surfactants, and polymer modifiers.[13]

Two large markets for Firestone's styrenic block copolymer SBS Stereon TPEs are: (1) impact modifiers (enhancers) for flame-retardant polystyrene and polyolefin resins and (2) PSA and hot-melt adhesives. Moldable SBS block copolymers possess high clarity, gloss, good flex cycle stability for "living hinge" applications, FDA-compliant grades for food containers, and medical/hospital products.[1] Typical mechanical properties are: 4600-lb/in^2 (31.7-MPa) tensile strength, 6000-lb/in^2 (41.4-MPa) flexural strength, and 200,000-lb/in^2 (1.4-GPa) flexural modulus.[1]

Stereon stereospecific butadiene styrene block copolymer is used as an impact modifier in PS, high-impact polystyrene (HIPS), polyolefin sheet and films, such as blown film grade linear low-density polyethylene (LLDPE), to achieve downgauging and improve tear resistance and heat sealing.[1] Blown LLDPE film modified with 7.5% stereospecific styrene block copolymers has a Dart impact strength of 185°F per 50 g, compared with 135°F per 50 g for unmodified LLDPE film. These copolymers also improve environmental stress crack resistance (ESCR), especially to fats and oils for meat/poultry packaging trays, increase melt flow rates, increase gloss, and meet U.S. FDA 21 CFR 177.1640 (PS and rubber modified PS) with at least 60% PS for food contact packaging.[1] When used with thermoformable foam PS, flexibility is improved without sacrificing stiffness, allowing deeper draws.[1] The stereospecific butadiene block copolymer TPEs are easily dispersed and improve blendability of primary polymer with scrap for recycling.

Vector SBS, SIS, and SB styrenic block copolymers are produced as diblock-free and diblock copolymers.[29] The company's process to make linear SBCs yields virtually no diblock residuals. Residual styrene butadiene and styrene isoprene require endblocks at both ends of the polymer in order to have a load-bearing segment in the elastomeric network.[29] However, diblocks are blended into the copolymer for certain applications.[29] Vector SBCs are injection-molded, extruded, and formulated into pressure-sensitive adhesives for tapes and labels, hot-melt product-assembly adhesives, construction adhesives, mastics, sealants, and asphalt modifiers.[29] The asphalts are used to make membranes for single-ply roofing and waterproofing systems, binders for pavement construction and repair, and sealants for joints and cracks.[29] Vector SBCs are used as property enhancers (additives) to improve the toughness and impact strength at ambient and low temperatures of engineering thermoplastics, olefinic and styrenic thermoplastics, and thermosetting resins.[29] The copolymers meet applicable U.S. FDA food additive 21 CFR 177.1810 regulations and United States Pharmacopoeia (USP) (Class VI medical devices) standards for health-care applications.[29]

The company's patented hydrogenation techniques are developed to improve SBC heat resistance as well as ultraviolet resistance.[29]

EniChem Europrene SOL T are styrene butadiene and styrene isoprene linear and radial block copolymers.[1] They are solution-polymerized using anionic type catalysts.[33] The molecules have polystyrene endblocks with central elastomeric polydiene (butadiene or isoprene) blocks.[33] The copolymers are $(S\text{-}B)_nX$ type where S = polystyrene, B = polybutadiene and polyisoprene, and X = a coupling agent. Both configurations have polystyrene (PS) endblocks, with bound styrene content ranging from 25 to 70% (wt).[1] Polystyrene contributes styrene hardness, tensile strength, and modulus; polybutadiene and polyisoprene contribute high resilience and flexibility, even at low temperatures.[1] Higher molecular weight (MW) contributes a little to mechanical properties, but decreases melt flow characteristics and processability.

The polystyrene and polydiene blocks are mutually insoluble, and this shows with two T_g peaks on a cartesian graph with tan δ (y axis) versus temperature (x axis): one T_g for the polydiene phase and a second T_g for the polystyrene phase. A synthetic rubber, such as SBR, shows one T_g.[33] The two phases of a styrenic TPE are chemically bound, forming a network with the PS domains dispersed in the polydiene phase. This structure accounts for mechanical/elastic properties and thermoplastic processing properties.[33] At temperatures up to about 167°F (80°C), which is below PS T_g of 203 to 212°F (95 to 100°C) the PS phase is rigid.[33] Consequently, the PS domains behave as cross-linking sites in the polydiene phase, similar to sulfur links in vulcanized rubber.[33] The rigid PS phase also acts as a reinforcement, as noted here.[33] Crystal PS, HIPS, poly-alpha-methylstyrene, ethylene vinyl acetate (EVA) copolymers, low-density polyethylene (LDPE), and high-density polyethylene (HDPE) can be used as organic reinforcements. $CaCO_3$, clay, silica, and silicates act as inorganic fillers, with little reinforcement, and they can adversely affect melt flow if used in excessive amounts.[33]

The type of PS, as well as its % content, affect properties. Crystal PS, which is the most commonly used, and HIPS increase hardness, stiffness, and tear resistance without reducing melt rheology.[1] High styrene copolymers, especially Europrene SOL S types produced by solution polymerization, significantly improve tensile strength, hardness, and plasticity, and they enhance adhesive properties.[33] High styrene content does not decrease the translucency of the compounds.[33] Poly-alpha-methylstyrene provides higher hardness and modulus, but abrasion resistance decreases.[33] EVA improves resistance to weather, ozone, aging, and solvents, retaining melt rheology and finished product elasticity. The highest Shore hardness is 90 A, the highest melt flow is 16 g/10 min, and specific gravity is 0.92–0.96.[1]

Europrene compounds can be extended with plasticizers which are basically a paraffinic oil containing specified amounts of naphthenic and aromatic fractions.[33] Europrenes are produced in both oil-extended and dry forms.[1] Oils were specially developed for optimum mechanical, aging, processing, and color properties.[33] Increasing oil content significantly increases melt flow properties, but it reduces mechanical properties. Oil extenders must be incompatible with PS in order to avoid PS swelling, which would decrease mechanical properties even more.[33]

The elastomers are compounded with antioxidants to prevent thermal and photooxidation which can be initiated through the unsaturated zones in the copolymers.[33] Oxidation can take place during melt processing and during the life of the fabricated product.[33] Phenolic, or phosphitic antioxidants, and dilauryldithiopropionate as a stabilizer during melt processing are recommended.[33] Conventional UV stabilizers are used such as benzophenone and benzotriazine.[33] Depending on the application, the elastomer is compounded with flow enhancers such as low MW polyethylene (PE), microcrystalline waxes or zinc stearate, pigments, and blowing agents.[33]

Europrene compounds, especially oil-extended grades, are used in shoe soles and other footwear.[1] Principal applications are: impact modifiers in PS, HDPE, LDPE, polypropylene (PP), other thermoplastic resins and asphalt; extruded hose, tubing, O-rings, gaskets, mats, swimming equipment (eye masks, snorkels, fins, "rubberized" suits) and rafts; and pressure-sensitive adhesives (PSA) and hot melts.[1] SIS types are used in PSA and hot melts; SBS types are used in footwear.[1]

The copolymer is supplied in crumb form, and mixing is done by conventional industry practices, with an internal mixer or low-speed room temperature premixing and compounding with either a single- or twin-screw extruder.[33] Low-speed premixing/extrusion compounding is the process of choice.

Europrenes have thermoplastic polymer melt processing properties and characteristics of TPEs. At melt processing temperatures, they behave as thermoplastics, and below the PS T_g of 203 to 212°F (95 to 100°C) the copolymers act as cross-linked elastomers, as noted earlier. Injection-molding barrel temperature settings are from 284 to 374°F (140 to 190°C). Extrusion temperature at the head of the extruder is maintained between 212 and 356°F (100 and 180°C).

3.2.2 Olefenics and TPO elastomers

Thermoplastic polyolefin (TPO) elastomers are typically composed of ethylene propylene rubber (EPR) or ethylene propylene diene "M" (EPDM) as the elastomeric segment and polypropylene thermoplastic segment.[18] LDPE, HDPE, and LLDPE; copolymers ethylene vinyl acetate (EVA), ethylene ethylacrylate (EEA), ethylene, methyl-acrylate (EMA); and polybutene-1 can be used in TPOs.[18] Hydrogenation of polyisoprene can yield ethylene propylene copolymers, and hydrogenation of 1,4- and 1,2-stereoisomers of S-B-S yields ethylene butylene copolymers.[1]

TPO elastomers are the second most used TPEs on a tonnage basis, accounting for about 25% of total world consumption at the close of the twentieth century (according to what TPOs are included as thermoplastic elastomers).

EPR and polypropylene can be polymerized in a single reactor or in two reactors. With two reactors, one polymerizes propylene monomer to polypropylene, the second copolymerizes polypropylene with ethylene propylene rubber (EPR) or EPDM. Reactor grades are (co)polymerized in a single reactor. Compounding can be done in the single reactor.

Montell's in-reactor Catalloy®* ("catalytic alloy") polymerization process alloys propylene with comonomers, such as EPR and EPDM, yielding very soft, very hard, and rigid plastics, impact grades or elastomeric TPOs, depending on the EPR or EPDM % content. The term *olefinic* for thermoplastic olefinic elastomers is arguable, because of the generic definition of olefinic. TPVs are composed of a continuous thermoplastic polypropylene phase and a discontinuous vulcanized rubber phase, usually EPDM, EPR, nitrile rubber, or butyl rubber.

Montell describes TPOs as flexible plastics, stating "TPOs are not TPEs."[32a] The company's Catalloy catalytic polymerization is a cost-effective process, used with propylene monomers which are alloyed with comonomers, including the same comonomers with different molecular architecture. Catalloy technology uses multiple gas-phase reac-

*Catalloy is a registered trademark of Montell North America Inc., wholly owned by the Royal Dutch/Shell Group.

tors that allow the separate polymerization of a variety of monomer streams.[32] Alloyed or blended polymers are produced directly from a series of reactors which can be operated independently from each other to a degree.[32]

Typical applications are: flexible products such as boots, bellows, drive belts, conveyor belts, diaphragms, keypads; connectors, gaskets, grommets, lip seals, O-rings, plugs; bumper components, bushings, dunnage, motor mounts, sound deadening; and casters, handle grips, rollers, and step pads.[9]

Insite®* technology is used to produce Affinity®† polyolefin plastomers (POP), which contain up to 20 wt % octene comonomer.[14] Dow Chemical's 8-carbon octene polyethylene technology produces the company's ULDPE "Attane" ethylene-octene-1 copolymer for cast and blown films. Alternative copolymers are 6-carbon hexene and 4-carbon butene, for heat-sealing packaging films. Octene copolymer POP has lower heat-sealing temperatures for high-speed form-fill-seal lines and high hot-tack strength over a wide temperature range.[14] Other benefits cited by Dow Chemical are: toughness, clarity, and low taste/odor transmission.[14]

Insite technology is used for homogeneous single-site catalysts which produce virtually identical molecular structure such as branching, comonomer distribution, and narrow molecular weight distribution (MWD).[14] Solution polymerization yields Affinity polymers with uniform, consistent structures, resulting in controllable, predictable performance properties.[14]

Improved performance properties are obtained without diminishing processability because the Insite process adds long-chain branching onto a linear short-chain, branched polymer.[14] The addition of long-chain branches improves melt strength and flow.[14] Long-chain branching results in polyolefin plastomers processing at least as smooth as LLDPE and ultra-low-density polyethylene (ULDPE) film extrusion.[14] Polymer design contributes to extrusion advantages such as enhanced shear flow, drawdown, and thermoformability.

For extrusion temperature and machine design, the melt temperature is 450 to 550°F (232 to 288°C), the feed zone temperature setting is 300 to 325°F (149 to 163°C), 24/1 to 32/1 *L/D*; for the sizing gear box, use 5 lb/h/hp (1.38 kg/h/kW) to estimate power required to extrude POP at a given rate; for single-flight screws, line draw over the length of the line, 10 to 15 ft/min (3 to 4.5 m/min) maximum. Processing conditions and equipment design vary according to the resin selection and finished product. For example, a melt temperature of 450 to 550°F (232 to 288°C)

---

*Insite is a registered trademark of DuPont Dow Elastomers LLC.

†Affinity is a registered trademark of The Dow Chemical Company.

applies to cast, nip-roll fabrication using an ethylene alpha-olefin POP,[14a] while 350 to 450°F (177 to 232°C) is recommended for extrusion/blown film for packaging, using an ethylene alpha-olefin POP.[14b]

POP applications are: sealants for multilayer bags and pouches to package cake mixes, coffee, processed meats and cheese, and liquids; overwraps; shrink films; skin packaging; heavy-duty bags and sacks; and molded storage containers and lids.[14]

Engage®*polyolefin elastomers (POE), ethylene octene copolymer elastomers, produced by DuPont Dow Elastomers, use Insite catalytic technology.[5] Table 3.2 shows their low density and wide range of physical/mechanical properties (using ASTM test methods).[5]

The copolymer retains toughness and flexibility down to −40°F (−40°C).[5] When cross-linked with peroxide, silane, or by radiation, heat resistance and thermal aging increase to >302°F (>150°C).[5] Cross-linked copolymer is extruded into covering for low- and medium-voltage cables. POE elastomers have a saturated chain, providing inherent UV stability.[5] Ethylene octene copolymers are used as impact modifiers, for example, in polypropylene. Typical products are: foams and cushioning components, sandal and slipper bottoms, sockliners and midsoles, swim fins, and winter and work boots; TPO bumpers, interior trim and rub strips, automotive interior air ducts, mats and liners, extruded hose and tube, interior trim, NVH applications, primary covering for wire and

*Engage is a registered trademark of DuPont Dow Elastomers LLC.

TABLE 3.2 Typical Property Profile for Engage Polyolefin Elastomers (Unfilled Polymer, Room Temperature Except Where Indicated)

Property	Values
Specific gravity, g/cm^3	0.857–0.913
Flexural modulus, lb/in^2 (MPa), 2% secant	435–27,550 (3–190)
100% modulus, lb/in^2 (MPa)	145–>725 (1–>5)
Elongation, %	700+
Hardness, Shore A	50–95
Haze, % 0.070 in (1.8 mm) Injection-molded plaque	<10–20
Low-temperature brittleness, °F (°C)	<−104 (<−76)
Melt flow index, g/10 min	0.5–30
Melting point, °F (°C)	91–225 (33–107)

cable voltage insulation (low and medium voltage), appliance wire, semiconductive shields, nonhalogen flame-retardant and low smoke emission jackets, and bedding compounds.[5]

Union Carbide elastomeric polyolefin flexomers combine flexibility, toughness, and weatherability, with properties midrange between polyethylene and EPR.

3.2.3 Polyurethane thermoplastic elastomers (TPUs)

TPUs are the third most used TPEs, accounting for about 15% of TPE consumption worldwide.

Linear polyurethane thermoplastic elastomers can be produced by reacting a diisocyanate [methane diisocyanate (MDI) or toluene diisocyanate (TDI)] with long-chain diols such as liquid polyester or polyether polyols, and adding a chain extender such as 1,4-butanediol.[17,18c] The diisocyanate and chain extender form the hard segment, and the long-chain diol forms the soft segment.[18c] For sulfur curing, unsaturation is introduced, usually with an allyl ether group.[17] Peroxide curing agents can be used for cross-linking.

The two principal types of TPUs are polyether and polyester. Polyethers have good low-temperature properties and resistance to fungi; polyesters have good resistance to fuel, oil, and hydrocarbon solvents.

BASF Elastollan®* TPU elastomer property profiles show typical properties of polyurethane thermoplastic elastomers (see Table 3.3).

Shore hardness can be as soft as 70 A and as hard as 74 D, depending on the hard/soft segment ratio. Specific gravity, modulus, compressive stress, load-bearing strength, and tear strength are also hard/soft ratio dependent.[18c] TPU thermoplastic elastomers are tough, tear resistant, abrasion resistant, and exhibit low-temperature properties.[4]

Dow Plastics Pellethane®† TPU elastomers are based on both polyester and polyether soft segments.[3]

Five series indicate typical applications:

1. Polyester polycaprolactones for injection-molded automotive panels, painted (without primer) with urethane and acrylic enamels, or water-based elastomeric coating
2. Polyester polycaprolactones for seals, gaskets, and belting
3. Polyester polyadipates extruded into film, sheet, and tubing

---

*Elastollan is a registered trademark of BASF Corporation.

†Pellethane is a registered trademark of The Dow Chemical Company.

TABLE 3.3 Mechanical Property Profile of Elastollan

Property	Value
Specific gravity	1.11–1.21
Hardness Shore	70 A–74 D
Tensile strength, lb/in^2 (MPa)	4600–5800 (31.7–40.0)
Tensile stress, lb/in^2 (MPa)	
100% elongation	770–1450 (5.3–10.0)
300% elongation	1300–1750 (9.0–12.0)
Elongation @ break, %	550–700
Tensile set @ break, %	45–50
Tear strength, Die C pli	515–770
Abrasion-resistance, mg loss (Tabor)	25

4. Polytetramethylene glycol ethers with excellent dielectrics for extruded wire and cable covering, and also for films, tubing, belting, and caster wheels
5. Polytetramethylene glycol ethers for healthcare applications[3]

Polyether-polyester hybrid specialty compounds are the softest nonplasticized TPU (Shore hardness 70 A), and are used as impact modifiers.

Polycaprolactones possess good low-temperature impact strength for paintable body panels, good fuel and oil resistance, and hydrolytic stability for seals, gaskets, and belting.[3] Polycaprolactones have fast crystallization rates, high crystallinity, and are generally easily processed into complex parts.

Polyester polyadipates have improved oil and chemical resistance, but slightly lower hydrolytic stability than polycaprolactones which are used for seals, gaskets, and beltings.

Polytetramethylene glycol ether resins for wire/cable covering have excellent resistance to hydrolysis and microorganisms, compared with polyester polyurethanes. Healthcare grade polyether TPUs are resistant to fungi, have low levels of extractable ingredients, excellent hydrolysis resistance, and can be sterilized for reuse by gamma irradiation, ethylene oxide, and dry heat but not with pressurized steam (autoclave).[3] Polyether TPUs are an option for sneakers and athletic footwear components such as outer soles.

Bayer Bayflex®* elastomeric polyurethane reaction injection molding (RIM) is a two-component diphenylmethane diisocyanate- (MDI)-based liquid system produced in unreinforced, glass-reinforced, and mineral-/microsphere-reinforced grades.[15] They possess a wide stiff-

*Bayflex is a registered trademark of Bayer Corporation.

ness range, relatively high impact strength, quality molded product surface, and can be in-mold coated. Room temperature properties are: specific gravity, 0.95 to 1.18; ultimate tensile strength, 2300 to 4200 lb/in^2 (16 to 29 MPa); flexural modulus, 5000 to 210,000 lb/in^2 (34 to 1443 MPa); and tear strength, Die C, 230 to 700 lb/in (40 to 123 kN/m).[15] Related Bayer U.S. patents are: TPU-Urea Elastomers, U.S. Patent 5,739,250 assigned to Bayer AG, April 14, 1998, and RIM Elastomers Based on Prepolymers of Cycloaliphatic Diisocyanates, U.S. Patent 5,738,253 assigned to Bayer Corp., April 14, 1998.

Representative applications are: tractor body panels and doors, automotive fascia, body panels, window encapsulation, heavy-duty truck bumpers, and recreation vehicle (RV) panels.

Bayer's Texin®* polyester and polyether TPU and TPU/polycarbonate (PC) elastomers were pioneer TPEs in early development of passenger car fronts and rear bumpers. PC imparts Izod impact strength toughness required for automotive exterior body panels. Extrusion applications include film/sheet, hose, tubing, profiles, and wire/cable covering. Hardness ranges from Shore A 70 to Shore D 75. Texin can be painted without a primer.

Morton International Morthane®† TPU elastomers are classified into four groups: polyesters, polyethers, polycaprolactones, and polyblends. Polyester polyurethanes have good tear and abrasion resistance, toughness, and low-temperature flexibility, and Shore hardness ranges from 75 A to 65 D. They are extruded into clear film and tubing and fuel line hose. Certain grades are blended with acrylonitrile butadiene styrene (ABS), styrene acrylonitrile (SAN), nylon, PC, polyvinyl chloride (PVC), and other thermoplastic resins. Polyethers possess hydrolytic stability, resilience, toughness, good low-temperature flexibility, easy processability, and fast cycles. They also have tensile strength up to 7500 lb/in^2 (52 MPa) and melt flow ranges from about 5 to 60 g/10 min. Hardnesses are in the Shore A range up to 90. Certain grades can be used in medical applications. Aliphatic polyester and polyether grades provide UV resistance for pipes, tubing, films, and liners. They can be formulated for high clarity.

The polyblends are polyester TPUs blended with ABS, SAN, PC, nylon, PVC, and other thermoplastics for injection molding and extrusion. A 10 to 20% loading into PVC compositions can increase mechanical properties 30 to 40%.[1] Although the elastomeric TPUs are inherently flexible, plasticizers may be recommended, for example, in films.

TPU elastomers are processed on rubber equipment, injection molded, extruded, compression molded, transfer molded, and calendered. In

*Texin is a registered trademark of Bayer Corporation.

†Morthane is a registered trademark of Morton International Inc.

order to be fabricated into products, such as athletic shoe outer soles, the elastomer and ingredients are mixed in conventional rubber equipment: two-roll mills, internal mixers, and compounded.[17] Subsequently, the compound is processed, for example, injection molded.[17]

Typical melt processing practices are described with Pellethane TPU (see Table 3.4). The moisture content is brought to <0.02% before molding or extruding.[3] Dessicant, dehumidifying hopper dryers that can produce a −40°F (−40°C) dew point at the air inlet are suggested. A dew point of −20°F (−29°C) or lower is suggested for TPU elastomers.[3] The suggested air-inlet temperature range is 180 to 230°F (82 to 110°C)[3]: 180 to 200°F (82 to 93°C) for the softer Shore A elastomers, to 210 to 230°F (99 to 110°C) for harder Shore D elastomers.

Drying time to achieve a given moisture content for resin used directly from sealed bags is shown in Fig. 3.3: about 4 h to achieve <0.02% moisture content @ 210°F (99°C) air-inlet temperature and −20°F (−29°C) dew point.[3] When TPU elastomers are exposed to air just prior to processing, the pellets are maintained at 150 to 200°F (65 to 93°C), a warmer temperature than the ambient air.[3] A polymer temperature that is warmer than the ambient air reduces the ambient moisture absorption.[3]

Melt temperature is determined by T_m of resin and processing and equipment specifications, including machine capacity, rated shot size, screw configuration (*L*/*D*, flight number, and design), part design, mold design (gate type and runner geometry), and cycle time.[3] Shear energy created by the reciprocating screw contributes heat to the melt, causing the actual melt temperature to be 10 to 20°F (6 to 10°C) higher than the barrel temperature settings.[3] Temperature settings should take shear energy into account. In order to ensure maximum product quality, the processor should discuss processing parameters, specific machinery, equipment, and tool and product data with the resin supplier. For example, if it is suspected that an improper screw design will be used, melt

TABLE 3.4 Typical Injection-Molding Settings* for Pellethane TPU

Temperature, °F (°C)	Shore A 80	Shore D 55
Melt temperature max	415 (213)	435 (224)
Cylinder zone		
Rear (feed)	350–370 (177–188)	360–380 (183–193)
Middle (transition)	360–380 (183–193)	370–390 (188–198)
Front (metering)	370–390 (188–199)	390–410 (199–210)
Nozzle	390–410 (199–210)	400–410 (204–210)
Mold temperature	80–140	(27–60)

*Typical temperature and pressure settings are based on Ref. 3. Settings are based on studies using a reciprocating screw, general-purpose screw, clamp capacity of 175 tons, and rated shot capacity of 10 oz (280 g). Molded specimen thicknesses ranged from 0.065 to 0.125-in (1.7 to 3.2-mm) thickness.

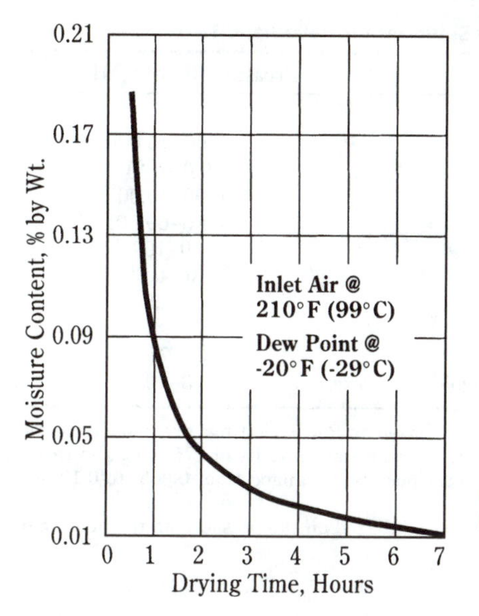

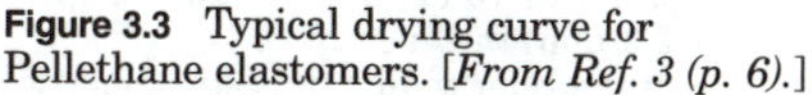
Figure 3.3 Typical drying curve for Pellethane elastomers. [*From Ref. 3 (p. 6).*]

temperature gradient may be reversed. Instead of increasing temperature from rear to front, it may be reduced from rear to front.[3]

A higher mold temperature favors a uniform melt cooling rate, minimizing residual stresses, and improves the surface finish, mold release, and product quality. The mold cooling rate affects finished product quality. Polyether type TPU can set up better and release better.

High pressures and temperatures fill a high surface-to-volume ratio mold cavity more easily, but TPU melts can flash fairly easily at high pressures (Table 3.5). Pressure can be carefully controlled to achieve a quality product by using higher pressure during quick-fill, followed by lower pressure.[3] The initial higher pressure may reduce mold shrinkage by compressing the elastomeric TPU.[3]

The back pressure ranges from 0 to 100 lb/in^2 (0 to 0.69 MPa). TPU elastomers usually require very little or no back pressure.[3] When additives are introduced by the processor prior to molding, back pressure will enhance mixing, and when the plastication rate of the machine is insufficient for shot size or cycle time, a back pressure up to 200 lb/in^2 (1.4 MPa) can be used.[3]

Product quality is not as sensitive to screw speed as it is to process temperatures and pressures. The rotating speed of the screw, together with flight design, affects mixing (when additives have been introduced) and shear energy. Higher speeds generate more shear energy (heat). A speed above 90 r/min can generate excessive shear energy, creating voids and bubbles in the melt, which remain in the molded part.[3]

TABLE 3.5 Typical Temperature Pressure Settings* for Pellethane TPU

	Pressure, lb/in^2† (MPa)
Injection pressure	
First stage	8000–15,000 (55.0–103)
Second stage	5000–10,000 (34.5–69.60)
Back pressure	0–100 (0–0.69)
Cushion, in/mm	0.25 (6.4)
Screw speed, r/min	50–75
Cycle time, s (injection, relatively slow to avoid flash, etc.)	3–10

*Typical temperature and pressure settings are based on Ref. 3. Settings are based on studies using a reciprocating screw, general-purpose screw, clamp capacity of 175 tons, and rated shot capacity of 10 oz (280 g). Molded specimen thicknesses ranged from 0.065- to 0.125-in (1.7- to 3.2-mm) thickness.

†U.S. units refer to line pressures; metric units are based on the pressure on the (average) cross-sectional area of the screw.

Cycle times are related to TPU hardness, part design, temperatures, and wall thickness. Higher temperature melt and a hot mold require longer cycles, when the cooling gradient is not too steep. The cycle time for thin-wall parts, <0.125 in (<3.2 mm), is typically about 20 s.[3] The wall thickness for most parts is less than 0.125 in (3.2 mm), and a wall thickness as small as 0.062 in (1.6 mm) is not uncommon. When the wall thickness is 0.250 in (6.4 mm), the cycle time can increase to about 90 s.[3]

Mold shrinkage is related to TPU hardness and wall thickness, part and mold designs, and processing parameters (temperatures and pressures). For a wall thickness of 0.062 in (1.6 mm) for durometer hardness Shore A 70, the mold shrinkage is 0.35%. Using the same wall thickness for durometer hardness Shore A 90, the mold shrinkage is 0.83%.[3]

Purging when advisable is accomplished with conventional purging materials, polyethylene or polystyrene. Good machine maintenance includes removing and cleaning the screw and barrel mechanically with a salt bath or with a high-temperature fluidized-sand bath.[3]

Reciprocating screw injection machines are usually used to injection mold TPU, and these are the preferred machines, but ram types can be successfully used. Ram machines are slightly oversized in order to avoid: (1) incomplete melting and (2) steep temperature gradients during resin melting and freezing. Oversizing applies especially to TPU durometers harder than Shore D 55.[3]

Molded and extruded TPU have a wide range of applications, including:

Automotive: body panels (tractors) and RVs, doors, bumpers (heavy-duty trucks), fascia, and window encapsulations

Belting

Caster wheels

Covering for wire and cable

Film/sheet

Footwear and outer soles

Seals and gaskets

Tubing

3.2.4 Copolyesters

Thermpolastic copolyester elastomers are segmented block copolymers with a polyester hard crystalline segment and a flexible soft amorphous segment with a very low T_g.[35] Typically, the hard segments are composed of short-chain ester blocks such as tetramethylene terephthalate, and the soft segments are composed of aliphatic polyether or aliphatic polyester glycols, their derivatives, or polyetherester glycols. The copolymers are also called thermoplastic etheresterelastomers (TEEEs).[35] The terms COPE and TEEE are used interchangeably (see Fig. 3.4).

TEEEs are typically produced by condensation polymerization of an aromatic dicarboxylic acid or ester with a low MW aliphatic diol and a polyalkylene ether glycol.[35] Reaction of the first two components leads to the hard segment, and the soft segment is the product of the diacid or diester with a long-chain glycol.[35] This can be described as a melt transesterification of an aromatic dicarboxylic acid, or preferably its dimethyl ester, with a low MW poly(alklylene glycol ether) plus a short-chain diol.[35]

An example is melt phase polycondensation of a mixture of dimethyl terephthalate (DMT) + poly(tetramethylene oxide) glycol + an excess of tetramethylene glycol. A wide range of properties can be built into the TEEE by using different mixtures of isomeric phthalate esters, different polymeric glycols, and varying MW and MWD.[35] Antioxidants, such as

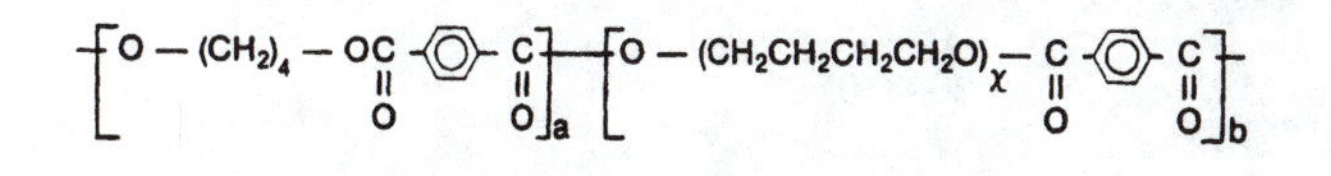

Hard Segment
Crystalline

Soft Segment
Amorphous

Figure 3.4 Structure of a commercial COPE TPE: a = 16 to 40, x = 10 to 50, and b = 16 to 40. (*Source: Ref. 10, p. 5.14.*)

hindered phenols or secondary aromatic amines, are added during polymerization, and the process is carried out under nitrogen because the polyethers are subject to oxidative and thermal degradation.[35]

Hytrel®* TEEElastomer block copolymers' property profile is given in Table 3.6.

Their mechanical properties are between rigid thermoplastics and thermosetting hard rubber.[35] Mechanical properties and processing parameters for Hytrel, and for a number of other materials in this chapter, can be found on the producers' Internet home page.

Copolymer properties are largely determined by the soft/hard segment ratio; and as with any commercial resin, properties are determined with compound formulations.

TEEEs combine flexural fatigue strength, low-temperature flexibility, good apparent modulus (creep resistance), DTUL and heat resistance,

*Hytrel is a registered trademark of DuPont for its brand of thermoplastic polyester elastomer.

TABLE 3.6 Typical Hytrel Property Profile[1]—ASTM Test Methods

Property	Value
Specific gravity, g/cm^3	1.01–1.43
Tensile strength @ break, lb/in^2 (MPa)	1,400–7,000 (10–48)
Tensile elongation @ break, %	200–700
Hardness Shore D	30–82
Flexural modulus, lb/in^2 (MPa)	
−40°F (−40°C)	9,000–440,000 (62–3,030)
73°F (22.8°C)	4,700–175,000 (32–1,203)
212°F (100°C)	1,010–37,000 (7.0–255)
Izod impact strength, ft-lb/in (J/m), notched	
−40°F (−40°C)	No break–0.4 (No break–20)
73°F (22.8°C)	No break–0.8 (No break–40)
Tabor abrasion, mg/1000 rev	
CS-17 wheel	0–85
H-18 wheel	20–310
Tear resistance, lb/in, initial Die C	210–1,440
Vicat softening temperature, °F (°C)	169–414 (76.1–212)
Melt point, °F (°C)	302–433 (150–223)

resistance to hydrolysis, and good chemical resistance to nonpolar solvents at elevated temperatures. A tensile stress/percent elongation curve reveals an initial narrow linear region.[19] COPEs are attacked by polar solvents at elevated temperatures. The copolymers can be completely soluble in meta-cresol which can be used for dilute solution polymer analysis.[19]

TEEEs are processed by conventional thermoplastic melt-processing methods, injection molding and extrusion, requiring no vulcanization.[35] They have sharp melting transitions, rapid crystallization (except for softer grades with higher amount of amorphous segment), and apparently melt viscosity decreases slightly with shear rate (at low shear rates).[35] The melt behaves like a Newtonian fluid.[35] In a true Newtonian fluid, the coefficient of viscosity is independent of the rate of deformation. In a non-Newtonian fluid, the apparent viscosity is dependent on shear rate and temperature.

TPE melts are typically highly non-Newtonian fluids, and their apparent viscosity is a function of shear rate.[10] TPE's apparent viscosity is much less sensitive to temperature than it is to shear rate.[10] The apparent viscosity of TPEs as a function of apparent shear rate and as a function of temperature are shown in Figs. 3.5 and 3.6.

TEEEs can be processed successfully by low-shear methods such as laminating, rotational molding, and casting.[35] Standard TEEElastomers are usually modified with viscosity enhancers for improved melt viscosity for blow molding.[35]

Riteflex®* copolyester elastomers have high fatigue resistance, chemical resistance, good low-temperature [−40°F (−40°C)] impact strength, and service temperatures up to 250°F (121°C).[1] Riteflex grades are classified according to hardness and thermal stability. The

*Riteflex is a registered trademark of Ticona.

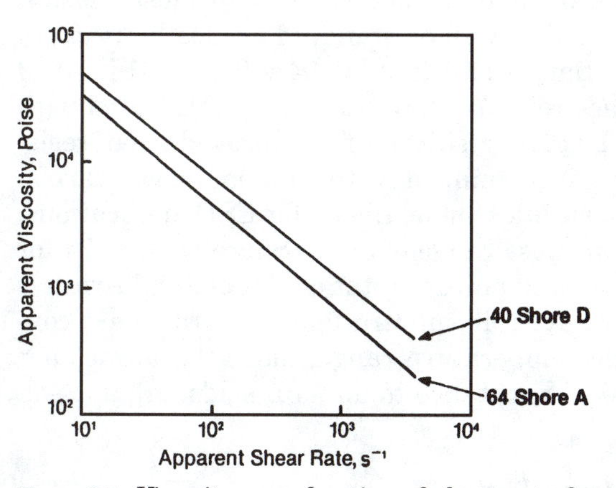

Figure 3.5 Viscosity as a function of shear rate for hard and soft TPEs. (*Source: Ref. 10, p. 5.31.*)

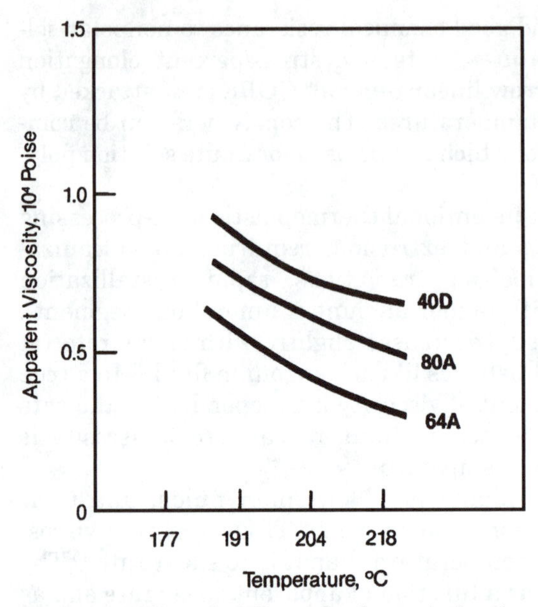

Figure 3.6 TPE (with different hardnesses) viscosity as a function of temperature. (*Source: Ref. 10, p. 5.31.*)

typical hardness range is Shore D 35 to 77. They are injection molded, extruded and blow molded. The copolyester can be used as a modifier in other polymer formulations. Applications for Riteflex copolyester and other compounds that use it as a modifier include bellows, hydraulic tubing, seals, wire coating, and jacketing; molded air dams, automotive exterior panel components (fender extensions, spoilers), fascia and fascia coverings, radiator panels; extruded hose, belting, and cable covering; and spark plug and ignition boots.

Arnitel®* TPEs are based on polyether ester or polyester ester, including specialty compounds as well as standard grades.[34] Specialty grades are classified as: (1) flame-retardant UL 94 V/0 @ 0.031 in (0.79 mm), (2) high modulus glass-reinforced, (3) internally lubricated with polytetrafluoro ethylene (PTFE) or silicone for improved wear resistance, and (4) conductive, compounded with carbon black, carbon fibers, nickel-coated fibers, stainless-steel fibers, for ESD applications. Standard grades have a hardness range of about Shore D 38 to 74 for injection molding, extrusion, and powder rotational molding.[1] Arnitels have high impact strength, even at subzero temperatures, near-constant stiffness over a wide temperature range, and good abrasion.[34] They have excellent chemical resistance to mineral acids, organic sol-

*Arnitel is a registered trademark of DSM.

vents, oils, and hydraulic fluids.[34] They can be compounded with property enhancers (additives) for resistance to oxygen, light, and hydrolysis.[34] Glass fiber–reinforced grades, like other thermoplastic composites, have improved DTUL, modulus, and coefficient of linear thermal expansion (CLTE).[34]

Typical products are: automotive exterior trim, fascia components, spoilers, window track tapes, boots, bellows, underhood wire covering, connectors, hose, and belts; appliance seals, power tool components, ski boots, and camping equipment.[1]

Like other thermoplastics, processing temperatures and pressures and machinery/tool designs are adjusted to the compound and application.

The following conditions apply to Arnitel COPE compounds, for optimum product quality: melt temperature range, 428 to 500°F (220 to 260°C); cylinder (barrel) temperature setting range, 392 to 482°F (200 to 250°C); mold temperature range for thin-wall products, 122°F (50°C) and for thick-wall products, 68°F (20°C).

Injection pressure is a function of flow length, wall thickness, and melt rheology, and it is calculated to achieve uniform mold filling. The Arnitel injection pressure range is <5000 to >20,000 lb/in^2 (<34 to >137 MPa). Thermoplastic elastomers may not require back pressure, and when back pressure is applied, it is much lower than for thermoplastics that are not elastomeric. Back pressure for Arnitel is about 44 to 87 lb/in^2 (0.3 to 0.6 MPa). Back pressure is used to ensure a homogeneous melt with no bubbles.

The screw configuration is as follows: thread depth ratio, approximately 1:2, and *L/D* ratio, 17/1 to 23/1 (standard three-zone screws: feed, transition or middle, and metering or feed zones).[34] Screws are equipped with a nonreturn valve to prevent backflow.[34] Decompression-controlled injection-molding machines have an open nozzle.[34] A short nozzle with a wide bore (3-mm minimum) is recommended in order to minimize pressure loss and heat due to friction.[34] Residence time should be as short as possible, and this is accomplished with barrel temperatures at the lower limits of recommended settings.[34]

Tool design generally follows conventional requirements for gates and runners. DSM recommends trapezoidal gates or, for wall thickness more than 3 to 5 mm, full sprue gates.[34] Vents approximately 1.5 × 0.02 mm are located in the mold at the end of the flow patterns, either in the mold faces or through existing channels around the ejector pins and cores.[34] Ejector pins and plates for thermoplastic elastomers must take into account the molded product's flexibility. Knock-out pins/plates for flexible products should have a large enough face to distribute evenly the minimum possible load. Prior to ejection, the part is cooled, carefully following the resin supplier's recommendation. The cooling system configuration in the mold base, and the cooling rate, are critical to optimum cycle time and product quality. The product is cooled as fast as possible without causing warpage. Cycle times vary from about 6 s for a wall thickness of 0.8 to 1.5 mm to 40 s for a wall thickness of about 5 to 6 mm. Drying temperatures and times range from 3 to 10 h at 194 to 248°F (90 to 120°C).

In general, COPEs can require drying for 4 h @ 225°F (107°C) in a dehumidifying oven in order to bring the pellet moisture content to 0.02% max.[1] The melt processing range is typically about 428 to 448°F (220 to 231°C); however, melt processing temperatures can be as high as 450 to 500°F (232 to 260°C). A typical injection-molding grade has a T_m of 385°F (196°C).[1] The mold temperature is usually between 75 and 125°F (24 and 52°C).

Injection-molding screws have a gradual transition (center) zone to avoid excess shearing of the melt and high metering (front) zone flight depths [0.10 to 0.12 in (2.5 to 3.0 mm)], a compression ratio of 3.0:1 to 3.5:1, and an *L/D* of 18/1 min (24/1 for extrusion).[18a] Barrier screws can provide more efficient melting and uniform melt temperatures for molding very large parts and for high-speed extrusions.[18a] When Hytrel is injection molded, molding pressures range from 6000 to 14,000 lb/in^2 (41.2 to 96.2 MPa). When pressures are too high, overpacking and sticking to the mold cavity wall can occur.[18a] Certain mold designs are recommended: large knock-out pins and stripper plates, and generous draft angles for parts with cores.[18a]

3.2.5 Polyamides

Polyamide TPEs are usually either polyester-amides, polyetherester-amide block copolymers, or polyether block amides (PEBA) (see Fig. 3.7). PEBA block copolymer molecular architecture is similar to typical block copolymers.[10] The polyamide is the hard (thermoplastic) segment, while the polyester, polyetherester, and polyether segments are the soft (elastomeric) segment.[10]

Polyamide TPEs can be produced by reacting a polyamide with a polyol such as polyoxyethylene glycol or polyoxypropylene glycol, a polyesterification reaction.[1] Relatively high aromaticity is achieved by esterification of a glycol to form an acid-terminated soft segment, which is reacted with a diisocyanate to produce a polyesteramide. The polyamide segment is formed by adding diacid and diisocyanate.[1] The chain extender can be a dicarboxylic acid.[1] Polyamide TPEs can be composed of lauryl lactam and ethylene-propylene rubber (EPR).

Polyamide thermoplastic elastomers are characterized by their high service temperature under load, good heat aging, and solvent resistance.[1] They retain serviceable properties >120 h @ 302°F (150°C) without adding heat stabilizers.[1] Addition of a heat stabilizer increases

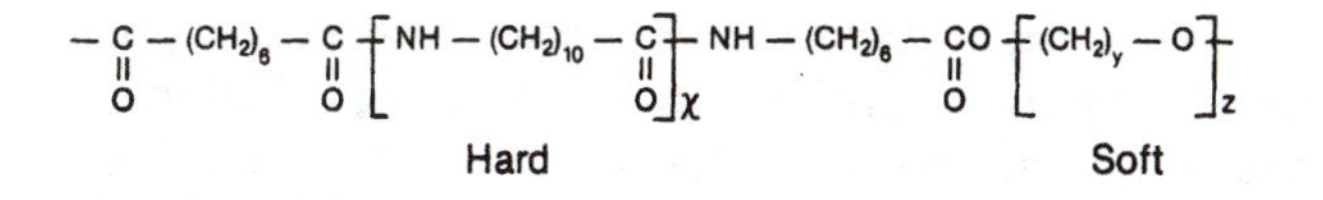

$$\left[(CH_2)_5 - \underset{\|}{C}(O) \right]_X - NH - B - NH\overset{}{C}(=O) - A - C(=O) - NH - B - NH -$$

$$\left[(CH_2)\right]_X O - C(=O) - (CH_2)_y - C(=O) - \left[NH - C_6H_4 - CH_2 - C_6H_4 - NH - A - C(=O) - CH_2 - C(=O) \right]_n - O -$$

Soft Hard

Where A = C_{19} to C_{21} dicarboxylic acid
B = $-(CH_2)_3 - O - \left[(CH_2)_4 - O\right]_b (CH_2)_3 -$

Figure 3.7 Structure of three PEBA TPEs. (*Source: Ref. 10, p. 5.17.*)

service temperature. Polyesteramides retain tensile strength, elongation, and modulus to 347°F (175°C).[1] Oxidative instability of the ether linkage develops at 347°F (175°C). The advantages of polyether block amide copolymers are their elastic memory which allows repeated strain (deformation) without significant loss of properties, lower hysteresis, good cold-weather properties, hydrocarbon solvent resistance, UV stabilization without discoloration, and lot-to-lot consistency.[1]

The copolymers are used for waterproof/breathable outerwear; air-conditioning hose; underhood wire covering; automotive bellows; flexible keypads; decorative watch faces; rotationally molded basket-, soccer-, and volley balls; and athletic footwear soles.[1] They are insert-molded over metal cores for nonslip handle covers (for video cameras) and coinjected with polycarbonate core for radio/TV control knobs.[1]

Pebax®* polyether block amide copolymers consist of regular linear chains of rigid polyamide blocks and flexible polyether blocks. They are injection molded, extruded, blow molded, thermoformed, and rotational molded.

The property profile is as follows: specific gravity about 1.0; Shore hardness range about 73 A to 72 D; water absorption, 1.2%; flexural modulus range, 2600 to 69,000 lb/in² (18.0 to 474 MPa); high torsional modulus from −40° to 0°C; Izod impact strength (notched), no break from −40 to 68°F (−40 to 20°C); abrasion resistance; long wear life; elastic memory, allowing repeated strain under severe conditions without permanent deformation; lower hysteresis values than many thermoplastics and thermosets with equivalent hardness; flexibility temperature range, −40 to 178°F (−40 to 81°C), and flexibility temperature range is achieved without plasticizer (it is accomplished by engineering the poly-

*Pebax is a registered trademark of Elf Atochem.

mer configuration); lower temperature increase with dynamic applications; chemical resistance similar to polyurethane (PUR); good adhesion to metals; small variation in electrical properties over service temperature range and frequency (Hz) range; printability and colorability; tactile properties, such as good "hand," feel; and nonallergenic.[1]

The T_m for polyetheresteramides is about 248 to 401°F (120 to 205°C) and about 464°F (240°C) for aromatic polyesteramides.[18b]

Typical Pebax applications are: one-piece, thin-wall soft keyboard pads; rotationally molded, high resiliency, elastic memory soccer-, basket-, and volley balls; flexible, tough mouthpieces for respiratory devices, scuba equipment, frames for goggles, and ski and swimming breakers; decorative watch faces; good adhesion to metal, nonslip for coverings over metal housings for hand-held devices such as remote controls, electric shavers, camera handle covers; coinjected over polycarbonate for control knobs; and films for waterproof, breathable outerwear.[1]

Polyamide/ethylene-propylene, with higher crystallinity than other elastomeric polyamides, has improved fatigue resistance and improved oil and weather resistance.[1] T_m and service temperature usually increase with higher polyamide crystallinity.[1]

Polyamide/acrylate graft copolymers have a Shore D hardness range from 50 to 65, and continuous service temperature range from −40 to 329°F (−40 to 165°C). The markets are: underhood hose and tubing, seals and gaskets, and connectors and optic fiber sheathing, snap-fit fasteners.[1] Nylon 12/nitrile rubber blends were commercialized by Denki Kagaku Kogyo, as part of the company's overall nitrile blend development.[1]

3.3 Melt Processable Rubber (MPR)

MPRs are amorphous polymers, with no sharp melt point,[1] which can be processed in both resin melt and rubber processing machines, injection molded, extruded, blow molded, calendered, and compression molded.[1] Flow properties are more similar to rubber than to thermoplastics.[1] The polymer does not melt by externally applied heat alone, but becomes a high-viscosity, intractable semifluid. It must be subjected to shear in order to achieve flowable melt viscosities, and shear force applied by the plasticating screw is necessary. Without applied shear, melt viscosity and melt strength increase too rapidly in the mold. Even with shear and a hot mold, as soon as the mold is filled and the plasticating screw stops or retracts, melt viscosity and melt strength increase rapidly.

Melt rheology is illustrated with Aclryn®.* The combination of applied heat and shear-generated heat brings the melt to 320 to

*Alcryn is a registered trademark of Advanced Polymer Alloys Division of Ferro Corporation.

330°F (160 to 166°C). The melt temperature should not be higher than 360°F (182°C). New grades have been introduced with improved melt processing.

Proponents of MPR view its rheology as a processing cost benefit by allowing faster demolding and lower processing temperature settings, significantly reducing cycle time.[1] High melt strength can minimize or virtually eliminate distortion and sticking and cleanup is easier.[1] MPR is usually composed of halogenated (chlorinated) polyolefins, with reactive intermediate-stage ethylene interpolymers that promote H^+ bonding.

Alcryn is an example of single-phase MPR with overall midrange performance properties, supplementing the higher-price COPE thermoplastic elastomers. Polymers in single-phase blends are miscible but polymers in multiple-phase blends are immiscible, requiring a compatibilizer for blending. Alcryns are partially cross-linked halogenated polyolefin MPR blends.[1] The specific gravity ranges from 1.08 to 1.35.[1] MPRs are compounded with various property enhancers (additives), especially stabilizers, plasticizers, and flame retardants.[1]

The applications are: automotive window seals and fuel filler gaskets, industrial door and window seals and weatherstripping, wire/cable covering, and hand-held power tool housing/handles. Nonslip soft-touch hand-held tool handles provide weather and chemical resistance and vibration absorption.[16] Translucent grade is extruded into films for face masks and tube/hosing and injection-molded into flexible keypads for computers and telephones.[1] Certain grades are paintable without a primer. Typical durometer hardnesses are Shore A 60, 76, and 80.

The halogen content of MPRs requires corrosion-resistant equipment and tool cavity steels along with adequate venting. Viscosity and melt strength buildup are taken into account with product design, equipment, and tooling design: wall thickness gradients and radii, screw configuration (flights, L/D, length), gate type and size, and runner dimensions.[1] The processing temperature and pressure setting are calculated according to rheology.[1]

In order to convert solid pellet feed into uniform melt, moderate screws with some shallow flights are recommended. Melt flow is kept uniform in the mold with small gates which maximize shear, large vents, and large sprues for smooth mold-filling.[1] Runners should be balanced and radiused for smooth, uniform melt flow.[1] Recommendations, such as balanced, radiused runners, are conventional practice for any mold design, but they are more critical for certain melts such as MPRs. Molds have large knockout pins or plates to facilitate stripping the rubbery parts during demolding. Molds may be chilled to 75°F (24°C). Mold temperatures depend on grades and applications; Hot molds are used for smooth surfaces and to minimize orientation.[1]

Similar objectives of the injection-molding process apply to extrusion and blow molding, namely, creating and maintaining uniform, homogeneous, and properly fluxed melt. Shallow-screw flights increase shear and mixing. Screws that are 4.5 in (11.4 cm) in diameter with *L/D* 20/1 to 30/1 are recommended for extrusion. Longer barrels and screws produce more uniform melt flux, but *L/D* ratios can be as low as 15/1. The temperature gradient is reversed. Instead of the temperature setting being increased from the rear (feed) zone to the front (metering) zone, a higher temperature is set in the rear zone and a lower temperature is set at the front zone and at the adapter (head).[1] Extruder dies are tapered, with short land lengths, and die dimensions are close to the finished part dimension.[1] Alcryns have low-to-minimum die swell.

The polymer's melt rheology is an advantage in blow molding during parison formation because the parison is not under shear, and it begins to solidify at about 330°F (166°C). High melt viscosity allows blow ratios up to 3:1 and significantly reduces demolding time.

MPRs are thermoformed and calendered with similar considerations described for molding and extrusion. Film and sheet can be calendered with thicknesses from 0.005 to 0.035 in (0.13 to 0.89 mm).

3.4 Thermoplastic Vulcanizate (TPV)

TPVs are composed of a vulcanized rubber component, such as EPDM, nitrile rubber, and butyl rubber, in a thermoplastic olefinic matrix. TPVs have a continuous thermoplastic phase and a discontinuous vulcanized rubber phase. TPVs are dynamically vulcanized during a melt-mixing process in which vulcanization of the rubber polymer takes place under conditions of high temperatures and high shear. "Static" vulcanization of thermoset rubber involves heating a compounded rubber stock under zero shear (no mixing), with subsequent cross-linking of the polymer chains.

Advanced Elastomer Systems' Santoprene®* thermoplastic vulcanizate is composed of polypropylene and finely dispersed, highly vulcanized EPDM rubber. Geolast®† TPV is composed of polypropylene and nitrile rubber, and the company's Trefsin®‡ is a dynamically vulcanized composition of polypropylene plus butyl rubber.

---

*Santoprene is a registered trademark of Advanced Elastomer Systems LP.

†Geolast is a registered trademark of Advanced Elastomer Inc. Systems LP.

‡Trefsin is a registered trademark of Advanced Elastomer Systems LP.

EPDM particle size is a significant parameter for Santoprene's mechanical properties, with smaller particles providing higher strength and elongation.[1] Higher cross-link density increases tensile strength and reduces tension set (plastic deformation under tension).[1] Santoprene grades can be characterized by EPDM particle size and cross-link density.[1]

These copolymers are rated as midrange with overall performance generally between the lower cost styrenics and the higher cost TPUs and copolyesters.[1] The properties of Santoprene, according to its developer (Monsanto), are generally equivalent to the properties of general purpose EPDM, and oil resistance is comparable to that of neoprene.[1] Geolast has higher fuel/oil resistance and better hot oil aging than Santoprene (see Tables 3.7, 3.8, and 3.9).

Tensile stress-strain curves for Santoprene at several temperatures for Shore 55 A and 50 D hardnesses are shown in Fig. 3.8.[8]

Generally, tensile stress decreases with temperature increase, while elongation at break increases with temperature. Tensile stress at a given strain increases with hardness from the softer Shore A grades to the harder Shore D grades. For a given hardness, the tensile stress-strain curve becomes progressively more rubberlike with increasing temperature. For a given temperature, the curve is progressively more rubberlike with decreasing hardness. Figure 3.9 shows dynamic mechanical properties for Shore 55 A and 50 D hardness grades over a wide range of temperatures.[8]

TPVs composed of polypropylene and EPDM have a service temperature range from −75 to 275°F (−60 to 135°C) for more than 30 days

TABLE 3.7 Santoprene Mechanical Property Profile—ASTM Test Methods—Durometer Hardness Range, Shore 55 A to 50 D

Property	Shore hardness		
	55 A	80 A	50 D
Specific gravity, g/cm^3	0.97	0.97	0.94
Tensile strength, lb/in^2 (MPa)	640 (4.4)	600 (11)	4000 (27.5)
Ultimate elongation, %	330	450	600
Compression set, %, 168 h	23	29	41
Tension set, %	6	20	61
Tear strength, pli			
77°F (25°C)	108 (42)	194 (90)	594 (312)
212°F (100°C)	42 (5.6)	75 (24)	364 (184)
Flex fatigue megacycles to failure	>3.4	—	—
Brittle point, °F (°C)	<−76 (<−60)	−81 (−63)	−29 (−34)

TABLE 3.8 Santoprene Mechanical Property Profile—Hot Oil Aging/Hot Air Aging*—Durometer Hardness Range Shore 55 A to 55 D[1]

Property	Shore hardness 55 A	80 A	50 D
Tensile strength, ultimate			
lb/in² (MPa)	470 (3.2)	980 (6.8)	2620 (18.10)
Percent retention	77	73	70
Ultimate elongation, %	320	270	450
Percent retention	101	54	69
100% modulus, lb/in² (MPa)	250 (1.7)	610 (4.2)	1500 (10.3)
Percent retention	87	84	91

*Hot oil aging (IRM 903), 70 h @ 257°F (125°C).

TABLE 3.9 Santoprene Mechanical Property Profile—Hot Oil Aging/Hot Air Aging*—Durometer Hardness Range Shore 55 A to 55 D[1]

Property	Shore hardness 55 A	80 A	50 D
Tensile strength, ultimate			
lb/in² (MPa)	680 (4.7)	1530 (10.6)	3800 (26.2)
Percent retention	104	109	97
Ultimate elongation, %	370	400	560
Percent retention	101	93	90
100% modulus, lb/in² (MPa)	277 (1.9)	710 (4.9)	1830 (12.6)
Percent retention	105	111	117

*Hot air aging, 168 h @ 257°F (125°C).

and 302°F (150°C) for short times (up to 1 week). Reference 8 reports further properties, including tensile and compression set, fatigue resistance, and resilience and tear strength. Polypropylene/nitrile rubber high/low service temperature limits are 257°F (125°C)/−40°F (−40°C).

Santoprene automotive applications include: air ducts, body seals, boots (covers), bumper components, cable/wire covering, weatherstripping, underhood and other automotive hose/tubing, and gaskets. Appliance uses include diaphragms, handles, motor mounts, vibration dampers, seals, gaskets, wheels, and rollers. Santoprene rubber is used in building/construction for expansion joints, sewer pipe seals, valves for irrigation, weatherstripping, and welding line connectors. Prominent electrical uses are in cable jackets, motor shaft mounts, switch boots, and terminal plugs. Business machines, power tools, and plumbing/hardware provide TPVs with numerous applications. In healthcare applications, it is used in disposable bed covers, drainage

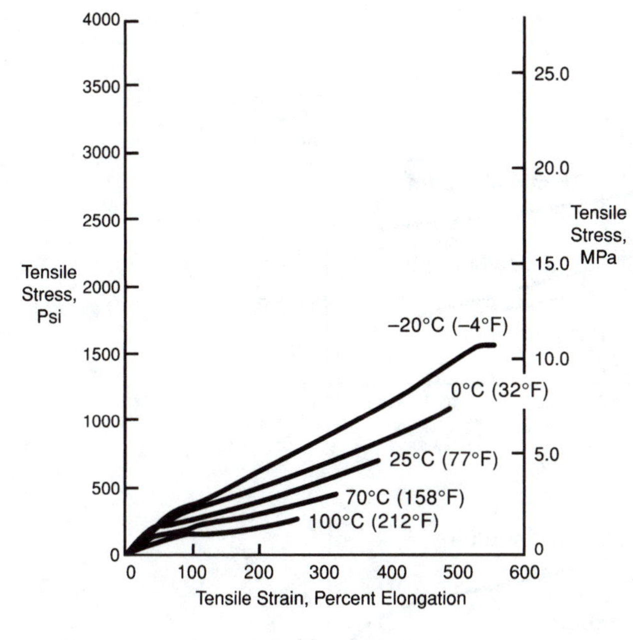

(a)

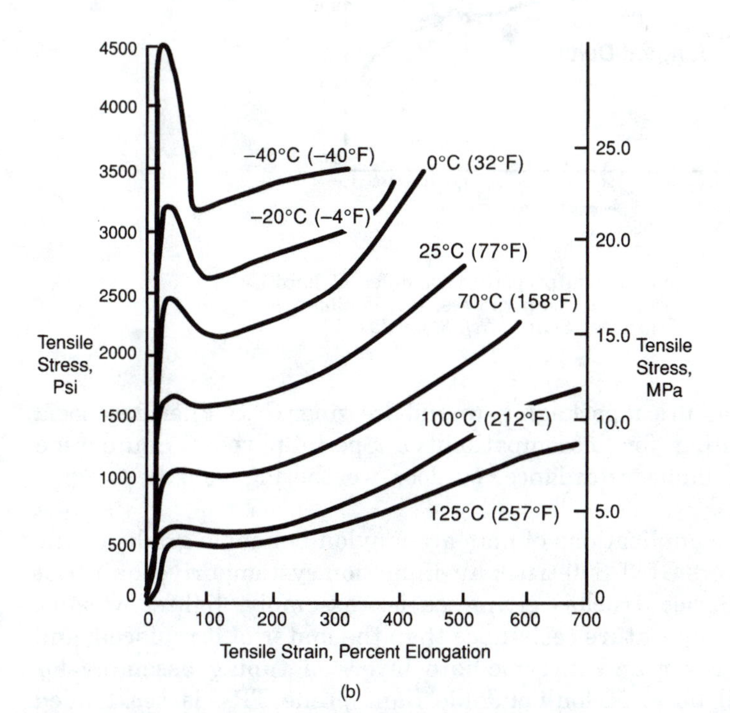

(b)

Figure 3.8 Tensile stress-strain curves for Santoprene at several temperatures for different hardness grades. (*a*) 55 Shore A grades (ASTM D 412); (*b*) 50 Shore D grades (ASTM D 412). (*Source: Ref. 8, pp. 3–4.)*

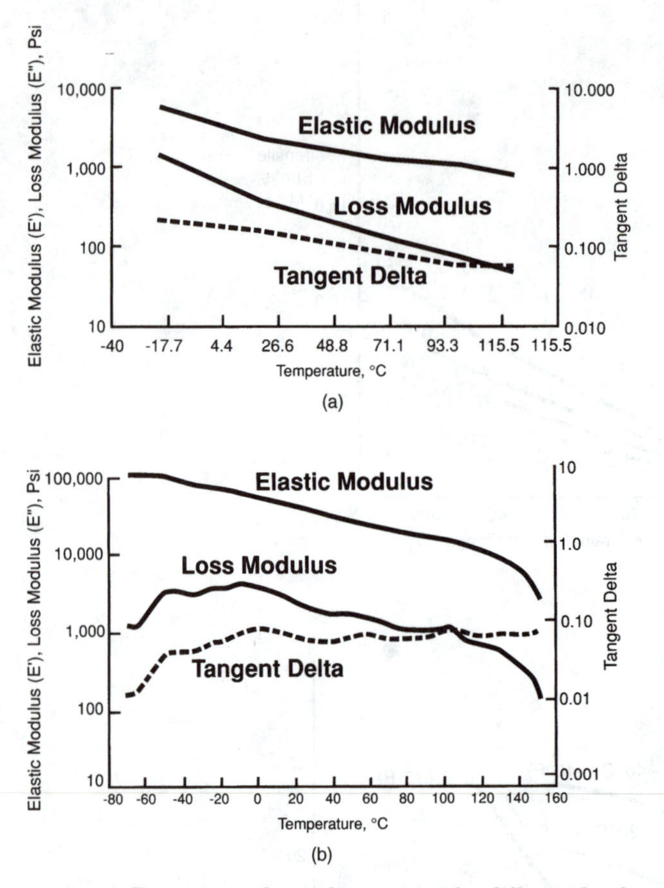

Figure 3.9 Dynamic mechanical properties for different hardness grades over a range of temperatures. (*a*) 55 Shore A grades; (*b*) 50 Shore D grades. (*Source: Ref. 8, pp. 12–13.*)

bags, pharmaceutical packaging, wound dressings (U.S. Pharmacopoeia Class VI rating for biocompatibility). Special-purpose Santoprene grades meet flame retardance, outdoor weathering, and heat aging requirements.

Santoprene applications of note are a nylon-bondable grade for the General Motors GMT 800 truck air-induction system; driveshaft boot in Ford-F Series trucks, giving easier assembly, lighter weight, and higher temperature resistance than the material it replaced; and Santoprene cover and intermediate layers of tubing assembly for hydraulic oil hose. Nylon-bondable Santoprene TPV is coextruded with an impact modified (or pure) nylon 6 inner layer.

Polypropylene/EPDM TPVs are hygroscopic, requiring drying at least 3 h at 160°F (71°C) and avoiding exposure to humidity.[1] They are not susceptible to hydrolysis.[1] Moisture in the resin can create voids, disturbing processing and finished product performance properties. Moisture precautions are similar to those for polyethylene or polypropylene.[1]

Typical of melts with a relatively low melt flow index (0.5 to 30 g/10 min for Santoprene), gates should be small and runners and sprues should be short; long plasticating screws are used with an *L/D* ratio typically 24/1 or higher.[1] The high viscosity at low shear rates (see Fig. 3.10) provides good melt integrity and retention of design dimensions during cooling.[1]

Similar injection-molding equipment design considerations apply to extrusion equipment such as long plasticating screws with 24/1 or higher *L/D* ratios and approximately 3:1 compression ratios.[1]

Equipment/tool design, construction, and processing of TPVs differ from that of other thermoplastics. EPDM/polypropylene is thermally stable up to 500°F (260°C) and it should not be processed above this temperature.[1] It has a flash ignition temperature above 650°F (343°C).

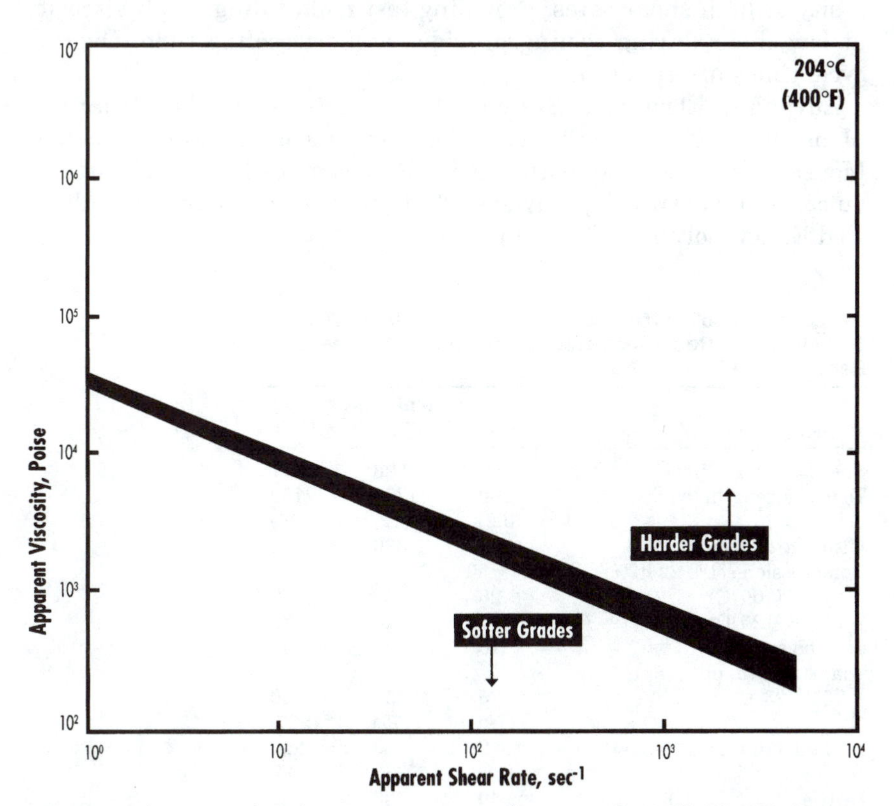

Figure 3.10 Apparent viscosity versus apparent shear rate @ 400°F (204°C). (*Source: Ref. 7, p. 36.*)

TPV's high shear sensitivity allows easy mold removal, thus sprays and dry powder mold release agents are not recommended.

Geolast TPVs are composed of polypropylene and nitrile rubber. Table 3.10 profiles the mechanical properties for these TPVs with Shore hardness range of 70 A to 45 D.

Geolast (polypropylene plus nitrile rubber) has a higher resistance than Santoprene (polypropylene plus EPDM) to oils (such as IRM 903) and fuels, plus good hot-oil/hot-air aging.[1] Geolast applications include: molded fuel filler gasket (Cadillac Seville), carburetor components, hydraulic lines, and engine parts such as mounts and tank liners.

Three property distinctions among Trefsin grades are: (1) heat aging; (2) high energy attenuation for vibration damping applications such as automotive mounts, energy absorbing fascia and bumper parts, and sound deadening; and (3) moisture and O_2 barrier. Other applications are: soft bellows; basket-, soccer-, footballs; calendered textile coatings; and packaging seals. Since Trefsin is hygroscopic, it requires drying before processing. Melt has low viscosity at high shear rates, providing fast mold filling. High viscosity at low shear during cooling provides a short cooling time. Overall, cycle times are reduced.

Advanced Elastomer Systems L.P. (AES) is the beneficiary of Monsanto Polymers' TPE technology and business, which included Monsanto's earlier acquisition of BP Performance Polymers' partially vulcanized EPDM/polypropylene (TPR), and Bayer's partially vulcanized EPDM/polyolefin TPEs in Europe.

TABLE 3.10 Geolast Mechanical Property Profile—ASTM Test Methods—Room Temperature—Durometer Hardness Range, Shore 70 A to 45 D

Property	Shore hardness		
	70 A	87 A	45 D
Specific gravity, g/cm^3	1.00	0.98	0.97
Tensile strength, lb/in^2 (MPa)	900 (6.2)	1750 (12)	2150 (15)
Ultimate elongation, %	265	380	350
Compression set @ 22 h, %			
212°F (100°C)	28	39	52
257°F (125°C)	37	48	78
Tension set (%)	10	24	40
Tear strength, pli			
73°F (23°C)	175 (79)	350 (177)	440 (227)
212°F (100°C)	52 (11)	150 (66)	220 (104)
Brittle point, °F (°C)	−40 (−40)	−33 (−28)	−31 (−36)

3.5 Synthetic Rubbers

A second major group of elastomers is that group known as synthetic rubbers. Elastomers in this group, discussed in detail in this section, are

Acrylonitrile butadiene copolymers (NBR)
Butadiene rubber (BR)
Butyl rubber (IIR)
Chlorosulfonated polyethylene (CSM)
Epichlorohydrin (ECH, ECO)
Ethylene propylene diene monomer (EPDM)
Ethylene propylene monomer (EPM)
Fluoroelastomers (FKM)
Polyacrylate (ACM)
Polybutadiene (PB)
Polychloroprene (CR)
Polyisoprene (IR)
Polysulfide rubber (PSR)
Silicone rubber (SiR)
Styrene butadiene rubber (SBR)

Worldwide consumption of synthetic rubber can be expected to be about 11 million metric tons in 2000 and about 12 million metric tons in 2003, based on earlier reporting (1999) by the International Institute of Synthetic Rubber Producers.[26] About 24% is consumed in North America.[1] Estimates depend on which synthetic rubbers are included and reporting sources from world regions.

New synthetic rubber polymerization technologies replacing older plants and increasing world consumption are two reasons new production facilities are being built around the world. Goodyear Tire & Rubber's 110,000-metric tons/y butadiene-based solution polymers goes onstream in 2000 in Beaumont Texas.[25] Goodyear's 18,200-metric tons/y polyisoprene unit went onstream in 1999 in Beaumont.[25] Sumitomo Sumika AL built a 15,000-metric tons/y SBR plant in Chiba, Japan, adding to the company's 40,000-metric tons/y SBR capacity at Ehime.[25] Haldia Petrochemical Ltd. of India is constructing a 50,000-metric tons/y SBR unit and a 50,000-metric tons/y PB unit using BASF technology.[25]

Bayer Corporation added a 75,000-metric tons/y SBR and PB capacity at Orange, Texas, in 1999, converting a lithium PB unit to produce solution SBR and neodymium PB.[25] Bayer AG increased SBR and PB capacity from 85,000 to 120,000 metric tons/y at Port Jerome, France, in 1999.[25] Bayer AG will complete a worldwide butadiene rubber capacity increase from 345,000 metric tons/y in 1998 to more than 600,000 metric tons/y by 2001.[25] Bayer AG increased EPDM capacity at Orange, Texas, using slurry polymerization and at Marl, Germany, using solvent polymerization in 1999.[25] Bayer Inc. added 20,000-metric tons/y butyl rubber capacity at Sarnia, Ontario, to the company's 70,000-metric tons/y butyl rubber capacity and 50,000-metric tons/y halo-butyl capacity at Sarnia. Bayer's 90,000-metric tons/y halo- or regular butyl capacity will be restarted in 2000.[25]

Mitsui Chemicals goes onstream with a 40,000-metric tons/y metallocene EPDM in Singapore in 2001.[25] The joint venture Nitrilo SA between Uniroyal Chemical subsidiary of Crompton & Knowles and Girsa subsidiary of Desc SA (Mexico) went onstream with a 28,000-metric tons/y NBR at Altamira, Mexico, in 1999.[25] Uniroyal NBR technology and Girsa process technology were joined.[25] Chevron Chemical went onstream in 1999 with a 60,000-metric tons/y capacity polyisobutylene (PIB) at Belle Chase, Louisiana, licensing technology from BASF.[25] BASF is adding a 20,000-metric tons/y medium MW PIB at its Lufwigshafen complex, which will double the unit's capacity to 40,000 metric tons/y. This addition will be completed in 2001. BASF has 70,000-metric tons/y low MW PIB capacity. The company is using its own selective polymerization technology which allows MW to be controlled.[25]

BST Elastomers, a joint venture of Bangkok Synthetics, Japan Synthetic Rubber (JSR), Nippon Zeon, Mitsui, and Itochu, went onstream in 1998–1999 with 40,000-metric tons/y PB capacity and 60,000-metric tons/y SBR at Map Ta Phut, Rayong, Thailand.[25] Nippon Zeon completed adding 25,000-metric tons/y SBR capacity to its existing 30,000-metric tons/y at Yamaguchi, Japan, in 1999, and the company licensed its solution polymerization technology to Buna Sow Leuna Olefinverbund (BSLO).[25] BSLO will start 60,000-metric tons/y SBR capacity at Schkopau, Germany, in 2000.[25]

Sinopec, China's state-owned petrochemical and polymer company, is increasing synthetic rubber capacity across the board, including butyls, SBRs, nitrile, and chloroprene. Sinopec is starting polyisoprene and EPR production, although the company did not produce polyisoprene or EPR prior to 1999.[25] Total synthetic rubber capacity will be 1.15 million metric tons/y by 2000. China's synthetic rubber consumption is forecast by the company to be almost 7 million metric tons/y in 2000.[25]

Synthetic rubber is milled and cured prior to processing such as injection molding. Processing machinery is designed specifically for synthetic rubber.

Engel (Guelph, Ontario) ELAST®* technology includes injection-molding machines designed specifically for molding cross-linked rubbers.[31] Typical process temperature settings, depending on the polymer and finished product, are 380 to 425°F (193 to 218°C). Pressures, which also depend on polymer and product, are typically 20,000 to 30,000 lb/in^2 (137 to 206 MPa). A vertical machine's typical clamping force is 100 to 600 U.S. tons, while a horizontal machine's typical clamping force is 60 to 400 U.S. tons. They have short flow paths, allowing injection of rubber very close to the cross-linking temperature.[31] The screw *L/D* can be as small as 10/1.[31] ELAST technology includes tiebarless machines for small and medium capacities and proprietary state-of-the-art computer controls.[31]

3.5.1 Acrylonitrile butadiene copolymers

Nitrile butadiene rubbers (NBRs) are poly(acrylonitrile-co-1,3-butadiene) copolymers of butadiene and acrylonitrile.[23] Resistance to swelling caused by oils, greases, solvents, and fuels is related to percent bound acrylonitrile (ACN) content, which usually ranges from 20 to 46%.[6] Higher ACN provides higher resistance to swelling but diminishes compression set and low temperature flexibility.[6] ACN properties are related to percent acrylo and percent nitrile content. Nitrile increases compression set, flex properties, and processing properties.[1] The rubber has good barrier properties due to the polar nitrile groups.[23] Continuous-use temperature for vulcanized NBR is up to 248°F (120°C) in air and up to 302°F (150°C) immersed in oil.[23]

NBR curing, compounding, and processing are similar to those for other synthetic rubbers.[6]

Fine-powder NBR grades are ingredients in PVC/nitrile TPEs and in other polar thermoplastics to improve melt processability; reduce plasticizer blooming (migration of plasticizer to the surface of a finished product); and improve oil resistance, compression set, flex properties, feel, and finish of the plastic product.[1] Chemigum®† fine powder is blended with PVC/ABS and other polar thermoplastics.[1] The powders are typically less than 1 mm in diameter (0.5 nominal diameter particle size), containing 9% partitioning agent. Partitioning agents may be SiO_2, $CaCO_3$, or PVC. Their structures may be linear, linear/cross-linked, and branched/cross-linked (see Table 3.11).

Nitrile rubber applications are: belting, sheeting, cable jacketing, hose for fuel lines and air conditioners, sponge, gaskets, arctic/aviation O-

*ELAST is a registered trademark of Engel Canada.

†Chemigum is a registered trademark of Goodyear Tire and Rubber Company.

TABLE 3.11 Typical Chemigum/PVC Formulations and Properties for General Purpose and Oil/Fuel-Resistant Hose[1]

	General-purpose	Oil/fuel resistant
Ingredient, parts by weight		
PVC	100	100
Chemigum	25	100
DOP plasticizer	78	40
$CaCO_3$	—	20
Epoxidized soya oil	3	5
Stabilizer, lubricant	2	3
Properties		
Specific gravity, g/cm^3	1.17	1.20
Tensile strength, lb/in^2 (MPa)	2566 (17.6)	1929 (13.3)
Elongation @ break, %	390	340
Hardness Shore A	57	73

rings and seals, precision dynamic abrasion seals, and shoe soles. Nitrile rubbers are coextruded as the inner tube with chlorinated polyethylene outer tube for automotive applications.[10d] Nitrile provides resistance to hydrocarbon fluids, and chlorinated polyethylene provides ozone resistance.[10d] Other automotive applications are: engine gaskets, fluid- and vapor-resistant tubing, fuel filler neck inner hose, fuel system vent inner hose, oil, and grease seals. Nitrile powder grades are used in window seals, appliance gasketing, footwear, cable covering, hose, friction material composites such as brake linings, and food contact applications.[1,6]

Blends based on nitrile rubbers are used in underground wire/cable covering; automotive weatherstripping, spoiler extensions, foam-integral skin core-cover armrests, and window frames; footwear; and flexible, lay-flat, reinforced, rigid, and spiral hose for oils, water, food, and compressed air.

3.5.2 Butadiene rubber (BR) and polybutadiene (PB)

Budene®* solution polybutadiene (solution polymerized) is cis-1,4-poly(butadiene) produced with stereospecific catalysts which yield a controlled MWD, which is essentially a linear polymer.[6] Butadiene rubber, polybutadiene, is solution-polymerized to stereospecific polymer configurations[10a] by the additional polymerization of butadiene monomer. The following cis- and trans-1,4-polybutadiene isomers can be produced: cis-1,4-polybutadiene with good dynamic properties, low

*Budene is a registered trademark of Goodyear Tire and Rubber Company.

hysteresis, and good abrasion resistance; trans isomers are tougher, harder, and show more thermoplasticity.[10,23] Grades are oil and nonoil extended and vary according to their cis-content.[6]

The applications are: primarily tire tread and carcass stock, conveyor belt coverings, V-belts, hose covers, tubing, golf balls, shoe soles and heels, sponges, and mechanical goods.[6]

They are blended with SBR for tire treads to improve abrasion and wear resistance.[10a] The tread is the part of the tire that contacts the road, requiring low rolling resistance, abrasion and wear resistance, and good traction and durability.[22]

Replacement passenger car shipments in the United States are expected to increase from 185.5 million units in 1998 to 199 million units in 2004, according to the Rubber Manufacturers Association. Synthetic rubber choices for tires and tire treads are related to tire design. Composition and design for passenger cars, sport utility vehicles (SUV), pickup trucks, tractor trailers, and snow tires are continually under development, as illustrated by the following sampling of U.S. Patents.[16] Tire Having Silica Reinforced Rubber Tread Containing Carbon Fibers to Goodyear Tire & Rubber, U.S. Patent 5,718,781, February 17, 1998; Silica Reinforced Rubber Composition to Goodyear Tire & Rubber, U.S. Patent 5,719,208, February 17, 1998; and Silica Reinforced Rubber Composition and Tire With Tread to Goodyear Tire & Rubber, U.S. Patent 5,719,207, February 17, 1998.

Other patents include: Ternary Blend of Polyisoprene, Epoxidized Natural Rubber and Chlorosulfonated Polyethylene to Goodyear Tire & Rubber, U.S. Patent 5,736,593, April 7, 1998, and Truck Tire With Cap/Base Construction Tread to Goodyear Tire & Rubber, U.S. Patent 5,718,782, February 17, 1998; Tread Of Heavy Duty Pneumatic Radial Tire to Bridgestone Corporation, U.S. Patent 5,720,831, February 24, 1998; Pneumatic Tire With Asymmetric Tread Profile to Dunlop Tire, U.S. Patent 5,735,979, April 7, 1998; and Tire Having Specified Crown Reinforcement to Michelin, U.S. Patent 5,738,740, April 14, 1998. Silica improves a passenger car's tread rolling resistance and traction when used with carbon black.[28] High dispersible silica (HDS) in high vinyl solution polymerized SBR compounds show improved processing and passenger car tread abrasion resistance.[28] Precipitated silica with carbon black has been used in truck tire tread compounds which are commonly made with NR.[28]

Modeling is the method of choice for analyzing passenger car cord-reinforced rubber composite behavior. Large scale three-dimensional finite element analysis (FEA) improves understanding of tire performance, including tire and tread behavior when "the rubber meets the road."

BR is extruded and calendered. Processing properties and performance properties are related to polymer configuration: cis- or transstereoisomerism, MW and MWD, degree of crystallization (DC), degree of branching, and Mooney viscosity.[23] Broad MWD and branched BR tend to mill and process more easily than narrow MWD and more linear polymer.[23] Lower Mooney viscosity enhances processing.[23] BR is blended with other synthetic rubbers such as SBR to combine BR properties with millability and extrudability.

3.5.3 Butyl rubber

Butyl rubber (IIR) is an isobutylene-based rubber which includes copolymers of isobutylene and isoprene, halogenated butyl rubbers, and isobutylene/*p*-methylstyrene/bromo-*p*-methylstyrene terpolymers.[22] IIR can be slurry polymerized from isobutylene copolymerized with small amounts of isoprene in methyl chloride diluent at −130 to −148°F (−90 to −100°C). Halogenated butyl is produced by dissolving butyl rubber in a hydrocarbon solvent and introducing elemental halogen in gas or liquid state.[23] Cross-linked terpolymers are formed with isobutylene + isoprene + divinylbenzene.

Most butyl rubber is used in the tire industry. Isobutylene-based rubbers are used in underhood hose for the polymer's low permeability and temperature resistance, and high damping, resilient butyl rubbers are used for NVH (noise, vibration, harshness) applications such as automotive mounts for engine and vehicle/road NVH attenuation.[22]

Butyl rubber is ideal for automotive body mounts which connect the chassis to the body, damping road vibration.[10d] Road vibration generates low vibration frequencies. Butyl rubber can absorb and dissipate large amounts of energy due to its high mechanical hysteresis over a useful temperature range.[10d]

Low MW "liquid" butyls are used for sealants, caulking compounds, potting compounds, and coatings.[23] Depolymerized virgin butyl rubber is high viscosity, and is used for reservoir liners, roofing coatings, and aquarium sealants.[10b] It has property values similar to conventional butyl rubber: extremely low VTR (vapor transmission rate); resistance to degradation in hot, humid environments; excellent electrical properties; and resistance to chemicals, oxidation, and soil bacteria.[10b]. In order to make high-viscosity depolymerized butyl rubber pourable, solvents or oil are added.[10b]

Chlorobutyl provides flex resistance in the blend chlorobutyl rubber/EPDM rubber/NR for white sidewall tires and white sidewall coverstrips.[22] An important application of chlorobutyl rubber in automotive hose is extruded air conditioning hose to provide barrier properties to reduce moisture gain and minimize refrigerant loss.[22] The polymer is used in compounds for fuel line and brake line hoses.[22] Brominated isobutylene-*p*-methylstyrene (BIMS) was shown to have better aging properties than halobutyl rubber for underhood hose and comparable aging properties to peroxide-cured EPDM, depending on compound formulations.[22] Bromobutyls demonstrate good resistance to brake fluids for hydraulic brake lines and to methanol and methanol/gasoline blends.[22]

3.5.4 Chlorosulfonated polyethylene (CSM)

Chlorosulfonated polyethylene is a saturated chlorohydrocarbon rubber produced from Cl_2, SO_2, and a number of polyethylenes, and contains about 20 to 40% chlorine and 1 to 2% sulfur as sulfonyl chloride.[23] Sulfonyl chloride groups are the curing or cross-linking sites.[23] CSM properties are largely based on initial polyethylene (PE) and percent chlorine. A free-radical-based PE with 28% chlorine and 1.24% S has a dynamic shear modulus range from 1000 to 300,000 lb/in^2 (7 MPa to 2.1 GPa).[23] Stiffness differs for free-radical-based PE and linear PE, with chlorine content: at about 30%, Cl_2 free-radical-based PE stiffness decreases to minimum value, and at about 35%, Cl_2 content linear PE stiffness decreases to minimum value.[23] When the Cl_2 content is increased more than 30 and 35%, respectively, the stiffness (modulus) increases.[23]

Hypalon®* CSMs are specified by their Cl_2, S contents, and Mooney viscosity.[23] CSM has an excellent combination of heat and oil resistance and oxygen and ozone resistance. CSM, like other polymers, is compounded to meet specific application requirements. Hypalon is used for underhood wiring and fuel hose resistance.

3.5.5 Epichlorohydrin (ECH, ECO)

ECH and ECO polyethers are homo- and copolymers, respectively: chloromethyloxirane homopolymer and chloromethyloxirane copolymer with oxirane.[23] Chloromethyl side chains provide sites for cross-linking (curing and vulcanizing). These chlorohydrins are chemically 1-chloro-2,3-epoxypropane. They have excellent resistance to swelling when exposed to oils and fuels; good resistance to acids, alkalis, water, and ozone; and good aging properties.[10a]. Aging can be ascribed to environments such as weathering (UV radiation, oxygen, ozone, heat, and stress).[10a]. High chlorine content provides inherent flame retardance,[10a] and, like other halogenated polymers, flame-retardant enhancers (additives) may be added to increase UL 94 flammability rating.

---

*Hypalon is a registered trademark of DuPont Dow Elastomers LLC.

ECH and ECO can be blended with other polymers to increase high- and low-temperature properties and oil resistance.[23] Modified polyethers have potential use for new, improved synthetic rubbers. ECH and ECO derivatives, formed by nucleophilic substitution on the chloromethyl side chains, may provide better processing.

3.5.6 Ethylene propylene copolymer (EPM) and ethylene propylene diene terpolymer (EPDM)

EPM [poly(ethylene-co-propylene)] and EPDM [poly(ethylene-co-propylene-co-5-ethylidene-2-norbornene)][23] can be metallocene catalyst polymerized. Metallocene catalyst technologies include: (1) Insite, a constrained geometry group of catalysts used to produce Affinity polyolefin plastomers (POP), Elite®* PE, Nordel®† EPDM, and Engage polyolefin elastomers (POP) and (2) Exxpol®‡ ionic metallocene catalyst compositions used to produce "Exact" plastomer octene copolymers.[24] Insite technology produces EPDM-based Nordel IP with property consistency and predictability[16] (see Sec. 3.2.2).

Mitsui Chemical reportedly has developed "FI" catalyst technology, called a phenoxycyimine complex, with 10 times the ethylene polymerization activity of metallocene catalysts, according to *Japan Chemical Weekly* (summer, 1999).[25]

EPM and EPDM can be produced by solution polymerization, while suspension and slurry polymerization are viable options. EPDM can be gas-phase 1,-4 hexadiene polymerized using Ziegler-Natta catalysts. Union Carbide produces ethylene propylene rubber (EPR) using modified Unipol low-pressure gas-phase technology.

The letter "M" designates that the ethylene propylene has a saturated polymer chain of the polymethylene type, according to the ASTM.[12] EPM (copolymer of ethylene and propylene) rubber and EPDM (terpolymer of ethylene, propylene, and a nonconjugated diene) with residual side chain unsaturation, are subclassified under the ASTM "M" designation.[12]

The diene ethylidiene norbornene in Vistalon®§ EPDM allows sulfur vulcanization (see Table 3.12).[12] 1,4-Hexadiene and dicyclopentadiene (DCPD) are also used as curing agents.[18] The completely saturated polymer "backbone" precludes the need for antioxidants which can

---

*Elite is a registered trademark of Dow Chemical Company.

†Nordel is a registered trademark of DuPont Dow Elastomers LLC.

‡Exxpol is a registered trademark of Exxon Chemical Company.

§Vistalon is a registered trademark of Exxon Chemical Company, Division of Exxon Corporation.

bleed to the surface (bloom) of the finished product and cause staining.[12] Saturation provides inherent ozone and weather resistance, good thermal properties, and a low compression set.[12] Saturation also allows a relatively high volume addition of low-cost fillers and oils in compounds, while retaining a high level of mechanical properties.[12] The ethylene/propylene monomer ratio also affects the properties.

EPM and EPDM compounds, in general, have excellent chemical resistance to water, ozone, radiation, weather, brake fluid (nonpetroleum based), and glycol.[12]

EPM is preferred for dynamic applications because its age resistance retains initial product design over time and environmental exposure.[12] EPDM is preferred for its high resilience.[12] EPM is resistant to acids, bases (alkalis), and hot detergent solution. EPM and EPDM are resistant to salt solutions, oxygenated solvents, and synthetic hydraulic fluids.[12] Properties are determined by the composition of the base compound. A typical formulation includes Vistalon EP(D)M, carbon black, process oil, zinc oxide, stearic acid, and sulfur.[12]

EPDM formulations are increasingly popular for medium voltage, up to 221°F (105°C) continuous-use temperature wire, and cable covering.[20] Thinner wall, yet lower (power) loss, and better production

TABLE 3.12 Typical Properties of EPM/EPDM Compounds[12] Based on Vistalon

Property	Value
Hardness Shore A	35–90
Tensile strength, lb/in^2 (MPa)	580–3200 (4–22)
Compression set (%), 70 h @ 302°F (150°C)	15–35
Elongation, %	150–180
Tear strength, lb/in (kN/m)	86–286 (15–50)
Continuous service temperature, °F (°C)	302 (150) max
Intermediate service temperature, °F (°C)	347 (175) max
Resilience (Yerzley), %	75
Loss tangent (15 Hz), % (dynamic)	0.14
Elastic spring rate, lb/in (kN/m) (15 Hz)	3143 (550)
Dielectric constant, %	2.8
Dielectric strength, kV/mm	26
Power factor, %	0.25
Volume resistivity, Ω-cm	1×10^{16}

rates are sought by cable manufacturers.[20] Low MW (Mooney viscosity, ML), high ethylene content copolymers, and terpolymers are used in medium-voltage cable formulations.[20] With ethylidiene norbornene or hexadiene, EPDMs are good vulcanizates, providing improved wet electrical properties.[20] When the diene vinyl norbornene is incorporated on the EPDM backbone by a gel-free process, a significantly improved EPDM terpolymer is obtained for wire/cable applications.[20] Other applications are automotive body seals, mounts, weatherstripping, roofing, hose, tubing, ducts, and tires. Molded EPDM rubber is used for bumpers and fillers to dampen vibrations around the vehicle, such as deck-lid over-slam bumpers, for its ozone and heat resistance.[10d] EPDM can be bonded to steels, aluminum, and brass, with modified poly(acrylic acid) and polyvinylamine water-soluble coupling agents.[27]

EPDM is a favorable selection for passenger car washer-fluid tubes and automotive body seals, and it is used for automotive vacuum tubing.[10d] EPDM has good water-alcohol resistance for delivering fluid from the reservoir to the spray nozzle and good oxygen and UV resistance.[10d] Random polymerization yields a liquid with a viscosity of 100,000 centipoise (cP) @ 203°F (95°C), room temperature–cured with para-quinone dioxime systems or two-component peroxide systems, or cured at an elevated temperature with sulfur.[10b] They can be used as automotive and construction sealants, waterproof roofing membranes, and for encapsulating electrical components.[10b]

3.5.7 Fluoroelastomers (FKM)

Fluoroelastomers can be polymerized with copolymers and terpolymers of tetrafluoroethylene, hexafluoroethylene, and vinylidene fluoride. The fluorine content largely determines chemical resistance and T_g, which increases with increasing fluorine content. Low-temperature flexibility decreases with increasing fluorine content.[1] The fluorine content is typically 57% wt.[11]

TFE/propylene copolymers can be represented by Aflas®* TFE, produced by Asahi Glass. They are copolymers of tetrafluoroethylene (TFE) + propylene, and terpolymers of TFE + propylene + vinylidene fluoride.

Fluoroelastomer dipolymer and terpolymer gums are amine- or bisphenol-cured and peroxide-cured for covulcanizable blends with other peroxide curable elastomers. They can contain cure accelerators for faster cures, and they are divided into three categories: (1) gums with incorporated cures, (2) gums without incorporated cures, and (3) specialty master batches used with other fluoroelastomers.

*Aflas is a registered trademark of Asahi Glass Company.

Aflas are marketed in five categories according to their MW and viscosities.[11] The five categories possess similar thermal, chemical, and electrical resistance properties but different mechanical properties.[11] The lowest viscosity is used for chemical process industry tank and valve linings, gaskets for heat exchangers and pipe/flanges, flue duct expansion joints, flexible and spool joints, and viscosity improver additives in other Aflas grades.[11]

The second lowest viscosity grade is high-speed extruded into wire/cable coverings, sheet, and calendered stock.[11] Wire and cable covering are a principal application, especially in Japan. The third grade, general purpose, is molded, extruded, and calendered into pipe connector gaskets, seals, and diaphragms in pumps and valves.[11] The fourth grade with higher MW is compression molded into O-rings and other seal applications.

The fifth grade, with the highest MW, is compression molded into oil field applications requiring resistance to high-pressure gas blistering.[11] It is used for down-hole packers and seals in oil exploration and production. Oilfield equipment seals are exposed to short-term temperatures from 302 to 482°F (150 to 250°C) and pressures above 10,000 lb/in^2 (68.7 MPa) in the presence of aggressive hydrocarbons H_2S, CH_4, CO_2, amine-containing corrosion inhibitors, and steam and water.[11]

Synthetic rubbers, EPM/EPDM, nitrile, polychloroprene (neoprene), epichlorohydrin, and polyacrylate have good oil resistance, heat stability, and chemical resistance. Fluoropolymers are used in oil and gas wells 20,000 ft (6096 m) deep. These depths can have pressures of 20,000 lb/in^2 (137.5 MPa) which cause "extrusion" failures of down-hole seals by forcing the rubber part out of its retaining gland. TFE/propylene jackets protect down-hole assemblies which consist of stainless steel tubes that deliver corrosion-resistant fluid into the well.

Aircraft jet engine O-rings require fluoropolymer grades for engine cover gaskets that are resistant to jet fuel, turbine lube oils, and hydraulic fluids.

Dyneon®* BREs (base-resistant elastomers) are used in applications exposed to automotive fluids such as ATF, gear lubricants, engine oils shaft seals, O-rings, and gaskets.

DuPont Dow Elastomers fuel-resistant Viton®† fluoroelastomers are an important source for the applications described previously. The company's Kalrez®‡ perfluoroelastomers with reduced contamination are widely used with semiconductors and other contamination-sensi-

---

*Dyneon is a registered trademark of Dyneon LLC.

†Viton is a registered trademark of DuPont Dow Elastomers LLC.

‡Kalrez is a registered trademark of DuPont Dow Elastomers LLC.

tive applications. Contamination caused by high alcohol content in gasoline can cause fuel pump malfunction. The choice of polymer can determine whether an engine functions properly.

The three principal Viton categories are: (1) Viton A dipolymers composed of vinylidene fluoride (VF_2) and hexafluoropropylene (HFP) to produce a polymer with 66% (% wt) fluorine content, (2) Viton B terpolymers of VF_2 + HFP + tetrafluoroethylene (TFE) to produce a polymer with 68% fluorine, and (3) Viton F terpolymers composed of VF_2 + HFP + TFE to produce a polymer with 70% fluorine.[36] The three categories are based on their resistance to fluids and chemicals.[36] Fluid resistance generally increases but low-temperature flexibility decreases with higher fluorine content.[36] Specialty Viton grades are made with additional or different principal monomers in order to achieve specialty performance properties.[36] An example of a specialty property is low-temperature flexibility.

Compounding further yields properties to meet a given application.[36] Curing systems are an important variable affecting properties. DuPont Dow Elastomers developed curing systems during the 1990s, and the company should be consulted for the appropriate system for a given Viton grade.

FKMs are coextruded with lower-cost (co)polymers such as ethylene acrylic copolymer.[10d] They can be modified by blending and vulcanizing with other synthetic rubbers such as silicones, EPR and EPDM, epichlorohydrin, and nitriles. Fluoroelastomers are blended with modified NBR to obtain an intermediate performance/cost balance. These blends are useful for underhood applications in environments outside the engine temperature zone such as timing chain tensioner seals.

Fluoroelastomers are blended with fluorosilicones and other high-temperature polymers to meet engine compartment environments and cost/performance balance. Fine-particle silica increases hardness, red iron oxide improves heat resistance, and zinc oxide improves thermal conductivity. Hardness ranges from about Shore 35 A to 70 A. Fluorosilicones are resistant to nonpolar and nominally polar solvents, diesel and jet fuel, and gasoline, but not to solvents such as ketones and esters.

Typical applications are: exhaust gas recirculating and seals for engine valve stems and cylinders, crankshaft, speedometers, and O-rings for fuel injector systems.

FKMs are compounded in either water-cooled internal mixers or two-roll mills. A two-pass mixing is recommended for internally mixed compounds with the peroxide curing agent added in the second pass.[11] Compounds press-cured 10 min @ 350°F (177°C) can be formulated to possess more than 2100-lb/in^2 (14.4-MPa) tensile strength, 380% elongation, 525% @ 100% modulus, and higher values when postcured 16 h @ 392°F (200°C).[11] Processing temperatures are >392°F (200°C).[30]

3.5.8 Polyacrylate acrylic rubber (ACM)

Acrylic rubber can be emulsion- and suspension-polymerized from acrylic esters such as ethyl, butyl, and/or methoxyethyl acetate to produce polymers of ethyl acetate and copolymers of ethyl, butyl, and methoxyl acetate. Polyacrylate rubber, such as Acron®* from Cancarb Ltd., Alberta, Canada, possesses heat resistance and oil resistance between nitrile and silicone rubbers.[23] Acrylic rubbers retain properties in the presence of hot oils and other automotive fluids, and resist softening or cracking when exposed to air up to 392°F (200°C). The copolymers retain flexibility down to −40°F (−40°C). Automotive seals and gaskets comprise a major market.[23] These properties and inherent ozone resistance are largely due to the polymer's saturated "backbone" (see Table 3.13).

Polyacrylates are vulcanized with sulfur or metal carboxylate, with a reactive chlorine-containing monomer to create a cross-linking site.[23]

Copolymers of ethylene and methyl acrylate, and ethylene acrylics, have a fully saturated "backbone," providing heat-aging resistance and inherent ozone resistance.[23] They are compounded in a Banbury mixer and fabricated by injection molding, compression molding, resin transfer molding, extrusion, and calendering.

3.5.9 Polychloroprene (neoprene) (CR)

Polychloroprene is produced by free-radical emulsion polymerization of primarily trans-2-chloro-2-butenylene moieties.[23] Chloroprene rubber possesses moderate oil resistance, very good weather and oil resistance, and good resistance to oxidative chemicals.[10a] Performance properties depend on compound formulation, with the polymer providing fundamental properties. This is typical of any polymer and its compounds. Chloride imparts inherent self-extinguishing flame retardance.

*Acron is a registered trademark of Cancarb Ltd.

TABLE 3.13 Property Profile of Polyacrylic Rubbers[23]

Property at room temperature*	Value
Tensile strength, lb/in^2 (MPa)	2212 (15.2)
100% modulus	1500 (10.3)
Compression set, % [70 h @ 302°F (150°C)]	28
Hardness Shore A	80

*Unless indicated otherwise.

Crystallization contributes to high tensile strength, elongation, and wear resistance in its pure gum state before CR is extended or hardened.[10a]

3.5.10 Polyisoprene (IR)

Polymerization of isoprene can yield high-purity cis-1,4-polyisoprene and trans-1,4-polyisoprene. Isoprene is 2-methyl-1,3-butadiene, 2-methyldivinyl, or 2-methylerythrene.[23] Isoprene is polymerized by 1,4 or vinyl addition, the former producing cis-1,4 or trans-1,4 isomer.[23]

Synthetic polyisoprene, isoprene rubber (IR), was introduced in the 1950s as odorless rubber with virtually the same properties as natural rubber. Isoprene rubber product and processing properties are better than natural rubber in a number of characteristics. MW and MWD can be controlled for consistent performance and processing properties.

Polyisoprene rubber products are illustrated by Natsyn®,* which is used to make tires and tire tread (cis isomer). Tires are the major cis-polyisoprene product. Trans-polyisoprene can be used to make golf ball covers, hot-melt adhesives, and automotive and industrial products.

Depolymerized polyisoprene liquid is used as a reactive plasticizer for adhesive tapes, hot melts, brake linings, grinding wheels, and wire and cable sealants.[10b]

3.5.11 Polysulfide rubber (PSR)

PSR is highly resistant to hydrocarbon solvents, aliphatic fluids, and aliphatic-aromatic blends.[10a] It is also resistant to conventional alcohols, ketones, and esters used in coatings and inks and to certain chlorinated solvents.[10a] With these attributes, PSR is extruded into hose to carry solvents and printing rolls, and due to its good weather resistance it is useful in exterior caulking compounds.[10a] Its limitations, compared with nitrile, are relatively poor tensile strength, rebound, abrasion resistance, high creep under strain, and odor.[10a]

Liquid PSR is oxidized to rubbers with service temperatures from −67 to 302°F (−55 to 150°C), excellent resistance to most solvents, and good resistance to water and ozone.[10b] It has very low selective permeability rates to a number of highly volatile solvents and gases and odors. Compounds formulated with liquid PSR can be used as a flexibilizer in epoxy resins, and epoxy-terminated polysulfides have better underwater lap shear strength than toughened epoxies.[10b] Other PSR applications are: aircraft fuel tank sealant, seals for flexible electrical connections, printing rollers, protective coatings on metals, binders in gaskets, caulking compound ingredient, adhesives, and to provide water and solvent resistance to leather.

---

*Natsyn is a registered trademark of Goodyear Tire and Rubber Company.

3.5.12 Silicone rubber (SiR)

Silicone rubber polymers have the more stable Si atom compared with carbon. Silicone's property signature is its combined: (1) high-temperature resistance [>500°F (260°C)], (2) good flexibility at <−100°F (−73°C), (3) good electrical properties, (4) good compression set, and (5) tear resistance and stability over a wide temperature range.[10a] When exposed to decomposition level temperature, the polymer forms SiO_2 which can continue to serve as an electrical insulator.[10a] Silicone rubber is used for high-purity coatings for semiconductor junctions, high-temperature wire, and cable coverings.[1]

RTV (room temperature vulcanizing) silicones cure in about 24 h.[10b] They can be graded according to their room temperature viscosities which range from as low as 1500 cP (general-purpose soft) up to 700,000 cP (high-temperature paste). Most, however, are between 12,000 and 40,000 cP.[10b] RTV silicone has a low modulus over a wide temperature range from −85 to 392°F (−65 to 200°C), making them suitable for encapsulating electrical components during thermal cycling and shock.[10b] Low modulus minimizes stress on the encapsulated electrical components.[10b]

High-consistency rubber (HCR) from Dow Corning is injection-molded into high-voltage insulators, surge arrestors, weather sheds, and railway insulators. The key properties are: wet electrical performance and high tracking resistance.

Liquid silicone rubbers (LSRs) are two-part grades which can be coinjection-molded with thermoplastics to make door locks and flaps for vents.[10b] LSRs can be biocompatible and have low compression set, low durometer hardness, and excellent adhesion.[10b] One-part silicones are cured by ambient moisture. They are used for adhesives and sealants with plastic, metal, glass, ceramic, and silicone rubber substrates.[10b] A solventless, clear silicone/PC has been developed which requires no mixing, and can be applied without a primer.[1]

3.5.13 Styrene butadiene rubber (SBR)

SBR is emulsion- and solution-polymerized from styrene and butadiene, plus small volumes of emulsifiers, catalysts and initiators, endcapping agents, and other chemicals. It can be sulfur-cured. SBR types are illustrated with Plioflex®* emulsion SBR (emulsion polymerized) and Solflex®† solution SBR.[6] Emulsion SBR is produced by hot polymerization for adhesives and by cold polymerization for tires and other molded automotive and industrial products.[6] Solution SBR is used for tires.

---

*Plioflex is a registered trademark of Goodyear Tire and Rubber Company.

†Solflex is a registered trademark of Goodyear Tire and Rubber Company.

SBR is a low-cost rubber with slightly better heat aging and wear resistance than NR for tires.[10a] SBR grades are largely established by the bound styrene/butadiene ratio, polymerization conditions such as reaction temperature, and auxiliary chemicals added during polymerization.

SBR/PVC blends with nitrile rubber (NBR) as a compatibilizer show improved mechanical properties at lower cost than NBR/PVC.[21] This was the conclusion of studies using a divinylbenzene cross-linked, hot-polymerized emulsion polymer with 30% bound styrene and a cold-polymerized emulsion polymer with 23% bound styrene; PVC with inherent viscosity from 0.86 to 1.4; NBR with Mooney viscosity from 30 to 86; acrylonitrile content of 23.5, 32.6, and 39.7%; and ZnO, stabilizers, sulfur, and accelerators.[21]

3.6 Natural Rubber (NR)

Natural rubber, the original elastomer, still plays an important role among elastomers. Worldwide consumption of NR in 2000 is expected to be about 7 million metric tons/y, based on earlier reporting by the International Rubber Study Group. Chemically, natural rubber is cis-1,4-polyisoprene and occurs in Hevea rubber trees. NR tapped from other rubber trees (gutta-percha and balata) is the trans isomer of polyisoprene.[23] NR's principal uses are automotive tires, tire tread, and mechanical goods. Automotive applications are always compounded with carbon black to impart UV resistance and to increase mechanical properties.[10d] Latex concentrate is used for dipped goods, adhesives, and latex thread.[23] Latex concentrate is produced by centrifuge-concentrating field latex tapped from rubber trees. The dry rubber content is subsequently increased from 30 to 40 to 60% minimum.[23]

Vulcanization is the most important NR chemical reaction.[23] Most applications require cross-linking via vulcanization to increase resiliency and strength. Exceptions are crepe rubber shoe soles and rubber cements.[23] There are a number of methods for sulfur vulcanization, with certain methods producing polysulfidic cross-linking and other methods producing more monosulfidic cross-links.[10d]

NR is imported from areas such as Southeast Asia to the world's most industrial regions, North America, Europe, and Japan, since it is not indigenous to these regions. The huge rubber trees require about 80 to 100 in/y (200 to 250 cm/y) rainfall, and they flourish at an altitude of about 1000 ft (300 m).[23] As long as NR is needed for tires, industrial regions will be import-dependent.

NR has good resilience; high tensile strength; low compression set; resistance to wear and tear, cut-through and cold flow; and good electrical properties.[10a] Resilience is the principal property advantage compared with synthetic rubbers.[10a] For this reason, NR is usually used for engine mounts because NR isolates vibrations caused when an engine is running. NR is an effective decoupler, isolating vibrations such as engine vibration from being transmitted to another location such as the passenger compartment.[10d] With decoupling, vibration is returned to its source instead of being transmitted through the rubber.[10d] Polychloroprene is used for higher underhood temperatures above NR service limits; butyl rubber is used for body mounts and for road vibration frequencies which occur less frequently than engine vibrations or have low energy; EPDM is often used for molded rubber bumpers and fillers throughout the vehicle, such as deck-lid over-slam bumpers.

Degree of crystallinity (DC) can affect NR properties, and milling reduces MW. MW is reduced by mastication, typically with a Banbury mill, adding a peptizing agent during milling to further reduce MW, which improves NR solubility after milling.[23] NR latex grades are provided to customers in low (0.20 wt %) and high (0.75 wt %), with ammonia added as a preservative.[23] Low NH_4 has reduced odor and eliminates the need for deammoniation.[23]

Properties of polymers are improved by compounding with enhancing agents (additives), and NR is not an exception. Compounding NR with property enhancers improves resistance to UV, oxygen, and ozone, but formulated ETPs and synthetic rubbers overall have better resistance than compounded NR to UV, oxygen, and ozone.[10a] NR does not have satisfactory resistance to fuels, vegetable, and animal oils, while ETPs and synthetic rubbers can possess good resistance to them.[10a] NR has good resistance to acids and alkalis.[10a] It is soluble in aliphatic, aromatic, and chlorinated solvents, but it does not dissolve easily because of its high MW. Synthetic rubbers have better aging properties; they harden over time, while NR soften over time (see Table 3.14).[10a]

There are several visually graded latex NRs, including ribbed smoked sheets (RSS) and crepes such as white and pale, thin and thick brown latex, etc.[23] Two types of raw NR are field latex and raw coagulum, and these two types comprise all NR ("downstream") grades.[23]

Depolymerized NR is used as a base for asphalt modifiers, potting compound, and cold-molding compounds for arts and crafts.[10b]

3.7 Conclusion

Producers can engineer polymers and copolymers, and compounders can formulate recipes for a range of products that challenges the designers' imaginations. Computer variable–controlled machinery, tools, and dies can meet the designers' demands. Processing elastomeric materials is not as established as the more traditional thermoplastic and thermosetting polymers. Melt rheology, more than just viscosity, is the central differentiating characteristic for processing

TABLE 3.14 Typical Thermal and Electrical Property Profile of NR[23]

Property	Value
Specific gravity	
@ 32°F (0°C)	0.950
@ 68°F (20°C)	0.934
T_g, °F (°C)	−98 (−72)
Specific heat	0.502
Heat of combustion, cal/g (J/g)	10,547 (44,129)
Thermal conductivity, (BTU-in)/(h-ft^2-°F)	0.90
W/(m·K)	0.13
Coefficient of cubical expansion, in^3/°C	0.00062
Dielectric strength, V/mm	3.937
Dielectric constant	2.37
Power factor @ 1000 cycles	0.15–0.20
Volume resistivity, Ω·cm	1015
Cohesive energy density, cal/cm^3 (J/cm^3)	64 (266.5)
Refractive index	
68°F (20°C) RSS*	1.5192
68°F (20°C) pale crepe	1.5218

*RSS = ribbed smoked sheet.

elastomeric materials. Processing temperature and pressure settings are not fixed ranges; they are dynamic, changing values from the hopper to the demolded product. Operators and management of future elastomeric materials processing plants will be educated to the finesse of melt processing these materials. Elastomeric materials industries, welcome to the twenty-first century.

References

1. James M. Margolis, Thermoplastic Elastomer Markets in 2000, Multiclient Report, 1996.
2. *Kraton Polymers and Compounds, Typical Properties Guide,* Shell Chemical Company, Houston, Texas, 1997.
3. *Products, Properties and Processing for PELLETHANE Thermoplastic Polyurethane Elastomers,* Dow Plastics, The Dow Chemical Company, Midland, Michigan, ca. 1997.
4. *Modern Plastics Encyclopedia* '99, McGraw-Hill, New York, 1999, pp. B-51, B-52.
5. Engage, A Product of DuPont Dow Elastomers, Wilmington, Delaware, December 1998.
6. *Product Guide,* Goodyear Chemical, Goodyear Tire & Rubber Company, Akron, Ohio, October 1996.
7. *Injection Molding Guide for Thermoplastic Rubber—Processing, Mold Design, Equipment,* Advanced Elastomer Systems LP, Akron, Ohio, 1997.
8. *Santoprene Rubber Physical Properties Guide,* Advanced Elastomer Systems LP, Akron, Ohio, ca. 1998.
9. Hifax MXL 55A01 (1998), FXL 75A01 (1997) and MXL 42D01 Developmental Data Sheets secured during product development and subject to change before final commercialization. Montell Polyolefins, Montell North America Inc., Wilmington, Delaware.

10. Charles B. Rader, "Thermoplastic Elastomers," in *Handbook of Plastics, Elastomers, and Composites,* 3d ed., Charles A. Harper, ed., McGraw-Hill, New York, 1996.
10*a*. Joseph F. Meier, "Fundamentals of Plastics and Elastomers," in *Handbook of Plastics, Elastomers, and Composites,* 3d ed., Charles A. Harper, ed., McGraw-Hill, New York, 1996.
10*b*. Leonard S. Buchoff, "Liquid and Low-Pressure Resin Systems," in *Handbook of Plastics, Elastomers, and Composites,* 3d ed., Charles A. Harper, ed., McGraw-Hill, New York, 1996.
10*c*. Edward M. Petrie, "Joining of Plastics, Elastomers, and Composites," in *Handbook of Plastics, Elastomers, and Composites,* 3d ed., Charles A. Harper, ed., McGraw-Hill, New York, 1996.
10*d*. Ronald Toth, "Elastomers and Engineering Thermoplastics for Automotive Applications," in *Handbook of Plastics, Elastomers, and Composites,* 3d ed., Charles A. Harper, ed., McGraw-Hill, New York, 1996.
11. Aflas TFE Elastomers Technical Information and Performance Profile Data Sheets, Dyneon LLC, A 3M-Hoechst Enterprise, Oakdale, Minnesota, 1997.
12. *Vistalon User's Guide, Properties of Ethylene-Propylene Rubber,* Exxon Chemical Company, Houston, Texas, Division of Exxon Corporation, ca. 1996.
13. Kraton Liquid L-2203 Polymer, Shell Chemical Company, Houston, Texas, 1997.
14. Affinity Polyolefin Plastomers, Dow Plastics, The Dow Chemical Company, Midland, Michigan, 1997.
14*a*. Affinity HF-1030 Data Sheet, Dow Plastics, The Dow Chemical Company, Midland, Michigan, 1997.
14*b*. Affinity PF 1140 Data Sheet, Dow Plastics, The Dow Chemical Company, Midland, Michigan, 1997.
15. Bayer Engineering Polymers Properties Guide, Thermoplastics and Polyurethanes, Bayer Corporation, Pittsburgh, Pennsylvania, 1998.
16. *Rubber World Magazine,* monthly, 1999.
17. Jim Ahnemiller, "PU Rubber Outsoles for Athletic Footwear," *Rubber World,* December 1998.
18. Charles D. Shedd, "Thermoplastic Polyolefin Elastomers," in *Handbook of Thermoplastic Elastomers,* 2d ed., Benjamin M. Walker and Charles P. Rader, eds., Van Nostrand Reinhold, New York, 1988.
18*a*. Thomas W. Sheridan, "Copolyester Thermoplastic Elastomers," *Handbook of Thermoplastic Elastomers,* 2d ed., Benjamin M. Walker and Charles P. Rader, eds., Van Nostrand Reinhold, New York, 1988.
18*b*. William J. Farrisey, "Polyamide Thermoplastic Elastomers," in *Handbook of Thermoplastic Elastomers,* 2d ed., Benjamin M. Walker and Charles P. Rader, eds., Van Nostrand Reinhold, New York, 1988.
18*c*. Eric C. Ma, "Thermoplastic Polyurethane Elastomers, in *Handbook of Thermoplastic Elastomers,* 2d ed., Benjamin M. Walker and Charles P. Rader, eds., Van Nostrand Reinhold, New York, 1988.
19. N. R. Legge, G. Holden, and H. E. Schroeder, eds., *Thermoplastic Elastomers, A Comprehensive Review,* Hanser Publishers, Munich, Germany, 1987.
20. P. S. Ravishanker, "Advanced EPDM for W & C Applications," *Rubber World,* December 1998.
21. Junling Zhao, G. N. Chebremeskel, and J. Peasley, "SBR/PVC Blends With NBR As Compatibilizer," *Rubber World,* December 1998.
22. John E. Rogers and Walter H. Waddell, "A Review of Isobutylene-Based Elastomers Used in Automotive Applications," *Rubber World,* February 1999.
23. *Kirk-Othmer Concise Encyclopedia of Chemical Technology,* John Wiley & Sons, New York, 1999.
24. *PetroChemical News (PCN),* weekly, William F. Bland Company, Chapel Hill, North Carolina, September 14, 1998.
25. *PetroChemical News (PCN),* weekly, William F. Bland Company, Chapel Hill, North Carolina, 1998 and 1999.
26. *PetroChemical News (PCN),* weekly, William F. Bland Company, Chapel Hill, North Carolina, February 22, 1999.

27. C. P. J. van der Aar, et al., "Adhesion of EPDMs and Fluorocarbons to Metals by Using Water-Soluble Polymers," *Rubber World,* November 1998.
28. Larry R. Evans and William C. Fultz, "Tread Compounds with Highly Dispersible Silica," *Rubber World,* December 1998.
29. Vector Styrene Block Copolymers, Dexco Polymers, A Dow/Exxon Partnership, Houston, Texas, 1997.
30. *Fluoroelastomers Product Information Manual* (1997), *Product Comparison Guide* (1999), Dyneon LLC, A 3M-Hoechst Enterprise, Oakdale, Minnesota, 1997.
31. Engel data sheets and brochures, Guelph, Ontario, 1998.
32. Catalloy Process Resins, Montell Polyolefins, Wilmington, Delaware.
32*a*. Catalloy Process Resins, Montell Polyolefins, Wilmington, Delaware, p. 7.
33. EniChem Europrene SOL T Thermoplastic Rubber, styrene butadiene types, styrene isoprene types, EniChem Elastomers Americas Inc., Technical Assistance Laboratory, Baytown, Texas.
34. "Arnitel Guidelines for the Injection Molding of Thermoplastic Elastomer TPE-E," DSM Engineering Plastics, Evansville, Ind., ca. 1998.
35. Correspondence from DuPont Engineering Polymers, July 1999.
36. Correspondence from DuPont Dow Elastomers, Wilmington, Delaware, August 1999.

Chapter

4

Plastic Additives

Lou Kattas

Project Manager

Fred Gastrock

Senior Research Analyst

Inessa Levin

Research Analyst

Allison Cacciatore

Research Analyst
TownsendTarnell, Inc.
Mount Olive, New Jersey

4.1 Introduction

Plastic additives represent a broad range of chemicals used by resin manufacturers, compounders, and fabricators to improve the properties, processing, and performance of polymers. From the earliest days of the plastics industry, additives have been used initially to aid these materials in processing and then to improve their properties. Plastics additives have grown with the overall industry and currently represent over $16 billion in global sales.

4.2 Scope

This chapter includes all of the major chemical additives for plastics that are consumed worldwide. Materials excluded from the scope of this chapter include fillers, reinforcements, colorants, and alloys.

4.2.1 Definitions

To ensure understanding we will define the terms additives and plastics.

Additives. *Plastic additives* are comprised of an extremely diverse group of materials. Some are complex organic molecules (antioxidants and light stabilizers for example) designed to achieve dramatic results at very low loadings. At the opposite extreme are a few commodity materials (talc and glyceryl monostearate) which also can impart significant property improvements.

Adding to this complexity is the fact that many varied chemical materials can, and frequently do, compete in the same function. Also, the same material type may perform more than one function in a host plastic. An example would include the many surfactant type materials based on fatty acid chemistry which could impart lubricant, antistatic, mold release, and/or slip properties to a plastic matrix, depending upon the materials involved, loading level, processing conditions, and application.

Given the range of materials used, plastic additives are generally classified by their function rather than chemistry.

Plastics. *Plastics* denotes the matrix thermoplastic or thermoset materials in which additives are used to improve the performance of the total system. There are many different types of plastics that use large volumes of chemical additives including (in order of total additive consumption): polyvinyl chloride (PVC), the polyolefins [polyethylene (PE) and polypropylene (PP)], the styrenics —[polystyrene (PS) and acrylonitrile butadiene styrene (ABS)], and engineering resins such as polycarbonate and nylon.

4.3 Antiblock and Slip Agents

4.3.1 Description

Antiblocking agents. *Antiblocking agents* function by roughening the surface of film to give a spacing effect. The inherent tack of linear low-density polyethylene (LLDPE) and low-density polyethylene (LDPE) is a detriment when used in film where self-adhesion is undesirable. An antiblock additive is incorporated by the compounder to cause a slight surface roughness which prevents the film from sticking to itself. Years ago, efforts were made to prevent this by dusting the surface with corn starch or pyrogenic silica. This process was abandoned because of potential health concerns. Antiblocking agents are now melt-incorporated into the thermoplastic either via direct addition or by use of a master batch.

Antiblocking agents are used in polyolefin films in conjunction with slip agents in such consumer items as trash bags, shipping bags, and a variety of packaging applications. The most common polymers extruded into film include LLDPE and LDPE. Lesser amounts of high-density polyethylene (HDPE) are used for these as well as other film applications. PE resins are used in film for their toughness, low cost and weight, optical properties, and shear sealability. Four criteria are used in the selection of an antiblocking agent, as shown in Table 4.1.

While both organic and inorganic materials are used as antiblocking agents, the inorganics make up the bulk of the market. The four major types of antiblocking agents are

- Diatomaceous earth
- Talc
- Calcium carbonate
- Synthetic silicas and silicates

The suppliers of inorganic additives to the plastics industry market their products primarily as fillers and extenders. While many of these products can also be used as antiblocking agents in polyethylene films, only a few suppliers actively market their products for this end use.

Slip agents. *Slip agents* or *slip additives* are the terms used by industry for those modifiers that impart a reduced coefficient of friction to the surface of finished products. Slip agents can significantly improve the handling qualities of polyolefins and, to a lesser extent, PVC, in film and bag applications. They help speed up film production and

TABLE 4.1 Criteria Used in Selection of an Antiblocking Agent

Specification	Function
Particle size distribution	Affects both the level of antiblock performance and the physical properties of the final film.
Surface area	Measured in square meters per gram. Affects the coefficient of friction of the film and level of wear on equipment.
Specific gravity	Indicates the relative weight of the product.
Density	Measures the mass/volume ratio. Affects the quality of the film.

ensure final product quality. Fatty acid amides, the primary chemical type used as slip agents, are similar to migratory antistatic agents and some lubricants with a molecule which has both a polar and non-polar portion. These additives migrate to the surface and form a very thin molecular layer that reduces surface friction.

Slip agents are typically employed in applications where surface lubrication is desired—either during or immediately after processing. To accomplish this, the materials must exude quickly to the surface of the film. To function properly they should have only limited compatibility with the resin. Slip agents, in addition to lowering surface friction, can also impart the following characteristics:

- Lower surface resistivity (antistatic properties)
- Reduce melt viscosity
- Mold release

Slip agents are often referred to as *lubricants.* However, they should not be confused with the lubricants which act as processing aids. While most slip agents can be used as lubricants, many lubricants cannot be used as slip agents since they do not always function externally.

The major types of slip agents include:

- Fatty acid amides (primarily erucamide and oleamide
- Fatty acid esters
- Metallic stearates
- Waxes
- Proprietary amide blends

Antiblock and slip agents can be incorporated together using combination master batches which give the film extruder greater formulation control.

4.3.2 Suppliers

Because of the different chemical composition of antiblocking and slip agents, few companies are involved in both. Table 4.2 presents a list of the selected global suppliers of antiblocking and slip agents.

4.3.3 Trends and forecasts

The trend toward downgauging in PE film has favorably affected the use of slip agents. Although the value of resin decreases as films are made thinner, surface area increases, therefore, requiring higher load-

TABLE 4.2 Selected Suppliers of Antiblocking and Slip Agents

Supplier	Type	
	Antiblock	Slip
Akcros	—	×
Akzo Nobel	—	×
AlliedSignal	—	×
American Ingredients	—	×
Asahi Denka Kogyo	—	×
Baerlocher	—	×
BASF	—	×
Cabot	×	—
Celite	×	—
Chemson	—	×
Clariant	—	×
Croda Universal	—	×
Cyprus Minerals	×	—
Degussa	×	—
Eastman Chemical	—	×
ECC	×	—
Ferro	—	×
Guangpin Chemical	×	—
Henkel	—	×
Huels	—	×
Idemitsu Kosan	—	×
J. M. Huber	×	—
Kao	—	×
Katsuta Kako	—	×
Kawaken Fine Chemical	—	×
Kawamura Kasei	—	×
Lion Akzo	—	×
Lonza	—	×
Matsumura Oil Research	—	×
Mitsui Petrochemical	—	×
Miyoshi Oil and Fat	×	—
New Japan Chemical	—	×
Nippon Fine Chemical	—	×
Nippon Kasei Chemical	—	×
Nippon Seiro	—	×
P T Sumi Asih	—	×
Petrolite	—	×
Pfizer	×	—
Sakai Chemical	—	×
Sankyo Organic Chemicals	—	×
Sanyo Chemical	—	×
Shinagawa Chemical	—	×
Struktol	—	×
Unichema	—	×
Witco	—	×
Yasuhara Chemical	—	×
Zeelan Industries	×	—

ings of slip agents. Both slip and antiblocking agents are expected to grow at a rate of about 4% annually over the next 5 years.

4.4 Antioxidants

4.4.1 Description

Antioxidants are used in a variety of resins to prevent oxidative degradation. Degradation is initiated by the action of highly reactive free radicals caused by heat, radiation, mechanical shear, or metallic impurities. The initiation of free radicals may occur during polymerization, processing, or fabrication.

Once the first step of initiation occurs, propagation follows. *Propagation* is the reaction of the free radical with an oxygen molecule, yielding a peroxy radical. The peroxy radical then reacts with an available hydrogen atom within the polymer to form an unstable hydroperoxide and another free radical. In the absence of an antioxidant, this reaction continues and leads to degradation of the polymer. Degradation is manifested either by cross-linking or chain scissoring. Cross-linking causes the polymer to increase in molecular weight, leading to brittleness, gellation, and decreased elongation. Chain scissoring decreases molecular weight, leading to increased melt flow and reduced tensile strength.

The function of an antioxidant is to prevent the propagation steps of oxidation. Products are classified as primary or secondary antioxidants depending on the method by which they prevent oxidation.

Primary antioxidants, usually sterically hindered phenols, function by donating their reactive hydrogen to the peroxy free radical so that the propagation of subsequent free radicals does not occur. The antioxidant free radical is rendered stable by electron delocalization. Secondary antioxidants retard oxidation by preventing the proliferation of alkoxy and hydroxy radicals by decomposing hydroperoxides to yield nonreactive products. These materials are typically used in synergistic combination with primary antioxidants.

Table 4.3 lists the chemical types of primary and secondary antioxidants and their major resin applications. Through the remainder of this chapter, antioxidants will be addressed by type based on overall chemistry. The class of antioxidant merely describes its mode of stabilization.

Amines. *Amines,* normally arylamines, function as primary antioxidants by donating hydrogen. Amines are the most effective type of primary antioxidant, having the ability to act as chain terminators and peroxide decomposers. However, they tend to discolor, causing staining, and, for the most part, lack FDA approval. For this reason, amines are found in pigmented plastics in nonfood applications. Amines are com-

TABLE 4.3 Antioxidants by Chemical Type with Major Resin Applications

Types	Major resins	Comments
Primary		
Amine	Rubber, some pigmented plastics, and polyurethane polyols	Arylamines tend to discolor and cause staining.
Phenolic	Polyolefins, styrenics, and engineering resins	Phenolics are generally stain resistant and include simple phenolics (BHT), various polyphenolics, and bisphenolics.
Metal salts	Polyolefin wire and cable	These are metal deactivators used in the inner coverings next to the metal.
Secondary		
Organophosphite	Polyolefins, styrenics, and engineering resins	Phosphites can improve color stability, and engineering resins but can be corrosive if hydrolyzed.
Thioester	Polyolefins and styrenics	The major disadvantage with thioesters is their odor which is transferred to the host polymer.

monly used in the rubber industry but also find minor use in plastics such as black wire and cable formulations and in polyurethane polyols.

Phenolics. The most widely used antioxidants in plastics are *phenolics*. The products generally resist staining or discoloration. However, they may form quinoid (colored) structures upon oxidation. Phenolic antioxidants include simple phenolics, bisphenolics, polyphenolics, and thiobisphenolics.

The most common simple phenolic is butylated hydroxytoluene (BHT) or 2,6-di-*t*-butyl-4-methylphenol. BHT possesses broad FDA approval and is widely used as an antioxidant in a variety of polymers. It is commonly called the “workhorse” of the industry but is losing ground to the higher molecular weight antioxidants which resist migration. The disadvantage of BHT is that it is highly volatile and can cause discoloration. Other simple phenolics include BHA (2- and 3-*t*-butyl-4-hydroxyanisole) which is frequently used in food applications.

Polyphenolics and bisphenolics are higher in molecular weight than simple phenolics and both types are generally nonstaining. The increased molecular weight provides lower volatility, but is generally more costly. However, the loading of polyphenolics is much less than that of the simple phenolics. The most commonly used polyphenolic is tetrakis(methylene-(3,5-di-*t*-butyl-4-hydroxyhydrocinnamate) methane or IRGANOX1010 from Ciba. Other important bisphenolics include: Cytec Industries' CYANOX 2246 and 425 and BISPHENOL A from Aristech, Dow, and Shell.

Thiobisphenols are less effective than hindered phenols in terminating peroxy radicals. They also function as peroxide decomposers (secondary antioxidants) at temperatures above 100°C. Typically, thiobisphenols are chosen for use in high-temperature resin applications. Users generally prefer hindered phenolics over thiobisphenols where high-temperature service is not involved.

Organophosphites. Acting as secondary antioxidants, *organophosphites* reduce hydroperoxides to alcohols, converting themselves to phosphonates. They also provide color stability, inhibiting the discoloration caused by the formation of quinoid reaction products which are formed upon oxidation of phenolics. Tris-nonylphenyl phosphite (TNPP) is the most commonly used organophosphite followed by tris(2,4-di-*tert*-butylphenyl)phosphite (for example, Ciba's IRGAFOS 168). The disadvantage of phosphites is their hygroscopic tendency. Hydrolysis of phosphites can ultimately lead to the formation of phosphoric acid, which can corrode processing equipment.

Thioesters. Derived from aliphatic esters of B-thio dipropionic acid, *thioesters* act as secondary antioxidants and also provide high heat stability to a variety of polymers. Thioesters function as secondary antioxidants by destroying hydroperoxides to form stable hexavalent sulfur derivatives. Thioesters act as synergists when combined with phenolic antioxidants in polyolefins. The major disadvantage of thioester antioxidants is their inherent odor which is transferred to the host polymer.

Deactivators. *Metal deactivators* combine with metal ions to limit the potential for chain propagation. Metal deactivators are commonly used in polyolefin inner coverings in wire and cable applications where the plastic comes in contact with the metal. In effect, the deactivator acts as a chelating agent to form a stable complex at the metal interface, thereby preventing catalytic activity. The most common deactivators contain an oxamide moiety that complexes with and deactivates the metal ions. A typical product is Ciba's IRGANOX MD-1024.

4.4.2 Recent developments

Some of the most significant new product development trends in antioxidants are as follows:

- "Lactone" stabilizers are a new class of materials that are reputed to stop the autoxidation process before it starts. These products, which are derivatives of the benzofuranone family, act as C-radical scavengers in combination with primary and secondary antioxidants. These blends (Ciba's HP) claim to be particularly effective in high-temperature and high-shear processing.

- A new phosphite secondary antioxidant, based on butyl ethyl propane diol, reputedly yields high activity, solubility, and hydrolytic stability in a range of polymers. This would allow the producer to use lower levels of additives to achieve similar results.
- Antioxidants (AO) in the form of pellets are challenging the granule forms. Advantages include low-dusting, easy flowing, and lower-cost systems. Most major AO suppliers are now marketing these product forms.
- Selected suppliers are promoting hindered amine light stabilizers for the combined use as antioxidants.

4.4.3 Suppliers

There are over 70 suppliers of antioxidants worldwide. Numerous suppliers offer both primary and secondary antioxidants to complete their product line. However, very few actually manufacture both primary and secondary antioxidants, since the products are based on different manufacturing routes, processes, and feedstock sources. As a result, it is quite common in this industry to resell products produced by another company. Table 4.4 displays selected suppliers of antioxidants by type.

4.4.4 Trends and forecasts

The overall growth of antioxidants in plastics will be influenced by the following factors:

- Growth of the polyolefin industry, especially polypropylene.
- Increased price competition as patents expire; this will force some suppliers to accept lower margins and/or to segment their customer base and concede lower margin accounts to selected competitors.
- Continued premiums will be possible for technical innovation where unique products bring value to the market. Examples include:

  Higher processing temperature performance
  New chemistry (for example, hydroxyl amines) replacing phenolic-based systems, avoiding potential toxicity and color issues
  Higher molecular weight AOs to reduce volatility during processing
  Better long-term stability

TABLE 4.4 Selected Antioxidant Suppliers

Supplier	Type					
	Amine	Phenolic	Organophosphite	Thioester	Metal deactivator	Other
3V Sigma	—	×	—	—	—	—
Akcros (Akzo)	—	×	×	×	—	—
Albemarle	—	×	×	—	—	—
Albright and Wilson	—	—	×	—	—	—
Asahi Denka Kogyo	—	×	×	×	—	—
Asia Stabilizer	—	×	×	—	—	—
Bayer	×	×	×	—	—	—
Cambrex	—	×	—	—	—	—
Chang-Chun Petrochemical	—	×	—	×	—	×
Ciba Specialty	—	×	×	×	×	×
Clariant	—	×	×	—	—	—
Coin Chemical Industrial	—	×	—	—	—	—
Cytec Industries	—	×	—	×	—	—
Dai-ichi Kogyo Seiyaku	—	×	—	—	—	×
Daihachi Chemical Ind.	—	—	×	—	—	—
Dongbo S.C.	—	—	×	×	—	—
Dover Chemical	—	—	×	—	—	—
Eastman Chemical Products	—	×	—	×	×	—
Everspring Chemical	—	×	—	—	—	—
Fairmount Chemical Company	—	×	—	—	×	—
Ferro Corporation	—	×	×	—	—	×
GE Specialty Chemical	—	—	×	—	—	×
Goodyear Tire and Rubber	×	×	—	—	—	—
Great Lakes Chemical	×	×	×	×	—	×
Hampshire Chemicals	—	—	—	×	—	—
Han Nong Adeka	—	×	×	—	—	—

Harwick Chemical Corporation	—	×	—	—	—	—
Honshu Chemical	—	×	—	—	—	—
Johoku Chemical	—	—	×	—	—	—
Kawaguchi Chemical Industry	—	×	×	—	—	—
Kolon Industries	—	—	×	—	—	—
Maruzen Petrochemical	—	×	×	—	—	—
Mayzo	—	×	—	×	—	—
Morton International	—	—	—	×	—	—
Musashina Geigy	—	×	×	—	—	—
Nan Ya Plastics	—	—	—	×	—	—
Nanjin Chemical Plant	—	×	×	—	—	—
Nippon Oil and Fats	—	×	—	×	—	—
Orient Chemical	—	×	—	—	—	—
Ouchi Shinko Chemical	—	×	×	—	—	—
PMC	—	×	—	—	—	—
R. T. Vanderbilt	×	×	—	×	—	—
Raschig Corporation	—	×	—	—	—	—
Reagens	—	—	×	×	—	—
Rhodia	—	×	—	—	—	—
Sakai Chemical Industry	—	—	×	—	—	—
Sankyo Chemical	—	×	×	—	—	—
Sanyo Chemical	—	×	—	—	—	—
Schenectady Chemicals	—	×	—	—	—	—
Seiko Chemical	—	×	—	—	—	—
Solutia	—	×	—	—	—	—
Song-Woun	—	×	—	×	—	—
Sumitomo Chemical	—	×	×	×	—	—
Taiwan Ciba Geigy	—	×	—	×	—	—
Tiyoda Chemical	—	×	—	—	—	—
Ueno Fine Chemicals	—	×	—	—	—	—
Uniroyal Chemical	×	×	×	—	×	—
UOP Biological & Food Products	—	×	—	—	—	—
Witco Corporation	×	×	×	×	—	—
Yoshitomi	—	×	×	×	—	—

Equal performance at lower loading levels
More economical product forms and blends

Over the next 5 years, consumption of antioxidants is expected to grow somewhat evenly around the world at a rate of about 5%/year.

4.5 Antistatic Agents

4.5.1 Description

Plastics are inherently insulative (typical surface resistivities in the range of 10^{12} to 10^{14} Ω/square) and cannot readily dissipate a static charge. The primary role of an *antistatic agent or antistat* is to prevent the buildup of static electrical charge resulting from the transfer of electrons to the surface. This static electricity can be generated during processing, transportation, handling, or in final use. Friction between two or more objects (for example, the passage of copy paper over a roller) is usually the cause of static electricity. Typical electrostatic voltages can range from 6000 to 35,000 V.

When the unprotected plastic is brought into contact with another material, loosely bound electrons pass across the interface. When these materials are then separated, one surface has an excess charge, while the other has a deficiency of electrons. In most plastics the excess charge will linger or discharge, causing the following problems:

- Fire and explosion hazards
- Poor mold release
- Damage to electrical components
- Attraction of dust

Antistats function to either dissipate or promote the decay of static electricity. Secondary benefits of antistat incorporation into polymer systems include improved processability and mold release, as well as better internal and external lubrication. Therefore, in certain applications, antistatic agents can also function as lubricants, slip agents, and mold release agents.

This discussion will focus on *chemical* antistats and excludes inorganic conductive additives such as carbon black, metal-coated carbon fiber, and stainless steel wire. Chemical antistatic additives can be categorized by their method of application (external and internal) and their chemistry. Most antistats are hydroscopic materials and function primarily by attracting water to the surface. This process allows the charge to dissipate rapidly. Therefore, the ambient humidity level plays a vital role in this mechanism. With an increase in humidity, the surface conductivity of the treated polymer is increased, resulting in a rapid flow of charge and better antistatic properties. Conversely, in dry ambient conditions, antistats which rely on humidity to be effective may offer erratic performance.

External antistats. *External, or topical, antistats* are applied to the surface of the finished plastic part through techniques such as spraying, wiping, or dipping. Since they are not subjected to the temperatures and stresses of plastic compounding, a broad range of chemistries is possible. The most common external antistatic additives are quaternary ammonium salts, or "quats," applied from a water or alcohol solution.

Because of low temperature stability and potential resin degradation, quats are not normally used as internal antistats. However, when topically applied, quats can achieve low surface resistivities and are widely used in such short-term applications as the prevention of dust accumulation on plastic display parts. More durable applications are not generally feasible because of the ease with which the quat antistat coating can be removed from the plastic during handling, cleaning, or other processes. For longer-term protection internal antistats are used.

Internal antistats. *Internal antistats* are compounded into the plastic matrix during processing. The two types of internal antistats are migratory, which is the most common, and permanent.

Migratory antistats (MAS). *Migratory antistats* have chemical structures that are composed of hydrophilic and hydrophobic components. These materials have limited compatibility with the host plastic and migrate or bloom to the surface of the molded product. The hydrophobic portion provides compatibility within the polymer and the hydrophilic portion functions to bind water molecules onto the surface of the molded part. If the surface of the part is wiped, the MAS is temporarily removed, reducing the antistat characteristics at the surface. Additional material then migrates to the surface until the additive is depleted. These surface-active antistatic additives can be cationic, anionic, and nonionic compounds.

Cationic antistats are generally long-chain alkyl quaternary ammonium, phosphonium, or sulfonium salts with, for example, chloride counterions. They perform best in polar substrates, such as rigid PVC and styrenics, but normally have an adverse effect on the resin's thermal stability. These antistat products are usually not approved for use in food-contact applications. Furthermore, antistatic effects comparable to those obtained from other internal antistats such as ethoxylated amines are only achieved with significantly higher levels, typically, five- to tenfold.

Anionic antistats are generally alkali salts of alkyl sulfonic, phosphonic, or dithiocarbamic acids. They are also mainly used in PVC and styrenics. Their performance in polyolefins is comparable to cationic antistats. Among the anionic antistats, sodium alkyl sulfonates have found the widest applications in styrenics, PVC, polyethylene terephthalate, and polycarbonate.

Nonionic antistats, such as ethoxylated fatty alkylamines, represent by far the largest class of migratory antistatic additives. These additives are widely used in PE, PP, ABS, and other styrenic polymers. Several types of ethoxylated alkylamines that differ in alkyl chain length and level of unsaturation are available. Ethoxylated alkylamines are very effective antistatic agents, even at low levels of relative humidity, and remain active over prolonged periods. These antistatic additives have wide FDA approval for indirect food contact applications. Other nonionic antistats of commercial importance are ethoxylated alkylamides such as ethoxylated lauramide and glycerol monostearate (GMS). Ethoxylated lauramide is recommended for use in PE and PP where immediate and sustained antistatic action is needed in a low-humidity environment. GMS-based antistats are intended only for static protection during processing. Even though GMS migrates rapidly to the polymer surface, it does not give the sustained antistatic performance that is obtainable from ethoxylated alkylamines or ethoxylated alkylamides.

The optimum choice and addition level for MAS additives depends upon the nature of the polymer, the type of processing, the processing conditions, the presence of other additives, the relative humidity, and the end use of the polymer. The time needed to obtain a sufficient level of antistatic performance varies. The rate of buildup and the duration of the antistatic protection can be increased by raising the concentration of the additive. Excessive use of antistats can, however, lead to greasy surfaces on the end products and adversely affect printability or adhesive applications. Untreated inorganic fillers and pigments like TiO_2 can absorb antistat molecules to their surface, and thus lower their efficiency. This can normally be compensated for by increasing the level of the antistat. The levels of antistat for food-contact applications are regulated by the U.S. Food and Drug Administration (FDA).

Permanent antistats. The introduction of *permanent antistats* is one of the most significant developments in the antistat market. These are polymeric materials which are compounded into the plastic matrix. They *do not* rely on migration to the surface and subsequent attraction of water to be effective. The primary advantages of these materials are

- Insensitivity to humidity
- Long-term performance

- Minimal opportunity for surface contamination
- Low offgassing
- Color and transparency capability

There are two generic types of permanent antistats: hydrophilic polymers and inherently conductive polymers. *Hydrophilic polymers* are currently the dominant permanent antistats in the market. Typical materials that have been used successfully are such polyether block copolymers as PEBAX from Atochem. Typical use levels for these materials are in excess of 10%. B.F. Goodrich is supplying compounds utilizing their permanent antistat additive, STAT-RITE. Office automation equipment, such as fax and copier parts, is the principal application for permanent antistats based on hydrophilic polymers. The most common resins are ABS and high-impact polystyrene (HIPS).

Another approach to achieving permanent antistatic properties is through the use of *inherently conductive polymers* (ICP). This technology is still in the early development stages. The potential advantages of ICP include achieving higher conductivity in the host resin at lower additive loading levels than can be achieved with hydrophilic polymers. The principal ICP technology to date is polyaniline from Zipperling-Kessler and Neste. This material is a conjugated polymer composed of oxidatively coupled aniline monomers converted to a cationic salt with an organic acid and is frequently described as an organic metal. Other approaches to ICPs include neoalkoxy zirconates from Kenrich Petrochemical and polythiophenes from Bayer. The issues to be resolved in achieving commercial success with these materials include improved stability at elevated temperatures and reduction in their relatively high cost. ICPs are not expected to compete with other chemical additives but primarily with carbon black or other conductive fillers.

Permanent antistatic properties can be readily obtained with such particulate materials as carbon black. However, these materials are inappropriate for applications where color and/or transparency capability is important. Also, particulate additives can negatively affect the physical properties of the final part and contribute to contamination in electronic applications also known as sloughing.

4.5.2 Suppliers

The antistatic additive market is served by fewer than 50 suppliers. The major suppliers include Akzo, Witco, Henkel, Elf Atochem, Kao, and Clariant. Table 4.5 lists some of the more prominent suppliers and the types of antistatic agents offered.

TABLE 4.5 Selected Suppliers of Antistatic Agents

Supplier	Type			
	Quats*	Amines	Fatty acid esters	Other†
Akzo Nobel	—	×	×	—
Bayer	—	—	—	×
Ciba Specialty Chemicals	—	—	—	×
Clariant	—	×	×	—
Cytec Industries	×	—	—	—
Elf Atochem	—	—	—	×
Henkel Corporation	—	—	×	—
ICI Americas	—	×	—	—
Kao Corporation	—	×	×	—
Lion Akzo	×	—	—	—
Lonza	×	—	×	—
NOF Corp.	—	—	×	—
Sanyo Chemical	×	—	—	—
Witco	×	×	×	—

*Quaternary ammonium compounds.
†"Other" category includes aliphatic sulfonates, fatty amides, and polymeric antistats.

4.5.3 Trends and forecasts

Continuing increases are expected in the markets for electronic components, devices, and equipment. Plant modernization activities will increase requirements for automated production machinery. Improvement in communication will continue to promote sales of items such as facsimile machines, personal computers, and cellular telephones. This will provide more opportunities for antistatic agents for static and electromagnetic interference control. Globally antistatic agents are expected to grow at a rate of 5 to 6%/year over the next 5 years.

4.6 Biocides

4.6.1 Description

Biocides are additives that impart protection against mold, mildew, fungi, and bacterial growth to materials. Without biocides, polymeric materials in the proper conditions can experience surface growth, development of spores causing allergic reactions, unpleasant odors, staining, embrittlement, and premature product failure. It is important to note that the biocide protects the material, not the user of the final product.

In general, in order for mold, mildew, and bacterial growth to develop, the end product must be in an environment that includes warmth, moisture, and food. Specifically, if the environment includes soil where microbes and bacteria abound, protection against bacterial growth is needed. If the end product has a water or moist environment, protection from fungi may be the most important feature. Environmental conditions overlap and many biocides are effective over a broad range.

Biocides, also referred to as antimicrobials, preservatives, fungicides, mildewcides, or bactericides, include several types of materials that differ in toxicity. OBPA (10, 10′-oxybisphenoxarsine) is the most active preservative of those commonly used for plastics. Amine-neutralized phosphate and zinc-OMADINE (zinc 2-pyridinethianol-1-oxide) have a lower activity level but are also effective. In the United States all biocides are considered pesticides and must be registered for specific applications with the U.S. Environmental Protection Agency (EPA).

The effectiveness of a biocide depends on its ability to migrate to the surface of the product where microbial attack first occurs. Most biocides are carried in plasticizers, commonly epoxidized soybean oil or diisodecyl phthalate, which are highly mobile and migrate throughout the end product. This mobility results in the gradual leaching of the additive. If significant leaching occurs, the product will be left unprotected. The proper balance between the rates of migration and leaching determines the durability of protection.

The majority of biocide additives are used in flexible PVC. The remaining portion is used in polyurethane foam and other resins. PVC applications using biocides include flooring, garden hoses, pool liners, and wall coverings, among others.

The use level of biocide additives depends on the efficacy of the active ingredient. OBPA, the most active, requires approximately 0.04% concentration in the final product. Less active ingredients, such as *n*-(trichloro-methylthio) phthalimide, require a loading of 1.0% in the final compound to achieve a similar level of protection.

Biocides are generally formulated with a carrier into concentrations of 2 to 10% active ingredient. They are available to plastics converters, processors, and other users in powder, liquid, or solid pellet form. The carrier, as noted previously, is usually a plasticizer, but it can also be a resin concentrate such as PVC/PVA (polyvinyl acetate) copolymer or polystyrene. For example, OBPA, the most common biocide active ingredient, is typically purchased as a dispersion in a plasticizer at a concentration of 2% active ingredient.

Of the hundreds of chemicals that are effective as biocides, only a few are used in plastic applications. After OBPA, the most common group of active ingredients are 2-*n*-octyl-4-isothiazolin-3-one, 4, 5-dichloro-2-*n*-octyl-4-isothiazolin-3-one (DCOIT), zinc OMADINE, trichlorophenoxyphenol (TCPP or TRICLOSAN), *N* trichloromethylthio-4-cyclohexene-1,2-dicarboximide (CAPTAN), and *N*-(trichloromethylthio) phthalimide (FOLPET).

4.6.2 Suppliers

There are two tiers of biocide suppliers to the plastics industry: those who sell active ingredients and those who provide formulated products, both of which are shown in Table 4.6. The active ingredient manufacturer typically does not produce formulated biocides and formulators do not typically synthesize active materials.

The major formulated plastic biocide suppliers are Akcros Chemicals (owned by Akzo) and Morton International. Other suppliers of formulated biocide products include Ferro, Huels, Olin, and Microban. Akzo-Nobel, Ciba, and Rohm and Haas are the major suppliers of active ingredients.

Among the industry leaders, Morton International offers one of the broadest ranges of formulated OBPA, TCPP, and isothiazole products.

4.6.3 Trends and forecasts

Biocides for plastics are growing at about 7%/year. OBPA, which currently holds the largest market share of all the biocides, is a mature market, growing at half that rate. Other biocides, such as isothiazolin and TCPP, will grow at a much faster rate than OBPA.

Most of this growth in biocides is attributed to increased consumer awareness. The end-use customers are now demanding that nontraditional biocide applications, like door handles, hospital chair rails, garden hoses, and blue ice packs, incorporate biocides to "protect" them from germs. Consumers seem, in some cases, to be misinformed about the true function of a biocide since it is intended to protect the plastic,

TABLE 4.6 Selected Suppliers of Active Ingredients and Formulated Biocides for Plastics

Supplier	Type	Active(A)/formulated(F)
Akcros (Akzo)	OBPA	F
Akzo Nobel	OBPA	A
Allied Resinous Products	Triclosan	F
Ciba Specialty	Triclosan	A
Creanova	Folpet, Captan	A,F
Ferro	Isothiazolin	F
Microban	Triclosan	F
Morton	OBPA, Isothiazolin, and Triclosan	F
Olin	Zinc OMADINE	A,F
Rohm and Haas	Isothiazolin	A
Sanitized, Inc.	Triclosan	F
Thomson Research	Triclosan	F
Witco	OBPA	F
Zeneca	Isothiazolin	F

not the consumer. Suppliers need to be cautious regarding product claims to avoid misinformation. However, this increased awareness does appear to be a long-term trend and not solely a fad.

4.7 Chemical Blowing Agents

4.7.1 Description

The term *blowing agent* in the broadest sense denotes an inorganic or organic substance used in polymeric materials to produce a foam structure. There are two major types of blowing agents: physical and chemical.

Physical blowing agents. *Physical blowing agents* are volatile liquids or compressed gases that change state during processing to form a cellular structure within the plastic matrix. The gases or low-boiling liquids that are dissolved in the resin, evaporate through the release of pressure or the heat of processing. The compounds themselves do not experience any chemical changes. Cell size is influenced by the pressure of the gas, the efficiency of dispersion, melt temperature, and the presence of nucleating agents. The most common gases used are carbon dioxide, nitrogen, and air. The liquid blowing agents are typically solvents with low boiling points, primarily aliphatic hydrocarbons and their chloro- and fluoro- analogs.

The blowing agents should be soluble in the polymer under reasonably achievable conditions but excessive solubility is not desirable. The permeability of the gas within the polymer is also significant, as is the volume of gas released per unit weight of agent. This latter measure is called the *blowing agent efficiency,* and is an important yardstick for all types of materials. Effective blowing agents should yield at least 150 to 200 cm^3 of gas (measured at standard temperature and pressure) per gram of agent.

Physical blowing agents comprise over 90% of the market. They are heavily used in thermoset foams, especially polyurethanes, polyesters, and epoxies. These additives also have some application in such low-density thermoplastics as polystyrene. Until recently, fluorocarbons had the highest consumption among the liquid physical blowing agents. Because of environmental concerns, the market is shifting to alternative blowing agents, primarily partially halogenated chlorofluorocarbons.

Chemical blowing agents. *Chemical blowing agents* (CBAs) are products that decompose at high temperature. At least one of the decomposition products is a gas, which expands the plastics material to give a foam structure. The amount and type of the blowing agent influence the density of the finished product and its pore structure. Two types of pore structures are possible: open and closed cell. *Closed-cell plastics* have discrete, self-contained pores which are roughly spherical. *Open-celled plastics* contain interconnected pores, allowing gases to pass through voids in the plastic.

Factors that determine the formation of a fine-celled plastic foam with a regular cell structure are the particle size of the blowing agent, dispersion properties of the plastics processing machine used, decomposition rate of the blowing agent, and the melt viscosity of the resin processed.

CBAs are mainly solid hydrazine derivatives. The gas formation must take place in a temperature range close to the processing temperature range of the polymer. In addition, the decomposition products must be compatible with the polymer. Typically, these additives decompose over a relatively narrow temperature range. CBAs can be mixed with the polymer at room temperature, requiring no special processing equipment. In most operations, they are self-nucleating and are stable under normal storage conditions. In addition, CBAs may be reformulated with such other additives as blowing agent catalysts or nucleating agents. Blowing agent catalysts lower the temperature of decomposition for the CBAs while nucleating agents provide sites for formation of a cell in the foamed plastic.

Blowing agents are used in plastics for several reasons: weight reduction, savings in cost and material, and achievement of new properties. The new properties include insulation against heat or noise, different surface appearance, improved stiffness, better quality (removal of sink marks in injection molded parts), and/or improved electrical properties.

CBAs may also be subdivided into two major categories, endothermic and exothermic. *Exothermic blowing agents* release energy during decomposition, while *endothermic blowing agents* require energy during decomposition. In general, endothermic CBAs generate carbon dioxide as the major gas. Commercially available exothermic types primarily evolve nitrogen gas, sometimes in combination with other gases. Nitrogen is a more efficient expanding gas because of its slower rate of diffusion through polymers compared to carbon dioxide.

Exothermic blowing agents. Once the decomposition of exothermic blowing agents has started, it continues spontaneously until the material has been exhausted. As a result, parts that are being foamed with this type of agent must be cooled intensely for long periods of time to avoid postexpansion.

Azodicarbonamide (AZ). The most widely used exothermic CBA is *azodicarbonamide.* In its pure state, this material is a yellow-orange powder, which will decompose at about 390°F. Its decomposition yields 220

cm^3/g of gas, which is composed mostly of nitrogen and carbon monoxide with lesser amounts of carbon dioxide and, under some conditions, ammonia. The solid decomposition products are off-white, which not only serves as an indicator of complete decomposition but also does not normally adversely affect the color of the foamed plastic. Unlike many other CBAs, AZ is not flammable. In addition, it is approved by the FDA for a number of food-packaging uses. AZ can be used in all processes and with most polymers, including PVC, PE, PP, PS, ABS, and modified polyphenylene oxide (PPO).

Modified AZ. *Modified AZ systems* have been developed which offer improved performance and increase versatility in a wide variety of applications. Each system has a formulated cell nucleation system (usually silica) and gas yield is approximately the same as unmodified AZ. Modified types are also available in several particle size grades.

The simplest form of modified AZ is a paste. It is composed of a plasticizer, which forms the liquid phase, and may also contain dispersing agents and catalysts. Its principal field of application is the expansion of PVC plastisols. The agents facilitate the dispersion of the blowing agent when it is stirred into the PVC plastisol, while catalysts lower the decomposition temperature.

Other modified AZs have been developed for the manufacture of integral-skin foams by extrusion and injection molding. These contain additives that modify the usual decomposition process of AZ and suppress the formation of cyanuric acid, which causes plateout on the surfaces of molds, dies, and screws. The additives used include zinc oxide and/or silicic acid (a colloidal silica) with a very low water content. The additives also act as nucleating agents, producing a cell structure that is both uniform and fine-celled.

There are also grades that have been flow-treated. This type contains an additive to enhance the flowability and dispersability of the powder. These grades are very useful in vinyl plastisols, where complete dispersion of the foaming agent is critical to the quality of the final foamed product.

Another method of modifying AZ is to mix it with such other CBAs as those from the sulfonyl hydrazide group. These "auxiliary" blowing agents decompose at lower temperatures than AZ, broadening the decomposition range.

Sulfonyl hydrazides. *Sulfonyl hydrazides* have been in use as CBAs longer than any other type. The most important sulfonyl hydrazide is 4,4′-oxybis (benzenesulfonyl hydrazide) (OBSH). OBSH is the preferred CBA for low-temperature applications. It is an ideal choice for the production of LDPE and PVC foamed insulation for wire where it does not interfere with electrical properties. In addition, it is capable of cross-linking such unsaturated monomers as dienes. Additional applications include PVC plastisols, epoxies, phenolics, and other thermosetting resins. Like AZ, it is approved by the FDA for food-packaging applications and is odorless, nonstaining, and nontoxic.

Sulfonyl semicarbazides. *Sulfonyl semicarbazides* are important CBAs for use in high-temperature applications. TSS (*p*-toluene sulfonyl semicarbazide) is in the form of a cream colored crystalline powder. Its decomposition range is approximately 440 to 450°F with a gas yield of 140 cm^3/g, composed mostly of nitrogen and water. TSS is flammable, burning rapidly when ignited and producing a large amount of residue. TSS is used in polymers processed at higher temperatures such as ABS, PPO, polyamide (PA), and HIPS.

Dinitropentamethylene tetramine (DNPT). *Dinitropentamethylene tetramine* is one of the most widely used CBAs for foamed rubber. Its use is limited in plastics because of its high decomposition temperature and the unpleasant odor of its residue. DNPT is a fine yellow powder that decomposes between 266 and 374°F, producing mainly nitrogen and a solid white residue.

Endothermic blowing agents. *Endothermic* CBAs are used primarily in the injection molding of foam where the rapid diffusion rate of carbon dioxide gas through the polymers is essential. This allows postfinishing of foamed parts right out of the mold without the need for a degassing period. Nucleation of physically foamed materials, especially those used for food packaging, has become a well-established application area for endothermic CBAs.

Sodium borohydride ($NaBH_4$). *Sodium borohydride* is an effective endothermic blowing agent because its reaction with water produces 10 to 20 times the amount of gas produced by other CBAs that give off nitrogen. Sodium borohydride must be blended with the polymer to be foamed to prevent reaction with water during storage.

Sodium bicarbonate ($NaHCO_3$). *Sodium bicarbonate* decomposes between 212 and 284°F giving off CO_2 and H_2O and forming a sodium carbonate residue. Its gas yield is 267 cm^3/g. At 287°F or higher, decomposition becomes more rapid, facilitating its use as a blowing agent for such higher-temperature thermoplastics as styrenic polymers.

Polycarbonic acid. *Polycarbonic acid* decomposes endothermically at approximately 320°F and gives off about 100 cm^3/g of carbon dioxide. Further heating will release even more gas. In addition to being used as the primary source of gas for foaming in some applications, this class of materials is frequently used as a nucleating agent for physical foaming agents.

4.7.2 Suppliers

There are fewer than 50 suppliers of primary chemical blowing agents worldwide. Most of the leading companies have built their chemical blowing agent business over at least 20 years of experience.

Many of the chemical blowing agents suppliers sell their complete product line in a single region and export only selected products. There are no suppliers of chemical blowing agent that have a leading position in all three major regions of the world. Many of the major chemical blowing agents producers are located in the Asia/Pacific region. There are a few dozen chemical blowing agent producers in China alone. Due to the poor logistics in China, the shipment of the chemicals is rather costly, so most of the companies there supply locally.

The leading supplier of chemical blowing agents in North America is Uniroyal Chemical. Bayer is the leading supplier of chemical blowing agents in Europe followed by Dong Jin. Asia/Pacific, the largest consuming region, has numerous suppliers, many selling only in that area of the world. The leading suppliers in this region typically manufacture in more than one country. For example, Dong Jin Chemical and Otsuka Chemical have primary manufacturing locations in Korea and Japan, respectively, but also produce in Indonesia through joint venture partnerships. A list of selected major suppliers of chemical blowing agents globally by type is shown in Table 4.7.

4.7.3 Trends and forecasts

A major concern for producers of AZ type blowing agents is the shortage of the raw material hydrazine. There are few companies globally that manufacture hydrazine and there is currently an insufficient supply to satisfy market demand. However, many leading suppliers like Bayer, Otsuka, and Dong Jin are planning to expand globally. For example, Bayer is doubling its capacity by the year 2000. Its big advantage over most of the leading suppliers, with the exception of Elf Atochem and Otsuka, is that it is backward integrated into hydrazine. Long term, the global expansion of backward integrated CBA suppliers should resolve the hydrazine supply issue.

The annual growth rate globally for chemical blowing agents over the next 5 years is in the 5%/year range.

4.8 Coupling Agents

4.8.1 Description

Coupling agents are additives used in reinforced and filled plastic composites to enhance the plastic–filler-reinforcement interface to meet

TABLE 4.7 Selected Suppliers of Chemical Blowing Agents

Supplier	Type				
	AZ*	TSS†	OBSH‡	DNPT§	Other
Bayer	×	—	—	—	—
Boehringer Ingelheim	—	—	—	—	—
Dong Jin Chemical	×	×	×	×	×
Eiwa Chemical Industry	×	×	×	×	×
Elf Atochem	×	—	—	—	—
Jiangmen Chemical Factory	×	—	—	—	—
Juhua Group	×	—	—	—	—
Kum Yang	×	×	×	—	—
Otsuka Chemical	×	—	—	—	—
Sankyo Kasei	×	×	×	×	—
Shanghai Xiangyang Chemical Industry Factory	×	—	—	×	—
Toyo Hydrazine Industry	×	—	—	—	×
Uniroyal Chemical (Crompton & Knowles)	×	×	×	×	—
Yonhua Taiwan Chemical	×	—	—	—	—
Zhenjiang Chemical Industry Factory	×	—	—	—	—
Zhuxixian Chemical Industry Factory	×	—	—	—	—

*AZ—azodicarbonamide.
†TSS—*p*-toluene sulfonyl semicarbazide.
‡OBSH—4,4′-oxybis (benzenesulfonyl hydrazide).
§DNPT—dinitropentamethylene tetramine.

increasingly demanding performance requirements. In general, there is little affinity between inorganic materials used as reinforcements and fillers and the organic matrices in which they are blended. With silicate reinforcements (glass fiber or wollastonite), silane coupling agents act by changing the interface between the dissimilar phases. This results in improved bonding and upgraded mechanical properties. By chemically reacting with the resin and the filler or reinforcement components, coupling agents form strong and durable composites. Coupling agents significantly improve mechanical and electrical properties for a wide variety of resins, fillers, and reinforcements. In addition, they act to lower composite cost by achieving higher mineral loading.

Fiberglass reinforcement for plastics is the major end use of coupling agents. Thermoset resins, such as polyester and epoxy, account for approximately 90% of coupling agent consumption. Kaolin clay, wollastonite, and glass fiber are the leading fillers or reinforcements chemically treated with coupling agents. Coupling agents are either purchased and applied by the glass fiber or inorganic filler manufacturer or by the compounder for incorporation into the composite system.

Another important market for silane coupling agents is in the cross-linking of polyolefins. In this market silanes are growing at the expense of organic peroxides. Silanes and titanates, along with several minor product types, make up the coupling agent market.

Silanes. *Silanes* comprise more than 90% of the plastic coupling agent market. They can be represented chemically by the formula $Y{-}Si(X)_3$ where X represents a hydrolyzable group such as ethoxy or methoxy and Y is a functional organic group which provides covalent attachment to the organic matrix. The coupling agent is initially bonded to the surface hydroxy groups of the inorganic component by the $Si(X)_3$ moiety—either directly or more commonly via its hydrolysis product, $Si(OH)_3$. The Y functional group (amino, methoxy, epoxy, etc.) attaches to the matrix when the silane-treated filler or reinforcement is compounded into the plastic, resulting in improved bonding and upgraded mechanical and electrical properties.

Table 4.8 lists four different silane chemistries and their related composite systems.

Titanates. *Titanates* are used primarily as dispersing aids for fillers in polyolefins to prevent agglomeration. Titanium-based coupling agents react with free protons at the surface of the inorganic material, resulting in the formation of organic monomolecular layers on the surface. Typically, titanate-treated inorganic fillers or reinforcements are hydrophobic, organophilic, and organofunctional and, therefore, exhibit enhanced dispersibility and bonding with the polymer matrix. When used in filled polymer systems, titanates claim to improve impact strength, exhibit melt viscosity lower than that of virgin polymer at loadings above 50%, and enhance the maintenance of mechanical properties during aging.

TABLE 4.8 Silane Chemistries and Related Composites

Silane type	Resin	Filler or reinforcement
Amino	Phenolic	Alumina
	Phenolic	Silicon carbide
	Acrylic	Clay
	Nylon	Clay
	Nylon	Wollastonite
	Furan	Sand
Epoxy	Epoxy	Alumina trihydrate
Methacrylate	Polyester	Mica
Vinyl	PVC	Clay
	PVC	Talc
	EPDM	Clay

4.8.2 Suppliers

Table 4.9 presents a list of selected suppliers of coupling agents. The two leading suppliers in North America are Witco and Dow Corning. Worldwide, Witco is the leading supplier with a strong presence in Europe, in Asia/Pacific (through a distribution agreement), as well as in North America.

4.8.3 Trends and forecasts

The coupling agent market follows the growth of its three major uses: fiberglass reinforced plastics, plastics compounding, and mineral filler pretreatment. The latter two markets, although smaller than the reinforced polyester area, are leading the growth, which is running at about 6%/year globally.

4.9 Flame Retardants

4.9.1 Description

Flame retardants are in a unique position among plastics additives in that they are both created by regulations and yet are threatened by other regulations. The huge $2.3 billion industry was created over the years by various industry, federal, and state statutes, which aimed to protect people from fire and smoke situations. Indeed, the Underwriters Laboratories (UL), whose standards are integral to the success or failure of flame retardants, were created by the insurance industry. Without these regulations, the plastics industry, which accounts for 85 to 90% ($2 billion) of the global sales of flame retardants, wouldn't use these products because they are expensive and lower the physical properties of the plastics in which they are

TABLE 4.9 Selected Suppliers of Coupling Agents

Supplier	Type		
	Silane	Titanate	Other
Aristech Chemical	—	—	×
Degussa	×	—	—
Dow Corning	×	—	—
Kenrich Petrochemicals	—	×	×
Nippon Unicar	×	—	—
PCR	×	—	—
Rhodia	×	×	×
Shin-Etsu Chemical	×	—	—
Sivento	×	—	—
Uniroyal	—	—	×
Witco	×	—	—

employed. On the other hand, environmental and toxicity concerns now have regulators looking at the important halogenated and antimony-based synergist flame retardants that have been developed over the years. Any regulations which limit the use of such products will again change the industry and force producers to develop a new generation of products.

Flame-retardant additives for plastics are essential safety materials. The transportation, building, appliance, and electronic industries use flame retardants in plastics to prevent human injury or death and to protect property from fire damage. Fundamentally, flame retardants reduce the ease of ignition smoke generation and rate of burn of plastics. Flame retardants can be organic or inorganic in composition, and typically contain either bromine, chlorine, phosphorus, antimony, or aluminum materials. The products can be further classified as being reactive or additive. *Reactive flame retardants* chemically bind with the host resin. *Additive* types are physically mixed with a resin and do not chemically bind with the polymer. Flame retardants are used at loading levels from a few percent to more than 60% of the total weight of a treated resin. They typically degrade the inherent physical properties of the polymer, some types significantly more than others.

Since flame retardants work by minimizing at least one of the requirements for a fire to exist, namely, fuel, heat energy, and oxygen, they also may be classified in another way as follows:

Char formers. Usually phosphorus compounds, which remove the carbon fuel source and provide an insulating layer against the fire's heat.

Heat absorbers. Usually metal hydrates such as aluminum trihydrate (ATH) or magnesium hydroxide, which remove heat by using it to evaporate water in their structure.

Flame quenchers. Usually bromine or chlorine-based halogen systems which interfere with the reactions in a flame.

Synergists. Usually antimony compounds which enhance performance of the flame quencher.

Resin formulators and compounders must select a flame retardant that is both physically and economically suitable for specific resin systems and the intended applications. It is common to formulate resins with multiple flame-retardant types, typically a primary flame retardant plus a synergist such as antimony oxide, to enhance overall flame-retardant efficiency at the lowest cost. Several hundred different flame-retardant systems are used by the plastics industry because of these formulation practices.

Flame retardants consumed in plastics are a diverse group of chemical types and are classified in the major groups, shown in Table 4.10.

Brominated hydrocarbons. *Brominated hydrocarbons* represent the highest dollar volume among all flame retardants used worldwide. The major additive types are decabromodiphenyl oxide (DBDPO) and derivatives of tetrabromobisphenol A (TBA). The major reactive type is TBA itself. Significant amounts of TBA are also used to make additive types. Typically, brominated compounds are used with a synergist

TABLE 4.10 Flame-Retardant Types and Typical Products

Types	Typical products
Brominated	Reactive
	Tetrabromobisphenol A (TBA)
	Brominated polyols
	Tetrabromophthalic anhydride
	Additive
	Decabromodiphenyl oxide (DBDPO)
	TBA derivatives
	Hexabromocyclodecane/dodecane
	Hexabromodiphenoxyethane
	Brominated polystyrene
Phosphate esters	Halogenated
	Pentabromodiphenyl oxide/phosphate ester mixtures
	Tris (chloropropyl) phosphate (TCPP)
	Tris (chloroethyl) phosphate (TCEP)
	Tridichloroisopropyl phosphate (TDCPP)
	Nonhalogenated
	Triaryl phosphates
	Alkyldiaryl phosphates
	Trialkyl phosphates
Chlorinated	Chlorinated paraffins—liquid
	Chlorinated paraffins—resinous
	DECHLORANE PLUS
	Chlorendic anhydride/HET Acid
	Bromochloroparaffins
Alumina trihydrate	—
Antimony oxides	Antimony trioxide
	Antimony pentoxide
	Sodium antimonate
Other flame retardants	Inorganic phosphorus
	Ammonium polyphosphate
	Red phosphorus
	Melamines
	Melamine crystal
	Melamine cyanurate
	Melamine phosphates
	Magnesium hydroxide
	Molybdenum compounds
	Zinc borate

such as antimony oxide in a 3:1 (brominated compound-synergist) ratio. A variety of plastic resins use brominated flame retardants, with HIPS, ABS, and PC being the most prominent resins using these additive types. Epoxies for microchips and circuit boards and unsaturated polyesters are the most important applications for reactives.

Phosphate esters. The *phosphate esters* are divided into halogenated and nonhalogenated types. The halogenated compounds, typically chloroalkyl esters, are used widely in polyurethane foam. The nonhalogenated products, with the triaryl phosphates being the most common, are used as flame retardants in engineering plastics and as flame retardants or plasticizers in PVC. There are also significant quantities of phosphate esters used outside of plastics in textile and lubricant applications. Confusion sometimes exists in the PVC category as to whether these products should be called plasticizers or flame retardants. Typically, phosphate esters are not used with a synergist.

Chlorinated hydrocarbons. Three major product types comprise the *chlorinated hydrocarbons.* The largest volume, but lowest performance category, is the chlorinated paraffins. These products, like the phosphate esters, are used as flame retardants or plasticizers for PVC and in polyurethane foams. There are both liquid and resinous types with the liquids being larger in the previous applications. The resinous types are used in polyolefins, unsaturated polyesters, and some HIPS. The second category is the DECHLORANE PLUS product sold by Occidental. This is a higher-performance product used primarily in polyolefin wire and cable and nylon. The third category is the chlorendic anhydride/acid (HET acid) reactive product which is used in unsaturated polyesters. Like the brominated products, the chlorinated products (other than the HET acid) are used with antimony oxide synergists.

Antimony oxide. A variety of antimony compounds, including antimony trioxide, antimony pentoxide, and sodium antimonate, are combined under the category of *antimony oxides.* These are synergists used in a 1:3 ratio with halogenated flame retardants in typical formulations.

Aluminum trihydrate. *Aluminum trihydrate* is a low-priced commodity that is used at high loadings (up to 50 to 60% on the plastic) as a flame-retardant filler. It is only effective in plastics processed at lower temperatures. Acrylics, polyolefins, PVC, and unsaturated polyesters are the major users. Fully one-third of all ATH is used outside of plastics as a flame retardant in elastomers, carpet backings, and textiles. The major aluminum companies make the basic white hydrate product and sell it to processors who tailor the product for the plastics industry.

Other flame retardants

Inorganic phosphates. *Inorganic phosphates* consist of ammonium polyphosphate and red phosphorus. The ammonium polyphosphate product is primarily used in intumescent coatings and rubber as well as plastics. Red phosphorus is used as a flame retardant in coatings and nylon.

Melamines. *Melamines* consist of melamine crystal, which is used to impart flame-retardant properties to flexible polyurethane foam in upholstery applications, and melamine salts, such as melamine phosphates and melamine cyanurates, used in intumescent coatings and some plastics.

Magnesium hydroxide. *Magnesium hydroxide* is finding increasing use as a replacement for ATH. It is a good smoke suppressant and its price is coming down relative to ATH.

Molybdenum compounds. *Molybdenum compounds* include such products as molybdic oxide or ammonium and metal molybdates and are used in PVC and carpet backings. These products are good smoke suppressants and have been looked at as replacements for antimony oxides.

Zinc borate. *Zinc borate* is the major boron compound used as a flame retardant in plastics. It competes with antimony oxide when antimony prices are high. The largest application for boron compounds as a flame retardant is in cellulose insulation. The flame-retardant categories and the major plastics where they are used are summarized in Table 4.11.

4.9.2 Driving forces

In addition to cost and performance demands, the plastics market for flame retardants is driven by a number of competing forces ranging from fire standard legislation and toxicity regulations to price situations, performance, and other market factors. These combined factors have resulted recently in significant shifts in demand for the major types of flame retardants. Further, large numbers of new flame retardants have emerged, designed for both traditional and specialty niche markets. Recent acquisitions, joint ventures, and alliances by flame-retardant producers have also created constant change in this market. The largest area of activity is in nonhalogenated flame retardants because of environmental concerns associated with the halogen-based products.

TABLE 4.11 Flame-Retardant Types—Major Plastics Applications

Type	Major resins	Comments
Brominated (additive types)	ABS, engineering resins, HIPS, urethane foam	Typically used with antimony synergist; DBDPO is most common product used.
Brominated (reactive types)	Epoxies, unsaturated polyesters	Major use in printed circuits and microchips; TBA is representative product.
Organic phosphate esters	Engineering resins, PVC, urethane foams	Halogenated types typically used in urethane foam; nonhalogenated types used in PVC and engineering resins; synergists not used with phosphate esters.
Chlorinated hydrocarbons	Engineering resins, polyolefins, PVC, urethane foams	Chlorinated paraffins used in PVC as plasticizer/flame retardant and in urethane foam; higher-performance types used in polyolefin wire/cable and engineering resins.
Antimony oxide	ABS, engineering resins, HIPS, polyolefins, PVC	Synergist used with brominated and chlorinated flame retardants.
Aluminum trihydrate (ATH)	Acrylic (counters and panels), polyolefins, PVC, unsaturated polyesters, urethane	Used at high loadings in plastics with low process temperatures; significant foams uses outside of plastics
Inorganic phosphates	Nylon, unsaturated polyesters, polyolefins	Consists of ammonium-phosphates and red phosphorus; uses outside of plastics in textiles and intumescent coatings.
Melamines	Nylon, polyolefins, urethane foam	Mainly melamine crystal for flexible urethane foam; some melamine salts (cyanurates, phosphates) used in plastics and intumescent coatings.
Magnesium hydroxide	Polyolefins, PVC	Replacement for ATH in wire/cable applications; good smoke suppressent.
Molybdenum compounds	PVC	High-priced replacement for antimony oxide; used in some PVC wire/cable.
Zinc borate	Cellulose insulation, miscellaneous other plastics	Replacement for antimony oxide.

Most flame-retardant suppliers, even those making halogenated types, are focusing their product research and development on nonhalogenated products. Since the impact of this research on the markets for halogenated products is still beyond 5 years, halogenated flame retardants are still expected to show a healthy growth pattern at least through 2005. New halogenated products are still being introduced.

The environmental scrutiny that has impacted halogenated flame retardants has primarily focused on brominated diphenyl oxides such as DBDPO. There is concern that these compounds release dioxins when burned. Activity has primarily been in Europe. Currently there are no legislative bans or limits on halogenated flame retardants anywhere in the world, and there are not any on the near-term horizon. However, there are some voluntary bans on selected brominated compounds (particularly DBDPO and related types) in some of the "green" countries of Europe. In many cases, these brominated products are replaced by other brominated products that are not under immediate suspicion.

This pressure is primarily political and is coming basically from the Green Parties in northern Europe (Scandinavia), Germany, and The Netherlands. In these countries, voluntary Eco-labels (ecology) have been developed for branded consumer products. In Scandinavia, the label is named White Swan, while in Germany and The Netherlands, the name Blue Angel is used. This trend has gained a substantial following from the environmental movement to eliminate chlorofluorocarbons (CFCs) and other chlorinated materials, augmented by the parallel movement against plasticizers in PVC cling film packaging.

These trends in Europe are translatable to the other parts of the world. There is some activity in Japan and Canada, and within a 5- to 10-years time span, some impact could be felt in the United States. Eventually, this movement could lead to regulations on halogenated products around the world.

Down the road, the need and the market exists for nonhalogenated approaches to the flame retarding of plastics. All the major flame-retardant companies, including those making halogenated types, are working in the area. Viable, nonhalogenated flame-retardant products do exist, but customers are reluctant to sacrifice the cost/performance advantage of brominated products. Organic phosphate, inorganic phosphorus, melamine salts, and inorganic metal hydrate approaches seem to be the major directions being followed to develop nonhalogenated alternatives.

4.9.3 Suppliers

There are at least 100 suppliers worldwide that are involved in some phase of the flame-retardant business. Most suppliers are involved in only one type of flame retardant, although in the past 2 years, the major brominated supplier (Great Lakes) and the major chlorinated supplier (Occidental) have both acquired an antimony oxide synergist supplier. Some of the major suppliers are basic in raw materials, such as bromine, phosphorus, or alumina, but others buy these materials. Backward integration into raw materials seems to be more prevalent on the brominated side with all three major producers (Great Lakes, Dead Sea Bromine, and Albemarle) integrated back to bromine. This is likely to be a criteria for long-term success on the brominated side.

Brominated and chlorinated flame retardants are sold throughout the world by the major producers or their affiliates. The phosphorus flame retardants are more likely to be sold through regional producers particularly in the Asia/Pacific region. Antimony oxide producers, other than Anzon, are regional, although most of the crude material is sourced from the same place, namely, China. ATH is produced by the major aluminum companies, but is upgraded and treated by other processors who sell to the plastics industry.

A global listing of the major flame-retardant suppliers is provided in Table 4.12. Included are the types that each supplies.

4.9.4 Trends and forecasts

The flame-retardant business has historically outpaced many segments of the plastics additives business as new regulations on fire protection were enacted. This trend will continue, especially in Asia/Pacific, Latin America, Africa, and the Middle East, since regulations regarding fire safety are in their infancy in these regions. Growth in North America, Europe, and Japan will still be healthy but lag behind the other regions. The global market for flame retardants in plastics should grow at a rate of 5%/year over the next 5 years. North America and Europe will see growth in the 3 to 4%/year range while Asia/Pacific (for example, Japan) and the rest of the world will grow at 2 or 3 times the North America/Europe rate.

In spite of continued commentary on the undesirability of halogenated compounds, brominated flame retardants are still expected to pace the growth of the overall flame-retardant market over the next 5 years. Other than in Europe, the growth rate should exceed 5% in all regions. Phosphate ester growth has slowed down in North America and Europe, but rapid growth in the other regions will push overall growth to about 4%/year. The chlorinated flame retardants are suffering in Europe and will likely grow slowly, if at all, worldwide during the period. Antimony oxide growth will not keep pace with brominated flame-retardant growth, but still should increase at a rate of 3%/year worldwide. ATH will show a healthy 5%/year growth as it replaces halo-

TABLE 4.12 Selected Suppliers of Flame Retardants in Plastics

Supplier	Type					
	Brominated hydrocarbons	Phosphate esters	Chlorinated hydrocarbons	Antimony oxides	ATH	Others
Ajinomoto Company	—	×	×	—	—	—
Akzo Nobel	—	×	—	—	—	×
Albemarle Corporation	×	—	—	—	—	—
Albright & Wilson	—	×	—	—	—	×
Alcan Chemicals	—	—	—	—	×	×
Alcoa	—	—	—	—	×	—
Aluchem	—	—	—	—	×	—
Aluminum Pechiney	—	—	—	—	×	—
Amspec Chemical	—	—	—	×	—	—
Asaha Glass	×	—	—	—	—	×
Asahi Denka Kogyo	—	—	×	—	—	—
Bayer	—	×	—	—	—	—
Campine	—	—	—	×	—	—
Clariant	—	—	—	—	—	×
Climax Performance Materials	—	—	—	—	—	×
Courtaulds	—	×	—	—	—	—
Custom Grinders	—	—	—	—	×	—
Daihachi Chemical	—	×	—	—	—	—
Daiichi Kogyo Seiyaku	×	—	—	—	—	—
Dainippon Ink & Chemicals	×	—	—	—	—	—
Dead Sea Bromine	×	—	—	—	—	×
Dover Chemical	×	—	×	—	—	—
Dow Chemical	—	—	—	—	—	×
DSM	—	—	—	—	—	×
Elf Atochem	×	—	—	—	—	—
Ferro Corporation	×	—	×	—	—	—
FMC Corporation	—	×	—	—	—	—

Great Lakes Chemical/ Anzon	×	×	—	×	—	
ICI	—	—	×	—	—	—
Industry Chimiche Caffaro	—	—	×	—	—	—
J.M. Huber	—	—	—	—	×	×
Kyowa Chemical Industry	—	—	—	—	—	×
Manac	×	—	—	—	—	—
Martin Marietta	—	—	—	—	—	×
Martinswerke (Lonza)	—	—	—	—	×	×
Melamine Chemicals	—	—	—	—	—	×
Mikuni Smelting & Refining	—	—	—	×	—	—
Mines de la Lucette	—	—	—	×	—	—
Mitsubishi Gas Chemicals	×	—	—	—	—	—
Morton International	—	—	—	—	—	×
Nabaltec	—	—	—	—	×	—
Nihon Seiko	—	—	—	×	—	—
Nippon Chemicals	×	—	—	—	—	—
Nyacol Products	—	—	—	×	—	—
Occidental/Laurel	—	—	×	×	—	—
Sherwin Williams	—	—	—	—	×	×
Showa Denko	—	—	×	—	×	—
Société Industrielle et Chimique	—	—	—	×	—	—
Solutia	—	×	—	—	—	×
Stibiox	—	—	—	×	—	—
Sumitomo Chemical	—	—	—	—	×	—
Sumitomo Metal Mining	—	—	—	×	—	—
Teijin Chemicals	×	—	—	—	—	—
Tosoh	×	—	×	—	—	—
U.S. Borax & Chemical	—	—	—	—	—	×
United States Antiomy Sales	—	—	—	×	—	—
Velsicol	—	—	×	—	—	—
Witco	—	—	×	—	—	—

gen-based products where it can. Look for some rapid growth in the other flame-retardant categories, particularly melamines, inorganic phosphates, and magnesium hydroxide. The latter is replacing ATH in some applications, and the melamine salts and inorganic phosphates are potential replacements down the road for halogenated compounds.

Technologically, efforts will continue to focus on halogen and antimony oxide replacements. Smoke suppression and higher heat stability are also hot topics in the general flame-retardant area. From a competitive standpoint, the acquisition of an antimony oxide business by two major halogenated flame-retardant suppliers is a trend worth watching. Historically, companies have been content to compete within one type, but now more horizontal integration, particularly among products which are used together, might be the way to go. Also, continued efforts by major suppliers to increase their operations in the growing markets of Asia/Pacific and the rest of the world is not to be discounted.

All in all, it will be another period of rapid change in the world of flame retardants over the next 5 years.

4.10 Heat Stabilizers

4.10.1 Description

Heat stabilizers are used to prevent the thermal degradation of resins during periods of exposure to elevated temperatures. Almost all heat stabilizers are used to stabilize PVC, polyvinylidene chloride (PVDC), vinyl chloride copolymers (for example, vinyl chloride/vinyl acetate), and PVC blends (for example, PVC and ABS). Thermal degradation is prevented not only during processing but also during the useful life of the finished products.

There are three major types of primary heat stabilizers, which include:

- Mixed metal salt blends
- Organotin compounds
- Lead compounds

Heat stabilizers belong to one of the two major classes: primary heat stabilizers and secondary heat stabilizers. When heated, chlorinated vinyl resins liberate HCl which causes further polymer degradation and discoloration. *Primary heat stabilizers* function both by retarding this dehydrochlorination and by reacting with liberated HCl to delay progressive degradation.

When mixed metal salts are used as primary heat stabilizers, metallic chlorides are formed by the reaction with labile Cl. These materials have a destabilizing effect that sometimes result in color formation in the resin. To prevent this, *secondary heat stabilizers or costabilizers* are used to scavenge liberated HCl from the PVC resin or to react with the metallic chloride by-products of the primary mixed metal stabilizers.

Of less importance are antimony mercaptides, which find occasional use as low-cost replacements for organotins. The organotin and lead stabilizers are usually present as the only heat stabilizers in the resin formulation. However, the mixed metal stabilizers are used in combination with secondary heat stabilizers. The secondary heat stabilizers are usually organophosphites and epoxy compounds, but polyols and beta diketones are also used. The major types of primary heat stabilizers, along with their end uses are summarized in Table 4.13.

Primary heat stabilizers

Mixed metal stabilizers. *Mixed metal stabilizers* are primarily used in flexible or semirigid PVC products. The most common are barium/zinc (Ba/Zn) metal salts. Typical liquid barium, cadmium, and zinc stabilizer products consist of such salts as octoates, alkylphenolates, neo decanoates, naphthenates, and benzoates. Typical solid barium, cadmium, and zinc stabilizer products consist of the salts of such fatty acids as stearates or laurates. Generally, Ba/Cd products provide the best thermal stability, followed by Ba/Zn and finally Ca/Zn. However, Ba/Cd stabilizers have come under increased environmental and Occupational Safety and Health Administration (OSHA) pressure and are being replaced by cadmium-free products that are usually Ca/Zn and Ba/Zn. Several Ca/Zn stabilizers have been approved by the FDA for use in food-contact applications.

Organotin heat stabilizers. *Organotin heat stabilizers* are used primarily for rigid PVC applications. Individual products usually consist of methyltin, butyltin and octyltin mercaptides, maleates, and carboxylates. Organotin stabilizers may be divided into sulfur-containing and sulfur-free products. Sulfur-containing products (mercaptides) provide excellent overall stabilization properties but suffer from odor and cross-staining problems. The nonsulfur organotins, such as the maleates, are less efficient heat stabilizers but do not suffer from odor problems and provide better light stability. Generally, butyl and methyltins have been used when toxicity is not a concern. Some octyltin mercaptoacetates and maleates, and to a lesser extent methyltin mercaptoacetates, have FDA approval for use in food-contact applications.

Lead heat stabilizers. *Lead heat stabilizers* are used primarily for wire and cable applications. Here they provide cost-effective stabilization while offering excellent electrical insulation properties. Most lead stabilizers are water-insoluble, an advantage in UL-approved electrical insulation applications. Lead stabilizers may be either organic- or inorganic-based products. Selected organic products consist of dibasic

TABLE 4.13 Major Primary Heat Stabilizers

Type	Major end use	Comments
Mixed metal		
Barium/cadmium	Flexible and semirigid PVC applications	Cadmium-based stabilizers are under pressure to be replaced because of toxicological problems.
Barium/zinc	Flexible and semirigid PVC applications	This is the most common type of heat stabilizer benefiting from the cadmium replacement.
Calcium/zinc	Flexible PVC—food-contact applications	Many of these products are sanctioned by the FDA under Title 21, *Code of Federal Regulations*. This will benefit from the trend away from lead.
Organotin		
Butyl	Rigid PVC	Provides excellent heat stability. Most versatile organotin stabilizer.
Methyl	Rigid PVC—particularly for pipe	Very effective stabilizer on a cost-performance basis. Some of these products are sanctioned by the DA under Title 21, *Code of Federal Regulations* for food-contact applications.
Octyl	Rigid PVC—food-contact applications	Several of these products are sanctioned by the FDA under Title 21, *Code of Federal Regulations*.
Lead	Wire and cable	Excellent insulation properties.

lead stearates and phthalates, while some inorganic lead products are tribasic lead sulfate, dibasic lead phosphite, and dibasic lead carbonate. There is increasing pressure to replace lead with other products. However, no suitable cost-effective replacement for lead stabilizers in primary cable insulation applications has been found.

Antimony. *Antimony compounds* are effective at low concentrations as primary heat stabilizers in rigid PVC applications. They have National Sanitation Foundation (NSF) acceptance for use in potable PVC water pipe. A disadvantage of antimony compounds is their poor light stability.

Secondary heat stabilizers

Alkyl/aryl organophosphites. *Alkyl/aryl organophosphites* are often used with liquid mixed metal stabilizers in the stabilization of PVC resin. They prevent discoloration by functioning as chelators of such by-products as barium chloride from the primary heat stabilizers. The use of phosphites as secondary heat stabilizers has many additional benefits. They reduce the melt viscosity, which contributes to smoother and easier processing, and also function as antioxidants. The liquid organophosphites are usually formulated with the liquid-metal stabilizers and sold as convenient one-package systems. Solid mixed metal stabilizers do not contain liquid organophosphites. Typical organophosphites used for heat stabilization include didecylphenyl, tridecyl, and triphenyl phosphites. A few organophosphite products have been given FDA approval for flexible and rigid vinyl applications. An example is tris (nonylphenyl) phosphite (TNPP).

Epoxy compounds. *Epoxy compounds* function both as plasticizers and stabilizers in flexible and semirigid PVC. As stabilizers, epoxies react with liberated HCl. In addition, they react with the polymer chain at labile-chlorine sites—either directly or catalytically by increasing the reactivity of the labile-chlorine site with metal salt stabilizers. Most epoxy stabilizers are derived from unsaturated fatty oils and fatty acid esters. Epoxidized soybean and linseed oils and epoxy tallate are commonly used products. Epoxy tallate also increases light stability. Epoxy compounds can be formulated with metallic liquid stearates and, thus, can be sold to compounders as a one-package system if a constant ratio of stabilizer-to-epoxy is acceptable. However, since these epoxy compounds are also plasticizers, the balance of the formulation must be adjusted for this effect.

Beta diketones. *Beta diketones* are secondary heat stabilizers used in combination with Ca/Zn and Ba/Zn metallic heat stabilizers to improve initial color. Beta diketones usually require the presence of other secondary heat stabilizers such as epoxidized oils and organophosphites.

Polyfunctional alcohols. *Polyfunctional alcohols* are secondary heat stabilizers used in combination with mixed metal products. They function by forming complexes that deactivate the metallic chloride by-products of the primary stabilizers.

4.10.2 Suppliers

There are over 100 suppliers of primary heat stabilizers. The majority of these companies use heat stabilizers as their core product and serve the PVC industry with other additives such as lubricants and organophosphite stabilizers. Many specialty suppliers sell their complete product line in a single region and export selected products. There are no suppliers of heat stabilizers that have leading positions in all three major regions of the world. A major change among suppliers of heat stabilizers took place recently with Witco's acquisition of Ciba's heat stabilizer business in exchange for Witco's epoxy and adhesives businesses. In addition, Akzo recently acquired the remaining half of the Akcros joint venture, a major heat stabilizer supplier. A global list of selected suppliers of heat stabilizers is shown in Table 4.14.

4.10.3 Trends and forecasts

There are several heat stabilizer products that have received environmental scrutiny in selected regions of the world. The European Directives banning the use of cadmium-based stabilizers, due to the effect on human health and the environment, has successfully limited their global use. This forced the industry to find cadmium-free alternatives. Ba/Zn and Ca/Zn are being substituted in the short term. The Ca/Zn material is much less effective but benefits from having two almost nontoxic components that have worldwide approvals. Organotins will experience long-term growth at the expense of Cd. Lead is being phased out in selected regions of the world. However, this will occur over a long period of time.

In response to the concerns regarding the use of heavy metals, producers are developing reduced metal and metal-free organic stabilizer systems. One reduced metal system is based on selected difunctional epoxides and zinc compounds and is reported to perform comparably to commercial lead-based systems.

Completely organic (metal-free) heat stabilizer systems are under development by all major producers. One system undergoing commercial testing is based on heterocyclic ketone compounds (the pyrimidindione ring) with HCl scavenging co-stabilizers. Although relatively insignificant in the present heat-stabilizer business, current environmental pressure might permit materials of this type to achieve 5 to 10% market penetration within the next 5 years.

The growth of heat stabilizers is dependent on PVC growth. Rigid PVC applications are expected to grow at a faster rate than flexible PVC applications worldwide. This indicates organotins will experience higher growth than mixed metals. Over the next 5 years, heat stabilizers are expected to grow at a rate of 6%/year paced by the Asia/Pacific and the developing regions of the world.

4.11 Impact Modifiers

4.11.1 Description

Impact modifiers are used in a wide variety of thermoplastic resins to absorb the energy generated by impact and dissipate it in a nondestructive fashion. The behavior and definition of impact modifiers are complex. The selection of an impact modifier is dependent on compatibility, physical solubility, impact performance, and cost.

Impact modifiers are primarily used in PVC, engineering resins, and polyolefins. The use levels of impact modifiers vary widely depending upon the modifiers, matrix type, and properties desired. The major types are shown in Table 4.15 along with the resins in which they are primarily used.

TABLE 4.14 Selected Heat Stabilizer Suppliers

Supplier	Type		
	Mixed metal	Organotin	Lead
Akzo	×	×	×
Asahi Denka Kogyo K.K.	×	×	—
Baerlocher	×	—	—
BASF	×	×	×
Cardinal Chemical	—	×	—
Chemson	×	×	×
Clariant/Hoechst	×	—	—
Dainippon Ink and Chemicals	×	×	—
Elf Atochem	×	×	—
Ferro	×	×	×
Hammond Lead	—	—	×
Kolon Chemical	×	—	×
Kyodo Chemical	×	×	—
Morton International	—	×	×
Nan Ya Plastics	×	×	×
Nanjing Chemical Factory	×	×	×
NOF	×	—	—
OMG	×	—	—
Reagens SpA	×	×	×
Sakai Chemical	×	×	×
Tokyo Fine Chemical	×	×	—
Witco	×	×	—

TABLE 4.15 Major Types of Impact Modifiers by Resin

Type	Resin					
	PVC*	PE**	PP†	PA‡	PET/PBT§	Other
MBS (methacrylate butadiene styrene)	×	—	—	—	—	—
MABS (methacrylate/acrylonitrile-butadiene-styrene)	×	—	—	—	×	—
ABS (acrylonitrile-butadiene-styrene)	×	—	—	—	×	—
CPE (chlorinated polyethylene)	×	×	×	—	—	—
EVA (ethylene vinyl acetate)	×	—	—	—	—	—
PMMA (polymethylmethacrylate)	×	—	—	—	—	×
EPDM (ethylene propylene diene monomer)	×	×	×	—	—	—
EPR (ethylene propylene rubber)	×	×	×	—	—	—
SBR (styrene butadiene rubber)	—	—	—	—	—	×
Others	—	—	—	—	—	—
Maleated EPDM	—	—	—	×	×	—
Maleated PP and PE	—	—	—	×	×	—
PUR (Polyurethane)	—	—	—	—	—	×
SAN-g-EPDM	—	—	—	—	×	—

*PVC—polyvinylchloride.
**PE—polyethylene.
†PP—polypropylene.
‡PA—polyamide.
§PET—polyethylene terephthalate; PBT—polybutylene terephthalate.

Methacrylate-butadiene-styrene (MBS). *Methacrylate-butadiene-styrene* represents the highest volume of the styrenic type impact modifiers. This modifier is used in transparent packaging applications due to its clarity. Rigid applications include film, sheet, bottles, credit cards, and interior profiles. MBS has limited use in exterior applications due to poor ultraviolet (UV) stability. Methacrylate/acrylonitrile-butadiene-styrene (MABS) is closely related to MBS, but has minor use in the industry and has been completely replaced by MBS in North America.

Acrylonitrile-butadiene-styrene (ABS). *Acrylonitrile-butadiene-styrene* is used in a variety of resins, with about 60% in PVC. The primary ABS applications are in automotive parts, credit cards, and packaging. ABS, like MBS, is not suitable for outdoor applications unless it is protected by a UV-resistant cap. ABS, although compatible with MBS, suffers from the disadvantage of not being regarded as an industry standard.

Acrylics. *Acrylics* are similar to MBS and ABS but have butyl acrylate or 2-ethyl-hexyl acrylate graft phases. Acrylics offer greater resistance to UV degradation and are used primarily in PVC siding, window profiles, and other applications calling for weather resistance. Due to growth in the building and construction industry, acrylics are experiencing the highest growth rate.

Chlorinated polyethylene (CPE). *Chlorinated polyethylene* modifiers are most commonly used in pipe, fittings, siding, and weatherable profiles. CPE modifiers compete primarily with acrylics in siding applications. CPE can be used in resins other than PVC, for example, PE and PP.

Ethylene vinyl acetate (EVA). *Ethylene vinyl acetate* modifiers have minor usage compared to other types of impact modifiers. EVA finds use in limited segments of the flexible PVC sheet business.

Ethylene propylene diene monomer (EPDM). *Ethylene propylene diene monomer* is used in thermoplastic olefin (TPO) for automotive bumpers and parts as well as scattered consumer durable markets.

Maleic anhydride grafted EPDM. *Maleic anhydride grafted EPDM* reacts with the matrix resin, typically nylon, to become its own compatibilizer. This type of modifier provides for excellent balance in impact, hardness, modulus, and tensile strength and is the major additive component of "super tough" nylon.

4.11.2 Suppliers

There are over 30 suppliers of impact modifiers worldwide. Most concentrate their efforts in one type of modifier as a result of their developed technologies and backward integration. Selected suppliers resell other producers' technologies in their home regions to broaden their product lines

Rohm and Haas, Kaneka, and Atochem are the leading suppliers of impact modifiers worldwide. Each has strong positions in both the acrylic and MBS-related modifiers. Elf Atochem is stronger in acrylics, while Kaneka is stronger in MBS types. Rohm and Haas, including its joint venture with Kureha in the Asia/Pacific region, has a more balanced position. Table 4.16 presents the major global suppliers of impact modifiers by type.

4.11.3 Trends and forecasts

The need for cost-effective materials that are strong, stiff, and ductile will continue to increase. In many cases the key to success will be the development of tailored impact modifier systems for specific resins.

The EPDM market will probably see a decline over the next couple of years due to the advent of reactor-generated polypropylene. This material incorporates the impact modifier in the polymer chain and does not require a secondary compounding operation.

The MBS market is decreasing partially due to PVC bottles being replaced by PET. This trend is more evident in Europe due to widespread use of water bottles. In contrast, the film and sheet market remain strong. Overall, MBS sales are heavily dependent on the future of PVC, particularly flexible PVC. Flexible PVC, comprising 15% of the total PVC market, is vulnerable to penetration by metallocene catalyzed polyolefins (for example, "super soft polypropylene").

Acrylic impact modifiers will continue to grow with the growth of rigid PVC in the construction market. Product development in this market will target improved low-temperature impact properties to reduce failures, lengthen the installation season, and lower cost.

A significant area for product development is the impact modification of engineering plastics. The replacement of such conventional materials as metal, glass, and wood by plastics has been underway for years. The applications are typically converted to engineering plastics and then lost to lower-cost polyolefins and/or vinyl type materials. Most of the "easy" applications have already converted to plastic. The remaining ones, particularly in durable goods, require new levels of strength and impact performance.

Consumption of impact modifiers worldwide is projected to grow at 5%/year over the next 5 years.

TABLE 4.16 Selected Impact Modifier Suppliers

Supplier	Type					
	Acrylic	ABS/MBS/ MABS	EVA	EPR/ EPDM	CPE	Other
Baerlocher	×	—	—	—	—	—
Bayer	—	×	×	—	—	—
Chisso	—	—	×	—	—	—
DSM Copolymer	—	—	—	×	×	—
Dupont/Dow Elastomers	—	—	×	×	×	—
Elf Atochem	×	×	—	—	—	×
Exxon	—	—	—	×	×	×
GE Specialty Chemicals	—	×	—	—	—	—
Huels	—	—	—	—	×	—
JSR	—	—	—	—	—	×
Kaneka	—	×	—	—	—	×
Kureha	—	—	—	—	—	×
Mitsubishi Rayon	—	—	—	—	—	×
Mitsui Petrochemical	—	—	—	—	—	×
Nippon Zeon	—	—	×	—	—	×
Osaka Soda	—	—	—	—	×	—
Polysar	—	—	—	×	—	—
Rohm and Haas	×	×	—	—	—	—
Shell	—	—	—	—	—	×
Showa Denko K.K.	—	—	—	—	×	—
Sumitomo Chemical	—	×	—	×	—	—
Toyo Soda	—	—	—	×	—	—
Ube Cycon	—	×	—	—	—	—
Uniroyal	—	—	—	×	—	×

4.12 Light Stabilizers

4.12.1 Description

Light stabilizers are used to protect plastics, particularly polyolefins, from discoloration, embrittlement, and eventual degradation by UV light. The three major classes of light stabilizers are UV absorbers, excited state quenchers, and free-radical terminators. Each class is named for the mechanism by which it prevents degradation. The major types included in each light stabilizer class may be categorized by their chemistries, as shown in Table 4.17.

Benzophenone. *Benzophenone UV absorbers* are mature products and have been used for many years in polyolefins, PVC, and other resins. These products also have wide use in cosmetic preparations as sunscreens and protectants.

Benzotriazole. *Benzotriazole UV absorbers* are highly effective in high-temperature resins such as acrylics and polycarbonate. They also find extensive use in areas outside plastics such as coatings.

Benzoates and salicylates. *Benzoates* and *salicylates* such as 3,5-di-*t*-butyl-4hydroxybenzoic acid *n*-hexadecyl ester, function by rearranging to 2-hydroxybenzophenone analogs when exposed to UV light to perform as UV absorbers.

Nickel organic complexes. *Nickel organic complexes* protect against degradation caused by UV light via excited state quenching. These deactivating metal ion quenchers stop the energy before it can break any molecular bonds and generate free radicals. Nickel complexes are primarily used in polyolefin fiber applications. Some examples of nickel complexes are nickel dibutyldithiocarbamate and 2,2′ thiobis (4-octylphenolato)-*n*-butylamine nickel II which are also used in agricultural film because of their resistance to pesticides.

Hindered amine light stabilizers (HALS). *Hindered amine light stabilizers* are the newest type of UV light stabilizer. They were introduced in 1975 by Ciba and Sankyo. HALS do not screen ultraviolet light, but stabilize the resin via free-radical termination. HALS are used at lower levels than benzophenones and benzotriazoles, and are widely used in polyolefins for their cost-effectiveness and performance. The successful growth of HALS has been directly related to their substitution for benzophenones and benzotriazoles in many applications as well as their blending with benzophenones.

TABLE 4.17 Major Types of Light Stabilizers

Type	Representative chemistry
UV light absorbers	
Benzophenone	2-hydroxy-4-methoxybenzophenone
	2-hydroxy-4-*n*-octoxybenzophenone
	2,4-dihydroxy-4-*n*-dodecycloxybenzophenone
Benzotriazole	2,2-(2-hydroxy-5-tert-octylphenyl) benzotriazole
	2-(3′-tert-butyl-2-hydroxy-5-methylphenyl)-5-chlorobenzotriazole
	2-(3′,5′-di-tert-butyl-2′-hydroxyphenyl)-5′-chlorobenzotriazole
	2-(2′hydroxy-3′-5′-di-tert amyl phenyl) benzotriazole
	2-(2-hydroxy-5-methylphenyl) benzotriazole
Phenyl esters	3,5-di-t-butyl-4hydroxybenzoic acid *N*-hexadecyl ester
Diphenylacrylates	Ethyl-2-cyano-3,3-diphenyl acrylate
	2-ethylhexyl-2-cyano-3,3-diphenyl acrylate
Excited state quenchers	
Nickel compounds	Nickel dibutyldithiocarbamate
	2,2′-thiobis (4-octylphenolato)-*n*-butylamine nickel II
Free-radical terminators	
Hindered amine light stabilizers (HALS)	Bis (2,2,6,6-tetramethyl-4-piperidinyl)
	N,N-bis(2,2,6,6-tetramethyl-4-piperidinyl)-1,6-hexane diamine polymer with 2,4,6-trichloro-1,3,5 triazine and 2,4,4-trimethyl-1,2-pentanamine

4.12.2 Suppliers

There are about 40 suppliers of light stabilizers worldwide. Some of these companies also produce antioxidants and PVC heat stabilizers. Of these 40 or so suppliers, only Ciba Specialty Chemicals is a significant player in every region of the world with the broadest product line of light stabilizers. Selected global suppliers of light stabilizers are given in Table 4.18.

4.12.3 Trends and forecasts

The entrance of Great Lakes into the European light stabilizer market with a series of acquisitions has been the most significant restructuring that has occurred in the light stabilizer market. This move has accelerated the trend toward a more competitive market in these materials.

Growth in the light stabilizer business is strongly dependent on the growth of the polyolefin applications. Polyolefins account for about three-quarters of the total global consumption of light stabilizers in plastics. Polyolefins, particularly PP, are replacing metals, engineer-

TABLE 4.18 Selected Light Stabilizer Suppliers

Supplier	Type			
	HALS	Benzotriazole	Benzophenone	Others
3V Sigma	×	—	×	—
Akcros (Akzo)	—	×	×	—
Asahi Denka Kogyo	×	×	×	—
Asia Stabilizer	—	×	×	—
BASF	×	—	×	×
BF Goodrich	×	—	—	—
Chemipro Kasei Kaisha	—	×	×	×
Ciba Specialty Chemicals	×	×	×	×
Clariant	×	×	×	×
Cytec Industries	×	×	×	—
Dai-ichi Chemical Industries	—	—	—	×
Dainippon Ink and Chemicals	—	—	—	×
Eastman Chemical	—	—	—	×
Elf Atochem	×	—	—	—
Everlight Chemical Industrial	×	×	×	—
Fairmount Chemical	—	×	—	—
Ferro	—	—	×	×
Great Lakes Chemical	×	×	×	—
Honshu Chemical	—	—	×	—
Iwaki Seiyaku	—	—	—	×
Johoku Chemical	—	×	—	—
Kolon Industries	—	—	×	—
Korea Fine Chemicals	—	—	×	—
Kyodo Chemicals	—	×	×	×
Liaoyang Organic Chemical	×	—	—	×
Mitsubishi Petrochemical	—	—	—	×
Musashino Geigy	×	×	—	—
Nissan Ferro Organic Chemical	—	—	—	×
Osaka Seika Chemical Ind.	—	×	×	—
Sakai Chemical Industry	—	—	—	×
Sankyo	×	—	×	—
Shipro Kasei	—	×	×	×
Shonan Kagaku Kogyo	—	—	×	—
Sumitomo Chemical	—	×	×	×
Witco	—	—	×	—
Yashiro Seiyaku	—	—	—	×
Yoshitomi Fine Chemicals	×	×	×	—

ing plastics, and styrenics in automotive and other applications, further increasing the volume of stabilizers consumed.

The use of nickel-containing stabilizers is decreasing in the marketplace, particularly in North America, due to potential toxicity concerns. In Europe, nickel continues to be used in agricultural film applications.

Design efforts are focusing on down-gauging of exterior plastic parts for weight and cost reduction. This will place increased value on light stabilization to maintain adequate performance at thinner wall sections.

HALS will experience the strongest growth due to their widespread use in polyolefins and their cost-effectiveness and performance. Benzotriazoles and benzophenones, however, are more effective than HALS in vinyl and engineering plastics.

Significant product development work is being done in HALS technology to produce higher-performance products in polyolefin systems. Low molecular weight alkoxy substituted amine systems and higher molecular weight HALS stabilizers significantly improve the performance of pigmented TPO parts with regard to color and gloss retention.

HALS are being promoted by selected suppliers as effective light stabilizer with excellent capabilities as antioxidants. In some cases, these materials are comparable to well-established antioxidant products such as Ciba's IRGANOX 1010.

Suppliers continue to improve on the physical forms of light stabilizers. For example, Cytec is introducing a flake form light stabilizer which reduces dusting and increases the shelf life of the products.

Consolidation is expected to continue due to margin pressures caused by regulatory issues such as FDA compliance, toxicological testing, environmental compliance, and the continual need for capital investment. This trend may be most apparent in the Asia/Pacific region where there are a large number of small suppliers.

Globally, light stabilizers should grow at a rate of 7%/year over the next 5 years, with the less developed regions in Asia/Pacific, Latin America, and Africa leading the way. This robust growth parallels the growth of polyolefins, particularly polypropylene/TPO, and engineering resins into more exterior applications replacing metal and painted plastic.

4.13 Lubricants and Mold Release Agents

4.13.1 Description

Lubricants. *Lubricants* represent a broad class of materials that are used to improve the flow characteristics of plastics during processing. Besides this primary task of improving flow properties, lubricants can act as melt promoters, antiblock, antitack, and antistatic agents as well as color and impact improvers. They can be used in conjunction with metal release agents and heat stabilizers. Lubricants are widely used in packaging film to prevent sticking to the metal processing equipment. Lubricants can improve efficiency by lowering the resin melt viscosity, resulting in reduced shear and equipment wear, increased rate of production, and decreased energy consumption.

Selection of lubricants is dependent upon the type of polymer as well as the process by which it is manufactured. The method of selection is easier when the manufacturing process is fully developed. Lubricant choices for new processes require careful experimentation.

The selection process is driven by the lubricant's compatibility with the hot resin, lack of adverse effects on polymer properties, good transparency, regulatory approval, and the balance of other additives in the polymer. The amount of lubricant used can also affect the final polymer properties. Overlubrication can cause excessive slippage and underlubrication can cause degradation and higher melt viscosities.

The two general classifications of lubricants are internal and external. *External lubricants* do not interact with the polymer but function at the surface of the molten polymer between the polymer and the surface of the processing equipment and are generally incompatible with the polymer itself. These lubricants function by coating the process equipment and reducing friction at the point of interface. They delay fusion and give melt control and the desired polymer flow to such applications as rigid PVC pipe, siding, and window frames.

Internal lubricants are usually chemically compatible with the polymer and act by reducing friction between polymer molecules. They reduce van der Waals forces, leading to lower melt viscosity and lowering energy input needed for processing.

Several chemicals are used as both internal and external lubricants since lubricants can function at several different points during polymer processing. When used during the blending portion of processing, they are usually waxy substances that coat the surface of resin pellets allowing easier movement through the cold portions of the processing equipment. As the polymer mix is heated, the lubricant softens, melts, and penetrates the polymer. The rate of penetration is dependent upon the solubility of the particular lubricant in the specific polymer.

Metallic stearates. *Metallic stearates* are the most widely used lubricants. They are utilized predominantly in PVC, but also find use in polyolefins, ABS, polyesters, and phenolics. The primary disadvantage of metallic stearates is their lack of clarity. Calcium stearate, the most common metallic stearate, is primarily used as an internal lubricant, but in PVC applications, it provides external lubricant and metal release characteristics while also acting as a heat stabilizer.

Esters. *Esters,* including fatty esters, polyol esters, and even wax esters, are reasonably compatible with PVC. They are also used in polystyrene and acrylic polymers. High molecular weight esters are used as external lubricants; conversely, low molecular weight esters are used as internal lubricants, although they are somewhat inefficient as either.

Fatty amides. *Fatty amides* possess unique mold release properties. Simple primary fatty amides are used as slip and mold release agents primarily in polyolefins but also in a variety of other polymers. The more complex bis-amides, such as ethylene bis-stearamide, offer mold release as well as internal and external lubricity functions in materials such as PVC and ABS.

Fatty alcohols. *Fatty alcohols* are used primarily in rigid PVC. Because of their compatibility and internal and external lubricant capabilities, they are chosen where clarity is important.

Waxes. *Waxes* are nonpolar and are, therefore, very incompatible with PVC which makes them excellent external lubricants for this material. Partially oxidized PE wax works well as an external lubricant for PVC by delaying fusion and is almost always combined with calcium stearate for melt flow control. Although the primary function of waxes, as well as metallic soaps, fatty acid esters, and amides is lubrication, they are in fact multifunctional, as noted previously, providing slip, antiblock, and mold release properties.

Mold release agents. When a plastic part tends to stick in the mold, a *mold release agent* is applied as an interfacial coating to lower the friction. Improper mold release can lead to long cycle times, distorted parts, and damaged tooling. The two types of mold release agents are internal and external.

Internal mold release agents are mixed directly into the polymer. These materials have minimal compatibility with the polymer. The additive either migrates to the surface of the polymer and sets up a thin barrier coating between the resin and mold cavity or is present in a sufficient quantity on the surface of the polymer to reduce adhesion to the mold cavity.

Traditionally, *external release agents* are applied by spraying or painting the surface of the mold with an aerosol, liquid, or by applying a paste. The solvent or water carrier then evaporates leaving a layer of release agent on the mold.

Mold release agents are used in a variety of applications, including fiber-reinforced plastics, castings, polyurethane foams and elastomers, injection-molded thermoplastics, vacuum-formed sheets, and extruded profiles. Because each application has its own plastic, mold material, cycle time, temperature, and final product use, there is no universal release agent. Mold release selection is dependent upon all of these conditions.

Release agents should ideally have high tensile strength so they are not worn by abrasive mineral fillers or glass fiber reinforcements. The agents should also be chemically resistant to decomposition and should stick to the mold to prevent interference with the final product. The major types of materials used as mold release agents are fatty acid esters and amides, fluoropolymers, silicones, and waxes.

Fatty acid esters and amides. *Fatty acid esters* and *amides* do not usually interfere with the secondary finishing operations and some have high-temperature stability making them well-suited for rotational mold resins and engineering plastics.

Fluoropolymers. *Fluoropolymers* form a monolayer providing easy application but are expensive.

Silicones. Although *silicones* are used as both external and internal mold release agents, the primary application is as the active ingredient in external release agents. The silicone is in a solution or aqueous dispersion that is sprayed intermittently into the mold cavity between shots. A disadvantage of silicones as internal release agents is their possible interference with painting and contamination of finish surfaces.

4.13.2 Suppliers

There are numerous suppliers of lubricants and mold release agents as a result of the variety of chemistries that perform the function of internal and external lubrication. The suppliers are generally large specialty chemical companies that sell the particular chemistry to a wide variety of end-use applications. The amount of material sold to function as a lubricant or mold release agent for plastics is typically small in comparison to each company's total sales. Table 4.19 shows the major global suppliers of lubricants and mold release agents by type.

4.13.3 Trends and forecasts

Other than plasticizers, lubricants come closest to being a commodity business within the plastic additives market. Since over 70% of lubricant consumption is directed at PVC for applications such as pipe, siding, and windows, demand will be highly dependent on the construction industry.

The use of lubricants with heat stabilizers, particularly lead types, in "one-pack" systems has not taken off in North America as it has in Europe. North America has focused more on the tin-based stabilizer systems, and customers still prefer buying the additives separately.

TABLE 4.19 Selected Suppliers of Lubricants and Mold Release Agents

Supplier	Type						
	Metallic stearates	Petroleum wax	Fatty amides	Fatty esters	PE wax	Fatty acid /alcohols	Silicones
Akcros (Akzo)	×	—	—	—	—	—	—
AlliedSignal	—	—	—	—	×	—	—
Baerlocher	×	—	—	—	—	—	—
BASF	—	—	—	—	×	—	—
Chemson	×	—	—	—	—	—	—
Clariant	—	×	—	×	×	—	—
Croda	—	—	×	—	—	×	—
Dow Corning	—	—	—	—	—	—	×
Eastman Chemical	—	—	—	×	×	—	—
Elf Atochem	×	—	—	—	—	—	—
Faci	×	—	×	—	—	—	—
Ferro	×	—	—	—	—	—	—
GE Specialty Chemicals	—	—	—	—	—	—	×
Henkel	×	—	—	×	—	×	—
Huels	—	—	—	—	×	—	—
ICI Specialty Chemicals	—	—	×	×	—	×	—
Lonza	—	—	×	×	—	—	—
Morton Plastic Additives	×	—	×	—	×	—	—
Olefina	×	—	×	×	—	—	—
Rhodia	—	—	—	—	—	—	×
Sogis	×	—	×	—	—	—	—
Wacker Silicones	—	—	—	—	—	—	×
Witco	×	×	×	×	—	×	—

Key technology trends in lubricants include the development of high-temperature lubricants and the continuing work on lubricants that are compatible with other additives and colors in the plastic.

Mold release agents are actually a different business than lubricants although there are some related chemistries at the lower end. These products are typically higher-priced formulations and are used primarily in thermoset urethanes, polyesters, and epoxies. The active ingredients are sold by silicone and fluorochemical producers such as Dow Corning, GE Silicones, Wacker, DuPont, and ICI.

Overall, the lubricant and mold release businesses are growing at 4 to 5%/year worldwide.

4.14 Nucleating Agents

4.14.1 Description

Nucleating agents are used in polymer systems to increase the rate of crystallization. These agents are added to partly crystalline polymers and change the polymer's crystallization temperature, crystal spherulite size, density, clarity, impact, and tensile properties. These intentional contaminates achieve these functions by acting as sites for crystalline formation.

Nucleating agents are typically added postreactor and are used primarily in injection molding applications. However, they can also be found in blow molding, sheet extrusion, and thermoforming. They are incorporated into materials such as nylon, PP, crystalline polyethylene terephthalate (CPET), and thermoplastic PET molding compounds at use levels typically below 1%, although CPET uses higher levels. The incorporation of these nucleating agents can be done in several ways, including powder mixtures, suspensions, solutions, or in the form of a masterbatch. Whichever method is used, good dispersion of the nucleating agent throughout the polymer must be achieved to provide the optimal effect. The addition of nucleating agents into polymers yields benefits such as higher productivity and improved optical properties.

Nucleating agents can shorten cycle time by reducing set-up time in the mold. Care must be taken to ensure that shrinkage and impact properties are not negatively affected. With some difficult-to-crystallize thermoplastics, such as partially aromatic polyamides or PET, nucleants are needed to obtain useful parts with reasonable cycle times and mold temperatures.

The optical benefits of nucleating agents are increased clarity and improved gloss. These properties improve because of an increase in the number of fine crystals. When crystals are smaller than the wavelength of visible light, the light is scattered at smaller angles, decreasing the hazy effect seen when nucleating agents are not used. When utilized to improve transparency in materials such as PP, these materials are referred to as *clarifiers or clarifying agents.* An example of how clarifiers work is depicted in Fig. 4.1.

Types. Several different types of nucleating agents are used in specific polymers, as shown in Table 4.20. The four major categories of chemical nucleating agents are substituted sorbitols, low molecular weight polyolefins, sodium benzoate, and ionomer resins. In addition, a variety of mineral fillers, reinforcements, and pigments are used in nylon and other polymers. These nonchemical nucleating agents are easily dispersed, inexpensive, and typically available "on-site" since they are commonly used for their primary reinforcing and filling function.

Substituted sorbitols. *Substituted sorbitols* are used in polyolefins, particularly PP, for nucleation and clarification purposes. They have varying degrees of miscibility in PP and different melting points and process temperatures as well as odor. Both homopolymers and random copolymers of PP use sorbitols. Use levels range from 0.1 to 0.3% on the polymer. The FDA has regulated the use of substituted sorbitols, but has given its approval for their use in PP. These materials are used in injection molded housewares, medical devices, and protective packaging. Smaller amounts are used in blow-molded bottles.

Low molecular weight polyolefins. *Low molecular weight polyolefins* are primarily used in CPET for rapid crystallization of otherwise amorphous material. These products are typically sold by the CPET suppliers in a package along with the base resin. Use levels are higher than with the sorbitols and average 1 to 3% of the resin. The major application is in

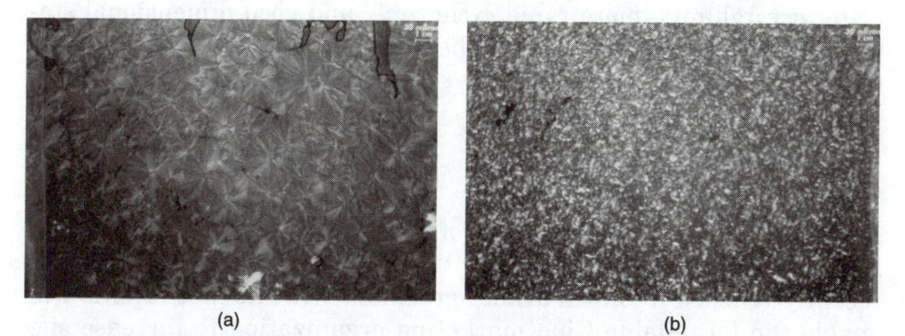

Figure 4.1 How clarifiers work: Conventional homopolymer PP (*a*) consists of large uneven "crystal" microstructures that refract light and increase opacity. Sorbitol clarifiers, (*b*) generate smaller, highly dispersed crystallites which are smaller than the wavelength of light. The result is a clarified PP in which the haze percentage falls; clarity and surface gloss are boosted. (*Courtesy Ciba Specialty Chemicals.*)

TABLE 4.20 Nucleating Agents Used in Specific Polymers

Polymer	Nucleating agents
Polyethylene terephthalate (PET/CPET)	Inert mineral fillers, chalk, clay, talc, silicates, carbonates, pigments
	Organic compounds, carboxylic acids, diphenylamine
	Polymers, mainly polyolefins, PE, PP, ethylene and styrene copolymers, ionomers
Polyamides (nylon)	Highly dispersed silica
	Sodium benzoate
	Talc
	Titanium dioxide
Polypropylene	Sodium benzoate
	Bis-benzylidene sorbitol
Polyethylene	Potassium stearate
	Nucleated PE or higher polyolefins

thermoformed dual-purpose food trays for conventional and microwave ovens. The nucleating agent promotes fast crystallization during the tray thermoforming process.

Sodium benzoate. *Sodium benzoate* is an inexpensive traditional nucleating agent used predominantly in nylon and PP homopolymer. Sodium benzoate has full FDA approval in PP and is used in food applications and pharmaceutical synthesis. Typical use levels of sodium benzoate as a nucleating agent in PP are lower than the sorbitols. The major application is in injection-molded packaging closures.

Ionomer resins. *Ionomer resins* are metal salts of ethylene/methacrylic acid copolymers and have a long chain semicrystalline structure. DuPont's SURLYN is the representative material. Ionomers are used as nucleating agents to control crystallization in PET molding resins. PET is processed at high mold temperatures. The ionomer provides faster crystallinity, more rapid cycle time, and good dimensional stability at elevated temperatures. The improvement rate in crystallization at lower temperatures allows the use of water-cooled molds. Typical use levels are below 1%.

4.14.2 Suppliers

Milliken is the leading producer of substituted sorbitol clarifiers in North America and Europe under the MILLAD trademark. Ciba has recently reached a joint market agreement with Roquette. This will enable the formidable Ciba marketing organization to increase significantly the market exposure of Roquette's sorbitol-based clarifiers. Significant amounts of sodium benzoate are sold to the plastics industry through distributors, who purchase from basic suppliers such as Kalama Chemical. The suppliers of low molecular weight polyolefins are the CPET resin producers such as Shell, Eastman, and ICI. AlliedSignal also offers related compounds. DuPont and others supply ionomer resins. A list of selected global suppliers can be seen in Table 4.21.

4.14.3 Trends and forecasts

PP, CPET, and PET molding resins, and, to some extent, nylon, account for most of the nucleating agent consumption. Approximately 10% of all PP and nearly 50% of the injection molding category is nucleated. Smaller percentages of the PP blow molding and extrusion categories use nucleating agents.

Improved clarity of PP has provided the ability for replacement of PVC with PP in applications such as blisterpacks for hardware. In addition, new PP resins are being developed that use single-site metallocene catalysts (mPP). While virtually no difference exists in the processing behavior or finished product properties between conventional PP and mPP, these new materials are easier to nucleate. The use of nucleated mPP provides for a product with the higher physical properties of PP homopolymer and the clarity of nucleated random PP copolymer.

There is continuing growth of nucleated PP, particularly in the blow molding and extrusion markets. CPET continues to expand in thermoforming applications, and PET molding compounds continue to pen-

TABLE 4.21 Selected Suppliers of Nucleating Agents

	Type			
Supplier	Sorbitols	Sodium benzoates	LMW polyolefins	Other
AlliedSignal	—	—	×	—
Ciba/Roquette	×	—	—	—
Clariant	—	—	—	×
Cytec Industries	—	—	—	×
DuPont	—	—	—	×
Eastman Chemical	—	—	×	—
FBC	—	×	—	—
ICI	—	—	×	—
Jarchem	—	×	—	—
Kalama Chemical	—	×	—	—
Milliken	×	—	—	—
Mitsui Toatsu Chemicals	×	—	—	—
New Japan Chemical	×	—	—	—
Shell	—	—	×	—
Witco	—	—	—	×

etrate electrical uses. Based on this activity, consumption of nucleating agents is likely to increase at a rate of about 6%/year globally over the next 5 years.

4.15 Organic Peroxides

4.15.1 Description

Organic peroxide initiators serve as sources of free radicals in the preparation of a variety of resins for plastics, elastomers, and coatings. Their usage in plastics processing can be divided into four functions:

- Polymerization of thermoplastic resins
- Curing for unsaturated polyester thermoset resins
- Cross-linking of polyethylene and various elastomers
- Visbreaking (rheology modification) of polypropylene

The peroxide group (—O—O—) contained in all organic peroxides is highly unstable. This instability eventually leads to homolytic cleavage. When the bond is broken between the two oxygen molecules, the peroxide decomposes and two free radicals are formed. The general formula for such compounds is R1—O—O—R2, whereby R1 and R2 either symbolize organic radicals or an organic radical and hydrogen atom.

Types. Organic peroxide initiators can be further classified by functional groups into seven major classes as follows:

- Dialkyl peroxides
- Diacyl peroxides
- Hydroperoxides
- Ketone peroxides
- Peroxydicarbonates
- Peroxyesters
- Peroxyketals

Each class denotes the varying chemistry of both substituent groups, R1 and R2. Figure 4.2 displays the general formulas of the major classes of these organic peroxides.

Dialkyl peroxides. *Dialkyl peroxides* can be further categorized depending on the two substituent groups. This class may contain two organic

Dialkyl peroxides	$R_1 - O - O - R_2$
Diacyl peroxides	$R_1 - C(=O) - O - O - C(=O) - R_2$
Hydroperoxides	$R - O - O - H$
Ketone peroxides	$H\text{-}O\text{-}O - C(R_1)(R_2) - O - O - C(R_1)(R_2) - O\text{-}O\text{-}H$
Peroxydicarbonates	$R_1 - O - C(=O) - O\text{-}O - C(=O) - O - R_1$

Peroxyesters

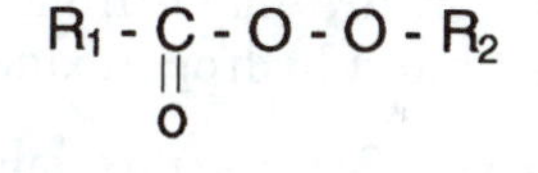

Peroxyketals

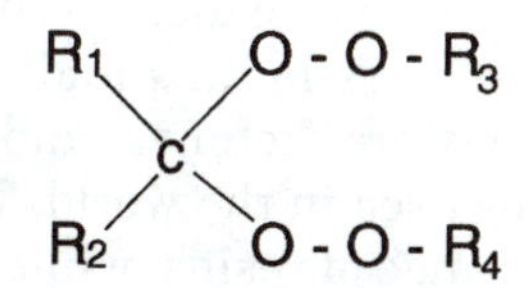

Figure 4.2 General chemical structures of organic peroxides by major class.

radicals which are wholly or partially aliphatic. Depending on this substitution, further categorizing may occur. For example, when both groups are aliphatic, it is known as a dialkyl peroxide. When both substituent groups are aromatic, the peroxide is known as a *diarylalkyl peroxide*. When the substituent groups are alkyl and aromatic, the peroxide is known as an *alkylaryl peroxide*. The workhorse product among the dialkyl peroxides is *dicumyl peroxide* which accounts for one-third of the worldwide volume for dialkyls.

Diacyl peroxides. *Diacyl peroxides* can be subdivided similarly to dialkyls, depending on the composition of the organic groups R1 and R2:

- Dialkanoyl peroxides
- Alkanoyl-aroyl peroxides
- Diaroyl peroxides

Benzoyl peroxide is the most common of the diacyl peroxides.

Hydroperoxides. *Hydroperoxides* are generally unsuitable for cross-linking and polymerization reactions since the possibility of a side reaction, such as ionic decomposition, is too great. They are used as a raw material to manufacture other organic peroxides. The most common hydroperoxides include cumene hydroperoxide and *t*-butyl hydroperoxide.

Ketone peroxides. *Ketone peroxides* are mixtures of peroxides and hydroperoxides that are commonly used during the room temperature curing of polyester. Methyl ethyl ketone peroxide (MEKP) is the major product.

Peroxydicarbonates. *Peroxydicarbonates,* such as di-(*n*-propyl) peroxydicarbonate and di-(sec-butyl) peroxydicarbonate, are relatively expensive products used largely to initiate polymerization of PVC.

Peroxyesters. *Peroxyesters,* such as *t*-butyl peroxybenzoate and *t*-octyl peroxyester, are made from the reaction of an alkyl hydroperoxide, such as *t*-butyl hydroperoxide, with an acid chloride.

Peroxyketals. *Peroxyketals,* such as *n*-butyl-4,4-di-(*t*-butylperoxy) valerate and 1,1-di-(*t*-butyl peroxy)-3,3,5-trimethylcyclohexane, are high-temperature peroxides used in selective applications for PE and elastomer cross-linking and in the curing of unsaturated polyester.

Peroxyesters, ketones, and dialkyls are the largest volume organic peroxides used in the world. The peroxyesters and dialkyls are used in a broad range of resins, while the ketones are the highest volume product used in the large unsaturated polyester market. Others, such as peroxydicarbonate types, are used in only one resin, in this case, PVC. The largest application globally for organic peroxides, based on tonnage, is in glass-reinforced unsaturated polyester resins. These resins represent about one-third of the total global organic peroxide consumption in plastics. Traditional high-pressure LDPE resins and PVC together account for another one-third of the tonnage, with ABS, cross-linked HDPE, PP, PS, and solid acrylics making up most of the remainder. Peroxides are also used in applications outside of plastics in elastomers and emulsion acrylics for coatings. A summary of organic peroxide types with primary uses is provided in Table 4.22.

Raw materials. The major raw materials for the organic peroxides are basic petrochemicals (propylene, benzene, and isobutane), organic intermediates (such as acid chlorides), and, in some cases, hydrogen peroxide or an inorganic peroxide salt. Diacyl peroxides may be manufactured by reacting hydrogen peroxide, or an alkali metal peroxide, with an acid chloride. Hydrogen peroxide is used to make ketone peroxides. Peroxyesters are made by reacting an alkyl hydroperoxide with an acylating agent such as acid chloride. A major class of peroxyesters is the *t*-butyl peroxyesters. The starting material, *t*-butyl hydroperoxide, is produced as an intermediate to manufacture *t*-butyl alcohol and propylene oxide from isobutane and propylene. Dicumyl peroxide, an important dialkyl peroxide, can be made from cumene hydroperoxide obtained from the oxidation of cumene.

4.15.2 Suppliers

There are about 30 major worldwide suppliers of organic peroxides. Most of these companies serve the plastics industry, and others produce hydroperoxides that are used as raw materials to produce other peroxides. Some of these companies also produce other plastics additives such as antioxidants, light stabilizers, PVC heat stabilizers, and flame retardants. Only three companies, namely, Akzo Nobel, Elf Atochem, and, to some extent, LaPorte, are significant suppliers of organic peroxides to the plastics industry in every region of the world. Important regional suppliers include Witco (North America) and Nippon Oil and Fats (Asia/Pacific). In North America, Hercules supplies dicumyl peroxide, while Aristech and Arco supply hydroperoxide raw materials. Norac makes a variety of peroxides for use in unsaturated polyesters. Selected global suppliers of organic peroxides are given in Table 4.23.

TABLE 4.22 Organic Peroxides Types and Functions

Type	Function
Dialkyl peroxides	Polyethylene cross-linking Initiator for polystyrene polymerization Polypropylene rheology modification
Diacyl peroxides	Initiator for polystyrene polymerization Unsaturated polyester curing
Hydroperoxides	Initiator for ABS polymerization Raw material for other organic peroxides
Ketone peroxides	Unsaturated polyester curing
Peroxydicarbonates	Initiator for PVC polymerization
Peroxyesters	Initiator for ABS polymerization Initiator for polystyrene polymerization Unsaturated polyester curing
Peroxyketals	Polyethylene cross-linking Unsaturated polyester curing

4.15.3 Trends and forecasts

The development of completely new organic peroxide chemicals continues to be limited by regulatory consent degrees, safety and health testing, and by threats from new technologies for manufacturing and modifying plastics.

The global producers of organic peroxides have been focusing on the following areas to solidify and expand their existing product offerings:

- Research and development efforts directed at formulation, blending, and mixing known peroxide components rather than developing new chemicals.
- Focus on reduction of safety and handling issues, including reduction of solvent-based carrying systems which generate emissions of volatile organic compounds (VOC).
- Development of new recyclable and returnable packaging systems.
- Continuing efforts on newer alternate technologies, such as single-site metallocene catalysis which have the potential of replacing organic peroxides in some polyolefin systems.

Concerns with VOCs and a consent decree relating to carcinogenity have limited development and, in most cases, changed the order of preference for organic peroxide products. For example, government regulations on styrene emissions from unsaturated polyester operations have increased the trend toward elevated closed molding operations and away from traditional open molding. This favors the use of peroxyester and peroxyketal types versus diacyl types in these operations.

The organic peroxide business historically has followed the growth patterns of the major resins. Over the next 5 years, the global market is expected to grow at 4%/year, paced by the Asia/Pacific and other developing regions, especially in the latter half of the period.

From a competitive standpoint, there will be continued efforts at consolidation, through joint ventures, alliances, and acquisitions as the majors look to the growing markets in Asia/Pacific, outside of Japan, and the developing countries. The remaining independent and regional producers of organic peroxides are largely located in countries such as Korea, Taiwan, China, and India, and this is where the action will be.

4.16 Plasticizers

4.16.1 Description

Plasticizers are the largest volume additives in the plastic industry. They are largely used to make PVC resin flexible and are generally regarded as commodity chemicals, although significant specialty nich-

TABLE 4.23 Selected Organic Peroxide Suppliers

Supplier	Type						
	Dialkyls	Diacyls	Hydro-peroxides	Ketones	Peroxy-dicarbonates	Peroxyesters	Peroxyketals
Akzo Nobel	×	×	×	×	×	×	×
Arco	—	—	×	—	—	—	—
Aristech	—	—	×	—	—	—	—
Central Chemicals	×	×	×	×	×	×	×
Chon Ya Fine Chemical	—	×	—	—	—	—	—
Coin Chemical Ind.	×	—	×	—	—	—	—
Concord Chem. Ind.	×	—	×	—	—	—	—
Hercules	×	—	×	—	—	—	—
Jain & Jain	×	×	×	×	—	×	—
Kawaguchi Chemical	—	×	—	—	—	—	—
Kayaku Akzo	×	×	×	×	×	×	×
LaPorte	×	×	×	×	×	×	×
Mitsui Petrochemical	×	—	—	—	—	—	—
NOF	×	×	×	×	×	×	×
Norac	—	×	—	×	—	×	—
Peroxide Catalysts	—	—	—	—	×	—	—
Peroxidos Organicos S.A.	×	×	×	×	×	×	—
Plasti Pigments	×	×	×	×	—	×	—
Seiki Chemical Ind.	—	—	—	×	×	×	—
Shandong Lauiu	—	×	—	×	—	×	—
Tianjin Akzo Nobel Peroxides	×	×	×	×	×	×	×
Tianjin Dongfang	—	×	—	—	—	—	—
Tung Hung Enterprise	—	—	—	×	—	—	—
Witco	×	×	×	×	×	×	×
Youngwoo Chemical	—	×	—	—	—	—	—
Yuh Tzong Enterprise	×	×	—	×	—	×	—

es exist. The primary role of a plasticizer is to impart flexibility, softness, and extensibility to inherently rigid thermoplastic and thermoset resins. Secondary benefits of plasticizers include improved processability, greater impact resistance, and a depressed brittle point. Plasticizers can also function as vehicles for plastisols (liquid dispersions of resins which solidify upon heating) and as carriers for pigments and other additives. Some plasticizers offer the synergistic benefits of heat and light stabilization as well as flame retardancy.

Plasticizers are typically di- and triesters of aromatic or aliphatic acids and anhydrides. Epoxidized oil, phosphate esters, hydrocarbon oils, and some other materials also function as plasticizers. In some cases, it is difficult to discern if a particular polymer additive functions as a plasticizer, a lubricant, or a flame retardant.

The major types of plasticizers are

- Phthalate esters
- Aliphatic esters
- Epoxy esters
- Phosphate esters
- Trimellitate esters
- Polymeric plasticizers
- Other plasticizers

There are a number of discrete chemical compounds within each of these categories. As a result, the total number of plasticizers available to formulators is substantial.

Phthalate esters. The most commonly used plasticizer types are *phthalate esters.* They are manufactured by reacting phthalic anhydride (PA) with 2 moles of alcohol to produce the diester. The most often used alcohols vary in chain length from 6 to 13 carbons. Lower-alcohol phthalate esters are also manufactured for special purposes. The alcohols may be either highly branched or linear in configuration. The molecular weight and geometry of the alcohol influences plasticizer functionality. The most frequently used alcohol is 2-ethylhexanol (2-EH). Other plasticizer alcohols include isooctanol, isononanol, isodecanol, tridecanol, and a variety of linear alcohols. The three major diester phthalate plasticizers are as follows:

- Dioctylphthalate or di-2-ethylhexyl phthalate (DOP or DEHP)
- Diisononyl phthalate (DINP)
- Diisodecyl phthalate (DIDP)

Aliphatic esters. *Aliphatic esters* are generally diesters of adipic acid, although sebacic and azelaic acid esters are also used. Alcohols employed in these esters are usually either 2-EH or isononanol. Higher esters of these acids are used in synthetic lubricants and other nonplasticizer materials. Lower esters are used as solvents in coating and other applications. Adipates and related diesters offer improved low-temperature properties compared with phthalates.

Epoxy ester. *Epoxy ester plasticizers* have limited compatibility with PVC. Therefore, they are used at low levels. Epoxidized soybean oil (ESO), the most widely used epoxy plasticizer, is also used as a secondary heat stabilizer. As a plasticizer, it provides excellent resistance to extraction by soapy water and low migration into adjoining materials that tend to absorb plasticizers. Other epoxy plasticizers include epoxidized linseed oil and epoxidized tall oils. Tall oils are prepared from tall oil fatty acids and C_5–C_8 alcohols.

Phosphate triesters. Phosphorous oxychloride can be reacted with various aliphatic and aromatic alcohols and phenols to yield *phosphate triesters*. Commercially, the trioctyl (from 2-EH) and triphenyl (from phenol) phosphates are often seen. Mixed esters are frequently encountered as well. Phosphate esters are considered to be both secondary plasticizers as well as flame retardants.

Trimellitates. *Trimellitates,* the esters of trimellitic anhydride (1,2,4-benzenetricarboxylic acid anhydride), are characterized by low volatility. This property increases the service life of a PVC compound subjected to elevated temperatures for long periods of time and reduces fogging. The most important trimellitates are trioctyl trimellitate (TOTM) and triisononyl trimelliate (TINTM). Trimellitates are most commonly used for PVC wire insulation, often in conjunction with phthalates.

Polymer plasticizers. Esterification of diols with dibasic acids yields high molecular weight (1000 to 3000) *polymeric plasticizers* that can plasticize PVC and other polymers. These polymerics are used in conjunction with phthalates to provide improved permanence and reduced volatility.

Other plasticizers. A number of other chemical compounds are employed in special cases to plasticize PVC and other polymers. These include benzoates, citrates, and secondary plasticizers.

Benzoates are esters of benzoic acid and various polyhydric alcohols and glycols. They are most often used in vinyl floor covering products because of their resistance to staining.

Citrates are plasticizer alcohol esters of citric acid. They are used in food-contact and medical applications due to their perceived low toxicity.

Other *secondary plasticizers* include various liquid aromatic and aliphatic hydrocarbons, oils, and esters. They are used in conjunction with such primary plasticizers as phthalates. While some offer particular functional benefits, secondaries are often chosen to lower formulation cost at the expense of other properties.

4.16.2 Suppliers

The general trend in plasticizer supply has been a consolidation among the leading plasticizer suppliers. Smaller suppliers are either vacating the business or focusing on selected specialty products. Although there are still a large number of suppliers, the majority of the market is held by the leading petrochemical companies of the world. The top three global plasticizer producers are Exxon, BASF, and Eastman, respectively. Table 4.24 lists selected global suppliers of plasticizers by type.

4.16.3 Trends and forecasts

Environmental concerns with PVC seem to have abated, although issues have arisen concerning alleged "hormone mimicking" properties of phthalate plasticizers. The industry has rigorously disputed these claims, but research into test materials is still going on. Although the industry is confident that there is no problem with the safety of phthalate plasticizers, alternatives to these materials are being developed. All in all, plasticizer usage is likely to follow flexible PVC growth with consumption increasing at about a 4%/year growth rate over the next 5 years.

4.17 Polyurethane Catalysts

4.17.1 Description

Polyurethanes are versatile polymers typically composed of polyisocyanates and polyols. By varying constituents, a broad range of thermosets and thermoplastics can be produced and used in different applications. Possible systems include high-strength, high-modulus, structural composites; soft rubbers; elastic fibers; and rigid or flexible foams. Although isocyanates have the ability to form many different polymers, very few types are used in actual production. The most common diisocyanates are methylene diphenylene diisocyanate (MDI) and toluene diisocyanate (TDI). Of these, TDI is the most commercially important dimer.

While polyurethanes can be formed without the aid of catalysts, the reaction rate increases rapidly when a suitable catalyst is selected. A

TABLE 4.24 Selected Suppliers of Plasticizers

Supplier	Type				
	Phthalate	Trimellitate	Polymeric	Adipate	Other
Aristech	×	×	—	×	—
BASF	×	×	—	—	—
Bayer	×	—	—	—	×
C.P. Hall	×	×	×	×	×
DuPont	—	—	—	×	×
Eastman Chemical	×	×	—	—	×
Elf Atochem	—	—	—	—	×
Exxon	×	×	—	×	×
Ferro	—	—	—	—	×
Huels	×	—	×	—	—
Kyowa Yuka	×	—	—	—	—
Mitsubishi Gas Chemical	×	×	—	—	—
Nan Ya Plastics	×	×	—	×	—
New Japan Chemical	×	×	—	—	—
Sekisui Chemical	×	—	—	×	—
Solutia	×	—	—	—	×
Union Petrochemical	×	—	—	—	×
Velsicol	—	—	×	—	×

well-chosen catalyst also secures the attainment of the desired molecular weight, strength, and, in the case of foams, the proper cellular structure. In some applications catalysts are used to lower the temperature of the polymerization reaction.

The major applications for polyurethane catalysts are in flexible and rigid foam, which account for over 80% of the catalyst consumption. Other applications are in microcellular reaction injection-molded (RIM) urethanes for automobile bumpers and a variety of noncellular end uses such as solid elastomers, coatings, and adhesives.

There are more than 30 different polyurethane catalyst compounds. The two most frequently used catalyst types are tertiary amines and organometallic salts which account for about equal shares of the market. The tertiary amine-catalyzed reaction causes branching and cross-linking and is used primarily for polyurethane foam formation. Organometallic salts, such as organotin catalysts, encourage linear chain extension and are used in flexible slabstock, rigid foam, and in a variety of noncellular elastomer and coating applications.

Tertiary aliphatic amines. The most common of the amine catalysts are *tertiary aliphatic amines,* and they are used to accelerate the isocyanate-hydroxyl reaction and give off carbon dioxide. Triethylenediamine, also known as diazabicyclooctane (DABCO), is the most prevalent of the tertiary amine catalysts used for polyurethane manufacture due to its high basicity and low steric hindrance which yields high catalytic activity. It should be noted that tertiary aliphatic amines can be discharged from fresh foams, causing unpleasant odor and potential skin irritation. Safety precautions are necessary when working with these materials to produce polyurethane foam.

Organometallic compounds. While *organometallic compounds* make excellent polyurethane catalysts, they affect the aging characteristics of the polymer to a higher degree than tertiary amines. Stannous octoate is the most broadly accepted catalyst of this type of polyurethane formation, although other organotins and potassium salts are also used. While minute quantities of the inorganic portion of these substances speed up polyurethane reactions during processing, residual amounts of metal from these catalysts can cause side reactions or change properties of the final product.

Different catalyst types can also be combined to obtain a desired effect. For example, polyurethane foam production can use both organotin and amine catalysts for a balance of chain extension and cross-linking.

4.17.2 Suppliers

Air Products is the major supplier of polyurethane catalysts in North America and one of the largest in Europe, making both amine and organometallic types. BASF is also active in both regions with amine types. Witco and Huntsman in North America and Goldschmidt in Europe are major regional suppliers. The Asia/Pacific market is served by a number of regional suppliers largely out of Japan. Selected global suppliers of polyurethane catalysts by type are listed in Table 4.25.

4.17.3 Trends and forecasts

As the guidelines for environmental safety become more stringent and chlorofluorocarbons (CFCs) gradually phase out as blowing agents for polyurethane foams, the demand for urethane catalysts will rise. Alternative blowing agents, such as methylene chloride, acetone, hydrochlorofluorocarbons, and carbon dioxide, are being introduced and, as a result, new catalyst technology is required to rectify problems caused by these new procedures. In addition, volatile organic compound (VOC) emissions are raising new concerns which are likely to propagate additional changes to adjust the viscosity and control the behavior of the polyurethane foam as well as its final properties.

TABLE 4.25 Selected Suppliers of Polyurethane Catalysts

Supplier	Type	
	Amines	Organometallics
Air Products and Chemicals	×	×
Akzo Nobel	—	×
BASF	×	—
Bayer	×	—
Cardinal Stabilizers	—	×
Ferro	—	×
Goldschmidt	×	—
Huntsman	×	—
Johoku Chemical	—	×
Kao Corporation	×	—
Kyodo Chemical	—	×
New Japan Chemical	×	—
Nitto Kasei Kogyo	—	×
Sankyo	×	×
Sanyo	×	—
Tosoh	×	—
Witco (OSI)	×	×
Yoshitimi Fine Chemicals	—	×

The global market for urethane catalysts is growing at a rate of approximately 4%/year. Growth is tied closely to the flexible and rigid foam markets. Rigid foam is growing at slightly above the average and flexible foam is growing at slightly below the average. The smaller automotive market in reaction injection molding urethanes is declining because thermoplastic polyolefins (TPO) are now the preferred materials over polyurethanes in bumpers.

The major driving forces, besides end-use growth, affecting urethane catalysts will be the continued phase-out of CFC blowing agents and the development of new blowing agent alternatives, along with the related concern over VOC emissions, which also affects blowing agent and catalyst choice. These forces will have more of an effect on catalyst mix than the overall volume of catalyst used.

Chapter

5

Processing of Thermoplastics

Carol M. F. Barry

Stephen A. Orroth

University of Massachusetts
Lowell, Massachusetts

McKelvey[1] defined plastics processing as "operations carried out on polymeric materials or systems to increase their utility." These types of operations produce flow, chemical change, and/or a permanent change in physical properties.[1] Plastics processing techniques can be grouped into three categories:

- Forming operations
- Bonding operations
- Modifications

Although forming operations always involve flow, thermoplastic processes, such as extrusion, thermoplastic injection molding, thermoforming, and rotational molding, produce physical changes in the polymer whereas chemical change occurs in the casting of liquid monomers. Reactive extrusion and thermoset injection molding induce both chemical and physical change in the plastics materials. Bonding operations join two or more materials by causing one or both joining surfaces to become molten or flow. The former occurs while laminating polyethylene to aluminum or paper, coating of polyvinyl chloride plastisols on fabric (to produce vinyl upholstery), and using adhesives to join materials. In contrast, both surfaces are molten during the heating sealing of polyethylene bags and ultrasonic welding of plastics parts.

Modifications include surface activation, mixing, and polymer modifications. Surface activation improves adhesion or printability of plastics materials. Typically, corona discharge and flame treatments oxidize the polymer surface to a depth of 1 nm; this oxidized surface is more compatible with polar inks and adhesives. Mixing reduces the nonuniformity of the polymer composition. While polymer processing uses many types of mixers, there are two types of mixing. The first, distributive or spatial mixing, causes randomization of a mixture but no physical changes. It takes place in drum tumblers and ribbon blenders. The second, dispersive mixing, involves heat and shear which reduce particle size and eliminate clumps or agglomerates. This mixing is observed in high-intensity mixers, Banbury mixers, two-roll mills, and extruders. Polymer modifications, such as annealing molded parts and radiation of plastics parts, change the amount of orientation, crystallinity, and/or cross-linking in the plastic.

This chapter focuses on the primary processing of thermoplastic materials. It begins with material concepts used in processing. The chapter continues with processing techniques, and each process is outlined, equipment requirements are specified, and processing parameters are discussed.

5.1 Material Concepts

Polymers are long-chain molecules with one or more repeat units called *mers.* The number of repeat units in a polymer, and thus the length of the polymer chain, can be varied during manufacture of the resin. The molecular weight of a polymer is a way of indicating chain length. The average molecular weight number is merely the molecular weight of the repeat unit multiplied by the number of repeat units. Since the molecular weight of the styrene repeat unit is 104 daltons, a polystyrene with 2500 mers would have a molecular weight of 260,000 daltons. However, not all polymer chains have the same length; some chains are short, while others are long. The average chain length is indicated by the number average molecular weight, but the spread or range of chains is given by the molecular weight distribution (MWD). As discussed later in this section, both molecular weight and molecular weight distribution significantly affect flow during polymer processing.

A homopolymer has one repeat unit while two or more mers polymerized together in a copolymer. The properties and processing characteristics of copolymers are often very different from those of the corresponding homopolymers. These characteristics also vary with the ratio of the components and their arrangement within the copolymer. As shown in Fig. 5.1*a*,[2] the repeat units of random copolymers are distributed randomly along the polymer chain. Thus, ethylene propylene rubber (EPR) is an elastomer, whereas polyethylene and polypropylene are plastics. In poly(styrene-co-acrylonitrile) (SAN), the small amount of acrylonitrile improves the heat resistance and increases processing temperatures when compared to polystyrene. Fluorinated ethylene propylene (FEP), unlike its "parent," polytetrafluoroethylene, is melt processible. Alternating copolymers (Fig. 5.1*b*[2]), in which every second repeat unit is the same, have, until recently, been laboratory curiosities. However, new catalysts may make these copolymers commercially viable. Block copolymers (Fig. 5.1*c*[2]) contain alternating segments of each repeat unit, but the segments are often several repeat units long. Such materials include polyetheramides, hard segment–soft segment polyurethanes, and butadiene-styrene elastomers (SEBS). Graft copolymers (Fig. 5.1*d*[2]) consist of a main chain containing only one repeat unit with side chains of the second mer. These copolymers link the two phases in high-impact polystyrene (HIPS) and acrylonitrile butadiene styrene (ABS).

Blends are physical mixtures of polymers rather than monomers. Like copolymers, properties and processing characteristics are often very different from those of the component polymers and also vary with the ratio of components. Unlike copolymers, blend properties can be sensitive to processing conditions. Miscible blends mix on a molecular level to produce a single phase and exhibit a single transition temperature that corresponds to the blend composition. The most important commercial miscible blend is polystyrene-polyphenylene

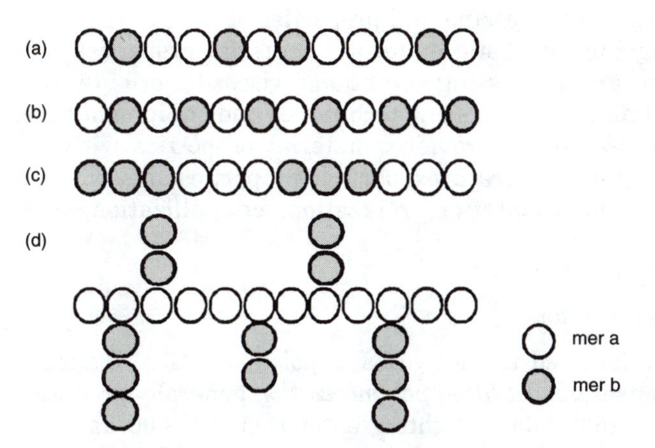

Figure 5.1 Types of copolymers (*Adapted from Ref. 2.*)

oxide (modified polyphenylene oxide). Partially miscible blends can exist as a single- or two-phase system, depending on composition and processing conditions. When partial miscible blends, such as polycarbonate/acrylonitrile butadiene styrene (PC/ABS), polycarbonate/polyethylene terephthalate (PC/PET), and polycarbonate/polybutylene terephthalate (PC/PBT), are modified so that the morphology is stable, they are often called *alloys*. Since immiscible blends cannot mix on a molecular level, they exist as two phases. These blends exhibit the transition temperatures of the component polymers, and the morphology and properties are very sensitive to processing conditions. Immiscible blends are the basis of many impact-modified plastics.

Plastics are usually not pure polymer, but contain the following substances:

- Fillers such as mica, talc, and calcium carbonate
- Fibers such as glass fibers and carbon fibers
- Plasticizers such as the dioctyl phthalate used in polyvinyl chloride (PVC)
- Flame retardants
- Heat stabilizers such as organo tin compounds used in PVC
- Antioxidants
- Ultraviolet (UV) stabilizers, which are added to polyolefins and other polymers
- Colorants
- Lubricants, which facilitate processing or performance
- Processing aids

All additives impact the processing and properties of the plastics.

When processing thermoplastic materials, material properties not only dictate drying and processing conditions, viscosity, orientation, and shrinkage, but also the processing techniques and equipment that can be used. The next section reviews material properties with an emphasis on processing temperatures, particle properties, melt viscosity and elasticity, and orientation, relaxation, crystallization, and shrinkage.

5.1.1 Processing temperatures

Polymers are manufactured using two basic polymerization methods: addition and condensation. *Addition polymerization* generally produces rapid chain growth, molecular weights greater than 100,000 daltons, and no by-products. In contrast, *condensation polymerization* provides for lower chain growth, typical molecular weights of 10,000 to 50,000 daltons, and by-products such as water. As a result, addition polymers are less susceptible to water absorption, and seldom depolymerize during processing. When these materials are dried prior to processing, it is usually to prevent foaming and surface defects such as splay. However, if poorly dried condensation polymers are melt processed, they tend to depolymerize. Since this reduces the molecular weight, material properties decrease. Consequently, condensation polymers are always dried prior to processing (although with special screws and vented processing equipment condensation polymers can be dried during processing). Polyethylene, polypropylene, polystyrene, impact-modified polystyrene, acrylonitrile-butadiene-styrene terpolymer, polymethylmethacrylate, poly(vinyl chloride), and polytetrafluoroethylene are addition polymers, whereas polyacetal, polycarbonate, polyamides, poly(ethylene terephthalate), poly(butylene terephthalate), polysulfones, polyetherimide, and polyetheretherketones (PEEK) are condensation polymers. Water absorption values, maximum water contents for molding, and suggested drying conditions are presented in Table 5.1.[3,4]

Processing temperatures (see Table 5.1) are associated with the transition temperatures of a polymer. The *glass transition temperature* (T_g) is the temperature at which the amorphous (unordered) region of a polymer goes from a glassy state to a rubbery state. In amorphous polymers, T_g is related to processing temperatures. As shown in Figure 5.2*a,* the modulus (stiffness) is relatively constant until the temperature rises above the T_g. The modulus then decreases gradually. When the polymer reaches its melt processing temperature, the polymer flows easily and can be extruded, injection molded, and extrusion blow molded. For polycarbonate, the difference between the softening temperature and processing temperature is about 140°C. Since this slow reduction in modulus over a wide temperature range facilitates stretching of the rubbery material, amorphous materials, such as polycarbonate, are easily thermoformed.

Figure 5.2*b* presents the modulus-temperature curve of a semicrystalline polymer, polypropylene. It exhibits a glass transition and a melting transition (T_m). The modulus of polypropylene, like other polymers with high levels of crystallinity, does not decrease substantially when the temperature is raised above the glass transition temperature. Thus, polypropylene remains relatively rigid until it reaches its T_m. At that point the crystallites (highly ordered regions) in semicrystalline polymers break up and the polymer begins to flow. Since all polymers contain amorphous regions, they do not have well-defined melting temperatures. Melt processing temperatures of semicrystalline polymers are usually less than 100°C above their melt temperatures.

TABLE 5.1 Suggested Drying Conditions for Generic Resins[3,4]

Material	Water absorption, %	Maximum water, %	$T_{extrusion}$, °C	$T_{inj.\ molding}$, °C	T_{drying}, °C	t_{drying}, h
Acrylonitrile butadiene styrene (ABS)	0.25–0.40	0.20	225	260	88	3–4
Acetal	0.25	—	—	200	93	1–2
Acrylic	0.20–0.30	0.08	190	235	82	1–2
Polyamide-6 (nylon) (PA-6)	1.60	0.15	270	290	82	4–5
Polyamide-6, 6 (nylon) (PA-6,6)	1.50	0.15	265	265	82	4–5
Polycarbonate (PC)	0.20	0.02	290	300	120	3–4
Polybutylene terephthalate (PBT)	0.08	0.04	—	240	125	2–3
Polyethylene terephthalate (PET)	0.10	0.005	250	255	160	4–5
Polyetherimide (PEI)	0.25	—	—	370	155	4–5
High-density polyethylene (HDPE)	<0.01	—	210	250	—	—
Low-density polyethylene (LDPE)	<0.01	—	180	205	—	—
Linear low-density polyethylene (LLDPE)	<0.01	—	260	220	—	—
Polyphenylene oxide (PPO)	0.07	—	250	275	100	2–3
Polypropylene (PP)	<0.01	—	235	255	—	—
Polystyrene (PS)	0.03	—	210	220	—	—
High-impact polystyrene (HIPS)	0.10	—	235	230	—	—
Polyphenylene sulfide (PPS)			—	330	140	2–3
Polysulfone (PSU)	0.30	0.05	345	360	135	3–4
Polyurethane (PU)	0.10	0.03	205	205	82	2–3
PU (elastomers)	0.07	0.03	200	205	100	2–3
r-PVC (polyvinyl chloride)	0.10	0.07	185	195	—	—
p-PVC (polyvinyl chloride)	0.02	—	175	150	—	—
Styrene acrylonitrile (SAN)	0.03	0.02	215	245	82	3–4

While thermoplastic polymers soften at T_g, and if semicrystalline, melt at T_m, cross-linked polymers do not melt and flow (Fig. 5.2c[5]). Lightly cross-linked polymers soften as the temperature exceeds T_g, but they remain rubbery solids until the polymer decomposes. Highly cross-linked polymers often do not even soften and retain a high modulus until reaching the decomposition temperature. Thermoset resins, like unsaturated polyester, epoxy, and polyurethanes, have varied levels of cross-linking. However, thermoplastic resins can be modified to contain few cross-links; lightly cross-linked polyethylene (XLPE) often improves the mechanical properties of rotomolded parts.

Some thermoplastics will decompose before they melt and flow. Extremely long polymer chains combined with intermolecular attractions prevent conventional melt processing of ultrahigh-molecular-

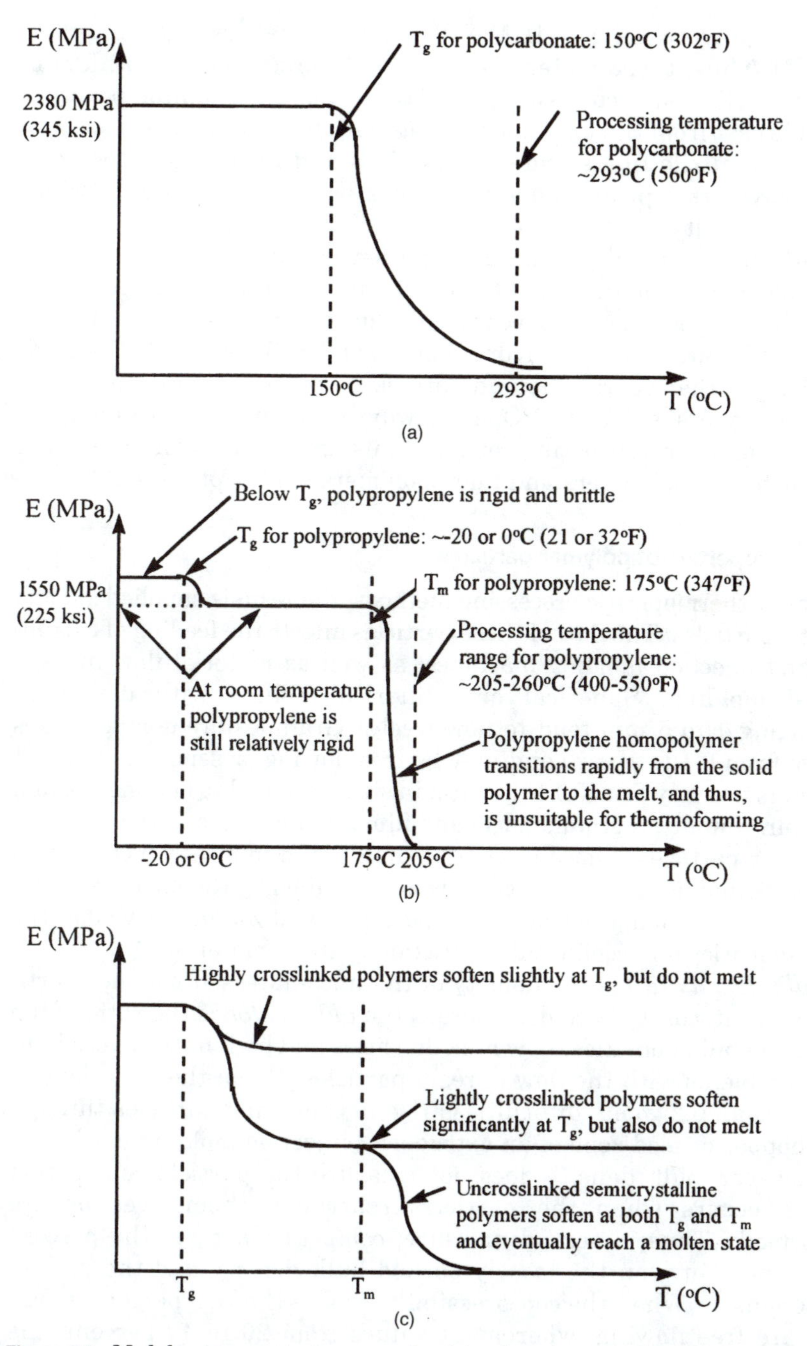

Figure 5.2 Modulus-temperature curves for (*a*) polycarbonate, (*b*) polypropylene, and (*c*) generic thermoplastic and cross-linked materials. (*Part c was adapted from Ref. 5.*)

weight polyethylene (UHMWPE) and polytetrafluoroethylene (PTFE). Thus, these materials are usually processed as powders or slurries. High-molecular-weight (MW > 1,000,000 daltons) acrylic sheet is cast from the monomers. Coagulation techniques are used for polyacrylonitrile fibers and cellulose film and fibers. High-temperature polymers typically have backbones with aromatic rings (which reduce mobility).

Certain functional groups are inherently stable or unstable. For example, chloropolymers and fluoropolymers degrade relatively easily. Dehydrohalogenation removes a chloride or fluorine atom and the adjacent hydogen to form hydrochloric or hydrofluoric acid. The acids catalyze further degradation and attack processing equipment. Thus, chloropolymers, such as PVC, are always compounded with heat stabilizers, and special coatings or materials are required for equipment in which chloropolymers and fluoropolymers will be processed.

5.1.2 Properties of polymer particles

For most thermoplastic processing methods, the resin is supplied as solid plastic particles. The flow of these particles affects the feeding of extruders and injection molding machines, as well as material flow in rotational molding. Spherical or cylindrical pellets with diameters averaging 1 to 5 mm tend to flow freely.[6] Granules are smaller (sizes range from 0.1 to 1 mm) and may be free flowing or semifree flowing.[6] Powders are very fine (0.1 to 100 μm in size) and tend to be cohesive and trap air.[6] Finally, reground parts and film and fiber scrap provide large, irregular particles (typically greater than 5 mm in size) which tend to interlock and not flow.[6] Particle flow is quantified by the measurements of particle size, shape, surface area, pore size, and volume, as well as the bulk densities and coefficients of friction of these materials.

Bulk density (ρ_b) is the density of the uncompressed polymer particles and interparticle voids, whereas the *bulk factor* is the ratio of the solid and bulk densities. Low bulk densities and high bulk factors indicate problems with the flow of resin particles. When the bulk density is less than 0.2 g/cm^3, difficulties arise in conveying particles through the hopper or feed zone of an extruder or injection molding machine.[7]

However, bulk density does not measure the particle compaction which occurs when the particles rearrange themselves or are deformed.[7] Thus, a second measure, *compressibility,* is the percent difference between the loosely packed bulk density and the packed bulk density. When the compressibility is less than 20 percent, particles are free flowing, whereas at values from 20 to 40 percent, the particles are prone to packing during storage.[7] Compressibilities in excess of 40 percent indicate that the material compacts easily and will probably not flow from the hopper without assistance.[7] In addition, water absorption by polymers causes agglomeration of particles and can reduce flowability.[6]

Two other measures of particle flow are pourability and angle of repose. *Pourability* is the time required for a standard quantity of particles to flow through a funnel with dimensions specified by American Society for Testing and Materials (ASTM) standard D1895.[8] Low pourability suggests problems with solids conveying. The *angle of repose* is the angle formed between a pile of material and the horizontal surface. When the angle of repose is greater than 45 percent, the material does not flow.[7]

The *coefficient of friction* affects feeding of polymer particles, and can be divided into two parts: internal and external. The *internal coefficient of friction* is the friction between the layers of polymer particles, while the *external coefficient of friction* occurs between the particles and other (metal) surfaces. In the transport of solid particles by plug flow (solids conveying in the feed zones of extruders), the external coefficient of friction is much greater than the internal coefficient of friction. However, when particles are deformed during flow, the internal coefficient of friction becomes important. While the coefficient of friction is influenced by temperature, sliding speed, contact pressure, metal surface conditions, polymer particle size, degree of compaction, time, and relative humidity,[9] these factors are not fully understood. Consequently, no equation of state describes the relationship between the external coefficient of friction and these factors. In addition, additives, such as lubricants, can change the frictional characteristics of polymers.[6]

5.1.3 Melt viscosity and elasticity

Viscosity is the resistance to flow. As shown in Table 5.2,[10] the viscosity of polymer melts is relatively high when compared to that of water. Thus, the polymer melts generally exhibit laminar flow, that is, the melt moves in layers. Plug flow results when these layers move at the same velocity. However, the layers typically flow at different velocities; for example, the melt at the center of a channel (Fig. 5.3*a*[11]) flows faster than the melt near the channel walls. If the geometry of the melt channel expands (Fig. 5.3*b*[11]) or contracts, the velocity of the melt decreases or increases, respectively. As the layers of melt move relative to each other, they produce shear, whereas a change in the velocity of all layers induces elongation or extension. These phenomena give rise to the terms defined in Table 5.3 and greatly affect melt viscosity.

In Newtonian fluids, such as water, alcohols and other solvents, oligomers, and low-molecular-weight polymers like hydraulic and mineral oils, viscosity is independent of shear rate. However, most polymer melts are pseudoplastic, that is, their viscosity decreases with increasing shear rate. As shown in Fig. 5.4,[12] the decrease in the viscosity of pseudo-

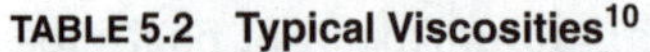

TABLE 5.2 Typical Viscosities[10]

Material	Viscosity, Pa·s
Air	10^{-5}
Water	10^{-3}
Polymer latexes	10^{-2}
Olive oil	10^{-1}
Glycerine	1
Polymer melts	10^{2}–10^{6}
Pitch	10^{9}
Plastics	10^{12}
Glass	10^{21}

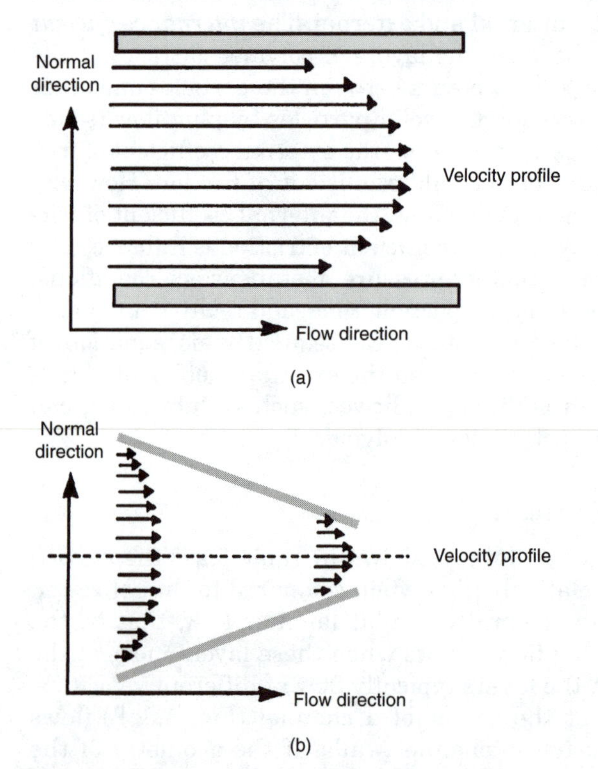

Figure 5.3 Velocity profiles: (*a*) shear flow and (*b*) elongational flow. (*Adapted from Ref. 11.*)

plastic fluids does not occur immediately. At low shear rates, the polymer molecules flow as random coils and the viscosity of the polymer melt is not affected by the increasing shear rate. The constant viscosity of this lower Newtonian plateau is called the zero shear-rate viscosity (η_0).

As the shear rate continues to increase, the polymer chains begin to align in the direction of flow. Since less force (stress) is required to move

TABLE 5.3 Terms Related to Shear Phenomena

Term	Symbol	Definition
Shear:		
Shear rate	$\dot{\gamma}$	Difference in velocity per unit normal distance
Shear stress	τ	Stress required to achieve a shearing deformation
Shear viscosity	η	Resistance to shear flow
Simple elongation:		
Hencky strain rate	$\dot{\epsilon}$	Rate of elongational deeformation
Tensile stress	σ_E	Stress required to achieve simple elongational deformation
Elongational viscosity	η_E	Resistance to elongational flow

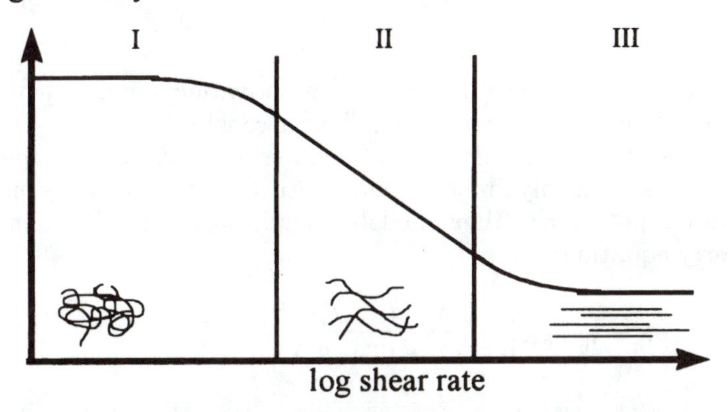

Figure 5.4 The effect of shear on the viscosity of polymeric materials: low newtonian plateau (I), power law region (II), and upper newtonian plateau (III). (*Adapted from Ref. 12.*)

the polymer melt, the viscosity decreases. Higher shear rates further align the polymer chains and the viscosity decrease eventually becomes proportional to the increase in the shear rate. In this power law region, the viscosity and shear rate are related by

$$\eta = k\dot{\gamma}^{n-1} \tag{5.1}$$

where k is the consistency index and n is the power law index. The *power law index* is an indicator of a material's sensitivity to shear (rate), or the degree of non-newtonian behavior. For Newtonian fluids $n = 1$, and for pseudoplastic fluids $n < 1$, with smaller values indicating greater shear sensitivity. At very high shear rates, the polymer chains are, in theory, fully aligned in the direction of flow. Thus, the viscosity cannot decrease further and is constant in the upper Newtonian plateau.

The shear rate varies considerably with processing method (see Table 5.4[13]). Therefore, the degree of alignment, shear thinning, and material relaxation varies considerably with the process. Compression and rotational molding typically induce very little alignment of the polymer chains, and, thus, produce low levels of orientation and retained stress. In contrast, the polymer chains are highly oriented during injection molding, and such parts exhibit high levels of residual stress.

Viscosity is also affected by temperature and pressure. Increasing the temperature increases the mobility of the polymer molecules, and, thus, reduces viscosity. One expression for the temperature dependence of viscosity is given by an Arrhenius equation:

$$\eta = A \exp\left(\frac{E_a}{RT}\right) \tag{5.2}$$

where A is a material constant, E_a is the activation energy (which varies with polymer and shear rate), R is a constant, and T is the absolute temperature. Since the activation energy varies with temperature, Eq. (5.2) is used only for short temperature ranges, typically only in the molten polymer. Other models, particularly the William's-Landel- Ferry equation:

$$\eta = \eta_r \exp\left[-\frac{C_1 (T - T_r)}{C_2 + (T - T_r)}\right] \tag{5.3}$$

where η_r is the viscosity at a reference temperature T_r and C_1 and C_2 are material constants and are used for wider temperature ranges. While pressure increases viscosity, the effects are relatively insignificant when the processing pressures are less than 35 MPa (5000 lb/in^2).[14] At higher pressures, the viscosity increases as an exponential function of pressure, expressed by

$$\eta = \eta_r \exp\left[\alpha_p (P - P_r)\right] \tag{5.4}$$

where η_r is the viscosity at a reference P_r and α_p is an empirical constant with values of 200 to 600 MPa^{-1} (Ref. 15).

TABLE 5.4 Typical Shear Rates for Selected Processes[13]

Process	Shear rate, s^{-1}
Compression molding	1–10
Calendering	10–100
Extrusion	100–1,000
Injection molding	1,000–100,000

Polymer type, molecular weight, molecular-weight distribution, and additives influence viscosity. As shown in Fig. 5.5, zero shear-rate viscosity increases with molecular weight ($\overline{M}_w$) as

$$\eta_0 = K\overline{M}_w^{\,a} \tag{5.5}$$

where K is an empirical constant and a is 1 when the molecular weight is less than the critical molecular weight (M_c) and about 3.4 above the critical value. Since thermoplastic resins typically have molecular weights in excess of the critical molecular weight, longer polymer chains decrease the processability of a material. For extremely long chains, as are present cast in acrylic sheet, conventional melt processing is not possible due to the excessively high viscosity. However, for lower molecular-weight polymers, the onset of shear thinning occurs at higher shear rates. This produces the extended lower newtonian plateau often observed with some grades of polycarbonate, poly(butylene terephthalate), and other materials.

A broad molecular-weight distribution (MWD) typically increases the shear and pressure sensitivity of polymer melts. However, the effect of MWD on zero shear-rate viscosity varies; a broad MWD increases the zero shear-rate viscosity of polyethylene and HIPS, but decreases η_0 for ABS and poly(vinyl chloride). The effect of MWD on temperature sensitivity also seems polymer-dependent.

In general, polymers with flexible chains, such as polypropylene and polyamide-6,6, flow easily, whereas those with rigid chains exhibit higher viscosity. Blending two polymers can significantly alter poly-

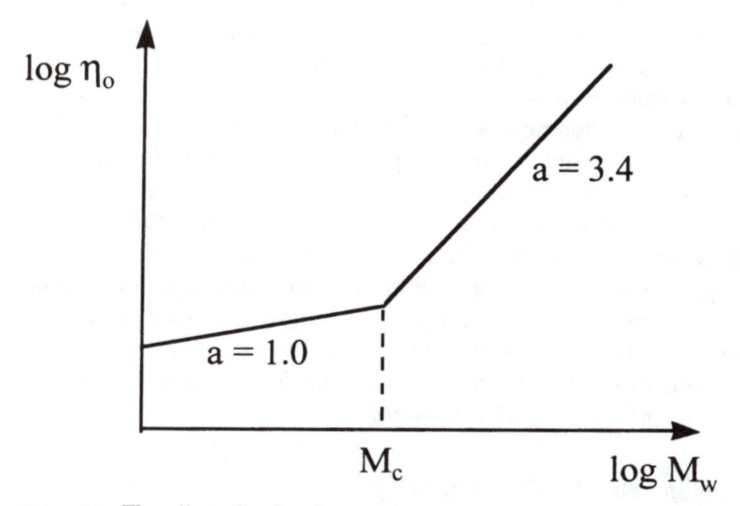

Figure 5.5 The effect of molecular weight on viscosity.

mer viscosity, but the effect depends on the two polymers. Additives also alter melt viscosity. Lubricants typically decrease viscosity and fillers and fibers increase viscosity. However, the effect of colorant varies with the color and dispersants, while the effect of impact modifiers varies with the base system and the impact modifier.

Melt viscosity, or flow, is typically measured using extrusion plastometers (or melt indexers), capillary rheometers, and parallel plate rheometers. The extrusion plastometer measures the flow of a polymer melt under conditions specified by ASTM standard D 1238.[16] This test yields a single, low-shear-rate value which is typically used to specify resins. Capillary rheometers determine viscosity over a range of shear rates in channel flow. While they are subject to error, these rheometers are still the only means of measuring viscosity at high shear rates (typically $\dot{\gamma} > 1000\ s^{-1}$). Parallel-plate rheometers also measure viscosity over a range of shear rates, but the maximum allowable shear rate is about 100 s^{-1}.

Extensional viscosity also varies with strain rate. At low strain rates, the extensional viscosity is given by

$$\eta_E = 3\eta_0 \tag{5.6}$$

where η_0 is the zero shear-rate viscosity. As the strain rate increases, elongation viscosity increases for materials which exhibit extensional thickening, while viscosity decreases polymers which show extensional thinning. The former behavior is typical of branched polymers, such as low-density polyethylene, which maintain a continuous cross section when stretched. The latter phenomenon in which the melt necks (that is, thins in the center) is common with linear polymers like linear low- and high-density polyethylene and polypropylene. Extensional viscosity is enhanced primarily by the longest polymer chains and by the branching of polymer chains.

Since extensional rheometers are limited to low strains and low strain rates, they cannot simulate plastics processing conditions. Thus, melt elasticity, the ability to recover from shear or extensional strain or flow, is used to assess the effects of stretching the polymer melt during plastics processing. It is related to extrudate (die) swell, bubble strength in blown film extrusion, and blow molding. Melt elasticity is usually determined from melt strength measurements in which a molten polymer strand is wound on a drum, and the drum speed increased, until the strand breaks. The force (stress) at which the extrudate breaks is the *melt strength.*

5.1.4 Orientation, relaxation, and shrinkage

Since plastic melts expand upon heating, the melt density (ρ_m) is related to the solid density by

$$\rho_m \cong 0.8\rho \tag{5.7}$$

Plastic melts are also compressible. Thus, a temperature and pressure change of 200°C and 50 MPa, respectively, causes a 10 to 20 percent difference in density, depending on whether the polymer is amorphous or semicrystalline.[17] *Specific volume* (ν), the inverse of density, is often used to relate density to temperature and pressure. As shown in the typical pressure-volume-temperature ($p\nu T$) curves presented in Fig. 5.6,[18] specific volume increases with temperature but decreases with increasing pressure. The isobars exhibit significant changes as the polymer passes through its transition temperature; this is T_g for amorphous

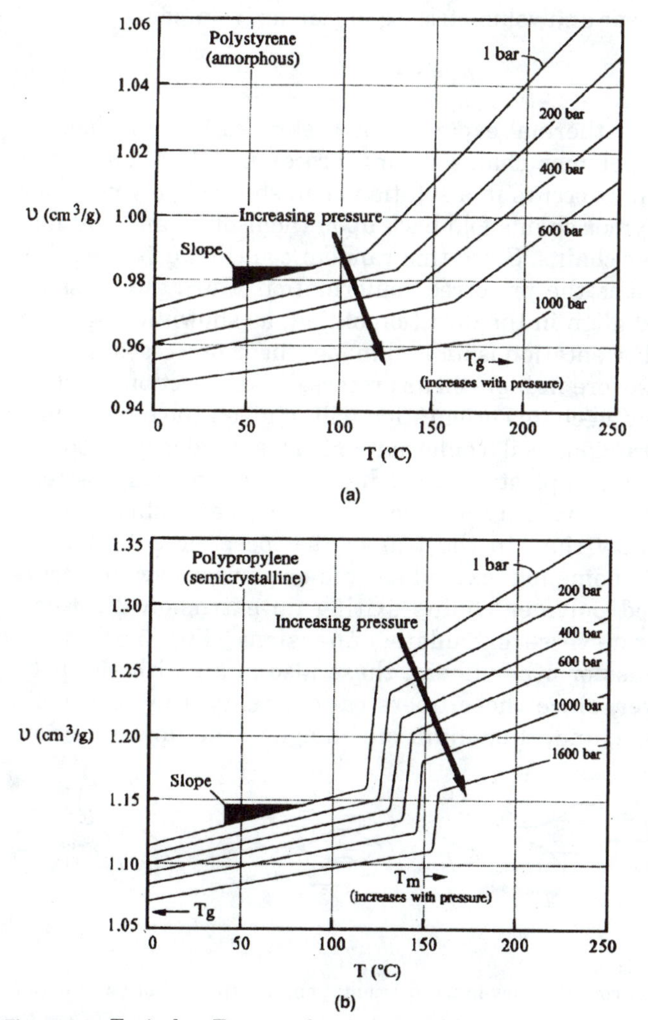

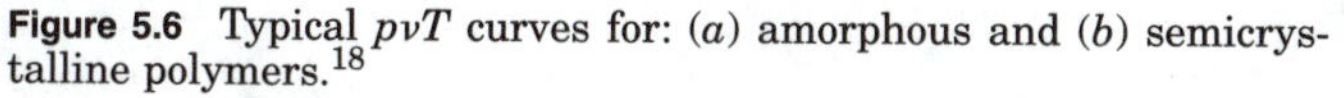

Figure 5.6 Typical $p\nu T$ curves for: (*a*) amorphous and (*b*) semicrystalline polymers.[18]

polymers and T_m for semicrystalline materials. When cooling semicrystalline polymers, the large reduction in specific volume at T_m is due to the formation of tightly packed regions (crystallites). Increasing pressure shifts the transitions shift to higher temperatures as the pressure increases, while rapid cooling not only increases T_g and T_m, but also produces higher specific volume below the transition temperature. With relaxation, molecular rearrangement decreases this specific volume. Consequently, annealing time and temperature affect density.

A polymer's dimensional changes are quantified using expansivity or coefficient of (volumetric) expansion (β or α_v) and the coefficient of linear thermal expansion (α or CLTE). Since volumetric expansion and shrinkage are typically determined from *pvT* data, they are relatively isotropic. Thus, in theory, linear and volumetric expansion are related by

$$\alpha = \frac{\beta}{3} \tag{5.8}$$

In practice, linear thermal expansion and shrinkage are influenced by orientation of polymer chains during processing. In their totally relaxed state, such as occurs in a solution or unsheared polymer melt, the individual polymer chains fold back upon themselves and entangle with other polymer chains. Since this random coil (Fig. 5.7*a*) requires the least energy, it is the preferred conformation. During processing, polymer molecules align in the direction of flow, as shown in Fig. 5.7*b*. This flow-induced orientation is limited in low-shear processes, such as rotomolding, but severe in high-shear processes like injection molding. However, the oriented chains orientation will, if given the opportunity, return to their random coil conformation. This tendency produces anisotropic shrinkage in plastics parts, die swell, and molded-in stress.

As highly oriented chains relax, they produce greater shrinkage in the direction of flow than in the transverse (perpendicular) direction. A similar phenomenon, extrudate (die or parison) swell, occurs when the oriented polymer chains exiting an extrusion die relax, increasing the transverse extrudate dimensions (Fig. 5.8). If the resin contains glass or other fibers, these also align with the polymer flow. However, since these fibers cannot relax, they constrain the polymer chains, and, thus, limit shrinkage in the flow direction.

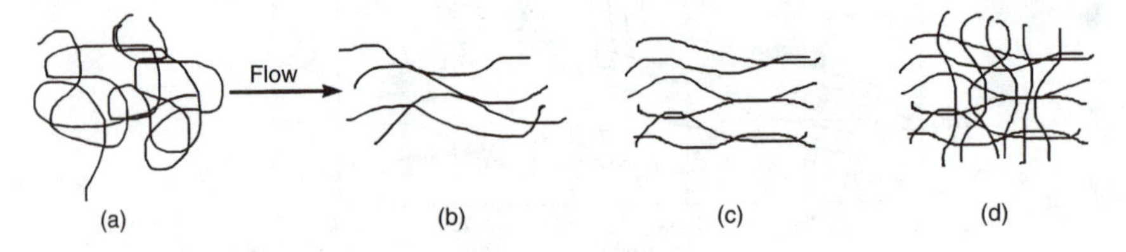

Figure 5.7 (*a*) Random coil. (*b*) Flow-induced orientation. (*c*) Uniaxial orientation. (*d*) Biaxial orientation.

In such materials, transverse shrinkage is usually greater than shrinkage in the flow direction. When the polymer is cooled so rapidly that the polymer chains do not have sufficient time to relax, this orientation produces retained or molded-in stress.

Alignment of the polymer chains improves plastics' properties. During extrusion, uniaxial orientation (Fig. 5.7*c*) is often increased by drawing the extrudate after it exits the die; this enhances mechanical strength in synthetic fibers and flat film. Biaxial orientation (Fig. 5.7*d*), which occurs in both the flow and transverse directions, produces high strength in blown film and stretch blow-molded bottles. The orientation and relaxation of polymer is also used for products such as strapping, heat-shrink tubing, and packaging film.

5.1.5 Crystallization

Whether oriented or in random coils, most molten polymers have no well-defined morphology. The exception is liquid crystalline polymers (LCP) which are rod shaped in the melt. As they are cooled, some polymers remain shapeless (amorphous), while segments of other polymer chains form well-defined structures called *crystallites*. Atactic (crystal) polystyrene, atactic polymethylmethacrylate, stereo-irregular aliphatic-aromatic polyamides, polysulfones, random copolymers such as EPR, and polymer blends like modified PPO are always amorphous because their irregular structures cannot be ordered. Semicrystalline polymers usually have a polymer's regular structures that can pack into tightly ordered crystallites. Thus, the polymer is a mixture of crystallites joined by amorphous regions. Since the level of crystallinity depends on processing conditions, these polymers can be completely amorphous.

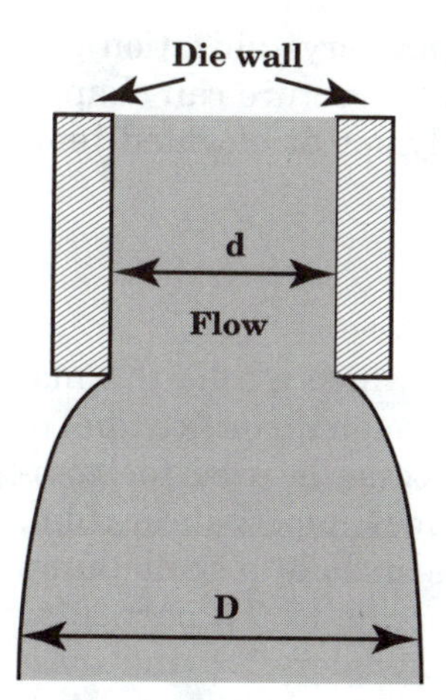

Figure 5.8 Die swell.

Crystallization is the competition between two processes: nucleation and crystal growth. *Nucleation* is the formation of small sites (nuclei) from which crystallites can grow. Primary nucleation creates the initial nuclei. Crystallites develop around these nuclei. Then in secondary nucleation, the surfaces of the crystallites are nucleated. More polymer chains diffuse to the crystallite surfaces and growth continues.

Crystallization is influenced by cooling rate, pressure, shear and orientation, and the structure and molecular weight of the polymer. For homogeneous or spontaneous nucleation, rapid cooling creates more local irregularities in the polymer melt, from which the crystals grow. When spontaneous nucleation does not provide sufficient nuclei, heterogeneous nucleation, the use of secondary phases like talc nucleating agents, facilitates nucleation. Since the growth of crystals requires chain mobility, it is favored by slow cooling and low pressures. Thus, the rate of crystallization depends on the available nuclei and rate of molecular transport.

The orientation and structure of the polymer melt affects the structure of the crystallites. With free crystallization (no stress or strain), crystals grow isotropically from the nuclei, but in oriented melts, crystals have anisotropic "row-nucleated structures."[19] Liquid crystal polymers form very different structures than flexible polymers. Polymers with flexible structures easily form crystallites rapidly, while more rigid polymers need more time. Thus, polyethylene, polypropylene, polyacetal, and polyamide-6,6 crystallize readily and to high levels. The maximum crystal growth rates are 30,000 and 3,000 nm/s for high-density polyethylene and polyamide-6,6, respectively, whereas it was 100 nm/s for the more rigid PET.[20] Polymers with very rigid structures, such as syndiotactic polystyrene, often crystallize so slowly (at 3 nm/s)[20] that crystallization occurs along shear bands within a plastic part.[21]

When the T_g is less than ambient temperature, crystallization can continue for weeks after processing and crystal structure can change after molding. Annealing (conditioning the polymer at elevated temperature) also permits further crystallization.

5.2 Extrusion

Extrusion is a polymer conversion operation in which a solid thermoplastic material is melted or softened, forced through an orifice (die) of the desired cross section, and cooled. The process is used for compounding plastics and for the production of tubes, pipes, sheet, film, wire coating, and profiles. All extrusion lines include a melt pump called an extruder, but other equipment is specific to the particular process. Although there are many types of extruders,[22–24] the most common types are single-screw extruders, intermeshing twin-screw extruders, and ram extruders for special processes.

5.2.1 Single-screw extruders

A single-screw extruder (Fig. 5.9[25]) consists of a screw in a metal cylinder or barrel. One end of the barrel is attached to the feed throat while the other end is open. A hopper is located above the feed throat and the barrel is surrounded by heating and cooling elements. The screw itself is coupled through a thrust bearing and gear box, or reducer, to a drive motor that rotates the screw in the barrel. A die is connected to the "open" end of the extruder with a breaker plate and screen pack (or a screen changer) forming a seal between the extruder and die.

During extrusion, resin particles are fed from the hopper, through the feed throat of the extruder, and into the extruder barrel. The resin falls onto the rotating screw and is packed in the first section or feed zone of the screw. The packed particles are melted as they travel through the middle section (transition or compression zone) of the screw, and the melt is mixed in the final section or metering zone. Pressure generated in the extruder forces the molten polymer through the die.

Extruder drive motors must turn the screw, minimize the variation in screw speed, permit variable speed control (typically 50 to 150 r/min), and maintain constant torque. In selecting drive motors, the three major factors are: (1) base speed variation, (2) the presence or absence of brushes, and (3) cost. The speed variation of a drive motor is based on the maximum speed available for the motor. Since this variation does not change when the speed is reduced, screw speed,

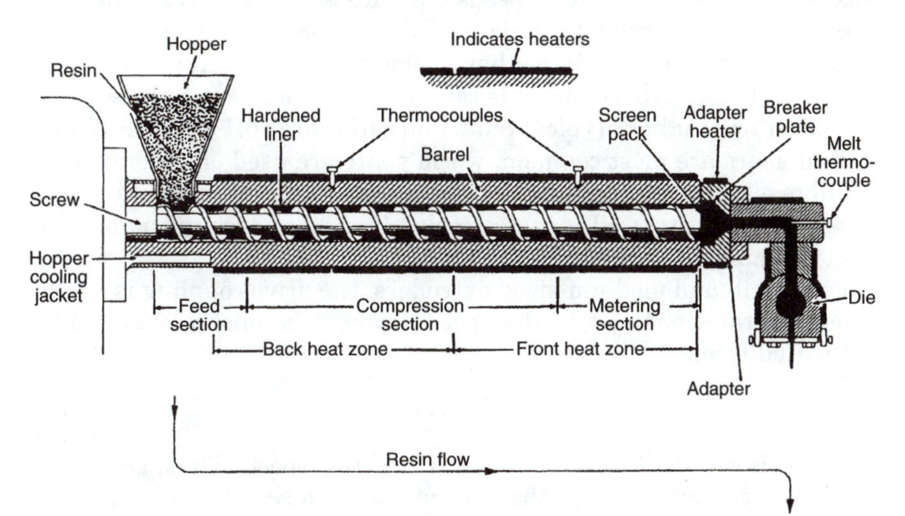

Figure 5.9 Single-screw extruder.[25]

which is generally 5 to 10 percent of the motor speed, varies more than the motor speed. For example, a 0.1 percent base-speed variation on a motor with a maximum speed of 1750 r/min produces a speed variation of ±1.75 r/min. If the maximum screw speed is 117 r/min (reduction ratio of 15:1), the screw speed variation is ±1.75 r/min or 1.5 percent. Brushes in the drive motor are subject to corrosion and may not be used with plastics like poly(vinyl chloride).

The three basic types of drives are alternating current (ac), direct current (dc), and hydraulic. While a number of drives have been used in extruders,[26,27] the most common are dc silicon control rectified (SCR) and ac adjustable frequency drives. A dc SCR drive is a solid-state dc rectifier connected to a dc motor. The base speed is about 1 percent, but reduces to 0.1 percent when a tachometer is added to the drive. These drives are very reliable, can handle high starting torques, can maintain a constant torque through a speed range of 20:1, and are relatively easy to maintain (that is, replace brushes). However, since the drives have brushes, they are limited to noncorrosive polymers.

In contrast, an ac adjustable frequency drive consists of a solid-state power supply connected to an ac "high-efficiency" or "vector" motor. The power supply converts three-phase ac line voltage to variable voltage dc power and then back to controlled ac frequency. Since the voltage-to-frequency ratio is adjusted to provide constant torque from the ac motor, speed-torque characteristics can be optimized by varying the voltage-to-frequency ratio. These motors provide constant torque up to base speed, "have operational ranges of 1000:1 with an encoder and 100:1 without one,"[27] give a base speed variation of 0.01 percent, have a higher power factor (than dc SCR) at low speeds, and are brushless. While they are usually more expensive than dc SCR drives, price reductions and the improved base-speed variation have made them competitive.[27]

The high-speed drive motor is coupled to the low-speed screw using a reducer or gear box. Typical reduction ratios are 15:1 or 20:1.[28] While helical gears are most common, worm gears are used on older or very small machines. A forced lubrication system allows oil to cool the bearings and gears; this oil is water-cooled by a heat exchanger in high-load machines.

For small- and medium-sized extruders, the drive coupling is a belt. This facilitates changes in the speed-torque relationship that can be estimated from

$$P = NT \tag{5.9}$$

where P is power, T is torque, and N is screw speed. Thus, switching to larger gears increases the torque but reduces the screw speed. However, to increase available power, the drive motor, and usually the gear box, must be replaced. For large drives ($P > 225$ kW), the drive motor is coupled directly to the screw.

As shown in Fig. 5.10,[29] a thrust bearing supports the screw and couples the gear box to the screw. The bearing sees high temperature, high pressure, and possible contamination. Although there are various bearing designs, the thrust bearing should last the life of the extruder (10 or more years). The life expectancy is given by the B-10 life, the number of hours where 10 percent of the bearings will fail when tested at a standard pressure and screw speed. If the operating head pressure and/or screw speed is greater than the "B-10 standard," then the bearing life is reduced. The new life expectancy can be estimated using

$$\text{B-10}(P, N) = \text{B-10}_{\text{Std}} \left(\frac{N_{\text{Std}}}{N}\right) \left(\frac{P_{\text{Std}}}{P}\right)^{K} \tag{5.10}$$

where N is the screw speed, P is the pressure, and K is a bearing constant.[30]

The feed throat fits around the first few flights of the screw and is usually separate from the barrel of the extruder. It is insulated from the barrel and cooled with water to prevent bridging and premature melting of the resin particles. The feed port is the opening in the feed throat. Standard feed ports (Fig. 5.11*a*[31,32]) are round or square and should match the geometry and size of the hopper opening. Although these ports are suitable for plastics pellets and some granules, specialized designs are employed with other materials. An undercut feed port (Fig. 5.11*c*[31,32]), which exposes the bottom of the screw, is used for rolls and strips of film or fiber, for film scrap, and for polymer melts. A sloped feed port (Fig. 5.11*b*[31]) is better suited to irregularly shaped particles, whereas a tangential feed port (Fig. 5.11*d*[32]) can be used for powders and regrind.

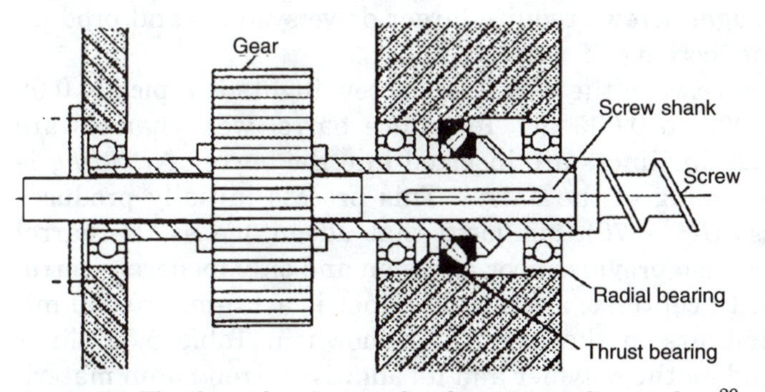

Figure 5.10 The thrust bearing couples the screw to the gear box.[29]

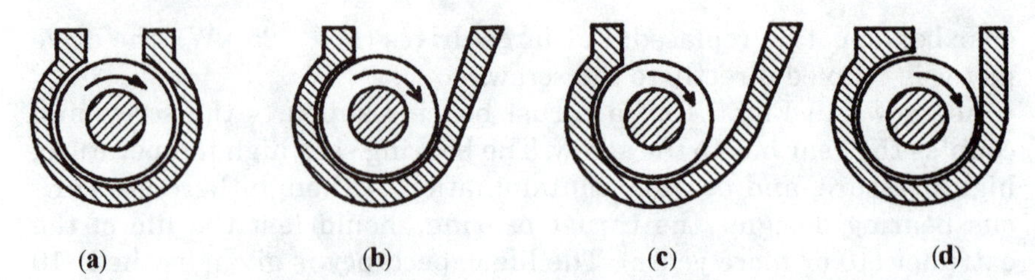

Figure 5.11 Feed port designs: (*a*) standard feed port (*adapted from Refs. 31 and 32*), (*b*) undercut feed port (*adapted from Refs. 31 and 32*), (*c*) sloped feed port (*adapted from Ref. 31*), and (*d*) tangential feed port (*adapted from Ref. 32*).

The feed hopper feeds material to the extruder. As shown in Fig. 5.12*a*,[33] single-screw extruders are usually fed gravimetrically through standard conical or rectangular hoppers. Although pellets and some granules flow smoothly in these hoppers, powders and other particles often require modifications for proper feeding. A spiral hopper (Fig. 5.12*b*) improves dry flow, while vibrating pads or hammers are sometimes attached to hoppers to break up bridges (blockages at the base of the hopper). Vacuum feed hoppers reduce the trapped air that hinders proper feeding. In crammer feeders (Fig. 5.12*c*[34]), an auger forces material into a barrel, whereas metered (starve) feeding (Fig. 5.12*d*[33]) uses an auger to feed a set amount of material to the barrel. Starve feeding not only minimizes bridging and air entrapment but also prevents vent flooding in vented barrel extruders.

The barrel is a metal cylinder that surrounds the screw. One end fastens to the feed throat and the opposite end connects directly to the die adapter. Since extruder barrels must withstand pressures up to 70 MPa (10,000 lb/in^2), they are usually made from standard tool steels, with special tool steels required for corrosive polymers (see Table 5.5[35]). Extruder barrels typically have length-to-diameter (L/D) ratios of 24:1 to 36:1, but they can be larger. Since melting occurs over a longer transition zone, longer barrels provide increased output. However, the longer screws require larger drive systems and produce greater screw deflection.

The clearance between the barrel and screw flights is typically 0.08 to 0.13 mm (0.003 to 0.005 in). To reduce barrel wear, barrels are nitrided or bimetallic liners are inserted into the barrel. *Nitriding* is the surface hardening of the barrel. This process initially produces higher hardness ($R_c \sim 70$), but loses that advantage as the barrel wears. Nitriding also provides poor abrasion and only moderate corrosion resistance. In contrast, a bimetallic liner is a 1.5-mm (0.060-in)-thick sleeve that fits in the barrel. As shown in Table 5.5,[35] liner materials depend on the polymer and its additives. Iron/boron materials are used as general-purpose liners, whereas nickel/cobalt liners

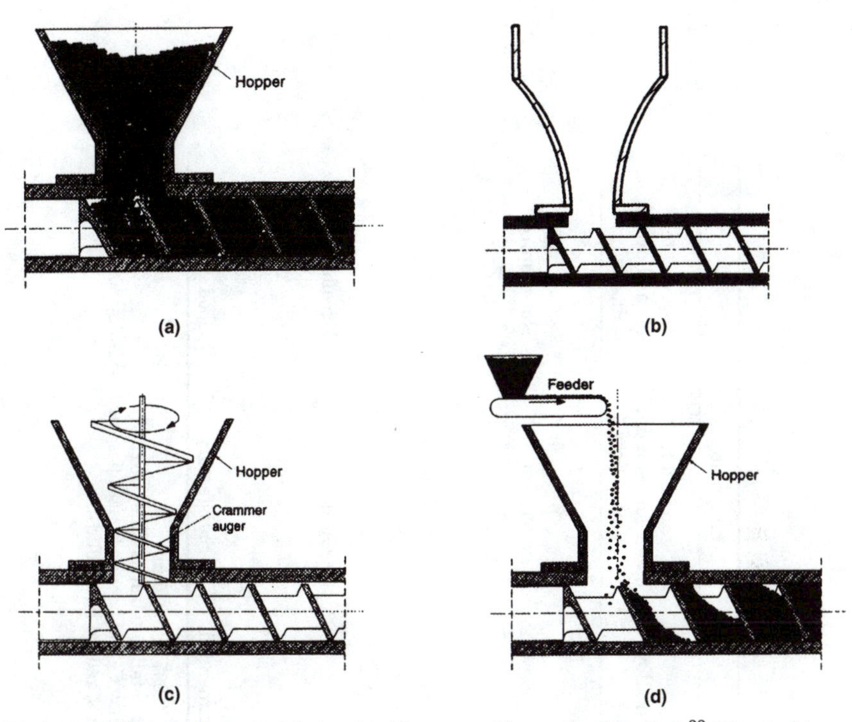

Figure 5.12 Hopper designs: (*a*) standard hopper with gravimetric feed,[33] (*b*) spiral hopper, (*c*) crammer feeder,[34] and (*d*) standard hopper with metered feed.[33]

increase chemical resistance at the expense of wear properties. Iron/boron carbide is used for high levels of abrasion resistance and vanadium/high carbon alloy steel provides the highest abrasion resistance. Bimetallic liners are favored for single-screw extruders, but nitriding is preferred for twin-screw extruders (due to the expense of the bimetallic sleeves).

Although barrel and barrel liners typically have smooth surfaces, a liner or barrel with axial grooves can be installed in the feed section of the extruder (Fig. 5.13[36]). Groove depth is greatest at the feed throat and gradually decreases with axial distance. Since the grooves increase shear, friction, and pressure, they improve extruder output. However, the feed section requires additional cooling, thermal insulation must be installed between the grooved section and the rest of the barrel, the feed zone must also be able to withstand higher pressures (typically 100 to 300 MPa), and significant wear occurs with abrasive materials. Grooved barrel liners were developed for low-bulk-density materials, but have increased output for regular pellets.[36–38]

The breaker plate (Fig. 5.14*a*) acts as a seal between the extruder barrel and the die adapter, thus preventing leakage of the melt. The

TABLE 5.5 Barrel Materials[35]

Resins being processed		Lining materials		
		Nitrided alloy steel*	Standard tool steels and standard bimetallics†	Special tool steels and premium bimetallics‡
Mild:	Acetates, polyethylenes, polypropylenes, polystyrenes, PET	Good	Excellent	Not required
Medium:	ABS, polyacetals, acrylics, polyamides, PVC, polcarbonates, polyesters, SAN	Satisfactory§	Excellent	Not required
Severe:	Resins containing up to 30% glass, mineral, flame-retardant and other filler material	Poor	Good	Excellent
Critical:	Fluoropolymers, phenolics, resins containing more than 30% glass, mineral, flame-retardant and other filler material	Not recommended	Satisfactory	Good

*AISI 4130-4140 nitrided to a depth of 0.18 to 0.38 mm (0.007 to 0.015 in).
†Includes D-2 tool steel and centrifugally cast bonded bimetallics (such as Xaloy 101).
‡Includes A-11 (CPM10V) tool steel, Inconel 718, Hastalloy C, and bonded bimetallics (such as Xaloy 800).
§ABS and polycarbonate require a standard tool steel or standard bimetallic liner.

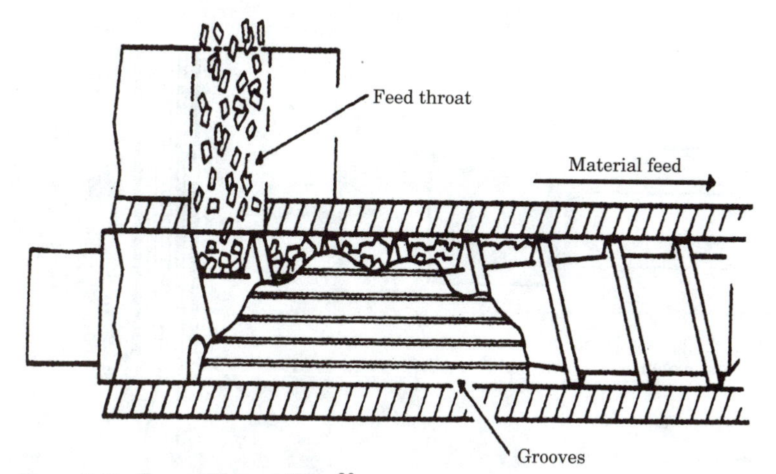

Figure 5.13 Grooved barrel liner.[36]

breaker plate also supports the screen pack, develops head pressure (restricts flow), and converts the rotational motion of the melt to axial motion. The screen pack filters melt for contamination and gel particles, generate head pressure, and minimize surging (pulsing of the melt). Five or more screens are used in a typical screen pack; screens are rated by the number of holes per millimeter (or inch). As shown in Fig. 5.14*b*, the screens become finer as they approach the breaker plate. A coarse screen next to the breaker plate supports the finer screens and prevents the melt pressure from forcing them through the breaker plate. Although the selection of screen sizes depends on the material and extrusion process, increasing the number of screens or the mesh size increases the pressure developed during extrusion.

Since screen packs become blocked by contaminants, they must be changed periodically. This requires separating the extruder and die adapter, removing the breaker plate, inserting a new breaker plate with fresh screens, and then reconnecting the extruder and die adapter. Thus, screen changers that switch screens without stopping the extrusion operation are often incorporated into extrusion lines. Discontinuous screen changers, such as the sliding plate, interrupt flow, whereas autoscreen and rotary screen changers permit continuous operation. The flow interruption limits the use of discontinuous screen changers with heat-sensitive materials. Autoscreen and rotary screen changers tend to leak, especially with low-viscosity melts like polyamide-6,6.

A sliding-plate screen changer (Fig. 5.15*a*[39]) has two breaker plates mounted on a support plate; a similar screen changer uses cartridge filters in place of the breaker plates. The breaker plates are manually or hydraulically switched when the pressure drop across the breaker plate exceeds a preset value. Then the dirty breaker plate and screens are

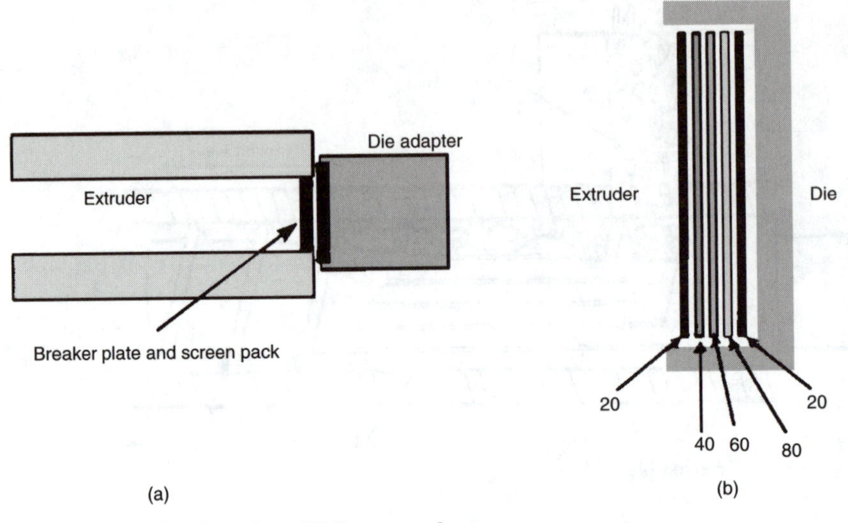

Figure 5.14 (*a*) Breaker plate. (*b*) Screen pack.

replaced. In an autoscreen system (Fig. 5.15*b*[40]), a steel screen is held against a fixed breaker plate. When the pressure drop exceeds the preset value, heaters melt the polymer at the dirty end of the screen and the screen, which is under tension, advances. With a clean screen in the melt stream, the heaters turn off and the polymer at the dirty end of the screen solidifies, preventing further screen movement. As shown in Fig. 5.15*c*,[41] the rotary screen is a wheel containing a series of crescent-shaped breaker plates. The wheel is moved in response to preset pressure limits and dirty breaker plates are replaced by fresh ones.

A rupture disk is located in the extruder barrel just before the breaker plate. When the extruder pressure exceeds the disk's rated value, the rupture disk opens, thereby reducing the pressure. Rupture disks are typically rated for 34.5, 51.7, and 70.0 MPa (5000, 7500, and 10,000 lb/in^2). They are required for operator safety.

Barrels, dies, and die adapters are heated to bring them to operating temperatures and to maintain set temperatures during operation. Although electrical-resistance heater bands are typically used for heating extruder barrels and dies, occasionally the barrel temperature is maintained by an oil jacket surrounding the barrel. Table 5.6[42] lists the four types of heater bands used in plastics processing. Mica heating bands are nichrome bands sandwiched between mica. While they provide rapid heating and cooling, they lose intensity (power) with time, burn out often, and must contact the metal surface for best performance. Thus, they are used primarily on dies and nozzles. With cast aluminum bands, nichrome heating elements are cast in aluminum

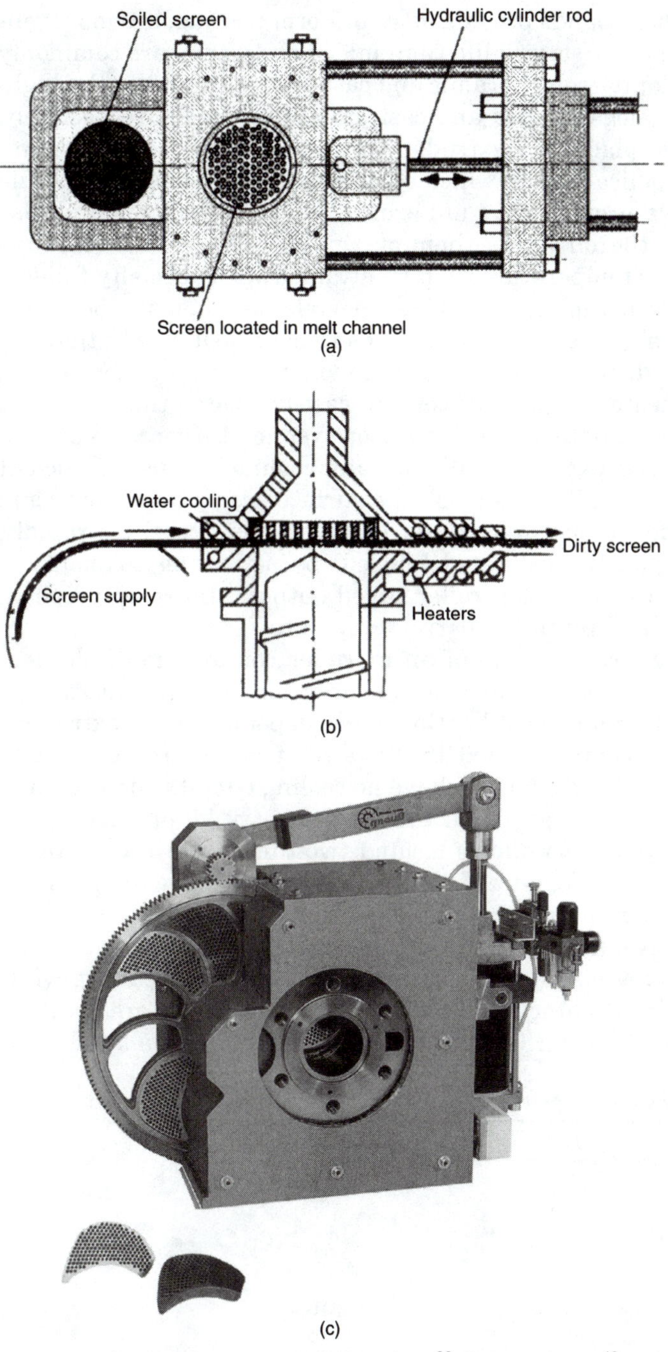

Figure 5.15 Screen changers: (*a*) sliding plate,[39] (*b*) autoscreen,[40] and (*c*) rotary screen.[41]

blocks. Since these bands provide more uniform heating, are more reliable, and give longer service life than mica bands, they are commonly used for extruder barrels. Ceramic and cast bronze bands are the high-temperature analogs of mica and cast aluminum bands, respectively. The former are placed on extruder barrels, whereas the latter are employed for applications in which ceramic bands are too bulky. While jacketed barrels provide very uniform barrel temperatures and also cool efficiently, the oil limits their maximum temperature to 250°C (although higher temperatures can be attained with specialty fluids).

To maintain constant temperatures, barrels must usually be cooled by fans (blowers) or water. Although fans remove heat slowly, they are inexpensive, and, thus, are the most commonly used. Water cooling requires cast heater bands that contain cast-in cooling tubes through which water is circulated. The closed-loop system also contains a tank, a pump, and a heat exchanger. While water cooling is twice as efficient as forced air cooling, it depends on the surface area of the tubing and the speed of the water. Direct water cooling tends to cool too rapidly, so water is typically modulated (pulsed) by activating solenoids in response to temperature-controller timed outputs. If set up properly, this provides better control of barrel temperature.

Each heating circuit (zone) of an extruder contains multiple heat bands, cooling fan(s), a temperature sensor, and a temperature controller. While the number of heating zones depends on the extruder's L/D, each zone contains 3500 to 4000 W of power. In contrast to extruders, dies and die adapters have no cooling circuits and the number of heating zones depends on the die geometry. Temperature sensors are usually thermocouples located two-thirds into the barrel or die thickness; these "deep-well" sensors provide accurate temperature readings and are relatively insulated from air currents.

Extruder screws fit into the barrel and are supported by the thrust bearing. The screw's shank length fits into the thrust bearing, while the flighted length contacts the plastic. Extruder screws are specified by their outside diameter (D) and the L/D, which is given by

TABLE 5.6 Heater Band Properties[42]

Heater band	T_{max}, °C	Power density, kW/m^2
Mica:		
Conventional	480	85
High performance	760	155^b
		356^n
Aluminum:		
319 or 356	370	310
443	425	310
Ceramic	~750	35
Bronze	870	310

$$L/D = \frac{L_{\text{flight}}}{D} \tag{5.11}$$

where L_{flight} is the length of the flighted section, with or without the feed pocket. Screw features, such as the flights and channel nomenclature are shown in Fig. 5.16.[43] *Pitch* (t) is the axial distance from center of one flight to the center of the next flight, whereas *lead* is the axial distance the screw moves in one full rotation. Since most extruders use square-pitched metering screws, both the pitch and lead equal the diameter, and the helix angle is 17.7°. The helix angle (ϕ) can also be calculated as

$$\phi = \tan^{-1}\left(\frac{t}{\pi D}\right) \tag{5.12}$$

In metering screws, the flighted section is divided into three zones: feed, transition or compression, and metering. The feed zone has a constant channel depth as does the metering zone. However, the channel depth gradually decreases in the transition zone. Since molten polymer requires less volume than the solid particles, the metering zone channel depth is shallower than the depth of the feed. The squeezing of the polymer is quantified by the compression ratio (CR):

$$\text{CR} = \frac{V_{\text{first channel}}}{V_{\text{last channel}}} \cong \frac{H_{\text{first channel}}}{H_{\text{last channel}}} \tag{5.13}$$

where V is the channel volume and H is the channel depth.* Low compression ratios do not fully pack the solid particles, and so the melt will contain air bubbles. In contrast, high-compression ratios deliver too much polymer to the metering zone, and with polyolefins, produce melting problems in the transition zone. Thus, typical metering screws have compression ratios of 1.5:1 to 4.5:1.

Since single-screw extruders provide relatively poor mixing, mixing elements are added to the screw's metering zone to improve mixing. Figure 5.17[44,45] presents a selection of mixing elements. Distributive mixing elements randomize the melt, while dispersive mixing elements provide shear (mechanical action) that reduces particle size. Although no element is purely distributive or dispersive, they are grouped according to the dominant mixing produced by the element.

As shown in Fig. 5.17[44,45] *mixing pins* are rows of metal pins inserted into the root diameter of a screw, while *slotted flights* are slots that are cut

*Since CR calculations ignore the screw radii, the volumetric compression ratio is always smaller than the depth compression ratio.

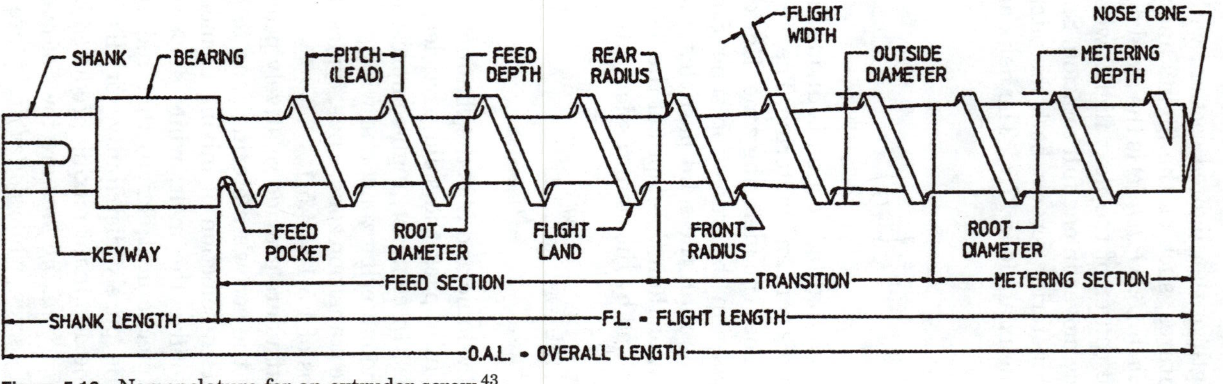

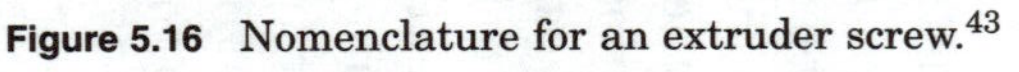
Figure 5.16 Nomenclature for an extruder screw.[43]

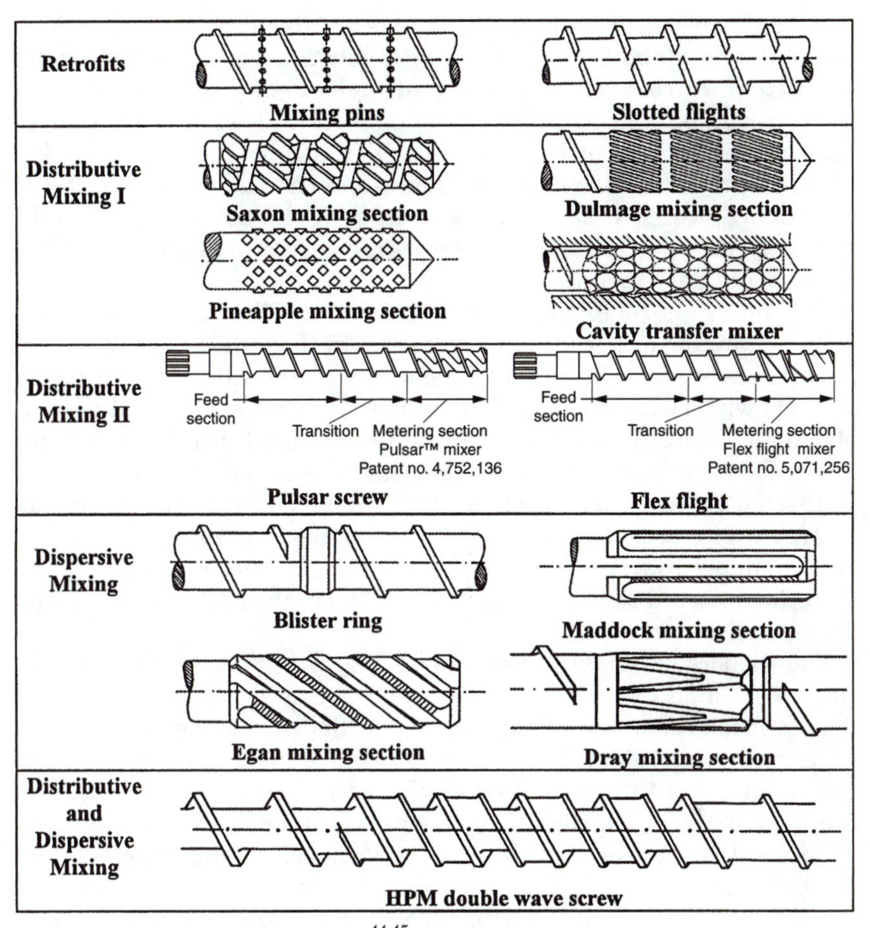

Figure 5.17 Selected mixing sections.[44,45]

in the flights of the metering section of a screw. Both are simple and easily adapted to existing screws, but provide the potential for material stagnation and a limited degree of mixing. In addition, material flows back through the slotted flights; this increases residence time and provides an opportunity for degradation of heat and shear-sensitive materials.

Some distributive mixing elements (Fig. 5.17[44,45]) effectively mix, but provide no forward conveyance of the melt. These elements are usually incorporated into the last three or more channels of a screw. In the dulmage mixing section, the polymer is divided into 10 to 12 narrow channels, recombined, and then divided again. This produces excellent mixing in foam screws and for other applications. The saxton mixing section contains many minor flights on a helix angle that differs from that of the primary flights. The flights create new channels which divide the flow, and flow is recombined between segments of the mixing section. In contrast, with the pineapple mixing section, the polymer stream is divided and recombined as it flows around pinlike obstructions in the mixing element. For a cavity transfer mixing (CTM) section, an entire section is added to the end of the screw and extruder. Since cavities are present in both the screw and "barrel," melt is divided and recombined as it is transferred from screw to barrel cavities and back again.

The third group of mixing elements in Fig. 5.17[44,45] provides both distributive mixing and forward conveyance. In the pulsar mixing section (screw), the metering section of the screw is divided into alternating sections with either deep or shallow root diameters. Material is tumbled as it is forced from one section to the other. Since this produces good mixing without excessive shearing, the screw can be used for heat-sensitive polymers such as PVC. A flex flight mixing section incorporates a second flight in the screw's metering zone; this creates two channels that vary in width. As channel width decreases, the material is forced over the second flight to produce both shearing and a tumbling action.

The dispersive mixing elements in Fig. 5.17[44,45] include a blister ring, the Union Carbide or Maddock mixing section, and variants of the Maddock section. A blister ring is a cylindrical screw section with a clearance of 0.50 to 0.76 mm (0.020 to 0.030 in) that is added to the root diameter of the screw. As melt flows through the tight clearance, it is only sheared. This breaks up gel particles, but provides no mixing or forward conveyance. In contrast, a Maddock mixing element consists of axial grooves that are alternately open to the upstream and downstream sections of the screw. The grooves are separated by mixing and wiping lands with clearances of 0.64 mm (0.025 in) and 0.013 (0.005 in), respectively. As shown in Fig. 5.18,[44] polymer flows into the former set of grooves. Since the axial downstream discharge is blocked, the polymer is forced over a mixing land and into the next channel. This channel is blocked upstream but open to downstream discharge. A wiping land ensures that the polymer exits the groove. Maddock mixing sections, which are usually inserted one-third of the way down or at the end of the metering zone, reduces the size of gel particles and breaks up clumps of filler and pigment. However, they provide too much shear for heat- or shear-sensitive polymers, tend to increase melt temperature in some polymers, and do not convey the melt.

The Egan and dray mixing sections represent improvements on the Maddock mixing section. In the Egan section, the grooves are placed at an angle to the screw axis, and the groove depth slowly decreases from the entry to the end of groove. While this reduces the pressure

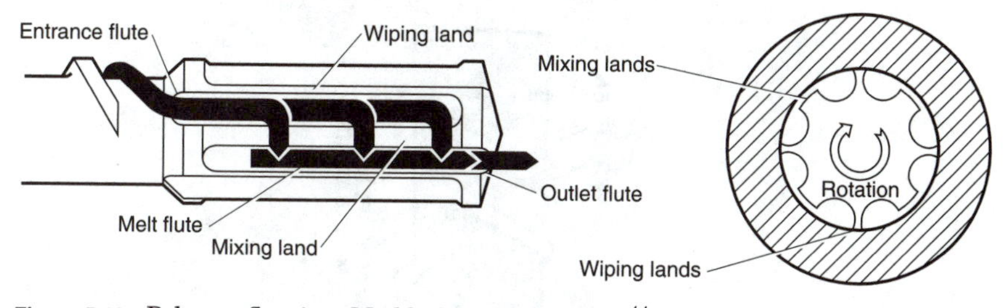

Figure 5.18 Polymer flow in a Maddock mixing section.[44]

drop typical of Maddock sections, an optimized section design can actually generate pressure (and so increase output). With the dray mixing section, outlet channels are open at the start of the mixing section. Thus, not all material is forced over the mixing land; some just flows through the section. This reduces the pressure drop observed in Maddock sections, but provides a nonuniform shear on the polymer melt.

The double wave screw (Fig. 5.17[44,45]) provides both distributive and dispersive mixing. This screw has a second flight in the metering zone. The depth of both channels alternately increases and decreases. When the channel depth is shallow, melt is sheared and can be transferred into the adjacent channel.

In addition to mixing sections incorporated into the screw, static mixing elements can be added to the end of extruder screws. They provide distributive mixing and can improve melt temperature uniformity. However, they severely restrict flow, producing large pressure drops, and also permit stagnation and possible degradation of the melt. Consequently, they are better suited to low-viscosity materials. A Kenics mixer is a helical spiral in which helix divides the melt stream, whereas a Ross mixer has multiple elements, each containing four holes. The holes divide melt stream, and, thus, mix the melt. In a Koch mixer, multiple elements of corrugated surfaces are placed at angles to each other.

Although most extruders use metering screws, other screws are available. Of these, the most significant are barrier screws and two-stage screws. Barrier screws (Fig. 5.19[46]) facilitate melting. While there are many designs, all are double-flighted in the transition zone, have primary (solids) channels with decreasing volumetric capacity, and a secondary (melt) channel with increasing volumetric capacity. Two-stage screws (Fig. 5.20[47]) are used in vented barrel extruders, single-screw extruders with a vent section. The screws have five zones: feed, transition, first metering, vent, and second metering. Polymer is conveyed and melted in the first three zones. Then the pressure is reduced in the vent section so that volatiles will exit the extruder. Finally, pressure is regenerated in the second metering zone. Pump ratio (PR)

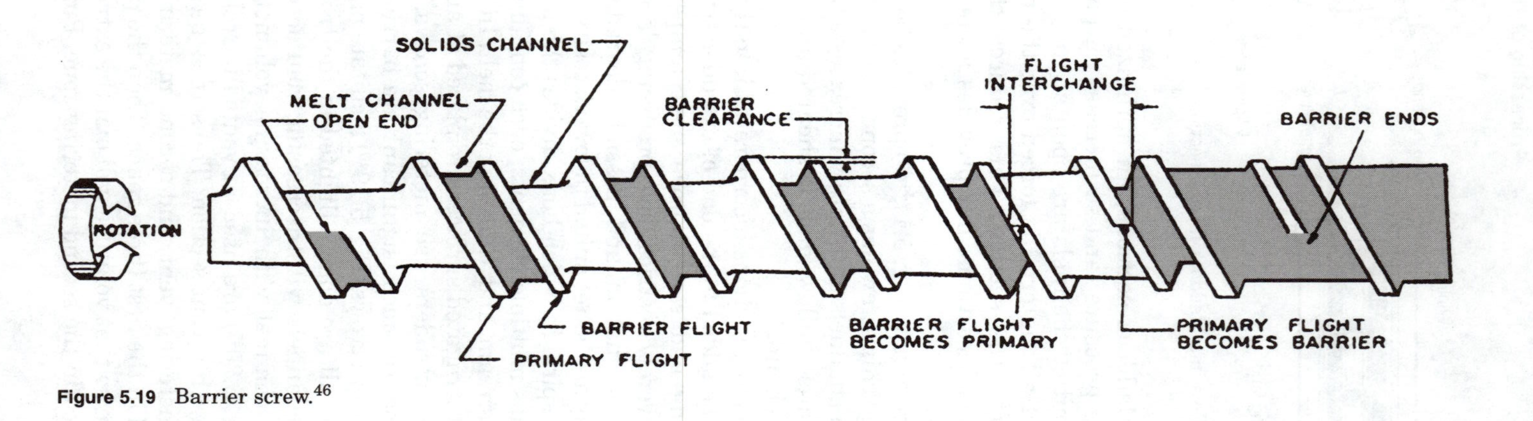

Figure 5.19 Barrier screw.[46]

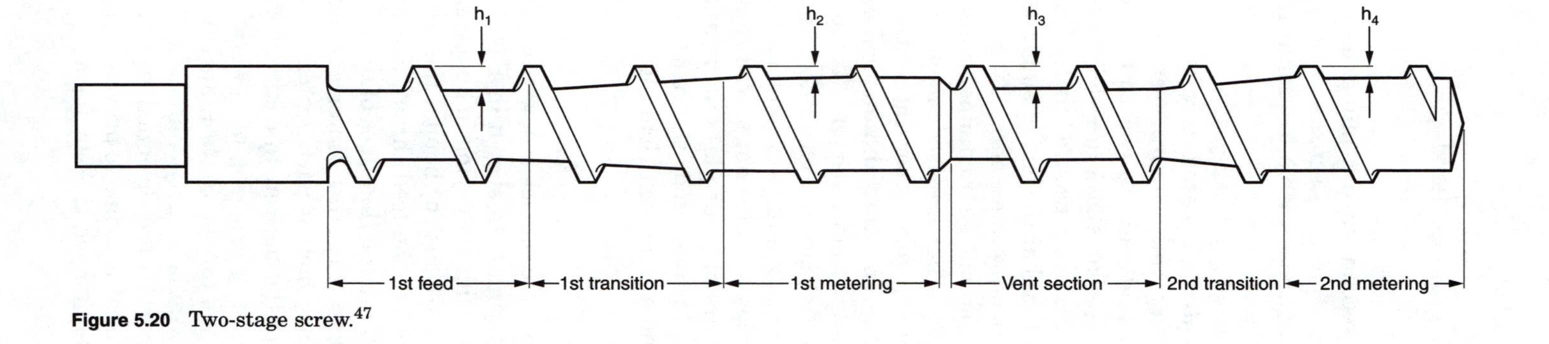

Figure 5.20 Two-stage screw.[47]

$$\mathrm{PR} = \frac{H_{\text{second metering}}}{H_{\text{first metering}}} \tag{5.14}$$

is the ability of the second stage to pump more material than the first stage delivers to it. Typically, screws with a high pump ratio will exhibit surging, while those with a low compression ratio will produce vent flow.

Screws are made from good tool-grade steels or stainless steel. They are then treated, coated, or plated to reduce the coefficient of friction between screw and polymer, improve chemical resistance, improve abrasion resistance, and reduce wear. In high-wear areas, particularly the flight lands, special alloys are welded to the base steel of the screw. Screw materials are presented in Table 5.7.[48]

Gear pumps are placed at the end of single- or twin-screw extruders to increase or stabilize pressure generation. The two counterrotating gears convey melt from one gear channel to the next until the material exits the gear pump. They are used on extruders with poor pressure-generating capabilities, such as corotating twin-screw extruders, and in processes, like fiber spinning and medical extrusion, where output stability must be better than 1 percent. The pumping is accompanied by viscous heating that can increase melt temperatures up to 20 to 30°C. Consequently, gear pumps cannot be used with heat-sensitive polymers. Gear pumps are also not suited to polymers with abrasive components. The gears degrade glass fibers and other such additives, while the abrasive additives erode the gears. Finally, gear pumps provide poor mixing and are not self-cleaning.

5.2.2 Extrusion

During extrusion, plastic particles flow from the hopper and into the feed throat of an extruder. As shown in Fig. 5.12*a*, gravity is usually the driving force for solids conveying in hoppers. With pellets and some granules, this produces mass or hopper flow in which all the particles flow down the hopper to the feed throat. However, when the particles pack easily, material around the edges of the hopper is stationary while particles flow through a narrow center channel or funnel, producing funnel flow. The mass flow of particles ceases if the particles form a bridge at the base of the hopper. In contrast, funnel flow is halted by piping in which the stagnant ring of material prevents any particles from flowing down the funnel. Since funnel flow, bridging, and piping are prevalent in low-bulk-density materials, the previously discussed hoppers and feed ports were designed to remedy such problems. In addition, the feed throat is cooled to prevent plastic particles from melting at the base of the hopper, thereby forming a bridge.

TABLE 5.7 Barrel Materials[47]

Condition	Example	Base screw materials			
		Alloy steel*	Nitrided steel†	Chrome plated‡	Special alloys§
Normal	Thermoplastics (except fluoropolymers) without fillers or flame retardants	Satisfactory	Good	Good	Excellent
Above normal wear	Resins containing up to 30% glass, mineral, or other abrasive fillers	Poor	Satisfactory	Satisfactory	Good
Severe wear	Resins containing more than 30% glass, mineral, or other abrasive fillers	Unsatisfactory	Poor	Poor	Satisfactory
Severe corrosion	Thermoplastics with certain flame retardants, all fluoropolymers	Unsatisfactory	Poor	Poor	Satisfactory

*Flame or induction hardened.

†AISI 4140 or a nitrided steel such as Nitralloy 135M or Crucible 135 modified.

‡Although chrome can be attacked by some acids, especially HCl, a generous plating (0.005 mm +) forms a very wear- and corrosion-resistant surface.

§Includes tool steels (D-2, H-13, A-11, CPM9V), various nickel alloys which are excellent for corrosion resistance but poor for wear, stainless steels in 400 series, coatings such as UCAR LW-IN30, Jet Kote, and plasma spray welded surfaces.

When single-screw extruders are starve fed (Fig. 5.12*d*), plastic particles do not immediately fill the screw channel. As a result, the first few channels of the feed zone lack the pressure required to compact the polymer particles. Particle conveyance in the unfilled channels is not as steady as transport with filled channels. Consequently, metered feeding is seldom used with single-screw extruders. Such feeding can be used to reduce the motor load,[49] limit temperature rises,[49] add several components through the same hopper,[49] improve mixing in single-screw extruders,[50] control flow into vented barrel extruders, and feed low-bulk-density materials.

Once the plastic particles reach the feed zone, they are compacted into a solid bed. The solid bed is conveyed down the extruder via viscous drag or drag flow. In this process, the solid bed adheres preferentially to the barrel wall and is pushed down that surface by the rotational motion of forward edge of the screw flights. As the solid bed moves down the feed zone, pressure on the solid bed rises gradually.

Developed by Darnell and Mol,[51] the prevailing feed zone theory assumes that the solid bed moves in plug flow. Solid conveyance, therefore, depends on the channel dimensions, screw speed, bulk density of the solid bed, coefficient of friction between the solid bed and barrel wall, coefficient of friction between the solid bed and screw, and the pressure gradient in the feed zone. Thus, solids transport increases with screw speed and, at least initially, with deeper channels. Further increases in channel depth increase the coefficient of friction between the solid bed and the screw, thereby reducing solids conveyance. Increasing pressure generation also slows the movement of the solid bed. However, solid conveyance is most sensitive to the coefficients of friction (COF) at the barrel wall and screw root. At low pressures, the coefficient of friction generally increases rapidly with temperature and then decreases.[52] With high pressures, the COF merely decreases with increasing temperature.[52] As a result, solids conveyance varies with the polymer, its additives, and the extrusion conditions.

Barrel temperature has a limited ability to improve frictional characteristics, and, therefore, solids conveyance. Moreover, as shown in Fig. 5.21,[53] the first few heating zones are usually cooler than the remaining zones. This prevents premature melting of the solid bed. While standard screw coatings prevent polymer from adhering to the screw root, low-friction coatings are also available. These include polytetrafluoroethylene (PTFE)-impregnated nickel plating, PTFE chrome plating, titanium nitride, boron-nitride, and tungsten-disulfide.[54] Heating, rather than cooling, the screw (Fig. 5.22[55]) alters the COF at the screw root, and, in some instances, can improve solids conveyance.[54] Finally, grooved barrels increase the friction between the solid bed and barrel wall, thereby increasing solids transport.

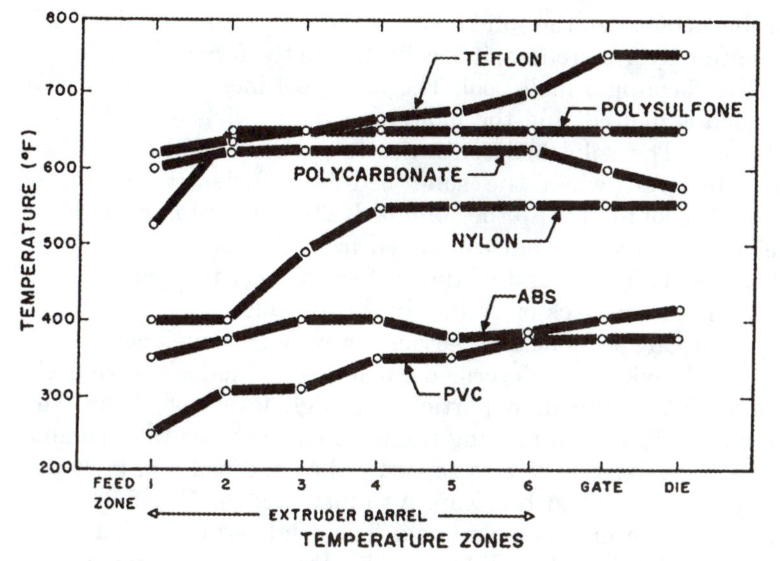

Figure 5.21 Barrel temperatures for selected materials.[51]

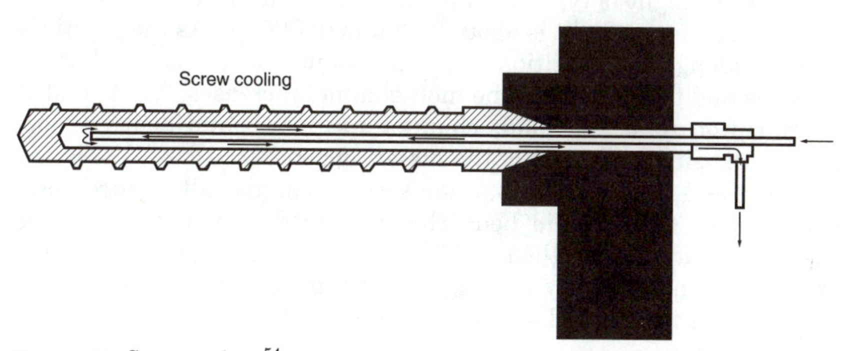

Figure 5.22 Screw cooling.[54]

While most melting or plastication occurs in the transition zone, the polymer can melt in the feed zone and may not be fully molten until well into the metering zone. Although the heater bands and cooling fans maintain the barrel at a set temperature profile, conduction from the barrel walls provides only 0 to 30 percent[56] of the energy required to melt the resin. The remainder of the energy is generated from the mechanical motion of the screw. This viscous dissipation is a far more efficient method for melting the plastic. Viscous drag also conveys the solid bed and melt in the transition zone.

The dominant melting model (Fig. 5.23[57]) was initially developed by Tadmor.[58] At the beginning of the transition zone, a layer of molten polymer forms between the solid bed and the barrel wall. As the thickness of molten layer increases, it is collected on the forward edge of the screw flight, forming a melt pool. The melt pool increases as more of the solid bed is melted, and the molten polymer swirls or vortexes in the melt pool. The solid bed continues to shrink and the melt pool expands. However, when the solid bed is sufficiently small, the swirling melt pool breaks up the solid bed. The remaining particles or gels are then melted by heat conducted from the molten polymer. If they do not melt by the end of the metering zone, the particles are caught by the screen pack or end up in the extrudate.

Maddock mixing sections and barrier screws were developed to combat solid-bed breakup. As described earlier, the Maddock mixing elements subject the unmelted particles to high levels of shear. This shear is more effective in melting the polymer particles than conduction. In contrast, barrier screws separate the solid bed and melt pool to help prevent solid-bed breakup. As illustrated in Fig. 5.24,[59] the transition zone contains two channels. The solid bed is formed in the primary channel. When the solid bed melts, the melt film is forced over a barrier flight and collected in the secondary channel. The clearance of the barrier flight is typically 0.50 mm (0.020 in), whereas the clearance of the primary flight is about 0.13 mm (0.005 in). As the plastic is conveyed along the transition zone, the volume of the solids channel decreases and the volume of the melt channel increases. At the end of the transition zone, the solids channel disappears and melt channel becomes the single channel of the metering zone.

While there are various barrier screw designs, all ensure complete melting of the solid bed. This prevents flow instabilities or surging produced when the solid bed breaks up. However, the solids channels are more likely to plug if melting cannot keep pace with the reduction in channel size in the screw.[60] The plugging, in turn, causes surging.

In the metering zone of an extruder, the polymer is usually molten. The rotating screw pushes material along the walls of the stationary barrel creating drag flow (Q_D). This drag flow provides the forward conveying action of the extruder, and, in the absence of a die, is effectively the only flow present. The addition of a die restricts the open discharge at the end of an extruder and produces a large pressure gradient

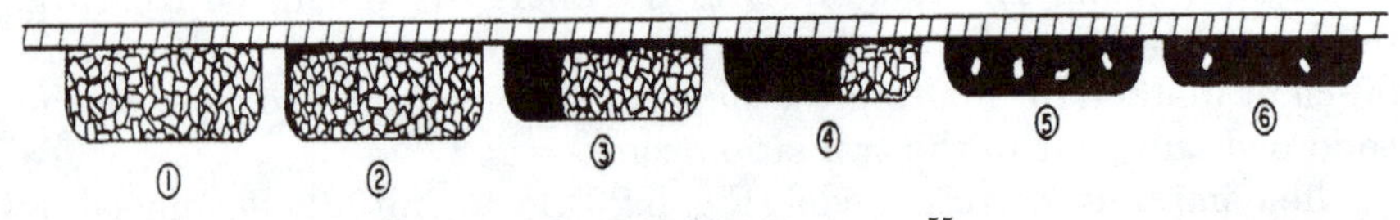

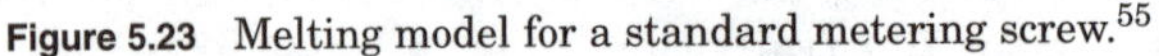

Figure 5.23 Melting model for a standard metering screw.[55]

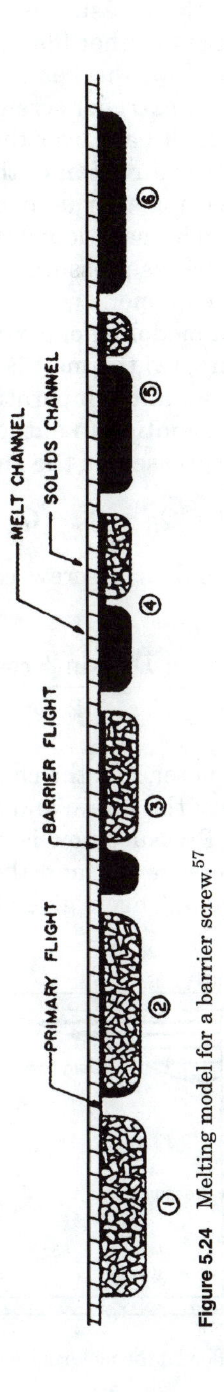

Figure 5.24 Melting model for a barrier screw.[57]

along the extruder. Since the pressure is greatest just before the die, this head pressure creates two other flows, pressure flow Q_P and leakage flow Q_L. In pressure flow, the head pressure forces the melt to rotate in the channels of the extruder screw. Leakage flow occurs when the head pressure forces melt back over the flights of the screw. Since they both counter the forward motion of the melt, pressure and leakage flow are often lumped together as back flow. As depicted in Fig. 5.25, during normal extruder operation, drag flow conveys the polymer along the barrel walls, whereas pressure flow forces the material near the screw back toward the hopper.

A simple mathematical modeling of extrusion assumes that: (1) the extruder is at steady state, (2) the melt is newtonian, (3) the extruder is isothermal (at a constant temperature), and (4) the metering zone makes the only contribution to output. Thus, the net output Q, of the extruder can be expressed as the sum of the three flows:

$$Q = Q_D - Q_P - Q_L \tag{5.15}$$

Drag flow is proportional to a screw constant (A) and the screw speed, expressed as

$$Q_D = AN = \frac{\pi^2 D^2 h \sin \phi \cos \phi}{2} N \tag{5.16}$$

where D is the screw diameter, h is the channel depth in the metering zone, ϕ is the helix angle of the screw, and N is the screw speed in revolutions per second (r/s). Pressure flow is related to a screw geometry constant B, the head pressure ΔP, and the apparent viscosity of the melt in the metering zone μ. This is given by

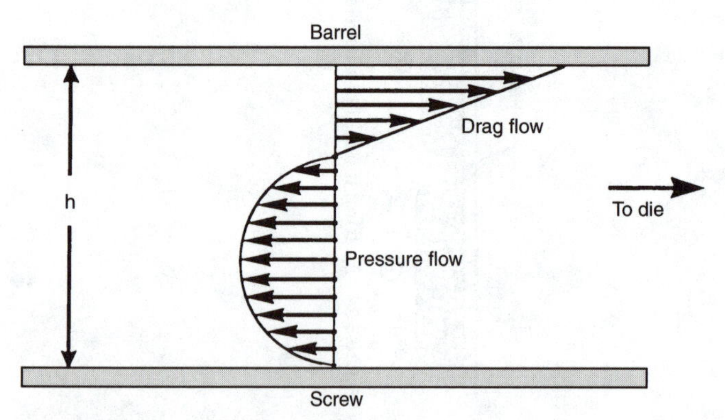

Figure 5.25 Drag and pressure flow in the metering zone of a single-screw extruder.

$$Q_P = \frac{B\,\Delta P}{\mu} = \frac{\pi D h^3 \sin^2 \phi}{12 L_m} \frac{\Delta P}{\mu} \tag{5.17}$$

where L_m is the length of the metering zone. Leakage flow is a function of a constant C, the head pressure ΔP, and the apparent viscosity in the flight clearance (μ). This is expressed by

$$Q_L = \frac{C\,\Delta P}{\mu} = \frac{\pi^2 D^2 \delta^3 \tan \phi}{12 e L_m} \frac{\Delta P}{\mu} \tag{5.18}$$

where δ is the flight clearance and e is the flight width. With similar assumptions, die output becomes:

$$Q_{\mathrm{Die}} = \frac{\Delta P}{K\mu} \tag{5.19}$$

where the die constant K varies with geometry. Selected constants are given in Table 5.8.[61]

When drag flow dominates, extruder output increases linearly with screw speed, and larger screws and deeper channels carry more melt. However, head pressure also rises with screw speed. Although this increases pressure flow, the actual effect depends heavily on melt viscosity. With high-viscosity melts, pressure flow may be minimal and have little effect on extruder output. In contrast, low-viscosity melts produce less head pressure but greater pressure flow. Thus, pressure flow will reduce the expected output. Deeper channels, neutral screws (Fig. 5.26), and shorter metering zones enhance these effects.

Leakage flow varies with the flight clearance. It is also enhanced by low-viscosity melts and high head pressures. With new screws and barrels, leakage flow is minor and has no apparent effect on extruder output. As the flight clearance increases, leakage flow rises, thereby reducing output. Consequently, the decrease in extruder output over time is used to monitor screw and barrel wear.

In contrast to extruder output, die output increases with head pressure (Fig. 5.27). Die output is also enhanced by low-viscosity melts and larger die gaps. The match between extruder and die output shifts with operating conditions. The simple die characteristic curve in Fig. 5.27 shows the optimized processing conditions. However, this curve does not consider extrudate quality. Other "lines" would be required to locate the onset of surface defects, such as melt fracture, and for incomplete melting.

Head (melt) pressure is measured at the end of the extruder. One pressure transducer is typically mounted just before the breaker plate while others may be placed in the die adapter or die itself. Pressure is monitored for safety purposes, product quality, research and develop-

TABLE 5.8 Selected Die Constants[61]

Die profile	Die constant, K
Circular	$128(L + 4D_c)/\pi D_c^4$
Slit	$12L/WH^3$
Annular	$12L/\pi D_m H^3$

Note: L = die land length, D_c = diameter of a circular die, D_m = mean diameter of an annular die, W = width of a slit die, and H = die opening (gap).

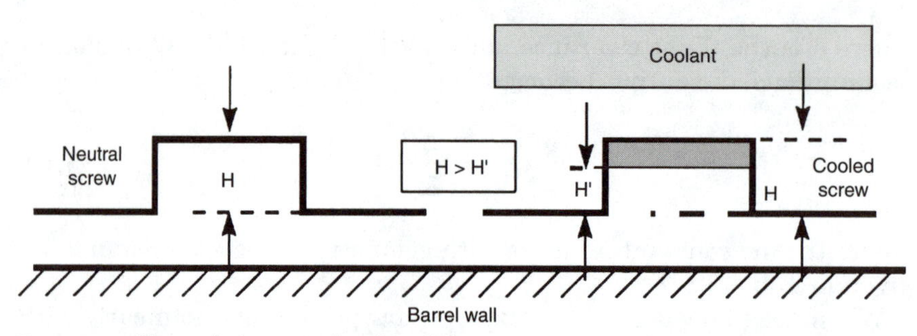

Figure 5.26 Channels in neutral and cooled screws.

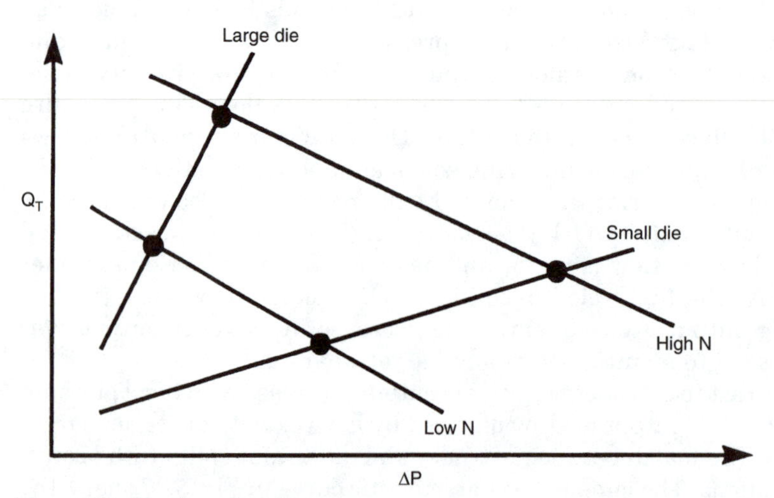

Figure 5.27 Die characteristic curve.

ment, screen pack or changer condition, process monitoring, and troubleshooting. Since head pressures can reach 69 MPa (10,000 lb/in²), pressure is monitored during extruder start-up and operation to adjust operating conditions or halt operation before the pressure opens the rupture disk. Variations in head pressure (surging) produce variations in output. These changes are used to track product quality and troubleshoot the extrusion process. Larger pressure increases also trigger the movement of screen changers or signal the need to replace screen packs and breaker plates.

Melt temperature is also monitored during extrusion. Melt temperature varies with the placement of the measuring device, material, and processing conditions. Thermocouples measure temperature at one point in the melt stream. As shown in Fig. 5.28,[62] melt temperature measured with a flush-mounted thermocouple is influenced by the barrel wall temperature. Protruding thermocouples interrupt flow and produce varying levels of shear heating. While straight protruding thermocouples measure more shear heating, they are more robust than upstream fixed or radially adjustable thermocouples. A bridge with multiple thermocouples measures melt temperature at several points in the melt stream. However, the bridge produces a greater interruption of the melt flow. Infrared sensors measure the average melt temperature and are more sensitive to temperature variations; these sensors are expensive and have limited availability.

A material's sensitivity to shear and temperature produces varying levels of shear. While melt travels fastest in the center of the channel, shear is highest near the wall. Cooling effects are also greatest near the wall, producing a melt temperature differential as great as 50°C in the melt channel.

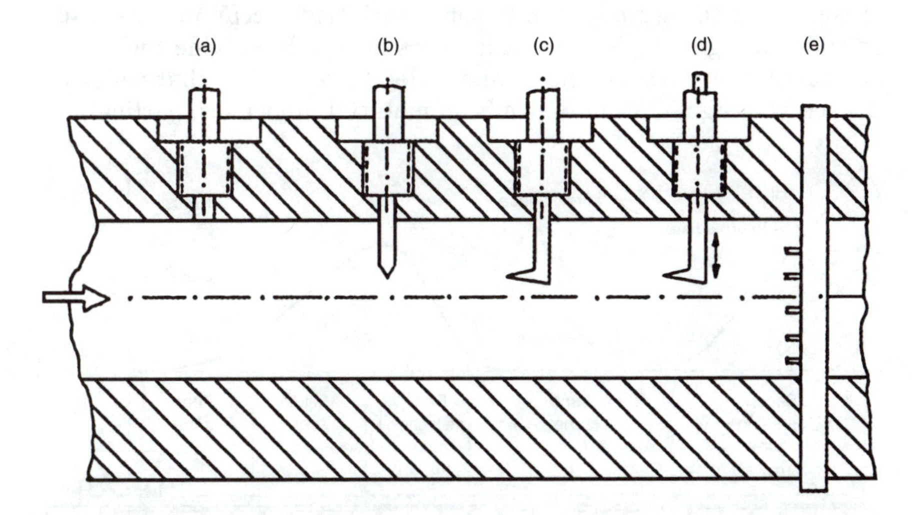

Figure 5.28 Various temperature sensor configurations: (*a*) flush-mounted, (*b*) straight protruding, (*c*) upstream fixed, (*d*) upstream radially adjustable, and (*e*) bridge with multiple probes.[62]

5.2.3 Vented extruders

Vented extruders (Fig. 5.29[63]) contain a vent port and require two-stage screws. As described previously, the screw has five zones: feed, transition (or melting), first metering, vent, and second metering. Material is fed, melted, and conveyed in the first three zones of the extruder. Melt pressure increases gradually as the plastic moves down the barrel. However, the channel depth increases abruptly in the vent zone. Since the thin layer of melt from the metering zone cannot fill this channel, the melt is decompressed and volatiles escape through the vent. The melt is repressurized in the second metering zone and this pressure forces the melt through the die.

While vented extruders are used for devolatilization, they can only handle materials with a volatiles content up to 5 percent.[64] They are also subject to vent flooding. If the die resistance is too high or the screen pack is clogged, the melt pressure will rise in the vent zone, causing vent flooding. In screws with high pump ratios or when the feeding rate is too high, the second zone cannot convey the material fed from the first metering zone and this floods the vent. Consequently, vented extruders are often starve fed and pressure is monitored carefully during operation.

5.2.4 Twin-screw extrusion

Single-screw extruders are relatively similar in design and function. All single-screw extruders convey the polymer to the die by means of viscous drag (drag flow). While some variations occur in screw and extruder design, single-screw extruders generally provide high head pressures, uncontrolled shear, and a degree of mixing that relies on the screw design. Output depends on material properties, particularly

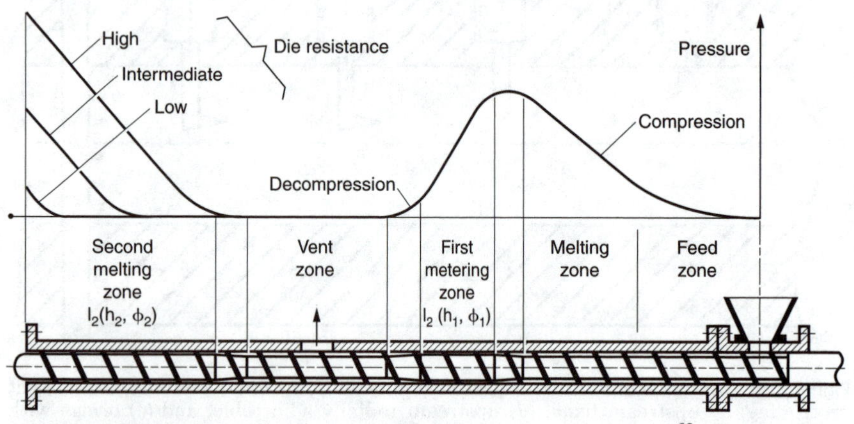

Figure 5.29 Pressure profile associated with a vented barrel extruder.[63]

the bulk properties (coefficient of friction, particle size, and particle-size distribution). In contrast, the design, principles of operation, and applications of twin-screw extruders vary widely. While the two screws are usually arranged side by side, the introduction of two screws produces different conveyance mechanisms, varied degrees of mixing, and controllable shear. The low head pressure generated by twin-screw extruders initially limited their use to processing of shear-sensitive materials, such as polyvinyl chloride, and to compounding. Although changes in design have permitted higher speeds and pressures,[65] the primary use of twin-screw extruders is still compounding. Twin-screw extruders are used in 10 percent of all extrusion.

The two screws are the key to understanding the conveyance mechanisms and probable applications of different twin-screw extruders. The screws may rotate in the same direction (corotating) or in opposite directions (counterrotating). In addition, the flights of the two screws may be separated, just touch (tangential), or intermesh to various degrees. The flights of partially intermeshing screws interpenetrate the channels of the other screw, whereas the flights of fully intermeshing screws completely fill (except for a mechanical clearance) the channels of the adjacent screw. While many configurations are possible, in practice the major designs are: (1) nonintermeshing, (2) fully intermeshing counterrotating, and (3) fully intermeshing corotating twin-screw extruders (Fig. 5.30[66]).

Nonintermeshing (separated or tangential) twin screws do not interlock with each other. The polymer is conveyed, melted, and mixing by drag flow. Since two corotating nonintermeshing screws would provide uncontrollable shear at the nip between two screw and little distributive mixing, they are not used commercially.[67] Counterrotating screws must rotate at the same rate to produce sufficient output. With matched flights, little plastic material is transferred between screws, however, substantial interscrew transfer occurs with staggered

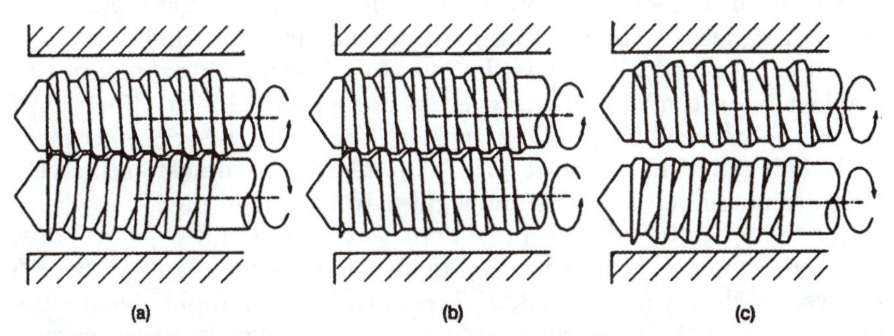

Figure 5.30 Twin-screw extruders: (*a*) counterrotating, fully intermeshing; (*b*) corotating, fully intermeshing; and (*c*) counterrotating, nonintermeshing.[66]

flights.[68] As a result, counterrotating nonintermeshing twin-screw extruders provide good distributive mixing but little shear. The screws of commercial counterrotating tangential (CRT) twin-screw extruders are either matched or one screw is longer than the other. With the latter configuration, the single screw at the end of the extruder improves pressure generation. Thus, counterrotating nonintermeshing twin-screw extruders have been used for devolatilization, coagulation, reactive extrusion, and halogenation of polyolefins.

With intermeshing twin-screw extruders, the flights of one screw fit into the channels of the other. Since the extruders are usually starve fed, the screw channels are not completely filled with polymer. By transferring some polymer from the channels of one screw to those of the other, the intermeshing divides the polymer in the channel into at least two flows. Thus, intermeshing twin-screw extruders provide positive conveyance of the polymer and improved mixing.

In counterrotating, intermeshing twin-screw extruders, a bank of material flows between the screws and the barrel wall. The remainder is forced between the two screws and undergoes substantial shear. With little intermeshing, drag flow between the screws is greater than that at the barrel walls. However, for the commercial fully intermeshing screws, most material flows along the screws in a narrow channel (C chamber) and is subject to relatively low shear. Consequently, the degree of mixing in counterrotating, intermeshing twin-screw extruders depends on the degree of intermeshing and screw geometry. Increasing the distance between the screws increases flow between the screws and permits effective distributive mixing. However, increased screw separation only decreases the shear rate in the nip, and hence reduces dispersive mixing. Since screw length and geometry are also used to prevent excessive shearing, melting in these extruders is limited, and most of the heat transferred to the polymer is conducted from the barrel. This mechanism provides very sensitive control over the melt temperature.

With good temperature control and the low shear, these extruders are well suited for compounding and for extrusion of rigid poly(vinyl chloride). Typically, high-speed (200- to 500-r/min[69]) extruders are employed for compounding, whereas low-speed (10- to 40-r/min[69]) machines are used for profile extrusion. Conical twin-screw extruders with their tapering screws (Fig. 5.31[70]) are utilized almost exclusively for chlorinated polyethylene and rigid poly(vinyl chloride).

Corotating fully intermeshing twin screws are self-wiping. Thus, they tend to move the polymer in a figure-eight pattern around the two screws, as shown in Fig. 5.32.[71] Typically, a screw flight pushes the material toward the point of intersection between the two screws. Material is then forced to change its direction through a large angle,

Figure 5.31 Screws for a conical twin-screw extruder.[70]

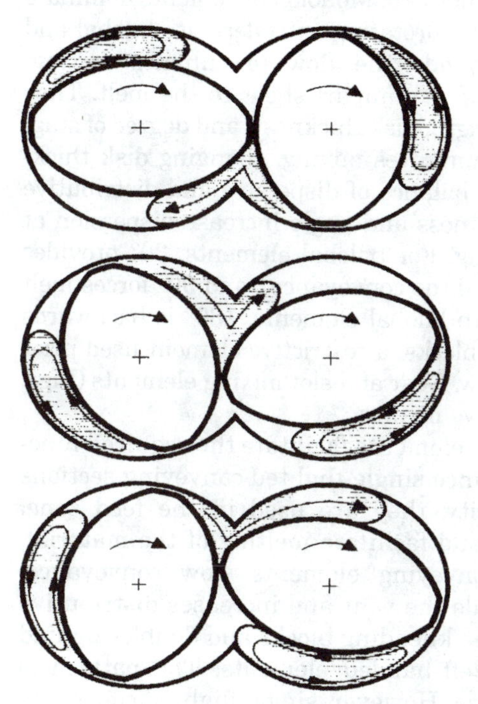

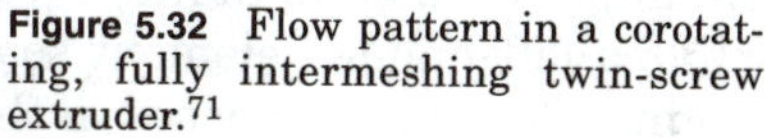

Figure 5.32 Flow pattern in a corotating, fully intermeshing twin-screw extruder.[71]

which mixes the material. Very little material is able to leak between the screws. Finally, the material is transferred from one screw to the next. The flow pattern provides a longer flow path for the material, and hence, the longer residence time of corotating extruders. Mixing elements, such as kneading blocks, are not fully self-wiping, but are usually incorporated to improve melting and mixing. However, unlike counterrotating screws, the shear between the corotating screws is relatively mild. Consequently, the combination of longer flow paths, more uniform shear, and self-wiping conveying elements make corotating intermeshing twin screws well suited to mixing and compounding applications.

The design of intermeshing twin-screw extruders also differs from that of single-screw extruders. With the exception of conical screws, the screws are usually not a single piece of metal, but two shafts onto which component screw elements are arranged (Fig. 5.33[72]). Thus, screw profiles may be "programmed" to impart specific levels of shear, mixing, and conveyance. Conveying elements (Figs. 5.34*a*[73] and *b*[73]) do not mix the plastic, but merely convey the material down the screw. Single-flighted conveying elements provide rapid transport, whereas triple-flighted elements impart shear to the plastic; the performance of double-flighted elements is intermediate to the other two. Monolobal elements dominate in counterrotating extruders while corotating extruders use bilobal and trilobal elements. The latter divide the flow to enhance mixing. Kneading blocks (Figs. 5.34*c*[73] and *d*[74]) impart shear to the melt. They have three critical dimensions: length, disk thickness, and degree of stagger. Although increasing length improves mixing, changing disk thickness and stagger angle alter the balance of dispersive and distributive mixing. Typically, increased thickness and angle increase dispersion at the expense of distributive mixing. For trilobal elements, 30° provides forward conveyance, 60° is neutral (no conveyance), and 90° forces melt backwards along the screws. With bilobal elements, 180° is backwards conveyance. Left-hand kneading blocks, a restrictive element used prior to vent ports, also induce back flow. Gear and slot mixing elements (Figs. 5.34*e*[73] and *f*[74]) provide distributive mixing.

When programming the screw, elements facilitate the required function of the screw (Fig. 5.35[75]). Since single-flighted conveying sections have a large volumetric capacity, they are used in the feed zone. Kneading blocks impart shear and facilitate melting of the material. Small-pitch, double-flighted conveying elements slow conveyance, while a left-handed element seals the vent and increases distributive mixing. Pressure increases in the kneading blocks and double-flighted elements and drops with the left-handed elements. The pattern is repeated for the second vent zone. However, single-flighted conveying elements increase the melt conveyance near the die, which facilitates the generation of pressure that forces the melt through the die. Small pitch is used to reduce conveyance and increase residence time in reactive extrusion.[76] With conventional twin-screw extruders, the time can be extended to 10 min, whereas special twin-screw extruders can provide residence times up to 45 min.[77] Narrow kneading blocks are used after fiber addition to prevent fiber degradation.[76]

The extruder barrels are also modular in design. Extruder L/D may be changed by adding or removing barrel segments. Features, such as multiple stages and venting sections, may also be added, subtracted, or moved. Finally, special barrel sections, such as those with expensive abrasion-resistant barrel liners, may be located after the appropriate feed ports.[78] Metered feeding of material is typically required to keep the channels of intermeshing twin-screw extruders from filling com-

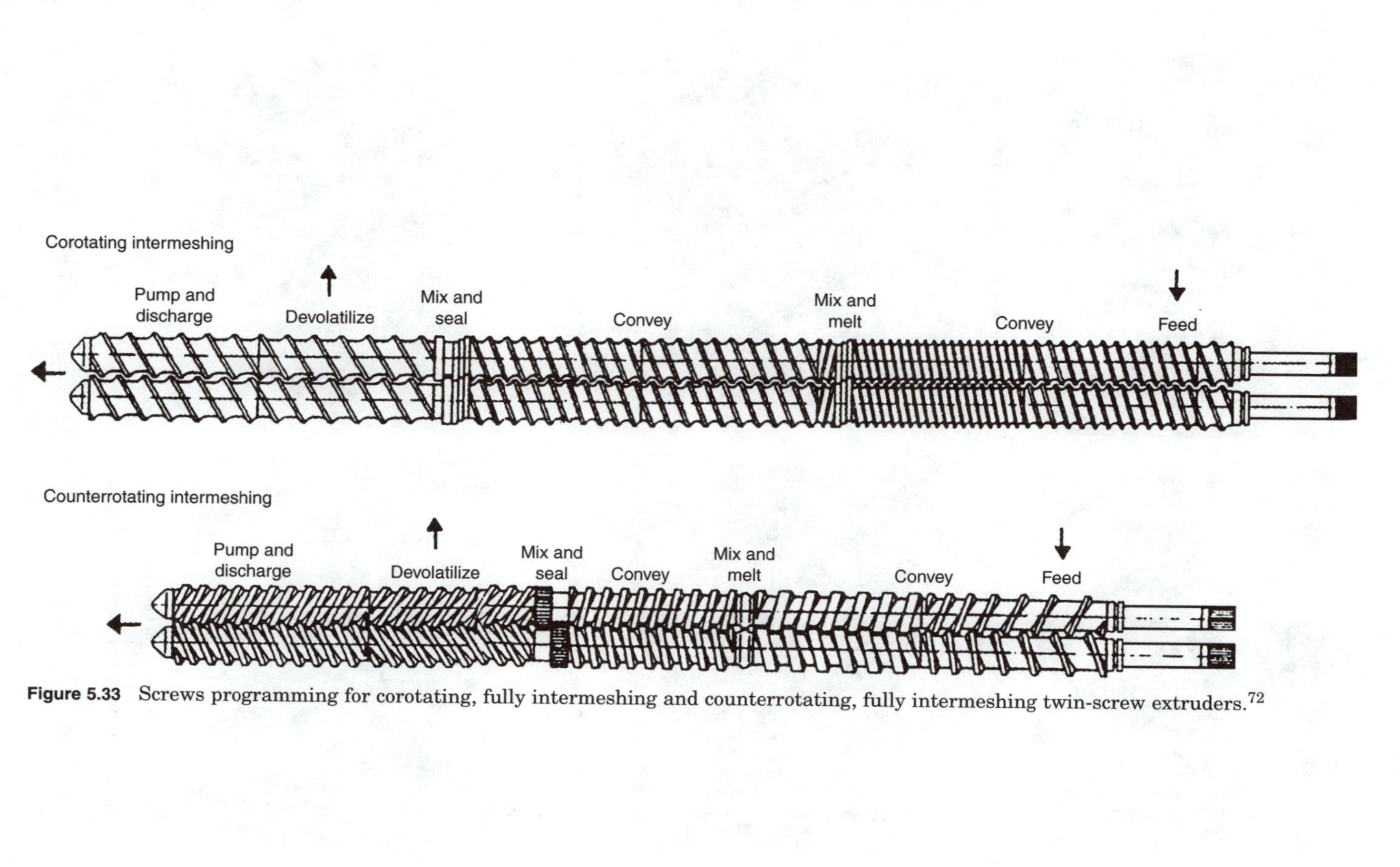

Figure 5.33 Screws programming for corotating, fully intermeshing and counterrotating, fully intermeshing twin-screw extruders.[72]

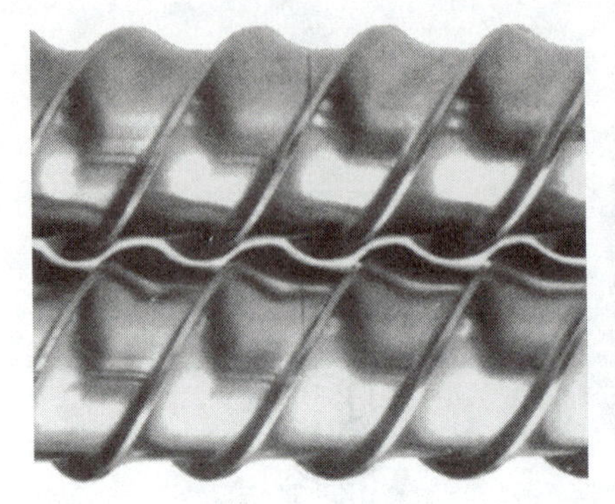

(a)Triple-flighted conveying element

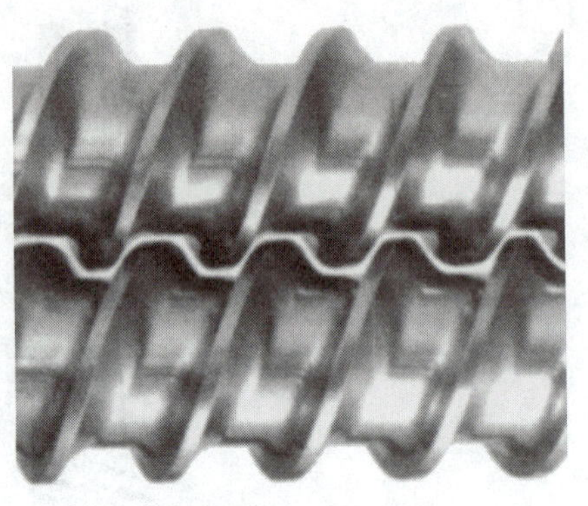

(b) Double-flighted conveying element

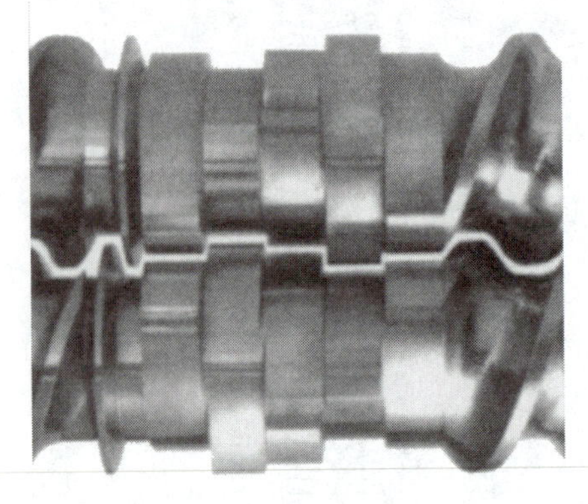

(c) Kneading blocks

(d) Kneading blocks

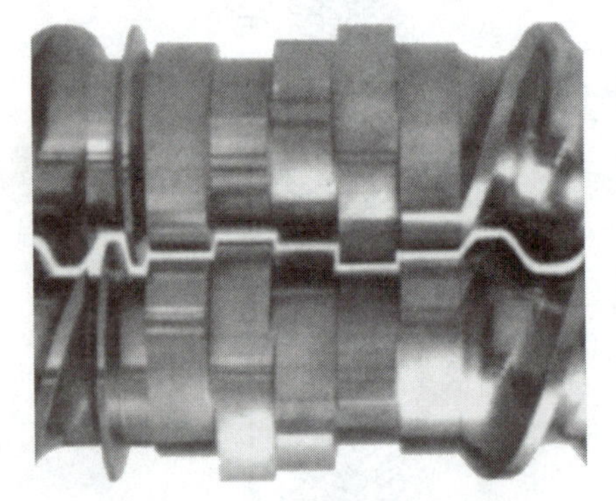

(e) Gear mixing elements

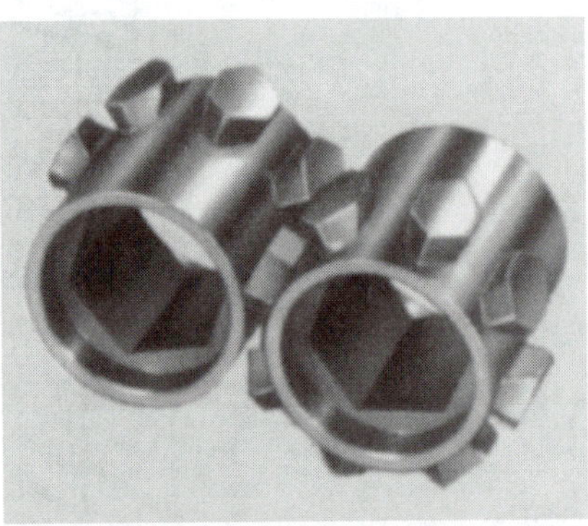

(f) Slot restrictor elements

Figure 5.34 Twin-screw elements.[73,74]

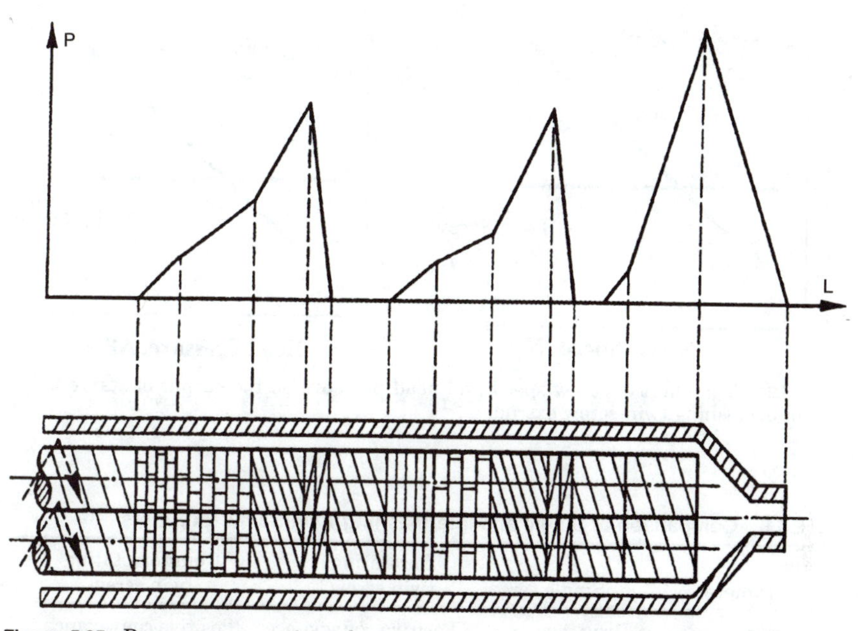

Figure 5.35 Pressure generation along a corotating, fully intermeshing twin-screw extruder.[75]

pletely. Drive motors can torque out when the screw channels are filled (except near die).

Recent changes in gear-box design have produced dramatic increases in the output of corotating twin-screw extruders. In these extruders, the high power-factor gear boxes[79] permit a 40 percent increase in torque and screw speeds up to 1200 r/min.[80] As a result, smaller-diameter extruders provide the output that once required much larger machines.

Finally, the effect of process variables, such as screw speed, differs with twin-screw extruders. Due to positive conveyance, the output of starve-fed twin-screw extruders (Fig. 5.36[81]) is independent of screw speed and, to some degree, head pressure. Head pressures and melt temperatures of intermeshing twin-screw extruders are also not as sensitive to screw speed as they are in single-screw extruders. In contrast, nonintermeshing twin-screw extruders exhibit output characteristics that are similar to those of single-screw extruders. A comparison of the characteristics of single and intermeshing twin-screw extruders is summarized in Table 5.9.

5.2.5 Ram extrusion

In ram extruders (Fig. 5.37[82]), a hydraulic ram forces the polymer through die. While high pressures ($\leq$ 345 MPa) can be achieved, the

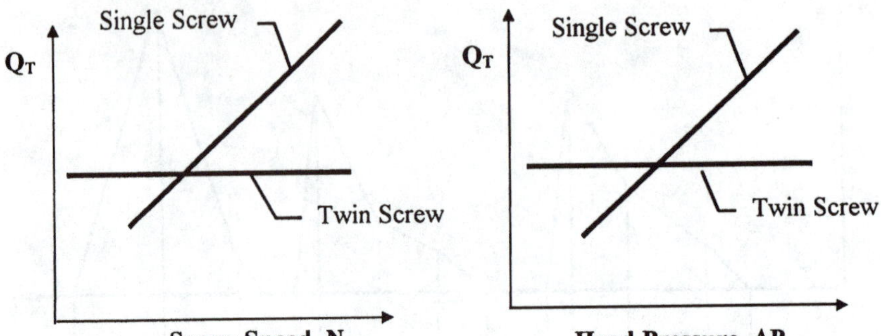

Figure 5.36 The effect of screw speed and head pressure on the output of starve-fed, fully intermeshing twin-screw extruders.[81]

TABLE 5.9 Comparison of Single and Intermeshing Twin-Screw Extruders

Parameter	Single screw	Corotating twin screw	Counterrotating twin screw
Conveyance	Drag flow	Positive conveyance	Positive conveyance
Mixing efficiency	Poor	Medium-high	Excellent
Shear	High (depends on N)	Screw design dependent	Screw design dependent
Self-cleaning	No	Yes	Partially
Energy efficiency	Low	Medium	High
Screw speed (r/min)	50–300	25–300	50–100

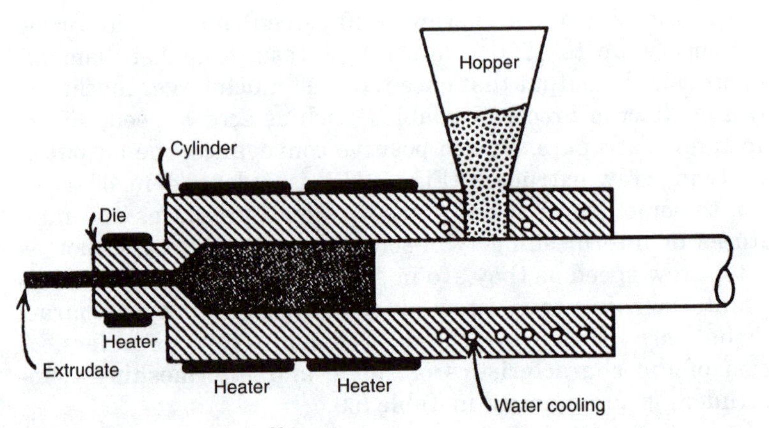

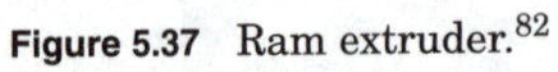
Figure 5.37 Ram extruder.[82]

process is discontinuous. All heating occurs by conduction from the barrel walls. The absence of shear gives ram extruders limited melting capacity and poor temperature uniformity.

Ram extrusion is generally used for specialty processing. When wet processing cellulosics and polytetrafluoroethylene, the polymer is softened with heat and solvent. Then the high-pressure ram forces this "slurry" through the die. Ultrahigh-molecular-weight polyethylene (UHMWPE) also requires the high pressure and low shear of ram extruders. Since the long chains of UHMWPE produce a melt temperature that is greater than the decomposition temperature, UHMWPE is also processed as slurry. However, the long chains are also sensitive to shear; thus, ram processed prevents the loss of mechanical properties that can occur when these materials are processed with screw extruders. The absence of shear is an advantage when processing thermoset materials and composites. With thermosets, the water or oil-heated barrel jacket barrel provides controlled temperature, whereas for composites, the ram does not degrade (break) fibers.

In solid-state extrusion, the high pressures of ram extruders form powdered polymer into solid objects. A variation of this technique is also used for large-diameter profiles. Polyamides and polypropylenes exhibit high levels of shrinkage when melt processed and this problem is enhanced for thick cross sections. Thus, the materials are processed below their crystalline melt temperature. While the material viscosity is high, the ram extruders provide sufficient pressure to extrude the profiles.

5.3 Extrusion Processes

While the extruder pumps the molten polymer, the die and downstream equipment determine the final form of the plastic. Blown film and flat film extrusion both produce plastic films, but require very different dies and take-off systems. Similarly, different extrusion lines are used for pipes, tubing, profiles, fibers, extrusion coating, and wire coating.

5.3.1 Blown film extrusion

Blown, or tubular, film extrusion is one of the major processes used for manufacturing plastic films. In this process, plastic pellets are fed into the hopper and melted in the extruder. After exiting the extruder barrel, the molten resin enters an annular die. The resin is forced around a mandrel inside the die, shaped into a sleeve, and extruded through the round die opening in the form of a rather thick-walled tube. The molten tube is then expanded into a "bubble" of the desired diameter and correspondingly lower film thickness (gauge) by the pressure of internal air, which is introduced through the center of the mandrel. Inside the bubble, air is maintained at constant pressure to ensure uniform film width and gauge. The inflated bubble is closed off at the bottom by the die and at the top by the nip rollers. While the nip rollers collapse the bubble, they also stretch the film and serve as a take-off device for the line. An air ring above the die cools the bubble so that the film is solid when it reaches the nip rollers. After it passes through the rollers, the collapsed film (or layflat) is wound up on a roll.

A blown film line (Fig. 5.38) consists of an extruder, annular die, cooling system, take-off tower, wind-up system, and auxiliary equipment such as a film gauge measuring system, surface treatment, sealing operation, and slitter. Extruders are typically single-screw extruders with L/D ratios of 24 to 30:1.[83,84] For polyolefins, screws often incorporate barrier flights in transition zone and dispersive mixing heads. Continuous or discontinuous screen changers may be used, but discontinuous units are preferred.[85]

Blown film lines use three major die designs: side fed, spider arm, and spiral flow. In side-fed dies (Fig. 5.39*a*[86]), the melt is fed into one side of the die body while air is introduced into the bottom of the mandrel. Since this produces a relatively low pressure drop, such dies are good for high-viscosity materials. However, the melt encircles the mandrel and joins in a single, relatively weak weld line. The pressure drop around mandrel deflects the mandrel to produce nonuniform flow.

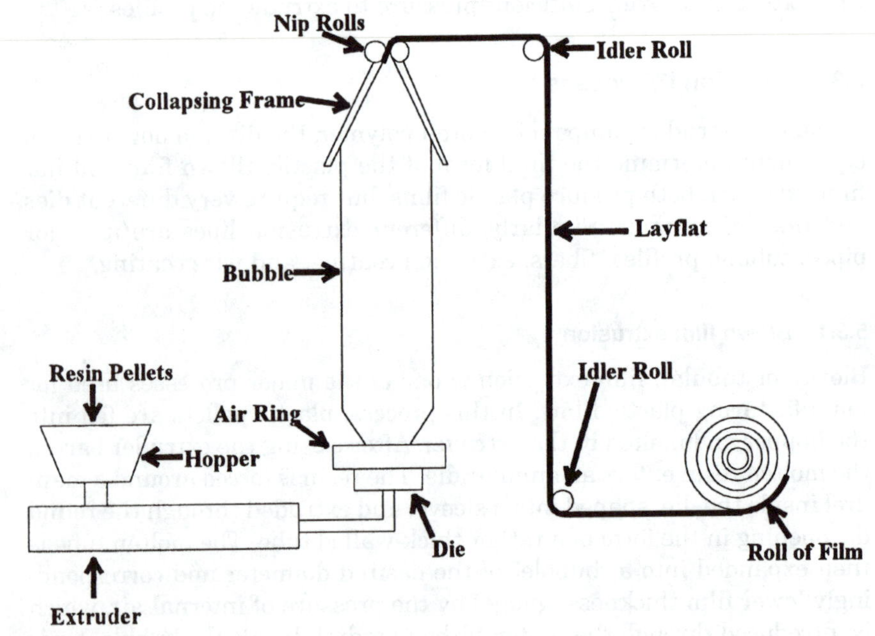

Figure 5.38 Blown film line.

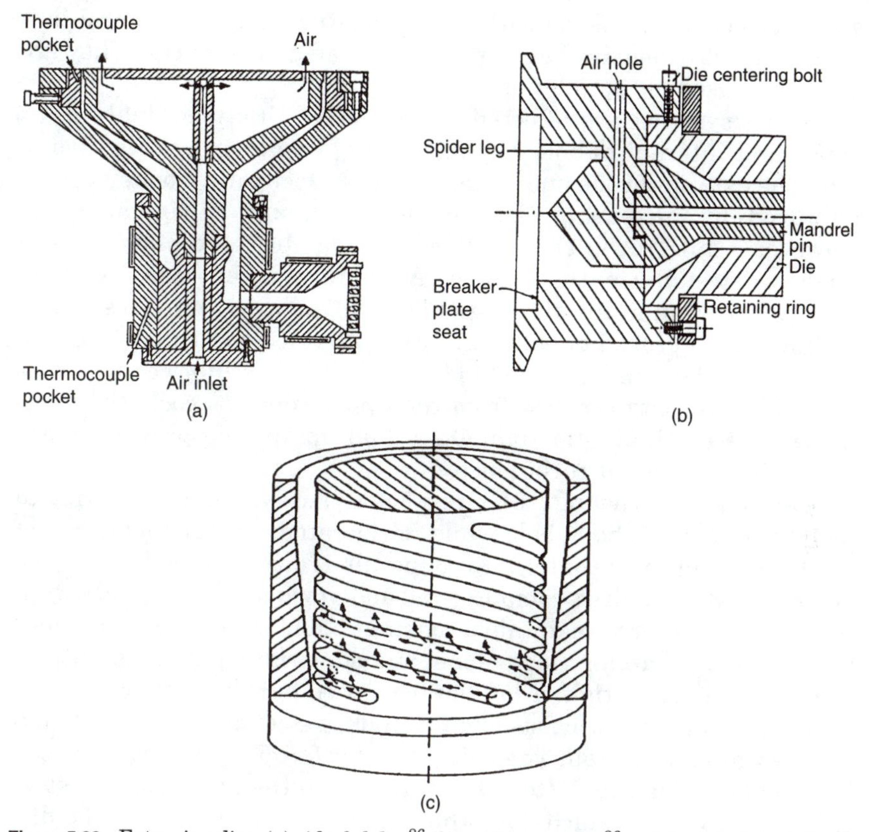

Figure 5.39 Extrusion dies: (*a*) side-fed die,[86] (*b*) spider-arm die,[86] and (*c*) spiral-flow die.[87]

Thus, film tends to be thicker on one side of the die. In addition, gauge control is poor, and so the die is seldom used in production lines.

With a spider-arm die (Fig. 5.39*b*[86]), polymer melt is fed through the bottom of the die while air is introduced through three or more "spider arms" that extend from the sides of the die to support the mandrel. This creates a relatively low pressure drop, and allows the die to be used with high-viscosity materials. The weld lines are stronger than the single-weld line produced with a side-fed die, but can result in weak points in the film. Spider-arm dies are used for poly(vinyl chloride) and other high-viscosity, heat-sensitive materials.

With a spiral-flow die (Fig. 5.39*c*[87]), melt is fed into the bottom of the die flows along the spiral channels of the die and also jumps from channel to channel. Air is introduced into the bottom of the die. This creates uniform flow and minimizes gauge bands (thick and thin areas in the film). However, high pressure drops [~35 MPa (5000 lb/in^2)] make the die unsuitable for high-viscosity materials. High levels of shear also cause problems with shear-sensitive materials. Spiral-flow dies are more expensive but provide improved gauge control. They are commonly used for polyolefins.

Cooling systems are affected by single-lip air rings, dual-lip air rings, and internal bubble cooling. A single-lip air ring (Fig. 5.40*a*) cools the exterior of the bubble using high-velocity air. Cooling can be improved by increasing the air flow or using refrigerated air. However, while turbulent air flow provides better cooling, it also tends to destabilize the bubble. As a result, the most common approach is a dual-lip air ring (Fig. 5.40*b*), which provides better cooling and improved bubble stability. Low-velocity air flow from the lower ring, Q_1, stabilizes the bubble and acts as a lubricant. In contrast, high-velocity air flow from the upper ring, Q_2, cools the melt. Since Q_2 is much greater than Q_1, a dual-lip air ring provides high-inlet velocity without turbulence.

Internal bubble cooling (IBC) (Fig. 5.40*c*[88]) uses a dual-lip air ring to cool the outside of the bubble while refrigerated air cools the inside of the bubble. Since the internal cooling air is introduced through the mandrel, IBC requires computerized monitoring of pressure within the bubble in order to maintain a constant bubble pressure. It provides better cooling than air rings alone, and so permits increased output, faster start-up, and tighter lay-flat (collapsed bubble) control.

The take-off tower consists of guide rolls, a steel nip (pinch) roll, and a rubber nip (pinch) roll. The guide rolls (or forming tent) collapse the bubble and guide the flattened film tube into the nip rolls. The steel roll is a driven roll which pulls the collapsed tube away from the die. The rubber rotates with the steel nip roll. Typical line speeds are 10 to 90 m/min (35 to 300 ft/min.).[89]

Although Fig. 5.38 shows upward extrusion, blown film is occasionally extruded downward or horizontally. The most common technique, upwards extrusion, provides more control over the amount of stretching, less machine vibration, and easier servicing of the extruder. However, the resin must have sufficient melt strength to support the bubble. Downward extrusion is used with lower melt-strength materials since gravity works with the flow. Heat transfer also facilitates handling the low melt-strength materials; heat from the extruder rises while cooling air is directed downward. The process is limited by scaffolding vibration, difficulties in serving the extruder, and limited extruder sizes. Horizontal extrusion is easy to start up, but uses a lot of floor space, requires high melt-strength materials, and limits the size of bubble diameter. Since the bubble requires supports, the bubble diameter is typically less than 50 mm (2 in).

The wind-up unit rolls up film as tube or flat film, provides constant tension, and produces a uniform wind-up rate. There are two types of

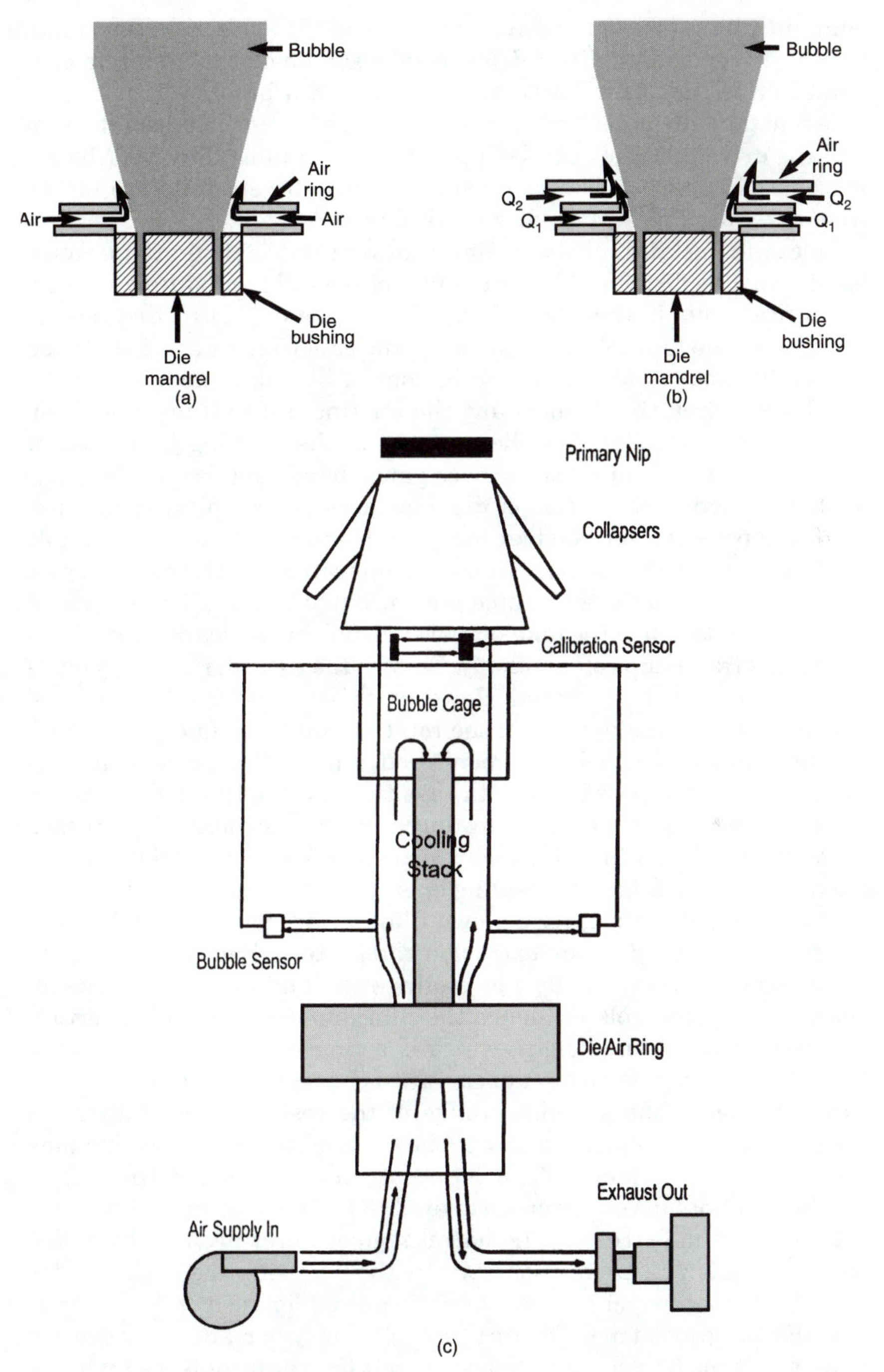

Figure 5.40 Cooling methods for blown film lines: (*a*) single-lip air ring, (*b*) dual-lip air ring, and (*c*) internal bubble cooling.[88]

wind-up units: (1) center-driven shafts and (2) surface-driven units. Center-driven shafts (direct drive) need a servomotor to maintain constant film tension as roll diameter increases. Surface-driven units provide constant surface speed, but require some friction between the film and the drive roll. This can be a problem when films have poor blocking and slip characteristics. However, center-driven units are not as frequently used as surface-driven wind-ups.

Since the thickness of blown films is never completely uniform, gauge bands must be distributed over the face of the roll to produce a smooth, cylindrical roll. If they are always in the same place, they create "bumps" in the film roll. Consequently, the gauge bands are distributed across the face of the roll by rotating one of five film line components: the die mandrel, the die bushing, the air ring, take-off tower and nip roll, and the extruder. The die mandrel or die bushing is rotated at about 0.3 r/min. While this removes gauge bands and is good for polyolefins, it increases the cost of die, increases die complexity, and provides the potential for polymer hang-up. Neither technique is suitable for heat-sensitive materials. Due to complications with the air hoses, the air ring is usually not rotated but oscillated by ±270°. The common technique, rotation of the take-off tower, usually involves oscillating the collapsing frame-nip roll assembly. Finally, the extruder can be mounted on a turntable and rotated. However, this method is expensive and its use is decreasing in favor of the rotating nip technique.

Film thickness is usually monitored during the process. A beta gauge detects the passage of beta rays through the film bubble, while capacitance gauges measure the increase in thickness as increased capacitance. Blown film lines may also include corona or flame treatment to improve adhesion, sealing operations for bags, and slitters.

The principal controls for a blown film line are barrel (cylinder) and die temperatures, die gap, extrusion rate, internal air pressure, bubble diameter, cooling air flow or cooling rate, and line speed (take-off speed). These controls influence the film dimensions and properties. The frost (freeze) line height, which is a ring-shaped zone where the bubble frequently begins to appear "frosty" because the film temperature falls below the softening range of the resin and crystallization occurs, is an indicator for many of these variables. The frost line may not be visible at times. When it is not, the zone where the bubble reaches its final diameter is considered to be the frost line.

In blown film extrusion, the barrel temperatures are relatively low, which permits fast production rates without raising the frost line too high. Sufficient barrel temperature is required for good optical properties, that is, to avoid melt fracture, unmelted polymer, and other defects. However, if the barrel temperature or melt temperature is too high, the viscosity becomes too low, and the bubble becomes unstable and may break. Thus, the temperatures should be as high as the resin and the cooling equipment permit. A lower die temperature may be used when starting up the film extruder, and then later the temperatures may be adjusted to optimize the film properties.

Die temperatures usually match the extruder's metering zone temperatures. Sufficient die temperatures also contribute to good optical properties. Elevated die lip temperatures are used to reduce or eliminate melt fracture. This requires dies with separate heater controls for the die lip temperature; these dies are commercially available.

Since the film thickness must be uniform, die gaps are usually adjustable. Improper adjustment of the die opening may cause variations in film gauge and, thus, nonuniform cooling and nonlevel frost line. Die gaps typically range from 0.70 to 2.55 mm (0.028 to 0.100 in).[90] Large die gaps increase output slightly, making gauge and frost-line control more difficult. They also promote film snap-off when the film is drawn down to small gauge [<0.13 mm (0.005 in[91])]. The die entry angle controls the pressure drop and, thus, shear stress and melt fracture or orange peel. Smaller entry angles and longer die land lengths permit more relaxation of the aligned polymer chains, thereby reducing melt fracture.

The extrusion rate is controlled by screw speed and head pressure. Since output increases with screw speed, films become thicker. Extruder size should also match the die size, as illustrated in Table 5.10.[92] Internal air is introduced through a 6.5- to 12.5-mm-(0.25- to 0.50-in-)[91] diameter hole in the die mandrel. The air pressure, typically 0.7 to 34 kPa (0.1 to 5.0 lb/in^2), is used to expand the bubble, but then held constant once the bubble diameter is fixed. This ensures uniform film width, uniform film thickness, and wind-up of wrinkle-free rolls.

After the molten tube exits the die gap, it travels upward before the internal air pressure expands the tube into a bubble. This upward distance, or stalk height, allows the melt to cool slightly and orient axially. As a result, the longer stalks increase film strength in both the flow, or machine direction (MD), and the perpendicular, or transverse direc-

TABLE 5.10 Extruder and Film Die Sizes[92]

Extruder diameter, mm	Blown film die diameter, mm	Cast film die width, mm	Extrusion coating die width, mm
38	<100	—	—
64	75–200	<900	—
89	150–380	600–1520	610–1220
114	≥220	900–1830	915–2290
152	≥220	1520–3050	1370–3550
203	—	—	≥4065

tion (TD). Stalk height depends on material properties and processing conditions. Typically, low-density polyethylene has a short stalk, whereas high-density polyethylene (HDPE) has long stalk.[93]

When the molten tube expands, it stretches the film in the transverse direction. A measure of this transverse stretching of the bubble is the blow-up ratio (BUR):

$$\text{BUR} = \frac{D_b}{D_d} = \frac{0.637W_{\text{LF}}}{D_d} \tag{5.20}$$

where D_b is the stabilized bubble diameter, D_d is the die diameter, and W_{LF} is the lay-flat width. The BUR is typically 2:1 to 4:1,[94,95] but can be as high as 7:1 for HDPE.[95] At constant take-off speed, increasing the BUR stretches the film, thereby increasing film width and biaxial orientation, reducing film thickness and promoting rapid cooling. As shown in Figs. 5.41*a* and *b*,[96] the thinner film has lower tensile and tear strength in both the machine and transverse directions. However, the increased biaxial orientation has aligned more polymer chains in the transverse direction. This decreases the tensile strength more in the machine direction while the tear strength reduces more in the transverse direction. With polyethylenes, a BUR of 2.5:1 provides equivalent orientation in both directions, making this "balanced" film

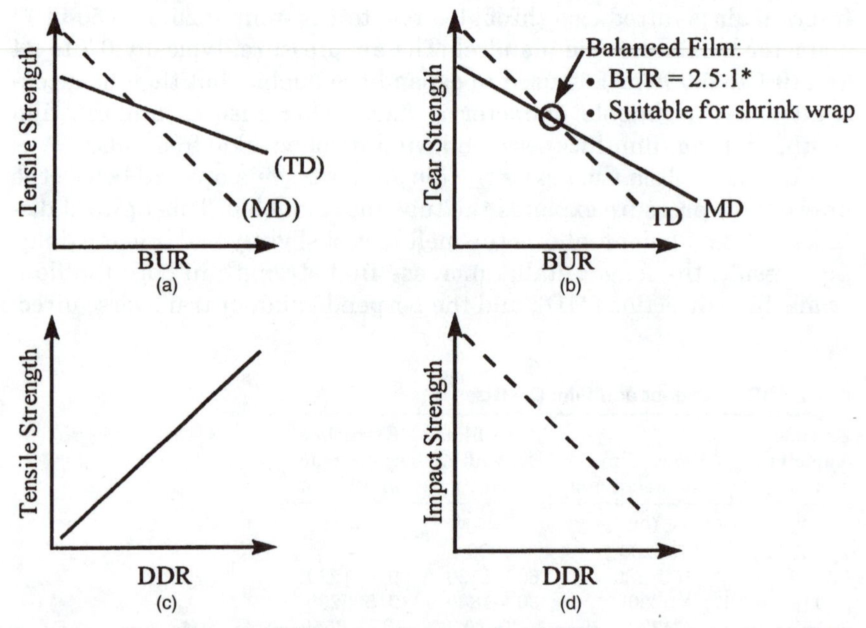

Figure 5.41 The effect of (*a*) BUR on tensile strength, (*b*) BUR on tear strength, (*c*) DDR on tensile strength, and (*d*) DDR on impact strength.[96]

suitable for shrink-wrap films. Impact strength is also enhanced by the biaxial orientation while any blocking tendency decreases due to the rapid cooling. Although stretching the film washes out defects, which improves optical properties and gloss, the larger, more unwieldy tube produced at higher BURs reduces bubble stability.

Increasing the take-off speed increases film length and uniaxial orientation but reduces film thickness. The machine-direction stretching of the film is gauged by the drawn-down ratio (DDR)

$$\mathrm{DDR} = \frac{W_d}{H_f\,\mathrm{BUR}} \tag{5.21}$$

where W_d is the die gap and H_f is the film thickness. Thus, the DDR is, effectively, the ratio of the take-off speed to the extruder speed. Since faster speeds are profitable, take-off speeds are as fast as possible. However, the cooling rate must match the take-off rate to prevent bubble instabilities and blocking. Increased uniaxial orientation also aligns the polymer chains in the machine direction. As shown in Figs. 5.41*c* and *d*,[96] this increases machine direction tensile strength, but reduces tear strength. The aligned chains are easily spread, and so the film impact strength is reduced.

The rate of bubble cooling is critical for obtaining the highest film quality and averting blocking in the nip rolls and on the wind-up roll. Generally, a large volume of low-pressure air is preferred to a smaller volume of higher-pressure air. Controlling the quantity and direction of this air is important because both are essential in gauge thickness control. The cooling rate also affects optical properties. Rapid cooling freezes in flaws and die lines, while slow cooling, which permits crystal growth, increases haze.

The frost-line height (FLH) is controlled by the cooling rate (preferred method), extrusion rate, take-off speed, and melt temperature. The recommended height is two to three die diameters [~200 to 450 mm (8 to 18 in)]. The frost-line height affects gauge control, and it becomes more critical with higher frost-line heights. Since higher frost-line heights also permit slower cooling and more crystal growth, they produce stiffer, more opaque films. However, high frost-line heights also wash out die lines and other defects, thereby improving gloss and the surface finish. Finally, high frost lines increase the tendency to block (external film surfaces adhere to each other, particularly in the film roll) and failure to slip (internal surface stick to each other).

Blown film extrusion is a scrapless operation with high outputs. The films are versatile; they can be used as tubes or slit to become flat film. Finally, the process inherently produces biaxial orientation. However, blown film extrusion requires high melt-strength materials, and so is limited to polyethylene, polypropylene, poly(ethylene-co-vinyl acetate) (EVA), flexible poly(vinyl chloride), and some polyamide and polycarbonate grades. The process provides slower cooling, and thus higher haze. Moreover, gauge control is difficult. Blown films are typically 0.0025 to 1.25 mm (0.0001 to 0.050 in) thick.[97]

5.3.2 Flat film and sheet extrusion

Another process for producing plastic films is flat, or cast, film extrusion. In this process the extruder pumps molten resin through a flat film or sheet die. The melt leaves the die in the form of a wide film, or sheet. This is typically fed into a chill-roll assembly which stretches and cools the film. After the film passes through the rolls, the film is wound up on a roll.

A film or sheet extrusion line consists of an extruder, film or sheet die, cooling system, take-off system, wind-up system, and auxiliary equipment such as film gauging systems, surface treatment, and slitters. Single-screw extruders with relatively long barrels (L/D = 27 to 33:1) are used for most resins, but polyvinyl chloride powders are often processed with twin-screw extruders.[98] As shown in Table 5.10,[92] extruder size is matched to the die width.

Film and sheet dies are wide, flat dies consisting of two pieces: a manifold and the die lip. The manifold distributes the melt across the width of the die, whereas the die lip controls molten film thickness. Three basic manifold designs are used in flat film dies. The T design (Fig. 5.42*a*[99]) is simple and easy to manufacture. Although it produces a nonuniform pressure drop across the die and thus causes nonuniform flow through the die lip, the distribution of the melt does not produce die distortion of clamshelling. This design is not suitable for high-viscosity or easily degradable melts, but can be used for extrusion coating.[100] In the coat-hanger manifold (Fig. 5.42*b*[99]) a channel distributes the melt, but flow is restricted in the preland. While the shape of the manifold channel compensates for the pressure drop, the placement of the die bolts permits die distortion that varies with melt viscosity and polymer flow rate. With a fishtail, the manifold, the entire land area, rather than a flow channel, changes to adjust the pressure drop. This gives better melt distribution, but the die is massive and contains a large mass of polymer. As a result, the fishtail manifold can create temperature nonuniformities and degradation problems. Since they provide uniform pressure drops with less thermal mass, coat-hanger manifolds are most commonly used. Fishtail manifolds are often employed in sheet extrusion,[100] and T manifolds are also becoming more popular due to the absence of clamshelling.

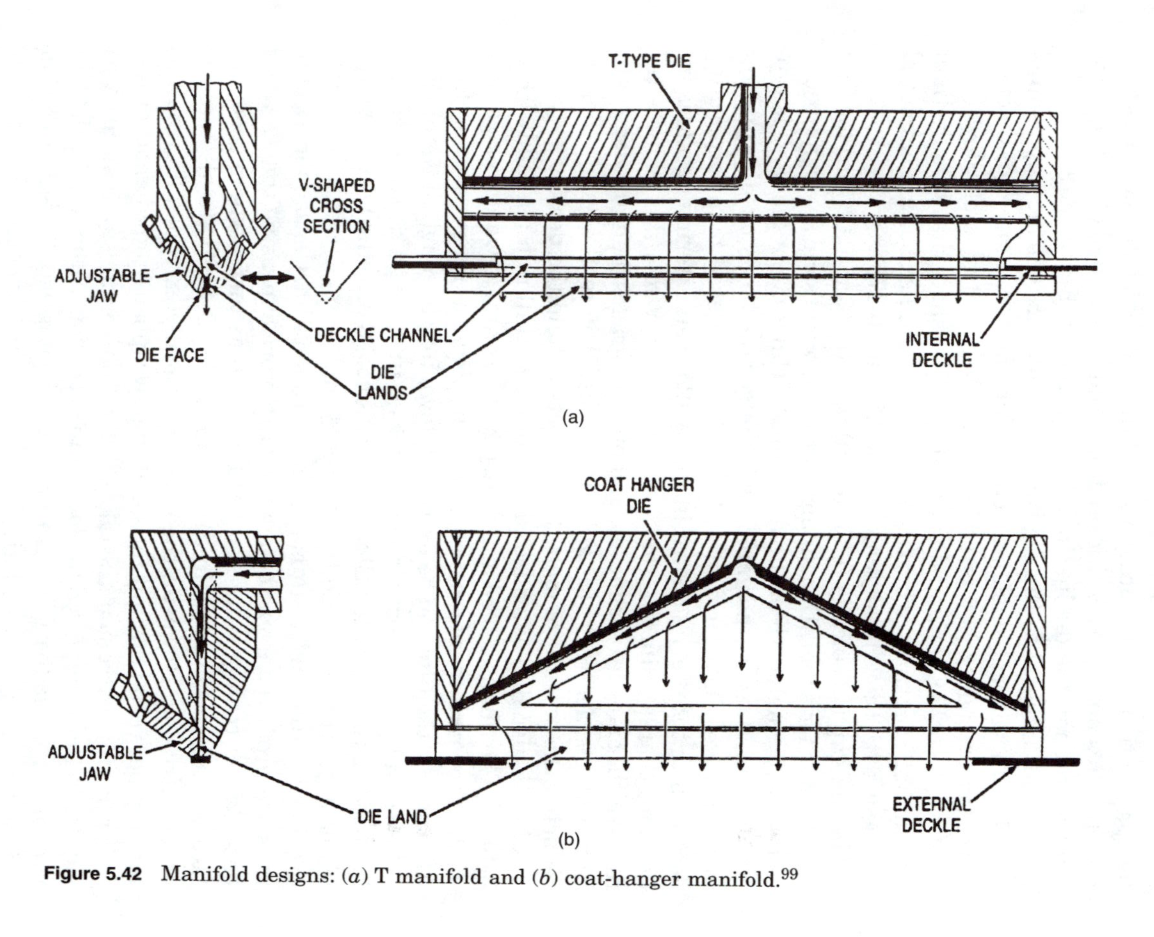

Figure 5.42 Manifold designs: (*a*) T manifold and (*b*) coat-hanger manifold.[99]

Three mechanisms produce the fine adjustment for flow through the die. In a flex lip die, a metal bar with a flexible hinge is machined into the upper part of the die. Bolts at an angle to the die lip adjust the die gap. For thin films [<0.25 mm (0.010 in)], the bolts only push the die lips together; the natural spring of the metal and the melt pressure force the lips apart. With thicker films and sheet, the bolts alternately push and pull the die lip. In a flex bar design (Fig. 5.43*a*[101]), the die lips are not integral to the die body. The thickness of the upper lip is varied with vertical push/pull bolts. For thicker films and sheet, a restrictor or choker bar is added to further control melt flow (Fig. 5.43*b*[101]). A choker bar and flex lip are used for sheet that is more than 6.4 mm (0.25 in) thick, whereas a choker bar and flex bar are employed for sheet with thickness greater than 9.5 mm (0.375 in).

The die gap is typically 0.25 to 0.50 mm (0.010 to 0.020 in) for film and greater for sheet. Depending on the material's die swell, gaps can be larger or smaller than the final film thickness. The die land is typically 10 to 20 times the gap,[102] with longer die lands used to prevent melt fracture. Sheet or film width can be decreased by incorporating deckles into the die. As shown in Fig. 5.42*b*,[99] an external deckle is little more than a shim inserted into the die lip. While this reduces the film width, material stagnates, and often degrades, behind the deckle. External deckles (Fig. 5.42*a*[99]) are placed before the land, thereby reducing the amount of melt that collects behind the deckle.

Film is usually extruded down onto a chill-roll assembly (Fig. 5.44[103]). The film is cooled and drawn by two or more water-cooled rolls while the surface of the chill roll imparts a finish to the film. Alternately, the film can be extruded down into a trough of water (Fig. 5.45[104]). In the water bath, the film passes under a guide shoe before it is raised out of the water. Since chill rolls cool more efficiently than the water bath, they increase transparency and gloss, decrease haze, and provide a better surface finish to the extruded film. Consequently, chill rolls are preferred even though they are more expensive than a water bath. Water cooling is limited to films, such as tarps, where good surface finishes are not required. Chill rolls (Fig. 5.46[105]) are also used in sheet extrusion. However, the sheet is typically extruded horizontally into nip between two chill rolls and then wrapped around another roll.

For flat film extrusion, the take-off and film-winding systems are very similar to those employed in blown film extrusion. Although the flat film cannot be rotated to randomize gauge bands, it is usually oscillated from side to side using an oscillating die, winder, or edge guide. Due to complications in extruding the film, the first method is seldom practiced. With the latter two systems, the film veers back and forth by 3.2 to 6.4 mm (0.125 to 0.250 in).[106] As illustrated in Fig. 5.44,[103] slitters (or trimmers) are also required to remove the edge bead in many film and sheet

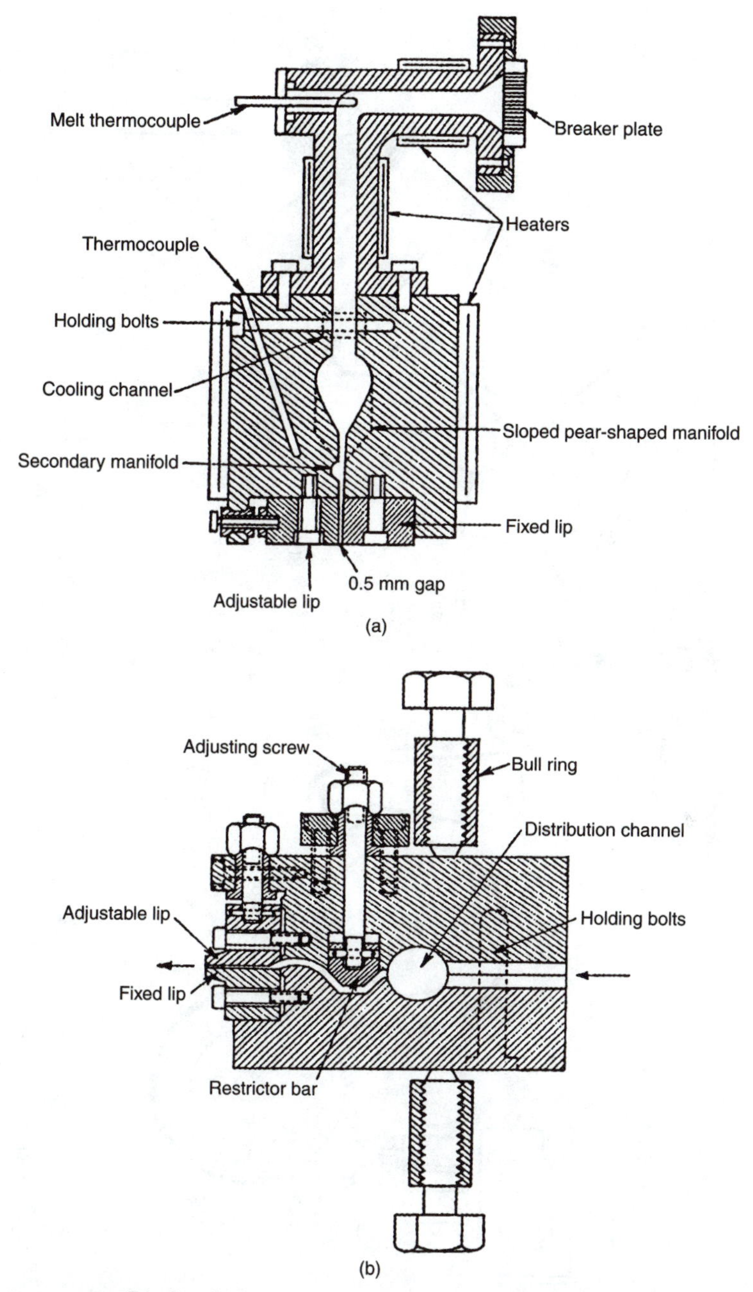

Figure 5.43 Die lip designs: (*a*) flex bar die and (*b*) flex bar die with a choker bar.[101]

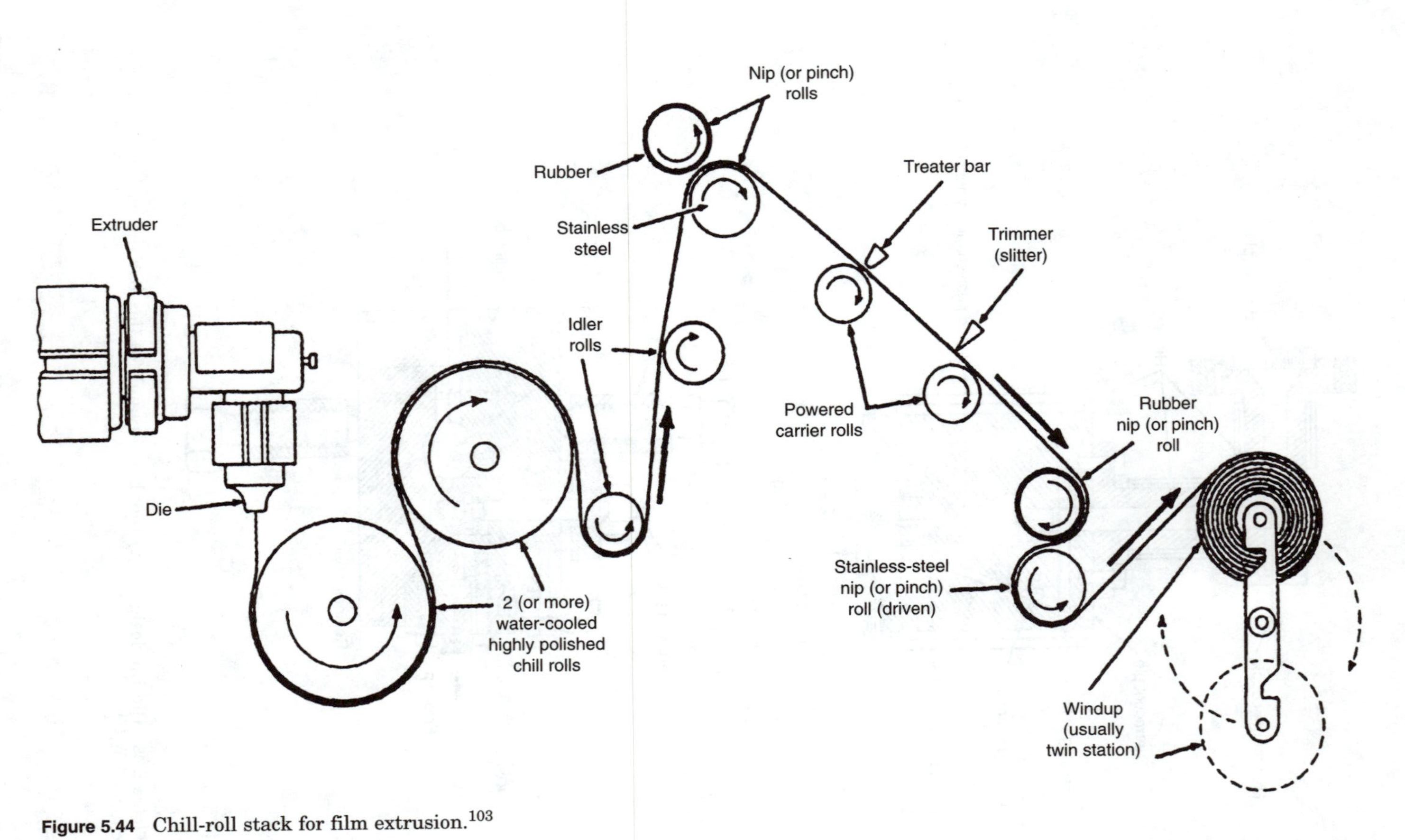

Figure 5.44 Chill-roll stack for film extrusion.[103]

lines. The edge bead (Fig. 5.47[107]) results when the flow of polymer at the ends of the die creates a thicker section on the edge of the film. In newer dies, design improvements and the incorporation of internal deckles reduce or eliminate this edge bead. Film gauging systems also located between the chill rolls and take-off rolls monitor film thickness. Since die lip bolts are placed closely along the width of the film line, local regulating of the die gap can accommodate small variations in output. Thus, the thickness values are continuously fed into manual or automatic die bolt adjustment systems.

The principal controls for a flat film or sheet line are barrel and die temperatures, die gap, extrusion rate, air gap, and chill-roll speed and temperature. Barrel and die temperatures influence melt temperature and, thus, melt viscosity. Increased melt temperature reduces viscosity which facilitates drawdown but enhances neck in (Fig. 5.47[107]) and decreases film (melt) strength. The result is a narrower film with improved uniaxial orientation. Higher melt temperatures reduce surface defects, thereby decreasing haze and increasing gloss. Since changes in melt temperature can alter the flow through the die, dies typically have multiple heating zones.

Initial die gaps are set to about 20 percent greater than the final film thickness,[108] and then adjusted to accommodate changes in polymer flow which are "resin and rate sensitive."[108] Higher screw speeds increase extruder output, overall film thickness, the tendency toward melt fracture, and may alter the flow pattern. Thus, extruder speed is not a recommended control. In contrast, increased chill-roll speeds decrease film thickness, reduce film width due to increased neck in, increase uniaxial orientation, and alter the optimum air gap or drawdown distance. The optimum air gap, which produces the best orientation, crystallization, and surface properties, depends on the material and chill-roll speed. At 23 to 30 m/min (75 to 100 ft/min), the air gap for low-density polyethylene is about 100 mm (4 in), but when the line speed increases, the air gap is found by trial and error.[109] Since the chill-roll speed controls film stretching, the take-off speed has little effect on the film dimensions.

Chill-roll or water-bath temperature influences film cooling. Increased temperature smooths surface defects but facilitates crystallization. Thus, film clarity and haze are a balance between surface properties and crystallinity. However, in general, films extruded onto chill rolls have better clarity than blown film. Hot chill rolls also reduce the film's tendency to pucker (produce nonuniformities in the film), but increase the tendency to block when the film is wound up.

Flat film and sheet extrusion produce uniaxially oriented film and sheet with improved clarity, better thickness uniformity, and a wide range of thicknesses. While the dies are very expensive and difficult to

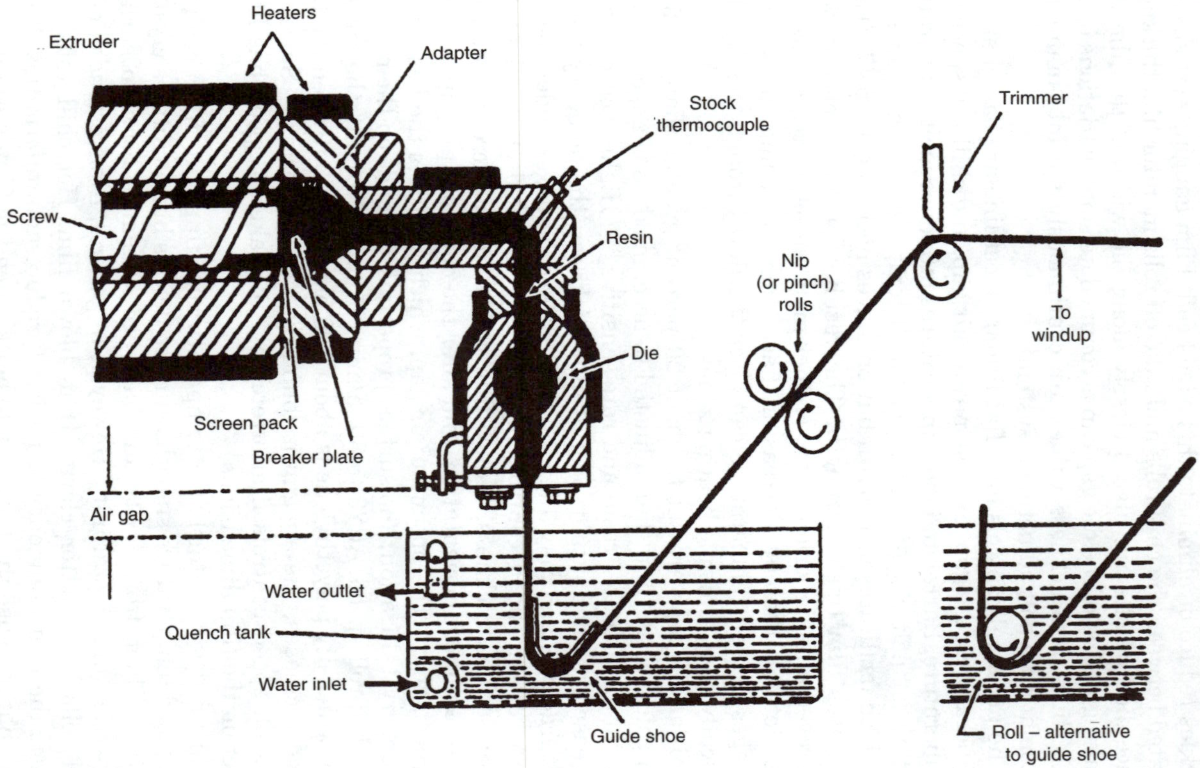

Figure 5.45 Water cooling of flat film.[104]

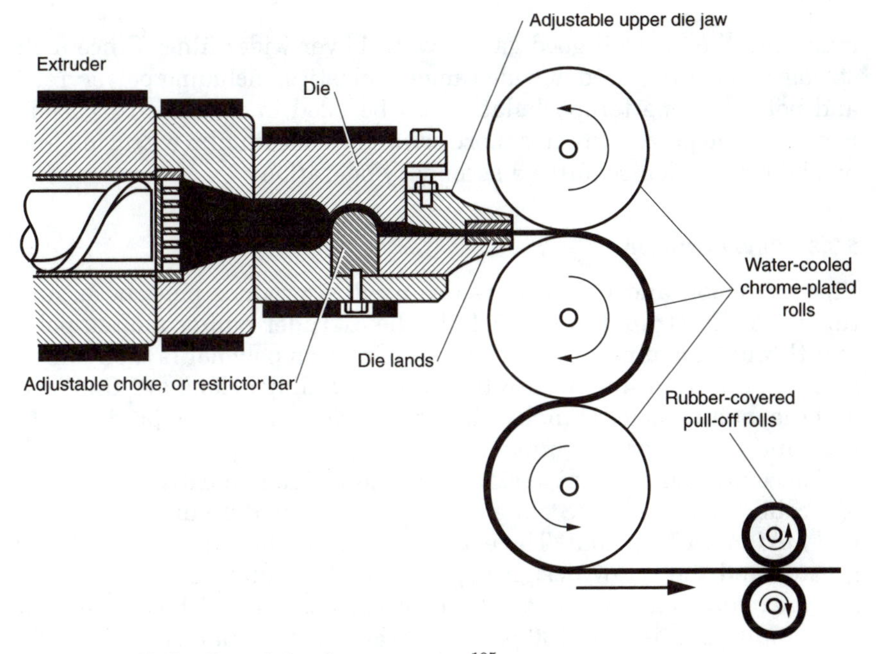

Figure 5.46 Chill-roll stack for sheet extrusion.[105]

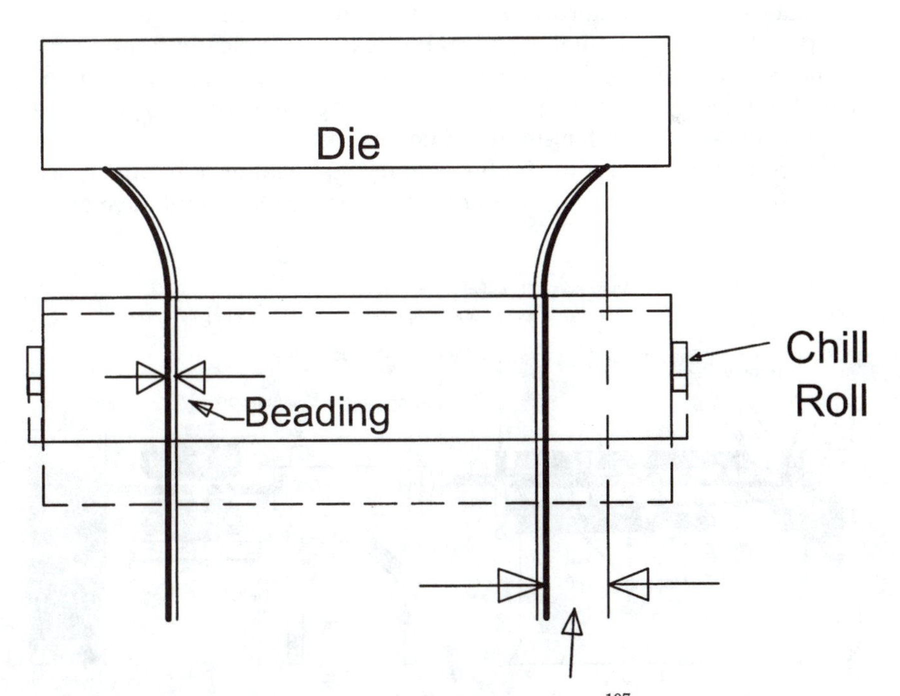

Figure 5.47 Neck in and edge beading in flat film extrusion.[107]

maintain, they permit good gauge control over wider films. Since melt strength is not required, a wide range of plastics, including polystyrene and poly(ethylene terephthalate), can be used in flat film extrusion. However, the process creates more scrap than blown film extrusion and biaxial orientation requires a secondary operation.

5.3.3 Pipe extrusion

Pipe extrusion and its variants are used to produce pipe, hose, and tubing. As illustrated in Fig. 5.48,[110] the extruder pumps molten polymer through an annular die. The tube of molten polymer is then pulled into a calibration unit where the extrudate dimensions are finalized and the melt is cooled. Finally, the puller draws the cooled product into the wind-up or stacking unit.

The components of a pipe extrusion line include an extruder, annual die, calibration system, cooling system, puller, wind-up unit or stacker, and auxiliary equipment. The extruder typically has an L/D of 24:1 or greater and the screw design depends on the material. Conical twin-screw extruders are often used for poly(vinyl chloride). The diameter of the annular dies is 25 to 100 percent of the extruder diameter. However, the output of a pipe extrusion line is limited by the ability of the downstream equipment to cool the product and hold tolerances. Since extruder capacity is often underutilized, the same die may be used for a range of pipes of differing diameter and wall thickness. Typically, a number of adapters and replaceable components are used with the same rear section of die. Dies with multiple outlets are employed when large-diameter extruders are used for small-diameter pipe.

Since poly(vinyl chloride) is the primary material used in pipe extrusion, spider-arm dies (Fig. 5.49*a*[111]) are the predominant design. In

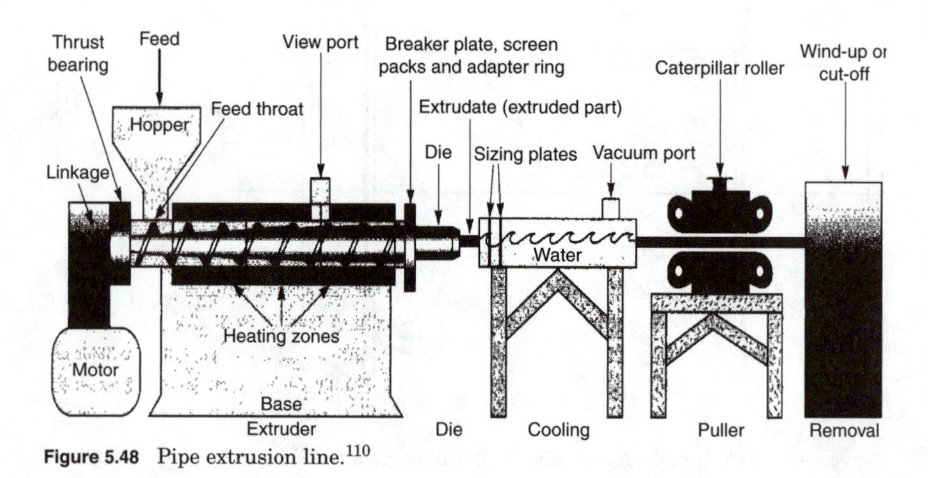

Figure 5.48 Pipe extrusion line.[110]

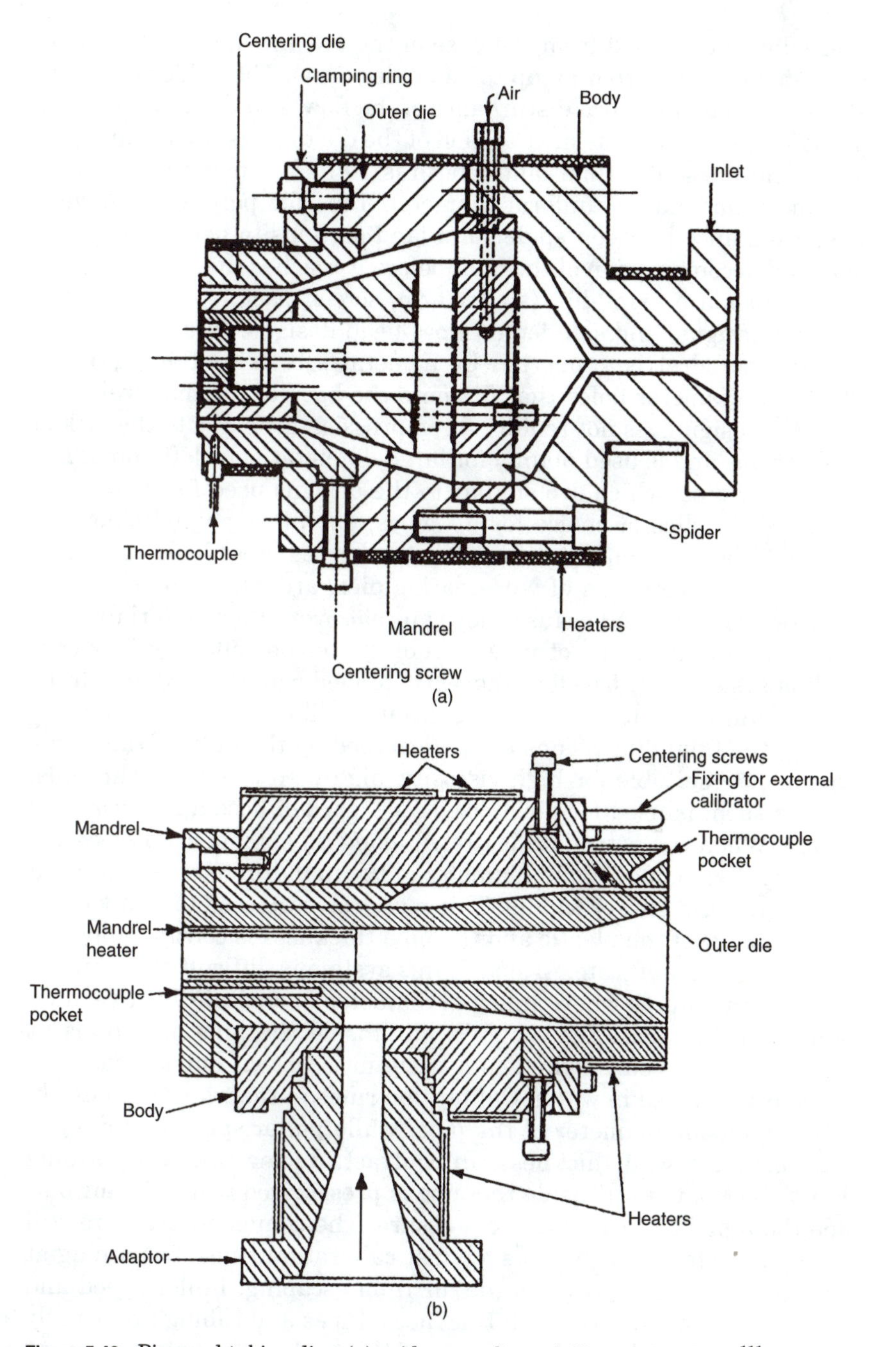

Figure 5.49 Pipe and tubing dies: (*a*) spider-arm die and (*b*) cross-head die.[111]

these dies, melt is fed from the base of the die and surrounds a mandrel which is supported by up to 24 spider arms. The spider arms are streamlined to minimize disturbance of the flow, and the channel cross section is reduced in the next section of the die to assist fusion and prevent weld lines. The final land length is 10 to 30 times the pipe wall thickness and has a uniform diameter. Dies for polyolefins have a longer distance between spider and land sections to permit compression and decompression of the melt before it reaches the land. The other pipe dies are axial dies (spider arm, screen pack, spiral flow, and helicoid designs) and side-fed or cross-head designs. In a screen pack die, the mandrel is supported by a metal screen containing 1-mm (0.040-in) diameter holes that fits onto the back of the mandrel. This compact design does not require a compression zone due to the lack of spider arms and is used mainly for large-diameter polyolefin pipe. The spiral flow dies, which are similar to the designs used for blown film dies, are the dies of choice for all pipes, except poly(vinyl chloride). Helicoid dies are employed for tubing, whereas cross-head dies (Fig. 5.49*b*[111]), an adaptation of wire-coating dies, are used extensively for catheters and tubing because they can coat over other materials.

The final dimensions of pipe or tubing are controlled by four calibration systems: (1) free forming, (2) extended water-cooled mandrels, (3) vacuum calibration, and (4) pressure calibration. In free-formed tubing, the final dimensions are determined by the pull-off rate. This type is only suitable for high-viscosity, high melt-strength materials, and, as such, is used for plasticized PVC garden hose and laboratory tubing. When the extrudate is drawn along an extended water-cooled mandrel (Fig. 5.50*a*[112]), the diameter of the mandrel fixes the internal diameter of the pipe or tubing. This mandrel is tapered to compensate for shrinkage of the plastic and the pipe thickness is controlled by the die gap and pull-off rate. Although this system is difficult to control, it is used in products where internal diameter is critical. With vacuum calibration systems (Fig. 5.50*b*[112]), air inside the pipe or tube is at atmospheric pressure, whereas a vacuum is drawn outside the pipe which is immersed in water. Calibration rings in the water bath establish the outside diameter of the pipe, while puller speed and die gap determine the wall thickness. In contrast, during pressure forming (Fig. 5.50*c*[112]), the air inside the pipe is pressurized while the air outside the pipe is at atmospheric pressure. The change in pressure still forces the exterior of the pipe against calibration rings, but a plug at the end of the pipe prevents the air from escaping. Puller speed and die gap again control the wall thickness. Pipes and tubing are usually cooled in a water bath that is part of the calibration unit and many lines require a second water bath for high throughputs or for profiling of the bath temperature. With large-diameter pipes, the dimensions

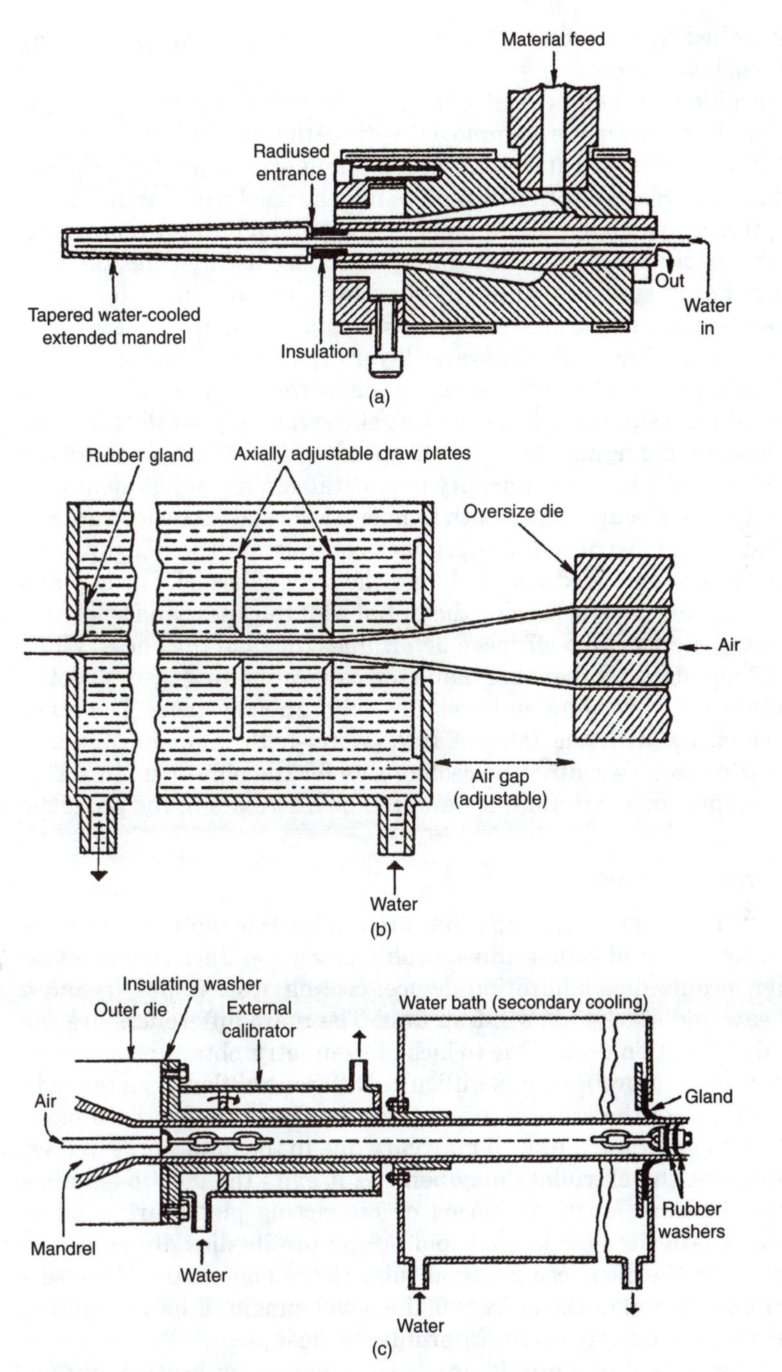

Figure 5.50 Calibration methods for pipe extrusion: (*a*) extended water-cooled mandrel, (*b*) vacuum forming, and (*c*) pressure forming.[112]

are controlled by pressure or vacuum calibration, and a water spray system cools the pipe.

A caterpillar haul-off is used when pipe or tubing is rigid, while servodrive puller systems are employed with catheters and medical tubing. Soft elements of caterpillar are a shallow V shape to grip the extrudate, whereas smooth belt pullers may be used with tubing. Since pulling forces are high, pullers must be long enough to prevent marking of the extrudate. Although wind-up units are used for flexible tubing and hose, stackers and cut-off saws are required for pipe. Four-sensor ultrasound measuring systems are usually located in or just after the cooling bath; the water layer between the pipe and ultrasound probes is beneficial for accurate measurement since it acts as a coupling fluid. Departure from the target average pipe-wall thickness is reduced by changing the take-off speed while deviation from the required wall-thickness uniformity necessitates a die adjustment.

While barrel, die, and water-bath temperatures and extrusion rate can be varied in pipe extrusion, the primary controls are the die gap, calibration rings or extended mandrel, and the take-off speed. Die gap and centering of the mandrel provide the initial wall thickness and thickness uniformity, whereas take-off speed determines the final thickness. While extended mandrels fix the inner diameter, calibration rings establish the outer diameter in vacuum and pressure calibration systems. The other diameter varies with the take-off rate. Water-bath temperatures and temperature profiles control the cooling rate for the pipe or tubing. This alters the shrinkage, crystallinity, and retained stresses in the products.

5.3.4 Profile extrusion

Profiles comprise cross sections that are not a circle, annulus, or wide sheet. Like pipe and tubing lines, profile extrusion lines consist of an extruder, profile die, calibration device, cooling system, puller, and a cut-off saw and stacker or wind-up unit. The main differences are the dies and calibration units. Due to lack of symmetry, obtaining a correct cross section in a profile die is difficult. Differential flow resistance in different parts of the cross section alters the flow rate for these parts of the die. In addition, die swell may vary due to the differences in flow. Consequently, the extrudate may bend as it exits the die. To equalize flow, the die land length is varied or restricting plates are used in channels where the flow is too rapid. Many profile dies are split into sections, with the die sliced perpendicular to the major axis. Thus, sections can be altered in the process of die development. Flow simulation software is particularly useful in profile die design.

A key component of a profile line is the vacuum calibration system. This technique involves the use of a calibration or sizing block that pulls the profile against its walls by a partial vacuum. The lengths of the calibration blocks are dependent on the thickness of the profile and the desired extrusion rate. The calibration blocks are temperature controlled and are coated or plated to reduce drag on the walls.

5.3.5 Fiber spinning

During the melt spinning of fibers such as polyamides, polyesters, and polypropylenes, the extruder pumps the melt or slurry through a spinneret die and into a containment box where the melt is cooled. The fiber is then drawn and wound on a bobbin. The process produces high levels of uniaxial orientation and facilitates crystallization of the fibers.[113]

Extruders for typical fiber lines are located on a platform that is 9 to 12 m (30 to 40 ft) above the floor and extrude downward. The die has three components: (1) a manifold, (2) multiple spinneret dies, and (3) a gear pump per spinneret. While the manifolds are similar to those used in flat film dies, the presence of many spinneret dies produces a large pressure drop. Thus, the gear pumps improve melt conveyance to compensate for this pressure drop. The spinneret dies contain many small circular holes [diameter is 0.13 mm (0.005 in[114])] and produce strands of extrudate. When these strands pass through the containment box, forced air cools the strands. Then the strands are gathered into bundles (tows) and stretched by a series of rollers. The oriented strands are wound on bobbins or occasionally the fiber is chopped into staple. Nonwoven fabrics are produced by extruding the strands onto a release sheet. Turbulence in the cooling air disturbs the flow of the strands, thus creating the random pattern of the nonwoven material. Since the strands are still molten when they reach the release paper, they bond together. For nonwovens, air and gravity provide the only drawing of the strands.

The melt spinning process is modified for polymers that are not melt processable. In wet spinning, the polymer is dissolved in a solvent, and the solution is forced through the spinneret die. However, the containment box is replaced by a chemical bath that solidifies or "regenerates" the fibers. With dry spinning, the polymer is also dissolved in a solvent and pumped through the die. When the fluid exits the die, the solvent evaporates leaving the fiber. During gel spinning, the material exiting the die forms a rubbery solid upon cooling. The filaments first pass through air and then through a liquid cooling bath. Wet spinning is used for rayon (cellulosic), acrylic, aramid, modacrylic, and spandex fibers, whereas acrylic, spandex, triacetate, and vinyon fibers can be dry spun.[115] Gel spinning is utilized for some high-strength polyethylene fibers.

5.3.6 Wire coating

In wire coating (Fig. 5.51[116]), metal wire is covered with plastic insulation. This process requires a pay-off (input) drum, input capstan, preheater, extruder, cross-head or offset die, cooling trough, spark tester, diameter and eccentricity gauges, output capstan, and wind-up drum. The pay-off drum and input capstan unwind wire at high speeds and provide constant speed and tension. In the preheating station, a gas burner or electrical resistance heater preheats the wire to improve adhesion, reduce plastic shrinkage, and remove any moisture or wire-drawing lubricant from the wire. As the wire passes through the die, it is coated with polymer melt. The coated wire is cooled in a series of water baths and then the coating is measured and tested prior to being wound up on the wind-up drum.

Wire coating uses two die designs: (1) pressure-coating dies and (2) tubing dies (tools). With the more commonly used pressure-coating die (Fig. 5.52*a*[117]), wire is fed into the die through a tapered guider. The space between the guider and the inside of the die is the "gum space," and can be adjusted using bolts at the back of the die. Melt flows around and coats the wire before it exits the die and melt pressure forces the melt against the wire. The clearance between wire and mandrel is less than 0.05 mm (0.002 in) to prevent back leakage of melt. A hardened insert used around clearance to prevent wear and a heart-shaped compensating plate distributes the melt. In contrast, a tubing die (tool) (Fig. 5.52*b*[118]) coats the wire as it exits the die. A low-level vacuum applied to rear of the die removes the air and pulls the melt toward the wire. Tubing tools are suited for larger wire and cable because excessive pressure produces back leakage of melt with larger wires.

As the wire passes through one or more cooling troughs, the wire speed and wall thickness determine the length of the cooling troughs. Gradual, controlled cooling is used for coatings thicker than 50 mm

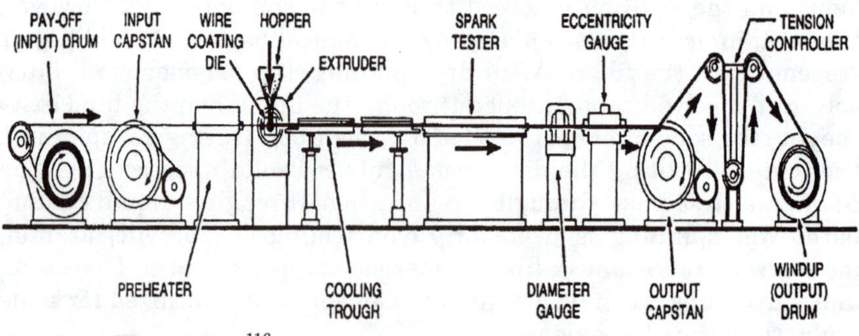

Figure 5.51 Wire coating.[116]

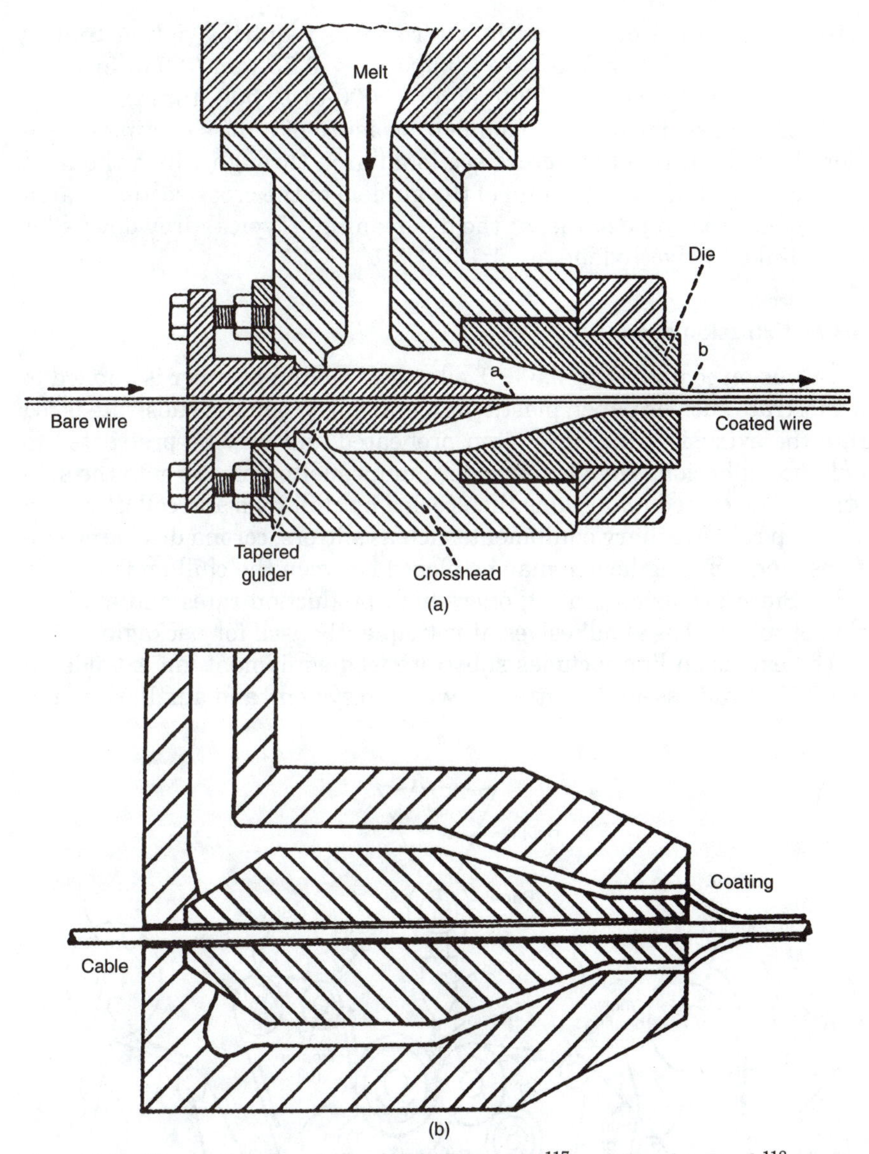

Figure 5.52 Wire-coating dies: (*a*) pressure-coating die[117] and (*b*) tubing tool.[118]

(0.020 in)[119] in order to avoid shrinkage in the resin wall next to the wire. Since quick cooling tends to harden the outer resin layers, this inner shrinkage produces voids along the wire, thereby reducing the integrity of the coating. A retention time of more than 1 min is required for thick coatings. Coating thickness is typically 0.13 to 13 mm (0.005 to 0.500 in).[119] Line speed is dictated by the speed of the wire and extruder output, with faster speeds requiring tighter control over the process. For fine wire [diameter < 0.75 mm (0.030 in)], the line speed can be 2100 to 3000 m/min (7000 to 10,000 ft/min).[120]

Coating drawdown is the common measure of a die's coating operation. It is the ratio of the cross-sectional area through which the melt is extruded (that is, at the end of the guider) to the cross-sectional area of the final coating (that is, at the die opening). Typical drawdowns for cross-linked polyethylene are 1:1 to 1.5:1.

5.3.7 Extrusion coating

In extrusion coating (Fig. 5.53[121]), a layer of molten polymer is applied to a substrate such as paper, plastic, or metal foil. When the substrate is fed into the extrusion line, it is often preheated and may be pretreated to enhance adhesion. Then an extruder pumps molten plastic onto the substrate. The coated material is then cooled by chill rolls and collected on a wind-up roll. Auxiliary equipment, such as slitters, corona discharge stations, and printing heads, may be placed between the chill and wind-up rolls. Since extrusion coating offers high production rates and elimination of solvent-based adhesives, it is frequently used for packaging.

The extrusion line includes substrate feed equipment, an extruder, a die, a chill-roll assembly, slitters, a wind-up system, and auxiliary equip-

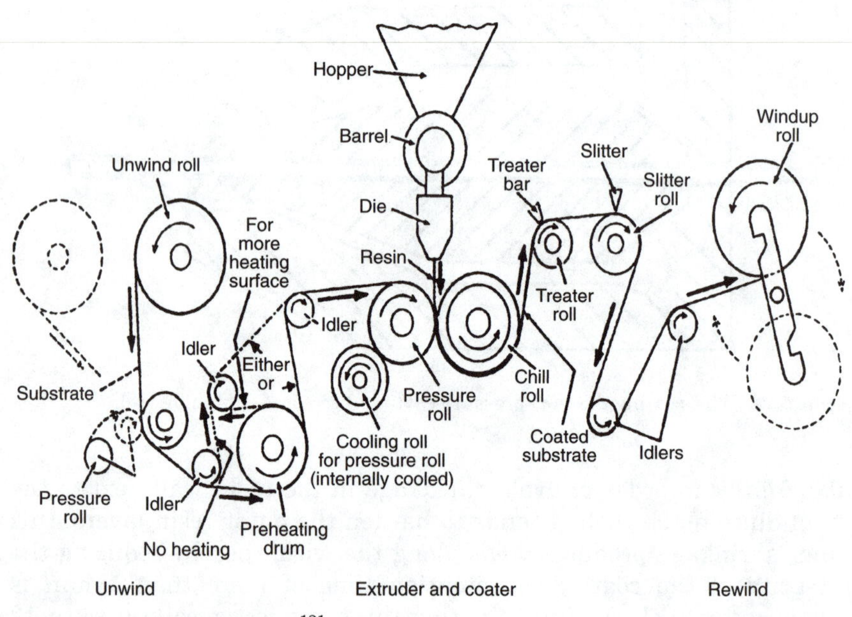

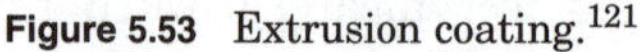
Figure 5.53 Extrusion coating.[121]

ment. Extrusion coating dies are similar to flat film dies, and as shown in Table 5.10,[92] the die size should match extruder size. The chill-roll design is also similar to chill rolls used for film and sheet lines. However, the roll contacting the polymer is steel, while the roll behind the substrate is rubber. Material is typically extruded downward into the nip between two rolls. The rubber roll is not cooled or heated directly, but as shown in Fig. 5.53, it is cooled by a second metal roll.

In extrusion coating, the primary controls are the barrel and die temperatures, substrate temperature, chill-roll temperatures, line speed, and die gap. The barrel and die temperatures are very high, producing a melt temperature that is usually just below the degradation temperature of the polymer. Extrusion coating resins usually have lower molecular weight, and so are more fluid than other extrusion-grade materials. Since the polymer melt has the lowest possible viscosity, the polymer melt flows onto and adheres to the substrate. The high temperatures also promote oxidation of the molten surface. This oxidation occurs in the air gap, the space between the die lip and nip of the chill and pressure rolls. Larger air gaps facilitate oxidation, but permit cooling of the polymer melt.

The substrate can be preheated, and paper can be dried, before it reaches the pressure roll and the chill rolls are also heated. The heated substrates and chill rolls allow the molten polymer to flow on the substrate and permit relaxation of the polymer chains. This not only facilitates mechanical interlocking of the coating and substrate, but also relieves stresses that would reduce adhesion of the coating to the substrate. The chill-roll temperature also affects the coating surface properties. Since rapid cooling freezes in coating defects, the coated surfaces are not as glossy.

Since higher screw speeds increase extruder output, they produce thicker coatings with decreased neck-in. In contrast, the chill rolls draw down the melt and thus reduce coating thickness and increase neck-in. Typically high drawdown is required for fast line speeds and thin coatings, whereas slower speeds are used for thicker coatings. Increasing the chill-roll speed also increases stress because melt drawing enhances uniaxial orientation, but the thinner coating freezes the polymer chains in their oriented state. The result is reduced adhesion. Thicker coatings also exhibit decreased adhesion due to the nonuniform stresses developed during the cooling of the coating. Two other measures affected by the chill-roll speed are coating weight per unit area and coating area output. The former decreases as the coating thins, while the latter increases as more polymer is spread over a larger area of substrate.

Larger die gaps also increase coating thickness but reduce neck-in. Larger air gaps increase neck-in since the longer distance between the die and chill roll's nip facilitates stretching of the molten polymer. The chill-roll surface affects the coating surface. Smooth rolls provide smooth, glossy coatings, whereas rough rolls are used for matte finishes. Greater roll pressures force the polymer melt and substrate into closer contact, thereby improving adhesion. Adhesion also increases with porous substrates, adhesion promoters, and priming of the substrate.

5.3.8 Coextrusion

Coextrusion permits multiple-layer extrusion of film, sheet, pipes, tubing, profiles, wire coating, and extrusion coating. It is used mostly in packaging applications to obtain desired barrier properties. The process eliminates the need for a laminator for plastic-plastic surfaces, is less expensive, and provides property enhancement.

Coextrusion is applicable to most extrusion processes. Multiple extruders feed into the process. Two techniques, feedblocks and multi-manifold dies, are used to achieve layering in the coextruded product. A feedblock, placed between the extruders and a conventional die, divides the polymer and programs the layers into the desired configuration. In modular feedblocks, the manifold is sliced into functional segments as shown in Fig. 5.54*a*.[122] The modules may be changed to accommodate more or fewer extruders, alter layer configurations, or adapt to different dies. Cloeren feedblocks (Fig. 5.54*b*[122]) have adjustable vanes in the flow channels. The position of these vanes determines the layer thickness and may be varied during extrusion. Since multimanifold dies (Fig. 5.54*c*[123]) are special dies which permit multiple-layer extrusion, the polymer feeds directly into the die. As summarized in Table 5.11, feedblocks are versatile and permit an unlimited number of layers, whereas multimanifold dies are dedicated to specific applications and restrict the number of layers that can be extruded. However, the polymer is delivered from the extruders directly to the multimanifold die. The shorter residence time hinders the mixing of layers and prevents degradation of heat-sensitive materials. As a result, multimanifold dies can coextrude resins with widely varying viscosity.

In general, extruders are located as close to the feedblock or multi-manifold die as possible to minimize the length of transfer lines (which connect the extruder to the feedblock or die) and so reduce the residence time of the melt. Typically, the extruder located closest to the feedblock contains the heat-sensitive polymer. Adhesive resins tend to cross-link if their residence time is too long. The barrier resin, poly(vinylidene chloride), is also heat sensitive.[122]

Adhesive resins or tie layers often connect two incompatible polymers such as a polyolefin and a barrier resin. These adhesive materials

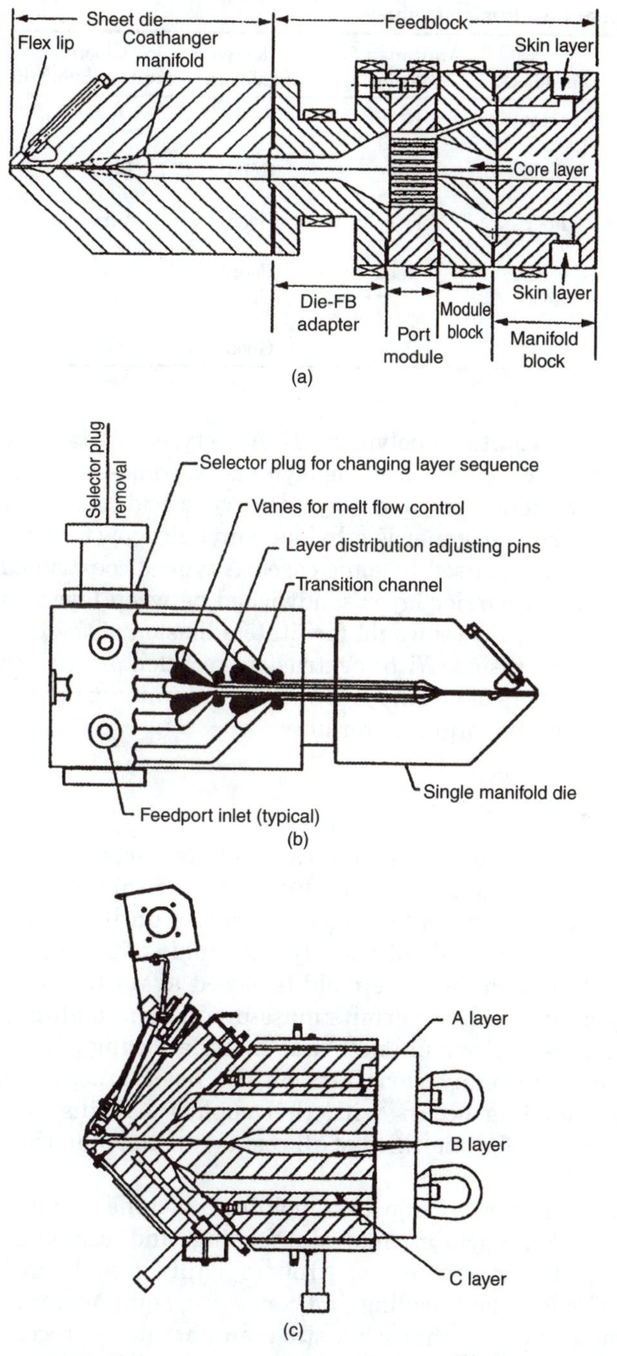

Figure 5.54 Coextrusion: (*a*) modular feedblock,[122] (*b*) Cloeren feedblock,[122] and (*c*) multimanifold die.[123]

TABLE 5.11 Comparison of Coextrusion Methods

	Feedblock	Multimanifold die	Cloeren feedblock	Cloeren multimanifold die
Equipment cost	Low	High	Low	High
Application to existing dies	Good	Poor	Good	Poor
Ability of extrudewide viscosity ratios	Poor	Very good	Good	Very good
Suitability for heat-sensitive materials	Poor	Good	Poor	Good
Ease of die adjustment	Good	Poor	Good	Fair
Flexibility in number and position of layers	Good	Poor	Good	Poor

include ethylene-vinyl acetate copolymer (EVA), styrene-butadiene (SBS), and styrene isoprene (SIS) block copolymers, styrene-ethylene-butylene-styrene (SEBS), ionomers, and ethylene acrylic acid copolymer (EEA). The principal barrier polymer is ethylene vinyl alcohol (EVOH), with poly(vinylidene chloride) used in some cases. A typical coextruded structure is comprised of a barrier layer sandwiched between layers of the structural resin; two tie layers would facilitate adhesion of the barrier layer to the structural resins. With coextrusion, recycled plastic can also be placed between layers of virgin resin so that contaminants in the recycle can be masked by the virgin material.

5.4 Injection Molding

Injection molding is a versatile process that can produce parts as small as a fraction of a gram and as large as 150 kg.[124] During this process, molten plastic is forced (injected) into a mold and cooled until the melt solidifies. When the part is cooled sufficiently, the mold is opened, the part is ejected from the mold, and the mold is closed again to repeat the cycle. Thus, injection molding permits mass-production, high-precision, and three-dimensional virtual netshape manufacturing of plastic parts. While there are many variations on the basic process, 90 percent of injection molding occurs with thermoplastic resins, and injection molding accounts for one-third of all resins consumed in thermoplastic processing.[124]

Injection molding requires an injection-molding machine, a mold, and ancillary equipment such as material-feeding and conveying equipment, dryers, mold temperature controllers, chillers, and robotics and conveyers. The material-feeding and conveying equipment and the dryers are common to most thermoplastic manufacturing processes, while the robotics and conveyers automate the molding process. Basic injection-molding machines are discussed in Sec. 5.4.1 while molds are examined in Sec. 5.4.2.

5.4.1 Injection-molding machines

Injection-molding machines have three components: the injection unit, the clamping unit, and the control system. The injection unit plasticates and injects the polymer melt while the clamping unit supports the mold, opens and closes the mold, and contains the part ejection system. Typically different injection units can be assembled with the same clamping unit.

The injection unit brings the nozzle into contact with the sprue bushing of the mold, generates contact pressure between the nozzle and sprue bushing, melts the plastic material, injects the molten material into the mold, and builds up packing and holding pressure. While several types of injection units have been used, the primary types are single-stage plunger units, two-stage screw-plunger machines, and single-stage reciprocating-screw units. Single-stage plunger units (Fig. 5.55[125]) melt the polymer by conduction from the barrel walls and then inject the melt using a hydraulic plunger. Since these units provide limited shot size [<435 cm^3 (16 oz[126]], poor shot size control, poor melting and mixing, nonuniform heating of the melt, long residence time, wide residence time distribution, and high pressure losses, they have been replaced in most applications by reciprocating screw units. However, single-stage plunger units are still used with bulk molding compounds[127] and for very small parts. Two-stage screw-plunger machines (Fig. 5.56[128]) use a screw to melt the plastic and a separate hydraulic ram to inject the molten plastic into the mold. Separation of the plasticating and injection functions permits improved melting and mixing of the polymer melt, large shot sizes, better shot-size control, lower pressure losses, and faster cycles. Two-stage screw-plunger machines are generally employed for fast cycles, large parts, and parts with long flow lengths. In single-stage reciprocating-screw injection units (Fig. 5.57[129]), the screw rotates to plasticate the polymer and moves linearly to inject the melt. As the single-stage reciprocating-screw units are the most common, the detailed discussion of injection units focuses on them.

As shown in Fig. 5.58,[130] the injection unit has two major components: the plasticating unit and the sled. The former melts and injects the resin, while the latter translates the plasticating unit. The plasticating unit consists of a hopper, feed throat, barrel, screw, screw drive motor, and nozzle. When plasticating the resin, single-stage reciprocating-screw injection units operate like extruders. Solid resin particles are fed from the hopper, through the feed throat, and onto a rotating screw. Although plastic particles may be metered onto the plasticating screw, the material is usually gravity fed into the feed zone of the barrel. The feed throat is water cooled to prevent the granulated plastic from melting (bridging) in the feed throat. A motor rotates the screw, conveying and melting the plastic as it travels

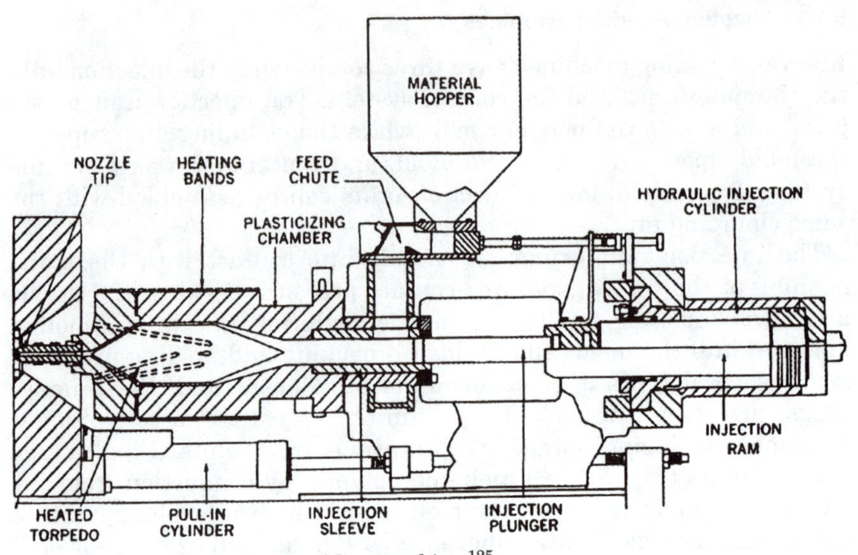

Figure 5.55 Plunger injection-molding machine.[125]

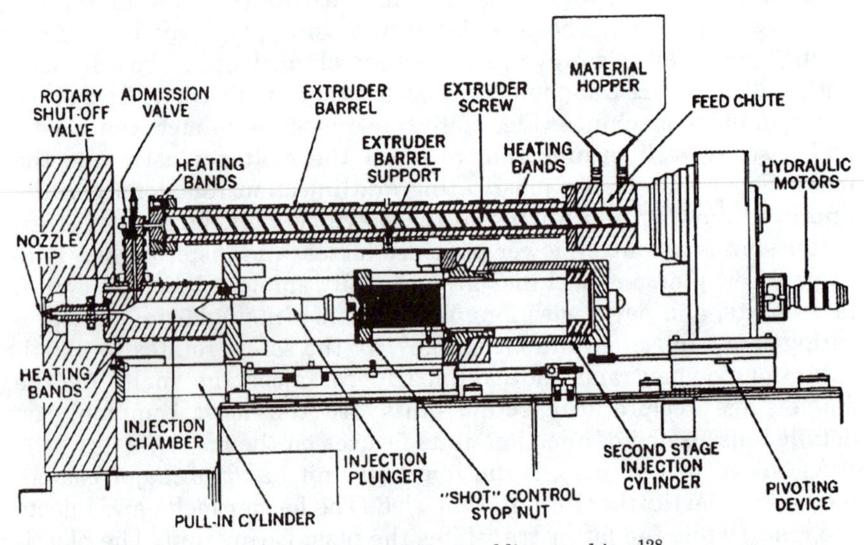

Figure 5.56 Two-stage screw-plunger injection-molding machine.[128]

through the barrel. However, injection-molding screws, unlike extrusion screws, end with a nonreturn valve. Once molten plastic flows over or through this valve, it cannot stream back toward the feed throat during the injection phase of the molding cycle. While the melt will flow out of an open nozzle, the nozzle is typically in contact with the mold. Thus, discharge from the nozzle is blocked by the sprue of

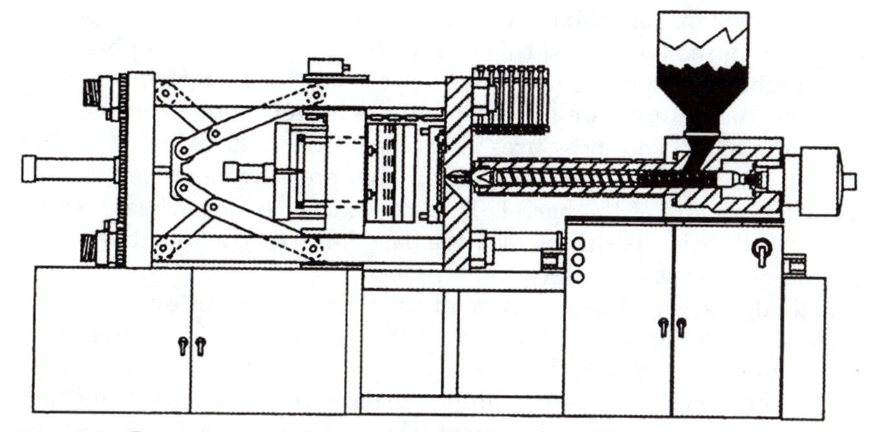
Figure 5.57 Reciprocating-screw injection-molding machine.[129]

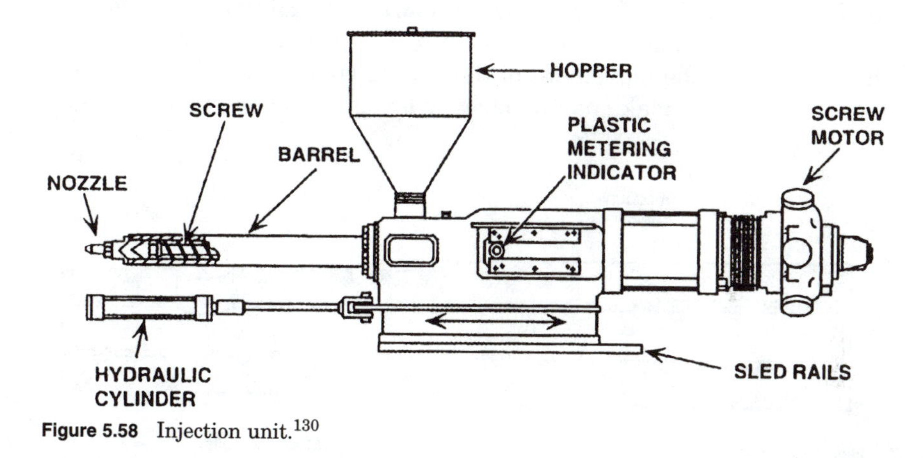

Figure 5.58 Injection unit.[130]

the cooling part and the melt collects between the nonreturn valve and the nozzle. The pressure developed by this melt forces the screw backward until the screw travels a predetermined distance. Typically, a limit switch is reached, and this triggers a signal that stops the screw's motor. This gives the desired shot size. The plastic metering or shot-size indicator indicates the linear position of the screw. This is used on simpler machines to set shot size (or maximum travel of the screw). During injection, the screw is driven forward by a hydraulic piston or electric motor. Melt is forced through the nozzle into the mold.

Injection-molding barrels (Fig. 5.59*a*[131]) are shorter than extruder barrels. The typical L/D ratio is 18:1 to 24:1 for conventional machines, 22:1 to 26:1 for "fast running machines,"[132] and 28:1 for vented barrel injection-molding machines. Unlike extruder barrels, the feed port is cut through the barrel and connects with a water-cooled feed housing. The discharge end of the barrel fastens directly to an end cap or nozzle adapter, the counterbore at the end of the barrel centers the end cap. Since the barrel sees pressures in excess of 138 to 207 MPa (20,000 to 30,000 lb/in^2),[133] "bell ends"[133] and/or a high-pressure sleeve are located at discharge end of the barrel. This sleeve is a "stronger heat-treated alloy steel"[133] that is shrunk over the barrel. Barrel treatments and liners are similar to those used in extruders.

Although screw designs are similar to those used in extrusion, an injection-molding screw (Fig. 5.59*b*[134]) does not end in a nose cone. Instead it has a counterbore that accepts a smear tip or nonreturn valve. During injection and holding, the nonreturn valve prevents melt from flowing back along the screw. There are two types of nonreturn valves: sliding rings and ball check valves. A sliding ring valve (Fig. 5.60*a*[135]) is pushed forward during plastication and is forced backward as injection begins. This provides more streamlined flow, is less restrictive in terms of materials, and produces a smaller pressure drop. However, the movement of the sliding ring causes wear and permits material to leak, particularly when glass fibers become stuck

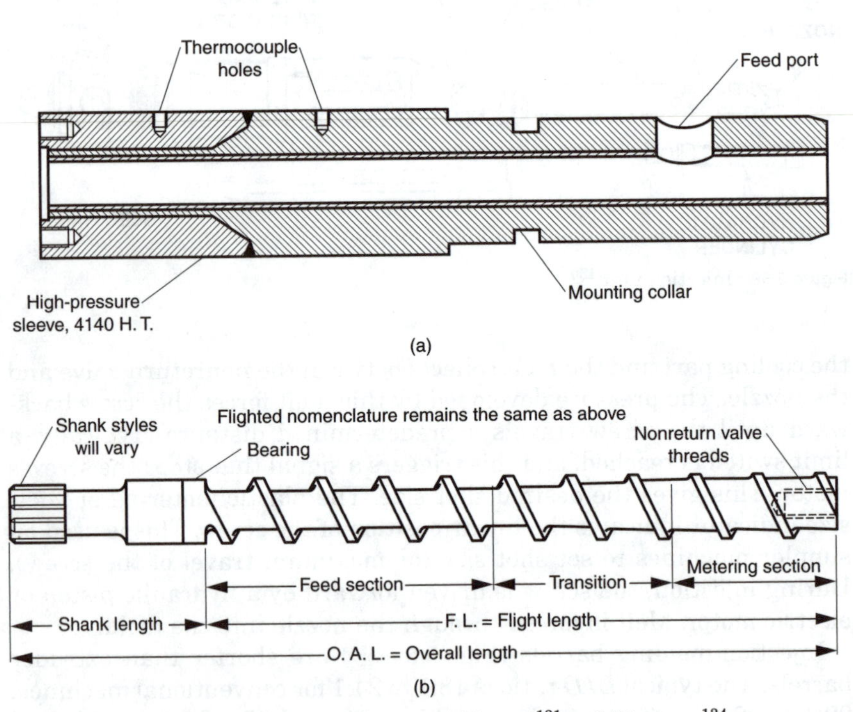

Figure 5.59 Components of an injection unit: (*a*) barrel[131] and (*b*) screw.[134]

under the ring. Sliding ring valves are used for high-viscosity materials and filled compounds and with vented injection molding. In a ball check valve (Fig. 5.60*b*[136]), a sphere is forced forward during plastication and backward for injection. While this produces more positive shutoff and better shot control, it is more restrictive to flow, provides a greater pressure drop, and causes more barrel wear than sliding rings. Ball check valves are available in front discharge and side charge designs. The front discharge valves are more difficult to clean and have a front angle that does not match the angle of the end cap or nozzle adapter. Side discharge valves eliminate these problems. Ball check valves are typically employed with unfilled, low-viscosity materials. A variation of the check valve is the Dray nonreturn valve that uses a sliding rod rather than a sphere. These provide more positive shutoff than sliding rings,[137] but are not as restrictive to flow as ball check valves. When no nonreturn valve is used, a smear tip (Fig. 5.60*c*[138]) is added to the end of the screw. Although these do not restrict flow, they permit back flow. Smear tips are used with high-viscosity and heat-sensitive materials such as rigid PVC.

The nozzle is the tip of the plasticating unit and provides a leak-proof connection from the barrel to the injection mold with minimum pressure loss. The radiused tip aligns the nozzle and the sprue bushing of the mold. SPI has two standard radii: 12.7 and 19.1 mm (0.5 and 0.75 in), and the nozzle tip opening should be 0.79 mm ($^1/_{32}$ in) smaller than the sprue bushing. Nozzles are also long enough to have heater bands and require their own heating zone(s). There are three types of nozzles: open channels, internally actuated shutoff nozzles, and externally actuated shutoff nozzles. In the common design (open channel), no mechanical valve is placed between the barrel and mold. This permits the shortest nozzle and unimpeded flow of the polymer melt. With highly fluid plastics, such as polyamide-6,6, the nozzle diameter becomes smaller and then is enlarged again before reaching the sprue bushing; this prevents drooling (of melt through the nozzle). Internally actuated shutoff nozzles are held closed by either an internal or an external spring. They are opened by the plastic injection pressure. Externally actuated shutoff nozzles are operated by external sources such as hydraulic or pneumatic pistons. While both types of shutoff nozzles are longer than open-channel nozzles, they eliminate drooling and permit plastication when the nozzle is not in contact with the sprue bushing.[139] Special cut-off nozzles are also used for injection molding of foams and for gas-assisted injection molding.

An injection-molding screw is rotated using an electric drive motor coupled with a reducer or gear box or a direct hydraulic drive. Electric drive motors are usually employed with larger hydraulic machines [clamp force $>$ 15,000 kN (1700 tons)][140] and in all-electric molding

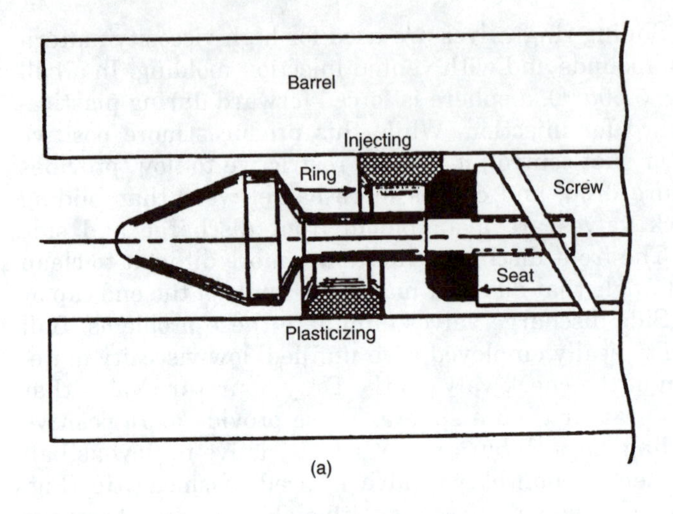

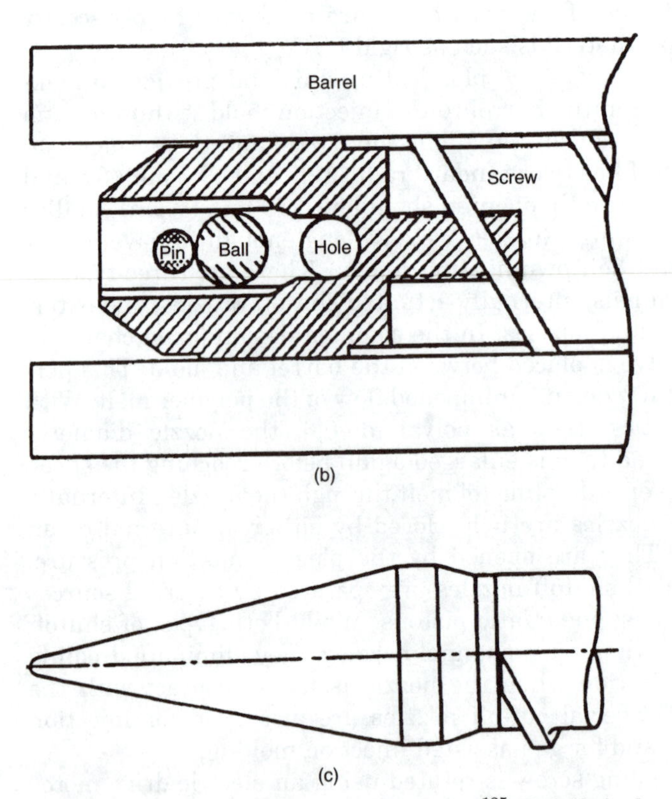

Figure 5.60 Nonreturn valves: (*a*) sliding ring,[135] (*b*) ball check valve,[136] and (*c*) smear tip.[138]

machines. While the electric motors can be started at the beginning of plastication, the screw is often connected to the running motor by an electromechanical coupling.[140] Since electric motors have high starting torques, the latter method prevents shearing of the screw. The dc SCR and ac adjustable frequency motors used for extruders are also suitable for injection molding. However, a mechanical brake or nonreverse lock may stop the motor at the end of plastication.

The former works well with machines that have no nonreturn valve, whereas the latter acts on the screw shank and is used when braking is not possible.[140] Various hydraulic drives are available, but vane motors, radial piston motors, and axial piston motors are preferred.[140] Radial piston motors are used for screws with diameters of 50 to 200 mm (~2 to 4 in)[140] because they provide smooth operation.

Screw drive motors can be positioned between the hydraulic piston and screw or at the far end of hydraulic piston.[140] Since direct hydraulic drives are lightweight, they can be placed in either position, even when the drive motor moves during injection. However, with electric motors coupled with reduction gears, the drive system is kept stationary so that the mass shifted during injection is small. The preferred design connects the screw with the stationary drive motor via a hollow shaft.[140]

The injection unit (or sled) retracts away from the mold along its rails. This permits purging of material from the barrel when changing the resin type used in the machine or getting rid of contaminated or degraded material. The injection unit's position is also adjustable to accommodate proper sprue bushing seat position for different mold bases and for different nozzles. The contact force of the injection unit prevents melt from leaking at the open interface between the nozzle and sprue bushing. As shown in Table 5.12,[141] the contact force increases with the size of the clamping unit. The unit is moved along rails or slides by a hydraulic cylinder or electric motor. Small- and medium-sized machines use guide bars (parallel to the machine axis) to steer the movement of the sled. The piston(s) can be combined with guide bars, mounted between the sled and stationary platen, either above or below the base, or mounted on either side of the sled. Since the two pistons used in the last method do not produce any lever action, they permit concentric positioning of the nozzle and sprue bushing. Due to the weight of the injection unit, large machines use slide ways backed up by adjustable guide shoes. One shoe supports and prevents tilting of the barrel.

Injection units are specified by shot size, maximum injection pressure, plasticating capacity and recovery rate, maximum injection velocity, and other less important factors. The shot size is the maximum weight or volume of plastic that can be injected in one shot. Injection

TABLE 5.12 Contact Force[141]

Clamp force, kN (tons)	Contact force, kN (tons)
500 (55)	50–60 (6–9)
1,000 (115)	60–90 (7–10)
5,000 (560)	170–220 (19–25)
10,000 (1,125)	220–280 (25–31)
15,000 (2,250)	250–350 (28–39)

unit sizes are specified by the maximum amount of plastic material they can dispense with one forward movement of the injection screw. For U.S. machines, the shot size is rated in ounces of general-purpose polystyrene (GPPS), whereas in European machines, the shot capacity is based on the volume (in cubic centimeters^3) displaced with 100-MPa injection pressure. The shot volume (size) should be at least 10 to 15 percent of the maximum[142] and no more than 75 to 90 percent of the maximum shot size.[142] Smaller shot volumes provide longer residence times, thereby producing greater variations in melt viscosity and the amount of material delivered to the mold. If the shop volume (size) is too large, the cycle time is dictated by the screw recovery time (discussed later), and the melt cushion may be insufficient for packing.

The *maximum injection pressure* is the maximum available pressure for injection. In hydraulic machines, the injection cylinders are rated for a maximum hydraulic pressure, typically 14 to 21 MPa (2000 to 3000 lb/in^2). Pressure on the melt increases because the screw diameter is usually smaller than the diameter of the injection cylinder. Thus, injection pressure P_{inj} is given by

$$P_{inj} = \frac{A_{hyd}}{A_{inj}} P_{hyd} \tag{5.22}$$

where A_{hyd} is the area of the hydraulic cylinder, A_{inj} is the cross-section area of the screw, and P_{hyd} is the hydraulic pressure. Typical intensification ratios, the area ratios of the hydraulic cylinder and screw, range from 10:1 to 16:1, but can be as high as 25:1.[143] With all-electric injection-molding machines, the injection force is obtained from a load cell, usually located behind the screw. This force is divided by the cross-sectional area of the screw to obtain the injection pressure. Standard machines have a maximum injection pressure of 138 MPa (20,000 lb/in^2), while standard machines with undersized barrels can achieve 205 MPa (30,000 lb/in^2) and special machines (for thin-wall molding) reach 310 MPa (45,000 lb/in^2).[124]

The plasticating capacity and recovery rate are standard tests developed by the Society of the Plastics Industry. Both rates are based on running polystyrene at 50 percent of maximum capacity. While the plasticating capacity provides a weight output in pounds per hour (kilograms per hour), the recovery rate gives the volumetric output in cubic inches per second (cubic centimeters per second). The *maximum injection velocity* is the maximum injection rate available in the machine; its units are in inches per second (millimeters per second). Standard machines typically have maximum injection velocities of about 150 to 250 mm/s (6 to 10 in/s), whereas thin-wall machines can reach 1500 mm/s (59 in/s).[124] Both plasticating capacity and maximum injection velocity are critical for thin-wall molding. Rapid injection fills the cavity before the melt solidifies, and high plasticating capacities build up the next shot within the very short cooling time required of thin-wall parts. Other less important specifications include screw diameter and L/D ratio, type of nonreturn valve, barrel material, screw material, barrel heater capacity, number of barrel/nozzle heating zones, screw speed range, torque available for screw drive, and power consumption.

The clamping unit supports the mold, holds the mold closed during injection, opens and closes the mold as rapidly as possible, provides for part ejection, and provides mold close protection. The four types of clamps are: hydraulic, hydraulically actuated toggle (mechanical), electrically actuated toggle, and hydromechanical. As shown in Fig. 5.61,[144] all clamps have a stationary and moving (or movable) platen. Since the stationary platen supports the core or A side of the mold, it contains a hole through which the nozzle contacts the sprue bushing. Typically, the sprue bushing is surrounded by a locating ring that aligns the mold with the nozzle. The stationary platen is often water cooled when the mold has a hot manifold. While the moving platen supports the B side of the mold, it moves horizontally to open and close the mold, applies clamping force to the mold, and houses the ejector system. The two platens align the two halves of the mold, thereby minimizing wear on contacting surfaces. These platens are usually supported and aligned by four tie bars. The moving platen is also guided by these tie bars when it travels to open or close the mold. Since the forces that hold the mold closed also stretch the tie bars, tie-bar adjustments are used to periodically realign the platens. In some machines, the four tie bars are replaced with slides or two tie bars (bottom). This supports and aligns the platens from the bottom only and facilitates mold changes and part removal.

A conventional hydraulic clamp (Fig. 5.61[144]) has one large cylinder in the center of the movable platen with no mechanical advantage applied. Thus, hydraulic fluid and pressure open and close the clamp. To move the clamp forward, hydraulic fluid is directed to a booster tube or cylinder and the prefill valve is opened. As the booster cylinder

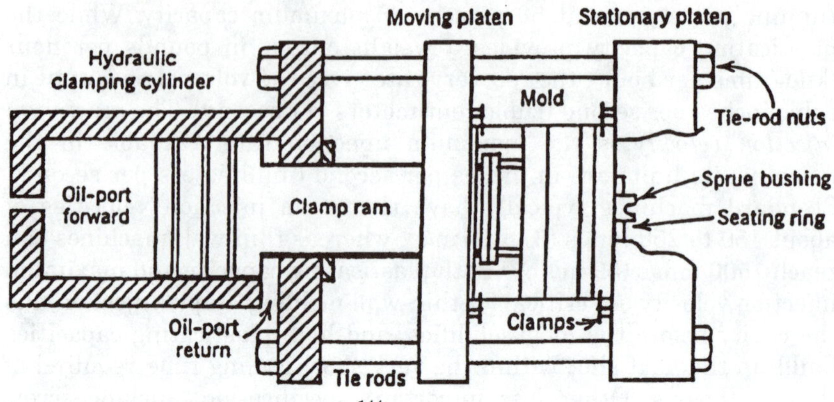

Figure 5.61 Hydraulic clamping unit.[144]

moves the clamp forward, a slight vacuum in the main clamp cylinder pulls oil from the tank, through the prefill valve, and into chamber of the main clamp cylinder. Once the clamp is closed, the prefill valve is closed. This traps oil in the main cylinder area. When the clamp is pressurized, high-pressure fluid moves the main cylinder forward, thereby compressing the oil in the main cylinder. To open the clamp, hydraulic fluid is directed to the pull-back side of the booster cylinder while the prefill valve is open. The backward motion of the cylinder forces the fluid through the prefill valve and back into the tank.

In hydraulic clamps, the clamp force is controlled by the pressure in the main cylinder. As a result, clamp force can be varied during the molding cycle. Typically, a higher clamp force is used during mold filling and packing while the force is reduced during cooling. The clamp force can be built up by increasing clamp pressure and maintaining maximum pressure using a high-pressure pump or a pressure pump intensifier or the clamp pressure can be raised with a high-pressure pump or a pressure intensifier and then maintained using a check valve in the return line.[145] The former techniques maintain a constant tie-bar deflection after the mold is closed and pressurized. However, the use of a check valve in the return line does not allow excess hydraulic fluid to leave the clamp, and so, the tie bars stretch further during injection of the melt into the die.[145] Die-height adjustment (setting the distance between the stationary and moving platens) occurs through travel of the main clamp cylinder.

While a hydraulic clamp machine requires a very large cylinder in the center of the platen to apply full clamp tonnage with hydraulic pressure, hydraulically actuated toggle clamps (Fig. 5.62[146]) use a small cylinder and a mechanical toggle. To close the clamp, the cylinder moves forward, extending the toggle links (Fig. 5.62*b*[146]). Clamp movement is rapid until the B side of the mold approaches the A side. At a predetermined distance, the clamp speed slows and continues its forward travel until the mold halves are joined. The low speed prevents damage to the mold, while the fully extended toggle links (Fig. 5.62*a*[146]), not the hydraulic cylinder, provide the clamping pressure. Retraction of the hydraulic cylinder opens the clamp. The opening stoke also starts slowly and then speeds up when the B side of the mold has cleared the leader (guide) pins on the A side of the mold.

Hydraulically actuated toggle clamps are the most common type of clamp. Single toggles are used for small machines [clamp force $\leq$500 kN (55 tons)[147]]. These require little space because the actuating cylinder is connected directly to the toggle with cross-links or it is pivoted at the tail stock platen (platen at the clamp cylinder end of the clamp unit) or machine support. Double toggles are employed when the clamp tonnage is about 1000 to 50,000 kN ($\sim$110 to 5620 tons).[147] As shown in Fig. 5.62,[146] most double toggles have one centrally located hydraulic cylinder. While double toggle clamps can have four- or five-point toggles, the

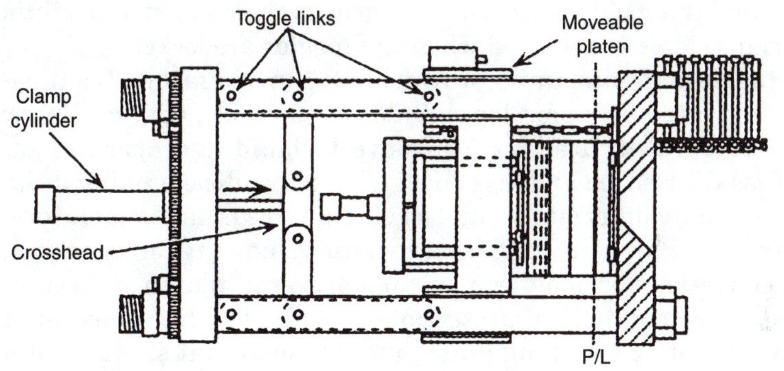

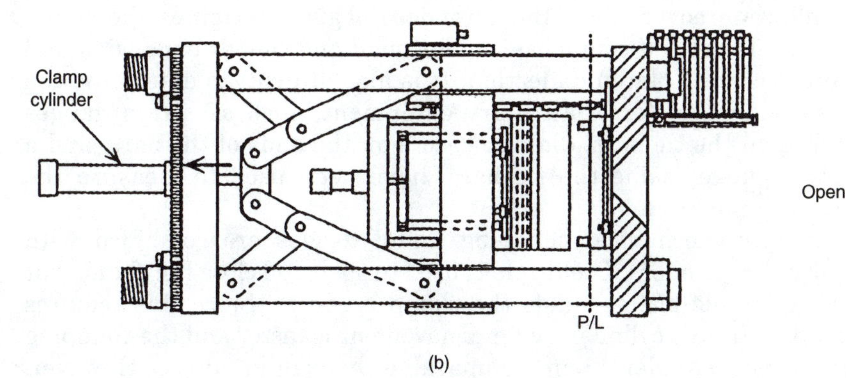

Figure 5.62 Toggle clamping unit[146]: (*a*) closed and (*b*) open.

five-point toggles are more common because they provide longer opening strokes and require less floor space.[147] These clamps provide relatively rapid and constant platen speed for most of the platen travel, whereas the platen speed slows as the moving platen approaches the stationary platen. Double toggles can be actuated laterally. This produces rapid clamp movement and requires minimal floor space, but opening strokes are relatively short and the clamp speed is not uniform. Since the maximum clamp speed occurs when the clamp is fully open, laterally actuated toggles are seldom used.[147]

In toggle clamps, die-height adjustment employs a clamp-adjusting ring gear (Fig. 5.63[148]). This motor-driven gear rotates all four tie-bar adjusting nuts simultaneously, thereby producing linear movement of the moving platen. While procedures for die-height adjustment vary, the moving platen is typically backed away from the stationary platen during mold changes. Thus, when the clamp toggles are fully extended, the moving platen does not touch the mold. The ring gear is then rotated until the moving platen contacts with the mold. After the B side of the mold is attached to the moving platen, the clamp toggle is retracted and the mold opens. The ring gear is then adjusted until the desired clamp force is achieved when the toggles are locked.

Since toggle clamping units provide a 50:1 mechanical advantage, they are activated by a smaller cylinder. As a result, they are faster by 10 to 20 percent, are less expensive to build and operate, and require less floor space and less energy. In hydraulic clamping units, the force is easily determined, and so clamp force is easily monitored and controlled. This also provides a constant clamping force, permits self-compensation for mold expansion, prevents platen deflection, and facilitates mold-close protection and die-height adjustment. However, hydraulic clamping units produce somewhat slower platen movement, are more expensive to manufacture and operate, and require more floor space. Toggles are subject to wear (around the toggle). Moreover, since the extended toggle determines the clamp force, it is not self-compensating for mold and tie-bar expansion and cannot correct platen deflection. It is also difficult to determine the exact clamping force; auxiliary equipment, such as strain gauges placed on the tie bars, dial indicators on the ends of the bars, and a strain gauge inside the tie bars, have been used to measure the exact clamp force.

In hydromechanical clamping units, toggles are combined with hydraulic cylinders. The toggle is used to open and close the clamp, but the hydraulic piston builds the clamp pressure. Since this requires small hydraulic cylinders, clamp movement is faster and the clamping units are smaller than comparable hydraulic units. However, hydraulic clamping provides better control of the clamp force.

Figure 5.63 Clamp-adjusting ring gear.[148]

An all-electric machine has no hydraulic pumps. The toggles in the clamping unit are extended and retracted by a servomotor and a reduction drive gear is used to achieve the required forces. The clamps are much more stable since they have no hydraulic system to generate heat and the servomotors provide extremely accurate movement of the machine components. It is also much cleaner to operate than the other types of molding machines. All-electric machines are the machine of choice for most medical products.

At the end of the molding cycle, the mold opens and cooled parts are ejected from the injection mold. This requires the ejection system shown in Fig. 5.64.[149] When the mold opens, the plastic part typically stays with the B side of the mold. Once the mold is opened, the hydraulic ejection cylinder extends, forcing the ejection platen forward (in all-electric machines, the cylinders are replaced by servomotors). Since the platen is usually tied to the mold's ejector plate by ejector rods, the ejector plate also moves forward. This forces the ejector pins (in the mold) forward

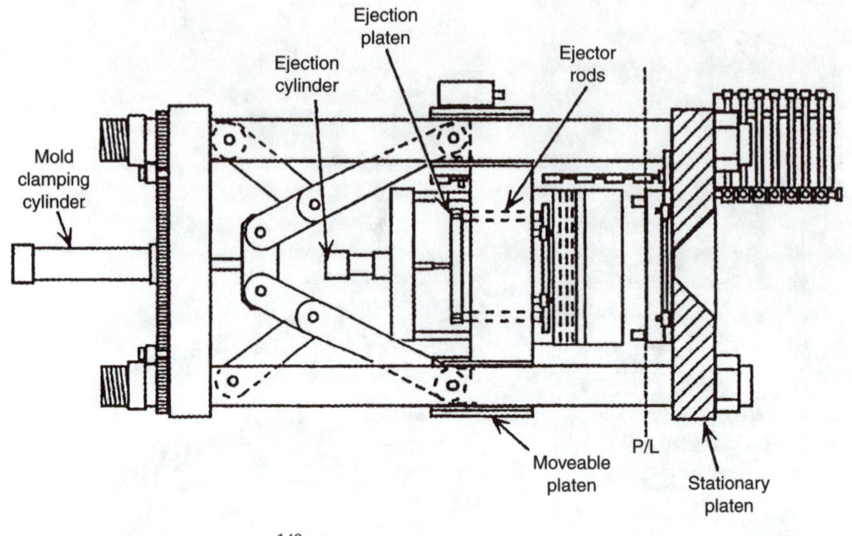

Figure 5.64 Ejection system.[149]

and they, in turn, push the plastic part out of the mold. Since movement of the ejection system is controlled separately from movement of the clamp, the timing of ejection can be modified. To speed up the molding cycle, the ejection system can be activated once the B side of the mold clears the A side. As a result, the parts are free of the mold by the time the clamp is fully open, and the clamp may close immediately. Ejection may also await for the positioning of robotics, thereby preventing damage during ejection or facilitating part assembly.

Clamp and ejection systems are specified jointly. The primary specification for a clamping unit is the *maximum clamp force,* the force needed to keep the mold closed during injection and packing. If the clamp force is not greater than the injection and packing forces, the mold can be forced open during injection or packing. This results in leakage of the molten polymer or flash. Clamp force F_c can be estimated from

$$F_c = f_\eta A \tag{5.23}$$

where f_η is an empirical pressure or viscosity factor and A is the projected area of the part (i.e., the area perpendicular to the sprue). For easy flow materials, f_η is 2 to 3 tons/in^2, while for viscous polymers, f_η is 4 to 5 tons/in^2. The viscosity factor can be replaced by the actual or estimated peak cavity pressure, P_{max}, to provide another estimate of clamp force:

$$F_c = P_{max} A \tag{5.24}$$

Clamp force can also be determined from

$$F_c \propto \frac{L}{H} \tag{5.25}$$

where L is the flow length and H is the part thickness. This method uses experimentally determined charts (Table 5.13[150]) to obtain clamp pressure or force.

Other clamp specifications (Fig. 5.65[151]) include the maximum daylight, clamp stroke, maximum and minimum mold height, distance between tie rods, tie-rod diameter, platen size, clamp speeds, ejector stroke, and ejector force. The *maximum daylight* is the maximum distance between the stationary and moving platens when the clamp is fully retracted. In hydraulic clamping units, this distance, minus the distance required for part ejection, determines the maximum depth for the mold. The mold opening must be at least twice the height of the core pin or the part cannot be ejected. With toggle machines, the extension of the toggle provides the clamp stroke (stroke is not specified for hydraulic clamps). Since the stroke limits the depth of the mold that can be mounted in the clamp, toggle machines specify a maximum mold height. As shown in Table 5.14,[152] the maximum daylight is the sum of the clamp stroke and the maximum mold height. The *minimum closed daylight, shut height,* or *mold height* is the minimum distance between platens when the clamp is closed. This dictates the shortest mold that may be run in the injection-molding machine. When the mold height is smaller than the minimum closed daylight, a spacer bar mounted to the platen provides a shorter minimum shut height. The distance between tie rods and tie-rod diameter limits the size of the mold that may be mounted in the clamp unit. While molds are typically hung vertically between the tie bars, small molds may be mounted horizontally. This restriction to mold size does not occur with tie-bar-less machines, but the mold size is limited by the platen size. If a mold is too small, the platens will bend, thereby damaging the clamping unit. Thus, the minimum mold size is typically one-half the distance between the tie rods,

TABLE 5.13 Clamping Pressure (MPa) for PE, PP, and PS[150]

	L/H				
H, mm	200	150	125	100	50
1.0	—	69	62	50	31
1.5	83	59	41	31	21
2.0	62	41	31	26	17
2.5	48	31	24	21	17
3.0	34	28	21	21	17
3.5	31	24	21	21	17

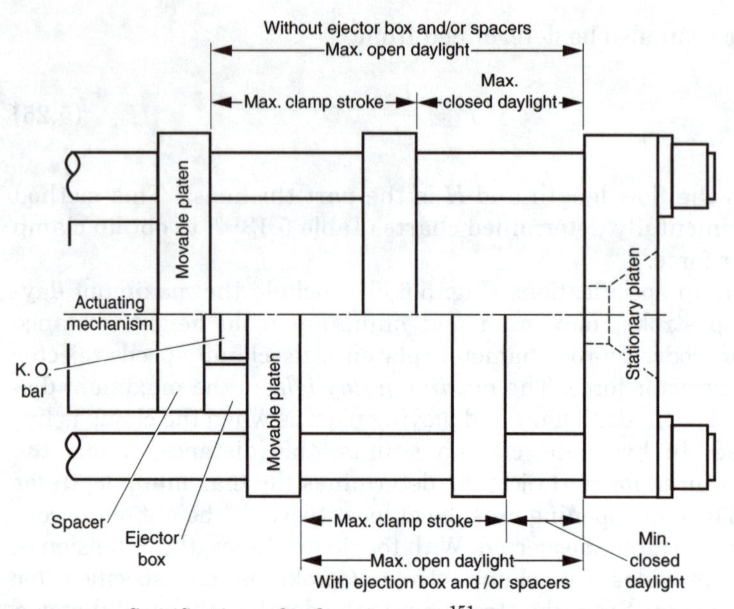

Figure 5.65 Specifications of a clamping unit.[151]

TABLE 5.14 Clamp Specifications[152]

Specification	Metric	English
Clamp force	110 tons	120 tons
Clamp stroke	350 mm	13.78 in
Maximum daylight	825 mm	32.48 in
Maximum mold height	475 mm	18.70 in
Minimum mold height	150 mm	5.9 in
Platen size ($H \times V$)	670 × 670 mm	26.38 × 26.38 in
Distance between tie rods ($H \times V$)	435 × 435 mm	17.13 × 17.13 in
Tie-rod diameter	80 mm	3.15 in
Ejector stroke	120 mm	4.72 in
Ejector force	3.6 tons	3.96 tons

both vertically (V) and horizontally (H). Although clamp opening and closing speeds contribute to overall cycle time, they are more important for thin-wall and other high-speed molding.

The ejector stroke and ejector force affect mold design. To properly eject a part, the ejector pins must clear the cavity. However, mold bases are typically available with standard stroke heights. Thus, part design plus machine stroke height will determine whether standard or custom-made mold bases can be used. The ejector force dictates the number and size of the ejector pins, with the surface area of the ejector pins being about one-one hundredth of the peripheral area of the part. When the ejection force is insufficient, the number and/or size of the pins must be increased, introducing new problems into the mold design. If the mold is run without enough ejection force, the ejector pins bend or break, or the part is stressed during ejection. The former produces a lot of down time and mold repairs, whereas the latter causes cosmetic defects such as stress whitening and other damage to the part.

5.4.2 Injection molds

Injection molds distribute the melt, give final shape to the part, cool the material, and eject the final part. They also must withstand the forces of injection and ejection, transfer motion, such as ejector platen (machine) to the ejector plate (mold), and guide the moving parts of the mold. A standard two-plate mold (Fig. 5.66[153]) has two halves: an A side and a B side. The division between these halves is called the *parting line*. The A side, or cavity half of the mold, contains the top clamp plate, A or cavity plate, sprue bushing, locating ring, and leader pins. The B side, or core half, includes the support plate, bottom clamp plate, B or core plate, leader bushings, and ejection system. Since the A and B plates define the shape of the molded part, they contain the machined mold cavity, parts of the delivery system, and the coolant lines. The top and bottom clamp plates support the mold and are used to attach the mold to the machine platens. The support plate backs up the B plate in the mold and melt is delivered from the nozzle to the sprue bushing. The locating ring aligns the mold with the nozzle, whereas the leader pins and bushings align the two sides

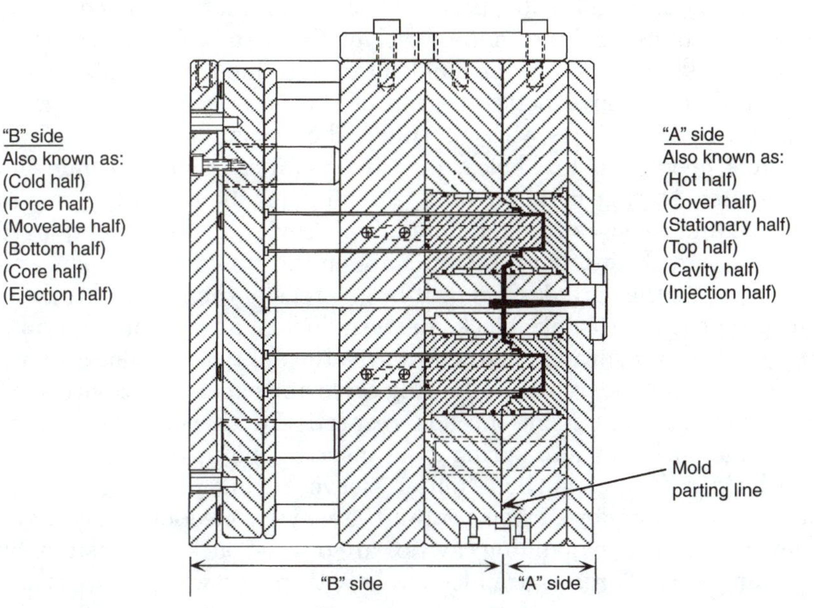

Figure 5.66 Standard mold base.[153]

of the mold. In the ejection system, the sprue puller and ejector pins are attached to the ejector plate, which is, in turn, connected to the machine's ejection platen by the ejector rod. When the mold is opened, the sprue puller draws the part to the B side of the mold, whereas the ejection pins push the part out of the mold during ejection. The ejection system also contains supports and guide systems for ejector plate movement.

Injection molds are often classified by the basic design of the mold. A standard two-plate mold (Fig. 5.67*a*[154]) opens in one direction and the part is demolded by gravity, ejector pins, or ejector sleeves. The sprue, runner, and parts are connected after ejection, with the sprue and runner(s) forming part scrap. Two-plate molds are used for all kinds of moldings that do not contain undercuts and provide the best overall part properties. A stripper plate mold is a two-plate mold in which the core and ejector plates are combined into a stripper plate. It is employed for thin-wall parts and parts with symmetry and no undercuts. A three-plate mold (Fig. 5.67*b*[154]) has two parting lines, thereby providing automatic separation of the parts and runner system. This does not work with all materials; brittle materials tend to fracture upon ejection. It also limits the selection of gate locations and does not eliminate scrap. A stack mold (Fig. 5.67*c*[154]) is used to mold two layers of parts without increasing clamp force. Since stack molds are as precise as two-plate molds, they are used for low-tolerance parts such as polystyrene drinking cups.

Parts containing undercuts are not usually ejected directly after the mold is opened. With flexible polymers, the part can be stripped off the core. For vinyls, polyurethanes, polyolefins, and elastomers, this requires an angle of 33° on the molded threads, and the threads should be interrupted to give better flexing of the outer wall.[124] Slide core molds are used for parts with internal and external undercuts such as ribs, gaps, openings, blind holes, and threads. These contain a slide and a cam lifter. During mold opening, the slide on the cam lifter moves sideward and releases the undercuts. Split cavity molds are employed when the threads or other undercuts are on the outside of the part and fast molding cycles are desired or if large areas of the parts have to be formed by the split cavity block. In these molds, an ejector rod (via a strap joint) or a hydraulic cylinder moves the split cavity block, thereby releasing the undercuts. Collapsing cores and expanding cavities are used for inside and outside threads, respectively. These are threads or undercuts machined into spring steel inserts. The inserts are held in place by a rod or expansion limiter sleeve during molding, but are released and, therefore, spring out of the way for ejection. During the closing of the mold, expanding cavities are pushed back into position by a striker insert. Threads are also made with unscrewing (or twist-off)

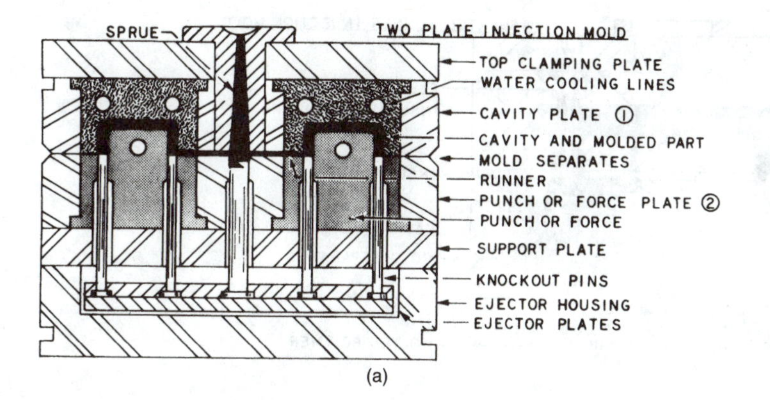

(a)

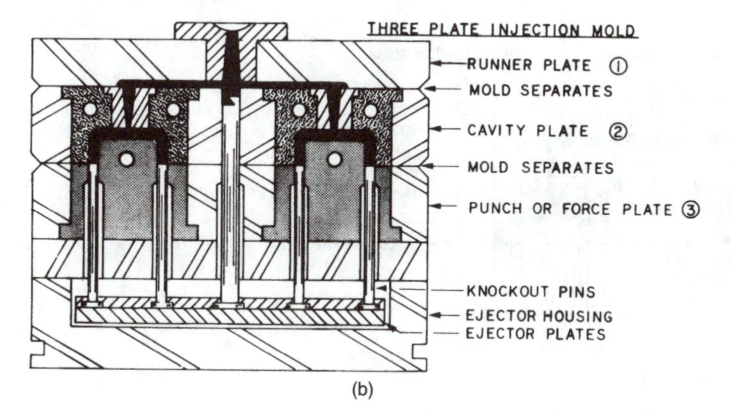

(b)

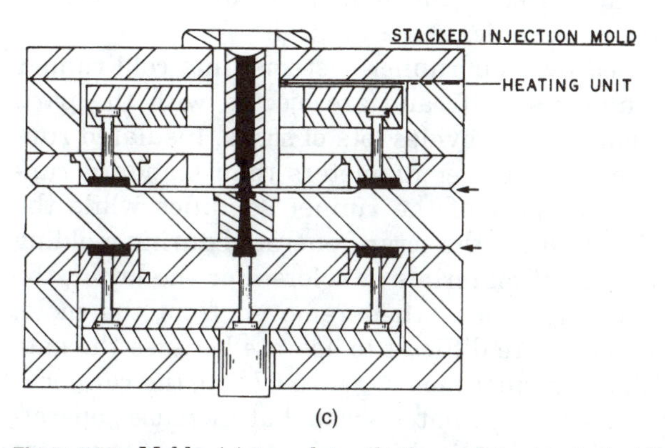

(c)

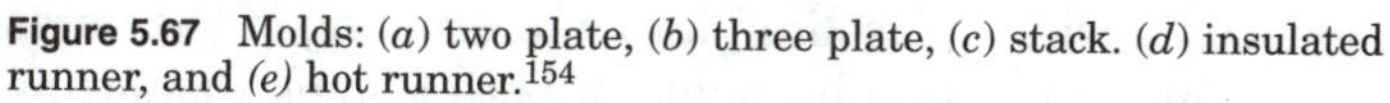
Figure 5.67 Molds: (*a*) two plate, (*b*) three plate, (*c*) stack. (*d*) insulated runner, and (*e*) hot runner.[154]

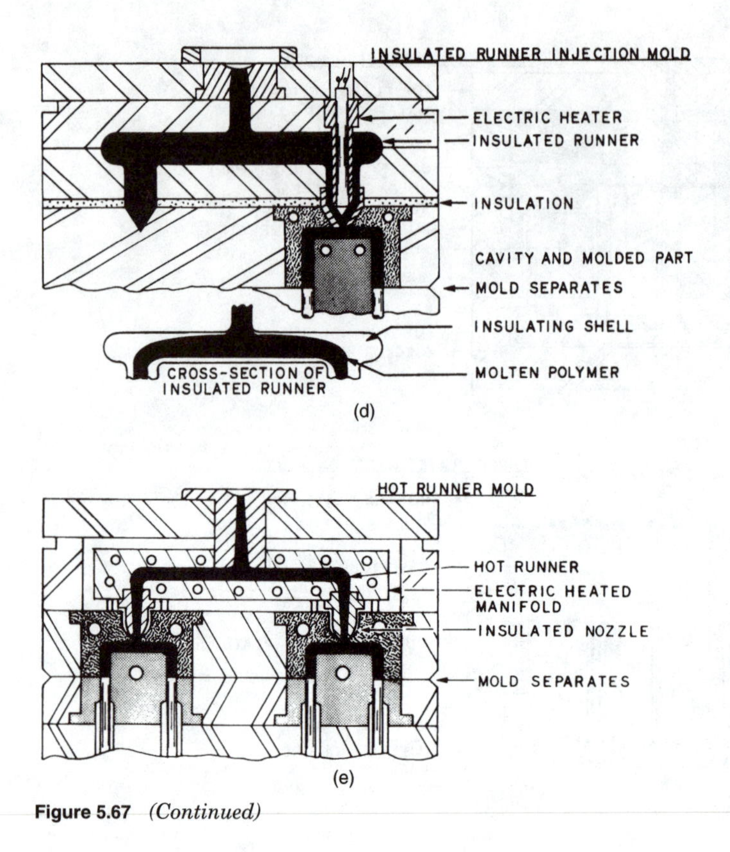

Figure 5.67 *(Continued)*

molds. The cavity elements unscrew during mold opening; the turning mechanism is hydraulic, pneumatic, or mechanical.

Molds are also classified by their runner systems. In a cold-runner mold, the sprue and runner solidify and are ejected with the part. Although this is fairly simple, it provides lots of scrap. Insulated runners (Fig. 5.67*d*[154]) have much larger diameters than standard runners. As a result, the outer part of the runner solidifies while the center remains fluid. This reduces the scrap generated during molding and facilitates the changing of materials or colors. For the latter, the runner is allowed to solidify and is then removed from the mold. Insulated runner molds are more difficult to operate because the gate tends to freeze off. With hot-runner molds (Fig. 5.67*e*[154]), the runner is always in the melt state. This eliminates scrap, but increases operating difficulties since freezing off (solidifying) and drooling of the gates must be balanced. Due to the requirements of better steel, insulating sheets, and electric controls, mold costs are typically 25 percent higher than comparable cold-runner molds. Both insulated and hot-runner molds have some restrictions on the materials they can process, but changing colors is a problem only in hot-runner molds.

Molds are also distinguished by the number of cavities cut in the mold. Single-cavity molds contain one cavity, while multiple-cavity molds can mold several parts simultaneously. In family molds (Fig. 5.68[155]), several different parts are molded at one time.

The basic parts of the mold are illustrated in Fig. 5.69.[156] The sprue receives the polymer melt from the nozzle and delivers it to the runners. The runners are as short as possible and provide minimum pressure and temperature drop from the sprue to the cavity. There are four basic types: round, half round, trapezoid, and modified trapezoid. Round runners are the most efficient type, but are the most expensive because they must be machined into both plates of the mold. Since half-round runners produce the highest pressure drops and greatest cooling of the melt, they are seldom recommended. Trapezoidal runners are a compromise between round and half round, and modified trapezoidal runners merely have rounded bottoms. These runners are machined into only one mold plate, but reduce the pressure and temperature losses observed with half-round runners. Runner sizes are determined from the part-wall thickness. The diameter of the runner connecting directly to the cavity is typically equal to the part thickness, and each upstream branch is increased in diameter as[157]

$$D_{\text{branch}} = n^{1/3} D \tag{5.26}$$

where D is the diameter of the runner, D_{branch} is the diameter of the upstream branch, and n is the number of branches. While runners should deliver melt to each cavity at the same time, they may be nat-

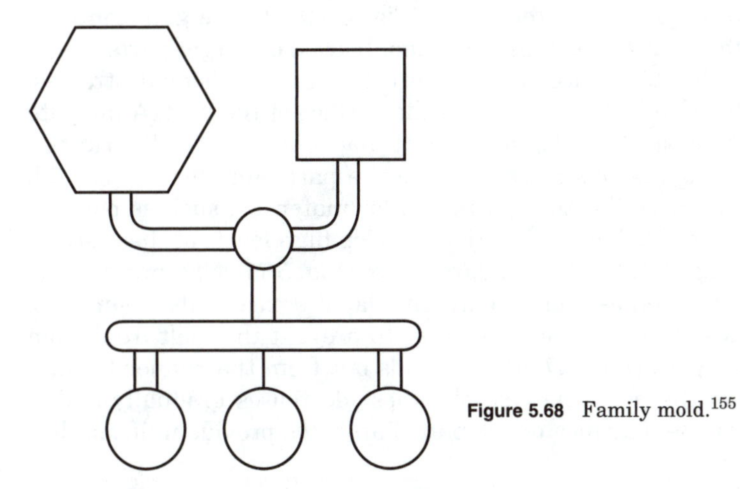

Figure 5.68 Family mold.[155]

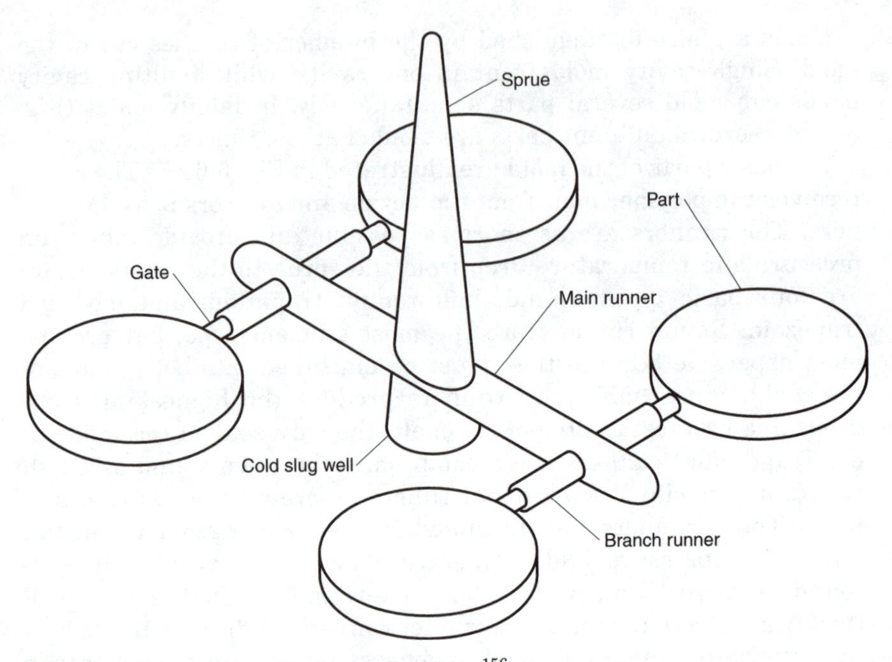

Figure 5.69 Sprue, runners, gates, and cavities.[156]

urally balanced (Fig. 5.70*a*[158]) or artificially balanced (Fig. 5.70*b*[158]). In naturally balanced runners, the distance from sprue to all gates is the same, whereas artificially balanced runners have the same pressure drop from sprue to all gates. Cold-runner molds also incorporate cold slug wells at each turn in the melt flow. These collect the cold melt from the sprue plug (melt frozen at the nozzle tip), thereby preventing this melt from entering the cavities.

As shown in Fig. 5.69*a*,[156] a gate connects the runner to each mold cavity. With a sprue (or direct) gate (Fig. 5.71*a*[159]), the gate connects directly to the cavity. This type is commonly used on larger parts in single-cavity molds to reduce the pressure losses and residual stresses. However, the gate leaves a mark on the surface of the part. A tab gate (Fig. 5.71*b*[159]) is an extension of the part that connects with the runner, thereby locating shear stresses outside the part. For this reason, tab gates are used with flat thin parts and for materials, such as polycarbonate, acrylics, SAN, and ABS, that develop high levels of shear stress during molding.[160] Edge (or standard) gates (Fig. 5.71*c*[159]) connect to the edge of a part, whereas overlap gates overlap the part walls or surfaces. Overlap gates often replace edge gates to prevent the melt front from jetting. A fan gate (Fig. 5.71*d*[159]) spreads out from the runner to fill a wider edge of the part; the gate thickness decreases gradually and is shallowest at the entrance to the part. Fan gates provide uniform flow into parts where warpage and dimensional stability are important.[160] In disk (or diaphragm) gates (Fig. 5.71*e*[159]) the melt flows from the runner into a thin disk before entering the mold. This produces uniform filling of cylindrical or round parts and prevents weld lines. Spoke (or spider) gates (Fig. 5.71*f*[161]) are also employed with cylindrical parts. While these gates are easier to remove than disk gates, they produce weld lines and do not permit perfect concentricity. Ring gates (Fig. 5.71*g*[159]) can also be used for cylindrical or round parts and, because it allows easy venting, is most effective when the part has a hole in the center. In ring gates, the melt fills a thick ring and then passes through thinner inner ring to reach the part. A film (or flash) gate (Fig. 5.71*h*[159]) permits uniform filling of flat parts. The melt fills a runner and then crosses a wide, but thin "land" to enter the cavity. Pin gates (Fig. 5.71*i*[159]) are very fine-diameter (typically 0.25- to 1.6-mm[160]) entrances to the cavity. They are used in three-plate molds or hot-runner molds to facilitate degating. A hot-probe (or hot-runner) gate (Fig. 5.71*j*[162]) delivers material through heated runners directly into the cavity. Submarine (tunnel or chisel) gates

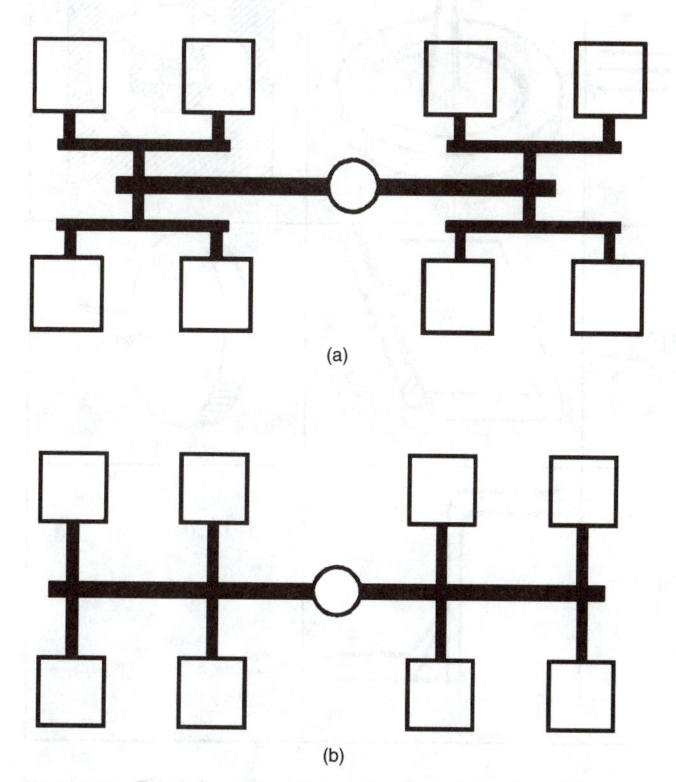

Figure 5.70 Runner systems: (*a*) naturally balanced and (*b*) artificially balanced.[158]

(Fig. 5.71*k*[163]) are fine-diameter (typically 0.25- to 2.0-mm[160]) gates that are angled from the parting line. Used in two-plate molds, they permit automatic degating of the part from the runner system.

Vents in the mold's cavities and runners permit air in the cavities to leave the mold ahead of the melt front. The vents are typically 0.01 to 0.06 mm (0.0005 to 0.0025 in) deep, 1 to 3 mm (0.04 to 0.12 in) long, and must be connected to the atmosphere.[164] The absence, poor loca-

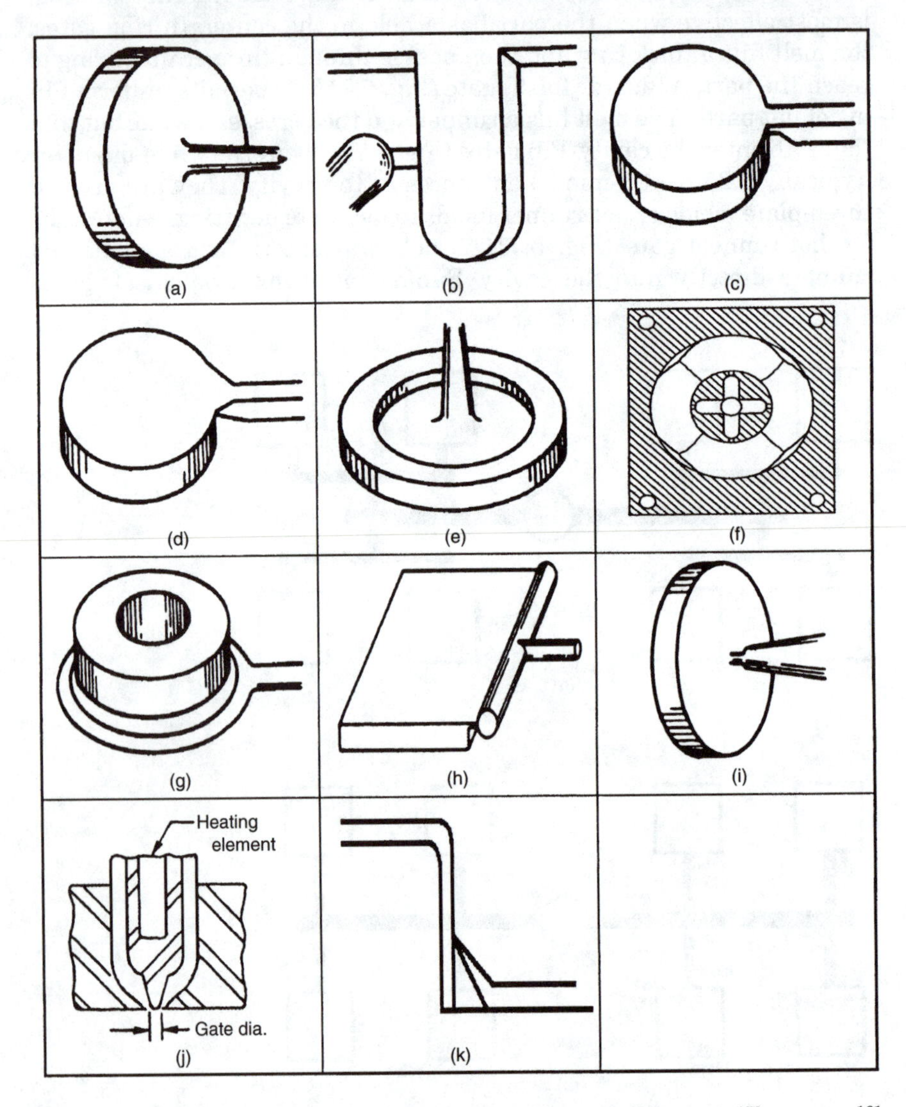

Figure 5.71 Gates: (*a*) sprue,[159] (*b*) tab,[159] (*c*) edge,[159] (*d*) fan,[159] (*e*) disk,[159] (*f*) spoke,[161] (*g*) ring,[159] (*h*) film,[159] (*i*) pin,[159] (*j*) hot probe,[162] and (*k*) submarine.[163]

tion, or clogging of vents can produce short shots (incomplete filling of the cavity) accompanied by char marks from burning of the trapped air (commonly referred to as *dieseling*). Since thin-wall molds are filled very rapidly, vacuum is applied to the vents to improve removal of air in the mold.

Injection molds are machined from a variety of tool steels and then hardened or in some cases plated with chromium, nickel, or proprietary materials. Large molds use prehardened tool steels because they cannot be hardened after machining.[124] Stainless steel is employed for some smaller molds, particularly those used for optical and medical parts and for corrosion resistance. Since they provide better heat transfer and, thus, shorter molding cycles, materials such as beryllium copper are used as inserts in critical areas. Injection molds are usually cooled or heated with water, although oil or electric heater cartridges are employed for high-mold temperatures. Mold-temperature controllers pump water into the manifolds and then into cooling line machine into the molds.

5.4.3 Injection molding process

An injection molding cycle has a number of steps, including mold close, injection, packing, holding, cooling, plastication, mold open, and part ejection. Injection molding machines are operated in three modes: manual, semiautomatic (single cycle), and automatic (continuous). With manual control, all steps of the molding cycle are performed manually; that is, the operator activates each part of the cycle with separate controls. Manual control is used for purging, machine and mold setup, and troubleshooting whereas semiautomatic or automatic molding cycles are used for actual molding. Semiautomatic mode initiates one cycle each time the gate is closed. In automatic mode, the machine cycles continuously unless the gate is opened or other interlocks (or protections) are tripped. The former mode is employed for process setup and when manual part removal is required. Automatic mode is used in production, with or without robots and other automatic part removal.

A typical molding cycle will not start until a number of interlock conditions are satisfied. While the interlock conditions vary with the machine, the machine will not enter semiautomatic or automatic mode in the absence of these interlock conditions. Generally, the gate (operator guard door) must be closed and the gate safety (interlock) switches tripped. In hydraulic machines, the front gate has three switches (electrical, mechanical, and hydraulic) whereas the rear gate usually has one or more electrical limit switches. The electrical interlocks are usually electrical limit switches. When tripped, the hydraulic interlock prevents flow of hydraulic oil. The mechanical interlock is often a notched safety bar that prevents or limits forward movement of the moving platen when the gate is open. Rear gate switches may be overridden for conveyors and robots, and the conveyors and robots may have interlocks that are tied to the machine. In addition to the gate interlocks, the purge guard (around the nozzle) must be closed, the clamp must be fully open, and the ejector platen (not the ejector pins) must be fully retracted. These conditions are determined by electrical limit switches or other sensors. Any mold side action may also have electrical limit switches. Often the injection unit must be forward (contacting stationary platen) and a shot must be built up in the injection unit.

Once the interlocks are satisfied, the mold closes. Then the injection unit occasionally moves slightly forward to counter a motion called sprue breakaway. With cold runner systems, the injection unit is moved slightly away from the sprue bushing to facilitate sprue removal and/or prevent nozzle freeze-off. Thus, injection unit must be moved forward to provide intimate contact between the nozzle and sprue bushing during injection. Sprue breakaway is not used with hot runner systems.

When the nozzle and sprue bushing are touching, injection (also called filling or first stage) begins. Linear (axial) movement of screw or ram forces plastic through nozzle and into mold. In hydraulic machines, the ram velocity is controlled by a flow control, proportional, or servo valve combined with a pressure control valve. Servo motors provide this motion in all-electric machines. The end of fill (or transfer) is determined by time, ram position, or pressure. At this point, the mold cavity is filled with plastic melt. After transfer, the packing stage begins and additional material flows into the mold to compensate for shrinkage of the cooling melt. Once packing is completed, the machine transfers to the holding stage, thereby applying pressure to the melt until the mold gates solidify. During packing and holding, movement of the plastic melt is determined by melt viscosity and pressure exerted on the ram. The duration of the packing or holding cycle is fixed by a timer. In some machines, packing and holding are combined into the second stage.

Cooling and plastication occur after the holding stage. If sprue breakaway is called for, the injection unit moves slightly away from the sprue bushing. The plastic in the mold is cooled until the part can be ejected. The length of the cooling cycle is determined by a timer whereas the mold temperature is set on the mold temperature controller. During cooling, the screw rotates and melt is built up for the next molding cycle. The rotational speed of the screw is determined by a flow control valve or servo motor, depending on the machine type. The duration of screw rotation is dictated by the shot size (axial travel of the screw). This is typically controlled by electrical limit switch or LVDT. Since plastication should end just prior to the end of the cooling cycle, extruder delay timers can delay the rotation of the screw (so that screw rotation does not have to be slowed). Additional mixing of the melt is controlled by the back pressure. In hydraulic machines, the flow of oil out of the injection cylinders is restricted, requiring more melt pressure to force the screw backwards; with all-electric machines, the axial screw motion is subject to a similar restriction. After the shot is built up, the screw can be retracted axially in a motion called decompression or suck back. This relieves the pressure on the melt, and thus prevents drooling. When cooling ends, the clamp motion opens the mold and the part is ejected, thereby completing the molding cycle.

Molding parameters are set up in five steps: shot size, injection velocity, injection pressure, injection time, and packing and holding parameters.[165] The initial shot size is the amount of material injected into the cavity during filling. About 95 to 98 percent of the cavity is typically filled during injection and the rest of the cavity during packing (or second stage). For thin-wall parts, the entire cavity is filled during injection because the part solidifies before packing can occur. The actual shot volume is determined at a high injection speed and pressure. The shot size is increased in increments until the desired fill is reached. Too little material produces a short shot (unfilled cavity) whereas too much material will force the mold to open and leak around the edges of the cavity, gate, and runners. This flash not only must be removed from molded parts, but it also damages the mold, particularly at the parting line.

After the shot size has been fixed, the injection velocity or ram speed is set. Typically, the injection pressure setting remains at a high or maximum value while the injection speed is set for the desired level. Then parts are molded and inspected for jetting, burning and discoloration at the gate, dieseling, and cold flow marks. Since any polymer melt is very viscous, it usually enters a mold as fountain flow. The melt flows into the cavity in waves (Fig. 5.72*a*[166]). The first material contacts the mold walls and freezes. Then new, hotter melt flows under the older material to continue filling the cavity. Thus, a cross section of the cavity (Fig. 5.72*b*[167]) shows the frozen layer at the mold walls with molten material near the center of the cavity. The thickness of the frozen layer varies with melt temperature, mold temperature, and injection velocity or time. When the cavity is filled rapidly, excessive shearing of the polymer melt increases the temperature of the flow (melt) front and reduces the frozen layer thickness. In contrast, slow injection into cold molds increases the thickness of the frozen layer. This increase or decrease in the frozen layer thickness also affects shear heating (viscous dissipation) and fill pressure. If the melt is injected into the cavity too rapidly, fountain flow does not occur.

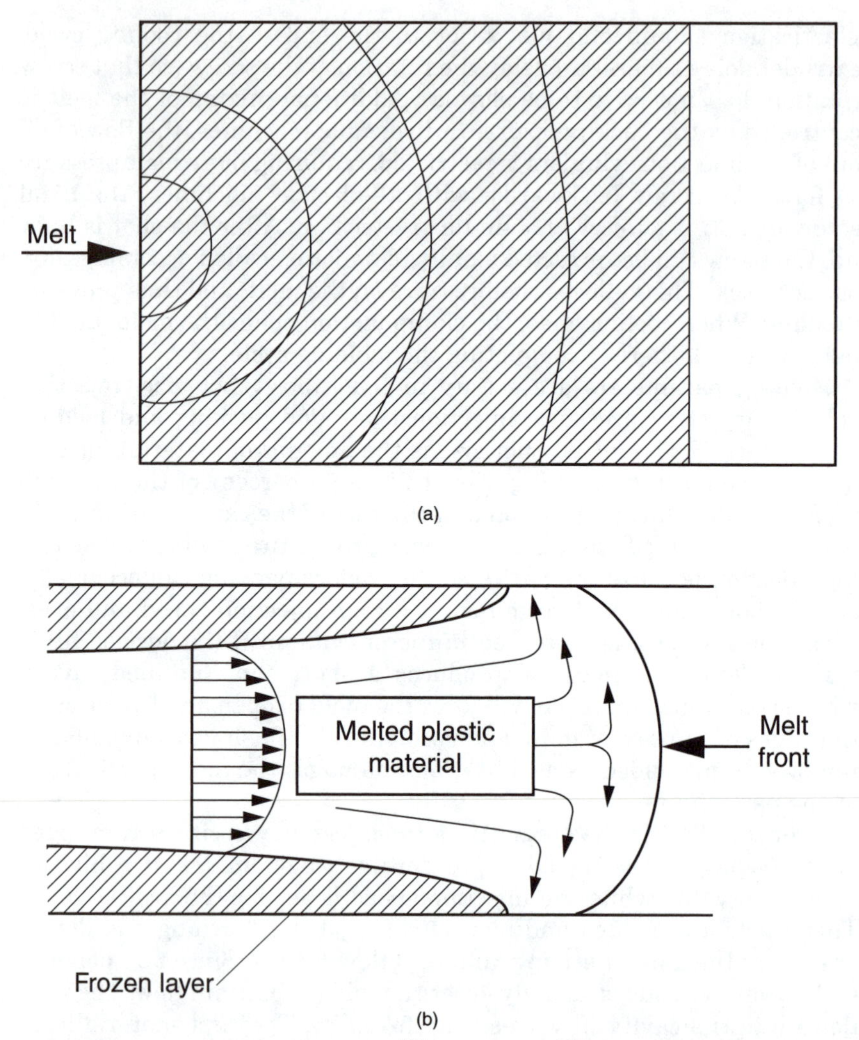

Figure 5.72 Fountain flow. (*a*) Top view;[166] (*b*) side view.[167]

Instead, a melt stream shoots straight into the cavity, only stopping when the stream contacts the end of the cavity. The remaining melt then fills the cavity as fountain flow. As shown in Fig. 5.73,[166] this jetting produces as weld line within the cavity. The jetting is not aesthetically pleasing, and the weld line is not as strong as the surrounding material. While jetting can be reduced or eliminated by reducing the velocity of the melt front, the effects of injection velocity are very complex and difficult to predict.[166] Typically, enlarging the gate and runner, reducing gate land length, and promoting mold-wall contact by locating the gate so that the flow is directed against a cav-

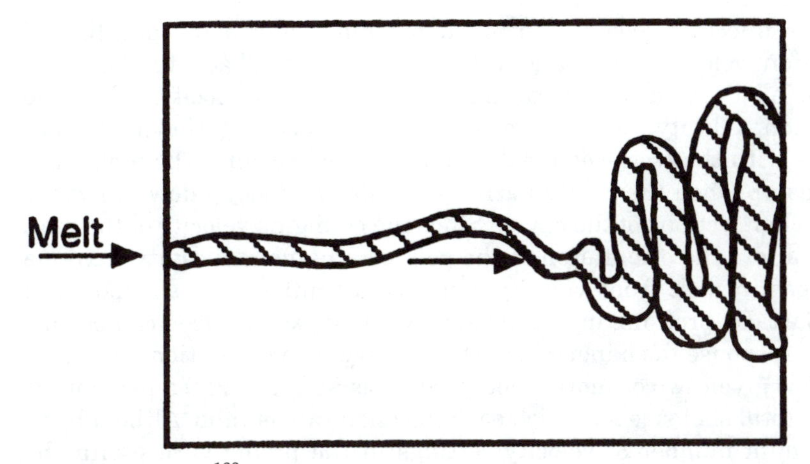

Figure 5.73 Jetting.[166]

ity wall are more effective in eliminating jetting. In contrast, when cold melt is slowly injected into the cavity, the melt freezes as it flows into the cavity. While this cold flow produces poor surface and part properties, it can be eliminated by increasing the injection speed, melt temperature, and/or mold temperature.

Rapid injection can also cause degradation of the polymer at the gate. Gate blush, a discoloration outside the gate, can be accompanied by a reduction in part properties, particularly impact strength.[168] Dieseling occurs when air trapped in the cavity burns. This produces discoloration or burning of parts, particularly at the end of the fill, and is often accompanied by short shots. Reducing injection speed, increasing melt temperature, and increasing gate size all tend to reduce or eliminate gate blush whereas remedies for dieseling include reducing the injection velocity, cleaning or enlarging mold vents, and redesigning venting in the mold.

Since hydraulic pumps maintain a constant or controlled flow of hydraulic oil, the flow to a particular hydraulic cylinder is regulated by either a flow control valve, a proportioning valve, or a servo valve. A flow control valve merely restricts the flow at a particular opening size until the valve's setting is changed. These values are typically used in "pressure-controlled' machines where the injection speed is a percentage of the valve's full open position. A proportional valve is a directional valve which adjusts the flow of oil in response to the position of an electric solenoid whereas a servo valve controls the flow of oil proportional to an electrical feedback signal. The latter system maintains a constant flow rate even with changes in the force against which the flow is working. Proportional valves or servo valves control fill for injection-velocity-controlled machines and allow profiling of injection velocity. With all-electric machines feedback to the servo motor also provides tight control of the injection velocity. When the injection velocity remains constant, the velocity of the melt front varies with the cross-sectional area of the mold cavity. The melt front speeds up when the cavity narrows or thins and slows down in wider or thicker sections of the cavity. Since the changing velocity of the melt front alters the orientation of the polymer chains in a cavity, extreme changes in melt front velocity can produce differential shrinkage in the part. By profiling injection velocity, the injection speed can be varied in response to changes in part geometry. The proportional or servo valve is given two or more velocity settings and (usually) a percentage of the total shot size for which each injection rate is valid. Although the maximum number of velocity settings in the profile varies with the machine manufacturer, velocity profiles are typically determined using flow analysis software.

The injection pressure setting is the maximum pressure that can develop in the hydraulic lines that feed the injection cylinder(s). Thus, injection pressure is typically read as hydraulic pressure and is the setting on a pressure control valve. The valve opens when the line pressure reaches the set value, and the remaining hydraulic oil is returned to the reservoir. For pressure-controlled machines, the injection pressure is typically a percentage of the maximum hydraulic pressure whereas with injection-velocity-controlled machines, the injection pressure (or fill pressure high limit) is read as a hydraulic pressure. The injection pressure is read directly for all-electric machines. For all machines, the injection velocity is controlled by the settings for injection velocity and injection pressure. In pressure-controlled machines, the injection speed is set using a flow control valve, but the desired injection speed is not achieved unless pressure is high enough. Thus, pressure is incremented until the entire shot size is injected into the mold, and the actual injection speed is not usually known. With injection-velocity-controlled machines, the hydraulic pressure at transfer (line pressure) is typically displayed on the control system. If the injection pressure (fill pressure high limit) is greater than the hydraulic pressure at transfer, the injection speed is usually constant for the entire injection time (or travel). Since the injection pressure is similarly displayed in all-electric machines, exceeding the pressure limit during filling also causes the injection speed to rapidly decrease.

The viscosity of the hydraulic oil can severely affect injection pressure. While hydraulic oils are fairly Newtonian, their viscosities will decrease with increasing temperature. Decreases in oil viscosity generally affect pressure settings; pressure develops more rapidly in the hydraulic lines, causing pressure control valves to open prematurely.

Consequently, many injection molding machines monitor and/or control oil temperature. Typically, hydraulics are run prior to molding to heat the oil to an acceptable temperature, and an oil temperature setting or window is often an interlock on the molding cycle. Oil is also cooled by water that is forced through a cooling manifold. This prevents degradation of the oil. Oil is also filtered to prevent wear in the hydraulic cylinders and lines.

Injection pressure is a primary factor affecting molding velocity and part quality. The pressure is influenced by part, gate, runner, and sprue dimensions, part surface area, melt temperature, mold temperature, injection velocity or injection time, and polymer viscosity. While pressure increases as part thickness and gate, runner, and sprue dimensions are decreased, part thickness and the pressure required to fill thin-wall parts is typically the most important factor in determining whether a part will fill.[169] Injection pressure also increases with part surface area since this increases the drag on the polymer melt. Although the pressure decreases with increased mold and melt temperatures, melt temperature has a greater effect than mold temperature. Finally, injection pressure is typically high at low injection times and high injection speeds because the polymer chains cannot orient in the direction of flow. At greater injection times or slower speeds, the required pressure decreases. However, if the injection time is too long or the velocity too slow, the polymer melt cools, thereby increasing melt viscosity and injection pressure.

Injection time, the maximum time for which injection can occur, is the setting on a timer. When this time is set for the transfer technique (transfer from fill to pack), it determines the time in which the cavity fills. However, if other transfer techniques are used, the time setting is merely a safety or default value. Thus, if transfer does not occur by the other technique, the machine switches to pack or second stage at the set time. When time is set for the transfer technique, injection time is incremented until the entire shot size is injected into the mold and the plastication begins immediately. When other transfer techniques are used, injection time is set 1 to 2 seconds higher than the time required for injection.

The switch from one part of the injection molding cycle to the next is called transfer. Although the transfer from fill to pack or from first stage to second stage, can occur using time, ram position, hydraulic (line) pressure, nozzle pressure, cavity pressure, tie bar force, and tie bar deflection techniques, the time is used for other stages of the molding cycle. Time is the oldest and easiest control for transfer in injection molding. The control device is merely an electrical timer. Thus, for timed transfer from fill to pack, the pressure and injection speed are maintained for a specified length of time. However, since the timer does not control the pressure/injection speed interaction during filling nor the viscosity of the resin, timed transfer is considered the least reproducible method for the transfer from fill to pack. Although not considered ideal, timed transfer still typically determines the pack/hold, hold/cooling, and cooling/mold open transitions. Ram position is commonly used for transfer from fill to pack in injection molding. For this, the machine can monitor the position of the ram using one or more transducers or a linear variable differential transformer (LVDT). When the set position is reached, the machine switches control. While such controls are typically used with injection velocity–controlled machines, position transfer can be attained when the shot size indicator (on the injection unit) trips an electrical microswitch. Unlike timed transfer, position transfer delivers a constant volume (shot) of melt to the mold. In conjunction with controlled ram velocity, such transfer permits uniform delivery of polymer melt to the mold cavities. Ram position has been suggested as a control for packing, but is not generally available on commercial machines.

Hydraulic pressure is available on most injection molding machines, but is not as commonly used as position for the transfer from fill to pack. In hydraulic pressure transfer, the machine switches from fill to pack when the pressure in the hydraulic lines behind the injection cylinders reaches the set position. With well-maintained machines and properly set controls, hydraulic pressure transfer is more consistent than ram position transfer. While position transfer delivers a constant volume of material, polymers expand upon heating, and the constant volume of melt does not always produce consistent part weights. Hydraulic pressure can compensate somewhat for changes in viscosity and for expansion of the melt. However, hydraulic pressure transfer is not easily set up as position transfer. Although cavity pressure transfer is considered the most accurate method for transfer because it measures both material changes (such as viscosity) and machine behavior, it is not commonly used for transfer. As shown in Fig. 5.74,[170] both cavity and hydraulic line pressure enable fairly accurate transfer from fill to pack by monitoring the sharp increase in pressure that occurs when the cavity is completely filled. Since only the cavity pressure technique can measure peak packing pressure, it permits an accurate switchover from pack to hold. For either method, the transfer from hold to cooling occurs at the time when a consistent part weight is reached. Since cavity pressure must be monitored in the mold, the expense and positioning of the pressure transducers is the major problem. Cavity pressure measurement is also highly dependent on the position of the pressure transducer. Ideally, a pressure sensor can be placed about one-third of the way into the cavity. However, multicavity molds, particularly family molds or those with artificially balanced runner systems require multiple transducers. Additionally, the transducers can be mounted flush with a cavity wall or behind ejector pins. The former is more accurate, whereas the latter allows the transducers to be used in other molds without disassembling the mold. Nozzle pressure transfer is a method for reducing the cost, but maintaining some of the advantages of cavity pressure transfer. For this, a pressure transducer is installed in the nozzle of the injection molding machine. Since this transducer is farther from the mold cavity, it is not as accurate as cavity pressure measurements are. Nozzle pressure is also not always useful for family molds or those with artificially balanced runner systems. Consequently, nozzle pressure is also not commonly used as a transfer method.

Tie bar deflection is relatively uncommon transfer technique. Typically a transducer or other device measures the strain in a tie bar. Since this strain changes with the cavity pressure, measuring tie bar deflection is a simpler (and less expensive) way to monitor the cavity pressure. This method has been shown to correlate with cavity pressure.[171,172] However, the relatively small changes in strain have produced practical problems in amplifying the signals from the transducers.

With a single gate, a cavity with a uniform wall thickness fills in the manner shown in Fig. 5.72*a*.[166] When the cavity has more than one gate, the polymer flows out from each gate and joins at a weld line.

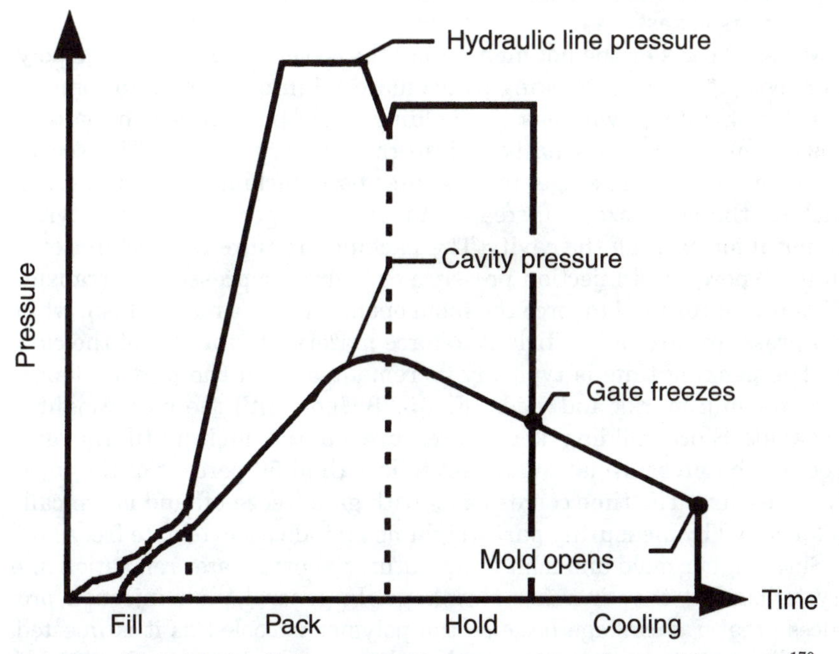

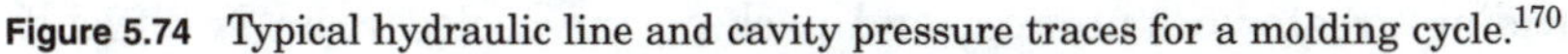

Figure 5.74 Typical hydraulic line and cavity pressure traces for a molding cycle.[170]

Since the polymer chains in the flow front are oriented perpendicular to the direction of flow, the chains diffuse across the weld line. Thus, weld lines are weak areas in the part. If the flow fronts meet at wider angle (>135°,[173] the polymer chains can flow together, thereby producing stronger weld lines called melt lines. As shown in Fig. 5.75,[173] weld and melt lines also occur when the melt must flow around an obstruction, such as a hole in the part. Weld lines are also produced when the part thickness is not uniform. In this case, the melt flows first through the thicker sections and then across the thinner sections of the part in an effect called race tracking (Fig. 5.76[174]). Since they are weak, weld lines are not typically located at critical areas in the part. They can be eliminated in multigated parts by the use of sequential valve gating, in which the next in a series of valve gates is opened when the flow front reaches the gate position.

During filling, the polymer melt freezes at the cavity wall to form the frozen layer. Melt adjacent to this is dragged along the frozen layer and then frozen. Since the frozen layer insulates the cavity, the melt in the center cools more slowly. This produces the orientation shown in Fig. 5.77*a*.[175] The melt frozen at the wall and the melt dragged along the frozen layer is highly oriented whereas the melt in the center has relaxed. This effect also depends on mold and melt temperature and the pressure on the melt. As illustrated in Fig. 5.77*b*,[175] the orientation is greater near the gate and lowest at the end of the fill (where cavity pressure is lowest).

Material cools as the hot melt enters the cooler mold, and as the polymer cools, it shrinks. Packing forces material into the mold to compensate for shrinkage whereas the holding stage pressurizes the gate to prevent melt from flowing back into the delivery system. The second stage (or the holding stage) incorporates both packing and holding. For packing the shot size is increased by 10 to 20 percent. This provides material for packing the cavity. The packing pressure is typically set at 50 to 60 percent of injection pressure or hydraulic pressure at transfer. High pressures tend to force the mold open, thereby causing flash, while low pressures are not sufficient to force material to the end of the cavity. The packing time is typically incremented until the part no longer exhibits sink marks and voids (Fig. 5.78[176]) or until the part weight is constant. Since holding keeps pressure on the melt until the gate freezes, the pressure is usually set to less than 50 percent of the injection pressure. The time corresponds with gate freeze-off and is typically determined by measuring part weight as an indication of gate freeze-off.

Since higher mold and melt temperatures allow greater relaxation and crystallization, they increase shrinkage. In general, slow injection produces greater shrinkage because the polymer is cooled as it is injected, providing greater orientation, and packing the cooled resin is difficult.

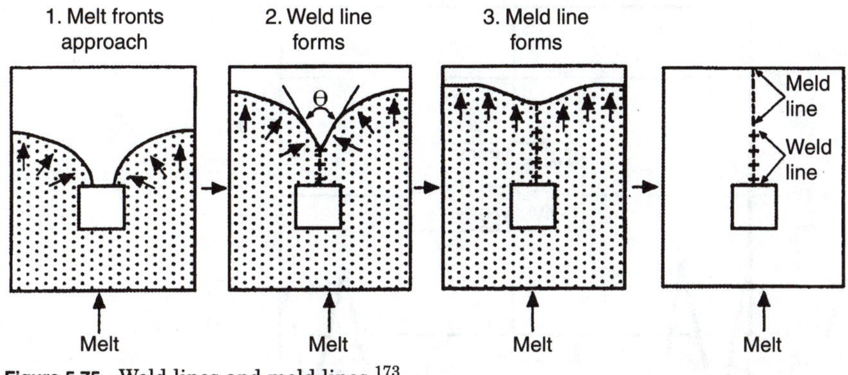

Figure 5.75 Weld lines and meld lines.[173]

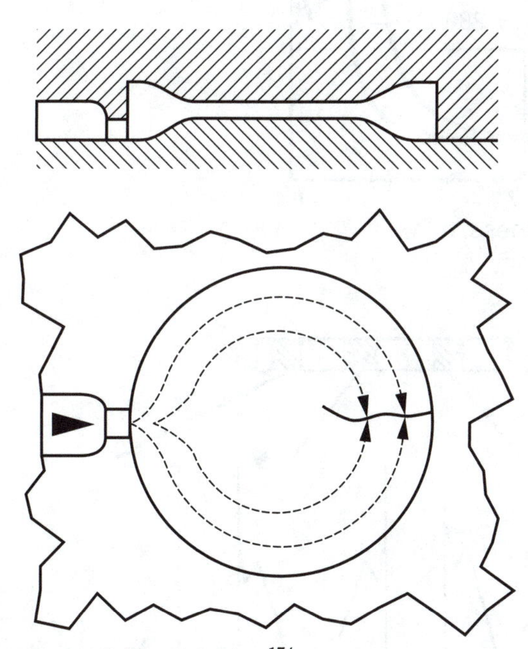

Figure 5.76 Race tracking.[174]

However, fast injection causes shear heating of the melt, thereby requiring the longer cooling times that facilitate relaxation and crystallization. Increased packing of the mold will reduce shrinkage, but this is limited by the gate freeze-off time. Molds with unbalanced filling will also exhibit over- and underpacking; this creates nonuniform shrinkage in the part.

Once the mold is filled and packed and the gate freezes off, the injection molding machine switches to the cooling stage. The amount of cooling is determined by the cooling time. While the melt in the mold cools to solid,

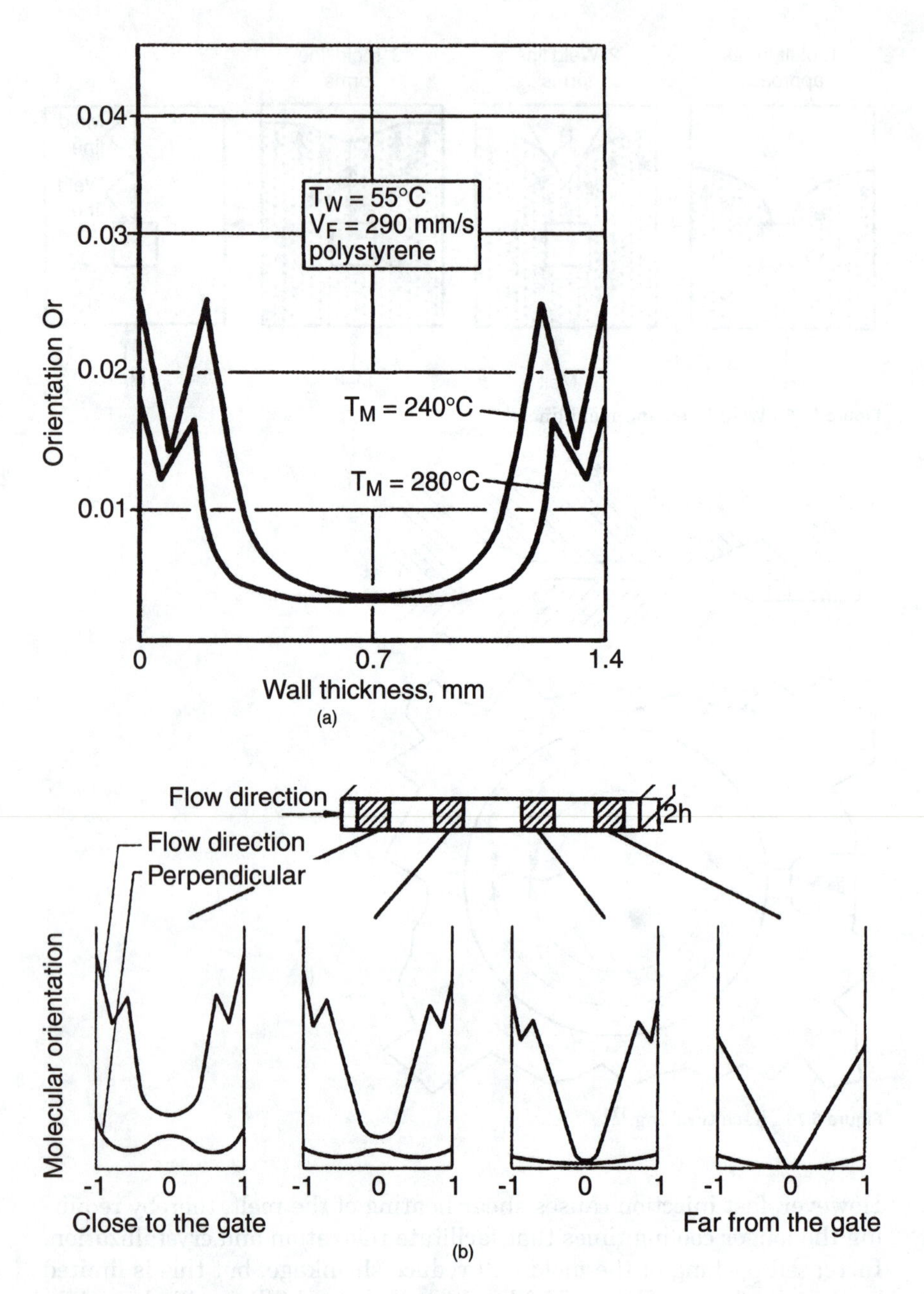

Figure 5.77 Orientation developed during filling. (*a*) Across the cavity thickness; (*b*) along the length of the cavity.[175]

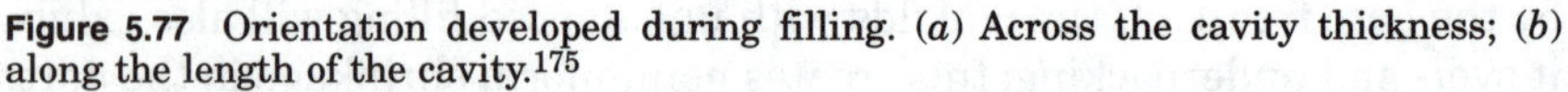

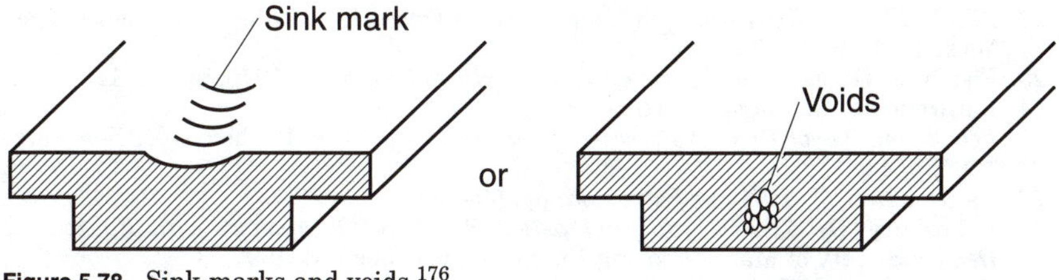

Figure 5.78 Sink marks and voids.[176]

the screw rotates and builds up a new shot. Cooling is usually the longest part of the molding cycle because the part is cooled until it can be ejected from the mold. The ejection temperature is estimated using the heat deflection temperature or Vicat softening temperature, but parts are ejected as soon as the ejection process does not damage the part. Parts ejected at high temperatures may warp due to stresses from ejection or due to the temperature of the surface onto which they fall.

References

1. McKelvey, J. M., *Polymer Processing,* John Wiley, New York, 1962, p. 1.
2. Michaeli, W., *Plastics Processing: An Introduction,* Hanser Publishers, New York, 1995, p. 19.
3. *Injection Molding Reference Guide,* 3d ed., Advanced Process Engineering, Seminole, Fla., 1996, p. 65.
4. "Plasticating Components Technology," Spirex Corporation, Youngstown, Oh, 1992, p. 65.
5. Deanin, R. D., *Polymer Structure, Properties and Applications,* Cahners Books, Boston, 1972, p. 333.
6. Rauwendaal, C., *Polymer Extrusion,* 2d ed., Hanser Publishers, New York, 1990, pp. 175–176.
7. Ibid., pp. 168–170.
8. American Society for Testing and Materials, ASTM D1895-89, "Test Methods for Apparent Density, Bulk Factor, and Pourability of Plastic Materials," 1992.
9. Rauwendaal, C., *Polymer Extrusion,* pp. 170–174.
10. Sperling, L. H., *Introduction to Physical Polymer Science,* 2d ed., John Wiley, New York, 1992, p. 487.
11. Rauwendaal, C., *Polymer Extrusion,* p. 177.
12. Ibid., p. 182.
13. Morton-Jones, D. H., *Polymer Processing,* Chapman & Hall, New York, 1989, p. 35.
14. Rauwendaal, C., *Polymer Extrusion,* p. 190.
15. Carreau, P. J., D. C. R. De Kee, and R. P. Chhabra, *Rheology of Polymeric Systems—Principles and Applications,* Hanser Publishers, New York, 1997, p. 52.
16. American Society for Testing and Materials, ASTM D1238-90b, "Test Method for Flow Rates of Thermoplastics by Extrusion Plastometer," 1992.
17. Tadmor, Z., and C. G. Gogos, *Principles of Polymer Processing,* Wiley, New York, 1979, pp. 135–137.
18. Malloy, R. A., *Plastics Product Design for Injection Molding,* Hanser Publishers, New York, 1994, p. 68.
19. Baird, D. G., and D. I. Collias, *Polymer Processing Principles and Design,* John Wiley, New York, 1998, p. 121.
20. Ibid., p. 123.
21. Ulcer, Y., M. Cakmak, J. Miao, and C.-M. Hsiung, ANTEC'95, 1995, p. 1788.

22. Bikales, N. M., *Extrusion and Other Plastics Operations,* Wiley-Interscience, New York, 1971, pp. 35–90.
23. Fisher, E. G., *Extrusion of Plastics,* John Wiley, New York, 1976, pp. 14–140.
24. Rauwendaal, C., *Polymer Extrusion,* pp. 23–48.
25. *Petrothene Polyolefins...A Processing Guide,* 3d ed., U. S. I. Chemicals, New York, 1965, p. 43.
26. Rauwendaal, C., *Polymer Extrusion,* pp. 50–59.
27. Lounsbury, D. C., *Proceedings: Plastics Extrusion Technology with Equipment Demos,* Society of Manufacturing Engineers, October 7–8, 1997.
28. Rauwendaal, C., *Polymer Extrusion,* p. 60.
29. Ibid., p. 19.
30. Ibid., p. 62.
31. Bernhardt, E. C., *Processing of Thermoplastic Materials,* Reinhold Publishing Corporation, New York, 1959, p. 157.
32. Rauwendaal, C., *Polymer Extrusion,* p. 65.
33. Ibid., pp. 117–118.
34. Ibid., p. 64.
35. *Cylinder and Screw Maintenance Handbook,* 2d ed., CAC Tool Corporation, Wichita, Kan., p. 16.
36. *Plasticating Components Technology,* pp. 22–23.
37. Kruder, G. A., and R. E. Nunn, “Applying Basic Solids Conveying Measurements to Design and Operation of Single-Screw Extruders,” *38th Annual Technical Conference of the Society of Plastics Engineers,* 1980, p. 62.
38. Wortberg, J., and R. Michaels, “Single-Screw Extruder Runs Diverse Range of Resins,” *Modern Plastics International,* vol. 28, no. 12, 1998, p. 93.
39. Rauwendaal, C., *Understanding Extrusion,* p. 13.
40. Ibid., p. 71.
41. Gneuß product literature, Matthews, N.C.
42. *Watlow Heaters,* Watlow, St. Louis, Mo., 1999.
43. *Plasticating Components Technology,* p. 6.
44. Ibid., pp. 10, 12–14.
45. Rauwendaal, C., *Polymer Extrusion,* pp. 432–433.
46. *Plasticating Components Technology,* p. 17.
47. Ibid., p. 5.
48. *Cylinder and Screw Maintenance Handbook,* p. 21.
49. Rauwendaal, C., *Understanding Extrusion,* p. 67.
50. Thompson, M. R., G. Donoian, and J. P. Christiano, “Examinations of Starve-Fed Single Screw Extrusion in Conventional and Barrier Feed Screws,” *57th Annual Technical Conference of the Society of Plastics Engineers,* 1999, p. 145.
51. Darnell, W. H., and E. A. J. Mol, “Solids Conveying in Extruders,” *Society of Petroleum Engineer Journal,* vol. 12, 1956, p. 20.
52. Rauwendaal, C., *Polymer Extrusion,* pp. 171–172.
53. Rosato, D. V., and D. V. Rosato, *Plastics Processing Data Handbook,* Chapman & Hall, New York, 1989, p. 93.
54. Rauwendaal, C., *Understanding Extrusion,* pp. 69–70.
55. Rosato, D. V., and Rosato, D. V., *Plastics Processing Data Handbook,* p. 105.
56. Steward, E. L., “Control of Melt Temperature on Single Screw Extruders,” *57th Annual Technical Conference of the Society of Plastics Engineers,* 1999, p. 195.
57. *Plasticating Components Technology,* p. 9.
58. Tadmor, Z., “Fundamentals of Plasticity Extrusion—I. A Theoretical Model for Melting,” *Polymer Engineering and Science,* vol. 6, 1966, p. 185.
59. *Plasticating Components Technology,* p. 16.
60. Rauwendaal, C., *Understanding Extrusion,* p. 75.
61. Glanvill, A. B., *The Plastics Engineer's Data Book,* Industrial Press, Inc., New York, 1974, p. 63.
62. Rauwendaal, C., *Polymer Extrusion,* p. 95.
63. Schenkel, G., *Plastics Extrusion Technology and Theory,* American Elsevier Publishing Company, New York, 1966, p. 45.
64. Rauwendaal, C., *Polymer Extrusion,* p. 26.
65. Kahns, A., *Almost Two Years in the Field and the ZSK Mega Compounder Speeds*

Ahead, Krupp Werner and Pfleiderer, Ramsey, N.J., 1998.
66. Baird, D. G., and D. I. Collias, *Polymer Processing,* p. 216.
67. Stevens, M. J., and J. A. Covas, *Extruder Principles and Operation,* 2d ed., Chapman & Hall, New York, 1995, p. 319.
68. White, J. L., *Twin-Screw Extrusion: Technology and Principles,* Hanser Publishers, New York, 1990, pp. 132–133.
69. Rauwendaal, C., *Understanding Extrusion,* pp. 3–4.
70. *Cincinnati Milicron,* Cincinnati Milicron Austria, Vienna, 1998.
71. White, J. L., *Twin-Screw Extrusion,* p. 229.
72. *Twin Screw Report,* American Leistritz Extruder Corp., Somerville, N.J., November 1993.
73. Baird, D. G., and D. I. Collias, *Polymer Processing,* p. 217.
74. *Leistritz Extrusionstechnik,* Leistritz Aktiengesellschraft, Nurnberg, Germany, 1998.
75. White, J. L., *Twin-Screw Extrusion,* p. 248.
76. Rauwendaal, C., *Plastics World,* vol. 50, no. 4, 1992, p. 68.
77. Rauwendaal, C., *Plastics Formulating and Compounding,* vol. 2, no. 1, 1996, p. 22.
78. Mielcarek, D. F., "Twin-Screw Extrusion," *Chemical Engineering Progress,* vol. 83, no. 6, 1987, p. 59.
79. *The MEGA Compounder: Productivity by Design,* Werner and Pfleiderer, Ramsey, N.J., 1997.
80. Callari, J., "Mega Machine is Fast, Productive," *Plastics World,* vol. 53, no. 11, 1995, p. 12.
81. Glanvill, A. B., *The Plastics Engineer's Data Book,* p. 89.
82. Bikales, N. M., *Extrusion and Other Plastics Operations,* p. 39.
83. Hensen, F., *Plastics Extrusion Technology,* 2d ed., Hanser Publishers, New York, 1997, p. 106.
84. *Petrothene® Polyolefins,* p. 49.
85. Hensen, F., *Plastics Extrusion Technology,* p. 107.
86. Fisher, E. G., *Extrusion of Plastics,* pp. 216–217.
87. Rauwendaal, *Polymer Extrusion,* p. 451.
88. Hellmuth, W., "Considerations for Choosing an Integrated Control System," *54th Annual Technical Conference of the Society of Plastics Engineers,* 1996, p. 2.
89. Petrothene® Polyolefins, p. 51.
90. Vargas, E., T. I. Butler, and E. W. Veazey, *Film Extrusion Manual: Process, Materials, Properties,* TAPPI Press, Atlanta, 1992, p. 18.
91. *Petrothene® Polyolefins,* p. 52.
92. Glanvill, A. B., *The Plastics Engineer's Data Book,* p. 96.
93. Berins, M. L., *Plastics Engineering Handbook of the Society of the Plastics Industry, Inc.,* 5th ed., van Nostrand Reinhold, New York, 1991, p. 104.
94. Ibid., p. 102.
95. *Petrothene® Polyolefins,* p. 54.
96. Glanvill, A. B., *The Plastics Engineer's Data Book,* p. 102.
97. Berins, M. L., *Plastics Engineering Handbook,* p. 105.
98. Hensen, F., *Plastics Extrusion Technology,* p. 163.
99. *Petrothene® Polyolefins,* p. 75.
100. Rauwendaal, *Polymer Extrusion,* p. 443.
101. Fisher, E. G., *Extrusion of Plastics,* pp. 222–223.
102. Berins, M. L., *Plastics Engineering Handbook,* p. 107.
103. *Petrothene® Polyolefins,* p. 62.
104. Ibid., p. 67.
105. Ibid., p. 68.
106. Vargas, E., *Film Extrusion Manual,* pp. 338–339.
107. *Petrothene® Polyolefins,* p. 78.
108. Vargas, E., *Film Extrusion Manual,* p. 247.
109. *Petrothene® Polyolefins,* p. 65.
110. Strong, A. B., *Plastics: Material and Processing,* Prentice-Hall, Englewood Cliffs, N.J., 1996, p. 258.
111. Fisher, E. G., *Extrusion of Plastics,* pp. 227–228.
112. Ibid., pp. 238–241.

113. Deanin, R. D., *Polymer Structure, Properties and Applications,* Cahners Books, Boston, 1972, p. 222.
114. Strong, A. B., *Plastics,* p. 295.
115. Ibid., pp. 296–297.
116. *Petrothene® Polyolefins,* p. 84.
117. Berins, M. L., *Plastics Engineering Handbook,* p. 117.
118. *Wire and Cable Coaters' Handbook,* DuPont, Wilmington, Del., 1968, p. 22.
119. *Petrothene® Polyolefins,* p. 85.
120. Berins, M. L., *Plastics Engineering Handbook,* p. 116.
121. *Petrothene® Polyolefins,* p. 73.
122. Finch, C., presentation notes.
123. Bezigian, T., *Extrusion Coating Manual,* 4th ed., TAPPI Press, Atlanta, 1999, p. 48.
124. Schott, N. R., private correspondence.
125. Berins, M. L., *Plastics Engineering Handbook,* p. 134.
126. Ibid., p. 136.
127. Wright, R. E., *Injection/Transfer Molding of Thermosetting Plastics,* Hanser Publishers, New York, 1995, pp. 39–40.
128. Berins, M. L., *Plastics Engineering Handbook,* p. 136.
129. Graham, L., *What Is a Mold?,* Tech Mold, Inc., Tempe, Ariz., 1993, p. 2-2.
130. Ibid., p. 2-6.
131. *Plasticating Components Technology,* Spirex Corporation, Youngstown, Ohio, 1992, p. 38.
132. Johannaber, F., *Injection Molding Machines: A Users Guide,* Hanser Publishers, New York, 1994, p. 51.
133. *Plasticating Components Technology,* p. 37.
134. Ibid., p. 6.
135. Berins, M. L., *Plastics Engineering Handbook,* p. 141.
136. Ibid., p. 142.
137. Lai, F., and J. N. Sanghavi, "Performance Characteristics of the Dray Non-Return Valve Using SPC/SQC in Injection Molding," *51st Annual Technical Conference of the Society of Plastics Engineers,* 1993, p. 2804.
138. Rosato, D. V., and D. V. Rosato, eds., *Injection Molding Handbook,* van Nostrand Reinhold, New York, 1986, p. 58.
139. Johannaber, F., *Injection Molding Machines: A Users Guide,* Hanser Publishers, New York, 1994, p. 78.
140. Ibid., pp. 45–47.
141. Ibid., p. 44.
142. Belofsky, H., *Plastics: Product Design and Process Engineering,* Hanser Publishers, New York, 1995, p. 289.
143. *Injection Molding Reference Guide,* 4th ed., Advanced Process Engineering, Corvallis, Oreg., 1997, p. 66.
144. Berins, M. L., *Plastics Engineering Handbook,* p. 145.
145. Johannaber, F., *Injection Molding Machines,* pp. 109–110.
146. Graham, L., *What Is a Mold?,* pp. 2–8.
147. Johannaber, F., *Injection Molding Machines,* pp. 93–94.
148. Ibid., p. 97.
149. Graham, L., *What Is a Mold?,* pp. 2–9.
150. Belofsky, H., *Plastics,* p. 288.
151. Berins, M. L., *Plastics Engineering Handbook,* p. 145.
152. *Cincinnati Milicron Vista Sentry—VSX User's Manual,* Cincinnati Milicron Marketing Company, Batavia, Ohio, 1997, p. II-22.
153. Graham, L., *What Is a Mold?,* pp. 4–10.
154. Rosato, D. V., and D. V. Rosato, eds., *Injection Molding Handbook,* 2d ed., Chapman and Hall, New York, 1995, pp. 222–223.
155. Belofsky, H., *Plastics,* p. 298.
156. *C-Mold Design Guide,* AC Technology, Ithaca, N.Y., 1994, p. 173.
157. *Injection Molding Reference Guide,* p. 17.
158. *C-Mold Design Guide,* p. 29.
159. McCrum, N. G., C. P. Buckley, and C. B. Bucknall, *Principles of Polymer*

Engineering, 2d ed., Oxford University Press, New York, 1997, p. 338.
160. *C-Mold Design Guide,* pp. 43–48.
161. Malloy, R. A., *Part Design for Injection Molding,* Hanser Publishers, New York, 1994, p. 16.
162. Rosato, D. V., and D. V. Rosato, eds., *Injection Molding Handbook,* p. 259.
163. Graham, L., *What Is a Mold?,* pp. 5-4.
164. *Injection Molding Reference Guide,* p. 16.
165. Nunn, R. E., *Short Shot Method,* internal documentation, University of Massachusetts, Lowell.
166. *C-Mold Design Guide,* AC Technology, Ithaca, N.Y., 1994, p. 134.
167. Graham, L., *What Is a Mold?,* pp. 1–3.
168. Yeager, M., Moldflow User's Group Meeting, Kalamazoo, Mich., 1999.
169. Coxe, M. M., C. M. F. Barry, D. Bank, and K. Nichols, "The Establishment of a Processing Window for Thin Wall Injection Molding of Syndiotactic Polystyrene," *ANTEXC 2000,* in press.
170. *Injection Molding Technology: Videocassette Education Course,* 2d ed., Workbook 1, Sessions 1-7, Paulson Training Programs, Inc., Cromwell Conn., 1983, p. 39.
171. Ulik, J., "Using Tie Rod Bending to Monitor Cavity Filling Pressure," *ANTEC'97,* 1997, p. 3659.
172. Mueller, N., private correspondence, 1999.
173. *C-Mold Design Guide,* p. 144.
174. Belofsky, H., *Plastics: Product Design and Process Engineering,* Hanser Publishers, New York, 1995, p. 304.
175. Potsch, G., and W. Michaeli, *Injection Molding: An Introduction,* Hanser Publishers, New York, 1995, pp. 116–117.
176. *C-Mold Design Guide,* p. 103.

Chapter

6

Processing of Thermosets

John L. Hull

President, Hull Corporation
Warminster, Pennsylvania

6.1 Introduction

Because thermoset plastics undergo an irreversible chemical reaction (cross-linking or polymerization) involving a time-temperature exposure, processing of such materials consists of: (1) getting the resin formulation to a liquid state, (2) causing the liquid to flow into a cavity of the desired configuration, (3) heating the liquid sufficiently to cause the chemical reaction to advance to the point where the plastic is essentially rigid (conforming to the cavity configuration), and (4) removing the rigid part from the cavity.

The other critical variable in most thermoset processing is pressure. Pressure on the molding compound is needed: (1) to create the flow of plastic, during its liquid state, into the interstices of each cavity, prior to the material becoming so viscous from the cross-linking reaction that it flows no more; and (2) to ensure that the plastic, during cross-linking, is kept at maximum density in order to obtain optimum physical properties of the molded part.

The several processing techniques described in this chapter represent the current methods for effecting the liquification of the formulation, the timely flow into the cavity, and the required heat and pressure to enable the chemical reaction to proceed to completion as rapidly as practical in order to achieve an acceptably short production cycle with full densification of the plastic. (See the time-temperature-viscosity curves in Fig. 6.1.)

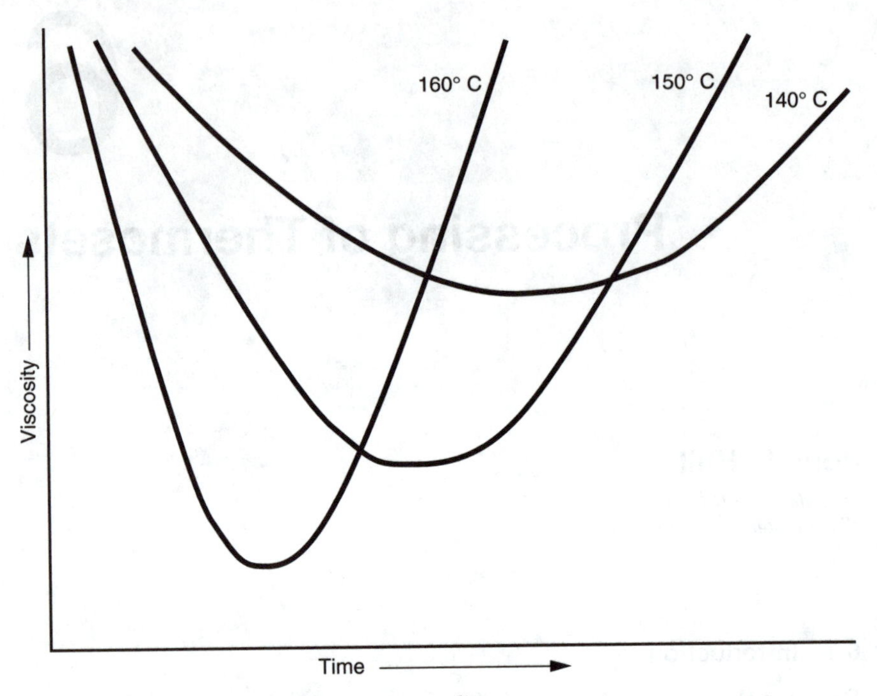

Figure 6.1 Viscosity-time–mold temperature diagram.

6.2 Molding Processes

6.2.1 Casting with liquid resins

For prototyping or for limited production runs, *liquid resin casting* offers simplicity of process, relatively low investment in equipment, and fast results. Thermoset casting resins may be epoxies, polyesters, phenolics, etc. The resins harden by a polymerization or cross-linking reaction. Such resins are often poured into open molds or cavities. Because pouring is done at atmospheric pressure, molds are simple, often made of soft metals (Fig. 6.2).

An example of liquid casting is in fabrication of design details such as scrolls and floral or leaf patterns for furniture decoration. Such parts are often made from filled polyester resins, and the parts, after curing, are simply glued to the wooden bureau drawer or mirror frame. The parts can readily be finished to look like wood. Molds for such parts are often made by casting an elastomeric material over a wood or plastic model of the part. When the elastomer is removed, it generally yields a cavity enabling very faithful reproduction of the original pattern.

Another widely used industrial application is casting with liquid resins to embed objects, such as electronic components or circuits, in

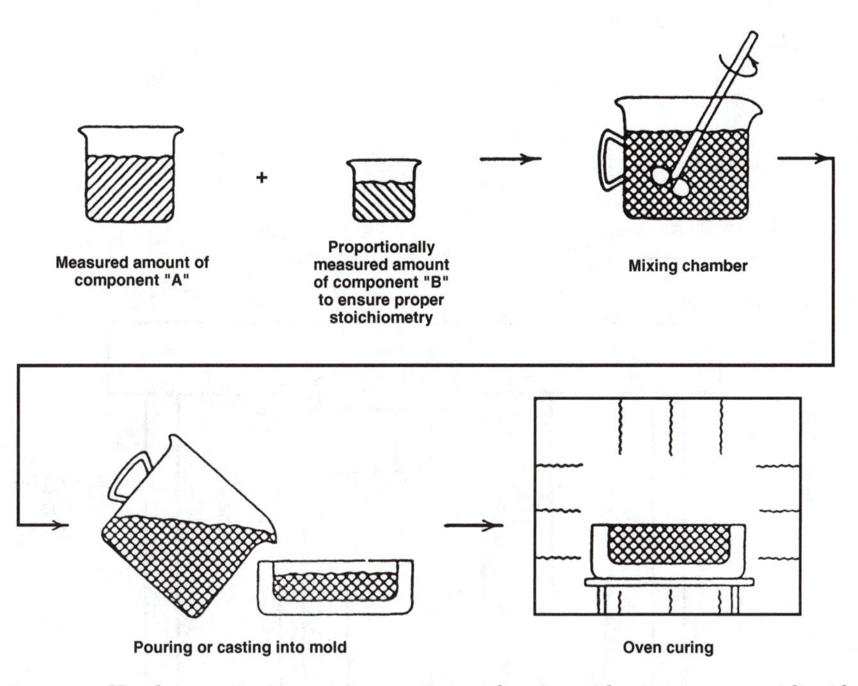

Figure 6.2 Hand proportioning, mixing, casting, and curing with a two-component liquid reactive thermosetting resin system.

plastic cups or cases or shells, giving the components mechanical protection, electrical insulation, and a uniform package. Such processing, when the case or housing remains with the finished part, is called *potting.*

When casting applications require fairly high-volume production, machines for mixing and dispensing the liquid plastics may be used for shorter production cycles, and curing ovens, conveyors, and other auxiliary capital equipment may be added. In short, the liquid resin casting operation may be a low-cost manual one, or it may be highly automated, depending on the nature of product desired and the quantities required (Fig. 6.3).

Potting of high reliability products is often done under vacuum to ensure void-free products, and may be followed with positive pressure in the casting chamber to speed the final penetration of casting formulation into the interstices of the component being embedded before the resin system hardens (Figs. 6.4 and 6.5).

6.2.2 Hand lay-up (composites)

When larger plastic parts are required, and often when such parts must be rigid and robust, a process referred to as *hand lay-up* is closely related to liquid resin casting.

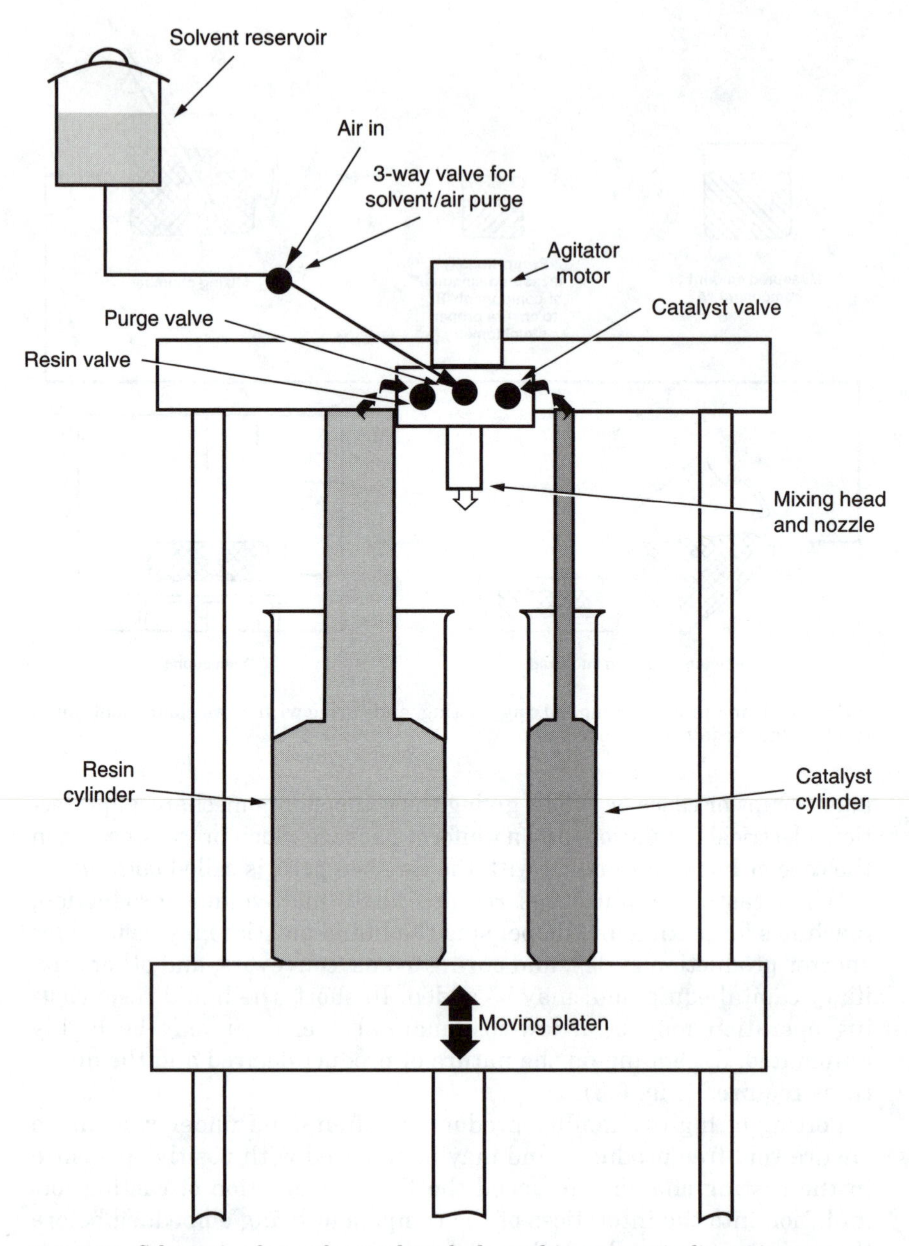

Figure 6.3 Schematic of one of several methods used in automatic dispensing systems for thermosetting resin systems.

Figure 6.4 Vacuum-pressure potter with planetary turntable. System enables resin degassing, addition of correct amount of catalyst under original vacuum, vacuum bake-out of products and molds before casting, and casting under original vacuum.

A reinforcing fabric or mat, frequently fiberglass, is placed into an open mold or over a form, and a fairly viscous liquid resin is poured over the fabric to wet it thoroughly and to penetrate into the weave, ideally with little or no air entrapment. When the plastic hardens, the object is removed from the mold or form, trimmed as necessary, and is then ready for use (Fig. 6.6). Many boats are produced using the hand lay-up process, from small sailing dinghies and bass boats, canoes, and kayaks to large sailboats, commercial fishing boats, and even military landing craft.

This basic process can be automated as required, with proportioning, mixing, and dispensing machines for liquid resin preparation;

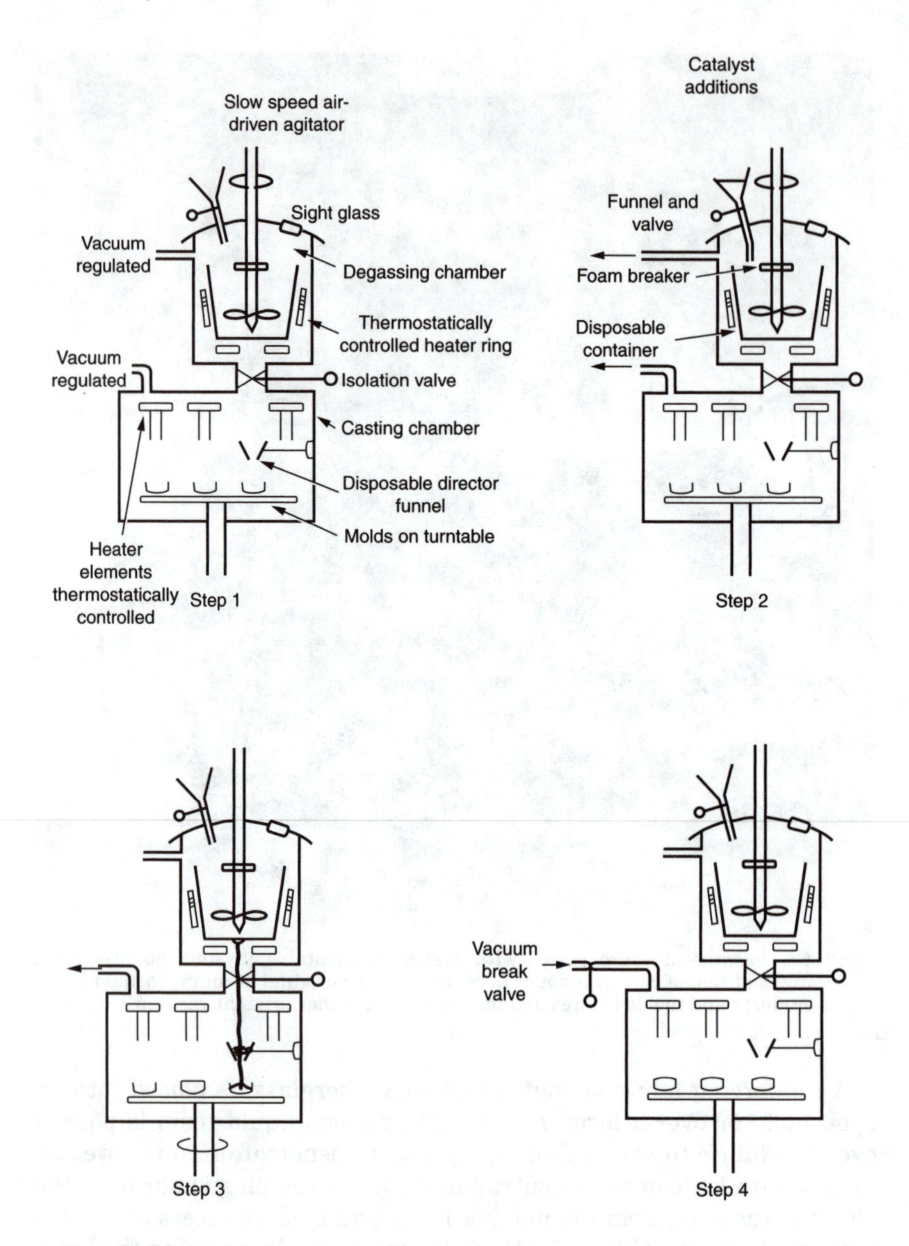

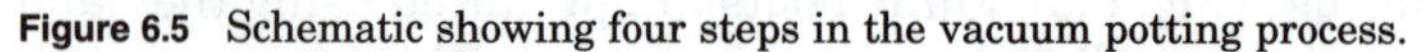

Figure 6.5 Schematic showing four steps in the vacuum potting process.

with matched molds (that is, with two mold halves closed after the reinforcing material has been impregnated with the liquid to produce a smooth uniform surface on both top and bottom of the part); and with conveyors, curing ovens, etc.

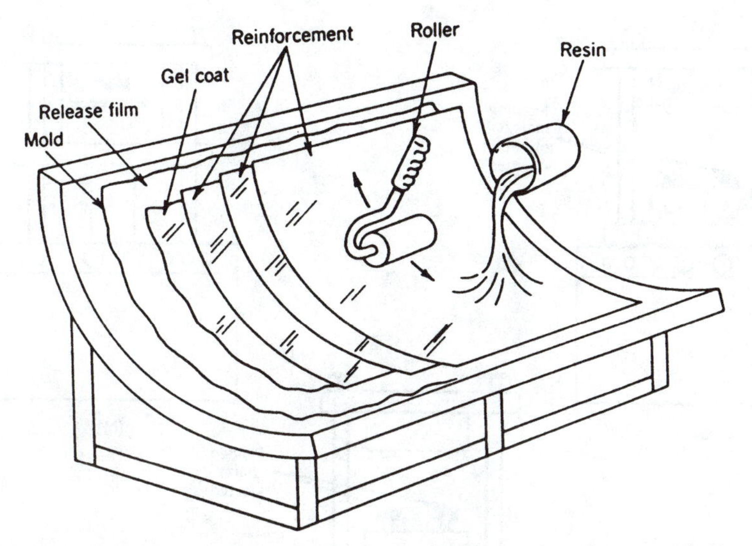

Figure 6.6 Hand lay-up process for producing highly reinforced parts with fiber matting and polyester, epoxy, or other thermosetting liquid resin systems.

6.2.3 Compression molding

Compression molding is a process that is very similar to making waffles. The molding compound, generally a thermosetting material such as phenolic, melamine, or urea, is placed in granular form into the lower half of a hot mold, and the heated upper mold half is then placed on top and forced down until the mold halves essentially come together, forcing the molding compound to flow into all parts of the cavity, where it finally "cures" or hardens under continued heat and pressure. When the mold is opened, the part is removed and the cycle is repeated.

The process can be manual, semi-automated, or fully automated (unattended operation), depending on the equipment. Molds are generally made of through-hardened steel and are highly polished and hard chrome plated, and the two mold halves, with integral electric, steam, or hot oil heating provisions, are mounted against upper and lower platens in a hydraulic press capable of moving the molds open and closed with adequate tonnage to make the plastic flow.

Molds may be single cavity or multiple cavity, and press tonnage must be adequate to provide as much as 300 kg/cm^2 for phenolics, less for polyesters, of projected area of the molded part or parts at the mold parting surfaces. Overall cycles depend on molding material, part thickness, and mold temperature, and may be about 1 min for parts of 3-mm thickness to 5 or 6 min for parts of 8-cm thickness (Fig. 6.7).

The process is generally used for high-volume production because the cost of a modern semi-automatic press of modest tonnage, say 50

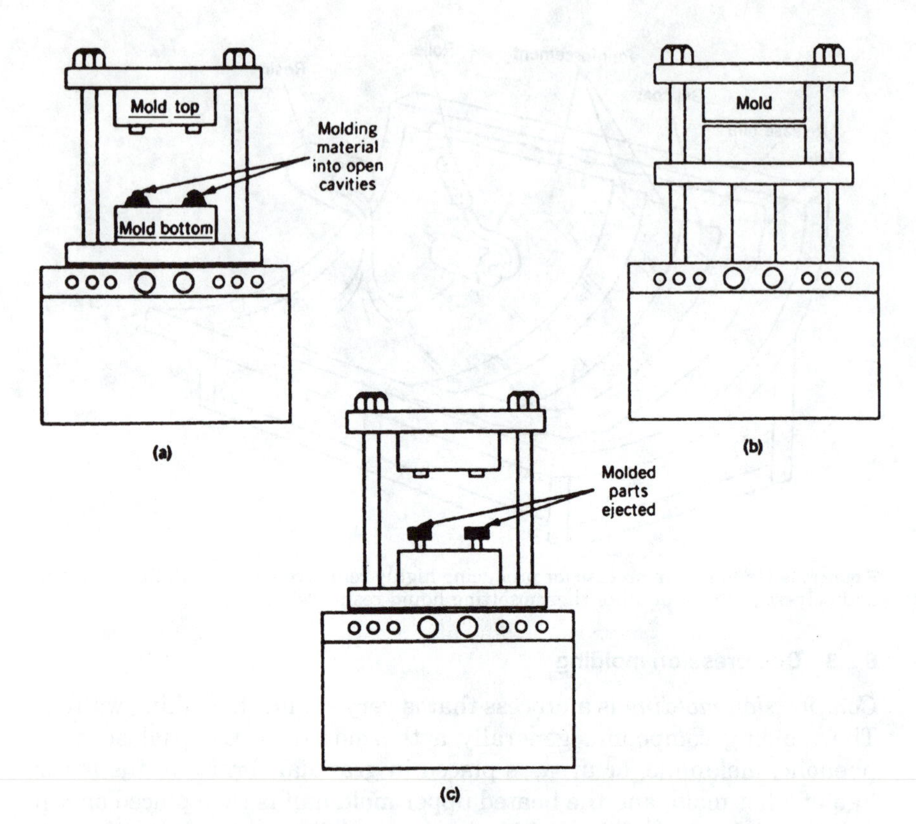

Figure 6.7 Compression molding sequence: (*a*) molding material is placed into open cavities; (*b*) the press closes the mold, compressing material in the hot mold for cure; and (*c*) the press opens and molded parts are ejected from the cavities.

tons clamp, may be as much as $50,000, and a moderately sophisticated self-contained multicavity mold may also cost $50,000 (Fig. 6.8).

Typical applications include melamine dinnerware; phenolic toaster legs; and pot handles, electrical outlets, wall plates, and switches—parts which require the rigidity, dimensional stability, heat resistance, or electrical insulating properties typical of thermosetting compounds.

To simplify feeding material into the mold, the molding compound is often precompacted into "preforms" or "pills" on a specially designed automatic preformer which compacts the molding compound at room temperature into cylindrical or rectangular blocks of desired weight of charge. And to reduce molding cycle time, the preform is often heated with high-frequency electrical energy in a self-contained unit called a *preheater,* which is arranged beside the press. The preform is manually placed between the electrodes of the preheater before each molding cycle, and heated throughout in as little as 10 to 15 s to about 90°C, at which

Figure 6.8 Fully automatic compression molding press with material feed to multiple cavities and removal of cured parts each cycle. (*Courtesy Hull Corporation.*)

temperature the plastic holds together but is slightly mushy. It is then manually placed in the bottom mold cavity and the molding cycle is initiated. Cure time may be cut in half through the use of preheating, a step which reduces mold wear and improves part quality (Figs. 6.9 and 6.10).

6.2.4 Transfer molding

A related process for high-volume molding with thermosetting materials is *transfer molding,* so called because instead of the material being placed between the two halves of an open mold, followed by closing the mold, to make the material flow and fill the cavity, the material is placed into a separate chamber of the upper mold half, called a *transfer pot,* generally cylindrical, which is connected by small runners and smaller openings called *gates* to the cavity or cavities.

Figure 6.9 High-frequency preheater with roller electrodes to raise temperature of preforms prior to placing in cavities of compression mold or in transfer pot of transfer mold. Preheating shortens cure time and minimizes mold wear.

In operation, the mold is first closed and held under pressure; the preheated preform is dropped into the pot; a plunger comes down into the pot where the material liquifies from the heat of the mold and the pressure of the plunger, and flows (is "transferred") through the runners and gates into the cavity or cavities. The plunger keeps pushing on the molding compound until the cavities are full and until the material cures. At that point, the mold is opened, the plunger is retracted, and the part or parts, runners, and cull (the material remaining in the pot, generally about $^1/_8$ in thick and the diameter of the pot and plunger) are removed. Because the gate is small, the runners and cull are readily separated from the molded parts at the surface of the parts, leaving a small and generally unobtrusive but visible "gate scar" (Figs. 6.11, 6.12, and 6.13).

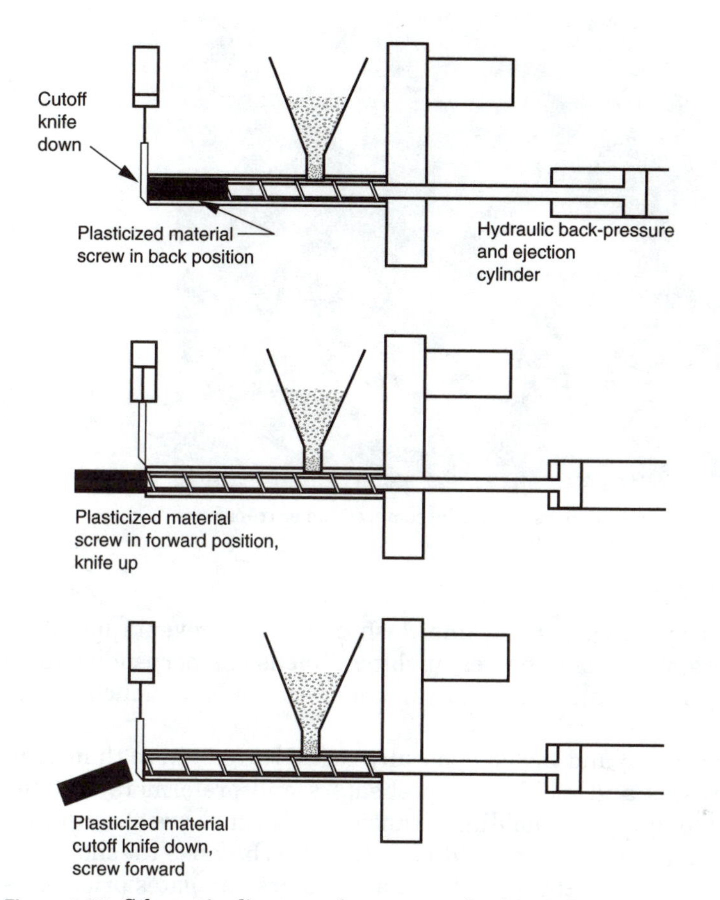

Figure 6.10 Schematic diagram of screw preplasticizer for preheating thermosetting molding compounds before molding. Preplasticizer may be integrated with molding press for fully automatic molding.

Transfer molding is often used when inserts are to be "molded in" the finished part, as, for example, contacts in an automobile distributor cap or rotor or solenoid coils and protruding terminals for washing machines. Whereas in compression molding, such inserts might be displaced during a compression molding cycle, in transfer molding the inserts are being surrounded by a liquid flowing into the cavity at controlled rates and pressures, and generally at a relatively low viscosity. The inserts are also generally supported by being firmly clamped at the mold parting line or fitted into close-toleranced holes of the cavity. Also, when dimensions perpendicular to the parting line or parting surfaces of the mold must be held to close tolerances, transfer molding is used because the mold is fully closed prior to being filled with plastic. With

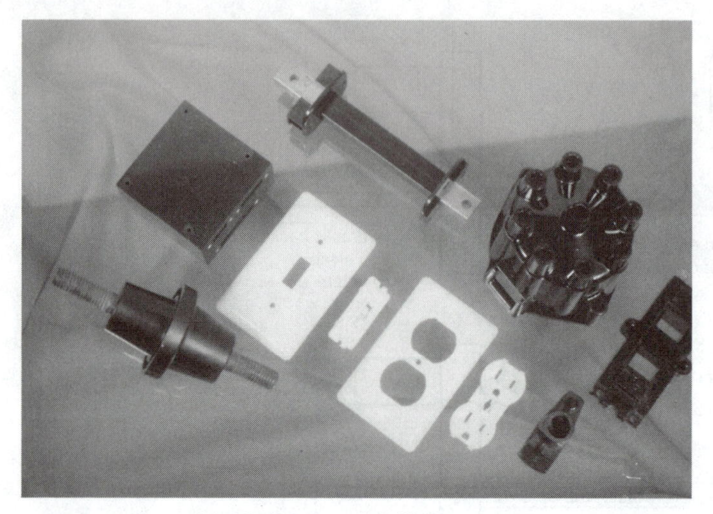

Figure 6.11 Several typical parts molded by compression or transfer or injection molding of thermosetting plastics.

compression molding, parting line flash generally prevents metal-to-metal closing of the mold halves, making dimensions perpendicular to the parting line greater by flash thickness—perhaps as much as 0.1 to 0.2 mm.

Transfer presses and molds generally cost 5 to 10% more than compression presses and molds, but preheaters and preformers cost the same as for compression molding. Transfer cycle times are often slightly shorter than compression molding cycle times because the motion of the compound flowing through the small runners and gates prior to its entering the cavity raises the compound temperature by frictional and mechanical shear, therefore accelerating the cure.

One highly significant application of transfer molding is for direct encapsulation of electronic components and semiconductor devices. Adaptation of the basic transfer molding process to enable successful molding around the incredibly fragile devices and whisker wires required, first, the development of very soft flowing materials, generally epoxies and silicones and then modifications to conventional transfer presses to enable sensitive low-pressure control and accurate speed control (both often programmed through several steps during transfer). Finally, new mold design and construction techniques were needed to ensure close tolerance positioning and holding of the components in the cavities prior to material entry. It can be fairly stated that the successful development of the transfer molding encapsulation process was a large factor in high-volume manufacture of low-cost transistors and integrated circuits (Figs. 6.14 to 6.18).

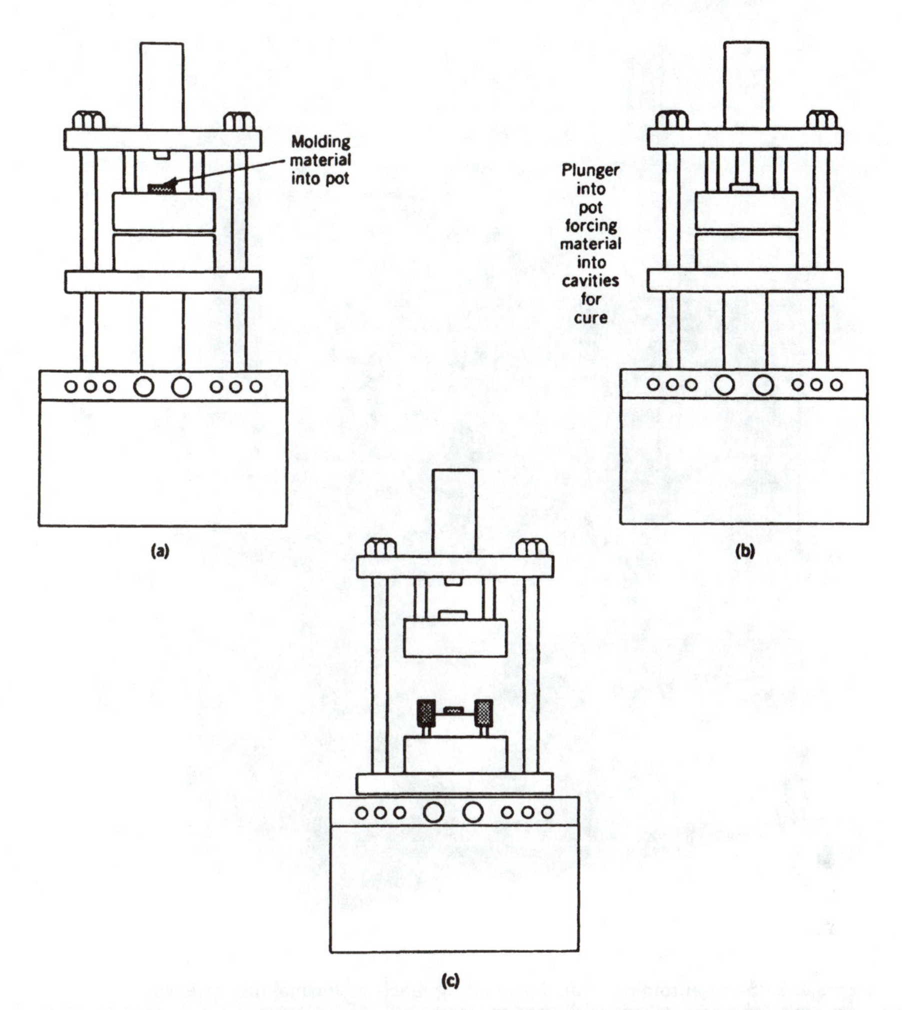

Figure 6.12 Transfer molding sequence: (*a*) the mold is closed and material is placed in the pot; (*b*) the plunger descends into the pot, causing material to melt and flow through runners into cavities; (*c*) after cure, the press opens, the plunger retracts, and the parts are ejected with cull and runners.

6.2.5 Injection molding of thermosets

Injection molding of thermosets is similar in many respects to transfer molding. The process is also a "closed mold" process, and the mold uses runners and gates leading to cavities in much the same way as does a transfer mold. But instead of a pot and plunger, the injection process generally uses an auger-type screw rotating inside a long cylindrical tube called a *barrel.* The barrel temperature is closely controlled, usually by hot-water jackets surrounding the barrel. The granular molding compound is fed by gravity from a hopper into the rear end of the

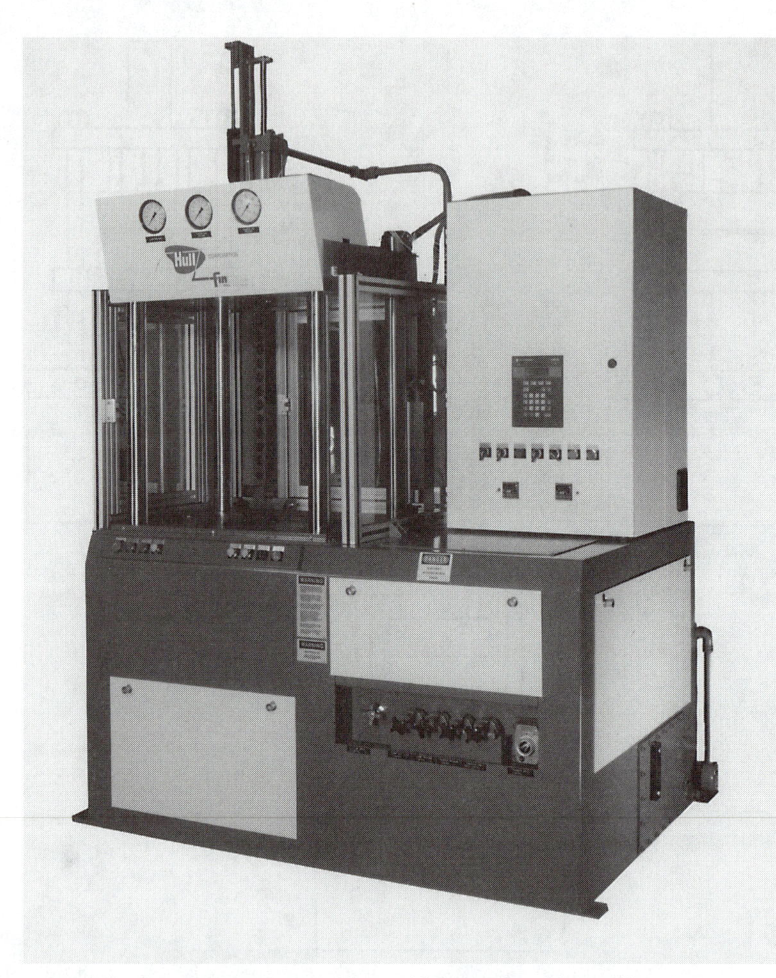

Figure 6.13 Semi-automatic transfer molding machine for molding or encapsulating with thermosetting molding compounds.

screw and barrel assembly. The front of the barrel narrows down to a small opening or nozzle which is held firmly against a mating opening in the center of one of the mold halves, called a *sprue hole,* which leads the fluid material into the runner system at the parting surfaces of the mold. The screw and barrel are generally positioned horizontally, and the press opens horizontally (as compared to the up-and-down movement that is traditional with compression and transfer presses), so the mold parting surface is in a vertical plane rather than a horizontal one (Fig. 6.19).

In operation, after the molded parts and runners have been removed from the open mold, the press closes the mold in preparation for the

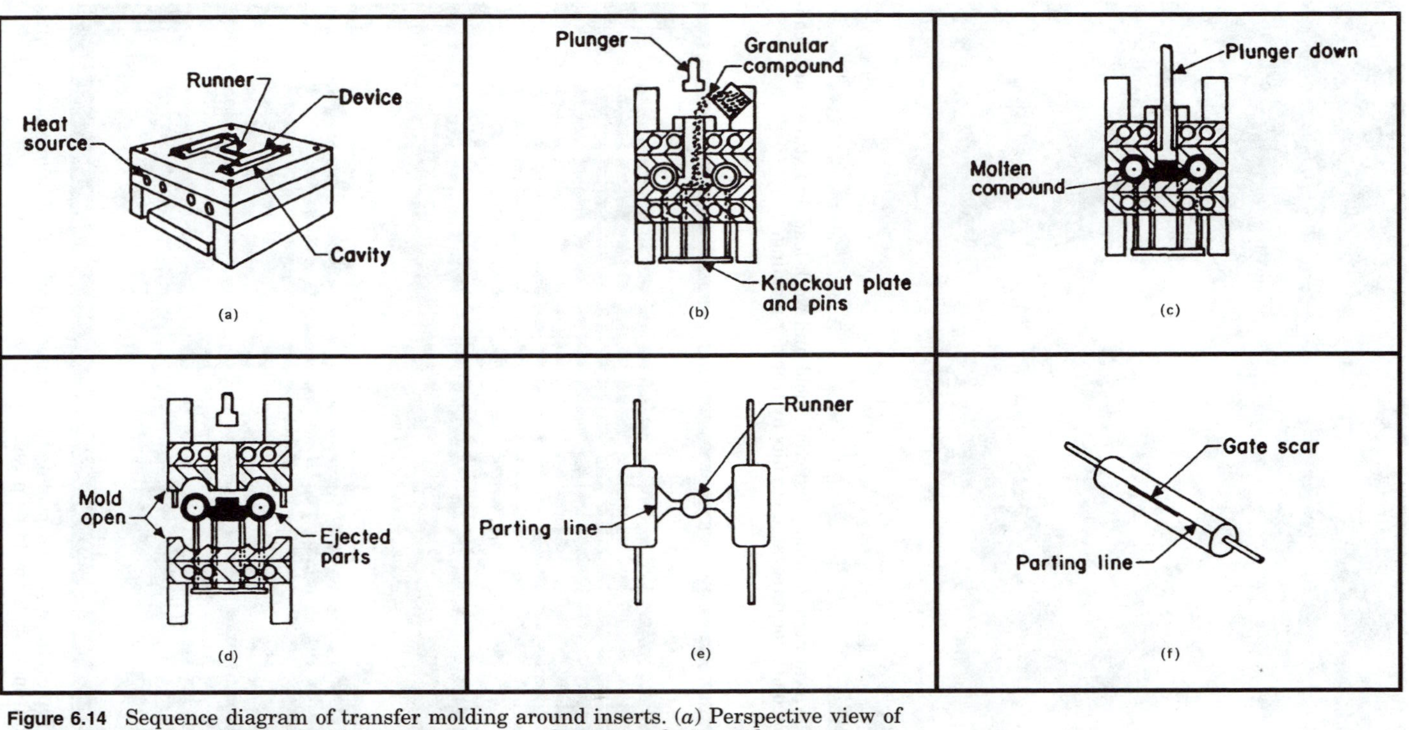

Figure 6.14 Sequence diagram of transfer molding around inserts. (*a*) Perspective view of transfer mold; (*b*) mold closed with inserts in position for encapsulations, plunger retracted, and granular or preformed compound fed into heating chamber; (*c*) plunger moves downward, forcing molten compound around devices in cavities; (*d*) mold opens following cure, and knockout pins eject encapsulated devices; (*e*) encapsulated devices as molded; (*f*) encapsulated device showing parting line and gate scar.

Figure 6.15 Typical high-volume production mold and work-loading fixtures for encapsulating semiconductor integrated circuits by transfer molding.

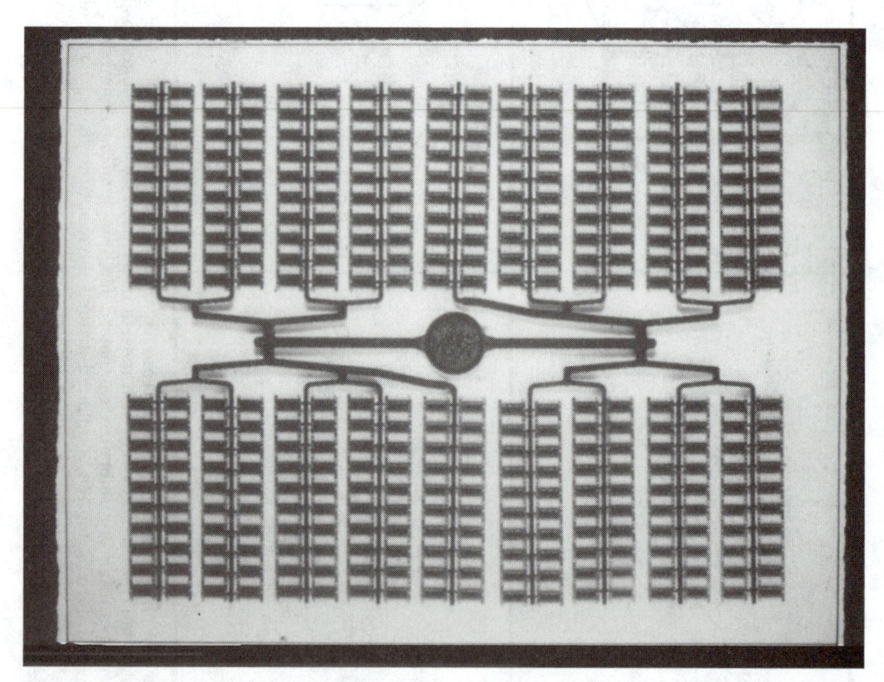

Figure 6.16 Typical "shot" of encapsulated integrated circuits (dual in-line packages) as it is removed from transfer molding press.

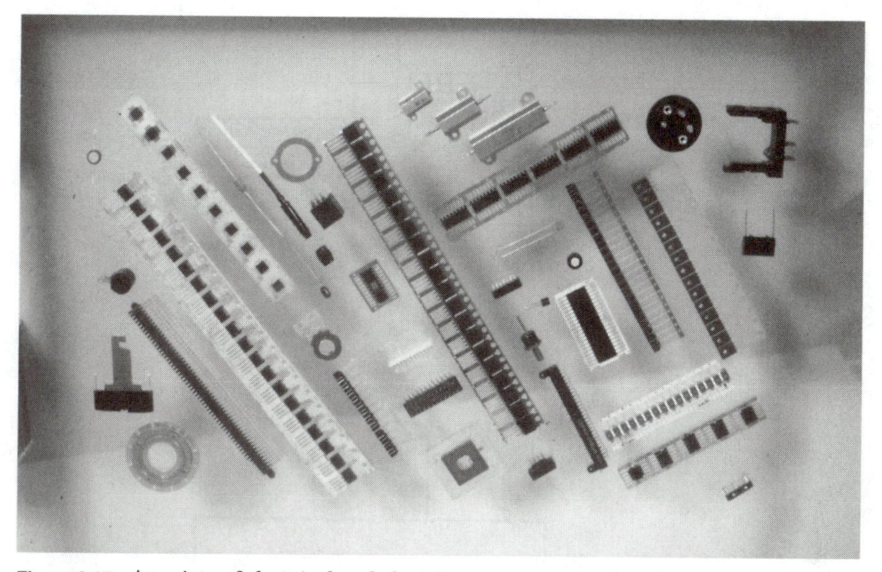

Figure 6.17 A variety of electrical and electric components produced by transfer molding with thermosetting compounds.

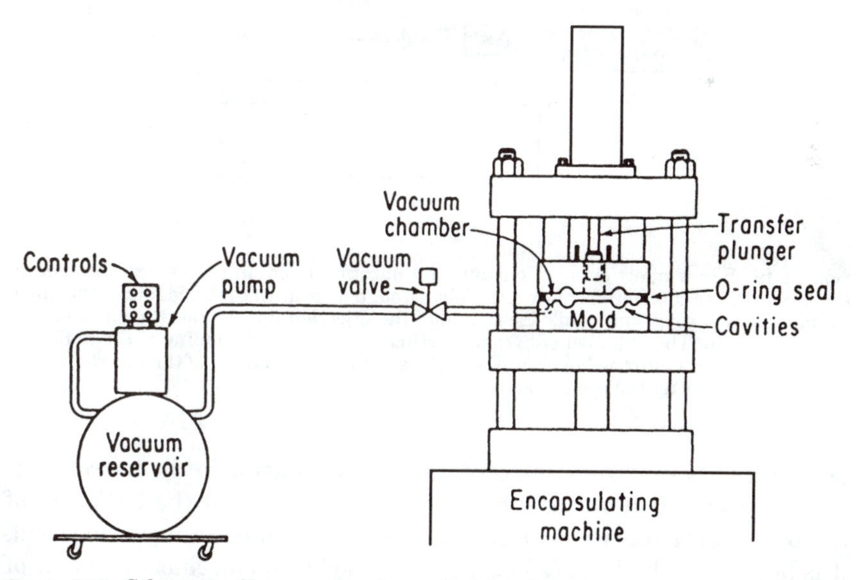

Figure 6.18 Schematic diagram showing principle of vacuum venting with transfer molding. The principle also is used with automatic transfer and screw injection molding where the cavity configuration precludes adequate parting line venting. (*Courtesy Hull Corporation.*)

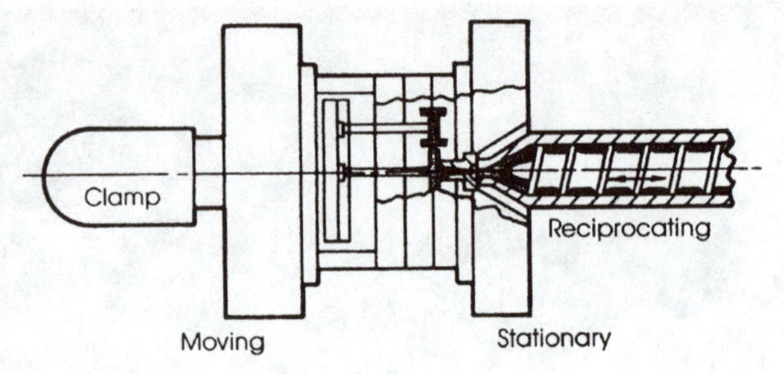

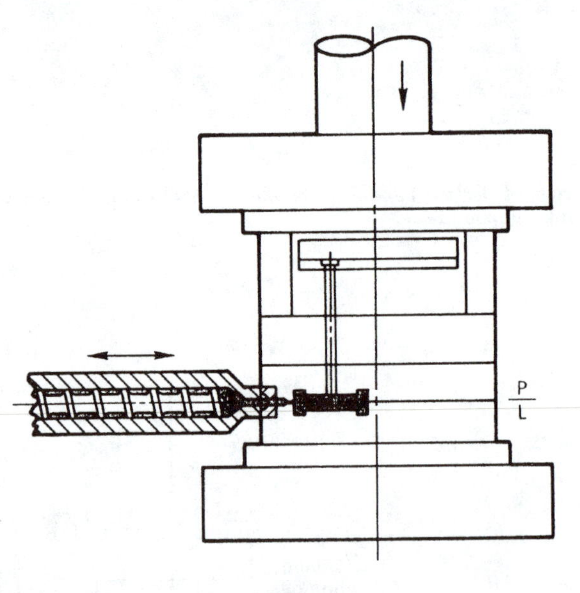

Figure 6.19 Screw injection. (*a*) Conventional injection mold, in-line. As can be seen in the drawing, the compound enters the mold through a sprue in the fixed half of the mold. (*b*) Parting line injection mold. The nozzle of the injection unit retracts upon the opening of the mold. The chief benefit from "parting line" injection molding is the ability to load metal inserts into the horizontal mold face without the danger of the inserts becoming dislodged during the mold closing.

next cycle. By this time, the screw has been rotating in the barrel, conveying granular material forward from the hopper at the back end of the barrel over the screw flights. As the material is conveyed forward, it is heated by the jacketed barrel and also by the mechanical shear of the screw rotation in the barrel and the constant forward motion of the material.

The material becomes a viscous paste-like fluid by the time it is delivered to the nozzle end. At this point there is not enough pressure for it to flow through the small nozzle opening, so it exerts a pressure against the front end of the screw, forcing the screw to move linearly back ("reciprocate") into the barrel against a piston in a hydraulic cylinder at the back end of the screw. As the plasticated material accumulates at the nozzle end of the receding screw, it finally reaches the required charge weight or volume for the mold, whereupon the screw "backward" motion is automatically detected by a limit switch or linear potentiometer, stopping further backward motion and further rotation.

The screw is now positioned in the barrel with the correct measured charge of material between the screw tip and the nozzle end of the barrel. This plasticizing step occurs automatically in the press cycle such that it is completed by the time the mold is closed and ready for another cycle.

When the injection molding machine senses that the mold is closed, and being held closed under full tonnage, the screw advances forward rapidly. During this stroke, it acts as a piston driving the plasticized charge of material through the nozzle and sprue and runners and gates to fill the mold cavities. Fill time is generally 1 to 3 s, depending on the charge mass, as compared to 10 to 30 s in a transfer molding filling cycle. Frictional heat from the high-velocity flow raises the molding compound temperature rapidly such that the material time-temperature experience assures a rapid cure in the cavity. Overall cycles of thermoset injection processes are often half those of comparable parts produced by the transfer molding process (Fig. 6.20).

Modern thermoset injection molding presses are usually fully automatic and produce parts at a high production rate. They are ideal for applications requiring a high volume of parts at a minimum cost. Machines cost about twice as much as comparable tonnage semi-automatic machines for transfer and compression molding. Mold costs are about the same as for transfer molding. No preforming or preheating is required, and the labor content of automatic injection molding is significantly lower than the labor content of semi-automatic transfer and compression molding.

To achieve maximum strength parts, a high concentration of glass or other reinforcing fibers may be mixed with the molding compound in this process. *Bulk molding compounds* (BMC) are often used, in which the formulation, generally polyester with glass reinforcing fibers up to 1 cm in length and uniformly distributed, is putty-like in consistency. Many electrical switchgear components are produced with BMC injection molding.

Figure 6.20 Injection machine for thermosetting plastics, available with screw or granular materials or plunger for bulk molding compounds. Machine also has an integral Rollaveyor tumble-blast deflasher inside the housing to deliver deflashed parts on a conveyor out through open end at left. (*Courtesy of Hull Corporation.*)

To minimize fiber breakage in BMC injection molding, the conventional barrel is often replaced with a special stuffing mechanism to load the BMC through an opening in the side of the nozzle end of the barrel. The conventional screw is replaced with a reciprocating plunger. As BMC is forced into the opening in the barrel during the barrel filling ("stuffing"), the plunger in the barrel is forced backward until the barrel is full of BMC and the plunger is at the back end of the barrel. Then, with each molding cycle, the plunger advances, pushing heated BMC through the nozzle into the mold, until the mold cavities are filled, and keeps pressure on the material until it has cured in the cavities. On the next cycle, the plunger advances further, delivering another charge to the mold. Thus, several charges are delivered until the barrel is nearly empty, at which time the stuffing mechanism refills the barrel through the side opening at the nozzle end of the barrel.

Screw injection and plunger injection molding may be used with vertical presses for insert molding. In such molding, the screw and barrel (or plunger, barrel, and stuffing mechanism) may be positioned vertically above the upward closing mold. After inserts are placed into the open bottom mold half, the press closes the mold, and the plastic is injected into the mold.

As an alternative, the screw and barrel (plunger, barrel, and stuffing mechanism) are horizontal, feeding into the horizontally positioned closed mold through a sprue bushing at the parting line of the closed mold (Fig. 6.21).

In recent years, injection molding with thermosets has been perfected using the "gas assist" more commonly used when injection molding thermoplastic parts that are fairly thick in cross section (say, 1 to 10

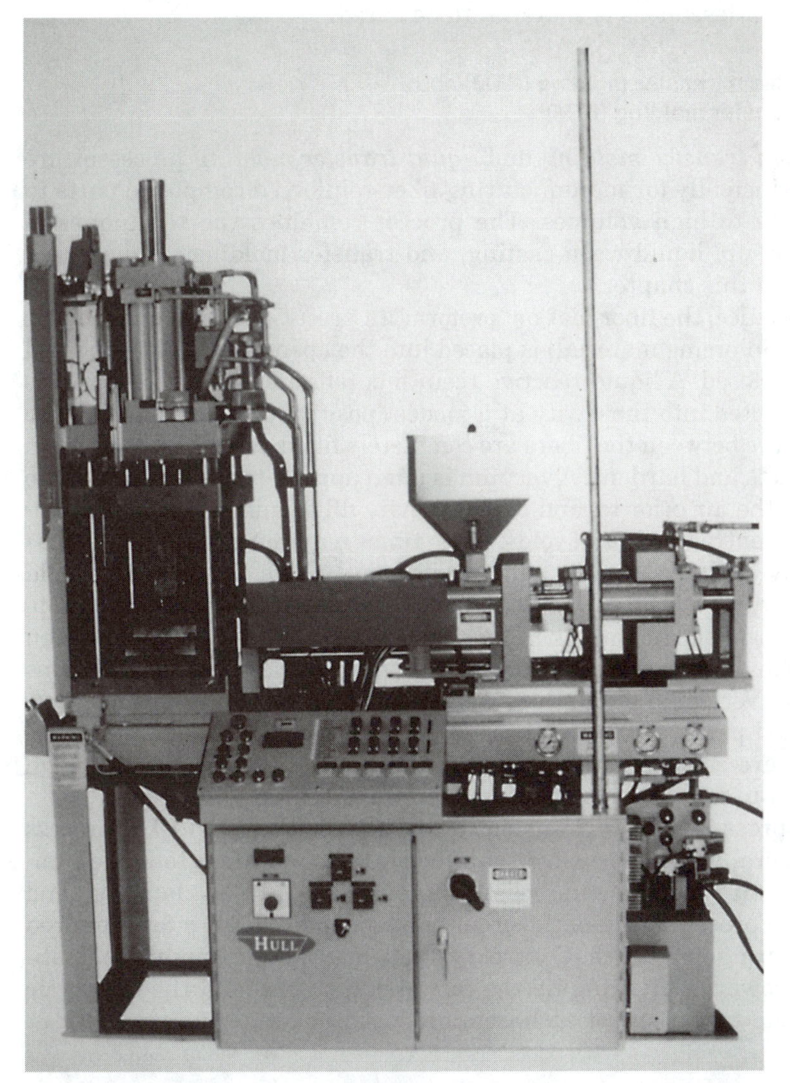

Figure 6.21 Parting line injection molding machine for semi-automatic or fully automatic molding of thermosetting plastic parts.

cm). In this process, an inert gas is introduced into the mold cavities shortly after the inflowing plastic has flowed past the gas inlet port of the cavity. The gas creates a bubble in the inflowing plastic, leaving a relatively uniform wall thickness of plastic for the length of the part, but with an open center. Such molded parts, properly designed and processed, are lighter in weight than the otherwise "solid" molded part, yet have the necessary strength and "rigidity" for the application. Material cost has been reduced and production cycles are shorter because of less mass of material to be cured.

6.2.6 Resin transfer molding (RTM) and liquid transfer molding (LTM)

The *resin transfer molding* and *liquid transfer molding* processes are used principally for manufacturing fiber-reinforced composite parts in moderate to high volumes. The process combines the techniques of hand lay-up, liquid resin casting, and transfer molding, as discussed earlier in this chapter.

In practice, the fiber mat or "preform" (a precut and preshaped insert of the reinforcing material) is placed into the open mold and the heated mold is closed. A liquid reactive resin mix (often epoxy or polyester) is then injected into the cavity at a modest positive pressure until all the interstices between the fibers are completely filled, after which the resin mix reacts and hardens. A vacuum is often applied to the mold cavity to remove the air prior to, and during, cavity fill to minimize the possibility of air entrapment and voids. Cycle times may run several hours, particularly for large aircraft and missile components. During cure, mold temperatures are often ramped up and down to achieve optimum properties of the plastic and, therefore, of the finished product. Fibers often used include glass, carbon, Kevlar (DuPont trademark), or combinations of these. Very high strength-to-weight ratios are achieved in such exotic molded parts, exceeding those of most metals, and complex configurations are achieved more easily than by machining and forming high strength-to-weight metals such as tantalum or aluminum.

The press and resin transfer equipment for the RTM/LTM process are generally less expensive than for an injection press to make comparable size components, partly because RTM/LTM clamping and injection pressures are of the order of magnitude of only a few hundred pounds per square inch. Control systems have become highly sophisticated, however, bringing up the cost of equipment, and the long cycle times add to the cost of each item produced.

6.2.7 Pultrusion

Pultrusion is a process somewhat similar to extrusion in concept, in that it produces continuous profiles by forcing the plasticized material through a precision die configured to the desired profile. The differences between the two processes, however, start with the fact that pultrusion uses thermoset plastics rather than thermoplastics, and the process *pulls* the resin and reinforcing fibers or woven web through the die instead of *pushing* it through. Because pultruded profiles are heavily reinforced, the pull is on the continuous reinforcing material (Fig. 6.22).

The steps in the process involve pulling the reinforcing matrix through a tank filled with liquid catalyzed resin, often polyester, where the fibers become totally saturated. The wet material is then passed through a stripper to squeeze out the excess liquid and sometimes to start shaping the material into a profile. The material is then pulled into a 20- to 30-in-long heated die, through which it is pulled at a rate controlled to ensure complete curing of the resin as it passes continuously from one end of the die to the other. The material may be partially brought up to temperature prior to entering the curing die by use of high-frequency heating of the saturated fibers or web between the stripper and the die, thereby speeding up the curing and enabling either faster pulling rates or shorter curing times, or both.

Perhaps the most crucial element in a pultrusion machine is the mechanism which grips the cured profile downstream of the curing die and continuously pulls at a carefully regulated rate to ensure total curing in the die, taking into account the degree of advanced cure of the catalyzed resin as it reaches the die.

Applications include electrical buss ducts, side rails for safety ladders, third-rail covers, walkways and structural supports for harsh chemical environments, and resilient items such as fishing rods, bike flag poles, and tent poles. Common profile sections generally fall between 4 in × 6 in and 8 in × 24 in.

6.2.8 Filament winding

For structural tubes up to several inches in diameter, and for tubes and tanks which may hold fluids or gases (especially corrosive fluids

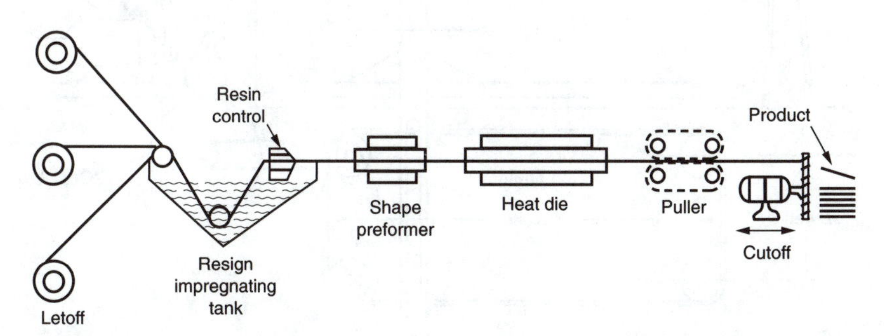

Figure 6.22 Schematic illustration of pultrusion process, using liquid polyester or other reactive thermosetting resin systems.

and gases) at elevated pressures, one or more continuous filaments of glass or other strong polymeric material are drawn through a liquid polyester or epoxy bath and then fed to a rotating mandrel or form, allowing the wetted filaments to closely wind onto the mandrel, often with layering in different directions or angles (Fig. 6.23).

When the predetermined number of turns has resulted in the desired thickness, heat is applied to the winding, creating a rigid tube or container, after which the mandrel or form is removed. The resulting tube or container has extremely high hoop strengths and is ideal for storing or transporting liquids or gases.

6.3 Techniques for Machining and Secondary Operations

6.3.1 Plastic part deflashing

Most thermoset molding operations result in some excess material, called *flash,* at the parting line and on inserts that protrude from the

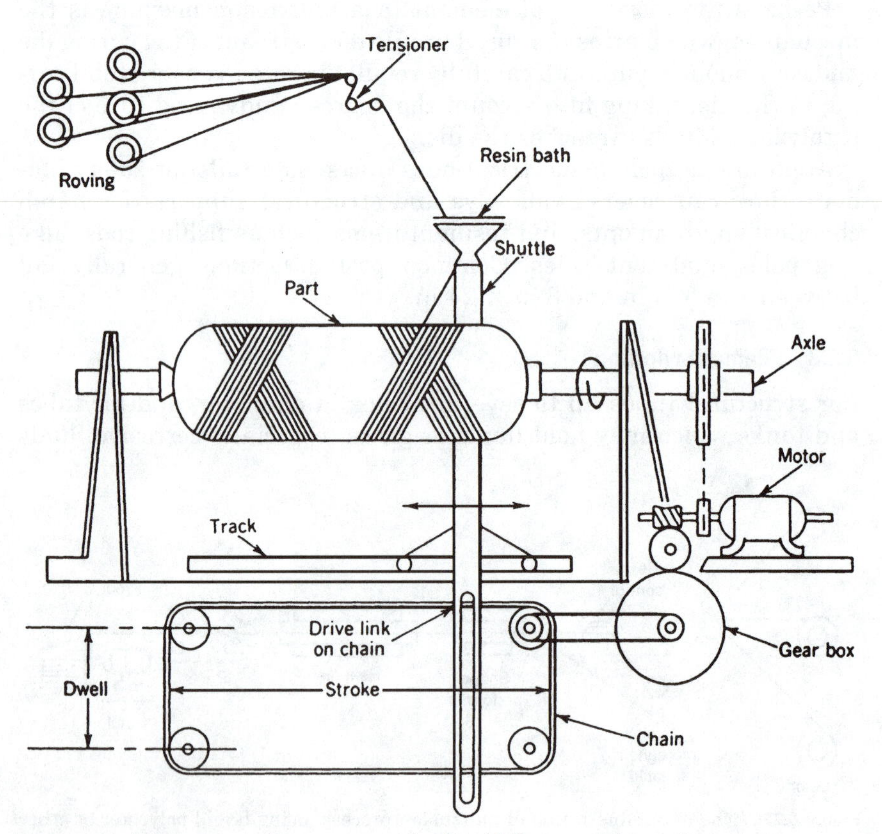

Figure 6.23 Filament winding process using liquid thermosetting resin systems.

molded part. It is generally necessary to remove this flash either for cosmetic reasons or, in the case of terminals or leads extending from an encapsulated electrical or electronic device, to ensure good electrical contact to the leads.

Robust parts may be randomly tumbled in a rotating container to remove the flash. Ten to 15 min of slow rotation of the container, with parts gently falling against one another, will usually suffice.

For more thorough flash removal, the tumbling action may be augmented by a blast of moderately abrasive material, or media, either an organic type (such as ground walnut shells or apricot pits) or a polymeric type (such as small pellets of nylon or polycarbonate) directed against the tumbling parts.

For more delicate parts, a deflashing system passing such components, on a conveyor that holds each part captive, past one or more directed blast nozzles generally proves practical. Such systems have been perfected for transfer molded electronic components, holding the lead frames captive, often temporarily masking the molded body as the devices pass through the blast area, and using as many as 20 individually positioned blast nozzles to ensure the complete removal of flash on a continuous basis. Such devices are magazine-fed and collected in magazines to maintain batch separation (Fig. 6.24).

Modern deflashers recycle the blast media and utilize dust collectors to minimize area pollution. The blasting chamber is effectively sealed

Figure 6.24 Fully automatic blast-type deflasher with magazine-to-magazine feed for thermoset molded parts. *(Courtesy Hull/Finmac Corporation.)*

with entry and exit ports designed to avoid the escape of dust and media.

Chemical deflashing using solvents to remove the flash from such components is also used. Additionally, water honing jet deflashing has been found successful for some types of devices.

Although very simple tumbling deflashers may be built by the processor, the sophisticated applications are best handled by commercial specialists who manufacture a wide variety of special and custom systems.

6.3.2 Lead trimming and forming

Most semiconductor devices encapsulated by transfer molding and a host of other high-volume small and fragile electronic components utilize lead frames as carriers during assembly and molding. These lead frames need to be trimmed off prior to testing, marking, and packing the devices.

Progressive trimming and forming presses and dies have been developed for this application, and are available both as manually fed and actuated systems and as fully automated magazine-to-tube carrier systems.

6.3.3 Adhesives in plastics assemblies

Bonding plastics to plastics has become a well-developed art through use of *adhesives.* For optimum strength in such bonding applications, the design of the joint where the two pieces are to be bonded must provide a physical means for the plastic parts to take the loads rather than the bonded surfaces themselves. Tongue-in-groove, or a molded boss fitting into a molded hole, or a stepped joint, with bonding on all the mating surfaces, ensures that so long as the two pieces remain together, shear or tension or torque loads will generally be transmitted from one piece to the other without stressing the bond itself.

6.3.4 Threaded joints

When two pieces of thermoset plastic must be joined together in service, but must also be suited to simple disassembly, threaded joints are possible. If metal screws are to be used, female threads can be molded in a plastic part to accept the metal screw. Metal female threaded inserts may also be molded into plastic parts to accept metal screws in a subsequent assembly operation.

It is important to locate holes for such threaded connections such that there is ample wall thickness to accommodate the concentrated stress in the immediate vicinity of the screw. Screwing a relatively thin plastic part to a more robust part could result in cracking or breaking of the thin material at the screw head if separating forces are severe. Designing a thicker section or using an ample washer or oversized screw head will yield more satisfactory results.

Screw joints require exceptional care with most thermosetting plastics. In such applications, slight overtightening of the screw may result in cracking the plastic or stripping the molded threads.

6.3.5 Machining

Machining of plastics is often necessary as, for example, in cutting and finishing Formica (melamine laminate) countertops or drilling structural pultruded shapes for attaching to supports, etc. While such operations are commonplace when working in metals or wood, special tools and practices are necessary when machining plastics.

Thermoset plastics tend to break or chip under concentrated loads such as those imposed by a saw tooth or a cutting edge of a drill bit. Thermosetting plastics are good thermal insulators, which means that the high energy imparted by cutting tools turns into frictional heat which, because it does not dissipate easily, quickly reaches the heat distortion point of some plastics and the burning point of others. Essentially, all plastics have heat distortion limits where the plastics lose rigidity and strength.

When conventional woodworking and metalworking cutting and grinding tools are used on plastics, therefore, the plastics often become gummy and sticky in the cutting area, binding the cutting tool and distorting the plastic. For machining operations on plastics, therefore, special cutting tools and special cutting techniques are necessary to achieve the desired results. Cutting tools that are suited to epoxies will not be appropriate for polyethylene, etc. When contemplating such machining operations, therefore, it is critical to contact the plastic supplier for specific recommendations as to cutting tool configurations, speeds, and procedures for cooling.

Mechanical finishing of thermosets. While the machining of thermoset plastics *must* consider the abrasive interaction with tools, it rarely involves the problems of melting from high-speed frictional heat. Although high-speed steel tools may be used, carbide and diamond tools will perform much better with longer tool lives. Higher cutting speeds improve machined finishes, but high-speed abrasion reduces tool life. Since the machining of thermosets produces cuttings in powder form, vacuum hoses and air jets should be used to remove the abrasive chips. To prevent grabbing, tools should have an O rake, which is similar to the rake of tools for machining brass. Adding a water-soluble coolant to the air jet will necessitate a secondary cleaning operation. The type of plastic will dictate the machining technique.

Drilling thermosets. Drills which are not made of high-speed steel or solid carbide should have carbide or diamond tips. Also, drills should have highly polished flutes and chrome-plated or nitrided surfaces. The drill design should have the conventional land, the spiral with regular or slower helix angle (15 to 30°), the positive angle rake (0 to +5°), the conventional point angle (90 to 118°), the end angle with conventional values (120 to 135°), and the lip clearance angle with conventional values (12 to 18°). Because of the abrasive material, drills should be slightly oversize, that is, 0.025 to 0.050 mm.

Taps for thermosets. *Solid carbide taps* and standard taps of high-speed steel with flash-chrome–plated or nitrided surfaces are necessary. Taps should be oversize by 0.05 to 0.075 mm and have two or three flutes. Water-soluble lubricants and coolants are preferred.

Machining operations will remove the luster from molded samples. Turning and machining tools should be high-speed steel, carbide, or diamond tipped. Polishing, buffing, waxing, or oiling will return the luster to the machined part, where required.

6.4 Postmolding Operations

6.4.1 Cooling fixtures

Thermoset molded parts are ejected from the mold while still warm. As cooling to room temperature takes place, parts may warp or deform due, in part, to internal stresses or to stresses created because of uneven cooling. Such changes in shape may be minimized by placing the parts in a restraining fixture which holds them to tolerance during the final stages of cooling.

Often the part design is at fault for such deformation. Thick sections in combination with thin sections, for example, experience faster cooling of the thinner sections, resulting in localized shrinkage which produces distortion, while the thick sections are still relatively soft. Designing parts with thin reinforcing ribs rather than thick sections often reduces or eliminates such distortion.

Another cause of internal stresses is the flow pattern as cavities are filled. In general, shrinkage is greater in a direction transverse to the flow line than in the direction of the flow. Part designers and mold designers need to agree on gate and vent locations to minimize such distortion due to flow direction. Other molding parameters, such as temperature and cavity fill pressures and rates, can be adjusted to minimize such distortion. But when all else fails, cooling fixtures may prove to be the final solution.

6.4.2 Postcure

Thermosetting plastics harden by cross-linking under heat and pressure. But at the time of ejection from cavities, especially in relatively short cycles for high production, parts have not fully cross-linked, and are to some extent "rubbery" at that stage of cure. They may be distorted from stresses created when ejector pins force parts from cavities or from parts piling on top of each other as they fall into a container or conveyor on being ejected. Longer cure times in the cavities may be necessary to lessen distortion from such conditions immediately following ejection.

Cross-linking, or polymerization, of thermosetting plastics is generally about 90% completed at the time of ejection, with the irreversible reaction continuing for minutes, possibly hours, and sometimes for days or months with certain formulations. Mechanical properties of such materials may be improved by a programmed cooling following ejection, a program generally providing staged cooling in ovens instead of conventional cooling in room temperature air. Molding compound formulators can recommend postcuring cycles where appropriate.

6.5 Process-Related Design Considerations

6.5.1 Flow

As thermosetting materials polymerize, changing from a viscous liquid to a shaped solid, the plastic parts often carry some history of their flow experiences preceding hardening. This history often adversely affects mechanical and electrical properties, dimensions, cosmetic appearance, and even the density of the finished parts.

Flow inside a mold cavity in injection, compression, or transfer molding ideally should be such that the cavity is completely filled while the material is still fluid. If such is the case, the curing under sustained cavity heat will proceed uniformly until the part is sufficiently rigid to withstand the rigors of ejection. But such is rarely the case.

Not infrequently, flow lines showing material hardened earlier adjacent to material hardened later will be visible. Material flowing around two sides of a boss or insert in the cavity may have partially hardened before coming together on the other side, showing "weld" marks or lines where the material actually failed to weld to the material coming from the other side. Material flowing through a thinner section may cool so rapidly that it hardens before it reaches the far side of this section.

A slow cavity fill, or a fill into a mold too hot, will cause the thermosetting plastic to cure before completely filling the cavity. Corners may not fill out, or detailed configurations in the cavity will not fill completely (Fig. 6.25).

If the material is reinforced, such as a glass fiber–filled phenolic or polyester, relatively thin passages in the cavity may cause the low-viscosity plastic to flow into the passage, but may "strain" out the glass fibers resulting in a resin-rich, but inadequately reinforced (and, therefore, weaker) area in the final part.

In thermosetting plastics molding, dynamic flow dichotomies occur because of cavity fill rate, obstructions in the cavity configuration, possible separation of resin from filler, and nonuniform cross-linking during flow and cavity fill, due to slightly differing time-temperature relationships causing precure or delayed cure—higher or lower viscosity—in various locations of the cavity during fill. Weld lines, incomplete fill, and resin-rich sections can result in reduced quality of parts and possibly lower yields of acceptable parts.

6.5.2 Cavity venting

Cavity venting—machined-in passages at one or more locations along the parting line or along ejector pins—is vital to obtaining complete cavity fill in the shortest possible time. Air in the cavity must have a

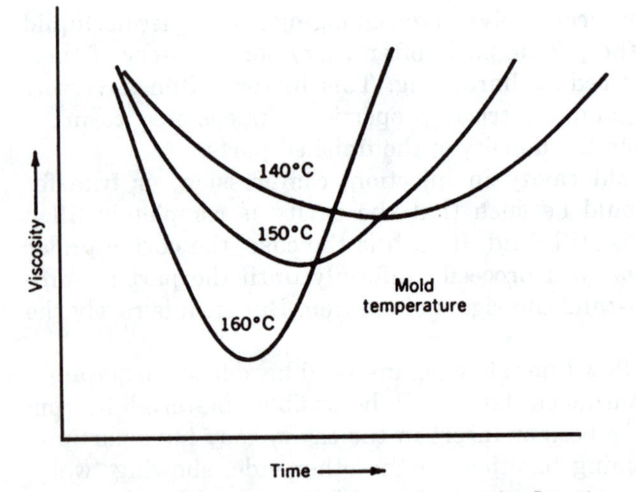

Figure 6.25 Higher mold temperatures shorten the flow time in the mold and accelerate cure of thermosetting resin systems.

clear and rapid escape route. If the air isn't eliminated either because it can't find a vent or because the vent is too restricted, a void will result, or possibly a burn mark on the finished part where the highly compressed unvented air overheats according to Boyle's law—the compression of gases creating very high temperatures, enabling diesel engines to ignite fuel, and enabling air in inadequately vented mold cavities to overheat and burn plastic.

Experienced part and mold designers need to understand the flow phenomena of various plastics during the molding process in order to minimize the occurrence of unwanted defects in the final product. But even with the "perfect" design of part, and the "perfect" design of mold, the processing parameters of temperatures (of fluid plastic and of mold), fill rate, fill pressures, and in-mold dwell are equally critical in achieving quality parts.

6.5.3 Final part dimensions

Physical dimensions of many processed parts must be held to fairly close tolerances to ensure proper assembly of parts into a complete structure, as, for example, molded fender panels bolted to steel chassis cars, plastic screw caps for glass jars, etc. In general, the final dimensions of the processed part will differ from the dimensions of the mold cavity or the pultrusion die. Such differences are somewhat predictable, but are usually unique to the specific material and to the specific process. The dimensions of a mold cavity for a phenolic part requiring close tolerances will often be different from dimensions of a cavity for an identical polyester part. Both the part designer and the mold or die designer must have a full understanding of the factors affecting final dimensions of the finished product, and often need to make compromises in tolerances of both part and cavity dimensions (or even in plastic material selection) in order to achieve satisfactory results with the finished product.

The following subsections will address the behavior characteristics of plastics that affect dimensional tolerances.

Mold shrinkage. Shrinkage of a plastic as it polymerizes is a fact of life, and is often specified as "parts per thousand." Such "mold shrinkage" is reasonably easy to compensate for by making the cavity proportionately larger in all dimensions as compared to the desired part dimensions.

But shrinkage in many materials is different when measured transverse to the material flow as when measured longitudinal to the flow. In reinforced or heavily filled materials, this difference is significant. Gate location and size, and multiple gates in some instances, must be considered for cavity and part design to minimize effects of such mold shrinkage.

Sink marks. Sink marks in a molded part often occur in relatively thick sections, usually reflecting progressive hardening of the molded part from the cavity wall to the inside area. The outside wall hardens while the mass of plastic in the thick section is still somewhat fluid. As this inside mass subsequently hardens (and shrinks, as most plastics will), the cured outer wall is distorted inwards, resulting in a "sink mark." The best way to avoid such deformation is to avoid thick sections wherever possible. Often one or more judiciously designed thin ribs in select locations will give a part adequate strength and thickness without the need for thick sections.

Nonuniform hardening of material. Nonuniform hardening of the material during residence in the mold generally produces internal stresses in a molded part which, after removal from the mold, may distort the part widely from the intended dimensions. Flat panels become concave, straight parts may curve, round holes may elongate, etc. Part and cavity design can generally accept some necessary compromises to accommodate such deformations, yet still yield a part meeting its functional requirements.

Deep molded parts. Deep molded parts may require design considerations to ensure minimum stresses during the ejection phase of the molding process. Imagine a straight-walled plastic drinking "glass" of internal and external diameters unchanging from top to bottom. As the part hardens in the cavity, it will tend to shrink around the force, or male part, of the cavity. When the mold opens, the part will stay with the force. To remove it from the force will require considerable pressure, either from ejector pins or air pressure coming out the end of the force against the bottom of the glass or from an "ejection ring" moving longitudinally from the inner end of the force. The pressure exerted by either of these ejection methods will be considerable until the open end of the "glass" finally slips off the end of the force.

To minimize such ejection stresses, forces for deep molded parts are designed with an appropriate "draft" or taper, up to 5° in some cases, such that very slight movement of the molded part with respect to the force will suddenly free the part from its strong grip on the force, and the remainder of the ejection stroke exerts almost no stress on the part. Such draft is advisable on all plastic parts, even those with depths of only 6 mm, to minimize ejection pressures and to prevent possible localized damage where the knock-out pins push against the not-yet fully hardened plastic.

Parting lines. Parting lines on molded parts require special consideration in part and mold design, especially where two molded parts must come together as, for example, on each half of a molded box with a hinged opening.

In compression molding or injection-compression of thermoset parts, the mold is fully closed only after the material has been placed into the cavity. More often than not, some material is forced out of the cavity onto the land area before the mold is fully closed, metal-to-metal. In effect, then, the mold closing is halted short of full close, perhaps by as much as 0.1 mm or more. Such overflow hardens, leaving "flash" on the molded part. Under these circumstances, the molded part dimension perpendicular to the mold parting surface will be 0.1 mm, or more, greater than intended. If the cavity and land area is so designed that such flash is perpendicular to the parting surfaces of the mold, the correct part dimension perpendicular to the mold parting surfaces may be achieved by removing such flash during a secondary operation of tumbling, blasting, or machining after the part has reached room temperature. More often, the tolerance of such compression molded parts is kept very wide in deference to the inherent characteristics of the process.

When an assembly of two molded parts is ultimately required, even if the materials and the molding processes are the same, it will be virtually impossible to achieve a perfect match where the parts come together. Slight variations in shrinkage or warpage will yield an easily noticeable or "feelable" mismatch. Intentionally designing mating surfaces with an overlap or a ridge enables ingenious camouflaging of the nonuniformity of mating areas in the final assembly of the two contacting parts.

Ejection of molded part. In designing molded parts, and the molds used to produce them, it is necessary to consciously determine how the part will be removed from the mold cavity or force and to maintain positive control of the part during mold opening, such that it is ejected as intended. This positive control is especially vital in automatic molding.

Assuming that the decision has been made that the molded part must be ejected from the moving half of the mold (as opposed to the fixed half), then it is necessary to make provisions such that the part will not remain in the fixed half of the mold during mold opening, but will invariably remain with the moving half.

One common way to accomplish such positive part control is to provide undercuts in the cavity or force of the moving half. These undercuts will enable plastic to flow into them and harden there before the mold is opened. Upon opening, the hardened plastic in the correctly designed and sized undercuts will hold the molded part in the moving mold half during the opening stroke. After mold opening, the ejector pins or mechanism in the moving half of the mold will then have to push hard enough to allow the molded part to distend sufficiently to be pushed off the undercuts and away from the moving half of the mold.

If undercuts are not practical, the fixed mold half may be provided with spring-loaded or mechanically actuated "hold-down pins," which are ejector pins that assist the molded part in leaving the fixed mold half and in following the moving mold half during "breakaway" and initial travel, perhaps 6 mm or more.

The part and the mold designer need to consider this aspect of the molding process, and to agree on how to ensure proper travel of the part to guarantee controlled ejection.

6.6 Mold Construction and Fabrication

This chapter is not intended to cover the broad field of mold and die design and construction, but the plastic part designer needs to be aware of the several aspects of mold design which can affect the cost of mold construction. With such knowledge, the part designer may be able to achieve the desired finished product with lower mold costs, faster deliveries, lower processing costs (shorter cycles and few, if any, postmolding costs), less stringent processing parameters, and minimum mold maintenance, as well as longer mold life.

6.6.1 Mold types

For the basic compression, transfer, and injection molding processes, a wide variety of mold types may be considered. Decisions as to the optimum type will often be based on the production volume anticipated and the allowable final part cost, including mold maintenance and amortization costs and hourly cost rates for molding machine and labor.

6.6.2 Molds for low production volumes

If production quantities are as low as a few hundred parts, *single-cavity molds* may be feasible, even recognizing the longer time period to produce parts one at a time and the increased labor and machine time to produce the required number of parts. Single-cavity molds for compression or transfer molding can be of the hand molding type, with no mechanical ejection mechanism, no heating provisions, but requiring a set of universal heating plates bolted into the press and between which the hand mold is placed and removed each cycle. Such hand molds may be of two-plate or three-plate construction, depending on the configuration of the parts to be molded. If the part is relatively small, hand molds can be multicavity, yet still be light enough that an operator can manually place them into and out of the press each cycle without physical strain. Although hand molds may be made of soft metals, such as aluminum or brass, and the cavities may produce acceptable parts, such metals quickly develop rough surfaces and rapidly lose their practicality after a few dozen cycles. It is best to use conventional metals as used for production molds for the respective process. Hand molds then may be used not only for prototyping, but also for modest production while waiting for full-size production tooling.

6.6.3 Production molds

Production molds are generally multicavity and have integral heating provisions and ejection systems. And when they follow single-cavity hand molds, cavity dimensions may be fine tuned, and vents and gates can be repositioned, based on experience with the single-cavity molds.

Family molds are multicavity molds that mold one or more sets of a group of parts that are required to make up a complete assembly of the finished product. A base, a cover, and a switch, for example, may be needed for a limit switch assembly. A family mold of, say, 36 cavities may be constructed which will yield parts for 12 assemblies each cycle.

Production molds may also be made using a standard mold base which will accept, say, 12 identically sized cavity inserts. If the component to be produced is a small box with a lid and attachable cover, in, say, 12 different sizes, each cavity insert could contain the cavities for one size box and cover. The complete mold would produce 12 boxes and covers each cycle. If sales of any one size require greater quantities than are required for one or more of the other sizes, a second cavity insert could be made to fit into the mold base, enabling twice as many of the faster-selling size each cycle. For family molds to be successful, all parts should have approximately the same wall thickness so that molding cycles may be optimum for all sizes.

6.6.4 Mold cavity removal

Some molded parts have configurations which require portions of the cavity to be removed in order for the part to be ejected. Solenoid coil bobbins, as an example, consist of a cylindrical body around which wire will be wound, with two large flat flanges at each end of the cylindrical body to keep the wire contained within the length of the body. Additionally, there is a hole through the length of the cylindrical body to accommodate the plunger of the solenoid assembly.

Such bobbins may be molded with the cylindrical body axis parallel to the parting surface of the mold. The cavity of the bottom half has the half-round shape to mold half of the body of the bobbin, as well as two thin slots at each end to mold half of each flange. The upper mold half is almost identical in order to produce the other half of the bobbin. To mold the hole through the body requires a cylindrical metal mandrel, with its outside diameter equal to the inside diameter of the coil body. This removable mandrel is manually placed in the matching half-round shape on the parting line. When the mold is closed, it effectively seals around the mandrel, leaving an open cavity into which the plastic will be injected in an injection or transfer molding machine. Following hardening of the plastic, the molded part with the mandrel is removed from the mold, the mandrel is pushed out of the bobbin with a simple manually actuated fixture, leaving the finished thin-walled bobbin intact. The mandrel is replaced in the mold for the next cycle.

For fully automatic cycles, molds for such a part may be constructed with cam-actuated or hydraulically actuated side cores which serve the same purpose as the previously described mandrel. In each cycle, after the mold is closed and prior to injection of material, the side core is automatically actuated into place. Following the cycle, the side core is retracted automatically prior to mold opening and part ejection (Fig. 6.26).

6.6.5 Molded-in inserts

Many plastic parts are produced with molded-in inserts such as a screw driver with a plastic handle. Molds for such items are designed

Figure 6.26 Open mold halves for injection molding coil bobbins and using vacuum venting to improve quality of thermosetting parts. The four angled cam pins enter the holes in the moving mold half to cause the mandrels to slide into place as the mold closes and to slide them out as the mold opens prior to bobbin ejection.

to accept and hold in the correct location the steel shaft with the flat blade or Phillips head end away from the cavity, and the other end, often knurled or with flats (to ensure that in use, the handle, when rotated, won't slip around the shaft) protruding into the handle cavity. After the mold is closed with the insert in place, plastic is injected or transferred into the cavity where it surrounds the shaft end and fills out the cavity to achieve its shape as a handle. After hardening of the plastic, the mold opens and the finished part is removed.

In many insert molding operations, fully automatic molding becomes possible when mechanisms are installed to put the insert into place before each cycle and to remove the insert with the molded part following each cycle.

When inserts are to be molded into plastic parts, close coordination is obviously needed between the part and mold designer and manufacturer to ensure the desired results.

6.7 Summary

Although this chapter has touched on many aspects of thermoset part design and processing, it has not covered a myriad of special considerations that may arise in the real world. The less-experienced part designer is advised to consult with others in the field as he or she develops part design, selects the material, and chooses the optimum process for the production requirements.

Chapter

7

Auxiliary Equipment

Section 1

Material Handling

William R. Lukaszyk

Universal Dynamics, Inc.
North Plainfield, New Jersey

7.1 Introduction

Correct selection of auxiliary equipment and utilization of efficient material handling systems can make the difference between a marginal plastic processing operation and a successful one. Introduction of a well thought out system, whether it is done all at once or in phases as capital budgets allow, results in savings and improvements in many areas including:

1. *Direct reduction in material cost.* When material is purchased in bulk and transported either by truck or rail, the cost per pound is significantly lower than the same material in bags or gaylords.
2. *Reduction in labor cost.* Automating a plant via bulk storage and pneumatic conveying systems eliminates costly manual transfer of gaylords or bags.
3. *Efficient space utilization.* Since almost all bulk storage is in outdoor silos, floor space can be devoted to production, assembly, and finished goods storage rather than raw material storage.
4. *Reduced spillage, contamination, and improved housekeeping.* Obviously, the less manual handling that occurs, the lesser the possibility that these types of problems will occur.

While the benefits of a properly designed system are many, so are the pitfalls that must be avoided by the system designer. An ill-conceived material flow pattern or poor choice of system elements can result in operational limitations or maintenance headaches which reduce the expected benefits. In this section, we will examine each element going into a system, such as silos, conveying systems, dryers, blenders, and then take an overview of how they fit together.

7.1.1 Factors to consider

First and foremost, when evaluating either an individual piece of equipment, such as a dryer or an entire system, we must focus on the nature of the specific material we will be working with. Plastic in the raw material state is either a pellet, granule, flake, or powder. The vast majority of resin is in pellet form, but even here wide variations in properties exist that must be taken into account. Some of these factors are

1. *Pellet size and shape.* This can affect material bulk density and flow properties, which in turn affect the sizing and geometry of components.

2. *Softness or hardness of material.* Extremely soft resins can present problems such as "massing up" or bridging in storage vessels or partial melting from frictional heat generated in conveying lines. Hard materials can cause accelerated wear on system elements such as conveying line elbows and flex, feed screws, grinder knives, and screens.

3. *Presence of fillers in material.* These can greatly change the properties of the base resin which they occur in. For example, the presence of carbon black in polyethylene can turn a resin, which is normally nonhygroscopic, into one which is very moisture sensitive. The addition of glass fillers in the range of 30 to 40% turns nylon from a nonabrasive to an extremely abrasive material, thus requiring very careful selection of equipment and system elements.

4. *Sensitivity to moisture.* Most commodity resins, such as polyethylene, polypropylene, and styrene, are classified as nonhygroscopic which means that they do not absorb water molecules into the pellets. Many engineering resins, such as acrylonitrile butadiene styrene (ABS), polycarbonate, and nylon, are, to varying degrees, hygroscopic and require drying to remove internal moisture before processing to assure that finished parts will have the desired properties.

5. *Dust considerations.* Some resins, such as unfilled polyethylene (PE) and polypropylene (PP), are relatively dust-free and do not require elaborate filtration systems. Other resins, such as some nylons and styrenes, can even have significant amounts of dust on virgin pellets. Dust and fines become an even greater consideration when regrind is introduced into the processing equation. In both selection of individual pieces of equipment or design of integrated systems, considerable attention must be given to dust containment and removal.

6. *Natural or precolored.* If virgin resin is delivered in its natural (uncolored) form, it typically requires equipment to add colorant prior to processing. If virgin material is delivered precolored, care has to be taken in equipment selection and system design to prevent cross-contamination.

7. *Granules, flakes, and powders.* Each has properties which require special consideration such as flow-inducing devices on storage and processing vessels or elaborate filtration systems for dust containment.

8. *Regrind considerations.* Most plastic processors generate some scrap material, which must be ground and reintroduced into the process. This is one of the most often overlooked aspects of plant-

systems layout and can cause no end of operational problems if not carefully thought through.

In addition to the preceding resin characteristics, other factors must be taken into account, including

1. *Throughput consideration.* The pound per hour consumption of individual machines and of a plant in total is a major determinant in the selection and sizing of almost all auxiliary equipment under discussion.
2. *Plant layout.* The actual physical layout and distances over which resin must be conveyed has a major impact on placement of individual equipment items and their integration into a system.
3. *Method of delivery.* Throughput and layout considerations generally dictate whether material will be delivered by bag, box, truck, or railcar. Each method has its own set of considerations, which we will examine.
4. *Economics.* Except where safety is concerned, economics is a primary consideration when evaluating auxiliary equipment and systems. Calculations of equipment capital and operating costs versus expected savings must be carefully evaluated to assure that they meet both operational needs and management's investment criteria.

7.2 Raw Material Delivery

Plastic is shipped in a number of different containers. The correct method of delivery for a particular operation is determined by material throughput and economics. The various methods of delivery are discussed in the following sections.

7.2.1 Bags

Each paper or plastic bag holds approximately 50 lb of resin, with 20 bags shrink-wrapped on a pallet. This method of delivery is appropriate for only the lowest throughput operation or where small quantities of a resin are required for test or prototype work. It is the most labor-intensive method in that each bag must be opened individually. Storage involves warehouse space, fork truck movement, and a great potential for spillage and contamination.

Some materials, such as fillers and additives used in compounding operations, are delivered *only* in bags. In cases such as this, an investment in automatic bag-handling equipment is justified. Such equipment slits bags open, empties the contents, contains dust, and compacts the empty bag.

7.2.2 Gaylord containers

These are essentially heavily constructed cardboard boxes, each holding approximately 1000 lb of resin. The box is usually provided with a heavy plastic liner, which offers some protection from contamination and moisture. The use of Gaylords eliminates much of the manual labor associated with bags, but they still take up warehouse space and require the use of fork trucks to move them about the plant. There are several items which greatly increase the efficiency of Gaylord containers:

1. *Gaylord tilters.* These devices hold a Gaylord on a platform which pivots automatically when the container is approximately half full. The tilting action is accomplished by either air cylinders or air bags and is governed by pressure regulators. By tilting the container, all material flows to the lowest point where it can be completely evacuated by a conveying system.
2. *Gaylord dumpers.* This piece of equipment lifts and tilts a Gaylord to dump its contents into a holding hopper, usually designed to hold at least 2000 to 3000 lb of material. From this surge hopper, material is typically conveyed pneumatically to various use points.

Both Gaylord tilters and dumpers, as illustrated in Fig. 7.1, can be supplied with a variety of accessories such as vibrators to induce material flow, low-level alarms, and dust containment systems.

One problem that is often overlooked when considering bags or Gaylords is the fact that after they are used, they must be disposed of. The solid waste disposal problem presents quite a housekeeping challenge and is very labor intensive. The problem grows in direct proportion to resin usage, but at some point, economics dictate the switch to bulk delivery of material, which eliminates the inefficiency associated with manual methods.

7.2.3 Bulk truck delivery

Specially designed trailers, each holding approximately 40,000 to 45,000 lb of resin, are common throughout the industry. Each trailer, in addition to being a transport vessel, is actually a delivery system. The tanks are rated as low-pressure vessels and are designed to work in conjunction with onboard blowers, piping, and valving. The driver makes flex-hose connections to a fill pipe on a storage silo or other receiving vessel, starts the blower, adjusts the valve, and unloads 40,000 to 45,000 lb of material in under 2 h. This method is very desirable because it requires only a

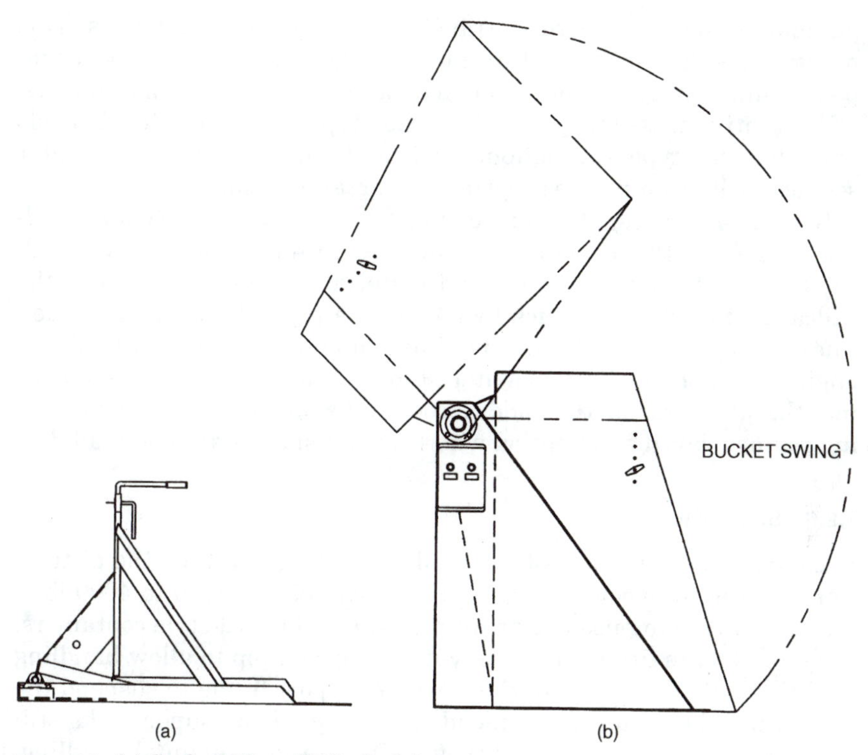

Figure 7.1 (*a*) Gaylord tilter; (*b*) Gaylord dumper.

simple silo to receive the material, with no unloading equipment required. A word of caution is necessary regarding this method of delivery for certain very soft resins such as low-density polyethylene. If the transfer rate and material velocities are too high, the frictional heat generated by the impact of pellets on the inside of the conveying line is sufficient to partially melt the pellets and coat the line with a film of material. This film later separates from the line and is conveyed downstream, where it collects in various choke points in the system such as silo takeoff compartments. This phenomenon is commonly referred to as *angel hair* or *streamers.* Some methods for minimizing the problem will be explored later in this section, but one of the most important precautions to take is to unload trucks at a relatively low speed to prevent the problem in the first place.

7.2.4 Bulk railcar

Whereas a truck holds 40,000 to 45,000 lb of resin, a railcar holds 180,000 to 220,000 lb. Delivery by this method generally yields the lowest price per pound. Unlike bulk trucks, railcars *do not* have any pumping system on board to transfer material to storage silos. This necessitates the use of a railcar unloading system, which can be a simple vacuum conveying unit or a much more elaborate combination negative-positive pressure system. The exact type of unit selected depends upon material type, throughput rates, and conveying distances. Later sections will examine these systems in greater detail.

In most instances, plants receiving resin via railcar have a rail siding on site, but this is not always the case. In some instances, the railcar is held at a local railyard and material is transferred out of the railcar and into a bulk truck by means of a portable railcar unloader. These units are typically mounted on trailers and powered by diesel engines. The truck shuttles material to the plant where it self-unloads into the silos. This method adds a cost to the operation, thereby reducing savings, but it is an option where on-site sidings are not available.

7.2.5 Super sacks

Recently, a novel method of material delivery has entered the picture. Large fabric transport–storage sacks, each holding 2000 to 4000 lb of material, are being used as an alternative to Gaylord-style containers. These sacks are fitted with heavy-duty loops on top to allow handling by fork truck. They require the use of a support frame to suspend the sack while the material is unloaded. Storage of the super sacks still requires warehouse space and they do require some manual handling, but their use should be considered.

7.3 Bulk Storage of Resin

Processing plants which receive resin in bulk shipments require a considerable storage capability to hold their raw material inventory. Silos have been widely adopted to fill this requirement. In this section, we will examine the two basic methods of construction (bolted and welded) and look at the accessories required for a workable tank installation.

7.3.1 Methods of construction

Welded silos are available in aluminum, stainless steel, and mild steel construction and are fabricated at the manufacturer's plant. The shell, deck, and hopper sections are all welded together into one structure and, on the mild steel tanks, both interior epoxy coating and exterior finish are applied before shipment. The epoxy coating provides a smooth corrosion- and abrasion-resistant lining for the tank interior surfaces which come in contact with the resin. Brackets are welded to the tank prior to finishing to allow attachment of various accessory items such as ladders, guardrails, fill lines, etc.

Welded silos are fabricated in both fully skirted and structural leg types. In the *fully skirted silo,* the tank side wall extends down to the foundation and provides the support for the structure. This enclosed area is extremely useful for locating items such as railcar unloader pumps, dehumidifiers and control panels, car unloading accessories, and flex hose.

An alternate method of construction is the *structural leg silo* where the side wall terminates at the hopper with the tank being supported by four I-beam legs welded to the tank.

Bolted steel silos are made up of a series of curved steel panels which are assembled in the field. Each panel is formed at the manufacturer's plant, at which time both the interior epoxy coating and exterior finish are applied. The tank panels (staves) are approximately 8 ft high, so bolted tanks are available in 8-ft increments. Depending on the tank diameter, silo height starts at 16 ft and progresses in 8-ft increments, that is, 24, 32, 40 ft, etc. The thickness of the staves varies from top to bottom, with the heavier staves being used on the bottom to support the weight of the tank and material above. Bolted tanks are almost always supplied in the fully skirted configuration. If a special application requires a leg type of installation, then the structure must be engineered and installed by a local contractor, as the tank manufacturer will make no design recommendations on this item.

Because of their structural integrity and ease of installation, welded tanks have become the preferred construction. The use of bolted tanks is limited to the following situations:

1. Installing additional tanks alongside existing ones.
2. Storage volume requirement is greater than can be provided by welded tanks.
3. Poor site access for welded silo delivery and erection.

An important consideration when specifying a bulk storage silo is the seismic zone in which it will be installed. These zones vary from zone 1, where there is no earthquake activity, to zone 4 in southern California where there is a serious threat. In zones 3 and 4, tank construction, foundation, and anchor bolt requirements are designed to resist the higher loads.

7.3.2 Methods of shipment

Welded steel silos are shipped by the manufacturer on specially designed trucks of their own. In almost all cases, these trucks are capable of unloading the tanks without a crane or lift truck. The only thing required is a clean, firm area as close to the silo foundation as possible. The truck rig is actually driven out from under the tank to accomplish unloading, so there must be a clear area at one end of the tank, slightly longer than the tank itself, to allow the truck to perform this maneuver. The accessories, such as ladders, guardrails, and crossovers, are shipped loose.

Bolted silos are shipped from the manufacturer's plant as a series of prefabricated steel panels. The curved shell panels are contained in special shipping racks, each of which holds 15 to 18 staves. The weight of one of these loaded racks can be well in excess of 3000 lb. The pie-shaped segments for the deck and hopper are shipped on skids with the accessories, along with the nuts, bolts, and hardware. The shipments are delivered either on truck or trailer and can be unloaded in several different ways. A small crane can be used to lift the racks from a flat-bed truck, while a heavy-duty fork lift can be used to unload the racks from either truck or boxcar.

Note: When lifting the racks with either a crane or fork lift, a spacer bar and top-lift connection should be used rather than sling straps below the load.

It is possible to disassemble the racks and unload the panels one at a time, but this does take considerably more time.

7.3.3 Methods of installation

Welded tanks require a clear site access for tank delivery and installation. The process is relatively simple, with the crane lifting the tank and setting it in place in the proper orientation. Ladders are usually installed prior to tank erection, while guardrails, crossovers, level switches, and other accessories are installed after erection in order to prevent their damage. A single welded steel silo can be erected, have its accessories attached, and be ready for filling in as little as 4 to 6 h.

Bolted steel silos require highly specialized installation techniques, which are best performed by experienced tank erectors.

In the most common method of fabricating, the staves are stacked in the center of the anchor bolt circle with the staves for the bottom ring being on top of the pile. The bottom ring is formed and the staves for the second ring are lifted one at a time until the second ring is formed. This procedure is repeated until the entire tank shell is complete. The roof brace, deck, and finally the hopper are set in place, after which time the accessories are attached. The seams between adjacent staves are sealed with neoprene gaskets. Proper alignment, gasketing, and tightening are necessary if a watertight tank is to result. In an average size bolted silo which is 12 ft in diameter × 40 ft high (see Fig. 7.2), there are nearly 5000 bolted connections, so the need for experienced installers is obvious.

The potential for a leak is inherent with a bolted tank simply because of its method of fabrication. It is recommended that bolted

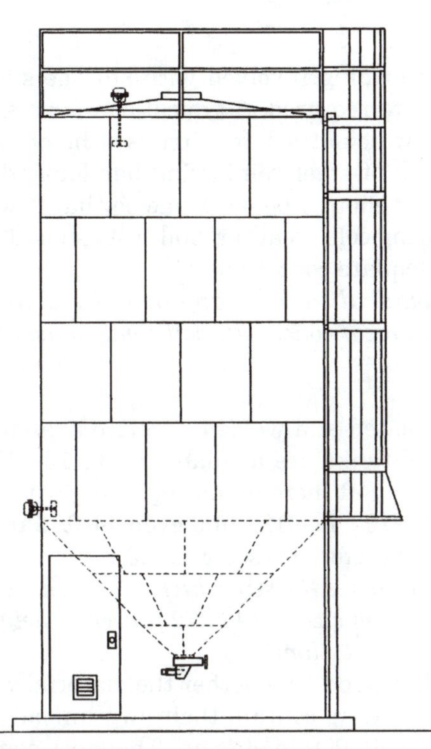

Figure 7.2 A 12-ft-diam bolted steel silo.

silos be leak-tested by running water hoses over seams and inspecting for signs of internal leaks. This should be done before the erection crew leaves. This procedure is really not necessary for welded tanks.

7.3.4 Tank finishing

Welded tanks may be finish painted at the factory with almost any desired color. All that is required in the way of field finishing is minor touch-up of scratches which may occur during erection.

Bolted tanks are supplied with either a prime painted exterior or a baked-on acrylic enamel (white) or light shades of green, blue, gray, or brown. If other colors are desired, then field painting after erection is required.

There are several choices available for tank interior coatings. Plants making food containers or medical items should consider Food and Drug Administration (FDA) approved coatings while standard coatings are appropriate for most other applications. Extremely abrasive materials may require special coatings to reduce wear.

7.3.5 Foundation

Because of the weight and wind loading involved with storage silos, heavy concrete slab foundations are required. A typical pad for a single silo contains the anchor bolts, which attach to stirrups or holddown rings on the tank, and two layers of steel reinforcing bar for added strength. The pad must extend far enough below grade so that it will not be affected by frost heave during cold weather. Soil tests should be conducted at the site to verify adequate conditions.

Note: Foundation designs should always be approved by a local architect or engineer familiar with local codes, site soil conditions, and seismic zone requirements.

Silo application guidelines. Both bolted and welded silos are available in a large number of sizes. Bolted tanks are available in 9-, 12-, 15-, and 18-ft diameters, with heights in 8-ft increments beginning with 24 ft. Welded silos are available in 9-, 10-, 12-, 13-, and even 14-ft diameters, with heights in roughly 6-ft increments (see Fig. 7.3).

Note: One limitation on welded tanks is the size which can be shipped over the road. Currently, the maximum size is 14 × 72 ft overall height (OAH). Bolted tanks have no such limitation.

The size a customer selects will depend on whether the material will be shipped by truck or in railcar. Trucks contain their own unloading equipment and hold approximately 40,000 lb of resin. The most popular size tanks for truck fill applications are

9 × 32* bolted or welded	65,000 lb
12 × 24 bolted or welded	75,000 lb
12 × 32 bolted or welded	108,000 lb
10 × 32 welded	77,000 lb

*That is, 9-ft diameter by 32-ft OAH.

Railcars hold up to 225,000 lb of resin and require separate unloading equipment to transfer resin to the silo. The more popular railcar silos are

12 × 56 welded	208,000 lb
12 × 60 welded	225,000 lb
15 × 40 bolted	215,000 lb
15 × 48 bolted	256,000 lb

All volumes shown take into account a 30° angle of repose common to most plastic material. Capacity is based on a 38-lb/ft^3 bulk density.

In any application, you must allow for a tank capacity equivalent to the volume of the carrier plus as large a safety margin as practical.

Another factor which must be taken into consideration when sizing tanks is local zoning ordinances, which may restrict the overall height.

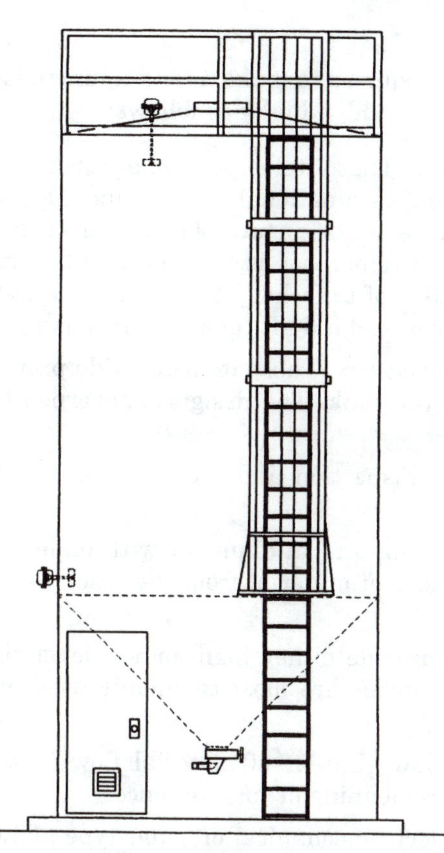

Figure 7.3 A 12-ft-diam welded steel silo.

If such a limitation exists, it may require going to a shorter tank of the next larger diameter.

At this point, the material flow characteristics must be taken into account. If the resin is a free-flowing pellet, a standard 45° hopper bottom is adequate. If the resin has poorer flow characteristics, such as polyvinyl chloride (PVC) dry blend, then a 60° hopper bottom should be specified. If the material is extremely difficult, a specially engineered hopper with air pads or other flow-inducing devices may be required.

Since almost all common resins have a bulk density of 40 lb/ft^3 or less, the standard tank design is based on 40-lb material with an adequate safety factor.

In simple clean pellet storage applications, a plain clam shell–type vent is adequate to allow air in and out of the tank. In very dusty or powder applications, either bag-type or continuous self-cleaning filters may be required.

7.3.6 Accessories

A number of accessory items are required to make a tank a functional piece of equipment. These are listed individually as follows:

1. *Ladder, cage, and rest platform.* These items provide access to the tank deck from ground level. Occupational Safety and Health Administration (OSHA) requirements dictate the design of this hardware. One requirement to remember is that a rest platform must be provided for every 30 ft of tank height. Therefore, a 32-ft tank will have one rest platform and a 64-ft tank will have two.
2. *Guardrails, toeboards, and crossovers.* These items provide protection to individuals on top of the tank. The design is governed by OSHA requirements.
3. *Manholes.* These items allow inspection and access to the upper area of the silo.
4. *Truck fill lines.* Aluminum tubing 4 in in diameter with male disconnect fittings to allow transfer of material from the truck to the silo.
5. *Level switches.* These items provide either high- or low-level signals. Rotating paddle-type switches are most commonly used but others are available.
6. *Slidegate shut-offs.* These allow shut-off of material flow in the event the take-off box requires cleaning or maintenance.
7. *Continuous level indicator.* Electromechanical or sonar-type allows remote readout of the silo level.

In addition to these, a wide variety of special accessories are used to meet specific material requirements.

Silo dehumidification systems. Under certain conditions, condensation forms on the interior surfaces of storage silos. This occurs mainly in the spring and fall when the daytime environment is warm and relatively humid followed by cool evenings. The moisture that has entered the tank during the day is condensed by the cooler temperatures and collects in the stored material. This problem can be particularly critical if moisture-sensitive materials or powders are stored in the silo. If hygroscopic resins are involved, the added moisture burden may exceed the ability of the drying systems to remove it. If powders are being stored, the moisture can cause them to behave more like wet sand than free-flowing material. In sensitive cases such as described, it is advisable to eliminate the buildup of condensation in silos by using dehumidifiers to blanket the upper volume, above material level, with a slight positive pressure of dry air. This arrangement is shown in Fig. 7.4*b*.

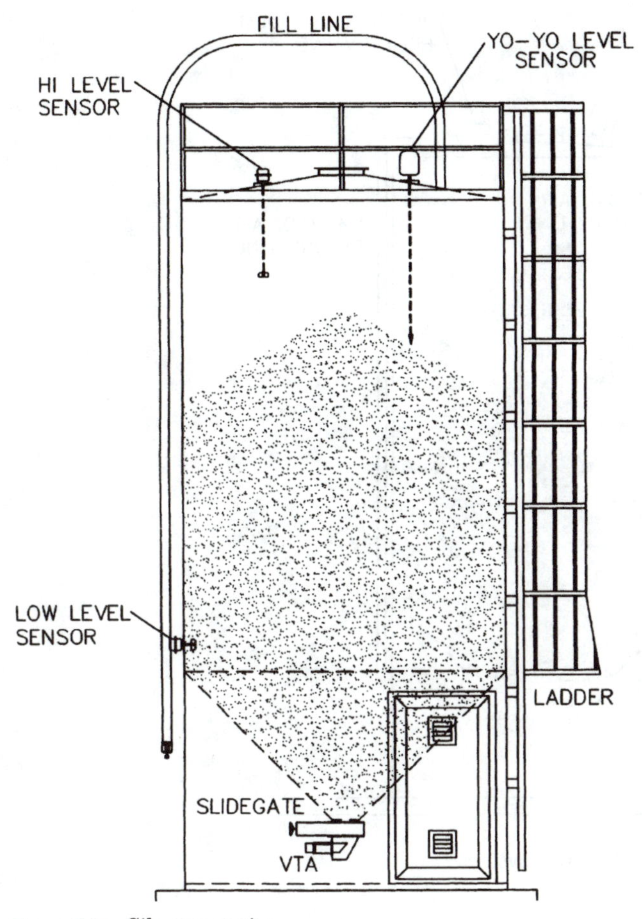

Figure 7.4*a* Silo accessories

When properly designed and installed, these systems eliminate condensation as a storage problem.

Flow-inducing devices. Plastic material stored in silos exhibits a wide variety of flow properties. Most pellets require no special devices. Notable exceptions to this are very soft materials such as ethylene vinyl acetate (EVA) and some rubber-modified polymers which require special devices.

Powders typically require, at a bare minimum, steep cone angles and, quite often, additional measures beyond that.

Some recycled plastic material, such as PET bottle flake, is shipped in bulk and requires special treatment. The following is a brief description of several methods widely used to promote reliable material discharge (see Fig 7.5.):

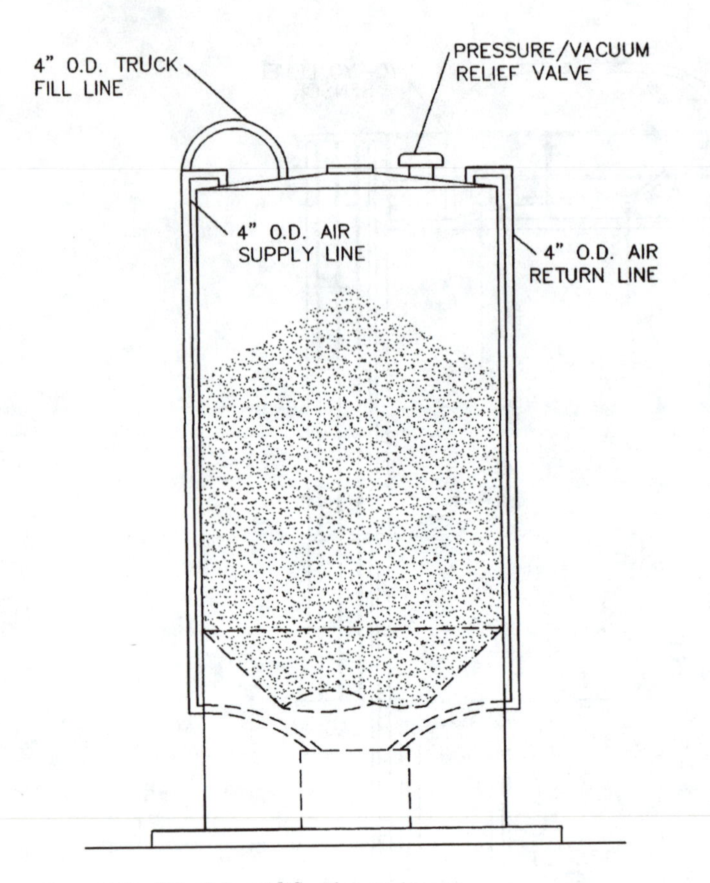

Figure 7.4*b* Silo dehumidification system.

1. *Cone angles.* Cone angles of 60 and even 70° are used to prevent bridging and promote mass flow patterns in silos.

2. *Mechanical agitators.* A wide variety of agitators are available. These are typically very low- (2- to 3-r/min) speed devices with shafts and breaker bars. Agitators typically cycle at the same time as silo unloading systems to break up bridging. It is generally not advisable to allow agitators to run continuously as they may cause fines in the material.

3. *Live bottom hoppers.* These devices replace the lower section of the silo discharge cone. Typically, they are approximately one-third of the silo diameter. They consist of an upper section bolted to the cone and a lower section secured to the upper by connecting links and flex collars. The lower section vibrates to break up bridging and aid product discharge.

4. *Pneumatic flow aids.* Air pads or nozzles are often installed in the cones of powder storage silos. When air is pumped through them,

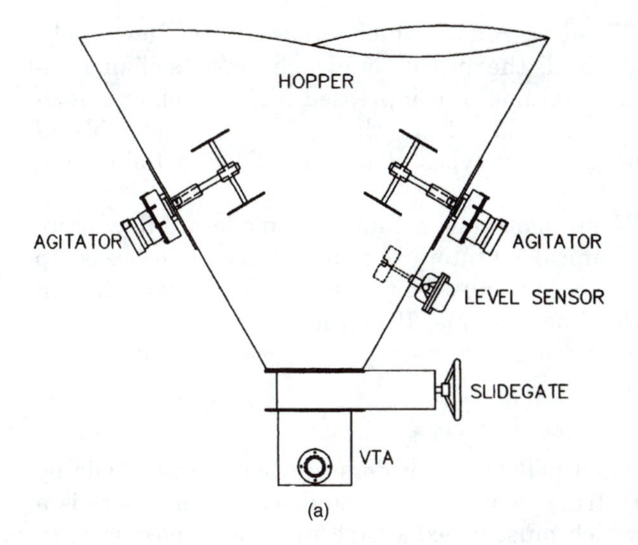

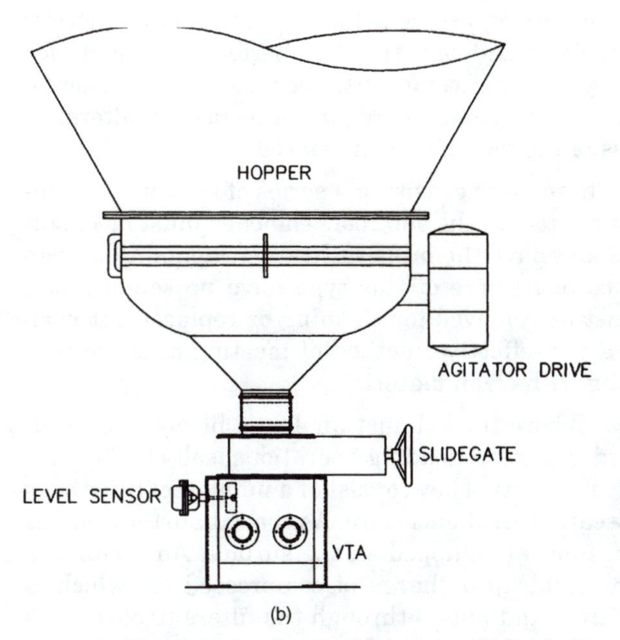

Figure 7.5 Flow-inducing devices. (*a*) Agitator; (*b*) live bottom.

the material is fluidized (aerated) which allows it to flow freely. Another pneumatic device is the air cannon which consists of an accumulator tank holding a volume of compressed air, a discharge solenoid, and nozzle. The air blast from these devices is capable of breaking up bridging in many types of non-free-flowing flakes and pellets.

5. *Bin vibrators.* Where materials exhibit only moderate flow problems, the use of mechanical vibrators is often sufficient to break up any bridging. Here again, care must be taken to cycle the vibrators in conjunction with unloading systems. If vibrators are allowed to run continuously, they may cause the material to pack up in discharge cones and worsen the situation.

Filtration systems. When pellets or powders are blown into a silo by either a bulk delivery truck or via a railcar unload system, there is a large volume of air which must be exhausted into the atmosphere. In the case of materials such as PE or PP pellets, the amount of fines contained in the exhaust is negligible while with other materials, especially powders, the dust loading is considerable and elaborate filtration is required. Several systems in widespread use are (see Fig. 7.6).

Weatherproof vent. This is commonly used with clean pellet materials that have little or no dust. There are large openings on the silo deck with clamshell covers and screens to keep out rain and insects. While from a functional standpoint they are perfectly adequate for clean materials, they may not conform to local code requirements. Many local and state governments require the use of filters on exhausts regardless of the material being stored.

Simple bag filters. These units consist of a series of bags in a weatherproof enclosure mounted on the silo roof. The bags must be of sufficient surface area and have the proper filtration capability to trap dust particles. Since bag filters of this type have no self-cleaning capability, they must be removed for cleaning or replacement periodically. They are a cost-effective method of meeting code requirements when working with clean materials.

Self-cleaning filters. Where the exhaust air has a heavy dust loading, such as in powder truck unloading operations, self-cleaning filters are an absolute necessity. They consist of a number of filter bags or cartridges in a weatherproof enclosure. As the exhaust air passes through the media, fines are trapped on the surface. An accumulator tank is used to build up a charge of compressed air which is released through valves and pulses through the filters to blow fines off of the surface and back into the tank.

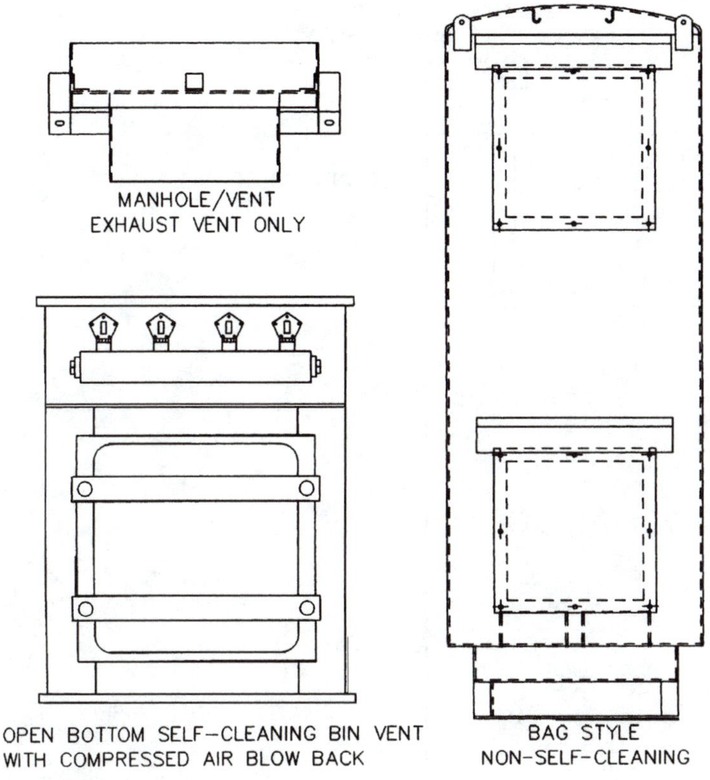

Figure 7.6 Filtration systems.

When considering the use of filters for dust-laden exhaust applications, some points to consider are

1. Micron size of dust particles
2. Volume of air (ft^3/min) to be filtered
3. Local, state, and national EPA requirements

Note: Whenever bin vent filters are used on a storage silo, the tank must also be provided with additional pressure-vacuum relief fittings. In the event the filter becomes clogged, they will allow the tank to vent, thereby eliminating the possibility of structural damage due to overpressurization.

7.3.7 Inventory measurement

Processors must have an accurate method for determining their raw material inventory. Several methods are in common use with plastic storage silos (see Fig. 7.7):

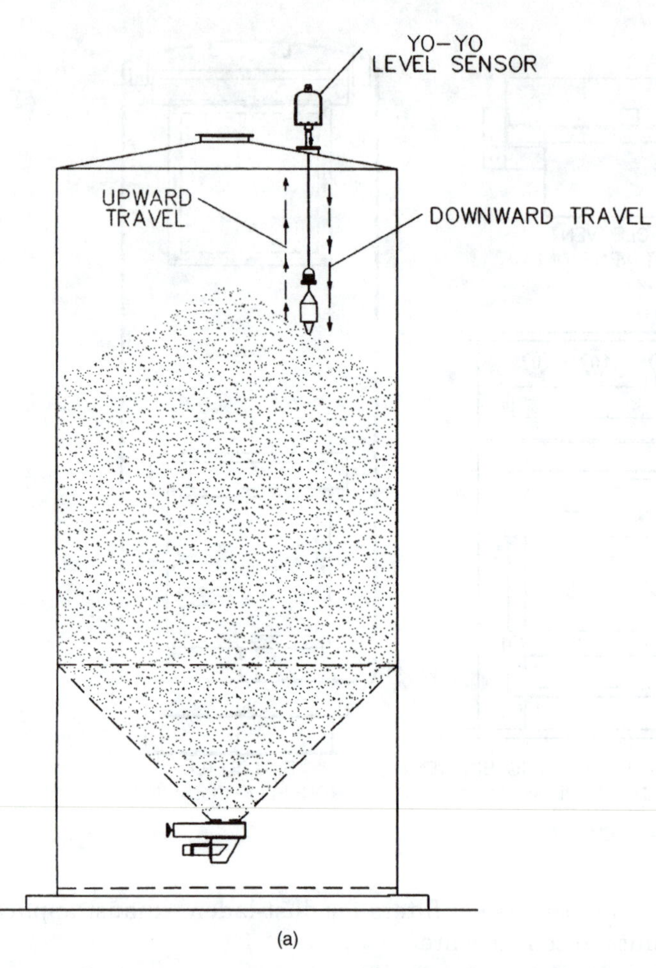

Figure 7.7 Inventory measurement. (*a*) Electromechanical; (*b*) "sonar" type.

Electromechanical devices. Commonly called *yo-yo devices,* they consist of a weight on a cable which, on command, is lowered into the silo. When the weight reaches the material, the cable goes slack and the weight is retracted. A readout on a remote panel indicates the level in the tank, which can then be translated to pounds of inventory via conversion tables.

Sonar systems. Instead of lowering a weight, these devices emit sound waves, which are reflected off the material surface and give a reading of the material level. Their accuracy is slightly better than the electromechanical system, but their primary advantage is that they give continuous readouts of material levels. Once installed and

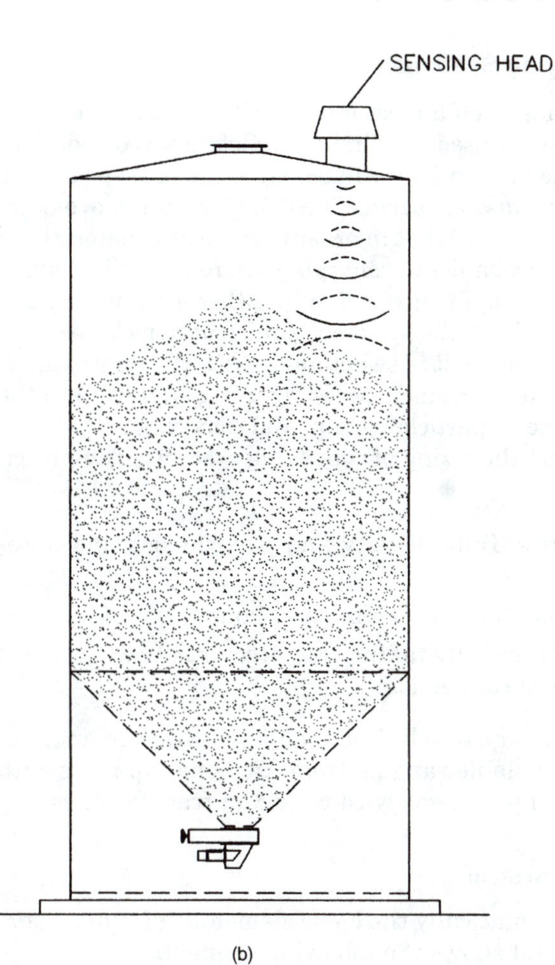

Figure 7.7 *(Continued)*

calibrated, they can be connected with displays which continuously display, both graphically and digitally, levels in multiple silos.

Both electromechanical and sonar systems are volumetric in that they measure height of material in a silo and require the use of conversion tables to determine actual weight. For a more direct method, another system is used.

Load cell systems. By mounting load cells either on the lower side wall section or structural leg supports of a silo, direct measurement of material weight can be taken. This method is accurate and the only choice where "certified" weight requirements exist.

7.4 Bulk Resin Conveying Systems

Pneumatic conveying lends itself almost ideally to the handling of plastic pellets, hence its universal use in the industry. Before we consider different types of pneumatic systems, we will examine the basic principles involved. The term *pneumatic conveying* itself implies air movement, and it must be clearly understood that in order to move any material, we first must move air. The mechanics of conveying are really quite simple. Air is caused to move through a transfer line by either a pump or blower. The air velocity at the inlet of the system is sufficient to pick up material and keep it in suspension as it is swept along with the airstream. If you simply think of a vacuum cleaner sweeping a rug, you can see the relationship of air movement, particle pickup, and conveying.

Numerous factors affect the sizing of conveying systems. The principal ones are

1. *Material characteristics.* Bulk density and particle size and abrasiveness.
2. *System capacity.* Throughput, in pounds per hour.
3. *Conveying distances.* Carefully taking into account elbows, vertical distances, and flex-hose connections.

These parameters can vary widely from one application to another, therefore, systems are available ranging from fractional hp units with 1 or $1^1/_4$-in lines to 100+ hp systems with 6- or 8-in transfer lines.

7.4.1 Vacuum conveying system

Figure 7.8 illustrates schematically the basic elements of a *simple vacuum conveying system,* and shows the following elements:

1. The vacuum power pack with motor/blower to provide air movement in the system.
2. The filter which protects both the pump and environment from contaminants.
3. The vacuum receiver which accumulates resin during the loading portion of the cycle. The bottom of the receiver is fitted with a "flapper" style of dump throat which closes to provide a vacuum seal during the load portion of the cycle and opens to allow material discharge during the dump portion of the cycle.
4. A level sensing device which signals to the control system.
5. A conveying line which routes the material from the source to the destination.
6. A pickup device, which is either fixed to the discharge of a piece of equipment, such as a silo or surge bin, or a suction lance to allow

pick up from Gaylord-style containers. The main consideration with pickup devices is that they must have the ability to vary the air inlet and, therefore, the air/material ratio in the system to optimize system performance.

A variation of a simple vacuum system is shown in Fig. 7.9. This is referred to as a *central vacuum system* in that it uses a single vacuum pump to draw material to multiple receivers on different machines by means of a common vacuum line with sequence valves which open when loading is required at a particular station.

Systems such as those described in Figs. 7.8 and 7.9 are the most widely used in the plastics industry and are available in a wide range of sizes.

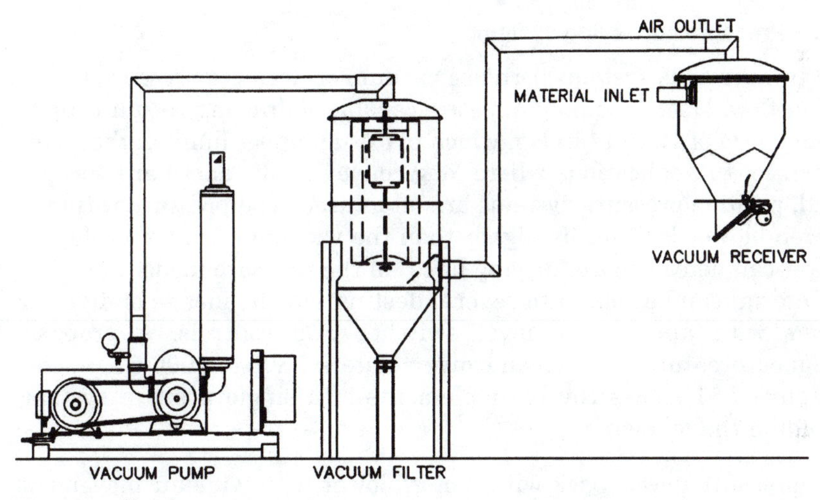

Figure 7.8 Simple vacuum system.

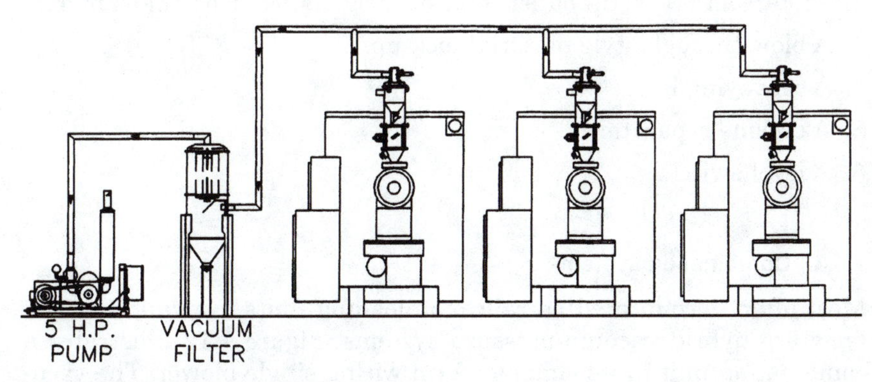

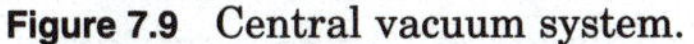

Figure 7.9 Central vacuum system.

Figure 7.8 illustrates what is often referred to as a *batch type of loading system* in that material is conveyed in discrete size batches, which fill the receiver and are then discharged into a hopper, with the load-dump cycle repeated until the level switch is satisfied. A variation of this system, which yields higher throughput, albeit at a higher cost, is a continuous vacuum system as shown in Fig. 7.10. With a continuous vacuum system, the flapper assembly is replaced by a rotary airlock. The airlock itself is a cast-iron housing with inlet and discharge and incorporates a cylindrical bore housing a rotor. The bore and rotor are precision machined with tolerances of 0.003 to 0.005 in between surfaces. In operation, the rotor is turning continuously. The tight tolerances provide a vacuum seal while the pockets fill with material from the upper section and discharge continuously from the lower.

7.4.2 Pressure conveying systems

The two previous systems both use vacuum or negative pressure to create air flow. Most vacuum pumps are capable of drawing vacuums up to a maximum of 12 to 14 in Hg, which places an upper limit on their performance. In applications where long conveying distances are encountered, positive pressure systems are often used. The pressure rating of a given blower is typically higher than the vacuum rating, therefore, a blower can deliver more "driving force" in the pressure mode.

Material can be routed to several destinations by means of diverter valves with appropriate level switches and controls. A properly designed pressure system can convey material 800 to 1000 ft.

Figure 7.11 shows the basic elements of a single positive system, including the following:

1. A pressure power pack with motor/blower to provide air movement in the system.
2. A rotary airlock on the outlet of the material source (silo or in-plant bin).
3. A blow-through style material pick up.
4. A conveying line.
5. A cyclone separator.
6. A level switch.

7.4.3 Combination systems

Most often encountered as railcar unloading units, *combination systems* are hybrid vacuum-pressure systems. Figure 7.12 illustrates a combination unit in its simplest form with a single blower. The vacuum side of the blower provides a continuous vacuum to draw material

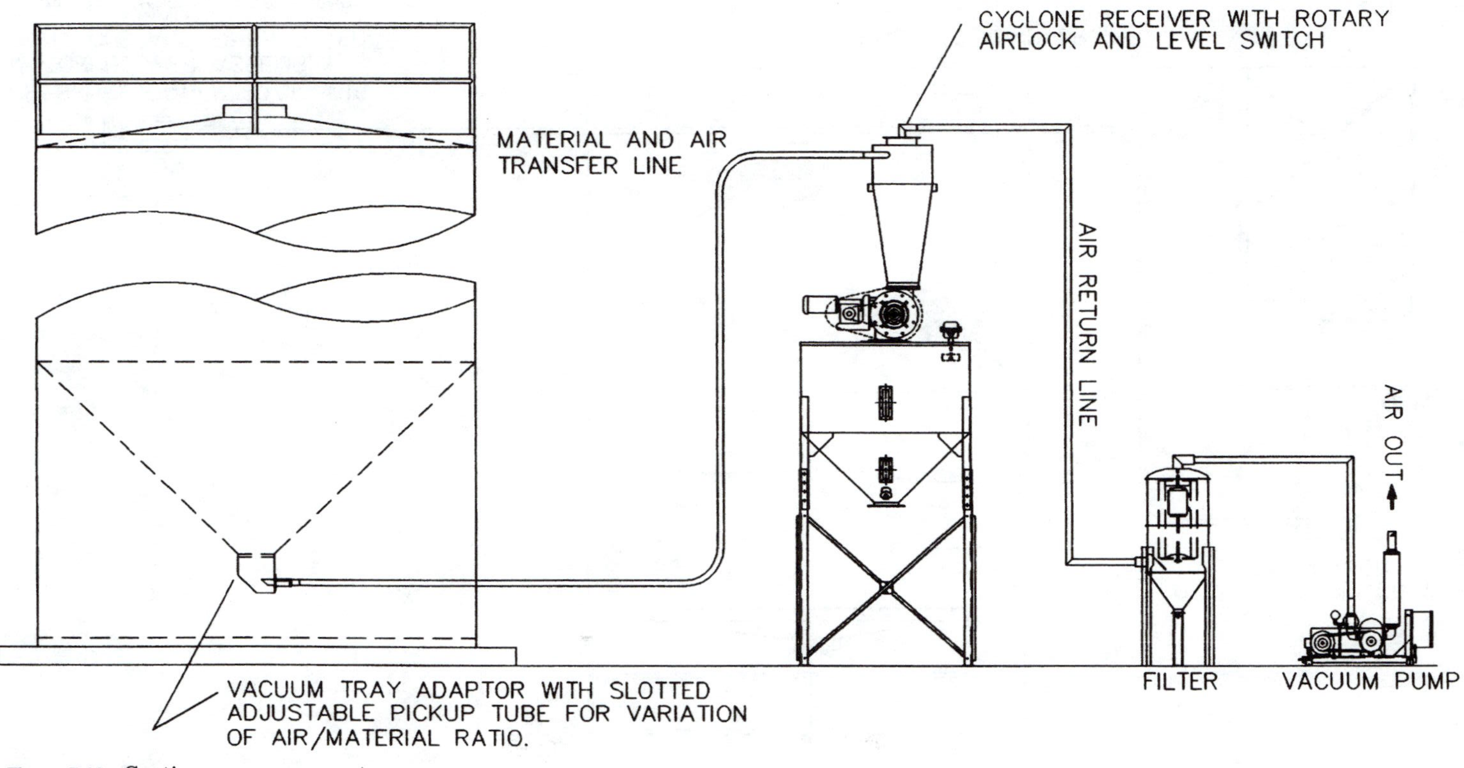

Figure 7.10 Continuous vacuum system.

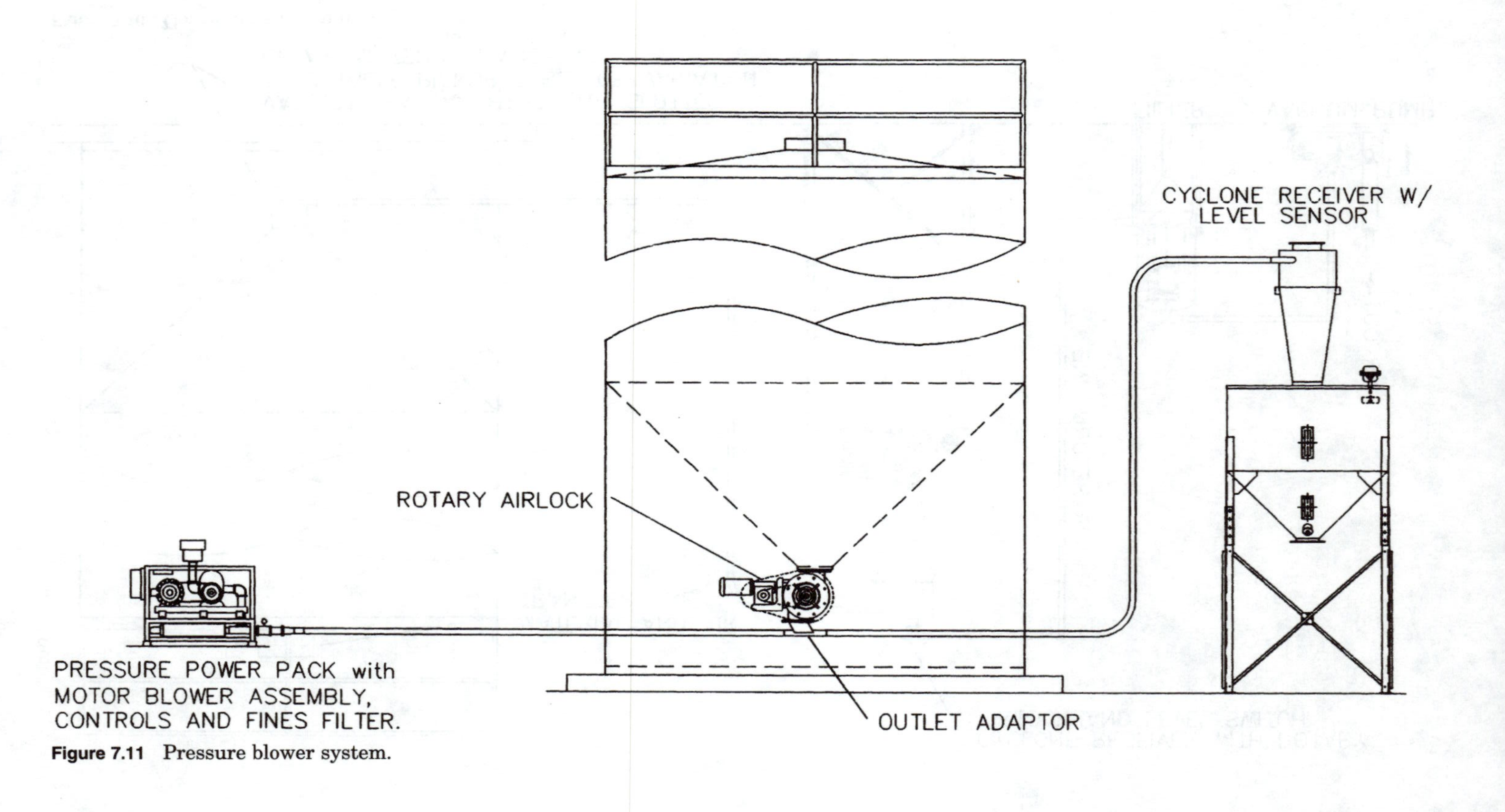

Figure 7.11 Pressure blower system.

from a railcar into a cyclone separator. The rotary airlock seals the vacuum from the pressure side of the system and also dumps material into the positive pressure airstream. Combination units are typically high-volume systems and in many cases are provided with two power packs, one for the vacuum side and one for the pressure side, as shown in Fig. 7.13.

7.4.4 Points to consider

All of the systems described utilize filters at some point in their pneumatic circuit. The type of filter selected depends on the nature of the material being conveyed and the dust loading of the air being filtered. Inadequate filtration is one of the most common reasons for system underperformance. Abrasive material may require special materials in the construction of receivers, cyclones, airlocks, and conveying lines. When sizing systems, future growth must be taken into account.

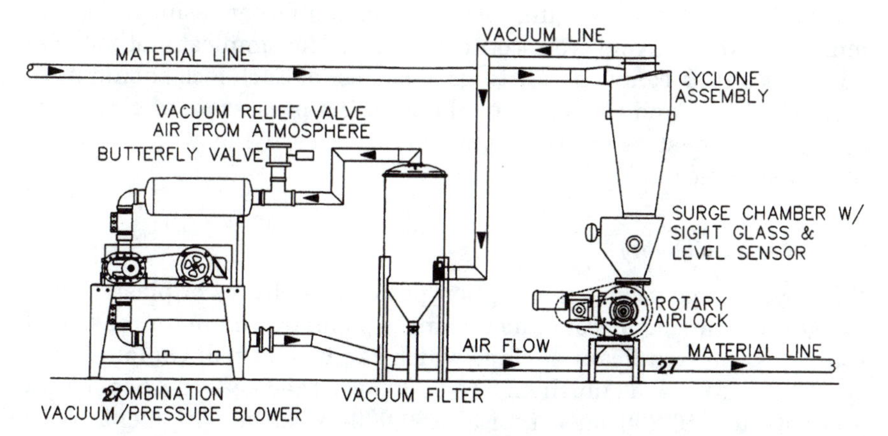

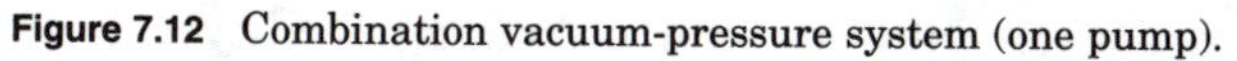

Figure 7.12 Combination vacuum-pressure system (one pump).

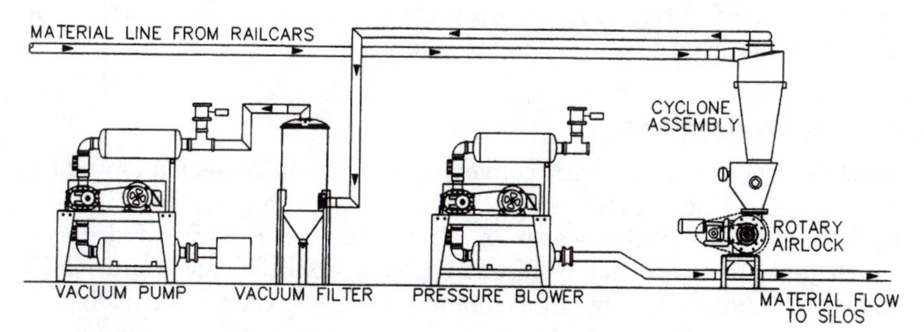

Figure 7.13 Combination vacuum-pressure system (two pump).

7.5 Bulk Delivery Systems

7.5.1 Truck delivery

Bulk shipment of resin is accomplished either by special bulk trucks or railcars dedicated to that service. As explained earlier, all that is necessary to receive material by bulk truck is a storage vessel large enough to hold the quantity delivered and a fill line for the truck to connect its delivery hose. Since trucks are equipped with their own blower systems, no additional conveying equipment is provided at the plant level (Fig 7.14). Several considerations are, however, worth noting.

It is important that silos be equipped with high-level switches and an alarm—either audible or visual—to alert personnel that the silo is full and avoid a situation where material backs up and clogs the fill line. It is, of course, always advisable to assure that the empty volume in a silo is sufficient to hold a truck load of resin prior to calling for delivery.

If multiple silos are on site, each holding a different resin, great care must be taken to avoid cross-contamination. The simplest method is to put locks on the connection fittings of each silo. Each lock should have a different key, and the keys should be in the possession of plant personnel who must select the right one after verifying which resin is being delivered.

7.5.2 Railcar delivery

With ever-rising resin consumption patterns, delivery of plastic resin by bulk railcar has become increasingly popular. The main reason, of course, is the significant savings which can be enjoyed. On low-density PE, this can be as much as 5 to 6¢/lb over the Gaylord price, which amounts to $10,800 on a typical 180,000-lb railcar shipment. With such savings, the cost of the storage and unloading facility can be recovered quite rapidly. Of almost equal importance is the fact that during a period of tight resin supply, the customer can be assured of a large raw material inventory in storage.

7.5.3 Types of systems

Two basic types of pneumatic conveying systems have been adopted for railcar unloading service: the vacuum system and the combination negative-positive pressure system. Each possesses certain advantages which make it useful for different types of applications. Figure 7.15 shows schematically a straight vacuum type of loader that utilizes a silo-mounted vacuum chamber. In this type system, the vacuum pump

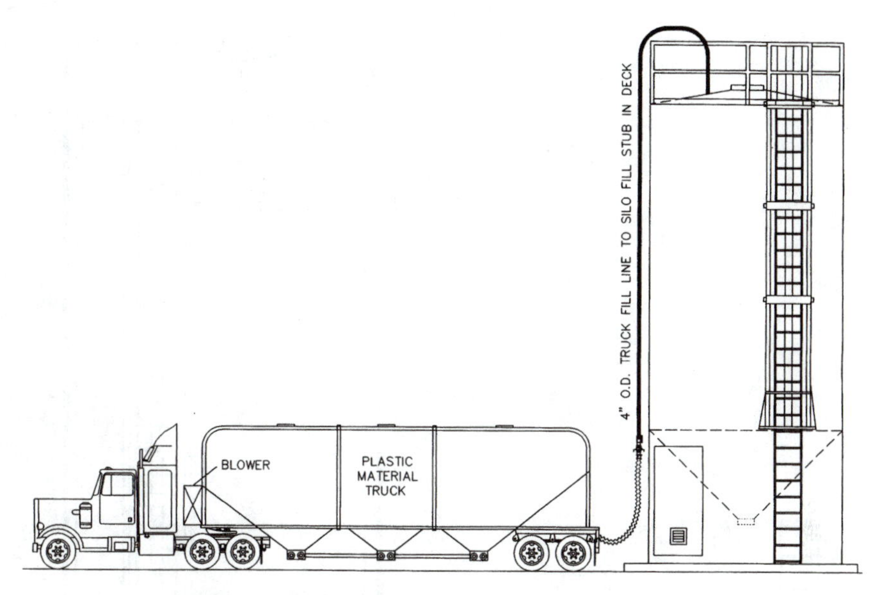

Figure 7.14 Silo fill line by truck (blower).

is generally located in the skirt of the silo. Air-return lines extend up to the vacuum hopper. On multiple silo installations, each vacuum hopper has a vacuum line extending to a central area near the pump. A manual flex hose switching station is utilized to selectively draw the vacuum on any silo loader. Fill lines extend from each vacuum chamber to a central area. The silo fill lines are always equipped with male disconnect fittings, while the air-return lines are fitted with female disconnects. By utilizing stainless-steel flex-hose connections, unloading manifolds, and accessories, the hookup to the railcar discharge can be accomplished. When the pump is started, the unit functions identically to smaller vacuum loaders in that it runs for a period of time until the chamber is full. It then allows the material to dump into the silo. This process is repeated until either the silo is full or the railcar compartment is empty. The vacuum hopper and extension are of welded aluminum construction. The hopper has approximately a 200-lb capacity of 38-lb/ft^3 material. The extension is equipped with a clean out door which allows inspection and maintenance of the flapper assembly.

Note: The chamber shown is used only for clean pellet applications. Where powder is to be handled, special chamber and filter designs are required.

The chamber is fitted with a pellet screen to prevent material being sucked back to the vacuum pump. In addition, the pellet screen

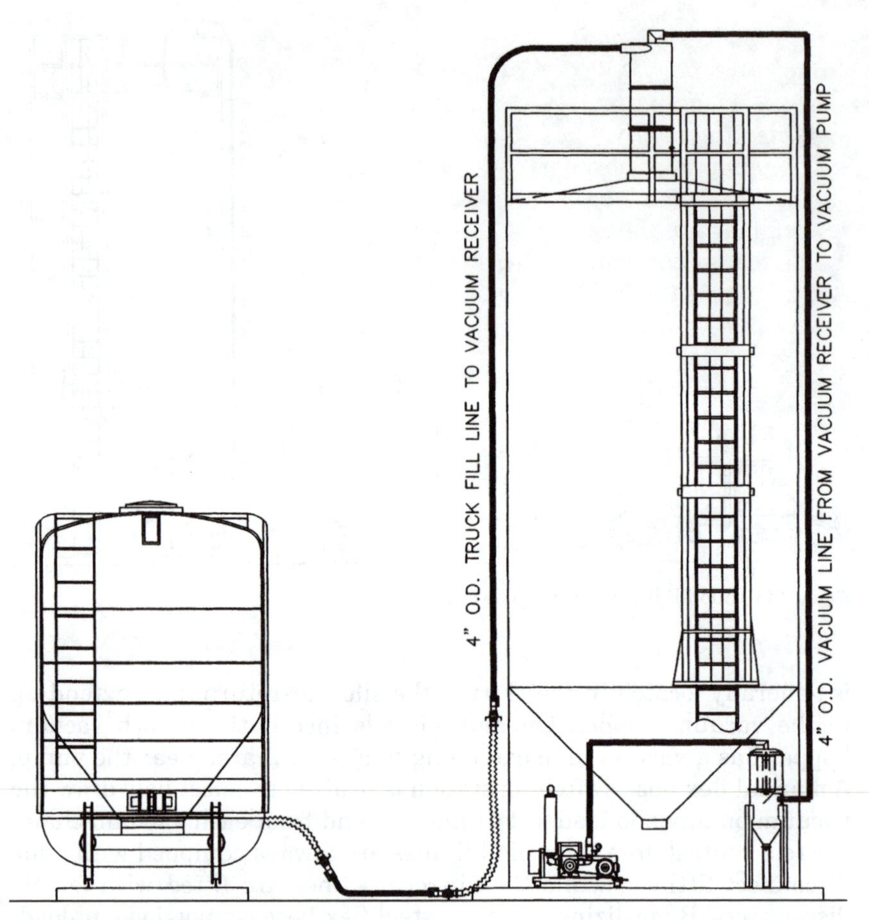

Figure 7.15 Silo fill line by railcar (vacuum).

is surrounded by a shroud which eliminates direct impingement of material on the screen. The entire hopper-extension assembly is bolted in place on the center dome of the silo.

A high-level switch must be utilized with this unloader. Most often, the rotating paddle type is used. The switches themselves are mounted in the silo deck with the paddles mounted on extension shafts. Extensions must be used because of the angle of repose the material adopts when loaded into the silo. When material contacts the paddle, the unit is shut off. The pumping system itself utilizes a positive displacement motor/blower assembly. The inlet of the pump is fitted with a two-stage secondary filter. The purpose of this filter is to prevent fines which may have passed the pellet screen from entering the pump.

Note: As a matter of routine maintenance, this filter should be inspected frequently.

The pump inlet also has a manual vacuum relief which, in the event of a material line blockage, will allow air to enter the system and prevent damage to the system.

The loader control panel, which may be mounted on the unit or at a remote point, contains motor starters, timers, high- and low-level switch lights and, on multiple silo systems, a selector switch to energize the proper high-level switch for automatic operation. When the operator wishes to change from loading one silo to another, the flex-hose connections to the material and vacuum line of the new tank must be changed and the tank selector switch on the control panel must be positioned to the new silo. The system is then ready to run.

A variation of the vacuum system just described utilizes a rotary airlock on the bottom of the hopper in place of the flapper assembly. The airlock seals the vacuum in the hopper while, at the same time, allows the material to be discharged into the silo. The primary advantage of this system is the fact that it conveys continuously rather than in intermittent batches. Its drawbacks are higher cost than the flapper discharge, airlocks tend to be a high-maintenance item located on top of the silos, and the tendency of some materials to break up or smear when passed through the rotor.

Figure 7.12 shows a typical combination negative-positive pressure vacuum system. Vacuum from the pump draws material into either a cyclone separator or filter receiver. The pellets are passed through a rotary airlock and enter the blower discharge airstream, which is at positive pressure. The air/material mixture is transferred to the silos via stainless-steel flex-hose and fill lines. Combination units tend to be high-capacity systems used primarily with multiple silo systems. One advantage is that they require no equipment whatsoever on top of the silo, only a simple fill line. All maintenance is performed at ground level.

7.5.4 Transfer rates

As with any type of pneumatic conveying system, the transfer rate depends upon material characteristics and distance. Railcar unloading has the added complication of always being a high-lift situation, anywhere from 40 to 70 ft vertical. A 25-hp unit with a 4-in conveying line moving polyethylene pellets approximately 100 ft horizontal and 60 ft vertical will maintain an approximate throughput of 10,000 lb/h.

Note: Final hookups to railcars and silo fill lines are always accomplished with flex hose. These connections should be kept as short and straight as possible, otherwise the conveying rate will be severely impaired.

Precautions must be taken to avoid cross-contamination. Fill lines can be equipped with locks, as described previously in the truck delivery section, or interlocking proof switches can be fitted to the fill lines so that the hose connection must agree with a selector switch setting on the control panel in order for the transfer to proceed.

7.5.5 Railcar connections and accessories

Figure 7.16 shows various details and layouts of commonly used railcar discharge systems. The cars themselves are divided into various compartments, each of which has its own outlet. These outlets can be adjusted for the air/material ratio and perform the same function as a vacuum tray adaptor on a dryer hopper. Three basic accessories are required for unloading:

1. Railcar adaptor
2. Air inlet filter
3. Hatch filter

The railcar adaptor slips over the end of the discharge tube and is held in place by set screws. The railcar discharge can be adjusted for the air/material ratio. Adjustments must be made to establish the proper ratio for optimal material conveying. The introduction of air on the opposite side of the car provides for a much smoother material pickup.

The adaptor is equipped with a female quick disconnect fitting which is the piece that makes the connection to the flex hose.

The air inlet filter fits on the far side of the discharge and prevents contaminants from entering the conveying airstream from that point.

The hatch on the compartment must be empty and open to allow air to take the place of the material being withdrawn. Contamination is prevented by placing the hatch filter over the opening.

Most railcar unloading installations are equipped with a manifold arrangement. This is a transfer tube that runs parallel to the siding and equipped with “Y” laterals and disconnect fittings. The disconnects are all capped, except for the one used for drawing material from the railcar. The use of a manifold in a system allows for variations

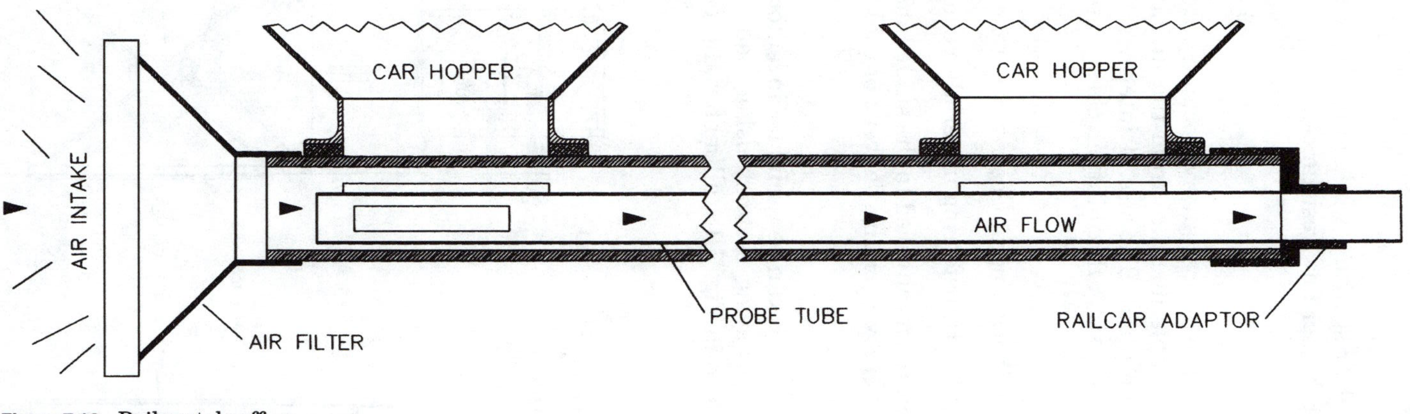

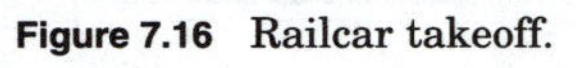
Figure 7.16 Railcar takeoff.

which occur in railcar spotting, while minimizing the use of flex hose which has a much higher flow resistance.

7.5.6 Surge bins

Resin storage silos are almost always located outdoors and, in many cases, at a considerable distance from the ultimate point of use of the material in the plant. The use of intermediate surge bins (see Fig. 7.17) enhances system flexibility in these cases by providing the following:

1. An internal distribution point closer to the resin use point.
2. A simple method of transferring material to drums or boxes.
3. In cold weather, resin conveyed to the surge bins will come to room temperature, thereby avoiding delivery of cold resin to processing machines.

In addition, surge bins can be useful for inventory control purposes. If equipped with conveying systems using load cell–weigh chamber technology, as shown in Fig. 7.18, all material brought into the plant

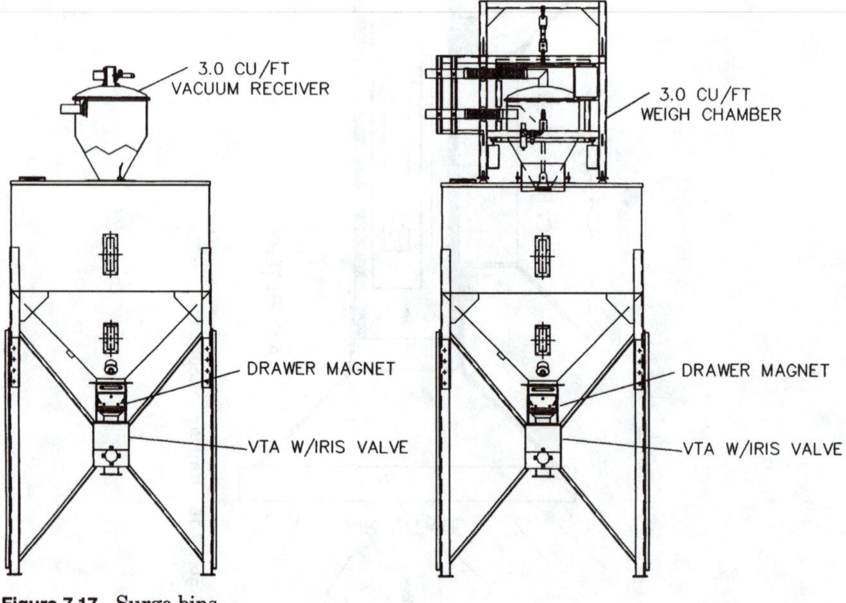

Figure 7.17 Surge bins.

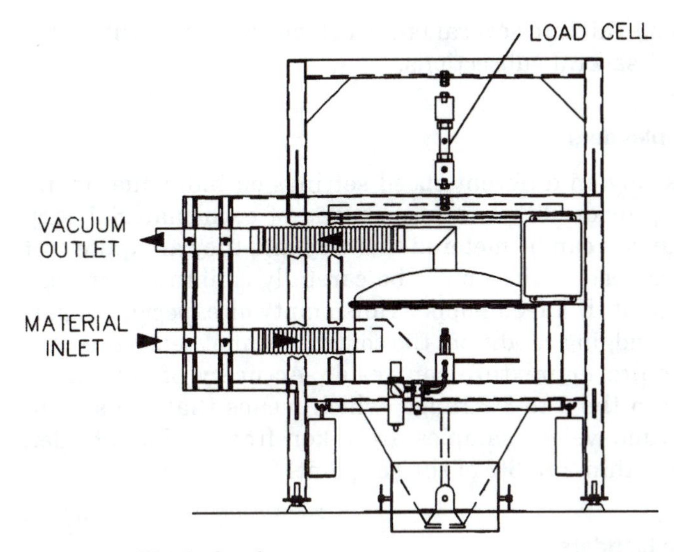

Figure 7.18 Weigh chamber.

is weighed and totalized. This greatly simplifies the tracking of resin usage patterns over a given period of time.

7.6 Blending Systems

Because of economic considerations, an increasing volume of resin is shipped in the natural or uncolored state. Color must be added at the plant level before the material is fed to the processing machine. This, along with the fact that most processing operations produce scrap which must be reintroduced into the material stream, has led to numerous systems for blending these various ingredients.

In their simplest form, all blenders contain the same elements:

1. Individual feed hoppers to contain each ingredient, typically virgin resin, regrind, and color pellets.
2. Metering devices, such as a feed screw, vibrator tray or air-operated slidegate, to regulate the flow of each ingredient.
3. A mixing section to homogenize the batch before leaving the blender.
4. A control section ranging from simple speed controllers to sophisticated microprocessors or loss in weight blenders.

Blending systems fall into several broad categories which will be discussed in the next several subsections.

7.6.1 Volumetric blenders

Volumetric units rely on different speed settings on individual ingredient feeders to provide different proportions of each material. Any number of ingredients can be metered this way and they are quite cost effective; however, feed devices must be carefully calibrated for each individual ingredient. If a feed hopper runs empty or experiences partial flow due to a bridging condition, the controls cannot sense this condition, and an improper mixture can result. Accuracy of volumetric units is typically in the 2 to 3% range, which means that if a significant number of time-weight samples are taken from a single feeder, they will cluster within 2 to 3% of the set point.

7.6.2 Gravimetric blenders

Gravimetric units offer more consistent and accurate performance at a very small premium. The simplest gravimetric blenders are so-called batch weighing systems which meter ingredients, one at a time, into a weigh chamber attached to a load cell. The load cell cuts off one feeder when its required weight is reached and then calls for the next ingredient and so on. The only calibration that is required is for the load cell itself, which usually takes less than 1 min. If bridging or an empty feed hopper occur, the load cell will sense this short weight and indicate an alarm condition, thereby minimizing the chance of incorrect mixes reaching the process. Accuracy of gravimetric units is typically greater than 1% of the batch weight size.

7.6.3 On-the-press blenders

Either volumetric or gravimetric units can be mounted directly on the feed throat of a processing machine. This arrangement brings all equipment and controls directly to the machine where it is easily monitored by the operator. This method is typically used where a plant has numerous machines, each using a different material-color combination. The main drawbacks to this approach are difficult access to the equipment on some large presses or extruders and the downtime required for cleanout when material changes are made. Examples of machine-mounted and central blenders are shown in Fig. 7.19.

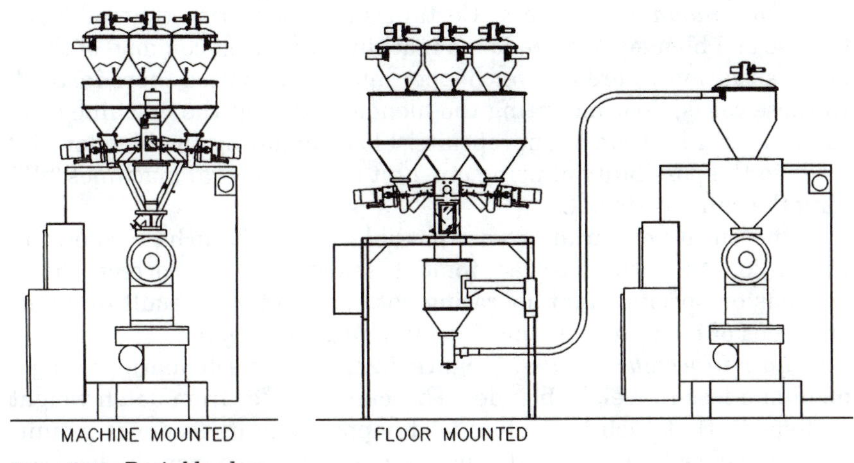

Figure 7.19 Basic blenders.

7.6.4 Central blenders

One blender can be mounted on a floor stand and feed the same material to a number of processing machines. In such cases, this reduces the number of blenders required and has the additional benefit of minimizing cleanout on the machine for color changes.

The choice of blender configuration depends on numerous factors which must be evaluated at the plant level. The following is a brief summary of different configurations widely used in the industry:

1. *Dual ratio receiver–color pellet feeder.* This arrangement uses a proportional valve arrangement to load virgin and regrind materials. As the processing machine screw rotates and material flows into the feed throat, the color is metered in by the feeder. The feeder typically has a variable speed drive to allow different feed rates for different coloring requirements.

2. *Dual-feeder arrangement.* By incorporating two separate feeders, one for regrind and one for color concentrate, a more precise control of proportions can be achieved.

Note: When using either the single- or dual-feeder arrangements described here, it is advisable to mount the riser section on all processing machines so that the feeder units can be relocated if necessary.

3. *Machine-mounted blenders.* By placing a volumetric or gravimetric blender directly on the feed throat of a machine, direct control is achieved over all ingredient feed rates. As with any press-mounted system, careful consideration must be given to allow access for maintenance, calibration, and cleanout.

4. *Floor-mounted blenders.* Certain situations arise that preclude the use of a blender mounted on a machine. The most common of these is the situation where a dryer hopper must be mounted there instead. In these cases, floor mounting the blender either at the machine or in some central location is appropriate. Floor mounting the blender also leads to the possibility of using one unit to feed several machines running the same material.

Numerous other variations are possible, some of which are shown in Fig. 7.20. As with most equipment selections, the correct choice depends on specific plant operating patterns and may result in different types of blending systems for different applications.

5. *Loss-in-weight blenders.* Figure 7.21 shows a schematic arrangement of a loss-in-weight blender. These units differ from batch weight systems in that each individual feed hopper with its feeder is mounted on a load cell. As material is metered out of the hopper, the load cell senses the loss in weight of the hopper and adjusts the feeder speed to maintain correct proportioning. Since all feeders run simultaneously, higher throughput rates are possible with loss-in-weight units. They are used mostly in sheet and film extrusion applications and are often interfaced with extruder or downstream drives to maintain precise gauge control.

7.7 Regrind Systems

Every plastic processing operation produces scrap material either generated directly from the production operation, such as runners and sprues or extrusion edge trim, or from reject finished parts. Whatever the source, economics dictate that the scrap be reintroduced into the material stream. The first step in the process is to grind the parts into a particle size small enough to be mixed with virgin pellets and flow through the various blenders, dryers, and loaders upstream of the injection press or extruder.

Figure 7.22 shows a simple upright or "beside-the-press" scrap grinder and identifies key elements of the unit.

Factors to consider when selecting a grinder are

1. Part size, shape, and wall thickness. These factors have a direct bearing on unit size, rotor, knife, and screen configuration.
2. Material. Hard or abrasive materials require one set of features while very soft materials may require their own option.
3. Throughput.

Many grinder variations exist, each tailored for a specific type of

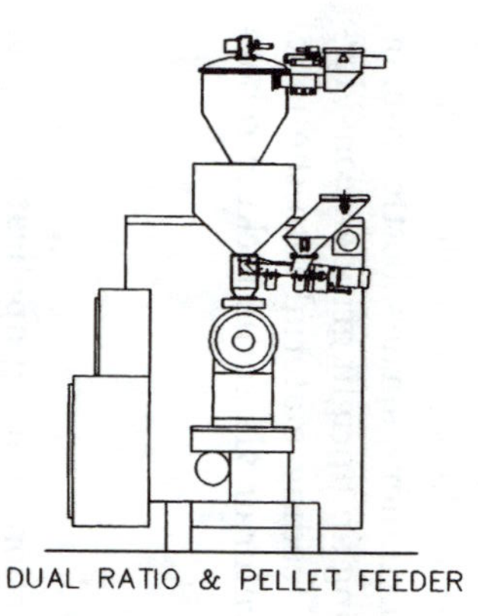

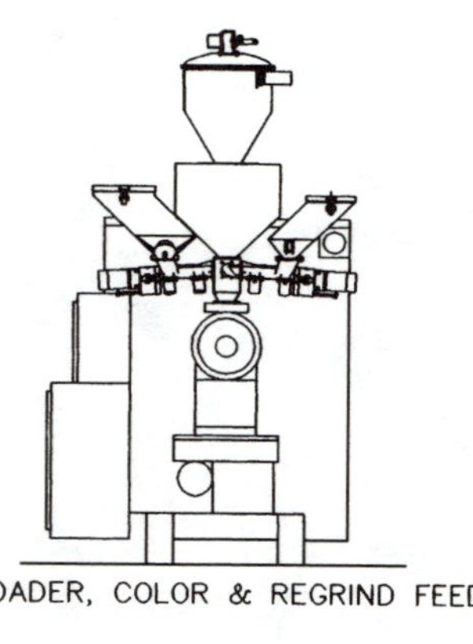

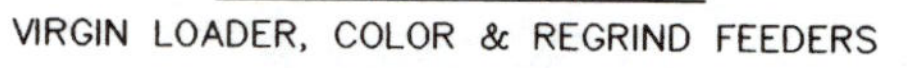

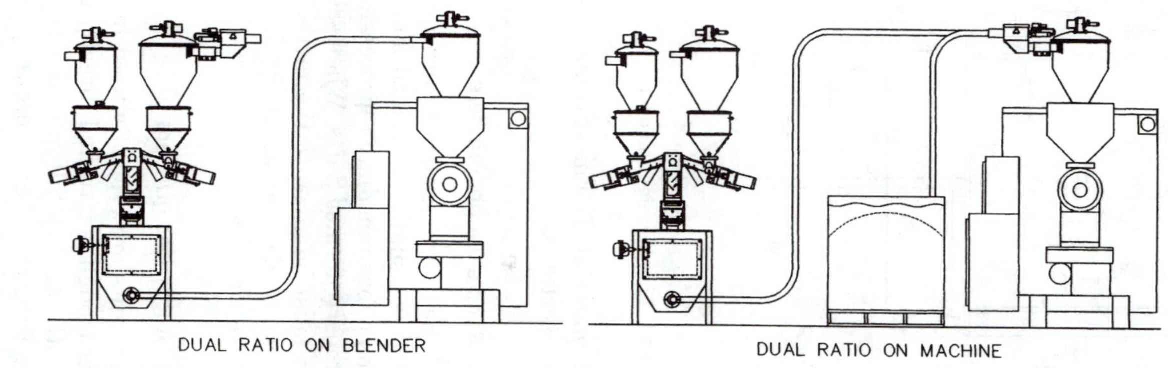

Figure 7.20 Blender options.

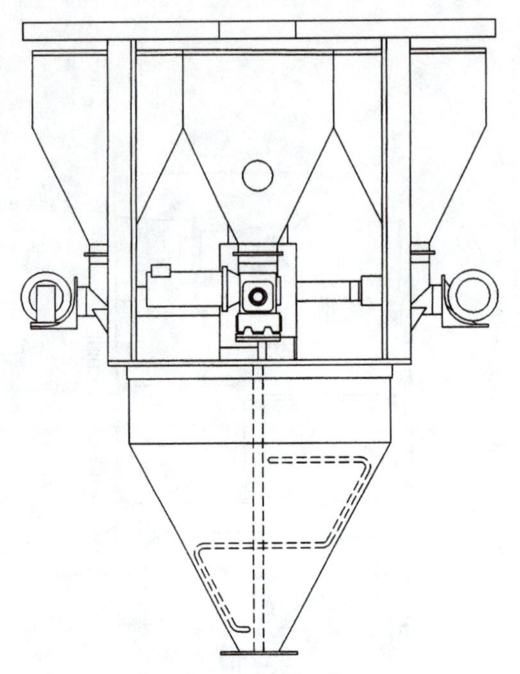

Figure 7.21 Loss-in-weight blender.

scrap material. The most common of these are discussed in the following subsections.

7.7.1 Simple upright grinders

This type of grinder has an arrangement similar to that shown in Fig. 7.22. It is usually positioned alongside an injection or blow molding machine for immediate reprocessing of runners, sprues, bottle trim, or improperly formed parts. When selecting an upright grinder, consideration must be given to the method of feeding material into the unit. This can be done manually by an operator or via sprue picking robots or conveyor belts.

7.7.2 Auger feed grinders

These units are typically used in injection molding applications. The auger trough is situated directly under the mold and catches runner networks as they fall from the mold. The auger pulls the runners into the cutting chamber where they are ground. Because of their low profile, they do not have a large storage capacity for ground material and must be continuously evacuated (see Fig. 7.23).

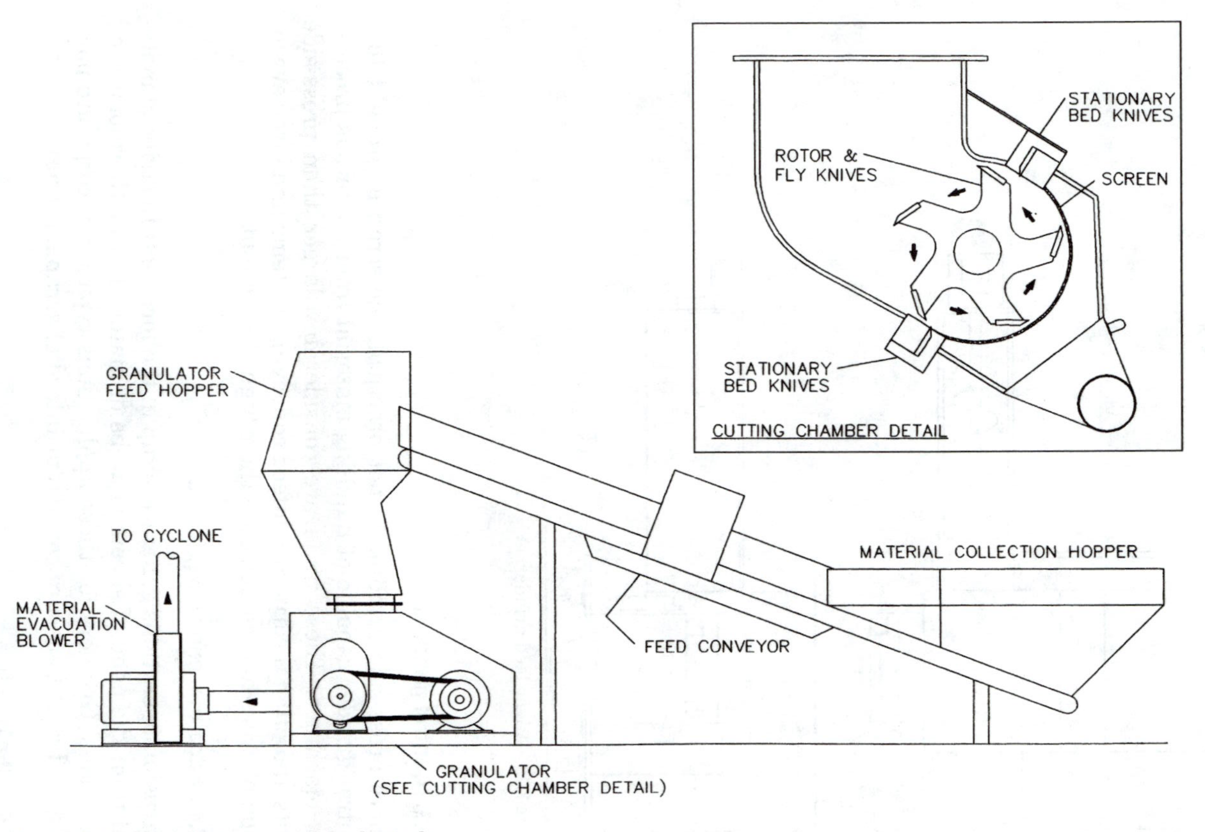

Figure 7.22 Upright tangential feed grinder.

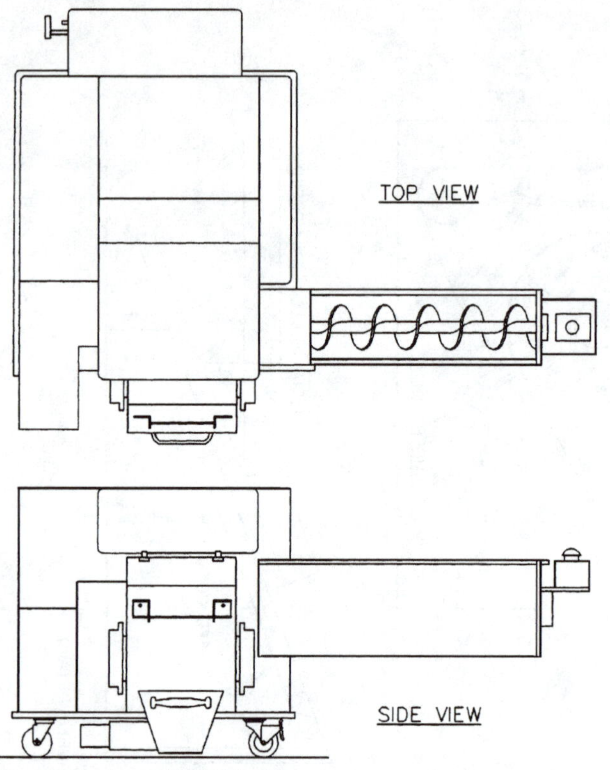

Figure 7.23 Auger feed grinder.

7.7.3 Central grinders

Where press-side grinding is not practical, the scrap is brought to a central grinder capable of handling material from numerous processing machines. These units are considerably larger than press-side units and often incorporate infeed conveying systems and large evacuation blowers to accommodate their high throughput.

7.7.4 Edge trim and web grinders

Extrusion operations generate scrap in the form of thin strips of material from the edge of sheets or webs of material after thermoforming operations. Grinders for these applications typically incorporate powered feed rolls to assure consistent infeed of scrap material.

7.7.5 Film grinders

Low-density PE blown film in the range of 1- to 5-mil thickness presents a unique set of problems for grinder design. Extremely tight knife tolerances are required, very small screen openings to assure reasonable bulk density of the regrind, and high-volume evacuation systems both to draw material through the chamber and provide a cooling action to prevent melting.

7.7.6 Combination shredder/granulators

For very large parts or extremely high volumes, a two-stage shredder/granulator process is often used. The first stage consists of multiple low revolutions-per-minute rotors which tear the parts into smaller pieces which then are fed into the grinder section.

7.7.7 Noise and safety considerations

By their very nature, scrap grinders generate very high noise levels. Sound treatments consisting of special acoustic enclosures to reduce noise levels are available for most units. All grinders must be equipped with safety switches and interlock circuits to assure that personnel cannot be exposed to dangerous conditions while maintaining or cleaning these units.

7.7.8 Grinder evacuation systems

Manual removal of regrind material is inefficient, dangerous, and messy; therefore, most grinders are equipped with evacuation blowers. In addition to continuously removing material, they have the added advantages of drawing a large quantity of air through the cutting chamber which helps cool the material, thus reducing degradation.

7.7.9 Size classification and fines removal

For elutriator systems, mechanical screen separators, and cyclone separators, various devices to perform the classification and removal function, include eludriator systems, mechanical system separators, and cyclone separators. Once parts have been ground, there is often a need to remove dust or oversize slivers of material from the regrind prior to its reintroduction into the process. Removal of fines can be accomplished by means of elutriation or air scalper systems which are shown in Fig. 7.24. Mechanical screen separators, such as Fig. 7.25, have the ability to remove both fines and oversize particles in one step.

7.8 Material Drying

Resin moving from storage to the molding machine often requires a drying stage due to hygroscopic properties of the material. Efficient

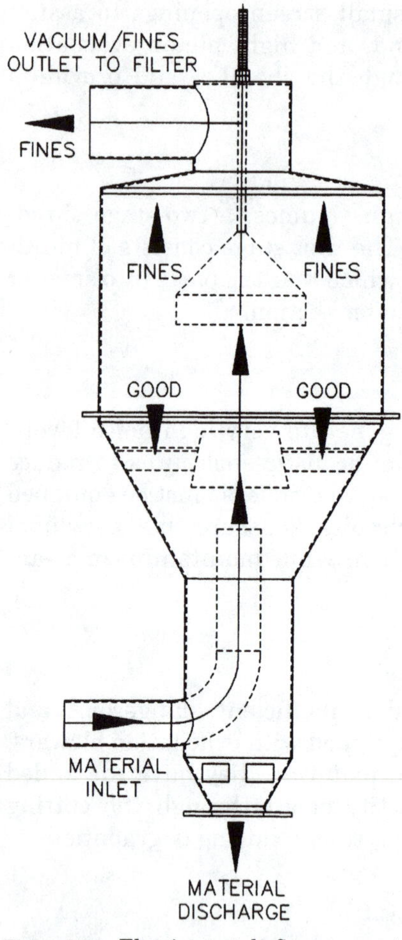

Figure 7.24 Elutriator style fines separator.

heating and drying of plastic resins can be important to the manufacture of a consistently acceptable product that meets quality requirements. The processor must pretreat plastic resins in strict adherence with the manufacturer's recommendations.

Plastic resins may be either hygroscopic or nonhygroscopic, depending on whether or not the resin will absorb or adsorb moisture. Nonhygroscopic resins collect moisture only on the pellet surface (adsorption), making it easy to remove. Hygroscopic resins, however, collect moisture throughout the pellet (absorption), making its removal more difficult. Wherever it is found, the presence of moisture presents potential problems from cosmetic surface blemishes to serious structural defects.

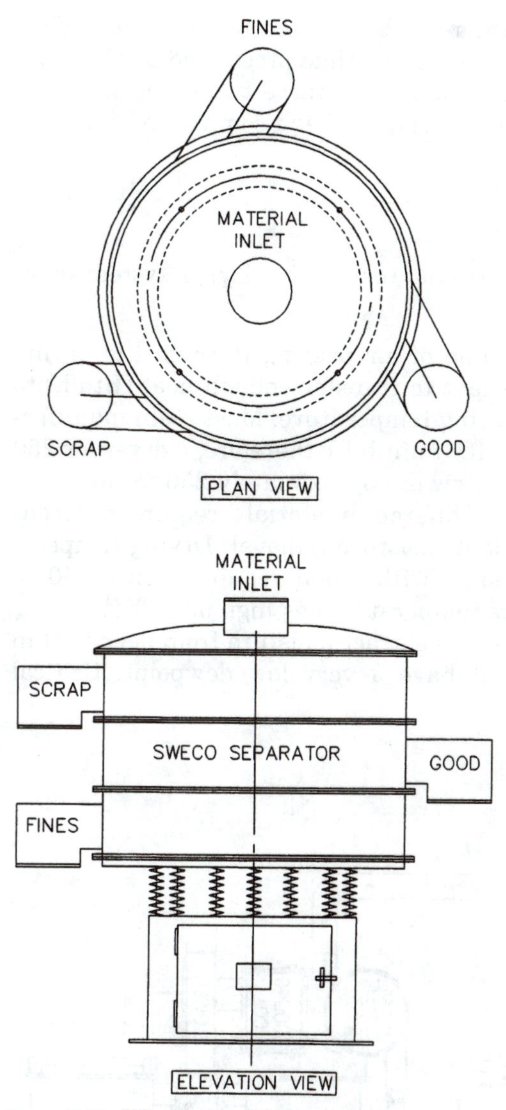

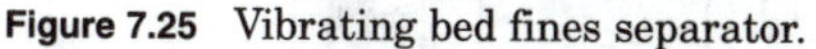

Figure 7.25 Vibrating bed fines separator.

Typical nonhygroscopic resins are PE, PP, polystyrene (PS), and PVC. Fillers such as talc, calcium carbonate, carbon black, and wood flour, when added to nonhygroscopic resins like PE or PP in sufficient quantity, can make the resin behave like hygroscopic resins.

Surface moisture on nonhygroscopic plastic pellets can be effectively removed using only heated ambient air, but removal of moisture collected within hygroscopic pellets requires dehumidified heated air.

Either system requires that the resin be exposed to adequate airflow heated to the proper temperature for the time prescribed by the resin manufacturer. The equipment required for these two types of drying systems varies considerably in terms of cost and complexity (see Figs. 7.26 and 7.27).

7.8.1 Drying system parameters

In order to properly dry hygroscopic resins, the drying system must provide the following parameters:

1. *Process airflow.* The volume of air passing through the drying vessel (ft^3/min) must be sufficient to transfer enough heat (Btu/h) to raise material to its proper drying temperature. Most resin manufacturers recommend airflow of 1 ft^3/(min·lb·h); therefore, a dryer for 250 lb/h of ABS should have an airflow of approximately 250 ft^3/min.

2. *Process air temperature.* Different materials require different drying temperatures for efficient moisture removal. Drying temperatures vary widely among resins, with nylon requiring only 140 to 160°F, while PET may require temperatures as high as 350°F.

3. *Low dewpoint air.* In order to extract moisture from deep within a pellet, the process air must have a very low dewpoint. Critical

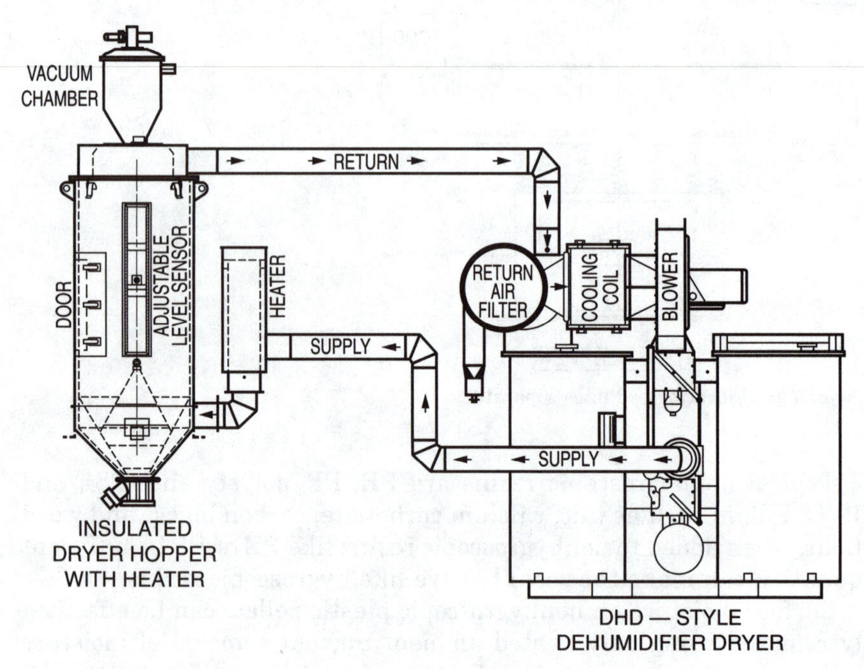

Figure 7.26 High-temperature material drying system.

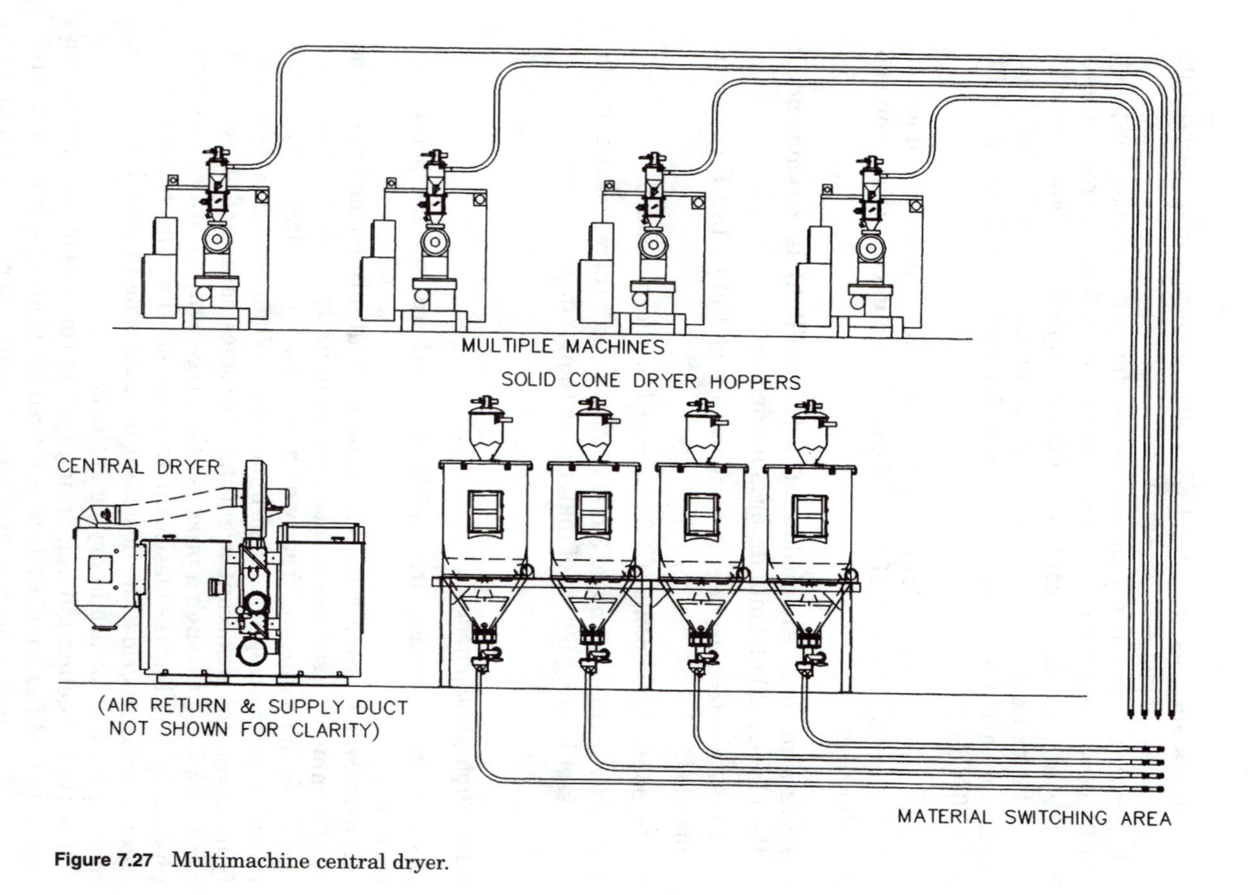

Figure 7.27 Multimachine central dryer.

applications such as molding PET bottle preforms may necessitate dewpoints as low as −50 or −60°F.

4. *Residence time.* Given proper airflow, temperature, and dewpoint, it still takes time for the moisture to be extracted from the resin, therefore, the drying vessel must have sufficient capacity to allow the material to be exposed to drying conditions for the recommended time. Times can vary from 1 to 2 h for removal of surface moisture from PE to 5 to 6 h for PET.

As an illustration of drying requirements, consider an injection molder running 200 lb/h of ABS. Material manufacturers recommend the following:

1. *Process airflow.* An airflow of 1 $ft^3/(min \cdot lb \cdot h)$ is recommended; therefore, a 200-ft^3/min dryer is needed.
2. *Process air temperature.* A temperature of 160 to 180°F is recommended.
3. *Dewpoint.* A dewpoint of −30 to −40°F is recommended.
4. *Residence time.* A time of 3 h is recommended; therefore at 200 lb/h, a vessel holding 600 lb of material is required.

7.8.2 Drying equipment

Desiccant-type dryers are widely used where low dewpoints are required.

A *desiccant* is a material with a natural affinity for moisture. The most commonly used desiccant is a synthetically produced crystalline metal, alumina-silicate, from which the water of hydration has been removed, permitting it to adsorb moisture. When used in a dehumidifying dryer, the desiccant eventually becomes saturated; however, it can be renewed through a process called *regeneration,* which is accomplished by heating the desiccant to drive off the collected moisture. After a cooling period, the desiccant is again able to adsorb moisture, making it ideal for use in drying systems.

The typical dehumidifying drying system consists of the process air fan, dehumidifier, an electric or gas-fired air heating system, a control system, the drying hopper, and a process air filter. The dehumidifier itself includes the desiccant regeneration system.

Single rotating bed. A single desiccant-coated honeycomb wheel rotates slowly, exposing part of the wheel to process air, part to regeneration, and part to cooling prior to returning to the process.

Multiple indexing bed. This is usually a three-bed arrangement, with one bed on process and one regenerating while the third is being cooled prior to going on line to the process.

Twin stationary beds. One bed is on process while the second is being regenerated and then cooled prior to going on to process (see Fig. 7.26). The twin stationary bed dehumidifier is simple, with relatively few moving parts, making it easy and inexpensive to maintain. The initial investment is usually lower than the other designs.

Selecting a bed design. Systems utilizing the twin-bed design are the most widely used, followed by the multiple indexing beds, and then the single rotating bed. All of these systems are designed to provide a continuous supply of dry air. Both the multiple indexing bed and single rotating bed work very effectively, although they tend to be mechanically complex, costly to maintain, and usually involve a higher initial investment.

Machine mounted. In many applications, the drying hopper is mounted directly on the feed throat of the processing machine. This arrangement eliminates conveying of material once it has been dried. Several disadvantages of this method are

1. Tall stack-up height, which may be a problem because of low ceilings or overhead cranes.
2. Lost production time due to hopper cleanout and predry when material changes are made.
3. Requires one dryer per machine, which may be inefficient if the plant is running a limited number of materials requiring drying.

These disadvantages of machine-mounted dryers have led to the increasing popularity of remote-monitored or central dryers.

Remote mounted. Putting dryer hoppers on floor stands at a remote location from the processing machines has several advantages:.

1. Head space requirements are minimized.
2. Multiple machines can be serviced by central drying systems, increasing energy efficiency and reducing changeover time. Because material must be conveyed after it is dried in a central drying system, this movement is often accomplished by using dry air supplied by a separate dehumidifier. Additionally, the conveying lines can be purged of material by means of appropriate valves and controls.

The starting point for any evaluation of central drying systems is the gathering of material throughput information by defining the number of materials requiring drying and accurate estimates of pound per hour throughput of each of the materials. This information leads to specifications regarding number of drying hoppers, size of hoppers, and cubic feet per minute of central dryers. Future growth in throughput and additional materials must be taken into account at this stage.

7.8.3 Dryer system controls

Almost all modern dryers utilize either microprocessor or programmable logic controllers which monitor process parameters such as temperature, airflow, dewpoint at numerous points in the system, and also monitor various machine functions such as heater amps, valve positions, and alarms for diagnostic and troubleshooting purposes. The most advanced dryers include energy-saving features, such as regeneration, based on process dewpoint versus time and an ability to protect material from degradation by reducing drying temperature when throughput is reduced or temporarily stopped.

7.9 Loading Systems

When designing a plantwide conveying system, throughput rates and conveying distances are two of the most important criteria to evaluate. Figure 7.28 represents a set of curves corresponding to system throughputs over a range of distances for conveying systems starting at 3-hp units with 2-in-OD lines to 25-hp pumps with 4-in-OD lines. This chart should be used as only a rough guide to system throughput, and the following notes regarding its use are important:

1. For multiple-station vacuum sequencing systems, you must use the total throughput for all loading stations serviced by a single pump.
2. The distance shown on the X axis of the graph represents total equivalent feet of the conveying system and is calculated as follows:

$$\text{Equivalent feet} = H + (E \times 15) + (V \times 2) = (F \times 3)$$

where H = horizontal material run
E = number of 90° elbows in the run
V = vertical material run
F = flex-hose run

When evaluating multistation systems, it is best to use the longest material run in the system.

3. Figure 7.28 applies to pellet conveying systems. If powder is being conveyed, the expected throughputs are approximately one-third lower than those indicated.

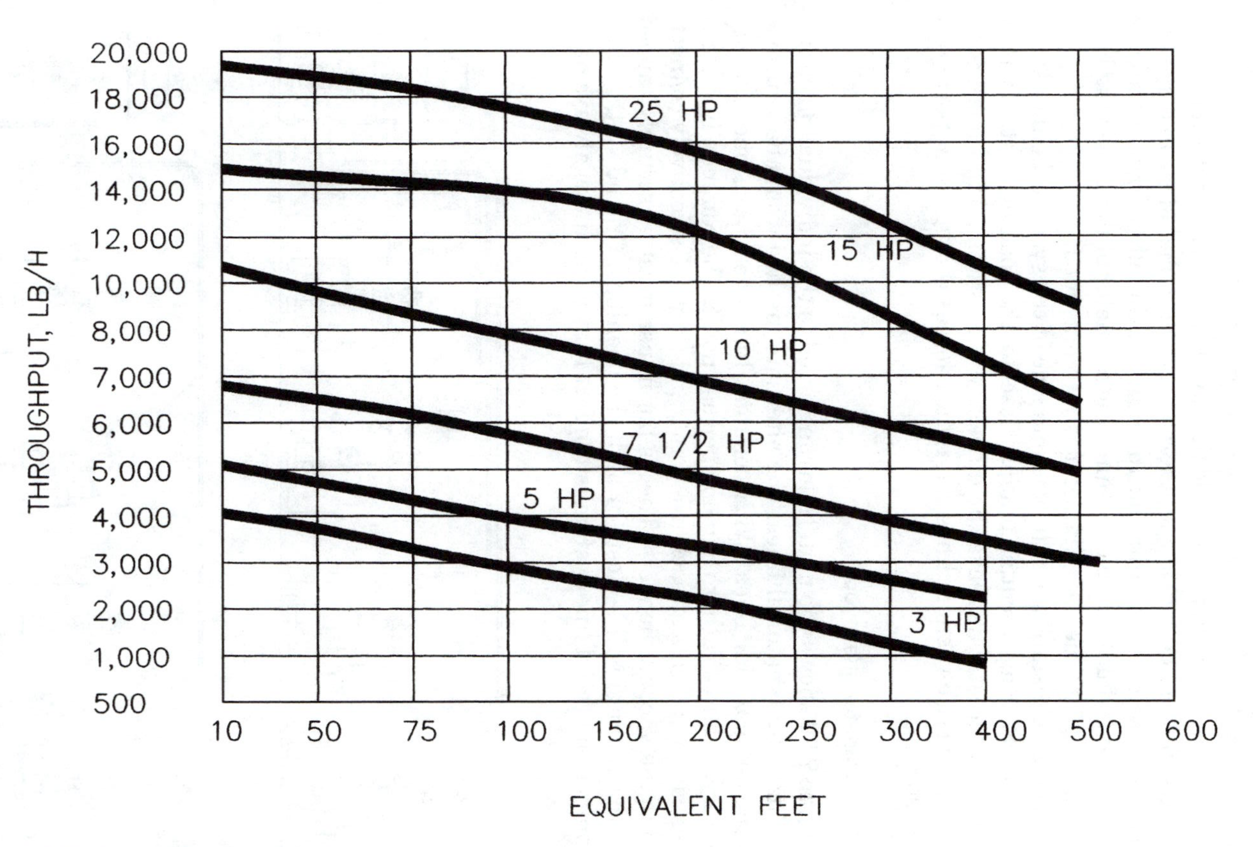

Figure 7.28 Loading system throughput curves.

Whenever possible, distances between various system elements, such as surge bins, blending stations, and processing machines, should be kept as short as possible given the site specifics.

From the equivalent feet calculation, it is evident that each elbow equals 15 ft of horizontal run; therefore, the number of els in a system should be held to a minimum. In addition, the elbows should have a centerline radius of at least 6 times the diameter of the conveying line. If possible, 10 times the diameter is recommended.

Excessive distances on the vacuum side of the system should also be avoided. Placing vacuum pumps close to the loading stations they serve can greatly increase the efficiency of a system. Additionally, the use of oversize vacuum lines can reduce losses through that portion of a system.

7.9.1 Transfer line layout

Figure 7.29 shows the material transferred to the individual machines by individual material lines. The alternate method would be to utilize common material lines. With common lines, you have one line per material source rather than one per use point. The common line is fitted with Y laterals at each use point, and they are selectively connected to the vacuum hoppers. The Ys not in use must be securely capped or the vacuum will be lost through the openings. One advantage to a common material line system is the fact that each line is always dedi-

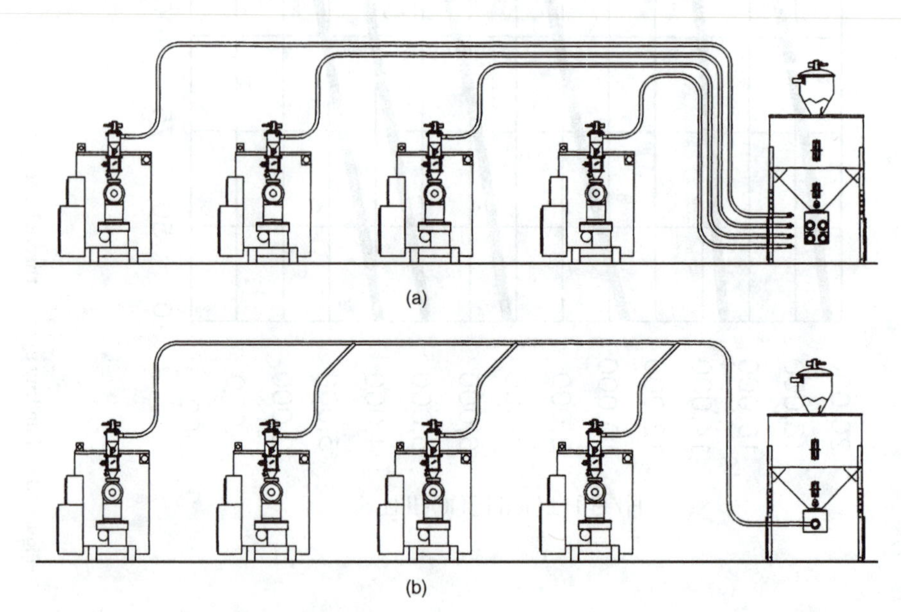

Figure 7.29 (*a*) Individual material transfer lines; (*b*) common material transfer lines.

cated to the same material, thereby eliminating the possibility of cross-contamination.

It is also possible to distribute the vacuum in a multiple-station system using either common or individual lines. Common vacuum lines are fitted with Ts at each station on which the vacuum sequencing valves are located. The common vacuum line has been almost universally adopted because of the savings in material and installation labor.

7.9.2 System hardware and installation

The principal point to remember when discussing material distribution systems is the fact that material lines *cannot* be run the way ordinary plumbing, compressed air, and electrical conduits are installed. Because you are moving particles of material in an airstream, every attempt must be made to keep the runs as straight as possible with a minimum of turns. Long lengths of flex hose should also be avoided because of their negative effect on conveying rates.

Almost all conveying systems installed in resin transfer systems utilize thin-wall aluminum tubing and elbows. The principal sizes are $1^1/_2$-, 2-, $2^1/_2$-, 3-, and 4-in OD. The tubing is available in 10- and 20-ft lengths, with the latter preferred because fewer couplings are required.

7.9.3 Material line elbow selection

Most wear in resin transfer lines occurs at elbows where the material impinges on the interior surface with maximum force. Various materials are used for conveying line elbows as the following table illustrates:

Elbow material	Application
Aluminum	Nonabrasive material such as PE, PP, soft PVCs
Stainless steel	Moderately abrasive materials such as PC, PET, and rigid PVCs
Ceramic-lined steel and glass elbows	Extremely abrasive materials such as 30 to 40% glass-filled nylons

When extremely soft materials are being conveyed, the use of elbows with internal spiral grooving is recommended. The grooving has a tendency to lessen the impact of pellets on the el's interior and also breaks up streamers into smaller flakes which are less prone to form blockages downstream.

Elbows are of the long radius type to assure smooth flow characteristics. The elbow centerline dimensions are listed in the following table:

Tube OD, in	Elbow centerline radius, in
$1^1/_2$	24
2	24
$2^1/_2$	24
3	30
4	36 or 48

The tubing elbows and other parts of the material line are held together by couplings, as illustrated in Fig. 7.30. This is a sleeve-type compression coupling with a neoprene gasket and stainless-steel ground strip to allow the dissipation of static electricity.

Connections between tubing and flex hose are accomplished with disconnect fittings which are also illustrated in Fig. 7.30.

Figure 7.31 shows a bulkhead-style switching station which is frequently used where multiple lines are to be brought in from silos through a block or masonry wall. The unit consists of an inner and outer aluminum plate mounted over the wall opening. Tube stubs are welded to the inner plate which is fitted with female disconnect fittings.

Once the lines are inside of the plant there are various ways of supporting them. The most common method involves the use of Uni-Strut bracket and clamps which are shown in Fig. 7.30. These brackets along with their accessory items can be used to fabricate crossovers to span tracks or roadways and to suspend tubing from roof trusses, columns, or other supports in the plant.

With the material and air-return lines thus run to the area of the vacuum hoppers, the final connections are generally made with flex

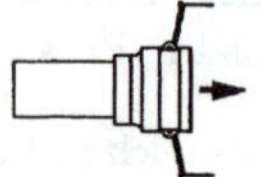

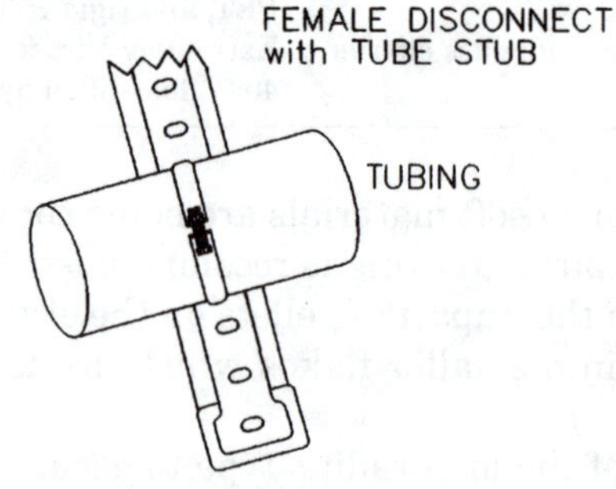

Figure 7.30 Conveying line hardware.

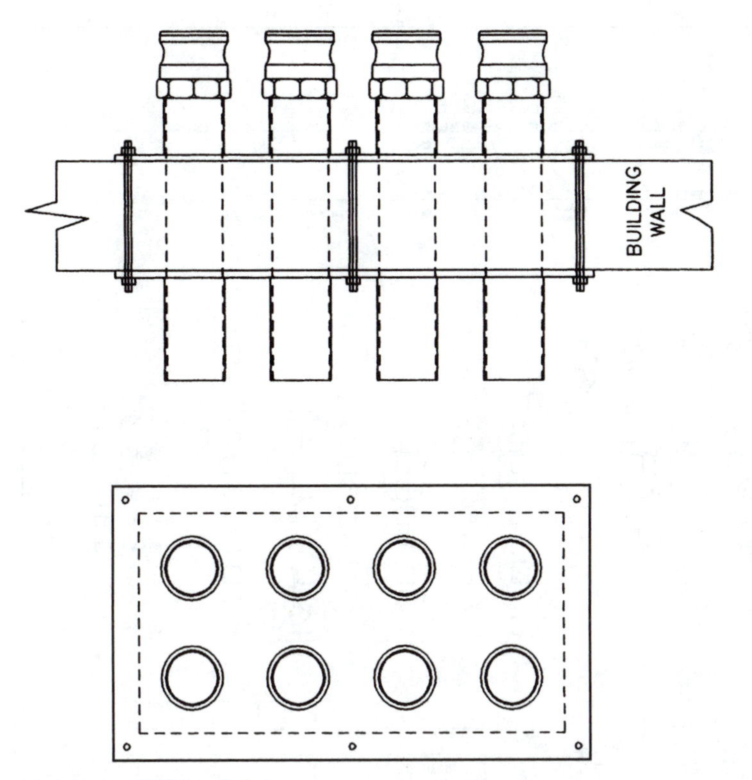

Figure 7.31 Bulkhead plate.

hose to allow easy movement of the hopper if cleaning or servicing becomes necessary.

7.9.4 Controls

All conveying systems require controls to monitor the status of level switches and other sensors and to control the sequence of operation of pumps, valves, filters, and other system elements (see Fig. 7.32). Basic types of loading system controls include:

1. Electromechanical panels that use relay logic to sequence operation.
2. Dedicated microprocessor panels tailored to specific applications.
3. Programmable logic controllers (PLCs), which can be reconfigured to accommodate changing system requirements.

When evaluating control systems for conveying systems, the following should be considered:

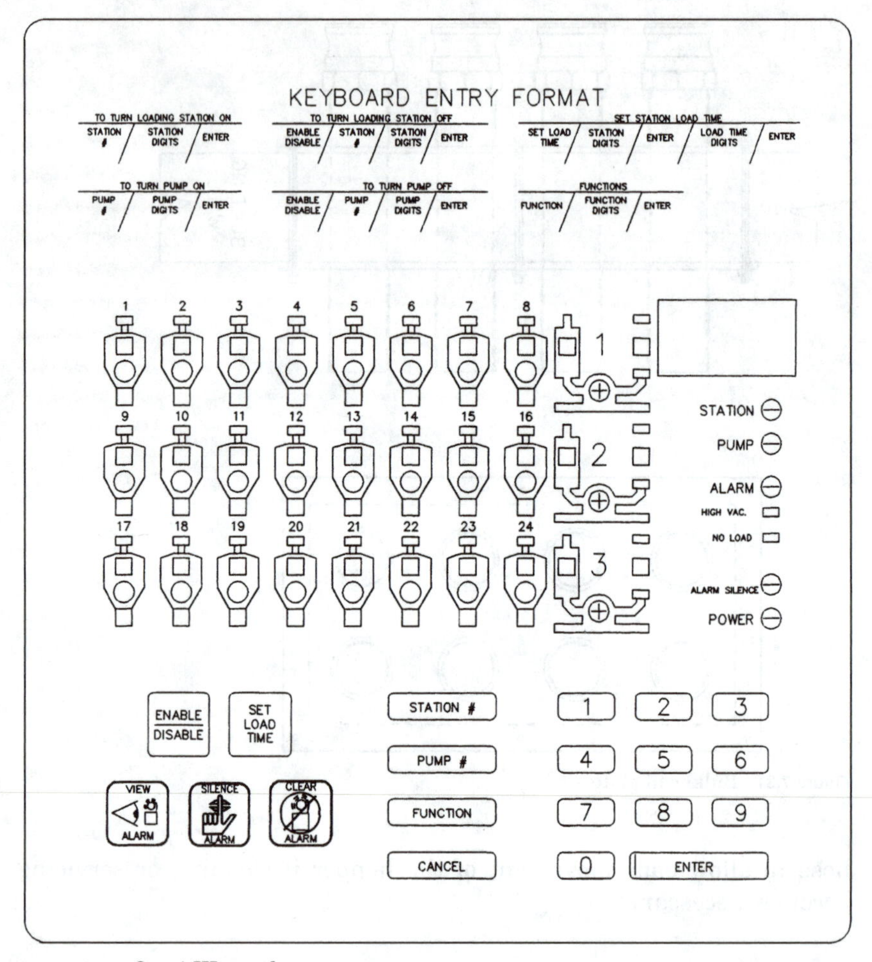

Figure 7.32 Omni-III panel.

1. In addition to performing control functions, the system should incorporate display and diagnostic capabilities to aid in setup procedures and pinpoint system problems when they arise.

2. *Central controls* versus distributed controls. Central control panels offer a convenient overview at a single location, but this can be a drawback when loading stations and other system elements are spread over a large area.

Distributed controls offer the ability to enter system commands and monitor system status at various points throughout the system. Historically, the drawback to distributed systems has been the high cost associated with wiring all system elements together, but new technology has addressed this issue and hybrid central/distributed systems are available which require a minimum of field wiring.

3. The ability to communicate with a plantwide monitoring computer is becoming a necessity.

With the increased popularity of central drying systems, special attention has been focused on conveying dried material downstream to the processing machines. Once a resin has been dried, the objective is to keep it dry by minimizing its exposure to ambient conditions. This is accomplished by the use of

Dry-air conveying

Line purging

Small-capacity receivers

Dry-air conveying utilizes a small, separate dryer to provide dehumidified air at the inlet of the pneumatic conveying system.

Line-purging systems use material control valves at the discharge of each dryer hopper. These valves open to allow material into the system during the conveying portion of a load cycle and then shift to cut off the material and allow the line to be purged completely of material by dry air.

Small-capacity hoppers, typically glass-tube receivers, are mounted directly on the throat of a processing machine and replace machine hoppers. They generally utilize an adjustable level switch which allows a minimum inventory of predried material to be maintained on the machine. Figure 7.33 illustrates a system incorporating all of these elements.

7.10 System Integration

Whenever auxiliary equipment requirements are discussed, it is advisable to adopt a systems perspective, that is to say, the engineer should evaluate not only the individual piece of equipment under consideration, but also the impact that unit will have on operations upstream and downstream of it in the plant's material flow. It is not uncommon to encounter situations where correcting a bottleneck in plant operations in one area creates another one downstream because those systems are inadequate for the increased volume or because existing systems upstream of the new equipment are inadequate to keep up with it.

Whenever an expansion of operations is planned, whether by expanding existing capacity or especially when contemplating an

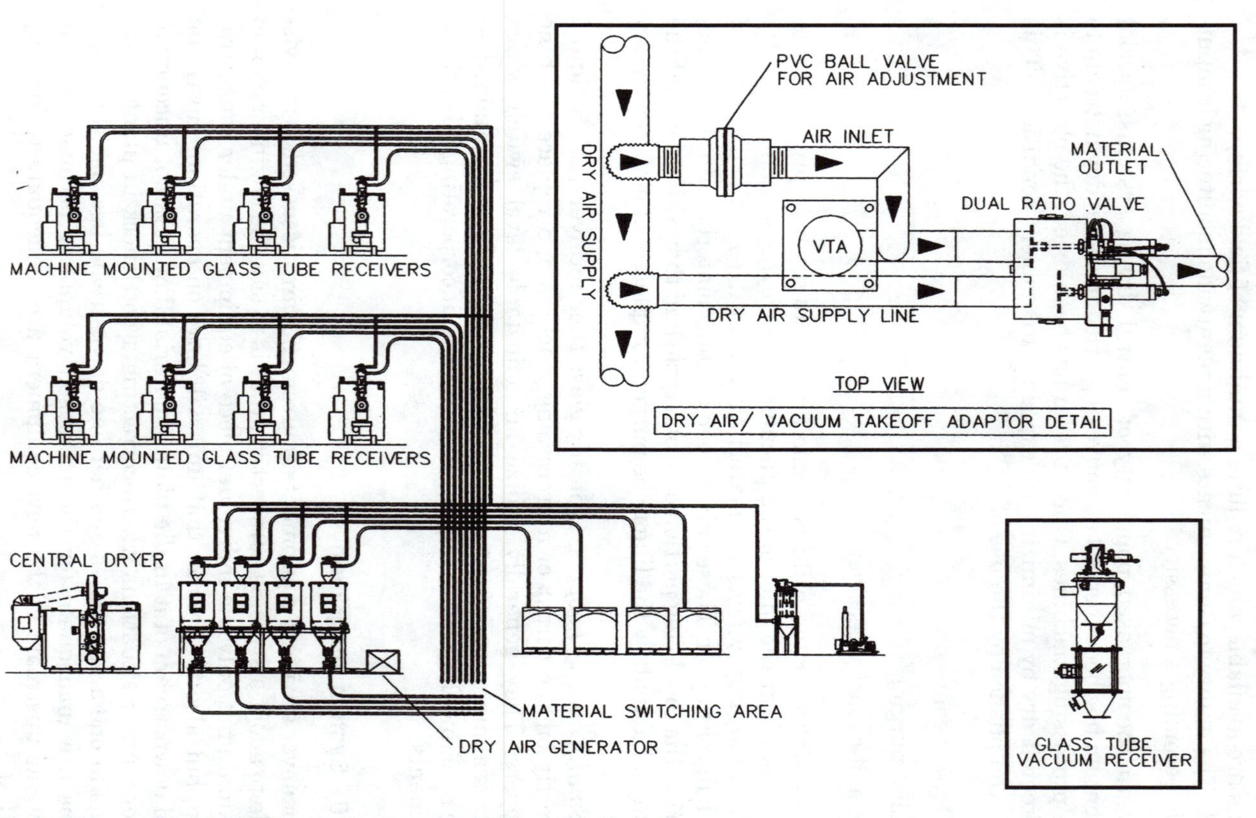

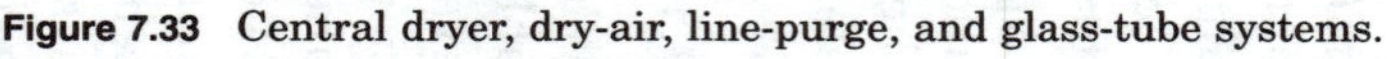

Figure 7.33 Central dryer, dry-air, line-purge, and glass-tube systems.

entirely new facility, a careful analysis of auxiliary needs must be conducted and a master plan must be developed. The analysis must involve all operations personnel who can provide input about operational requirements for each auxiliary subsystem. Such input early in the project can help avoid costly errors further along. Outside sources, such as consultants and vendors, can provide new and innovative ideas for either increasing capacity or reducing project costs.

Each of the areas we have discussed in this section, including bulk delivery, storage requirements, conveying, blending, grinding, and drying, must be considered along with others such as tower and chilled water systems, mold temp control units, robotic equipment, downstream parts handling, compressed air systems, and numerous others.

The analysis of needs in these areas must take into account initial plant capacity as defined by number of production units and throughput requirements and also provide for planned future growth. The analysis of needs will lead to the development of a master plan which should include material flow diagrams, detailed equipment specifications, plant layout drawings, along with accurate cost and time estimates for the various phases of the project. It is quite normal for a master plan to be reviewed and revised several times before it is adopted for a project. Once again, the more input at this stage of the project, the better. Master plans often call for the phased introduction of auxiliary items as planned production expansion takes place over a number of budget cycles. At each phase of the expansion, the original assumptions and specifications must be re-evaluated to see if expense or operational changes dictate modifications.

The following list summarizes questions to be considered when evaluating auxiliary systems for plant expansion or construction:

1. Resin delivery: Bags or boxes? Bulk truck? Railcar?
2. Number of different materials?
3. Desired inventory levels?

Also at this point, take into account whether resin is ready for use as delivered or if further conditioning or additives will be required downstream.

7.10.1 Blending systems

the following questions should be considered for blending systems:

Blend on the machine or central blending?

Number of components?

Throughputs?

Accuracy requirements?

Recordkeeping requirements?

Dedicated or subject to material changes?

7.10.2 Drying systems

The following questions should be considered for drying systems:

Drying on the machine or central drying?

Specific resin requirements such as drying temperature, dewpoint, residence time?

Throughputs?

Dedicated or subject to material change?

7.10.3 Regrind systems

The following questions should be considered for regrind systems:

Grinding at the press versus central grinding?

Part size and shape?

Material characteristics?

Combination of press side for mold room or blow molding trim combined with central grinder for reject parts?

Storage of regrind material if short-term situation provides more regrind than can be immediately re-introduced into the process?

7.10.4 Resin distribution systems

Decisions made in each of the previous areas influence the layout and sizing of the resin distribution system. It must accommodate all of the requirements of the storage, blending, drying, and scrap recovery operations while taking into account resin characteristics and plant layouts. It is typically one of the final items designed after many other system decisions have been made.

Since resin distribution systems are spread out throughout an entire facility, the controls of such systems become a major concern. The trend within the plastics industry is toward distributed control systems, which allow an operator to enter commands at a point of use rather than having to repeatedly go back to one central panel location.

7.10.5 Controls and monitoring systems

Most auxiliary equipment comes with controllers that not only perform the normal control functions, but also provide diagnostic information to help optimize operations or analyze problems when they arise. It is also extremely desirable that controls include a communications ability to allow their monitoring by a central computer which can

1. Record operational performance. This is particularly important for quality control and ISO requirements.
2. Keep track of equipment alarms and malfunctions. This is useful for preventive maintenance programs.
3. Provide material inventory and throughput data via silo measurement systems, weigh chamber, and blender throughput data.
4. Print out change reports indicating when an operator enters a command to change a setting or parameter on any piece of equipment monitored.

Central monitoring systems as outlined here are becoming less of a luxury and more of a necessity as greater operating efficiency and higher quality standards are called for. Any auxiliary system analysis must make provisions for such monitoring capability, if not initially, then as part of a planned enhancement at a future date.

Drying and Dryers

Pete Stoughton
Global Accounts Manager, Conair
Pittsburgh, Pennsylvania

Without careful drying and handling before processing, plastic polymers may process inconsistently. Residual moisture on or within the pellet can have adverse effects on product quality and performance

At temperatures above the melting point, water reacts rapidly with certain polymers such as nylon, polycarbonate, polybutylene terephthalate (PBT), and polyethylene terephthalate (PET). This reaction results in a decrease in the molecular weight. At the same time, absorbed water can form steam that results in surface roughness, splay, and internal bubbles. The reaction between water and the molten polymer is accelerated by prolonged exposure to temperatures above the melting point.

Processing problems due to moisture can be avoided by carefully following the proper handling and drying procedures outlined by your material supplier and auxiliary equipment supplier.

7.11 Why Do We Dry Plastic Materials?

Plastics, when exposed to the atmosphere, will pick up moisture. How much will depend on the type of polymer, the humidity of the surrounding air, and several other factors. If the amount of moisture pick up is excessive, it may affect the performance of the processing machine, the cosmetic qualities of the product being produced, or even the physical and structural properties of the product. It is, therefore, generally understood that the processing of thermoplastics that contain high concentrations of moisture normally results in the production of unacceptable products.

7.11.1 Effects of unremoved moisture

The effects of processing a polymer that contains excess moisture manifest themselves in a variety of ways, depending on the type of

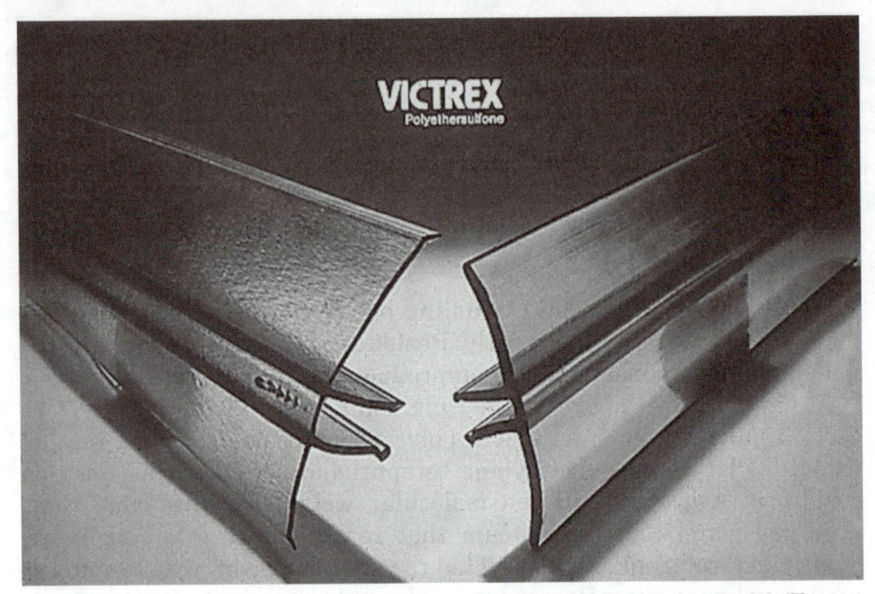

Figure 7.34 The profile on the right is extruded from properly dried material. The one on the left shows a trail of gas bubbles and surface roughness caused by excess moisture.

polymer, the process, and the product being produced. Figure 7.34 shows typical moisture-related flaws in extruded products.

In most materials, small amounts of moisture will cause a change in the polymer's melt viscosity which, in turn, will affect the way it processes. When polymers, such as polycarbonate and polyester, are heated above the melt temperature, small amounts of moisture in the pellets or on the surface will cause a chemical reaction. This reaction can degrade the polymer, changing its molecular weight, melt viscosity, and mechanical strength. Larger amounts of moisture may result in a rough and scaly surface finish and even bubbles and voids in the product.

Certain moisture effects are specific to the process involved. For instance:

Injection molding. High moisture concentrations in thermoplastics that are injection molded can result in nozzle drool, foamy melt, lower melt viscosity, flow lines and silver streaks in the gate area, rough or scaly part surface, voids in the part, poor dimensional control, and reduced mechanical properties.

Extrusion. The effects of excess moisture during extrusion may include a foamy melt, lower melt viscosity, bubbles in the extrudate, surging that results in arrow heads and wave forms in the extrudate, surface roughness, and reduced mechanical properties.

7.12 Hygroscopic and nonhygroscopic polymers

Moisture can accumulate on the surface of any plastic material, and some polymers actually absorb moisture into the pellet itself. Materials that absorb moisture—usually engineering-grade materials like nylon or polycarbonate—are called *hygroscopic.* Those that do not have the same attraction or affinity for moisture, and do not absorb it, are called *nonhygroscopic polymers.* They include commodity polymers such as polyolefins.

However, even nonhygroscopic polymers may pick up small amounts of surface moisture when exposed to atmospheric air. Although it is a good practice to remove any surface moisture that is present prior to processing, very small amounts of surface moisture may not have a detrimental affect on the products being produced. Just how much moisture will cause a problem depends on the process and final product being produced.

Nonhygroscopic polymers. Generally, these materials pick up surface moisture if they are handled carelessly. For instance, if a polymer is taken out of a cool, unheated storage area and moved into a warm processing area, moisture from the warm air in the processing area may condense onto the comparatively cold polymer surface. This is the same as condensation that occurs when a cold soft drink is taken out of the refrigerator on a warm summer day. Moisture from the humid ambient air quickly condenses on the surface of the soft drink container. To avoid processing problems, surface moisture on nonhygroscopic polymers can be removed easily by a simple hot-air dryer with a single-pass air circuit.

If, however, a nonhygroscopic polymer is compounded with a hygroscopic pigment or additive, the resultant compound will be hygroscopic and may need to be dried differently. For more detail on drying nonhygroscopic polymers, refer to Sec. 7.15.1.

Hygroscopic polymers. On the other hand, most engineering-grade polymers have a chemical nature that causes them to attract and absorb moisture from the surrounding ambient air. Most engineering polymers are hygroscopic polymers.

When a hygroscopic polymer is exposed to the atmosphere, water vapor migrates into the pellet where some of the water molecules become bound to the resin's polymer chains by intermolecular forces. These forces are what makes it so difficult to dry hygroscopic polymers.

Moisture absorption takes place over time, and eventually slows and stops when the pellet's moisture content reaches equilibrium with that of the surrounding air. Some hygroscopic polymers have a greater affinity for moisture than others. Therefore, under identical conditions, one type of polymer may absorb moisture faster or slower than another type of polymer. One type of polymer may also reach its equilibrium moisture content at a higher or lower level than another polymer when exposed to identical conditions (see Fig. 7.35).

To summarize, the moisture content of a hygroscopic polymer will vary, depending on

- The type of polymer
- The length of time the polymer is exposed to the atmosphere
- The humidity of the atmosphere
- The temperature

Effective control of moisture in hygroscopic polymers almost always requires a dehumidifying air dryer (Fig. 7.36), and the processor should be armed with a good basic understanding of how the dehumidifying drying system functions, in order to keep it in good repair.

7.13 Drying Hygroscopic Polymers

A basic understanding of the relationships between water vapor and the polymer, and water vapor and the surrounding air, is a prerequisite to understanding the mechanics of drying. A hygroscopic polymer will absorb moisture from the air or give up some of its absorbed moisture to the air, depending on the temperature of the polymer and the air's relative humidity.

As noted, a dry hygroscopic polymer exposed to the atmosphere will absorb water vapor until it reaches equilibrium with the surrounding air. This process may take several minutes, or several days, depending on the type of polymer and the relative humidity of the air. Even after reaching equilibrium at, say, 50% relative humidity (RH), the polymer will absorb more moisture if it is later exposed to 80% RH. Or, if the RH were reduced to 30%, some of the absorbed moisture would be given up until it again reaches equilibrium.

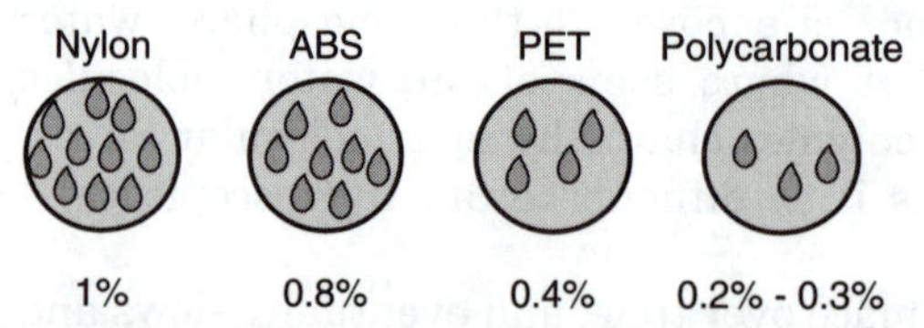

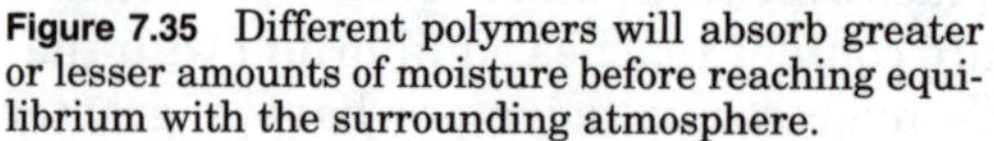

Figure 7.35 Different polymers will absorb greater or lesser amounts of moisture before reaching equilibrium with the surrounding atmosphere.

Figure 7.36 This bank of desiccant dryers is part of a central materials handling and conditioning system installed at a plastics molding plant.

Thus, the process of moisture absorption is completely reversible and is governed by these fundamental parameters:

- Polymer temperature
- Relative humidity/dewpoint of the air surrounding the polymer
- Elapsed time at the prescribed temperature and humidity conditions
- Airflow in the hopper

7.13.1 Polymer temperature

Polymer temperature is the most important consideration in most drying applications. It has a strong effect on the diffusion rate of water molecules through a hygroscopic polymer, so the rate of moisture loss is largely dependent on the polymer temperature. As the temperature of the polymer is increased, the molecules move about more vigorously and the attraction between the polymer chains and the water molecules is greatly reduced. This allows the water molecules to escape from the polymer chains. In fact, heating a hygroscopic polymer is a prerequisite to drying.

Generally, the higher the drying temperature, the faster the polymer will dry. There is, however, a practical temperature limit. If the polymer is exposed to a drying temperature that is too high, the pellets may stick together and bridge in the drying hopper. Some polymers, such as nylon, may oxidize and discolor at drying temperatures above those recommended by the material supplier.

Conversely, there is also a practical limit to how low drying temperatures can be and still be effective. The lower the drying temperature is, the longer it will take for a polymer to give up its moisture. Figure 7.37 illustrates how temperature controls the rate of drying.

7.13.2 Relative humidity/dewpoint

To reach a lower moisture level, the polymer must be exposed to a dry air environment. Thus, relative humidity and dewpoint comprise the second fundamental drying parameter. *Relative humidity* defines the amount of moisture a sample of air holds, compared to the amount it could hold at saturation. *Dewpoint* expresses how much moisture is in an air sample when it is at 100% relative humidity.

One way to increase the air's drying "ability" is to increase its temperature. As ambient air is heated, its relative humidity will decrease. The amount of moisture in the air does not change, but the total amount it can hold actually increases. If you think of the air as a sponge with a certain amount of water in it, then raising its temperature doesn't change the amount of water; rather, it increases the size of the sponge, making it "thirstier." For example, heating ambient air with a relative humidity of 50% (half saturated) to 190°F will reduce the relative humidity of the air to 2% (or 98% *un*saturated).

That is the principle by which common dryers—your home clothes dryer and hair dryer—operate. It is also the principle behind the hot-air dryer used in drying nonhygroscopic polymers. Although a hot-air

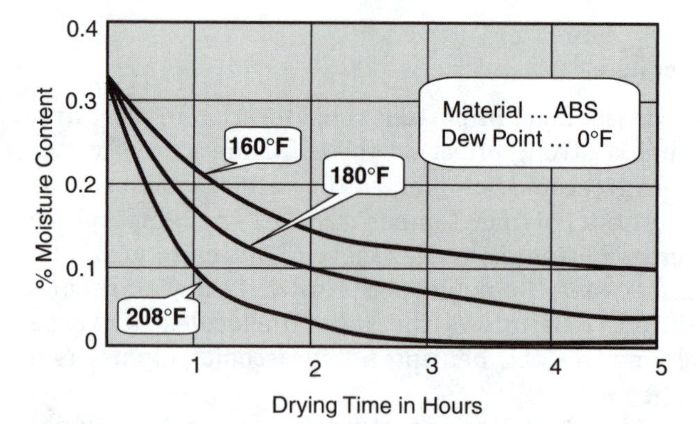

Figure 7.37 The drying temperature controls the rate of drying and, to some degree, the ultimate moisture content achievable.

dryer is capable of doing an excellent job removing surface moisture from polymers, it is not capable of producing air that is dry enough to remove moisture from hygroscopic polymers year-round.

For that, you need a dehumidifying dryer, which actually removes moisture from the drying air before heating it. The actual "dryness" of the air is expressed in terms of its "dewpoint." The *dewpoint* of air is the temperature at which the air's relatively humidity reaches 100% and moisture begins to condense. On a humid summer night, moisture in the air condenses onto surfaces when the temperature drops to the "dew" point. As the temperature drops, the relative humidity increases until it reaches saturation or 100%. When humidity is high, the dewpoint is high and dew may form at 65 or 75°F. When the air is dryer, moisture may not condense until the air temperature drops to freezing or below. *Remember,* the amount of moisture contained in the air does not change with temperature, but the air's ability to hold moisture does.

Dehumidifying dryers, then, actually remove moisture from the drying air using a molecular sieve desiccant (see Sec. 7.15.2). By removing moisture, the desiccant lowers the dewpoint of the drying air. The actual dewpoint is less important (see Fig. 7.38) than the extremely low relative humidity that is ultimately achieved by heating the drying air, but the industry has generally settled on −40°F as a desirable dewpoint for drying air and most dryers can deliver it.

When −40°F dewpoint air is heated to 200°F, for instance, its relative humidity drops to just thousandths of a percent. It is extremely thirsty air, and it will take up moisture much more readily than either low-dewpoint air at room temperature or moist air at a high temperature. Thus, most resin dryers use a desiccant to remove moisture from

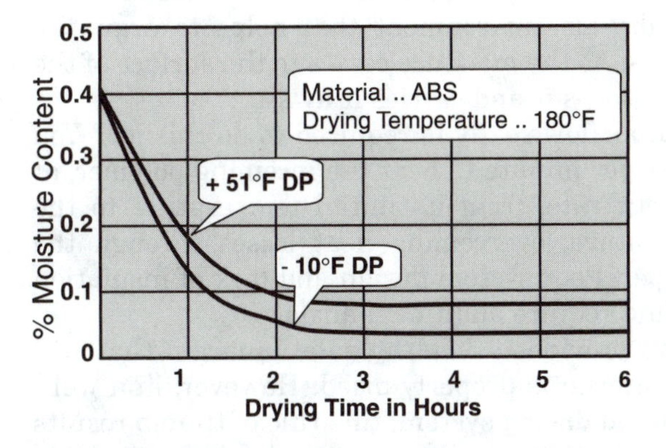

Figure 7.38 Dewpoint is important, but not critical, to final moisture level. Widely different dewpoints can result in similar ultimate moisture content.

the air to lower its dewpoint before heating it to reach the minuscule relative humidity percentage required for optimum drying efficiency.

7.13.3 Drying time

Time is critical to the drying process because plastic pellets do not dry instantaneously. Once the pellets are surrounded with heated, low-dewpoint air in a dehumidifying dryer, sufficient time must be provided to allow the heat from the surrounding air to migrate to the center of the pellets. As the temperature inside the pellet increases, the moisture diffuses and migrates toward the hot, dry air surrounding the surface of the pellet. When the water molecule reaches the pellet's surface, the hot, dry air will carry it away.

Resin manufacturers have defined how long this process takes for their particular type and grade of polymer (see Table 7.1). However, it is important to note that *effective drying time* is the time the pellets are exposed to the hot, dry air. The time the polymer is in the drying hopper at anything less than the recommended temperature and dewpoint cannot be considered drying time.

7.13.4 Airflow

In a dehumidifying air dryer it is the drying air that delivers the heat to the pellets in the hopper, and it is the drying air that strips the moisture off the surface of the pellets and carries it away. So how much airflow—cubic feet per minute—is required to do this drying job? Using an old rule of thumb which calls for 1 ft^3/min/lb/h, you would need a 100-ft^3/min dryer to dry a polymer at the rate of 100 lb/h.

However, let's consider the function the airflow plays. The air heats the pellets in the hopper to the setpoint temperature. The air also provides a low dewpoint, dry-air environment that helps to draw the moisture out of the pellets. As the moisture comes to the surface of the pellets, the air quickly absorbs it and carries it away.

But, again, how much airflow really is required to do this job? You need sufficient cubic feet per minute to heat the incoming polymer, at the applicable throughput rate, from its initial temperature to the desired setpoint temperature, overcoming heat losses through the system's hoses and hopper. Poor system design and lack of insulation will increase heat loss and require additional airflow.

The old rule of thumb provides a healthy safety margin that will normally ensure that material is properly dried. However, in a well-designed and well-insulated drying system, this rule of thumb results in wasted energy. There is economically priced technology available today to allow more scientific regulation of airflow and this will be covered later (see Sec. 7.17.3).

TABLE 7.1 Drying Times for Typical Plastics

Material	Drying temperature, F°	Drying time, h
Acrylonitrile butadiene styrene (ABS)	180	3–4
Acetal	210	2
Acrylic	160–180	2
Acrylonitrile	160	6
Cellulosics	160	6
Ionomer	150	8
Nylon	160	6
Polycarbonate (w/ 40% black)	250	3–4
Polyethylene	195	3
Polyethylene terephthalate (PET)	325–375	4–6
Polybutylene terephthalate (PBT)	250	2–3
PETG	160	3–4
Polyamide	250	2
Polyester elastomer	225	3
PEM	300	4
Polyethersulfone (PES)	300	4
Polyphenylene sulfide (PPS)	300	6
Polypropylene	195	1
GP polystyrene	180	1
High impact PS	180	1.5
Polysulfone	250	4
Polyurethane	180	3
Polyphenylene oxide (PPO)	255	2
Rynite	250	2
Styrene acrylonitrile (SAN)	180	2
Vinyls	160	1

7.14 How Physical Characteristics of Plastics Affect Drying

The size of the pellets and regrind particles being dried will have a significant affect on drying performance. Remember that polymer temperature is probably the most important consideration in most drying applications and that plastic pellets must be thoroughly heated in order for the water molecules to diffuse from the polymer chains and migrate to the surface of the pellet.

Remember also that plastic is not a good conductor of heat. It takes time for the heat from the drying air to migrate to the center of the pellet. Only when the center of the pellet is heated will the water molecules in that area be released and allowed to migrate to the pellets' surface. It is easy to see, then, that drying larger pellets requires more time for the heat to travel the longer distance to the pellet's center. Once heated, the larger pellets require more time for the water molecules to travel the greater distance to the surface of the pellet. Therefore, larger pellets may require an inordinate amount of time to dry.

If larger pellets dry slowly for the reasons mentioned here, it stands to reason that smaller pellet will dry faster for those same reasons. This is a fact. Smaller pellets will dry faster than larger pellets. However, keep in mind that we are working with a dehumidifying *air* dryer, which uses air to move the heat from the heating elements in the dryer to the pellets in the drying hopper. It also uses air to strip the moisture off the surface of the pellet and carry it away.

A hopper full of smaller pellets will have less free air space between those pellets. Less free air space between the pellets presents a restriction that *reduces* airflow. Reduced airflow will limit the amount of heat that can be delivered to the pellets in the hopper, thus reducing drying capacity. The end result is the same as if a filter becomes blinded and restricts airflow.

Pellet size is a major consideration of the material suppliers in the manufacture of hygroscopic polymers. It should also be a major consideration when reclaiming scrap regrind materials.

Virgin pellets may come in a variety of shapes, including cylindrical, round, oval, and square cut, depending on the type of material and pelletizer they were produced on. Although pellet shape may vary from one material type to another, once a material and supplier are selected, the pellets are usually of a uniform size and shape. From the previous discussion of how pellet size and shape affect drying, it is reasonable to conclude that uniform pellet size and shape are necessary for consistent drying system performance.

Irregularly shaped pellets or particles will also affect the material flow characteristics. Uniformly shaped pellets tend to be more free flowing. Free-flowing materials perform well in a mass-flow drying hopper. Irregularly shaped pellets or particles may not be free flowing and may result in bridging or other nonuniform flow through the drying hopper.

Regrind poses several problems in this regard. Excess material fines in the drying hopper will sift down through the free air space between the pellets. The fines, by their presence, will reduce the amount of free air between the pellets, causing a restriction to the drying air. Restricted airflow means reduced airflow which results in reduced drying capacity.

Regrind from flat-sheet or thin-wall bottles may tend to nest and layer as it passes through a drying hopper. This may result in reduced airflow through the drying hopper or it could force all the drying air to channel through one area of the drying hopper. Either occurrence, however, will result in a degradation of drying system performance.

7.15 How Dryers Work

Anyone who has a clothes dryer in the laundry room, or a hair dryer in the bathroom, already has a basic understanding of how heated air will pick up and carry away excess moisture.

The same principles are used in a pellet dryer. The difference is the volume of moisture involved. With a clothes dryer or a hair dryer, the starting moisture level is usually "sopping wet" and the final moisture level desired is "dry to the touch." When drying plastic pellets, the starting moisture level is usually less than 1%—already dry to the touch—and the final moisture level desired may range from a high of 0.25% for some nylons to a low of 0.003% for some grades of PET

The simplest drying system for plastic materials is the hot-air dryer. In operation, they are not much different from the laundry dryer and they are used primarily for two purposes:

1. To preheat the material, eliminating any temperature variation of the material in-feed
2. To remove surface moisture from nonhygroscopic polymers

A hot-air dryer may also be used successfully to remove some of the absorbed moisture from a mildly hygroscopic polymer, provided the application is not too moisture sensitive and the ambient relative humidity is not excessive.

7.15.1 Hot-air drying

A *hot-air dryer* is a single-pass, ambient air dryer that consists of an inlet air filter, blower, heater, drying hopper, and exhaust air filter. The blower draws ambient air into the system via the inlet air filter, and across a bank of electrical heating elements, where the air is brought up to the selected drying temperature. After being heated, the now-thirsty drying air enters the drying hopper where it comes into contact with the pellets being dried. As the temperature of the pellets increases, they give up moisture to the heated, low-RH drying air. Sufficient drying time must be provided to thoroughly heat the pellets and achieve a state of moisture equilibrium between the surface of the pellets and the drying air. The moisture-laden air exiting the drying hopper is filtered of any airborne fines before it is exhausted into the atmosphere.

Some hot-air dryers may utilize natural gas energy to heat the drying air. In these systems heating should be via an indirect heat exchanger. Direct heating would allow the emissions created during combustion to come into contact with the polymer being dried, risking contamination.

Unlike a dehumidifying dryer, a hot-air dryer *does not* remove moisture from the drying air. Its performance depends entirely on reducing the relative humidity of the drying air by heating. Remember that heating the air reduces its relative humidity, thereby greatly increasing its ability to absorb moisture off the surface of the pellets and carry it away.

For example, atmospheric air at a temperature of 70°F, with a relative humidity of 50%, contains 4.032 grains of water per cubic foot. Heating that air to 200°F *will not* reduce its moisture content, but it will greatly increase its ability to hold additional water. Just by raising the temperature of our sample of air from 70 to 200°F, while maintaining the water content at 4.032 grains per cubic foot, will reduce the relative humidity to less than 2%.

7.15.2 Dehumidifying drying

A *dehumidifying dryer* is used to remove absorbed moisture from hygroscopic polymers. While a hot-air dryer heats the drying air to lower its relative humidity, a typical dehumidifying dryer first removes as much moisture as possible from the drying air by means of a desiccant, and then heats the air to obtain an extremely low relative humidity.

Most dehumidifying dryers used in the plastics industry today employ a molecular sieve desiccant. A *molecular sieve* is a synthetic zeolite, which has a very strong attraction for water across a certain temperature range. When the temperature of the molecular sieve is maintained at a reasonably low level (under 150°F), such as when it is in the on-stream drying position, it will adsorb moisture readily. When the molecular sieve temperature is increased above 400°F, as is the case during regeneration, it will release its adsorbed moisture. Figure 7.39 shows how temperature dramatically affects desiccant's ability to hold moisture.

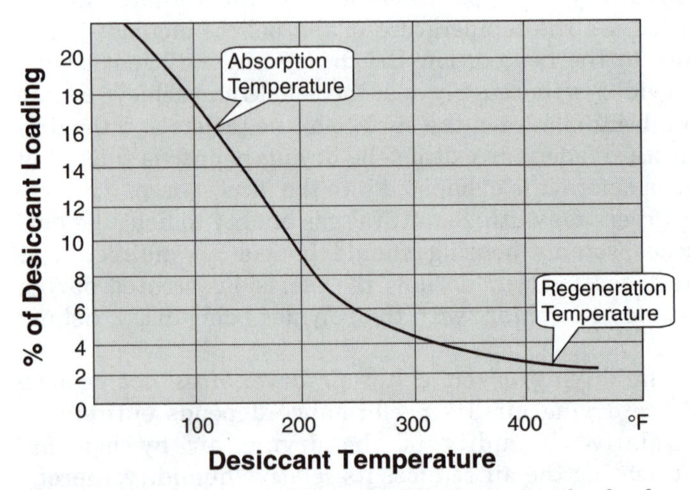

Figure 7.39 The moisture holding (adsorption) capacity of molecular sieve desiccant drops quickly as the temperature rises.

A hot-air dryer uses a single-pass drying air circuit, exhausting the spent drying air and removed moisture to the atmosphere, while a dehumidifying dryer employs a closed-loop air circuit, returning the moisture-laden, spent air to the dehumidifier, where the moisture is removed.

Figure 7.40 shows the major components of a dehumidifying drying system. The process sequence is as follows:

1. Material is supplied to the insulated drying hopper (1) on demand.
2. Heated, dehumidified air from the dryer enters the drying hopper near the bottom (2). The air flows evenly up through the material, heating it to a temperature prescribed by the resin manufacturers. At the proper drying temperature, the molecular attraction between the polymer and the moisture weakens, and the released water molecules diffuse and migrate to the pellet surface where the dry air absorbs them and carries them to the return-air outlet (3) at the top of the hopper.
3. Moisture-laden air passes through a filter (4) to remove fines and other potential contaminants.

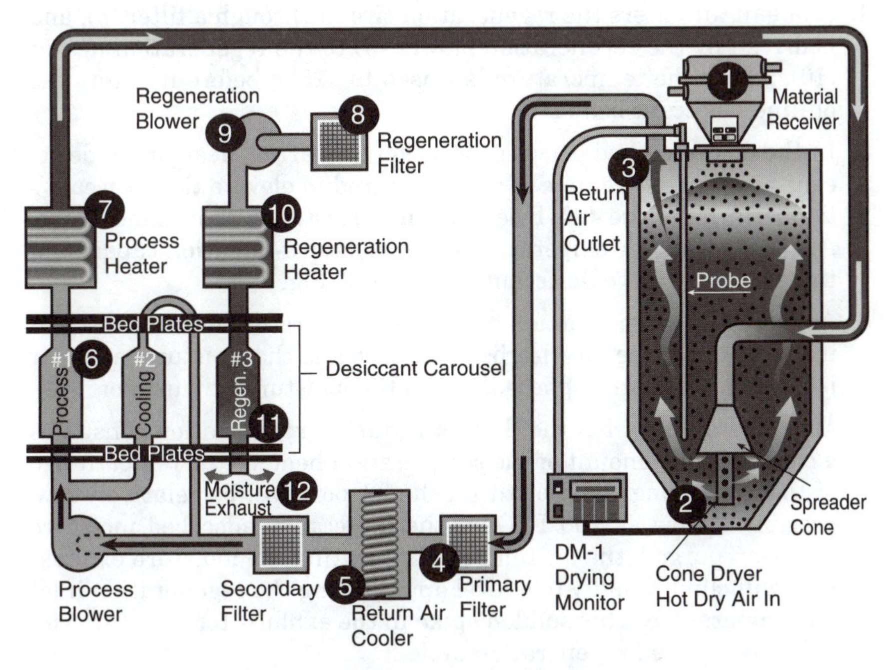

Figure 7.40 Carousel dehumidifying dryer airflow diagram.

4. If necessary, return air is passed through a heat exchanger (5) to reduce its temperature to under 150°F before it enters the desiccant bed (6).
5. The relatively cool, relatively moist return air flows through the desiccant, where it gives up its moisture.
6. A process-air blower forces the dried air through a heater (7) and back into the drying hopper to start the cycle over again.

During the drying cycle, the desiccant will continue to adsorb moisture. If the on-stream desiccant is not replaced with fresh, thirsty desiccant periodically, it will eventually become saturated and no longer remove moisture from the air. Saturated desiccant will result in high dewpoint levels in the drying air. In order to maintain the low dewpoint levels desired, the on-stream desiccant must be removed before it becomes saturated and must be replaced with freshly regenerated desiccant. Desiccant bed switching may be initiated from a timing circuit or with an integrated dewpoint meter that monitors the process air leaving the dryer.

Desiccant regeneration is similar to the process used in drying nonhygroscopic polymers with a single-pass hot-air dryer, and is best illustrated by referring to Fig. 7.40:

1. Ambient air enters the regeneration circuit through a filter (8), and is driven by the regeneration blower (9) to the regeneration heater (10), where the temperature is raised to 425°F before it enters the off-line desiccant bed (11).
2. In the desiccant bed, the regeneration air carries heat to the desiccant. A sufficient volume of air is required to elevate the desiccant's temperature to the 425°F setpoint in a relatively short time frame. At these elevated temperatures, the moisture-retention capacity of the molecular sieve desiccant is very low (Fig. 7.39).
3. As the hot regeneration air raises the temperature of the desiccant, it begins to release its adsorbed moisture and the moisture-laden air is purged to the atmosphere through the moisture exhaust port (12).
4. Moisture driven from the desiccant during regeneration consumes a considerable amount of the regeneration heat, so the temperature of the air exiting the moisture exhaust port will be relatively low, possibly as low as 200°F. Once the desiccant's adsorbed moisture has been purged, the air temperature exiting the moisture exhaust port will climb rapidly to a level approaching the regeneration inlet air temperature. This sudden spike in the exhaust temperature signals a completed regeneration cycle.

The regeneration air exhaust temperature can signal problems in the dryer. For instance, if the exhaust temperature climbs almost immediately to a level approaching the regeneration inlet temperature of 425°F, it indicates that there is no adsorbed moisture in the desiccant. This condition will result if

- The material being dried is extremely dry and therefore has not given up any moisture to be adsorbed by the desiccant.
- The desiccant has been severely contaminated and, therefore, has not adsorbed any moisture.

Once the regeneration cycle is complete, the desiccant temperature is elevated and the residual moisture is at a very low level. Now, the desiccant must be cooled before being moved in to the process airstream. This should be accomplished by diverting dry process air through the hot, regenerated desiccant bed. If the desiccant is not cooled, it will not adsorb moisture, and moist air returning from the drying hopper will pass through the hot desiccant and back into the process circuit, resulting in an undesirable dewpoint spike in the drying air.

A second problem associated with moving a hot desiccant bed back on-line is that it may raise the drying air temperature above the setpoint. If the dryer setpoint is in the lower end of the operating range (140 to 180°F) and temperature-sensitive materials are being processed, even a small spike in the delivery air temperature could be catastrophic. Surlyn, cyclohexylenedimethylene terephthalate (PETG), and acrylonitrile butadiene styrene (ABS), for example, will agglomerate when exposed to drying temperatures even slightly above the drying temperature suggested by their manufacturer. Nylon will oxidize and yellow when exposed to higher than suggested drying temperatures.

7.16 Critical Dryer Components

To function properly, a drying system must be sized for the type and amount of material being processed. Each of the components in the system deserves special attention.

7.16.1 The drying hopper

At first glance, a drying hopper may appear to be a fairly mundane piece of equipment (Fig. 7.41). But don't be misled. The drying hopper

Figure 7.41 Typical insulated drying hoppers. The two hoppers on the right are stainless-steel construction. The hopper on the left is painted carbon steel.

is a critical component of a dehumidifying drying system, and a poorly designed and manufactured drying hopper can spell disaster in a critical drying application.

More often than not, a poorly performing drying hopper will result in drying problems that tend to come and go. Many times, the problem will seem to disappear by itself, before the cause can be pinpointed. Or, deficiencies of a poorly performing drying hopper often can be overlooked during periods when the ambient relative humidity (and, therefore, the polymers initial moisture content) is fairly low. This is often the case during the winter months when a dehumidifying dryer usually doesn't need to perform to its peak. But it is usually a different story when the hot, humid, dog days of June, July, and August arrive and a dehumidifying drying system must perform to the peak of its ability.

Therefore, a well-designed and manufactured drying hopper is critical for year-round consistency in product drying. The important features of a well-designed drying hopper are

- Mass material flow
- Even air distribution
- Material heat retention
- Operator safety

The distinguishing characteristic of mass flow is that each and every pellet and particle within the drying hopper will move uniformly and steadily down through the hopper when material is discharged. A drying hopper exhibits mass flow when all of the material flows through the hopper at the same rate of speed.

Mass material flow through the drying hopper is the key to achieving consistent drying time. Mass flow is contrasted with funnel flow in Fig. 7.42. Funnel flow, the opposite of mass flow, occurs when material flows faster through the center of the drying hopper than it does

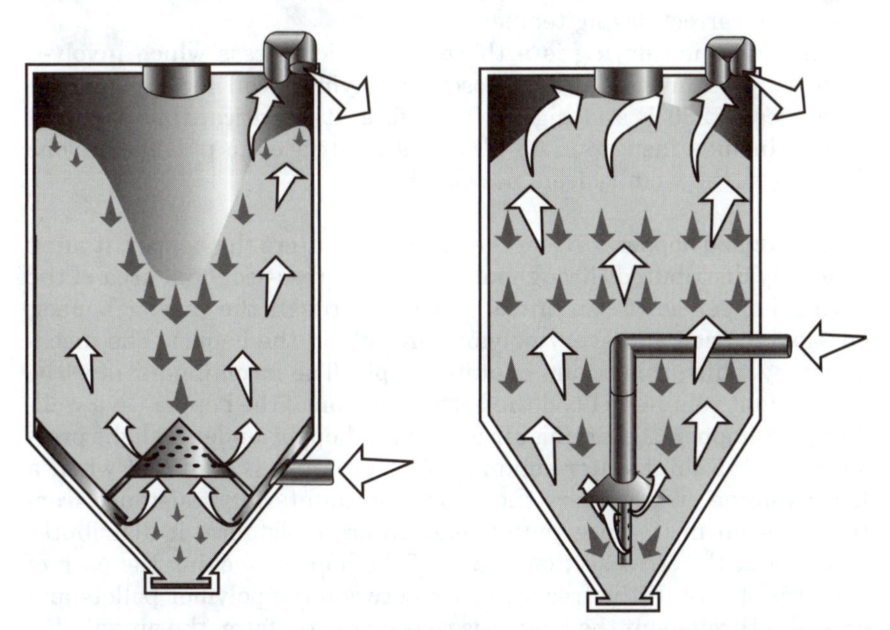

Figure 7.42 Schematics contrast funnel flow (left) and mass flow (right) in material hoppers. In a properly designed hopper, mass flow ensures that resin moves through the hopper evenly, so each pellet is exposed to the same drying conditions.

along the side wall of the hopper. Funnel flow results in inconsistent drying time in the drying hopper. The material that flows quickly through the center of the drying hopper is exposed to the drying air for a shorter period of time. It may not be exposed to the drying air long enough. The material that flows slowly along the side wall of the hopper, on the other hand, is exposed to the drying air for a longer period of time. If exposure time is too long, then overdrying or thermal degradation may occur.

Even air distribution throughout the cross-sectional area of the drying hopper is required to envelope each and every pellet with the hot, dry air needed for drying. As noted previously, heat and a dry-air environment are fundamental to successful drying. If the drying air tends to channel through the drying hopper, neglecting one side of the hopper, while favoring the other, uneven, inconsistent drying will result.

Because polymer temperature is so important, heat retention in the drying hopper is of paramount concern. Well-designed and well-insulated drying hoppers will minimize heat loss and maximize dryer performance, while reducing operating costs and airflow requirements. Most important, a well-insulated hopper will ensure that even the pellets that come in contact with the metal side wall of the drying hopper are at the correct drying temperature.

Dehumidifying drying is a thermodynamic process which involves high temperatures. Many hygroscopic polymers call for drying temperatures of 250°F and higher. Therefore, it is critical that a drying hopper include insulation on all exposed surfaces to prevent anyone working in the vicinity from possible burns.

Sizing a drying hopper. When the drying air enters the hopper, it must be evenly distributed throughout the entire cross-sectional area of the drying hopper so it can travel upward through the drying hopper, enveloping each and every polymer granule in the hopper. The mechanism by which this occurs is quite simple. The incoming air must be of sufficient volume to flood the bottom section of the hopper. In a well-designed hopper, the drying air enters the hopper under a slight pressure (5 to 15 in of water column). This pressure is achieved when a large volume of air enters the hopper, encountering resistance from the material in it. Once the air encounters resistance, it distributes throughout the cross-sectional area of the hopper, seeking the path of least resistance. If the free air space between the polymer pellets and granules throughout the hopper is reasonably uniform, the air will disperse itself evenly throughout the hopper.

The air *volume* passing through the drying hopper is also critical for obtaining and maintaining adequate drying performance. The dehumidifier, with its fixed volume of airflow, must be matched properly to the drying hopper size with its fixed diameter. A mismatched dryer-hopper combination will not produce the drying performance expected.

For instance, if the dehumidifier's air volume is too low for the hopper size, the drying air may not develop sufficient pressure as it enters the bottom of the drying hopper. This will result in poor air distribution throughout the cross-sectional area of the drying hopper, which will allow the drying air to channel through the hopper. If, on the other hand, the dehumidifier has too much air volume, the airflow will interfere with the mass flow of material through the hopper. Excessive air volume may even suspend material and cause bridging in the cone section of the hopper.

Cone drying. Molders and extruders who frequently change the type of materials they are processing may find that material in the lower portion of the drying hopper is not properly conditioned during static predrying. Because drying air is not always forced to the very bottom of the drying hopper, material at the bottom of the hopper may not be exposed to the drying air when the hopper is first filled with the wet material.

The most common solution to this problem has been to drain the undried material from the bottom of the hopper after predrying is complete, being careful to drain all the material that was not exposed to the drying air. This was an acceptable solution in the past, but in today's environment this wasteful procedure is no longer an acceptable practice. Another solution is the cone dryer.

The cone dryer connects the return line of the dryer (suction) through a hand-operated valve to a specially designed hopper material discharge assembly. When the valve is opened during predrying, a small quantity of hot drying air is drawn from the hopper air inlet, through the hopper cone, down to the discharge assembly, and into the dryer return line. Once predrying is completed and there is no longer a need to draw hot drying air down through the cone, the hand-operated valve is closed and drying continues conventionally.

7.16.2 Air-handling systems

In a dehumidifying dryer, unlike a hot-air dryer, airflow is a closed-loop process (see Fig. 7.40). Hot, dry air is delivered to the drying hopper where it picks up moisture from the resin before returning to the desiccant bed where the moisture is removed. Air is then reheated on its way back to the hopper for another trip through the system.

Air delivery circuit. As the name implies, this circuit delivers hot, dry air from the dryer to the drying hopper. As noted previously, the drying air delivered to the hopper must be of a sufficient volume to

- Flood the drying hopper, enveloping each and every polymer granule
- Carry enough Btus to develop and maintain the desired temperature profile through the vertical height of the drying hopper
- Overcome heat losses through the delivery hose and hopper side walls

In smaller dryers, mounted on casters for mobility, a flexible hose is used to move air to and from the drying hopper. On larger, permanently installed dryers, both rigid and flexible tubing may be used in the air-handling circuits.

The air delivery circuit must be airtight, since any air leaks will result in reduced air volume to the drying hopper, wasted heat energy, loss of drying efficiency, and dewpoint deterioration. The delivery air piping should always be well insulated. Any heat loss will result in wasted energy and increased operating costs. The routing of the delivery air circuit—like all system piping—should be arranged so as to prevent damage. It should be situated out of reach of lift trucks, and placed so it is not a convenient foothold for anyone climbing on the equipment.

The preset drying temperature should be monitored and controlled at the inlet of the drying hopper to compensate for any heat loses in the delivery air piping.

Return air circuit. The reason all dehumidifying polymer dryers employed in the plastics industry use a closed-loop drying air circuit is that the moisture content of the air exiting the drying hopper is always drier than the ambient air. The amount of moisture being driven from the polymer in a typical drying application is very low—always less than 1% by weight. This small amount of moisture, when added to the drying air as it travels through the drying hopper, does not raise the return air dewpoint temperature above that of the ambient air. Therefore, it makes economic sense to salvage and reuse the moisture-laden (but still relatively dry) air exiting the hopper, rather than to dehumidify the much wetter ambient air.

The return air circuit must be airtight to prevent high-humidity ambient air from entering the dryer and prematurely overloading the molecular sieve desiccant. But, the return air line is *not normally insulated,* since heat loss from the return air will actually improve dehumidifier performance.

When the drying air exits the top of the drying hopper, it has already given up a substantial amount of its heat to the polymer in the hopper, picking up moisture in exchange. Low return air temperatures are good because the lower the temperature of the air returning to the dehumidifier, the more efficiently the desiccant in the dehumidifier is going to operate. High return air temperatures, on the other hand, decrease desiccant efficiency. In fact, if the return air reaches 180°F, the absorption capacity of the molecular sieve is reduced by over 50%. These same high temperatures can also result in premature blower failure. As a general rule, the temperature of the return air entering the dehumidifier should not exceed 150°F.

Any time return air temperatures above 150°F are expected, a return air heat exchanger needs to be included in the system design. For instance, in a lower-temperature drying application (drying temperature between 140 and 250°F), the hopper exit temperature can be expected to range from 100 to 150°F, which is in the acceptable operating range of the molecular sieve desiccant. If, however, a change in the system operating conditions (a reduction in material throughput rate, for instance) causes return air temperatures to rise, a heat exchanger will be required to reduce return temperatures to under 150°F.

In a higher-temperature system, drying at 250 to 350°F, hopper exit temperatures can be expected to range from 150 to 250°F, and a return air heat exchanger will always be required.

Volatile traps. Occasionally, a hygroscopic polymer may contain components or additives, such as plasticizers, and fire retardants, that will vaporize at less than the recommended drying temperature. These volatiles or vapors, once driven from the polymer, will be carried into the return air circuit and will pass through the return air filter in the gaseous state. However, as the return air makes its way back to the dehumidifier and its temperature is lowered, the volatiles condense, usually into a waxy, oily substance that contaminates the desiccant.

There are several common methods for contending with volatiles:

- Lower the drying temperature to minimize the amount of volatiles being driven from the polymer. This will reduce the volatiles but it will also diminish the performance of the drying system.
- Add a volatile trap to the return air circuit. A *volatile trap* is a heat exchanger that lowers the return-air temperature and promotes condensation of the volatiles before they enter the dehumidifier. The volatile trap will collect the condensate, which is usually a waxy, oily substance, so it can be disposed of in a controlled manner.
- Add a demister to the return air circuit. A demister is made up of a heat exchanger and a coalescing filter. The return air is first cooled to near the condensing temperature of the volatile. It is then passed through the coalescing filter where the volatile condenses and is filtered from the air.

Desiccant bed. The heart of a dehumidifying dryer, the desiccant, removes moisture from the drying air. Without the desiccant, the dehumidifying dryer would be no more than a hot-air dryer.

There are several basic approaches to the design of desiccant beds used in plastics dryers:

- Twin-tower dryers employ two large cylindrical "towers" of loose desiccant material. While one is on-line removing moisture from process air, the second tower is in either the regeneration or the cooling cycle. Before the on-line desiccant becomes saturated, process airflow is diverted to the regenerated tower and the saturated desiccant is regenerated. Twin-tower dryers can be relatively inexpensive, but they require large quantities of desiccant and long regeneration cycles.
- Carousel dryers (Fig. 7.43) use densely packed cartridges of desiccant that are sized to match the dryer capacity rating. Cartridges are cycled through drying, regeneration, and cooling stations. The small amount of desiccant in each cartridge requires less air and less heat for regeneration and faster cycling tends to minimize temperature and dewpoint spikes. During service, the self-contained cartridges can be replaced easily, eliminating the need to handle loose desiccant.

Filters. Many plastic materials—particularly reclaimed scrap—contain fines that can accumulate over time and blind the filters. If the filter fails and allows the fines to pass, they will contaminate the molecular sieve desiccant. Therefore, a dryer should be equipped with well-maintained filters on the return air line and on the regeneration air inlet. As a general rule, the process air filter will be the most critical because of the quantity of fines in the material being dried. But under dusty plant conditions, the regeneration filter is also important and must be maintained periodically.

These filters are the most maintenance- and service-intensive components of a dehumidifying dryer. They must be kept relatively clean so they do not restrict and reduce air volume, and they must be maintained in good condition since any ruptures will allow fines to pass into the dehumidifier, risking desiccant contamination. Depending on the amount of fines in the material being dried, the return air filter may require service as frequently as every shift or as little as every several months. The only prudent way to determine a reasonable filter service schedule is to monitor the filter condition on a daily basis for a period of time. If the amount of fines in a particular application is extreme, it may be advisable to add a secondary return air filter, such as a cyclone separator or free-standing bag house type filter.

Figure 7.43 Carousel dryers use densely packed cartridges of desiccant that are sized to match the dryer capacity rating. Cartridges are cycled through drying, regeneration, and cooling stations (see also Fig. 7.40).

7.17 Monitoring Drying Conditions

To effectively control the drying process, it is necessary to monitor and control the four fundamental drying parameters: temperature, dewpoint, airflow, and time.

7.17.1 Drying temperature

It is imperative that the drying air entering the hopper be maintained at the correct temperature. Make sure the dryer's temperature controller is adjusted to the correct setpoint temperature for the material being processed. Then, monitor the air temperature entering the drying hopper with an independent, hand-held pyrometer to verify the accuracy of the dryers' controller. Pay particular attention to the drying air temperature immediately after a freshly regenerated and cooled desiccant bed cycles to the on-stream position. If the freshly regenerated desiccant has not been properly cooled prior to being brought on-line, the hot desiccant may cause an unacceptable temperature spike in the drying air.

Double check the temperature of the air returning to the dehumidifying dryer, because its temperature will have a strong effect on the moisture-loading capacity of the molecular sieve desiccant and, therefore, on the dryers' dewpoint temperature capability. The temperature of the air returning to the dryer should be in the range of 120 to 130°F for optimum efficiency of operation, although return air temperatures of up to 150°F are considered acceptable. Above 150°F, the moisture-loading capacity of the desiccant deteriorates quickly, and corrective action should be taken to lower it into an acceptable range.

If, on the other hand, return air temperatures drop below 120°F, it will marginally increase the desiccant's moisture loading capacity, but it will also limit the high end drying temperature the dryer is able to achieve. Reducing the return air temperature to below 120°F is not normally considered cost effective.

7.17.2 Dewpoint

To get a complete picture of a dryer's dewpoint performance, the dewpoint must be monitored continuously throughout the entire dryer cycle to ensure it remains relatively constant. The dewpoint should be monitored closely when a fresh desiccant tower or cartridge is first brought on-line. If the freshly regenerated desiccant has not been properly cooled, the dewpoint will spike to an unacceptable level and will remain there until the desiccant has cooled to a temperature at which the desiccant can adsorb the moisture from the return air.

Most dryers are designed to produce a drying air dewpoint temperature in the −40°F range under normal drying conditions. This will usually be more than adequate, since most moisture-sensitive polymers dry very well with a dewpoint temperature in the 0 to −20°F range.

Many of today's dehumidifying dryers have dewpoint instruments integrated into the dryers control, but it is also a good idea to have a hand-held dewpoint instrument to periodically check the accuracy of the integrated instrument. However, the temperature of the air sample will affect the accuracy of most dewpoint instruments, so it is advisable to add a cooling coil, consisting of several feet of metal tubing, to a hand-held dewpoint instrument. This will cool the air before it reaches the temperature-sensitive sensor in the dewpoint instrument.

As long as your dryer is producing an acceptable drying air dewpoint temperature, there is no need to monitor the *return* air dewpoint. However, if you are having trouble maintaining an acceptable drying air dewpoint, then monitoring the return air dewpoint can be very helpful in determining what the problem is.

7.17.3 Drying time and airflow

You might think it is easy to monitor drying time. If a given material requires 4 h of drying time, the hopper capacity is 500 lb, and the machine throughput rate is 100 lb/h, then keeping the hopper at least four-fifths full at all times should ensure adequate drying when the correct air temperature and dewpoint settings are maintained.

Remember, though, that "time in the hopper" is not necessarily "effective drying time." If dryer airflow is not sufficient to keep the correct amount of material (400 lb in the previous example) at the proper temperature, then the material will not be adequately dried. Airflow, then, is just as important as temperature and dewpoint. But direct measurement of a dryer's airflow can be tricky at best, an indirect measurement—monitoring the vertical temperature profile of the material—in the drying hopper is recommended (see Fig. 7.44).

This is accomplished by placing a temperature probe directly in the hopper and monitoring material temperatures from the bottom of the hopper to the top. The temperature of the pellets at different levels throughout the vertical height of the hopper (see Fig. 7.45) will provide a good indication of whether airflow is too low, too high, or just right.

In the drying example given previously, if the air volume flowing through the hopper is restricted or inadequate for any reason, the dryer will not be able to heat 4 h worth of material (400 lb) up to the correct drying temperature. If the entire hopper of 500 lb of material is at the correct drying temperature, the indication is excessive airflow. Excessive airflow wastes energy and results in high dryer return air temperatures, which will effect the dryer's desiccant capacity.

A perfect balance of airflow and energy use would be achieved when the dryer heats the specified 400 lb of material up to the correct drying temperature and the temperature of the remaining 100 lb of material in the top of the hopper drops sharply.

Figure 7.44 Drying monitor provides operator with temperature readings from inside the drying hopper, which, along with residence time, provides the best indication that material is being properly dried.

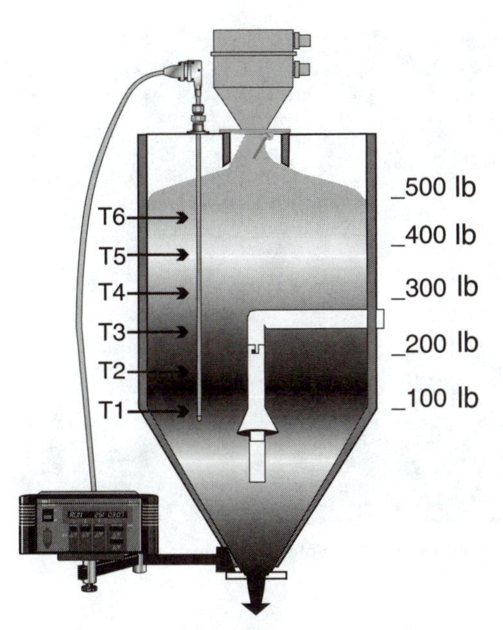

Figure 7.45 Drying monitor probe reads temperature at six different points in the hopper. Maintaining the correct drying temperature profile through the hopper results in proper drying. A change in the profile can warn of problems.

7.18 Drying System Configurations

Any dehumidifying drying system will include the dryer itself, the hopper in which the material is actually dried, and a means to deliver dry material to the processing machine feed throat. Beyond that, there are several different ways to configure these system components, and each has definite advantages and disadvantages.

7.18.1 On-machine drying

The original drying system configuration, this setup incorporates a machine-mounted hopper and drying unit (Fig. 7.46). Although manual material loading is possible, raw material is usually loaded into the drying hopper by vacuum conveying and then, once dry, flows by gravity directly into the processing machine. The dryer is sized to match the hopper requirements and normally sits next to the machine. Among the advantages of this system are

- *Flexibility.* Temperature and dewpoint conditions in the hopper can be set to exactly match the material being processed.

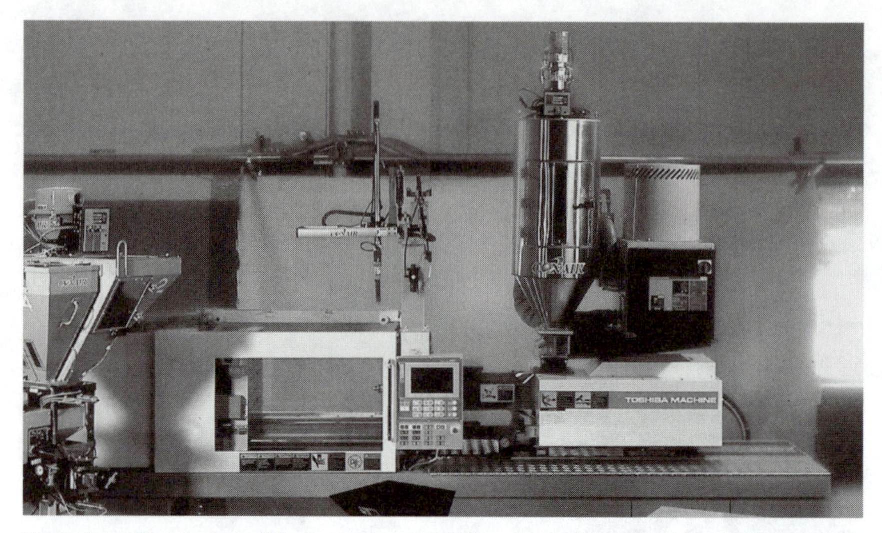

Figure 7.46 The dryer and hopper are mounted on the molding machine feed throat.

- *Simplicity.* Each drying unit is self-contained and can be placed on a machine as needed or removed when not needed. No piping or long-distance conveying required.
- *Material control.* Because dried material flows directly into the machine, there is no need to be concerned about moisture regain after drying.

The dedicated machine-mounted drying arrangement has some definite disadvantages too:

- *Difficult material changes.* The machine-mounted hopper holds several hundred pounds of material and it will need to be completely emptied, cleaned, and refilled whenever a material change is required.
- *Slow startup.* Once a new material is loaded, it will need to dry for several hours before it can be processed. During that time, the machine will be down and unproductive.
- *Floor space.* Space around the processing machine must be available for the dryer.

7.18.2 Mobile drying systems

An excellent alternative to machine-mounted drying, mobile systems (Fig. 7.47) are comprised of a dedicated dehumidifying dryer and hopper, mounted on a sturdy frame with casters for mobility. This system has

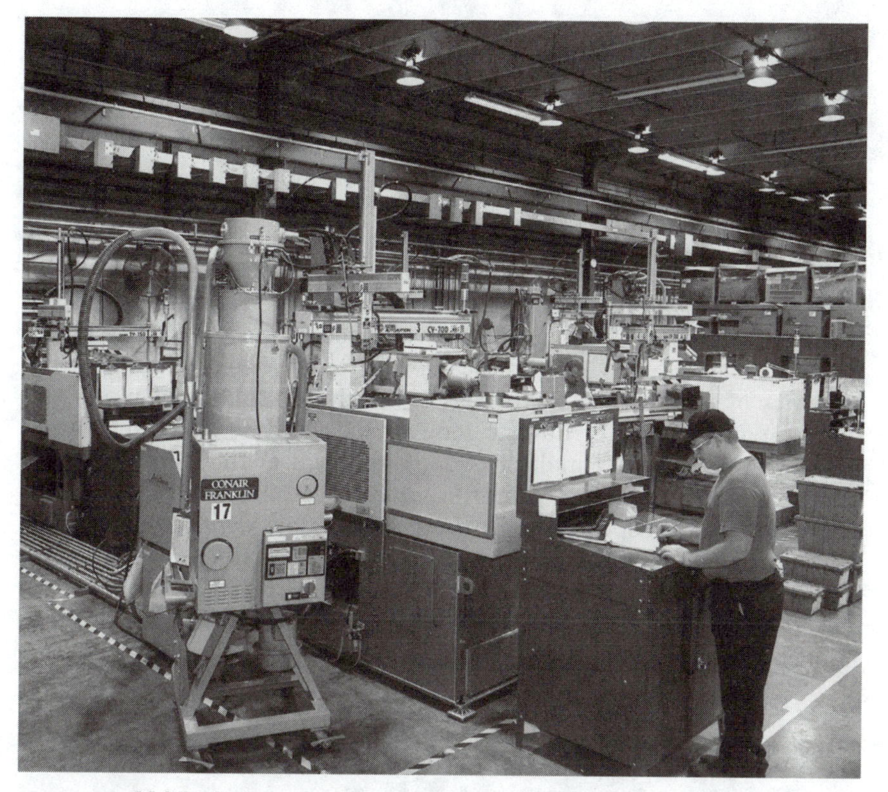

Figure 7.47 Mobile drying-conveying systems like this one feature a dedicated dehumidifying dryer and hopper, and vacuum or compressed air loader, mounted on a sturdy frame with casters for mobility.

most of the advantages of a machine-mounted system and it eliminates some of the disadvantages.

- *Flexibility.* Drying conditions can be tailored to the specific material being dried.
- *Easy material changes.* Material can be loaded and predried away from the machine and then wheeled into place for immediate startup.
- *Economy.* In a plant where material and color changes are frequent, a mobile dryer that can be used to predry material remotely is required only rarely; one mobile drying unit can be scheduled to service several different machine, saving capital costs.
- *Safety.* All drying equipment and material are off the machine within easy reach and do not interfere with other work on the processing machine.

The chief disadvantage of mobile drying is

- *Floor space.* Space around the processing machine must be available for the dryer. However, the mobility of these units means that they can be stored out of the way when not in use.

7.18.3. Central drying systems

For a processor that uses a lot of hygroscopic material, a central system (Fig. 7.48) may offer significant advantages. In fact, it is easy to see how a large-volume, continuous run processor could justify a central system. Generally, these processors have the need to handle large amounts of similar materials, fed to machines that may make the same product day after day for long periods of time. Custom processors, however, usually make frequent material changes and are less likely to need a central drying system. And yet, even these short-run processors tend to specialize, running lots of similar parts using similar materials. Even if they cannot standardize production plantwide, it may be possible to create discrete manufacturing cells within their plant and centralize drying with each cell. Thus almost any processor can achieve the following:

Figure 7.48 A central material drying-distribution system includes one or more dryers, serving multiple hoppers, dedicated to drying different materials. A manifold system allows dried material to be conveyed wherever it is needed.

- *Energy efficiency.* One large dryer can efficiently serve multiple hoppers. Booster heaters located at the inlet to each hopper make it possible to exactly match the temperature requirements of each resin and avoid heat losses between the dryer and the hopper.
- *Safety.* Drying takes place in one area, away from the processing machine.
- *Easy material changes.* Switching from one material to another may be as simple as switching conveying tubes at a central materials distribution manifold. New materials can be loaded and predried without affecting on-going production.
- *Floor space and manpower.* No machine-side floor space is required. A central system requires less floor space per machine because one dryer can serve several hoppers and one hopper can serve several machines.
- *Material control.* With a well-organized, well-controlled central system there is less chance of contamination and improper blending. Pocket conveying meters material into the distribution box under the drying hopper so that only small amounts of material are conveyed and no extra material remains in the conveying lines to absorb moisture or contaminate subsequent material lots.

The only real disadvantages to a central system are

- *Capital cost.* A central system will always require more up-front expenditures and costs associated with central material conveying systems required to get material from the dryers to the processing machines. System expansion can be expensive too.
- *Material control.* This is both an advantage and a disadvantage. An error or miscalculation in material control can be extremely costly because of the amount of material involved and the number of processing machines served by a central system.

7.19 Gas or Electric?

Plastics processors, historically, have been very dependent on electric power for most of their processing needs. And, because of its dependence on heat, the resin drying process can be a heavy consumer of electrical power. It's not surprising, then, that processors are showing increasing interest in natural gas as an alternative energy source for drying. In fact, a gas dryer can provide energy cost savings of up to 70% (see Fig. 7.49).

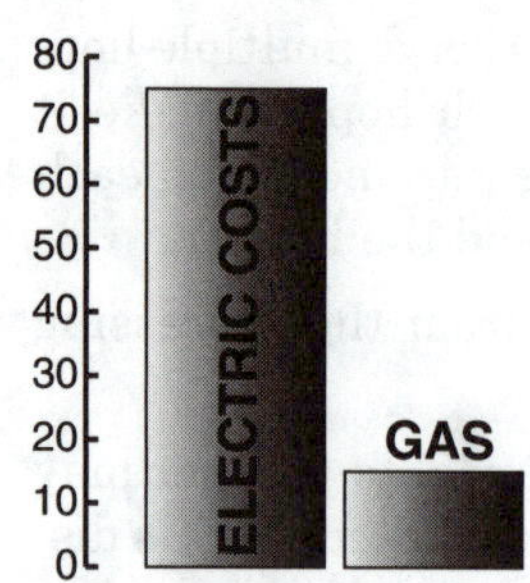

Figure 7.49 Bar chart illustrates the savings that are possible when using gas instead of electricity for drying.

Today's gas dryers incorporate the most advanced gas burners available, featuring a ceramic-metal fiber matrix firing surface that provides flameless, efficient radiant gas heat, with low emissions. Gas-fired dryers are available in both large central systems and machine-side portable units. In addition, process air heaters are available to convert installed dryers from electrical heating to gas (Fig. 7.50).

As an alternate heat source, natural gas has these advantages:

1. The cost per btu is approximately one-quarter of the cost of electric per Btu.
2. Natural gas appliances have a long and proven track record as a safe heat source.
3. By reducing electric consumption during peak hours, companies whose rates are determined by peak usage can qualify for a lower overall electric rate.
4. Retrofitting existing dryers with a gas process air heater can free up existing electric switch gear for use on other new machinery.
5. Gas process heaters require less maintenance than comparable electrical units.

7.20 Handling Dried Material

Once plastic materials have been properly dried, it is imperative that they be protected from moisture regain prior to molding. As noted in Sec. 7.18, each drying system will approach the problem in a slightly different way.

A drying hopper that is mounted directly to the machine throat, for instance, will require no special accommodations because the resin

Figure 7.50 Here, a gas-fire process air heater has been used to convert an existing electric dryer to clean, economical, and safe natural gas.

goes directly from the hot, dry environment of the hopper to the processing machine, eliminating the possibility of moisture regain. When dried material needs to be conveyed from hopper to machine, however, precautions need to be taken.

The key is to prevent material from coming into contact with moisture-laden ambient air for any appreciable length of time. Some materials regain moisture slowly and will stay dry enough to process for 2 or 3 h after exposure to ambient air, while others will regain an unacceptable amount of moisture in a matter of minutes.

Many processors choose to use a dry air generator to produce conveying air, thus avoiding exposing dried resin to ambient air.

However, the cost associated with adding another piece of equipment to the system may be completely avoidable if precautions can be taken to limit the amount of time the material is exposed to ambient air. A better approach is to keep material out of the conveying lines as much as possible. That calls for *pocket conveying,* a technique in which a discrete amount of dried material is dispensed into a closed chamber (see Fig. 7.51) under the drying hopper. That small "pocket" of material is then vacuum conveyed, using ambient air, to a small hopper loader (Fig. 7.52) on the processing machine. The quantities being conveyed are so small that they can be processed within minutes of leaving the drying hopper. In addition, the supply lines are constantly purged of material, so nothing is left behind to become separated or to pick up moisture.

For the ultimate protection against moisture regain, processors can combine pocket conveying with dry-air conveying.

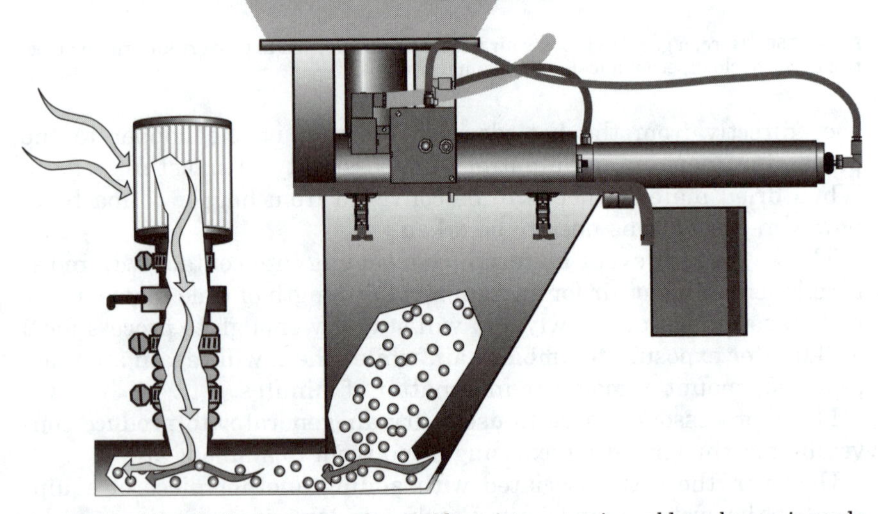

Figure 7.51 A "pocket" conveying valve avoids moisture regain problems by moving only small amounts of dried material and preventing material from remaining in conveying lines.

Figure 7.52 To minimize the time dried material is outside the drying hopper before molding, and thus prevent moisture regain, mini hoppers, like the one in the foreground, are used.

Section 3

CAD, CAM, and CAE

Peter Kennedy

Moldflow Corporation, Lexington, Massachusetts

7.21 Introduction

We begin with some definitions of the acronyms used in this section. The term *computer-aided drafting* (CAD) is part of common language today and refers to the use of computers for drafting and modeling of product designs. In a sense, CAD is the technological backbone that provides opportunities for concurrent engineering and for the subsequent use of CAE and CAM. The CAD industry has seen tremendous growth since its inception in the 1970s and continues to grow with advances in both technology and integration of CAE and CAM products. We will not discuss CAD specifically in this section but will mention how trends in the CAD industry are impacting plastics CAE. CAD has been embraced by many companies and plays a central role in CAM and CAE. Using a CAD system, the designer creates a representation of the part to be manufactured. An application is a component to be made from plastic or the mold to produce the part. The CAD system creates a representation of the component's geometry. This representation may be used for a variety of downstream operations including rapid prototyping, CAE analysis, numerical machining, mold building, or tolerance and assembly checking.

Computer-aided manufacturing (CAM) refers to the production or alteration of control data for manufacturing. Often the term is used to specifically refer to computer numerically controlled (CNC) machine tooling. With regard to the plastics industry, CAM generally refers to the generation of CNC cutter paths for the production of molds and dies. More recently, plastics CAM has been extended by the availability of "smart" controllers for injection-molding machines. This is an important development and will be discussed in detail subsequently.

The term *computer-aided engineering* (CAE) describes the use of computers for analysis of a particular design. Often the design is a new product, but in the context of plastics it could be a cooling circuit layout for an injection mold or even the mold itself. Frequently, the term CAE is used to embrace both CAM and CAD though, strictly speaking, it refers to the analysis stage only. There are many types of analyses available these days. Typical examples include

- Structural analysis for determination of deflections and stresses in a design subject to applied load.
- Thermal analysis in which the temperature distribution is calculated.

- Flow analysis in which the flow of a material through a defined region is calculated.
- Mechanical analysis where motion of a linkage or mechanical system is determined.

Regardless of the type of analysis, all CAE involves the use of a mathematical model that simulates the physical process or conditions to which the design is exposed.[1] The mathematical model is typically a set of equations, usually involving partial derivatives and suitable boundary conditions to ensure a unique solution. In order to implement the mathematical model in computer software, we need to use appropriate numerical methods for the solutions of the equations forming the mathematical model. One of the features of all numerical methods is that the problem must be discretized. For CAE this means creating a set of points, which are called *nodes,* at which the quantities (e.g., temperature, pressure, stress) of interest will be calculated. One of the most popular methods is the finite element method.[2] In addition to the generation of points, the finite element method requires that the points be arranged to form geometrical entities called *elements.* The combination of nodes and elements is called a *mesh.* For two-dimensional problems, the mesh generally consists of triangular or quadrilateral elements. In three dimensions, the elements are usually tetrahedral or hexahedral in shape. Thus, the discretization step in finite element analysis involves the generation of a mesh which represents the region in which a solution to the problem will be sought. These ideas are illustrated in Fig. 7.53. It is common to refer to meshed objects as models for analysis. These models, not to be confused with the mathematical model mentioned earlier, are abstractions of the component under consideration and provide information for the analysis in a form understood by the computer. In the plastics industry, and for the purposes of this section, CAE describes the simulation of a particular process, e.g., extrusion, injection molding, film blowing, etc. Generally, this will involve use of a computer code, input of material properties, definition of the region in which calculations are to be carried out, and input of processing conditions. We will see later that generation of a mesh is an important part of the process and can represent a significant part of the total time involved in CAE.

In this section we focus on injection molding. For us, CAE will involve the simulation of the injection-molding process. Injection molding is the most mature area of the plastics industry with respect to utilization of CAD/CAM and CAE. Moreover, developments in the injection-molding field represent the state of the art. Of course, many of the general ideas of simulation of injection molding may also be applied to other plastic part production methods.

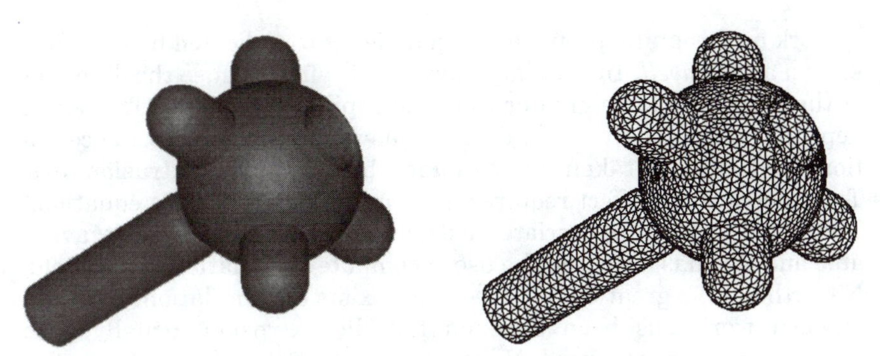

Figure 7.53 Solid model geometry on left and meshed model on right.

7.22 Simulation and Polymer Processing

All major manufacturing processes for plastic products have a common feature, namely, the melting and subsequent solidification of the material. The notion that processing has a dramatic effect on the properties of the manufactured article has been known since plastic processing began. In practice, the relationship between process variables and article quality is extremely complex. It is very difficult to gain an understanding of the relationship between processing and part quality by experience alone. It is for this reason that simulation was born, but it is interesting to note that CAE has been much more successful in injection molding than in other areas. In this section we review why this is the case and discuss some aspects of simulation for other processes.

The major processes encountered today are

- Extrusion
- Blown film extrusion
- Blow molding
- Vacuum forming
- Injection molding and its variants such as injection compression molding, gas injection, and co-injection.

7.22.1 Extrusion

Much effort has been spent on studying extrusion, as the plastication stage is applicable to other processes. Plastication models were given in Ref. 3. However, the plastication process is still actively researched today, particularly with regard to mixing and screw efficiency.[4,5]

Work on designing profile extrusion dies is complicated by the effect known as *die swell.* In capillary flow, elastic effects cause the diameter of the extrudate to be greater than the capillary diameter. This effect depends on the length of the capillary as well as the processing conditions and must be taken into account when designing extrusion dies. To model such an effect requires a viscoelastic constitutive equation.[6] There is a lack of appropriate models for which data is readily available and this has hindered the use of computer simulation in this field. Nevertheless, a great deal of literature exists on simulation.[7]

Much work has been done on flat die extrusion, usually with Newtonian or generalized Newtonian material models.[8] When flat extrusion dies are operated at high pressure, the deflection of the die itself can be significant and may need to be accounted for.[9]

7.22.2 Blown film extrusion

This process has been studied for some time[10] and the effect of viscoelasticity on bubble shape, velocity, and stress are still being explored.[11] Some simulation of the flow in the spiral mandrel has also been performed.[8]

7.22.3 Blow molding

Blow molding is complicated by the complex stress field set up in the materials when the parison is inflated. This amounts to a biaxial stretching of the molten polymer and it is difficult to obtain material data under these conditions so that simulation may be performed. Despite this, much work on the inflation stage has been done, mostly with the aim of determining the final thickness distribution.[12] Recently parison inflation has been simulated using three-dimensional finite elements[13] and with remeshing of the parison as it inflates to minimize error from element distortion.[14]

7.22.4 Thermoforming

In principle, thermoforming is quite similar to the parison inflation stage of blow molding.[12] A complication is the use of plugs to assist forming. The physics of the interaction between the molten material and the plug is not well understood and is difficult to simulate. As a result, there are some limitations on what can be simulated today.

7.22.5 Injection molding

Injection molding and its variants have been by far the most successful area of simulation and many codes are available. Reasons for this were mentioned in the introduction but may be reduced to three key facts:

1. The process may be represented by a relatively simple material model, namely, the generalized Newtonian fluid which allows the viscosity of the fluid to be a function of the rate of deformation.
2. The governing equations may be reduced to a simple form that is suitable for solution on ordinary computers.
3. Injection-molding simulation has a high return on investment.

The last point requires some explanation. Previously in this section we considered a number of common production processes for polymers. Of these, the majority were continuous processes. For these processes, although the process physics may be complex, the die is generally quite simple and inexpensive to make. Moreover, the processes allow considerable flexibility in changing process conditions. Of the two noncontinuous processes mentioned earlier, namely, blow molding and thermoforming, the cost of tooling in these industries is also relatively inexpensive. In fact, the cost of a blow-molding mold can be as low as one-tenth that of an injection mold for a similar article.[15] Moreover, blow-molding machines provide the operator with enormous control so problems can generally be solved on the factory floor. In contrast, in injection molding, problems experienced in production may not be fixed by varying process conditions as with other processes. While there is scope to adjust process conditions to solve one problem, often the change introduces another. For example, increasing the melt temperature and so decreasing the viscosity of the melt may cure a mold that is difficult to fill and which is flashing slightly. The increase in temperature may, however, cause gassing or degradation of the material possibly resulting in unsightly marks on the product. The fix may be to increase the number of gates or mold the part on a larger machine. Both of these are economically unfavorable—the first, involving significant retooling—is also costly in terms of time and the second will erode profit margins as quotes for the job were based on the original machine which would be cheaper to operate. Simulation, on the other hand, can be performed relatively cheaply in the early stages of part and mold design and offers the ability to evaluate different design options in terms of both part and mold design.

In short, compared to other production processes, injection molding demands more of part and mold designers—experimentation after the mold is built is expensive in terms of time and money. Injection-molding simulation is relatively inexpensive in terms of project cost and offers great benefits to those using it early in the manufacturing process.

7.23 The Injection-Molding Process

Injection molding is a seemingly simple process. A mold is created to form the shape of the component to be made and molten plastic is injected into it and then ejected when sufficiently cool. Despite this apparent simplicity, there are factors which complicate the process significantly. They are

- Nature of injection molding, in particular the basic physics of the process
- Material properties
- Geometric complexity of the mold
- Process optimization/stability

7.23.1 Physics of injection molding

In injection molding there are two major heat transfer mechanisms—convection and conduction.

During mold filling, the molten material enters the mold and heat transfer, due to convection of the melt, is the dominant mechanism. Due to the rapid speed of injection, heat may also be generated by viscous dissipation. Viscous dissipation depends on the viscosity and rate of deformation of the material. It most often occurs in the runner system and gates where flow rates are highest, however, it can also occur in the cavity if flow rates are sufficiently high or the material is very viscous. Finally, the mold acts as a heat sink and heat is removed from the melt by conduction through the mold wall and out to the cooling system. As a result of these heat-transfer mechanisms, a thin layer of solidified material is formed as the melt contacts the mold wall. Depending on the local flow rate of the melt, this "frozen layer" may rapidly reach equilibrium thickness or continue to grow, thereby restricting the flow of the incoming melt. This has a significant bearing on the pressure required to fill the mold. When filling is complete, pressure is maintained on the melt and the packing phase begins. The purpose of the packing stage is to add further material to compensate for the shrinkage of material as it cools in the cavity. Since the cavity is full, mass flow rate into the cavity is much smaller than during filling and, consequently, both convection and viscous dissipation are minor effects. During packing, conduction becomes the major heat-transfer mechanism and the frozen layer continues to increase in thickness until such time as the component has sufficient mechanical stiffness to be ejected from the mold.

7.23.2 Material complexity

Polymers for injection molding generally can be classified as semicrystalline or amorphous. Both have complex thermorheological behavior that has an enormous bearing on the molding process. Thermoplastics typically have viscous behavior that exhibits shear thinning and a dependence on pressure and temperature (Fig. 7.54). In addition, their thermal properties are temperature dependent and, for the case of semicrystalline materials, many properties also depend on the rate of temperature change. An extra complexity in injection-molding simulation is the need to incorporate an equation of state to calculate density variation as a function of temperature and pressure. The equation of state relates the material's specific volume (inverse of density), pressure, and temperature. This is referred to as the material's *pvT* characteristic. It, too, is complex and varies depending on the type of material (Fig. 7.55).

7.23.3 Geometric complexity

Injection-molded parts are generally thin-walled structures and may be extremely complex in shape. The combination of thin walls and rapid injection speeds leads to significant flow rates and shear rates and these, coupled with the material's complex viscosity characteristics, lead to large variations in material viscosity and so variation in

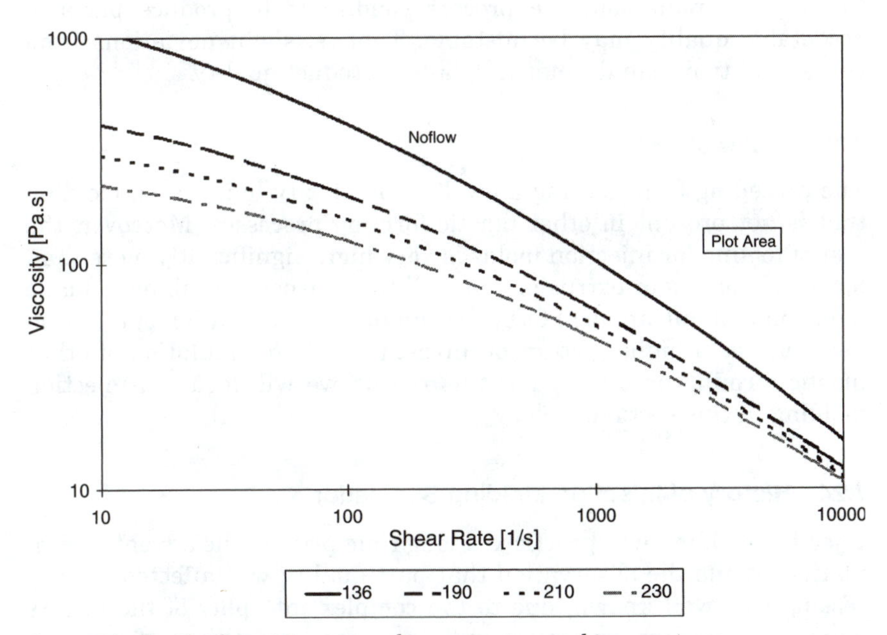

Figure 7.54 Graph of viscosity versus shear rate at several temperatures.

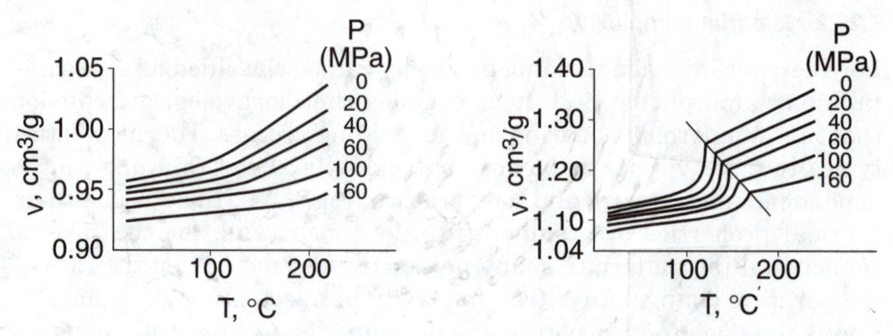

Figure 7.55 Graphs of *pvT* data for amorphous (left) and semicystalline (right) polymer.

fill patterns. The mold has two main functions in injection molding. The first is to form the shape of the part to be manufactured and the second is to remove heat from the mold as quickly as possible. Frequently the injection mold is a complex mechanism with provision for moving cores and ejection systems. This complexity influences the position of cooling channels which, in turn, can lead to variations in mold temperature. These variations, in turn, affect the material viscosity and so the final flow characteristics of the material.

7.23.4 Process stability

Finally, in production, the process conditions to produce parts of acceptable quality may be unstable. That is, slight deviations from these conditions can dramatically affect product quality.

7.23.5 Value of simulation

The preceding factors bring a level of complexity to injection molding that is not present in other plastic forming processes. Moreover, the cost of tooling for injection molds is very high, significantly more than for blow molding or extrusion dies. All these aspects combine to make injection molding an ideal focus for simulation. Simulation of injection molding has a higher return on investment than simulation of other plastic forming processes. For this reason we will focus on injection molding in this section.

7.24 History of Injection-Molding Simulation

Injection molding was practiced a long time prior to the advent of simulation. While the observation that part quality was affected by processing was well known, due to the complex interplay of the factors mentioned earlier, injection molding was something of an art.

Experience was the only means of dealing with problems encountered in the process. An overview of this approach is given in Ref. 16. The bibliography of Ref. 16 cites hundreds of empirical studies, each contributing to the relationship between processing and part quality. Demand for increased quality of molded parts in the 1970s saw an increased interest in mathematical modeling of the injection-molding process. During this time, many pioneering studies were published (see, for example, Refs. 17 to 23). These focused on rather simple geometries and, while of academic interest, offered little assistance to engineers involved with injection molding. Nevertheless, these studies provided a scientific base for simulation.

7.24.1 Early simulation of filling

In 1978, Moldflow introduced commercial software on a worldwide computer time-sharing system. This software enabled users to determine process conditions (melt temperature, mold temperature, and injection time) and to balance flow in cavities and runner systems. Although accepted at the time, this software was difficult to use as it required the user to produce a "layflat."[24]

The layflat was a representation of the part under consideration that reduced the problem of flow in a three-dimensional geometry to flow in a plane. For example, consider an open box with a thickened lip at the open end. If the box is to be injected at the center of its base, a potential problem could arise from polymer flowing around the rim of the box and forming an airtrap (Fig. 7.56). The layflat of the box is shown in Fig. 7.57. As can be seen, the box has been "folded out" to form the layflat. Analysis could now be done on the various flow paths on the layflat. For example, the results of such analysis could be used to thicken the sections shown to promote flow and so prevent the flow of material around the thickened rim. While this type of analysis was undoubtedly useful, it did require considerable skill on the part of the user to produce the layflat and optimize the various flow paths.

Hieber and Shen[25] made a significant breakthrough with the introduction of finite element analysis to the injection-molding process. Despite the fact that computing technology was not up to the demands of finite element analysis for injection molding, their use of the technique indicated the advantage of the approach, namely, that the model for analysis was created in a form that resembled the part geometry and that results could be shown on the part representation. In 1983, Moldflow introduced a finite element filling analysis program that found ready acceptance in the market. For the first time, analysts could model the part under consideration in a form that resembled the component and display the results of analysis on the part.

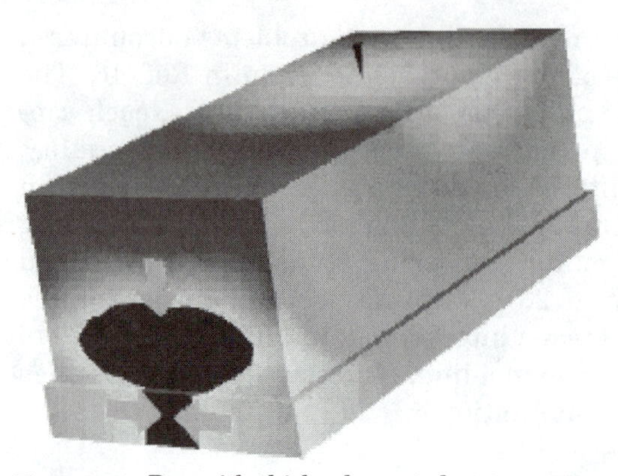

Figure 7.56 Box with thick edge gated at top center may have gas trap (dark region) and weld line due to flow "race tracking" around thick edge.

Since finite element was introduced, the CAE industry has developed rapidly. By 1987 there were no less than six companies involved in the development of software to simulate injection molding.[26]

7.24.2 Simulation of packing, cooling, and variations of injection molding

Over the past decade, development has been rapid with simulation now available for the packing and cooling phases of the injection-molding process. Also, in the 1990s some variants of injection molding have been introduced. These include

- Co-injection, in which a charge of material is first injected and then followed by a second charge. This results in the second material forming the core of the part which is encapsulated with the first material.
- Gas injection, in which after introduction of some polymer into the mold, a charge of inert gas is introduced to force the polymer to the extremities of the mold and so produce a hollow component in the areas penetrated by the gas.
- Injection compression molding in which the mold is initially open and is then closed after the polymer is injected or closed while polymer is being injected.

We will not discuss the details of simulation for these processes, as many of the fundamental ideas are common to injection-molding simulation.

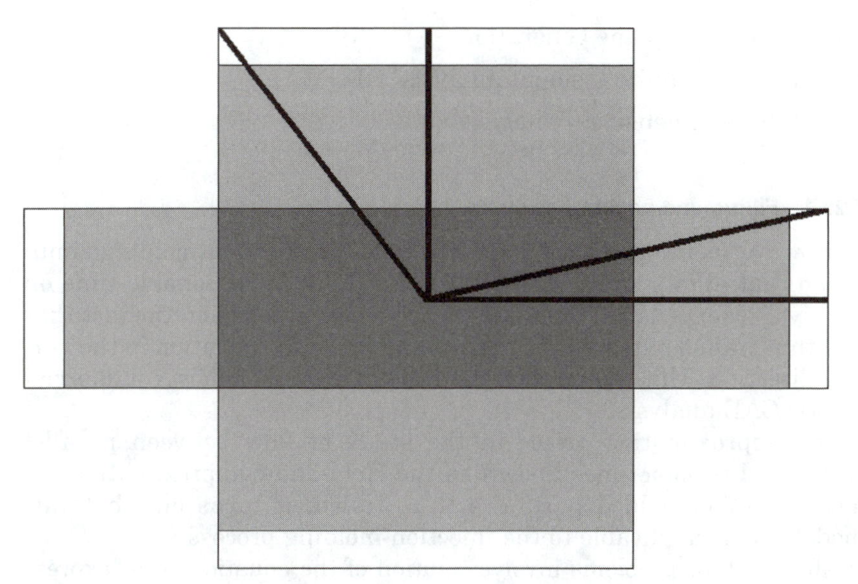

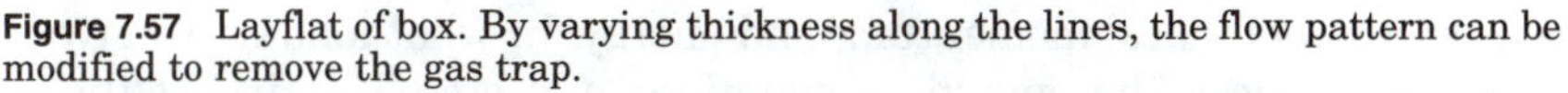

Figure 7.57 Layflat of box. By varying thickness along the lines, the flow pattern can be modified to remove the gas trap.

7.24.3 Simulation of warpage

Moldflow introduced software for prediction of shrinkage and warpage in 1990. Since this time, analysis of fiber-reinforced materials has been introduced in which the orientation of the fibers in the molding are calculated. With the distribution of fibers known, it is possible to calculate mechanical properties of the composite material in principal directions. This data can then be used in warpage calculations or used for structural analysis of the resulting part.

7.25 Current Technology for Injection-Molding Simulation

In this section we review the current state of technology for simulation of injection molding. In particular, we consider analysis of

- Filling and packing phases
- Mold cooling
- Fiber orientation distribution
- Warpage

In addition, we discuss modeling for simulation and review recent developments such as

- Automatic midplane generation
- Dual domain finite element analysis (DD/FEA)*
- Full three-dimensional analysis

7.25.1 Filling and packing analysis

Earlier we mentioned that a model exists for injection-molding simulation that allows simulation to be carried out in reasonable time on relatively inexpensive computers. This model is based on the fact that in a thin-walled part we may neglect any pressure variation in the normal direction. This assumption underlies the basis of most commercial plastic CAE analysis.

The approximation arises in the study of flow between parallel plates and is sometimes known as the Hele-Shaw approximation. As most injection-molded parts are thin walled, it turns out that this model is also applicable to the injection-molding process.

All fluid flow problems involve solution of the equations that express conservation of mass, momentum, and energy. In what follows we give a brief summary of the governing equations and simplification of them. Details may be found in Ref. 27. The conservation equations take the form

$$\text{Mass:} \quad \frac{\partial \rho}{\partial t} + (\nabla \cdot \rho \nu) = 0 \tag{7.1}$$

$$\text{Momentum:} \quad \rho \frac{\partial \nu}{\partial t} = -\nabla \rho + [\nabla \cdot \eta \, \dot{\underline{\gamma}}\,] - \rho\, [\nu \cdot \nabla \nu] \tag{7.2}$$

$$\text{Energy:} \quad \rho c_p \left(\frac{\partial T}{\partial t} + \nu \cdot \nabla T \right) = \beta T \left(\frac{\partial p}{\partial t} + \nu \cdot \nabla p \right) + \eta\, \dot{\gamma}^2 + \nabla \cdot (k \nabla T) \tag{7.3}$$

where ρ is the material density, t is time, ν is the velocity of the melt, p is pressure, η is the melt viscosity, $\dot{\underline{\gamma}}$ is the rate of deformation tensor, c_p is the specific heat of the material, T is temperature, β is the coefficient of thermal expansion, $\dot{\gamma}$ is the shear rate, and k is the thermal conductivity. It should be noted that the viscosity of polymers depends on pressure, temperature, and shear rate, and this dependency must be incorporated in the simulation.

Adopting a cartesian coordinate system and assuming the cavity thickness is small compared to the other dimensions, the mass and momentum equations may be reduced to the single equation

---

*DD/FEA is a registered trademark of Moldflow Corporation.

$$\frac{\partial}{\partial x}\left(S\frac{\partial p}{\partial x}\right)+\frac{\partial}{\partial y}\left(S\frac{\partial p}{\partial y}\right)=\frac{1}{2}\int_{-H}^{H}\left\{\kappa\frac{\partial p}{\partial t}-\frac{\beta}{\rho c_p}\left[\eta\dot{\gamma}^2+\frac{\partial}{\partial z}\left(k\frac{\partial T}{\partial z}\right)\right]\right\}dz \tag{7.4}$$

$$\text{where } S=\frac{1}{2}\left\{\int_{h^-}^{h^+}\frac{z^2}{\eta}\,dz-\frac{\left(\int_{h^-}^{h^+}\frac{z}{\eta}\,dz\right)^2}{\int_{h^-}^{h^+}\frac{1}{\eta}\,dz}\right\}$$

h^+ and h^- are, respectively, the upper and lower z coordinates of the frozen layer position, κ is the coefficient of compressibility, and H is half the wall thickness.

With the additional assumption that convection in the z direction may be ignored, the energy equation takes the form

$$\rho c_p\left(\frac{\partial T}{\partial t}+v_x\frac{\partial T}{\partial x}+v_y\frac{\partial T}{\partial t}\right)=\beta T\frac{\partial T}{\partial y}+\eta\dot{\gamma}^2+\frac{\partial}{\partial z}\left(k\frac{\partial T}{\partial z}\right) \tag{7.5}$$

The left-hand side of the energy equation represents the rate of change of temperature and convection, while the terms on the right-hand side account for heat of expansion/compression, viscous dissipation, and conduction to the mold, respectively.

These equations, and their respective boundary conditions, are generally solved with a hybrid approach introduced in Ref. 25. The pressure solution is found from Eq. (7.4) using finite elements and the temperature field is obtained from Eq. (7.5) using the finite difference method.[28]

7.25.2 Cooling analysis

The cooling of the mold is a key factor in efficient production and quality. Generally, cooling systems are given too little priority in the mold design process. This is unfortunate as careful attention to cooling system design can reduce cycle time and so increase the productivity of the mold/machine combination.

Injection-molding cooling systems can be analyzed quite readily today. Generally, two analyses are done. The first concerns the flow of the coolant in the cooling system; the second concerns the heat transfer from the part, through the mold, and into the coolant.

The flow of the coolant can be handled using conventional hydraulics theory. Results include the flow rates, required pressure, and the Reynold's number for the coolant. The latter is a measure of the degree of turbulence achieved. In general, a turbulent flow increases the heat transfer from the mold to the coolant. Beyond a certain Reynold's number, however, the heat transfer increases only marginally while the power required to pump the fluid increases significantly. So while it is desirable to achieve turbulence, it is inefficient to use too high a pump capacity.

Heat transfer from the plastic to the mold can be calculated using finite difference schemes for thin-walled, shell-like parts or finite element methods for true three-dimensional analysis. Heat transfer through the mold is generally performed using the boundary element method (BEM). As the mold is three-dimensional, use of the BEM permits a heat-transfer analysis through the mold using only a surface mesh.[29] This is far easier than meshing the mold steel internally as would be required for a finite element analysis and accounts for the popularity of the method for this application. Results from the cooling analysis provide the temperature distribution on the surface of the mold at particular points of time during the cycle. These temperature distributions enable mold designers to position cooling lines so as to minimize temperature variations over the mold surface and, in particular, from one side of the mold to the other. Temperature variations from side to side are a frequent cause of part warpage and should be avoided if possible.

The temperature field calculated by the cooling analysis, while of interest in its own right, may also be used as a boundary condition for the flow analysis. That is, the cooling analysis is used to define the mold temperature for the flow analysis. This type of coupling between flow and cooling analysis most accurately describes the real process.

7.25.3 Fiber orientation analysis

In many engineering applications short glass fiber reinforcement is added to the material. This has the advantage of increasing the strength and modulus of the material. As the material flows into the mold, the fluid deformation and interaction with other fibers alters the orientation of the fibers. The final orientation state will depend on the processing history of the material and may lead to highly anisotropic material properties.

In simplest terms, fibers tend to align in the flow direction when the flow is converging and align transverse to flow where the flow diverges (Fig. 7.58). With multiple injection points and complicated geometry, the final orientation distribution can be extremely complex. Simulation can assist here by first calculating the final orientation state and then using this information to derive the thermomechanical

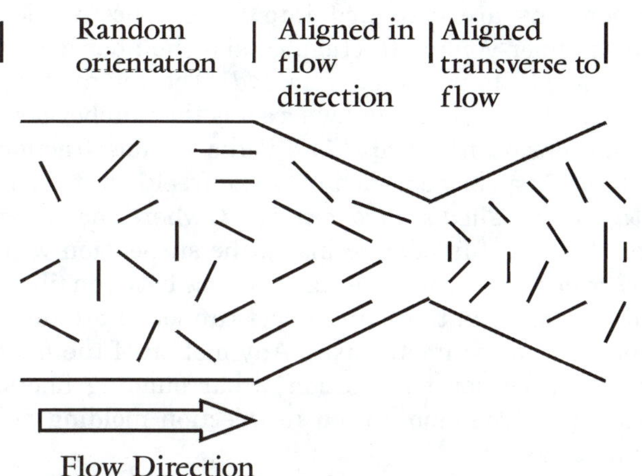

Figure 7.58 Fibers tend to align in the flow direction in converging flow and transverse to the flow in diverging flow.

properties of the material such as elastic moduli, Poisson's ratio, and linear coefficients of expansion. These derived properties may be used for subsequent structural analysis of the part. They may also be used for determining shrinkage and warpage of the component.

Predicting fiber orientation. Isotropic constitutive models are not valid for injection-molded fiber-reinforced composites. Unless the embedded fibers are randomly oriented, they introduce anisotropy in the thermomechanical properties of the material. The fiber orientation distribution is induced by kinematics of the flow during filling and, to a lesser extent, packing. An extensive literature deals with flow-induced fiber orientation while much other work has been devoted to micromechanical models which estimate anisotropic elastic and thermal properties of the fiber-matrix system from the properties of the constituent fiber and matrix materials based on given microstructures. Comprehensive reviews of both research areas have been given in two recent books edited, respectively, by Advani[30] and by Papathanasiou and Guell[31] where many references can be found.

Analysis of the final properties of injection-molded short-fiber composite parts requires accurate prediction of flow-induced fiber orientation. Several different fiber suspension theories and numerical methods are available for the calculation of the motion of fibers during flow. Only the works of Folgar and Tucker[32] and Fan[33] are briefly reviewed here since they are the most relevant. The reader is referred to Phan-Thien and Zheng[34] for additional information for other constitutive theories of fiber suspensions. In what follows we assume the fibers are rigid rods of circular cross section.

Usually, fiber suspensions are classified into three concentration regimes according to the fiber volume fraction, ϕ, and the fiber aspect ratio a_R (defined by the length-to-diameter ratio, L/d). The volume fraction satisfies $\phi = n\pi d^2 L/4$ for rodlike fibers where n is the number density of the fibers. A suspension is called *dilute* if the volume fraction satisfies $\phi a_R^2 < 1$. In dilute suspensions, each fiber can freely rotate. The region in which $1 < \phi a_R^2 < a_R$ is called *semiconcentrated*, where each fiber has only two rotating degrees-of-freedom. Finally, the suspension with $\phi a_R > 1$ is called *concentrated,* where the average distance between fibers is less than a fiber diameter, and, therefore, fibers cannot rotate independently except around their symmetry axes. Any motion of the fiber must necessarily involve a cooperative motion of surrounding fibers. Most commercial composites commonly used in injection molding fall into the semi- or highly concentrated regimes.

There are several choices available to model the motion of the fibers. One is to track a large number of fibers and explicitly determine their interaction and motion. This can be done by attaching a unit vector $\mathbf{p}$ along the axis of each fiber and track its evolution with time. While possible, this method is not really practical for complex models. An alternative is to use a probability distribution function, $\Psi(\mathbf{p}) = \Psi(\theta, \phi)$, whose value represents the probability of finding a fiber between the angles ϕ and $\phi + d\phi$ and θ and $\theta + d\theta$ (see Fig. 7.59). Given such a distribution, we assume that one end of the fiber is indistinguishable from the other and so, $\Psi(\theta, \phi) = \Psi(\pi - \theta, \phi + \pi)$. Also, the integral of the distribution over all possible directions must be one. That is,

$$\int_0^{2\pi} \int_0^{\pi} \Psi(\theta, \phi) \sin\theta \, d\theta \, d\phi = 1$$

Although the distribution function completely describes the state of orientation, it is still too difficult to implement in simulation.

A far more compact method involves the use of orientation tensors.[35] Orientation tensors are defined as even-ordered tensors by relationships such as

$$a_{ij} = \Psi(\mathbf{p})\, p_i p_j \, d\mathbf{p}$$

Similarly, it is possible to define a fourth-order tensor

$$a_{ijkl} = \int \Psi(\mathbf{p})\, p_i p_j p_k p_l \, d\mathbf{p}$$

and so on.

The starting point for using an orientation tensor to predict fiber orientation is the evolution equation of Jeffery[36] for the motion of an isolated fiber in a Newtonian fluid. Jeffery's equation is valid for dilute fiber suspensions where there are no fiber-fiber interactions.

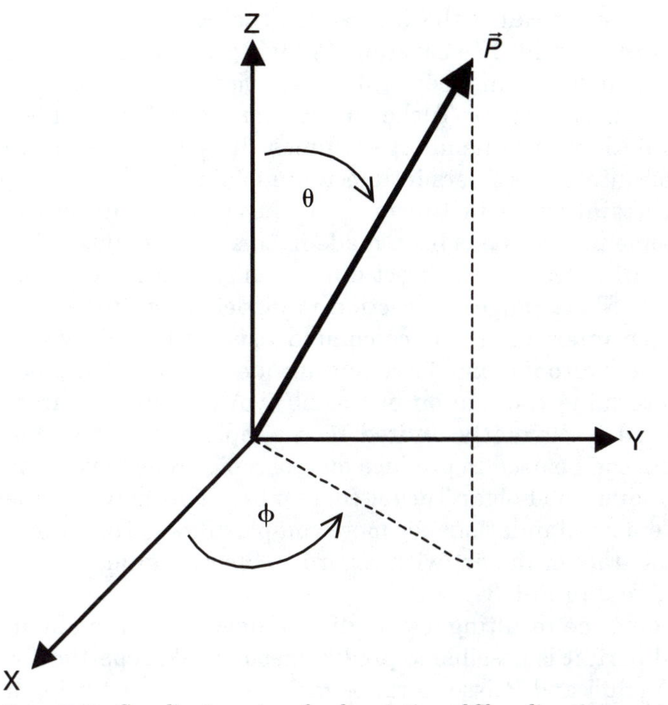

Figure 7.59 Coordinate system for description of fiber direction.

For an individual fiber, with a unit vector, **p**, directed along its length, Jeffery's equation for the time evolution of the fiber is as follows:

$$\dot{p}_i = -\tfrac{1}{2}\,\omega_{ij}p_j + \tfrac{1}{2}\,\lambda\,(\dot{\gamma}_{ij}p_j - \dot{\gamma}_{kl}p_k p_l p_i)$$

where
$$\gamma_{ij} = \frac{\partial v_j}{\partial x_i} + \frac{\partial v_i}{\partial x_j} \qquad \omega_{ij} = \frac{\partial v_j}{\partial x_i} - \frac{\partial v_i}{\partial x_j}$$

v_i is the fluid velocity, and λ is a constant that depends on the shape of the particle and is approximately 1 for slender rods.

Jeffery's equation was extended to concentrated solutions by Folgar and Tucker who added a diffusion term to account for the fiber-fiber interaction. In terms of the orientation tensor, the Tucker-Folgar equation has the form

$$\frac{Da_{ij}}{Dt} = \tfrac{1}{2}\,\lambda\,(\dot{\gamma}_{ik}a_{kj} + a_{ik}\dot{\gamma}_{kj} - 2\dot{\gamma}_{kl}a_{ijkl})$$

$$-\tfrac{1}{2}\,\lambda\,(\omega_{ik}a_{kj} - a_{ik}\omega_{kj}) + 2C_I\dot{\gamma}\,(\delta_{ij} - 3a_{ij})$$

An empirical constant called the interaction coefficient C_I is introduced in the diffusion term. The constant C_I for a given suspension is assumed to be isotropic and independent of the orientation state, as a first approximation. The Folgar-Tucker model has extended the fiber orientation simulations into nondilute regimes. It is widely used to determine the orientation of fibers in injection molding.

The main uncertainty in using this model is the chosen value for the coefficient C_I. Some progress has been made in this area recently. Fan et al.[33] have recently presented a direct numerical simulation of fiber-fiber interactions. Short-range interaction is modeled by lubrication forces. Long-range interaction was calculated using a boundary element method. The hydrodynamic force and torque on each fiber were calculated to determine the motion of the fiber. Although the direct simulation method is currently limited to a simple shear flow, the numerical results can be used to produce macroscopic properties of the suspension, including the Folgar-Tucker constant C_I. This may then be used in finite element simulations for more complex flows. This represents the current state of the art with regard to fiber orientation, and details may be found in Ref. 37.

After determining the resulting distribution of fiber orientation in an injection-molded part, it is possible to predict mechanical properties for the composite. Moduli and Poisson's ratios may be determined using a variety of mechanical theories for composite materials. These properties may then be used in structural analysis or warpage analysis.

7.25.4 Warpage analysis

Warpage is a common problem in injection molding. It is evidenced by deformation of the part such that assembly is difficult or the part is not fit for its purpose. Warpage is a complex phenomenon and a direct consequence of processing effects on the material. Warpage is caused by variations in shrinkage of the material. These variations are of three types:

- *Variation from point to point on a part.* This type of variation is frequently caused by variation in the density distribution which is caused by variation in the pressure/temperature history experienced by the material.
- *Variation of shrinkage in different directions.* This type of variation, which also varies from point to point in the molding, is due to anisotropic shrinkage. This, in turn, arises from anisotropy in the thermomechanical properties of the material due to molecular orientation and the morphology of semicrystalline materials.

Shrinkage anisotropy is a common problem when using fiber-filled materials as the orientation of the fibers leads to quite extreme anisotropy in thermomechanical properties.

- *Variation of shrinkage from one side of the molding to the other.* This type of variation arises from asymmetry in the flow and temperature fields of the molding. It is mostly due to temperature variation from one side of the mold to the other.

There are two main approaches to analysis of warpage—strain-based methods and residual stress methods. Strain-based methods are among the earliest and owe their existence to the difficulty of accurately determining residual stresses. Essentially, the idea is to predict the shrinkage strain experienced by the material.[38] These strains are then input to a structural analysis program that determines the overall part shrinkage and the deformation of the part. Residual stress methods are more directly linked to the physics of the process but require accurate material characterization to perform well. The idea is to predict the residual stress distribution in the injection-molded material while it is in the mold. This stress distribution is then used as input to a structural analysis that determines the shrinkage and deformation of the part. Early calculations considered only thermally induced stresses[39] caused by the material cooling while in the mold. This was later extended to models incorporating both thermally induced stress and the stress induced in solidified material by the pressure exerted on it by the melt.[40] Regardless of the method used, the prediction of warpage requires accurate prediction of the filling, packing, and cooling phases of the injection-molding process.

Results from warpage simulation include the deformed shape of the part; part shrinkage; deflections in the x, y, and z directions; and residual stresses and strains. The deformed shape and its accompanying stress distribution may be subsequently used for structural analysis.

7.25.5 Optimization

One of the most exciting possibilities for simulation is the potential to optimize part designs and manufacturing processes.

Simulation has always been used to improve part designs and while this is a form of optimization, in the current context, optimization refers to an automated procedure whereby part designs can be improved or an optimum set of manufacturing conditions is obtained.

Initial efforts in optimizing part designs have focused on sizing runners for family and multicavity tools. Here a series of flow analyses are run, usually subject to the constraint that the cavities all fill at the same time. Runner dimensions are defined to lie in a range and the optimization algorithm varies the diameters of the runners subject to the constraints until a satisfactory solution is obtained.

More recently, there has been interest in optimizing the manufacturing conditions used to mold the part. Much of the motivation for this work stems from the possibility of using these results to provide information to the injection-molding machine. We return to the topic of machine interaction in a later section. For now, we consider the scope of optimization of processing conditions. For this purpose it is convenient to consider the molding process as consisting of the filling phase in which the velocity of the ram is controlled as a function of position and the packing phase in which pressure as a function of time is the main variable.

For the filling phase it is known that many surface defects may be caused by sudden changes in the flow front velocity. Hence a common constraint is that the flow front velocity be constant. Of course in attempting to achieve this, one must be careful to ensure that the material shear stress limits are not exceeded and that the temperature of the material is within permissible limits.

By considering these factors, optimization algorithms can define a ram profile that meets these conditions that can be entered on the molding machine directly. The advantage of simulation here is that it effectively allows you to view the plastic in the mold—something that machine setters cannot do.

The packing phase is known to have a dramatic affect on part quality—particularly part weight, dimensional tolerance, and warpage. In the packing phase, material is subjected to high pressure while it cools in the mold. The pressure temperature history determines the ultimate density of the molded material. A common goal of the optimization is, therefore, to minimize density variation throughout the part. Alternatively, it is possible to minimize variation in linear shrinkage in directions along, and transverse to, the flow direction. The result of these operations is a pressure profile with pressure varying as a function of time. Figure 7.60 gives an example of the variables and the type of output that can be achieved today. Here the optimization algorithm determines the initial pack pressure, P_c, the time of application t_c, the level to which pressure should decay P_d and the time to do this, t_{cl1}, the time to decay to zero pressure t_{cl2}, and the cooling time t_{cool}. Such a profile is capable of improving part quality dramatically. It is almost impossible for a molder to define an optimum profile such as this since there is no way of quantifying the effect of changing pressures and rates of pressure decay.

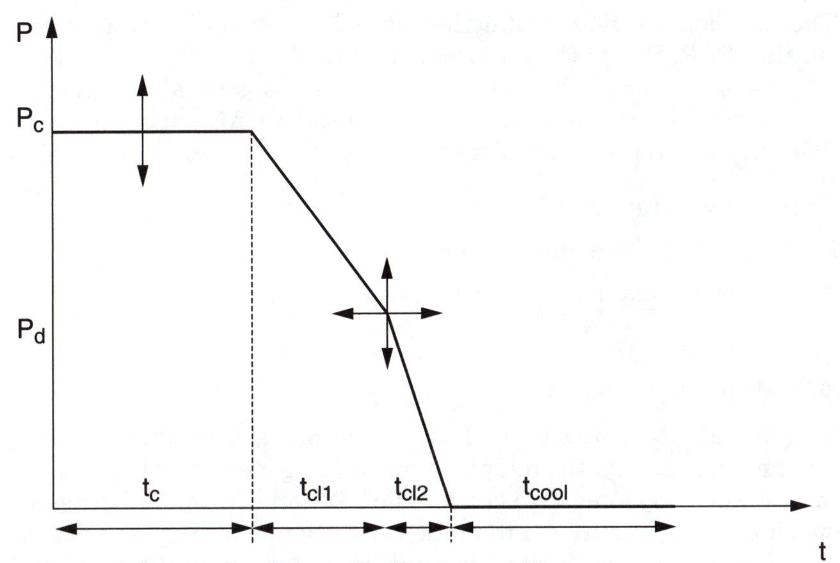

Figure 7.60 Optimized packing profile showing variables.

7.25.6 Modeling for CAE analysis

Of particular importance is the assumption of thin-walled geometry. From Eq. (7.3) we see that the pressure is independent of the z coordinate. Consequently, the finite element utilized for pressure calculation need have no thickness. That is, the element is a plane shell—generally a triangle or quadrilateral. This has great implications for users of plastics CAE. It means that a finite element model of the component is required that has no thickness. In the past this was not a problem. Almost all common CAD systems were using surface or wireframe modeling and thickness was never shown explicitly. The path from the CAD model to the FEA model was clear and direct.

In recent years, tremendous development has occurred in CAD using solid models. With solid modeling the component is represented faithfully. All details are shown and the model is photo-realistic. Initially, solid modelers were not so popular for plastic designers. This was because injection-molded parts were thin walled. Advances in modeling technology in the area of "shelling" now mean that the designer can produce thin-walled models very easily. Solid modeling is fast becoming the norm, due to the realization that the solid model can be the master for design as well as all downstream operations such as rapid prototyping, assembly analysis, tolerancing analysis, and mold making. The increased interest in solid modeling is evidenced by the number of products available now and the reduction in price due to competition.[41]

The adoption of solid modeling introduced a problem for many users of plastics CAE. Due to the requirement of having an FEA model of no thickness, solid models had to be "midplaned" to generate an appropriate model. To overcome this problem, plastic CAE suppliers have developed three recent technologies:

- Midplane generation
- Dual domain finite element analysis
- Three-dimensional finite element analysis

7.25.7 Midplane generation

This is the direct approach in which a solid model is read into a program and an automatic midplane mesh is generated (Fig. 7.61). Details of such a system were given in Ref. 42 and will not be discussed here. In practice, the fully automatic generation of a midplane model is difficult. In many cases there is some need to clean up the resulting model before analysis. Nevertheless, midplane generation can save an enormous amount of time for many types of part.

7.25.8 Dual domain finite element analysis

Dual domain finite element analysis (DD/FEA) is a method that enables analysis on the solid model. It uses a surface mesh on the solid geometry

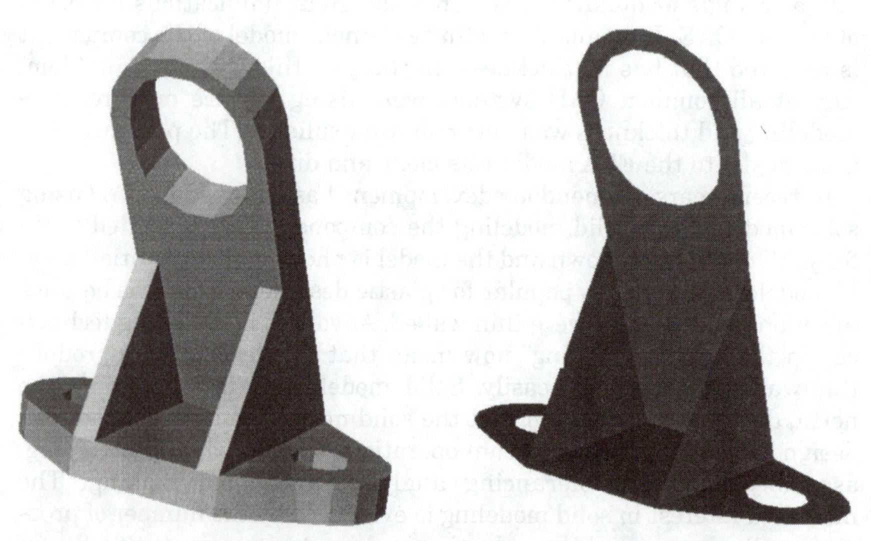

Figure 7.61 Automatic midplane generation seeks to transform three-dimensional solid geometry (left) to midplane shell representation (right).

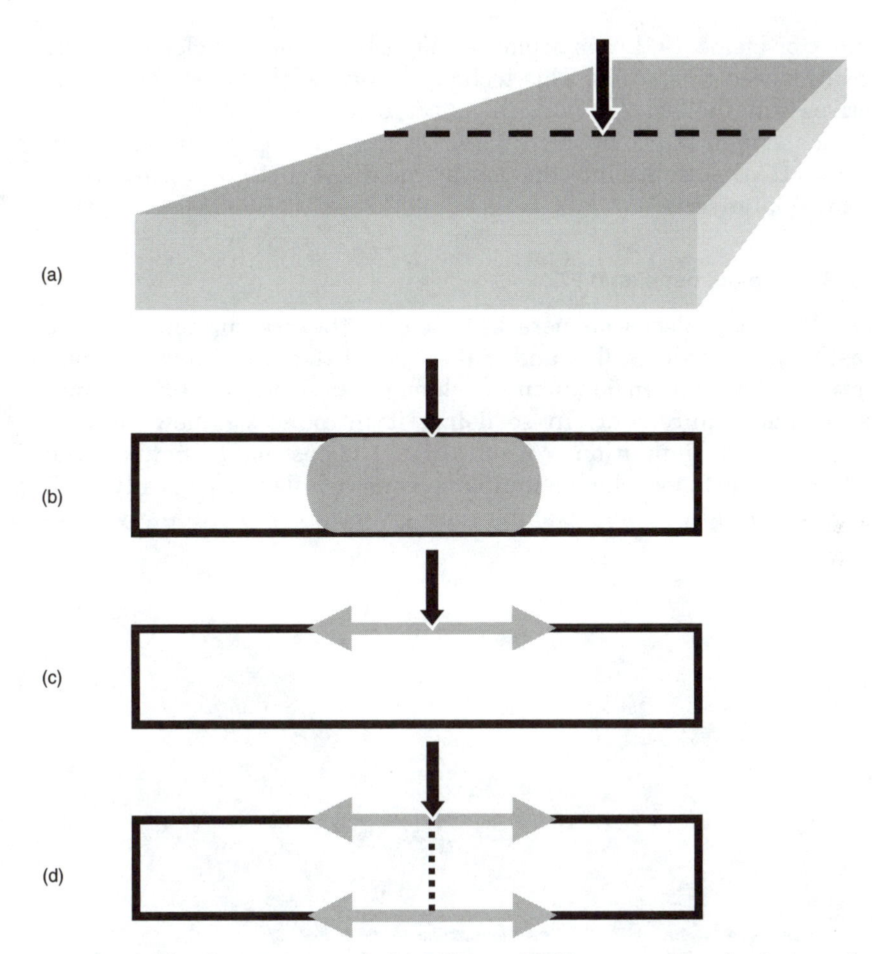

Figure 7.62 Flow in a center-gated plate. If normal FEA were performed using a surface mesh, the flow would run along top surface only (*c*) and not match the physical reality (*a*) and (*b*). Dual domain finite element analysis uses a connector element to synchronize flows on opposite surfaces (*d*).

and then inserts extra elements to ensure that the filling pattern is physically sensible. Figure 7.62 gives the idea. Imagine injecting a polymer into a rectangular plate as shown. If we simply used the surface mesh on the plate with a conventional FEA, the flow pattern would be physically incorrect. The material would flow out from the injection point across the top of the plate and then down the edges and finally along the bottom surface. However, by inserting an extra element from the injection point though the thickness to the other side, we can obtain a satisfactory result using a conventional solver based on the Hele-Shaw approximation. For more complex geometry, it is necessary to insert more connector elements. In particular, at any rib a connector element must be introduced (Fig. 7.63). This technique is called DD/FEA. The name derives from the fact that you are, in fact, doing two FEA analyses—one on each side of the component. DD/FEA has had a striking impact on plastics CAE since its introduction by Moldflow in 1997. We return to this in a later section.

7.25.9 Three-dimensional FEA

All technology discussed here is based on the assumption that the plastic part is thin walled and makes use of the Hele-Shaw approximation. Three-dimensional finite element analysis (3D/FEA) eliminates this requirement. In so doing, it introduces a new class of components to simulation. With 3D/FEA it is possible to simulate the molding of parts for which a midplane is not available. Typically, such parts are chunky—some examples are given in Fig. 7.64. Many parts

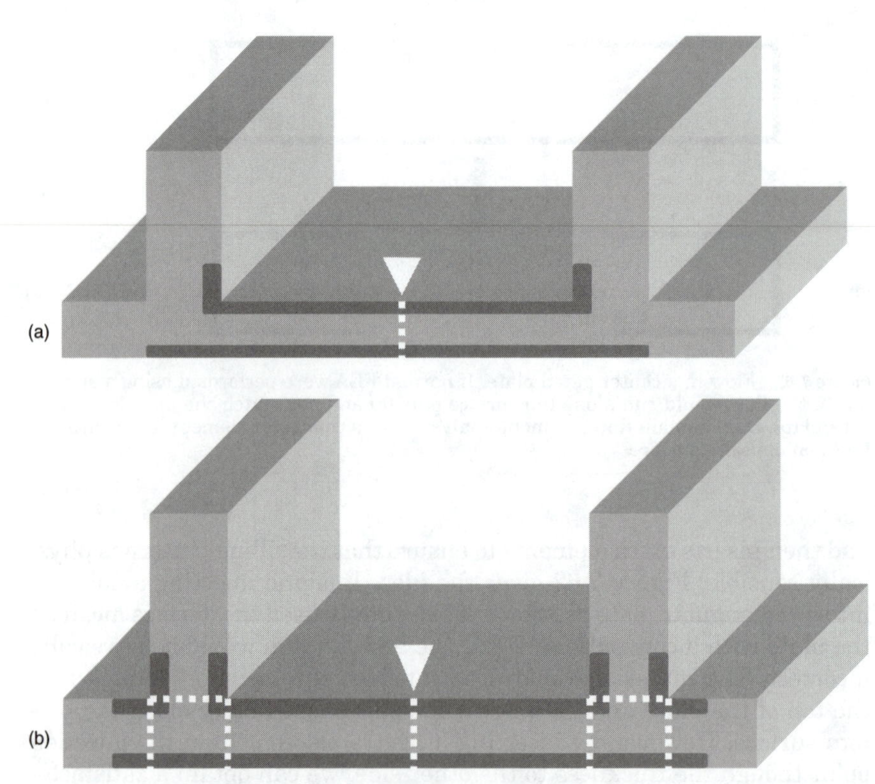

Figure 7.63 For ribs, additional connector elements must be inserted. Without connector elements (*a*) the flow is unrealistic. Connector elements (*b*) are introduced to ensure the flow is physically realistic.

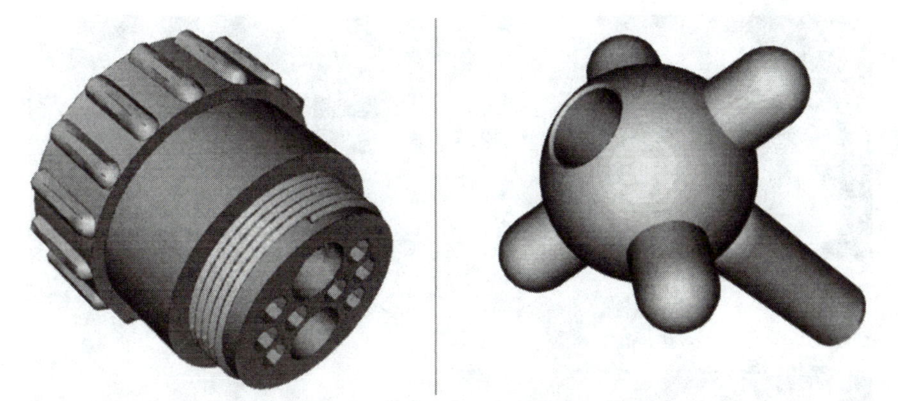

Figure 7.64 Parts such as these do not possess a midplane and so cannot be analyzed with conventional CAE. Such parts require full 3D FEA analyses.

contain inserts, either metal or some other material, and it can be difficult to analyze these using conventional shell-based analysis. These parts are also amenable to 3D/FEA.

In addition to broadening the range of parts that can be simulated, 3D/FEA also couples well with solid modeling. A particular advantage is that the model for analysis is an unambiguous representation of the real part geometry. Sometimes it is difficult to achieve this with shell-based modeling.

Three-dimensional FEA solves the conservation equations, Eqs. (7.1) to (7.3), discussed earlier with fewer assumptions. Generally, inertial and gravitational terms are omitted, as viscous forces are dominant. Importantly, 3D/FEA has a pressure gradient in the thickness direction and so there is explicit calculation of convection of the melt from the midstream to the wall at the flow front. The phenomena of "fountain flow" is thus accounted for in 3D/FEA.

Another advantage in three-dimensional calculations is that the model for analysis unambiguously represents the part. For example, the modeling of the gate regions of some parts is quite complex and is better handled in three dimensions. Figure 7.65 shows a gate region comprising a runner of circular cross section, a conical feed to the part gate that is of square cross section. This feed configuration would be impossible to model accurately with shells and beams.

7.26 The Changing Face of CAE

In an earlier section we mentioned the dramatic effect of dual domain finite element analysis on the CAE industry. In this section we elaborate on this.

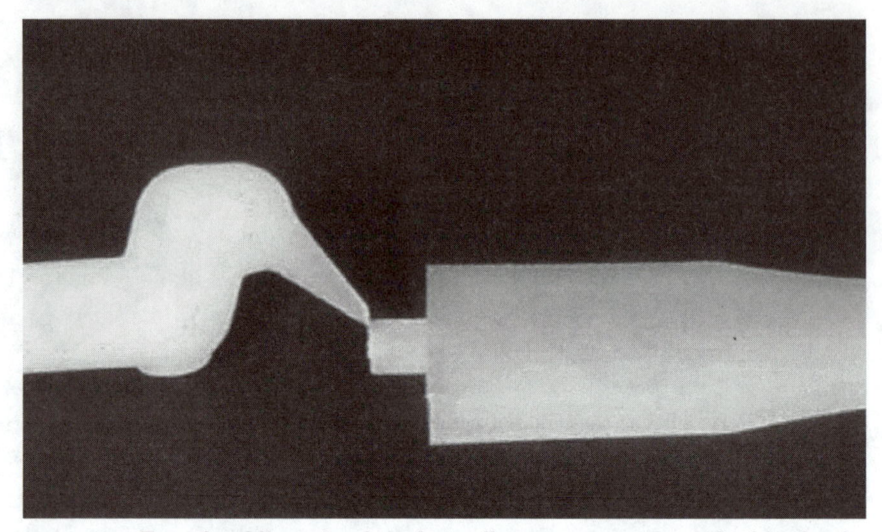

Figure 7.65 True three-dimensional analysis allows users to accurately describe the geometry at complex gate regions. This provides greater accuracy in fluid and heat-transfer calculation.

Throughout its relatively short history, plastics CAE has been seen as a specialist activity. Analysis was typically performed by dedicated staff who were expert users of software and had a good understanding of injection-molding theory. While it has always been recognized that simulation performed early in the design stage provides more benefit, the lack of direct interfaces between the CAD systems on which parts were designed and CAE software meant analysis was outside the design environment. DD/FEA provides a solution to this problem. For the first time, it is possible to closely couple plastics CAE with solid modeling. In doing this, the use of CAE by plastic designers rather than dedicated analysts is possible.

Several solids-based CAD systems now offer products with DD/FEA technology to facilitate analysis at an early stage. This type of technology redefines the use of plastic CAE analysis by enabling nonspecialists to perform analysis very early in the design stage. Accordingly, special attention was paid to results presentation. Whereas the traditional outputs have been pressure and temperature distributions, new display technology has been introduced in response to the fact that the designer may have little previous experience in plastics CAE. For example, in order to choose the number of gates, wall thickness, and resin type, a key variable is the pressure required to fill the part. This, in turn, depends on the temperature of the material which, in turn, depends on the processing conditions, locations of gates, and the part geometry. Determination of the pressure required to fill is therefore a multidimensional task that requires simultaneous interpretation of pressure and temperature distributions. To simplify the interpretation of results, the pressure and temperature distributions are processed to produce a single plot called "confidence of fill." This is displayed by overlaying the colors red, yellow, and green over the part geometry in areas that have low, medium, and high probabilities of filling, respectively. Figure 7.66 gives an example of such a plot (red shows here as black, yellow as light gray, and green as gray. The gate position is at the end of the part. Most of the part has a high confidence of fill. However regions far from the gate are shown in yellow, while areas at the end of the part are shown in red. Such a plot is far easier to understand than the simultaneous pressure and temperature distributions.

DD/FEA enables plastic designers to begin analysis very early in the design phase. The information gleaned here is valuable only in so far as it can be rapidly communicated to other people involved in the process. For example, material suppliers, mold designers, and the molder can all benefit from this early knowledge of how the part will fill. The rapid development of the Internet has been adopted by CAE suppliers as the way to facilitate communication among team members. Latest products are now offering report writers in which analysis images and notes can be linked together into a format that can be viewed on a browser or sent as a message on the Internet.

DD/FEA provides the means to take analysis into the design stage. As well as its appearance in CAD systems, DD/FEA-based advanced

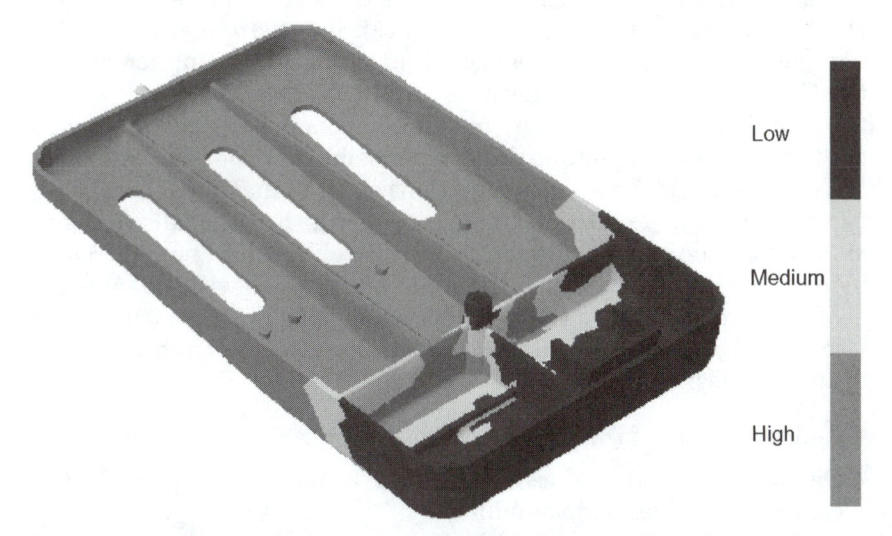

Figure 7.66 Simplified results interpretation in DD/FEA technology. Regions that are unlikely to fill have low confidence of fill shown here as black, regions of medium confidence are in light gray while regions with high confidence are in gray.

analysis modules are also available. This means that designers who encounter problems can send their models to analysts armed with advanced products for detailed analysis. In this way DD/FEA has provided a link between the traditional CAE user and part designers. The result is that more designs are subject to analysis and parts can be modified, if necessary, at a stage where change is least expensive.

7.27 Machine Control

While simulation is of great benefit, it is aimed at the part design and mold design areas of injection molding. Of course the actual processing of the material has a dramatic effect on the quality of the component, and much effort has been devoted to controlling the injection-molding machine. Much of this effort has been focused on ensuring that the molding machine is capable of repeating a particular cycle. While this is certainly important, the part quality is affected by the polymer flow. Much of the focus in machine control has been on making the machine respond rather than concern for the melt. Moreover, injection-molding machine controllers do not provide systematic tools for optimization of the molding cycle.

Advances in simulation technology have led to several attempts to link the results of simulation to the injection-molding machine. One of the difficulties in doing this is that injection-molding machines have characteristics that are not easily accounted for in simulation. For example, the ability of the machine to respond to a desired change in process conditions is not known nor is the performance of the flow check valve. Nevertheless, simulation can be used to get somewhere near an optimum set of process conditions. The setup can then be fine-tuned to optimum performance.

It is a fact of life that many plastic components are designed without the benefit of simulation. For such a mold, how does one determine the optimal set of processing conditions so as to maximize part quality and minimize production time? In general, the task falls to highly skilled machine setters who are, unfortunately, in short supply. What's more, there is only so much a machine setter can do without some level of instrumentation on the mold, for example, a pressure transducer. Use of in-mold sensors while providing some information brings some disadvantages:

- Additional cost of the pressure transducers
- Damage to the transducer, wiring, or connectors in a production environment due to mishandling
- Need for an operator to interpret the information from the transducer and adjust the machine control

The preceding considerations led to an endeavor to try and bridge simulation with injection-molding machine controllers. The aims of this endeavor are twofold:

- To ensure that the results of simulation are able to be used in production.
- To ensure that if simulation has not been done, there is a systematic way to optimize a given mold, material, and machine combination.

In October 1998, Moldflow released a family of products called MPX (Moldflow Plastics Xpert). The Moldflow Xpert Series is the downstream realization of Moldflow's Process Wide Plastics Simulation Strategy to link design, CAE, and the shop floor. Resulting from several years' research and development with industry and academia, the Plastics Xpert has been designed to assist the user in setting up and optimizing injection-molding machines to ensure the production of consistently good quality parts. There are three modules in the product, each dealing with one of the following tasks:

- Initial setting up of the injection-molding machine
- Defining the molding window
- Monitoring production

One of the main features of this product is that it utilizes the existing instrumentation on the molding machine.[43] For example, in its most basic configuration, all that is required is the hydraulic pressure and screw displacement. The system is capable of using additional input such as cavity pressure and mold and melt temperatures but this is not necessary.

7.27.1 Setting up the molding machine

This module, called Setup Xpert, enables systematic mold setup independent of operator and location. Mold setup can be done via

- A setup wizard that takes the user though a defined set of questions to generate the initial velocity and pressure profiles.
- Profiles obtained by CAE analysis. These are velocity versus screw displacement or time for the filling phase and pressure versus time for the holding phase.

The module determines the best optimization route for each setup process, then starts the automated machine optimization process. Using information from the machine transducers, it automatically

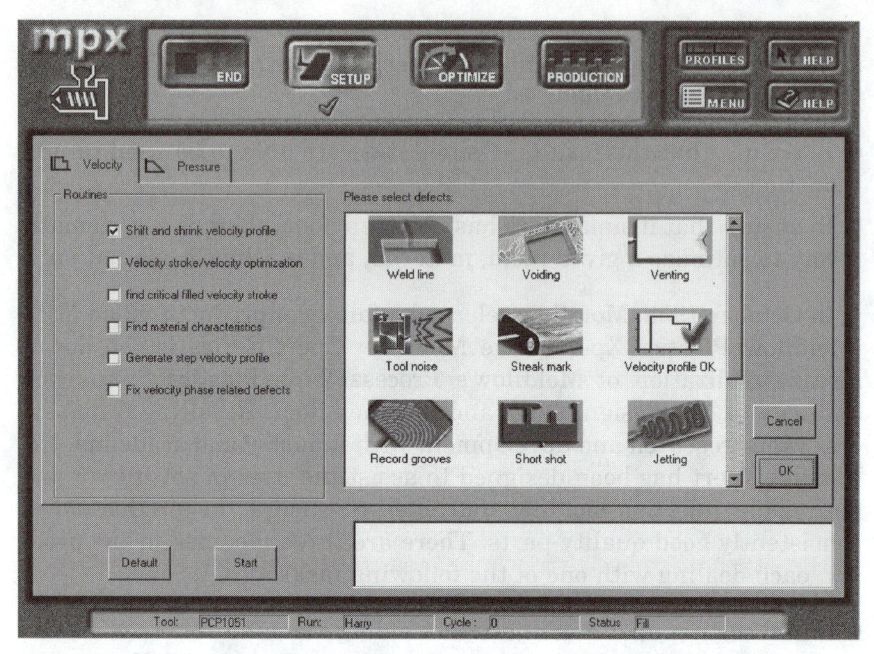

Figure 7.67 User interface for new generation of control technology. Molding problems are listed and the machine settings requiring adjustment are automatically changed. The same interface appears to all users regardless of machine type.

determines ram velocity and velocity stroke, velocity profile, gate freeze time, pressure profile, and cooling time.

A special feature of the interface is that the user is not required to understand the details of what is being changed on the machine. The interface presents a number of faults to the user (see Fig. 7.67). Faults currently supported are

- Weld line visibility and voiding
- Venting and burn marks
- Tool noise and flashing
- Streak marks and record grooves
- Short shots and jetting
- Gloss and delamination

These faults are then interpreted by an expert system to define the necessary changes to be made to the machine to bring about a reduction in the fault. In this way the user need not have a detailed understanding of the relationship between process parameters and injection-molding faults. Another advantage is that the user is presented with the same user interface regardless of machine type. In summary then, Setup Xpert allows the operator to produce a good part without an in-depth knowledge of the individual machine or process.

7.27.2 Defining the molding window

While an optimized set of process conditions is desirable, it is also important that the set conditions are stable to small perturbations around the set point. This is achieved by automatically performing an experimental design. Prior to this, however, the module automatically determines shot to shot variability of the machine. With the variability known, design of experiment technology is used to define a space of molding parameters. For each set of conditions, a series of shots are determined and the operator is asked for feedback on part quality. Once the extremes of the space are determined, a set of conditions at the center of the space is selected as the most robust operating point.

7.27.3 Monitoring production

This module, called Production Xpert, graphically monitors variables *specific* to the injection-molding process and can automatically determine the quality control limits. The key advantage of the Production Xpert is its ability to automatically detect process variations and drift and it can either suggest how to correct the process or make the necessary changes itself.

7.28 Future Trends

The future will see increased use of three-dimensional modeling, increased flexibility in choosing software due to improved interoperability, and increased use of three-dimensional analysis technology.

7.28.1 Increased use of three-dimensional modeling

Today, a lot of design is still performed in two dimensions. Decreasing costs of high-performance computers, the relatively low cost of RAM, and falling prices of solid modeling systems will combine to increase the use of solids modeling in all facets of the plastic industry. Increased use of solids modeling will lead to increased use of 3D CAE and interfacing to CAM.

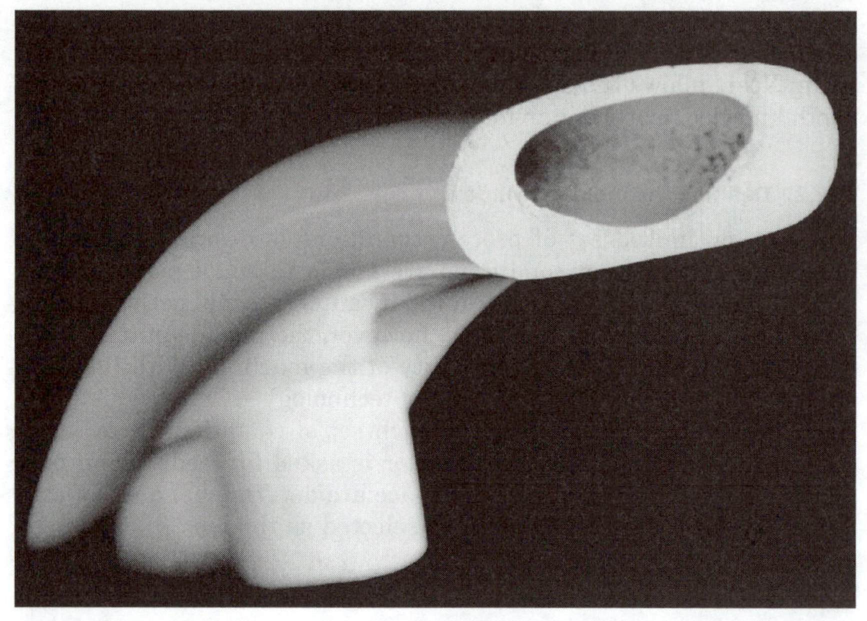

Figure 7.68 A gas-injected part. Such a complex shape requires three-dimensional analysis to accurately simulate the gas-injection process.

7.28.2 Flexibility and interoperability

Windows NT is emerging as the preferred operating system for many of the new vendors of solid modeling CAD systems. These newer companies are offering many downstream applications such as CAE analysis and NC machining with their products. Frequently, these downstream products are produced by different vendors and are well integrated with the CAD application. This will allow users to mix and match their CAD/CAE/CAM systems in future.

The CAD vendors who have been offering three-dimensional modeling systems for the past decade are tending to offer more integrated environments with downstream modules of their own making as well as some integration from other vendors. The resulting environments are extremely well integrated with seamless transmission of data.

The Internet is also having an impact on CAD and CAE products. Already many companies are linking to the web to facilitate report generation and communication as well as product data management (PDM).

7.28.3 Extension of three-dimensional analysis

From the analysis side, there is no doubt that the introduction of three-dimensional techniques will increase. While existing technology for 3D/FEA is notable for its ability to deal with thick components, it is also a platform technology for future developments in simulation. Considerable improvement in simulation for co-injection, gas injection, and warpage are possible with a three-dimensional approach. For example, the gas-injected part shown in Fig. 7.68 is very difficult to model with shell elements. The initial shape really requires a three-dimensional description, as does the hollow core formed by the gas.

Increased computer speed will also enable CAE vendors to offer optimization codes. Coupled with three-dimensional analysis, a real possibility exists of providing a virtual prototyping capability. The visualization capabilities of tomorrow's computers are also key to achieving this goal. We can expect in the near future to be able to link simulation results with visual defects such as splay marks and gloss variations.

References on CAD, CAM, CAE

1. C. L. Tucker III, ed., *Computer Modeling for Polymer Processing,* Hanser, New York, 1989.
2. J. N. Reddy, *An Introduction to the Finite Element Method,* McGraw-Hill, New York, 1993.
3. Z. Tadmor and I. Klein, *Engineering Principles of Plasticating Extrusion,* Van Nostrand Reinhold Company, New York, 1970.
4. D. H. Sebastian and R. Rakos, "Extrusion Process Analysis with PASS," in K. T. O'Brien, ed., *Computer Modeling for Extrusion and Other Continuous Polymer Processes,* Hanser, New York, 1992.
5. N. S. Rao, K. T. O'Brien, and D. H. Harry, "Designing Polymer Machinery with Computer Programs," in K. T. O'Brien, ed., *Computer Modeling for Extrusion and Other Continuous Polymer Processes,* Hanser, New York, 1992.
6. M. J. Crochet et al. "POLYFLOW: A Multi-Purpose Finite Element Program for Continuous Polymer Flow," in K. T. O'Brien, ed., *Computer Modeling for Extrusion and Other Continuous Polymer Processes,* Hanser, New York, 1992.
7. K. T. O'Brien, ed., *Computer Modeling for Extrusion and Other Continuous Polymer Processes,* Hanser, New York, 1992.
8. J. Vlcek, J. Perdikoulas, and J. Vlachopoulos, "Extrusion Die Flow Simulation and Design with FLATCAD, COEXCAD and SPIRALCAD," in K. T. O'Brien, ed., *Computer Modeling for Extrusion and Other Continuous Polymer Processes,* Hanser, New York, 1992.
9. W. A. Gifford, "A Three-Dimensional Analysis of the Effect of Die Body Deflection in the Design of Extrusion Dies," *Polymer Engineering Science,* vol. 38, 1998, pp. 1729–1739.
10. C. J. S. Petrie, "Mathematical Modeling and the Systems Approach in Plastics Processing: the Blown Film Process," *Polymer Engineering Science,* vol. 15, 1975, pp. 708–724.
11. J. M. André et al., "Numerical Modelling of the Film Blowing Process," in J. Huétink and F. P. T. Baaijens, eds., *Simulation of Materials Processing: Theory, Methods and Applications,* A. A. Balkema, Rotterdam, 1998.
12. H. DeLorenzi and H. F. Nied, "Finite Element Simulation of Thermoforming and Blow Molding," in A. Isayev, ed., *Modeling of Polymer Processing,* Hanser, Munich, 1991.
13. D. Laroche and F. Erchiqui, "3D Modeling of the Blow Molding Process," in J. Huétink and F. P. T. Baaijens, eds., *Simulation of Materials Processing: Theory, Methods and Applications,* A. A. Balkema, Rotterdam, 1998.
14. M. Bellet, A. Rodriguez-Villa, and J. F. Agassant, "Finite Element and Automatic Remeshing Methods for the Simulation of Complex Blow Molded Polymer

Components," in J. Huétink and F. P. T. Baaijens, eds., *Simulation of Materials Processing: Theory, Methods and Applications,* A. A. Balkema, Rotterdam, 1998.
15. A. Garcia-Rejon, "Advances in Blow Moulding Process Optimization," *Rapra Review Reports,* vol. 7, no. 10, Rapra Technology Ltd., 1995.
16. I. I. Rubin, *Injection Molding, Theory and Practice,* John Wiley & Sons, New York, 1972.
17. I. T. Barrie, "Understanding How an Injection Mold Fills," *SPE Journal,* August 1971, pp. 64–69.
18. E. Broyer, C. Gutfinger, and Z. Tadmor, "A Theoretical Model for the Cavity Filling Process in Injection Molding," *Transactions of the Society of Rheology,* vol. 19, 1975, pp. 423–444.
19. Z. Tadmor, E. Broyer, and C. Gutfinger, "Flow Analysis Network (FAN)—A Method for Solving Flow Problems in Polymer Processing," *Polymer Engineering Science,* vol. 14, 1974, pp. 660–665.
20. G. Williams and H. A. Lord, "Mold Filling Studies for the Injection Molding of Thermoplastic Materials Part I: The Flow of Plastic Materials in Hot and Cold Walled Circular Channels," *Polymer Engineering Science,* vol. 15, 1975, pp. 553–568.
21. H. A. Lord and G. Williams, "Mold Filling Studies for the Injection Molding of Thermoplastic Materials Part II: The Transient Flow of Plastic Materials in the Cavities of Injection Molding Dies," *Polymer Engineering Science,* vol. 15, 1975, pp. 569–582.
22. J. F. Stevenson et al., "An Experimental Study and Simulation of Disk Filling by Injection Molding," *Society of Plastics Engineers Technical Papers,* vol. 22, 1976, pp. 282–288.
23. K. K. Kamal and S. Kenig, "The Injection Molding of Thermoplastics Part 1: Theoretical Model," *Polymer Engineering Society,* vol. 12, 1972, pp. 295–308.
24. C. A. Austin, "Filling of Mold Cavities," in E. C. Bernhardt, ed., *Computer Aided Engineering for Injection Molding,* Hanser, New York, 1983.
25. C. A. Hieber and S. F. Shen, "A Finite Element/Finite Difference Simulation of the Injection Molding Filling Process," *Journal of Non-Newtonian Fluid Mech.,* vol. 7, 1980, pp. 1–32.
26. L. T. Manzione, *Applications of Computer Aided Engineering in Injection Molding,* Hanser, New York, 1987.
27. P. K. Kennedy, *Flow Analysis of Injection Molds,* Hanser, New York, 1995.
28. S. Güçeri, "Finite Difference Solution of Field Problems," in C. L. Tucker III, ed., *Computer Modeling for Polymer Processing,* Hanser, New York, 1989.
29. M. Rezayat and T. E. Burton, "A Boundary Integral Formulation for Complex Three-dimensional Geometries," *International Journal of Numerical Methods in Engineering,* vol. 29, 1990, pp. 263–273.
30. S. G. Advani, ed., *Flow and Rheology in Polymer Composites Manufacturing,* Elsevier Science, Amsterdam, 1994.
31. T. D. Papathanasiou and D. C. Guell, eds., *Flow-induced Alignment in Composite Materials,* Woodhead, Cambridge, England, 1997.
32. F. P. Folgar and C. L. Tucker, "Orientation Behaviour of Fibres in Concentrated Suspensions," *Journal of Reinforced Plastics and Composites,* vol. 3, 1984, pp. 98–119.
33. X. J. Fan, N. Phan-Thien, and R. Zheng, "A Direct Simulation of Fibre Suspensions," *Journal of Non-Newtonian Fluid Mechanics,* vol. 74, 1998, pp. 113–136.
34. N. Phan-Thien and R. Zheng, "Macroscopic Modelling of the Evolution of Fibre Orientation during Flow," in T. D. Papathanasiou and D. C. Guell, eds., *Flow-induced Alignment in Composite Materials,* Woodhead, Cambridge, England, 1997.
35. S. G. Advani and C. L. Tucker III, "The Use of Tensors to Describe and Predict Fiber Orientation in Short Fiber Composites," *Journal of Rheology,* vol. 31, 1987, pp. 751–784.
36. G. B. Jeffery, "The Motion of Ellipsoidal Particles Immersed in Viscous Fluid," *Proceedings of the Royal Society of London A,* vol. 102, 1922, pp. 161–179.
37. R. Zheng et. al., "Thermoviscoelastic Simulation of Thermally and Pressure Induced Stresses in Injection Moulding for the Prediction of Shrinkage and Warpage for

Fibre-Reinforced Thermoplastics," *Journal of Non-Newtonian Fluid Mechanics,* vol. 84, 1999, pp. 159–190.
38. S. F. Walsh, "Shrinkage and Warpage Prediction for Injection Moulded Components," *Journal of Reinforced Plastics and Composites,* vol. 12, 1993.
39. M. Rezayat, "Numerical Computation of Cooling Induced Residual Stress and Deformed Shape for Injection-Molded Thermoplastics," *Society of Plastics Engineers Technical Papers,* vol. 35, 1989, pp. 341–343.
40. F. P. T. Baaijens, "Calculation of Residual Stress in Injection Moulded Products," *Rheol. Acta,* vol. 30, 1991, pp. 284–299.
41. R. Mills, "The Advancing State of Solid Modeling," *Computer-Aided Engineering,* vol. 17, no. 9, 1998, pp. 56–66.
42. P. Kennedy and H. Yu, "Plastic CAE Analysis of Solid Geometry," *Society of Plastics Engineers, Technical Papers,* vol. 43, 1997.
43. R. G. Speight et. al., "In-line Process Monitoring for Injection Molding," *Proc. of the I. MechE., Part E, J. Process Mechanical Engineering,* vol. 211, 1996.

Chapter

8

Design of Plastic Products

Jordan I. Rotheiser

Rotheiser Design, Inc., Highland Park, Illinois

8.1 Fundamentals

8.1.1 Introduction

The times we live in are often known as the "Computer Age." It could also be referred to as the "Plastics Age," as the production of plastics has exceeded that of steel (by volume) since 1979. In fact, the volume of plastics produced has more than doubled in the last 20 years. Nonetheless, most students who graduate from the major engineering universities are generally unprepared to design in plastics. Thus, it is left to the individual engineer to learn plastics engineering on his or her own, often by trial and error.

Unfortunately, this type of education can come at great cost—both to the company and the career of the individual. First, second, and even third and fourth efforts can be disastrous. That is because plastics design is more complicated, and more time consuming, than designing with metals. There are several reasons for this and they center around the processes which are used to manufacture plastic parts, the tooling used for those processes, and the nature of the plastics themselves.

Unlike metals, the properties of most plastics vary considerably within normal operating temperatures. A particular acrylonitrile butadiene styrene (ABS) whose tensile strength is 5500 lb/in^2 at room temperature can drop to 2800 lb/in^2 at 125°F. Other properties are also affected. For example, brittleness increases as the temperature drops, etc.

What does this mean to the design engineer? Basically, it means there will be more work to do. Various pertinent properties will need to be examined at both extremes of the service range. Furthermore, the design parameters must be explored more fully. It cannot be assumed that the product will survive the temperatures endured in cleaning, shipping, or storage unscathed.

Other exposures can cause problems with plastic components as well. Ultraviolet light causes or catalyzes chemical degradation in many resins. Plastics are vulnerable to attack from many chemicals, particularly in heavy concentrations. Some are even affected by water and there is one, polyvinyl acetate, that actually dissolves in water (think of soap packets).

Engineers will want to calculate the various performance values in the traditional fashion. Unfortunately, these computations cannot be regarded as valid. It is not that the laws of physics are different for plastics, it is simply that the data employed is not reliable for calculations.

That is not to suggest any skullduggery on the part of the test engineers; it is simply that the standard test sample and conditions are narrowly defined and likely to be significantly different from those to be endured by any specific product. The values obtained for most plastics will vary according to the process, gating, wall thickness, rate of loading, etc. It should be noted that there is some latitude within the test procedures themselves which can affect results. Most plastics engineers use the data sheets principally for the purposes of comparison in material selection.

It is important that the design engineer be cognizant of this when employing the data in an empirical fashion. Generous safety factors are customarily used. When greater precision is required, the proposed material may be independently tested under conditions more appropriate to the actual application. When the anticipated market is sufficient, the resin manufacturer may be inclined to perform these tests at the company's expense. Otherwise, it will be the obligation of the product manufacturer to bear the cost.

The general stiffness range of most plastics, combined with the general effort to use the thinnest possible wall thickness, means the geometry has a pronounced effect. Other than through comparison to similar constructions, the stiffness of the actual part is difficult to predict in a precise fashion. Although the traditional equations will produce approximate results, stiffness remains a question until the first part is molded. (Finite element analysis results are vulnerable to the many variables involved.) Fortunately, there are so many compound variations available within a given resin, it is usually possible to adjust stiffness within a reasonable range. This has been the saving grace of many an engineer. Also, many plastics engineers withhold the placement of ribbing until the first parts are tested.

Even if the material maintained its properties throughout the product's temperature range and the data was perfectly reliable, the product's performance could still vary. That is because the plastics processes are subject to tooling quality and process parameter variations.

Nonetheless, the fact that plastic parts can be successfully designed is attested to by the wide variety of products in the marketplace. It is clearly, however, more work to design in plastic and it is virtually impossible to perfectly predict the initial results. That is the reason prototypes are frequently made.

It is tempting to test a fabricated sample before constructing tooling. However, it should be noted that the final part is likely to produce substantially different test results than one fabricated by some other process. Furthermore, sophisticated plastics engineers will often deliberately underdesign a product, adding a little material at a time in a prototype mold to determine the thinnest acceptable wall. This approach also permits the testing of the complete assembly whereby the walls of the mating parts reinforce each other to produce a stronger overall structure.

8.1.2 The holistic design approach

There are four principle elements to a successful plastic product: material selection, part design, tooling, and processing. Typically, product designers are effective part designers but have limited background in the other disciplines. This leads to products which are more expensive than necessary and difficult to manufacture—which also increases the cost. Many companies have solved this problem through the use of multidisciplinary design teams. However, team members report that such teams can be dysfunctional, often due to the fact that team members' schedules are difficult to synchronize or a lack of availability of required skills.

Ideally, the part designer would know enough about these other disciplines to be able to design with them in mind. That utopian situation could create the ultimate in efficient product design—*the holistic design approach.*

Bottle cap example. An example of this type of thinking would be the case of a closure for a bottle containing cosmetics. In order to protect the contents, the closure must seal the opening. Typically, that seal is created with a seal ring or liner which is clamped down with force provided by screw threads, thus creating what is generally known as a *bottle cap.*

Process selection is simplified because the process of choice for most bottle cap applications is injection molding. Compression and transfer molding are possibilities, however, they are slower than injection molding and are rarely used for thermoplastic materials. Therefore, they would only be considered if the material of choice turned out to be a thermoset.

In the case of a cosmetic bottle cap, the most extreme temperatures it is likely to encounter will be in transit or washing prior to application. This limits the range to that which is readily accommodated by most thermoplastics without deformation. However, thermal expansion will need to be considered as the cap cannot loosen enough to break the seal at elevated temperatures nor contract enough to crack at low temperatures. Furthermore, it must not fail due to stress relaxation over time nor impart an odor of its own to the contents. For a cosmetic application, there may also be an appearance requirement of a high-gloss surface. Most importantly, it must withstand the chemical attack of the contents. While resin manufacturers typically perform limited tests (more on this topic will be discussed later) for resistance to various chemical compounds, they cannot do this when the composition of the exposure is a secret such as with a cosmetic. The manufacturer is expected to conduct such tests privately.

There is another element of material selection for a bottle cap which involves the tool building and processing disciplines. It derives from the fact that the decision must be made as to whether the threads are to be stripped off the core, turned off the core or the core is to be collapsed to permit ejection. The tool for the former will be far less expensive and the mold will operate at a much faster rate of speed. However, the material must be one which is flexible enough to strip off the mold, yet rigid enough to perform its other functions.

The problem is created by the fact that the formation of a thread creates plastic at a point inside the largest diameter of the hardened core as shown in Fig. 8.1*a*. As the force of ejection pushes on the base of the cap to remove it from the core, it must be flexible enough to expand off the core as illustrated in Fig. 8.1*b*. The part becomes more rigid as it cools in the mold. Even an essentially rigid material might be successfully ejected from the mold if this function is performed while the part is still soft. However, the part must also be rigid enough to withstand the force of ejection without enduring permanent distortion.

The point at which the cap is cool enough to eject, yet warm enough to strip off the core, will vary according to the means of ejection employed. Ejector pins provide very localized forces at the base of the cap. An ejector plate creates an ejection force which is distributed uniformly across the base of the cap. Therefore the cap can be ejected in a softer condition with the use of a stripper plate. That results in a cycle

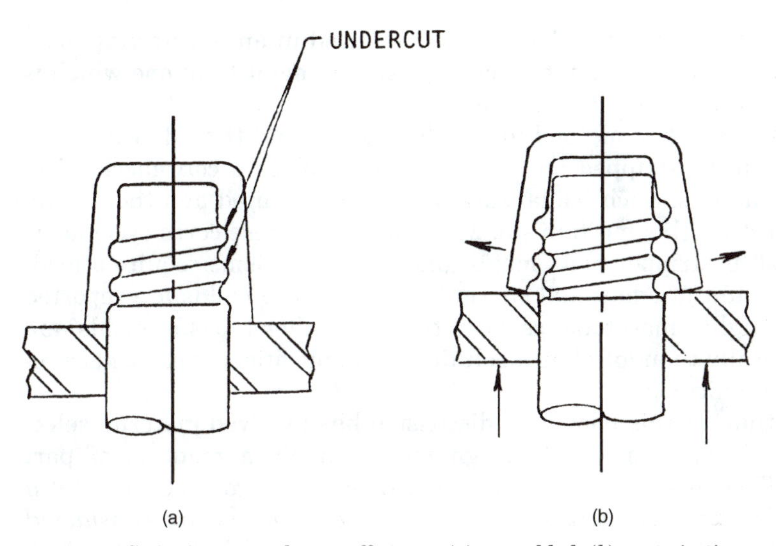

Figure 8.1 Stripping an undercut off a core: (*a*) as molded; (*b*) part ejection.

reduction on the order of 30%, however, the stripper plate adds a significant increment of cost to the tooling.

The amount of force required to eject the part can also be attained through the use of interrupted threads on the bottle cap. By breaking the continuity of the thread, the amount of material which must be stretched to permit removal of the cap from the mold is significantly reduced.

The determining factor in how deep a thread can be stripped is its strain rate. For the sake of this bottle cap example, it will be presumed that it must be made of acrylic for appearance reasons because this polymer can provide a very high level of gloss. *Acrylic* is an amorphous thermoplastic with a very low rate of strain. In this case, too low to permit the part to be stripped off the mold. Therefore, the cap would need to be turned off the core of the mold.

In order to turn the part off the core, an unscrewing mechanism must be employed. There are several ways to go about this, however, all of them incur significant additional cost. Furthermore, the space required for the mechanism limits the number of cavities which can be placed within a mold base. If the platens of the molding machine have sufficient space, a larger mold base can be used. However, if the mold was already sized to the limits of the platen, the number of cavities will need to be reduced or a larger molding machine will be required. Either way, fewer bottle caps will be produced with each molding cycle and the machine cost for each cap will be increased, thus reducing the efficiency of the production. The machine cost will also be increased by the longer cycle necessitated by the time required for the unscrewing mechanism to function. Thus, a cap produced in an unscrewing mold will always have a greater machine cost increment than one which is stripped off a core—all other elements being equal.

There is another method for producing internal threads which are too rigid to be stripped off a mold. That involves a core mechanism which collapses. Such "collapsing cores" are patented and there is an added cost for this mechanism. Molds utilizing these cores cycle nearly as fast as stripper plate molds and the mechanisms require a moderate amount of additional space. However, these molds are reported to have higher maintenance costs than the other types of molds and are generally thought of as a solution for applications with lower production quantities.

Note that, at this point, the discussion has involved material selection, processing, and tool design with scarcely a mention of part design. That is the whole point of the holistic design concept—that *a higher level is achieved when all of these elements are considered simultaneously.* The details of the part design are dependent on the decisions reached as a whole. Clearly the thread design will be dependent on the determination of whether the cap is to be turned or stripped off the core or whether a collapsing core is to be used instead. When appearance is of greater importance than molding efficiency, esthetic requirements may be the determining factor.

8.1.3 Basic design considerations

In order to avoid unpleasant surprises which can cause a design to fail, it is necessary to know everything possible about the conditions which the product will be exposed to in its lifetime. Armed with that information, the plastics designer can determine if the design, material, process, and tooling are appropriate for the application. That is, at least to the limits of the available information. A certain degree of risk is inherent in plastics design because the cost in time and resources is too great to permit the accumulation of enough information to eliminate that risk. Higher levels of risk are acceptable where tooling investment is low and where product failure results only in very low levels of property loss. As the cost of failure increases, more resources are devoted to risk reduction and greater safety factors are used. When product failure could result in serious injury or loss of life, exhaustive testing and greater safety factors are employed.

Most product structural failures result from conditions the designer did not anticipate. Thus, the first order of business is to establish the design parameters. This is achieved by the rather tedious process of considering all the conditions the product will be exposed to. A checklist is a useful means of reminding the designer of all these conditions

(Table 8.1). This is a general checklist, and the individual designer will no doubt find it necessary to make additions appropriate to his or her specific product area.

Note 2 of the checklist in Table 8.1 serves to remind the designer that not all of the conditions encountered by the product occur in its use. The highest temperature the product may be called upon to endure may actually occur in a boxcar in the desert in midsummer, in cleaning, or in a mandated test procedure. Decorating, cleaning, or assembly may result in unforeseen chemical exposures. The author recalls one project where he was "blindsided" by the discovery, long after production had commenced, that the product was washed in an acid solution which was followed by a base solution out in the field. Fortune smiled on that occasion and the material used happened to be one which could withstand those exposures.

Impacts may also occur under odd circumstances. One time, the author found that the sales department of a company had brought a baseball bat to the product's introduction and was merrily inviting customers to take a swing at the housing to demonstrate its superiority over the competitors' metal housing. The engineers joined in a silent nondenominational prayer that the housing would be able to withstand 3 days of that kind of treatment.

Apparently their prayer was answered and the product survived the onslaught. However, there is a well-known story in the automobile industry about a bumper mold which had to be rebuilt because the engineers did not account for the heat of the oven which dried the paint. The original mold had been built for a material which would not withstand the oven temperatures and the resin with a higher temperature resistance had a significantly different mold shrinkage. The mold could not be salvaged and a new one had to be made.

Fear often strikes the heart of those who are called upon to fill out a checklist, and they find themselves specifying higher levels of performance than is really required. Performance has a price and that practice leads to unnecessary cost, particularly when the designer incorporates a safety factor to ensure that the required standard is met. (Safety factors are added in the design stage.) In some cases, requirements are raised to a level which causes a more expensive material or process to be used. Occasionally, they are lifted to a level which cannot be reached using plastics.

Section A of Table 8.1, "Physical Limitations," sets the basic parameters of the design. The sizes alone are enough to eliminate some of the processes. Section B, "Mechanical Requirements," will reveal some interesting aspects of the design not often considered. For example, there is the question of whether it is more important to have a long functional life, a static life, or a shelf life.

TABLE 8.1 Product Design Checklist

NOTES:
1. Items left blank presume "does not apply."
2. Include conditions encountered in use, cleaning, shipping, assembly, testing, and decorating.
3. Remember: Overspecification results in unnecessary cost.

Part Name and Description of Application: ____________

A. Physical Limitations

Length: ______ Width: ______ Height: ______ Weight: ______ Density: ______

Mating parts: ______ Mating fitments: ______

B. Mechanical Requirements

Functional (dynamic) life: ______	Nonfunctional (static) life: ______	Shelf life: ______
Tensile strength: −40°F ______	72°F ______	140°F ______
Flexural strength: −40°F ______	72°F ______	140°F ______
Compressive strength: −40°F ______	72°F ______	140°F ______
Flexural modulus (stiffness): ______	Hardness: ______	
Maximum allowable creep: ______	Deflection: ______	
Impact strength: −40°F ______	72°F ______	140°F ______
Shear strength: ______	Abrasion resistance: ______	

C. Environmental Limitations

Chemical resistance: Continuous: ______	Intermittent: ______	Occasional: ______
Temperature: Maximum + Duration: ______	Minimum + Duration: ______	Operating + Duration: ______
Ultraviolet light: ______	Water immersion: ______	Moisture vapor transmission: ______
Radiation: ______	Flammability: ______	

Electrical requirements

Volume resistivity: ____________ Surface resistivity: ____________ Dielectric constant: ____________
Dissipation factor: ____________ Dielectric loss: ____________ Arc resistance: ____________
Electrical conductance: ____________ Microwave transparency: ____________ EMI/RFI shielding: ____________

E. Appearance Requirements

Surface finish: Inside (SPI#): ________ Outside (SPI#): ________ Texture No: ____________ Depth: ____________
Match to: ____________ Color: ____________
Color maintenance: ____________ Transparency: ____________ Translucency: ____________
Metallizing: ____________ Decoration: ____________
Identification: Model: ____________ Production date: ____________ Recycling: ____________
Warnings: ____________ Nameplates: ____________ Instructions: ____________
Manufacturing limits: Flash :____________ Mismatch: ____________
Gate location: ____________ Ejector location: ____________

F. Assembly Requirements

Parts to be assembled to, method prepared for (screws, solvents, etc.), and type (permanent, serviceable, occasionally reopenable, water-tight, or hermetic).

1. ____________ 2. ____________
3. ____________ 4. ____________

G. Other Design Parameters

Anticipated volume: Annual: ____________ Order: ____________
Anticipated production date: Start: ____________ Volume: ____________
Unusual legal exposure: Liability: ____________ Patents: ____________
Regulatory approvals required: 1.________ 2.________ 3.________ 4.________ 5.________ 6.________

Specifications: Military: ____________ Building codes: ____________
Foreign production: ____________ Foreign sales areas: ____________

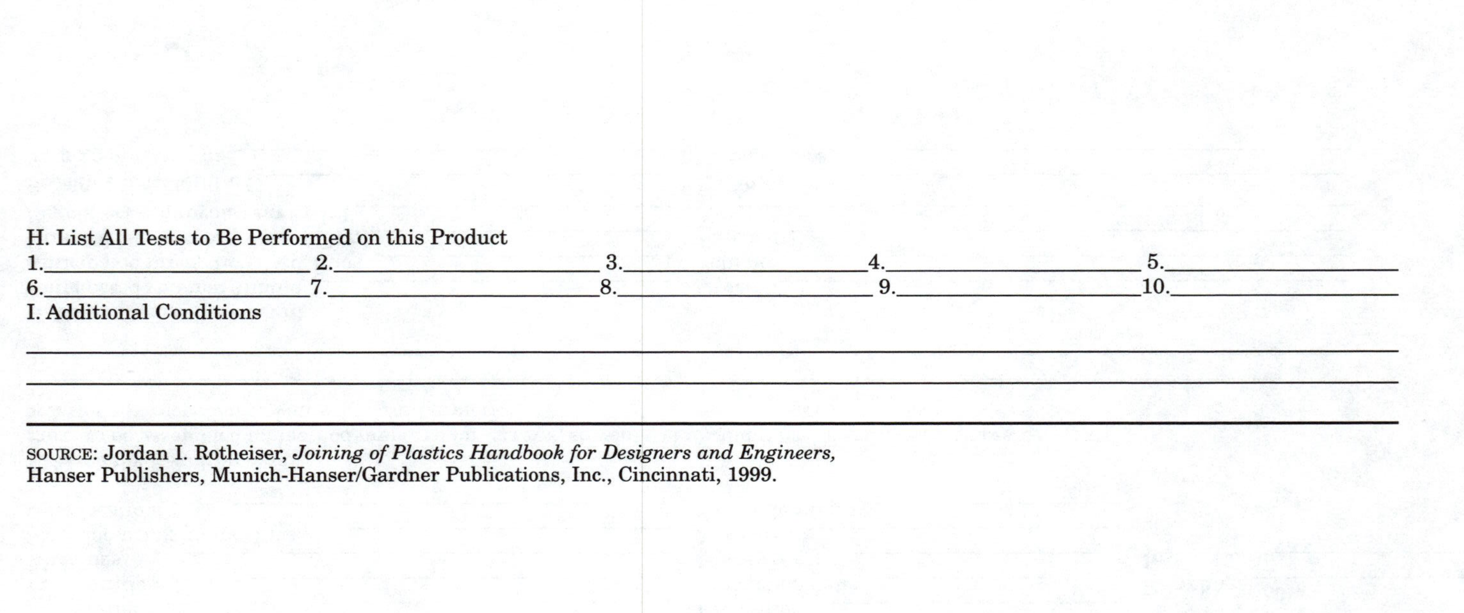

H. List All Tests to Be Performed on this Product

1.________ 2.________ 3.________ 4.________ 5.________

6.________ 7.________ 8.________ 9.________ 10.________

I. Additional Conditions

SOURCE: Jordan I. Rotheiser, *Joining of Plastics Handbook for Designers and Engineers,* Hanser Publishers, Munich-Hanser/Gardner Publications, Inc., Cincinnati, 1999.

Sometimes one type of performance must be sacrificed to another. An example of this type of phenomenon might be a shaft seal in place on a product. Its *shelf life* would be the length of time the product could remain unused and still have the seal function properly when used. Its *static life* would be the length of time it would be able to function after it had been first used. Finally, its *functional life* would be the period of time or number of rotations the seal could withstand while the shaft was rotating. Bear in mind that the temperatures would rise when the shaft was turning and abrasion would also be taking place. It is, therefore, quite likely that the ideal polymer for one circumstance might be less favorable for another. Decisions must often be made.

Which of the product requirements in this section is most likely to result in a structural failure? Of course any of them can or they would not be listed. However, in the author's experience, the most common failures (short of gross design errors) occur due to weakening of the material at elevated temperature, impact failure at low temperature, or creep failure over time.

Designers often forget that plastics are, themselves, chemical compounds. Hence they are particularly vulnerable to environmental exposures. Section C in Table 8.1, "Environmental Limitations," attempts to reveal the exposures which could result in failure. The chemical exposures are the most difficult to discover because there are so many of them and they are often hidden from view. Many common cleaning, cosmetic, and food preparations contain unusual and secret ingredients. Nonetheless, we identify what we can and list the commercial names for the others. In some cases, resin manufacturers have actually tested their polymers in these commercial preparations.

Often these exposures take place in combination as in products to be used out of doors. In fact, design for products to be used outdoors is practically a category in itself as ultraviolet light significantly affects most plastics over time. In a few cases, even indoor applications have been severely affected.

Section D in Table 8.1, "Electrical Requirements," is largely limited to electrical applications. However, Section E, "Appearance Requirements," affects nearly every application to some degree. Designers often overlook these details in the early stages of the project because they are preoccupied with the structural aspects of the design. Nonetheless, appearance requirements can lead to expensive mold changes. For example, failure to select the proper mold finish can require a change in the draft angle. That can make the part larger, affecting the fitments and so on. Section F in the table deals with assembly requirements.

There are some other design parameters as indicated in Section G. For example, the anticipated volume will often dictate the process to be used. Section H, "List All Tests to Be Performed on This Product," may reveal test protocols more stringent than the demands on the product in the field, particularly if they are archaic in nature.

8.1.4 Material selection

The early chapters of this book cover the plastics materials in depth. For a thorough discussion of each polymer, the author recommends that the reader refer to these chapters. This section looks at material selection from the designer's perspective.

It is said that there are some 30,000 to 35,000 plastic compounds on the market as this is being written, with more being added all the time. That number is enough to stagger the mind of the designer trying to make a material selection. Fortunately, only a small percentage of them are actually serious contenders for any given application. Some of them were developed specifically for a single product, particularly in the packaging industry. Others became the material of choice for certain applications because of special properties they offer which are required for that product or process. For example, the vast majority of roto-molded parts are made of polyethylene, while glass fiber–reinforced polyester is the workhorse of the thermoset industry. A bit of research should reveal if there is a material of choice for any given product application.

First, a bit of a review of the basic categories of plastics materials. In general, they fall into one of two categories: thermosets and thermoplastics. *Thermosets* undergo a chemical reaction when heated and cannot return to their original state. Consequently, they are chemical resistant and do not burn. Cross-linked plastics are an example of thermosets. *Thermoplastics* constitute the bulk of the polymers available. Although some degradation does occur, they can be remelted. Most are readily attacked by chemicals and they burn readily.

Thermoplastics can also be broken down into two basic categories: amorphous and semicrystalline (hereafter referred to as crystalline). The names refer to their structures; *amorphous* have molecular chains in random fashion and *crystalline* have molecular chains in a regular structure. Polymers are considered semicrystalline because they are not completely crystalline in nature. Amorphous resins soften over a range of temperatures whereas crystallines have a definite point at which they melt. Amorphous polymers can have greater transparency and lower, more uniform postmolding shrinkage. Chemical resistance is, in general, much greater for crystalline resins than for amorphous resins, which are sufficiently affected to be solvent welded. The triangle illustrated in Fig. 8.2 provides an easy way to categorize the thermosplastics.

The cost of plastics, generally, increases with a corresponding improvement in thermal properties (other properties, typically, go up as well). The lowest cost plastics are the most widely used. The triangle in Fig. 8.2 is organized with the least temperature-resistant plastics at the base and those with the highest temperature resistance at the top. Therefore, the plastics designated "Standard" at the base of the triangle, often referred to as *commodity* plastics, are the lowest in cost and most widely used. They can be used in applications with temperatures up to 150°F. (Note that these are very loose groupings and the precise properties of a specific resin must be evaluated before specifying it.)

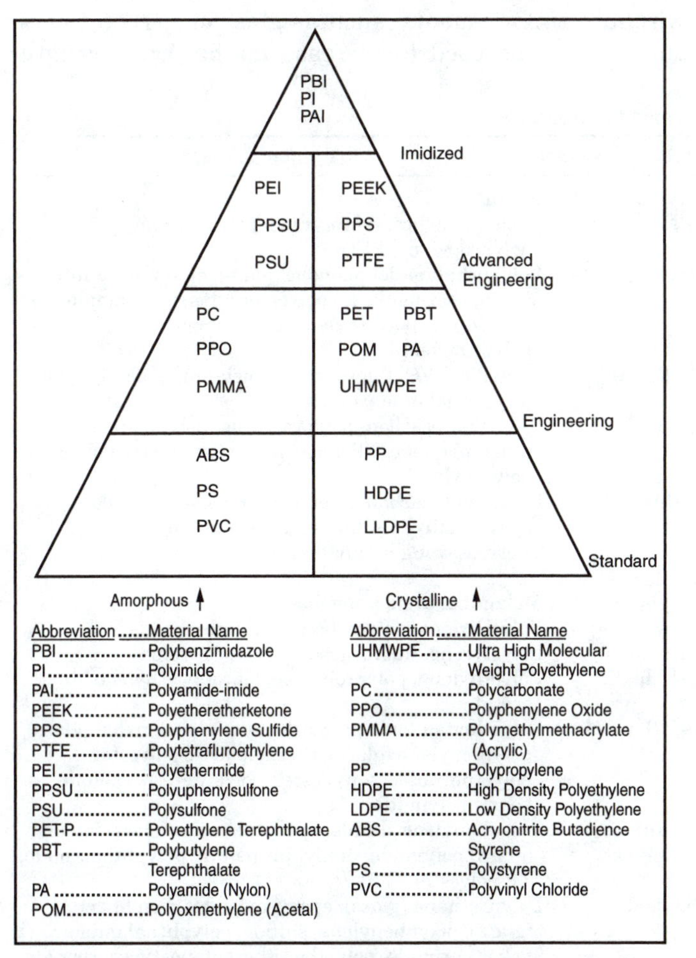

PBI
PI
PAI
Imidized
PEI
PPSU
PSU
PEEK
PPS
PTFE
Advanced Engineering
PC
PPO
PMMA
PET PBT
POM PA
UHMWPE
Engineering
ABS
PS
PVC
PP
HDPE
LLDPE
Standard
Amorphous
Crystalline
Abbreviation......Material Name
PBI..................Polybenzimidazole
PI....................Polyimide
PAI..................Polyamide-imide
PEEK..............Polyetheretherketone
PPS................Polyphenylene Sulfide
PTFE..............Polytetrafluroethylene
PEI..................Polyethermide
PPSU..............Polyuphenylsulfone
PSU................Polysulfone
PET-P..............Polyethylene Terephthalate
PBT.................Polybutylene Terephthalate
PA...................Polyamide (Nylon)
POM................Polyoxmethylene (Acetal)
Abbreviation......Material Name
UHMWPE.........Ultra High Molecular Weight Polyethylene
PC..................Polycarbonate
PPO................Polyphenylene Oxide
PMMA.............Polymethylmethacrylate (Acrylic)
PP...................Polypropylene
HDPE..............High Density Polyethylene
LDPE..............Low Density Polyethylene
ABS................Acrylonitrite Butadience Styrene
PS...................Polystyrene
PVC................Polyvinyl Chloride

Figure 8.2 Classification of thermoplastics. (*Source: Laura Pugliese, Defining Engineering Plastics, Plastics Machining and Fabrication, January-February, 1999, Courtesy DSM Engineering Plastic Products.*)

The next level are the "Engineering" plastics which can be used for applications ranging up to 250°F. ABS is often considered an engineering plastic for its other properties although it cannot withstand this temperature level. For applications requiring temperature resistance up to 450°F, the next step, "Advanced Engineering," is appropriate. The amorphous plastics at this level are often used in steam environments and the crystalline plastics have improved chemical resistance. The top level, the "Imidized" plastics, can withstand temperatures up to 800°F and have excellent stress and wear properties as well.

Table 8.2 lists many of the principle properties and some of the polymers which are noted for those properties. While incomplete, this table should at least provide a beginning. They are primarily listed in their natural state without reinforcements, such as glass or carbon fibers. These reinforcements can be used to increase mechanical strength,

TABLE 8.2 Recommended Materials

Property	Recommendation
Abrasion, resistance to (high)	Nylon
Cost:weight (low)	Urea, phenolics, polystyrene, polyethylene, polypropylene, PVC
Compressive strength	Polyphthalamide, phenolic (glass), epoxy, melamine, nylon, thermoplastic polyester (glass), polyimide
Cost:volume (low)	Polystyrene, polyethylene, urea, phenolics, polypropylene, PVC
Dielectric constant (high)	Phenolic, PVC, fluorocarbon, melamine, alkyd, nylon, polyphthalamide, epoxy
Dielectric strength (high)	PVC, fluorocarbon, polypropylene, polyphenylene ether, phenolic, TP polyester, nylon (glass), polyolefin, polyethylene
Dissipation factor (high)	PVC, fluorocarbon, phenolic, TP polyester, nylon, epoxy, diallyl phthalate, polyurethane
Distortion, resistance to under load (high)	Thermosetting laminates
Elastic modulus (high)	Melamine, urea, phenolics
Elastic modulus (low)	Polyethylene, polycarbonate, fluorocarbons
Electrical resistivity (high)	Polystyrene, fluorocarbons, polypropylene
Elongation at break (high)	Polyethylene, polypropylene, silicone, ethylene vinyl acetate
Elongation at break (low)	Polyether sulfone, polycarbonate (glass), nylon (glass), polypropylene (glass), thermoplastic polyester, polyetherimide, vinyl ester, polyetheretherketone, epoxy, polyimide
Flexural modulus (stiffness)	Polyphenylene sulfide, epoxy, phenolic (glass), nylon (glass) polyimide, diallyl phthalate, polyphthalamide, TP polyester
Flexural strength (yield)	Polyurethane (glass), epoxy, nylon (carbon fiber) (glass), polyphenylene, sulfide, polyphthalamide, polyetherimide, polyetheretherketone, polycarbonate (carbon fiber)

TABLE 8.2 Recommended Materials (*Continued*)

Property	Recommendation
Friction, coefficient of (low)	Fluorocarbons, nylon, acetal
Hardness (high)	Melamine, phenolic (glass) (cellulose), polyimide, epoxy
Impact strength (high)	Phenolics, epoxies, polycarbonate, ABS
Moisture resistance (high)	Polyethylene, polypropylene, fluorocarbon, polyphenylene sulfide, polyolefin, thermoplastic polyester, polyphenylene ether, polystyrene, polycarbonate (glass or carbon fiber)
Softness	Polyethylene, silicone, PVC, thermoplastic elastomer, polyurethane, ethylene vinyl acetate
Tensile strength, break (high)	Epoxy, nylon (glass or carbon fiber), polyurethane, thermoplastic polyester (glass), polyphthalamide, polyetheretherketone, polycarbonate (carbon fiber), polyetherimide, polyether-sulfone
Tensile strength, yield (high)	Nylon (glass or carbon fiber), polyurethane, thermoplastic polyester (glass), polyetheretherketone, polyetherimide, polyphthalamide, polyphenylene sulfide (glass or carbon fiber)
Temperature, heat deflection	Phenolic, epoxies, polysulfone, thermoset polyesters, polyether sulfone, polyimide (glass)
Temperature (maximum use)	Fluorocarbons, phenolic (glass), polyphthalamide, polyimide thermoplastic polyester (glass), melamine, epoxy, nylon (glass or carbon fiber), polyetheretherketone, polysulfone, polyphenylene sulfide
Thermal conductivity (low)	Polypropylene, PVC, ABS, polyphenylene oxide, polybutylene, acrylic, polycarbonate, thermoplastic polyester, nylon
Thermal expansion, coefficient of (low)	Polycarbonate (carbon fiber or glass), phenolic (glass), nylon (carbon fiber or glass), thermoplastic polyester (glass), polyphenylene sulfide (glass or carbon fiber), polyetherimide, polyetheretherketone, polyphthalamide, alkyd, melamine
Transparency, permanent (high)	Acrylic, polycarbonate
Weight (low)	Polypropylene, polyethylene, polybutylene, ethylene vinyl acetate, ethylene methyl acrylate
Whiteness retention (high)	Melamine, urea

maximum use temperature, impact resistance, stiffness, mold shrinkage, and dimensional stability.

Generally, the resin prices increase with improved mechanical and thermal properties. When there is no clear-cut material of choice, plastics designers generally follow the practice of looking for the lowest-cost material which will meet the product's requirements. If there is a reason that polymer is not acceptable, they start working up the cost ladder until they find one that will fulfill their needs. In thermoplastics, there are the so-called *commodity resins.* These are the low-cost resins used in great volume for housewares, packaging, toys, and so on. This group is made up of polyethylene, polypropylene, polystyrene, and PVC. Reinforcements can improve the properties of these resins at moderate additional cost. A lower-priced resin with reinforcement will often provide properties comparable to a more expensive resin. Table 8.3 is a list of the approximate cost of a number of plastics in increas-

TABLE 8.3 Approximate,* in Dollars per Cubic Inch of Plastics

Plastic	$/in^3
Polypropylene	0.010
Polyethylene (HD)	0.010
Polyvinylchloride	0.013
Polyethylene (LD)	0.014
Polystyrene	0.015
ABS	0.032
Acrylic	0.034
Phenolic	0.038
Thermoplastic elastomer	0.039
Styrene acrylonitrile	0.043
Polyester (TS)	0.043
Melamine	0.049
Polyphenylene ether	0.050
Polyphenylene oxide	0.050
Urea	0.053
Polyester (PET)	0.055
Nylon	0.058
Styrene maleic anhydride	0.059
Vinyl ester	0.060
Polyurethane (TS)	0.062
Alkyd	0.068
Polycarbonate	0.073
Polyurethane (TP)	0.074
Acetal	0.074
Cellulosics	0.075
Polycarbonate/ABS	0.078
Epoxy	0.081
Polyphenylene sulfide	0.119
Dially phthalate	0.154
Polysulfone	0.201
Polyetherimide	0.253
Liquid crystal polymer	0.434
Fluorpolymer	0.481
Polyamideimide	0.949
Polyetheretherketone	1.720

*These values are very approximate. They were arrived at by multiplying the average density by the average price at the time this was being written. In many cases, the range from which the average was taken was quite wide.

ing order of cost per cubic inch. This is regarded as a more useful figure than cost per pound in selecting a plastic material.

Thermosets usually provide higher mechanical and thermal properties at a lower material cost than do thermoplastics, "more bang for the buck" so to speak. However, most of the processes used to fabricate thermoset parts are slower and more limited in design freedom than the thermoplastic processes. Furthermore, the opportunity to utilize 100% of the material which thermoplastics provide is simply not available with thermosets because the regrind cannot be reused. Recycling possibilities are far more limited for thermosets for the same reason. Nonetheless, glass fiber–reinforced thermoset polyester is the material of choice for many outdoor applications in a severe environment such as boats and truck housings.

About the data. Comparison of resins is usually done with data sheets supplied by the resin manufacturers. It is extremely important that the plastics design engineer understand the limitations of this data. Since the properties of polymers change with temperature, the data sheet does not provide the total picture of a given compound. Instead, think of it as a "snapshot" of the material taken at 72°F. As the temperature goes down from this point, the material becomes harder and more brittle. Increasing the temperature makes the polymer softer and more ductile. These are general statements and the effect of temperature will vary widely between resins. For one material, tensile strength at 140°F may be only half that at 72°F. For another polymer, it may change only slightly.

The graph depicted in Fig. 8.3 illustrates this phenomenon. The upper curve indicates that the value at 0°F is 14,000 lb/in^2. At 72°F, it has dropped to around 12,000 lb/in^2. By the time it reaches 140°F, the tensile yield strength is approximately 7000 lb/in^2. This data is for nylon, a polymer particularly affected by moisture. The lower curve illustrates the effect of 2.5% moisture. In the range of temperatures between 30 and 100°F, the tensile yield strength appears to be about 20% lower for the moist material. Note that the curves begin to run together beyond 150°F as most of the water has been driven off by that point.

Time is also a significant factor. Figure 8.4 illustrates the effect of time on the stress-strain relationship of Delrin, an acetal polymer, at room temperature. Note that the strain rate increases with time. Figures 8.5 and 8.6 demonstrate how the strain rate of this material increases as the temperature rises.

Most of the physical properties, even properties such as the coefficient of friction or the coefficient of linear thermal expansion, can change significantly with changes in temperature, although not necessarily to the degree depicted in the graphs for these materials.

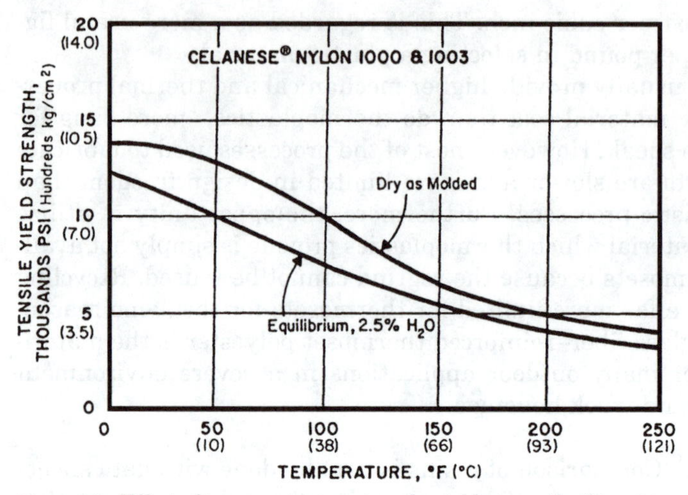

Figure 8.3 Effect of temperature on tensile yield strength. (*Courtesy of Ticona.*)

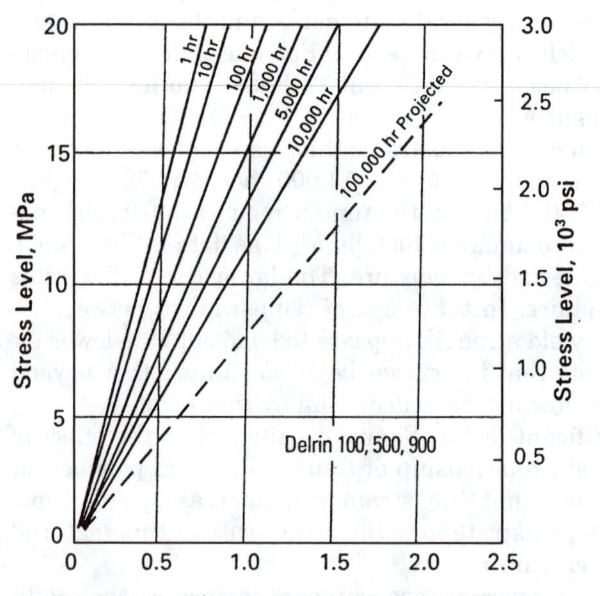

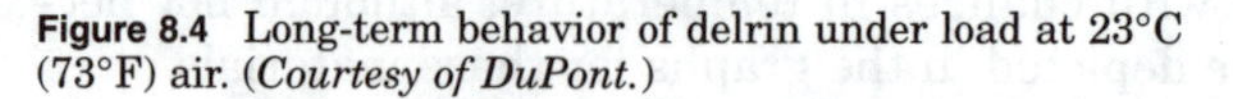

Figure 8.4 Long-term behavior of delrin under load at 23°C (73°F) air. (*Courtesy of DuPont.*)

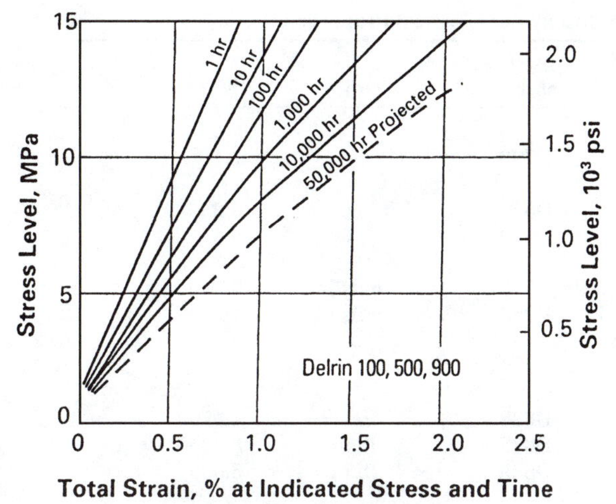

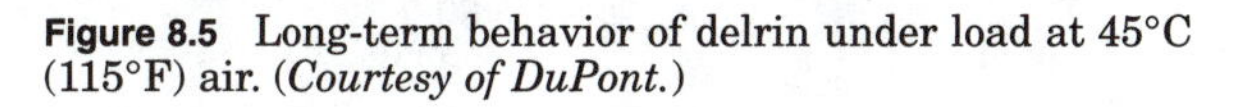

Figure 8.5 Long-term behavior of delrin under load at 45°C (115°F) air. (*Courtesy of DuPont.*)

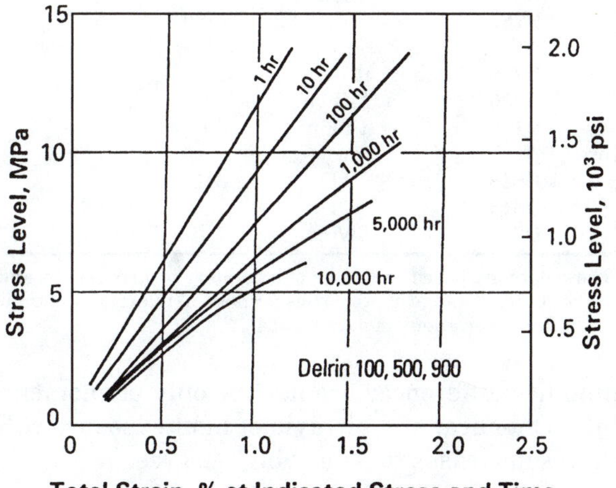

Figure 8.6 Long-term behavior of delrin under load at 85°C (185°F) air. (*Courtesy of DuPont.*)

TABLE 8.4 Celcon™ Acetal Copolymer—Typical Properties

Property	ASTM test method	Units	M90™
Physical:			
Special gravity	D792	—	1.41
Water absorption: 24 h			
@73°F	D570	%	0.22
@equilibrium	D570	%	0.8
Mold shrinkage:			
Flow direction	D955	mils/in	22
Transverse direction	D955	mils/in	18
Mechanical and thermal:			
Tensile strength @ yield			
−40°F	D638	lb/in^2	13,700
73°F	D638	lb/in^2	8,800
160°F	D638	lb/in^2	5,000
Elongation @ break:			
−40°F	D638	%	20
73°F	D638	%	60.0
160°F	D638	%	>250
Flexural stress @ 5% deformation	D790	lb/in^2 × 10^3	13.0
Flexural modulus:			
73°F	D790	lb/in^2 × 10^4	37.5
160°F	D790	lb/in^2 × 10^4	18.0
220°F	D790	lb/in^2 × 10^4	10.0
Fatigue endurance (limit @ 10^7 cycles)	D671	lb/in^2	4,100
Compressive strength:			
@ 1% deflection	D695	lb/in^2	4,500
@ 10% deflection	D695	lb/in^2	16,000
Rockwell hardness	D785	—	M80
Izod impact strength:			
−40°F (notched)	D256	ft-lb/in	1.0
73°F (notched)	D256	ft-lb/in	1.3
Tensile impact strength	D1822	ft-lb/in^2	70
Heat deflection temperature:			
@ 66 lb/in^2	D648	°F	316
@ 264 lb/in^2	D648	°F	230
Shear strength: 73°F	D732	lb/in^2	7,700

SOURCE: Courtesy Ticona. These data are based on testing of laboratory test specimens and represent data that fall within the standard range of properties for natural material. Colorant or other additives may cause significant variations in data values.

Temperature and humidity differences are not the only phenomena which can affect the data. Chemical and ultraviolet light exposure will also affect most resins, in some cases to a very high degree.

Interpreting the test data. This section discusses plastics test data from the point of view of the plastic product designer. For a far deeper discussion of the topic, see Chap. 11.

It is important to recognize that the test data represent a very precise set of circumstances, those established by the test protocol. Product designers must be aware of exactly how the test is performed in order to determine how well the results relate to the conditions experienced by the product under development. The following sections discuss the test procedures for the mechanical and thermal properties most commonly required. Table 8.4 represents a typical property sheet as supplied by the resin manufacturer. The material is Celcon® M90™ acetal copolymer (polyoxymethylene), a high-performance, engineering thermoplastic. For the plastic part designer to use this data effectively, he or she must thoroughly understand exactly what information it provides and the limits of that information. The following discussion will proceed to review the properties provided with a brief description of the test procedure and a few comments from the author.

Specific gravity—ASTM D792. This is a simple test in which a piece of the material is weighed, submerged in water, and weighed again to determine the difference in its weight. This test is not normally performed until 24 h after molding to permit most of the postmolding shrinkage to take place. It is important to consider the specific gravity (or density) in establishing the actual cost of a resin. It is also used to determine the extent of packing in molded parts. The value of this resin, 1.41, is in the upper range for unfilled thermoplastics.

Water absorption: 24 h—ASTM D570. The specimens for this test are 0.125 in thick and 2.00 in in diameter for molding materials. The material is submerged in water and the increase in weight is measured. This test can be performed with several procedures ranging from 1 h in boiling water to 24 h in water at 73.4°F. The physical and electrical properties of plastics can be affected by moisture absorption. The value for this resin, 0.22%, is quite low, too low to have a significant effect on properties.

Mold shrinkage—ASTM D955. The *mold shrinkage* is the difference between the mold dimensions and the molded part. This is established by measuring the part after molding it according to a prescribed set of molding parameters and cooling it for a short period of time. It does not account for all of the shrinkage because shrinkage can continue for up to 48 h for some resins. Furthermore, shrinkage is significantly affected by the wall thickness, shape, and size of the part in addition to the molding temperature, cycle, nozzle size, and packing. The values for this resin, 0.022 in/in in the flow direction and 0.018 in/in in the transverse direction, are typical for unfilled crystalline thermoplastics.

Tensile test—ASTM D638. The first mechanical property most product designers look for in evaluating a potential material is its *strength,* and by this they mean its *tensile strength at yield or break*. Therefore, it is often found at the top of the data sheet. The principal test for this property is ASTM D638; it calls for a "dog bone"–shaped specimen 8.50 in long by 0.50 in wide. The gripping surfaces at the ends are 0.55 in wide, giving it its characteristic shape. The test protocol permits the thickness to range from 0.12 to 0.55 in, and the rate at which the stress is applied from 0.5 to 20 in/min. This test is also used to obtain the percentage elongation at break and produce the stress-strain curve, from which the modulus of elasticity is derived. The values for this resin are upper midrange for unfilled thermoplastics, but much higher values can be achieved with the use of glass or carbon fiber fillers.

Flexural properties of plastics—ASTM D790. This test is performed by suspending a specimen between supports and applying a downward load at the midpoint between them. The specimen is a 0.50- by 5.00-in rectangular piece. Thickness can vary from 0.06 to 0.25 in, however, 0.125 in is the most commonly used. The distance between the supports is 16 times the specimen thickness. The load is applied at rates defined by the specimen size until fracture occurs or until the strain in the outer fibers reaches 5%. (Most thermoplastics do not break in this test.) The *flexural strength* is the flexural stress at 5% strain. In the event of failure before that point, the flexural strength is the tensile stress in the outermost fibers at the break point. The flexural modulus is the ratio of stress to strain within the elastic limit of the material. It is the primary means of measuring the stiffness of a material. Acetals have a wide range of flexural stress values. This one is about upper midrange for acetals and for unfilled thermoplastics in general; however, its flexural modulus is on the high side.

Fatigue endurance—ASTM D671. The *fatigue endurance* of a material is its ability to resist deterioration due to cyclic stress. The test covers determination of the effect of repeated flexural stress of the same magnitude with a fixed-cantilever apparatus designed to produce a constant amplitude of force on the plastic test specimen. The results are suitable for application in design only when all of the application parameters are directly comparable to those of the test. Consequently, this test is primarily used for comparison purposes.

Compressive strength—ASTM D695. Except for foams, plastic products rarely fail from compression alone. Consequently, the compressive strength is of limited value. The apparatus for this test resembles a C-clamp with the specimen compressed between the jaws of the apparatus, which close at the rate of 0.05 in/min until failure occurs. A wide range of specimen sizes is permitted for this test. The values for this resin are in the upper range for unfilled thermoplastics.

Rockwell hardness—ASTM D785. Hardness of plastics is difficult to establish and compare because there is an enormous range and there is an elastic recovery as well. However, the *Rockwell hardness* is useful in determining the relative indentation hardness between plastics. An indenter is placed on the surface of the test specimen and the depth of the impression is measured as the load on the indenter is increased from a fixed minimum value to a higher value and then returned to the previous value. A number of different diameter steel balls and a diamond cone penetrator are used. The Rockwell scale refers to a given combination of indenter and load. M80 places this resin in the upper middle range of hardness for plastics. A number of scales are used within the plastics industry. Figure 8.7 illustrates the relationship between them.

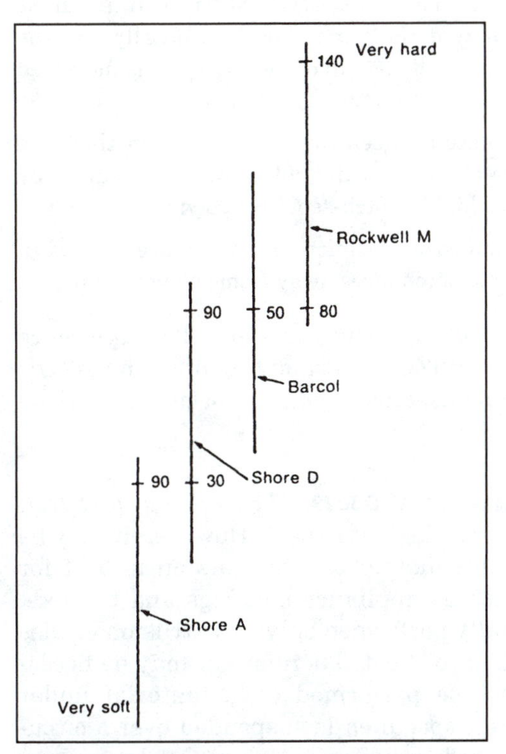

Figure 8.7 Range of hardness common to plastics. (*Source: Dominick V. Rosato, Rosato's Plastics Encyclopedia and Dictionary, Carl Hanser Verlag, Munich, 1993.*)

The IZOD impact test—ASTM D256. D256, the IZOD impact test, is the most common variety of impact test. It is not regarded as a reliable indicator of overall toughness or impact strength, however, it is a reasonable measure of a plastic's notch sensitivity. It is a pendulum test with the pendulum dropping from the 12 o'clock position to hit a sample held in a clamp at the 6 o'clock position. The pendulum breaks the sample and the distance it travels beyond the specimen is a measure of the energy absorbed in breaking the sample. The value calculated from this test is usually expressed in foot-pounds per inch of sample width.

The plastics design engineer must be wary of the fact that there are five different methods of performing this test, and the results will vary with each method. The four used by design engineers are as follows:

Method A. The specimen for this method is 2.50 in long by 0.50 in thick. There is a 45° included angle notch at midpoint which is 0.10 in deep and has a 0.01-in radius at the V. The notch faces the pendulum. The impact point is just above the notch.

Method B. This procedure is also known as the "Charpy" test. It is similar to the previous method, except that the bar is laid horizontally and the impact is directly behind the notch. The length of the specimen is increased to 5.0 in for this method.

Method D. The principal difference between this method and method A is that it permits a larger radius at the V of the notch, which substantially affects the results. It is used for highly notch-sensitive polymers.

Method E. In this case, the same size specimen and procedure applied in method A is used, except that the notch faces away from the pendulum.

This resin is notch sensitive and its values are low. The difference between the results of the notched IZOD test and the unnotched IZOD test (when available) can also be used as a measure of notch sensitivity for a given material.

The falling dart (tup) impact test—ASTM D3029. The *falling dart test,* ASTM D3029, is not on this particular data sheet. However, it may be more appropriate to reveal the behavior of materials on impact for many product applications such as appliance housings and the like. Unfortunately, this test is usually performed only for extrusion grades of resins which are to be made into sheet. Therefore, it may be necessary to request that this test be performed on a material under consideration. For testing, a flat specimen is suspended over a circular opening below a graduated column with a cantilever arm attached. A *weight,* also known as a *dart or tup,* is attached to the arm, from which it is released to strike the sample. The arm can be raised or lowered and the weight of the tup varied until 50% of the sample quantity fails the test. Method A of this test calls for an opening below the sample which is 5.00 in in diameter. For method B, it is 1.50 in in diameter.

The tensile impact strength—ASTM D1822. The *tensile impact test* eliminates the notch-sensitivity aspect of the IZOD impact test and is a more reliable indicator of impact strength for many applications. This test uses an apparatus very similar to that used for the IZOD tests except that, in this case, the specimen is attached to the pendulum on one end and has a T bar attached to the other end. When the pendulum drops, the T bar catches on the apparatus at its base causing the specimen to undergo tensile impact. For this test, the specimen is 2.50 in long and necks to 0.125 in at the center. The thickness can vary. The gripping surfaces at the ends are 0.50 in wide. This test is typically performed on materials which are too elastic to fail in the IZOD test and is normally found on data sheets. It can be performed on request if it best represents the product's performance requirements.

Heat deflection temperature—ASTM D648. This data is dangerous in the respect that it often is the only temperature data provided on a resin data sheet, which leaves the impression that it is a reliable indicator of the limit to which the product can be used (see the section on "Relative Temperature Index" later in this chapter). It is really nothing more than the temperature at which a given load (66 or 264 lb/in^2) will deflect a specimen an arbitrary amount. Other temperature tests are also used, and some are described in the following sections. Results of those tests are usually available from the resin manufacturer.

The apparatus for this test resembles somewhat that of the flexural test in that the specimen is suspended between two supports 4 in apart with a downward load at the midpoint. However, in this case, the entire structure is immersed in a liquid whose temperature is increased at the rate of 2°C/min. Two loadings are used, 66 and 264 lb/in^2. Consequently, the plastics engineer must be careful to compare values for the same loading. The *heat deflection temperature* is the temperature at which the specimen deflects 0.010 in. The specimen for this test is 0.50 in wide by 5.00 in long. Thicknesses vary from 0.125 to 0.50 in. The values for this resin are fairly high for an unfilled thermoplastic.

Different sample thicknesses and processes can produce significant differences in values. The author recalls a project where the field had been narrowed to two competing materials. One of them had a 15% heat deflection temperature advantage at a slightly higher cost. However, the other was produced by a long-standing supplier and, before taking the business away from that vendor, it seemed only fair to call and ask if a comparable resin were available which was not in the current brochure. The discussion revealed that the competing material was, in fact, the very same resin which the competitor bought from our supplier and resold as the company's own brand. Why then the difference in test values? Further research revealed that one supplier had used an injection-molded sample 0.125 in thick and the other had tested an extruded sample 0.50 in thick, which resulted in higher values.

Vicat softening point—ASTM D1525. The *Vicat softening point* is not provided on this particular data sheet, however, it is a method of determining the softening point of plastics which have no definite melting point. A 1000-g load is placed on a needle with a 0.0015-in^2 circular or square cross section. The softening point is taken as the point where the needle penetrates the specimen to a depth of 1 mm.

Glass transition temperature—ASTM D3418. Thermoplastics exhibit a characteristic whereby they change from a material which behaves like glass (strong, rigid, but brittle) to one with generally reduced physical properties (weaker and more ductile). This is known as the *glass transition temperature* (T_g) and is actually a range of temperatures as the value is different for each property and is significantly affected by variations in the test protocol. Usually a single value is provided, therefore, it should be treated as an approximation.

Relative temperature index—UL746B. Underwriter's Laboratories' Inc. (UL) has devised a thermal aging test protocol whereby a subject material is tested in comparison with a material with an acceptable service experience and correlates numerically with the temperatures above which the material is likely to degrade prematurely. The end of life of a material is regarded as the point where the value of the critical properties have dropped to half their original values. The resin manufacturer must submit his material to UL to have it tested. When this has not been done, the designer can use the generic value for the polymer, which is usually regarded as conservative.

Shear strength—ASTM D732. *Shear strength* is rarely a factor in molded and extruded plastic products due to their relatively thick wall sections, however, it can be important in film and sheet products. In this test, 2-in-diameter or 2-in-square specimens, ranging in thickness from 0.005 to 0.500 in are placed in a punch-type shear fixture. Pressure is applied to the punch at the rate of 0.005 in/min until the moving part of the sample clears the stationary part. The force divided by the area sheared determines the shear strength.

There are a number of other properties commonly used to evaluate plastic materials. Space limitations prevent a detailed description of the tests used to establish values for these properties. However, a listing of them is provided for the reader to research independently in Table 8.5, and data is often available from the resin manufacturer. More information about plastics tests can be found in the *Plastics Engineering Handbook* and can be obtained from the American Society

TABLE 8.5 Other Tests of Interest

Test	Standard
Coefficient of linear thermal expansion	ASTM D696, E228
Creep	
Stress relaxation	ASTM D2991
Tensile, compressive, and flexural creep and creep-rupture	ASTM D2990
Crystallization, heat of	ASTM D3417
Cure kinetics of thermosets	ASTM D4473
Deformation under load	ASTM D621
Dimensional stability	ASTM D756
Electrical properties	
Arc resistance	ASTM D495
Dielectric constant	ASTM D150
Insulation resistance	ASTM D257
Friction, coefficient of	ASTM D1894
Fatigue	
Flexural fatigue	ASTM D671
Tension-tension fatigue	ASTM D3479
Flammability	
Density of smoke from burning	ASTM D2843
Flammability	UL94
Flooring radiant panel test	ASTM E684
Oxygen index	ASTM D2863
Smoke emission	ASTM E662
Steiner 25-ft tunnel test	ASTM E84
Vertical test for cellular plastics	ASTM D3014
Flowability of thermosets	ASTM D3123
Fracture toughness	ASTM D5045
Moisture vapor transmission	ASTM D675
Optical properties	
Color	ASTM E308
Haze	ASTM D1003
Specular gloss	ASTM D253
Shear modulus	ASTM D5279
Viscosity-shear rate	ASTM D3835
Wear	
Abrasion resistance	ASTM D1242
Mar resistance	ASTM D673
Weathering	
Accelerated	ASTM G23, G26, G53
Light and water exposure	ASTM D1499, D2565
Outdoor	ASTM D1435, E838

for Testing Materials, 100 Barr Harbor Drive, West Conshohoken, PA 19428-2959.

8.1.5 Process selection

Selection criteria. A brief description of each process is found further in this chapter in Sec. 8.3. For a thorough discussion of each plastic process, refer to Chaps. 1 or 6. This section looks at process selection from the designer's perspective.

Plastics product designers are primarily interested in the ability of a given process to produce the shape they require. Therefore, the processes have been grouped according to their ability to produce a given shape. The groupings are

Thermoplastic open shapes

Thermoset open shapes

Hollow parts

Profiles

Ultra high strength

Within each shape category, the plastic product designer must be concerned with the ability of the process to produce the level of detail necessary for the application. A further discussion of the ability of each process to produce design details is also found in Sec. 8.3.

Beyond the fundamental design requirements, cost becomes the most significant factor in selecting the optimum process for the application. Product cost has three interrelated components: part cost, labor cost, and tooling amortization. Labor is related to process selection because some plastics processes permit the combining of parts to eliminate labor cost. For example, the cost of blow molding a hollow container must be compared to the cost of injection molding two halves and assembling them. Tooling amortization and piece part cost are directly related to anticipated annual volume, which is often difficult to forecast for new products.

As a broad statement, processes that require a higher initial investment in tooling produce parts at a lower cost. This is largely due to the fact that the reduced part cost is the product of faster molding cycles. Faster cycles require pressure on the plastic to reduce the time required to fill the mold cavity. The greater the pressure, the stronger the tooling and the more sophisticated the processing equipment must be. Both of these are factors which increase the initial investment.

The product design engineer is, therefore, keenly interested in the volume at which the additional investment would be justified by reduced part cost. It would simplify the decision making tremendously if it were possible to determine that point in terms of a given volume for each process. That might be feasible if all parts were identical in shape and size. One may presume that the larger and/or more complex the part, the greater the investment will be. As the investment grows, the production volume must be greater in order to pay the difference with lower piece part costs within an acceptable time period. That period also varies considerably between companies.

In the following discussion, the processes are ranked in order of increased tooling expense, which is usually commensurate with decreasing piece part cost. The first topics to be addressed are the thermoplastic processes. These easily comprise the largest volume of applications.

Thermoplastic shapes

1. Machining
2. Thermoforming and pressure thermoforming
3. Structural foam molding and coinjection molding
4. Injection molding and gas-assisted injection molding

The thermoplastic process pecking order begins with *machining*. While this is a common means of producing prototypes, the volume does not have to climb very high before other processes become more cost-effective.

Thermoforming and pressure thermoforming become cost-effective at very low volumes. The only problem with that equation lies in minimum sizes. Parts under roughly 1 ft^2 in area become difficult to justify for these processes. Thus, we really need to develop several pecking orders which are size dependent.

Therefore, for small parts, *injection molding* is the next logical process to consider. Fortunately, the injection molding industry has developed relatively inexpensive methods of dealing with low-volume small parts through the use of such tooling devices as insert and family molds. There are, however, a limited number of molders interested in such work.

As the part increases in size beyond 1 ft^2, other processing options beyond thermoforming, such as *structural foam and coinjection molding,* come into the pecking order. These processes can compete with injection molding to higher volumes as the size of the part grows because the cost differential with the injection mold grows disproportionately with size.

Actually, depending on the part configuration (such as a requirement for internal details), structural foam or coinjection molding may be the preferable starting process (after machining). As a low-pressure process, structural foam molds have traditionally cost significantly less than injection molds (although more than thermoforming dies). While the structural foam cycle is much longer than injection molding, the mold cost differential can cover quite a volume before injection molding becomes more competitive. However, as product designers create structural foam designs with detail equivalent to injection-molded parts or require gas counterpressure molds, that advantage has diminished. When the cost of a structural foam mold is nearly that of an injection mold, the process will be selected for its unique attributes. Such features as weight reduction, filling of hollows, and the ability to make larger parts (due to the low pressure) become the principal selection criteria.

For some parts, there might be one more stop on the pecking order before going to injection molding, that is, *gas-assisted injection molding*. This process provides a rigid wall with a hollow channel within its thick sections and a solid wall elsewhere. Molding pressures can be as low as 15% of injection molding pressures with the resultant increase in the size of the part which can be molded in a given molding machine. Gas-assisted injection molding is of particular value in the elimination of assembly operations used to fabricate box structures. There have been successful applications in the elimination of parts in automobile door frames where the gas has been used to create the box structures within the frame.

For most parts, the ultimate in production volume economies can be reached with injection molding. At a size of about 5 ft^2 of surface area, however, injection molding drops out of the competition because molding machines which can handle that size are very scarce and are usually reserved for proprietary applications. In that case, structural foam molding may be the most efficient process.

Unfortunately, structural foam molding can handle a much more limited palette of materials than injection molding can. The materials that are readily molded with structural foam are ABS, ionomers, polycarbonate, polyethylene, and polypropylene. Most of the remainder can only be handled with difficulty and some, like styrene and acrylic, cannot be processed at all. The only high-strength and temperature material in the readily processed group is polycarbonate. For cost or chemical compatibility reasons, polycarbonate may not be acceptable. That means that the next step up in volume economies for a part which is too large for injection molding and which requires higher-temperature resistance than the structural foam materials can inexpensively offer may be a thermoset process like compression molding.

Machining. With the ready availability of most machining equipment and little or no tooling required, machining is the *ideal* process for very low-volume applications, provided the shape is one which can be readily machined. Piece part cost will be highest with machining as each piece must be processed individually. Machining also is the process which permits the closest tolerances for extremely critical applications where the cost can be justified. The high cost is due to the fact that the work must be done slowly to avoid thermal expansion or melting which would result in imprecision. It should be noted that some plastics are too flexible to be readily positioned for machining or may deflect away from the cutting tool when pressure is applied.

Machining is also used in combination with other processes. Some processes require that gates be removed or parts trimmed by mechanical means before they are ready to be used. Machining may be used to create details the process cannot create. An example of this type of application would be holes in thermoformed parts. Finally, machining may be used because the product's production volume is too low to warrant the additional tooling cost necessary to mold the detail into the part. An example of this type of application would be a hole in an injection-molded part parallel to the parting line that would require an expensive side action. The additional amortization cost could exceed the cost of drilling the hole for low volumes.

Many plastic prototypes are machined and fabricated. However, this process is only useful to resolve fitment, appearance, and ergonomic issues. That is because the strength, internal stresses, and environmental behavior characteristics of the part made in this fashion will be significantly different from the properties of the same part made by the ultimate production process. Furthermore, many plastic parts are in the strength and stiffness range where the characteristics of the assembly are different than those of the individual parts. Therefore, testing should be performed on the completed assembly made with production parts. Even tests performed on parts made by the prototyping method for a given process cannot be completely reliable because the processing conditions would be different and plastics are process-sensitive.

Thermoforming and pressure thermoforming. Thermoforming has the reputation of being the process to use for applications requiring large parts with thin walls that have low tooling budgets. Tooling for thermoforming is relatively inexpensive because this is an "open-mold" process, meaning that a mold with only one side is used. Thus, it is the first step up in tooling cost from simple machining. Product piece parts are typically less costly for thermoformed parts than for machined parts. However, this is a three-stage process because there are two additional stages besides the thermoforming stage. First, resin must be extruded into sheet before it can be thermoformed and, after thermoforming, the excess sheet (offal) must be trimmed and any holes or openings cut in the parts. These factors can result in fairly high piece part costs. Therefore, as the volume of a given product grows, it may reach a point where the additional cost of injection molding tooling is more than offset by the reduction in piece part cost.

Thermoforming is a stretching process which permits thinner wall thicknesses than machining, for which walls must withstand machining forces and heat, or the molding processes that require wall thickness thick enough to permit melt flow. For products that do not require thicker walls for strength, this can result in lower piece part prices. There are actually several thermoforming process variations, and they can be considered according to the thickness of the sheet used ("thin gauge" or "heavy gauge" thermoforming), the manner in which it is supplied (roll or sheet), or the contour of die used (male or female).

Thin-gauge (thickness less than 0.060 in) thermoforming uses material supplied in a roll. It is the high-volume variety of this process and is generally associated with packaging. With the exception of disposables, such as cups and plates, it is not generally used for product manufacturing.

Heavy-gauge (thickness greater than 0.060 in) material is supplied in sheet form. However, it should be noted that gauges less than 0.060 in are sometimes used in sheet thermoforming. This is the variety of thermoforming that product designers are normally concerned with. It can produce relatively flat parts with rounded corners in its simplest form, known as "vacuum forming." However, its "pressure forming" version can produce detail that can rival injection molding. Heavy-gauge thermoforming tends to lose its competitive edge over other processes when the part size falls below 1 ft^2. Conversely, it has a very large part capability. This allows the combination of several parts into one in many cases, thus eliminating some assembly operations completely.

Pressure thermoforming uses air pressure, often with the addition of a plug assist, to force the material deeper into the mold cavity. With pressure, the fine detail and surface finishes associated with injection molding can be achieved. However, injection molding usually requires much higher annual volume than thermoforming to be economically feasible. Thus, this development increases the capabilities of the product designer by extending the range of products which can use such detail to those with lower annual volumes. Injection molding becomes a more serious contender for the application as its volume increases. Then the piece part cost differential can be applied to the substantially greater cost of the injection molds.

The cost of the tooling for injection molding rises substantially with increasing size, and the payoff volume, the point at which the additional tooling cost is offset by piece part savings, goes up accordingly. Thermoforming can, however, make substantially larger parts than can injection molding.

Structural foam molding, gas counterpressure structural foam molding, and co-injection molding. The high cost of tooling is the factor which governs access to the injection molding process. This cost can be reduced through the use of related processes which require less molding pressure, such as structural foam molding, gas counterpressure structural foam molding, and co-injection-molding. Lower pressure allows for a less substantial, and therefore less costly, mold.

The part produced by structural foam molding is not solid like the part produced by thermoforming or injection molding. Within the outer skin, there is a cellular structure with the cell size increasing toward the center. The part must have a wall thickness of 0.187 in in order for any significant amount of foaming to take place. Thus, the structural foam part may actually require more resin than the equivalent part made by either thermoforming or injection molding if the wall thickness must be increased to accommodate the foaming process. In addition, the molding cycle for structural foam molding is much longer than that of injection molding due to the time required for gas expansion. For these reasons, the piece part cost will be greater for a part produced by structural foam molding than for a similar part made by injection molding. The reduced molding pressure does permit larger parts to be molded in a given molding machine, at least to the limits of the machine platens.

Parts made by the structural foam molding process have a characteristic swirled surface. Gas counterpressure structural foam molding and co-injection molding are variations of the process which can produce a solid, nonswirled surface. Depending on the application, the additional mold cost for the gas counterpressure feature reduces the tooling savings over the injection molding alternative. Coinjection molding permits a solid material to be used for the outer skin and a foamed material for the inner structure which can also be a less expensive material. This process requires sophisticated equipment.

These three processes use closed molds and low pressures. The closed molds are more costly than the open molds used in thermoforming because there are two halves instead of one. However, the low pressure keeps the tool cost significantly lower than traditional injection molding. Unfortunately, it is that high pressure associated with injection molding that permits its fast cycles. Thus, piece part prices are higher for these methods than they are for injection.

As in thermoforming, these processes are most competitive with injection molding for large parts. The larger the part, the greater the mold cost advantage over injection molding. Unless, of course, one designs parts of such intricacy that this advantage is negated.

Piece part costs are a different matter. As a broad statement, parts from these processes will be less costly than those from thermoforming, but more costly than those from injection molding. However, there are other reasons for using them besides simple piece part cost.

One purpose could be to change the “feel” or “heft” of a product such as a steering wheel or a vacuum cleaner handle. That permits the handle to be made in one piece instead of the older method which consisted of two injection-molded halves usually assembled with screws and nuts. These foam molding processes eliminate the need for assembly and the cost of the fasteners. This results in a lower product cost even though the injection molding process has a faster molding cycle.

Injection molding. While the extrusion process is the highest-volume process, injection molding is used for the greatest number of product design applications, largely because it provides the lowest piece part cost for volume applications. Injection molding can accommodate a range of applications from huge parts which require a cycle of several minutes to high-volume bottle caps that have been molded at the rate of 288 parts every 4 s.

Injection molding gets its name from the fact that plastic is injected into a mold at very high pressure. That gives the process high output capability plus the ability to produce fine detail and tight tolerances. However, the high pressure at which the plastic is injected into the mold requires sturdy, robust steel molds that are inherently expensive for that reason alone. Added to that is the fact that the process is used for the most precise, demanding piece part designs which also require expensive core and cavity details. Finally, the low piece part cost provided by the injection molding process is often obtained through the use of multiple cavities. Hence, the cost of each cavity is multiplied by the number of cavities in the mold.

Size is probably the major limitation to injection molding. As the parts grow larger, the cost of tooling becomes prohibitive for many applications and the number of molding machines available that are large enough to make them diminishes significantly. Many of the very largest injection molding machines have been manufactured for special applications, are owned by proprietary manufacturers, and are not available for custom molding. Gas-assisted injection molding significantly increases the size of the part which can be molded in a given size molding machine because of its much lower pressure.

Thermoset shapes

1. Machining
2. Casting
3. Lay-up or spray-up
4. Cold-press molding
5. Resin transfer molding
6. Reaction injection molding

7. Compression molding, bulk molding compound (BMC), sheet molding compound (SMC), low-pressure molding compound (LPMC), transfer molding, and thermoset injection molding

Since many of the thermoset processes are best suited to large parts, two pecking orders exist in this area as well. Up to 1 ft^2, the next step after machining and casting would, depending on part configuration, be either compression molding, transfer molding, or injection molding. All of these are processes which require expensive tooling, however, the alternatives do not lend themselves to small parts very well. For large parts, the pecking order would include the full gamut of thermoset processes with the exception of transfer or injection molding.

Machining. As with thermoplastics, thermosets can also be machined. In fact, most of the thermosets are somewhat easier to machine than the thermoplastics in that the melting problem is less of a factor. While localized heat at the machining surface can still create difficulties, the temperatures are much higher and charring or burning is the likely result.

A considerable amount of machining is done with thermosets because, with the exception of a couple of methods, the mechanical removal of molding flash is necessary in thermoset processes. Drilling holes and cutting openings is also commonplace in some processes because it is difficult, bordering on impossible, to mold them in for many designs.

Casting. Casting is a low-pressure closed-mold process; however, mold costs are kept low because it is often possible to cast the mold directly from the model. It is difficult to create fine detail to close tolerances with this process. Casting cycles are long and, consequently, piece part prices are high.

Casting is often used to enclose an object, usually an electrical component, in order to protect it. Casting applications also include furniture and decorative objects where fine detail is required or simulation of wood is desired in relatively low volume.

Lay-up and spray-up. The largest of parts (mine sweeper hulls) can be made by lay-up and spray-up. However, the ability to create fine detail is limited and close tolerances are not possible. Machining is necessary to create holes and trim the parts. The construction is laminated with polyester with glass reinforcement, and it is the reinforcement application method which defines the name of the process. Open molds are used; therefore, one side of each part is rough and unfinished. Mold costs are low; however, they only last for a small number of parts. Thus many molds would be required for high volumes, although the pattern needs to be made only once. Since these are very slow processes, piece part costs are high. However, robotics have been used to reduce labor costs.

Cold molding or cold-press molding. This process is a step up from the open-mold processes previously discussed in that it is a closed-mold process. Therefore, the parts are finished on both sides. However, the process does not quite produce the surface quality required by the transportation industry. Hence, cold-press–molded parts are more likely to be used for interior parts. To a limited degree, this process can provide a part with some inside structure such as ribs, etc. However, it has poor tolerance control and, therefore, somewhat limited application.

Cold-press molds are plastic, which makes them comparatively inexpensive. Relatively long cycle times and postmolding machining result in parts which are more expensive than compression-molded parts, but less costly than parts produced by the lay-up or spray-up processes.

Resin transfer molding (RTM). Resin transfer molding can produce parts of higher finish, greater complexity, and wall thickness consistency than cold-press molding; however, the part cost can also be higher. Therefore, the processes are sometimes used together with external parts made by the resin transfer molding process using cold-press–molded parts for internal supports. Closed plastic molds are used which are sometimes plated for longer life and better finish. These molds are considerably less expensive than compression molds, thus considerable volume is required before the additional cost of tooling can be offset by the lower piece part cost of compression molding. Parts as large as truck hoods, small boat hulls, and car body halves have been produced using this process.

Reaction injection molding (RIM). RIM is a low-pressure process using closed molds. However, it has a much shorter cycle than resin transfer molding, which results in lower piece part costs. Unfortunately, very few materials are available. Thermosetting polyurethane is the principle material available, with epoxy, nylon, and polyester also available but to a limited extent. Reinforced reaction injection molding (RRIM) is available with chopped glass used as the reinforcement. This process is most widely used in the automotive field, although there have been other applications where large parts or limited volumes are required.

Compression molding, BMC, SMC, LPMC, transfer molding, and thermoset injection molding. Compression molding is the primary thermoset process; the other processes in this group, with the exception of thermoset injection molding, are derived from it. The term *compression molding* also includes BMC, SMC, and LPMC which actually describe the molding compound used. BMC and SMC refer to the type of reinforcement and the manner in which the resin is prepared for molding.

Compression molding, transfer molding (not to be confused with resin transfer molding), and thermoset injection molding compete for short fiber–reinforced or –unreinforced applications with fine detail and close tolerances. Molds are expensive, but piece part costs are low. For a product which can be manufactured by all three processes, compression molding will have the lowest tooling cost and highest piece part cost. Injection molding will have the lowest production cost with a higher tooling cost, and transfer molding is somewhere in between them (the use of preheated resin allows transfer molding cycles to approach injection molding cycles). That is a gross generalization because part design details will likely favor one process or the other in most cases.

In BMC, long strands of reinforcement ($^1/_4$ to $^1/_2$ in) are placed in the material along with other additives. A ball, slab, or log of this mixture is formed and placed in the mold. This method of reinforcement is less expensive than SMC, however, it is not as strong. SMC fibers are spread into a resin paste to form a sheet. Reinforcement fibers can range from the very smallest to those of indefinite length, although they usually do not exceed 3 in. However, SMC is generally known as a long-fiber process; it is used for higher strength applications such as truck tractor hoods and fenders.

LPMC material is prepared like SMC but is formulated to permit molding at a lower pressure. The reduction in pressure results in less expensive molds, often constructed of aluminum. The additional cost of tooling requires a high production level for the piece part savings to justify a change of process from cold-press molding or resin transfer molding to one of the compression processes. However, the use of LPMC can make compression molding competitive at relatively low levels.

Through their large-part capability, the compression molding processes permit the combination of parts with the resultant assembly savings. They are also processes which offer economies of high volume. Thus, they are the processes of choice for large-part, high-volume applications.

Hollow part processes. Hollow products can be made either by assembling parts made by most of the processes (even extrusions can be capped) or by one of the hollow part processes. Which is most cost effective varies considerably according to the application and to the state of the art of the techniques being applied. The savings associated with molding the part in one piece may be offset by the use of robotics in automating the assembly process coupled with a more cost-effective molding process.

For thermosets, there really is only one hollow part process—*filament winding.* That process is very limited in shape and structure such that, except for pressure applications, most thermoset hollow products are assembled.

There are three hollow part processes for thermoplastics and they do have a pecking order of sorts. When proceeding with the development of thermoplastic hollow shapes which do not lend themselves well to assembly, the following sequence should be used, starting with the lowest volume requirements:

1. Rotational molding or twin-sheet thermoforming (application dependent)
2. Blow molding

Processes for hollow shapes have a much simpler pecking order. The ideal shape for *rotational molding* is a sphere. The ideal shape for *twin-sheet thermoforming* is a flat panel. The selection of which process to begin with would be based on the shape of the particular part to be manufactured, although it should be noted that rotational molding can make parts with sections as thin as 1 in. Additional factors would be the size of the part (rotational can go larger) and the selected material, since each of these processes favors different materials.

Depending on the size and shape of the part, *blow molding* may become competitive as the volume grows. However, there are size limits to which this step can be taken as the other two processes can make larger parts than blow molding.

Rotational molding and blow molding make similar hollow parts. In some cases, such as automotive ducts, the ends are removed from a hollow shape to create the final part. Large containers can be molded so they are integral with their covers which are cut off to form the two parts. Other shapes can also be molded as one hollow piece and cut apart to make multiple parts (usually two). Structural components are usually molded as double-walled parts which can be filled with foam for greater strength and rigidity.

Rotational molding, blow molding, and twin-sheet thermoforming can also make large, double-walled parts which are relatively flat (such as pallets). Twin-sheet thermoforming is best suited to the flattest of such parts.

Rotational molding. The *rotational molding process,* sometimes referred to as *rotomolding,* can produce parts ranging in size from small balls to enormous containers. The principal material used in rotational molding is polyethylene, however, some nylon parts are also made. For simple parts with low appearance requirements, inexpensive sheet metal molds can be used. More costly cast molds are used for parts of a higher level of finish and complexity. Even these molds are comparatively low in cost as rotational molding is a low-pressure process which does not require heavy steel molds. Piece part price can be quite high due to the long cycle time. In addition, most openings in the part must be machined as a secondary operation. For small products, such as balls, a large number of cavities can be used to offset the low output. Rotational molding is well suited to large parts and, for the largest of parts, it has no real competition in thermoplastic processes.

Twin-sheet thermoforming. *Twin-sheet thermoforming* is generally used to produce large, flat, double-walled parts. Nearly all of the thermoplastics can, at least theoretically, be thermoformed. However, the bulk of parts made by this process are usually polystyrene or ABS. The highest-temperature plastic thermoformed in any significant amount is polycarbonate.

Blow molding. With few exceptions, plastic bottles produced in volume are *blow molded.* Less well known are the other blow-molded hollow shapes such as children's toys, storage sheds, etc. Typically, these are double-walled. For high-volume molding of hollow shapes, blow molding is the only option if the shape and material requirements are suitable.

Profiles. Extrusion and pultrusion are the only profile processes, although injection molding can compete with extrusion for short pieces such as rulers. Tooling, although made of hardened steel, tends to be relatively inexpensive. Neither extrusion nor pultrusion lends itself well to short runs due to the lengthy setup required. Therefore, there is no pecking order by volume. Instead, the profile processes are discussed according to their applications.

Extrusion is the principal process for producing film and sheet. It is also used to create open (weather stripping) and closed (soda straws) profiles. Dual extruders can be used to produce laminates, a process called *coextrusion.* This is often done to provide barriers to ultraviolet light or moisture vapor. Two colors can also be extruded side-by-side for interesting effects. All thermoplastics can, theoretically, be extruded including foamed and reinforced materials. However, in practice, the highest melt temperature thermoplastic extruded in any volume is polycarbonate. Thermosets can also be extruded on a limited basis as they require special equipment and are slow to process.

Pultrusion is the principal means of producing reinforced profiles. It is more expensive than extrusion and is usually reserved for demanding structural applications such as light poles, wind turbine blades, and structural beams. There are some consumer applications as well and they include pole vault poles, fishing rods, flag poles, and ski poles. The process is noted for its ability to produce a part with extraordinary resilience which can sustain considerable deflection and return to its original shape. While pultrusion is not limited to thermoset materials, its principal resin is thermoset polyester.

If there is to be any preference between extrusion and pultrusion, it would be based on design freedom and strength. Extrusion can produce a wider range of shapes and has a wider availability of materials; however, pultrusion can create the stronger part. Most applications clearly favor one process over the other for these reasons. When both processes can make the part, extrusion usually has the edge because it processes at a higher rate of speed. For higher-strength applications, however, that advantage can be offset by the higher cost of reinforced thermoplastics over the reinforced polyester used in pultrusion.

Ultra high strength. Filament winding is a process which stands alone. It is an ultra high-strength process which defines its own applications. Its principle competition is metal fabrication, over which it has significant cost and weight advantages when employed.

Filament winding is the process associated with such ultra high-strength requirements as rocket motor casings, rocket tubes, helicopter blades, automobile leaf springs, and aircraft parts. A variety of winding configurations are used to wind resin-impregnated or bath-dipped glass filaments around a mandrel, which is often retained as part of the end product. Using this process, small-diameter pipe rated at 2000 lb/in^2 has been created for the chemical industry. Tooling investment for this process is low to moderate, however, piece part costs are relatively high.

Additional process selection considerations. Material limitations will affect the process selection because none of the processes will accept all of the plastics. Therefore, the optimum process may not accept the desired material. When this problem occurs, one of the two will need to be changed. Table 8.6 lists the acceptable materials for the principal processes. Bear in mind that both resin and equipment manufacturers are continuously working to enhance their products and a list such as this can be made obsolete at any time. If the desired process does not indicate that the preferred polymer can be used, it may be worthwhile to investigate further.

One additional determining factor in the selection of a process is the size of the part to be manufactured. Each process is limited in the size of parts it can handle. Table 8.7 gives some indication of the size range that each process is capable of. Note that, on the high side, two sizes are indicated. The column, "largest known," lists the largest size the

TABLE 8.6 Plastics Available for Processes

Process	Materials
Blow molding	ABS, acrylic, cellulosics, nylon, polycarbonate, polyester (thermoplastic), polyethylene, polypropylene, polystyrene, polysulfone, PVC, SAN
Casting	Acrylic (thermoset), alkyd, epoxy, nylon, phenolic, polyester (thermoset), polyurethane (thermoset), silicone
Cold molding	Epoxy, phenolic, polyester (thermoset), polyurethane (thermoset)
Compression molding (including BMC and SMC)	Alkyd, allyl, amino, epoxy, fluorocarbons, phenolic, polyester (thermoset), polyurethane (thermoset), silicone
Extrusion	ABS, acetal, acrylic, cellulosics, liquid crystal polymer, nylon, polycarbonate, polyester (thermoplastic), polyethylene, polyphenylene oxide, polypropylene, polystyrene, polysulfone, polyurethane (thermoplastic), PVC, SAN
Filament winding	Epoxy, polyester (thermoset)
Gas-assisted injection molding	ABS, acetal, acrylic, cellulosics, nylon, polycarbonate, polyester (thermoplastic), polyethylene, polyphenylene oxide, polypropylene, polystyrene, PVC, SAN
Injection molding	ABS, acetal, acrylic, alkyd,* allyl,* amino,* cellulosics, epoxy,* fluorocarbons,* liquid crystal polymer, nylon, phenolic,* polycarbonate, polyester (thermoplastic), polyester (thermoset),* polyethylene, polyphenylene oxide, polypropylene, polystyrene, polysulfone, polyurethane (thermoplastic), PVC, SAN
Lay-up and spray-up	Epoxy, polyester (thermoset)
Pultrusion	Epoxy, polyester (thermoset), silicone
Reaction injection molding	Nylon, polyurethane (thermoset), epoxy, polyester
Resin transfer molding	Epoxy, polyester (thermoset), silicone
Rotational molding	Acetal, acrylic, cellulosics, fluorocarbons, nylon, polyester (thermoplastic), polyethylene, polypropylene, polystyrene, polyurethane (thermoplastic), PVC
Structural foam molding	ABS, acetal, nylon, polycarbonate, polyethylene, polyphenylene oxide, polypropylene, polystyrene, polysulfone, SAN
Thermoforming	ABS, acrylic, cellulosics, polycarbonate, polyethylene, polypropylene, polystyrene, polysulfone, PVC, SAN
Transfer molding	Alkyd, allyl, amino, fluorocarbons, phenolic, polyester (thermoset), polyurethane (thermoset), silicone

*Special equipment required.

TABLE 8.7 Approximate Part Size Ranges for the Principal Processes

Process	Smallest known	Largest commercial	Largest known
Blow molding	⅜ in deep × 1¼ in long	9½ ft long × 1 ft deep × 4 in thick, 28 in deep × 44 in long	1320-gallon tank
Casting	No limit	Limited only by physical ability to handle molds and moldments	Limited only by physical ability to handle molds and moldments
Coinjection molding	¼ × ¼ × ¼ in	2 × 5 × 5 ft	2½ × 4 × 10 ft
Cold molding	1½ × 1½ ft	10 × 10 × 1½ ft	14-ft boat hull
Compression molding	¼ × ¼ × 1/16 in	4 × 5 × 8 ft	1½ × 4½ × 14 ft
Extrusion	No limit	12 × 12 in	42 in deep
Filament winding	4 ft deep × 8 in long	13 ft deep × 60 ft long	10-ft high × 82½ ft deep
Injection molding	0.008 × 0.020 × 0.020 in	2½ deep × 3 ft	4 ft × 4ft 6 in × 7 ft
Lay-up and spray-up	¼ × 6 × 6 in	150-ft minesweeper	Continuous roadway
Machining	No limit	10 ft wide or 15 in deep	Limited by size of stock available
Pultrusion	1/16 in deep	12 × 12 in	15 × 100 in
Reaction injection molding	4 × 12 in	3 × 4 × 10 ft	10 × 10 ft
Resin transfer molding	1 in × 3 in × 2 ft	16 in × 4 ft × 8 ft	4 ft × 8 ft × 28 in
Rotational molding	½-in -diam sphere	6 ft deep × 18 ft long	12 ft deep × 30 ft long
Structural foam molding	¼ × ¼ × ¼ in	2½ × 6½ × 6½ ft	2½ × 4 × 10 ft
Transfer molding	⅛ × ⅛ × 1/16 in	2 × 1 × ½ ft	2½ × 1 ft × 9 in
Thermoforming: thin gauge	¼ × 1 × 1 in	3 × 3 × 3 ft	3 × 3 × 3 ft
Thermoforming: heavy gauge	6 × 6 × 6 in	3 × 10 × 12 ft	2 × 6 × 20 ft
Twin-sheet thermoforming	6 × 6 × 6 in	6 in × 3 ft × 6 ft	6 in × 3 ft × 6 ft

SOURCE: Jordan I. Rotheiser, "The Bigger Picture," *Plastics Engineering Magazine,* January 1997.

author has ever heard of. That size is not realistic for general use as not all materials can be run at the limits for a given process and the equipment that made it may not be available for custom projects. Indeed, such equipment has typically been built for some company's proprietary needs. The column, "largest commercial," lists sizes at which a reasonable selection of competing bids might be obtained. However, when working near the limits of a given process, it behooves the savvy product designer to be certain that the equipment is available for the project before proceeding with the design.

8.1.6 Tooling selection

Tooling is a critical aspect of the total plastic product design picture because a plastic part can be no better than the tooling that created it. While it is not necessary for the product designer to be able to design a tool, a fundamental knowledge of tooling is essential—not merely for the design of the part, but for the process selection as well. The cost of tooling for a process and the type of design detail it is capable of producing are among the principal determining factors in the selection of the process. These two factors are determined by the amount of pressure required for the process and the amount of tooling that has to be built. For the purposes of the tooling discussion, the processes have been broken up according to the amount of pressure required with the open-mold and profile processes discussed separately.

Basic tooling construction. In general, the open-mold processes can be expected to have lower tooling costs than any of the closed-mold processes for parts of equivalent complexity because only one side of a mold has to be built. Other types of molding require both a male and female half (core and cavity) or two female halves (split cavity for hollow parts); however, open molds require either a male or a female section. This is illustrated in Fig. 8.8, which depicts a female mold which forms the outside of the part as in Fig. 8.8*a* and a male mold which forms the inside of the part as in Fig. 8.8*b*.

Open-mold processes have little or no molding pressure. In general, the tooling costs for the closed-mold processes increase according to the amount of molding pressure required for the process. The greater the pressure in the tool, the stronger (and more expensive) it must be because higher pressures require stronger molds which need to be made from stronger materials that cost more and require more time to machine. However, the greater pressure permits the molding of finer detail and results in shorter molding cycles. Therefore, it is generally axiomatic that the piece part price decreases as the cost of tooling increases.

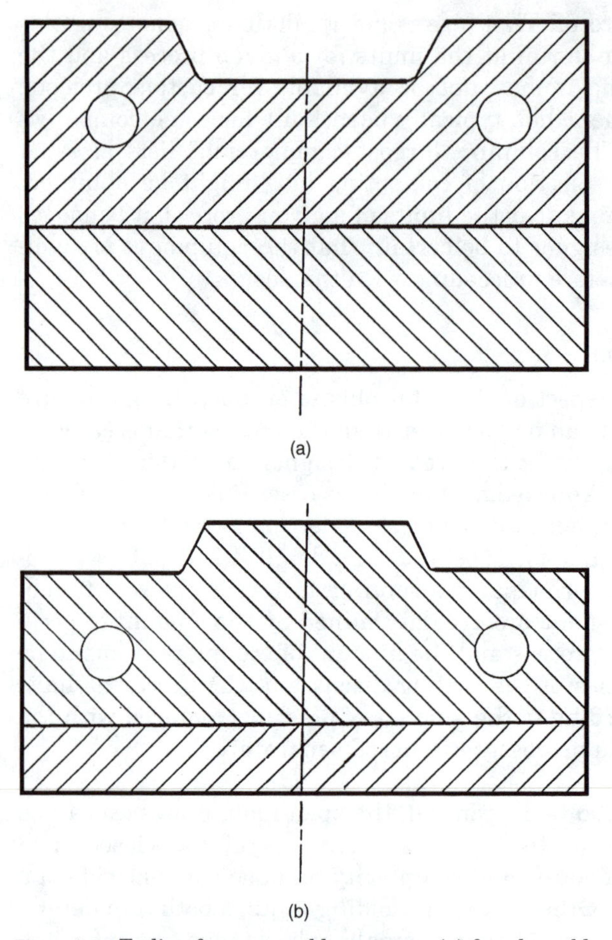

Figure 8.8 Tooling for open-mold processes: (*a*) female mold and (*b*) male mold.

Closed-molds consist of either a core and a cavity as illustrated in the injection mold in Fig. 8.9*a* or, in the case of hollow processes (blow or rotational molding), a split cavity as shown in Fig. 8.9*b*. Where the two (or more) parts of the mold meet is called the *parting line*. To prevent plastic from leaking from the mold, the two halves must match perfectly. When the parting line is flat and on one plane, it is relatively easy to align the two halves. However, some designs require the mold to have a contoured parting line. This is referred to as a *broken parting line* and requires the halves of the mold to be machined with great care. Thus, a mold with a broken parting line is considerably more expensive than one with a straight parting line.

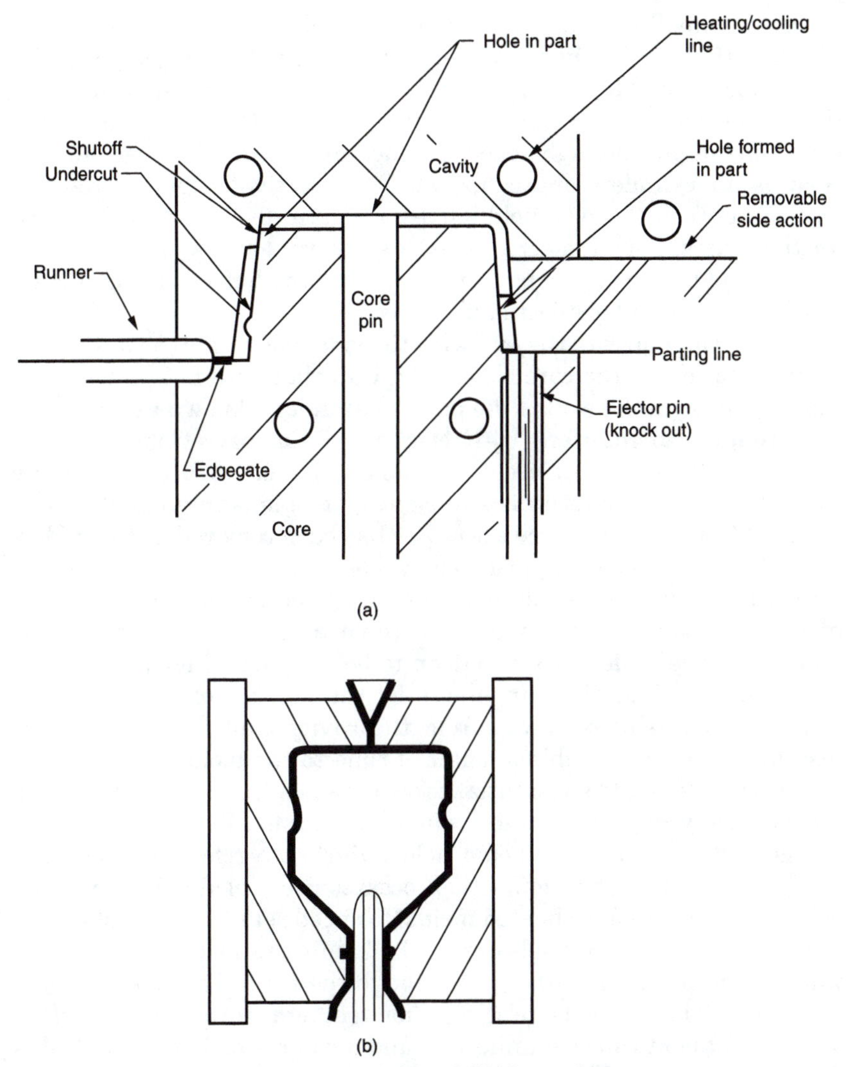

Figure 8.9 Closed molds: (*a*) injection mold and (*b*) blow mold.

Openings in the part are created when a portion of one half of the mold closes against the other half such that plastic cannot flow through that area. When the hole is to be in a plane parallel to the parting line, this is called a *shut-off*. It is normally the core which shuts off against the cavity (see Fig. 8.9*a*) because the shrinkage causes the part to grip to the sides of the shut-off and it is preferable for that to occur on the core side of the mold. For odd-shaped or very large openings, the raised portion is usually cut in the solid core, however, round holes are most economically made from a core pin as illustrated in Fig. 8.9*a*.

Note in this illustration that there is a hole in the side of the part. The core that makes that hole would interfere with the ejection of the part from the mold. Therefore, that core is placed in a removable section which has to be replaced after every cycle. This procedure is too time consuming for high-speed production, and mechanical devices, such as air cylinders and cams, which are activated by the relative motion between the two mold halves, are used to achieve higher production rates. These mechanisms are referred to as *slides* or *side actions* and they can also be used to operate the halves of a split cavity when it is used for injection molding.

On the wall opposite the one with the side hole in Fig. 8.9*a,* there is an indentation on the core side of the wall that does not protrude all the way through the wall of the part. That indent, known as an *undercut,* would also interfere with ejection of the part from the core. However, when the undercut is shallow, its edges are adequately radiused and the material is sufficiently flexible, an undercut can be stripped from the mold once the cavity has been removed. If the undercut is too deep to be stripped, removable core inserts or a collapsing core will be required. Undercuts can be placed in the cavity as well; however, a split cavity, slides, or removable core inserts will be required if the undercut is too deep to be stripped. Cavity undercuts can be deeper than core undercuts by the amount of the shrinkage since the part shrinks away from the cavity wall. Removable core inserts add a considerable amount of time to the molding cycle; split cavities or slides add substantial cost to the mold; and collapsing cores are, generally, expensive to tool when they are feasible at all.

Figure 8.9*a* also shows some holes which are referred to as *cooling/heating lines.* Thermoplastic processes use cooled molds for most polymers (some require heated molds) to speed set up of the moldment. Therefore, cooling channels are drilled into the mold to allow cool water (sometimes chilled) to run through the mold. The objective is to achieve a cool mold at the precise, same temperature across its surface so the moldment can cool uniformly and avoid distortion. This ideal is difficult to achieve, but, depending on the contours of the design, very high levels have been attained. Often, elaborate cooling systems are employed to do this and that adds cost to the mold. Frequently, the difference in cost between sources can be attributed largely to the amount of cooling they have included in their quote.

Thermosets do not have to be cooled because they cross-link during molding. Consequently, they are run hot and the channels are used for heating media like steam or oil.

Tooling for the open-mold processes. The open-mold processes are thermoforming, lay-up, and spray-up. Since open molds form only one side of the part, they permit design details only on that side. Tooling for the open-mold processes tends to be very simple. Cooling systems are employed for thermoforming and undercuts are readily created using break-away or removable sections which come out of the mold with the part and which are replaced before the next cycle. In high-volume applications, springs or air cylinders can be used to actuate these sections. Mechanical ejection is sometimes employed using springs or mechanical devices when there is an undercut to be stripped.

In general, any openings need to be cut as secondary operations since there is no way to close off the mold faces to create holes within the part. For the lay-up and spray-up processes, a very large opening like the center of a window frame could be included in the shape of the mold. Openings which need to be cut may call for additional tooling fixtures, but these are relatively inexpensive. Because they are secondary operations which must be performed after the moldment is created, openings do add to the cost of the part. Fine detail, which requires pressure to create, cannot be part of the design because the molding pressure is nearly nonexistent for the open-mold processes. There is, however, a variety of thermoforming, known as *pressure thermoforming,* that can produce fine detail using special thermoforming equipment.

Tooling for the low- to moderate-pressure processes. This category covers those processes that require closed molds which do not have to withstand high molding pressures that need hardened steel tooling. Processes that fall into this category are casting, cold-press molding, resin transfer molding (RTM), reaction injection molding (RIM), low-pressure compression molding (LPMC), plus the hollow part processes, rotational molding, and blow molding. The lower molding pressures used for these processes permit the use of a variety of techniques for mold construction, most of which involve machining a softer metal or casting from a pattern. A softer metal, such as aluminum, machines much faster than hardened steel and reduces the cost of the tool.

The lowest-cost tools are made by creating a pattern of the part and casting the tool from that pattern, usually in epoxy and glass. The patterns are often made from wood, however, they can be made from a computer-driven method, such as *stereolithography,* providing the configuration and size are suitable. Such tools tend to wear out within a limited number of parts, in which case a new mold is cast from the pattern. The number of parts that a cast mold can produce before it must be replaced varies from a dozen to several thousand depending on the process and the complexity of the design. In general, the more complex the part design and the higher the molding pressure, the shorter the life of the mold. Complex designs require fine detail which is more susceptible to damage and higher pressure is more likely to cause such damage. In some cases, epoxy molds can be nickel-plated to provide a better finish and prolong the life of the mold. The most common materials used for the tooling of the various plastics processes are indicated in Table 8.8.

The materials used for most of the low- to moderate-pressure process tooling do not lend themselves well to mechanical removal of cores. Therefore, holes are machined as secondary operations or cores are manually operated, removed, and emplaced by hand. This procedure takes more time than mechanically operated cores. However, since these are not generally high-speed processes, the additional time is not normally a problem.

Tooling for the high-pressure processes. Injection and compression molding develop the highest molding pressures, and they require hardened steel molds for large-scale production. However, the types of mold construction discussed for the low- to moderate-pressure processes are often used for prototypes and short runs for these processes. It is important to note that these methods have significant limitations when used for processes which develop higher pressures. That is because engineers are accustomed to incorporating the fine detail and surface finish these processes are capable of into their designs without taking into consideration that these prototype and short-run molds cannot withstand the pressures necessary to create them for very many shots. Just how many shots a mold will withstand is design dependent. Nonetheless, it is obvious that the harder the mold material is, the longer the tool will last without significant repair or replacement.

TABLE 8.8 Mold Materials for Plastics Process

Process	Mold material
Casting	Epoxy, silicone
Lay-up and spray-up	Epoxy (sometimes plated), aluminum, steel (occasionally)
Cold molding	Epoxy, glass
RTM	Epoxy, glass, aluminum
RIM	Epoxy, glass, aluminum, steel, kirksite
Compression and transfer	Steel (SMC, BMC), aluminum (LPMC)
Rotational molding	Sheet steel, stainless steel, aluminum, cast aluminum, machined metal, electroformed nickel and copper
Structural foam molding	Prehardened steel, aluminum with beryllium copper
Thermoforming	Aluminum (wood, epoxy, or polyester for prototypes)
Blow molding	Aluminum, beryllium copper, zinc alloy, brass, stainless steel
Extrusion	Steel
Pultrusion	Steel
Vacuum and bag molding	Epoxy, glass

When necessary, inserts made of a harder material, such as steel, may be used.

Molds made for prototyping and short runs often have other cost saving simplifications such as manual cores, manual ejectors, and limited, or nonexistent, cooling. In general, the molding cycle is largely determined by the length of time required to cool the part enough for it to be rigid enough to be ejected from the mold. Therefore, a significant difference in the mold construction which affects these parameters will affect the strength and shape of the part. As a consequence, it is necessary to construct a single cavity with the identical cooling, coring, and ejection of the production mold if a precise prototype of the production part is required. This is often done when a large, expensive, multicavity mold is to be built.

Tooling for the profile processes. Extrusion and pultrusion, the profile processes, require steel dies because they build up back pressures behind the die. However, the tooling costs for these processes are relatively low because the shapes manufactured with them are normally comparatively simple.

8.2 Design Fundamentals for Plastic Parts

The basic engineering formulas for structural design can be applied to plastic part design, within the limitations of the data available for the material properties, as previously discussed. This information is widely available and is not particular to plastic part design. Therefore, it will not be covered in this chapter. It is, however, found in *Machinery's Handbook,* 24th Edition, (Industrial Press Inc., New York) as well as in some of the other references listed at the end of this chapter.

A plastic part is like a chain in that it will fail first at its weakest link. Repair that, and it will fail at its next weakest link and so on. Therefore, the objective in plastic part design is to see to it that no weak point exists such that the part will fail below its design limit. This requires careful attention to a number of critical areas.

The majority of differences between the design of plastic parts and the design of parts made of other materials are in some way heat or pressure related. Heat and pressure are used to create the parts and their effects on the part plus the effects of the subsequent cooling of the parts to room temperature require consideration in the design phase.

8.2.1 Cooling effects and the need for uniform wall thicknesses

Most of the processes require the plastic to be heated to elevated temperatures. When the part cools, it shrinks away from the cavity and onto the core, if there is one. Clearly, the thicker the nominal wall of the plastic part, the longer it will take to cool. When the thickness of the wall varies within the part, the thinner portion of the moldment solidifies before the thicker segment. Nonuniform cooling leads to problems of voids, warpage, and sink, as illustrated in Fig. 8.10.

The extent of these problems varies considerably with the process and material (worse in higher shrinking materials) and is affected by the use of fillers in the resin. Consequently, plastic designers need to make a concerted effort to keep the wall thickness uniform throughout the part. In corners, this is accomplished by making the outside radius (OR) equal to the inside radius plus the wall thickness (IR + W). When the wall thickness must vary, the transition must be gradual as illustrated in Fig. 8.11. Every increment of wall thickness variation (T) should take place over a distance at least 3 times as great ($3T$). In no case should the thicker section exceed the thinner section (W) by more than 25% ($1.25W$). Not only will there be increased internal stresses, but there will likely be a sink with most unfilled resins. When a vertical edge is required, perhaps to position the part indicated by the dashed lines on the left side of Fig. 8.11, a rib can be used in place of a large block of material.

Excessive variation in wall thickness is not the only cause of nonuniform cooling of the part. It can also result from a mold which is not adequately cooled. Ideally, the mold temperature would be maintained at a constant temperature across the molding surface. This ideal is virtually impossible to attain because the shape of the part usually limits the mold designer's freedom to place water lines where they can best perform their function. An example of this type of situation is a box configuration. The core side of the box corner receives heat from three directions (both sides and the top), whereas the core which forms the side of the box is heated only by the melt on that side. For there to be no temperature differential between those two parts of the core, the corner must receive more coolant than the side. This could require a very expensive cooling system which the budget cannot support.

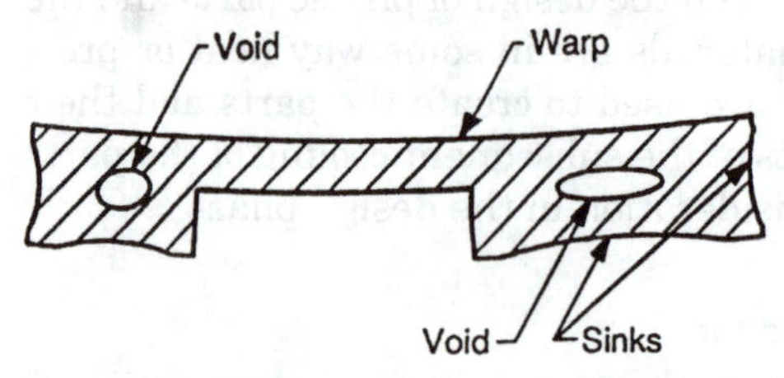

Figure 8.10 Voids, warpage, and sink in plastic parts. (*Source: Jordan I. Rotheiser, Joining of Plastics Handbook for Designers and Engineers, Hanser Publishers, Munich-Hanser/Gardner Publications, Inc., Cincinnati, 1999.*)

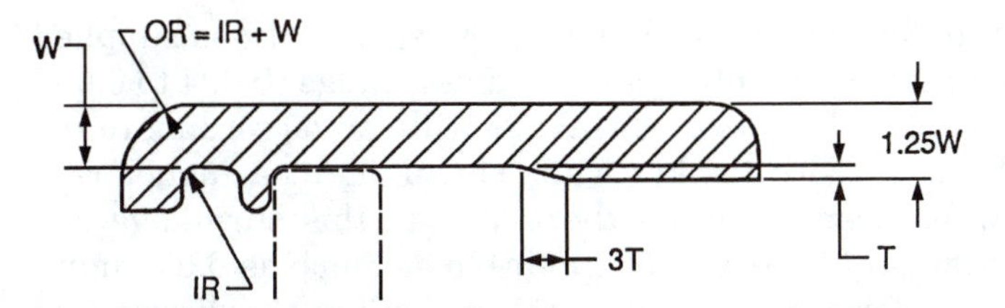

Figure 8.11 Nominal wall thicknesses in plastic parts. (*Source: Jordan I. Rotheiser, Joining of Plastics Handbook for Designers and Engineers, Hanser Publishers, Munich-Hanser/Gardner Publications, Inc., Cincinnati, 1999.*)

Furthermore, the cooling lines must avoid the ejection system as well. The mold designer must often choose between an ejector and a cooling line. Fewer ejectors creates more pressure per ejector which can also result in distorted parts or an extended molding cycle.

The shape of the part will also dictate gate location. Ideally, the gates should be located where they permit the cavity to be filled uniformly. If one part of the cavity fills before the rest, it will begin to cool immediately, resulting in differential cooling through the part. This phenomenon can also result in premature gate freeze-off which prevents adequate packing of the part. Improper gate location or inadequate number of gates are other ways to cause nonuniform cooling and part distortion.

8.2.2 Shrinkage and the use of draft in plastic parts

The *shrink rate* is the amount the part will reduce in size as it cools from processing temperature to room temperature. It is described in terms as the amount of shrinkage per inch of part size. Thus, a piece part 1 in long made of a material with a shrinkage rate of 0.005 in/in will require a mold 1.005 in long if the final part is to wind up a true 1 in in length. Mold dimensions for other sizes would be determined by using 1.005 as a multiplier. Thus, the mold for a 1.500-in dimension would be determined by multiplying 1.500 by 1.005 (1.508 in). Normally, the mold designer performs these computations and this is not a concern for the part designer. However, the part designer can help reduce mold costs by specifying part dimensions which take the shrinkage rate into consideration. Thus, a part specification of a 0.995-in diameter, instead of a 1.000-in diameter, would permit the use of a standard 1.000-in-diameter core pin which would require significantly less work for the moldmaker than the larger diameter. The savings might not be great for one cavity, but could be considerable for a multicavity mold.

The shrink rate for plastics as found on the specifications supplied by the resin manufacturer is often indicated as a range (0.004 to 0.006 in/in). That is because the rate of shrinkage will vary according to the thickness of the wall, with thicker walls shrinking at a higher rate than thinner walls. Depending on the material, this shrinkage can range from very slight (0.001 to 0.002 in/in) to as much as 0.050 in/in. The shrinkage rate for some resins is uniform both in the direction of flow from the gate and in the transverse direction (isotropic shrinkage). The shrinkage for other resins varies according to the direction of flow of the resin in the cavity (anisotropic shrinkage). In most cases, it is greater in the transverse direction.

The use of fillers can significantly alter both the amount of shrinkage and its direction. For example, glass fibers align in the flow direction which causes lower shrinkage in that direction. However, there is little or no reduction in shrinkage in the transverse direction. The result is in an increase in differential shrinkage and a greater tendency to warp.

In the injection molding process, the shrinkage will be affected by the density of the plastic in the part which can be controlled by "packing" more resin into the mold. The processor is the person most familiar with the behavior of a given resin in the equipment and it is the processor who determines what shrink factors will be employed in the construction of the mold, often after consulting the resin manufacturer.

Compression molds, which are not cooled, operate at elevated temperatures which cause the molds to expand. If this expansion is greater than the shrinkage of the material, the mold may be actually built smaller than the intended piece part to compensate for this expansion.

When the part shrinks onto a core, it grips the core very tightly and cannot be removed without considerable force. That force will tend to distort the part; therefore, the cycle must be extended long enough for the part to be removed from the mold without damage. The force can be reduced by placing a slight angle on the walls of the part perpendicular to the parting line, as indicated in Fig. 8.12. This angle is referred to as *draft* and is common to all molded plastic parts.

In the past, we followed the general rule that 1° per side of draft is ideal and $^1/_2$° per side is a bare minimum. However, economic pressures require ever shorter cycles, and improvements in equipment, materials, and tooling allow this demand to be met. However, in order to permit parts to be removed from the mold in less time, they must be ejected at higher temperatures when they are softer. Therefore, greater draft is required. Thus, the use of 2° and 3° per side drafts have become commonplace. For deep parts (beyond 6 in), a 3° per side draft is definitely called for and more would be better.

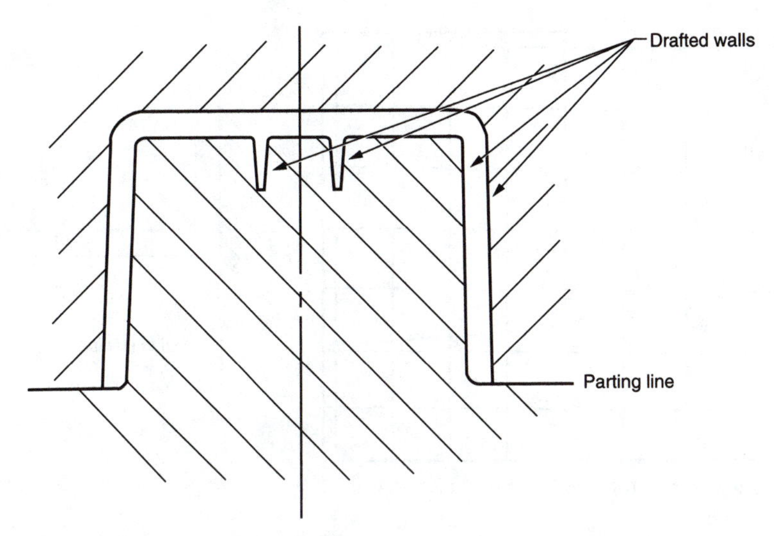

Figure 8.12 Draft in plastic molds.

Note that there are two ribs in Fig. 8.12. When the part shrinks, multiple ribs create a locking effect which causes the part to grip the core with greater force. Such ribs need to be well drafted to avoid this phenomenon. When multiple ribs are external to the part, they can cause the part to adhere to the cavity—an occurrence known as a *cavity hang-up*. Since most molds are built with their ejection mechanisms in the core side, a cavity hang-up leaves the molder with no convenient way to remove the part from the mold. The usual remedy for this problem is to add tiny undercuts to the core side until the cavity hang-up is eliminated. However, this changes the design and adds additional stresses to the part.

When texturing is added to the surface, the draft must be increased because, microscopically, the texture is composed of thousands of tiny crevices which must clear each other on removal from the mold to avoid scrape marks. The amount of draft required varies a bit depending on the texture, however, the general rule for textures is $1^1/_2$° per side plus $1^1/_2$° per side for each 0.001 in of texture depth. In tight situations, this can be reduced to 1° per side plus 1° per side for each 0.001 in of texture depth for some applications.

8.2.3 Stress concentrations in inside corners

Plastics are particularly vulnerable to the concentration of stresses on inside corners as most of them tend to be notch sensitive. The graph in

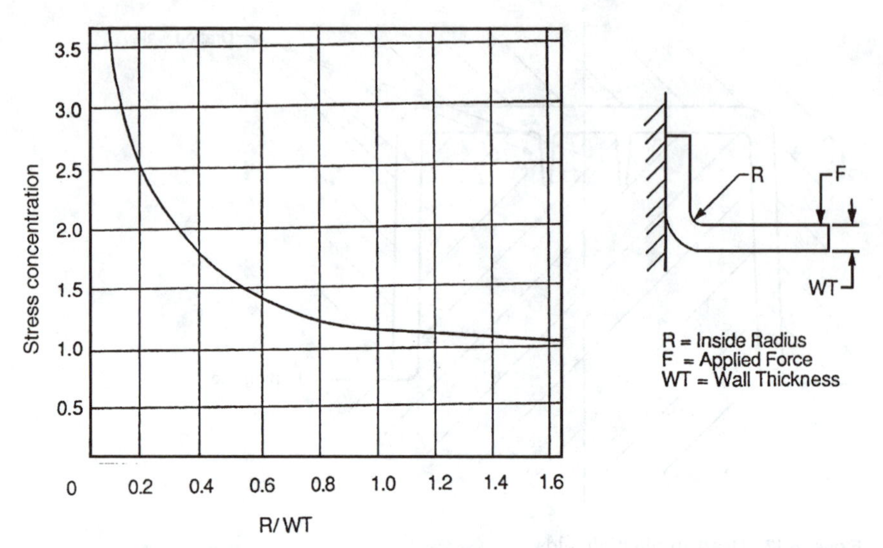

Figure 8.13 Relationship of inside corner radius to the stress concentration factor. (*Source: Jordan I. Rotheiser, Joining of Plastics Handbook for Designers and Engineers, Hanser Publishers, Munich-Hanser/Gardner Publications, Inc., Cincinnati, 1999.*)

Fig. 8.13 illustrates the manner in which stress concentrates in inside corners by plotting the ratio of the inside radius (R) to the wall thickness (WT) against the stress concentration factor. At ratios below 0.5, the stress concentration factor rises dramatically. At the ratio of 0.25, it has reached 2.25. That should be regarded as an absolute minimum. Below 0.25, the stress concentration factor reaches toward astronomical levels. Even a little radius is better than none at all. In addition to the reduction in stress concentration factor, radiused corners also improve the flow of plastic in the mold resulting in a more uniform melt and a shorter molding cycle.

8.2.4 Rib and post design

The elimination of sharp inside radii is important to the design of ribs as well. For free-standing ribs, that raises a paradox in that it becomes difficult to provide generous radii at the base of the rib and still maintain the nominal wall rule of not increasing the wall thickness by more than 25%. The effect of changing the inside radii from $0.25W$ to $0.5W$ is demonstrated in Fig. 8.14. The 25% increase in wall thickness is represented by the circle which is $1.25W$. In Fig. 8.14a, the radii at the base of the rib are $0.25W$, and they are $0.5W$ in Fig. 8.14b. Note how much thicker the rib is in Fig. 8.14a (Y_1) than it is in Fig. 8.14b (Y_2). There is a danger that the tip of the rib will become too thin when draft is taken into consideration; 0.040 in should be regarded as a

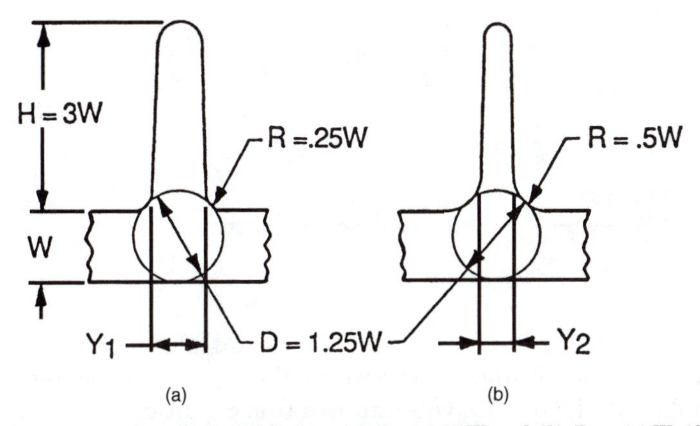

Figure 8.14 Standing rib design: (*a*) $R = 0.25W$ and (*b*) $R = 0.5W$. (*Source: Jordan I. Rotheiser, Joining of Plastics Handbook for Designers and Engineers, Hanser Publishers, Munich-Hanser / Gardner Publications, Inc., Cincinnati, 1999.*)

minimum. If the rib is used for locating another part or is lightly loaded, the use of inside corner radii in the $0.25W$ range is perfectly adequate. However, for load-bearing ribs, the greater the inside radius, the safer the design. The point of diminishing return is $0.8W$.

There is a point of diminishing returns for the height of the rib as well—3 times the wall thickness ($3W$). For corner gussets, that height is twice the wall thickness ($2W$), as shown in Fig. 8.15. The corner angle is effective at 30° from the vertical. If necessary, corner gussets and ribs can be as close to each other as twice the wall thickness ($2W$), however, that may be excessive for many applications considering the additional gripping force to the core that results. External ribs and gussets require a generous draft to prevent them from causing the part to stick in the cavity when the mold opens.

Posts are designed with the same cross section as a rib and use the same design parameters. However, free-standing posts are difficult to

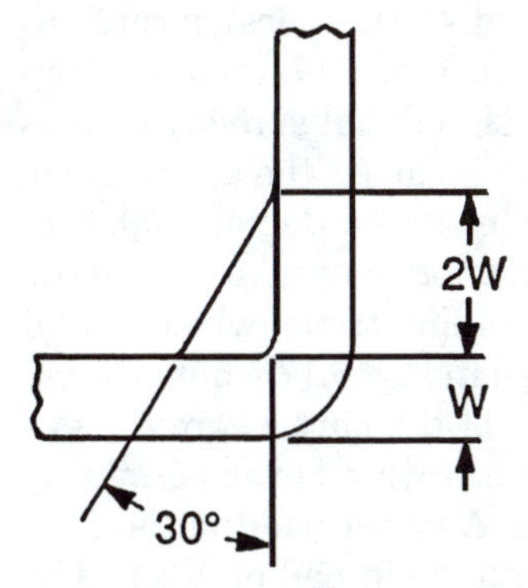

Figure 8.15 Design of corner gussets. (*Source: Jordan I. Rotheiser, Joining of Plastics Handbook for Designers and Engineers, Hanser Publishers, Munich-Hanser / Gardner Publications, Inc., Cincinnati, 1999.*)

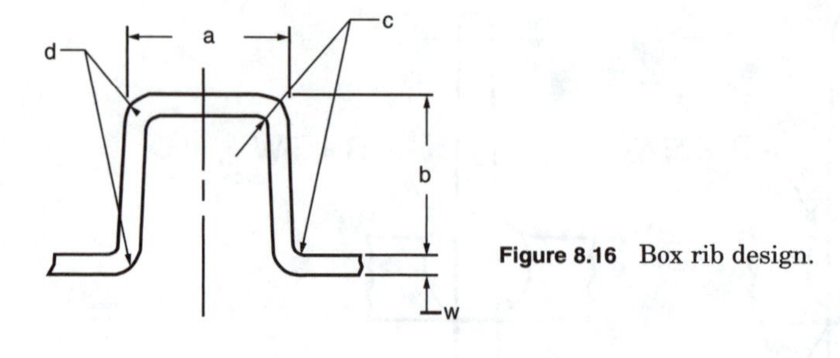

Figure 8.16 Box rib design.

cool uniformly and have no support. Therefore, they have a strong tendency to mold distorted. Gussets will improve this condition. A better solution, however, is to increase the diameter, hollow out the center, and create a small boss. That configuration is stronger, more controllable, and easier to design without sinks and molded-in stress. Boss design details are discussed in Sec. 8.2.7.

Certain processes, such as blow molding, lay-up, rotational molding, spray-up, and thermoforming, cannot create standing ribs. For these processes, the box rib illustrated in Fig. 8.16 can be used. For most applications, the width of the rib (a) can be 4 times the nominal wall thickness ($4W$), and the height of the rib (b) can be 4 to 5 times the nominal wall thickness. Inside corner design criteria (c, d) are the same as for standing ribs. Some large parts are difficult to adequately reinforce with integral ribbing which will not create sinks even when produced by a process which can create them. Box ribs (known as *hatbox ribs*) can be manufactured independently and attached to the large part, usually with adhesives.

8.2.5 Gas traps

The interior of a closed mold is not empty before it is filled with plastic. The space is occupied by a gas, air to be precise. The plastic displaces the gas, chasing the gas before it as it fills the mold. Certain design configurations, such as the free-standing rib shown in Fig. 8.17, create a trap with nowhere for the gas to escape. Such a design detail is referred to as a *gas trap* and the compression of the gas into it causes the gas to burn. The moldmaker can place very small vents (on the order of 0.0005 in deep by 0.25 to 0.50 in wide) at the parting line for the gas to escape the mold. These are often added after the first molding trials, which reveal the gas traps. A better design is also illustrated in Fig. 8.17 where the rib is extended until it attaches to the outer wall of the part where the gas is able to escape. Plastic part designers must be aware of the possibility of creating gas traps and avoid such designs. Another solution is illustrated in Fig. 8.17*b*, that is, the use of an ejector pin pad on a rib. The

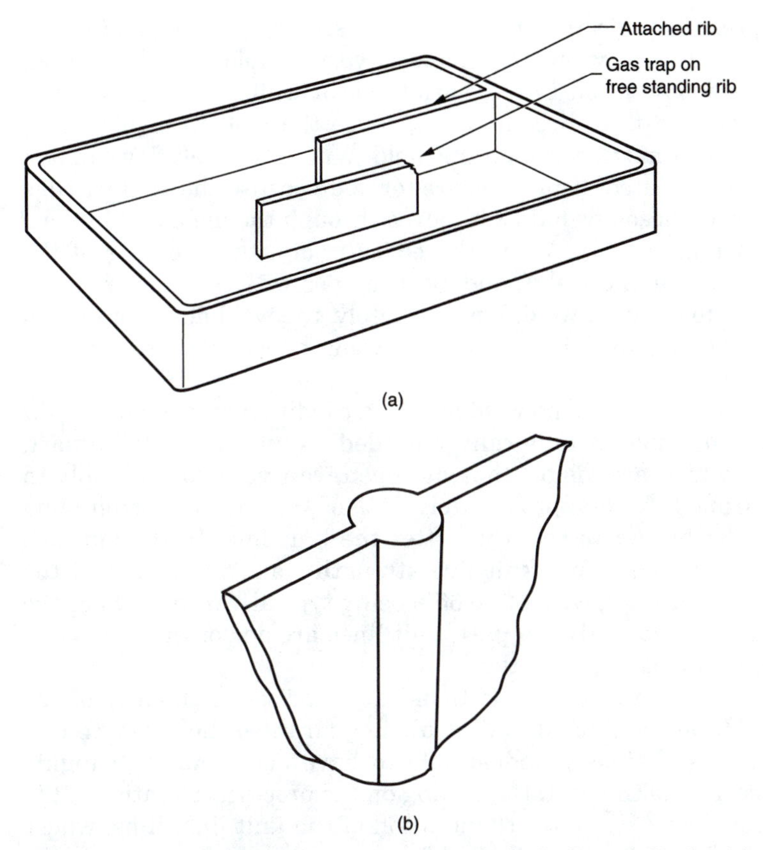

Figure 8.17 Gas traps: (*a*) free-standing and attached ribs and (*b*) rib with ejector pin pad. (*Source: Jordan I. Rotheiser, Joining of Plastics Handbook for Designers and Engineers, Hanser Publishers, Munich-Hanser/Gardner Publications, Inc., Cincinnati, 1999.*)

vent is then ground in the side of the ejector pin. The constant action of the ejector pin keeps the vent clean.

The problem of gas traps also exists in posts and bosses. In the case of a post which is trapping gas, the usual solution is to place an ejector pin on the post similar to the one used for the ejector pin pad on the rib. Bosses use a core pin that is stationary to create the hole. While the core pin can be modified to create a vent, that vent will be stationary and can plug with debris in time. That can require a more expensive solution: the use of an ejector sleeve around the core pin which then has to be fixed to the bottom plate of the mold base.

8.2.6 Knit or weld lines

Knit lines, also known as *weld lines,* occur in parts made of the processes in which the plastic fills out a mold (cold press, compression, injection, reaction injection, resin transfer, structural foam, and transfer molding), except when the shape is very simple and there is an unobstructed flow path. Cores, variations in wall thickness, depressions, changes in flow direction, etc., cause the melt to divide (often repeatedly) as it moves through the mold. Where the melt flows rejoin, knit lines are created. Knit lines are formed because the leading edge of each flow cools somewhat as it moves through the mold and forms a partially solidified skin. When the edges rejoin, the pressure of the melt bursts through the skin and intermingles with the melt from the joining flow to create a weld. Unfortunately, the weld does not extend through the total wall thickness at full strength, even if there is no visible knit line.

In some cases, there is no weld at all—a condition known as an *open knit line* and which is uniformly regarded as an unacceptable part. Knit lines which are visible, but not open, can vary considerably in strength from 10% to approximately 75% of that of the surrounding material; the harder to see, the better the knit line. In the author's experience, the maximum knit line strength was 85% of that of the surrounding material, with 50 to 65% being typical. Since they are the weakest link in the wall of a part, knit lines are one of the most common causes of plastic part failure.

Adjustments to molding conditions can improve the quality of the knit line. However, since it is difficult to guarantee the strength of a knit line, it is obviously undesirable to have one occur in a highly stressed section of the part. Depending on the process, relocation of the gate or the charge can alter the location of the knit line. Ribs, whose sole purpose is to provide a channel for molten material to speed to the desired area (path of least resistance) or flow interrupting depressions can also be used to relocate a knit line. The use of multiple gates will change the flow within the mold as well, however, an additional knit line is added for each new gate. Computer simulations are available which can forecast the locations of knit lines with reasonably good accuracy.

Openings in the part that are not to be cut into it require the use of cores. When the melt encounters a core, it divides and passes around it. Therefore, there is always a knit line around a hole or boss on the side opposite from the direction of flow. For relatively simple parts, that would be the direction of the gate. It is also difficult for fillers to reach these same areas and they may contain little or no filler—a condition known as *resin rich.* If knit lines or lack of filler result in a loss of strength beyond what the application can withstand, it may be advisable to cut the holes instead, even though the cost is higher. With the proper equipment, machining the holes can also be more precise than molding them.

8.2.7 Holes and Bosses

For those processes in which the plastic fills out a mold (cold press, compression, injection, reaction injection, resin transfer, structural foam, and transfer molding), openings in the part can be molded into the part through the use of cores. Holes in plastic parts are normally cut with a router (or drilled if round) for the remaining processes (although large openings can be molded into parts to be rotational or blow molded in some cases). When holes are created with cores, there will always be a knit line on one side. If the hole is for a fastening device, such as a screw, it is important to locate the knit line away from the side of the hole that the force is applied to.

When a boss is used, it will be formed by the melt running both into the cavity and around the sides simultaneously until it fills the cavity completely. Figure 8.18 depicts a partially formed boss in which the melt has not completely closed around the core (known as a *short shot*). Clearly, there will be a knit line along one side of the boss. Every boss will have such a knit line and it is rare that a boss fails other than at its knit line.

There are two criteria for the design of bosses: the appearance criteria and the strength criteria. The appearance criteria is illustrated in Fig. 8.19*a*. In this situation the 25% maximum increase in wall thickness rule has been applied ($D = 1.25W$) in order to reduce the molded-in stress and avoid sink. When the draft is taken into account, the wall thickness at the open end of the boss can become quite thin. It is generally unwise to permit it to become thinner than 0.040 in. Bosses of this type are usually used for location or as stand-offs with very little load, therefore, the inside radii can be $0.25W$, which will result in a boss wall thickness of approximately $0.5W$ when the rule is applied.

The boss in Fig. 8.19*a* will not supply the strength necessary for structural applications like threaded metal inserts and self-tapping screws. Structural applications require much stronger boss walls like those (*f*) in Fig. 8.19*b*. As the circle in the lower left corner illustrates, the 25% maximum wall increase rule is exceeded by a considerable

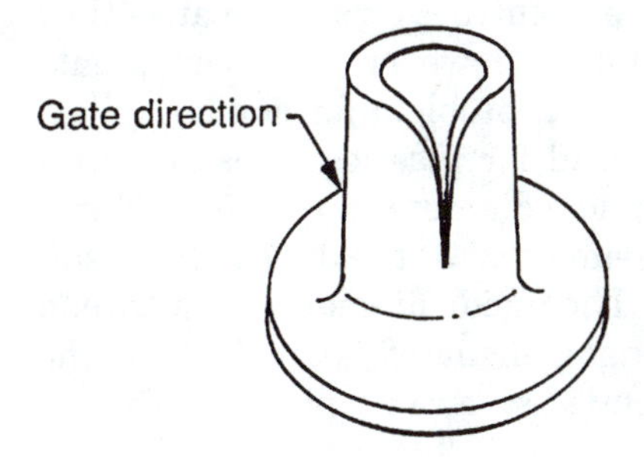

Figure 8.18 The formation of a molded boss. (*Source: Jordan I. Rotheiser, Joining of Plastics Handbook for Designers and Engineers, Hanser Publishers, Munich-Hanser / Gardner Publications, Inc., Cincinnati, 1999.*)

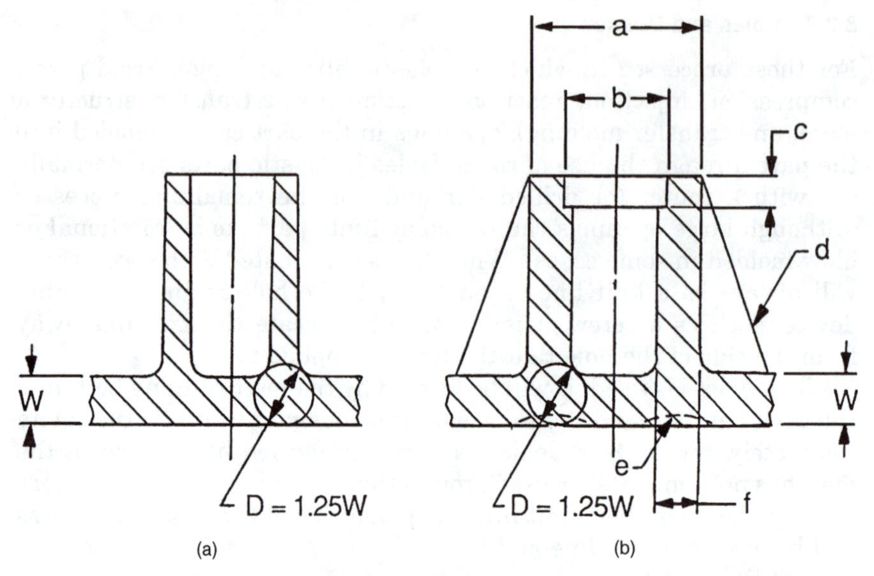

Figure 8.19 Molded boss configurations: (*a*) appearance criteria and (*b*) strength criteria. (*Source: Jordan I. Rotheiser, Joining of Plastics Handbook for Designers and Engineers, Hanser Publishers, Munich-Hanser / Gardner Publications, Inc., Cincinnati, 1999.*)

amount. This will result in a sink *(e)*, indicated by the dashed line, beneath each wall at the base of the boss. The injection molding process will result in the greatest amount of sink, however fillers can be used to reduce it. Foaming will also result in a reduction in sink, but there is a significant drop in physical properties when foam is used and the surface appearance is affected. Surface textures, labels, and some hot stamping patterns will also tend to hide sink. The other filling processes demonstrate sink to a lesser extent.

Bosses for self-tapping screws. The inside diameter of the boss for self-tapping screws is dependent on the size and type of the screw. Thread forming screws are generally used for plastics with a flexural modulus below 400,000 lb/in^2, however they can be used for stiffer plastics in some cases. They are preferable to thread cutting screws because they do not break the skin of the moldment. However, they can create greater stress in the boss which may be a problem for some applications. Thread cutting screws can be used for plastics whose flexural modulus exceeds 200,000 lb/in^2. The inside diameter for a thread forming screw boss should allow for engagement of 70 to 90% (with consideration for draft) of the thread depth. The inside diameter for a thread cutting screw boss should permit engagement of 50 to 70% of the thread depth. The generally recommended screws for plastics are type

AB for thread forming screws, type BF or BT/25 for thread cutting screws. These screws provide for more plastic between each metal thread than most of the other types. Specially designed screws for plastics are also available.

Once the inside diameter of the boss is established, the remaining dimensions can be readily determined. The length of thread engagement should be 2 to 3 times the screw diameter and the outside diameter (*a*) of the boss should be 2 to 3 times the inside diameter. (The outside draft is added to that.) The lead diameter (*b*) should just clear the outside diameter of the screw and the depth (*c*) should be enough to permit one full turn of thread. For screws 0.250 in in depth, the outside gussets should be used. The gusset design can follow the same criteria indicated in Sec. 8.2.3.

The two most common modes of failure for self-tapping screw bosses are boss cracking and low stripping torque. Poor weld lines are an obvious cause of boss cracking; however, too small a hole diameter, too large a screw diameter, too great a flexural modulus, and too long a thread engagement are also causes. Unfortunately, the remedy for one mode of failure may be the cause of the other as low stripping torque can be the result of too large a hole size, too small a screw diameter, too low a flexural modulus, and a length of engagement which is too short.

Bosses for threaded inserts. Bosses for threaded inserts are very similar to those for self-tapping screws. The inside diameter is determined by the diameter of the insert; the exact size for heat, press-in, or ultrasonic postmolded inserts are those recommended by the insert manufacturer for a given plastic who will also provide the lead diameter. (For self-tapping inserts, follow the same inside diameter criteria as for self-tapping screws.) In this case, however, the lead is an 8° included angle instead of a step. The outside diameter of the boss is 2 to 3 times the diameter of the insert and the minimum depth of the hole is the insert length plus 0.030 in.

8.2.8 Design for multiple part assemblies

Thus far, the design discussion has centered around the design of individual piece parts. However, most products require multiple part assemblies, often consisting of parts made of different materials. The first step is to ensure that the parts fit together properly—not merely at room temperature, but at the temperature extremes of what the product may be expected to encounter. (For example, a force fitment that works perfectly at room temperature may loosen at elevated temperatures or fracture at low temperatures.) That involves the determination of the fitments after the relationship of the parts to each other has changed due to differences in the coefficient of linear thermal expansion. Thus, the establishment of acceptable dimensional limits, generally known as *tolerances,* for the fitment dimensions is of critical importance.

Dimensional control. The next major consideration is the fact that plastic part quality is "process sensitive." That means that the part's size and shape can vary according to variations in process parameters. The thermoplastic processes generally operate with a cool mold, with the moldment remaining in the tool until the part is rigid enough to withstand the forces of ejection. If the part is ejected while it is too hot, it can be distorted and dimensional control lost. Furthermore, a mold core can act as a shrink fixture if the part is left in the tool beyond that point. Also, the temperature of the melt has an effect as a hotter melt is less viscous and can be injected into a mold at a higher speed. However, a hotter melt can lead to greater shrinkage, more distortion, and take longer to cool.

Plastic molding cost is also process sensitive because machine cost is a major expense component. The faster the mold is run, the lower the part cost. However, since running the mold fast requires higher temperatures and pressures which result in poorer dimensional control, it costs more to mold tight tolerances—often a great deal more. For this reason, molding quotes are typically tied to a drawing which has tolerances. The tighter the tolerances, the higher the part cost. Piece part quotes are not final until there is agreement that the part can be produced to the desired tolerance with the available tooling. If the molds are not capable of producing the parts within the required tolerances at the quoted prices, the molds will need to be upgraded or the piece part prices increased accordingly. For these reasons, it is sometimes said that "the money is in the tolerances" in plastics molding.

In the event of a dispute, the tolerances indicated on the drawing will determine legal liability. Therefore, it is important for the engineer to understand the establishment of plastics tolerances completely, for they are more complicated than for other materials. Plastics tolerances are divided between the toolmaker (one-third) and the processor (two-thirds). Therefore, a dimensional tolerance of ±0.006 in would require the toolmaker to build the tool to within ±0.002 in. If it is further presumed that the plastic has a shrinkage rate of 0.005 in/in, it is clear that a 1.000 ±0.005-in dimension requires a precision mold and very careful molding.

A shrinkage rate of 0.005 in/in is not really very high for plastic materials as some can reach as high as 0.050 in/in. What might be regarded as a "commercial" tolerance for one resin could be considered "fine" for another and "impossible" for a third. As a consequence, the Society of the Plastics Industry has prepared a series of charts for the principal plastics indicating what can be regarded as the "fine" and "commercial" tolerance ranges for each resin. These are available from:

The Society of the Plastics Industry, Inc.
Literature Sales Department
1801 K Street NW
Washington, DC 20006

Processes also vary in their ability to produce fine tolerances. Injection and transfer molding can produce the finest tolerances (with the proper resin). Processes like rotational molding and thermoforming require tolerances on the order of 0.010 in/in and can produce toleranced dimensions only on the mold side. Lay-up and spray-up cannot produce high-toleranced parts.

Casual requests for ultra-tight tolerances are widespread. When processors and toolmakers observe such tolerances, they will ask which are "critical tolerances," that is, those which really must be held. This practice undermines the validity of the entire tolerancing system since all tolerances should be held and no unnecessary ones should be specified. Since the drawing is part of the purchase contract, what is written on it is what will count the most.

Deviations from drawing tolerances can be approved if they turn out to be excessively tight when the actual parts are available. When parts are accepted with deviations from the contract drawing, a written record should be retained and the drawing should be altered accordingly to reflect the newly approved tolerance.

Regardless of how it is specified, the objective remains the same, namely, that the parts must fit together readily and stay together within acceptable parameters.

Establishing tolerances. A drawing or CAD (computer-aided design) file without tolerances is like a time bomb waiting to explode and destroy the project. Prior to the development of CAD, the engineer and a checker would examine the drawing before releasing it for tool construction and would accept responsibility for its accuracy. This system of checks and balances served industry well for many years. The purchase order referred to the drawing and the toolmaker and processor were responsible for meeting the stated tolerances if they accepted the purchase order. In the event of a dispute, the issue was clear-cut.

When a CAD file is produced without dimensions or tolerances, it is very time consuming and tedious to check. Hence, this step is frequently bypassed. The toolmaker will build the tool taking the file directly into the machining program without deviation. The responsibility for any errors in the tool reverts back directly to the individual who created the CAD file (unless it can be proven that the file was not correctly translated). The lack of tolerances practically gives the toolmaker and processor free rein to be off the mark without responsibility. The buyer will be obligated to pay for any corrections to the tool necessary to make the parts functional, assuming the tool can be corrected. Therefore, it is essential that a drawing be prepared with all fitment dimensions and tolerances. Broad blanket tolerances can be used for nonessential dimensions.

Blanket tolerances should not be used for critical dimensions. Not only is this poor engineering practice, but it betrays an ignorance of plastics processes, tooling, and good plastics design practice. Each fitment tolerance must be determined individually and, since plastics tolerances are expensive, should be no tighter than necessary to assure proper function. There is a methodology for determining tolerances as illustrated in Fig. 8.20.

A typical shaft-to-hub fitment is depicted in Fig. 8.20*a* with a shaft diameter established as 1.000 in and a trial tolerance of ±0.005 in. The hub diameter will be determined using the methodology illustrated in Fig. 8.20*b*. Adjusting the shaft diameter for the tolerance results in a low limit of 0.995 in and a high limit of 1.005 in. For this theoretical application, it is critical that the fitment be in clearance under the worst conditions. Since neither of the materials is flexible, a minimum clearance of 0.002 in is selected. Therefore, the low side limit for the hub is 1.007 in. Again, using a trial tolerance of ±0.005 in, the nominal diameter for the hub becomes 1.012 in and the high limit becomes 1.017 in. Subtracting that from the low limit for the shaft of 0.995 in, it is apparent that a 0.022-in gap can appear under the most extreme conditions.

If there were two such fitments on the part as illustrated in Fig. 8.20*c* with a center-to-center distance of 3.000 in and a ±0.005-in tolerance, an additional clearance of 0.010 in would have to be added for a total clearance of 0.032 in. That is enough clearance for the fitments to be quite loose. Little comfort can be found in the concept that the bell curve can be applied to a sampling of actual parts and that very few will actually occur at the tolerance extremes because, as previously discussed, processing conditions can result with the entire population of parts being at the extremes. Furthermore, if the shaft mold is machined to the low side and the hub mold is cut to the high side, there is no possibility of a bell curve resulting from the sampling.

If one of the resins were flexible, or if the wall of one of the parts were thin enough to allow some flexure, the maximum gap could be reduced by shifting the other limits in a modest interference of 0.001 to 0.003 in.

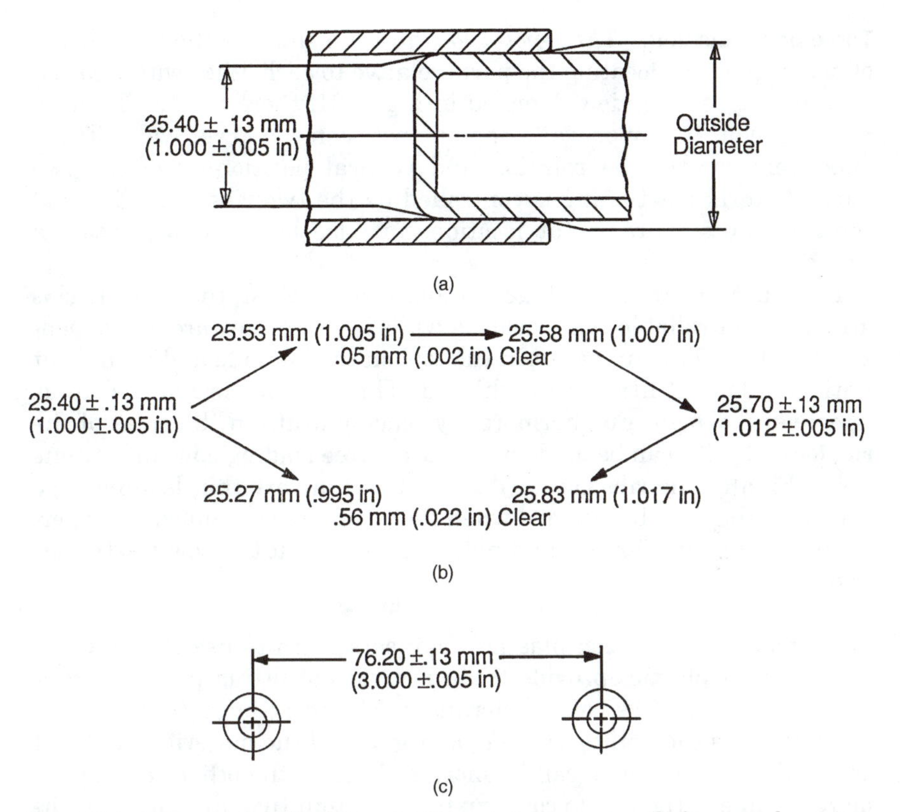

Figure 8.20 Tolerancing methodology: (*a*) fitment starting point; (*b*) fitment tolerancing; and (*c*) center-to-center tolerancing. (*Source: Jordan I. Rotheiser, Joining of Plastics Handbook for Designers and Engineers, Hanser Publishers, Munich-Hanser/Gardner Publications, Inc., Cincinnati, 1999.*)

The simplistic solution would be to tighten the tolerances. However, that would increase the processing cost and, possibly, the tooling cost, particularly if the material were one with a high shrinkage rate. It might even result in a change to a much more costly material.

One must also bear in mind the effect of differences in the coefficient of linear thermal expansion between the two parts. If the coefficient of linear thermal expansion is greater for the shaft than it is for the hub, the gap will tend to close at elevated temperatures and become greater at low temperatures. Whether or not this is desirable depends on whether the joint is intended to be fixed or sliding at elevated temperatures. In the case of a fixed joint, it may be desirable to elevate the temperature of the hub (only) for assembly in order to assure that the joint remains fixed at elevated temperatures. However, this practice can lead to stress cracking of the hub at low temperatures if it is not strong enough to withstand the added stress under such conditions.

Three-point location. *Three-point location* can simplify the tolerancing of plastic parts by locating the parts relative to each other with locating ribs. In the section view depicted in Fig. 8.21, there are two locating ribs between the upper walls and one between the lower walls. These effectively control the rotation and vertical positioning of the two parts. Lateral positioning is controlled by the two ribs providing end contact points. Three-point location can also be used to position a cylinder.

There are several advantages to this system. First, the locating ribs are more controllable than locating walls because they are less expensive to alter. They can be cut in the tool on the low side and then recut until the desired fitment is achieved. This practice is referred to as "steel safe" design. Furthermore, by placing draft on the sides of the rib, locating ribs can be made with a draft-free leading edge in extreme cases. Finally, the ribs can be designed with a very thin leading edge such that they can be crushed when the parts are assembled, thus permitting some interference and reducing the gap at the worst extremes of both tolerances.

Crush ribs. Experienced plastics engineers learn to use the inherent benefits that plastics provide to devise ways of fitting parts together using looser and less costly tolerances. The most important of these advantages is the ability to alter the rigidity of the material such that one of the mating parts can be more rigid than the other, forcing the more flexible of the two to conform to its contour (usually the outer fitment becomes the more flexible). This can be done by changing the

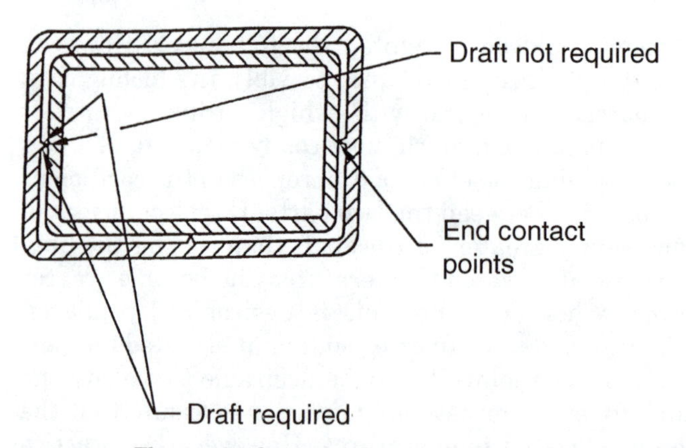

Figure 8.21 Three-point location. (*Source: Jordan I. Rotheiser, Joining of Plastics Handbook for Designers and Engineers, Hanser Publishers, Munich-Hanser/Gardner Publications, Inc., Cincinnati, 1999.*)

grade of the resin for one of the parts or even substituting another plastic.

Forcing the more flexible of two parts to conform to the other without changing the material can be accomplished by thinning the wall of one of the parts until it is flexible enough to do this. It may not be possible to reach this goal without weakening that part to an unacceptable level, and the cost of tool modifications may be unacceptable. Therefore, another approach is to accomplish the fitment with a series of tapered ribs which can be crushed as the parts are pressed together. These are generally referred to as *crush ribs.*

Crush ribs are much more controllable in plastic parts than are outer walls because they are less costly to alter and can be adjusted with little effect on the part weight or wall thickness. Most important, they permit the mating part to be located while also maintaining uniform wall thicknesses and without increasing the wall for locating purposes.

In Fig. 8.22*a,* the crush ribs are illustrated in a side cross-sectional view. The crush ribs are not in section and are located at the top and bottom of the cap tapering inward from the open end. As the shaft is pressed into the cap, it encounters the ribs with increasing force until it bottoms and the cap is fully emplaced on the shaft. The point of first contact and the amount of final interference depends on the actual sizes of the two parts. However, the ability of the ribs to crush over permits wider tolerances on their mating dimensions. The amount of taper varies with the application. However, the author usually begins trials with a 2° angle/side which results in reduction in the diameter of 0.035 in per side over a 1-in depth.

Figure 8.22*b* is a cross section taken through the end view which illustrates the use of eight crush ribs. The number, size, location, and included angle of crush ribs is dependent on the application. The author tends to keep them at a height of around 0.031 in with as small an included angle as possible (30°). The angle can be increased in-

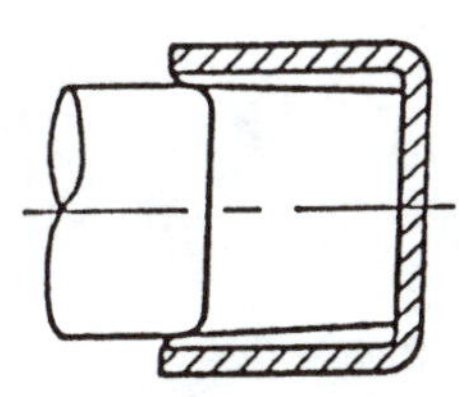

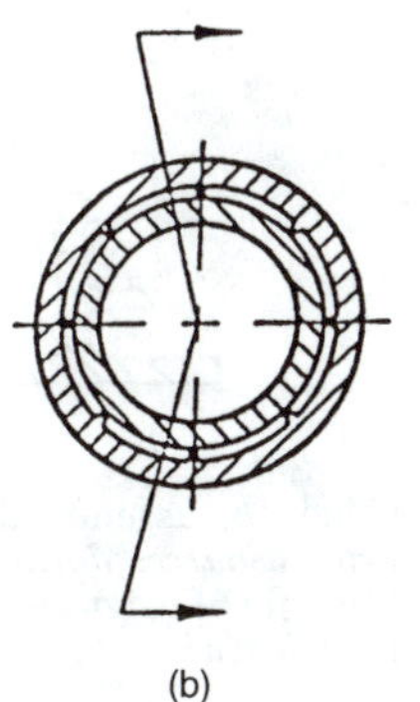

Figure 8.22 Crush ribs: (*a*) side section and (*b*) plan section. (*Source: Jordan I. Rotheiser, Joining of Plastics Handbook for Designers and Engineers, Hanser Publishers, Munich-Hanser/Gardner Publications, Inc., Cincinnati, 1999.*)

expensively in the tool, but decreasing it requires welding at a much greater cost. The tip of the rib needs to be called out as sharp because it is the very tip that crushes over, and a tip radius will increase the stiffness significantly.

It is difficult to precisely determine the effect of shrinkage on the dimension of many parts and the amount of deformation that will result from the thinned wall. Therefore, it is wise to build a prototype single-cavity mold designed with steel safe fitment dimensions and close in on the final dimensions by removing metal from the mold based on results from initial trials.

Flex ribs. The crush ribs described in the previous section will provide stiffness and accommodate small variations in size. They are excellent for centering one part to the other. However, flexible ribs may be required to provide a greater range of dimensional deviation. For this purpose, the design depicted in Fig. 8.23 may be used. The unassembled state is shown in Fig. 8.23*a*. The thickness and the angle of the rib may have to be experimentally determined as is often the case with flex and crush designs. For rigid plastics, the author usually begins with a rib angle 30° from the plane of the wall it is attached to and as thin a rib as possible with an included angle of 20°. This design is

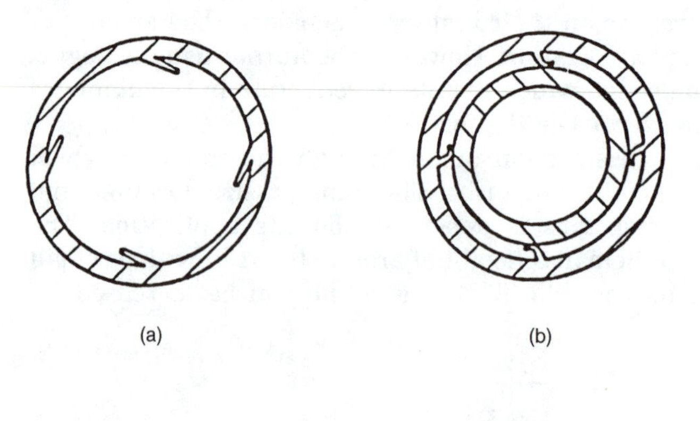

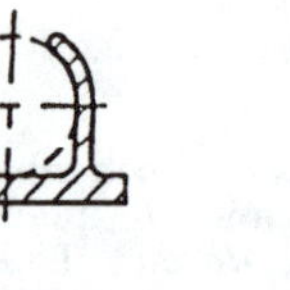

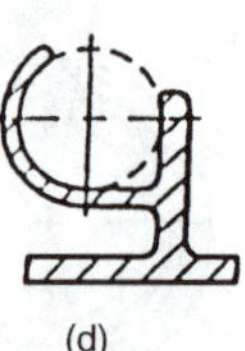

Figure 8.23 Flexible ribs: (*a*) unassembled; (*b*) assembled; (*c*) dual flex arms; and (*d*) single flex arm. (*Source: Jordan I. Rotheiser, Joining of Plastics Handbook for Designers and Engineers, Hanser Publishers, Munich-Hanser/Gardner Publications, Inc., Cincinnati, 1999.*)

shown assembled in Fig. 8.23*b* with the ribs deformed by their interference with the mating part.

There are other ways to use the flexible rib concept such as those illustrated in Figs. 8.23*c* and 8.23*d*. In both cases, much larger arms which can achieve a great deal more flexure are used. In the design in Fig. 8.23*c*, both arms flex somewhat to accept the mating part whereas only the left arm flexes in the design in Fig. 8.23*d*. The design of flexible arms such as these is application dependent.

Inside/outside fitments. The tolerance extremes can also be accommodated through the use of an additional fitment to the outside of the shaft fitment. This outside fitment will ensure a good joint when the shaft is small and the hub is large, as shown in Fig. 8.24*a*. The inside fitment tightens when the shaft is large and the hub is small as in Fig. 8.24*b*. The parts can be designed to create a wedging effect as depicted in Fig. 8.24*c*. Although the inside/outside fitment requires a greater tooling expense and is practically limited to round parts, it can be very effective for critical applications and tends to hide deformation. This

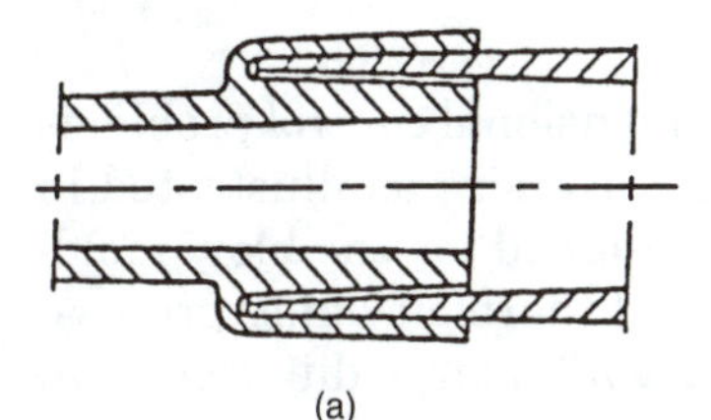

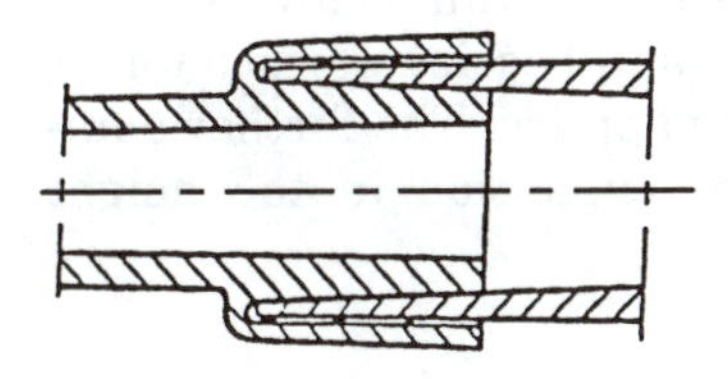

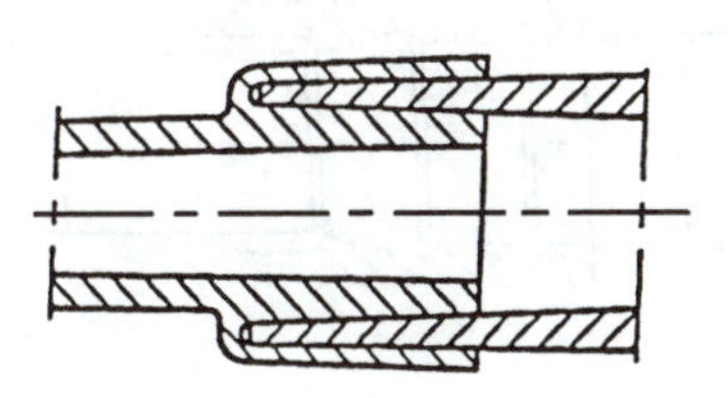

Figure 8.24 Inside/outside fitment: (*a*) small shaft-large hub; (*b*) large shaft-small hub; and (*c*) wedged joint. (*Source: Jordan I. Rotheiser, Joining of Plastics Handbook for Designers and Engineers, Hanser Publishers, Munich-Hanser/Gardner Publications, Inc., Cincinnati, 1999.*)

approach has been used successfully on ultrasonic shear joints which had to withstand severe drop tests.

Step fitments. *Step fitments* are another way to reduce dependency on tight tolerances provided one of the parts is more flexible than the other. They can be used externally or internally depending on whether the more flexible part is the internal or external fitment. Both styles require more expensive tooling. The external variety shown in Fig. 8.25*a* calls for a split cavity and the internal version in Fig. 8.25*b* needs a collapsing core unless the undercuts are small enough for the part to be stripped off the core.

Drilling in place. One way to eliminate the tolerancing problem between two parts is to locate them next to each other and drill in place. The holes then match perfectly and the need to hold expensive, tight tolerances is eliminated. The disadvantage is that a replacement part would not be interchangeable and might need to be drilled in the field or the original parts returned to the factory. This solution is normally employed for very large parts in which holes are not easily molded or when the volume is very low.

Oversize holes. Another approach to the dimensional control problems associated with plastics is to make the hole oversize as illustrated in Fig. 8.26*a*. This is often done with parts designed for the blow molding, lay-up, rotational molding, spray-up, and thermoforming processes and is also an excellent means of dealing with a large differential in coefficient of linear thermal expansion between the two parts. An elastomeric sleeve, such as the one in the illustration, can be used for this purpose. In order to compensate for the loss in bearing surface under the head of the screw, a washer is typically applied. The further addition of a spring washer will account for expansion in the height

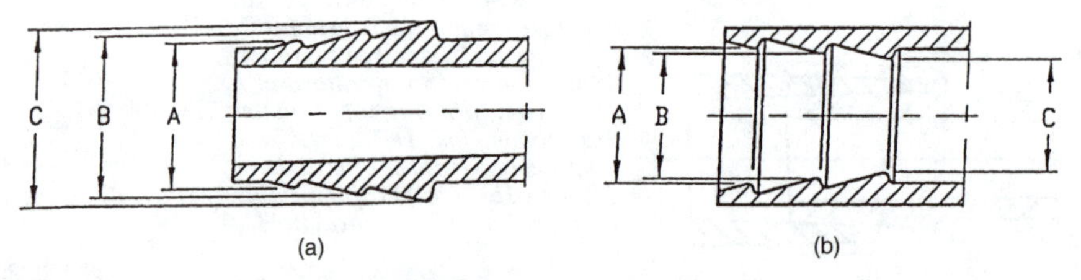

Figure 8.25 Step fitments: (*a*) external step fitment and (*b*) internal step fitment. (*Source: Jordan I. Rotheiser, Joining of Plastics Handbook for Designers and Engineers, Hanser Publishers, Munich-Hanser / Gardner Publications, Inc., Cincinnati, 1999.*)

direction; however, the spring washer should not be used without a flat washer because of the notch sensitivity of many plastics.

In some cases, there are dimensional variations greater than those which can be accommodated by oversize holes. A solution to that problem is the use of criss-cross slots (Fig. 8.26*b*) in the moldments in conjunction with a nut, bolt, and washers. It should be noted that criss-cross holes are more expensive to tool or machine than round holes.

Separation of functions. The assembly illustrated in Fig. 8.27*a* requires the bosses to perform both the function of location and that of joining. The location function requires the processor to maintain close dimensional control on one side of each boss while the joining function requires close dimensional control on the other side plus a good knit line. It may be necessary to increase the temperature of the melt to achieve an acceptable knit line, however close dimensional control may be more readily attained with a cooler melt. Thus, the processor

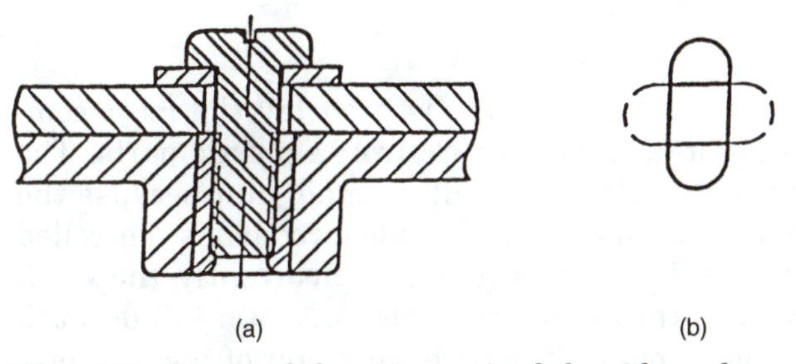

Figure 8.26 Oversize holes: (*a*) oversize hole with washer and (*b*) criss-cross holes. (*Source: Jordan I. Rotheiser, Joining of Plastics Handbook for Designers and Engineers, Hanser Publishers, Munich-Hanser / Gardner Publications, Inc., Cincinnati, 1999.*)

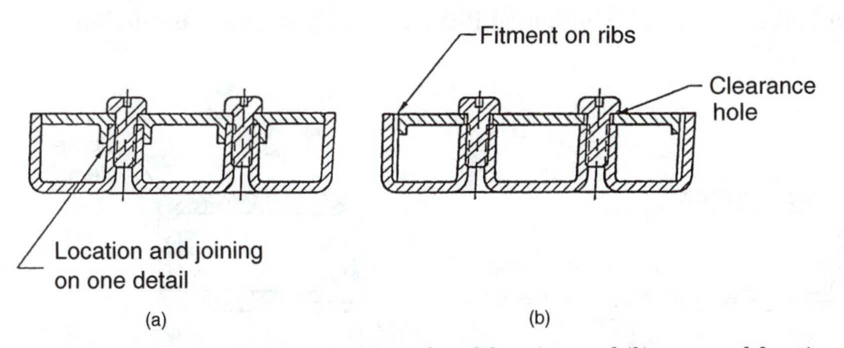

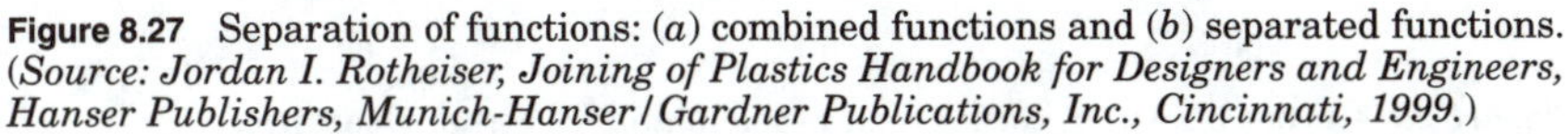

Figure 8.27 Separation of functions: (*a*) combined functions and (*b*) separated functions. (*Source: Jordan I. Rotheiser, Joining of Plastics Handbook for Designers and Engineers, Hanser Publishers, Munich-Hanser / Gardner Publications, Inc., Cincinnati, 1999.*)

winds up in the paradox where the attainment of one feature of a part results in the loss of another.

The solution may lie in the separation of the functions as illustrated in Fig. 8.27*b*. Clearance holes have replaced the locator bosses in the top piece, thereby removing the locating function from that design detail. Thus, the molder can concentrate on the integrity of the boss knit line itself. Location has been removed to the outer walls where the top is located to the base on ribs that can be adjusted independently without interfering with the strength of the bosses.

Corner binds. *Corner bind* is a condition where the corners of the parts, often the stiffest portions of a part, bind and prevent assembly of the parts. This can occur even if the sides of the parts warp inward, as illustrated in Fig. 8.28*a,* which is a common problem in plastic part assembly. The solution is to radius the corners of the part, a practice which should be followed to reduce the stress concentration factor in any case. However, the radii tolerances can still cause an interference. This problem can be avoided by designing the part so the inside radius on the outer part ($R1$) is smaller than the outside radius on the inner part ($R2$).

Semidovetail joint. The *semidovetail joint* is probably the most common joint used for position and contour control in plastic parts. The semidovetail joint is used in place of a full dovetail joint because the latter is not required since location in the inside direction is controlled by the joint on the other side of the part. More importantly, the semidovetail joint requires two-thirds the wall thickness of a full dovetail joint. A semidovetail joint around the entire perimeter of the part provides location, tends to mask minor warpage and debris from joining devices (surplus adhesive, solvent, welding flash, etc.), and is reasonably tolerant of dimensional variations. In addition, if designed steel-safe, it can be readily adjusted if the molded parts turn out to have too

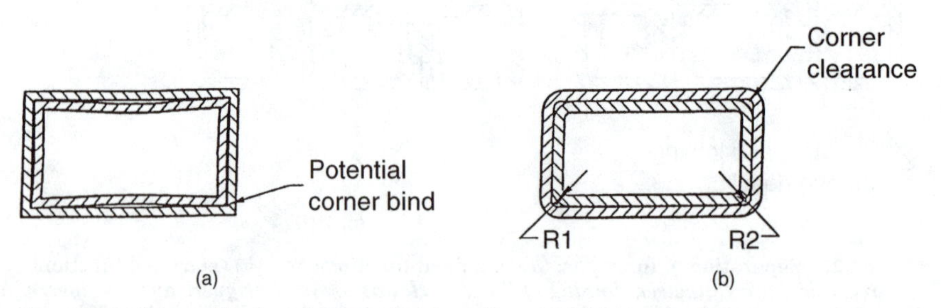

Figure 8.28 Corner binds: (*a*) corner bind problem and (*b*) corner clearance.

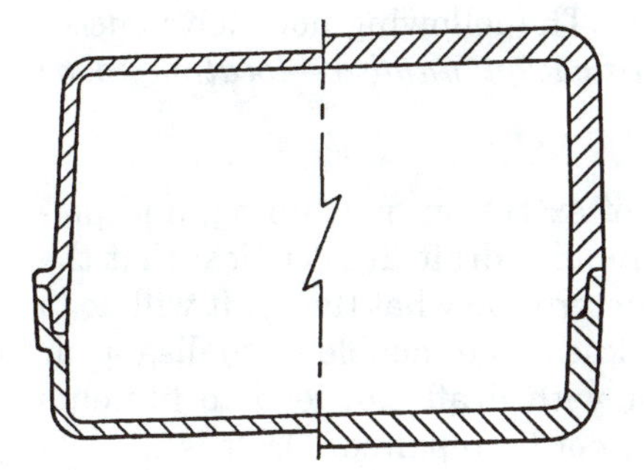

Figure 8.29 Semidovetail joint.

great a gap in the joint area. When the wall requires control in both directions, ribs can be added to the inside of the semidovetail joint.

Two semidovetail joints are illustrated in Fig. 8.29. The standard version is shown on the right and the thin-wall design is on the left. There are design concerns related to the drop in wall thickness in the joint area for the standard semidovetail joint. The arm of the joint must have a minimum wall thickness for rigidity, usually 0.040 to 0.050 in. The lower part of the joint shown can have any nominal wall equal to or greater than the thickness of the arm. However, for the outside wall to be flush, the upper part must have a nominal wall at least double that of the arm. Thus, the thickness of the joint dictates the thickness of the nominal wall.

If the outside surface does not need to be flush, the thin-wall semidovetail joint shown on the left side of Fig. 8.29 can be used. This design resolves the nominal wall problem inherent in the standard version by going to a uniform wall thickness throughout. Both versions of the semidovetail joint are vulnerable to sharp inside corners at the base of the arm and these are the most likely sources of failure. A minimum radius of 0.010 in should be used at those locations.

8.2.9 Plastics specifications and drafting practices

There are several aspects of plastics specifications and drawings (or files) which are particular to this family of materials and processes and which deal with realities which must be addressed. The following sections will discuss them in detail.

Cavity identification. When more than one cavity is built, they must be identified in order to evaluate their adherence to quality specifications. Each will vary in actual dimension due to the effects of location in the mold base on melt flow and to construction differences. The number or letter should be located in a place where it will not interfere with a fitment or the appearance of the part. The following notation is used: *Each cavity must contain an identification number—location to be approved by engineering.*

Draft indication. With the exception of extrusion, machining, and pultrusion, all of the plastics processes require draft. It is critical that the part be drawn with draft in order to determine what the draft will do to the design. Draft can cause wall thicknesses to double or to disappear altogether. Failure to draw the part with draft can lead to fitments which do not fit and molds which cannot be repaired. Draft is usually specified as $+x°$/side (or $-x°$/side) and placed on the dimension taken from that point ($1.000 + 1°/S$). Thus, the designation $+1°/S$ indicates that a 1° draft is intended to increase from the point of the dimension so indicated. Conversely, the designation $-1°/S$ would indicate that the draft decreases from the point dimensioned by $1°/S$.

The draft specification controls the way the mold is built as the direction of draft normally indicates the direction of draw (removal of the part from the tool) since reverse draft would be an undercut condition. For injection molding, the mold designer will attempt to locate the core on the back, or movable, half of the mold. That is the side where the molding machine's ejector bars are located, and locating the core on that half makes the ejector mechanism available to push the part off the core. Part designs with cores from both sides of the part may require slight undercuts to keep them on the side of the mold which has the ejector mechanism. The draft specification will also control the location of the parting line, which should be indicated.

Ejector locations. Ejection devices for plastic parts can range from screwdrivers used to pry parts out of a hand mold to mechanized stripper plates and elaborate mechanisms which also retract collapsible cores. All of them share one common characteristic: they exert pressure on a newly formed part. That pressure can distort the part to the point of disturbing its function or appearance if it occurs while the part is still too soft to withstand it. Therefore, the processor must delay ejection until the moldment can endure it. The more ejectors there are, the more ejection surface there is to distribute that pressure and the sooner the part can be removed from the mold, thereby shortening the molding cycle. However, ejectors cost money and leave marks on the surface of the moldment. Therefore, there is a mold cost associated with a faster molding cycle. (Differences between bidders on a project are often based on variations in cooling and ejection systems.) Additional ejectors leave more marks on the surface and their number and location may be limited by functional and appearance concerns.

Neither the molder nor the moldmaker is intimately familiar with the product or its application. Therefore, ejector location should be controlled with the following notation: *Ejector locations must be approved..*

The *stripper plate* is a variety of ejection system which need not leave a mark on the part. It also permits ejection of the part in a much warmer state because it distributes the force of ejection uniformly around the parting line of the part. This system can reduce cycle times by as much as 35%; however, there is a considerably greater tooling cost for a stripper plate and not all designs permit its use.

Flash specification. Flash consists of small amounts of plastic which adhere to a molded part beyond its desired contour, usually where wear occurs like the fitments around side actions, ejector pins, and at the parting line. In small amounts, it can be removed with a variety of cutting and scraping devices ranging from clippers to a tile knife. For large amounts, typical of compression-molded parts, a router is used.

Regardless of how well a tool is built or a part is molded, there will always be some flash, particularly as the mating parts wear in time. To demand "no flash" betrays an ignorance of the processes. That is not to suggest that it cannot be done if the purchaser is prepared to pay the price for the necessary secondary operations. A preferred practice is to design the part so the critical fitment dimensions do not fall in an area where flash can occur, such as a parting line. The most practical procedure is to determine how much flash is acceptable (0.010 in is reasonable for most applications) and provide the following specifications: *Maximum allowable flash is .XXX in.*

Flatness control. The very nature of plastics processing makes absolute flatness a virtual impossibility. Therefore, flatness specifications must take this factor into account. However, most plastics are flexible to some degree, thus they will conform to the shape of the mating part, making absolute flatness unnecessary. Flatness recommendations for each material are provided in the tolerance charts discussed in Sec. 8.2.8 under "Dimensional Control." Nonetheless, plastic parts are process sensitive and there must be limits or the parts can go out of control and proper fitments can be jeopardized. The following flatness specification is recommended: *Part must be flat within .XXX in.* (or within .XXX in/in).

Gate location specification. Gates can interfere with the function or the appearance of the part. Therefore, the engineer or designer must approve their location. Furthermore, they should specify surfaces where gates may not be placed (or where they must be trimmed flush) so the mold designer can lay out the mold accordingly. This issue can normally be addressed with the following specification: *Gate location must be approved by engineering.*

Knit or weld lines specification. As previously discussed (Sec. 8.2.6), parts from processes in which material flows in the mold nearly always have knit lines and these knit lines will be the weak points in the part. Therefore, a specification regarding the acceptability of knit lines is necessary.

For most applications, it is enough to indicate where knit lines are unacceptable. Knit lines which cannot be seen with the naked eye are always regarded as good; however, a good knit line can sometimes be faintly visible on a glossy surface. Thus the use of the specification which permits *no visible knit line* requires highly subjective decisions on the part of those charged with quality assurance. The only sure way of learning the strength of a weld line involves destructive testing. To reduce the level of subjectivity, weld-line examples can be tested and limit samples provided for quality assurance. In that case, the specification should read: *Limit samples for acceptable weld lines.*

Material specification. The material specification is, perhaps, the most critical of all the specifications. A material deviation can lead to a variety of problems in the molding of the part, its properties, and its performance, both short and long term. That is true, not only of the resin itself, but of the other additives, such as fire retardants, ultra-violet inhibitors, fillers, lubricants, and pigments as well. For applications with low safety factors, no material substitutions should be permitted without prior testing. Strange things do happen, even with applications which appear to be significantly overspecified. The following material specification is recommended: *Material is to be (name of manufacturer) (exact number of resin). Part is to include XX% additive (name of manufacturer) (exact number of additive). No substitutions permitted without written authorization.*

The pressure for substitutions arises from market conditions. Material shortages and price increases force processors to seek means of relief. They will sometimes offer an “equivalent” material. Equivalent is, however, an ambiguous term when referring to plastics. It cannot mean precisely the same resin because resins are covered by patents and, therefore, each one is somewhat different than the others in its behavior and properties. Therefore, whether the alternative resin is close enough to the specified material to be acceptable depends on the application. Product engineers find it difficult to rely on processors’ recommendations because they are primarily concerned with producing the part to a level which will achieve acceptance by the purchaser and are usually innocent of the product's performance requirements and characteristics, particularly in the long term. It is, therefore, wise practice for the end user to test at least two alternatives for each of the resins and additives. This should be done in advance of the need because there is usually no time for testing when the need for an alternative arises. Finally, the material supplier may not be the actual resin producer. In fact, the material supplier may use several sources. That needs to be established and resin from all possible sources tested for critical applications.

Mismatch specification. Closed molds are subject to misalignment when the parts are put together. This condition is referred to as "mismatch" and should be controlled according to how much it affects appearance or function in order to avoid any kind of dispute as to whether any misalignment is or is not included in the tolerance. When misalignment occurs between a core and cavity, a variation in the wall thickness between the two sides results. The flow of the melt will be altered since the cavity space is different between the two sides of the mold. This could affect the strength and location of a weld line. Misalignment between core and cavity is also controlled by placing a tolerance on the wall thickness. The following specification can be used to control mismatch: *Maximum allowable mismatch is .XXX in.*

Regrind limitation. Thermoplastics are capable of nearly 100% material utilization by regrinding sprues, runners, and reject parts and mixing that "regrind" in with virgin resin. However, the material suffers some degradation when it is raised to elevated temperatures. Regrind, therefore, reduces some of the physical properties of the material; the degree of drop varies between plastics and according to molding conditions. It is necessary to perform actual tests to determine the degree of property loss; however, readily visible signs of degradation are an increase in brittleness and a yellowing or darkening of color.

Physical properties drop according to the amount of regrind permitted. This is magnified by the multiplier effect which takes place. For example, if 20% regrind is used, then the batch will contain 4% (20% of 20%) which has been through twice, 0.8% of which has been through 3 times, 0.16% which has been through 4 times, etc. The processor may presume the use of all the regrind is acceptable unless it is controlled with a specification like the following: *XX% regrind acceptable.*

The engineer should be alert to the fact that the use of regrind runs the risk of contamination of the resin from a variety of sources up to and including cigarette butts. If a contaminant will pose a significant risk for the application (medical product) or if the need for all of the physical properties is critical, it may be necessary to prohibit the use of regrind. It is still possible to obtain near 100% material utilization in a closed runner system (hot or insulated runner system), which results in a significant increase in tooling cost.

Surface finish and textures. Surface finishes are generally indicated according to the Society of the Plastics Industry standard finishes, a plaque with samples of which is available from the same address indicated in Sec. 8.2.8, "Dimensional Control." The surfaces are created as indicated in Table 8.9. These finishes begin with A-1 as the highest polish and D-3 as the roughest surface. The highest diamond buff finishes are the most expensive and are rarely used for anything but lens applications and the like. Finish A-3 would be the level used when high gloss or transparency is required for the product. Finish B-3 is a level that resembles lightly brushed aluminum and is typically used for industrial products where high gloss is not required. It is also used as the working surface for texturing. An inexpensive finish often used for core surfaces that are not visible is C-3. The D finishes are blasted finishes which create a matte surface that should be well drafted. Esthetically, they resemble degrees of frosted glass. Functionally, they are used to create writing surfaces, absorb stray light, and provide a roughened surface to enhance adhesion. Surface finishes should be indicated with the following notation: *SPI Finish No. X on all outside surfaces except as noted. SPI Finish No. Y on all inside surfaces.*

Textures are available in an infinite variety of standard patterns from several texturizers. However, they are quite capable of custom reproducing virtually any texture the customer desires. In either case, the texture identification is substituted for the SPI finish in the surface finish notation. The area which is to receive the texture should be clearly indicated on the piece part drawing (with tolerances). As indicated previously, the general rule on draft for textures is $1^1/_2{}^\circ$ per side plus $1^1/_2{}^\circ$ per side for each 0.001 in of texture depth. In tight situations, this can be reduced to 1° per side plus 1° per side for each 0.001 in of texture depth for some applications.

TABLE 8.9 SPI Mold Finishes

A-1	Grit #3 diamond buff	C-1	600 stone
A-2	Grit #6 diamond buff	C-2	400 stone
A-3	Grit #15 diamond buff	C-3	320 stone
B-1	600 grit paper	D-1	Dry glass bead #11
B-2	400 grit paper	D-2	Dry blast #240 oxide
B-3	320 grit paper	D-3	Dry blast #24 oxide

SOURCE: M. L. Berins, ed., *Plastics Engineering Handbook,* 5th ed., The Society of the Plastics Industry, Chapman & Hall, New York, 1991.

Tolerance notation. Dimensional control and tolerances have been extensively discussed in Secs. 8.2.8 "Dimensional Control" and "Establishing Tolerances." For noncritical tolerances, a general *tolerance notation,* or a *tolerance box,* is a convenience which can save a considerable amount of drafting time when properly employed. Unfortunately, more often than not, it is not properly employed and it leads to excessively tight tolerances. That is because it is easier for the designer to use the overall tolerance than it is to work out the tolerance for each location. When the range of tolerances used is great, it may be more useful to use the note: *Dimensions reference if not toleranced.* (Reference means for informational purposes only—usually abbreviated to "Ref.") The following tolerance notation can be used for tolerances which are not critical to the function of the product: *Tolerances ±.XXX if not otherwise specified.*

8.3 Design Details Specific to Major Processes

Plastic part design varies considerably according to the process which will be used to manufacture the part. However, similarities exist for many processes which can be categorized according to the manner in which the part is formed and this fact can be used to better understand the nature of the part. The processes can be categorized according to whether the plastic is built up in layers, stretched into shape, or a mold is filled with liquid or melted resin.

The build-up processes are lay-up, spray-up, and rotational molding. Lay-up and spray-up are thermoset processes for open parts while rotational molding is a thermoplastic process (see Chaps. 1 and 6 for further explanation of the processes). Since build-up is accomplished without pressure, these processes produce parts which are relatively stress free. Indeed, they can produce the very largest parts (see Table 8.7).

The thinnest nominal walls are created by the stretching processes: thermoforming and blow molding. In each case, the material is first processed by some other method before it is shaped. Thermoforming uses extruded sheet which is then heated and stretched over a tool. Blow molding uses an extruded tube or injection moldment which is then blown into a mold. Plastic thins as it is stretched, and therefore, the effect of this thinning must be carefully considered in the part design.

Except for the profile processes (extrusion and pultrusion), the remaining processes are filling processes. For cold press and compression molding, the material is placed in the mold and the mold is closed on it. That eliminates the stress that results from forcing the material through a small opening (gate) under pressure, known as *gate stress.* However, the pressure processes are capable of producing the finest detail which is needed for many applications.

The following sections will be devoted to design details particular to each process. If the information provided does not resolve the reader's particular design issues, references which provide more information on these topics are listed at the end of the chapter. The processes will be addressed in alphabetical order.

8.3.1 Blow molding

This process is similar, in concept, to glass blowing. For *blow molding,* a preform, known as a *parison,* is either injection molded or extruded as a tube. The injection-molded preform then moves to another station where it is blown into a final shape as defined by the mold. For an extruded parison, two halves of the mold close on the tube and air is blown in. The air pressure causes the soft plastic tube to expand to the contour of the mold.

Blow molding can be used in several ways to create plastic parts. The process is best known for creating hollow shapes like bottles. However, this capability can also be used to create other hollow parts, such as large containers, which can be molded integrally with their covers that are cut off to form two parts. In another example, a hollow part could be cut lengthwise to create two opposite hand parts, however only their outer surfaces would be finished. The ends of the moldment could be removed from a hollow shape to create the final part such as is done with automotive air ducts. The walls can be molded close together to form a double-walled part for structural applications. For greater strength and rigidity, the space between the walls can be filled with foam.

Blow molding is a stretching process in that the material, which is first injection molded or extruded into a parison, is expanded to fill out the cavity. The ratio of the surface area of the expanded surface to that of its original surface is referred to as the *blow ratio,* which can be determined by

$$\mathrm{BR} = \frac{H}{D} \tag{8.1}$$

where BR = blow ratio

H = maximum height of projection perpendicular to the mold parting line (see Fig. 8.30)

D = smallest dimension at the base of the projection in the plane parallel to the parting line (see Fig. 8.30)

Since the part grows larger without the addition of any new material, the part wall thins in proportion to the expansion. Therefore, if a 1-in^2 area which is 0.100 in thick is expanded into a projection with a sur-

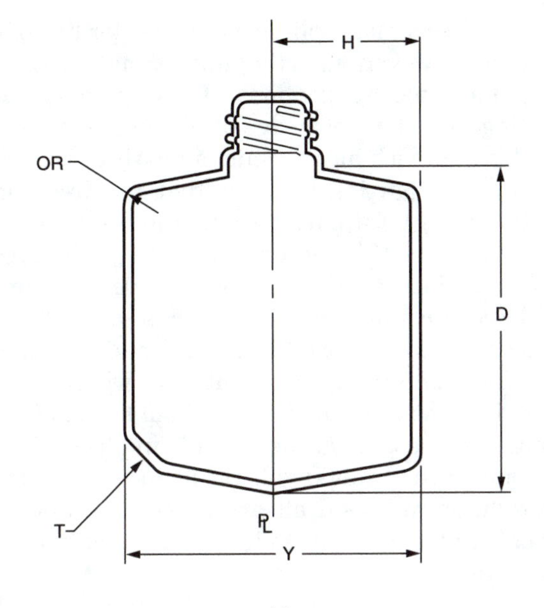

(a)

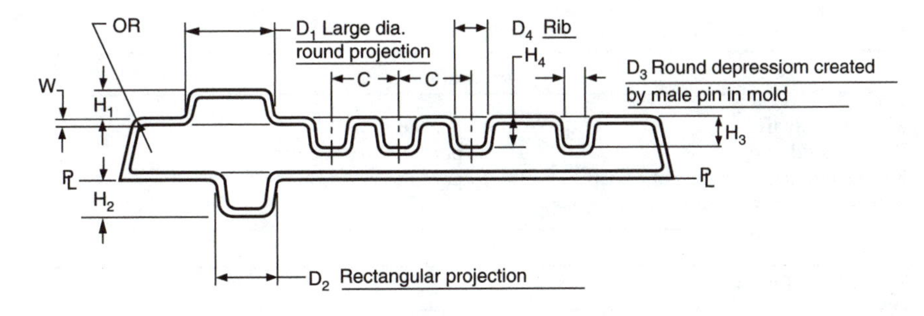

(b)

Figure 8.30 Design details for hollow parts: (*a*) center parting line and (*b*) edge parting line.

face area of 4 in², its wall will thin to an average of 0.025 in. The thinning is not uniform, since it is closer to the original, or parison, thickness where little stretching occurs and greatest where there is the most stretching. Therefore, both part projections and part recesses will have their thinnest sections at their outermost corners.

There are limits to how much the plastic can be stretched. For engineered parts, $1\frac{1}{2}$ to 2 is a comfortable blow ratio with a maximum of 4 to 5 depending on the material. However, much greater blow ratios are used for blown bottles, particularly those stretch blown from injection-molded parisons in thermoplastic polyester. The lowest blow ratios are required for parts that use a mold split down the center line of the moldment as in Fig. 8.30*a*. However, a parting line at the center of the part is not appropriate for some applications. Those products are better served with a parting line along one edge of the moldment as illustrated in Fig. 8.30*b*. This approach also permits a nearly sharp edge (0.020 to 0.040-in radius) at the parting line, but that requires higher blow ratios to achieve the same total part thickness and will result in wall thicknesses proportionately thinner on the blown side than on the unblown (parting line) side. Returning to Fig. 8.30*a*, the outside radii (OR), which will have the thinnest wall sections, are affected by the blow ratios. For average blow ratios, OR are acceptable in the range of 0.060 to 0.200 in. However, for high blow ratios or where the cross-sectional thickness (*Y*) is large, the outside radii should be 0.200 in or higher. Placing a chamfer at the edge of the part (*T*) in place of a radius (of the same dimensions as the radius) will result in reduced thinning of the wall. Corner radii as well as draft angles are also affected by the overall thickness of the part. Recommended draft angles and radii based on part thickness criteria are indicated in Table 8.10.

Referring to Fig. 8.30*b*, the proportions of the projection also affect the determination of the radii. For a round projection with a large

TABLE 8.10 Blow Molding Draft and Radii

(a) Draft

Minimum draft	1° per side
Recommended draft	2° per side (no texture)
For each 0.001-in depth of texture	1° per side additional
Increase draft as blow ratio increases	

(b) Radii

Typical Radius Values

Type	Description	Value, in (mm)
Tight	Smallest obtainable radii; possible only where little or no polymer stretching occurs. Generally, these radii are found at the parting line. May need zero radius in mold to obtain these values on part.	0.020–0.40 (0.51–1.02)
Average	Most radii will fall into this range. Use with average blow ratios.	0.060–0.200 (1.5–5.08)
Large	Used on tanks, boxes, and other components where cross-sectional part thickness is large. Also used with aggressive blow ratio geometries to minimize polymer thinning.	Over 0.200 (5.08)

SOURCE: J. Lincoln, *Engineered Blow Molding Part Design*, GE Plastics, 1989.

diameter, a blow ratio of 0.33 is the recommended maximum. Thus, the relationship between the height (H_1) and the diameter dimension (D_1) of the projection would be $H_1 \leq D_1/3$ or $D_1 \geq 3H_1$. Therefore, the relationship between the outside radius and the other dimensions becomes OR $\geq 0.1D_1$ or OR $\geq 0.3H_1$.

If the projection is rectangular instead of round, a blow ratio of 1.0 is the recommended maximum. Then the relationship between the height (H_2) of the projection and the dimension (D_2) across the short side of the rectangle is $H_2 \leq D_2$. In that case, the outside radius (OR) relationship becomes OR $\geq 0.1D_2$ or OR $\geq 0.1H_2$.

For round depressions, which are created by a male pin in the mold, larger blow ratios can be used. If the pin is to be stationary, the maximum recommended blow ratio is 2. Therefore, the relationship between the diameter (D_3) of the depression and its depth (H_3) is $D_3 \geq 0.5H_3$ or $H_3 \leq 2D_3$. For retractable pins, the blow ratio can be increased to 3, resulting in $D_3 \geq H_3/3$ or $H_3 \leq 3D_3$.

If the depression is a rib instead of a round, the recommended maximum blow ratio is 2.5, in which case the equations become $D_4 \geq 0.4 H_4$ or $H_4 \leq 2.5D_4$. The center-to-center (C) distance between the ribs should be greater or equal to the height of the rib (H_4) added to the width of the rib (D_4), or $C = H_4 + D_4$. The relationship between the rib dimensions and internal radius (IR) at the depth of the rib are as follows: IR $\geq 0.1D_4$ or IR $\geq 0.04H_4$.

Ribs, gussets, and crowning of the surface are commonly used to provide strength and stiffness to an engineered blow moldment as there is a tendency for flat sections to buckle inward or "oil can." Additional strength can also be achieved by having opposing rounds or ribs meet at the center of the part. This device is known as a *tack-off* and is illustrated in Fig. 8.31 at points *P, Q,* and *R*. The resulting combined wall thickness (S) should be 1.75W, but can range from 0.5W to 2.0W.

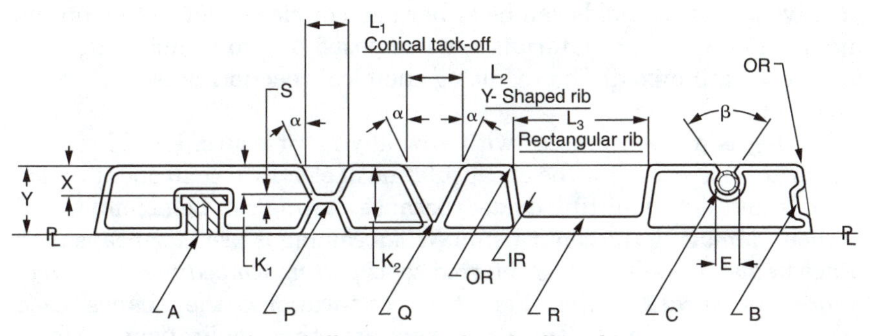

Figure 8.31 Design details for hollow parts with tack-offs and inserts.

The tack-off at P is a conical tack-off. Conical tack-offs provide compressive strength improvement and modest increases in stiffness and feeling of solidity. The cone angle (α) should be in the 25 to 30° range and the diameter (L_1) of the cone should be 1.2 to 1.4 times greater than its depth (K_1).

For greater rigidity and strength, rib tack-offs are preferred. Continuous ribs are very effective to prevent bending in one direction. Crossed ribs, staggered ribs, and various other configurations are also used depending on the application. The rib at Q is a v-shaped rib with the same side angle (α) as the conical tack-off, 25 to 30°. For an overall thickness (Y) of 0.75 in, the width (L_2) of the rib is 1.2 times greater than its depth (K_2) and the radius at the base of the rib would be 0.25 in.

For a rectangular rib like the one at R and for the same overall thickness, a 3.00-in rib width (L_3) would be typical with outside radii (OR) of 0.100 in and inside radii (IR) of 0.125 in. The sides of the rib would be drafted at 3 to 5° per side.

Metal reinforcement can be employed using stiffeners such as the one in detail C. In this case, the outside radius (OR) would be $0.1E$, the diameter of the reinforcing rod. The open angle (ß) would be approximately 70°. Molded-in inserts (A) and undercuts (B) are also possible, permitting the use of snap fits in some cases. Finally, hinge details can be compression molded at the parting line and hinged parts can also be created by drilling holes for the hinge pins.

Foam can be placed within the walls for further strength and stiffness. Foam densities control the property enhancement applied to the part, which is now a composite, and can range from 0.5 to 60 lb/ft^3. However, the usual range is 2 to 3 lb/ft^3.

8.3.2 Casting

Casting is a traditional process which served as the forerunner to many of the more sophisticated molding processes used in the plastics industry today. Since casting is a process which is free of negative or positive pressure, molds can be either open or closed depending on the application. Casting materials are composed of two components that are inert until mixed. The resulting chemical reaction causes the plastic to set.

Casting is a fill-in process with virtually no pressure. Liquid casting resins are often used for the encapsulation of electrical components, such as coils and light emitting diodes, because casting can be accomplished without damaging the components. When casting is used to encapsulate other items, it is sometimes referred to as *potting, embediment, impregnation,* or *encapsulation.* For those applications, the shapes used are usually very simple. However, there are other applications, such as furniture components and other decorative objects, which require very complex shapes with fine detail and varying wall thicknesses.

Heat is generated because the reaction of the resin and the catalyst is an exothermic reaction until cross-linking is complete. In some cases, additional heat is applied to speed the reaction. This heat results in shrinkage during curing, beginning where the heat of exotherm first generates and ending where the last area cures. This shrinkage is significant and difficult to control. Therefore, it is difficult to hold close tolerances with cast parts. The larger the parts are and the thicker the walls, the more difficult it is to hold tolerance. The shrinkage can be great enough to exert sufficient pressure on delicate electronic components to damage them or their wiring. On decorative parts, this shrinkage may result in warpage, sink marks, and internal stresses. In addition, shrinkage may not be uniform.

For nonencapsulation applications, the general rules regarding uniform wall thicknesses and the avoidance of thick sections and sharp corners apply. One characteristic unique to casting would be surface cracking due to very high localized exothermic heat. Furthermore, most of these materials are very brittle and ultra-thin sections can result in breakage due to handling. Also, the part must be designed so the mold can be built such that air does not become trapped which would result in a surface void. The special design details for casting would be the same as those used for compression molding.

Electrical encapsulation is practically a field unto itself with some very special design concerns. First, locating the component to be encapsulated such that it does not contact the outer wall is important as the area which made contact will be exposed. In some cases, it may be necessary to place a molded shell in the mold to be encapsulated to ensure a minimum protective wall. Also, the parts to be encapsulated may be contaminated with a material that could volatize with heat. Therefore, they should be cleaned before casting. It may be necessary to employ a vacuum bake for a period long enough to completely remove these volatiles. Finally, casting under vacuum may be necessary to prevent voids which could cause problems with high voltages or frequencies.

The requirement for assembly of castings is limited since casting can make complex shapes. The most common methods are adhesives and fasteners. Self-tapping screws can be used with most of the materials used for casting. Threaded inserts can be molded in place, however they are difficult to salvage in the event of a reject moldment because the material is a thermoset and the inserts cannot be melted out. Threaded inserts can also be glued or tapped in place.

Cast materials will usually need to have the gate or sprue removed by hand or machining methods. For appearance parts, additional surface finishing may be required. It may also be necessary to fill in surface voids. Decorative parts are often sprayed to represent a wood finish and can be marked with hot stamping, pad printing, silk screening, and heat transfers.

8.3.3 Co-injection molding

See Sec. 8.3.15.

8.3.4 Cold-press molding or cold molding

In the *cold-press molding* process, resin and reinforcement are placed in the mold and it is then closed, developing a small amount of pressure (on the order of 50 lb/in^2). The resin and catalyst react to set the plastic. This results in an exothermic reaction. Cold-press molding is a closed-mold fill-in process that is limited to shapes which are not too complex. It is used with a long fiberglass resin in the range of 15 to 25% glass content. It provides a modestly smooth surface which can be gel coated; however, considerable surface finishing may be necessary for an appearance surface without gel coating.

Thermoformed decorative inserts and labels have been molded into cold-press molded parts; however, threaded metal inserts cannot be molded-in. Threaded inserts can be glued or tapped in place, but not into bosses, which are impractical. Molded-in holes, core pulls, slides, undercuts, and ribs are, likewise, unfeasible. Holes need to be drilled and their location can be held to ±0.030 in. The same assembly methods used for parts made from compression molding can be used for those made from this process.

Parts which are cold-press molded cannot be trimmed in-mold. They are generally trimmed with a router to a minimum tolerance of ±0.060 in, although water-jetted trimlines can be toleranced to ±0.040 in. However the part is trimmed, it is expensive to request a tolerance tighter than is required for the application.

The minimum inside radius for cold-press molded parts is 0.25 in. The range of wall thicknesses it can create is 0.080 to 0.500 in, with a normal wall thickness variation of ±0.020 in. A 100% wall thickness variation can be tolerated. For parts under 6 in in depth, a 2° per side draft is required (1.5° minimum); beyond that depth, the draft should be 3 to 5° per side.

8.3.5 Compression molding (preforms, BMC, SMC, LPMC), transfer molding, and thermoset injection molding

Compression molding is a closed-mold filling process in that the resin is placed in the mold in the form of powder, pellets, preforms, bulk molding compound (BMC), sheet-molding compound (SMC), or low-pressure molding compound (LPMC). Pressure is applied to the resin with heated molds causing the now molten resin to fill out the mold cavity. Compression molds are built of steel because molding pressures range from 900 lb/in^2 up for all the systems except LPMC, for which molding pressures can be as low as 200 lb/in^2. LPMC molds are often made of aluminum. In either case, the closed mold permits both sides of the part to be finished.

There is a major division in applications between the processes which use short-fiber–reinforced resins (pellets) and those which use long-fiber–reinforced resins (preforms, BMC, SMC, and LPMC). For long-fiber–reinforced materials, the compression-molding processes are usually referred to by the type of reinforcement used. Design details differ somewhat between them. For unreinforced or short-fiber–reinforced (0.125 in or less) resins, the compression molding process finds its largest market in products such as electrical components and temperature-resistant housewares (plates and ashtrays). When fine detail is required, a variant known as *transfer molding* can be used in which the resin is first placed in a transfer pot and then forced through a sprue and runner system into the cavity much like injection molding. Transfer molding must compete with thermoset injection molding for its applications. Both are limited to short-fiber–reinforced materials.

Design details for short-fiber molding. Short- (0.125 in and under) fiber–reinforced resins are used with compression, transfer, and thermoset injection molding. The shorter fibers permit finer detail and thinner walls than can be molded with the long-fiber resins, although compression is more limited than injection or transfer molding. Permissible wall thicknesses also vary with the material. Table 8.11 provides general recommendations for the most common short-fiber molding materials. However, these recommendations can be stretched a bit for special circumstances. With reference to Fig. 8.32, wall thickness (W) can range from 0.015 in (for short distances) to 1.500 in (resulting in very long molding cycles); the usual range for most applications is 0.060 to 0.125 in. The wall thickness may be expected to vary by 5%, however, it is more difficult to control in the direction the mold opens. Therefore, the minimum tolerance for a wall in a plane parallel to the parting line would be ±0.005 in, whereas the minimum tolerance for a wall perpendicular to the parting line would be ±0.003 in. In general, these processes can handle the tightest tolerances of any of the plastics processes.

Thermosets do sink, however, they have much less tendency to do so than thermoplastics. Therefore, the rib thickness (B) for these processes can be equal to the nominal wall thickness (W). Rib heights (C) can

TABLE 8.11 Suggested Wall Thickness for Thermosetting Molding Materials

Thermosetting materials	Minimum thickness, in	Average thickness, in	Maximum thickness, in
Alkyd—glass filled	0.040	0.125	0.500
Alkyd—mineral filled	0.040	0.187	0.375
Diallyl phthalate	0.040	0.187	0.375
Epoxy glass	0.030	0.125	1.000
Melamine—cellulose filled	0.035	0.100	0.187
Urea—cellulose filled	0.035	0.100	0.187
Phenolic—general purpose	0.050	0.125	1.000
Phenolic—flock filled	0.050	0.125	1.000
Phenolic—glass filled	0.030	0.093	0.750
Phenolic—fabric filled	0.062	0.187	0.375
Phenolic—mineral filled	0.125	0.187	1.000
Silicone glass	0.050	0.125	0.250
Polyester premix	0.040	0.070	1.000

SOURCE: R. D. Beck, *Plastic Product Design,* Van Nostrand Reinhold Company, Inc., New York, 1980.

be 2 to 3 times the nominal wall thickness, however, 2 is better from a molding standpoint. Inside radii (IR) should be 25 to 50% of the nominal wall thickness, but can be as low as 10% in extreme conditions where stresses are low. The outside radius (OR) should be equal to the IR + *W*.

Posts and bosses can also have a wall thickness at the base equal to the nominal wall thickness, and their height dimension can be permitted to reach 4 times the wall thickness. These processes can make hollow bosses, blind holes, and through holes both parallel and perpendicular to the parting line. However, the knit line side of the hole is likely to have less reinforcement, a condition known as "resin rich." It is best to avoid placing molded-in holes within a two diameter distance of the outside edge of the part. If holes are needed in those

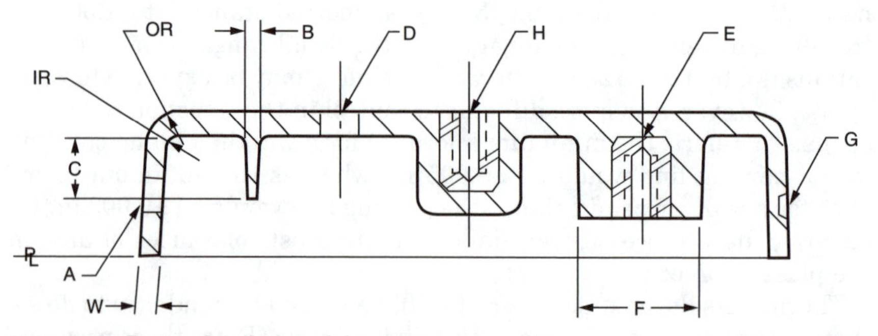

Figure 8.32 Design details for solid parts.

locations, it is best to drill them as a secondary operation. The solid material thickness between two holes should not be less than one diameter (the largest).

Resin transfer and injection molding are both closed-mold processes, so their edges do not require trimming in most cases. However, compression-molded parts may or may not be trimmed in-mold depending on the mold and the application. If not trimmed in-mold, the parts will have to be deflashed and, for some appearance applications, machined to a clean surface. Draft should be 3 to 5° per side, but the process can handle 1° per side for shallow depths. If the use of automatic molding equipment is essential to the economics of the project, the 3-in part depth limitation for these machines may be significant. The length-to-diameter ratio for cores should not exceed 2:1 for compression molding or 4:1 for injection and transfer molding.

It is possible to strip undercuts for parts molded of these processes in some cases and threads can be molded-in. However, side actions or split cavities, as would be required for details *A* and *G*, are more commonplace. Threaded inserts, like those at *E* and *H*, can also be molded-in. However, it may be more advantageous to emplace them after molding with adhesives since the molding cycle will be shorter and the risk of one misplaced insert ruining an entire moldment is eliminated. The other common methods of assembling compression, transfer, and thermoset injection-molded parts are self-tapping screws (holes can be tapped), adhesives, snap-fits, press-fits, and the usual screw and bolt techniques. None of the plastics welding techniques can be used with thermoset parts.

Design details for long-fiber molding. Whereas short-fiber molding is best suited to smaller, complicated shapes, long-fiber molding applications tend to be large, simple parts. Fiber lengths of 0.125 in and over can be regarded as long. Compression molding with preforms, sheet-molding compound (SMC), low-pressure molding compound (LPMC), and bulk molding compound (BMC) produce long-fiber–molded parts. Preform glass content ranges from 15 to 50% with fiber lengths from 0.125 to 3.00 in. Glass content usually ranges from 25 to 35% for SMC and LPMC, however it can reach 60 to 80% in some cases with the usual range in fiber length of 1.00 to 2.00 in. BMC uses a mixture of glass fibers from 0.125 to 0.500 in in length in the 10 to 20% glass content range. It has the lowest strength because it has the lowest glass content.

Preform molding has the highest strength of these three systems. It is best for uniform wall designs requiring deep draws. SMC is better able to handle ribs, bosses, and thinner walls. LPMC is a variety of SMC which is modified to require less molding pressure, enabling it to use softer tools (aluminum) at a lower cost. It follows the same design rules as SMC. BMC can also mold ribs and bosses, however, it is best for thick-walled parts which do not exceed 2 ft in length.

As a closed-mold process, both surfaces of compression-molded parts can be finished. Draft for long-fiber compression-molded parts should be 3° per side, but less can be used for depths under 6 in—down to 1° per side in extreme cases. Molded-in holes are feasible in the plane of the parting line as are bosses, corrugated sections, molded-in labels, and raised lettering. Molded-in holes within a two-diameter distance from the outside edge of the part will need to be drilled as a secondary operation. The solid material thickness between two holes should not be less than one diameter (the largest). Core pulls, slides, split molds (for external undercuts), and metal inserts are recommended only for the SMC and BMC varieties.

With respect to Fig. 8.32, the nominal wall thickness (*W*) can vary from 0.045 to 0.250 in for preform-molded parts, 0.080 to 0.500 in for SMC-molded parts, and 0.060 to 1.000 in for BMC parts. The process can handle variations of 100% for preform molded parts, even more for BMC and SMC (300%); however, the transition between wall thicknesses must be gradual. Wall thickness variations of ±0.010 in can be anticipated. In general, these processes can handle tolerances nearly as tight as those for the short-fiber processes.

Ribs (*B*) and metal inserts (*E, H*) are not recommended for preforms, but they can be used with the other versions. The minimum inside radius (IR) is 0.125 in for preform-molded parts and 0.062 in for SMC- and BMC-molded parts. Long-fiber compression-molded parts can be trimmed in the mold. For Class A finishes in SMC parts, the inside radius at the base of a rib should be 0.020 in; otherwise it should not be less than 0.060 in. The thickness at the base of the rib (*B*) can be equal to the nominal wall thickness (*W*), however, it must not exceed 0.75*W* under Class A surfaces. The height of the rib (*C*) should be 2 to 3 times the nominal wall thickness (*W*) and the draft should not be less than 1° per side except for ribs under Class A surfaces (which can handle the cost premium) where it can go down to 0.5° per side. These same rib design rules apply to boss wall thicknesses for these parts. In-mold coating can be used for high-class finishes.

The same assembly methods as described for short-fiber compression-molded parts apply to long-fiber compression-molded parts. However, since long-fiber–molded parts are usually much larger than short-fiber–molded parts, it is more difficult to use molded in metal inserts and tapped holes. Therefore, metal plates with tapped holes or studs, known as *tapping* or *stud plates,* are sometimes placed behind the wall to provide a means of fastening (with the plastic part sandwiched between). Rivets and U or J nuts can also be placed in recesses or on flanges. The SMC Automotive Alliance recommends "Hi-Lo" screws for SMC applications.

8.3.6 Extrusion and co-extrusion

An *extruder* consists of a heated chamber (barrel) with a screw located inside of it. At one end of the chamber, resin is fed into a hopper which deposits into the chamber. As the screw turns, the resin moves toward the other end of the chamber melting as it goes. There, the material moves through the screens and the breaker plate to the die. There is a die at the end of the chamber which causes the melt to take its shape as it passes through it. Beyond the die is a cooling system composed mainly of a tank of cooled water through which the extruded shape travels, hardening as it goes. When it is sufficiently cooled, it is either cut to shape, if rigid, or wound on a reel, if flexible. Extrusions can be created in the form of film, sheet, tubing, gasketing, profiles, etc.

When two extruders are used in tandem, two compatible resins (or two colors of the same resin) can be extruded together, either as a coating of one over the other, as for a laminate, or side by side, as for a two-color sheet. In practice, this is done at a stage where the two resins can be melted together to form a weld. This technique is known as *co-extrusion* and is often used to create barrier layers.

When not being used in conjunction with another process, like thermoforming or blow molding, profile extrusions are the variety most often used by product designers. They can be either solid or hollow and, for the most part, are made of thermoplastic materials (although thermoset materials can be extruded with special equipment). Therefore, the effect of nonuniform cooling is precisely the same for extrusion as it is for the molding processes: sink and distortion.

There are three designs illustrated in Fig. 8.33. The design in Fig. 8.33*a* is a very poor one. Wall thickness B is more than double the nominal wall A. That will result in a sink such as that represented by the dashed lines. The maximum it should be is 25% greater than the nominal wall ($B \leq 1.25A$). Furthermore, the inside radius C_1 is nonexistent. The minimum practical inside radius is 0.010 in; however, it really should be 25 to 50% of the nominal wall as discussed in Sec. 8.2.3. As for the outside radius E_1, the minimum practical outside radius for extrusion is 0.016 in.

If the requirement of the application is such that dimension B must be maintained, the design can be improved as shown in Fig. 8.33*b*. Here, the large center block of material has been cored out to alleviate the sink condition by creating a wall variation that does not exceed 25% of the nominal wall ($D \leq 1.25A$). This rule has been used both to create the

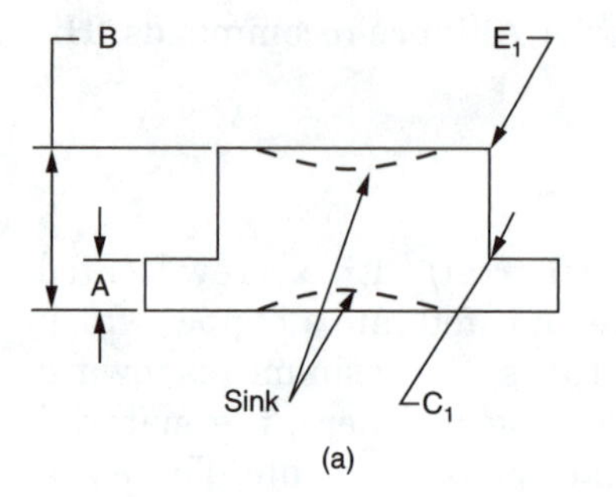

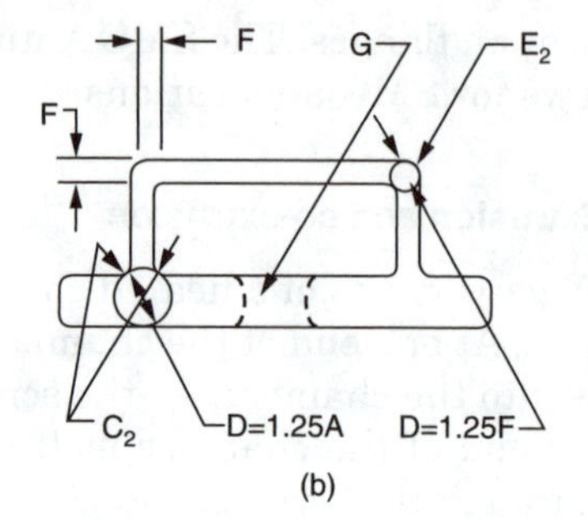

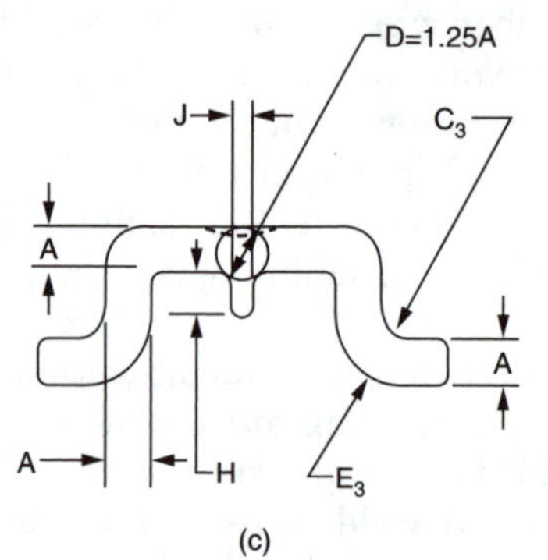

Figure 8.33 Design details for extrusions and pultrusions: (*a*) poor design; (*b*) improved design; and (*c*) best design.

wall thickness F and the new outside radius, E_2 ($D \leq 1.25F$). The inside radii at C_2 have been increased to 25% of the nominal wall (A).

The design in Fig. 8.33*b* still leaves much to be desired. The cored-out hole in the center is created by a mandrel which requires supports known as *spiders.* The melt recloses beyond the spiders, but knit lines are created which result in weak areas in the extrusion. That problem can be resolved by creating an opening (G) in the surface as represented by the dashed lines in the bottom of the part.

The design in Fig. 8.33*c* is an improvement on the concepts illustrated in Fig. 8.33*b*. Here the bottom cross leg has been removed. That permits the upper portions to become nominal wall thickness (A). The inside radii (C_3) have been increased to equal the nominal wall thickness, which is the ideal configuration. The outside radii are the inside radii plus a wall thickness ($E_3 = C_3 + A$). A stiffening rib has been added at the center of the part. Its height should not exceed the nominal wall thickness ($H \leq A$). Its thickness (J) should be designed using the same parameters as elsewhere where there are material build-ups ($D = 1.25A$). If the rib is greater than described by this equation, a sink will appear where there is a dashed line. This can be masked by a designed recess located where the sink will appear, a rib on the opposite side, or a series of serrations.

Tolerances are difficult to maintain with the extrusion process because the part must be supported with fixtures to retain its shape while it cools. Rigid resins, like rigid vinyl and polystyrene, hold tolerances better than softer materials, such as polyethylene. Table 8.12 provides a good basis for tolerancing extrusions.

Co-extrusion can be used to create products with areas exhibiting significantly different characteristics. Examples would be a rigid strip with a soft edge, a polyethylene hose with a soft vinyl exterior, or a rigid vinyl bumper strip with a flexible bumper surface in another color. Hollow cross sections are much easier to co-extrude than they are to extrude in the normal fashion since one side can be made of another material, eliminating the need for a mandrel supported by spiders. The rigid segment of a co-extrusion must be at least 0.020 in thick, whereas the flexible portion can be as thin as 0.005 in. The maximum thickness for flexible sections is 0.250 in. For rigid materials, sections thicker than 0.250 in can be made, however such sections create a flow

TABLE 8.12 Tolerance Guide for Plastic Profile Extrusions

	Material					
	Rigid vinyl	Poly-styrene	ABS, PPO, poly-carbonate	Poly-propylene	EVA, flexible vinyl	Poly-ethylene*
Wall thickness, % ±	8	8	8	8	10	10
Angles, (degrees ±	2	2	3	3	5	5
Profile dimensions, inches ±						
To 0.125	0.007	0.007	0.010	0.010	0.010	0.012
0.125 to 0.500	0.010	0.012	0.020	0.015	0.015	0.025
0.500 to 1	0.015	0.017	0.025	0.020	0.020	0.030
1 to 1.5	0.020	0.025	0.027	0.027	0.030	0.035
1.5 to 2	0.025	0.030	0.035	0.035	0.035	0.040
2 to 3	0.030	0.035	0.037	0.037	0.040	0.045
3 to 4	0.045	0.050	0.050	0.050	0.065	0.065
4 to 5	0.060	0.065	0.065	0.065	0.093	0.093
5 to 7	0.075	0.093	0.093	0.093	0.125	0.125
7 to 10	0.093	0.125	0.125	0.125	0.150	0.150

*Low density and regular grades.

SOURCE: R. D. Beck, *Plastic Product Design,* Van Nostrand Reinhold Company, Inc., New York, 1980.

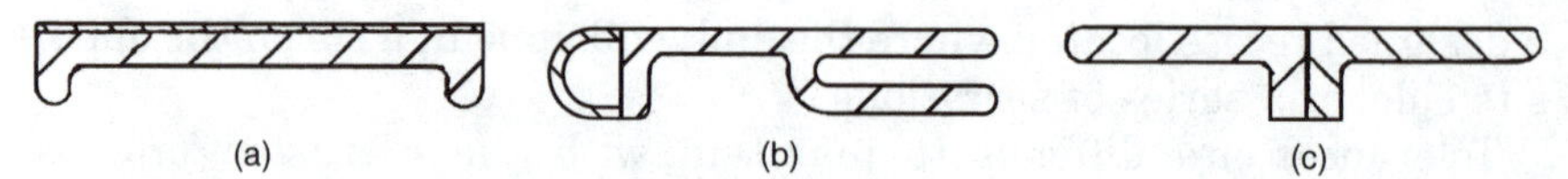

Figure 8.34 Co-extrusion design examples: (*a*) cap sheet; (*b*) hollow bumper; and (*c*) two-color strip.

imbalance during extrusion. Figure 8.34 depicts several examples of co-extrusion designs.

8.3.7 Gas-assisted injection molding

Gas-assisted injection molding is essentially similar to injection molding, except that in gas-assisted injection molding, a gas is injected into the center of the wall as the material enters the mold. A hollow core in the center of the thick sections is formed, thus creating a box-like geometry in cross section with much higher strength than a thinner solid wall using the same amount of material. This results in a much stronger part than a solid walled part of the same weight. It also permits a far wider variation in wall thickness than traditional injection molding and a thinner wall thickness than any of the structural foam processes. Figure 8.35 shows examples of gas-assisted injection molding channels.

The diameter (D) of a round gas channel should not exceed 1 in ($D \leq 1$ in). Gas channels in wall sections are determined experimentally and are very much a function of the wall thickness. The height (H) of the gas channel should not exceed 3 to 4 times the nominal wall (T) of the part ($H \leq 3$ to $4T$). The width (W) should not exceed 2.5 to 3 times the nominal wall (T) ($W \leq 2.5$ to $3T$). The basic rules regarding uniform wall and inside radii outlined in Sec. 8.2 apply to gas-assisted injection molded parts, except for the gas channels. However, careful consideration must be given to gate location, gas nozzle location, and channel path location to assure proper application of the principle. The prospective molder should be consulted during the design phase to get it right.

8.3.8 Injection molding

The design rules discussed in this section refer to the injection molding of thermoplastics. For the design rules for thermoset injection molding, please refer to Sec. 8.3.5.

In its basic form, a reciprocating screw injection molding machine (the most common variety) consists of a heated cylinder with a reciprocating screw inside and a large clamp into which the mold is mount-

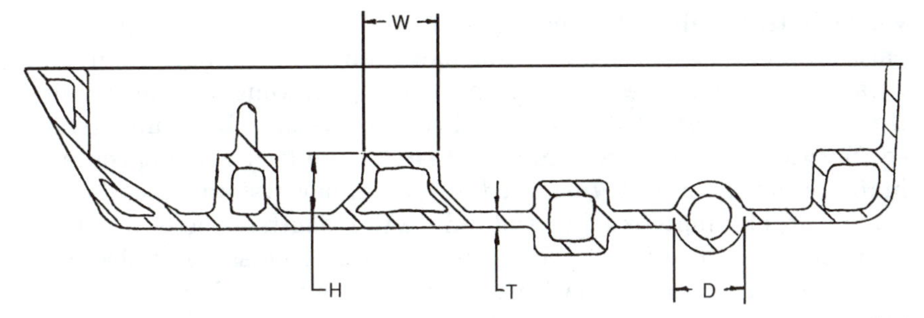

Figure 8.35 Examples of gas-assisted injection molding channels.

ed. Resin is fed into one end of the heated cylinder from a hopper. As the screw turns, the resin travels through the cylinder melting as it goes. When it reaches the nozzle end, the screw travels forward injecting a shot of melted resin into the mold. The clamp then opens, exposing the newly molded part which is adhered to the mold core. Ejector bars on the molding machine then move forward actuating an ejector mechanism (of which there are several types) which pushes the parts off the core into a basket located under the mold. The mold then closes and the cycle is repeated.

Injection molding is capable of accommodating the broadest range of materials. Nearly all the thermoplastics can be injection molded and, with special equipment, even many thermoset materials can be used. The mold capabilities of the process are also a major asset. Moldmakers have succeeded in constructing incredibly complex molds using side actions to create holes perpendicular to the parting line, split cavities for unusual shapes, and cores which collapse to permit withdrawal from undercuts.

The use of high molding temperatures and injection pressures with resins which are heat sensitive requires that the part designer be familiar with the parameters of the process. During processing, the material will degrade slightly under normal conditions. Recycling of sprues, runners, and reject parts, known as *regrind,* means that a portion of the material will have been through the process previously. When the amount of regrind is small, the effect of accumulated degradation will be negligible unless high temperatures and pressures are used. High temperatures increase the amount of degradation and high pressures can increase the amount of shear heat generated as the melt passes through constricted areas. The need for temperature and pressure increases is determined by the viscosity of the melt and the flow path, which extends from the nozzle of the cylinder to the furthest reaches of the furthest cavity. The melt cools as it passes along the flow path and an outer skin of hardened material causes a further restricted flow channel. Sharp turns and thin walls result in pressure drops which, in turn, call for higher temperatures and pressures. In part, the part designer controls this phenomenon by the configuration and details of the part design. (The mold designer controls the balance through the design of the sprue and runner system.) It should be noted that the use of regrind can also permit the introduction of contaminants into the process. Insulated or hot runner systems, installed at extra cost, eliminate regrind and reduce the length of the flow path.

Molders also use increased temperature and pressure to decrease the cycle and reduce the machine cost portion of the molding cost. High temperatures and pressures can work for a part perfectly designed for injection molding and in a mold with excellent cooling characteristics, particularly if the part has no tight tolerances. However, the smallest deviations from ideal part design can result in severely distorted parts. When this occurs, the molder must reduce these parameters and extend the cycle. For very fine tolerances slow cycling would be required for parts of the best design. Thus, tight tolerances and improper part design result in more expensive piece parts.

From the previous discussion, it should be clear that injection molding is process sensitive. This means that critical dimensions with tight tolerances can go out of tolerance during a molding run if proper control is not maintained. It also requires that quality be checked on a regular basis.

Standard injection molding. The design rules described in Sec. 8.2 apply to injection-molded parts. Referring once again to Fig. 8.32, the nominal wall (W) may range from a minimum of 0.040 in to a maximum of 0.500 in. Even thinner walls, down to 0.010 in, have been molded, but only for very short distances in the easiest flowing materials (like nylon). The typical nominal wall, however, will be in the range of 0.060 to 0.125 in. Nominal walls are commonly held to ±0.005 in in injection molding. Suggested wall thicknesses for thermoplastic molding materials are shown in Table 8.13.

Injection molding is a closed-mold process, therefore, both sides of the part are finished. Textures, raised letters and numbers, and molded-in labels are all done frequently. Ribs (B) and bosses (F) are feasible and metal inserts (E) can be molded-in, although it is often more efficient to add them after molding. While the rib configurations are discussed in detail in Sec. 8.2.4, the usual starting point for the base of the rib (B) to be set is at one-half the nominal wall thickness ($B = 0.5W$) with a rib height (C) of 3 times the wall thickness ($C = 3W$) and an inside radius at the base of the rib of $0.25W$. We then check to be certain the draft has not reduced the wall thickness at the tip of the rib to an unacceptable level and adjust accordingly if necessary. Hole and boss design are discussed in Sec. 8.2.7.

TABLE 8.13 Suggested Wall Thicknesses for Thermoplastic Molding Materials

Thermoplastic materials	Minimum, in	Average, in	Maximum, in
Acetal	0.015	0.062	0.125
ABS	0.030	0.090	0.125
Acrylic	0.025	0.093	0.250
Cellulosics	0.025	0.075	0.187
FEP fluoroplastic	0.010	0.035	0.500
Nylon	0.015	0.062	0.125
Polycarbonate	0.040	0.093	0.375
Polyester TP	0.025	0.062	0.500
Polyethylene (LD)	0.020	0.062	0.250
Polyethylene (HD)	0.035	0.062	0.250
Ethylene vinyl acetate	0.020	0.062	0.125
Polypropylene	0.025	0.080	0.300
Polysulfone	0.040	0.100	0.375
Noryl (modified PPO)	0.030	0.080	0.375
Polystyrene	0.030	0.062	0.250
SAN	0.030	0.062	0.250
PVC—rigid	0.040	0.093	0.375
Polyurethane	0.025	0.500	1.500
Surlyn (ionomer)	0.025	0.062	0.750

SOURCE: R. D. Beck, *Plastic Product Design,* Van Nostrand Reinhold Company, Inc., New York, 1980.

Draft angles are often specified as 1° for depths up to 6 in; however, more draft (2 to 3°) will permit more economical molding and should be used whenever possible. For greater depths, a draft of 3° + 1° for every additional 6 in is recommended. When the surface perpendicular to the parting line is to be textured, the preferred draft is 1.5° + 1.5° for every 0.001 in of texture depth. However, this can be reduced to 1° + 1° for every 0.001 in of texture depth in some cases.

Thin-wall molding. The molding of plastic parts with very thin walls is not really a new phenomenon. Small parts have been molded with thin walls for many years. However, thin-wall molding over large areas, which is a recent development in injection molding, has altered the design rules somewhat for that special area. This technique has been driven by the race to reduce the weight and size of cell phones and laptop computers. *Thin-wall molding* is defined as molding of wall thicknesses less than 0.060 in, although wall thicknesses as low as 0.040 in have been molded. The principal materials used are amorphous thermoplastics such as polycarbonate, modified PPO, and ABS. The operating concept is to inject the resin at a very high speed so that it can fill out the cavity before the solidified skin can form to restrict the flow path. Hot runner systems are used to bring the hottest possible material to the cavity and multiple gating is employed. A vacuum is used to evacuate the air from the cavity and eliminate the resistance the air could create, and slip coatings are used in the mold. This technique requires the use of molds which are much hotter than normal injection molding.

It is important to note that the standard published shrink rates do not apply for these very thin-wall thicknesses. The normal published shrink rate for polycarbonate, a material frequently used for thin-wall molding, is 0.005 to 0.007 in/in. However, published shrink rates are typically based on a 0.125-in wall thickness. As the wall thickness decreases, the shrinkage also decreases and can reach levels of 0.001 to 0.003 in/in. Also, engineers must be aware that there will be some grades of a given resin which are better suited to thin-wall molding than others. The resin supplier and the anticipated injection molder should be consulted before attempting to design a thin-wall moldment for the first time.

With wall thicknesses in the range of 0.040 to 0.060 in, it is clearly impossible to adhere to all the fundamental rules of proper injection-molded part design (rib thicknesses will be greater than 70% of the nominal wall). This means that surface finish problems will result and texture is required to mask these flaws. In any case, wall thickness variations should not exceed 25%, inside radii should not be permitted to go below 0.020 in, and design details that restrict flow should be avoided. Furthermore, the thinner parts are less rigid at the time of ejection from the mold. Thus, much larger ejector pins must be used than would be used for normal injection molding and there would be more of them.

Thin-wall molding has special requirements and not all molders are aware of them. Reject parts can be the result of poor molding technique. High temperatures can cause degradation or high stress that can result in cracking after assembly. Overdrying of the resin or excessive residence time (too long in the hot cylinder) can cause a color shift between parts. Velocity-related rejects include variations in surface sheen that can be the result of thick to thin areas, texture smears that can be due to overpacking of the material or thin to thick wall thickness transitions, and sink marks resulting from ribs at the end of the flow path and thin to thick wall thickness transitions.

Molding of filled materials. Fillers and reinforcements are commonplace in materials used for injection-molded parts. Calcium carbonate, talc, and glass fibers are materials often used for this purpose. There are other additives used for ultra-violet light protection, flame retardance, ease of molding, coloring, etc., however they typically constitute small portions of the total compound and do not significantly affect the part design. (That is not to suggest they have absolutely no effect.)

The important thing to remember is that the filler replaces plastic with another material (a 30% glass-filled resin is only 70% plastic). That material will have physical properties which are different from the plastic. It will not heat or cool at the same rate as the plastic nor will it have the same coefficient of linear thermal expansion. Fillers, such as calcium carbonate or talc, tend to decrease the shrink rate significantly, however they also are likely to result in a lower tensile strength. This knowledge can be used to advantage when, for example, a part is undersize, as the addition of talc will result in less shrinkage and a larger part without changing the mold. This reduction in shrinkage rate can also be used to eliminate an unsightly sink where the wall thickness build-ups have led to nonuniform cooling.

Glass fiber reinforcement will have a similar effect, but it is too expensive to use simply for that purpose. It is typically used to increase strength and stiffness. However, glass fibers cool at a significantly different rate than plastics and this can result in greater part distortion. Furthermore, the glass fibers do not always flow well into ribs, bosses, and to the side of a molded-in hole opposite the direction of flow. This results in sections of the part which are not as strong as the remainder of the part. Finally, glass-filled materials do not flow in the mold as readily as unfilled resins.

8.3.9 Lay-up or spray-up

Lay-up and *spray-up* are open-mold processes generally used to produce large thermoset parts in low volume (automobile bodies and boat hulls). In essence, they are really one process, with the names referring to the manner in which the reinforced resin is applied. The very large size capability of this process permits the economy of many parts being molded as one.

The process consists of a thin, unreinforced outer coat called a *gel coat.* The reinforcement (normally glass fibers) mixed with resin (usually polyester) is then applied by hand (lay-up) or via a spray gun (spray-up). When the desired amount of resin and reinforcement is in the mold, the air bubbles are pressed out with a roller. The part is then left to set up which, depending on the size of the part and the amount of resin, can take more than an hour. Consequently, many molds are used when higher production volumes are required. The molds themselves are made of plastic cast from a pattern. As such, they have a very short life and must be replaced frequently. Fortunately, they are inexpensive; unfortunately, they are not capable of producing fine detail or close tolerances. Since the mold has only one side, only that side of the part has a finished surface; the other side is very rough. The mold can be textured and the part will reproduce the textured surface. Raised numbers and letters can also be molded-in and labels can be molded into the surface.

Referring again to Fig. 8.32, the nominal wall (*W*) can range from a bare minimum of 0.060 to over 1.500 in (there really is no technical limit) with a variation of ±0.020 in. Holes (*A, D*) are not economically feasible to mold-in and are usually cut in as a postmolding operation, but large openings can be molded-in. Parts cannot be trimmed in the mold, but cores can be placed in the mold and removed from the part by hand after molding. Metal inserts (*H*) and bosses (*F*) can also be molded into the part, but only from the outside surface. External ribs and internal wood or metal stiffeners can also be molded-in. When needed, internal ribs, bosses, and other structures are usually glued in place or corrugated structures are used to increase strength and stiffness. The use of reinforcement mats strategically located in the wall may eliminate the need for structural details. Outside undercuts (*G*) are possible using slides or a split mold. Although it is possible to mold some parts with no draft, at least a 1 to 2° draft is desirable for economical molding.

8.3.10 Machining

Plastics can be machined just like other materials and, often, with less scrap because the stock is more readily formed to a size closer to the finished shape. *Machining* is the process which can provide the tightest tolerances because of the shrinkage inherent in the other processes. The use of CNC machining centers makes reproducibility very high. However, one-off prototypes are most economically made with traditional machine tools because the setup cost is lower.

The greatest limitation to the use of machining is the lack of a wide selection of materials available. Stock must be extruded or cast from the basic resin, and this is only done for a limited number of resins and rarely for more than one grade of each resin. A limited number of sizes are processed in this fashion, and only the most popular ones are stocked by most distributors. Consequently, it may be necessary both to order a quantity greater than required and to wait for it to be shipped to you. Thus, the cost of the stock can become a disproportionate part of the cost of the finished piece.

The largest volume of machining of plastics is done as secondary operations for other processes. With the exception of the more precise molding processes, such as injection molding, most of the plastics processes must have their openings machined (usually routed) and their excess material trimmed off. Even injection molding requires gates to be trimmed for the majority of applications. For applications of moderate volume, it may be most cost efficient, when mold amortization is taken into consideration, to mold only the outer shape and machine-in holes and other details.

Machining plastic is somewhat different than machining other materials, largely because of the effect of the heat generated during the process. The temperature of the plastic can rise high enough to melt or cause degradation of thermoplastics (thermosets are affected to a lesser degree). Melting can cause the molten plastic to adhere to the cutting tool causing it to gouge more material from the stock, thereby causing the loss of dimensional control. Degradation can cause the loss of physical properties which can result in premature failure of the part. Even when temperatures are not permitted to rise that high, they can still cause the stock to expand. Upon cooling, the part is out of tolerance. Material suppliers provide data on speeds and feeds to be used with their resin along with the ideal cutting tool configuration. These should be followed precisely. Also, the use of coolants may be restricted to those which will chemically react with the plastic. Finally, annealing, both before and after machining, is recommended to relieve the stresses created in forming the stock and in the machining process itself.

One of the problems associated with machining plastics is that the wall thickness may be too thin, or the material too soft, to withstand the pressure of the clamp without distortion or disfigurement. There are several potential remedies in use. One is to construct a special fixture to support the stock during machining. Another is to adhere the material to a block of sturdier material (wood) which can be machined with the plastic and discarded upon completion. Plastics can also be frozen to increase rigidity for machining. However, the heat of machining does soften the stock and the contraction due to the drop in temperature must be taken into consideration. Surface finish can be protected through the use of pads on the clamp, although some clear materials (acrylic) are available with paper adhered to the surface.

Machined thermoplastics are readily joined with fasteners, solvents, adhesives, and a variety of welding techniques. Machined thermosets are assembled much the way wood parts are assembled. Screws and adhesives are the principle means. Ultrasonic welding, spin welding, vibration welding, hot plate welding, induction/electromagnetic heating, and hot gas welding are not usable for thermosets since they are all methods which require the plastic to be melted. Self-tapping screws are available as are metal inserts, and they can be either the expansion type or the type that can be glued in place.

8.3.11 Pultrusion

The shapes produced by pultrusion are essentially similar to those created by extrusion. Whereas the plastic is pushed through a die in extrusion, it is pulled through a resin bath and then a heated die in the pultrusion process by means of glass fiber strands, which provide the reinforcement. As in extrusion, they are then cut to length.

Pultrusion is generally specified for high-strength applications, such as I-beams, using thermoset materials. However, thermoplastic materials can be used and pultrusion does compete with extrusion for medium-strength applications.

As shown in Fig. 8.33, the minimum wall thickness (*A*) is 0.080 in and the maximum can range to 0.500 in and beyond. The wall thickness can be held to a tolerance of ±0.010 in and it can vary by 100%. The minimum inside radius (*C*) is 0.060 in. Ribs (*H*) are feasible and both sides of the part can be finished.

8.3.12 Reaction injection molding (RIM)

Reaction injection molding creates its own resin by mixing two highly reactive liquid components which react with each other by high-pressure impingement. The reaction is completed shortly after the resin is injected into the mold. This is accomplished at a pressure of 100 lb/in^2, which permits the use of electro-formed and vapor-formed nickel lined epoxy, aluminum, and kirksite tools at a considerably reduced tooling cost. (Machined and plated cast steel molds can also be used.) As a closed-mold process, both sides of the part are finished and can have textures, raised numbers, and raised letters, but molded-in labels are not feasible. The finished part can be trimmed in the mold.

The principal material used for RIM is polyurethane, however nylon, polyethylene, epoxy, and polyester can also be used. Chopped glass fibers and inorganic fillers have been added to these materials to increase strength and stiffness. The process then becomes known as *reinforced reaction injection molding* (RRIM).

Referring again to Fig. 8.32, the nominal wall (*W*) can range from 0.080 to 2 in, with wall variations as low as ±0.005 in (for thinner walls); however, the larger wall thicknesses require very long curing periods (over 5 min). Large wall thickness variations can be tolerated without sink marks. The optimum wall thickness is in the range of 0.375 to 0.500 in. Holes (*A, D*) can be molded both parallel and perpendicular to the parting line. The rib thickness (*B*) can be as large as the nominal wall (*W*), however generous draft is called for to ease demolding. Inside corners (IR) should be a minimum 0.125 in, with outside corners (OR) equal to the inside corner plus a wall thickness (OR = IR + *W*). Inserts (*E, H*) can be molded-in as can bosses (*F*). Bosses should be connected to walls because air pockets can form during foaming and the air needs a way to escape. Draft of 1 to 3° is required for depths up to 6 in and 3° beyond 6 in. Undercuts (*G*) are feasible with the use of slides.

The principal means of assembling polyurethane RIM parts are fasteners and adhesives. For self-tapping screws, the pilot holes can be molded-in, providing dense skin surface for a better grip. The thermoplastic materials can be joined by fasteners and adhesives as well as all the welding techniques.

8.3.13 Resin transfer molding (RTM)

Resin transfer molding is a low-pressure molding process in which resin is pumped into a closed mold where a vacuum has been drawn. Molding cycles are very long and multiple molds may be required to meet production requirements. However, molds are relatively inexpensive (nickel-plated epoxy). Reinforcement, usually in the form of mats, can be placed in the mold to provide added strength and avoid the use of internal structures such as ribs. Structural reinforcing members are sometimes joined to the inside structure with adhesives. Since the mold is closed, both sides of the part can be finished to a good surface quality. Molded-in labels, raised numbers, and letters are feasible. The process is capable of parts as large as small car bodies and boat hulls. Parts cannot be trimmed in the mold.

Referring again to Fig. 8.32, the nominal wall (*W*) can range from 0.080 to 1.000 in, with wall variations as low as ±0.010 in for thin walls and ±0.020 in for heavy walls. However, the wall thickness can drop to 0.060 in when the glass fiber content is less than 20%. Large variations in wall thickness can be tolerated. Inside radii (IR) should be a minimum of 0.250 in. Draft should be 1 to 3° to a depth of 6 in, 3° from a depth of 6 to 12 in, and an additional 1° for every 6 in beyond that.

Holes (*A, D*) are possible both perpendicular and parallel to the parting line if they are large. Small holes must be drilled or routed. Metal inserts (*E, H*) can be molded into the part, however bosses (*F*), while possible, can be difficult. Cores are possible to a limited degree as they must be pulled by hand before demolding. Undercuts (*G*), however, can be performed with slides in the mold or with a split mold. Corrugated sections and ribs (*B*) are possible. The principal means of assembling RTM parts are fasteners and adhesives.

8.3.14 Rotational molding

Rotational molding utilizes a closed female mold. The mold is opened and filled with a predetermined weight of powder or liquid. It is then swung into the oven section of the equipment and rotated around its primary and secondary axes. The oven heats the mold which, in turn, heats the powder. As the powder melts, it deposits uniformly on the interior walls of the mold. When all the material is deposited on the walls, the mold is then rotated to the cooling chamber. There it is cooled by moving air, water fog, or water spray. When this stage is completed, the mold is then swung back to its original position where the part is removed and a new charge of powder is placed in the mold. Polyethylene is the workhorse of rotational molding with the bulk of its applications in that material. PVC, polypropylene, nylon, polycarbonate, and polybutylene make up most of the balance of its commercial applications; however, in theory, all thermoplastics can be rotationally molded.

Rotational molding and blow molding make similar hollow parts. However, the ideal rotational molded part is a sphere, whereas the ideal blow molded part is more elongated, like the part shown in Fig. 8.30*a*. Both processes, however, can make fairly flat parts such as those illustrated in Figs. 8.30*b* and 8.31. In some cases, such as automotive ducts, the ends are removed from a hollow shape to create the final part. Large containers can be molded integrally with their covers which are cut off to form the two parts. Other shapes can also be molded as one hollow piece and cut apart to make multiple parts (usually two). Structural components are usually molded as a double-walled part which can be filled with foam for greater strength and rigidity. This manner of construction is also used for twin-sheet thermoformed parts.

One advantage of rotational molding is that it is extremely easy to change the wall thickness of the part without changing the mold by simply varying the charge of material that is placed in the mold. Ideal wall thicknesses vary according to the material. Table 8.14 provides a guideline to follow. Rotational molded walls will be essentially uniform. The normal commercial tolerance is ±20%, but a tolerance of ±10% can be obtained at additional cost. It is difficult to maintain flatness on large roto-molded areas. Therefore, they are usually crowned at the rate of 0.010 in/in. Smaller flat areas can be maintained within a tolerance of ±0.020 in/in at a commercial cost and ±0.010 in/in at a significant additional cost. However, nylon and polycarbonate can be held to the closer tolerances of ±0.005 in/in commercially or ±0.003 in/in at higher cost. Ideally, the tolerance for these materials would still be ±0.010 in/in.

Referring again to Fig. 8.31, the minimum wall separation (X) should be no less than 5 times the wall thickness (W) ($X \geq 5W$), except for extreme situations when −3 times the wall thickness can be used. However, tack-offs (P, Q, R) used to stiffen the part are feasible. The combined wall thickness (S) should be 1.75 times the wall thickness ($S = 1.75W$). Inserts, such as the one at A, can also be used in rotational molded parts. The recommended draft angles will vary according to the material as indicated in Table 8.15.

The permissible inside and outside radii (IR and OR in Fig. 8.36) for rotational molded parts varies according to the material. Table 8.16 provides a handy reference.

Rotational molding cannot produce inside structures such as ribs; however, hollow ribs and multiple hollow ribs (corrugated sections)

TABLE 8.14 Recommended Wall Thicknesses for Commonly Rotationally Molded Materials

	Ideal, mm (in)		Possible, mm (in)	
Plastic Material	Minimum	Maximum	Minimum	Minimum
Polyethylene	1.50 (0.06)	12.70 (0.50)	0.50 (0.02)	50.80 (2.00)
Polypropylene	1.50 (0.06)	6.40 (0.25)	0.75 (0.03)	10.16 (0.40)
Polyvinyl Chloride	1.50 (0.06)	10.16 (0.40)	0.25 (0.01)	25.40 (1.00)
Nylon	2.50 (0.10)	20.32 (0.80)	1.50 (0.06)	31.75 (1.25)
Polycarbonate	2.00 (0.80)	10.16 (0.40)	1.50 (0.06)	12.70 (0.50)

SOURCE: Glenn L. Beall, *Rotational Molding—Design, Materials, Tooling, and Processing,* Hanser Publishers, Munich-Hanser/Gardner Publications, Inc., Cincinnati, 1998.

TABLE 8.15 Recommended Draft Angles for Commonly Molded Materials, Degrees per Side

	Inside surfaces		Outside surfaces	
Plastic material	Minimum	Better	Minimum	Better
PE	1.0°	2.0°	0.0°	1.0°
PP	1.5°	3.0°	1.0°	1.5°
PVC	1.0°	3.0°	0.0°	1.5°
Nylon	1.5°	3.0°	1.0°	1.5°
PC	2.0°	4.0°	1.5°	2.0°

SOURCE: Glenn L. Beall, *Rotational Molding—Design, Materials, Tooling, and Processing,* Hanser Publishers, Munich-Hanser/Gardner Publications, Inc., Cincinnati, 1998.

which must show from the outside can be made provided they do not create an undercut which prevents the mold from opening. The width of the hollow rib (X) should be at least 5 times the wall thickness (W) ($X \geq 5W$) and the rib height (H) should be 4 times the wall thickness (W) ($H \geq 4W$). Spacing between the ribs (V) should not be less than 3 times the wall thickness ($V \geq 3W$), although 5 times the wall thickness is better. Corner angles (Z), ideally, should not be less than 45° for parts molded of polyethylene, PVC, and nylon. However, corners as tight as 30° have been molded in polyethylene and PVC and 20° in nylon which flows easiest. Polycarbonate cannot handle tight corners as well; 45° is the absolute minimum, but larger is better.

Indentations in the surface are not undercuts if they are perpendicular to the parting line because they are in the direction of the draw. Therefore, it is desirable to set the parting line such that indentations are in that plane. In Fig. 8.36, the corner Z and the indentation at U are undercuts because they are in the direction parallel to the parting line. The undercut at U could be stripped from the mold; however, the one formed by the corner Z would require a removable section in the tool. Parts which are rotationally molded are often quite large and the materials used (polyethylenes) have high shrinkage rates. It is often possible to strip the undercut if the part will shrink more than the

TABLE 8.16 Recommended Radius Size for Commonly Molded Materials

Plastic material	Outside radii, mm (in)		Inside radii, min (in)	
	Minimum	Better	Minimum	Better
PE	1.52 (0.060)	6.35 (0.250)	3.20 (0.125)	12.70 (0.500)
PP	6.35 (0.250)	12.70 (0.500)	6.35 (0.250)	19.05 (0.750)
PVC	2.03 (0.080)	6.35 (0.250)	3.20 (0.125)	9.53 (0.375)
Nylon	4.75 (0.187)	12.70 (0.500)	6.35 (0.250)	19.05 (0.750)
PC	6.35 (0.250)	19.05 (0.750)	3.20 (0.125)	12.70 (0.500)

SOURCE: Glenn L. Beall, *Rotational Molding—Design, Materials, Tooling, and Processing,* Hanser Publishers, Munich-Hanser/Gardner Publications, Inc., Cincinnati, 1998.

undercut. If the undercut is too deep to be stripped, a side core which must be removed before demolding can be used.

Holes can be molded-in rotational molded parts by using core pins to which the resin does not adhere. Bosses can be molded by molding a raised cylinder such as that at *T.* The tip of the cylinder is then cut off to leave an opening. The diameter of this type of hole should be at least 5 times the nominal wall thickness ($D \geq 5W$). Threads, both inside and outside, such as the one at *S,* are readily rotational molded. Recommended tolerances are provided in Table 8.17 (see Fig. 8.37).

8.3.15 Structural foam molding, gas counter pressure structural foam molding, and co-injection molding

Structural foam molding is a low-pressure, closed-mold process which provides a finished surface on both sides of the part. In structural foam

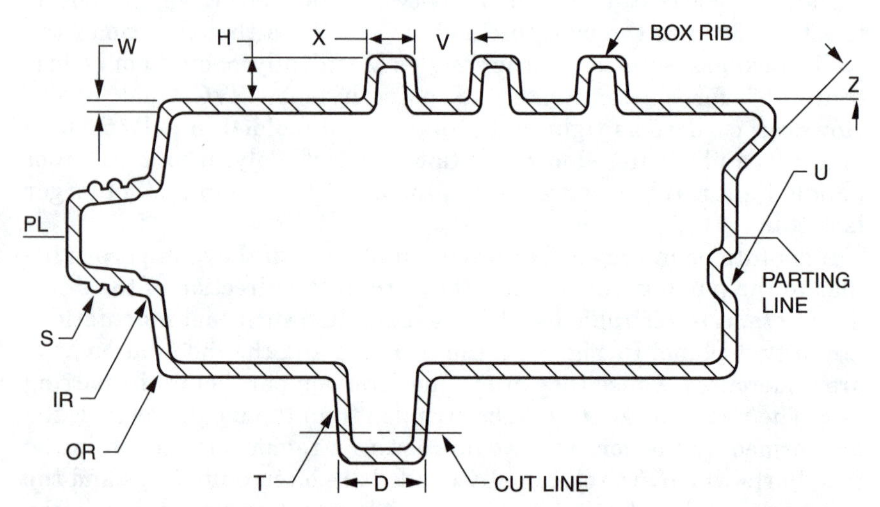

Figure 8.36 Design details for hollow rotational molded parts.

TABLE 8.17 Industry-Standard Dimensional Tolerances, ±cm/cm (±in/in)

Plastic material		A	B	C	D*	E	F
PE	Ideal	0.020	0.020	0.020	0.020	0.015	0.010
	Commercial	0.010	0.010	0.010	0.010	0.008	0.008
	Precision	0.005	0.005	0.005	0.005	0.004	0.004
PP	Ideal	0.020	0.020	0.020	0.020	0.015	0.010
	Commercial	0.010	0.010	0.010	0.010	0.008	0.008
	Precision	0.005	0.005	0.005	0.005	0.004	0.004
PVC	Ideal	0.025	0.025	0.025	0.025	0.015	0.015
	Commercial	0.020	0.020	0.020	0.020	0.010	0.010
	Precision	0.010	0.010	0.010	0.010	0.005	0.005
Nylon	Ideal	0.010	0.010	0.010	0.010	0.008	0.008
	Commercial	0.006	0.006	0.006	0.006	0.005	0.005
	Precision	0.004	0.004	0.004	0.004	0.003	0.003
PC	Ideal	0.008	0.008	0.008	0.008	0.005	0.005
	Commercial	0.005	0.005	0.005	0.005	0.003	0.003
	Precision	0.003	0.003	0.003	0.003	0.002	0.002

NOTE: This table refers to Fig. 8.37.
NOTE: Does not include cavity tolerance.
*Plus 0.250 cm for parting-line variation.
SOURCE: Glenn L. Beall, *Rotational Molding—Design, Materials, Tooling, and Processing*, Hanser Publishers, Munich-Hanser/Gardner Publications, Inc., Cincinnati, 1998.

molding, a mixture of plastic and blowing agent is injected into the mold, simultaneously resulting in a lower-pressure fill than injection molding. The material does not completely fill the mold until the gas expansion takes place. The result is a part with a skin of relatively rigid material with cells of increasing size toward the center of the wall. Due to the space occupied by the gas, the part will weigh less than an equivalent part of the same wall thickness. However, this process requires a minimum wall thickness of 0.187 in for there to be much in the way of gas expansion. If the wall thickness must be increased in order to utilize this process, this advantage is lost.

There are some special considerations in performing product strength computations on structural foam parts. The most significant of these is the basic fact that, since the center of the wall is not solid material, the data provided on the resin suppliers' material specification will not be directly applicable. Tensile strength will be reduced, the amount being dependent on the thickness of the wall and the amount of foaming that has occurred. Countering that is the effect of the greater wall thickness in creating increased rigidity as the rigidity of the part will increase by the cube of its relative wall thicknesses. (A 0.250-in wall thickness is 4 times as stiff as a 0.150-in wall.) The result is a more rigid part with a lower tensile strength (for the same weight of material).

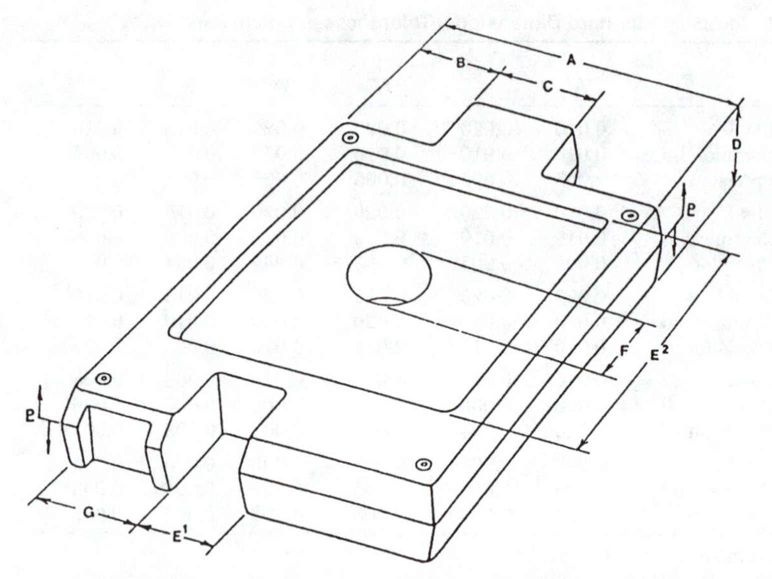

Figure 8.37 Dimensional references for Table 8.17. (*Source: Glenn L. Beall, Rotational Molding—Design, Materials, Tooling and Processing, Hanser Publishers, Munich-Hanser/Gardner Publications, Inc., Cincinnati, 1998.*)

Structural foam molding results in broken cells that create a swirled surface which, for appearance products, must be painted. Gas counterpressure structural foam molding and co-injection molding will both provide a swirl-free surface. Surface foaming can be prevented by preinjecting gas into a sealed mold. Then the gas pressure is released and the foaming of the core is permitted to take place. This process is referred to as gas counterpressure structural foam molding and it does allow solid skins to be created similar to those of conventional injection molding. However, the extra cost of sealing the mold reduces the savings in tooling cost (over injection molds) to be gained from structural foam molding.

A swirl-free surface can also be achieved by molding two different, but compatible, resins. This process is known as co-injection molding. Two joined heating cylinders are used with a different resin in each of them. The first resin to be injected into the mold forms an outer skin against the surface of the cavity. The second material creates a foamed inner core. Through co-injection molding, the outer skin can be a more costly, solid plastic to provide good surface finish and strength and the inner core can be a less expensive foamed resin.

Gas counterpressure structural foam molding and co-injection structural foam molding generally follow the same design rules as structural foam molding, although co-injection molding differs somewhat and behaves more like standard injection molding for certain details. Referring again to Fig. 8.32, the recommended nominal wall (W) increases with the distance the material must flow. For a flow path up to 12 in, the recommended wall thickness is 0.187 in; from 12 to 20 in, the wall thickness should be 0.250 in; and from 20 to 28 in, a wall thickness of 0.312 in is desirable. In general, if the wall thickness is less than 0.187 in, very little foaming will take place and what results is an unnecessarily expensive thick-walled solid part. There are, however, some specially formulated resins which purport to permit foamed wall thicknesses as low as 0.125 in. Flow paths greater than 28 in are not recommended. Structural foam parts should be designed to be gated in a thin section and fill into a thicker section as this practice requires less press than gating into a thick section. (Note that this is contrary to standard injection molding practice.)

Structural foam is more tolerant of nonuniform wall thicknesses than solid injection molding and variations of 75 to 100% are acceptable for standard foam and co-injected walls. Thin foamed walls (0.156 in) should not vary more than 50 to 75%. The change in wall thickness should be very gradual and abrupt wall thickness changes must be avoided to reduce stress concentrations. Wall thickness tolerances of ±0.005 in can be held. Draft should be specified according to the recommendations outlined in Sec. 8.2.2.

Overall tolerances in structural foam molding vary with the size of the part. For dimensions of less than 4 in, tolerances of ±0.005 in can be held. From 4 to 10 in, the minimum tolerance should be ±0.010 in. From 10 to 20 in, tolerances of ±0.015 in should be regarded as a minimum and beyond 20 in the tolerance should be ±0.050 in. Co-injection structural foam and thin-wall structural foam tolerances would be similar to those for injection molding.

The guidelines for inside radii (IR), Sec. 8.2.3, apply to structural foam-molded parts. The ideal inside radius is 0.5 to 0.6 times the wall thickness (IR = 0.5 to 0.6 W). The minimum inside radius should be 0.093 in for most parts, 0.060 for thin wall (0.156-in) parts. The same recommendations apply for co-injection molded parts. The outside radius should be equal to the inside radius plus a wall thickness (OR = IR + W).

Structural foam is also more generous in freedom of rib design. The thickness of the base of the rib (B) can follow the specification indicated in Table 8.18. However, the recommendations in Sec. 8.2.4 should be followed for co-injection-molded parts. The height of the rib (C) should not exceed 3 times the nominal wall (W) ($C \leq 3W$) since the law of diminishing returns sets in at that point and taller ribs, while feasible up to 8 times the wall thickness, tend to become too thin to foam due to the reduction in thickness resulting from draft. Besides, the

TABLE 8.18 Recommended Rib Thickness for Structural Foam Parts

Nominal wall thickness, W in	Rib thickness at base, B
0.156–0.175	$\leq 0.6W$
0.175–0.215	$\leq 0.7W$
0.215–0.250	$\leq 0.8W$
0.250–0.300	$\leq 1.0W$
>0.300	$\leq 1.2W$

strength of the structural foam rib is significantly increased by the geometry of its greater thickness. Ideally, the radius at the base of the rib should be 0.5 to 0.6W, however, 0.25W can be regarded as a minimum to be used when larger radii would increase the material buildup to unacceptable levels. Corrugating can also be used to strengthen walls where other design considerations preclude the use of ribs.

Metal inserts can be molded into structural foam parts and bosses and holes can be readily formed. In general, the recommendations outlined in Sec. 8.2.7 should be followed. Holes can be molded both in the plane parallel to the parting line (A) and perpendicular to the parting line (D). For structural foam, holes should be spaced at least two wall thicknesses (2W) from the edge of a part. Structural foam can tolerate the diameter of the boss to be a bit larger than these recommendations in extreme cases. Bosses have been successfully molded with the outside diameter reaching as much as 3 times the inside diameter for small holes and low wall thickness (0.156 to 0.187 in). With reference to Fig. 8.19, the circle drawn at the base of the boss should not generally be permitted to exceed 1.5W. Gussets (d) should be used to support the boss if its wall thickness is less than 0.120 in.

Snap fits can be molded by the structural foam processes. They can also be joined by the plastic welding processes and with solvents and adhesives. Fasteners can be used, but care must be taken to avoid generating pressures which crack through the rigid skin. Self-tapping screws have only the skin to bite into and should not be used in drilled holes because the screw threads will be in the foamed area. It is often more cost effective to install inserts after molding. Since the strength of the material in the structural foam part is reduced due to the foaming, the published sizes for insert holes will be incorrect and the insert supplier should be consulted for the correct size (it is usually a bit smaller). Hinges can be molded-in (but not living hinges).

Structural foam molds can be textured and have molded-in letters and numbers. When the surface perpendicular to the parting line is to be textured, the preferred draft is 1.5° + 1.5° for every 0.001 in of texture depth. However, in some cases this can be reduced to 1° + 1° for every 0.001 in of texture depth.

8.3.16 Thermoforming, pressure thermoforming, and twin-sheet thermoforming

Thermoforming is a stretching process which uses sheet or film created by the extrusion process. The sheet is first placed in clamps and then heated until it is soft enough to be stretched in all the thermoforming processes. In male or "drape" thermoforming, the softened sheet is lowered over the die and a vacuum draws the softened sheet down on to the die. Having taken the shape of the die, the part is then allowed to cool until it becomes rigid enough to permit removal. The part is then trimmed to size. Female thermoforming, the more common variety, and male thermoforming are similar, the identification referring to the shape of the die. More sophisticated thermoforming, such as pressure thermoforming, also uses positive pressure and mechanical pressure in the form of a plug assist or matched die. Twin-sheet thermoforming uses two molds that form two single-sheet parts. The parts are then pressed together while they are still hot enough for the surfaces in contact to weld together. The result is a two-sided part with a cavity between the sides.

Thermoformings are made from a solid sheet, so the spare material (offal) must be trimmed from the exterior. Holes and other openings must be machined since they cannot be formed in this process. While traditional cutting tool methods are commonplace, high-pressure water jets, hot wires, thermal laser, and robotic cutters are often used. The most recent trend has been the introduction of one or another of the automated devices as thermoformers strive to be competitive.

The removal of so much material results in a high scrap rate for thermoforming, often as high as 50%. This plastic is recycled as regrind, and thermoformers customarily presume its use in their quotes. The loss of physical properties from extensive use of regrind can have a significant effect on the performance of a product used under severe conditions. However, restrictions on, or elimination of, the use of regrind can have a significant effect on the cost of the parts. Trial parts typically have little or no regrind because there was no offal from previous runs to use for that purpose. For a critical application with very little safety margin, the trial parts should be tested with exactly the same percentage of regrind as the maximum permissible in production.

Design details for thermoforming and pressure thermoforming. Except for twin-sheet thermoforming, the thermoforming processes use open molds. This means that only one side of the part is finished. Decoration of that surface using a prepatterned sheet is commonplace. Lettering, numbers, and fine detail can be accomplished only with pressure thermoforming. Design details are also limited to the side of the part which contacts the tool, therefore thermoforming cannot produce internal projections like ribs, pins, and bosses. Furthermore, only that tool side surface can be dimensionally controlled. To help understand the design discussion, see Fig. 8.38.

In thermoforming, the term *nominal wall* is often replaced by the term *starting gauge* (T_1), based on the thickness of sheet stock introduced to the process. Since the sheet will have the smallest surface in its original flat state and no more material is introduced, it must be stretched to assume the contour of the mold. When it is so stretched, it thins out. The starting gauge of the original sheet divided by the final thickness of the piece part is referred to as the *stretch ratio.* The stretch ratio is determined by the following formula:

$$\mathrm{SR} = 1 + \frac{2D\,(L + W)}{LW}$$

where SR = stretch ratio
D = depth of the finished part
L = length of the finished part
W = width of the finished part

A stretch ratio of 2 means that the final wall thickness is one-half of the starting gauge; a stretch ratio of 3 means it is one-third, etc. A stretch ratio of 2 is regarded as conservative, whereas a stretch ratio of 3 is probably the maximum for good design. Starting gauges for sheet thermoforming begin around 0.050 in and can go to 0.500 in. The thinner gauges are best suited to vacuum thermoforming, whereas pressure thermoforming is better suited to the thicker gauges. The preferred range of starting gauge for pressure thermoformed parts is 0.125 to 0.250 in.

The stretch ratio is not to be confused with the depth of draw ratio. The *depth of draw ratio* is the ratio of the maximum depth of the part to the minimum distance across the face of the die at any location on

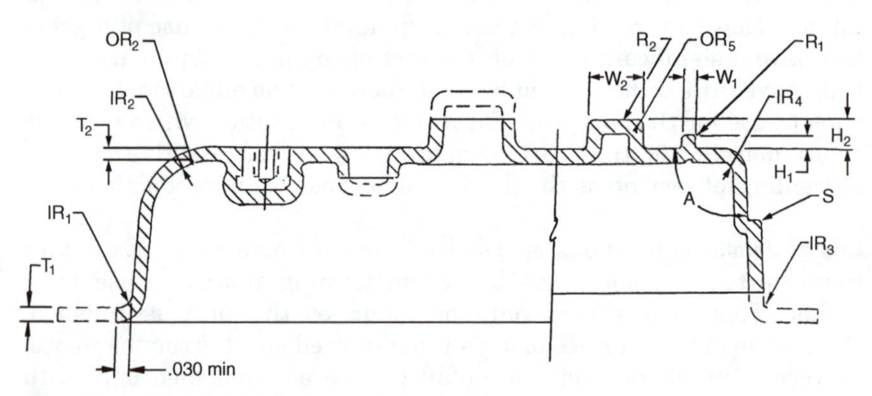

Figure 8.38 Design details for thermoformed parts.

the die. A 14-in-deep part which is 14 in across has a depth of draw ratio of 1:1. That is ideal for a pressure thermoformed part. Parts with depth of draw ratios of 2:1 and greater (3:1) have been thermoformed with methods such as plug-assist, snap-back, and billow forming.

In male thermoforming, the first part of the mold to contact the sheet is the highest point on the form. In most cases, this will be the thickest point on the part (T_2). As the sheet is stretched over the mold, it thins down. Therefore, the thinnest part of a male thermoforming will be along its edges (IR_3). The reverse is true for a female mold where the outer edges of the piece part make first contact (T_1). The thinnest wall will, therefore, be at the bottom of the mold (T_2). Any sharp corners (IR_2) at this level will be particularly thin.

The amount of draft required varies according to the type of die being used. The majority of thermoformed products are made on female molds because that provides a finished outer surface. In a female die, the part shrinks away from the wall of the mold. While a generous draft of 3 to 5° is desirable for economical molding, parts have been made with draft as low as 0 to 1° per side. Fairly deep undercuts have been molded in female molds for applications where a combination of thin walls and material flexibility permit the part to be pulled out of the mold. (Split cavities or pulls are required for stiffer parts.) Expect to pay extra for parts with little or no draft. It is also important to use materials like nylon or polyethylene which have high shrinkage rates and a slick surface and avoid those like acrylic or styrene which shrink at a much lower rate. Male thermoformed parts need more draft because, with a male die, the part shrinks down on the mold and grips it tightly. For these parts, a draft of 3 to 5° is really needed.

The ease with which a part can be removed from the mold is also affected by the corner radii as larger radii grip the mold less than small radii. However, in the stretching process, large radii thin out less than sharp radii. The significance of this thinning depends on the location of the radius and the type of mold used. Male thermoformed parts naturally thin the most at the extremities of the part and need to have the most generous radii at those locations. Female thermoformed parts are thinnest at the inside of the part and need the most generous radii in the inner corners. The ability of the process to create sharp corners is dependent on the thickness of the wall. Pressure thermoforming is capable of forming much sharper corners than vacuum thermoforming. Figure 8.39 illustrates the relative tightness of formed radius each process is capable of for a given material thickness.

The general rule for vacuum thermoforming is that it should have a minimum inside radius (IR_1) equal to the starting gauge ($IR_1 \geq T_1$) increasing to 4 times the starting gauge (T_1)($IR_2 \geq 4T_1$) at its thinnest

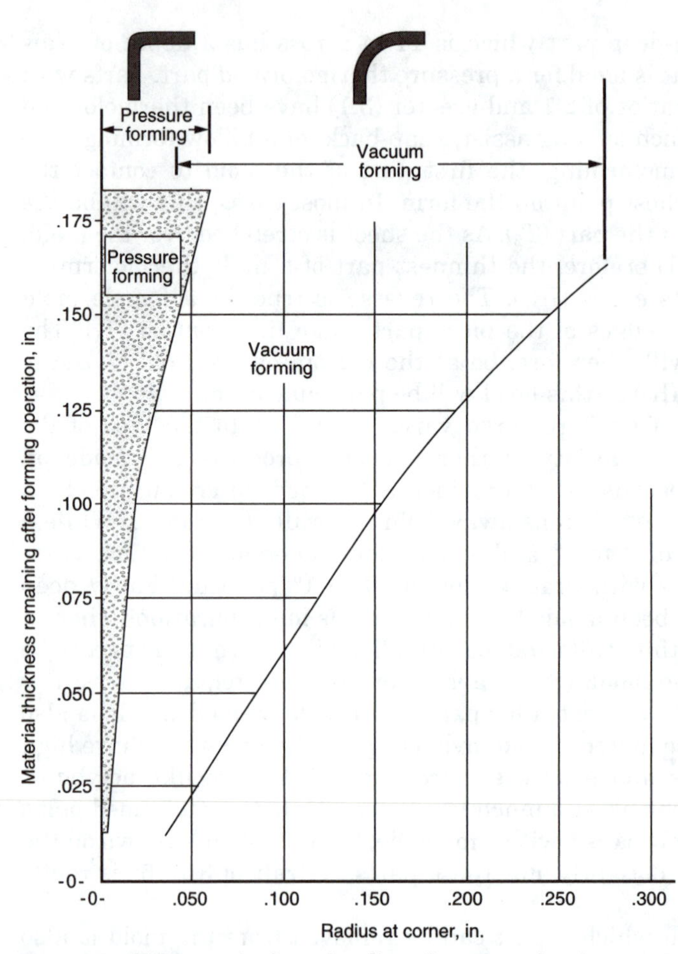

Figure 8.39 Difference in formed radius of pressure forming versus vacuum forming. (*Source: Designer's Guide to Pressure Forming, Arrem Plastics, Inc., 1981.*)

point. Pressure thermoforming can handle much tighter inside radii of 0.25 the starting gauge (T_1)($IR_3 \geq 0.25T_1$) at its thickest point increasing to $0.75T_1$ in its thinnest areas ($IR_4 \geq 0.75T_1$), the latter being ideal for pressure thermoforming. (For uncontrolled radii, such as IR_2 and IR_4 in Fig. 8.38, assume the wall thickness to be the starting gauge.) The discussion in Sec. 8.2.3 applies to thermoformed parts.

The inside angle of the corner created by the two intersecting walls (A) is also significant with respect to the capabilities of the thermoforming processes. The 90° corner illustrated in Fig. 8.38 is commonly done in thermoforming. Wider ($A = 120°$) is better. Narrower is more costly to thermoform and 60° ($A \geq 60°$) is just about the minimum.

In female thermoforming, the outside radius is determined by the mold. For vacuum thermoforming it would be quite large, equal to the inside radius (IR_2) plus the thickness of the sheet (T_2) at that point on the part ($OR_2 = IR_2 + T_2$). However, pressure thermoforming is capable of forcing material deep into a cavity and can readily create outside radii (OR_5) as small as 0.015 in. Even sharper radii (0.005 in) have been formed for special requirements like energy directors for ultrasonic welding. Lettering, numbers, fine detail, and textured areas can also be created with pressure thermoforming to the point of creating parts which are difficult to distinguish from those which have been injection molded.

Pressure thermoforming can create free-standing ribs (R_1) for the purpose of stiffening the part to a limited extent. For a rib height (H_1) no greater than the wall thickness (T_2), a rib as narrow as one wall thickness can be created (for $H_1 \leq T_2$, $W_1 \geq T_2$). Stiffening of thermoformed parts is usually created with design features such as hollow ribs (R_2) or shoulders (S). The width (W_2) of a hollow rib should be at equal or greater than 1.75 times its height (H_2) ($W_2 \geq 1.75H_2$). Corners, such as the one at the right edge of the part (IR_3), also provide stiffening. Care must be taken to leave at least 0.030 in of the flange in place. The trim on the vertical wall at the right side of the part will cost more to cut and have considerably less stiffness than the one at the left edge of the part because of the stiffness provided by the corner. Additional stiffness can be attained by adding a step or shoulder in the wall such as the one shown at S. If the design does not permit the addition of stiffening ribs or shoulders, large flat areas can be domed to prevent buckling. A rise of 0.015 in/in is desirable, however, 0.010 in/in will do if the part must be kept as flat as possible.

Most piece parts sheet thermoformed are fairly large. That can be a problem in attempting to use some of the welding processes. Hot gas welding can be used and vibration welding is also feasible. Hot plate welding is theoretically feasible, but the size and lack of stiffness of thermoformed parts usually precludes its use. Most thermoformed parts are also too large for ultrasonic welding except for the practice of spot welding on the flanges. Pressure thermoforming can form ultrasonic energy directors.

The use of adhesive and solvent joining is widespread with thermoformed parts; however, care must be taken to avoid selection of an adhesive which creates sufficient exothermic heat to distort the part. While external metal inserts can be molded into the part, internal ones must be placed in glued-on bosses which require postmolding operations at additional cost. For light-duty applications, self-tapping screws can be used. The flanges commonly found on thermoformed parts make an ideal location for fasteners and holes are often drilled in place (the parts are positioned together and the hole drilled through both at the same time so the holes are aligned). In general, tolerances for the machining work (like the drilled holes) can be held much tighter than for the thermoforming itself, particularly when computer-controlled machine tools are used. Clearance holes should be 0.015 to 0.030 in oversize to avoid inducing additional stress. Because they are conveniently applied to the flanges, Tinnerman™ type fasteners are often used.

Force fits and snap fits have been used for thermoformed parts, however the difficulty of holding tight tolerances limits their utility for these applications. Outside walls can be held to ±0.015 in for the first 1 in plus ±0.002 in/in thereafter. Tolerances of ±0.010 in plus ±0.001 in can be held for projections such as the rib in Fig. 8.39.

Design details for twin-sheet thermoforming. The design guidelines for twin-sheet thermoforming essentially follow those for pressure thermoforming. Depending on the material, both inside and outside radii can be as low as 0.060 in, however that is without taking into account internal stress considerations. All openings in the part must be machined. In addition, the outer flange must be cut from the original sheet.

The tooling is closed, but there is an uncontrolled internal space between the walls. Thus, all ribs and bosses must be hollow. Additional structural support can be attained by designing the walls or bosses from each side to meet. These contact surfaces will weld together and are known as *tack-offs.* Male and female threads and threaded inserts can be molded in from the outside surfaces.

References

Alvord, Lincoln J., "Engineered Blow Molding Part Design," GE Plastics, Pittsfield, Mass., 1989.

Beall, Glenn L., "Designing for Structural Foam," Cashiers Structural Foam Division, Cashiers, N. C., 1984.

Beall, Glenn L., *Glenn Beall Design Guide II for Pressure Formed Plastic Parts,* Arrem Plastics, Inc., Technical Bulletin, Addison, Ill., 1985.

Beall, Glenn L., *Rotational Molding, Design Materials, Tooling and Processing,* Hanser Publishers, Munich-Hanser/Gardner Publications, Inc., Cincinnati, 1998.

Beck, Ronald D., *Plastic Product Design,* Van Nostrand Reinhold Company Inc., New York, 1980.

Berins, Michael L., *Plastics Engineering Handbook,* 5th ed., The Society of the Plastics Industry, Inc., Chapman & Hall, New York, 1991.

Brockhaus, J., "Design Considerations for Optimum Cold Molded Fiberglas Parts," ETM Company Technical Bulletin, Grand Ledge, Mich., 1996.

Design Handbook for DuPont Engineering Polymers, Module 1, General Design Principles, Du Pont Polymers Technical Bulletin 201742B, Wilmington, Del., 1992.

Designer's Guide to Pressure Forming, Arrem Plastics, Inc., Technical Bulletin AP8349, Addison, Ill., 1981.

Designer's Guide to Vacuum, Pressure, and Twin Sheet Forming, Profile Plastics Corporation Technical Bulletin, Lake Bluff, Ill., 1992.

Designing with Fiberglas Reinforced Plastics, Molded Fiberglas Company, Ashtabula, Ohio, 1989.

Designing with Plastic—The Fundamentals, Hoechst Celanese Technical Bulletin 93-320/10M/0294, Chatham, N.J., 1994.

Engineering Structural Foam, GE Plastics Technical Bulletin SFR-2B, Pittsfield, Mass. 1989.
Fassett, J., "Thin Wall Molding," GE Plastics, Pittsfield, Mass., April 21, 1998.
Galli, Ed, "Designing for Structural RIM," Plastics Design Forum, November/December, 1989.
Green, R.E. ed., *Machinery's Handbook, 24th ed.,* Industrial Press, Inc., New York, 1992.
Malloy, R. A., *Plastic Part Design for Injection Molding,* Hanser Publishers, Munich-Hanser/Gardner Publications, Inc., Cincinnati, 1994.
Mulcahy, Charles M. and Berns, Evan M., "Thermoforming Takes on More Engineering Applications," *Plastics Engineering,* Vol. 46, no. 1, January, 1990, pp. 21-25.
Pugliese, Laura, "Defining Engineering Plastics," *Plastics Machining and Fabrication,* January/February, 1999. (Courtesy DSM Engineering Plastic Products.)
Pultrusions, Polygon Company Technical Bulletin, Walkerton, Ind.
Rosato, Dominick V., *Rosato's Plastics Encyclopedia and Dictionary,* Carl Hanser Verlag, Munich, 1993.
Sayer, Mathew, "Application of Gas Injection Technology," Plastics Product Design & Development Forum, Product Design and Development Division of the Society of Plastics Engineers, Inc., Schaumberg, Ill., June 1998.
SMC Design Manual / Exterior Body Panels, SMC Automotive Alliance Catalog No. AF-180, The Composites Institute of the Society of the Plastics Industry, Inc., Bloomfield Hills, Mich., 1991.
Throne, James L., *Thermoforming,* Hanser Publishers, Munich-Hanser/Gardner Publications, Inc., Cincinnati, 1987.
"Understanding Thermoplastic Part Warpage," *LNP Technical Times,* LNP Engineering Plastics, Inc., Exton, Pa., Spring 1995, pp. 1–3.
Van Doren, Robert, "Pultrusion Speeding in the 1990's," *Job Shop Technology,* March 1996, pp. 1-3.

Chapter

9

Finishing, Assembly, and Decorating

Edward M. Petrie

Electric Systems Technology Institute
ABB Power T&D Company, Inc.
Raleigh, North Carolina

9.1 Introduction

This chapter will provide practical information and guidance on several important plastic processes that occur only *after* the part has been formed. The following processes are sometimes referred to as postprocessing or secondary operations:

- Machining and finishing
- Assembly
- Decorating

They are common operations essential to producing practical commercial products from plastic materials.

Fortunately, many of the processes and tools that are satisfactory for working with metals, wood, and other common engineering materials also apply to plastics. Although there are similarities in these processes, there are also some critical differences that must be considered due to the unique nature of polymeric materials. The material properties of the polymeric resins will dictate the processing parameters that can be used. There are also certain unique assembly and finishing opportunities available for the designer because the material is a plastic. It is these material properties that must occupy the designer's early attention, and this chapter will elaborate on these.

The reader should consider the individual plastic supplier as an excellent source of information on fabricating and finishing processes for specific types of plastic materials. Generally, this information is readily available because the plastic resin producers benefit by providing the most complete and up-to-date information on how their materials can reliably and economically produce commercial products. Because of the tremendous number of plastic materials available, their many forms, and the possible finishing and fabrication processes, it would be difficult to include comprehensive information covering all product possibilities in this chapter. Thus, this chapter is a guide that will help direct the reader to more complete information if required and provide sufficient information for most common complications.

9.2 Machining and Finishing

This section will discuss the major machining and finishing operations commonly used in the manufacture of plastic products:

- Deflashing
- Smoothing and polishing
- Sawing and cutting
- Filing, grinding, and sanding
- Routing, milling, and turning
- Drilling
- Tapping and threading
- Cleaning
- Annealing

Almost all molded articles require finishing to some extent, if only to remove flash and gates from the molding operation.

Machining of plastic parts is normally employed only if there is no other way of designing the functional requirement (hole, rounded edge, etc.) into the primary molded part. Some finishing operations on molded articles can be avoided or reduced by careful design of the part and the mold. For example, placing the flash lines and gates in selected areas could greatly reduce filing needed to remove excess material, and holes molded into the part could eliminate the need for postmold drilling.

Machining requires skilled labor, equipment, and other production resources. However, in certain cases, it can prove to be the least costly production method. If the total number of parts to be produced is small, machining of stock shapes is generally less expensive than investing in expensive and complex molding.

There are several factors to consider when machining plastics: (1) the material's physical properties, such as toughness or modulus; (2) the material's thermal properties, such as thermal expansion coefficient, thermal conductivity, and glass transition temperature; and (3) stress effects on the plastic, either internal (molded-in) stress or external stress from a postprocessing operation.

Brittle plastics tend to break or chip when under a concentrated load such as a drill bit. Softer plastics tend to tear when local load concentrations occur. The modulus of elasticity of plastics is 10 to 60 times smaller than for metals, and this resilience permits much greater deflection of the work material during cutting. Thermoplastics, especially, must be held and supported firmly to prevent distortion, and sharp tools are essential to keep cutting forces to a minimum.

Plastics have a relatively high coefficient of thermal expansion, and they recover elastically during and after machining so that drilled or tapped holes often end up tapered or of smaller diameter than the tool. Turned diameters can end up larger than the dimensions measured immediately after the finishing cut. Plastic materials are also very good thermal insulators. Heat is difficult to conduct away from a plastic surface undergoing a machining operation. This results in the plastic expanding due to the heating and, perhaps, even deforming or melting due to the built-up frictional heat. Sufficient clearances must be provided on cutting tools to prevent rubbing contact between the tool and the work. Tool surfaces that meet the plastic part must be polished to reduce frictional drag and resulting temperature increase.

Internal stresses that are molded into the plastic part can be released when in contact with heat, chemicals, or during cutting. These stresses, when released on the surface of the part, cause stress cracking or crazing. They are noticeable as microcracks or fractures on the surface of the sections under highest stress. When large internal stresses are released by heating through a finishing operation, it could result in warpage and deformation. Like some metals, plastics may need to be annealed before finishing to avoid warpage or undesirable stress relief during postprocessing. Plastic resin suppliers can provide specific annealing instructions for individual plastics.

Some important guidelines should be remembered when machining plastics:

- Tools must be extremely sharp (tungsten carbide or diamond bit tools are recommended for long production runs).
- Adequate tool clearance is essential.

- Compression, thermal expansion, etc., can result in poor dimensional measurements.
- Heat buildup must be minimized (coolants are usually recommended).
- Fast tool speed and slow material feed are generally recommended.
- Parts may need annealing to relieve internal stress.

Since the many types of plastics have a wide range of elastic modulus and heat resistance, each material will have a certain set of machining parameters that are optimal. Therefore, it is critical to contact the plastic supplier for specific recommendations as to cutting tool configurations, working speeds, and cooling procedures. Table 9.1 provides plastic machining guidelines for several common plastics.

9.2.1 Deflashing

Many plastic parts formed in molding operations will have some excess material, called *flash,* at the parting lines and on molded-in inserts. *Gates* are also excess resin material resulting from the flow path of the plastic into the mold. It is usually necessary to remove this excess material for cosmetic or functional reasons.

No single method is universally applicable for the removal of flash and gate material. Each part geometry and material will have its specific requirements and individual problems. The techniques generally used are hand deflashing (filing, sanding, and machining) and tumbling. For removal of flash, tumbling is generally preferred. However, the shape, size, or contour of the article may require filing to remove heavy flash, gate sections, or burrs that may be left by machining operations such as cutting and drilling. Other methods of deflashing parts that are less common are chemical deflashing and water honing.

Hand deflashing. The ease of filing will depend on the type of file chosen. The file characteristics must be carefully matched to the plastic's properties (hardness, brittleness, flexibility, and heat resistance). The size, shape, and contour of the article being filed determine the size and shape of the file to use. For removal of flash, files should have very sharp, thin-topped teeth that will hold their edge, well-rounded gullets to minimize the tendency to clog, and the proper rake for clearing of the chips.

Thermoplastics that can be filed include those that are relatively soft as well as those that are hard. However, some materials (for exam-

TABLE 9.1 Machining Guidelines for Common Plastics

	Sawing (circular)	Sawing (band)	Lathe (turn)	Lathe (cutoff)	Drilling	Milling	Reaming
Acetals:							
Speed, ft^2/min	4,000 to 6,000	600 to 2,000	450 to 600	600	300 to 600	1,000 to 3,000	350 to 450
Feed, in/rev	Fast, smooth	Fast, smooth	0.0045 to 0.010	0.003 to 0.004	0.004 to 0.015	0.004 to 0.016	0.0055 to 0.015
Tool	HSS*, carbide	HSS	HSS, carbide	HSS, carbide	HSS, carbide	HSS, carbide	HSS, carbide
Clearance, deg	20 to 30		10 to 25	10 to 25	10 to 25	10 to 20	
Rake, deg	0	0 to 15 (positive)	0 to 5	0 to 15 (positive)	0 to 10 (positive)	0 to 5 (negative)	0 to 10 (positive)
Set	Slight	Slight					
Point angle, deg					90 to 118		
Cooling	Dry, air jet, vapor	Dry, air jet	Dry, air jet, vapor	Dry, air jet, vapor	Dry, air jet, vapor	Dry, air jet, vapor	Dry, air jet, vapor
Acrylics:							
Speed, ft^2/min	8,000 to 12,000	8,000 to 12,000	300 to 600	450 to 500	200 to 400	300 to 600	250 to 400
Feed, in/rev	Fast, smooth	Fast, smooth	0.003 to 0.008	0.003 to 0.004	Slow, steady	0.003 to 0.010	0.006 to 0.012
Tool	HSS, carbide	HSS	HSS, carbide	HSS, carbide	HSS, carbide	HSS, carbide	HSS, carbide
Clearance, deg	10 to 20		10 to 20	10 to 20	12 to 15	15	
Rake, deg	0 to 10 (positive)	0 to 10 (positive)	0 to 5	0 to 15 (negative)	0 to 5 (negative)	0 to 5 (negative)	0 to 10 (negative)
Set	Slight	Slight					
Point angle, deg					118		
Cooling	Dry, air jet, vapor	Dry, air jet, vapor	Dry, air jet, vapor	Dry, air jet, water solution	Dry, air jet, vapor	Dry, air jet, vapor	Dry, air jet, vapor
Fluorocarbons:							
Speed, ft^2/min	8,000 to 12,000	5,000 to 7,000	400 to 700	425 to 475	200 to 500	1,000 to 3,000	300 to 600
Feed, in/rev	Fast, smooth	Fast, smooth	0.002 to 0.010	0.003 to 0.004	0.002 to 0.010	0.004 to 0.016	0.006 to 0.015
Tool	HSS, carbide	HSS	HSS, carbide	HSS, carbide	HSS, carbide	HSS, carbide	HSS, carbide
Clearance, deg	20 to 30		15 to 30	10 to 25	20	7 to 15	10 to 20
Rake, deg	0 to 5 (positive)	0 to 10 (positive)	0 to 5	3 to 15 (positive)	0 to 10 (negative)	3 to 15 (positive)	0 to 10 (negative)

TABLE 9.1 Machining Guidelines for Common Plastics (*Continued*)

	Sawing (circular)	Sawing (band)	Lathe (turn)	Lathe (cutoff)	Drilling	Milling	Reaming
Fluorocarbons (*Continued*):							
Set	Heavy	Heavy					
Point angle, deg					90 to 118		
Cooling	Dry, air jet, vapor	Dry, air jet	Dry, air jet, vapor	Dry, air jet, vapor	Dry, air jet, vapor	Dry, air jet, vapor	Dry, air jet, vapor
Nylons:							
Speed, ft^2/min	4,000 to 6,000	4,000 to 6,000	500 to 700	700	180 to 450	1,000 to 3,000	300 to 450
Feed, in/rev	Fast, smooth	Fast, smooth	0.002 to 0.016	0.002 to 0.016	0.004 to 0.015	0.004 to 0.016	0.005 to 0.015
Tool	HSS, carbide	HSS	HSS, carbide	HSS, carbide	HSS, carbide	HSS, carbide	HSS, carbide
Clearance, deg	20 to 30		5 to 10	7 to 15	10 to 15	7 to 15	
Rake, deg	15 (positive)	0 to 15 (positive)	0 to 5	0 to 5 (positive)	0 to 5 (positive)	0 to 5 (negative)	0 to 10 (positive)
Set	Slight	Slight					
Point angle, deg					90 to 110 under $^1/_2$ in 118 over $^1/_2$ in		
Cooling	Dry, air jet, vapor	Dry, air jet	Dry, air jet, vapor	Dry, air jet, vapor	Dry, air jet, vapor	Dry, air jet, vapor	Dry, air jet, vapor
Polyolefins:							
Speed, ft/min	1,650 to 5,000	3,900 to 5,000	600 to 800	425 to 475	200 to 600	1,000 to 3,000	280 to 600
Feed, in/rev	Fast, smooth	Fast, smooth	0.0015 to 0.025	0.003 to 0.004	0.004 to 0.020	0.06 to 0.020	0.006 to 0.012
Tool	HSS, carbide	HSS, carbide	HSS, carbide	HSS, carbide	HSS, carbide	HSS, carbide	HSS, carbide
Clearance, deg	15		15 to 25	15 to 25	10 to 20	10 to 20	10 to 20
Rake, deg	0 to 8 (positive)	0 to 10 (positive)	0 to 15	3 to 15 (positive)	0 to 5 (positive)	0 to 10 (positive)	0 to 10 (negative)
Set	Heavy	Heavy					
Point angle, deg					90 to 118		
Cooling	Dry, air jet, vapor	Dry, air jet, vapor	Dry, air jet, vapor	Dry, air jet, vapor	Dry, air jet, vapor	Dry, air jet, vapor	Dry, air jet, vapor

*High-strength steel.

NOTE: This information is designed as a guideline and is not to be construed as absolute. Because of the variety of work and diversity of finishes required, it may be necessary to depart from the suggestions in the table. A good practice to follow is to run a test workpiece before starting a production run.

ple, nylon), because of their toughness and abrasion resistance, are not easily filed. Flash from thermoset materials should be filed off in such a way as to break it toward a solid portion of the part rather than away from the main body to prevent chipping. The file is pushed with a firm stroke to break off the flash close to the body, and then filing is continued to smooth the surface.

Tumbling. *Tumbling* is a simple, high-volume process to deflash rigid plastics. The tumbler is usually a metal drum or cylinder that is perforated in such a way that the plastic parts will not fall through the perforations (Fig. 9.1*a*). The tumbler is then placed on a rotating mechanism. The plastic parts are loaded in the tumbler so that at least one-half of the tumbler space is empty. The tumbler is set in motion allowing the parts to slowly roll and impact against one another. The constant impacting removes the brittle, external flash in a relatively fast time (10 to 15 min). The limitations of tumbling are that the parts

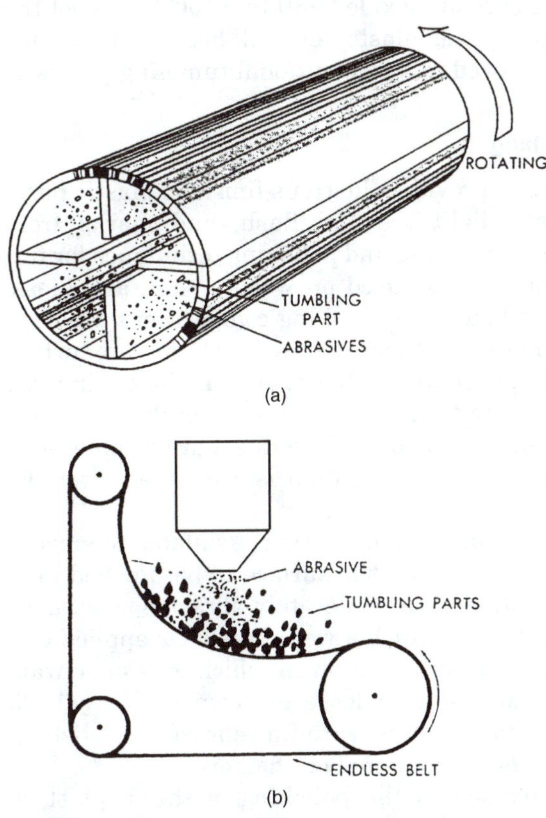

Figure 9.1 Two tumbling methods: (*a*) parts being tumbled in a revolving drum and (*b*) parts being tumbled on an endless revolving belt.[2]

must be robust; they must be molded from rigid, brittle material; and only external flash will be removed. Tumbling can also be used to round corners and apply a surface finish to plastic parts. Tumbling does not produce as high a finish as polishing, but for many articles a very high polish is not necessary and is not worth the higher costs involved.

For more thorough flash removal, moderately abrasive material is added to the tumbling method described previously. This "media deflashing" allows both internal and external flash to be removed. The parts to be deflashed are loaded into a sealed container (Fig. 9.1*b*). Once the tumbling starts, an abrasive media material (ground walnuts, hard nylon, polycarbonate pellets, etc.) is directed at the parts under air pressure and at high velocity. Modern deflashers recycle the blast media and use dust collectors to minimize air pollution.

Cryogenic deflashing employs cold temperatures to make the plastic rigid and brittle. Essentially, it uses a sealed tumbler with either liquid nitrogen or expanding carbon dioxide (−40 to −100°F) to cool the parts. At these temperatures, even elastomers will become rigid, thus allowing the flash to be removed in a conventional tumbling process.

9.2.2 Smoothing and polishing

Smoothing and polishing can provide a lustrous finish on plastic parts or remove surface defects, light residual flash, and marks from machining operations. The smoothing and polishing techniques for finishing plastics are similar to those used on wood, metal, and glass. Lathers, buffing wheels, and suitable polishing compositions are used in these processes. Automatic machines are available for production buffing of articles with simple contours, but they must be engineered for the particular job. All dry buffing and polishing operations require an efficient exhaust system since many of the dusts are combustible. Usually the smoothing or polishing operation is completed after the part has undergone rough grinding or sanding.

There are several types of finishing operations available to smooth plastic surfaces. All consist of holding the surface to be finished next to a soft rotating wheel containing a moderately abrasive substance. *Ashing* is a finishing operation in which a wet abrasive is applied to a loose muslin wheel. *Buffing* is an operation in which grease or wax-filled abrasive cakes are applied to a loose or sewn muslin wheel. *Polishing* employs wax compounds filled with fine abrasives. Polishing wheels are generally made of loose flannel or chamois.

Overheating must be avoided in the polishing of thermoplastics. Thus, it is necessary to avoid exceedingly hard buffing wheels and excessive speeds or pressures. For the buffing of thermoplastics, fine silica powders in special grease binders are available. For thermosetting plastics, a polishing operation using a greaseless compound on a more rigid buffing wheel is recommended. There are a large variety of polishing waxes available. It is necessary to determine experimentally, which wax provides the best results.

Plastics should never be finished on polishing wheels used for metal. Small metal particles may be left on the wheel and damage the plastic surface. All rotating machines should be electrically grounded because the frictional movement of the wheel against the plastic surface readily generates static electricity.

Polishing of certain plastic parts may also be accomplished by dipping the part in a solvent to dissolve surface irregularities. Cellulosic and acrylic parts may be polished in this manner. The parts are either dipped or sprayed with a selected solvent for approximately 1 min. The plastic supplier should recommend specific solvents. Solvents are sometimes also used to polish edges or drilled holes. Annealing of the part may be necessary before solvent dipping to prevent crazing. Surface coatings may be used on most plastics to produce a high surface gloss. This operation may be less costly than other finishing operations. Flame polishing may also be used to polish several plastics. This procedure is much like the flame treatment for surface preperation of plastics prior to adhesive bonding.

9.2.3 Sawing and cutting

Nearly all types of saws have been adapted to cutting plastics. In sawing plastic sheet, there is likely to be concentrated heat buildup in the saw blade because of poor heat conductivity of the material. To allow for this, the blade should be selected according to the gauge of the plastic material. Finer pitch blades should be used for thinner material. The saw blade for cutting thicker material should be heavier and should be hollow ground.

Circular saws have unlimited use in cutting thermoplastics. With thermosets above $^{3}/_{4}$-in gauge, the cut should be made on a band saw. Circular saws provide a better finish, but band saws run smoother and are generally preferred. The band saw will cut through thick plastic with less heating than a circular saw because the length of blade allows the saw teeth to cool between cuts. Band saw cutting speeds range from 500 to 3000 ft/min. In cutting thermoplastics, the material should be fed from 2 to 20 ft/min depending on thickness. Feed rates must be slow with thick plastic. For band sawing thin plastic, it is best to use a fine pitch blade; thick plastics require a coarse blade. The number of teeth per inch may vary from 8 to 22. Table 9.1 shows circle and band sawing parameters for various plastic materials.

For frequent cutting of reinforced, filled thermoplastics, or many thermosetting plastics, carbide-tipped blades will give accurate cuts and reasonably long blade life. Abrasive or diamond-tipped blades may also be used; however, a liquid coolant is recommended. All cutting tools should have protective shields and safety devices.

Shearing can be used for cutting light-gauge thermoplastic sheets. All shearing should be accomplished with the material to be sheared at least at room temperature. Shearing of plastic at low temperatures may crack the part. Thermoplastics may be heated to a more flexible condition to aid the shearing action.

A CO_2 laser can be used to punch intricate holes and cut delicate patterns in plastics. The laser can be directed to etch the plastic surface barely or actually vaporize and melt it. There is no physical contact between the plastic and the laser equipment, and no dust or drill chips are produced.

High-pressure water can also be used to water-jet cut many plastic materials. A fine jet of water (either plain or with abrasive added) is directed at the part at pressures of 60,000 lb/in^2 from a fine nozzle. Automated robotic equipment is available to water-jet cut intricate patterns. This method is very attractive for many hard to cut materials such as aramid fiber-reinforced composites. The water jet cuts most materials quickly without a burr and does not introduce stresses in the material.

9.2.4 Filing, grinding, and sanding

Thermosetting plastics are relatively hard and brittle and, as such, are relatively easy to file, grind, and sand. Filing removes material in the form of a light powder. Aluminum type A, shear tooth, or other files that have coarse, single-cut teeth with an angle of 45° are preferred. The deep angle file teeth enable the file to clear itself of plastic chips.

On the other hand, many thermoplastics are tough and flexible materials. These have a tendency to clog files. Curved tooth files, like those used in auto body shops, are good because they clear themselves of plastic chips. Specially designed files for plastics should be kept clean and not used for filing metals.

Grinding is generally not done on plastic parts unless open grit wheels with a coolant are used. Plastics are generally ground more slowly than metals, and the grinding is kept to a minimum to avoid unnecessary polishing later. Thermosets can be ground at speeds of 4 to 8 ft/s after a rough belt grinding. Thermoplastics are preferably ground wet because of better cooling at speeds of 3 to 5 ft/s, using a thicker application of grinding media.

Thermoplastics can be ground or sanded to a finished edge for fabrication fit-up or for close dimensional tolerance. The work is generally done on a machine (belt, disk, or band) by wet sanding using water cooling for removing the heat buildup. High tool speed and light feed or pressure is recommended. This will prevent softening of the plastic and subsequent buildup of viscous material on the grinding wheel. Wet grinding is used to square edges and to touch up angles, arcs, or curves for fit-up. Beveling can be done on a tilt-bed wet sander.

A number 80 grit silicon-carbide abrasive is recommended for rough sanding. Progressively finer abrasives are then used in finishing. In any machine sanding, light pressure is used to prevent overheating. Disk sanders should be operated at speeds of 1750 r/min and belt sanders at a surface speed of 3600 ft/min. After sanding, further finishing operations can be completed.

9.2.5 Milling, turning, and routing

For all operations such as milling, turning, and routing, carbide-tipped cutters and high speeds are used. Adequate support and suitable feed rates are very important. For many plastics a surface speed of 500 ft/min with feeds (depth of cut) of 0.002 to 0.005 in/rev will produce good results. The feeds and speeds should be similar to machining of brass or aluminum. Table 9.1 gives milling, turning, and reaming data for various plastic materials. Higher cutting speeds improve machined finish but also reduce the life of the cutting tool.

Table 9.2 gives side and end relief angles and back rake angles for cutting tools used on various plastics. Figure 9.2 illustrates the cutting characteristics of tools used to cut plastics. To prevent grabbing, tools should have an O-rake, which is similar to the rake of tools for machining brass. For thermoplastics, a diamond-tipped cutter should be used for a mirrorlike finish. The spiral plastic cuttings that leave the cutting edge can be removed by air jets or vacuum attachments directing the strips away from the cutter. The machining of thermosets generally produces cuttings in powder form.

Reaming may be used to size holes accurately, but diameters produced may also be affected by thermal expansion. Pilot holes may help when the hole is to be reamed or counterbored. Reamer speeds should approximate those used for drilling. Reaming can be done dry, but water-soluble coolants will produce better finishes. For thermoplastics it is recommended that a reamer 0.001 to 0.002 in larger than the desired hole size be used. Tolerances as close as ±0.0005 in can be held in holes of $^1/_4$-in diameter.

9.2.6 Drilling

Thermoplastics and thermosets may be drilled with any standard drill used for metals. However, there are drills specifically designed to drill

TABLE 9.2 Geometries for Cutting and Drill Tools[2]

Drill Geometry				
Material	Rake angle, degrees	Point angle, degrees	Clearance, degrees	Rake, degrees
Thermoplastic:				
Polyethylene	10–20	70–90	9–15	0
Rigid polyvinyl chloride	25	120	9–15	0
Acrylic (polymethyl methacrylate)	25	120	12–20	0
Polystyrene	40–50	60–90	12–15	0 to −5
Polyamide resin	17	70–90	9–15	0
Polycarbonate	25	80–90	9–15	0
Acetal resin	10–20	60–90	10–15	0
Fluorocarbon TFE	10–20	70–90	9–15	0
Thermosetting:				
Paper or cotton base	25	90–120	10–15	0
Fibrous glass or other fillers	25	90–120	10–15	0

Cutting Tool Geometry			
Work material	Side relief angle, degrees	End relief angle, degrees	Back rake angle, degrees
Polycarbonate	3	3	0–5
Acetal	4–6	4–6	0–5
Polyamide	5–20	15–25	−0.5–0
TFE	5–20	0.5–10	0–10
Polyethylene	5–20	0.5–10	0–10
Polypropylene	5–20	0.5–10	0–10
Acrylic	5–10	5–10	10–20
Styrene	0–5	0–5	0
Thermosets:			
Paper or cloth	13	30–60	−0.5–0
Glass	13	33	0

plastic materials. Drills that are not made of high-speed steel or solid carbide should be carbide or diamond tipped. Figure 9.2 shows important parts of a common twist drill with tapered shank. For drilling plastics, drills should have highly polished flutes and chrome-plated or nitrided surfaces. Drills should be ground with a 70 to 120° point angle and a 10 to 25° lip clearance angle. The rake angle on the cutting edge should be zero or several degrees negative. Highly polished, large, slowly twisting flutes are the most desirable for good chip removal. Table 9.2 provides common drill geometries for various plastic materials.

Normal feed rates are in the range of 0.001 to 0.012 in/rev for holes of $^{1}/_{16}$- to 2-in diameter with speeds of 100 to 250 ft/min. Lower speeds should be used for deep and blind holes. Table 9.3 provides suggested drill sizes and speeds for thermosets and thermoplastics.

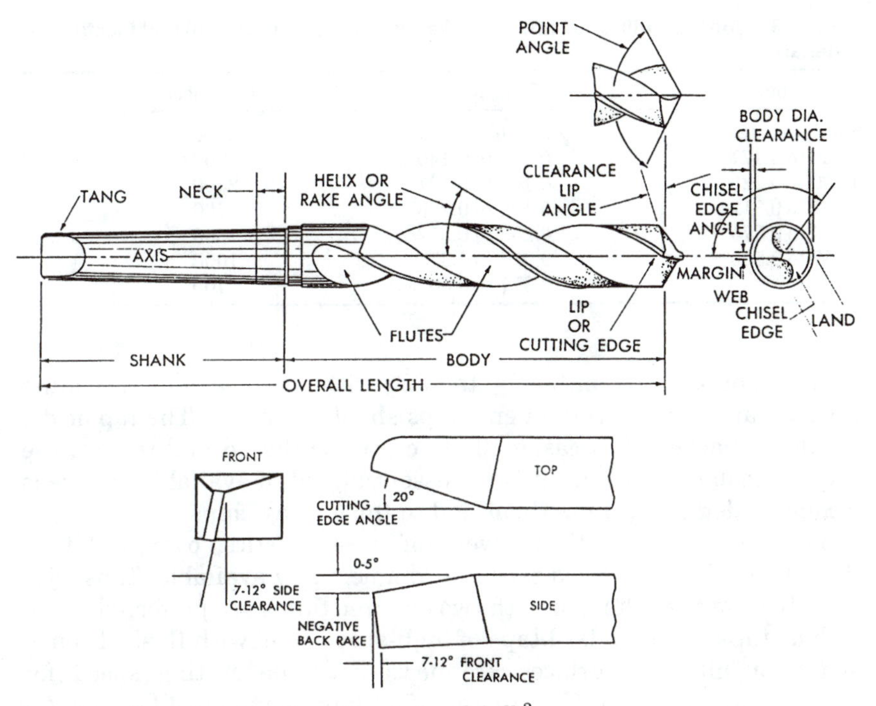

Figure 9.2 Characteristics of a cutting tool and drill.[2]

Holes drilled in most thermoplastics and some thermosets are generally 0.002 to 0.004 in undersized due to the thermal expansion effects. To alleviate this effect, drilling could be accomplished in two stages. A small pilot drill should be used first followed by the required size drill. The drill should be lifted frequently to prevent overheating. Thermoplastics may require external cooling to reduce frictional heat and gumming. Compressed air blown into the hole will help remove chips and provide some cooling effect. The part must be held firmly during drilling to counter the tendency for the tooling to grab and spin the work. Circle cutters are often preferred for making holes in thin materials.

9.2.7 Tapping and thread cutting

Many different threaded fasteners can be used with plastics. (These are described in the following section.) Sometimes threads must be machined into the part to provide for fasteners. Threads can be cut in thermoplastics and thermosets by the same methods used with metals and generally with the same ease of operation. Plastics may also be tapped and threaded on conventional lathes and screw machines.

TABLE 9.3 Suggest Drill Sizes (r/min) for Various Drill Diameters (in) and Plastic Materials[3]

Thermoplastics	Thermosets	Drill speed
Small to 0.125	Small to 0.093	5000
0.126 to 0.177	0.094 to 0.140	3000
0.180 to 0.228	0.141 to 0.191	2500
0.232 to 0.312	0.192 to 0.250	1700
0.315 to 0.375	0.252 to 0.312	1300
0.377 to 0.500	0.315 to 0.375	1000
	0.377 to 0.500	700

In cutting threads, either by tapping with serial taps or threading with die and die stock, frequent stops should be made. The tap or die should be backed off occasionally to clear the thread and remove the chips. A coolant or lubricant is not essential, but, if available, its use is recommended to minimize heat and to flush away chips.

Because of the elastic recovery of most plastics, oversized taps should be used. Oversized taps are commercially available. Taps with slightly negative rake and with two or three flutes are preferred. Solid carbide taps and standard taps of high-speed steel with flash chrome-plated or nitrided surfaces are necessary. The cutting speed for machine tapping should be less than 50 ft/min. Tapping of filled materials is done at 25 ft/min. These speeds should be further reduced for deep or blind holes and when the percentage of thread is greater than 65 to 75%. Sharp V-threads are not recommended.

9.2.8 Cleaning

Once machined, the plastic must be cleaned of any finishing lubricants and contaminants before other processes, such as decorating or assembly, can be considered. Simple cleaning of plastic parts is generally accomplished using solvent or alkaline cleaners. However, chemical treatments may be necessary to prepare the surface of the part for bonding, painting, or similar processes. These chemical treatment processes are described in Sec. 9.5 of this chapter.

Solvent cleaning is generally the first (and often the only) surface preparation applied to plastic parts. Solvent cleaning removes organic release agents, such as silicone, that may coat the part during molding, and any machine oil transferred to the part during finishing. Solvents, such as methyl ethyl ketone, toluene, naphtha, acetone, trichloroethylene, and perchloroethylene, are used for degreasing. Selection of the solvent should depend on the plastic being cleaned. Certain solvents will aggressively attack certain polymeric surfaces causing the surface to deform or cause stress crazing. Substrates can be wiped or sprayed with degreasing solvents, or they can be immersed in an agitated bath for about 5 min. Vapor degreasing equipment using trichloroethylene vapor is very effective in eliminating mold release. This operation, however, is performed at a high temperature, thus requiring the substrate to be thermally stable.

Alkaline cleaning is effective for removing the water dissolvable contaminants that may be difficult to remove with solvent wiping. Substrate immersion in an alkaline bath sometimes follows solvent degreasing. Parts are continuously agitated in a water solution maintained at 77 to 93°C and at an alkaline pH. As dirt, metal chips, and other contaminants are removed, more alkaline solution must be added to maintain the pH, and the bath must be replaced if contaminant levels become too high. After alkaline cleaning, substrates are rinsed in water to remove all traces of the alkaline solution. The solution should not be allowed to dry on the part for cosmetic reasons or because it could result in a weak boundary layer in a painting or bonding operation. Some plastics are hygroscope or absorb water, and a washing process, especially at elevated temperatures, could degrade the properties of the plastic. These plastics include nylons, acrylonitrile butadiene styrene (ABS), polycarbonate, and thermoplastic polyesters.

The use of detergent-free ultrasonic washing may be the best method to clean small plastic parts. In this type of washer, clean water is placed in a tank that is subjected to ultrasonic (high-frequency) vibrations. Thousands of small bubbles form from the ultrasonic agitation. These bubbles scrub the part's surface and clean it without the use of heavy detergents or solvents.

Plastic parts will often attract contaminants because of a static charge. Such static charging can also be a safety concern. The charges can be large enough to be painful and result in bodily damage along a high-speed production line. Electrostatic charge can also cause circuit failure if near an integrated circuit. To render a plastic part less susceptible to static buildup, the plastic has to be made less of a dielectric insulator and more conductive. Three ways of doing this are by internal antistatic agents, external antistatic agents, and surface discharging. External antistatic agents are sprayed, wiped, or rubbed onto the plastic part to provide a short-term increase in surface conductivity. Water is the most common antistatic agent.

9.2.9 Annealing

During the molding operation or during various finishing processes, the plastic part may develop internal stresses. Exposure to chemicals either in secondary processing or in the service environment may sensitize the stressed plastic and cause crazing. Stress crazing is evident by microfractures on the surface of the part near the areas of greatest stress. Crazing can be eliminated or greatly reduced by thermal annealing of the plastic part. *Annealing* consists of prolonged heating of the part at temperature lower than molding temperatures to relieve the internal stresses. The parts are then slowly cool after annealing. Annealing is a common procedure if the part is known to have internal stress and it will see harsh chemical environments either for the purpose of assembly (such as solvent welding) or in service. Acrylic is a common plastic that is susceptible to stress crazing.

9.3 Assembly of Plastic Parts—General Considerations

The next four sections of this chapter provide practical information and guidance on how to join parts made from plastics to themselves and to other substrates. Several processes are available for joining these materials:

- Mechanical fastening
- Adhesive bonding
- Thermal welding
- Solvent cementing

Solvent cementing and thermal welding use the resin in the part itself as the "fastener" to hold the assembly together. Adhesive bonding and mechanical fastening use another substance as the "fastener." The design engineer must determine the joining method that best suits the purpose. The choice will often depend on the type of plastic, the service environment, economic and time constraints, and production parameters. The designer should not force an assembly method on a plastic product originally designed for another assembly method. Usually, parts must be specifically designed to an assembly method. In fact, certain plastic materials are specifically chosen for an application because of their capability for assembly. For instance, in the automotive industry, plastics are often chosen based on the fact that they can be assembled with very fast processes such as ultrasonic welding.

It is important that the designer realize the unique opportunities and problems posed by each method of assembly. To do this well, he or she must have an understanding of materials science, chemistry, surface science, physics, mechanics, and industrial engineering since all of these disciplines will come into play. Even with this background, final selection of the most desirable assembly method involves some trial and error that can become costly and time consuming. The purpose of these next sections is to give the designer a foundation for finding the right assembly system for any particular combination of plastic material, part design, service environment, and production constraint.

The joining of plastics is generally more difficult than joining of other substrates because of their low surface energy, poor wettability, and presence of mold release agents and other contaminants that can create a weak interface. The relative differences in thermal expansion coefficient and elastic modulus also make joining of plastics to nonplastic materials difficult. They may also cause very high loads or loose fitting fasteners in parts assembled with mechanical fasteners. The nature of the polymeric material could also change with the service environment. Parts may swell in solvent, become brittle when exposed to UV, lose plasticizer on aging, gain a plasticizer (water) during exposure to humidity, and go through many other changes. All of these will have an effect on the joint.

With plastic materials, the designer also has a greater choice of assembly techniques than with many other materials. Thermosets must be adhesively bonded or mechanically joined, but most thermoplastics can be joined by solvent or heat welding. Plastic parts can also be designed for assembly by means of molded-in snap-fit, press-fit, pop-on, and threaded fasteners, so that no additional fasteners, adhesives, solvents, or special equipment is required.

Table 9.4 describes various joining methods for plastics and the advantages and disadvantages of each. Table 9.5 indicates which joining methods are appropriate for common plastic substrates. The plastic manufacturer is generally the leading source of information on the proper methods for joining a particular plastic.

9.4 Methods of Mechanical Joining

There are basically two methods of mechanical assembly for plastic parts. The first uses fasteners such as screws or bolts; the second uses an interference fit such as a press-fit or snap-fit and is primarily used in thermoplastic applications. This latter method of fastening is also called *design for assembly* or *self-fastening.*

If possible, the designer should try to design the entire product as a one-part molding because it will eliminate the need for a secondary assembly operation. However, mechanical limitations often will make it necessary to join one part to another using a fastening device. Fortunately, there are a number of mechanical fasteners designed for metals that are also generally suitable with plastics. There are also many fasteners specifically designed for plastics. Typical of these are thread-forming screws, rivets, threaded inserts, and spring clips.

TABLE 9.4 Bonding or Joining Plastics: What Techniques Are Available and What Do They Offer[4]

Technique	Description	Advantages	Limitations	Processing considerations
Solvent cementing and dopes	Solvent softens the surface of an amorphous thermoplastic; mating takes place when the solvent has completely evaporated. Bodied cement with small percentage of parent material can give more workable cement, fill in voids in bond area. Cannot be used for polyolefins and acetal homopolymers.	Strength, up to 100% of parent materials, easily and economically obtained with minimum equipment requirements.	Long evaporation times required; solvent may be hazardous; may cause crazing in some resins.	Equipment ranges from hypodermic needle or just a wiping media to tanks for dip and soak. Clamping devices are necessary, and air dryer is usually required. Solvent-recovery apparatus may be necessary or required. Processing speeds are relatively slow because of drying times. Equipment costs are low to medium.
Thermal Bonding				
Ultrasonics	High-frequency sound vibrations transmitted by a metal horn generate friction at the bond area of a thermoplastic part, melting plastics just enough to permit a bond. Materials most readily weldable are acetal, ABS, acrylic, nylon, PC, polyimide, PS, SAN, phenoxy.	Strong bonds for most thermoplastics; fast, often less than 1 s. Strong bonds obtainable in most thermal techniques if complete fusion is obtained.	Size and shape limited. Limited applications to PVCs, polyolefins.	Converter to change 20 kHz electrical into 20 kHz mechanical energy is required along with stand and horn to transmit energy to part. Rotary tables and high-speed feeder can be incorporated.
Hot-plate and hot-tool welding	Mating surfaces are heated against a hot surface, allowed to soften sufficiently to produce a good bond, then clamped together while bond sets. Applicable to rigid thermoplastics.	Can be very fast, for example, 4 to 10 s in some cases; strong.	Stresses may occur in bond area.	Use simple soldering guns and hot irons, relatively simple hot plates attached to heating elements up to semiautomatic hot-plate equipment. Clamps needed in all cases.
Hot-gas welding	Welding rod of the same material being joined (largest application is vinyl) is softened by hot air or nitrogen as it is fed through a gun that is softening part surface simultaneously. Rod fills in joint area and cools to effect a bond.	Strong bonds, especially for large structural shapes.	Relatively slow; not an "appearance" weld.	Requires a hand gun, special welding tips, an air source, and welding rod. Regular hand-gun speeds run 6 in/min; high-speed hand-held tool boosts this to 48 to 60 in/min.

Spin welding	Parts to be bonded are spun at high speed, developing friction at the bond area; when spinning stops, parts cool in fixture under pressure to set bond. Applicable to most rigid thermoplastics.	Very fast (as low as 1 to 2 s); strong bonds.	Bond area must be circular.	Basic apparatus is a spinning device, but sophisticated feeding and handling devices are generally incorporated to take advantage of high-speed operation.
Dielectric	High-frequency voltage applied to film or sheet causes material to melt at bonding surfaces. Material cools rapidly to effect a bond. Most widely used with vinyls.	Fast seal with minimum heat applied.	Only for film and sheet.	Require rf generator, dies, and press. Operation can range from hand-fed to semiautomatic with speeds depending on thickness and type of product being handled. Units of 3 to 25 kW are most common.
Induction	A metal insert or screen is placed between the parts to be welded, and energized with an electromagnetic field. As the insert heats up, the parts around it melt, and when cooled form a bond. For most thermoplastics.	Provides rapid heating of solid sections to reduce chance of degradation.	Since metal is embedded in plastic, stress may be caused at bond.	High-frequency generator, heating coil, and inserts (generally 0.02 to 0.04 in thick). Hooked up to automated devices, speeds are high. Work coils, water cooling for electronics, automatic timers, multiple-position stations may also be required.
		Adhesives*		
Liquids solvent, water base, anaerobics	Solvent- and water-based liquid adhesives, available in a wide number of bases—for example, polyester, vinyl—in one- or two-part form fill bonding needs ranging from high-speed lamination to one-of-a-kind joining of dissimilar plastics parts. Solvents provide more bite, but cost much more than similar base water-type adhesive. Anaerobics are a group of adhesives that cure in the absence of air.	Easy to apply; adhesives available to fit most applications.	Shelf and pot life often limited. Solvents may cause pollution problems; water-base not as strong; anaerobics toxic.	Application techniques range from simply brushing on to spraying and roller coating-lamination for very high production. Adhesive application techniques, often similar to decorating equipment, from hundreds to thousands of dollars with sophisticated laminating equipment costing in the tens of thousands of dollars. Anaerobics are generally applied a drop at a time from a special bottle or dispenser.

TABLE 9.4 Bonding or Joining Plastics: What Techniques Are Available and What Do They Offer[4] (*Continued*)

Technique	Description	Advantages	Limitations	Processing considerations
		Adhesives (*Continued*)		
Pastes, mastics	Highly viscous single- or two-component materials which cure to a very hard or flexible joint depending on adhesive type.	Does not run when applied.	Shelf and pot life often limited.	Often applied via a trowel, knife, or gun-type dispenser; one-component systems can be applied directly from a tube. Various types of roller coaters are also used. Metering-type dispensing equipment in the $2500 range has been used to some extent.
Hot melts	100% solids adhesives that become flowable when heat is applied. Often used to bond continuous flat surfaces.	Fast application; clean operation.	Virtually no structural hot melts for plastics.	Hot melts are applied at high speeds via heating the adhesive, then extruding (actually squirting) it onto a substrate, roller coating, using a special dispenser or roll to apply dots or simply dipping.
Film	Available in several forms including hot melts, these are sheets of solid adhesive. Mostly used to bond film or sheet to a substrate.	Clean, efficient.	High cost.	Film adhesive is reactivated by a heat source; production costs are in the medium to high range depending on heat source used.
Pressure-sensitive	Tacky adhesives used in a variety of commercial applications. Often used with polyolefins.	Flexible.	Bonds not very strong.	Generally applied by spray with bonding effected by light pressure.
Mechanical fasteners (staples, screws, molded-in inserts, snap-fits, and variety of proprietary fasteners)	Typical mechanical fasteners are listed on the left. Devices are made of metal or plastic. Type selected will depend on how strong the end product must be, appearance factors. Often used to join dissimilar plastics or plastics to nonplastics.	Adaptable to many materials; low to medium costs; can be used for parts that must be disassembled.	Some have limited pull-out strength; molded-in inserts may result in stresses.	Nails and staples are applied by simply hammering or stapling. Other fasteners may be inserted by drill press, ultrasonics, air or electric gun, hand tool. Special molding—that is, molded-in-hole—may be required.

*Typical adhesives in each class are: Liquids: 1. Solvent—polyester, vinyl, phenolics acrylics, rubbers, epoxies, polyamide; 2. Water—acrylics, rubber-casein; 3. Anaerobics—cyanoacrylate; mastics—rubbers, epoxies; hot melts—polyamides, PE, PS, PVA; film—epoxies, polyamide, phenolics; pressure, sensitive—rubbers.

TABLE 9.5 Assembly Methods for Plastics[5]

Plastic	Common assembly methods							
	Adhesives	Dielectric welding	Induction bonding	Mechanical fastening	Solvent welding	Spin welding	Thermal welding	Ultrasonic welding
Thermoplastics:								
ABS	X	…	X	X	X	X	X	X
Acetals	X	…	X	X	X	X	X	X
Acrylics	X	…	X	X	X	X	…	X
Cellulosics	X	…	…	…	X	X		
Chlorinated polyether	X	…	…	X				
Ethylene copolymers	…	X	…	…	…	…	X	
Fluoroplastics	X							
Ionomer	…	X	…	…	…	…	X	
Methylpentene	…	…	…	…	…	…	…	X
Nylons	X	…	X	X	X	…	…	X
Phenylene oxide–based materials	X	…	…	X	X	X	X	X
Polyesters	X	…	…	X	X	X	…	X
Polyamide-imide	X	…	…	X				
Polyaryl ether	X	…	…	X	…	…	…	X
Polyaryl sulfone	X	…	…	X	…	…	…	X
Polybutylene	…	…	…	…	…	…	…	X
Polycarbonate	X	…	X	X	X	X	X	X
Polycarbonate/ABS	X	…	…	X	X	X	X	X
Polyethylenes	X	X	X	X	…	X	X	X
Polyimide	X	…	…	X				
Polyphenylene sulfide	X	…	…	X				
Polypropylenes	X	X	X	X	…	X	X	X
Polystyrenes	X	…	X	X	X	X	X	X
Polysulfone	X	…	…	X	X	…	…	X
Propylene copolymers	X	X	X	X	…	X	X	
PVC/acrylic alloy	X	…	X	X	…	…	…	X
PVC/ABS alloys	X	X	…	…	…	…	X	
Styrene acrylonitrile	X	…	X	X	X	X	X	X
Vinyls	X	X	X	X	X	…	X	

TABLE 9.5 Assembly Methods for Plastics[5] (*Continued*)

Plastic	Common assembly methods							
	Adhesives	Dielectric welding	Induction bonding	Mechanical fastening	Solvent welding	Spin welding	Thermal welding	Ultrasonic welding
Thermosets:								
Alkyds	X							
Allyl diglycol carbonate	X							
Diallyl phthalate	X	...	...	X				
Epoxies	X	...	...	X				
Melamines	X							
Phenolics	X	...	...	X				
Polybutadienes	X							
Polyesters	X	...	...	X				
Silicones	X	...	...	X				
Ureas	X							
Urethanes	X	...	...	X				

As in other fabrication and finishing operations, special considerations must be given to mechanical fastening because of the nature of the plastic material. Care must be taken to avoid overstressing the parts. Mechanical creep can result in loss of preload in poorly designed systems. Reliable mechanically fastened plastic joints require

- A firm, strong connection
- Materials that are stable in the environment
- Stable geometry
- Appropriate stresses in the parts, including a correct clamping force

In addition to joint strength, mechanically fastened joints should prevent slip, separation, vibration, misalignment, and wear of parts. Well-designed joints provide all of these without being excessively large or heavy or without burdening assemblers with bulky tools. Designing plastic parts for mechanical fastening will depend primarily on the particular plastic being joined and the functional requirements of the application.

Mechanical fasteners and the design of parts to accommodate them are covered in detail in later sections of this chapter. The following section describes the use of press-fit and snap-fit designs that are integrated into the molded part to achieve assembly.

9.4.1 Design for self-fastening

It is often possible, and desirable, to incorporate fastening mechanisms in the design of the molded part itself. The two most common methods of doing this are by interference fit (including press-fit or shrink-fit) and by snap-fit. Whether these methods can be used will depend heavily on the nature of the plastic material and the freedom one has in part design.

Press-fit. In press-fits or interference-fits, a shaft of one material is joined with the hub of another material by a dimensional interference between the shaft's outside diameter and the hub's inside diameter. This simple, fast assembly method provides joints with high strength and low cost. Press-fitting is applicable to parts that must be joined to themselves or to other plastic and nonplastic parts. The advisability of its use will depend on the relative properties of the two materials being assembled. When two different materials are being assembled, the harder material should be forced into the softer. For example, a metal shaft can be press-fitted into plastic hubs. Press-fit joints can be made by simple application of force or by heating or cooling one part relative to the other. Press-fitting produces very high stresses in the plastic parts. With brittle plastics, such as thermosets, press-fit assembly may cause the plastic to crack if conditions are not carefully controlled.

Where press-fits are used, the designer generally seeks the maximum pullout force using the greatest allowable interference between parts that is consistent with the strength of the plastic. General equations for interference-fits (when the hub and shaft are made of the same materials and for when they are a metal shaft and a plastic hub) are given in Fig. 9.3. Safety factors of 1.5 to 2.0 are used in most applications. Figures 9.4 and 9.5 show calculated interference limits at room temperature for press-fitted shafts and hubs of Delrin acetal resin and Zytel nylon resin. These represent the maximum allowable interference based on yield point and elastic modulus data.

For a press-fit joint, the effect of thermal cycling, stress relaxation, and environmental conditioning must be carefully evaluated. Testing of the factory-assembled parts under expected temperature cycles, or under any condition that can cause changes to the dimensions or modulus of the parts, is obviously indicated. Differences in coefficient of thermal expansion can result in reduced interference due either to one material shrinking or expanding away from the other, or it can cause thermal stresses as the temperature changes. Since plastic materials will creep or stress-relieve under continued loading, loosening of the press-fit, at least to some extent, can be expected during service. To counteract this, the designer can knurl or groove the parts. The plastic will then tend to flow into the grooves and retain the holding power of the joint.

Snap-fit. In all types of snap-fit joints, a protruding part of one component, such as a hook, stud, or bead, is briefly deflected during the joining operation, and it is made to catch in a depression (undercut) in the mating component. This method of assembly is uniquely suited to thermoplastic materials due to their flexibility, high elongation, and ability to be molded into complex shapes. However, snap-fit joints cannot carry a load in excess of the force necessary to make or break the snap-fit. Snap-fit assemblies are usually employed to attach lids or covers which are meant to be disassembled or which will be lightly loaded. The design should be such that after the assembly, the joint will return to a stress free condition.

The two most common types of snap-fits are those with flexible cantilevered lugs (Fig. 9.6) and those with a full cylindrical undercut and mating lip (Fig. 9.7). Cylindrical snap-fits are generally stronger but require deformation for removal from the mold. Materials with good recovery characteristics are required. In order to obtain satisfactory results, the undercut design must fulfill certain requirements:

General equation for interference

$$I = \frac{S_d D_s}{W}\left[\frac{W\mu_h}{E_h} + \frac{1-\mu_s}{E_s}\right]$$

in which

$$W = \frac{1+\left(\frac{D_s}{D_h}\right)^2}{1-\left(\frac{D_s}{D_h}\right)^2}$$

I = Diametral interference, mm (in.)

S_d = Design stress limit or yield strength of the polymer, generally in the hub, MPa (psi) (A typical design limit for an interference fit with thermoplastics is 0.5% strain at 73°C.)

D_h = Outside diameter of hub, mm (in.)

D_s = Diameter of shaft, mm (in.)

E_h = Modulus of elasticity of hub, MPa (psi)

E_s = Elasticity of shaft, MPa (psi)

μ_h = Poisson's ratio of hub material

μ_s = Poisson's ratio of shaft material

W = Geometric factor

If the shaft and hub are of the same material, $E_h = E_s$ and $\mu_h = \mu_s$. The above equation simplifies to:

Shaft and hub of same material

$$I = \frac{S_d D_s}{W} \times \frac{W+1}{E_h}$$

If the shaft is a high modulus metal or other material, with $E_s > 34.4 \times 10^3$ MPa, the last term in the general interference equation is negligible, and the equation simplifies to:

Metal shaft, plastic hub

$$I = \frac{S_d D_s}{W} \times \frac{W+\mu_h}{E_h}$$

Figure 9.3 General calculation of interference-fit between a shaft and hub.[6]

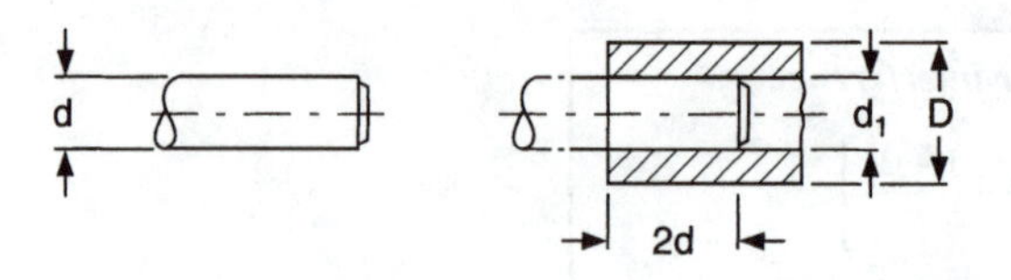

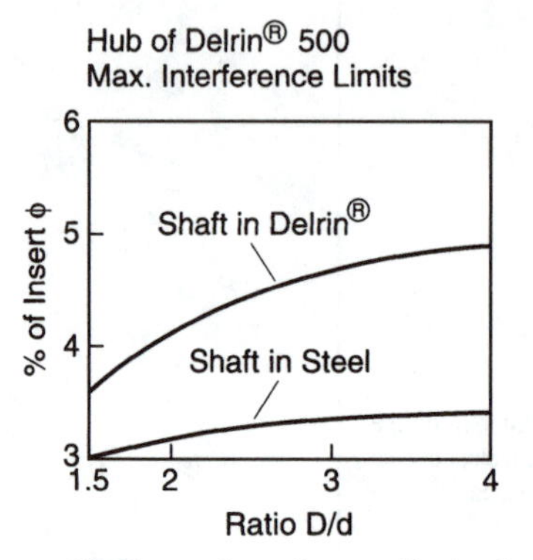

Figure 9.4 Maximum interference limits for Delrin acetal.[7]

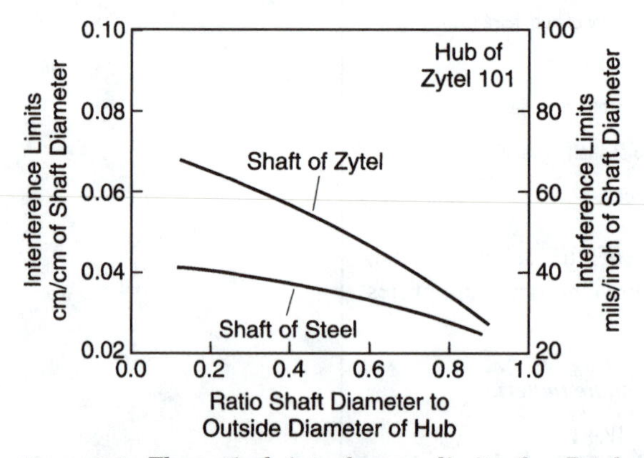

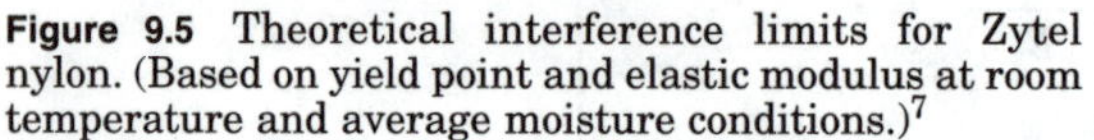

Figure 9.5 Theoretical interference limits for Zytel nylon. (Based on yield point and elastic modulus at room temperature and average moisture conditions.)[7]

- The wall thickness should be kept uniform.
- The snap-fit must be placed in an area where the undercut section can expand freely.
- The ideal geometric shape is circular.
- Ejection of an undercut core from the mold is assisted by the fact that the resin is still at relatively high temperatures.
- Weld lines should be avoided in the area of the undercut.

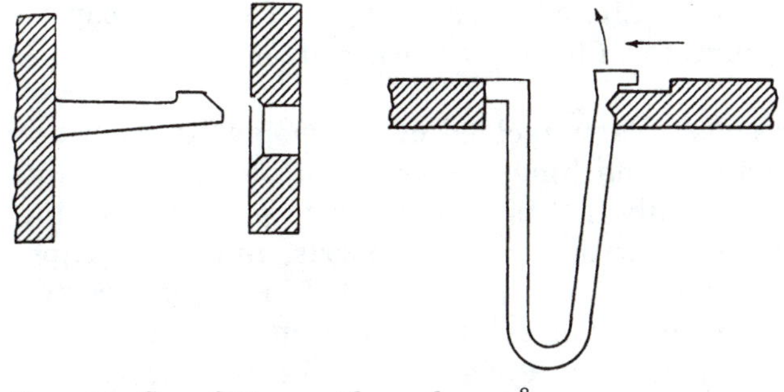

Figure 9.6 Snap-fitting cantilevered arms.[8]

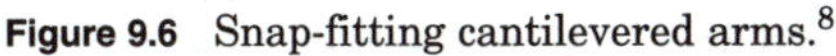

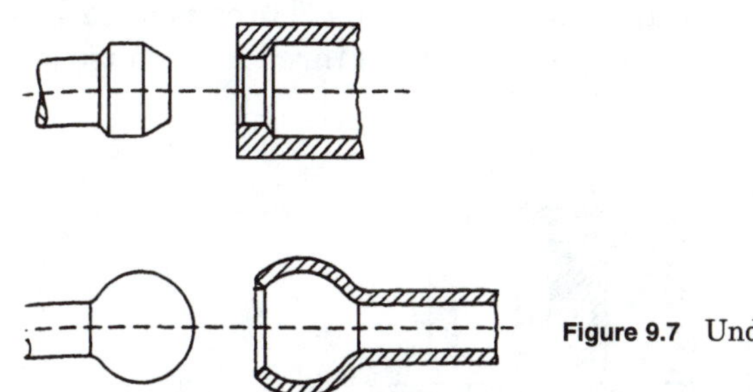

Figure 9.7 Undercuts for snap joints.[8]

In the cantilevered snap-fit design, the retaining force is essentially a function of the bending stiffness of the resin. Cantilevered lugs should be designed in a way so as not to exceed allowable stresses during assembly. Cantilevered snap-fits should be dimensioned to develop constant stress distribution over their length. This can be achieved by providing a slightly tapered section or by adding a rib. Special care must be taken to avoid sharp corners and other possible stress concentrations.

Many more designs and configurations can be used with snap-fit configurations. The individual plastic resin suppliers should be contacted for design rules and guidance on specific applications.

9.4.2 Mechanical fasteners

A large variety of mechanical fasteners can be used for joining plastic parts to themselves and to other materials. These include machine screws, self-tapping screws, rivets, spring clips, and nuts. In general, when repeated disassembly of the product is anticipated, mechanical fasteners are used.

Metal fasteners of high strength can overstress plastic parts, so torque-controlled tightening or special design provisions are required.

Where torque cannot be controlled, various types of washers can be used to spread the compression force over larger areas.

Machine screws and bolts. Parts molded of thermoplastic resin are sometimes assembled with machine screws or with bolts, nuts, and washers (Fig. 9.8), especially if it is a very strong plastic. Machine screws are generally used with threaded inserts, nuts, and clips. They are rarely used in pretapped holes. Molded-in inserts provide very high-strength assemblies and relatively low unit cost. However, molded-in inserts could increase cycle time while the inserts are manually placed in the mold. When the application involves infrequent disassembly, molded-in threads can be used successfully. Coarse threads can be molded into most materials. Threads of 32 or finer pitch should be avoided, along with tapered threads, because of

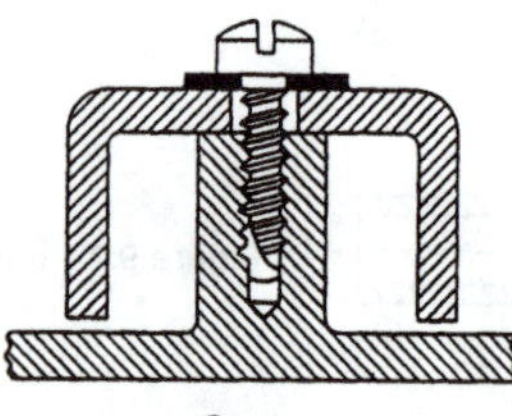

(a)

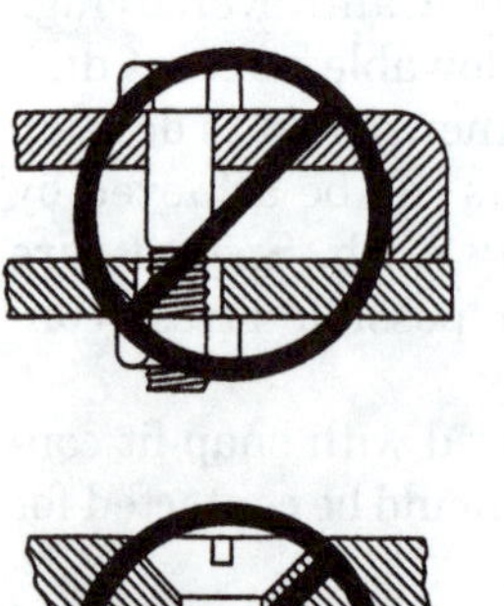

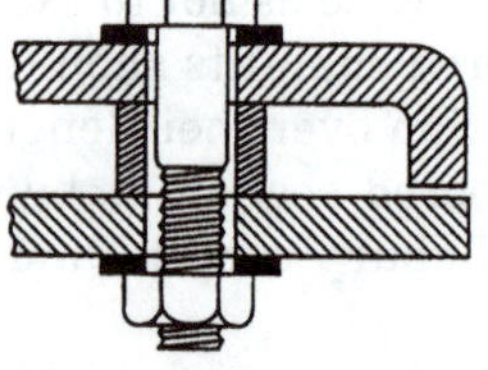

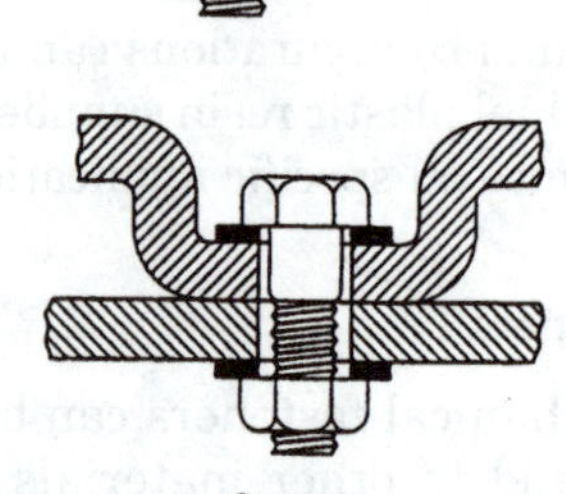

(b)

Figure 9.8 Mechanical fastening with: (*a*) self-taping screws and (*b*) bolts, nuts, and washers.[8]

excessive stress. If the mating part is metal, overtorque will result in part failure.

Postmolded inserts come in four types: press-in, expansion, self-tapping, and thread forming, and inserts that are installed by some method of heating (for example, ultrasonic). Metal inserts are available in a wide range of shapes and sizes for permanent installation. Inserts are typically installed in molded bosses, designed with holes to suit the insert to be used. Some inserts are pressed into place and others are installed by methods designed to limit the stress and increase strength. Generally, the outside of the insert is provided with projections of various configurations that penetrate the plastic and prevent movement under normal forces exerted during assembly.

Particular attention should be paid to the head of the fastener. Conical heads, called *flat heads,* produce undesirable tensile stresses and should not be used. Bolt or screw heads with a flat underside, such as pan heads, round heads, and so forth (Fig. 9.9), are preferred because the stress produced is more compressive. Flat washers are also suggested and should be used under both the nut and the fastener head. Sufficient diametrical clearance for the body of the fastener should always be provided in the parts to be joined. This clearance can nominally be 0.25 mm (0.010 in).

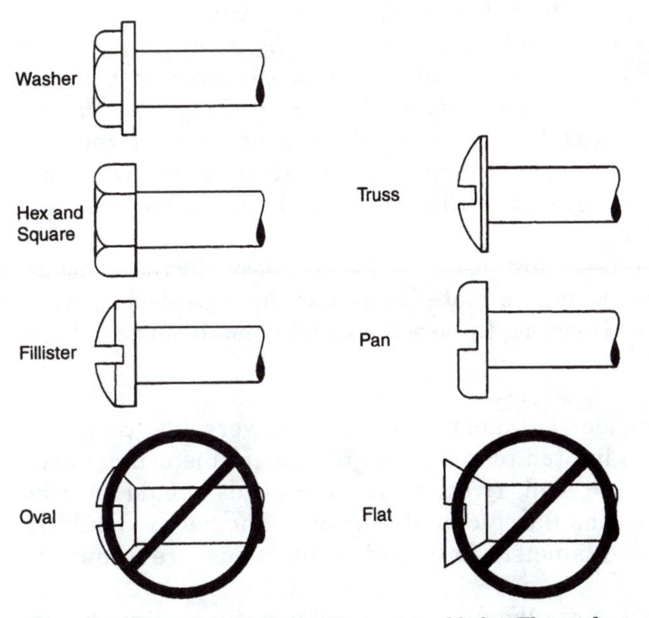

Figure 9.9 Common head styles of screws and bolts. Flat underside of head is preferred.[8]

Self-threading screws. Self-threading screws can either be thread-cutting or thread-forming. To select the correct screw, the designer must know which plastic will be used and its modulus of elasticity. The advantage of using these types of screws are

- Generally off-the-shelf items
- Low cost
- High production rates
- Minimum tooling investment

The principal disadvantage of these screws is limited reuse; after repeated disassembly and assembly, these screws will cut or form new threads in the hole, eventually destroying the integrity of the assembly.

Thread-forming screws are used in the softer, more ductile plastics with moduli below 1380 MPa (200,000 lb/in^2). There are a number of fasteners specially designed for use with plastics (Fig. 9.10). Thread-forming screws displace plastic material during the threading operation. This type of screw induces high stress levels in the part and is not recommended for parts made of weak resins.

Assembly strengths using thread-forming screws can be increased by reducing the hole diameter in the more ductile plastics, by increasing the screw thread engagement, or by going to a larger-diameter screw when space permits. The most common problem encountered with these types of screws is boss cracking. This can be minimized or eliminated by increasing the size of the boss, increasing the diameter of the hole, decreasing the size of the screw, changing the thread configuration of the screw, or changing the part to a more ductile plastic.

Thread-cutting screws are used in harder, less ductile plastics. Thread-cutting screws remove material as they are installed, thereby avoiding high stress. However, these screws should not be installed and removed repeatedly.

Rivets. Rivets provide permanent assembly at very low cost. The clamp load must be limited to low levels to prevent distortion of the part. To distribute the load, rivets with large heads should be used with washers under the flared end of the rivet. The heads should be 3 times the shank diameter. Standard rivet heads are shown in Fig. 9.11.

Riveted composite joints should be designed to avoid loading the rivet in tension. Generally, a hole $^1/_{64}$ in larger than the rivet shank is satisfactory for composite joints. A number of patented rivet designs

Blunt-tip fasteners are suitable for most commercial plastics. Harder plastics require a fastener with a cutting tip. Hardest plastics require both a piercing and drilling tip, as in these fasteners.

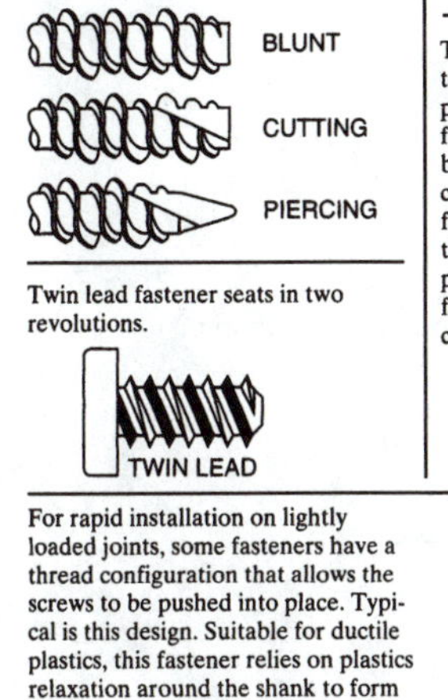

Twin lead fastener seats in two revolutions.

Reverse saw-tooth edges bite into the walls of the plastic.

Triangular configuration is another technique for capturing large amounts of plastic. After insertion, the plastic cold-flows or relaxes back into the area between lobes. The Trilobe design also creates a vent along the length of the fastener during insertion, eliminating the "ram" effect. In some ductile plastics, pressure builds up in the hole under the fastener as it is inserted, shattering or cracking the material.

Dual-height thread design boosts holding power by increasing the amount of plastic captured between threads.

HI-LO

Some specials have thread angles smaller than the 60° common on most standard screws. Included angles of 30 or 45° make sharper threads that can be forced into ductile plastics more readily, creating deeper mating threads and reducing stress. With smaller thread angles, boss size can sometimes be reduced.

For rapid installation on lightly loaded joints, some fasteners have a thread configuration that allows the screws to be pushed into place. Typical is this design. Suitable for ductile plastics, this fastener relies on plastics relaxation around the shank to form threads. The thread is helical so that it can be unscrewed, but reuse is limited.

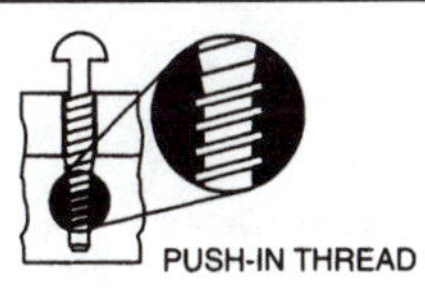

Barbs provide holding power.

BARBED

Pushtite fastener is pushed into place and can be screwed out.

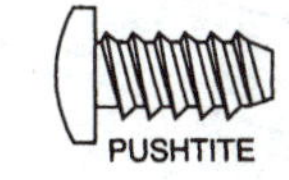

Figure 9.10 Thread forming fasteners for plastics.[9]

are commercially available for joining aircraft or aerospace structural composites.

Spring steel fasteners. Push-on spring steel fasteners (Fig. 9.12) can be used for holding light loads. Spring steel fasteners are simply pushed on over a molded stud. The stud should have a minimum 0.38-mm (0.015-in) radius at its base. Too large a radius could create a thick section, resulting in sinks or voids in the plastic molding.

9.4.3 Special consideration for composites

The efficiency of a composite structure is established, with very few exceptions, by its joints, not by its basic structure. The selection of a joining method for composites is as broad as with metals. Riveting, bolting, pinning, and bonding are all possible. Only welding and brazing cannot be applied to thermoset composites. However, thermoplastic and metallic matrix composites can be joined by welding or brazing. Composites can be mechanically fastened in a manner similar to metals. Parts are drilled, countersunk, and joined with a fastener.

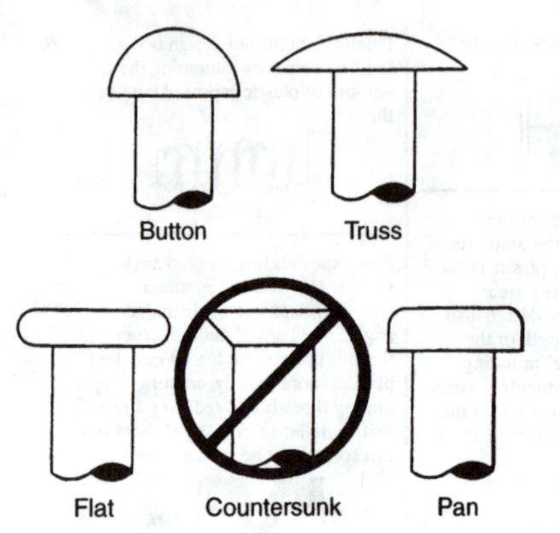

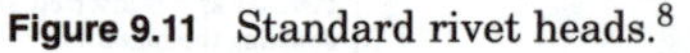
Figure 9.11 Standard rivet heads.[8]

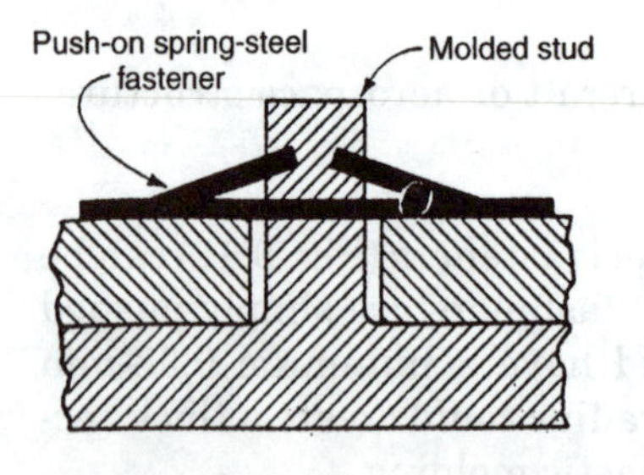

Figure 9.12 Push-on spring steel fasteners.[8]

Rivets, pins, two-piece bolts, and blind fasteners made of titanium, stainless steel, and aluminum are all used for composites. Several factors should be considered:

1. Differential expansion of the fastener in the composite
2. The effect of drilling on the structural integrity of the composite, as well as delamination caused by the fastener under load
3. Water intrusion between the fastener and composite
4. Electrical continuity of the composite and arching between fasteners
5. Possible galvanic corrosion at the composite joint
6. Weight of the fastening system
7. Environmental resistance of the fastening system

Table 9.6 provides a performance comparison for some of the mechanical fasteners most commonly used to join composite materials.

Aluminum and stainless fasteners expand and contract when exposed to temperature extremes, as in aircraft applications. In carbon-fiber composites, contraction and expansion of such fasteners can cause changes in clamping load. Pressure within the joint is often critical.

Drilling and machining can damage composites. Several techniques exist for producing quality holes in composites. Carbon, aramid, and boron fiber-reinforced materials each require different drilling methods and tools. When composites are cut, fibers are exposed. These fibers can absorb water, which weakens the material. Sealants can be used to prevent moisture absorption. Sleeved fasteners can provide fits that reduce water absorption as well as provide tightness (Fig. 9.13).

Fastener holes should be straight and round within limits specified. Normal hole tolerance is 0.075 mm (0.003 in). Interference-fits may cause delamination of the composite. Holes should be drilled perpendicular to the sheet within 1°. Special sleeved fasteners can limit the

TABLE 9.6 Fasteners for Advanced Composites[9]

		Suggested application				
Fastener type	Fastener material	Surface coating	Epoxy/ graphite composite	Kevlar	Fiberglass	Honeycomb
Blind rivets[a]	5056 Al	None	NR	E }[h]	E }[h]	}[e]
	Monel	None	G }[h]	E	E	
	A-286	Passivated	G	E	E	
Blind bolts[b]	A-286	Passivated	E[h]	E }[h]	E	}
	Alloy steel	Cadmium	NR	E	E	
Pull-type lockbolts	Titanium	None	E[d]	E[c]	E[c]	G or NR[f]
Stump-type lockbolts	Titanium	None	E[d]	E[c]	E[c]	G or NR[f]
Asp fasteners	Alloy steel	Cad/Nickel	G[g]	E	E	E
Pull-type lockbolts	7075 Al	Anodize	NR	E	E	NR

[a]Blind rivets with controlled shank expansion.
[b]Blind bolts are not shank expanding.
[c]Fasteners can be used with flanged titanium collars or standard aluminum collars.
[d]Use flanged titanium collar.
[e]Performance in honeycomb should be substantiated by installation testing.
[f]Depending on fastener design. Check with manufacturer.
[g]Nickel plated Asp only.
[h]Metallic structure on backside.
Note: NR = not recommended; E = excellent; G = good.

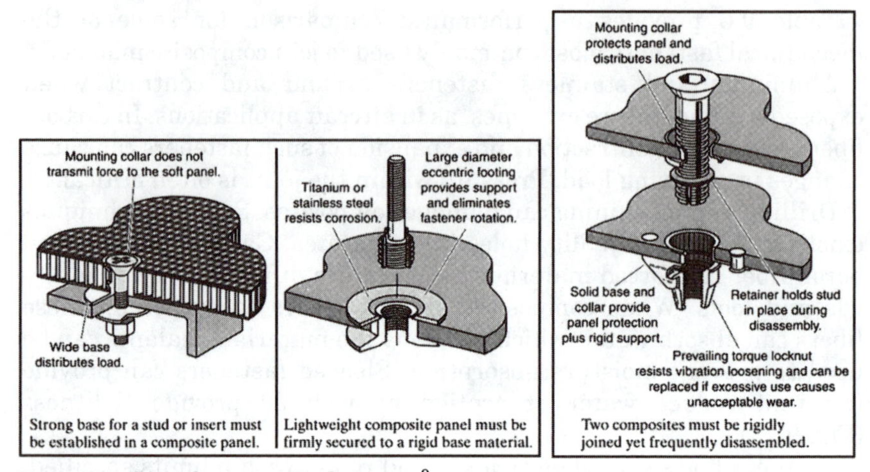

Figure 9.13 Specialty composite fasteners.[9]

chances of damage in the clearance hole and still provide an interference-fit. Fasteners can also be bonded in place with adhesives to reduce fretting.

Additionally, galvanic corrosion may occur in carbon fiber composites if aluminum fasteners are used. This occurs due to the chemical reaction of the aluminum with the carbon. Coating the fasteners guards against corrosion, but adds cost and time to assembly. Aluminum fasteners are often replaced by more expensive titanium and stainless steel fasteners in carbon fiber composite joints.

When joining composites with mechanical fasteners, special consideration must be given to creep. There are two kinds of creep: creep of the fastener hole and long-term material compression. The greater the material modulus, the lower the creep. There are mechanical ways to reinforce the hole or distribute the load so that the creep problem is minimized. For fasteners that rely on inserts, the ability of the composite to retain the fastener must be considered.

Like mechanically fastened metal structures, composites exhibit failure modes in tension, shear, and bearing. However, because of the complex failure mechanisms of composites, two further modes are possible, namely, cleavage and pullout. Figure 9.14 shows the location of each of these modes. Environmental degradation of a bolted joint, after exposure to a hot, wet environment is most likely to occur in the shear and bearing strength properties. The evidence shows that, for fiber-reinforced epoxies, temperature has a more significant effect than moisture, but in the presence of both at 127°C, a strength loss of 40% is possible. Evidence suggests that the failure behavior of thermoplas-

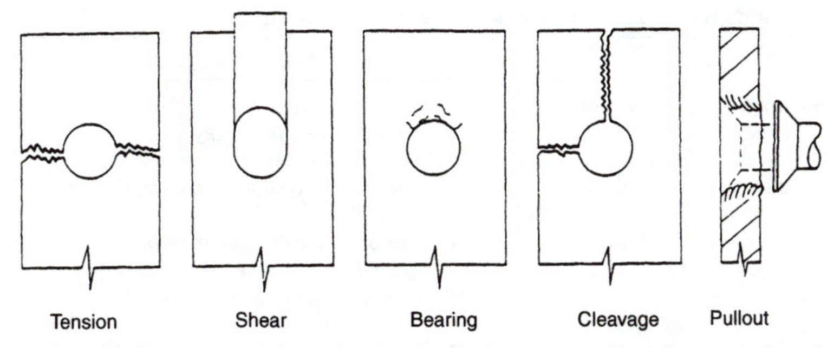

Figure 9.14 Modes of failure for mechanical joints in fiberglass reinforced plastics.[10]

tics is much the same as for thermoset composites. High joint efficiencies can be obtained with suitable consideration to the joint design, fastener type, and environmental factors.

9.5 Adhesive Bonding

Adhesive bonding presents several distinct advantages over other methods of fastening plastic substrates. These are summarized in Table 9.7. Solvent and heat welding may be considered, as a type of adhesive bonding process where the adhesive is actually part of the substrate itself. Bonding, as a method of assembly, is often preferred when different types of substrates (for example, metals to plastics) need to be joined, when high-volume production is necessary, or when the design of the finished part prohibits the use of mechanical fasteners.

Although there are various ways of joining plastics to themselves or to other materials, adhesive bonding has often proved to be the most effective assembly method. In many applications the use of adhesives rather than metal fasteners reduces product cost and the weight of the assembly, and, in some cases, provides longer service life. Adhesive bonding can also be used very effectively in prototypes and with large or intricate assemblies that for economic or design reasons cannot be molded or processed as a single part.

However, the joining of plastics with adhesives can be made difficult because of their low surface energy, poor wetting, and presence of contaminants such as mold release agents, low molecular weight internal components (for example, flexibilizers, UV inhibitors, and processing aids), and possible susceptibility to moisture and other environmental factors. Fortunately, numerous adhesives and processing methods are available for the joining of plastic materials, and they have been suc-

TABLE 9.7 Advantages and Disadvantages of Adhesives Bonding

Advantages	Disadvantages
1. Provides large stress-bearing area.	1. Surfaces must be carefully cleaned.
2. Provides excellent fatigue strength.	2. Long cure times may be needed.
3. Damps vibration and absorbs shock.	3. Limitation on upper continuous operating temperature (generally 350°F).
4. Minimizes or prevents galvanic corrosion between dissimilar metals.	4. Heat and pressure may be required.
5. Joins all shapes and thicknesses.	5. Jigs and fixtures may be needed.
6. Provides smooth contours.	6. Rigid process control usually necessary.
7. Seals joints.	7. Inspection of finished joint difficult.
8. Joins any combination of similar or dissimilar materials.	8. Useful life depends on environment.
9. Often less expensive and faster than mechanical fastening.	9. Environmental, health, and safety considerations are necessary.
10. Heat, if required, is too low to affect metal parts.	10. Special training sometimes required.
11. Provides attractive strength-to-weight ratio.	

cessfully used in many applications. Many of these processes and applications are described in articles and handbooks on the subject. The plastic resin manufacturer is generally the leading source of information on the proper methods of joining a particular plastic.

9.5.1 Theories of adhesion

Various theories attempt to describe the phenomena of adhesion. The adhesion theories that are applicable to plastic substrates are provided in the following sections.

Mechanical theory. The surface of a solid material is never truly smooth but consists of a maze of microscopic peaks and valleys. According to the mechanical theory of adhesion, the adhesive must penetrate the cavities on the surface and displace the trapped air at the interface. Some mechanical anchoring appears to be a prime factor in bonding many rough or porous substrates. Adhesives also frequently bond better to abraded surfaces than to natural surfaces (although this is sometimes not true of certain low-surface energy plastics). *Mechanical abrasion* is a popular surface preparation step prior to adhesive bonding. The beneficial effects of surface roughening may be due to

- Mechanical interlocking
- Formation of a clean surface
- Formation of a more reactive surface
- Formation of a larger surface area

Adsorption theory. The *adsorption theory* states that adhesion results from molecular contact between two materials and the surface forces that develop. The process of establishing intimate contact between an adhesive and the adherend is known as *wetting.* Figure 9.15 shows examples of good and poor wetting of a liquid adhesive spreading over a substrate surface. For an adhesive to adequately wet a solid surface, the adhesive should have a lower surface tension than the solid's critical surface tension:

Adhesive surface tension << Substrate's critical surface tension

Tables 9.8 and 9.9 list surface tensions of common adherends and liquids used as adhesives. Most liquid adhesives easily wet metallic solids because of the high surface tension of most metals. However, many solid organic substrates have surface tensions less than those of common adhesives.

From Tables 9.8 and 9.9, it can be forecast that epoxy adhesives will wet clean aluminum or copper surfaces. However, epoxy resin will not wet a substrate having a critical surface tension significantly less than 47 dyn/cm. Epoxies will not, for example, wet either a metal surface contaminated with silicone oil or a clean polyethylene substrate. For wetting to occur, the substrate surface has to be chemically or physically altered by some mechanism to raise its surface energy. This is why there are so many prebond surface treatments for plastic substrates.

TABLE 9.8 Critical Surface Tensions of Common Plastics and Metals

Materials	Critical surface tension, dyne/cm
Acetal	47
Acrylonitrile-butadiene-styrene	35
Cellulose	45
Epoxy	47
Fluoroethylene propylene	16
Polyamide	46
Polycarbonate	46
Polyethylene	31
Polyethylene terephthalate	43
Polyimide	40
Polymethylmethacrylate	39
Polyphenylene sulfide	38
Polystyrene	33
Polysulfone	41
Polytetrafluoroethylene	18
Polyvinyl chloride	39
Silicone	24
Aluminum	~500
Copper	~1000

TABLE 9.9 Surface Tension of Common Adhesives and Liquids

Material	Surface tension, dyne/cm
Epoxy resin	47
Fluorinated epoxy resin*	33
Glycerol	63
Petroleum lubricating oil	29
Silicone oils	21
Water	73

*Experimental resin; developed to wet low-energy surfaces. (Note low surface tension relative to most plastics.)

After intimate contact is achieved between adhesive and adherend through wetting, adhesion results primarily through the forces of molecular attraction. The adhesion between adhesive and adherend is believed to be primarily due to the van der Walls forces of attraction.

Electrostatic and diffusion theories. The *electrostatic theory* states that electrostatic forces in the form of an electrical double layer are formed at the adhesive-adherend interface. These forces account for resistance to separation. The electrostatic theory of adhesion is not generally applicable to common production assembly.

The fundamental concept of the *diffusion theory* is that adhesion occurs through the interdiffusion of molecules in the adhesive and

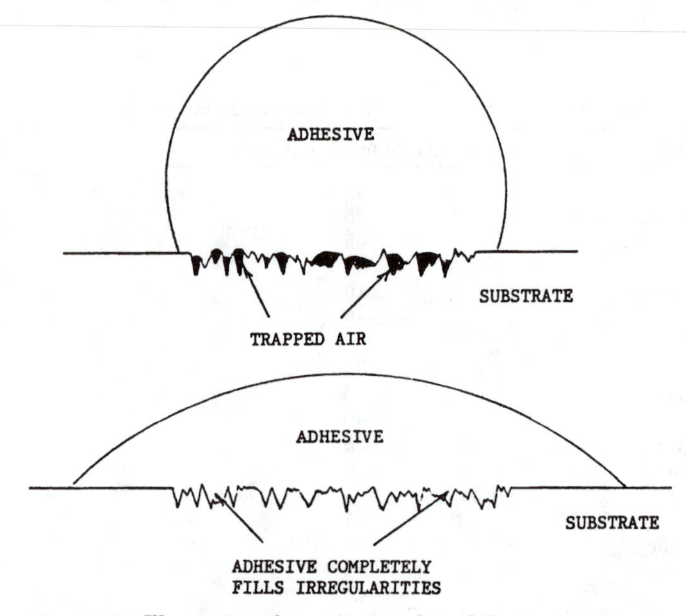

Figure 9.15 Illustration of poor (top) and good (bottom) wetting by an adhesive spreading over a substrate surface.

adherend. The diffusion theory is primarily applicable when both the adhesive and adherend are polymeric. For example, bonds formed by solvent or heat welding of thermoplastics result from the diffusion of molecules.

Weak boundary layer theory. According to the *weak boundary layer theory,* when bond failure seems to be at the interface, usually a cohesive break of a weak boundary layer is the real event. Weak boundary layers can originate from the adhesive, the adherend, the environment, or a combination of any of the three. When bond failure occurs, it is the weak boundary layer that fails, although failure seems to occur at the adhesive-adherend interface. Figure 9.16 shows examples of certain possible weak boundary layers for a metallic substrate. For plastic substrates there are many more opportunities for weak boundary layers, such as mold release, plasticizer migration, and moisture migrating to the interface. Certain weak boundary layers can be removed or strengthened by various surface treatments.

9.5.2 Requirements for a good bond

The basic requirements for a good bond to plastic substrates are surface cleanliness, wetting of the surface by the adhesive, solidification of the adhesive, and proper selection of adhesive and joint design. These requirements will be briefly defined here and then more thoroughly discussed in the following sections.

To achieve an effective adhesive bond, one must start with a clean surface. Foreign materials, such as dirt, oil, mold release agents, etc., must be removed from the surface or else the adhesive will bond to these weak boundary layers rather than to the actual substrate. To have acceptable joint strength, many plastic materials require treatment prior to bond-

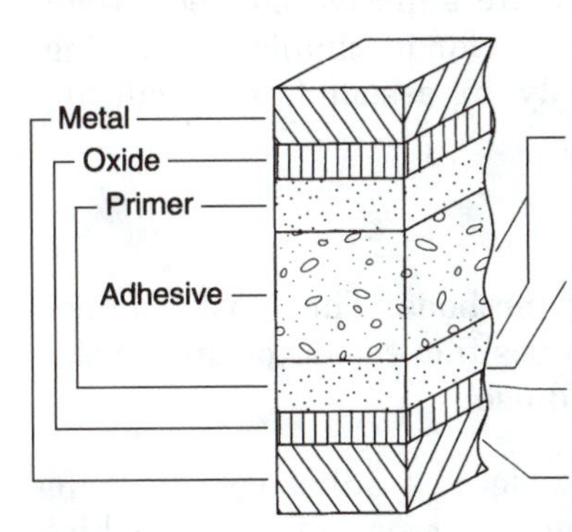
Metal
Oxide
Primer
Adhesive
Moisture permeates the adhesive or primer and causes debonding followed by corrosion or adhesive deterioration.
Moisture causes a reaction between the primer and oxide, which leads to debonding.
With aging, hydration or dissolution of the oxide occurs.
Oxide breaks down, allowing corrosive attack.

Figure 9.16 Examples of weak boundary layers in a metal adhesive joint.[9]

ing to provide one or a combination of the following: surface roughness, raised surface energy, removal of contaminants or weak boundary layers, and strengthening of boundary layers. At the minimum, the surface must be made free of all contaminants such as oil, grease, mold release agents, water, or polishing compounds. Solvents or detergents are generally used to clean the plastic parts prior to bonding.

While it is in the liquid state, the adhesive must wet the substrate. The result of good wetting is greater contact area between adherend and adhesive over which the forces of adhesion may act. If the surface is one of low energy, wetting cannot be obtained without some sort of surface modification. For example, polytetrafluoroethylene, polyethylene, and certain other polymeric materials are completely unsuitable for adhesive bonding in their natural state. For these surfaces, mechanical abrasion will only provide unwettable air pockets at the interface, resulting in lower bond strength. Thus, for surfaces where wetting is difficult, mechanical abrasion is not recommended as a surface preparation. Plastics with lower surface energy may need to be chemically or physically treated prior to bonding.

The liquid adhesive, once applied, must be capable of conversion into a solid. The process of solidifying can be completed in different ways. Adhesives are generally solidified by

- Chemical reaction by any combination of heat pressure and curing agent
- Cooling from a molten liquid to a solid state
- Drying due to solvent evaporation

The main areas of concern when selecting an adhesive are the materials to be bonded, service requirements, production requirements, and overall cost.

The adhesive joint should be designed to optimize the forces acting on and within the joint. Although adequate adhesive bonded assemblies have been made from joints designed for mechanical fastening, maximum benefit can be obtained only in assemblies specifically designed for adhesive bonding.

9.5.3 Basic adhesive materials

Adhesives may be classified by many methods. The most common methods are by function, chemical composition, mode of application and setting, physical form, cost, and end use.

Function. The *functional classification* defines adhesives as being structural or nonstructural. *Structural adhesives* are materials of high strength and permanence. They are generally represented by thermosetting adhesives with shear strengths greater than 1000 lb/in^2. Their primary function is to hold structures together and be capable of resisting high loads.

Nonstructural adhesives are not required to support substantial loads but merely to hold lightweight materials in place. Nonstructural adhesives are sometimes referred to as *holding adhesives.* They are generally represented by pressure-sensitive, contact, and hot-melt adhesives. Sealants usually have a nonstructural function. They are principally intended to fill a gap between adherends to provide a seal without having high degrees of adhesive strength.

Chemical composition. The *composition classification* describes synthetic adhesives as thermosetting, thermoplastic, elastomeric, or combinations of these. They are described generally in Table 9.10.

Thermosetting adhesives are materials that cannot be heated and softened repeatedly after the initial cure. Adhesives of this sort cure by chemical reaction at room or elevated temperatures, depending on the type of adhesive. Substantial pressure may also be required with some thermosetting adhesives, and others are capable of providing strong bonds with only contact pressure. Thermosetting adhesives are sometimes provided in a solvent medium to facilitate application. However, they are also commonly available as solventless liquids, pastes, and solids.

Thermosetting adhesives may be sold as multiple and single-part systems. Generally, the single-part adhesives require elevated-temperature cure, and they have a limited shelf life. Multiple-part adhesives have longer shelf lives, and some can be cured at room temperature or more rapidly at elevated temperatures. But they require metering and mixing before application. Once the adhesive is mixed, the working life is limited. Because molecules of thermosetting resins are densely cross-linked, their resistance to heat and solvents is good, and they show little elastic deformation under load at elevated temperatures.

Thermoplastic adhesives do not cross-link during cure, so they can be resoftened with heat. They are single-component systems that harden upon cooling from a melt or by evaporation of a solvent or water vehicle. Hot-melt adhesives commonly used in packaging are examples of a solid thermoplastic material that is applied in a molten state, and adhesion develops as the melt solidifies during cooling. Wood glues are thermoplastic emulsions that are a common household item. They harden by evaporation of water from an emulsion.

Thermoplastic adhesives have a more limited temperature range than thermosetting types. It is not suggested to use thermoplastic adhesives over 150°F. Their physical properties vary over a wide range because many polymers are used in a single adhesive formulation.

TABLE 9.10 Adhesives Classified by Chemical Composition[11]

Classification	Thermoplastic	Thermosetting	Elastomeric	Alloys
Types within group	Cellulose acetate, cellulose acetate butyrate, cellulose nitrate, polyvinyl acetate, vinyl vinylidene, polyvinyl acetals, polyvinyl alcohol, polyamide, acrylic, phenoxy	Cyanoacrylate, polyester, urea formaldehyde, melamine formaldehyde, resorcinol and phenol-resorcinol formaldehyde, epoxy, polyimide, polybenzimidazole, acrylic, acrylate acid diester	Natural rubber, reclaimed rubber, butyl, polyisobutylene, nitrile, styrene-butadiene, polyurethane, polysulfide, silicone, neoprene	Epoxy-phenolic, epoxy-polysulfide, epoxy-nylon, nitrile-phenolic, neoprene-phenolic, vinyl-phenolic
Most used form	Liquid, some dry film	Liquid, but all forms common	Liquid, some film	Liquid, paste, film
Common further classifications	By vehicle (most are solvent dispersions or water emulsions)	By cure requirements (heat and/or pressure most common but some are catalyst types)	By cure requirements (all are common); also by vehicle (most are solvent dispersions or water emulsions)	By cure requirements (usually heat and pressure except some epoxy types); by vehicle (most are solvent dispersions or 100% solids); and by type of adherends or end-service conditions
Bond characteristics	Good to 150 to 200°F; poor creep strength; fair peel strength	Good to 200 to 500°F; good creep strength; fair peel strength	Good to 150 to 400°F; never melt completely; low strength; high flexibility	Balanced combination of properties of other chemical groups depending on formulation; generally higher strength over wider temp range
Major type of use	Unstressed joints; designs with caps, overlaps, stiffeners	Stressed joints at slightly elevated temp	Unstressed joints on lightweight materials; joints in flexure	Where highest and strictest end-service conditions must be met; sometimes regardless of cost, as military uses
Materials most commonly bonded	Formulation range covers all materials, but emphasis on nonmetallics—especially wood, leather, cork, paper, etc.	For structural uses of most materials	Few used “straight” for rubber, fabric, foil, paper, leather, plastics films; also as tapes. Most modified with synthetic resins	Metals, ceramics, glass, thermosetting plastics; nature of adherends often not as vital as design or end-service conditions (that is, high strength and temperature)

Elastomeric-type adhesives are based on polymers having great toughness and elongation. These adhesives may be supplied as solvent solutions, latex cements, dispersions, pressure-sensitive tapes, and single- or multiple-part solventless liquids or pastes. The curing requirements vary with the type and form of elastomeric adhesives. These adhesives can be formulated for a wide variety of applications, but they are generally noted for their high degree of flexibility and good peel strength.

Adhesive alloys or hybrids are made by combining thermosetting, thermoplastic, and elastomeric adhesives. They utilize the most useful properties of each material. However, the adhesive alloy is usually never better than at its weakest constituent. For example, higher peel strengths are generally provided to thermosetting resins by the addition of thermoplastic or elastomeric materials, although usually at the sacrifice of temperature resistance. Adhesive alloys are commonly available in solvent solutions and as supported or unsupported film.

Mode of application and setting. Adhesives are often classified by their mode of application. Depending on viscosity, adhesives can be sprayed on, brushed on, or applied with a trowel. Heavily bodied adhesive pastes and mastics are considered extrudable; they are applied by syringe, caulking gun, or pneumatic pumping equipment.

Another distinction between adhesives is the manner in which they flow or solidify. Some adhesives solidify simply by losing solvent while others harden as a result of heat activation or chemical reaction. Pressure-sensitive systems flow under pressure and are stable when pressure is absent.

Physical form. Another method of distinguishing between adhesives is by physical form. The physical state of the adhesive generally determines how it is to be applied. Liquid adhesives lend themselves to easy handling via mechanical spreaders or spray and brush. Paste adhesives have high viscosities to allow application on vertical surfaces with little danger of sag or drip. These bodied adhesives also serve as gap fillers between two poorly mated substrates. Tape and film adhesives are poor gap fillers but offer a uniformly thick bond line, no need for metering, and easy dispensing. Adhesive films are available as a pure sheet of adhesive or with cloth or paper reinforcement. Another form of adhesive is powder or granules that must be heated or solvent-activated to be made liquid and applicable.

Cost. The cost of fastening with adhesives must include the material cost of the adhesive, the cost of labor, the cost of equipment, the time required to cure the adhesive, and the economic loss due to rejects of defective joints. Often the actual material cost of the adhesive is rather minor compared to the total assembly cost per unit.

Adhesive material cost should be calculated on a cost per bonded area basis. Since many adhesives are sold as dilute solutions, a cost per unit weight or volume basis may lead to erroneous comparisons.

Adhesive price is dependent on development costs and volume requirements. Adhesives that have been specifically designed to be resistant to adverse environments are more expensive than general-purpose adhesives. Adhesive prices range from pennies a pound for inorganic and animal-based systems to hundreds of dollars per pound for certain heat-resistant synthetic adhesives. Adhesives in film or powder form require more processing than liquid or paste types and are more expensive.

9.5.4 Joint design

Types of stress. In order to effectively design joints for adhesive bonding, it is necessary to understand the types of stress that are common to bonded structures. Four basic types of loading stress are common to adhesive joints: tensile, shear, cleavage, and peel. Any combination of these stresses, illustrated in Fig. 9.17, may be encountered in an adhesive application.

Tensile stress develops when forces acting perpendicular to the plane of the joint are distributed uniformly over the entire bonded area. Adhesive joints show good resistance to tensile loading because all the adhesive contributes to the strength of the joint. In practical applications, unfortunately, loads are rarely axial, and cleavage or peel stresses tend to develop. Since adhesives have poor resistance to cleavage and peel, joints designed to load the adhesive in tension should have physical restraints to ensure axial loading.

Shear stresses result when forces acting in the plane of the adhesive try to separate the adherends. Joints dependent upon the adhesive's shear strength are relatively easy to make and are commonly used. Adhesives are generally strongest when stressed in shear because all the bonded area contributes to the strength of the joint.

Cleavage and peel stresses are undesirable. *Cleavage* occurs when forces at one end of a rigid, bonded assembly act to split the adherends apart. *Peel stress* is similar to cleavage but applies to a joint where one or both of the adherends are flexible. Joints loaded in peel or cleavage provide much lower strength than joints loaded in shear because the

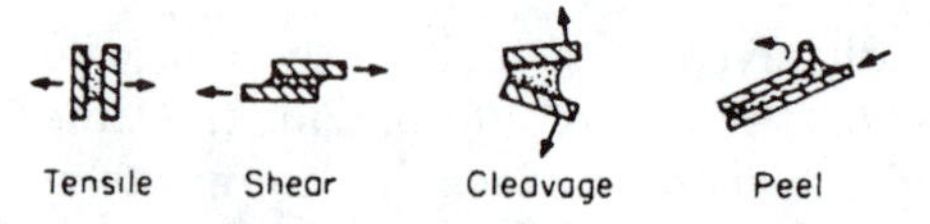

Figure 9.17 Four basic types of adhesive stress.

stress is concentrated on only a very small area of the total bond. Peel stress, in particular, should be avoided where possible, since the stress is confined to a very thin line at the edge of the bond (Fig. 9.18). The remainder of the bonded area makes no contribution to the strength of the joint.

Maximizing joint efficiency. Although adhesives have often been used successfully on joints designed for mechanical fastening, the maximum efficiency of bonded joints can be obtained only by designing the joint specifically for adhesive bonding. To avoid concentration of stress, the joint designer should take into consideration the following rules:

1. Keep the stress on the bond line to a minimum.
2. Design the joint so that the operating loads will stress the adhesive in shear.
3. Peel and cleavage stresses should be minimized.
4. Distribute the stress as uniformly as possible over the entire bonded area.
5. Adhesive strength is directly proportional to bond width. Increasing width will always increase bond strength; increasing the depth does not always increase strength.
6. Generally, rigid adhesives are better in shear, and flexible adhesives are better in peel.

Brittle adhesives are particularly weak in peel because the stress is localized at only a thin line, as shown in Fig. 9.18. Tough, flexible adhesives distribute the peeling stress over a wider bond area and show greater resistance to peel.

For a given adhesive and adherend, the strength of a joint stressed in shear depends primarily on the width and depth of the overlap and

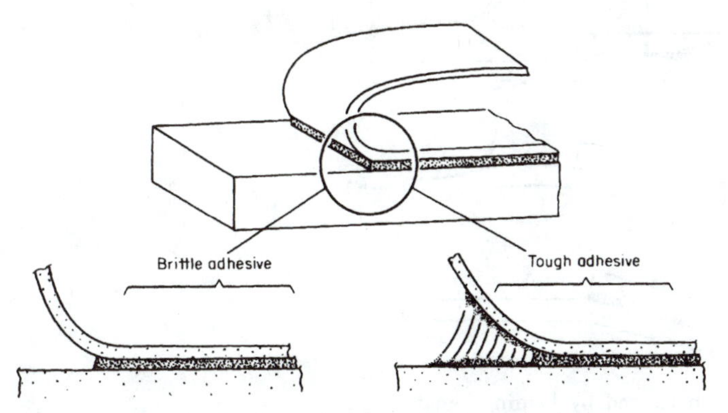

Figure 9.18 Tough, flexible adhesives distribute peel stress over a larger area.[12]

the thickness of the adherend. Adhesive shear strength is directly proportional to the width of the joint. Strength can sometimes be increased by increasing the overlap depth, but the relationship is not linear. Since the ends of the bonded joint carry a higher proportion of the load than the interior area, the most efficient way of increasing joint strength is by increasing the width of the bonded area.

In a shear joint made from thin, relatively flexible adherends, there is a tendency for the bonded area to distort because of eccentricity of the applied load. This distortion, illustrated in Fig. 9.19, causes cleavage stress on the ends of the joint, and the joint strength may be considerably impaired. Thicker adherends are more rigid, and the distortion is not as much a problem as with thin-gauge adherends. Since the stress distribution across the bonded area is not uniform and depends on joint geometry, the failure load of one specimen cannot be used to predict the failure load of another specimen with different joint geometry.

The strength of an adhesive joint also depends on the thickness of the adhesive. Thin adhesive films offer the highest shear strength provided that the bonded area does not have "starved" areas where all the adhesive has been forced out. Excessively heavy adhesive-film thickness causes greater internal stresses during cure and concentration of stress under load at the ends of a joint. Optimum adhesive thickness for maximum shear strength is generally between 2 and 8 mils. Strength does not vary significantly with bond-line thickness in this range.

Joint geometry. The ideal adhesive-bonded joint is one in which under all practical loading conditions the adhesive is stressed in the direction

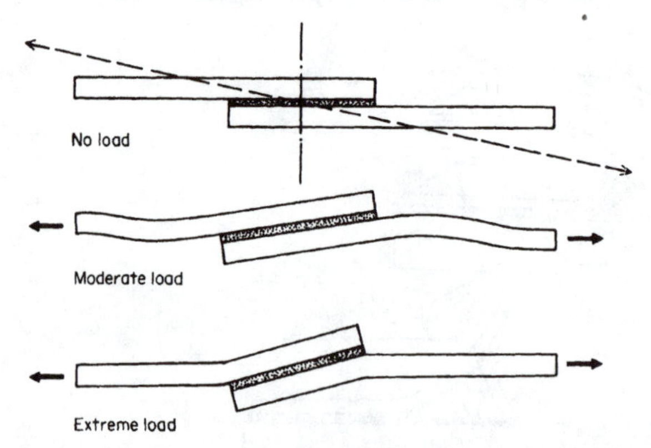

Figure 9.19 Distortion caused by loading can introduce cleavage stresses and must be considered in the joint design.[11]

in which it most resists failure. A favorable stress can be applied to the bond by using proper joint design. Some joint designs may be impractical, expensive to make, or hard to align. The design engineer will often have to weigh these factors against optimum adhesive performance.

Joints for flat adherends. The simplest joint to make is the plain *butt joint.* Butt joints cannot withstand bending forces because the adhesive would experience cleavage stress. If the adherends are too thick to design simple overlap-type joints, the butt joint can be improved by redesigning in a number of ways, as shown in Fig. 9.20. All the modified butt joints reduce the cleavage effect caused by side loading. Tongue-and-groove joints also have an advantage in that they are self-aligning and act as a reservoir for the adhesive. The scarf joint keeps the axis of loading in line with the joint and does not require a major machining operation.

Lap joints are the most commonly used adhesive joint because they are simple to make, are applicable to thin adherends, and stress the adhesive in shear. However, the simple lap joint can cause stresses other than shear. In this design, the adherends are offset and the shear forces are not in line, as was illustrated in Fig. 9.19. This factor results in cleavage stress at the ends of the joint, which seriously impairs its efficiency. Modifications of lap-joint design include:

1. Redesigning the joint to bring the load on the adherends in line
2. Making the adherends more rigid (thicker) near the bond area
3. Making the edges of the bonded area more flexible for better conformance, thus minimizing peel

Modifications of lap joints are shown in Fig. 9.21.

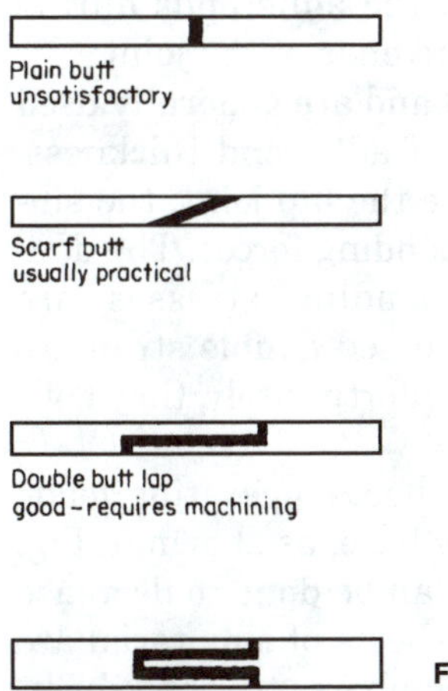

Figure 9.20 Butt connections.

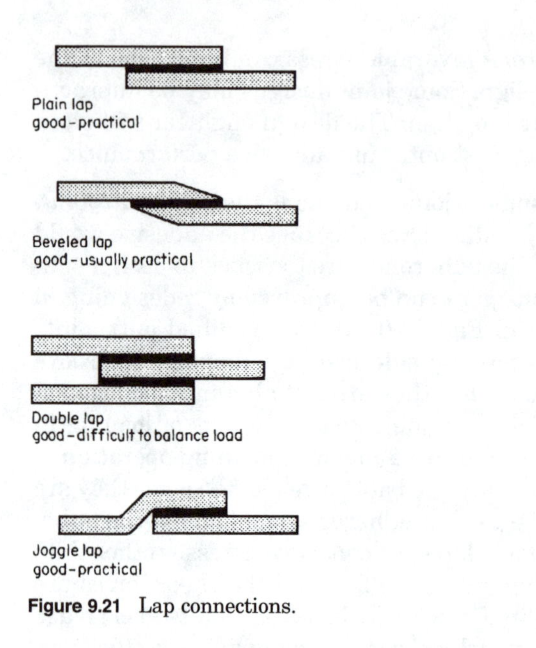

Figure 9.21 Lap connections.

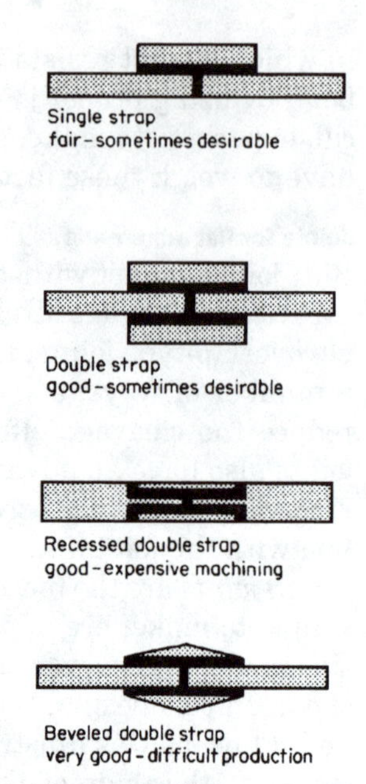

Figure 9.22 Strap connections.

The joggle-lap-joint design is the easiest method for bringing loads into alignment. The joggle lap can be made by simply bending the adherends. It also provides a surface to which it is easy to apply pressure. The double lap joint has a balanced construction, which is subjected to bending only if loads on the double side of the lap are not balanced. The beveled lap joint is also more efficient than the plain lap joint. The beveled edges allow conformance of the adherends during loading, thereby reducing cleavage stress on the ends of the joint.

Strap joints keep the operating loads aligned and are generally used where overlap joints are impractical because of adherend thickness. Strap-joint designs are shown in Fig. 9.22. Like the lap joint, the single strap is subjected to cleavage stress under bending forces. The double strap joint is more desirable when bending stresses are encountered. The beveled double strap and recessed double strap are the best joint designs to resist bending forces. Unfortunately, they both require expensive machining.

When thin members are bonded to thicker sheets, operating loads generally tend to peel the thin member from its base, as shown in Fig. 9.23. The subsequent illustrations show what can be done to decrease peeling tendencies in simple joints. Often thin sheets of a material are made more rigid by bonding stiffening members to the sheet.

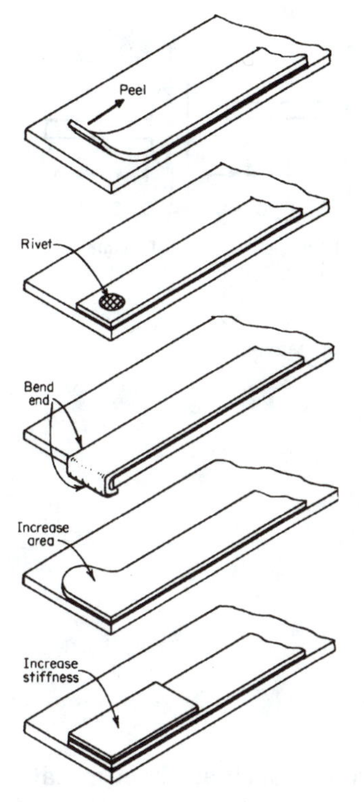

Figure 9.23 Minimizing peel in adhesive joint.[13]

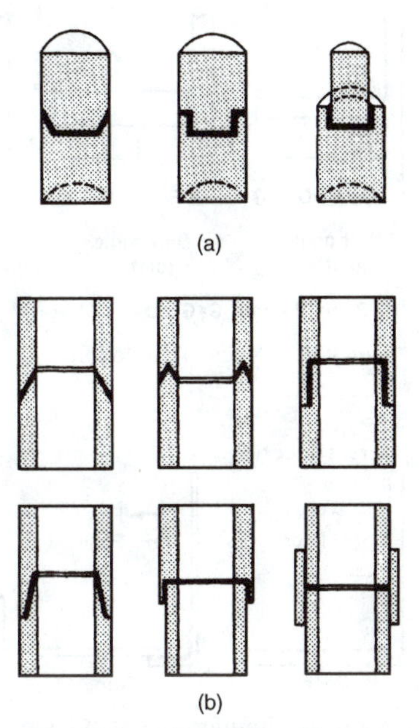

Figure 9.24 Recommended designs for rod and tube joints: (*a*) three joint designs for adhesive bonding of round bars and (*b*) six joint configurations useful in adhesive bonding cylinders or tubes.[14]

Resistance to bending forces is also increased by extending the bond area and increasing the stiffness of the base sheet.

Cylindrical-joint design. Several recommended designs for rod and tube joints are illustrated in Fig. 9.24. These designs should be used instead of the simpler butt joint. Their resistance to bending forces and subsequent cleavage is much better, and the bonded area is larger. Unfortunately, most of these joint designs require a machining operation.

Angle- and corner-joint designs. A butt joint is the simplest method of bonding two surfaces that meet at an odd angle. Although the butt joint has good resistance to pure tension and compression, its bending strength is very poor. Dado, L, and T angle joints, shown in Fig. 9.25, offer greatly improved properties. The T design is the preferable angle joint because of its large bonding area and good strength in all directions.

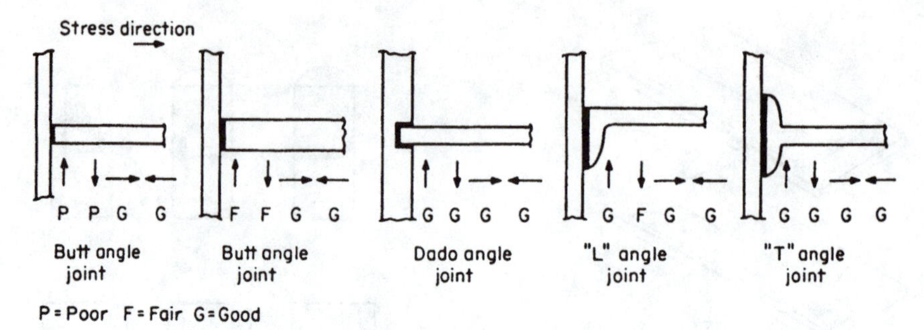

Figure 9.25 Types of angle joints and methods of reducing cleavage.[13]

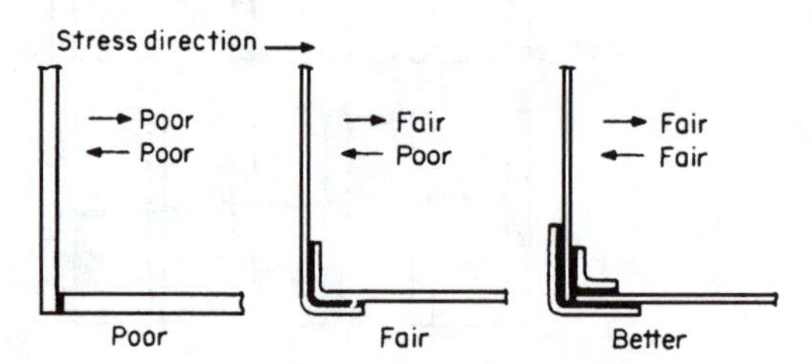

Figure 9.26 Reinforcement of bonded corner joints.

Corner joints for relatively flexible adherends, such as sheet metal, should be designed with reinforcements for support. Various corner-joint designs are shown in Fig. 9.26. With very thin adherends, angle joints offer low strengths because of high peel concentrations. A design consisting of right-angle corner plates or slip joints offers the most satisfactory performance. Thick, rigid members, such as rectangular bars and wood, may be bonded with an end lap joint, but greater strengths can be obtained with mortise and tenon. Hollow members, such as extrusions, fasten together best with mitered joints and inner splines.

Flexible plastics and elastomers. Thin or flexible polymeric substrates may be joined using a simple or modified lap joint. The double strap joint is best, but also the most time-consuming to fabricate. The strap material should be made out of the same material as the parts to be joined, or at least have approximately equivalent strength, flexibility, and thickness. The adhesive should have the same degree of flexibility as the adherends.

If the sections to be bonded are relatively thick, a scarf joint is acceptable. The length of the scarf should be at least 4 times the thickness; sometimes larger scarves may be needed.

When bonding elastic material, forces on the elastomer during cure of the adhesive should be carefully controlled, since excess pressure will cause residual stresses at the bond interface. Stress concentrations may also be minimized in rubber-to-metal joints by elimination of sharp corners and using metal thick enough to prevent peel stresses that may arise with thinner-gauge metals.

As with all joint designs, polymeric joints should avoid peel stress. Figure 9.27 illustrates methods of bonding flexible substrates so that the adhesive will be stressed in its strongest direction.

Rigid plastic composites. Reinforced plastics are often anisotropic materials. This means their strength properties are directional. Joints made from anisotropic substrates should be designed to stress both the adhesive and adherend in the direction of greatest strength. Laminates, for example, should be stressed parallel to the laminations. Stresses normal to the laminate may cause the substrate to delaminate.

Single and joggle lap joints are more likely to cause delamination than scarf or beveled lap joints. The strap-joint variations are useful when bending loads may be imposed on the joint.

9.5.5 Surface preparation

Many plastics and plastic composites can be treated prior to bonding by simple mechanical abrasion or alkaline cleaning to remove surface contaminants. In some cases it is necessary that the polymeric surface be physically or chemically modified to achieve acceptable bonding. This applies particularly to crystalline thermoplastics such as the polyolefins, linear polyesters, and fluorocarbons. Methods used to improve the bonding characteristics of these surfaces include

1. Oxidation via chemical treatment or flame treatment
2. Electrical discharge to leave a more reactive surface
3. Plasma treatment (exposing the surface to ionized inert gas)
4. Metal-ion treatment (for example, sodium naphthalene process for fluorocarbons)

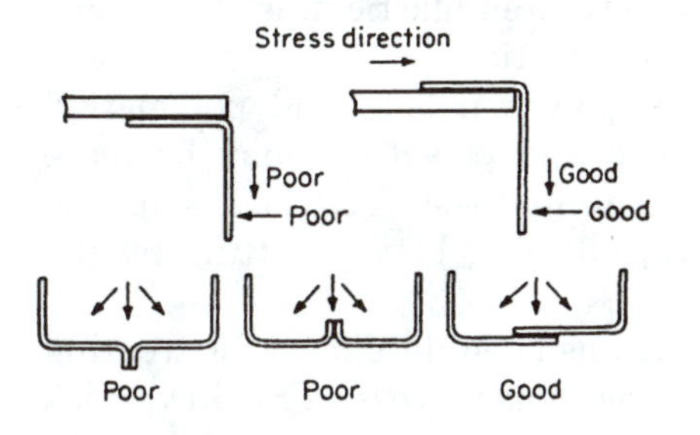

Figure 9.27 Methods of joining flexible rubber or plastic.[13]

Table 9.11 lists common recommended surface treatments for plastic adherends. These treatments are necessary when plastics are to be joined with adhesives. Solvent and heat welding are other methods of fastening plastics that do not require chemical alteration of the surface. Welding procedures will be discussed in another section of this chapter. The effects of plastic surface treatments decrease with time. It is necessary to prime or bond soon after the surfaces are treated. Some common plastic materials that require special physical or chemical treatments to achieve adequate surfaces for adhesive bonding are listed in the following sections.

Fluorocarbons. Fluorocarbons, such as polytetrafluoroethylene (TFE), polyfluororethylene propylene (FEP), polychlorotrifluoroethylene (CFE), and polymonochlorotrifluoroethylene (Kel-F), are notoriously difficult to bond because of their low surface tension. However, epoxy and polyurethane adhesives offer moderate strength if the fluorocarbon is treated prior to bonding.

The fluorocarbon surface may be made more "wettable" by exposing it for a brief moment to a hot flame to oxidize the surface. The most satisfactory surface treatment is achieved by immersing the plastic in a sodium-naphthalene dispersion in tetrahydrofuran. This process is believed to remove fluorine atoms, leaving a carbonized surface that can be wet easily. Fluorocarbon films treated for adhesive bonding are available from many suppliers. A formulation and description of the sodium-naphthalene process may be found in Table 9.11. Commercial chemical products for etching fluorocarbons are also listed.

Polyethylene terephthalate (Mylar). A medium-strength bond can be obtained with polyethylene terephthalate plastics and films by abrasion and solvent cleaning. However, a stronger bond can be achieved by immersing the surface in a warm solution of sodium hydroxide or in an alkaline cleaning solution for 2 to 10 min.

Polyolefins. These materials can be effectively bonded only if the surface is first oxidized. Polyethylene and polypropylene can be prepared for bonding by holding the flame of an oxyacetylene torch over the plastic until it becomes glossy or else by heating the surface momentarily with a blast of hot air. It is important not to overheat the plastic because it causes deformation. The treated plastic must be bonded as quickly as possible after surface preparation.

Polyolefins, such as polyethylene, polypropylene and polymethyl pentene, as well as polyformaldehyde and polyether, may be more effectively treated with a sodium dichromate–sulfuric acid solution. This treatment oxidizes the surface, allowing better wetting by the adhesive.

Another process, *plasma treatment,* has been developed for treating hard-to-bond plastics such as polyolefins. This process works in various

TABLE 9.11 Surface Preparation Methods for Plastics

Adherend	Degreasing solvent	Method of treatment	Remarks
Acetal (copolymer)	Acetone	1. Abrasion. Grit or vapor blast, or medium-grit emery cloth followed by solvent degreasing 2. Etch in the following acid solution: *Parts by wt.* Potassium dichromate 75 Distilled water 120 Concentrated sulfuric acid (96%, sp. gr. 1.84) 1500 for 10 s at 25°C. Rinse in distilled water, and dry in air at RT	For general-purpose bonding For maximum bond strength. ASTM D 2093
Acetal (homopolymer)	Acetone	1. Abrasion. Sand with 280A-grit emery cloth followed by solvent degreasing 2. "Satinizing" technique. Immerse the part in *Parts by wt.* Perchloroethylene 96.85 1,4-Dioxane 3.00 *p*-Toluenesulfonic acid 0.05 Cab-o-Sil (Cabot Corp.) 0.10 for 5–30 s at 80–120°C. Transfer the part immediately to an oven at 120°C for 1 min. Wash in hot water. Dry in air at 120°C	For general-purpose bonding For maximum bond strength. Recommended by DuPont
Acrylonitrile butadiene styrene	Acetone	1. Abrasion. Grit or vapor blast, or 220-grit emery cloth, followed by solvent degreasing 2. Etch in chromic acid solution for 20 min at 60°	Recipe 2 for methyl pentane
Cellulosics: Cellulose, cellulose acetate, cellulose acetate butyrate, cellulose nitrate, cellulose propionate, ethyl cellulose	Methanol, isopropanol	1. Abrasion. Grit or vapor blast, or 220-grit emery cloth, followed by solvent degreasing 2. After procedure 1, dry the plastic at 100°C for 1 h, and apply adhesive before the plastic cools to room temperature	For general bonding purposes
Diallyl phthalate, diallyl isophthalate	Acetone, methyl ethyl ketone	Abrasion. Grit or vapor blast, or 100-grit emery cloth, followed by solvent degreasing	Steel wool may be used for abrasion

TABLE 9.11 Surface Preparation Methods for Plastics (*Continued*)

Adherend	Degreasing solvent	Method of treatment	Remarks
Epoxy resins	Acetone, methyl ethyl ketone	Abrasion. Grit or vapor blast, or 100-grit emery cloth, followed by solvent degreasing	Sand or steel shot are suitable abrasives
Ethylene vinyl acetate	Methanol	Prime with epoxy adhesive and fuse into the surface by heating for 30 min at 100°C	
Furane	Acetone, methyl ethyl ketone	Abrasion. Grit or vapor blast, or 100-grit emery cloth, followed by solvent degreasing	
Ionomer	Acetone, methyl ethyl ketone	Abrasion. Grit or vapor blast, or 100-grit emery cloth, followed by solvent degreasing	Alumina (180-grit) is a suitable abrasive
Melamine resins	Acetone, methyl ethyl ketone	Abrasion. Grit or vapor blast, or 100-grit emery cloth, followed by solvent degreasing	For general-purpose bonding
Methyl pentene	Acetone	1. Abrasion. Grit or vapor blast, or 100-grit emery cloth, followed by solvent degreasing 2. Immerse for 1 h at 60°C in *Parts by wt.* Sulfuric acid (96% sp. gr. 1.84) 26 Potassium chromate 3 Water 11 Rinse in water and distilled water. Dry in warm air 3. Immerse for 5–10 min at 90°C in potassium permanganate (saturated solution), acidified with sulfuric acid (96%, sp. gr. 1.84). Rinse in water and distilled water. Dry in warm air 4. Prime surface with lacquer based on urea-formaldehyde resin diluted with carbon tetrachloride	Coatings (dried) offer excellent bonding surfaces without further pretreatment
Phenolic resins phenolic melamine resins	Acetone, methyl ethyl ketone detergent	1. Abrasion. Grit or vapor blast, or abrade with 100-grit emery cloth, followed by solvent degreasing 2. Removal of surface layer of one ply of fabric previously placed on surface before curing. Expose fresh bonding surface by tearing off the ply prior to bonding	Steel wool may be used for abrasion. Sand or steel shot are suitable abrasives. Glass-fabric decorative laminates may be degreased with detergent solution

TABLE 9.11 Surface Preparation Methods for Plastics (*Continued*)

Adherend	Degreasing solvent	Method of treatment	Remarks
Polyamide (nylon)	Acetone, methyl ethyl ketone, detergent	1. Abrasion. Grit or vapor blast, or abrade with 100-grit emery cloth, followed by solvent degreasing	Sand or steel shot are suitable abrasives
		2. Prime with a spreading dough based on the type of rubber to be bonded in admixture with isocyanate	Suitable for bonding polyamide textiles to natural and synthetic rubbers
		3. Prime with resorcinol-formaldehyde adhesive	Good adhesion to primer coat with epoxy adhesives in metal-plastic joints
Polycarbonate, allyl diglycol carbonate	Methanol, isopropanol, detergent	Abrasion. Grit or vapor blast, or 100-grit emery cloth, followed by solvent degreasing	Sand or steel shot are suitable abrasives
Fluorocarbons: Polychloro-trifluoroe-thylene, polytetra-fluoro-ethylene, polyvinyl fluoride, polymono-chlorotri-fluoro-ethylene	Trichloro-ethylene	1. Wipe with solvent and treat with the following for 15 min at RT: Naphthalene (128 g) dissolved in tetrahydrofuran (1 l) to which is added sodium (23 g) during a stirring period of 2 h. Rinse in deionized water, and dry in water air	Sodium-treated surfaces must not be abraded before use. Hazardous etching solutions requiring skillful handling. Proprietary etching solutions are commercially available (see 2). PTFE available in etched tape. ASTM D 2093
		2. Wipe with solvent and treat as recommended in one of the following commercial etchants: Bond aid (W.S. Shamban and Co.) Fluorobond (Joelin Mfg. Co.) Fluoroetch (Action Associates) Tetraetch (W. L. Gore Associates)	
		3. Prime with epoxy adhesive, and fuse into the surface by heating for 10 min at 370°C followed by 5 min at 400°C	
		4. Expose to one of the following gases activated by corona discharge: Air (dry) for 5 min Air (wet) for 5 min Nitrous oxide for 10 min Nitrogen for 5 min	Bond within 15 min of pretreatment
		5. Expose to electric discharge from a tesla coil (50,000 V ac) for 4 min	Bond within 15 min of pretreatment

TABLE 9.11 Surface Preparation Methods for Plastics (*Continued*)

Adherend	Degreasing solvent	Method of treatment	Remarks
Polyesters, polyethylene terephthalate (Mylar)	Detergent, acetone, methyl ethyl ketone	1. Abrasion. Grit or vapor blast, or 100-grit emery cloth, followed by solvent degreasing 2. Immerse for 10 min at 70–95°C in *Parts by wt.* Sodium hydroxide 2 Water 8 Rinse in hot water and dry in hot air	For general-purpose bonding For maximum bond strength. Suitable for linear polyester films (Mylar)
Chlorinated polyether	Acetone, methyl ethyl ketone	Etch for 5–10 min at 66–71°C in *Parts by wt.* Sodium dichromate 5 Water 8 Sulfuric acid (96%, sp. gr. 1.84) 100 Rinse in water and distilled water Dry in air	Suitable for film materials such as Penton. ASTM D 2093
Polyethylene, polyethylene (chlorinated), polyethylene terephthalate (see polyesters), polypropylene, polyformaldehyde	Acetone, methyl ethyl ketone	1. Solvent degreasing 2. Expose surface to gas-burner flame (or oxyacetylene oxidizing flame) until the substrate is glossy 3. Etch in the following: *Parts by wt.* Sodium dichromate 5 Water 8 Sulfuric acid (96% sp. gr. 1.84) 100 Polyethylene and polypropylene 60 min at 25°C or 1 min at 71°C Polyformaldehyde 10 s at 25°C 4. Expose to following gases activated by corona discharge: Air (dry) for 15 min Air (wet) for 5 min Nitrous oxide for 10 min Nitrogen for 15 min 5. Expose to electric discharge from a tesla coil (50,000 V ac) for 1 min	Low-bond-strength applications For maximum bond strength. ASTM D 2093 Bond within 15 min of pretreatment. Suitable for polyolefins. Bond within 15 min of pretreatment. Suitable for polyolefins.
Polymethyl methacrylate, methacrylate butadiene styrene	Acetone, methyl ethyl ketone, detergent, methanol, trichloroethylene, isopropano	Abrasion. Grit or vapor blast, or 100-grit emery cloth, followed by solvent degreasing	For maximum strength relieve stresses by heating plastic for 5 h at 100°C

TABLE 9.11 Surface Preparation Methods for Plastics (*Continued*)

Adherend	Degreasing solvent	Method of treatment	Remarks
Poly-phenylene	Trichloro-ethylene	Abrasion. Grit or vapor blast, or 100-grit emery cloth, followed by solvent degreasing	
Poly-phenylene oxide	Methanol	Solvent degrease	Plastic is soluble in xylene and may be primed with adhesive in xylene solvent
Polystyrene	Methanol, isopropanol, detergent	Abrasion. Grit or vapor blast, or 100-grit emery cloth, followed by solvent degreasing	Suitable for rigid plastic
Polysulfone	Methanol	Vapor degrease	
Polyurethane	Acetone, methyl ethyl ketone	Abrade with 100-grit emery cloth and solvent degrease	
Polyvinyl chloride, polyvinylidene chloride, polyvinyl fluoride	Trichloro-ethylene, methyl ethyl ketone	1. Abrasion. Grit or vapor blast, or 100-grit emery cloth followed by solvent degreasing	Suitable for rigid plastic. For maximum strength, prime with nitrile-phenolic adhesive
		2. Solvent wipe with ketone	Suitable for plasticized material
Styrene acrylonitrile	Trichloro-ethylene	Solvent degrease	
Urea for-maldehyde	Acetone, methyl ethyl ketone	Abrasion. Grit or vapor blast, or 100-grit emery cloth, followed by solvent degreasing	

SOURCE: Based on the following: N. J. DeLolis, *Adhesives for Metals Theory and Technology*, Industrial Press, New York, 1970; C. V. Cagle, *Adhesive Bonding Techniques and Applications*, McGraw-Hill, New York, 1968; W. H. Guttmann, *Concise Guide to Structural Adhesives*, Reinhold, New York, 1961; "Preparing the Surface for Adhesive Bonding," Bulletin G1-600, Hysol Division, Dexter Corporation; A. H. Landrock, *Adhesive Technology Handbook*, Noyes Publications, Park Ridge, N. J., 1985; and J. Schields, *Adhesive Handbook*, CRC Press, Boca Raton, Fla., 1970.

ways depending on the type of plastic being treated. For most polyolefins, plasma treatment cross-links the polymeric surface by exposing it to an electrically activated inert gas such as neon or helium. This forms a tough, cross-linked surface that wets easily and is adequate for printing and painting as well as bonding.

Table 9.12 shows the tensile-shear strength of bonded polyethylene pretreated by these various methods.

Elastomeric adherends. Vulcanized-rubber parts are often contaminated with mold release and plasticizers or extenders that can migrate to the surface. Solvent washing and abrading are common treatments for most elastomers, but chemical treatment is required for maximum properties. Many synthetic and natural rubbers require "cyclizing" with concentrated sulfuric acid until hairline fractures are evident on the surface.

Fluorosilicone and silicone rubbers must be primed before bonding. The primer acts as an intermediate interface, providing good adhesion to the rubber and a more wettable surface for the adhesive.

9.5.6 Adhesives selection

Factors most likely to influence adhesive selection are listed in Table 9.13. However, thermosetting adhesives, such as epoxies, polyurethanes, or acrylics, are commonly used for structural application. The adhesive formulations are generally tough, flexible compounds that can cure at room temperature. The reasons that these adhesives have gained the most popularity in bonding of plastics are summarized in this section.

The physical and chemical properties of both the solidified adhesive and the plastic substrate affect the quality of the bonded joint. Major elements of concern in selecting an adhesive for plastic parts are the thermal expansion coefficient and glass transition temperature of the

TABLE 9.12 Effect of Surface Treatments on Polyethylene[15]

	Relative bond strength*			
Polymer	Control	Plasma	Abrasion	Chemical
High-density polyethylene	1.0	12.1		5.1
Polypropylene	1.0	221		649
Valox 310 polyester (General Electric Company)	1.0	18.9	2.9	1.0
Silicone rubber	1.0	>20	4.7	

*Results normalized to the control for each material.
SOURCE: Branson International Plasma Corporation.

TABLE 9.13 Factors Influencing Adhesive Selection

Stress	Tension
	Shear
	Impact
	Peel
	Cleavage
	Fatigue
Chemical factors	External (service-related)
	Internal (effect of adherend on adhesives)
Exposure	Weathering
	Light
	Oxidation
	Moisture
	Salt spray
Temperature	High
	Low
	Cycling
Biological factors	Bacteria or mold
	Rodents or vermin
Working properties	Application
	Bonding time and temperature range
	Tackiness
	Curing rate
	Storage stability
	Coverage

substrate relative to the adhesive. Special consideration is also required of the polymeric surface which can change during normal aging or exposure to operating environments.

Significant differences in the thermal expansion coefficient between substrates and the adhesive can cause serious stress at the plastic's joint interface. These stresses are compounded by thermal cycling and low-temperature service requirements. Selection of a resilient adhesive or adjustments in the adhesive's thermal expansion coefficient via filler or additives can reduce such stress.

Structural adhesives must have a glass transition temperature higher than the operating temperature to avoid a cohesively weak bond and possible creep problems. Modern engineering plastics, such as polyimide or polyphenylene sulfides, have very high glass transition temperatures. Most common adhesives have a relatively low glass transition temperature so that the weakest thermal link in the joint may often be the adhesive.

Use of an adhesive too far below its glass transition temperature could result in low peel or cleavage strength. Brittleness of the adhesive at very low temperatures could also manifest itself in poor impact strength.

Plastic substrates could be chemically active, even when isolated from the operating environment. Many polymeric surfaces slowly undergo chemical and physical change. The plastic surface, at the time of bonding, may be well suited to the adhesive process. However, after aging, undesirable surface conditions may present themselves at the interface, displace the adhesive, and result in bond failure. These weak boundary layers may come from the environment or from within the plastic substrate itself.

Moisture, solvent, plasticizers, and various gases and ions can compete with the cured adhesive for bonding sites. The process by which a weak boundary layer preferentially displaces the adhesive at the interface is called *desorption.* Moisture is the most common desorbing substance, being present both in the environment and within many polymeric substrates.

Solutions to the desorption problem consist of eliminating the source of the weak boundary layer or selecting an adhesive that is compatible with the desorbing material. Excessive moisture can be eliminated from a plastic part by postcuring or drying the part before bonding. Additives that can migrate to the surface can possibly be eliminated by reformulating the plastic resin. Also, certain adhesives are more compatible with oils and the plasticizer than others. For example, the migration of the plasticizer from flexible polyvinyl chloride can be counteracted by using nitrile-based adhesives. Nitrile adhesives resins are capable of absorbing the plasticizer without degradation.

9.5.7 Equipment for adhesive bonding

After the adhesive is applied, the assembly must be mated as quickly as possible to prevent contamination of the adhesive surface. The substrates are held together under pressure and heated, if necessary, until cure is achieved. The equipment required to perform these functions must provide adequate heat and pressure, maintain constant pressure during the entire cure cycle, and distribute pressure uniformly over the bond area. Of course, many adhesives cure with simple contact pressure at room temperature, and extensive bonding equipment is not necessary.

Pressure equipment. Pressure devices should be designed to maintain constant pressure on the bond during the entire cure cycle. They must compensate for thickness reduction from adhesive flow-out or thermal expansion of assembly parts. Thus, screw-actuated devices like C clamps and bolted fixtures are not acceptable when constant pressure is important. Spring pressure can often be used to supplement clamps and compensate for thickness variations. Dead-weight loading may be applied in many instances; however, this method is sometimes impractical, especially when heat cure is necessary.

Pneumatic and hydraulic presses are excellent tools for applying constant pressure. Steam or electrically heated platen presses with hydraulic rams are often used for adhesive bonding. Some units have multiple platens, thereby permitting the bonding of several assemblies at one time.

Large bonded areas, such as on aircraft parts, are usually cured in an autoclave. The parts are mated first and covered with a rubber blanket to provide uniform pressure distribution. The assembly is then placed in an autoclave, which can be pressurized and heated. This method requires heavy capital equipment investment.

Vacuum-bagging techniques can be an inexpensive method of applying pressure to large parts. A film or plastic bag is used to enclose the assembly, and the edges of the film are sealed airtight. A vacuum is drawn on the bag, enabling atmospheric pressure to force the adherends together. Vacuum bags are especially effective on large areas because size is not limited by equipment.

Heating equipment. Many structural adhesives require heat as well as pressure. Most often the strongest bonds are achieved by an elevated temperature cure. With many adhesives, trade-offs between cure times and temperature are permissible. But generally, the manufacturer will recommend a certain curing schedule for optimum properties.

If, for example, a cure of 60 min at 300°F is recommended, this does not mean that the assembly should be placed in 300°F for 60 min. Total oven time would be 60 min plus whatever time is required to bring the adhesive up to 300°F. Large parts act as a heat sink and may require substantial time for an adhesive in the bond line to reach the necessary temperature. Bond line temperatures are best measured by thermocouples placed very close to the adhesive. In some cases, it may be desirable to place the thermocouple in the adhesive joint for the first few assemblies being cured.

Oven heating is the most common source of heat for bonded parts, even though it involves long curing cycles because of the heat-sink action of large assemblies. Ovens may be heated with gas, oil, electricity, or infrared units. Good air circulation within the oven is mandatory to prevent nonuniform heating.

Heated platen presses are good for bonding flat or moderately contoured panels when faster cure cycles are desired. Platens are heated with steam, hot oil, or electricity and are easily adapted with cooling water connections to further speed the bonding cycle.

Induction and dielectric heating are the fastest heating methods because they focus heat at or near the adhesive bond line. Workpiece heating rates greater than 100°F/s are possible with induction heating. For induction heating to work, the adhesive must be filled with metal particles or the adherend must be capable of conducting electricity or being magnetized. Dielectric heating is an effective way of curing adhesives if at least one substrate is a nonconductor. Metal-to-metal joints tend to break down the microwave field necessary for dielectric heating. This heating method makes use of the polar characteristics of the adhesive materials. Both induction and dielectric heating involve relatively expensive capital equipment outlays, and the bond area is limited. Their most important advantages are assembly speed and the fact that an entire assembly does not have to be heated to cure only a few grams of adhesive.

Adhesive-thickness control. It is highly desirable to have a uniformly thin (2- to 10-mil) adhesive bond line. Starved adhesive joints, however, will yield exceptionally poor properties. Three basic methods are used to control adhesive thickness. The first method is to use mechanical shims or stops which can be removed after the curing operation. Sometimes it is possible to design stops into the joint.

The second method is to employ a film adhesive that becomes highly viscous during the cure cycle, preventing excessive flow-out. With supported films, the adhesive carrier itself can act as the "shims." Generally, the cured bond line thickness will be determined by the original thickness of the adhesive film. The third method of controlling adhesive thickness is to use trial and error to determine the correct pressure-adhesive viscosity factors that will yield the desired bond thickness.

9.5.8 Quality control

A flowchart of a quality-control system for a major aircraft company is illustrated in Fig. 9.28. This system is designed to ensure reproducible bonds and, if a substandard bond is detected, to make suitable corrections. Quality control should cover all phases of the bonding cycle from inspection of incoming material to the inspection of the completed assembly. In fact, good quality control will start even before receipt of materials.

Prehandling conditions. The human element enters the adhesive bonding process more than in other fabrication techniques. An extremely high percentage of defects can be traced to poor workmanship. This generally prevails in the surface preparation steps but may also arise in any of the other steps necessary to achieve a bonded assembly. This problem can be largely overcome by proper motivation and education. All employees—from design engineer to laborer to quality-control inspector—should be somewhat familiar with adhesive bonding technology and be aware of the circumstances that can lead to poor joints. A great many defects can also be traced to poor design engineering.

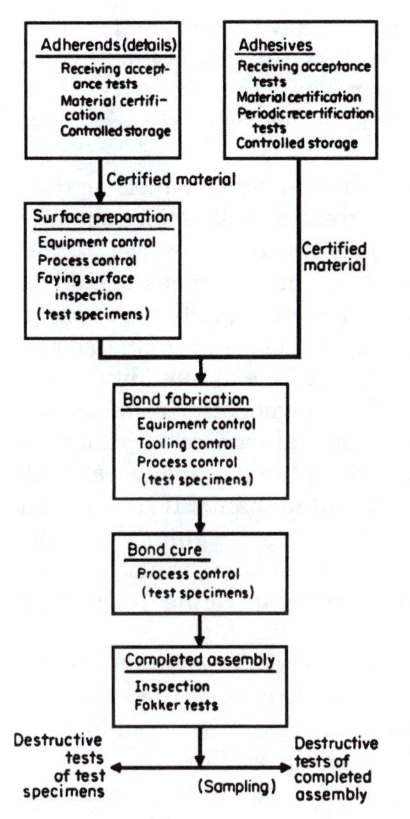

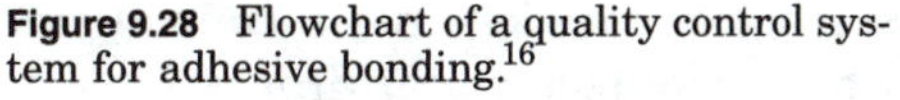
Figure 9.28 Flowchart of a quality control system for adhesive bonding.[16]

The plant's bonding area should be as clean as possible prior to receipt of materials. The basic approach for keeping the assembly area clean is to segregate it from the other manufacturing operations by placing it either in a corner of the plant or in isolated rooms. The air should be dry and filtered to prevent moisture or other contaminants from gathering at a possible interface. The cleaning and bonding operations should be separated from each other. If mold release is used to prevent adhesive flash from sticking to bonding equipment, it is advisable that great care be taken to assure that the release does not contaminate the adhesive or the adherends. Spray mold releases, especially silicone release agents, have a tendency to migrate to undesirable areas.

Quality control of adhesive and surface treatment. Acceptance tests on adhesives should be directed toward assurance that incoming materials are identical from lot to lot. The tests should be those which can quickly and accurately detect deficiencies in the adhesive's physical or chemical properties. A number of standard tests for adhesive bonds and for adhesive acceptance have been specified by the American Society for Testing and Materials (ASTM). Selected ASTM Standards are presented in Table 9.14. The properties usually reported by adhesive suppliers are ASTM tensile-shear and peel strength.

Actual test specimens should also be made to verify the strength of the adhesive. These specimens should be stressed in directions that are representative of the forces which the bond will see in service, that is, shear, peel, tension, or cleavage. If possible, the specimens should be prepared and cured in the same manner as actual production assemblies. If time permits, specimens should also be tested in simulated service environments, for example, high temperature and humidity.

Surface preparations must be carefully controlled for reliable production of adhesive-bonded parts. If a chemical surface treatment is required, the process must be monitored for proper sequence, bath temperature, solution concentration, and contaminants. If sand or grit blasting is employed, the abrasive must be changed regularly. An adequate supply of clean wiping cloths for solvent cleaning is also mandatory. Checks should be made to determine if cloths or solvent containers may have become contaminated.

The specific surface preparation can be checked for effectiveness by the water break-free test. After the final treating step, the substrate surface is checked for a continuous film of water that should form when deionized water droplets are placed on the surface.

After the adequacy of the surface treatment has been determined, precautions must be taken to assure that the substrates are kept clean and dry until bonding. The adhesive or primer should be applied to the treated surface as quickly as possible.

Quality control of the bonding process. The adhesive metering and mixing operation should be monitored by periodically sampling the mixed adhesive and testing it for adhesive properties. A visual inspection can also be made for air entrapment and degree of mixing. The quality-control engineer should be sure that the oldest adhesive is used first and that the specified shelf life has not been exceeded.

During the actual assembly operation, the cleanliness of the shop and tools should be verified. The shop atmosphere should be controlled as closely as possible. Temperature is in the range of 65 to 90°F and relative humidity from 20 to 65% is best for almost all bonding operations.

The amount of the applied adhesive and the final bond line thickness must also be monitored because they can have a significant effect on joint strength. Curing conditions should be monitored for heat-up rate, maximum and minimum temperature during cure, time at the

TABLE 9.14 ASTM Adhesive Standards Test Methods*

Aging
Resistance of Adhesives to Cyclic Aging Conditions, Test for (D 1183) Bonding Permanency of Water- or Solvent-Soluble Liquid Adhesives for Labeling Glass Bottles, Test for (D 1581) Bonding Permanency of Water- or Solvent-Soluble Liquid Adhesives for Automatic Machine Sealing Top Flaps of Fiber Specimens, Test for (D 1713) Permanence of Adhesive-Bonded Joints in Plywood under Mold Conditions, Test for (D 1877) Accelerated Aging of Adhesive Joints by the Oxygen-Pressure Method, Practice for (D 3632)
Amylaceous Matter
Amylaceous Matter in Adhesives, Test for (D 1488)
Biodeterioration
Susceptibility of Dry Adhesive Film to Attack by Roaches, Test for (D 1382) Susceptibility of Dry Adhesive Film to Attack by Laboratory Rats, Test for (D 1383) Permanence of Adhesive-Bonded Joints in Plywood under Mold Conditions, Test for (D 1877) Effect of Bacterial Contamination of Adhesive Preparations and Adhesive Films, Test for (D 4299) Effect of Mold Contamination on Permanence of Adhesive Preparation and Adhesive Films, Test for (D 4300)
Blocking Point
Blocking Point of Potentially Adhesive Layers, Test for (D 1146)
Bonding Permanency
(See Aging)
Chemical Reagents
Resistance of Adhesive Bonds to Chemical Reagents, Test for (D 896)
Cleavage
Cleavage Strength of Metal-to-Metal Adhesive Bonds, Test for (D 1062)
Cleavage/Peel Strength
Strength Properties of Adhesives in Cleavage Peel by Tension Loading (Engineering Plastics-to-Engineering Plastics), Test for (D 3807) (See also Peel Strength)
Corrosivity
Determining Corrosivity of Adhesive Materials, Practice for (D 3310)
Creep
Conducting Creep Tests of Metal-to-Metal Adhesives, Practice for (D 1780) Creep Properties of Adhesives in Shear by Compression Loading (Metal-to-Metal), Test for (D 2293) Creep Properties of Adhesives in Shear by Tension Loading, Test for (D 2294)

TABLE 9.14 ASTM Adhesive Standards Test Methods* (*Continued*)

Cryogenic Temperatures
Strength Properties of Adhesives in Shear by Tension Loading in the Temperature Range from −267.8 to −55°C (−450 to −67°F), Test for (D 2557)
Density
Density of Adhesives in Fluid Form, Test for (D 1875)
Durability (Including Weathering)
Effect of Moisture and Temperature on Adhesive Bonds, Test for (D 1151) Atmospheric Exposure of Adhesive-Bonded Joints and Structures, Practice for (D 1828) Determining Durability of Adhesive Joints Stressed in Peel, Practice for (D 2918) Determining Durability of Adhesive Joints Stressed in Shear by Tension Loading, Practice for (D 2919) (See also Wedge Test)
Electrical Properties
Adhesives Relative to Their Use as Electrical Insulation, Testing (D 1304)
Electrolytic Corrosion
Determining Electrolytic Corrosion of Copper by Adhesives, Practice for (D 3482)
Fatigue
Fatigue Properties of Adhesives in Shear by Tension Loading (Metal/Metal), Test for (D 3166)
Filler Content
Filler Content of Phenol, Resorcinol, and Melamine Adhesives, Test for (D 1579)
Flexibility
(See Flexural Strength)
Flexural Strength
Flexural Strength of Adhesive Bonded Laminated Assemblies, Test for (D 1184) Flexibility Determination of Hot Melt Adhesives by Mandrel Bend Test Method, Practice for (D 3111)
Flow Properties
Flow Properties of Adhesives, Test for (D 2183)
Fracture Strength in Cleavage
Fracture Strength in Cleavage of Adhesives in Bonded Joints, Practice for (D 3433)
Gap-Filling Adhesive Bonds
Strength of Gap Filling Adhesive Bonds in Shear by Compression Loading, Practice for (D 3931)

TABLE 9.14 ASTM Adhesive Standards Test Methods* (*Continued*)

High-Temperature Effects
Strength Properties of Adhesives in Shear by Tension Loading at Elevated Temperatures (Metal-to-Metal), Test for (D 2295)
Hydrogen-Ion Concentration
Hydrogen Ion Concentration, Test for (D 1583)
Impact Strength
Impact Strength of Adhesive Bonds, Test for (D 950)
Light Exposure
(See Radiation Exposure)
Low and Cryogenic Temperatures
Strength Properties of Adhesives in Shear by Tension Loading in the Temperature Range from −267.8 to −55°C (−450 to 67°F), Test for (D 2557)
Nonvolatile Content
Nonvolatile Content of Aqueous Adhesives, Test for (D 1489) Nonvolatile Content of Urea-Formaldehyde Resin Solutions, Test for (D 1490) Nonvolatile Content of Phenol, Resorcinol, and Melamine Adhesives, Test for (D 1582)
Odor
Determination of the Odor of Adhesives, Test for (D 4339)
Peel Strength (Stripping Strength)
Peel or Stripping Strength of Adhesive Bonds, Test for (D 903) Climbing Drum Peel Test for Adhesives, Method for (D 1781) Peel Resistance of Adhesives (T-Peel Test), Test for (D 1876) Evaluating Peel Strength of Shoe Sole Attaching Adhesives, Test for (D 2558) Determining Durability of Adhesive Joints Stressed in Peel, Practice for (D 2918) Floating Roller Peel Resistance, Test for (D 3167)
Penetration
Penetration of Adhesives, Test for (D 1916)
pH
(See Hydrogen-Ion Concentration)
Radiation Exposure (Including Light)
Exposure of Adhesive Specimens to Artificial (Carbon-Arc Type) and Natural Light, Practice for (D 904) Exposure of Adhesive Specimens to High-Energy Radiation, Practice for (D 1879)
Rubber Cement Tests
Rubber Cements, Testing of (D 816)

TABLE 9.14 ASTM Adhesive Standards Test Methods* (*Continued*)

Salt Spray (Fog) Testing
Salt Spray (Fog) Testing, Method of (B 117)
Modified Salt Spray (Fog) Testing, Practice for (G 85)
Shear Strength (Tensile Shear Strength)
Shear Strength and Shear Modulus of Structural Adhesives, Test for (E 229)
Strength Properties of Adhesive Bonds in Shear by Compression Loading, Test for (D 905)
Strength Properties of Adhesive in Plywood Type Construction in Shear by Tension Loading, Test for (D 906)
Strength Properties of Adhesives in Shear by Tension Loading (Metal-to-Metal), Test for (D 1002)
Determining Strength Development of Adhesive Bonds, Practice for (D 1144)
Strength Properties of Metal-to-Metal Adhesives by Compression Loading (Disk Shear), Test for (D 2181)
Strength Properties of Adhesives in Shear by Tension Loading at Elevated Temperatures (Metal-to-Metal), Test for (D 2295)
Strength Properties of Adhesives in Two-Ply Wood Construction in Shear by Tension Loading, Test for (D 2339)
Strength Properties of Adhesives in Shear by Tension Loading in the Temperature Range from −267.8 to −55°C (−450 to −67°F), Test for (D 2557)
Determining Durability of Adhesive Joints Stressed in Shear by Tension Loading, Practice for (D 2919)
Determining the Strength of Adhesively Bonded Rigid Plastic Lap-Shear Joints in Shear by Tension Loading, Practice for (D 3163)
Determining the Strength of Adhesively Bonded Plastic Lap-Shear Sandwich Joints in Shear by Tension Loading, Practice for (D 3164)
Strength Properties of Adhesives in Shear by Tension Loading of Laminated Assemblies, Test for (D 3165)
Fatigue Properties of Adhesives in Shear by Tension Loading (Metal/Metal), Test for (D 3166)
Strength Properties of Double Lap Shear Adhesive Joints by Tension Loading, Test for (D 3528)
Strength of Gap-Filling Adhesive Bonds in Shear by Compression Loading, Practice for (D 3931)
Measuring Strength and Shear Modulus of Nonrigid Adhesives by the Thick Adherend Tensile Lap Specimen, Practice for (D 3983)
Measuring Shear Properties of Structural Adhesives by the Modified-Rail Test, Practice for (D 4027)
Specimen Preparation
Preparation of Bar and Rod Specimens of Adhesion Tests, Practice for (D 2094)
Spot-Adhesion Test
Qualitative Determination of Adhesion of Adhesives to Substrates by Spot Adhesion Test Method, Practice for (D 3808)
Spread (Coverage)
Applied Weight per Unit Area of Dried Adhesive Solids, Test for (D 898)

TABLE 9.14 ASTM Adhesive Standards Test Methods* (*Continued*)

Spread (Coverage) (*Continued*)
Applied Weight per Unit Area of Liquid Adhesive, Test for (D 899)
Storage Life
Storage Life of Adhesives by Consistency and Bond Strength, Test for (D 1337)
Strength Development
Determining Strength Development of Adhesive Bonds, Practice for (D 1144)
Stress-Cracking Resistance
Evaluating the Stress Cracking of Plastics by Adhesives Using the Bent Beam Method, Practice for (D 3929)
Stripping Strength
(See Peel Strength)
Surface Preparation
Preparation of Surfaces of Plastics Prior to Adhesive Bonding, Practice for (D 2093) Preparation of Metal Surfaces for Adhesive Bonding, Practice for (D 2651) Analysis of Sulfochromate Etch Solution Used in Surface Preparation of Aluminum, Methods of (D 2674) Preparation of Aluminum Surfaces for Structural Adhesive Bonding (Phosphoric Acid Anodizing), Practice for (D 3933)
Tack
Pressure Sensitive Tack of Adhesives Using an Inverted Probe Machine, Test for (D 2979) Tack of Pressure-Sensitive Adhesives by Rolling Ball, Test for (D 3121)
Tensile Strength
Tensile Properties of Adhesive Bonds, Test for (D 897) Determining Strength Development of Adhesive Bonds, Practice for (D 1144) Cross-Lap Specimens for Tensile Properties of Adhesives, Testing of (D 1344) Tensile Strength of Adhesives by Means of Bar and Rod Specimens, Method for (D 2095)
Torque Strength
Determining the Torque Strength of Ultraviolet (UV) Light-Cured Glass/Metal Adhesive Joints, Practice for (D 3658)
Viscosity
Viscosity of Adhesives, Test for (D 1084) Apparent Viscosity of Adhesives Having Shear-Rate-Dependent Flow Properties, Test for (D 2556) Viscosity of Hot Melt Adhesives and Coating Materials, Test for (D 3236)
Volume Resistivity
Volume Resistivity of Conductive Adhesives, Test for (D 2739)
Water Absorptiveness (of Paper Labels)
Water Absorptiveness of Paper Labels, Test for (D 1584)

TABLE 9.14 ASTM Adhesive Standards Test Methods* (*Continued*)

Weathering
(See Durability)
Wedge Test
Adhesive Bonded Surface Durability of Aluminum (Wedge Test) (D 3762)
Working Life
Working Life of Liquid or Paste Adhesive by Consistency and Bond Strength, Test for (D 1338)

*The latest revisions of ASTM standards can be obtained from the American Society for Testing and Materials, 100 Barr Harbor Drive, West Conshohocken, Pa.

required temperature, and cool-down rate.

After the adhesive is cured, the joint area can be inspected to detect gross flaws or defects. This inspection procedure can be either destructive or nondestructive in nature. Destructive testing generally involves placing samples of the production run in simulated or accelerated service and determining if it has similar properties to a specimen that is known to have a good bond and adequate service performance. The causes and remedies for faults revealed by such mechanical tests are described in Table 9.15.

Nondestructive testing (NDT) is far more economical, and every assembly can be tested if desired. However, there is no single nondestructive test or technique that will provide the user with a quantitative estimate of bond strength. There are several ultrasonic test methods that provide qualitative values. However, a trained eye can detect a surprising number of faulty joints by close inspection of the adhesive around the bonded area. Table 9.16 lists the characteristics of faulty joints that can be detected visually. The most difficult defect to be found by any way are those related to improper curing and surface treatments. Therefore, great care and control must be given to surface preparation procedures and shop cleanliness.

9.6 Welding

Certain thermoplastic substrates may be joined by methods other than mechanical fastening or adhesive bonding. By careful application of heat or solvent to a thermoplastic substrate, one may liquefy the surface resin and use it to form the bond. With thermal or solvent welding, surface preparation is not as critical as with adhesive bonding. The bond strength is determined by diffusion of polymer from one surface into another instead of by the wetting and adsorption of an adhesive layer.

TABLE 9.15 Faults Revealed by Mechanical Tests

Fault	Cause	Remedy
Thick, uneven glue line	Clamping pressure too low	Increase pressure. Check that clamps are seating properly
	No follow-up pressure	Modify clamps or check for freedom of moving parts
	Curing temperature too low	Use higher curing temperature. Check that temperature is above the minimum specified throughout the curing cycle
	Adhesive exceeded its shelf life, resulting in increased viscosity	Use fresh adhesive
Adhesive residue has spongy appearance or contains bubbles	Excess air stirred into adhesive	Vacuum-degas adhesive before application
	Solvents not completely dried out before bonding	Increase drying time or temperature. Make sure drying area is properly ventilated
	Adhesive material contains volatile constituents	Seek advice from manufacturers
	A low-boiling constituent boiled away	Curing temperature is too high
Voids in bond (that is, areas that are not bonded), clean bare metal exposed, adhesive failure at interface	Joint surfaces not properly treated	Check treating procedure; use clean solvents and wiping rags. Wiping rags must not be made from synthetic fiber. Make sure cleaned parts are not touched before bonding. Cover stored parts to prevent dust from settling on them
	Resin may be contaminated	Replace resin. Check solids content. Clean resin tank
	Uneven clamping pressure	Check clamps for distortion
	Substrates distorted	Check for distortion; correct or discard distorted components. If distorted components must be used, try adhesive with better gap-filling ability
Adhesive can be softened by heating or wiping with solvent	Adhesive not properly cured	Use higher curing temperature or extend curing time. Temperature and time must be above the minimum specified throughout the curing cycle. Check mixing ratios and thoroughness of mixing. Large parts act as a heat sink, necessitating larger cure times

TABLE 9.16 Visual Inspection for Faulty Bonds

Fault	Cause	Remedy
No appearance of adhesive around edges of joint or adhesive bond line too thick	Clamping pressure too low	Increase pressure. Check that clamps are seating properly
	Starved joint	Apply more adhesive
	Curing temperature too low	Use higher curing temperature. Check that temperature is above the minimum specified
Adhesive bond line too thin	Clamping pressure too high	Lessen pressure
	Curing temperature too high	Use lower curing temperature
	Starved joint	Apply more adhesive
Adhesive flash breaks easily away from substrate	Improper surface treatment	Check treating procedures; clean solvents and wiping rags. Make sure cleaned parts are not touched before bonding
Adhesive flash is excessively porous	Excess air stirred into adhesive	Vacuum-degas adhesive before application
	Solvent not completely dried out before bonding	Increase drying time or temperature
	Adhesive material contains volatile constituent	Seek advice from manufacturers
Adhesive flash can be softened by heating or wiping with solvent	Adhesive not properly cured	Use higher curing temperature or extend curing time. Temperature and time must be above minimum specified. Check mixing

However, with welding some form of pretreatment may still be necessary. Certainly, the parts should be clean, and all mold release and contaminants must be removed by standard cleaning procedures. It may also be necessary to dry certain polymeric parts, such as nylon and polycarbonate, before welding so that the inherent moisture in the part will not affect the overall quality of the bond. It may also be necessary to thermally anneal parts, such as acrylic, before solvent welding to remove or lessen internal stresses caused by molding. Without annealing, the stressed surface may crack or craze when in contact with solvent.

9.6.1 Thermal welding

Welding by application of heat or *thermal welding* provides an advantageous method of joining many thermoplastics that do not degrade rapidly at their melt temperature. It is a method of providing fast, relatively easy, and economical bonds that are generally 80 to 100% the strength of the parent plastic.

The thermal welding process can be of two kinds: direct and indirect. With *direct welding,* the heat is applied directly to the substrate in the form of either a heated tool or hot gas. *Indirect welding* occurs when some form of energy other than thermal is applied to the joint. The applied energy, which causes heating at the interface or in the plastic as a whole, is generally in the form of friction, high-frequency electrical fields, electromagnetic fields, or ultrasonic vibration. Because the heating is localized at the bonding surface, indirect heating processes are very energy efficient, generally resulting in bonds that are stress free and of higher strength than those made by direct welding methods.

Heated tool welding. With the *heated tool welding* method, the surfaces to be fused are heated by holding them against a hot-metal surface (450 to 700°F); then the parts are brought into contact and allowed to harden under slight pressure (5 to 15 lb/in^2). Electric strip heaters, soldering irons, hot plates, and resistance blades are common methods of providing heat. One production technique involves butting flat plastic sheets on a table next to a resistance heated blade that runs the length of the sheet. Once the plastic adjacent to the blade begins to soften, the blade is raised, and the sheets are pressed together and held under pressure while they cool. The heated metal surfaces are usually coated with a high-temperature release coating, such as polytetrafluoroethylene, to discourage sticking to the molten plastic.

Successful heated tool welding depends on the temperature of the heated tool surface, the amount of time the plastic adherends are in contact with the hot tool, the time lapse before joining the substrates, and the amount and uniformity of pressure that is held during cooling. Heated tool welding can be used for structural plastic parts, and heat sealing can be used for plastic films. With heat sealing, the hot surface is usually hot rollers or a heated rotating metal band commonly used to seal plastic bags. Table 9.17 offers heat welding temperatures for a number of common plastics and films.

Resistance wire welding is also a type of heated tool welding. This method employs an electrical resistance heating wire laid between mating substrates to generate the heat of fusion. When energized, the wire undergoes resistance heating and causes a melt area to form around the adjacent polymer. Pressure on the parts during this process causes the molten material to flow and act as a hot-melt adhesive for the joint. After the bond has been made, the exterior wire is cut off and removed. Resistance wire welding can be used on any plastic that can be joined effectively by heated tool welding. The plastic resin manufacturer should be contacted for details concerning the specific parameters of this process.

TABLE 9.17 Hot-Plate Temperatures to Weld Plastics and Plastic Films[17]

Plastic	Temperature, °F	Film	Temperature, °F
ABS	450	Coated cellophane	200–350
Acetal	500	Cellulose acetate	400–500
Phenoxy	550	Coated polyester	490
Polyethylene			
LD	360	Poly(chlorotrifluoroethylene)	415–450
HD	390	Polyethylene	250–375
Polycarbonate	650	Polystyrene (oriented)	220–300
PPO	650	Poly(vinyl alcohol)	300–400
Noryl*	525	Poly(vinyl chloride) and copolymers (nonrigid)	200–400
Polypropylene	400	Poly(vinyl chloride) and copolymers (rigid)	260–400
Polystyrene	420	Poly(vinyl chloride)—nitrile rubber blend	220–350
SAN	450	Poly(vinylidene chloride)	285
Nylon 6, 6	475	Rubber hydrochloride	225–350
PVC	450	Fluorinated ethylene-propylene copolymer	600–750

*Trademark of General Electric Company.

Hot-gas welding. An electrically or gas-heated welding gun with an orifice temperature of 425 to 700°F can be used to bond many thermoplastic materials. The pieces to be bonded are beveled and positioned to form a V-shaped joint as shown in Fig. 9.29. A welding rod, made of the same plastic that is being bonded is laid into the joint, and the heat from the gun is directed at the interface of the substrates and the rod. The molten product from the welding rod then fills the gap. A strong filet must be formed, the design of which is of considerable importance.

A large difference between the plastic melting temperature and the decomposition temperature of the plastic is necessary for consistent, reliable hot-gas welding results. Usually the hot gas can be common air. However, for polyolefins and other easily oxidized plastics, the heated gas must be inert or nitrogen, since air will oxidize the surface of the plastic.

After welding, the joint should not be stressed for several hours. This is particularly true for polyolefins, nylons, and polyformaldehyde. Hot-gas welding is not recommended for filled materials or substrates less than $^{1}/_{16}$ in in thickness. Applications are usually large structural assemblies. The weld is not cosmetically attractive, but tensile strengths that are 85% of the parent materials are easily obtained.

Friction or spin welding. *Spin welding* uses the heat of friction to cause fusion at the interface. One substrate is rotated very rapidly while in

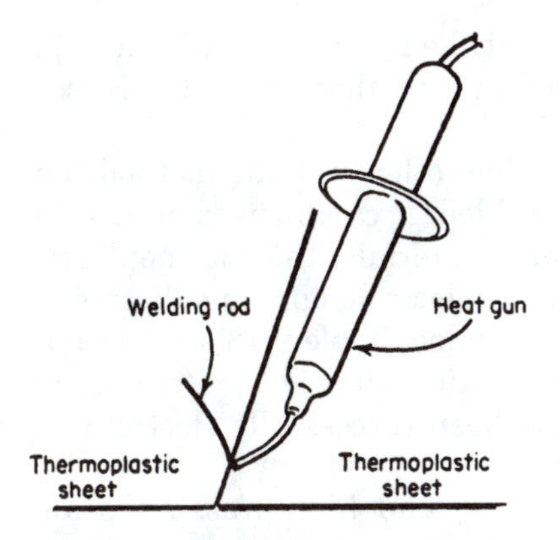

Figure 9.29 Hot-gas welding apparatus.

touch with the other stationary substrate so that the surfaces melt without damaging the part. Sufficient pressure is applied during the process to force out excess air bubbles. The rotation is then stopped, and pressure is maintained until the weld sets. Rotation speed and pressure are dependent on the thermoplastics being joined. The main process parameters are the spin of rotation, weld or axial pressure, and weld time. The equipment necessary depends upon production requirement, but spin welding can be adapted to standard shop machinery such as drill presses or lathes.

In commercial spin welding machines, rotational speeds can range from 200 to 14,000 r/min. Welding times (heating and cooling) can range from less than 1 to 20 s, with typical times being several seconds.

A wide variety of joints can be made by spin welding. Since the outer edges of the rotating substrate move considerably faster than the center, joints are generally designed to concentrate pressure at the center. A shallow tongue-and-groove type of joint design is useful to index the opposite parts and provide a uniform bearing surface. Spin welding is a popular method of joining large-volume products, packaging, and toys.

Induction heating. An electromagnetic induction field can be used to heat a metal grid or an insert placed between mating thermoplastic substrates. When the joint is positioned between energized induction coils, the hot insert material responds to the high-frequency ac source, causing the plastic surrounding it to melt and fuse together. Slight pressure is maintained as the induction field is turned off and the joint hardens.

In addition to metal inserts, electromagnetic adhesives can be used to form the joint. Electromagnetic adhesives are made from metal or ferromagnetic particle-filled thermoplastics. These adhesives can be shaped into gaskets or film that can easily be applied and will melt in an induction field. The advantage of this method is that stresses caused by large metal inserts are avoided.

Induction welding is less dependent than other welding methods on the properties of the materials being welded. It can be used on nearly all thermoplastics. In welding different materials, the thermoplastic resin enclosing the metal particles in the electromagnetic adhesive is made of a blend of the materials being bonded. Table 9.18 shows compatible combinations for electromagnetic adhesives. Reinforced plastics with filler levels over 50% have been successfully electromagnetically welded.

Strong and clean structural, hermetic, and high-pressure seals can be obtained from this process. Important determinants of bond quality in induction welding are the joint design and induction coil design. With automatic equipment, welds can be made in less than 1 s.

Ultrasonic and vibration welding. During *ultrasonic welding,* a high-frequency (20- to 40-kHz) electrodynamic field is generated that resonates a metal horn. The horn is in contact with one of the plastic parts and the other part is fixed firmly. The horn and the part to which it is in contact vibrates sufficiently fast to cause great heat at the interface of the parts being bonded. With pressure and subsequent cooling, a strong bond can be obtained with many thermoplastics. Rigid plastics with a high modulus of elasticity are best. Excellent results can be obtained with polystyrene, SAN, ABS, polycarbonate, and acrylic plastics.

TABLE 9.18 Compatible Plastic Combinations for Bonding with Electromagnetic Adhesives[18]

	ABS	Acetal	Acrylic	Nylon	PC	PE	PP	PS	PVC	SAN
ABS	X		X				X	X		
Acetal		X								
Acrylic	X		X				X	X		
Nylon				X						
Polycarbonate					X					
Polyethylene						X				
Polypropylene							X			
Polystyrene	X		X					X	X	
Polyvinyl chloride	X		X					X	X	
SAN										X

X = compatible combinations.

The basic variables in ultrasonic bonding are amplitude, air pressure, weld time, and hold time. The desired joint strength can be achieved by altering these variables. Increasing weld time generally results in increasing bond strength up to a point. After that point, additional weld time does not improve the joint and can even degrade it. Average processing times, including welding and cooling, are less than several seconds.

Typical ultrasonic joint designs are shown in Fig. 9.30. Often an energy director, or small triangular tip in one of the parts, is necessary. All of the ultrasonic energy is concentrated on the tip of the energy director and this is the area of the joint that then heats, melts, and provides the material for the bond. Ultrasonic welding is considered a faster means of bonding than direct heat welding.

Ultrasonic welding of parts fabricated from ABS, acetals, nylon, PPO, polycarbonate, polysulfone, and thermoplastic polyesters should be considered as early in the design of the part as possible. Very often minor modifications in part design will make ultrasonic welding more convenient. Best joint design and ultrasonic horn design can be recommended by the plastic resin manufacturer or ultrasonic equipment supplier.

Like resin materials, such as ABS to ABS, are the easier to weld ultrasonically; some unlike resins *may* be bonded provided they have similar melt temperatures, chemical composition, and modulus of elasticity. Generally, amorphous resins (ABS, PPO, and polycarbonate) are also easier to weld ultrasonically than crystalline resins (nylon, acetal, and thermoplastic polyester).

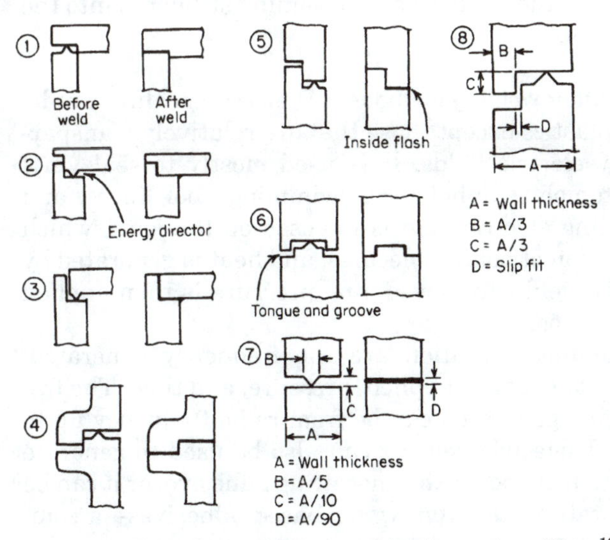

Figure 9.30 Typical joint designs used in ultrasonic welding.[19]

Ultrasonic equipment can also be used for mechanical fastening operations. Ultrasonic energy can be used to apply threaded inserts to molded plastic parts and to heat stake plastic studs.

Vibration welding is similar to ultrasonic welding, except that it uses a lower frequency (120 to 240 Hz) of vibration. In this way, very large parts can be bonded. The process parameters affecting the strength of the resulting weld are the amplitude and frequency of motion, weld pressure, and weld time. There are two types of vibration welding: linear, in which friction is generated by a linear motion of the parts, and orbital, in which one part is vibrated using circular motion in all directions.

Vibration welding has been used on large thermoplastic parts such as canisters, pipe sections, and other parts that are too large to be excited with an ultrasonic generator. An advantage of vibration welding over ultrasonic welding is that it can provide gas-tight joints in structures with long bond lines. Ultrasonic welding is basically a spot weld technique limited by the size of the horn.

Total process time for vibration joining is generally between 5 and 15 s. This is longer than spin or ultrasonic welding but much shorter than direct heat welding, solvent cementing, or adhesive bonding. Vibration welding can be applied to ABS, acetal, nylon, PPO, thermoplastic polyesters, and polycarbonate. For vibration welding, hydroscopic resins, such as nylon, do not have to be dried as is necessary with ultrasonic welding. Joint designs do not require an energy director, as in ultrasonic joint designs, but the joint area must be strong enough to resist the forces of operation without deformation. This often requires thickening the bond area or designing stiffeners into the part near the joint areas.

Dielectric welding and other welding methods. Dielectric sealing can be used on most thermoplastics except those that are relatively transparent to high-frequency electric fields. It is used mostly to seal vinyl sheeting such as automobile upholstery, swimming pool liners, and rainwear. An alternating electric field is imposed on the joint, which causes rapid reorientation of polar molecules, and heat is generated by molecular friction. The field is removed, and pressure is then applied and held until the weld cools.

Variable in the bonding operation are the frequency generated, dielectric loss of the plastic, power applied, pressure, and time. The frequency of the field being generated can be from radio frequency up to microwave frequency. Dielectric heating can also be used to generate the heat necessary for curing polar, thermosetting adhesive, or it can be used to quickly evaporate water from water based adhesives—a common application in the furniture industry.

Other thermal welding processes that are less common than those described previously are: extrusion welding, electrofusion welding, infrared welding, and laser welding. These are generally used in specialty processes or with applications that require unique methods of heating because of the joint design or nature of the final product.

9.6.2 Solvent welding

Solvent welding or cementing is the simplest and most economical method of joining many noncrystalline thermoplastics. Solvent-cemented joints are less sensitive to thermal cycling than joints bonded with adhesives, and they are as resistant to degrading environments as their parent plastic. Bond strength equal to 85 to 100% of the parent plastic can be obtained. The major disadvantage of solvent cementing is the possibility of stress cracking or crazing of the part and the possible hazards of using low vapor point solvents. When two dissimilar plastics are to be joined, adhesive bonding is generally desirable because of solvent and polymer compatibility problems.

Solvent cements should be chosen with approximately the same solubility parameter as the plastic to be bonded. Table 9.19 lists typical solvents used to bond major plastics. It is common to use a mixture of a fast-drying solvent with a less volatile solvent to prevent crazing. The

TABLE 9.19 Typical Solvents for Solvent Cementing of Plastics[4]

Plastic	Solvent
ABS	Methyl ethyl ketone, methyl isobutyl ketone, tetrahydrofuran, methylene chloride
Acetate	Methylene chloride, acetone, chloroform, methyl ethyl ketone, ethyl acetate
Acrylic	Methylene chloride, ethylene dichloride
Cellulosics	Methyl ethyl ketone, acetone
Nylon	Aqueous phenol, solutions of resorcinal in alcohol, solutions of calcium chloride in alcohol
PPO	Trichloroethylene, ethylene dichloride, chloroform, methylene chloride
PVC	Cyclohexane, tetrahydrofuran, dichlorobenzene
Polycarbonate	Methylene chloride, ethylene dichloride
Polystyrene	Methylene chloride, ethylene ketone, ethylene dichloride, trichloroethylene, toluene, xylene
Polysulfone	Methylene chloride

Note: These are solvents recommended by the various resin suppliers. A key to the selection of solvents is how fast they evaporate: a fast-evaporating product may not last long enough for some assemblies; too slow evaporation could hold up production.

solvent cement can be bodied to 25% by weight with the parent plastic to fill gaps and reduce shrinkage and internal stress during cure.

The parts to be bonded should be unstressed and, if necessary, annealed. The surfaces should mate well and have a clean, smooth surface. A V-joint or rounded butt joint are generally preferred for making a solvent butt joint. Scarf joints and flat butt joints are difficult to position and to apply pressure during the solvent evaporation phase of the process.

The solvent cement is generally applied to the substrate with a syringe or brush. In some cases the surface can be immersed in the solvent. After the area to be bonded softens, the parts are mated and held under pressure until dry. Pressure should be low and uniform so that the finished joint will not be stressed. After the joint hardens, the pressure is released, and an elevated temperature cure may be necessary, depending on the plastic and desired joint strength. The bonded part should not be packaged or stressed until the solvent has adequate time to escape from the joint.

9.7 Recommended Assembly Processes for Common Plastics

When decisions are to be made relative to assembly methods (mechanical fastening, adhesive bonding, thermal welding, or solvent cementing), special considerations must be taken because of the nature of the substrate and possible interactions with the adhesive or the environment. The following sections identify some of these considerations and offer an assembly guide to the various methods of assemblies that have been found appropriate for specific plastics.

9.7.1 Acetal homopolymer and acetal copolymer

Parts made of acetal homopolymer and copolymer are generally strong and tough, with a surface finish that is the mirror image of the mold surface. Acetal parts are generally ready for end use or further assembling with little or no postmold finishing.

Press fitting has been found to provide joints of high strength at minimum cost. Acetal copolymer can be used to provide snap-fit parts. Use of self-tapping screws may provide substantial cost savings by simplifying machined parts and reducing assembly costs.

Epoxies, isocyanate-cured polyester, and cyanoacrylates are used to bond acetal copolymer. Generally, the surface is treated with a sulfuric–chromic acid treatment. Epoxies have shown 150- to 500-lb/in^2 shear strength on sanded surfaces and 500 to 1000 lb/in^2 on chemically treated surfaces. Plasma treatment has also shown to be effective on acetal substrates. Acetal homopolymer surfaces should be chemically treated prior to bonding. This is accomplished with a sulfuric–chromic acid treatment followed by a solvent wipe. Epoxies, nitrile, and nitrile-phenolics can be used as adhesives.

Thermal welding and solvent cementing are commonly used for bonding this material to itself. Heated tool welding produces exceptionally strong joints with acetal homopolymers and copolymers. With the homopolymer, a temperature of the heated surface near 550°F and a contact time of 2 to 10 s are recommended. The copolymer can be hot-plate–welded from 430 to 560°F. It is claimed that annealing acetal copolymer joints will strengthen them further. Annealing can be done by immersing the part in oil heated to 350°F. Acetal resin can be bonded by hot-wire welding. Pressure on the joint, duration of the current, and wire type and size must be varied to achieve optimum results. Shear strength on the order of 150 to 300 lb/in or more have been obtained with both varieties, depending on the wire size, energizing times (wire temperature), and clamping force.

Hot-gas welding is used effectively on heavy acetal sections. Joints with 50% of the tensile strength of the acetal resin have been obtained. Conditions of joint design and rod placement are similar to those presented for ABS. A nitrogen blanket is suggested to avoid oxidation. The outlet temperature of the welding gun should be approximately 630°F for the homopolymer and 560°F for the copolymer. For maximum joint strength both the welding rod and parts to be welded must be heated so that all surfaces are melted.

Acetal components can easily be joined by spin welding, which is a fast and generally economical method to obtain joints of good strength. Spin-welded acetal joints can have straight 90° mating surfaces, or surfaces can be angles, molded in a V-shape, or flanged.

Although not common practice, acetal copolymer can be solvent-welded at room temperature with full-strength hexafluoroacetone sesquihydrate (Allied Chemical Corporation). The cement has been found to be an effective bonding agent for adhering to itself, nylon, or ABS. Bond strengths in shear are greater than 850 lb/in^2 using "as-molded" surfaces. Hexafluoroacetone sesquihydrate is a severe eye and skin irritant. Specific handling recommendations and information on toxicity should be requested from Allied Chemical Corporation. Because of its high solvent resistance, acetal homopolymer cannot be solvent cemented.

9.7.2 Acrylonitrile butadiene styrene (ABS)

ABS parts can be designed for snap-fit assembly using a general guideline of 5% allowable strain during the interference phase of the assembly. Thread-cutting screws are frequently recommended for nonfoamed

ABS, and thread-forming screws for foamed grades. Depending on the application, the use of bosses and boss caps may be advantageous.

The best adhesives for ABS are epoxies, urethanes, thermosetting acrylics, nitrile-phenolics, and cyanoacrylates. These adhesives have shown joint strength greater than that of the ABS substrates being bonded. ABS substrates do not require special surface treatments other than simple cleaning and removal of possible contaminants.

ABS can also be bonded to itself and to certain other thermoplastics by either solvent-cementing or any of the heat welding methods. For bonding ABS to itself, it is recommended that the hot-plate temperatures be between 430 and 550°F. Lower temperatures will result in sticking of the materials to the heated platens, while temperatures above 550°F will increase the possibility of thermal degradation of the surface. In joining ABS the surfaces should be in contact with the heated tool until they are molten, then brought carefully and quickly together, and held with minimum pressure. If too much pressure is applied, the molten material will be forced from the weld and result in poor appearance and reduced weld strength. Normally, if a weld flash greater than $^1/_8$ in occurs, too much joining pressure has been used.

Hot-gas welding has been used to join ABS thermoplastic with much success. Joints with over 50% of the strength of the parent material have been obtained. The ABS welding rod should be held approximately at a 90° angle to the base material; the gun should be held at a 45° angle with the nozzle $^1/_4$ to $^1/_2$ in from the rod. ABS parts to be hot-gas–welded should be bonded at 60° angles. The welding gun, capable of heating the gas to 500 to 600°F, must be moved continually in a fanning motion to heat both the welding rod and bed. Slight pressure must be maintained on the rod to ensure good adhesion.

Spin-welded ABS joints can have straight 90° mating surfaces, or surface can be angled, molded in a V-shape, or flanged. The most important factor in the quality of the weld is the joint design. The area of the spinning part should be as large as possible, but the difference in linear velocity between the maximum and minimum radii should be as small as feasible.

One of the fastest methods of bonding ABS and acetal thermoplastics is induction welding. This process usually takes 3 to 10 s, but can be done in as little as 1 s. During welding, a constant pressure of at least 100 lb/in^2 should be applied on the joint to minimize the development of bubbles; this pressure should be maintained until the joint has sufficiently cooled. When used, metal inserts should be 0.02 to 0.04 in thick. Joints should be designed to enclose the metal insert completely. Inserts made of carbon steel require less power for heating although other types of metal can be used. The insert should be located as close as possible to the electromagnetic generator coil and centered within the coil to assure uniform heating.

The solvents recommended for ABS are methyl ethyl ketone, methyl isobutyl ketone, tetrahydrofuran, and methylene chloride. The solvent used should be quick drying to prevent moisture absorption, yet slow enough to allow assembly of the parts. The recommended cure time is 12 to 24 h at room temperature. The time can be reduced by curing at 130 to 150°F. A cement can be made by dissolving ABS resin in a solvent of up to 25% solids. This type of cement is very effective in joining parts that have irregular surfaces or areas that are not readily accessible. Because of the rapid softening actions of the solvent, the pressure and amount of solvent applied should be minimal.

9.7.3 Cellulosics (cellulose acetate, cellulose acetate butyrate, cellulose nitrate, ethyl cellulose, etc.)

Cellulosic materials can be mechanically fastened by a number of methods. However, their rigidity and propensity to have internal molding stresses must be carefully considered. The adhesives commonly used are epoxies, urethanes, isocyanate-cured polyesters, nitrile-phenolic, and cyanoacrylate. Only cleaning is required prior to applying the adhesive. A recommended surface cleaner is isopropyl alcohol. Cellulosic plastics may contain plasticizers. The extent of plasticizer migration and the compatibility with the adhesive must be evaluated. Cellulosics can be stress cracked by uncured cyanoacrylate and acrylic adhesives. Any excess adhesive should be removed from the surface immediately.

Cellulosic materials can also be solvent cemented. Where stress crazing is a problem, adhesives are a preferred method of assembly.

9.7.4 Fluorocarbons (PTFE, CTFE, FEP, etc.)

Because of the lower ductility of the fluorocarbon materials, snap-fit and press-fit joints are seldom used. Rivets or studs can be used in forming permanent mechanical joints. These can be provided with thermal techniques on the melt processable grades. Self-tapping screws and threaded inserts are used for many mechanical joining operations. In bolted connections some stress relaxation may occur the first day after installation. In such cases, mechanical fasteners should be tightened; thereafter, stress relaxation is negligible.

The combination of properties that makes fluorocarbons highly desirable engineering plastics also makes them nearly impossible to heat or solvent weld and very difficult to bond with adhesives without proper surface treatment. The most common surface preparation for fluorocarbons is a sodium naphthalene etch, which is believed to remove fluorine atoms from the surface to provide better wetting properties. A formulation and description of the sodium naphthalene process can be found in Table 9.11. Commercial chemical products for etching fluorocarbons are also listed.

Another process for treating fluorocarbons, as well as some other hard to bond plastics (notably polyolefins), is *plasma treating.* Plasma surface treatment has been shown to increase the tensile shear strength of Teflon bonded with epoxy adhesive from 50 to 1000 lb/in^2. The major disadvantage of plasma treating is that it is a batch process, which involves large capital equipment expense, and part size is often limited because of available plasma treating vessel volume. Epoxies and polyurethanes are commonly used for bonding treated fluorocarbon surfaces.

Melt processable fluorocarbon parts have been successfully heat welded, and certain grades have been spin welded and hermetically sealed with induction heating. However, because of the extremely high temperatures involved and the resulting weak bonds, these processes are seldom used for structural applications.

Fluorocarbon parts cannot be solvent welded because of their great resistance to all solvents.

9.7.5 Polyamide (nylon)

Due to their toughness, abrasion resistance, and generally good chemical resistance, parts made from polyamide or resin (or nylon) are generally more difficult to finish and assemble than other plastic parts. However, nylons are used in virtually every industry and market. The number of chemical types and formulations of nylon available also provide difficulty in selecting fabrication and finishing processes.

Nylon parts can be mechanically fastened by most of the methods described in this chapter. Mechanical fastening is usually the preferred method of assembly because adhesives bonding and welding often show variable results, mainly due to the high internal moisture levels in nylon. Nylon parts can contain a high percentage of absorbed water. This water can create a weak boundary layer under certain conditions. Generally, parts are dried to less than 0.5% water before bonding.

Some epoxy, resorcinal formaldehyde, phenol resorcinol, and rubber-based adhesives have been found to produce satisfactory joints between nylon and metal, wood, glass, and leather. The adhesive tensile shear strength is about 250 to 1000 lb/in^2. Adhesive bonding is usually considered inferior to heat welding or solvent cementing. However, priming of nylon adherends with compositions based on resorcinol formaldehyde, isocyanate modified rubber, and cationic surfactant have been reported to provide improved joint strength.

Elastomeric (nitrile and urethane), hot-melt (polyamide and polyester), and reactive (epoxy, urethane, acrylic, and cyanoacrylate) adhesives have been used for bonding nylon.

Induction welding has also been used for nylon and polycarbonate parts. Because of the variety of formulations available and their direct effect on heat welding parameters, the reader is referred to the resin manufacturer for starting parameters for use with these welding methods. Both nylon and polycarbonate resins should be predried before induction welding.

Recommended solvent systems for bonding nylon to nylon are aqueous phenol, solutions of resorcinol in alcohol, and solutions of calcium chloride in alcohol. These solvents are sometimes bodied by adding nylon resin.

9.7.6 Polycarbonate

Polycarbonate parts lend themselves to all mechanical assembly methods. Polycarbonate parts can be easily joined by solvents or thermal welding methods; they can also be joined by adhesives. However, polycarbonate is soluble in selected chlorinated hydrocarbons. It also exhibits crazing in acetone and is attacked by bases.

When adhesives are used, epoxies, urethanes, and cyanoacrylates are chosen. Adhesive bond strengths with polycarbonate are generally 1000 to 2000 lb/in^2. Cyanoacrylates, however, are claimed to provide over 3000 lb/in^2 when bonding polycarbonate to itself. No special surface preparation is required of polycarbonate other than sanding and cleaning. Polycarbonates can stress crack in the presence of certain solvents. When cementing polycarbonate parts to metal parts, a room temperature curing adhesive is suggested to avoid stress in the interface caused by differences in thermal expansion.

Polycarbonate film is effectively heat sealed in the packaging industry. The sealing temperature is approximately 425°F. For maximum strength the film should be dried at 250°F to remove moisture before bonding. The drying time varies with the thickness of the film or sheet. A period of approximately 20 min is suggested for a 20-mil-thick film and 6 h for a $^1/_4$-in-thick sheet. Predried films and sheets should be sealed within 2 h after drying. Hot-plate welding of thick sheets of polycarbonate is accomplished at about 650°F. The faces of the substrates should be butted against the heating element for 2 to 5 s or until molten. The surfaces are then immediately pressed together and held for several seconds to make the weld. Excessive pressure can cause localized strain and reduce the strength of the bond. Pressure during cooling should not be greater than 100 lb/in^2.

Polycarbonate parts with thicknesses of at least 40 mils can be successfully hot-gas–welded. Bond strengths in excess of 70% of the parent resin have been achieved. Equipment capable of providing gas temperature of 600 to 1200°F should be used. As prescribed for the heated tool process, it is important to adequately predry (250°F) both the polycarbonate parts and welding rods. The bonding process should occur within minutes of removing the parts from the predrying oven.

For spin welding, tip speeds of 30 to 50 ft/min create the most favorable conditions to get polycarbonate resin surfaces to their sealing temperature of 435°F. Contact times as short as $^1/_2$ s are sufficient for small parts. Pressures of 300 to 400 lb/in^2 are generally adequate. For the best bonds parts should be heat treated for stress relief at 250°F for several hours after welding. However, this stress relief step is often unnecessary and may lead to degraded impact properties of the parent plastics.

Methylene chloride is a very fast solvent cement for polycarbonate. This solvent is recommended only for temperature climate zones and on small areas. A mixture of 60% methylene chloride and 40% ethylene chloride is slower drying and the most common solvent cement used. Ethylene chloride is recommended in very hot climate. These solvents can be bodied with 1 to 5% polycarbonate resin where gap-filling properties are important. A pressure of 200 lb/in^2 is recommended.

9.7.7 Polyethylene, polypropylene, and polymethyl pentene

Because of their ductility, polyolefin parts must be carefully assembled using mechanical fasteners. These assembly methods are normally used on the materials with higher moduli such as high molecular weights of polyethylene and polypropylene.

Epoxy and nitrile-phenolic adhesives have been used to bond these plastics after surface preparation. The surface can be etched with a sodium sulfuric-dichromate acid solution at an elevated temperature. Flame treatment and corona discharge have also been used. However, plasma treatment has proven to be the optimum surface process for these materials. Shear strengths in excess of 3000 lb/in^2 have been reported on polyethylene treated for 10 min in an oxygen plasma and bonded with an epoxy adhesive. Polyolefin materials can also be thermally welded, but they cannot be solvent cemented.

Polyolefins can be thermally welded by almost any technique. However, they cannot be solvent welded because of their resistance to most solvents.

9.7.8 Polyethylene terephthalate and polybutylene terephthalate

These materials can be joined by mechanical self-fastening methods or by mechanical fasteners.

Polyethylene terephthalate (PET) and polybutylene terephthalate (PBT) parts are generally joined by adhesives. Surface treatments recommended specifically for PBT include abrasion and solvent cleaning with toluene. Gas plasma surface treatments and chemical etch have been used where maximum strength is necessary. Solvent cleaning of PET surfaces is recommended. The linear film of polyethylene terephthalate (Mylar®) surface can be pretreated by alkaline etching or plasma for maximum adhesion, but often a special treatment is unnecessary. Commonly used adhesives for both PBT and PET substrates are isocyanate-cured polyesters, epoxies, and urethanes. Polyethylene terephthalate cannot be solvent cemented or heat welded.

Ultrasonic welding is the most common thermal assembly process used with polybutylene terephthalate parts. However, heated tool welding and other welding methods have provided satisfactory joints when bonding PET and PBT to itself and to dissimilar materials.

Solvent cementing is generally not used to assembly PET or PBT parts because of their solvent resistance.

9.7.9 Polyetherimide (PEI), polyamide-imide, polyetheretherketone (PEEK), polyaryl sulfone, and polyethersulfone (PES)

These high-temperature thermoplastic materials are generally joined mechanically or with adhesives. The high modulus, low creep strength, and superior fatigue resistance make these materials ideal for snap-fit joints.

They are easily bonded with epoxy or urethane adhesives; however, the temperature resistance of the adhesives do not match the temperature resistance of the plastic part. No special surface treatment is required other than abrasion and solvent cleaning. Polyetherimide (ULTEM®), polyamide-imide (TORLON®), and polyethersulfone can be solvent cemented, and ultrasonic welding is possible.

These plastics can also be welded using vibration and ultrasonic thermal processes. Solvent welding is also possible with selected solvents and processing conditions.

9.7.10 Polyimide

Polyimide parts can be joined with mechanical fasteners. Self-tapping screws must be strong enough to withstand distortion when they are inserted into the polyimide resin which is very hard. Polyimide parts can be bonded with epoxy adhesives. Only abrasion and solvent cleaning is necessary to treat the substrate prior to bonding. The plastic part will usually have a higher thermal rating than the adhesive. Thermosetting polyimides cannot be heat welded or solvent cemented.

9.7.11 Polymethylmethacrylate (acrylic)

Acrylics are commonly solvent cemented or heat welded. Because acrylics are a noncrystalline material, they can be welded with greater ease than semicrystalline parts. Ultrasonic welding is the most popular process for welding acrylic parts. However, because they are relatively brittle materials, mechanical fastening processes must be carefully chosen.

Epoxies, urethanes, cyanoacrylates, and thermosetting acrylics will result in bond strengths greater than the strength of the acrylic part. The surface needs only to be clean of contamination. Molded parts may stress crack when they come in contact with an adhesive containing solvent or monomer. If this is a problem, an anneal (slightly below the heat distortion temperature) is recommended prior to bonding.

9.7.12 Polyphenylene oxide (PPO)

Polystyrene modified polyphenylene oxide can be joined with almost all techniques described in this chapter. Snap-fit and press-fit assemblies can be easily made with this material. A maximum strain limit of 8% is commonly used in the flexing member of PPO parts. Metal screw and bolts are commonly used to assemble PPO parts or for attaching various components.

Epoxy, polyester, polyurethane, and thermosetting acrylic have been used to bond modified PPO to itself and other materials. Bond strengths are approximately 600 to 1500 lb/in^2 on sanded surfaces and 1000 to 2200 lb/in^2 on chromic acid etched surfaces.

Polystyrene modified polyphenylene oxide or Noryl® can be hot-plate–welded at 500 to 550°F and 20 to 30 s contact time. Unmodified PPO can be welded at hot-plate temperatures of 650°F. Excellent spin-welded bonds are possible with modified polyphenylene oxide because the low thermal conductivity of the resin prevents head dissipation from the bonding surfaces. Typical spin welding is done at a rotational speed of 40 to 50 ft/min and a pressure of 300 to 400 lb/in^2. Spin time should be sufficient to ensure molten surfaces.

Polyphenylene oxide joints must mate almost perfectly; otherwise, solvent welding provides a weak bond. Very little solvent cement is needed. Best results are obtained by applying the solvent cement to only one substrate. Optimum holding time has been found to be 4 min at approximately 400 lb/in^2. A mixture of 95% chloroform and 5% carbon tetrachloride is the best solvent system for general purpose bonding, but very good ventilation is necessary. Ethylene dichloride offers a slower rate of evaporation for large structures or hot climates.

9.7.13 Polyphenylene sulfide (PPS)

Since PPS is a semicrystalline thermoplastic, it is not ideally suited to ultrasonic welding. Because of its excellent solvent resistance, PPS cannot be solvent cemented. PPS assemblies can be made by a variety of mechanical fastening methods as well as by adhesives bonding.

Adhesives recommended for polyphenylene sulfide include epoxies and urethanes. Joint strengths in excess of 1000 lb/in^2 have been reported for abraded and solvent-cleaned surfaces. Somewhat better adhesion has been reported for machined surfaces over as-molded surfaces. The high heat and chemical resistance of polyphenylene sulfide plastics make them inappropriate for solvent cementing or heat welding.

Polyimide and polyphenylene sulfide resins present a problem in that their high temperature resistance generally requires that the adhesive have similar thermal properties. Thus, high-temperature epoxies adhesive are most often used with polyimide and PPS parts. Joint strength is superior (greater than 1000 lb/in^2) but thermal resistance is not better than the best epoxy systems (300 to 400°F continuous).

9.7.14 Polystyrene

Polystyrene parts are conventionally solvent cemented or heat welded. However, urethanes, epoxies, unsaturated polyesters, and cyanoacrylates will provide good adhesion to abraded and solvent-cleaned surfaces. Hot-melt adhesives are used in the furniture industry. Polystyrene foams will collapse when they come in contact with certain solvents. For polystyrene foams, a 100% solids adhesive or a water-based contact adhesive is recommended.

Polystyrene can be joined by either thermal or solvent welding techniques. Preference is generally given to ultrasonic methods because of speed and simplicity. However, heated tool welding and spin welding are also commonly used.

9.7.15 Polysulfone

Polysulfone parts can be joined with all the processes described in this chapter. Because of their inherent dimensional stability and creep resistance, polysulfone parts can be press fitted with ease. Generally, the amount of interference will be less than that required for other thermoplastics. Self-tapping screws and threaded inserts have also been used.

Urethane and epoxy adhesives are recommended for bonding polysulfone substrates. No special surface treatment is necessary. Polysulfones can also be easily joined by solvent cementing or thermal welding methods.

Direct thermal welding of polysulfone requires a heated tool capable of attaining 700°F. Contact time should be approximately 10 s, and then the parts must be joined immediately. Polysulfone parts should be dried 3 to 6 h at 250°F before attempting to heat seal them. Polysulfone can also be joined to metal since polysulfone resins have good adhesive characteristics. Bonding to aluminum requires 700°F. With cold-rolled steel the surface of the metal first must be primed with 5 to 10% solution of polysulfone and baked for 10 min at 500°F. The primed piece then can be heat welded to the polysulfone part at 500 to 600°F.

A special tool has been developed for hot-gas welding of polysulfone. The welding process is similar to standard hot-gas welding methods but requires greater elevated temperature control. At the welding temperature great care must be taken to avoid excessive application of heat, which will result in degradation of the polysulfone resin.

For polysulfone a 5% solution of polysulfone resin in methylene chloride is recommended as a solvent cement. A minimum amount of cement should be used. The assembled pieces should be held for 5 min under 500 lb/in^2. The strength of the joint will improve over a period of several weeks as the residual solvent evaporates.

9.7.16 Polyvinyl chloride (PVC)

Rigid polyvinyl chloride can be easily bonded with epoxies, urethanes, cyanoacrylates, and thermosetting acrylics. Flexible polyvinyl chloride parts present a problem because of plasticizer migration over time. Nitrile adhesives are recommended for bonding flexible polyvinyl chloride because of compatibility with the plasticizers used. Adhesives that are found to be compatible with one particular polyvinyl chloride plasticizer may not work with another formulation. Solvent cementing and thermal welding methods are also commonly used to bond both rigid and flexible polyvinyl chloride parts.

9.7.17 Thermoplastic polyesters

These materials may be bonded with epoxy, thermosetting acrylic, urethane, and nitrile-phenolic adhesives. Special surface treatment is not necessary for adequate bonds. However, plasma treatment has been reported to provide enhanced adhesion. Solvent cementing and certain thermal welding methods can also be used with thermoplastic polyester.

Thermoplastic polyester resin can be solvent cemented using hexafluoroisopropanol or hexafluoroacetone sesquihydrate. The solvent should be applied to both surfaces and the parts assembled as quickly as possible. Moderate pressure should be applied as soon as the parts are assembled. Pressure should be maintained for at least 1 to 2 min; maximum bond strength will not develop until at least 18 h at room temperature. Bond strengths of thermoplastic polyester bonded to itself will be in the 800- to 1500-lb/in^2 range.

9.7.18 Thermosetting plastics (epoxies, diallyl phthalate, polyesters, melamine, phenol and urea formaldehyde, polyurethanes, etc.)

Thermosetting plastics are joined either mechanically or by adhesives. Their thermosetting nature prohibits the use of solvent or thermal welding processes; however, they are easily bonded with many adhesives.

Abrasion and solvent cleaning are generally recommended as the surface treatment. Surface preparation is generally necessary to remove any contaminant, mold release, or gloss from the part surface. Simple solvent washing and abrasion is a satisfactory surface treatment for bonds approaching the strength of the parent plastic. An adhesive should be selected that has a similar coefficient of expansion and modulus as the part being bonded. Rigid parts are best bonded with rigid adhesives based on epoxy formulations. More flexible parts should be bonded with adhesives that are flexible in nature after curing. Epoxies, thermosetting acrylics, and urethanes are the best adhesives for the purpose.

9.8 Decorating Plastics

Several decorating processes can be accomplished with plastic parts—either during processing of the part, directly afterward, or before final assembly and packaging. The most inexpensive method of providing decorative designs on plastics is to incorporate the design into the mold or to apply the decoration as part of the molding operation. However, often this is not possible either because of mold complexity, the need to apply decorations to customers' specifications, or other reasons.

This section will discuss common decorative processes that are generally used with plastic parts. The most widely used decorating processes in the plastic industry are

- Painting
- Hot decorating (hot stamping, in-mold decorating, heat transfer)
- Plating (electroless, electrolytic, vacuum metallizing)
- Printing
- Application of labels, decals, etc.

As with adhesive bonding, surface treatment and cleanliness is of primary importance when decorating plastic parts. Prior to decoration, the surface of the plastic part must be cleaned of mold release, internal plastics lubricants, and plasticizers. Plastic parts can also become electrostatically charged and attract dust. This could disrupt the even flow of a coating or interfere with adhesion. Solvent or destaticizers may be used to clean and eliminate static from plastic parts prior to decorating. Cleaning of the plastic part requires an understanding of the plastic material to be cleaned and the effect of the solvents and processes on that plastic.

Some plastics may need to be roughened or chemically treated to promote adhesion of the decorating medium. Molded and extruded plastic parts tend to have glossy resin-rich surfaces. This is desirable if the part is used in the as-molded condition, but the glossy surface may require abrading or etching to hold paint or printing media or prior to application of an adhesive. Polyolefins, polyacetals, polyamides, fluorocarbons, and other surfaces that are difficult to wet might require special surface treatments before decorating processes can be completed satisfactorily. Many of the cleaning methods and surface treatments that are described in the previous sections of this chapter for surface treatment prior to bonding are also applicable prior to decorating.

9.8.1 Painting

Painting of plastic parts is fully addressed in another chapter of this book, and the reader is referred there for details on the painting processes and materials used. In this section, some of the fundamentals of painting will be reviewed from a perspective of decorating the plastic part. Small-volume processes will be emphasized because these processes are usually most conducive to the flexibility (changes in color or design, short lead time, etc.) needed in decorating plastic parts.

Painting is a method of providing a specific color to a plastic part. The most convenient and economical way to color plastics is to blend the color pigments into the base resin. However, often this does not provide the flexibility desired either to change colors, match colors, or to apply multiple colors to a single part. The painting processes used in decorating plastics are similar to those used with other substrates, including spray painting, electrostatic spraying, dip coating, fill-in marking, screen painting, and roller coating. In all of these painting processes the type of paint, especially the nature of the solvent, and the temperature of the paint-drying process must be chosen carefully for the specific plastic being coated.

Several categories of paint are generally used for painting on plastic. Table 9.20 shows paints that are recommended for various plastics.

TABLE 9.20 Recommended Paints for Plastics[20]

Plastic	Urethane	Epoxy	Polyester	Acrylic lacquer	Acrylic enamel	Waterborne
ABS	R	R	NR	R	R	R
Acrylic	NR	NR	NR	R	R	R
PVC	NR	NR	NR	R	R	NR
Styrene	R	R	NR	R	R	R
PPO/PPE	R	R	R	R	R	R
Polycarbonate	R	R	R	R	R	R
Nylon	R	R	R	NR	NR	NR
Polypropylene	R	R	R	NR	NR	NR
Polyethylene	R	R	R	NR	NR	NR
Polyester	R	R	R	NR	NR	NR
RIM	R	NR	NR	NR	R	R

Note: R = recommended; NR = not recommended.

Water-based acrylic paints provide no harmful solvents. They are easy to apply and a wide variety of colors is available. Lacquers contain solvents that evaporate allowing the paint coating to dry without chemical reaction. The lacquer must be carefully chosen so the solvent is compatible with the plastic being painted. Enamels are based on a chemical reaction that causes the coating film to form and harden. The effect of this reaction on the parent plastic part must be carefully noted. Certain plastics may require a primer to help improve the adhesion of the paint to the plastic. After coating, a top coat may also be required to improve the environmental or wear resistance of the paint. Two part paints, such as epoxy or urethane coatings, provide excellent adhesion qualities and do not require a top coat.

Masking is required when areas of the part are not to be painted. Paper-backed masking tape, polyvinyl alcohol coating masks, and electroformed metal masks have been all used with plastic parts. The plastic part can be designed to better accommodate the mask. The part may be designed with some depression or boss that will allow the mask to be properly registered on the part being sprayed so it may do an effective masking job.

In the design of the part for painting, certain features should be avoided. Depressions, which have to be painted, should be relatively wide and shallow. If they are narrow and deep, they are likely to catch too much paint and, thus, will then take much longer to dry. For coating depressions, a technique called *wipe-in or fill-in painting* can be used as shown in Fig. 9.31*a*. Here the paint is applied to the depressions and surrounding surfaces. The excess paint on the surrounding surfaces is then removed by wiping or by using a solvent-dampened pad.

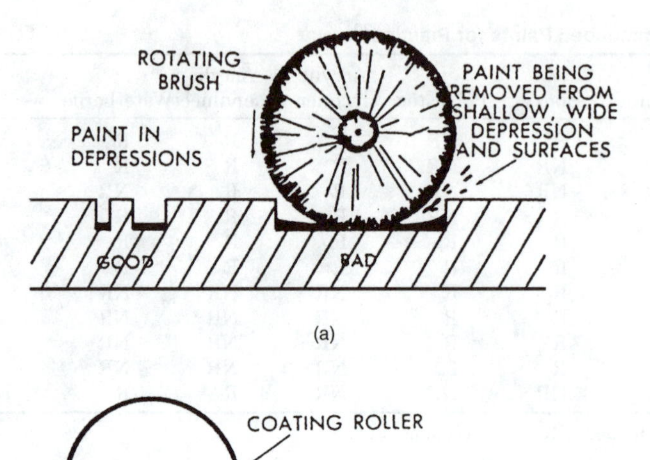

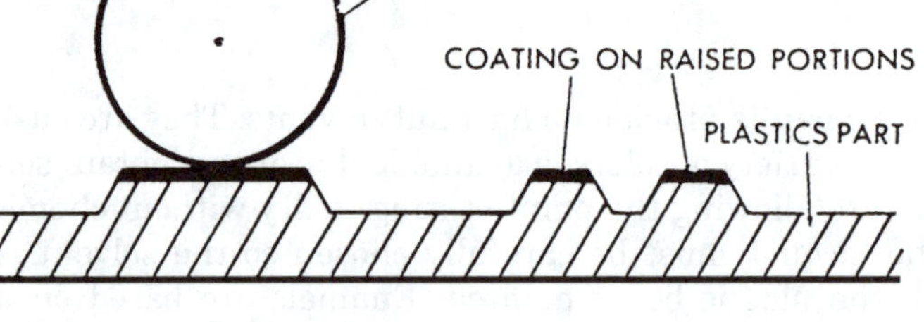

Figure 9.31 Roller coating of plastic parts: (*a*) fill-in method and (*b*) roller coating of raised portions.[21]

Painting on raised surfaces also requires special considerations. Trying to spray paint only on raised surfaces (that is, letter or numbers) is difficult to mask. However, roller coating (Fig. 9.31*b*) may be a method of obtaining good detail on raised sections without masks. Large, flat, smooth areas in which runs or drips would be apparent should also be avoided.

Molded letter, figures, and decorated designs are often used on molded parts. Letter designs can be raised or depressed. Raised lettering is less expensive since it merely required engraved letters on the mold surface. Depressed lettering on the part, however, requires raised lettering on the mold, and this is generally an expensive proposition. The depressed letters are filled with paint, and the surface is then wiped. Raised or depressed lettering usually does not exceed 0.03 in.

9.8.2 Hot decorating processes

Several decorating processes use heat to transfer a decoration or a printed image to the plastic part surface. These techniques usually use a carrier element, the decorative media, and an adhesive. In certain cases, the adhesive is the molten surface of the part that the decoration comes into contact with. The primary processes for hot decorating are hot stamping, in-mold decorating, and heat transfer.

Since these technologies employ no wet inks or paints, there are no offensive odors, environmental concerns over volatile emission, or storage problems. In addition, no ink mixing is necessary, and color or design changes involve merely changing a roll of dry printed foil, which minimizes setup time.

Hot stamping. The hot-stamping process is the most widely used in plastics decorating because of its convenience, versatility, and performance. The process of hot stamping consists of placing the part to be decorated under a hot-stamping die. The hot die then strikes the surface of the part through a metallized or painted roll-leaf carrier. The paint carried by the roll is fused into the impression made by the stamp. The stamp is programmed to be pressed down on the thermoplastic surface at a controlled pressure and temperature and set to penetrate a controlled distance.

Hot stamping is best suited for flat surfaces, although contoured dies could provide hot stamping for curved surfaces. The fit between the plastic part and the contoured die face must be very close to get uniform hot stamping. For the most part, hot stamping is used to apply small decorations and explanatory lettering, logos, etc., to plastic moldings. Hot stamping may be used to place gold, silver, or other metal foils (leaf), as well as paint pigments, onto plastic parts. Because these foils and pigments are dry, they are easy to handle and may be placed over painted surfaces. No masking is required and the process may be automatic or hand operated. Although hot stamping is used primarily on thermoplastics, the process is also applicable to thermosets.

Figure 9.32*a* shows a typical metallized hot-stamping foil. The carrier film supports the decorative coatings until they are pressed on the plastics part. A lacquer coating is passed over the releasing layer to provide protection for the metal foil. If paints are used, the lacquer and paint pigment are combined into one layer. The bottom layer functions as a heat- and pressure-sensitive hot-melt adhesive. Heat and pressure must have time to penetrate the various film coating and layers to bring the adhesive to a liquid state. Before the carrier film is stripped away, a short cooling time is required to ensure that the adhesive is solidified.

Typical hot-stamping processes are shown in Figure 9.33. The major components of a hot-stamping installation include the equipment, tooling, and hot-stamp foil. The hot-stamp die is generally either metal or silicone rubber–mounted to a heater head. The die is heated to a tem-

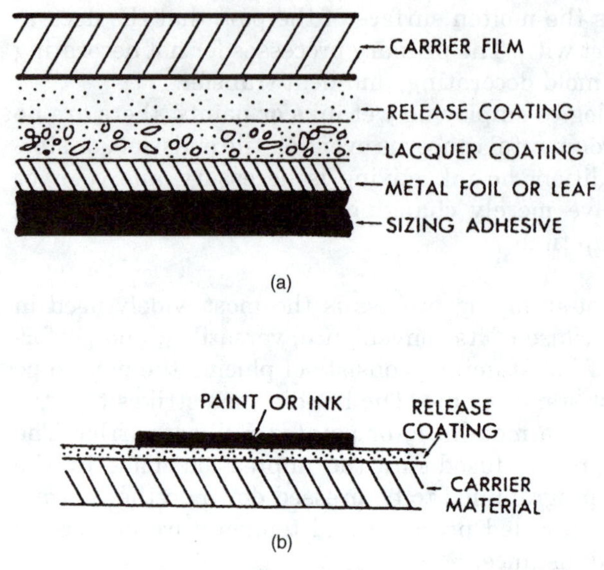

Figure 9.32 (*a*) Diagram of a typical hot-stamping foil and (*b*) structure of a heat-transfer decorating stock.[21]

perature near the melting point of the plastic substrate. A typical application is an automotive tail light trim piece, where decorative bright silver graphics are stamped onto the exterior plastic surface.

A wide variety of hot-stamping equipment is available. Sizes range from hand-held heated roller dies to continuous 4-ft-wide rolls for hot stamping ABS. Equipment is also available to do border stamping such as decorating perimeters of rollers or dials, like instrumentation dials.

In-mold decorating and coating. Much like hot stamping, the in-mold decorating process uses an overlay or a coated foil to become part of the molded product. The decorative image and, if possible, the film carrier are made of the same materials as the part to be molded. The in-mold overlay is placed in the molding cavity either before the molding operation or when the material is partially cured (thermosets). The molding cycle is then completed, and the decoration becomes an integral part of the product.

The design of the mold should take into consideration placement of the gates to prevent wrinkled or washed overlays. The overlay may be held in place in the mold by cutting it so that it fits snugly or by electrostatic means. Blow-molded parts can also be printed or decorated by in-mold processes. The ink or paint image is placed on a carrier film or paper and placed in the mold. As the hot plastics expands, filling the

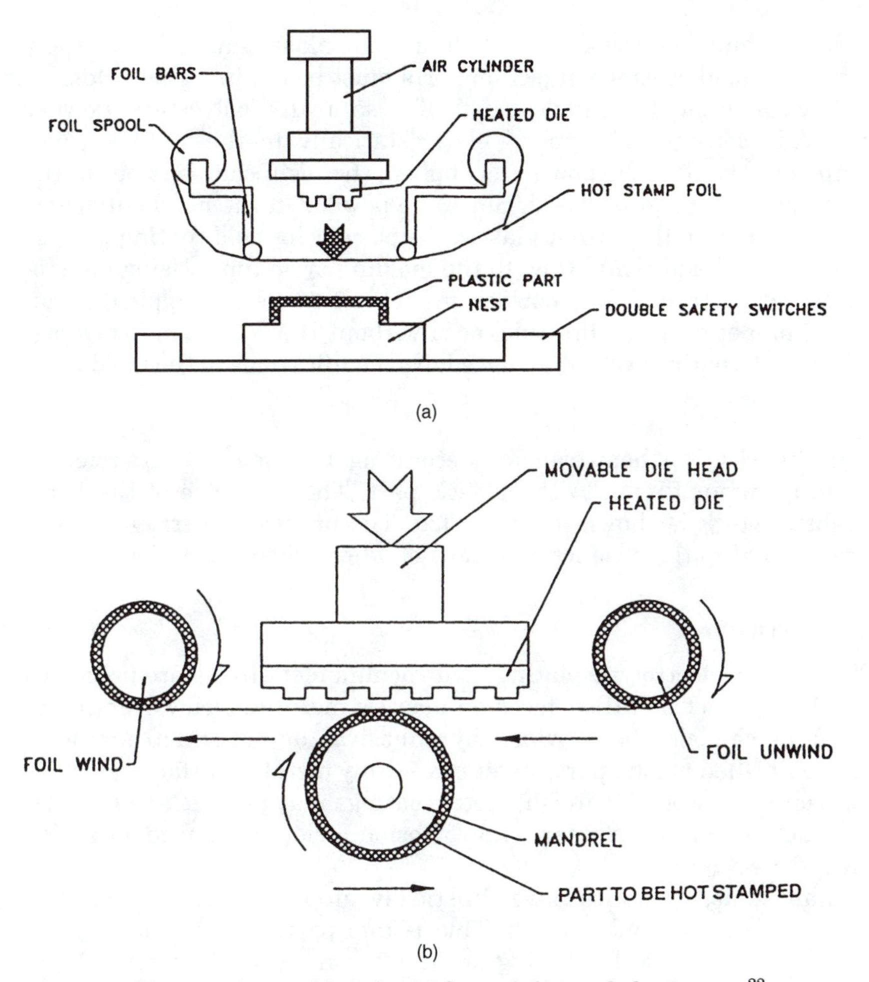

Figure 9.33 Hot-stamping machines for: (*a*) flat and (*b*) round plastic parts.[22]

mold cavity, the image is transferred from the carrier to the hot plastic surface.

In-mold coatings with either liquid or powder coating materials are becoming increasingly popular as primers and finish coats for sheet molding compounds and reaction injection–molded materials. In addition to providing decoration and color, in-mold coatings eliminate surface porosity and provide a hard, durable surface.

Liquid in-mold coatings are primarily two-component urethane systems. The liquid coating is injected into the mold at high pressure. The coating will cure somewhat faster than the parent plastic. The extent of the spread of the coating is dependent on the pressure of injection of the liquid in-mold coating. Film thickness will be dependent on the viscosity and compressibility of the parent plastic. In order to apply liquid in-mold coatings, injection ports must be cut into the molds.

Powder in-mold coating consists of unsaturated polyesters. Powder in-mold coatings are applied electrostatically to the mold surface. Immediately after the powder is applied, the resin charge is put in the mold and the mold is closed. Since the powder in the mold cures at a faster rate than the parent plastic, the powder in-mold coating process does not add additional time to the molding cycle time. Using electrostatic spray techniques, powder in-mold coatings are applied in an even, predetermined film thickness without the hazards of solvents. Very thick coatings can be achieved. No modifications to the mold itself are required.

Heat transfer. In heat-transfer decorating the image is transferred from a carrier film onto the plastic part. The structure of the heat-transfer stock is shown in Fig. 9.32*b*. The preheated carrier stock is transferred to the product by means of a heated rubber roller.

9.8.3 Plating

Electroless, electrolytic plating, and vacuum metallizing are processes used to deposit metal surfaces on plastics materials. However, metal surfaces can also be provided by adhesives or hot-stamp methods. Some finished plastic parts must have shiny metallic surfaces. Besides providing a decorative finish, metal coatings may provide an electrical conducting surface, a wear- and corrosion-resistant surface, or added heat deflection.

Many plated plastics have drastically altered physical properties compared to as-molded parts. This is due partly to the metal plate itself and partly to the plating process. For a part to be electroplated the part should be smooth and blemish-free. There should be no weld lines or sink marks on the part. Any surface irregularity will be magnified by the metallic coating. There should be very little surface stress, as this will affect adhesion of the coating to the part.

Several plastics can be plated on a commercial scale as shown in Table 9.21. Some thermoplastics that can be readily etched and plated include ABS, modified PPO, and modified polypropylene and polysulfone. Some grades of foamed plastics can also be plated. The greatest volume of plated plastic is represented by ABS and modified PPO.

Most of the plated plastics market is for automotive parts. However, a significant number of nonautomotive parts are plated and used for marine hardware, plumbing fixtures, packaging, and appliance and furniture hardware. One of the most important applications for electroless-plated plastics parts and all forms of metallized plastics is the

TABLE 9.21 Commercially Platable Plastics[23]

ABS*	Epoxy/glass†	Polyacetal
PEC*	Polyimide†	Polypropylene
ABS/polycarbonate*	Nylon	Polysulfone
Polystyrene	Polyester	Teflon/glass†
Phenolics†	Polycarbonate	

*Highest volume for plating on plastics.
†Typically plated for printed circuits.

radio frequency interference (RFI) and electromagnetic interference (EMI) markets.

Electroless plating. *Electroless plating* is a surface treatment process that does not require electric current, as does electrolytic plating. Electroless plating deposits a dissolved metal, such as copper or nickel, on the surface of a plastic part through the use of a chemical solution (Fig. 9.34*a*). Also, any plastic part to be plated with the electrolytic process must first be electroless plated to create a conductive surface.

Prior to electroless plating plastic parts, the surfaces have to be treated to ensure good adhesion. A flow diagram of the electroless plating process is shown in Fig. 9.35*a*. The etching process usually involves the use of a chromic acid solution to provide a microscopically roughened surface on the plastic part. The catalytic process is sometimes referred to as *seeding*. Here very small particles of an inactive noble metal catalyst, normally palladium, are deposited into the microcracks created during the etching process. The palladium will act as active catalyst sites for chemical reduction of the electroless copper or electroless nickel. The electroless metal layer formed is usually pure copper or a nickel/phosphorous alloy, depending upon the corrosion requirements of the plated part. In an exterior environment electroless copper has been shown to have better corrosion resistance than electroless nickel. Usually the thickness of the electroless metal is approximately 0.5 mils.

Plastics that have been plated with electroless metal can then be given any electrolytic metal coating. As a result, the design engineer has more possibilities regarding metal coating type and thicknesses.

Electrolytic process. Many plastic products require only electroless plating, but electrolytic plating can be added to the process to provide a metallized surface that is thicker and more like typical metal surfaces. The electrolytic plating process starts at the completion of the electroless process. To electrolytically deposit metal onto a plastic, the plastic must be made electrically conductive and then grounded. The

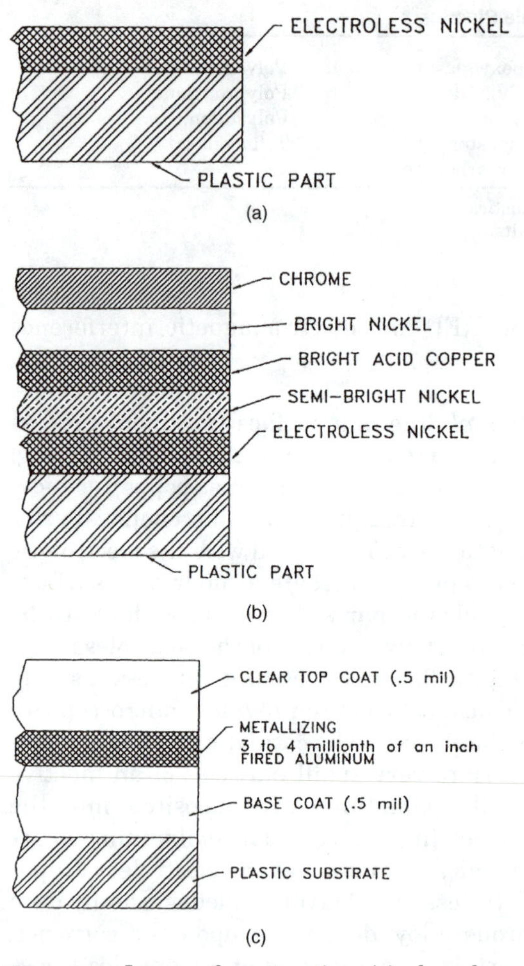

Figure 9.34 Layers of construction: (*a*) electroless plating, (*b*) electrolytic plating, and (*c*) vacuum metallizing.[22]

metal to be plated is positively charged, thus allowing the positively charged metal atoms to precipitate onto the negatively charged plastic.

Figure 9.35*b* is a flow diagram that shows the electrolytic plating process. First, the electroless plated plastic is etched once again, this time usually in sulfuric acid. The part then undergoes a semibright nickel deposition. This layer helps limit and control the amount of current that is exposed to the plated surface and thus protect it from burning off. A mild sulfuric acid bath follows. Bright acid copper is electrolytically deposited on the surface followed by bright nickel. The

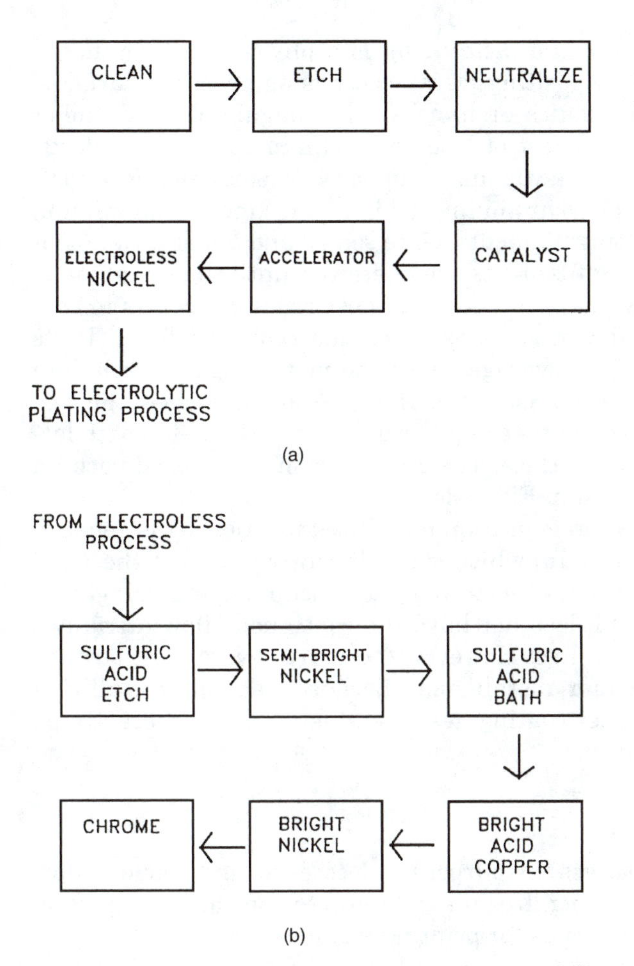

Figure 9.35 Flow diagram of: (*a*) electroless plating and (*b*) electrolytic plating processes.[22]

nickel layer is the immediate layer onto which the chrome will be deposited. The final electrolytically deposited layer is usually chrome but many other finishes are possible such as bright brass, satin nickel, silver, black chrome, and gold.

Figure 9.34*b* shows the various layers in an electrolytic coating. Electrolytic plated parts are nearly indistinguishable from metal parts, and due to the plastic body, the plated plastic parts are lighter, less expensive, and can be used in a wider range of designs than their metal counterparts. To provide a standard rating for plated plastic, the American Society of Electroplated Plastic (ASEP) has developed a service condition code for common applications.

Vacuum metallizing. *Vacuum metallizing* is a physical, rather than a chemical, process for depositing metal coatings on prepared surfaces. Vacuum metallizing plastics or films are thoroughly cleaned and/or etched and given a base coat of lacquer to eliminate surface defects. The parts are then put in a vacuum chamber with small pieces of coating metal (for example, chromium, gold, silver, zinc, or aluminum) placed on special heating filaments. Once sealed and under about 0.5 μ of vacuum, the heating filaments are energized until the metal melts and vaporizes. The vaporized metal coats everything it comes in contact with in the chamber, condensing on the cooler surfaces. Parts must be rotated for full coverage, since the metal vapor travels in a line-of-sight from its origin. Once the plating is accomplished, the vacuum is released and the parts are removed and gnarly coated with lacquer to help protect its surface. The cross section of a typical vacuum metallized part is shown in Fig. 9.34*c*.

The metal thickness on a vacuum metallized plastic part is generally 3 to 4 Å (1 to 4 $\times$ 10^{-6} in) which is significantly less than the 1-mil thickness for conventional electroplating. Vacuum metallizing is a batch-type process and does not have the continuous flow attributes that electrolytic plating has. However, there are several cost savings associated with vacuum metallizing. Because vacuum metallizing inherently has a thinner coating, less metal is used and there are no toxic chemicals involved.

9.8.4 Printing

Printing on plastics serves the purpose of decoration and placing information, such as bar coding, freshness dates, etc., on the part surface. The conventional techniques for printing include

- Stamp printing
- Silk screening
- Noncontact methods

Other methods of printing, such as hot stamping or heat transfer, have already been discussed in this section. The surface condition of the part must be considered before any attempts are made for printing on plastic. Surface cleaning and treating methods, described under the section on adhesive bonding, are applicable also for treating these surfaces prior to printing.

Stamp printing produces code or lettering by using a metal or rubber die exposed to an ink. Letterpress, letterflex, gravure, and flexography are methods of applying the ink via various types of dies. For lay-flat film products gravure and flexographic printing are most commonly used.

Letterpress printing is a method where raised, rigid printing plates are inked and pressed against the plastics part. The raised portion of the plate transfers the image. *Letterflex* is similar to letterpress, except that flexible printing plates are used. Flexible plates may transfer their designs to irregular surfaces.

Gravure (also called *rotogravure*) transfers ink for cells etched or engraved on a copper- and chrome-plated cylinder to the material being printed. The cylinder rotates in an ink fountain and the cells pick up the liquid ink. A doctor blade wipes excess ink from the roller, leaving ink only in the cells. The low-viscosity liquid inks used in this process contain a volatile solvent. Gravure is mainly used when the design requires good continuity of sharp pattern details and rich color effects. It is a high-speed process; however, it can be expensive for short runs because of the cost of the cylinders.

Pad transfer printing is a gravure-like process that offers versatility for printing on a variety of shapes and surfaces. The image is transferred using a special silicone pad instead of the etched plate used in gravure. Table 9.22 compares pad printing to other commonly used decorating methods. Pad printing can reproduce fine-line engraving, and it can apply wet ink on wet ink. Pad printing is generally most useful for printing small areas. Pad printers are generally more expensive than hot stamping or screen printing. The ink costs, however, are much less than the cost of hot-stamping foil.

Flexographic printing is more widely used for short and medium runs. This method prints by transforming images from a flexible raised printing palate directly to the material. A uniform film of ink is applied to the raised portion of the printing plate by bringing it in contact with a form roller. The plate continues to rotate until it contacts the substrate and transfers the image to it. Individual colors are applied by a series of separate printing stations.

Silk screening is the process where ink or paint is forced by a rubber squeegee through a fine metallic or fabric screen onto the product. The screen is blank or blocked off in areas where no ink is wanted. Printing from flat-bed or rotary systems makes it possible to decorate large or complex parts. However, only one color can be applied per pass. Image sharpness is of lower quality than with other printing processes. The major uses for screen printing have been short runs on molded products with complex contours, on flat substrates, or to get special color effects.

Noncontact methods of printing are electrostatic, ink jet, and laser etching. *Electrostatic printing* applies dry ink to areas by a difference in electrical potential. *Ink jet printing* is a direct adaptation of the ink jet printer used to print on paper. *Laser etching printing* requires that a laser beam be aimed via a series of computer-controlled mirrors. The

TABLE 9.22 Various Decorating Methods[24]

	Pad printing	Mask spray painting	Screen printing	Hot stamping
Comparative equipment costs	P	E	G	F
Comparative tool costs	G	P	E	F
Ink costs	E	F	G	P
Ease of setup	F	E	G	P
Suitability for various jobs:	E	P	G	E
Small, flat areas				
Fine detail	E	P	G	G
Large areas	P	E	E	P
Opaque coverage	P	E	E	G
Convex surfaces	E	G	F	F
Concave surfaces	E	P	P	P
Single, multicolor application	E	P	P	G
Low-volume application	F	G	E	F
High-speed production	E	F	E	G

Note: P = poor, F = fair, G = good, E = excellent.

laser burns the surface of the plastic, thus creating the printed image. Noncontact printing methods generally allow the designer to print on most plastics regardless of the shape of the surface.

9.8.5 Labels, decals, etc.

There are numerous miscellaneous decorating methods, including pressure-sensitive labels, decals, and flocking. Pressure-sensitive labels are easy to use and apply. The designs or printing is generally printed on an adhesive-backed foil or film, a pressure-sensitive adhesive is applied, and a release material covers the adhesive. To apply, the user must simply remove the release material and press the label to the part. Of course, the part must be clean and free of any weak boundary layers so that adequate adhesion may develop. The stripping off of the release layer and application of the label can be accomplished either by hand or by automatic mechanical means.

Decals are also a means of transferring a design or printed message to a plastic part. With decals the decoration is placed on a decorative adhesive-backed film and then on a paper backing. The adhesive is such that it activates when in contact with water. To apply, the decal is moistened in water and the film is placed on the plastic surface. The process is not widely used because of the difficulty in accurately placing the decal in a specific location on the surface.

Flocking is a process of placing a velvetlike finish on virtually any surface. On plastics flocking is applied generally by mechanical or electrostatic means. The process consists of coating the part with an adhesive and placing the flocking fibers on the adhesive areas before the adhesive sets.

Decorative wood-grained finishes are also possible using a number of processes. Some are accomplished by heat stamping or rolling an engraved plate over the surface of the part and then applying a coating. Wood-grained decorative designs can be applied to hot-stamp films. Various decorative wood-grained laminates and clad coatings are also applied with adhesives.

References

1. Cadillac Plastic & Chemical Co., Troy, Mich.
2. Terry A. Richardson, "Machining and Finishing," *Modern Industrial Plastics,* Howard W. Sams & Co., New York, 1974.
3. John L. Hull, "Design and Processing of Plastic Parts," *Handbook of Plastics Elastomers and Composites,* 2d ed., Charles A. Harper, ed., McGraw-Hill, New York, 1992.
4. J. O. Trauernicht, "Bonding and Joining, Weigh the Alternatives, Part 1, Solvent Cements, Thermal Welding," *Plastics Technology,* August 1970.
5. "Engineer's Guide to Plastics," *Materials Engineering,* May 1972.
6. "Mechanical Fastening," *Handbook of Plastics Joining,* Plastics Design Library, Norwich, NY, 1997.
7. "General Design Principals", *Design Handbook for DuPont Engineering Polymers,* DuPont Polymers, Wilmington, Del.
8. "Engineering Plastics," *Engineered Materials Handbook,* vol. 2, ASM International Materials Park, Ohio, 1988.
9. Fastening, Joining, and Assembly Reference Issue, *Machine Design,* November 17, 1988.
10. "Joining of Composites," in A. Kelley, ed., *Concise Encyclopedia of Composite Materials,* The MIT Press, Cambridge, 1989.
11. J. C. Merriam, "Adhesive Bonding," *Materials Design Engineering,* September 1959.
12. D. K. Rider, "Which Adhesives for Bonded Metal Assembly," *Product Engineering,* May 25, 1964.
13. G. W. Koehn, "Design Manual on Adhesives," *Machine Design,* April 1954.
14. "Adhesive Bonding Alcoa Aluminum," Aluminum Company of America, Pittsburgh, Pa., 1967.
15. "Surface Preparation of Plastics," in *Adhesives and Sealants,* vol. 3, *Engineered Materials Handbook,* H. F. Binson, ed., ASM International, Materials Park, Ohio, 1990.
16. D. F. Smith, and C. V. Cagle, "A Quality Control System for Adhesive Bonding Utilizing Ultrasonic Testing," in *Structural Adhesive Bonding,* M. J. Bodnar, ed., Interscience, New York, 1966.
17. D. F. Gentle, "Bonding Systems for Plastics," in *Aspects of Adhesion,* vol. 5, D. J. Almer, ed., University of London Press, London, 1969.
18. "Electromagnetic Bonding—It's Fast, Clean, and Simple," *Plastics World,* July 1970.
19. "How to Fasten and Join Plastics," *Materials Engineer,* March 1971.
20. R. C. Van Har, "Spray Painting Plastics," *Plastics Finishing and Decoration,* D. Satas, ed., Van Nostrand Reinhold Co., New York, 1986.
21. Terry A. Richardson, "Decorating Processes," *Modern Industrial Plastics,* Howard W. Sams & Co., New York, 1974.
22. Edward A. Muccio, "Finishing and Decorating Plastic Parts," *Plastic Part Technology,* ASM International, Materials Park, Ohio, 1991.
23. G. A. Krulik, "Electrolytic Plating," *Plastics Finishing and Decoration,* D. Satas, ed., Van Nostrand Reinhold Co., New York, 1986.
24. D. Satas, "Pad Printing," *Plastics Finishing and Decoration,* D. Satas, ed., Van Nostrand Reinhold Co., New York, 1986.

Chapter

10

Coatings and Finishes

Carl P. Izzo

Industrial Paint Consultant
Export, Pennsylvania

10.1 Introduction

Almost from the beginning, humans have used coatings and finishes. Biblical history reports that Noah received divine instructions to paint his ark with pitch. Amber was mixed with oils to make decorative varnishes by the Egyptians. White lead was mixed with amber and with pitch by the Romans. From cave dwellers decorating their walls with earth pigments ground in egg whites to factory workers protecting products with E-coat primers and urethane acrylic enamels, these coatings are still composed of film-forming vehicles, pigments, solvents, and additives.[1]

The first 50 years of the twentieth century were the decades of discovery. Significant changes were made in the *vehicles,* which are the liquid portions of the coatings composed of binder and thinner.[2] Since the 1900s and the introduction of phenolic synthetic resin vehicles, coatings have been designed to increase production and meet performance requirements at lower costs. These developments were highlighted by the introduction of nitrocellulose lacquers for the automotive and furniture industries; followed by the alkyds, epoxies, vinyls, polyesters, acrylics, and a host of other resins; and finally the polyurethanes.

In the 1950s, the decade of expansion, manufacturers built new plants to supply the coatings demand for industrial and consumer products. Coatings were applied at low solids using inefficient conventional air-atomized spray guns. The atmosphere was polluted with volatile organic compounds (VOCs), but no one cared as long as finished products were shipped out of the factories. Coatings suppliers were fine-tuning formulations to provide faster curing and improved performance properties. In 1956, powder coatings were invented. By 1959, there were several commercial conveyorized lines applying powder coatings by the fluidized bed process.

In the 1960s, the decade of technology, just as coatings were becoming highly developed, another variable—environmental impact—was added to the equation. Someone finally noticed that the solvents, which were used in coatings for viscosity and flow control, and evaporate during application and cure, were emitted to the atmosphere. Los Angeles County officials, who found that VOC emissions were a major source of air pollution, enacted Rule 66 to control the emission of solvents that cause photochemical smog. To comply with Rule 66, the paint industry reformulated its coatings using exempt solvents, which presumably did not produce smog. California's Rule 66 was followed by other local air-quality standards and finally by the establishment of the U.S. Environmental Protection Agency (EPA), whose charter, under the law, was to improve air quality by reducing solvent emissions. During this decade, there were three notable developments which would eventually reduce VOC emissions in coating operations, electrocoating, electrostatically sprayed powder coatings, and radiation curable coatings.

In the 1970s, the decade of conservation, the energy crisis resulted in shortages and price increases for solvents and coatings materials. Also affected was the distribution of natural gas, the primary fuel for curing ovens, which caused shortages and price increases. In response to those pressures, the coatings industry developed low-temperature curing coatings in an effort to reduce energy consumption. Of greater importance to suppliers and end users of coatings and finishes was the establishment of the Clean Air Act of 1970. An important development during this period was radiation curable coatings, mostly clears, for flat line applications using electron beam (EB) and ultraviolet (UV) radiation sources.

In the 1980s, the decade of restriction, the energy crisis was ending and the more restrictive air-quality standards were beginning. However, energy costs remained high. The importance of *transfer efficiency,* the percentage of an applied coating that actually coats the product, was recognized by industry and the EPA. This led to the development and use of coatings application equipment and coating methods that have higher transfer efficiencies. The benefits of using higher transfer efficiency coating methods are threefold: lower coating material usage, lower solvent emissions, and lower costs. The automotive industry switched almost exclusively to electrocoating for the application of primers. Powder coating material and equipment suppliers worked feverishly to solve the problems associated with automotive topcoats. Other powder coating applications saw rapid growth. Radiation curable pigmented coatings for three-dimensional products were developed.

In the 1990s, the decade of compliance, resin and coating suppliers developed compliance coatings—electrocoating, high solids, powder, radiation curable, and waterborne. Equipment suppliers developed devices to apply and cure these new coatings. These developments were in response to the 1990 amendments to the Clean Air Act of 1970. The amendments established a national permit program which made the law more enforceable, ensuring better compliance and calling for nationwide regulation of VOC emissions from all organic finishing operations. The amendments also established Control Technique Guidelines to allow state and local governments to develop Attainment Rules. Electrocoating was the process of choice for priming many industrial and consumer products. Powder coatings were used in a host of applications where durability was essential. UV curable coatings were applied to three-dimensional objects. Equipment suppliers developed more efficient application equipment.

In the 2000s, the beginning of the green millennium, coatings and equipment suppliers' investments in research and development will pay dividends. Improvements in coating materials and application equipment have enabled end users to comply with the air-quality regulations. Primers are applied by electrocoating. One-coat finishes are replacing two coats in many cases. High-solids and waterborne liquid coatings are replacing conventional solvent-thinned coatings. Powder coatings usage has increased dramatically. Radiation-cured coatings are finding more applications. Coatings and solvent usage, as well as application costs, are being reduced. Air-quality standards are being met. Coating and equipment suppliers, as well as end users, are recognizing the cost savings bonus associated with attainment. Compliance coatings applied by more efficient painting methods will reduce coatings and solvent usage, thereby effecting cost savings.

Coatings today are considered engineering materials. Their performance characteristics must not only match service requirements, they must also meet governmental regulations. In the past, the selection of a coating depended mainly on the service requirements and application method. Now, more than ever before, worker safety, environmental impact, and economics must be considered. For this reason, compliance coatings—electrocoating, high-solids, powders, radiation cured, and waterborne coatings—are the most sensible choices.

Coatings are applied to most industrial products by spraying. Figure 10.1 shows a typical industrial spray booth. In 1890 Joseph Binks invented the cold-water paint-spraying machine, the first airless sprayer, which was used to apply whitewash to barns and other building interiors. In 1924 Thomas DeVilbiss used a modified medical atomizer, the first air atomizing sprayer, to apply a nitrocellulose lacquer on the Oakland automobile. Since then, these tools have remained virtually unchanged, and, until the enactment of the air-quality standards, they were used to apply coatings at 25 to 50% volume solids at transfer efficiencies of 30 to 50%. Using this equipment, the remainder of the nonvolatile material, the overspray, coated the floor and walls of spray booths and became hazardous or nonhazardous waste, while the solvents, the VOCs, evaporated from the coating during application and cure to become air pollutants. Today, finishes are applied by highly transfer-efficient application equipment.

Even the best coatings will not perform their function if they are not applied on properly prepared substrates. For this reason surfaces must first be cleaned to remove oily soils, corrosion products, and particulates, and then pretreated before applying any coatings and finishes.

After coatings are applied, they form films and cure. Curing mechanisms can be as simple as solvent evaporation or as complicated as free-radical polymerization. Basically, coatings can be classified as baking or air drying, which usually means room temperature curing. The curing method and times are important in coating selection because they must be considered for equipment and production schedules choices.

Figure 10.1 A typical industrial spray booth used for applying industrial coatings. (*Courtesy of AFC Inc.*)

The purposes of this chapter are threefold: (1) to stress the importance of environmental compliance in coating operations; (2) to acquaint the reader with surface preparation, coating materials, application equipment, and curing methods; and (3) to aid in the selection of coating materials.

10.2 Environment and Safety

In the past, changes in coating materials and coating application lines were discussed only when lower prices, novel products, new coating lines, or new plants were considered. Today, with rising material costs, rising energy costs, and more restrictive governmental regulations, they are the subject of frequent discussions. Coating material and solvent costs, which are tied to the price of crude oil, have risen since the 1970s, as has the cost of natural gas, which is the most frequently used fuel for coatings bake ovens. The EPA has imposed restrictive air-quality standards. The Occupational Safety and Health Act (OSHA) and the Toxic Substances Control Act (TSCA) regulate the environment in the workplace and limit workers' contact with hazardous materials. These factors have increased coating costs and the awareness of product finishers. To meet the challenge, they must investigate and use alternative coating materials and processes for compliance and cost effectiveness.

Initial attempts to control air pollution in the late 1940s resulted in smoke control laws to reduce airborne particulates. The increased use of the automobile and industrial expansion during that period caused a condition called *photochemical smog* (smog created by the reaction of chemicals exposed to sunlight in the atmosphere) in major cities throughout the United States. Los Angeles County officials recognized that automobile exhaust and VOC emissions were major sources of smog, and they enacted an air pollution regulation called Rule 66. Rule 66 forbade the use of specific solvents that produced photochemical smog and published a list of exempt solvents for use in coatings. Further study by the EPA has shown that, if given enough time, even the Rule 66 exempt solvents will produce photochemical smog in the atmosphere.

The Clean Air Act of 1970 and its 1990 amendments, formulated by the EPA, established national air-quality standards that regulate the amount of solvents emitted. The EPA divided the 50 states into 250 air-quality regions, each of which is responsible for the implementation of the national air-quality standards. It is important to recognize that many of the local standards are more stringent than the national. For this reason, specific coatings that comply with the air-quality standards of one district may not comply with another's. Waterborne, high-solids, powder, electrophoretic, and radiation-cured coatings will comply. The use of precoated metal can eliminate all the compliance problems.

Not only because the EPA mandates the reduction of VOC emissions, but also because of economic advantages, spray painting, which is the most used application method, must be done more efficiently. The increased efficiency will reduce the amount of expensive coatings and solvents used, thereby reducing production costs.

10.3 Surface Preparation

The most important step in any coating operation is surface preparation, which includes cleaning and pretreatment. For coatings to adhere, surfaces must be free from oily soils, corrosion products, and loose particulates. New wood surfaces are often coated without cleaning. Old wood and coated wood must be cleaned to remove oily soils and loose, flaky coatings. Plastics are cleaned by using solvents and chemicals to remove mold release. Metals are cleaned by media blasting, sanding, brushing, and by solvents or aqueous chemicals. The choice of a cleaning method depends on the substrate and the size and shape of the object.

After cleaning, pretreatments are applied to enhance coating adhesion and, in the case of metals, corrosion resistance. Some wood surfaces require no pretreatment, while others require priming of knots and filling of nail holes. Cementitious and masonry substrates are pretreated, using acids, to remove loosely adhering contaminants and to passivate the surfaces. Metals, still the most common industrial substrates, are generally pretreated using phosphates, chromates, and oxides to passivate their surfaces and provide corrosion resistance. Plastics, second only to steel, are gaining rapidly in use as industrial substrates. Some are paintable after cleaning to remove mold release and other contaminants, while others require priming, physical treatments, or chemical etching to ensure coating adhesion. Since most of the industrial substrates coated are metals and plastics, their cleaning and pretreatment are described in the next sections. Because of their complexities, detailed descriptions of cleaning and pretreatment processes are beyond the scope of this chapter. Enough detail will be given to allow the reader to make a choice. As with the choice of a cleaning method, the choice of a pretreatment method depends on the composition, size, and shape of the product.

10.3.1 Metal surface cleaning

Oily soils must be removed before any other surface preparation is attempted. Otherwise these soils may be spread over the surface. These soils can also contaminate abrasive cleaning media and tools. Oily soils can be removed faster using liquid cleaners that impinge on the surface or in agitated immersion baths. It is often necessary to heat liquid cleaners to facilitate soil removal.

Abrasive cleaning. After removal of the oily soils, surfaces are abrasive-cleaned to remove rust and corrosion by media blasting, hand or power sanding, and hand or power brushing. Media blasting consists of propelling materials, such as sand, metallic shot, nut shells, plastic pellets, and dry ice crystals, by gases under pressure, so that they impinge on the surfaces to be cleaned. High-pressure water-jet cleaning is similar to media blasting.

Alkaline cleaning. To remove oily soils, aqueous solutions of alkaline phosphates, borates, and hydroxides are applied to metals by immersion or spray. After cleaning, the surfaces are rinsed with clear water to remove the alkali. These materials are not effective for removing rust and corrosion.

Detergent cleaning. Aqueous solutions of detergents are used to remove oily soils in much the same way as alkaline cleaners. Then they are rinsed with cold water to flush away the soils.

Emulsion cleaning. Heavy oily soils and greases are removed by aqueous emulsions of organic solvents such as mineral spirits and kerosene. After the emulsified solvent has dissolved the oily soils, they are flushed away using a hot-water rinse. Any remaining oily residue must be removed using clean solvent, alkaline, or detergent cleaners.

Solvent cleaning. Immersion, hand wiping, and spraying using organic solvents are effective methods for removing oily soils. Since these soils will contaminate solvents and wipers, it is important to change them frequently. Otherwise, oily residues will remain on substrates. Safe handling practices must be followed because of the hazardous nature of most organic solvents.

Manual spray cleaning. For large products, detergent and alkaline cleaners applied using steam cleaners are a well-known degreasing method. In addition to oily soils, the impingement of the steam and the action of the chemicals will dissolve and flush away heavy greases and waxes. Hot-water spray cleaning using chemicals is nearly as effective as steam cleaning.

Vapor degreasing. Vapor degreasing has been a very popular cleaning method for removing oily soils. Boiling solvent condenses on the cool surface of the product and flushes away oily soils, but does not remove particulates. Since this process uses chlorinated solvents, which are under regulatory scrutiny by government agencies, its popularity is declining. However closed-loop systems are still available.

10.3.2 Metal surface pretreatment

Cleaning metals will remove oily soils but will generally not remove rust and corrosion from substrates to be coated. Abrasive cleaning will remove corrosion products, and for this reason it is also considered a pretreatment because the impingement of blasting media and the action of abrasive pads and brushes roughen the substrate and therefore enhance adhesion. The other pretreatments use aqueous chemical solutions, which are applied by immersion or spray techniques. Pretreatments for metallic substrates used on industrial products are discussed in this section. Because they provide corrosion protection to ferrous and nonferrous metals, chromates are used in pretreatment stages and as conversion coatings. They are being replaced by nonchromate chemicals.

Aluminum. Aluminum is cleaned by solvents and chemical solutions to remove oily soils and corrosion products. Cleaned aluminum is pretreated using chromate conversion coating and anodizing. Phosphoric acid–activated vinyl wash primers, which are also considered pretreatments, must be applied directly to metal and not over other pretreatments.

Copper. Copper is cleaned by solvents and chemicals and then abraded to remove corrosion. Bright dipping in acids will also remove corrosion. Cleaned surfaces are often pretreated using chromates and vinyl wash primers.

Galvanized steel. Galvanized steel must be cleaned to remove the oil or wax that is applied at the mill to prevent white corrosion. After cleaning, the surfaces are pretreated using chromates and phosphates. Vinyl wash primer pretreatments can also be applied on galvanized steel surfaces having no other pretreatments.

Steel. Steel surfaces are cleaned to remove oily soils and, if necessary, pickled in acid to remove rust. Clean steel is generally pretreated with phosphates to provide corrosion resistance. Other pretreatments for steel are chromates and wash primers.

Stainless steel. Owing to its corrosion resistance, stainless steel is usually not coated. Otherwise the substrate must be cleaned to remove oily soils and then abraded to roughen the surface. Wash primers will enhance adhesion.

Titanium. Cleaned titanium is pretreated like stainless steel.

Zinc and cadmium. Zinc and cadmium substrates are pretreated like galvanized steel.

10.3.3 Plastic surface cleaning

Alkaline cleaning. Aqueous solutions of alkaline phosphates, borates, and hydroxides are applied to plastics by immersion or spray to remove oily soils and mold release agents. After cleaning, the surfaces are rinsed with clear water to remove the alkali.

Detergent cleaning. Aqueous solutions of detergents are used to remove oily soils and mold release agents in much the same way as with alkaline cleaners. Then they are rinsed with cold water to flush away the soils.

Emulsion cleaning. Heavy, oily soils, greases, and mold release agents are removed by aqueous emulsions of organic solvents such as mineral spirits and kerosene. After the emulsified solvent has dissolved the oily soils, they are flushed away using a hot-water rinse. The remaining oily residue must be removed using clean solvent, alkaline, or detergent cleaners.

Solvent cleaning. Immersion, hand wiping, and spraying, using organic solvents, are effective methods for removing oily soils and mold release agents. Since these soils will contaminate solvents and wipers, it is important to change them frequently. Otherwise, oily residues will remain on substrates. Compatibility of cleaning solvents with the plastic substrates is extremely important. Solvents that affect plastics are shown in Table 10.1. Suppliers of mold release agents are the best source for information on solvents which will remove their materials. Safe handling practices must be followed because of the hazardous nature of most organic solvents.

Manual spray cleaning. Detergent and alkaline cleaners applied using steam and hot-water spray cleaners are a well-known degreasing method. It can also be used for removing mold release agents. The impingement of the steam and hot water and the action of the chemicals will dissolve and flush away the contaminants. Manual spray cleaning is used for large products.

TABLE 10.1 Solvents That Affect Plastics

Resin	Heat-distortion point, °F	Solvents that affect surface
Acetal	338	None
Methyl methacrylate	160–195	Ketones, esters, aromatics
Modified acrylic	170–190	Ketones, esters, aromatics
Cellulose acetate	110–209	Ketones, some esters
Cellulose propionate	110–250	Ketones, esters, aromatics, alcohols
Cellulose acetate butyrate	115–227	Alcohols, ketones, esters, aromatics
Nylon	260–360	None
Polyethylene:		
High density	140–180	
Medium density	120–150	None
Low density	105–121	
Polypropylene	210–230	None
Polycarbonate	210–290	Ketones, esters, aromatics
Polystyrene (GP high heat)	150–195	Some aliphatics, ketones, esters, aromatics
Polystyrene (impact, heat-resistant)	148–200	Ketones, esters, aromatics, some aliphatics
Acrylonitrile butadiene styrene (ABS)	165–225	Ketones, esters, aromatics, alcohol

10.3.4 Plastic surface pretreatment

Cleaning will remove oily soils and mold release agents but additional pretreatment may be needed on certain plastic surfaces to ensure adhesion. Many of the plastic substrates are chemically inert and will not accept coatings because of their poor wettability. Depending on their chemical composition, they will require mechanical, chemical, and physical pretreatment or priming to enhance coating adhesion. Since mechanical pretreatment consists of abrasion, its effect on the substrate must be considered. Chemical pretreatments involve corrosive materials which etch the substrates and can be hazardous. Therefore, handling and disposal must be considered. Physical pretreatments consist of plasma, corona discharge, and flame impingement. Process control must be considered.

Abrasive cleaning. After removal of the oily soils, surfaces are abrasive-pretreated to roughen the substrate by media blasting, hand or power sanding, and hand or power brushing. Media blasting consists of propelling materials, such as sand, metallic shot, nut shells, plastic pellets, and dry ice crystals, by gases under pressure, so that they impinge on the surfaces to be pretreated.

Chemical etching. Chemical pretreatments use solutions of corrosive chemicals, which are applied by immersion or spray techniques, to etch the substrate.

Corona discharge. During corona discharge pretreatment the plastic is bombarded by gasses directed toward its surface.

Flame treating. During the flame pretreatment, an open flame impinging on the surface of the plastic product causes alterations in the surface chemistry.

Plasma pretreatment. Low-pressure plasma pretreatment is conducted in a chamber while atmospheric plasma pretreatment is done in the open. In both cases, ablation alters the surface chemistry and causes changes in surface roughness.

Laser pretreatment. Laser pretreatment ablates the plastic substrate causing increased surface roughness and changes in the surface chemistry.

10.3.5 Priming

Priming involves the application of a coating on the surface of the plastic product to promote adhesion or to prevent attack by the solvents in a subsequent protective or decorative coating. In some cases priming can be done after cleaning. In others, it must be done after pretreatment.

10.4 Coatings Selection

To aid in their selection, coatings will be classified by use in finish systems, physical state, and resin type. Coatings are also classified by their use as electrical insulation. It is not the intention of this chapter to instruct the reader in the chemistry of organic coatings but rather to aid in selection of coatings for specific applications. Therefore, the coating resin's raw materials feed stock and polymerization reactions will not be discussed. On the other hand, generic resin types, curing, physical states, and application methods are discussed.

10.4.1 Selection by finish systems

Finish systems can be one coat or multicoat that use primers, intermediate coats, and topcoats. Primers provide adhesion, corrosion protection, passivation, and solvent resistance to substrates. Topcoats provide weather, chemical, and physical resistance, and generally determine the performance characteristics of finish systems. Performance properties for coatings, formulated with the most commonly used resins, are shown in Table 10.2.

In coatings selection, intended service conditions must be considered. To illustrate this point, consider the differences between service conditions for toy boats and for battleships. Table 10.3 shows the use of coating finish systems in various service conditions.

10.4.2 Selection by physical state. A resin's physical state can help determine the application equipment required. Solid materials can be applied by powder coating methods. Table 10.4 lists resins applied as powder coatings. Liquids can be applied by most of the other methods, which are discussed later. Many of the coating resins exist in several physical states. Table 10.5 lists the physical states of common coating resins.

10.4.3 Selection by resin type

Since resin type determines the performance properties of a coating, it is used most often. Table 10.6 shows the physical, environmental, and film-forming characteristics of coatings by polymer (resin) type. It is important to realize that in selecting coatings, tables of performance properties of generic resins must be used only as guides because coat-

TABLE 10.2 Performance Properties* of Common Coating Resins[3]

Resin type	Humidity resistance	Corrosion resistance	Exterior durability	Chemical resistance	Mar resistance
Acrylic	E	E	E	G	E
Alkyd	F	F	P	G	G
Epoxy	E	E	G	E	E
Polyester	E	G	G	G	G
Polyurethane	E	G	E	G	E
Vinyl	E	G	G	G	G

Note: *E—excellent; G—good; F—fair; P—poor.

TABLE 10.3 Typical Industrial Finish Systems[3]

Service conditions	Primer	One-coat enamel	Intermediate coat	Topcoat
Interior				
Light duty		X		
Heavy duty	X			X
Exterior				
Light duty		X		
Heavy duty	X			X
Extreme duty	X		X	X

TABLE 10.4 Plastics Used in Powder Coatings

Resin	Fluidizing conditions			Fluidized bed powder		
	Preheat temperature, °F	Cure or fusion		Maximum operating temperature, °F	Adhesion	Weather resistance
		Temperature, °F	Time, min			
Epoxy	250–450	250–450	1–60	200–400	Excellent	Good
Vinyl	450–550	400–600	1–3	225	Poor	Good
Cellulose acetate butyrate	500–600	400–550	1–3	225	Poor	Good
Nylon	550–800	650–700	1	300	Poor	Fair
Polyethylene	500–600	400–600	1–5	225	Fair	Good
Polypropylene	500–700	400–600	1–3	260	Poor	Good
Penton	500–650	450–600	1–10	350	Poor	Good
Teflon	800–1000	800–900	1–3	500	Poor	Good

TABLE 10.5 Physical States of Common Coating Resins[3]

Resin type	Convenietal solvent	Waterborne	High-solids	Powder coating	100% solids liquid	Two-component liquid
Acrylic	X	X	X	X		
Alkyd	X	X	X			
Epoxy	X	X	X	X	X	X
Polyester	X		X	X	X	X
Polyurethane	X	X	X	X	X	X
Vinyl	X	X	X	X	X	

ings of one generic type, such as acrylic, epoxy, or polyurethane, are often modified using one or more of the other generic types. Notable examples are acrylic alkyds, acrylic urethanes, acrylic melamines, epoxy esters, epoxy polyamides, silicone alkyds, silicone epoxies, silicone polyesters, vinyl acrylics, and vinyl alkyds. While predicting specific coating performance properties of unmodified resins is simple, predicting the properties of modified resins is difficult, if not impossible. Parameters causing these difficulties are resin modification percentages and modifying methods such as simple blending or copolymerization. The performance of a 30% copolymerized silicone alkyd is not necessarily the same as one which was modified by blending. These modifications can change the performance properties subtly or dramatically.[3]

There are more than 1200 coatings' manufacturers in the United States, each having various formulations that could number in the hundreds. Further complicating the coatings selection difficulty is the well-known practice of a few coatings' manufacturers who add small amounts of a more expensive, better performing resin to a less expensive, poorer performing resin and call the product by the name of the former. An unsuspecting person, whose choice of such a coating is based on proper-

TABLE 10.6 Properties of Coatings by Polymer Type

Coating type	Electrical properties: Volume resistivity, Ω-cm (ASTM D 257)	Dielectric strength, V/mil	Dielectric constant	Dissipation factor	Physical characteristics: Maximum continuous service temperature, °F	Adhesion to metals	Flexibility	Approximate Sward hardness (higher number is harder)	Abrasion resistance
Acrylic	10^{14}–10^{15}	450–550	2.7–3.5	0.02–0.06	180	Good	Good	12–24	Fair
Alkyd	10^{14}	300–350	4.5–5.0	0.003–0.06	200 250 TS	Excellent	Fair to good Low temperature—poor	3–13 (air dry) 10–24 (bake)	Fair
Cellulosic (nitrate butyrate)		250–400	3.2–6.2		180	Good	Good Low temperature—poor	10–15	
Chlorinated polyether (Penton*)	10^{15}	400	3.0	0.01	250	Excellent	Good		
Epoxy-amine cure	10^{14} at 30°C 10^{10} at 105°C	400–550	3.5–5.0	0.02–0.03 at 30°C	350	Excellent	Fair to good Low temperature—poor	26–36	Good to excellent
Epoxy-anhydride, dicy		650–730	3.4–3.8	0.01–0.03	400	Excellent	Good to excellent Low temperature—poor	20	Good to excellent
Epoxy-polyamide	10^{14} at 30°C 10^{10} at 105°C	400–500	2.5–3.0	0.008–0.02	350	Excellent	Good to excellent Low temperature—poor	20	Fair to good
Epoxy-phenolic	10^{12}–10^{13}	300–450			400	Excellent	Good Low temperature—fair		Good to excellent
Fluorocarbon TFE	10^{18}	430	2.0–2.1	0.0002	500	Can be excellent; primers required	Excellent		
FEP	10^{18}	480	2.1	0.0003–0.0007	400				
CTFE	10^{18}	500–600	2.3–2.8	0.003–0.004	400	Can be excellent; primers required	Excellent		
Parylene (polyxylylenes)	10^{16}–10^{17}	700	2.6–3.1	0.0002–0.02	240°F (air) 510°F (inert atmosphere)	Good	Good		
Phenolics	10^{9}–10^{12}	100–300	4–8	0.005–0.5	350	Excellent	Poor to good Low temperature—poor	30–38	Fair

Phenolic-oil varnish					250	Excellent	Good Low temperature—fair		Poor to fair
Phenoxy	10^{13}–10^{14}	500	3.7–4.0	0.001	180	Excellent	Excellent		
Polyamide (nylon)	10^{13}–10^{15}	400–500	2.8–3.6	0.01–0.1	225–250	Excellent			
Polyester	10^{12}–10^{14}	500	3.3–8.1	0.008–0.04	200	Good on rough surfaces; poor to polished metals	Fair to excellent	25–30	Good
Chlorosulfonated (polyethylene Hypalon)†			6–10	0.03–0.07	250	Good	Elastomeric	Less than 10	
Polyimide	10^{16}–10^{18}	400 3000 (10 mil)	3.4–3.8	0.003	500	Good	Fair to excellent		Good
Polystyrene	10^{10}–10^{19}	500–700	2.4–2.6	0.0001–0.0005	140–180		Poor to fair		
Polyurethane	10^{12}–10^{13}	450–500 3800 (1 mil)	6.8 (1 kHz) 4.4 (1 MHz)	0.02–0.08	250	Often poor to metals (excellent to most nonmetals)	Good to excellent Low temperature—poor	10–17 (castor oil) 50–60 (polyester)	
Silicone	10^{14}–10^{16}	550	3.0–4.2	0.001–0.008	500	Varies, but usually needs primer for good adhesion	Excellent Low temperature—excellent	12–16	Fair to excellent
Vinyl chloride (poly-)	10^{11}–10^{15}	300–800	3–9	0.04–0.14	150	Excellent, if so formulated	Excellent Low temperature—fair to good	5–10	
Vinyl chloride (plastisol, organisol)	10^{9}–10^{16}	400	2.3–9	0.10–0.15	150	Requires adhesive primer	Excellent Low temperature—fair to good	3–6	
Vinyl fluoride	10^{13}–10^{14}	260 1200 (8 mil)	6.4–8.4	0.05–0.15	300	Excellent, if fused on surface	Excellent Low temperature—excellent		
Vinyl formal (Formvar‡)	10^{13}–10^{15}	850–1000	3.7	0.007–0.2	200	Excellent			

*Trademark of Hercules Powder Co., Inc., Wilmington, Del.
†Trademark of E. I. du Pont de Nemours & Co., Wilmington, Del.
‡Trademark of Monsanto Co., St. Louis, Mo.

TABLE 10.6 Properties of Coatings by Polymer Type *(Continued)*

Coating type	Resistance to environmental effects				Flammability	Repairability	Film formation		Application method	Typical uses
	Chemical and solvent resistance	Moisture and humidity resistance	Weatherability	Resistance to micro-organisms			Method of cure	Cure schedule		
Acrylic		Good	Excellent resistance to UV and weather	Good	Medium	Remove with solvent	Solvent evaporation	Air dry or low-temperature bake	Spray, brush, dip	Coatings for circuit boards. Quick dry protection for markings and color coding.
Alkyd	Solvents—poor Alkalies—poor Dilute acids—poor to fair	Poor	Good to excellent	Poor	Medium	Poor	Oxidation or heat	Air dry or baking types	Most common methods	Painting of metal parts and hardware.
Cellulosic (nitrate butyrate)	Solvents—good Alkalies—good Acids—good	Fair		Poor to good	High	Remove with solvents	Solvent evaporation	Air dry or low-temperature bake	Spray, dip	Lacquers for decoration and protection. Hot-metal coatings.
Chlorinate polyether (Penton*)		Good			Low		Powder or dispersion fuses	High-temperature fusion	Spray, dip, fluid bed	Chemically resistant coatings.
Epoxy-amine cure	Solvents—good to excellent Alkalies—good Dilute acids—fair	Good	Pigmented—fair; clear—poor (chalks)	Good	Medium	No	Cured by catalyst reaction	Air dry to medium bake	Spray, dip, fluid bed	Coatings for circuit boards. Corrosion-protective coatings for metals.
Epoxy-anhydride, Dicy	Solvents—good Alkalies— Dilute acids—	Good		Good	Medium	No	Cured by chemical reaction	High bakes 300 to 400°F	Spray, dip, fluid bed, impregnate	High-bake, high-temperature-resistant dielectric and corrosion coatings.
Epoxy-polyamide	Solvents—fair Alkalies—good Dilute acids—poor	Good		Good	Medium	No	Cured by coreactant	Air dry or medium bake	Spray, dip	Coatings for circuit boards. Filleting coating.
Epoxy-phenolic	Solvents—excellent Alkalies—fair Dilute acids—good	Excellent	Pigmented—fair; clear—poor	Good	Medium	No	Cured by coreactant	High bakes 300 to 400°F	Spray, dip	High-bake solvent and chemical resistant coating.
Fluorocarbon TFE	Solvents—excellent Alkalies—good Dilute acids—excellent	Excellent		Good	None	No	Fusion from water or solvent dispersion	Approximately 750°F	Spray, dip	High-temperature resistant insulation for wiring.
FEP CTFE		Excellent		Good	None	No	Fusion from water or solvent dispersion	500–600°F	Spray, dip	High-temperature-resistant insulation. Extrudable.
Parylene (polyxylylenes)		Excellent			None		Vapor phase deposition and polymerization requiring special license from Union Carbide.			Very thin, pinhole-free coatings, possible semiconductable coating.
Phenolics	Solvents—good to excellent Alkalies—poor Dilute acids—good to excellent	Excellent	Fair	Poor to good	Medium	No	Cured by heat	Bake 350–500°F	Spray, dip	High-bake chemical and solvent-resistant coatings.

Phenolic-oil varnish	Solvents—poor Alkalies—poor Dilute acids—good to excellent	Good	Good	Poor, unless toxic—additive	Medium	Poor	Oxidation or heat		Spray, brush, dip-impregnate	Impregnation of electronic modules, quick protective coating.
Phenoxy		Good		Good			Cured by heat			Chemical resistant coating.
Polyamide (nylon)		Fair				Fairly solderable				Wire coating.
Polyester	Solvents—poor Alkalies—poor to fair Dilute acids—good	Fair	Very good	Good	Medium	Poor	Cured by heat or catalyst	Air dry or bake 100–250°F	Spray, brush, dip	
Chlorosulfonated polyethylene (Hypalon†)		Good		Good	Low		Solvent evaporation	Air dry or low-temperature bake	Spray, brush	Moisture and fungus proofing of materials.
Polyimide	Solvents—excellent Alkalies—poor to fair Dilute acids—good	Good	Good	Good	Low	Poor	Cured by heat	High bake	Dip, impregnate, wire coater	Very high temperature resistant with insulation.
Polystyrene		Good		Good	High	Dissolve with solvents	Solvent evaporation	Air dry or low bake	Spray, dip	Coil coating, low dielectric constant, low loss in radar uses.
Polyurethane	Solvents—good Dilute alkalies—fair Dilute acids—good	Good		Poor to good	Medium	Excellent; melts, solder-through properties	Coreactant or moisture cure	Air dry to medium bake	Spray, brush, dip	Conformal coating of circuitry, solderable wire insulation.
Silicone	Solvents—poor Alkalies—good (dilute) poor (concentrated) Dilute acids—good	Excellent	Excellent	Good	Very low (except in O_2 atmosphere)	Fair to excellent. Cut and peel	Cured by heat or catalyst	Air dry (RTV) to high bakes	Spray, brush, dip	Heat-resistant coating for electronic circuitry. Good moisture resistance.
Vinyl chloride (poly-)	Solvents—alcohol, good Alkalies—good	Good	Pigmented—fair to good Clear—poor	Poor to good (depends on plasticizer)	Very low	Dissolve with solvents	Solvent evaporation	Air dry or elevated temperature for speed	Spray, dip, roller coat	Wire insulation. Metal protection (especially magnesium, aluminum).
Vinyl chloride (plastisol, organisol)		Good		Poor to good (depends on plasticizer)	Low	Poor	Fusion of liquid to gel	Bake 250–350°F	Spray, dip, reverse roll	Soft-to-hard thick coatings, electroplating racks, equipment.
Vinyl fluoride		Good	Excellent	Good	Very low	Poor	Fusion from solvent dispersion	Bake 400–500°F	Spray, roller coat	Coatings for circuitry. Long-life exterior finish.
Vinyl formal (Form-var‡)		Good		Good	Medium	Poor	Cured by heat	Bake 300–500°F	Roller coat, wire coater	Wire insulation (thin coatings) coil impregnation.

*Trademark of Hercules Powder Co., Inc., Wilmington, Del.
†Trademark of E. I. du Pont de Nemours & Co., Wilmington, Del.
‡Trademark of Monsanto Co., St. Louis, Mo.
SOURCE: This table has been reprinted from *Machine Design*, May 25, 1967. Copyright, 1967, by The Penton Publishing Company, Cleveland, Ohio.

ties of the generic resin, can be greatly disappointed. Instead, selections must be made on the basis of performance data for specific coatings or finish systems. Performance data are generated by the paint and product manufacturing industries when conducting standard paint evaluation tests. Test methods for coating material evaluation are listed in Table 10.7.

10.4.4 Selection by electrical properties

Electrical properties of organic coatings vary by resin (also referred to as polymer) type. When selecting insulating varnishes, insulating enamels, and magnet wire enamels, the electrical properties and physical properties determine the choice.

Table 10.8 shows electric strengths, Table 10.9 shows volume resistivities, Table 10.10 shows dielectric constants, and Table 10.11 shows dissipation factors for coatings using most of the available resins. Magnet wire insulation is an important use for organic coatings. National Electrical Manufacturer's Association (NEMA) standards and manufacturers' trade names for various wire enamels are shown in Table 10.12. This information can be used to guide the selection of coatings. However, it is important to remember the aforementioned warnings about blends of various resins and the effects on performance properties.

10.5 Coating Materials

Since it is the resin in the coating's vehicle that determines its performance properties, coatings can be classified by their resin types. The most widely used resins for manufacturing modern coatings are acrylics, alkyds, epoxies, polyesters, polyurethanes, and vinyls.[3] In the following section, the resins used in coatings are described.

10.5.1 Common coating resins

Acrylics. Acrylics are noted for color and gloss retention in outdoor exposure. Acrylics are supplied as solvent-containing, high-solids, waterborne, and powder coatings. They are formulated as lacquers, enamels, and emulsions. Lacquers and baking enamels are used as automotive and appliance finishes. Both these industries use acrylics as topcoats for multicoat finish systems. Thermosetting acrylics have replaced alkyds in applications requiring greater mar resistance such as appliance finishes. Acrylic lacquers are brittle and therefore have poor impact resistance, but their outstanding weather resistance allowed them to replace nitrocellulose lacquers in automotive finishes

TABLE 10.7 Specific Test Methods for Coatings

Test	ASTM	Federal STD. 141a, method	MIL-STD-202, method	Federal STD. 406, method	Others
Abrasion	D 968	6191 (Falling Sand) 6192 (Taber)		1091	Fed. Std. 601, 14111
Adhesion	D 2197	6301.1 (Tape Test, Wet) 6302.1 (Microknife) 6303.1 (Scratch Adhesion) 6304.1 (Knife Test)		1111	Fed. Std. 601, 8031
Arc resistance	D 495		303	4011	
Dielectric constant	D 150		301	4021	Fed. Std. 101, 303
Dielectric strength (breakdown voltage)	D 149 D 115			4031	Fed. Std. 601, 13311
Dissipation factor	D 150			4021	
Drying time	D 1640 D 115	4061.1			
Electrical insulation resistance	D 229 D 257		302	4041	MIL-W-81044, 4.7.5.2
Exposure (exterior)	D 1014	6160 (On Metals) 6161.1 (Outdoor Rack)			
Flash point	D 56, D 92 D 1310 (Tag Open Cup)	4291 (Tag Closed Cup) 4294 (Cleveland Open Cup)			Fed. Std. 810, 509
Flexibility	D1737 D522	6221 (Mandrel) 6222 (Conical Mandrel)		1031	Fed. Std. 601, 11041

TABLE 10.7 Specific Test Methods for Coatings (*Continued*)

Test	ASTM	Federal STD. 141a, method	MIL-STD-202, method	Federal STD. 406, method	Others
Fungus resistance	D 1924				MIL-E-5272, 4.8 MIL-STD-810, 508.1 MIL-T-5422, 4.8
Hardness	D3363 D 1474	6211 (Print Hardness) 6212 (Indentation)			
Heat resistance	D 115 D 1932	 6051			
Humidity	D 2247	6071 (100% RH) 6201 (Continuous Condensation)	103 106A		MIL-E-5272, Proc. 1 Fed. Std. 810, 507
Impact resistance	D2794	6226 (G. E. Impact)	1074		
Moisture-vapor permeability	E 96 D 1653	 6171	 7032		
Nonvolatile content		4044			
Salt spray (fog)	B 117	6061	101C	6071	MIL-STD-810, 509.1 MIL-E-5272, 4.6 Fed. Std. 151, 811.1 Fed. Std. 810, 509
Temperature-altitude					MIL-E-5272, 4.14 MIL-T-5422, 4.1 MIL-STD-810, 504.1
Thermal conductivity	D 1674 (Cenco Fitch) C 177 (Guarded Hot Plate)				MIL-I-16923, 4.6.9

Thermal shock			107		MIL-E-5272, 4.3 MIL-STD-810, 503.1
Thickness (dry film)	D 1005 D 1186	6181 (Magnetic Gage) 6183 (Mechanical Gage)		2111, 2121, 2131, 2141, 2151	Fed. Std. 151, 520, 521.1
Viscosity	D 1545 D 562 D 1200 D 88	4271 (Gardner Tubes) 4281 (Krebs-Stormer) 4282 (Ford Cup) 4285 (Saybolt) 4287 (Brookfield)			
Weathering (accelerated)	D 822	6151 (Open Arc) 6152 (Enclosed Arc)		6024	

Note: A more complete compilation of test methods is found in J. J. Licari, *Plastic Coatings for Electronics,* McGraw-Hill, New York, 1970.

The major collection of complete test methods for coatings is *Physical and Chemical Examination of Paints, Varnishes, Lacquers, and Colors,* by Gardner and Sward, Gardner Laboratory, Bethesda, Md. This has gone through many editions.

TABLE 10.8 Electric Strengths of Coatings

Material	Dielectric strength, V/mil	Comments*	Source of information
Polymer coatings:			
Acrylics	450–550	Short-time method	[a]
	350–400	Step-by-step method	[a]
	400–530		[b]
	1700–2500	2-mil-thick samples	Columbia Technical Corporation, Humiseal Coatings
Alkyds	300–350		[b]
Chlorinated polyether	400	Short-time method	[a]
Chlorosulfonated polyethylene	500	Short-time method	[a]
Diallyl phthalate	275–450		[b]
	450	Step-by-step method	[a]
Diallyl isophthalate	422	Step-by-step method	[a]
Depolymerized rubber (DPR)	360–380		H. V. Hardman Company, DPR Subsidiary
Epoxy	650–730	Cured with anhydride–castor oil adduct	Autonetics, Division of North American Rockwell
Epoxy	1300	10-mil-thick dip coating	
Epoxies, modified	1200–2000	2-mil-thick sample	Columbia Technical Corporation, Humiseal Coatings
Neoprene	150–600	Short-time method	[a]
Phenolic	300–450		[b]
Polyamide	780	106 mil thick sample	
Polyamide-imide	2700		
Polyesters	250–400	Short-time method	[a]
	170	Step-by-step method	[a]
Polyethylene	480		[b]
	300	60-mil-thick sample	[c]
	500	Short-time method	[a]
Polyimide	3000	Pyre-ML, 10 mils thick	[d]
	4500–5000	Pyre-ML (RC-675)	[e]
	560	Short-time method, 80 mils thick	
Polypropylene	750–800	Short-time method	[a]
Polystyrene	500–700	Short-time method	[a]
	400–600	Step-by-step method	[a]
	450	60-mil-thick sample	[c]
Polysulfide	250–600	Short-time method	[a]
Polyurethane (single component)	3800	1-mil-thick sample	[f]
Polyurethane (two components)/castor oil cured	530–1010		[g]
Polyurethane (two components, 100% solids)	275	125-mil-thick sample	Products Research & Chemical Corporation (PR-1538)
	750	25-mil thick sample	
Polyurethane (single component)	2500	2-mil-thick sample	Columbia Technical Corporation, Humiseal 1A27
Polyvinyl butyral	400		
Polyvinyl chloride	300–1000	Short-time method	
	275–900	Step-by-step method	[b]
Polyvinyl formal	860–1000		
Polyvinylidene fluoride	260	Short-time, 500-V/s, $^1/_8$-in sample	[h]
	1280	Short-time, 500-V/s, 8-mil sample	[h]
	950	Step by step (1-kV steps)	[h]
Polyxylylenes:			
Parylene N	6000	Step by step	Union Carbide Corporation
	6500	Short time	Union Carbide Corporation
Parylene C	3700	Short time	Union Carbide Corporation
	1200	Step by step	Union Carbide Corporation
Parylene D	5500	Short time	Union Carbide Corporation
	4500	Step by step	Union Carbide Corporation
Silicone	500	Sylgard 182	Dow Corning Corporation
Silicone	550–650	RTV types	General Electric & Stauffer Chemical Company bulletins
Silicone	800	Flexible dielectric gel	Dow Corning Corporation
Silicone	1500	2-mil-thick sample	Columbia Technical Corporation, Humiseal 1H34
TFE fluorocarbons	400	60-mil-thick sample	[c]
	480	Short-time method	[a]
	430	Step by step	[a]
Teflon TFE dispersion coating	3000–4500	1- to 4-mil-thick sample	E. I. du Pont de Nemours & Company
Teflon FEP dispersion coating	4000	1.5-mil-thick sample	E. I. du Pont de Nemours & Company

TABLE 10.8 Electric Strengths of Coatings *(Continued)*

Material	Dielectric strength, V/mil	Comments*	Source of information
Other materials used in electronic assemblies:			
Alumina ceramics	200–300		b
Boron nitride	900–1400		
Electrical ceramics	55–300		b
Forsterite	250		b
Glass, borosilicate	4500	40-mil sample	c
Steatite	145–280		b

*All samples are standard 125 mils thick unless otherwise specified.
[a]*Insulation,* Directory Encyclopedia Issue, no. 7, June–July 1968.
[b]*Material Engineering,* Materials Selector Issue, vol. 66, no. 5, Chapman-Reinhold Publication, mid-October 1967–1968.
[c]W. H. Kohl, *Handbook of Materials and Techniques for Vacuum Devices,* p. 586, Reinhold Publishing Corporation, New York, 1967, p. 586.
[d]J. R. Learn and M. P. Seegers, "Teflon-Pyre-M. L. Wire Insulation System," *13th Symposium on Technical Progress in Commun. Wire and Cables,* Atlantic City, NJ, December 2–4, 1964.
[e]J. T. Milek: Polyimide Plastics: A State of the Art Report, *Hughes Aircraft Report,* S-8, October, 1965.
[f]*Hughson Chemical Co.* Bulletin 7030A.
[g]*Spencer-Kellogg* (Division of Textron, Inc.) Bulletin TS-6593.
[h]*Pennsalt Chemicals Corp. Prod. Sheet* KI-66a, Kynar Vinylidene Fluoride Resin, 1967.

for many years. Acrylic and modified acrylic emulsions have been used as architectural coatings and also on industrial products. These medium-priced resins can be formulated to have excellent hardness, adhesion, abrasion, chemical, and mar resistance. When acrylic resins are used to modify other resins, their properties are often imparted to the resultant resin system.

Uses. Acrylics, both lacquers and enamels, were the topcoats of choice for the automotive industry from the early 1960s to the middle 1980s. Thermosetting acrylics are still used by the major appliance industry. Acrylics are used in electrodeposition and have largely replaced alkyds. The chemistry of acrylic-based resins allows them to be used in radiation curing applications alone or as monomeric modifiers for other resins. Acrylic-modified polyurethane coatings have excellent exterior durability.

Alkyds. Alkyd resin–based coatings were introduced in the 1930s as replacements for nitrocellulose lacquers and oleoresinous coatings. They offer the advantage of good durability at relatively low cost. These low- to medium-priced coatings are still used for finishing a wide variety of products, either alone or modified with oils or other resins. The degree and type of modification determine their performance properties. They were used extensively by the automotive and appliance industries through the 1960s. Although alkyds are used in outdoor applications, they are not as durable in long-term exposure, and their color and gloss retention is inferior to that of acrylics.

Uses. Once the mainstay of organic coatings, alkyds are still used for finishing metal and wood products. Their durability in interior exposures is generally good, but their exterior durability is only fair. Alkyd resins are used in fillers, sealers, and caulks for wood finishing

TABLE 10.9 Volume Resistivities of Coatings

Material	Volume resistivity at 25°C, Ω-cm	Source of information
Acrylics	10^{14}–10^{15}	[a]
	>10^{14}	[b]
	7.6×10^{14}–1.0×10^{15}	Columbia Technical Corporation, Humiseal
Alkyds	10^{14}	[a]
Chlorinated polyether	10^{15}	[b]
Chlorosulfonated polyethylene	10^{14}	[b]
Depolymerized rubber	1.3×10^{13}	H. V. Hardman, DPR Subsidiary
Diallyl phthalate	10^{8}–2.5×10^{10}	[a]
Epoxy (cured with DETA)	2×10^{16}	[c,d]
Epoxy polyamide	1.1–1.5×10^{14}	
Phenolics	6×10^{12}–10^{13}	[a]
Polyamides	10^{13}	
Polyamide-imide	7.7×10^{16}	[e]
Polyethylene	>10^{16}	
Polyimide	10^{16}–10^{18}	[f]
Polypropylene	10^{10}–>10^{16}	[a]
Polystyrene	>10^{16}	[b]
Polysulfide	2.4×10^{11}	[g]
Polyurethane (single component)	5.5×10^{12}	[h]
Polyurethane (single component)	2.0×10^{12}	[h]
Polyurethane (single component)	4×10^{13}	Columbia Technical Corporation
Polyurethane (two components)	1×10^{13}	Products Research & Chemical
	5×10^{9}(300°F)	Corporation (PR-1538)
Polyvinyl chloride	10^{11}–10^{15}	[b]
Polyvinylidene chloride	10^{14}–10^{16}	[a]
Polyvinylidene fluoride	2×10^{14}	[h]
Polyxylylenes (parylenes)	10^{16}–10^{17}	Union Carbide Corporation
Silicone (RTV)	6×10^{14}–3×10^{15}	Stauffer Chemical Company, Si-O-Flex SS 831, 832, & 833
Silicone, flexible dielectric gel	1×10^{15}	Dow Corning Corporation
Silicone, flexible, clear	2×10^{15}	Dow Corning Corporation
Silicone	3.3×10^{14}	Columbia Technical Corporation Humiseal 1H34
Teflon TFE	>10^{18}	[b]
Teflon FEP	>2×10^{18}	[b]

[a]*Materials Engineering,* Materials Selector Issue, vol. 66, no. 5, Chapman-Reinhold Publication, mid-October 1967–1968.
[b]*Insulation,* Directory Encyclopedia Issue, no. 7, June–July 1968.
[c]H. Lee and K. Neville, *Epoxy Resins,* McGraw-Hill, New York, 1966.
[d]Tucker, Cooperman, and Franklin, Dielectric Properties of Casting Resins, *Electronics Equipment,* July 1956.
[e]J. H. Freeman, "A New Concept in Flat Cable Systems," *5th Annual Symposium on Advanced Technology for Aircraft Electrical Systems,* Washington, D.C., October 1964.
[f]J. T. Milek, "Polyimide Plastics: A State of the Art Report," *Hughes Aircraft Rep.* S-8, October 1965.
[g]L. Hockenberger, *Chem.-Ing. Tech.,* vol. 36, 1964.
[h]*Hughson Chemical Company Technical Bulletin* 7030A; *Pennsalt Chemicals Corporation Product Sheet* KI-66a, Kynar Vinylidene Fluoride Resin, 1967.

TABLE 10.10 Dielectric Constants of Coatings

Coating	60–100 Hz	10^6 Hz	$>10^6$ Hz	Reference source
Acrylic		2.7–3.2		[a]
Alkyd			3.8 (10^{10} Hz)	[b]
Asphalt and tars			3.5 (10^{10} Hz)	[b]
Cellulose acetate butyrate		3.2–6.2		[a]
Cellulose nitrate		6.4		[a]
Chlorinated polyether	3.1	2.92		[a]
Chlorosulfonated polyethylene (Hypalon)	6.19 7–10 (10^3 Hz)	~5		E. I. du Pont de Nemours & Company
Depolymerized rubber (DPR)	4.1–4.2	3.9–4.0		H. V. Hardman, DPR Subsidiary
Diallyl isophthalate	3.5	3.2	3 (10^8 Hz)	[a,c]
Diallyl phthalate	3–3.6	3.3–4.5		[a,c]
Epoxy-anhydride—castor oil adduct	3.4	3.1	2.9 (10^7 Hz)	Autonetics, Division of North American Rockwell
Epoxy (one component)	3.8	3.7		Conap Inc.
Epoxy (two components)	3.7			Conap Inc.
Epoxy cured with methyl nadic anhydride (100:84 pbw)	3.31			[d]
Epoxy cured with dodecenylsuccinic anhydride (100:132 pbw)	2.82			[d]
Epoxy cured with DETA	4.1	4.2	4.1	[e]
Epoxy cured with *m*-phenylenediamine	4.6	3.8	3.25 (10^{10} Hz)	[e]
Epoxy dip coating (two components)	3.3	3.1		Conap Inc.
Epoxy (one component)	3.8	3.5		Conap Inc.
Epoxy-polyamide (40% Versamid* 125, 60% epoxy)	3.37	3.08		[e]
Epoxy-polyamide (50% Versamid 125, 50% epoxy)	3.20	3.01		[e]
Fluorocarbon (TFE, Teflon)	2.0–2.08	2.0–2.08		E. I. du Pont de Nemours & Company
Phenolic		4–11		[a]
Phenolic	5–6.5	4.5–5.0		[c]
Polyamide	2.8–3.9	2.7–2.96		
Polyamide-imide	3.09	3.07		
Polyesters	3.3–8.1	3.2–5.9		[c]
Polyethylene		2.3		[a]
Polyethylenes	2.3	2.3		[c]
Polyimide-Pyre-M.L.† enamel	3.8	3.8		[f]
Polyimide-Du Pont RK-692 varnish	3.8			[g]
Polyimide-Du Pont RC-B-24951	3.0(10^3 Hz)			[g]
Polyimide-Du Pont RC-5060	2.8(10^3 Hz)			[g]
Polypropylene		2.1		[a]

TABLE 10.10 Dielectric Constants of Coatings *(Continued)*

Coating	60–100 Hz	10^6 Hz	>10^6 Hz	Reference source
Polypropylene	2.22–2.28	2.22–2.28		c
Polystyrene	2.45–2.65	2.4–2.65	2.5(10^{10} Hz)	b
Polysulfides	6.9			h
Polyurethane (one component)	4.10	3.8		Conap Inc.
Polyurethane (two components—castor oil cured)	6.8(10^3 Hz)	2.98–3.28		i
Polyurethane (two components)		4.4		Products Research & Chemical Corporation (PR-1538)
Polyvinyl butyral	3.6	3.33		a
Polyvinyl chloride	3.3–6.7	2.3–3.5		a
Polyvinyl chloride—vinyl acetate copolymer	3–10			a
Polyvinyl formal	3.7	3.0		a
Polyvinylidene chloride		3–5		a
Polyvinylidene fluoride	8.1	6.6		i
Polyvinylidene fluoride	8.4	6.43	2.98(10^9 Hz)	k
p-Polyxylylene:				
Parylene N	2.65	2.65		Union Carbide Corporation
Parylene C	3.10	2.90		Union Carbide Corporation
Parylene D	2.84	2.80		Union Carbide Corporation
Shellac (natural, dewaxed)	3.6	3.3	2.75(10^9 Hz)	l
Silicone (RTV types)	3.3–4.2	3.1–4.0		General Electric and Stauffer Chemical Companies
Silicone (Sylgard‡ type)	2.88	2.88		Dow Corning Corporation
Silicone, flexible	3.0			Dow Corning dielectric gel Corporation
FEP dispersion coating	2.1(10^3 Hz)			E. I. du Pont de Nemours & Company
TFE dispersion coating	2.0–2.2(10^3 Hz)			E. I. du Pont de Nemours & Company
Wax (paraffinic)	2.25	2.25	2.22(10^{10} Hz)	l

*Trademark of General Mills, Inc., Kankakee, Ill.
†Trademark of E. I. du Pont de Nemours & Company, Wilmington, Del.
‡Trademark of Dow Corning Corporation, Midland, Mich.
[a]*Materials Engineering,* Materials Selector Issue, vol. 66, no. 5, Chapman-Reinhold Publication, mid-October 1967–1968.
[b]M. C. Volk, J. W. Lefforge, and R. Stetson, *Electrical Encapsulation,* Reinhold Publishing Corporation, New York, 1962.
[c]*Insulation,* Directory Encyclopedia Issue, no. 7, June–July 1968.
[d]C. F. Coombs, ed., *Printed Circuits Handbook,* McGraw-Hill, New York, 1967.
[e]H. Lee and K. Neville, *Handbook of Epoxy Resins,* McGraw-Hill, New York, 1967.
[f]J. R. Learn and M. P. Seeger, Teflon-Pyre-M.L. Wire Insulation System, *13th Symposium of Technical Progress in Communications Wire and Cable,* Atlantic City, NJ, Dec. 2–4, 1964.
[g]*Du Pont Bulletin* H65-4, Experimental Polyimide Insulating Varnishes, RC-B-24951 and RC-5060, January 1965.
[h]L. Hockenberger, *Chem.-Ing. Tech.* vol. 36, 1964.
[i]*Spencer-Kellogg* (division of Textron, Inc.) Bull. TS-6593.
[j]W. S. Barnhart, R. A. Ferren, and H. Iserson, 17th ANTEC of SPE, January 1961.
[k]*Pennsalt Chemicals Corporation Product Sheet* KI-66a, Kynar Vinylidene Fluoride Resin, 1967.
[l]A. R. Von Hippel, ed., *Dielectric Materials and Applications,* Technology Press of MIT and John Wiley & Sons, Inc., New York, 1961.

TABLE 10.11 Dissipation Factors of Coatings

Coating	60–100 Hz	10^6 Hz	$>10^6$ Hz	Reference source
Acrylics	0.04–0.06	0.02–0.03		*a*
Alkyds	0.003–0.06			*a*
Chlorinated polyether	0.01	0.01		*a*
Chlorosulfonated polyethylene	0.03	0.07(10^3 Hz)		*b*
Depolymerized rubber (DPR)	0.007–0.013	0.0073–0.016		H. V. Hardman, DPR Subsidiary
Diallyl phthalate	0.010	0.011	0.011	*b*
Diallyl isophthalate	0.008	0.009	0.014(10^{10} Hz)	*b*
Epoxy dip coating (two components)	0.027	0.018		Conap Inc.
Epoxy (one component)	0.011	0.004		Conap Inc.
Epoxy (one component)	0.008	0.006		Conap Inc.
Epoxy polyamide (40% versamid 125, 60% epoxy)	0.0085	0.0213		
Epoxy polyamide (50% Versamid 115, 50% epoxy)	0.009	0.0170		
Epoxy cured with anhydride–castor oil adduct	0.0084	0.0165	0.0240	Autonetics, North American Rockwell
Phenolics	0.005–0.5	0.022		*a*
Polyamide	0.015	0.022–0.097		
Polyesters	0.008–0.041			*a*
Polyethylene (linear)	0.00015	0.00015	0.0004(10^{10} Hz)	*d*
Polymethyl methacrylate	0.06	0.02	0.009(10^{10} Hz)	*d*
Polystyrene	0.0001–0.0005	0.0001–0.0004		
Polyurethane (two component, castor oil cure)		0.016–0.036		
Polyurethane (one component)	0.038–0.039	0.068–0.074		Conap Inc.
Polyurethane (one component)	0.02			Conap Inc.
Polyvinyl butyral	0.007	0.0065		
Polyvinyl chloride	0.08–0.15	0.04–0.14		*a*
Polyvinyl chloride, plasticized	0.10	0.15	0.01(10^{10} Hz)	*d*
Polyvinyl chloride-vinyl acetate copolymer	0.6–0.10			

TABLE 10.11 Dissipation Factors of Coatings *(Continued)*

Coating	60–100 Hz	10^6 Hz	$>10^6$ Hz	Reference source
Polyvinyl formal	0.007	0.02		
Polyvinylidene fluoride	0.049	0.17		[f]
	0.049	0.159	0.110	[g]
Polyxylylenes:				
Parylene N	0.0002	0.0006		Union Carbide Corporation
Parylene C	0.02	0.0128		Union Carbide Corporation
Parylene D	0.004	0.0020		Union Carbide Corporation
Silicone (Sylgard 182)	0.001	0.001		Dow Corning Corporation
Silicone, flexible dielectric gel	0.0005			Dow Corning Corporation
Silicone, flexible, clear	0.001			Dow Corning Corporation
Silicone (RTV types)	0.011–0.02	0.003–0.006		General Electric
Teflon FEP dispersion coating	0.0002–0.0007			E. I. du Pont de Nemours & Company
Teflon FEP	<0.0003	<0.0003		[a]
Teflon TFE	<0.0003			[a]
Teflon TFE	0.00012	0.00005		Union Carbide Corporation
Other materials:				
Alumina (99.5%)		0.0001		[h]
Beryllia (99.5%)		0.0003		[h]
Glass silica	0.0006	0.0001	0.00017(10^{10} Hz)	[d]
Glass, borosilicate		0.013–0.016		Corning Glass Works
Glass, 96% silica		0.0015–0.0019		Corning Glass Works

[a] *Machine Design,* Plastics Reference Issue, vol. 38, no. 14, Penton Publishing Co., 1966.
[b] *Insulation,* Directory, Encyclopedia Issue, no. 7, June–July 1968.
[c] H. Lee and K. Neville, *Handbook of Epoxy Resins,* McGraw-Hill, New York, 1967.
[d] K. Mathes, Electrical Insulation Conference, 1967.
[e] *Spencer-Kellogg* (Division of Textron, Inc.) *Technical Bulletin* TS-6593.
[f] W. S. Barnhart, R. A. Ferren, and H. Iserson, 17th ANTEC of SPE, January 1961.
[g] *Pennsalt Chemicals Corporation Product Sheet* KI-66a, Kynar Vinylidene Fluoride Resin, 1967.
[h] *Machine Design,* Design Guide, September 28, 1967.

because of their formulating flexibility. Alkyds have also been used in electrodeposition as replacements for the oleoresinous vehicles. They are still used for finishing by the machine tool and other industries. Alkyds have also been widely used in architectural and trade sales coatings. Alkyd-modified acrylic latex paints are excellent architectural finishes.

Epoxies. Epoxy resins can be formulated with a wide range of properties. These medium- to high-priced resins are noted for their adhesion, make excellent primers, and are used widely in the appliance and automotive industries. Their heat resistance permits them to be used for electrical insulation. When epoxy top coats are used outdoors, they tend to chalk and discolor because of inherently poor ultraviolet light resistance. Other resins modified with epoxies are used for outdoor exposure as topcoats, and properties of many other resins can be improved by their addition. Two-component epoxy coatings are used in environments with extreme corrosion and chemical conditions. Flexibility in formulating two-component epoxy resin–based coatings results in a wide range of physical properties.

Uses. Owing to their excellent adhesion, they are used extensively as primers for most coatings over most substrates. Epoxy coatings provide excellent chemical and corrosion resistance. They are used as electrical insulating coatings because of their high electric strength at elevated temperatures. Some of the original work with powder coating was done using epoxy resins, and they are still applied using this method. Many of the primers used for coil coating are epoxy resin based.

Polyesters. Polyesters are used alone or modified with other resins to formulate coatings ranging from clear furniture finishes—replacing lacquers—to industrial finishes—replacing alkyds. These moderately priced finishes permit the same formulating flexibility as alkyds but are tougher and more weather resistant. There are basically two types of polyesters: two-component and single-package. *Two-component polyesters* are cured using peroxides which initiate free-radical polymerization, while *single-package polyesters,* sometimes called *oil-free alkyds,* are self-curing, usually at elevated temperatures. It is important to realize that, in both cases, the resin formulator can adjust properties to meet most exposure conditions. Polyesters are also applied as powder coatings.

Uses. Two-component polyesters are well known as gel coats for glass-reinforced plastic bathtubs, lavatories, boats, and automobiles. Figure 10.2 shows tub and shower units using a polyester gel coat.

TABLE 10.12 NEMA Standards and Manufacturers' Trade Names for Magnet Wire Insulation

Manufacturer	Plain enamel	Polyvinyl formal	Polyvinyl formal modified	Polyvinyl formal with nylon overcoat	Polyvinyl formal with butyral overcoat	Polyamide	Acrylic	Epoxy
Thermal class NEMA Standard*	105°C MW 1	105°C MW 15	105°C MW 27	105°C MW 17	105°C MW 19	105°C MW 6	105°C MW 4	130°C MW 9
Anaconda Wire & Cable Company	Plain enamel	Formvar	Hermetic Formvar	Nyform	Cement-coated Formvar	....	....	Epoxy, epoxy-cement-coated
Asco Wire & Cable Company	Enamel	Formvar	....	Nyform	Formbond	Nylon	Acrylic	Epoxy
Belden Manufacturing Company	Beld enamel	Formvar	....	Nyclad	....	....	....	Epoxy
Bridgeport Insulated Wire Company	....	Formvar	....	....	Quickbond	Quick-Sol	....	....
Chicago Magnet Wire Corporation	Plain enamel	Formvar	....	Nyform	Bondable Formvar	Nylon	Acrylic	Epoxy
Essex Wire Corporation	Plain enamel	Formvar	Formetex	Nyform	Bondex	....	Ensolex/ ESX	Epoxy
General Cable Corporation	Plain enamel	Formvar	Formetic	Formlon	Formese	....	Solder-able acrylic	Epoxy
General Electric Corporation	....	Formex	....	....	....	Nylon	....	....
Haveg-Super Temp Division	....	....	....	....	....	....	....	....
Hitemp Wires Company Division Simplex Wire & Cable Company	....	....	....	....	....	....	....	....
Hudson Wire Company	Plain enamel	Formvar	....	Nyform	Formvar AVC	Ezsol	....	....
New Haven Wire & Cable, Inc.	Plain enamel	....	....	....	....	....	....	....
Phelps Dodge Magnet Wire Corporation	Enamel	Formvar	Hermeteze	Nyform	Bondeze	....	....	....
Rea Magnet Wire Company, Inc.	Plain enamel	Formvar	Hermetic Formvar special	Nyform	Koilset	Nylon	....	Epoxy
Viking Wire Company, Inc.	Enamel	Formvar	....	Nyform	F-Bondall	Nylon	....	....

*National Electrical Manufacturers Association.
SOURCE: Courtesy of Rea Magnet Wire Company, Inc.

Teflon	Polyurethane	Polyurethane with friction surface	Polyurethane with nylon overcoat	Polyurethane with butyral overcoat	Polyurethane with nylon and butyral overcoat	Polyester	Polyester with overcoat	Polyimide	Polyester polyimide	Ceramic, ceramic-Teflon, ceramic-silicon
200°C MW 10	105°C MW 2	105°C	130°C MW 28	105°C MW 3 (PROP)	130°C MW 29 (PROP)	155°C MW 5	155°C MW 5	220°C MW 16	180°C	180°C+ MW 7
....	Analac	....	Nylac	Cement-coated analac	Cement-coated nylac	....	Anatherm D Anatherm 200	AL 220 ML	Anatherm N	
....	Poly	....	Nypol	Asco bond-P	Asco bond	Ascotherm	Isotherm 200	ML	Anamid M (amide-imide),	
....	Beldure	....	Beldsol	....	....	Isonel	Polythermalese	ML	Ascomid	
....	Polyurethane	Uniwind	Polynylon	Polybond	....	Isonel 200				
....	Soderbrite	....	Nysod	Bondable polyurethene	....	Polyester 155				
....	Soderex	....	Soderon	Soderbond	Soderbond N	Thermalex F	Polythermalex/ PTX 200	Allex		
....	Enamel "G"	....	Genlon	....	....	Gentherm	Polythermaleze 200			
....	....	....	....	....	....	Alkanex				
Teflon	....	....	....	....	....	Isonel	....			
Temprite										
....	Hudsol	Gripon	Nypoly	Hudsol AVC	Nypoly AVC	Isonel 200	Isonel 200-A	ML	Isomid	
....	Impsol	....	Impsolon	....	....	....	Imp-200	....	....	
....	Sodereze	Gripeze	Nyleze	....	S-Y Bondeze	Polythermaleze 200 II	ML			
....	Solvar	....	Nylon solvar	Solvar Koilset	....	Isonel 200	Polythermaleze 200	Pyre ML	Isomid	Ceroc
....	Polyurethane	....	Polynylon	P-Bondall	....	Isonel 200	Iso-poly	ML	Isomid Isomid-P	

High-quality one-package polyester finishes are used on furniture, appliances, automobiles, magnet wire, and industrial products. Polyester powder coatings are used as high-quality finishes in indoor and outdoor applications for anything from tables to trucks. They are also used as coil coatings.

Polyurethanes. Polyurethane resin–based coatings are extremely versatile. They are higher in price than alkyds but lower than epoxies. Polyurethane resins are available as oil-modified, moisture-curing, blocked, two-component, and lacquers. Table 10.13 is a selection guide for polyurethane coatings. Two-component polyurethanes can be formulated in a wide range of hardnesses. They can be abrasion-resistant, flexible, resilient, tough, chemical-resistant, and weather-resistant. Abrasion resistance of organic coatings is shown in Table 10.14. Polyurethanes can be combined with other resins to reinforce or adopt their properties. Urethane-modified acrylics have excellent outdoor weathering properties. They can also be applied as air-drying, forced-dried, and baking liquid finishes as well as powder coatings.

Uses. Polyurethanes have become very important finishes in the transportation industry, which includes aircraft, automobiles, railroads,

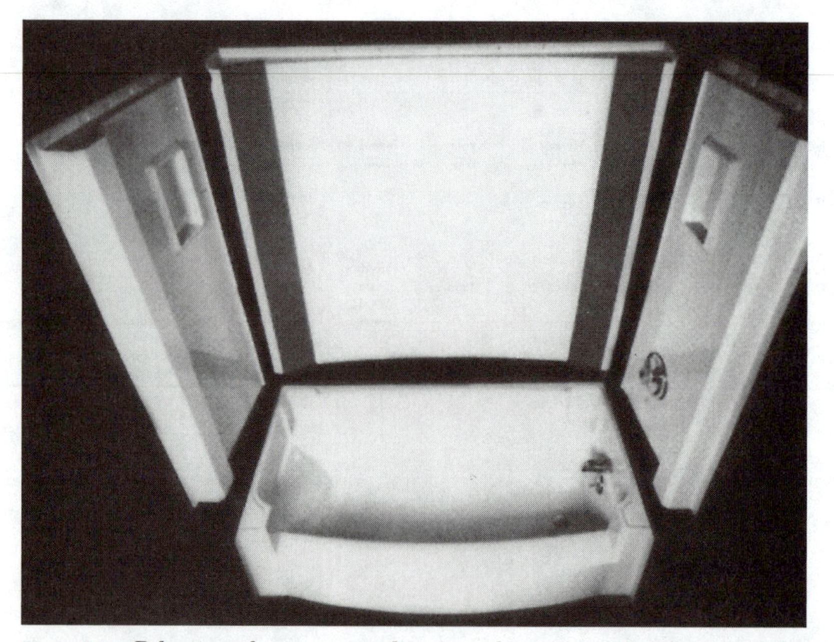

Figure 10.2 Polyester gel coats are used to give a decorative and protective surface to tub shower units which are made out of glass fiber–reinforced plastics. (*Courtesy of Owens-Corning Fiberglas Corporation.*)

TABLE 10.13 Guide to Selecting Polyurethane Coatings

Property	One-component			Two-component	Lacquer
	Urethane oil	Moisture	Blocked		
Abrasion resistance	Fair–good	Excellent	Good–excellent	Excellent	Fair
Hardness	Medium	Medium–hard	Medium–hard	Soft–very hard	Soft–medium
Flexibility	Fair–good	Good–excellent	Good	Good–excellent	Excellent
Impact resistance	Good	Excellent	Good–excellent	Excellent	Excellent
Solvent resistance	Fair	Poor–fair	Good	Excellent	Poor
Chemical resistance	Fair	Fair	Good	Excellent	Fair–good
Corrosion resistance	Fair	Fair	Good	Excellent	Good–excellent
Adhesion	Good	Fair–good	Fair	Excellent	Fair–good
Toughness	Good	Excellent	Good	Excellent	Good–excellent
Elongation	Poor	Poor	Poor	Excellent	Excellent
Tensile	Fair	Good	Fair–good	Good–excellent	Excellent
Weatherability					
Aliphatic	Good	Poor–fair		Good–excellent	Good
Conventional	Poor–fair	Poor–fair	Poor–fair	Poor–fair	Poor
Pigmented glass	High	High	High	High	Medium
Cure rate	Slow	Slow	Fast	Fast	None
Cure temp	Room temperature	Room temperature	300–390°F	212°F	150–225°F
Work life	Infinite	1 y	6 months	1 s–24 h	Infinite

TABLE 10.14 Abrasion Resistance of Coatings

Coating	Taber ware index, mg/1000 rev
Polyurethane type 1	55–67
Polyurethane type 2 (clear)	8–24
Polyurethane type 2 (pigmented)	31–35
Polyurethane type 5	60
Urethane oil varnish	155
Alkyd	147
Vinyl	85–106
Epoxy-amine–cured varnish	38
Epoxy-polyamide enamel	95
Epoxy-ester enamel	196
Epoxy-polyamide coating (1:1)	50
Phenolic spar varnish	172
Clear nitrocellulose lacquer	96
Chlorinated rubber	200–220
Silicone, white enamel	113
Catalyzed epoxy, air-cured (PT-401)	208
Catalyzed epoxy, Teflon-filled (PT-401)	122
Catalyzed epoxy, bake-Teflon–filled (PT-201)	136
Parylene N	9.7
Parylene C	44
Parylene D	305
Polyamide	290–310
Polyethylene	360
Alkyd TT-E-508 enamel (cured for 45 min at 250°F)	51
Alkyd TT-E-508 (cured for 24 h at room temperature)	70

trucks, and ships. Owing to their chemical resistance and ease of decontamination from chemical, biological, and radiological warfare agents, they are widely used for painting military land vehicles, ships, and aircraft. They are used on automobiles as coatings for plastic parts and as clear topcoats in the basecoat–clearcoat finish systems. Low-temperature baking polyurethanes are used as mar-resistant finishes for products that must be packaged while still warm. Polyurethanes are used in an increasing number of applications. They are also used in radiation curable coatings.

Polyvinyl chloride. *Polyvinyl chloride (PVC) coatings,* commonly called *vinyls,* are noted for their toughness, chemical resistance, and durability. They are available as solutions, dispersions, and lattices. Properties of vinyl coatings are listed in Table 10.15. They are applied as lacquers, plastisols, organisols, and lattices. PVC coating powders have essentially the same properties as liquids. PVC organisol, plastisol, and powder coatings have limited adhesion and

TABLE 10.15 Properties of Vinyl Coatings

Coating type	Outstanding characteristics	Mechanical properties[a]					Color and gloss[a]			Weathering[a] properties	
		Hardness	Abrasion resistance	Adhesion	Flexibility	Toughness	Film color	Color retention	Gloss	Weather resistance	Gloss retention
Solution[b]	Excellent color, flexibility, chemical resistance; tasteless, odorless	F	E	F to G	E	E	E[f]	E	G	E[f]	E
Plastisol[c]	Toughness; resilience; abrasion resistance; can be applied without solvents	F	E	E to cloth[e]	E	E	E[f]	E	F	E[f]	F to G
Organosol[d]	High solids content; excellent color, flexibility; tasteless, odorless	F	E	E to cloth[e]	E	E	E[f]	E	P to G	E[f]	G

[a]E = excellent; G = good; F = fair; P = poor.
[b]Vinyl chloride acetate copolymers; resins vary widely in compatibility with other materials.
[c]Vinyl chloride acetate copolymer and vinyl chloride resins.
[d]Vinyl chloride acetate copolymers; require grinding for good dispersions.
[e]Requires primer for use on metal.
[f]Pigmented.

require primers.

Uses. Vinyls have been used in various applications, including beverage and other can linings, automobile interiors, and office machine exteriors. They are also used as thick-film liquids and as powder coatings for electrical insulation. Owing to their excellent chemical resistance, they are used as tank linings and as rack coatings in electroplating shops. Typical applications for vinyl coatings are shown in Fig. 10.3. Vinyl-modified acrylic latex trade sale paints are used as trim enamels for exterior applications and as semigloss wall enamels for interior applications.

10.5.2 Other coating resins

In addition to the aforementioned materials, there are a number of other important resins used in formulating coatings. These materials, used alone or as modifiers for other resins, provide coating vehicles with diverse properties.

Aminos. Resins of this type, such as urea formaldehyde and melamine, are used in modifying other resins to increase their durability. Notable among these modified resins are the superalkyds used in automotive and appliance finishes.

Uses. Melamine and urea formaldehyde resins are used as modifiers for alkyds and other resins to increase hardness and accelerate cure.

Figure 10.3 Vinyl plastisols and organisols are used extensively for dip coating of wire products. The coatings can be varied from very hard to very soft. (*Courtesy of M & T Chemicals.*)

Cellulosics. Nitrocellulose lacquers are the most important of the cellulosics. They were introduced in the 1920s and used as fast-drying finishes for a number of manufactured products. Applied at low solids using expensive solvents, they will not meet air-quality standards. By modifying nitrocellulose with other resins such as alkyds and ureas, the VOC content can be lowered and performance properties can be increased. Other important cellulosic resins are cellulose acetate butyrate and ethyl cellulose.

Uses. Although no longer used extensively by the automotive industry, nitrocellulose lacquers are still used by the furniture industry because of their fast-drying and hand-rubbing properties. Cellulose acetate butyrate has been used for coating metal in numerous applications. In 1959 one of the first conveyorized powder coating lines in the United States coated distribution transformer lids and hand-hole covers with a cellulose acetate butyrate powder coating.

Chlorinated rubber. Chlorinated rubber coatings are used as swimming pool paints and traffic paints.

Fluorocarbons. These high-priced coatings require high processing temperatures and therefore are limited in their usage. They are noted for their lubricity or nonstick properties due to low coefficients of friction, and also for weatherability. Table 10.16 gives the coefficients of friction of typical coatings.

Uses. Fluorocarbons are used as chemical-resistant coatings for processing equipment. They are also used as nonstick coatings for cookware, friction-reducing coatings for tools, and dry lubricated surfaces in many other consumer and industrial products, as shown in Fig. 10.4. Table 10.17 compares the properties of four fluorocarbons.

Oleoresinous. Oleoresinous coatings, based on drying oils such as soybean and linseed, are slow curing. For many years prior to the introduction of synthetic resins, they were used as the vehicles in most coatings. They still find application alone or as modifiers to other resins.

Uses. Oleoresinous vehicles are used in low-cost primers and enamels for structural, marine, architectural, and, to a limited extent, industrial product finishing.

Phenolics. Introduced in the early 1900s, phenolics were the first commercial synthetic resins. They are available as 100% phenolic baking resins, oil-modified, and phenolic dispersions. Phenolic resins, used as modifiers, will improve the heat and chemical resistance of

TABLE 10.16 Coefficients of Friction of Typical Coatings

Coating	Coefficient of friction, μ	Information source
Polyvinyl chloride	0.4–0.5	*a*
Polystyrene	0.4–0.5	*a*
Polymethyl methacrylate	0.4–0.5	*a*
Nylon	0.3	*a*
Polyethylene	0.6–0.8	*a*
Polytetrafluoroethylene (Teflon)	0.05–0.1	*a*
Catalyzed epoxy air-dry coating with Teflon filler	0.15	*b*
Parylene N	0.25	*c*
Parylene C	0.29	*c*
Parylene D	0.31–0.33	*c*
Polyimide (Pyre-ML)	0.17	*d*
Graphite	0.18	*d*
Graphite-molybdenum sulfide:		
Dry-film lubricant	0.02–0.06	*e*
Steel on steel	0.45–0.60	*e*
Brass on steel	0.44	*e*
Babbitt on mild steel	0.33	*e*
Glass on glass	0.4	*e*
Steel on steel with SAE no. 20 oil	0.044	*e*
Polymethyl methacrylate to self	0.8 (static)	*e*
Polymethyl methacrylate to steel	0.4–0.5 (static)	*e*

[a]F. P. Bowder, *Endeavor,* vol. 16, no. 61, 1957, p. 5.
[b]Product Techniques Incorporated, Bulletin on PT-401 TE, October 17, 1961.
[c]Union Carbide data.
[d]*DuPont Technical Bulletin* 19, Pyre-ML Wire Enamel, August 1967.
[e]Electrofilm, Inc. data.

Figure 10.4 Nonstick feature of fluorocarbon finishes makes them useful for products such as saws, fan and blower blades, door-lock parts, sliding- and folding-door hardware, skis, and snow shovels. (*Courtesy of E. I. Du Pont de Nemours & Company.*)

TABLE 10.17 Properties of Four Fluorocarbons

Property	Polyvinyl fluoride (PVF) $(CH_2\text{-}CHF)_n$	Polyvinyl-idene fluoride (PVF-2) $(CH_2\text{-}CF_2)_n$	Polytrifluoro-chloroethylene (PTFC1) $(CC1F\text{-}CF_2)_n$	Polytetra-fluoroethylene (PTFE) $(CF_2\text{-}CF_2)_n$
Physical properties:				
Density	1.4	1.76	2.104	2.17–2.21
Fusing temperature, °F	300	460	500	750
Maximum continuous service and temperature, °F	225	300	400	550
Coefficient of friction	0.16	0.16	0.15	0.1
Flammability	Burns	Nonflammable	Nonflammable	Nonflammable
Mechanical properties:				
Tensile strength, lb/in^2	7000	7000	5000	2500–3500
Elongation, %	115–250	300	250	200–400
Izod impact, ft-lb/in		3.8	5	3
Durometer hardness		80	74–78	50–65
Yield strength at 77°F, lb/in^2	6000	5500	4500	1300
Heat-distortion temperature at 66 lb/in^2 °F	NA	300	265	250
Coefficient of linear expansion	2.8×10^5	8.5×10^5	15×10^5	8×10^5
Modulus (tension) $\times 10^5$ lb/in^2	2.5–3.7	1.2	1.9	0.6
Electrical properties:				
Dielectric strength, V/mil		260 (0.125)	500 (0.063)	600 (0.060)
Short time, V/mil, in	3400 (0.002)			
Dielectric constant, 10^3 Hz	8.5	7.72	2.6	2.1
Arc resistance (77°F) ASTM D 495	NA	60	300	300
Volume resistivity Ω-cm at 50% relative humidity, 77°F	10^{12}	10^{14}	10^{16}	10^{18}
Dissipation factor, 100 Hz	1.6	0.05	0.022	0.0003

other resins. Baked phenolic resin–based coatings are well known for their corrosion, chemical, moisture, and heat resistance.

Uses. Phenolic coatings are used on heavy-duty air-handling equipment, chemical equipment, and as insulating varnishes. Phenolic resins are also used as binders for electrical and decorative laminated plastics.

Polyamides. One of the more notable polyamide resins is nylon, which is tough, wear-resistant, and has a relatively low coefficient of friction. It can be applied as a powder coating by fluidized bed, electrostatic spray, or flame spray. Table 10.18 compares the properties of three types of nylon polymers used in coatings. Nylon coatings generally require a primer. Polyamide resins are also used as curing agents for two-component epoxy resin coatings. Film properties can be varied widely by polyamide selection.

Uses. Applied as a powder coating, nylon provides a high degree of toughness and mechanical durability to office furniture. Other polyamide resins are used as curing agents in two-component epoxy resin–based primers and topcoats, adhesives, and sealants.

Polyolefins. These coatings, which can be applied by flame spraying, hot melt, or powder coating methods, have limited usage.

TABLE 10.18 Properties of nylon coatings

	Nylon 11	Nylon 6/6	Nylon 6
Elongation (73°F), %	120	90	50–200
Tensile strength (73°F), lb/in^2	8,500	10,500	10,500
Modulus of elasticity (73°F), lb/in^2	178,000	400,000	350,000
Rockwell hardness	R 100.5	R 118	R 112–118
Specific gravity	1.04	1.14	1.14
Moisture absorption, %, ASTM D 570	0.4	1.5	1.6–2.3
Thermal conductivity Btu/(ft^2)(h)(°F/in)	1.5	1.7	1.2–1.3
Dielectric strength (short time), V/mil	430	385	440
Dielectric constant (10 Hz)	3.5	4	4.8
Effect of:			
Weak acids	None	None	None
Strong acids	Attack	Attack	Attack
Strong alkalies	None	None	None
Alcholos	None	None	None
Esters	None	None	None
Hydrocarbons	None	None	None

Uses. Polyethylene is used for impregnating or coating packaging materials such as paper and aluminum foil. Certain polyethylene-coated composite packaging materials are virtually moisture-proof. Table 10.19 compares the moisture vapor transmission rates of various coatings and films. Polyethylene powder coatings are used on chemical processing and food-handling equipment.

Polyimides. Polyimide coatings have excellent long-term thermal stability, wear, mar and moisture resistance, as well as electrical properties. They are high in price.

TABLE 10.19 Moisture-Vapor Transmission Rates per 24-h Period of Coatings and Films in g/(mil)(in^2)

Coating or film	MVTR	Information source
Epoxy-anhydride	2.38	Autonetics data (25°C)
Epoxy-aromatic amine	1.79	Autonetics data (25°C)
Neoprene	15.5	Baer (39°C)
Polyurethane (Magna X-500)	2.4	Autonetics data (25°C)
Polyurethane (isocyanate-polyester)	8.72	Autonetics data (25°C)
Olefane,* polypropylene	0.70	Avisun data
Cellophane (type PVD uncoated film)	134	DuPont
Cellulose acetate (film)	219	DuPont
Polycarbonate	10	FMC data
Mylar†	1.9	Baer (39°C)
	1.8	DuPont data
Polystyrene	8.6	Baer (39°C)
	9.0	Dow data
Polyethylene film	0.97	Dow data (1-mil film)
Saran resin (F120)	0.097–0.45	Baer (39°C)
Polyvinylidene chloride	0.15	Baer (2-mil sample, 40°C)
Polytetrafluoroethylene (PTFE)	0.32	Baer (2-mil sample 40°C)
PTFE, dispersion cast	0.2	DuPont data
Fluorinated ethylene propylene (FEP)	0.46	Baer (40°C)
Polyvinyl fluoride	2.97	Baer (40°C)
Teslar	2.7	DuPont data
Parylene N	14	Union Carbide data (2-mil sample)
Parylene C	1	Union Carbide data (2-mil sample)
Silicone (RTV 521)	120.78	Autonetics data
Methyl phenyl silicone	38.31	Autonetics data
Polyurethane (AB0130-002)	4.33	Autonetics data
Phenoxy	3.5	Lee, Stoffey, and Neville
Alkyd-silicone (DC-1377)	6.47	Autonetics data
Alkyd-silicone (DC-1400)	4.45	Autonetics data
Alkyd-silicone	6.16–7.9	Autonetics data
Polyvinyl fluoride (PT-207)	0.7	Product Techniques Incorp.

*Trademark of Avisun Corporation, Philadelphia, Pa.
†Trademark of E. I. du Pont de Nemours & Company, Wilmington, Del.

Uses. Polyimide coatings are used in electrical applications as insulating varnishes and magnet wire enamels in high-temperature, high-reliability applications. They are also used as alternatives to fluorocarbon coatings on cookware, as shown in Fig. 10.5.

Silicones. Silicone resins are high in price and are used alone or as modifiers to upgrade other resins. They are noted for their high temperature resistance, moisture resistance, and weatherability. They can be hard or elastomeric with baking or room temperature curing.

Uses. Silicones are used in high-temperature coatings for exhaust stacks, ovens, and space heaters. Figure 10.6 shows silicone coatings on fireplace equipment. They are also used as conformal coatings for printed wiring boards, moisture repellants for masonry, weather-resistant finishes for outdoors, and thermal control coatings for space vehicles. The thermal conductivities of coatings are listed in Table 10.20.

10.6 Application Methods

The selection of an application method is as important as the selection of the coating itself. Basically, the application methods for industrial liquid coatings and finishes are dipping, flow coating, and spraying, although some coatings are applied by brushing, rolling, printing, and silk screening. The application methods for powder coatings and finishes are fluidized beds, electrostatic fluidized beds, and electrostatic spray outfits. In these times of environmental awareness, regulation, and compliance, it is mandatory that coatings be applied in the most efficient manner.[3] Not only will this help meet the air-quality standards, but it will also reduce material costs. The advantages and disadvantages of various coating application methods are given in Table 10.21.

Liquid spray coating equipment can be classified by its atomizing method: air, hydraulic, or centrifugal. These can be subclassified into air atomizing, airless, airless electrostatic, air-assisted airless electrostatic, rotating electrostatic disks and bells, and high-volume, low-pressure types. While liquid dip coating equipment is usually simple, electrocoating equipment is fairly complex using electrophoresis as the driving force. Other liquid coating methods include flow coating, which can be manual or automated, roller coating, curtain coating, and centrifugal coating. Equipment for applying powder coatings is not as diversified as for liquid coatings. It can only be classified as fluidized bed, electrostatic fluidized bed, and electrostatic spray.

It is important to note that environmental and worker safety regulations can be met, hazardous and nonhazardous wastes can be reduced, and money can be saved by using compliance coatings (those

Figure 10.5 Polyimide coating is used as a protective finish on the inside of aluminum, stainless steel, and other cookware. (*Courtesy of Mirro Aluminum Company.*)

Figure 10.6 Silicone coatings are used as heat-stable finishes for severe high-temperature applications such as fireplace equipment, exhaust stacks, thermal control coatings for spacecraft, and wall and space heaters. (*Courtesy of Copper Development Association.*)

TABLE 10.20 Thermal Conductivities of Coatings

Material	*k* value,* cal/(s)(cm^2) (°C/cm) × 10^4	Source of information
Unfilled plastics:		
Acrylic	4–5	*a*
Alkyd	8.3	*a*
Depolymerized rubber	3.2	H. V. Hardman, DPR Subsidiary
Epoxy	3–6	*b*
Epoxy (electrostatic spray coating)	6.6	Hysol Corporation, DK-4
Epoxy (electrostatic spray coating)	2.9	Minnesota Mining & Manufacturing, No. 5133
Epoxy (Epon† 828, 71.4% DEA, 10.7%)	5.2	
Epoxy (cured with diethylenetriamine)	4.8	*c*
Fluorocarbon (Teflon TFE)	7.0	DuPont
Fluorocarbon (Teflon FEP)	5.8	DuPont
Nylon	10	*d*
Polyester	4–5	*a*
Polyethylenes	8	*a*
Polyimide (Pyre-ML enamel)	3.5	*e*
Polyimide (Pyre-ML varnish)	7.2	*f*
Polystyrene	1.73–2.76	*g*
Polystyrene	2.5–3.3	*a*
Polyurethane	4–5	*n*
Polyvinyl chloride	3–4	*a*
Polyvinyl formal	3.7	*a*
Polyvinylidene chloride	2.0	*a*
Polyvinylidene fluoride	3.6	*h*
Polyxylylene (Parylene N)	3	Union Carbide
Silicones (RTV types)	5–7.5	Dow Corning Corporation
Silicones (Sylgard types)	3.5–7.5	Dow Corning Corporation
Silicones (Sylgard varnishes and coatings)	3.5–3.6	Dow Corning Corporation
Silicone (gel coating)	3.7	Dow Corning Corporation
Silicone (gel coating)	7 (150°C)	Dow Corning Corporation
Filled plastics:		
Epon 828/diethylenetriamine = A	4	*b*
A + 50% silica	10	*b*
A + 50% alumina	11	*b*
A + 50% beryllium oxide	12.5	*b*
A + 70% silica	12	*b*
A + 70% alumina	13	*b*
A + 70% beryllium oxide	17.8	*b*
Epoxy, flexibilized = B	5.4	*i*
B + 66% by weight tabular alumina	18.0	*i*
B + 64% by volume tabular alumina	50.0	*i*
Epoxy, filled	20.2	Emerson & Cuming, 2651 ft
Epoxy (highly filled)	15–20	Wakefield Engineering Company
Polyurethane (highly filled)	8–11	International Electronic Research Company

TABLE 10.20 Thermal Conductivities of Coatings *(Continued)*

Material	*k* value,* cal/(s)(cm^2) (°C/cm) × 10^4	Source of information
Other materials used in electronic assemblies:		
Alumina ceramic	256–442 (20–212°F)	*a*
Aluminum	2767–5575	*a*
Aluminum oxide (alumina), 96%	840	*i*
Beryllium oxide, 99%	5500	*i*
Copper	8095–9334	*a*
Glass (Borosill, 7052)	28	*k*
Glass (pot-soda-lead, 0120)	18	*k*
Glass (silica, 99.8% SiO_2)	40	*l*
Gold	7104 (20–212°F)	*a*
Kovar	395	*m*
Mica	8.3–16.5	*a*
Nichrome‡	325	*m*
Silica	40	*k*
Silicon nitride	359	*m*
Silver	9995 (20–212°F)	*a*
Zircon	120–149	*a*

*All values are at room temperature unless otherwise specified.

†Trademark of Shell Chemical Company, New York, N.Y.

‡Trademark of Driver-Harris Company, Harrison, N.J.

[a]*Materials Engineering,* Materials Selector Issue, vol. 66, no. 5, Chapman-Reinhold Publication, mid-October 1967.

[b]D. C. Wolf, *Proceedings, National Electronics and Packaging Symposium,* New York, June 1964.

[c]H. Lee and K. Neville, *Handbook of Epoxy Resins,* McGraw-Hill, New York, 1966.

[d]R. Davis, *Reinforced Plastics,* October 1962.

[e]*DuPont Technical Bulletin* 19, Pyre-ML Wire Enamel, August 1967.

[f]*DuPont Technical Bulletin* 1, Pyre-ML Varnish RK-692, April 1966.

[g]W. C. Teach and G. C. Kiessling, *Polystyrene,* Reinhold Publishing Corporation, New York, 1960.

[h]W. S. Barnhart, R. A. Ferren, and H. Iserson, 17th ANTEC of SPE, January 1961.

[i]A. J. Gershman and J. R. Andreotti, *Insulation,* September 1967.

[j]*American Lava Corporation* Chart 651.

[k]E. B. Shand, *Glass Engineering Handbook,* McGraw-Hill, 1958.

[l]W. D. Kingery, "Oxides for High Temperature Applications," *Proceedings, International Symposium,* Asilomar, Calif., October 1959, McGraw-Hill, New York, 1960.

[m]W. H. Kohl, *Handbook of Materials and Techniques for Vacuum Devices,* Reinhold Publishing Company, New York, 1967.

[n]*Modern Plastics Encyclopedia,* McGraw-Hill, New York, 1968.

TABLE 10.21 Application Methods for Coatings

Method	Advantages	Limitations	Typical applications
Spray	Fast, adaptable to varied shapes and sizes. Equipment cost is low.	Difficult to completely coat complex parts and to obtain uniform thickness and reproducible coverage.	Motor frames and housings, electronic enclosures, circuit boards, electronic modules.
Dip	Provides thorough coverage, even on complex parts such as tubes and high-density electronic modules.	Viscosity and pot life of dip must be monitored. Speed of withdrawal must be regulated for consistent coating thickness.	Small- and medium-sized parts, castings, moisture and fungus proofing of modules, temporary protection of finished machined parts.
Brush	Brushing action provides good "wetting" of surface, resulting in good adhesion. Cost of equipment is lowest.	Poor thickness control; not for precise applications. High labor cost.	Coating of individual components, spot repairs, or maintenance.
Roller	High-speed continuous process; provides excellent control on thickness.	Large runs of flat sheets or coil stock required to justify equipment cost and setup time. Equipment cost is high.	Metal decorating of sheet to be used to fabricate cans, boxes.
Impregnation	Results in complete coverage of intricate and closely spaced parts. Seals fine leaks or pores.	Requires vacuum or pressure cycling or both. Special equipment usually required.	Coils, transformers, field and armature windings, metal castings, and sealing of porous structures.
Fluidized bed	Thick coatings can be applied in one dip. Uniform coating thickness on exposed surfaces. Dry materials are used, saving cost of solvents.	Requires preheating of part to above fusion temperature of coating. This temperature may be too high for some parts.	Motor stators; heavy-duty electrical insulation on castings, metal substrates for circuit boards, heat sinks.

Screen-on	Deposits coating in selected areas through a mask. Provides good pattern deposition and controlled thickness.	Requires flat or smoothly curved surface. Preparation of screens is time-consuming.	Circuit boards, artwork, labels, masking against etching solution, spot insulation between circuitry layers or under heat sinks or components.
Electrocoating	Provides good control of thickness and uniformity. Parts wet from cleaning need not be dried before coating.	Limited number of coating types can be used; compounds must be specially formulated ionic polymers. Often porous, sometimes nonadherent.	Primers for frames and bodies, complex castings such as open work, motor end bells.
Vacuum deposition	Ultrathin, pinhole-free films possible. Selective deposition can be made through masks.	Thermal instability of most plastics; decomposition occurs on products. Vacuum control needed.	Experimental at present. Potential use is in microelectronics, capacitor dielectrics.
Electrostatic spray	Highly efficient coverage and use of paint on complex parts. Successfully automated.	High equipment cost. Requires specially formulated coatings.	Heat dissipators, electronic enclosures, open-work grills and complex parts.

that meet the VOC emission standards) in equipment having the highest transfer efficiency (the percentage of the coating used which actually coats the product and is not otherwise wasted). The theoretical transfer efficiencies (TE) of coating application equipment are indicated in the text and in Table 10.22, where they are listed in descending order.[4] The aforementioned transfer efficiencies are meant to be used only as guidelines. Actual transfer efficiencies are dependent on a number of factors that are unique to each coating application line.

In the selection of a coating method and equipment, the product's size, configuration, intended market, and appearance must be considered. To aid in the selection of the most efficient application method, each will be discussed in greater detail.

10.6.1 Dip coating

Dip coating (95 to 100% TE) is a simple coating method where products are dipped in a tank of coating material, withdrawn, and allowed to drain in the solvent-rich area above the coating's surface and then allowed to dry. The film thickness is controlled by viscosity, flow, percent solids by volume, and rate of withdrawal. This simple process can also be automated with the addition of a drain-off area, which allows excess coating material to flow back to the dip tank.

Dip coating is a simple, quick method that does not require sophis-

TABLE 10.22 Theoretical Transfer Efficiencies TE of Coating Application Methods

Coating method	TE, %
Autodeposition	95–100
Centrifugal coating	95–100
Curtain coating	95–100
Electrocoating	95–100
Fluidized-bed powder	95–100
Electrostatic fluidized-bed powder	95–100
Electrostatic-spray powder	95–100
Flow coating	95–100
Roller coating	95–100
Dip coating	95–100
Rotating electrostatic disks and bells	80–90
Airless electrostatic spray	70–80
Air-assisted airless electrostatic spray	70–80
Air electrostatic spray	60–70
Airless spray	50–60
High-volume, low-pressure spray	40–60
Air-assisted airless spray	40–60
Multicomponent spray	30–70
Air-atomized spray	30–40

ticated equipment. The disadvantages of dip coating are film thickness differential from top to bottom, resulting in the so-called wedge effect; fatty edges on lower parts of products; and runs and sags. Although this method coats all surface areas, solvent reflux can cause low film build. Light products can float off the hanger and hooks and fall into the dip tank. Solvent-containing coatings in dip tanks and drain tunnels must be protected by fire extinguishers and safety dump tanks. The fire hazard can be eliminated by using waterborne coatings.

10.6.2 Electrocoating

Electrocoating (95 to 100% TE) is a sophisticated dipping method commercialized in the 1960s to solve severe corrosion problems in the automotive industry. In principle, it is similar to electroplating, except that organic coatings, rather than metals, are deposited on products from an electrolytic bath. Electrocoating can be either anodic (deposition of coatings on the anode from an alkaline bath) or cathodic (deposition of coatings on the cathode from an acidic bath). The bath is aqueous and contains very little VOCs. The phenomenon called *throwing power* causes inaccessible areas to be coated with uniform film thicknesses. Electrocoating has gained a significant share of the primer and one-coat enamel coatings market.

Advantages of the electrocoating method include environmental acceptability owing to decreased solvent emissions and increased corrosion protection to inaccessible areas. It is less labor intensive than other methods, and it produces uniform film thickness from top to bottom and inside and outside on products with a complex shape. Disadvantages are high capital equipment costs, high material costs, and more thorough pretreatment. Higher operator skills are required.

10.6.3 Spray coating

Spray coating (30 to 90% TE), which was introduced to the automobile industry in the 1920s, revolutionized industrial painting. The results of this development were increased production and improved appearance. Electrostatics, which were added in the 1940s, improved transfer efficiency and reduced material consumption. Eight types of spray-painting equipment are discussed in this section. The transfer efficiencies listed are theoretical. The actual transfer efficiency depends on many variables, including the size and configuration of the product and the airflow in the spray booth.

Rotating electrostatic disks and bell spray coating. *Rotating spray coaters* (80 to 90% TE) rely on centrifugal force to atomize droplets of liquid as they leave the highly machined, knife-edged rim of an electrically charged rotating applicator. The new higher-rotational speed applicators will atomize high-viscosity, high-solids coatings (65% volume solids and higher). Disk-shaped applicators are almost always used in the automatic mode, with vertical reciprocators, inside a loop in the conveyor line. Bell-shaped applicators are used in automated systems in the same configurations as spray guns and can also be used manually.

An advantage of rotating disk and bell spray coating is its ability to atomize high-viscosity coating materials. A disadvantage is maintenance of the equipment.

High-volume, low-pressure spray coating. *High-volume, low-pressure (HVLP) spray coaters* (40 to 60% TE) are a development of the early 1960s that has been upgraded. Turbines rather than pumps are now used to supply high volumes of low-pressure heated air to the spray guns. Newer versions use ordinary compressed air. The air is heated to reduce the tendency to condense atmospheric moisture and to stabilize solvent evaporation. Low atomizing pressure results in lower droplet velocity, reduced bounce back, and reduced overspray.

The main advantage of HVLP spray coating is the reduction of overspray and bounce back and the elimination of the vapor cloud usually associated with spray painting. A disadvantage is the poor appearance of the cured film.

Airless electrostatic spray coating. The *airless electrostatic spray coating method* (70 to 80% TE) uses airless spray guns with the addition of a dc power source that electrostatically charges the coating droplets.

Its advantage over airless spray is the increase in transfer efficiency due to the electrostatic attraction of charged droplets to the product.

Air-assisted airless electrostatic spray coating. The *air-assisted airless electrostatic spray coating method* (70 to 80% TE) is a hybrid of technologies. The addition of atomizing air to the airless spray gun allows the use of high-viscosity, high-solids coatings. Although the theoretical transfer efficiency is in a high range, it is lower than that of airless electrostatic spray coating because of the higher droplet velocity.

The advantage of using the air-assisted airless electrostatic spray method is its ability to handle high-viscosity materials. An additional advantage is better spray pattern control.

Air electrostatic spray coating. The *air electrostatic spray coating method* (60 to 70% TE) uses conventional equipment with the addition of electrostatic charging capability. The atomizing air permits the use of most high-solids coatings.

Air electrostatic spray equipment has the advantage of being able to handle high-solids materials. This is overshadowed by the fact that it has the lowest transfer efficiency of the electrostatic spray coating methods.

Airless spray coating. When it was introduced, *airless spray coating* (50 to 60% TE) was an important paint-saving development. The coating material is forced by hydraulic pressure through a small orifice in the spray gun nozzle. As the liquid leaves the orifice, it expands and atomizes. The droplets have low velocities because they are not propelled by air pressure as in conventional spray guns. To reduce the coating's viscosity without adding solvents, in-line heaters were added.

Advantages of airless spray coating are that less solvent used, less overspray, less bounce back, and compensation for seasonal ambient air temperature and humidity changes. A disadvantage is its slower coating rate.

Multicomponent spray coating. *Multicomponent spray coating equipment* (30 to 70% TE) is used for applying fast-curing coating system components simultaneously. Since they can be either hydraulic or air-atomizing, their transfer efficiencies vary from low to medium. They have two or more sets of supply and metering pumps to transport components to a common spray head.

Their main advantage, the ability to apply fast-curing multicomponent coatings, can be overshadowed by disadvantages in equipment cleanup, maintenance, and low transfer efficiency.

Air-atomized spray coating. *Air-atomized spray coating equipment* (30 to 40% TE) has been used to apply coatings and finishes to products since the 1920s. A stream of compressed air mixes with a stream of liquid coating material, causing it to atomize or break up into small droplets. The liquid and air streams are adjustable, as is the spray pattern, to meet the finishing requirements of most products. This equipment is still being used.

The advantage of the air-atomized spray gun is that a skilled operator can adjust fluid flow, air pressure, and coating viscosity to apply a high-quality finish on most products. The disadvantages are its low transfer efficiency and ability to spray only low-viscosity coatings, which emit great quantities of VOCs to the atmosphere.

10.6.4 Powder coating

Powder coating (95 to 100% TE), developed in the 1950s, is a method for applying finely divided, dry, solid resinous coatings by dipping products in a fluidized bed or by spraying them electrostatically. The fluidized bed is essentially a modified dip tank. When charged powder particles are applied during the electrostatic spraying method, they adhere to grounded parts until fused and cured. In all cases the powder coating must be heated to its melt temperature, where a phase change occurs, causing it to adhere to the product and fuse to form a continuous coating film. Elaborate reclaiming systems to collect and reuse oversprayed material in electrostatic spray powder systems boost transfer efficiency. Since the enactment of the air-quality standards, this method has grown markedly.

Fluidized bed powder coating. *Fluidized bed powder coating* (95 to 100% TE) is simply a dipping process that uses dry, finely divided plastic materials. A *fluidized bed* is a tank with a porous bottom plate, as illustrated in Fig. 10.7. The plenum below the porous plate supplies low-pressure air uniformly across the plate. The rising air surrounds and suspends the finely divided plastic powder particles, so the powder-air mixture resembles a boiling liquid. Products that are preheated above the melt temperature of the material are dipped in the fluidized bed, where the powder melts and fuses into a continuous coating. Thermosetting powders often require additional heat to cure the film on the product. The high transfer efficiency results from little dragout and consequently no dripping. This method is used to apply heavy coats, 3 to 10 mil, in one dip, uniformly, to complex-shaped products. The film thickness is dependent on the powder chemistry, preheat temperature, and dwell time. It is possible to build a film thickness of 100 mil using higher preheat temperatures and multiple dips. An illustration of film buildup is presented in Fig. 10.8.

Advantages of fluidized bed powder coating are uniform and reproducible film thicknesses on all complex-shaped product surfaces. Another advantage is a heavy coating in one dip. A disadvantage of this method is the 3-mil minimum thickness required to form a continuous film.

Electrostatic fluidized bed powder coating. An *electrostatic fluidized bed* (95 to 100% TE) is essentially a fluidized bed with a high-voltage dc grid installed above the porous plate to charge the finely divided particles. Once charged, the particles are repelled by the grid and repel each other, forming a cloud of powder above the grid. These electrostatically charged particles are attracted to and coat products that are at ground potential. Film thicknesses of $1^1/_2$ to 5 mil are possible on cold parts, and 20 to 25 mil are possible on heated parts.

The advantage of the electrostatic fluidized bed is that small products, such as electrical components, can be coated uniformly and quickly. The disadvantages are that the product size is limited and inside corners have low film thicknesses owing to the well-known fara-

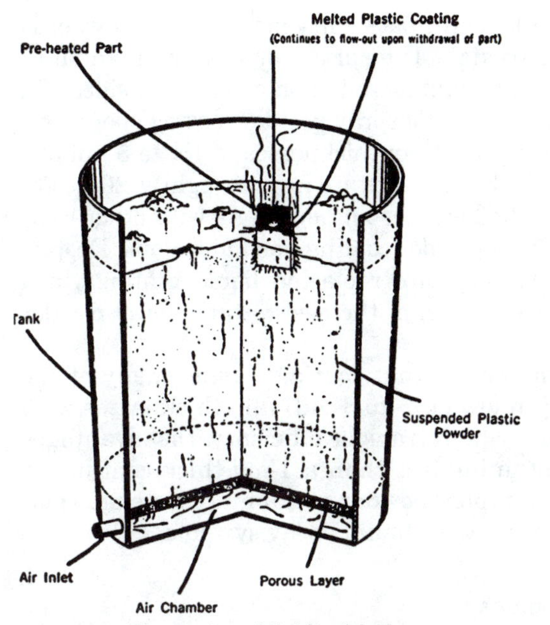

Figure 10.7 Illustration of fluidized-bed process principle.

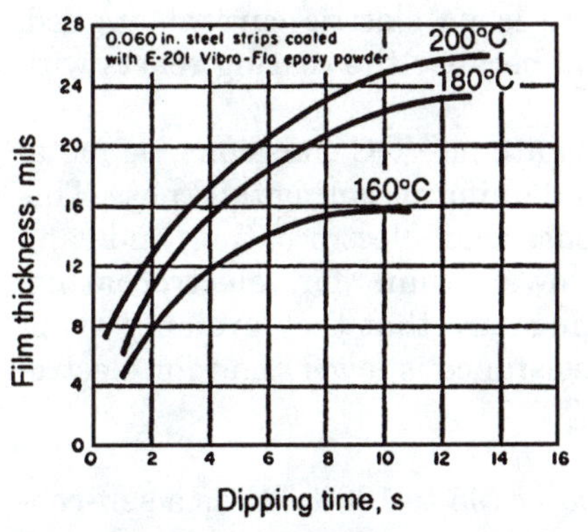

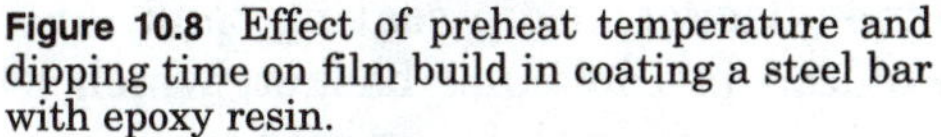

Figure 10.8 Effect of preheat temperature and dipping time on film build in coating a steel bar with epoxy resin.

day cage effect.

Electrostatic spray powder coating. *Electrostatic spray powder coating* (95 to 100% TE) is a method for applying finely divided, electrostatically charged plastic particles to products that are at ground potential. A powder-air mixture from a small fluidized bed in the powder reservoir is supplied by a hose to the spray gun, which has a charged electrode in the nozzle fed by a high-voltage dc power pack. In some cases the powder is electrostatically charged by friction. The spray guns can be manual or automatic, fixed or reciprocating, and mounted on one or both sides of a conveyorized spray booth. Electrostatic spray powder coating operations use collectors to reclaim overspray. Film thicknesses of $1^1/_2$ to 5 mil can be applied on cold products. If the products are heated slightly, 20- to 25-mil-thick coatings can be applied on these products. As with other coating methods, electrostatic spray powder coating has limitations. Despite these limitations, powder coatings are replacing liquid coatings in a growing number of cases. A variation of the electrostatic spray powder coater is the electrostatic disk.

The advantage of this method is that coatings, using many of the resin types, can be applied in low ($1^1/_2$- to 3-mil) film thicknesses with no VOC emission at extremely high transfer efficiency. Disadvantages include the difficulty in obtaining less than a 1-mil-thick continuous coating and, owing to the complex powder reclaiming systems, color changes are more difficult than with liquid spray systems.

10.6.5 Other application methods

Autodeposition coating. *Autodeposition* (95 to 100% TE) is a dipping method where coatings are applied on the product from an aqueous solution. Unlike electrocoating, there is no electric current applied. Instead, the driving force is chemical because the coating reacts with the metallic substrate.

Advantages of autophoretic coating are no VOC emissions, no metal pretreatment other than cleaning, and uniform coating thickness. This technique requires 30% less floor space than electrocoating, and capital equipment costs are 25% lower than for electrocoating. Disadvantages of autophoretic coatings are that they are only available in dark colors, and corrosion resistance is lower than for electrocoated products.

Centrifugal coating. A *centrifugal coater* (95 to 100% TE) is a self-contained unit. It consists of an inner basket, a dip coating tank, and exterior housing. Products are placed in the inner basket, which is dipped into the coating tank. The basket is withdrawn and spun at a speed high enough to remove the excess coating material by centrifugal force. This causes the coating to be flung onto the inside of the exterior housing, from which it drains back into the dip coating tank.

The advantage of centrifugal coating is that large numbers of small parts can be coated at the same time. The disadvantage is that the appearance of the finish is a problem because the parts touch each other.

Flow coating. In a *flow coater* (95 to 100% TE), the coating material is pumped through hoses and nozzles onto the surfaces of the product, from which the excess drains into a reservoir to be recycled. Flow coaters can be either automatic or manual. Film thickness is controlled by the viscosity and solvent balance of the coating material. A continuous coater is an advanced flow coater using airless spray nozzles mounted on a rotating arm in an enclosure.

Advantages of flow coating are high transfer efficiency and low volume of paint in the system. Products will not float off hangers, and extremely large products can be painted. As with dip coating, the disadvantages of flow coating are coating thickness control and solvent refluxing.

Curtain coating. *Curtain coating* (95 to 100% TE), which is similar to flow coating, is used to coat flat products on conveyorized lines. The coating falls from a slotted pipe or flows over a weir in a steady stream or curtain while the product is conveyed through it. Excess material is collected and recycled through the system. Film thickness is controlled by coating composition, flow rates, and line speed.

The advantage of curtain coating is uniform coating thickness on flat products with high transfer efficiency. The disadvantage is the inability to uniformly coat three-dimensional objects.

Roller coating. *Roller coating* (95 to 100% TE), which is used mainly by the coil coating industry for prefinishing metal coils that will later be formed into products, has seen steady growth. It is also used for finishing flat sheets of material. There are two types of roller coaters—direct and reverse—depending on the direction of the applicator roller relative to the direction of the substrate movement. Roller coating can apply multiple coats to the front and back of coil stock with great uniformity.

The advantages of roller coating are consistent film thickness and elimination of painting operations at a fabricating plant. The disadvantages are limited metal thickness, limited bend radius, and corrosion of unpainted cut edges.

10.7 Curing

No dissertation on organic coatings and finishes is complete without mentioning film formation and cure. It is not the intent of this chapter to fully discuss the mechanisms, which are more important to researchers and formulators than to end users, but rather to show that differences exist and to aid the reader in making selections. Most of the organic coating resins are liquid, which cure or dry to form solid films. They are classified as thermoplastic or thermosetting.

Thermoplastic resins dry by solvent evaporation and will soften when heated and harden when cooled. Thermosetting resins will not soften when heated after they are cured. Another classification of coatings is by their various film-forming mechanisms, such as solvent evaporation, coalescing, phase change, and conversion. Coatings are also classified as *room temperature curing,* sometimes called *air drying*; or *heat curing,* generally called *baking,* or *force drying,* which uses elevated temperatures to accelerate air drying. Thermoplastic and thermosetting coatings can be both air drying and baking.

10.7.1 Air drying

Air drying coatings will form films and cure at room or ambient temperatures (20°C) by the mechanisms described in this section.

Solvent evaporation. Thermoplastic coating resins that form films by solvent evaporation are shellac and lacquers such as nitrocellulose, acrylic, styrene-butadiene, and cellulose acetate butyrate.

Conversion. In these coatings, films are formed as solvents evaporate, and they cure by oxidation, catalyzation, or cross-linking. Thermosetting coatings cross-link to form films at room temperature by oxidation or catalyzation. Oxidative curing of drying oils and oil-modified resins can be accelerated by using catalysts. Monomeric materials can form films and cure by cross-linking with polymers in the presence of catalysts, as in the case of styrene monomers and polyester resins. Epoxy resins will cross-link with polyamide resins to form films and cure. In the moisture curing polyurethane resin coating systems, airborne moisture starts a reaction in the vehicle, resulting in film formation and cure.

Coalescing. Emulsion or latex coatings, such as styrene-butadiene, acrylic ester, and vinyl acetate acrylic, form films by coalescing and dry by solvent evaporation.

10.7.2 Baking

Baking coatings will form films at room temperature, but require elevated temperatures (150 to 200°C) to cure. Most coatings are baked in gas-fired ovens, although oil-fired ovens are also used. Steam-heated and electric ovens are used on a limited basis. Both electric and gas-fired infrared elements are used as heat sources in paint bake ovens.

Conversion. The cure of many oxidative thermosetting coatings is accelerated by heating. In other resin systems, such as thermosetting acrylics and alkyd melamines, the reactions do not occur below temperature thresholds of 135°C or higher. Baking coatings (those that require heat to cure) are generally tougher than air-drying coatings. In some cases the cured films are so hard and brittle that they must be modified with other resins.

Phase change. Thermoplastic coatings that form films by phase changes, generally from solid to liquid then back to solid, are polyolefins, waxes, and polyamides. Plastisols and organisols undergo phase changes from liquid to solid during film formation. Fluidized bed powder coatings, both thermoplastic and thermosetting, also undergo phase changes from solid to liquid to solid during film formation and cure.

10.7.3 Radiation curing

Films are formed and cured by bombardment with ultraviolet and electron beam radiation with little increase in surface temperature. Infrared radiation, on the other hand, increases the surface temperature of films and is therefore a baking process. The most notable radiation curing is UV curing. This process requires the use of specially formulated coatings. They incorporate photoinitiators and photosensitizers which respond to specific wavelengths of the spectrum to cause a conversion reaction. Curing is practically instantaneous with little or no surface heating. It is, therefore, useful in coating temperature-sensitive substrates. Since the coatings are 100% solids, there are no VOCs. Although most UV coating are clears (unpigmented), paints can also be cured. Figure 10.9 shows radiation curing equipment.

10.7.4 Force drying

In many cases the cure rate of thermoplastic and thermosetting coatings can be accelerated by exposure to elevated temperatures that are below those considered to be baking temperatures.

10.7.5 Vapor curing

Vapor curing is essentially a catalyzation or cross-linking conversion method for two-component coatings. The product is coated with one component of the coating in a conventional manner. It is then placed in an enclosure filled with the other component, the curing agent, in vapor form. It is in this enclosure that the reaction occurs.

10.7.6 Reflowing

Although not actually a curing process, certain thermoplastic coating films will soften and flow to become smooth and glossy at elevated

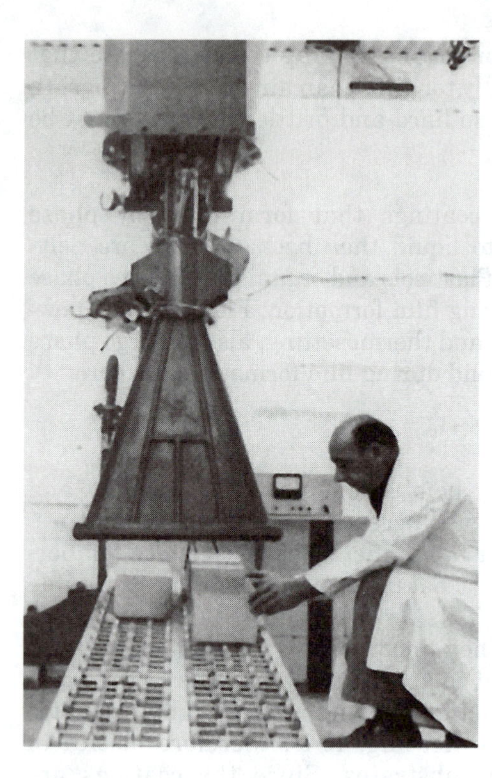

Figure 10.9 Radiation curing is fast, allowing production-line speeds of 2000 ft/min. This technique takes place at room temperature, and heat-sensitive wooden and plastic products and electronic components and assemblies can be given the equivalent of a baked finish at speeds never before possible. (*Courtesy of Radiation Dynamics.*)

temperatures. This technique is used on acrylic lacquers by the automotive industry to eliminate buffing.

10.8 Summary

The purpose of this chapter is to aid readers in selecting surface preparation methods, coating materials, and application methods. It also acquaints them with curing methods and helps them to comply with environmental regulations.

Coating selection is not easy, owing to the formulating versatility of modern coating materials. This versatility also contributes to one of their faults, which is the possible decline in one performance property when another is enhanced. Because of this, the choice of a coating must be based on specific performance properties and not on generalizations. This choice is further complicated by the need to comply with governmental regulations. Obviously, the choice of compliance coatings, electrocoating, waterborne, high solids, powder, radiation curable, and vapor cure is well advised.

To apply coatings in the most effective manner, the product's size, shape, ultimate appearance, and end use must be addressed. This chapter emphasizes the importance of transfer efficiency in choosing a coating method.

To meet these requirements, product finishers have all the tools at their disposal. They can choose coatings that apply easily, coat uniformly, cure rapidly and efficiently, and comply with governmental regulations at lower costs. By applying compliance coatings using methods that have high transfer efficiencies, finishers will not only comply with air- and water-quality standards, but they will also provide a safe workplace and decrease the generation of hazardous wastes.[4] Reducing hazardous wastes, using less material, and emitting fewer VOCs can also effect significant cost savings.

References

1. C. P. Izzo, "Today's Paint Finishes: Better than Ever," *Product Finishing Magazine,* vol. 51, no. 1, October 1986, pp. 48–52.
2. *Coatings Encyclopedic Dictionary,* Federation of Societies for Coatings Technology, 1995.
3. C. Izzo, "How Are Coatings Applied," in *Products Finishing Directory,* Gardner Publications, Cincinnati, Ohio, 1991.
4. C. Izzo, "Overview of Industrial Coating Materials," in *Products Finishing Directory,* Gardner Publications, Cincinnati, Ohio, 1991.

Chapter

11

Plastics Testing

Dr. Ranganath Shastri

The Dow Chemical Company
Midland, Michigan

11.1 Introduction

Although the terms "testing" and "characterization" are used interchangeably to describe evaluation of various properties of plastics, there is a subtle difference between the two terms. *Characterization* often refers to evaluation of the molecular or structural characteristics of plastics while *testing* is used to refer to evaluation of behavior of plastics in response to the applied external loads, environment, etc.

The testing is not limited to plastics resin form only. Besides the resin itself, very often testing the fabricated part in its final form needs to be an essential part of the design validation step to ensure that end-use performance requirements are adequately realized. Such part tests will have to be application specific and often need to involve testing under actual or simulated service conditions employing specialized and nonstandard methods/procedures.

11.2 The Need for Testing Plastics

It is simple to understand the rationale for testing plastics. For a resin supplier, plastics testing is an integral part of new product development, application development, and quality control. Specifically, the key reasons are

1. Assessment of the material's behavior in order to determine its suitability for desired application(s)

2. Evaluation of whether it meets various compliance requirements, for example, U.S. FDA and UL
3. Evaluation of lot-to-lot quality and consistency of the product being manufactured for quality control purposes
4. Specific evaluations of the material for customer support

The assessment of material's behavior is inclusive of understanding structure-property relationships as well as comparison of the performance characteristics of one product against another. In the case of new product development, understanding the structure-property relationship, thereby enabling the performance envelope of the product decisions for further development effort to be made. Such understanding is crucial to facilitate material design and predict how the product may perform under actual service conditions during use. Intermaterial comparison is essential in exploring substitution opportunities in existing applications.

Quality control of the resins supplied to the customer is an important aspect that the plastics manufacturer/supplier addresses on a regular basis to ensure the quality of the product and to demonstrate lot-to-lot consistency for compliance with the sales specifications. Since it is not practical to test every property for each lot of the material produced, specific properties that are more sensitive to swings in production fluctuations are usually chosen for quality control purposes. These could include

- Melt flow rate (MFR)
- Chemical composition by infrared spectroscopy
- Molecular weight and molecular-weight distribution (MWD) by gel permeation chromatography
- Tensile properties
- Notched impact strength, etc.

It is not too uncommon for internally developed, nonstandard test procedures to be employed for quality control purposes. Both resin and part testing are often carried out by the resin supplier.

The end user's reasons for testing plastics are essentially twofold. The primary reason is to verify the quality of the incoming material to assure that it is within the specifications. The second reason is to validate the part design, which may involve testing the resin as well as the final part.

11.3 Diverse Types of Testing

Testing of plastics is directly linked to the type of data that needs to be evaluated, and the data fall under several categories:

1. Analytical data associated with structural features
2. Data needed to aid in material preselection
3. Data needed for computer-aided engineering (CAE) and computer-aided design (CAD)
4. Data needed to understand processing behavior
5. Data needed for regulatory compliance/approval
6. Data needed to satisfy original equipment manufacturer (OEM) specification
7. Miscellaneous data

11.3.1 Analytical data associated with structural features

Analytical data associated with structural features includes molecular parameters like molecular weight, molecular weight distribution, tacticity, branching or chemical heterogeneity, morphological features like degree of crystallinity, etc. Such data are intended for material design and customer support and typically not shared broadly due to the proprietary nature of the data. Characterization of the analytical data is essential for establishing the structure-property relationships, which is useful in tailoring products designed to yield desired performance characteristics.

11.3.2 Data needed to aid in material preselection

Invariably every resin supplier provides a technical datasheet for each product that is commercially offered. These datasheets typically list a general set of 10 to 15 properties that are intended to represent the types of characteristics in each product, as illustrated in Fig. 11.1.

Since the primary purpose of the datasheet is a sales tool, the data reported in the datasheets generally focus on single-point data only, which is adequate for initial screening or intermaterial comparison. *By their very nature, such data are inadequate for design and engineering analysis, as they bear little relevance to the end-use performance of the product and provide very little insight into how well the plastics will perform in service.*[1–3]

11.3.3 Data needed for CAE and CAD

When plastics are considered for load-bearing applications involving complex shapes, CAD/CAE, utilizing finite element analysis (FEA) techniques, become powerful tools for design engineers in performing engineering analysis to predict the performance. Unfortunately, there are

Product Information

QUESTRA* WA 202

Crystalline Polymers – 30 Percent Glass-Filled, Moderate Heat Polymer for Automotive Electronic Systems

Properties(1)	Test Method	English Units	S. I. Units
Physical			
Specific Gravity	ASTM D 792	1.22	1.22
Water Absorption (24 hrs., 50% RH)	ASTM D 570	0.01%	0.01%
Mold Shrinkage	ASTM D 955	0.003 - 0.004 in/in (0.3 - 0.4%)	0.003 - 0.004 cm/cm (0.3 - 0.4%)
Mechanical (1)			
Tensile Strength			
@ Yield	ASTM D 638	13,400 psi	92 MPa
@Break	ASTM D 638	13,400 psi	92 MPa
Tensile Modulus	ASTM D 638	1,100,000 psi	7,580 MPa
Tensile Elongation @ Break	ASTM D 638	2.0%	2.0%
Flexural Strength	ASTM D 790	23,000 psi	158 MPa
Flexural Modulus	ASTM D 790	1,150,000 psi	7,930 MPa
Notched Izod Impact			
@ 73°F (23°C)	ASTM D 256	1.8 ft-lb/in	96 J/m
@ 0°F (-18°C)	ASTM D 256	1.6 ft-lb/in	85 J/m
Thermal(2)			
Vicat Softening Point	ASTM D 1525	505°F	263°C
Deflection Temp. Under Load			
@ 284 psi (1.82 MPa)	ASTM D 648	400°F	204°C
@ 66 psi (0.45 MPa)	ASTM D 648	500°F	260°C
Coefficient of Linear Thermal Expansion			
@ 72 -171°F (22 - 77°C), (×10⁶)	ASTM D 696	23.5 in/in/°F	42.3 cm/cm/°C
Electrical			
Dielectric Constant	ASTM D 150	3.1	3.1
Dielectric Strength	ASTM D 149	550 V/mil	21.8 kV/mm
Dissipation Factor	ASTM D 150	0.001	0.001

(1) Typical property values not to be construed as sales specifications.
(2) Tests conducted on 0.125 in (3.2 mm) specimen unannealed.

– See "Handling Considerations" reverse side.

Dow Plastics, a business group of The Dow Chemical Company and its subsidiaries Form No. 301-02731-1197X SMG CH 265-001-E-1197X

Figure 11.1 Example of a typical technical datasheet available from resin suppliers.

misconceptions about which properties are essential for the design process. (See also Chap. 7, Sec. 3.)

Whether the CAE tools are used or not, the data required for product design go beyond the single-point data. To effectively design with plastics, more comprehensive data, such as the following, are essential:

- Isothermal stress-strain curves at temperatures and strain rates reflective of service conditions
- Temperature dependency of dynamic modulus
- Isochronous creep curves at ambient and elevated temperatures and several stress levels
- Impact behavior at ambient and subambient temperatures
- Effect of exposure on the behavior of plastics to environments the product is typically exposed to during its service life
- Effect of anisotropy on the material performance
- Viscosity-shear rate data

With operating conditions varying over a rather broad range of temperatures, loading history and environments over the life cycle of the product, using the instantaneous material properties at 23°C and 50% relative humidity reported in the datasheet for design would be a gross mistake. Yet design engineers rely heavily on the datasheet properties because the datasheet is readily available.

The specific material properties relevant for plastics product design are adequately addressed[4–6] and a comprehensive new ISO guide for design data on plastics is currently under development.[7] In the interest of making the information readily available to readers, relevant information is culled here from these works.

The testing involved for generating design data naturally depends on the type of CAE analysis being carried out. The three main types of engineering analysis generally carried out are

1. Structural analysis
2. Manufacturability assessment
3. Part assembly design

Structural analysis is employed in assessing the structural integrity of the designed part over its useful life or in determining the required geometry of the part to ensure part functionality. As functional requirements are often specific to each application, the material properties essential for structural analysis can be classified into two categories: those that are somewhat application-specific and those that are generic in nature. Whether the individual property is application-specific or generic, certain properties are directly employed in design calculations while others are employed more or less for verification of design limits. For example, although parts may fail in service under multiaxial impact loading conditions, the impact energy data can only be used in design verification, at best. Additional examples of properties that are useful only for design verification include fatigue (*S-N*) curves, wear factor, PV limit, retention of properties following exposure to chemicals and solvents, and accelerated aging or ultraviolet (UV) exposure/outdoor weathering.

The essential properties needed for various types of structural design calculations (beam or plate, buckling, pipe, bearings analysis, or some combination thereof) are compiled in Table 11.1. The key design parameters in all the previous analyses, except in the case of bearings, are maximum deflection and maximum or critical stress, in order to determine appropriate part thickness to provide sufficient rigidity and strength to the part. The specific material properties employed in the calculations of each design parameter involved in these analyses are summarized in Table 11.2. The required modulus type for these calculations is determined by the stress type and duration. Tensile or secant modulus is adequate where loading is limited to short duration, with secant modulus justified when the stress-strain behavior exhibited by the plastic is nonlinear. Tensile creep modulus is required where stress is encountered over extended periods of time. Shear modulus is needed where torsional loads are involved.

Processing simulations are employed in assessing the manufacturability of parts from the plastics. Among the various plastics processing methods, injection molding is the most dominant method in practice. As such, the CAE tools for simulation of the injection-molding process are more advanced in terms of number of CAE programs available and their sophistication. Recently, greater emphasis seems to be given to the development of CAE tools for simulation of other processing methods such as extrusion, blow molding, and thermoforming.

Most of the commercially available injection-molding simulation programs allow a two-dimensional (2-D) analysis, incorporating temperature distribution through the thickness dimension. Enhancements touting full three-dimensional (3-D) simulation have only been introduced recently. Both types of simulation programs are rather complex in nature, utilizing a quite rigorous definition of the part geometry and incorporating various viscosity models to describe the flow behavior of polymer melts. Some expertise is required to use these programs. Simple two-dimensional programs, which do not involve such rigorous analysis, are also currently on the market.

The main objective of these methods is to simulate the part filling and post filling steps in order to optimize the manufacturability of the part.

TABLE 11.1 Material Properties Needed for Structural Design Calculations

Plate or beam analysis	Pipe analysis	Buckling analysis	Bearing analysis
Tensile modulus	Tensile modulus	Compressive modulus	PV limit*
Secant modulus	Tensile creep modulus	Secant modulus	Wear factor
Tensile creep modulus	Poisson's ratio	Tensile creep modulus	Coefficient of friction
Shear modulus	Critical stress intensity factor K_{1c}	Shear modulus	Coefficient of thermal expansion
Poisson's ratio	Tensile stress at yield	Poisson's ratio	
Tensile stress at yield	Tensile creep rupture stress	Compression strength	
Tensile creep rupture stress		Tensile stress at yield	
Shear strength			

*No ASTM or ISO standards exist today.

TABLE 11.2 Relevance of Material Properties in Structural Design Calculations

Type of analysis	Design parameter	Relevant properties needed
Plate analysis	Maximum deflection	Modulus (tensile, secant, tensile creep) Poisson's ratio
	Maximum stress	Tensile stress at yield Tensile creep rupture stress
Beam analysis	Maximum deflection	Modulus (tensile, secant, creep, shear) Poisson's ratio
	Maximum stress	Tensile stress at yield Tensile creep rupture stress Shear strength
Buckling analysis	Critical stress	Modulus (compressive, tensile, secant) Poisson's ratio Compression strength Tensile stress at yield
Pipe analysis	Hydrostatic design stress	Tensile stress at yield Tensile creep rupture stress Compression strength Critical stress intensity factor K_{1c}
	Radial displacement	Modulus (tensile, tensile creep) Poisson's ratio
Bearings analysis	PV value	PV limit Coefficient of friction
	Volumetric wear	Wear factor Coefficient of thermal expansion

There are three main types of analysis in injection-molding simulation:

1. Simple mold filling analysis to determine the ability to fill the mold cavity and to assess the pressure requirements.
2. Advanced mold filling, packing, and cooling analysis, which is carried out to optimize the processing conditions or to evaluate design alternatives such as number of gates, proper gate size, its location, etc.

TABLE 11.3 Data Needed for Injection-Molding Simulation—Simple Mold Filling Analysis of Thermoplastics and Thermoplastic Elastomers

Property	Variables
Melt viscosity	Temperature, shear rate
Melt density	
Thermal conductivity	
Specific heat	
Solidification temperature*	
Ejection temperature*	

*Reference temperatures defined by simulation software.

TABLE 11.4 Data Needed for Injection-Molding Simulation—Advanced Mold Filling, Packing, and Cooling Analysis of Thermoplastics and Thermoplastic Elastomers

Property	Variables
Melt viscosity	Temperature, shear rate
Specific volume	Pressure, temperature, cooling rate*
Thermal conductivity	Temperature
Specific heat	Temperature
Solidification temperature†	Pressure, cooling rate
Ejection temperature†	
Crystallization temperature (semicrystalline materials)	Pressure, cooling rate
Enthalpy of crystallization (semicrystalline materials)	Cooling rate

*Predicted by compensating for crystallization kinetics at different cooling rates using DSC.
†Reference temperatures defined by simulation software.

3. Shrinkage and warpage analysis to satisfy tolerances and predict dimensional stability of the manufactured part.

The material properties needed for simple mold filling simulation are listed in Table 11.3. The material properties essential for advanced filling, packing, and cooling simulations are listed in Tables 11.4 through 11.6. The requirements are essentially the same for thermoplastics and thermoplastic elastomers, while in the case of reactive materials, such as thermosets, the main differences are the reactive polymer viscosity in place of melt viscosity data and reaction kinetics data.

The simulation of extrusion generally includes consideration of the melting of the polymer in the barrel, flow of the melt in the die, and the cooling of the extruded shape. There are several simulation packages on the market, employing different viscosity models to describe the flow characteristics of the polymer melt. The material properties needed for simulation of extrusion process are listed in Table 11.7.

The material properties needed for simulation of blow molding, blown film extrusion, and thermoforming are listed in Table 11.8. The

TABLE 11.5 Data Needed for Molding Simulation—Mold Filling, Packing, and Cooling Analysis of Reactive Materials Including Thermosets

Property	Variables
Reactive viscosity	Temperature, time, shear rate
Density, reacted	
Thermal conductivity	Temperature
Specific heat	Temperature
Heat of reaction	Temperature
Isothermal induction time	Temperature
Gelation conversion	Temperature
Reaction kinetics	Temperature, conversion, heating rate

TABLE 11.6 Data Needed for Injection-Molding Simulation—Additional Data for Shrinkage and Warpage Analysis

Property	Variables
Molding shrinkage, parallel	Thickness, processing parameters* p_H, t_H
Molding shrinkage, normal	Thickness, processing parameters* p_H, t_H
Crystallinity (semicrystalline materials)	Cooling rate
Crystallization kinetics (semicrystalline materials)	Temperature, cooling rate
Tensile modulus, parallel	
Tensile modulus, normal	
Poisson's ratio	
In-plane shear modulus	
Coefficient of linear thermal expansion, parallel	Temperature, thickness
Coefficient of linear thermal expansion, normal	Temperature, thickness

*p_H = cavity pressure and t_H = hold time.

common denominator in these processes is the biaxial orientation step involved. While in blow molding and blown film extrusion the biaxial orientation is induced in the melt state, in thermoforming it is induced in the softened state.

The part assembly design addresses the ability to join/assemble the component parts. Where the components are assembled with adhesives, it is important to know the compatibility and strength of adhesion to dissimilar substrates, in addition to the chemical compatibility of the plastic with the specific adhesive and its constituents. If melt bonding methods, like ultrasonic, vibration, or spin-welding processes, are employed, thermal compatibility aspects have to be taken into account. The broad possibilities of mechanical assembly methods include snap fits, press fits, bolts, and threads. The material properties needed for each of these design calculations are listed in Table 11.9.

TABLE 11.7 Material Properties Needed for Simulation of Extrusion

Property	Variables
Melting temperature (semicrystalline materials)	
Glass transition temperature (amorhous materials)	
Enthalpy of fusion (semicrystalline materials)	
Coefficient of friction	Pressure, temperature, slip velocity
Thermal conductivity	Temperature
Specific heat	Temperature
Melt viscosity	Temperature, shear rate pressure
First normal stress difference	Temperature, shear rate
Uniaxial extensional viscosity	Temperature, time, strain rate
Crystallization temperature (semicrystalline materials)	Pressure, cooling rate
Enthalpy of crystallization (semicrystalline materials)	Cooling rate
Crystallization kinetics (semicrystalline materials)	Temperature, cooling rate

TABLE 11.8 Material Properties Needed for Simulation of Blow Molding, Blown Film Extrusion, and Thermoforming

Property	Variables
Uniaxial extensional viscosity	Temperature, time, strain rate
Biaxial extensional viscosity	Temperature, time, strain rate
First normal stress difference	Temperature, shear rate
Thermal conductivity	Temperature
Specific heat	Temperature
Crystallization temperature (semicrystalline materials)	Pressure, cooling rate
Enthalpy of crystallization (semicrystalline materials)	Cooling rate
Crystallization kinetics (semicrystalline materials)	Temperature, cooling rate

The specific material properties employed in the computation of each design parameter involved in these analyses are summarized in Table 11.10. Similar to design calculations in the case of structural analyses, the modulus type suitable for these calculations is determined by the applied stress type and duration.

It is worth noting that, although the exhaustive compilation of data identified here will suffice in most cases, commercial CAE tools often insist on a significantly large volume of additional data that are not

TABLE 11.9 Material Properties Needed for Part Assembly Design

Snap fit analysis:	*Bolt analysis:*
Tensile modulus	Compressive strength
Secant modulus	Tensile creep rupture stress
Creep modulus	Tensile creep strain
Shear modulus	Stress relaxation*
Poisson's ratio	*Thread analysis:*
Tensile strength at yield	Shear strength
Coefficient of friction	Coefficient of friction
Press fit analysis:	Tensile strength at yield
Tensile modulus	*Weldability:*
Compression modulus	Shear strength
Creep modulus	Density
Poisson's ratio	Coefficient of friction
Compression strength	Thermal conductivity
Coefficient of friction	Specific heat
Tensile strength at yield	Crystalline melting temperature
Tensile creep rupture stress	
Stress relaxation*	

*No ASTM or ISO standards exist today.

covered here. This is attributed to many of the models available in the software tools for structural analysis like ABAQUS, MSC/PATRAN, DYNA3D, etc., and processing simulation software tools like MOLDFLOW, C-MOLD, CADMOULD, TM Concept, POLYFLOW, etc.

11.3.4 Data needed to understand processing behavior

In order to assist customers in fabrication of useful articles from plastics, it is important to understand the processability of the resin in terms of maximum processing temperature to avoid degradation and any decomposition products, that is, off gases given off to ensure safety during processing. In addition, processing simulation tools require information on maximum shear stress and shear rate allowable.

11.3.5 Data needed for regulatory compliance

In many applications, where consumer safety is of paramount importance, the product performance criteria may stipulate compliance with various regulatory requirements. The specific requirements are, of course, dependent on the targeted application and should be determined for each application.

TABLE 11.10 Relevance of Material Properties in Structural Design Calculations

Type of analysis	Design parameter	Relevant material properties
Snap fit analysis:		
Cantilever	Maximum deflection	Modulus (tensile, secant, tensile creep) Poisson's ratio
	Deflection force	Modulus (tensile, secant, tensile creep)
	Mating force	Modulus (tensile, secant, tensile creep) Tensile stress at yield Coefficient of friction
Cylindrical	Maximum interference	Modulus (tensile, secant, tensile creep) Poisson's ratio
	Interference stress	Tensile stress at yield
	Deflection force	Modulus (tensile, secant, tensile creep) Poisson's ratio
	Mating force	Modulus (tensile, secant, tensile creep) Coefficient of friction
Torsional	Deflection force	Modulus (tensile, secant, shear) Poisson's ratio
	Permissible shear	Poisson's ratio
Press fit analysis	Allowable interference	Modulus (tensile, secant, compression) Poisson's ratio Coefficient of linear thermal expansion
	Maximum stress in hub	Tensile stress at yield Tensile creep rupture stress Compression strength
Thread analysis	Stripping torque	Shear strength Tensile stress at yield Coefficient of friction
	Stripping stress	Shear strength
	Pullout force	Tensile stress at yield Shear strength
Bolt analysis	Preload stress	Tensile stress at yield Tensile creep rupture stress Compression strength Stress relaxation
	Torque	Tensile stress at yield Coefficient of friction

Targeted application	Industry focus	Regulatory body	Material property
Electrical components (connectors, housings, lighting components)	Information technology Lighting Computer and business equipment	UL, CE	Relative thermal index (RTI) Glow wire temperature
Nondisposable products	Healthcare	U.S. FDA (medical devices)	Biocompatibility Extractables
Applications involving direct food contact	Packaging	U.S. FDA	Extractables

11.3.6 Data needed to satisfy OEM specification

Contractual agreements with major OEMs often require that the material under transaction meet the specifications set forth by the OEM for the incoming material, as part of the quality assurance effort. The resin supplier is required to demonstrate that the shipped material is within the specification. The testing involved is dependent on the OEM and targeted application, to provide proof of compliance.

11.3.7 Miscellaneous data

This category typically includes nonstandard or application-specific properties such as environmental effects on properties of plastics.

Targeted application	Industry focus	Material property
Components involving outdoor or UV exposure	Lawn and garden	UV stability
	Automotive exterior	Gasoline resistance
	Automotive interior (above belt line)	Fogging resistance
Food containers	Packaging	Gas barrier
Disposable components	Healthcare	Autoclaveability Radiation sterilizability
Enclosures for electronic units	Information technology	Electromagnetic interference/radio frequency (EMI/RF) shielding effectiveness

The environmental effects are not routinely reported primarily because of the unique and complex nature of the surrounding environments associated with each application.

11.4 Test Methods for Acquisition and Reporting of Property Data

For testing plastics, a wide spectrum of national standards have been practiced worldwide—for instance, American Society for Testing and Materials (ASTM) standards in the United States, Deutsches Institut für Normung (DIN) in Germany; British Standards Institution (BSI) standards in the United Kingdom, Association Française de Normalisation (AFNOR) standards in France, and Japanese Industrial Standards (JIS) in Japan (Fig. 11.2).

In theory, any of these national standards could achieve global acceptance. However, in reality, none of them is in contention for universal acceptance worldwide because of each's national identity. By contrast, test methods developed by the Geneva-based International

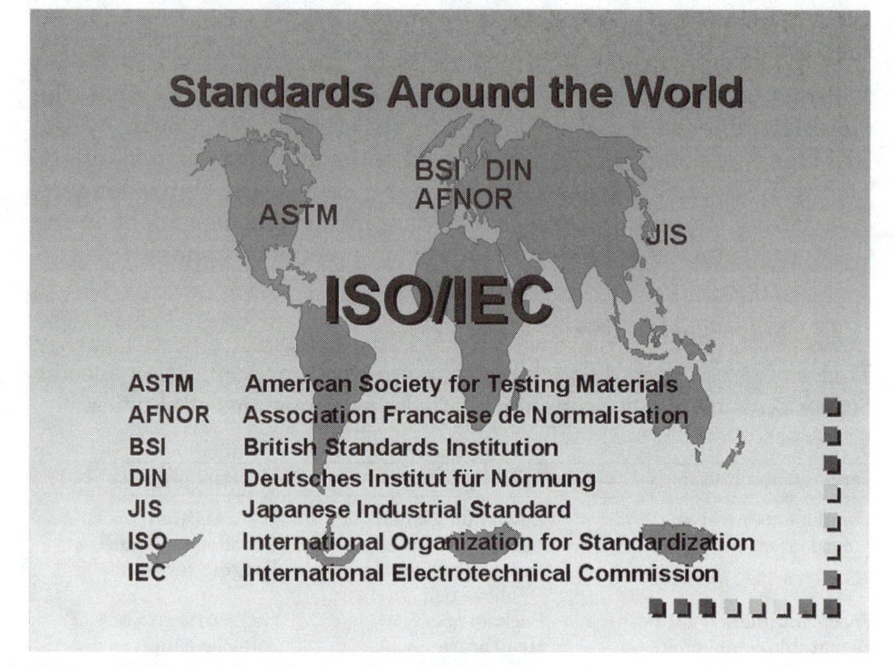

Figure 11.2 Plastics test standards around the world.

Organization for Standardization (ISO), which have been derived and from those developed by ASTM, DIN, BSI, and others, have the greatest chance to provide the basis for consensus on single set of global standards. Test methods developed by International Electrotechnical Commission (IEC) fulfill a similar role where electrical properties are concerned.

Table 11.11 lists the test methods commonly employed for determining the properties of plastics reported in datasheets.

For the design data, the relevant properties that need to be evaluated along with the applicable ISO and ASTM standard methods are summarized in Tables 11.12 and 11.13, respectively. For some of the properties like pvT data, no-flow temperature, ejection temperature, and fatigue in tension, etc., no national or international standards exist today. Efforts are under way to develop industrywide standards for these properties. Suggested test conditions in Tables 11.12 and 11.13 are intended to serve as a guide in establishing specific test conditions for the purpose of developing comparable data.

It should be emphasized that the specific operating conditions (temperature, applied stress or strain, environment, etc., and corresponding duration of such exposures) often vary significantly from one

TABLE 11.11 Key Properties of Plastics Reported in Datasheets and Common Test Methods

Property	Test method in accordance with	
	ISO methods	ASTM methods
Specific gravity/density	ISO 1183	D 792 D1505 (polyolefins)
Water absorption	ISO 62	D 570
Melt flow rate (MFR)	ISO 1133	D 1238
Mold shrinkage	ISO 294-4 (thermoplastics) ISO 2577 (thermosets)	D 955
Tensile properties	ISO 527-1 and 2	D 638
Flexural properties	ISO 178	D 790
Notched Izod impact strength	ISO 180	D 256
Instrumented dart impact strength	ISO 6603-2	D 3763
Deflection temperatures under load	ISO 75-1 and 2	D 648
Vicat softening temperature	ISO 306	D 1525
Coefficient of thermal expansion	ISO 11359-2	E 831

application to another. It, therefore, is critical for the design engineer or material specifier to test plastics under actual conditions of use to determine the suitability of the plastic for the intended application.

ISO brings together the interests of producers, users, governments, and the scientific community in the development of standards which can be accepted as international standards by consensus among the participating countries. The scope of ISO covers standardization in all fields except electrical and electronics engineering, including some areas of telecommunications, which are the responsibility of IEC. The objective of ISO and IEC is to promote the standardization and related activities in the world to facilitate the international exchange of goods and services and to develop cooperation in the spheres of intellectual, scientific, technological, and economic activity.

ISO, founded in 1947 with headquarters in Geneva, Switzerland, is a worldwide federation of 110 national standards bodies, represented, at present, by one "official" member for each participating country. A member body of ISO is the national body "most representative of standardization in its country." The American National Standards Institute (ANSI) is the U.S. member body. ISO has about 200 established technical committees (TC) to address specific areas; the technical committee which represents the field of plastics and semifinished plastic products is TC61 on Plastics. ISO TC61 currently has 26 participating (P) member bodies (Table 11.14). In addition, there are 42 observer (O) member bodies. IEC, founded in 1906, is now comprised of about 50 national electrotechnical committees.

TABLE 11.12 Material Property Testing for Structural Analysis

	Test Method		
	ISO	ASTM	Suggested conditions
Properties in tension	527-1,2, and 4	D 638	At 23°C, at least three elevated temperatures, and one temperature below standard laboratory conditions at standard strain rate; at three additional strain rates at 23°C.
Poisson's ratio	527-1, 2	D 638	At 23°C, at least one elevated temperature, and one temperature below standard laboratory conditions.
Properties in compression	604	D 695	At 23°C, two additional elevated temperatures, and one temperature below standard laboratory conditions.
Shear modulus	6721-2, 5	D 5279	−150°C to T_g + 20°C or T_m + 10°C @ 1 Hz.
Creep in tension	899-1	D 2990	At 23°C and at least two elevated temperatures for 1000 h at three stress levels.
Fatigue in tension			$S-N$ curves at 3 Hz at 23°C; 80, 70, 60, 55, 50, and 40% of tensile stress at yield; $R = 0.5$; 1 million cycles run out. $a-N$ curves at 3 Hz at 23°C; single edge notched specimens; three stress levels; $R = 0.5$.
Coefficient of friction	8295	D 3028	At 23°C against itself and steel.
Application-specific:			
Creep in bending	899-2	D 2990	At 23°C and at least two elevated temperatures for 1000 h and at least three stress levels.
Creep in compression		D 2990	At 23°C and at least two elevated temperatures for 1000 h and at least three stress levels.
Fatigue in bending*			At 23°C; fully reversed; 80, 70, 60, 55, 50, and 40% of tensile stress at yield @ 3 Hz.
Fracture toughness	13586-1	D 5045	

*No ASTM or ISO standard exists today.
†Although the test frequency is restricted to 30 Hz only, D 671 may, in principle, be used.

TABLE 11.13 Material Property Characterizations for Processing Simulations

	Test Method		
	ISO	ASTM	Suggested conditions
Melt viscosity–shear rate data	11443	D 3835	At three temperatures, over shear rate range 10–10000 s^{-1}
Reactive viscosity of thermosets	6721-10		A slit die rheometer according to ISO 11443 can also be used.
Uniaxial extensional viscosity*			
Biaxial extensional viscosity*			
First normal stress difference	6721-10		
Melt density		D 3835	At 0 MPa and processing temperature.
Bulk density	61	D 1895	
Density-reacted system	1183	D 792	
pvT data			At a cooling rate of 2.5°C/min at 40, 80, 120, 160, and 200 MPa with an estimation at 1 MPa.
Thermal conductivity		D 5930	23°C to processing temperature.
Specific heat	11357-4	D 3418	DSC cooling scan @ 10°C/min from processing temperature to 23°C.
No-flow temperature*			DSC cooling scan @ 10°C/min from processing temperature to 23°C (11357-3).
Ejection temperature*			DSC cooling scan @ 10°C/min from processing temperature to 23°C (11357-3).
Glass transition temperature	11357-2	D 3418	DSC cooling scan @ 10°C/min from processing temperature to 23°C.
Crystallization temperature	11357-3	D 3418	DSC cooling scan @ 10°C/min from processing temperature to 23°C.
Degree of crystallinity	11357-3	D 3418	DSC cooling scan @ 10°C/min from processing temperature to 23°C.
Enthalpy of fusion	11357-3	D 3417	DSC heating scan @ 10°C/min from 23°C to processing temperature.
Enthalpy of crystallization	11357-3	D 3417	Cooling scan @ 10, 50, 100, and 200°C/min from processing temperature to 23°C.
Crystallization kinetics	11357-7	D 3417	Isothermal scans at different cooling rates at three temperatures in the crystallization range.
Heat of reaction of thermosets	11357-5	D 4473	Heating scan @ 10°C/min from 23°C to reaction temperature.
Reaction kinetics of thermosets	11357-5	D 4473	Isothermal DSC runs at three temperatures in the reaction temperature range.
Gelation conversion	11357-5		Heating scan @ 10°C/min.
Isothermal induction time			
Coefficient of linear thermal expansion	11359-2	E 831	With specimens cut from ISO 294-3 plate over the range −40 to 100°C.
Mold shrinkage:			
Thermoplastics	294-4	D 955	At 1, 1.5, and 2 mm thickness with cavity pressures of 25, 50, 75, and 100 MPa.
Thermosets	2577		
In-plane shear modulus	6721-2 or 7		

*No ASTM or ISO standard exists today.

TABLE 11.14 Participating (P) Members of ISO TC61 on Plastics

Country	National standards osrganization
Belgium	Institut Belge de Normalisation (IBN)
Canada	Standards Council of Canada (SCC)
People's Republic of China	China State Bureau of Technical Supervision (CSBTS)
Colombia	Instituto Colombiano de Normas Técnicas (ICONTEC)
Czech Republic	Czech Office for Standards, Metrology and Testing (COSMT)
Finland	Finnish Standards Association (SFS)
France	Association Française de Normalisation (AFNOR)
Germany	Deutsches Institut für Normung (DIN)
Hungary	Magyar Szabványügyi Hivatal (MSZH)
India	Bureau of Indian Standards (BIS)
Islamic Republic of Iran	Institute of Standards and Industrial Research of Iran (ISIRI)
Italy	Ente Nazionale Italiano di Unificazione (UNI)
Japan	Japanese Industrial Standards Committee (JISC)
Republic of Korea	Korean Industrial Advancement Administration (KIAA)
Netherlands	Nederlands Normalisatie-Instituut (NNI)
Philippines	Bureau of Product Standards (BPS)
Poland	Polish Committee for Standardization (PKN)
Romania	Institutul Roman de Standardizare (RS)
Russian Federation	Committee of the Russian Federation for Standardization, Metrology and Certification (GOST R)
Slovakia	Slovak Office of Standards, Metrology and Testing (UNMS)
Spain	Associación Espanola de Normalización y Certificación (AENOR)
Sweden	Standardiseringen i Sverige (SIS)
Switzerland	Schweizerische Normen-Vereinigung (SNV)
United Kingdom	British Standards Institution (BSI)
United States	American National Standards Institute (ANSI)

11.4.1 Impact of globalization

The elimination of trade barriers around the world through negotiations among countries involved in major international trade accords, such as the World Trade Organization (WTO), General Agreement on Tariffs and Trade (GATT), North American Free Trade Agreement (NAFTA), the European Union (EU), Asia Pacific Economic Cooperation (APEC), and MERCOSUR, etc., is changing the world into a single—global—marketplace. This trend has provided a new meaning to the term "globalization." While some OEMs have seized this opportunity to establish a strong global presence, other OEMs have streamlined their product development process by relying on global sourcing and leveraging of resources across the continents. Business leaders around the globe have come to recognize the strategic importance of international standards and their implications in world trade to efficiently design, manufacture, and deliver the same products to virtually any location in the world and the competitive disadvantage unless the industry adapts to the wave of global changes.

In the European Union (EU) and most other European nations, ISO/IEC test methods, where they exist, are being adopted as common European standards by the Committee for European Normalization (CEN) and the Committee for European Normalization of Electrotechnical testing (CENELEC). All major European standards organizations, such as DIN, BSI, and AFNOR, are accepting these international standards as their national standards conforming to the Vienna Agreement between ISO and Committee for European Normalization (CEN). The Japan Plastics Industry Federation (JPIF) is involved in an aggressive 3-year project to bring Japanese industrial standards into compliance with ISO/IEC standards. This project is in response to the decision by the Japanese government to accelerate the compliance plan to demonstrate its commitment to the deregulation policy. Japan's serious commitment to convert to ISO/IEC test methods is underlined by its aggressive goal of completing the adoption of ISO/IEC standards for plastics by 2001. Most industrialized countries around the world have also adopted these universal standards outright or are using them as the basis for national standards.

In the United States, in order to achieve greater uniformity in their operations worldwide and to stay competitive in a global economy, the Big Three U.S. automakers have jointly instituted a strategic standardization initiative to develop plastics material specifications, based on ISO methodology, through the U.S. Council for Automotive Research (USCAR) consortium. The recently published Society for Automotive Engineers (SAE) specifications J1639[8] for nylon, and J1685[9] for ABS and ABS + PC, are the first two in a series of documents that are being adopted by the Big Three automakers. Currently in development are 10 new SAE specifications (Table 11.15).

The implication of this commitment from the Big Three automakers is that anyone supplying materials to the automotive industry will be required to report data based on ISO methodology as defined in the new SAE protocols in the very near future. A significant impact of this strategic move by the automotive industry is the drive to switching from current practices based on ASTM test methods to uniform global testing protocols based on ISO test standards. One can expect that most of the engineering thermoplastics (ETP) and polypropylene (PP) producers will be routinely reporting only ISO test data.

TABLE 11.15 SAE Material Specifications Currently in Development

SAE Standard	Title
J 1686	Classification System for Automotive Polypropylene (PP) Plastics[10]
J 1687	Classification System for Automotive Thermoplastics Elastomeric Olefins (TEO)[11]
J 2250	Classification System for Automotive Poly(Methyl methacrylate) (PMMA) Plastics[12]
J 2273	Classification System for Automotive Polyester Plastics[13]
J 2274	Classification System for Automotive Acetal (POM) Plastics[14]
J 2323	Classification System for Automotive Polycarbonate Plastics[15]
J 2324	Classification System for Automotive Polyethylene Plastics[16]
J 2325	Classification System for Automotive Poly(Phenylene ether) (PPE) Plastics[17]
J 2326	Classification System for Automotive S/MA (Styrene-Maleic Anhydride) Plastics[18]
J 2460	Classification System for Automotive Thermoplastic Elastomeric Polyesters[19]

OEMs in the computer and business machines, appliance, healthcare, and electronics industries and multinationals, who prefer to reduce the amount of resources and effort allocated for dual testing (separate testing by ISO/IEC and ASTM standards), are indicating an interest in ISO test methods. Xerox Corporation has already instituted multinational material specifications based on ISO/IEC test methods. Even the U.S. government, particularly the U.S. Department of Commerce, continues to encourage adoption of ISO test methods. The U.S. government's standards policy to encourage the development of standards in recognized organizations, such as ISO and IEC, and the subsequent use of these standards in the United States has not changed.

In late 1992, The Society of the Plastics Industry, Inc (SPI)'s Polymeric Material Producers Division (PMPD) recommended that its member companies begin to convert to the use of internationally accepted standards developed by ISO and IEC for determining the properties of plastics and to routinely supply data on product datasheets and advertisements, using the preferred ISO/IEC standards, by June 1994. This strategic move was in response to the growing needs of various market sectors, led by the automotive industry.

SPI recognized the considerable amount of confusion created by the conflicting messages and general misinformation that appeared in trade literature. In November 1993, SPI formed an ad hoc ISO Communications Committee under the auspices of the International Technical and Standards Advisory Committee (ITSAC), to provide a formal, coordinated response that adequately represents the interests of the resin producers and customers within the U.S. plastics industry. The main charter of this committee is to help SPI lead an industrywide effort to promote and educate the U.S. plastics industry on those issues surrounding the implementation of ISO/IEC test standards in accordance with the 1992 resolution.

During NPE '94, the ISO committee organized an industry forum with a roundtable panel discussion and issued a call for uniform global testing standards for resins.*[20] In early 1996, the committee also developed a Technical Primer[21] illustrating the similarities and differences between the ISO/IEC test methods and current U.S. practices.* This detailed primer targeted at the technical community also describes the essential steps involved in the conversion to ISO/IEC test methods. By mid-1996, the committee developed a Management Primer[22] to promote the benefits of converting to uniform global testing standards for the U.S. industry leaders.*

11.4.2 Need for uniform global standards in testing plastics

Access to reliable and, most importantly, comparable property data is essential in material selection for any application, without which any attempt to compare properties among similar resins from different suppliers or from different sources, is apt to become an exercise in futility. This is primarily due to the fact that search for the most likely candidates invariably involves screening among available grades in the market on the basis of the properties that are related to the end-use performance requirements of the application.

At the outset, this appears straightforward and quite simple. Unfortunately, however, the material selection process often turns out to be an ordeal for anyone involved in this exercise. This is largely attributed to a combination of factors such as inconsistent test methods, different test specimen geometry as well as specimen molding conditions, flexibility in the choice of test conditions, and also lack of uniform reporting format associated with the current practices.[23–25] The wide latitude for variability allowed today in specimen preparation and test conditions in data acquisition makes it difficult to meaningfully compare resin properties from different suppliers and even from various global manufacturing sites of the same supplier. The published deflection temperature under load (DTUL) of several ABS grades in Table 11.16 best demonstrates the variability arising from using specimens of varying thickness, different specimen preparation methods, and pretreatment, if any. If one is only focusing on the DTUL value, ignoring these important variables, the result could be disastrous. Similarly, multiple test standards often employed to determine the impact behavior (Table 11.17) add further confusion.

*Copies may be obtained from The Society of The Plastics Industry, Inc., 1801 K Street, N. W., Suite 600, Washington, D.C. 20005.

With more than 15,000 grades of resins to choose from in the United States alone, over 6000 grades in Europe, and nearly 10,000 grades in Japan, it is not difficult to appreciate the magnitude of this problem. It is further compounded by the fact that the data for many of the products often lacks sufficient information regarding test conditions, specimen details, etc., in commercial databases or even resin manufacturers' product literature.

Such lack of uniformity in the acquisition and data reporting, added to lot-to-lot variability and interlaboratory variations, contributes to more frustration among the material specifiers and designers. Adoption of uniform test standards on a global basis would alleviate this ordeal and facilitate true comparability, that is, an "apples-to-apples" comparison. The benefits of adopting one set of test standards worldwide are

TABLE 11.16 Comparison of Deflection Temperature under Load (DTUL) at 1.8 MPa for ABS Resins from Different Suppliers

	Specimen thickness, mm	Specimen preparation method	Annealed	DTUL @ 1.8 MPa, °C
Supplier A	3.2	Injection-molded	No	76
	3.2	Injection-molded	Yes	100
	3.2	Compression-molded	Yes	102
Supplier B	12.7	Injection-molded	No	85
	12.7	Injection-molded	Yes	93
Supplier C	12.7	?	No	92
	12.7	?	Yes	100
Supplier D	?	?	?	106

TABLE 11.17 Multiple Test Standards Employed in Reporting Impact Strengths of Polypropylene (PP) Homopolymer

Test standard	Method	Specimen	Notch form	Impact strength
ISO 180/1A	Izod	80 × 10 × 4 mm	V	6 kJ/m^2
ISO 180/1R	Izod	80 × 10 × 4 mm	U	50 kJ/m^2
ISO 179-1/1A	Charpy	80 × 10 × 4 mm	V	8 kJ/m^2
ISO 179-1/1U	Charpy	80 × 10 × 4 mm	U	NB
ASTM D 256	Izod	63 × 12.7 × 3.2 mm	V	0.5 ft·lb/in
ASTM D 4812	Cantilever beam impact	63 × 12.7 × 3.2 mm	U	1068 J/m
ASTM D 3029/G	Gardner	50 mm in diameter × 3.2 mm	U	<10 in·lb
ASTM D 3763	Instrumented dart impact	100 mm in diameter × 3.2 mm		U <2 ft·lb

NOTE: U = unnotched.

- *Long-term cost savings.* Multinational companies stand to save costs in the long run from elimination of the need to retest or investing time and effort in comparing test methods.
- *Increased opportunities for access to international markets.* Adoption of uniform global testing protocols equate to "speaking the same language with customers around the globe," facilitating greater access to international markets that were not possible earlier.
- *Easier procurement of materials worldwide.* In the growing global manufacturing environment, use of one set of universally accepted test standards would facilitate easier procurement of materials against uniform global specifications, regardless of where in the world they are manufactured or needed.
- *Greater consistency in data.* The adoption of more stringent, consistent, and uniform methodology in generation of material property data by resin suppliers will reduce the large variability associated with the data prevalent today.
- *Improved communication.* Communication between manufacturers and resin suppliers worldwide is expected to improve significantly by having comparable data. In the case of multinational companies, internal communication between their manufacturing sites around the world will be improved as well.

11.4.3 Uniform global testing protocols

In order to produce truly comparable data, use of uniform standards, uniform test specimens, standard molds, narrow specimen molding conditions specifically defined for each resin family, and uniform test conditions, that is, identical, reproducible conditions is vital. Simply providing detailed information about the specimen geometry, preparation, conditioning, and test procedure is not sufficient to allow true comparability.

Until recently, the lack of uniform testing protocols for plastics posed a major hurdle. Fortunately, three international standards—ISO 10350-1,[26] ISO 11403-1,[27] and ISO 11403-2[28]—were specifically developed by an international collaborative effort to address these issues. ISO 10350-1 forms the basic framework for testing and reporting of single-point data on plastics by designating specific test procedures and conditions that are specified in other ISO test standards, and when used in conjunction with the relevant ISO material standards, it defines rigid guidelines for the choice of specimen geometry, mold design, specimen preparation conditions, and test conditions. ISO 10350-1 clearly indicates which test specimens should be used for each test and how the specimens should be prepared. For example, to determine the tensile modulus of an ABS resin, ISO 10350-1 specifies using the 4-mm-thick, ISO 3167[29] multipurpose specimen, molded using the balanced mold design with gating as specified in ISO 294-1[30] at conditions specified in the ABS material document ISO 2580-2,[31] and tested according to the procedures described in ISO 527-2[32] at a specified test speed of 1 mm/min. The end result of such a comprehensive approach is a reduction in variables associated with testing, which yields more reliable, reproducible, and, above all, comparable data. The test methods recommended in ISO 10350-1 are shown in Table 11.18.

To complement these "single-point" data with data representing the time- and temperature-dependent behavior of plastics useful in product design, a similar document has been developed which deals with the acquisition and presentation of comparable multipoint data. It has three parts: ISO 11403 -1 which deals with mechanical properties; ISO 11403-2, which addresses the thermal and processing properties; and ISO 11403-3,[33] which focuses on environmental influences on properties. Similar to ISO 10350-1, the multipoint data standards ISO 11403-1 and -2 also define the types of specimens for testing, how the tests should be conducted, and provide a technically sound framework for acquisition of multipoint data.

11.4.4 Comparison between ISO/IEC methods as specified in ISO 10350-1 and ASTM approaches

To better understand the similarities and differences between the ISO 10350-1 and ASTM approaches, a detailed examination of three key aspects of data acquisition are in order:

1. Choice of the test specimen
2. Test specimen preparation
3. Test conditions

The key documents which define the appropriate tests for a specific type of material in both ISO/IEC and ASTM approaches are material standards. These material documents within ISO and ASTM are similar in that they both contain specimen preparation conditions, conditioning, and specific testing parameters for that material. The corresponding material standards for common polymer families are listed in Table 11.19.

Choice of test specimen. The most significant difference between the ISO 10350-1 and ASTM approaches is in the test specimen used to determine mechanical properties. The ISO 3167 multipurpose specimen, required by ISO 10350-1 (Fig. 11.3), has the dimensions 165 × 10 × 4 mm and must be prepared in end-gated balanced molds to achieve uniform

TABLE 11.18 Test Conditions and Format for Presentation of Single-Point Data According to ISO 10350-1: 1999

Property	Standard	Specimen type, mm	Unit	Test conditions and supplementary instructions
		Rheological Properties		
Melt mass-flow rate Melt volume-flow rate	ISO 1133	Molding compound	g/10 min cm^3/10 min	At test conditions for temperature and load specified in Part 2 of appropriate material standards.
Molding shrinkage	ISO 2577		%	Thermosetting materials only in parallel and normal directions.
	ISO 294-4	60 × 60 × 2 (ISO 294 3 type D2)		Thermoplastic materials only in parallel and normal directions.
		Mechanical Properties (At 23°C/50% RH, Unless Noted)		
Tensile modulus			MPa	Test speed 1 mm/min; between 0.05% to 0.25% strain.
Stress at yield Strain at yield Nominal strain at break Stress at 50% strain Stress at break Strain at break	ISO 527-1 and 527-2	ISO 3167	MPa % % MPa MPa %	Test speed 50 mm/min if strain at yield or break > 10%. Test speed 50 mm/min if no yield at > 50 Strain and 5 mm/min if strain at break ≤ 0%.
Tensile creep modulus Tensile creep modulus	ISO 899-1	ISO 3167	MPa MPa	At 1 h; Strain < 0.5% At 1000 h
Flexural modulus Flexural strength	ISO 178	80 × 10 × 4	MPa MPa	Test speed 2 mm/min
Charpy impact strength	ISO 179-1 or	80 × 10 × 4	kJ/m^2	At +23°C and −30°C; edgewise impact.
Charpy notched impact strength	ISO 179-2	80 × 10 × 4 Machined V-notch with r = 0.25		At +23°C and −30°C; edgewise impact.
Tensile impact strength	ISO 8256	80 × 10 × 4 Machined double V-notch with r = 1	kJ/m^2	At +23°C; record if fracture cannot be observed with notched Charpy test.
Puncture impact behavior	ISO 6603-2	60 × 60 × 2		Record maximum force and energy at 50% decrease in force after the maximum; Striker velocity 4.4 m/s; striker diameter 20 mm; specimen clamped sufficiently to prevent any out-of-plane movement; striker lubricated.
		Thermal Properties		
Melting temperature	ISO 11357-3	Molding compound	°C	Record peak melting temperature; at 10°C/min.
Glass transition temperature	ISO 11357-2	Molding compound	°C	Record midpoint temperature; at 10°C/min
Temperature of deflection under load	ISO 75-1 and 75-2	80 × 10 × 4	°C °C °C	0.45 MPa for less rigid materials. 1.8 MPa for both soft and rigid materials. 8.0 MPa for rigid materials only.
Vicat softening temperature	ISO 306	≥10 × 10 ×4	°C	Heating rate 50°C/h. Load 50 N.
Coefficient of linear thermal expansion	ISO 11359-2	Prepared from ISO 3167	1/K	Mean secant value over the temperature range 23 55°C in parallel and normal directions.
Burning behavior	ISO 1210	125 × 13 × 3 Additional thickness		Record one of the classifications V-0, V-1, V-2, HB4 HB75 or N.
	ISO 10351	≤150 × ≤150 × 3 Additional thickness		Record classifications 5VA, 5VB, or N.
Oxygen Index	ISO 4589-2	80 × 10 × 4	%	Use procedure A (top surface ignition).
		Electrical Properties		
Relative permittivity Dissipation factor	IEC 60250	≥60 × ≥60 × 2		At 100-Hz and 1-MHz frequency; compensate for electrode edge effects.
Volume resistivity Surface resistivity	IEC 60093		Ω · cm Ω	Use contacting line electrodes 1 to 2 mm wide, 50 mm long, and 5 mm apart; voltage 500V.
Electric strength	IEC 60243-1	≥60 × ≥60 × 1 and ≥60 × ≥60 × 2	kV/mm	Use 20-mm-dia spherical electrodes; immerse in transformer oil in accordance with IEC 60296; Use a voltage application of 2 kV/s.
Comparative tracking index	IEC 60112	≥15 × ≥15 × 4		Use solution A.
		Other Properties		
Water absorption	ISO 62	Thickness ≥1	%	Saturation value in water at 23°C and equilibrium value at 23°C/50%RH.
Density	ISO 1183	Use part of the center of the multipurpose test specimen	kg/m^3	

TABLE 11.19 ISO and ASTM Material Standards for Common Polymer Families

Family	ISO standards*		ASTM standards
ABS	2580 - 1 : 99	2580 - 2 : 94	D 4673 - 98
Styrene acrylonitrile	4894 - 1 : 97	4894 - 2 : 94	D 4203 - 95
Polystyrene, crystal	1622 - 1 : 94	1622 - 2 : 94	D 4549 - 98
Polystyrene, high impact	2897 - 1 : 97	2897 - 2 : 94	D 4549 - 98
Polypropylene	1873 - 1 : 95	1873 - 2 : 97	D 4101 - 98a
Polyethylene	1872 - 1 : 93	1872 - 2 : 97	D 4976 - 98
Polyvinyl chloride (PVC), plasticized	2898 - 1 : 97	2898 - 2 : 97	D 2287 - 96
Polyvinyl chloride (PVC), unplasticized	1163 - 1 : 95	1163 - 2 : 91	D 1784 - 97
PMMA	8257 - 1 : 98	8257 - 2 : 98	D 788 - 96
Polycarbonate	7391 - 1 : 95	7391 - 2 : 95	D 3935 - 94
Acetals	9988 - 1 : 98	FDIS 9988 - 2 : 99	D 4181 - 98
Polyamides	1874 - 1 : 96	1874 - 2 : 95	D 4066 - 98
Thermoplastic polyester	7792 - 1 : 98	7792 - 2 : 98	D 5927 - 97
Polyketone	FDIS 15526 - 1 : 99	FDIS 15526 - 2 : 99	D 5990 - 96
PPE	FDIS 15103 - 1 : 99	FDIS 15103 - 2 : 99	D 4349 - 96
Thermoplastic polyester elastomer	14910 - 1 : 97	14910 - 2 : 98	D 4550 - 92

*The Part 1 of each ISO material document addresses the "designatory properties."

orientation in test specimens with high reproducibility. With this approach, the end result is a reduction in variables typically associated with specimen preparation, thereby ensuring more reliable, reproducible, and comparable data. The specimen of choice for testing plastics with ASTM methods is often the ASTM D 638 type 1 specimen (Fig. 11.4) with dimensions of 165 × 12.7 × 3.2 mm. The cross-sectional areas of both specimens are nearly the same—40 mm² versus 40.6 mm²—a difference of only 1.5%. However, the thickness differences between the two test specimens *is* significant—4 mm versus 3.2 mm—a 20% difference.

Test specimen preparation. Although often overlooked, one of the most critical parameters in testing plastics is how the test specimen is prepared. The recommended specimen preparation conditions for some common polymer families, according to ISO material standards, are summarized in Table 11.20 and corresponding ASTM guidelines are listed in Table 11.21. A quick comparison of the two tables reveals slight differences, in some cases, in melt temperature recommendations between the two approaches. More often, the recommended mold temperatures are somewhat different between the two approaches, notably in the case of polypropylene, acetal copolymer, and ABS resin.

Test procedures. A detailed comparison of the specific tests recommended in ISO 10350-1 for the single-point data with the corresponding ASTM test methods is summarized in Table 11.22. Compilations of

(*text continues on page 11.66*)

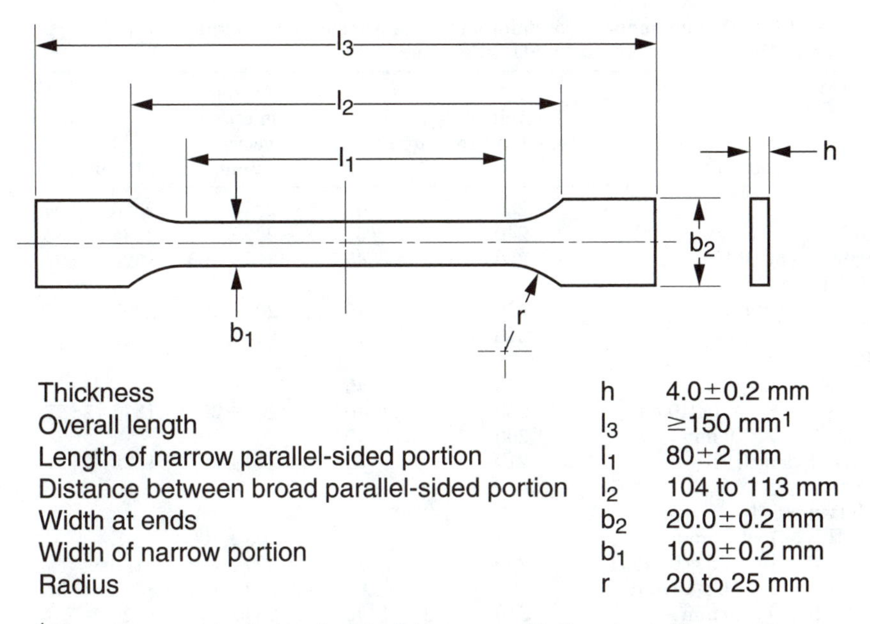

Thickness	h	4.0±0.2 mm
Overall length	l_3	≥150 mm[1]
Length of narrow parallel-sided portion	l_1	80±2 mm
Distance between broad parallel-sided portion	l_2	104 to 113 mm
Width at ends	b_2	20.0±0.2 mm
Width of narrow portion	b_1	10.0±0.2 mm
Radius	r	20 to 25 mm

[1] For some materials, the length of the tabs may need to be extended to prevent breakage or slippage in the jaws of the testing machine.

Figure 11.3 ISO 3167 multipurpose test specimen.

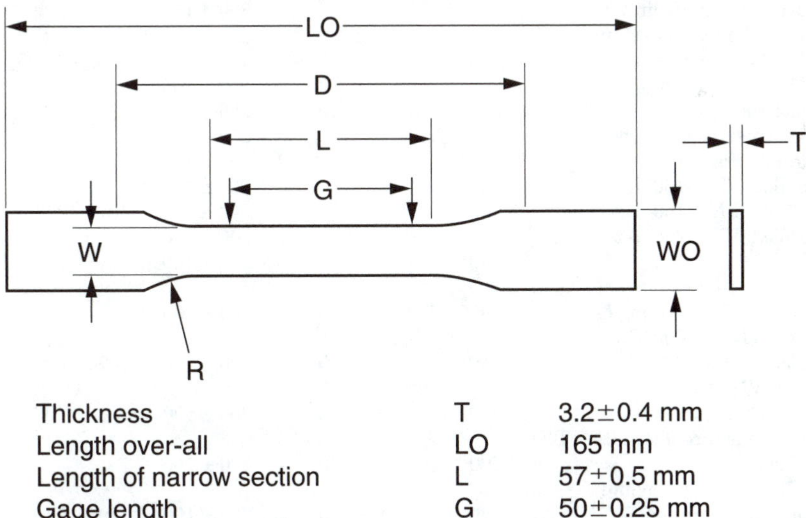

Thickness	T	3.2±0.4 mm
Length over-all	LO	165 mm
Length of narrow section	L	57±0.5 mm
Gage length	G	50±0.25 mm
Distance between grips	D	115±5 mm
Width over-all	WO	19 mm (+6.4, 0)
Width of narrow section	W	13±0.5 mm
Radius of fillet	R	76 ±1 mm

Figure 11.4 ASTM D 638 type I specimen.

TABLE 11.20 Recommended Conditions for Test Specimen Preparation from Common Materials According to ISO Guidelines

Material	Melt temperature, °C	Mold temperature, °C	Average injection velocity, mm/s	Reference
ABS	250	60	200 ± 100	2580 - 2 : 94
SAN	240	60	200 ± 100	4894 - 2 : 94
Polystyrene (PS)	220	45	200 ± 100	1622 - 2 : 94
PS - I				
General purpose	220	45	200 ± 100	2897 - 2 : 94
Flame-retarded	210	45	200 ± 100	
PP				
MFR $<$ 1.5 g/10 min	255	40	200 ± 20	
MFR $> 1.5 \leq 7$ g/10 min	230	40	200 ± 20	1873 - 2 : 97
MFR $>$ 7 g/10 min	200	40	200 ± 20	
Polyethylene (PE)	210	40	100 ± 20	1872 - 2 : 97
PC				
Unreinforced				
MFR $>$ 15 g/10 min	280	80	200 ± 100	
MFR $> 10 \leq 15$ g/10 min	290	80	200 ± 100	
MFR $> 5 \leq 10$ g/10 min	300	80	200 ± 100	7391 - 2 : 95
MFR $\leq$ 5 g/10 min	310	90	200 ± 100	
Glass fiber reinforced	300	110	200 ± 100	
Acetals				
Homopolymer				
MFR $\leq$ 7 g/10 min	215	90	140 ± 100	
MFR $\geq$ 7 g/10 min	215	90	300 ± 100	
Homopolymer, impact modified				
MFR $\leq$ 7 g/10 min	215	60	140 ± 100	9988 - 2 : 99
MFR $\geq$ 7 g/10 min	215	60	300 ± 100	
Copolymer	205	90	200 ± 100	
Copolymer, impact-modified	205	80	200 ± 100	
Polyamide (PA)6				
Unfilled, VN $\leq$160 mg/L	250	80	200 ± 100	
Unfilled, VN $\geq$160–$\leq$200mg/L	260	80	200 ± 100	1874 - 2 : 95
Unfilled, VN $\geq$200–$\leq$240 mg/L	270	80	200 ± 100	
Filled	290	80	200 ± 100	
PA66				
Unfilled, VN $\leq$200 mg/L	290	80	200 ± 100	
Filled, VN $\leq$200 mg/L, glass $>$10 to $\leq$50%	290	80	200 ± 100	1874 - 2 : 95
Filled, VN $\leq$200 mg/L, glass $>$50 to $\leq$70%	300	100	200 ± 100	
Polybutylene terephthalate (PBT)				
Unfilled	260	80	200 ± 100	
Unfilled, impact-modified or flame-retarded	250	80	200 ± 100	7792 - 2 : 98
Filled	260	80	200 ± 100	
Filled, impact-modified and flame-retarded	250	80	200 ± 100	

TABLE 11.21 Recommended Conditions for Test Specimen Preparation from Common Materials According to ASTM Guidelines

Material	Melt temperature, °C	Mold temperature, °C	Average injection velocity, mm/s	Reference
ABS:				
General purpose	250 ± 5	55 ± 5	200 ± 100	
Flame-retarded	220 ± 5	55 ± 5	200 ± 100	D 4673 - 98
High heat grade	255 ± 5	55 ± 5	200 ± 100	
Filled	255 ± 5	55 ± 5	200 ± 100	
SAN	Molding conditions shall be as specified in practice D 3641 unless otherwise agreed by the user and the supplier.			D 4203 - 95
PS	220 ± 10	50 ± 10		D 4549 - 98
PS - I	220 ± 10	50 ± 10		D 4549 - 98
PP				
MFR 1.0–1.5 g/10 min	250 ± 3	60 ± 3	The injection speed is set to produce equal weight parts (that is, part weight not varying by more than ± 2% regardless of material flow rates) and to minimize sink and flash.	D 4101 - 98a
MFR 1.6–2.5 g/10 min	240 ± 3	60 ± 3		
MFR 2.6–4.0 g/10 min	230 ± 3	60 ± 3		
MFR 4.1–6.5 g/10 min	220 ± 3	60 ± 3		
MFR 6.6–10.5 g/10 min	210 ± 3	60 ± 3		
MFR 10.6–17.5 g/10 min	200 ± 3	60 ± 3		
MFR 17.6–30.0 g/10 min	190 ± 3	60 ± 3		
PE	Unless otherwise specified, test specimens shall be compression-molded in accordance with procedure C of practice D 1928.			D 4976 - 98
PC				
Unfilled, MFR < 8 g/10 min	290 - 345	80 - 115		
Unfilled, MFR > 8 g/10 min	275 - 290	70 - 95		D 3935 - 94
Filled and reinforced	300 - 350	80 - 115		
High heat copolymer, unfilled	330 - 375	80 - 100		
High heat copolymer, filled/reinforced	Consult manufacturer for recommended molding conditions.			
Acetals:				
Homopolymer	215 ± 5	90 ± 10	200 ± 100	D 4181 - 98
Homopolymer, modified	210 ± 5	90 ± 10	200 ± 100	
Copolymer	195 ± 5	85 ± 5	200 ± 100	
PA6				
Unfilled	260	80	200 ± 100	D 4066 - 98
Filled	290	80	200 ± 100	
PA66				
Unfilled	260	80	200 ± 100	D 4066 - 98
Filled	290	80	200 ± 100	
PBT	260 ± 3	80 ± 5	200 ± 100	D 5927 - 97

TABLE 11.22 Comparison of Test Methods Between ISO 10350-1 and ASTM Approaches*

Property	ISO/IEC methods as specified by ISO 10350-1:99	ASTM methods
	Rheological Properties	
Melt flow rate (MFR), melt volume rate (MVR), and flow rate ratio (FRR):		
Standard	ISO 1133:97	D 1238-98
Specimen	Powder, pellets, granules, or strips of films.	Powder, pellets, granules, strips of films, or molded slugs.
Conditioning	In accordance with the material standard, if necessary.	Check the applicable material specification for any conditioning requirements before using this test. See practice D 618 for appropriate conditioning practices.
Apparatus	Extrusion plastometer with a steel cylinder 115 – 180 mm (L) × 9.55 ± 0.025 mm (ID), and a die with an orifice of 8.000 ± 0.025 mm (L) × 2.095 ± 0.005 mm (ID).	Extrusion plastometer with a steel cylinder 162 mm (L) × 9.55 ± 0.008 mm (ID), and a die with an orifice of 8.000 ± 0.025 mm (L) × 2.0955 ± 0.0051 mm (ID).
Test procedures	Test temperature and test load as specified in Part 2 of the material designation standards, or as listed in ISO 1133 Table A.2. Some examples from Table A.2: PC (300°C/1.2 kg) ABS (220°C/10 kg) PS (200°C/5 kg) PS-HI (200°C/5 kg) SAN (220°C/10 kg) PP (230°C/2.16 kg) PE (190°C/2.16 kg) POM (190°C/2.16 kg) PMMA (230°C/3.8 kg)	Test temperature and test load as specified in the applicable material specification, or as listed D1238 Table 1. Some examples from Table 1: PC (300°C/1.2 kg) ABS (230°C/10 kg) PS (200°C/5 kg) PS-HI (200°C/5 kg) SAN (220°C/10 kg) PP (230°C/2.16 kg) PE (190°C/2.16 kg) POM (190°C/2.16 kg) Acrylics (230°C/3.8 kg)
	Charge—within 1 min	Charge—within 1 min
	Preheat—4 min	Preheat—6.5 min
	Test time—last measurement not to exceed 25 min from charging Procedure A—manual operation using the mass and cut-time intervals shown here:	Test time—7.0 ± 0.5 min from initial charging Procedure A—manual operation using the mass and cut-time intervals shown here:
	Melt flow range / Mass / Cut-time interval 0.1 to 0.5 g/10 min / 3–5 g / 4 min >0.5 to 1 g/10 min / 4–5 g / 2 min >1 to 3.5 g/10 min / 4–5 g / 1 min >3.5 to 10 g/10 min / 6–8 g / 30 s >10 g/10 min / 6–8 g / 5–15 s	Melt flow range / Mass / Cut-time interval 0.15 to 1 g/10 min / 2.5–3 g / 6.00 min >1 to 3.5 g/10 min / 3–5 g / 3.00 min >3.5 to 10 g/10 min / 4–8 g / 1.00 min >10 to 25 g/10 min / 4–8 g / 30 s >25 g/10 min / 4–8 g / 15 s

	Procedure B—automated time or travel indicator is used to calculate the MFR (MVR) using the mass as specified previously in Procedure A for the predicted MFR.	Procedure B—MFR (MVR) is calculated from automated time measurement based on specified travel distances: <10 MFR travel distance is 6.35 ± 0.25 mm >10 MFR travel distance is 25.4 ±0.25 mm and using the mass as specified above for the predicted MFR.
Values and units	MFR (g/10 min) MVR (cm^3/10 min)	MFR (g/10 min) MVR (cm^3/10 min) FRR [ratio of the MFR (190/10) by MFR (190/2.16) (*used specifically for polyethylenes*)].
Injection molded shrinkage:		
Standard	ISO 294-4:97	D 955-89 (reapproved 1996)
Specimen	60- × 60- × 2-mm plate with specified fan runner of 66 × 25-30 × 4.0 mm and a low tolerance gate with dimensions of 60 × 4.0 × 1.50 mm. (Refer to ISO 294-3 Type D mold.)	127- × 12.7- × 3.2-mm bar with an end gate of 6.4 × 3.2 mm, or for diametral shrinkage at 102-mm-dia × 3.2-mm disk with a radial gate of 12.7 × 3.2 mm placed on the edge of the disk.
Conditioning	At 23 ± 2°C between 16 and 24 h. Materials which show marked difference in mold shrinkage if stored in a humid or dry atmosphere must be stored in dry atmosphere.	At 23 ± 2°C and 50 ± 5% relative humidity for 1–2 h for "initial molding shrinkage" (optional), 16–24 h for "24-h shrinkage" (optional), and 40–48 h for "48-h or normal shrinkage."
Test procedures	Mold at least five specimens, using a two-cavity ISO 294-3 type D2 mold, equipped with cavity pressure sensor.	Mold at least five specimens. No mold is specified and no cavity sensor is required.
	Molding equipment complies with the relevant 4.2 clauses in ISO 294-1 and ISO 294-3. In addition, accuracy of the cavity pressure sensor must be ±5%. The machine is operated such that the ratio of the molding volume to the screw-stroke volume is between 20–80%, when using the injection-molding conditions specified in Part 2 of the relevant material standard.	Molding in accordance with the practice D 3641 such that the molding equipment is operated without exceeding 50–75% of its rated shot capacity.
	Perform mold shrinkage measurements on specimens which have been molded such that one or more of the preferred "cavity pressure at pressure at hold (pch) of 20, 40, 60, 80 and/or 100 MPa is achieved.	No cavity pressure requirements are given.

*Information in this table is accurate as of June 1, 1999. ISO and ASTM standards have mandatory 5-y revisions; however the standards can be revised as needed.

TABLE 11.22 Comparison of Test Methods Between ISO 10350-1 and ASTM Approaches* (*Continued*)

Property	ISO/IEC methods as specified by ISO 10350-1:99	ASTM methods
Injection molded shrinkage (*Continued*):		
	Allow molded specimens to cool to room temperature by placing them on a material of low thermoconductivity with an appropriate load to prevent warping. Any specimen that has warpage >3% of its length is discarded.	Allow molded specimens to cool at 23 ± 2°C and 50 ± 5% relative humidity. No warpage limits are specified.
	Measure the length and width of the cavity and the corresponding molded specimens to within 0.02 mm at 23 ± 2°C.	Measure the length or diameter (both parallel and normal to the flow) of the cavity and the corresponding molded specimens to within 0.02 mm. Temperature requirement of the mold while measuring the cavity dimensions is not specified.
Values and units	Molding shrinkage (16–24 h): %* **Reported as mean value of the five specimens measured.*	Initial molding shrinkage: mm/mm (optional)* 24-h shrinkage: mm/mm (optional)* 48 h or normal shrinkage: mm/mm* **Reported as mean value of the five specimens measured.*
Mechanical Properties		
Tensile properties:		
Standard	ISO 527-1:93 and 527-2:93	D 638-98
Specimen	ISO 3167 (type A or B*) multipurpose test specimens (Figure 11.3). *Type A is recommended for directly molded specimens, so the 80- × 10- × 4-mm specimens required for most tests in ISO 10350-1 can be cut from the center of these specimens. Type B is recommended for machined specimens. **Dimensions for ISO 3167 specimens are:** Overall length >150 mm Width 10 mm Thickness 4 mm Fillet radius 20–25 mm (type A) or >60 mm (type B) Length of parallel narrow section 80 mm (type A) or 60 mm (type B)	For rigid/semirigid plastics: D 638 Type I specimens (Figure 11.4) are the preferred specimen and shall be used when sufficient material having a thickness of 7 mm or less is available. **Dimensions for D 638 Type I specimens are:** Overall length 165 mm Width 12.7 mm Thickness 3.2 mm Fillet radius 76 mm Length of parallel narrow section 57 mm

Conditioning	Specimen conditioning, including any postmolding treatment, shall be carried out at 23°C ± 2°C and 50 ± 5% RH for a minimum length of time of 88 h, except where special conditioning is required as specified by the appropriate material standard.	At 23 ± 2°C and 50 ± 5% relative humidity for not less than 40 h prior to testing in accordance with D 618 Procedure A for those tests where conditioning is required. For hygroscopic materials, the material specification takes precedence over the preceding routine preconditioning requirements.
Test procedures	A minimum of five specimens shall be prepared in accordance with the relevant material standard. When none exists, or unless otherwise specified, specimens shall be directly compression or injection molded in accordance with ISO 293 or ISO 294-1.	A minimum of five test specimens shall be prepared by machining operations or die cutting the materials in sheet, plate, slab, or similar form. Specimens can also be prepared by injection or compression molding the material to be tested.
	Test speed for ductile failure (defined as yielding or with a strain at break >10%) is 50 mm/min and for a brittle failure (defined as rupture without yielding or strain at break <10%) is 5 mm/min. For modulus determinations the test speed is 1 mm/min.	Test speed is specified in the specification for the material being tested. If no speed is specified, then use the lowest speed given in Table 1 (5, 50, or 500 mm/min) which gives rupture within 0.5 to response and resolution are adequate.
	Extensometers are required for determining strain at yield and tensile modulus. The specified initial gauge length is 50 mm. The extensometer shall be essentially free of inertia lag at the specified speed of testing and capable of measuring the change in gauge with an accuracy of 1% of the relevant value or better. This corresponds to ±1 μm for the measurement of modulus on a gauge length of 50 mm.	Extensometers are required for determining strain at yield and tensile modulus. The specified initial gauge length is 50 mm. For modulus determinations, an extensometer which meets Class B-2 (Practice E-38) is required, for low extensions (<20%) the extensometer must at least meet Class C (Practice E38) requirements, for high extensions (>20%) any measurement technique which has an error no greater than ±10% can be used.
	The reported tensile modulus is a chord modulus determined by drawing a straight line that connects the stress at 0.05% strain and the stress at 0.25% strain. There is no requirement for toe compensation in determining a corrected zero point, if necessary.	Tangent modulus is determined by drawing a tangent to the steepest initial straight line portion of the load-deflection curve and then dividing the difference in stress on any section of this line by the corresponding difference in strain.
		Secant modulus is the ratio of stress to corresponding strain at any given point on the stress-strain curve, or the slope of the straight line that joins the zero point or corrected zero point and the selected point corresponding to the strain selected on the actual stress-strain curve. Toe compensation, if applicable as defined, is mandatory.

TABLE 11.22 Comparison of Test Methods Between ISO 10350-1 and ASTM Approaches* (*Continued*)

Property	ISO/IEC methods as specified by ISO 10350-1:99	ASTM methods
Tensile properties (*Continued*):		
Values and units	For ductile materials: Stress at yield (MPa) Strain at yield (%) Stress at 50% strain* (MPa) Nominal strain at break** (%) Tensile modulus (MPa) *If the material does not yield before 50% strain, report stress at 50% strain. **Nominal strain at break based on initial and final grip separations, if rupture occurs above 50% nominal strain. One can either report the strain at break or simply >50%.	For ductile materials: Stress at yield (MPa) Strain at yield (%) Stress at break (MPa) Strain at break (%) Tangent modulus (MPa) Secant modulus (MPa)
	For brittle materials: Stress at break (MPa) Strain at break (%) Chord modulus (0.5–0.25% strain) (MPa)	For brittle materials: Stress at break (MPa) Strain at break (%) Modulus (MPa)
Tensile creep modulus:		
Standard	ISO 899-1:93	D 2990-95
Specimen	ISO 3167 type A specimen	D 638 type I specimens may be prepared by injection or compression molding or by machining from sheets or other fabricated forms.
Conditioning	Specimen conditioning, including any postmolding treatment, shall be carried out at 23°C ±2 °C and 50 ± 5% RH for a minimum length of time of 88 h, except where special conditioning is required as specified by the appropriate material standard.	At 23 ± 2°C and 50 ± 5% relative humidity for not less than 40 h, prior to testing in accordance with D 618 Procedure A. The specimens shall be preconditioned in the test environment for at least 48 h prior to testing. Those materials whose creep properties are suspected to be affected by moisture content shall be brought to moisture equilibrium appropriate to the test conditions prior to testing.
Test procedures	Conduct the test in the same atmosphere as used for conditioning, unless otherwise agreed upon by the interested parties, for example, for testing at elevated or low temperatures.	For material characterization, select two or more test temperatures to cover the useful temperature range. For simple material comparisons, select the test temperatures from the following: 23, 50, 70 90, 120, and 155°C.

	Select appropriate stress levels to produce data for the application requirements. Where it is necessary to preload the test specimen prior to loading, preloading shall not be applied until the temperature and humidity of the test specimen (finally gripped in the testing apparatus) correspond to the test conditions, and the total load (including preload) shall be taken as the test load.	For simple material comparisons, determine the stress to produce 1% strain in 1000 h. Select several loads to produce strains in the approximate range of 1% strain and plot the 1000-h isochronous stress-strain curve* from which the stress to produce 1% strain may be determined by interpolation. *Since only one point of an isochronous plot is obtained from each creep test, it is usually necessary to run at least three stress levels (preferably more) to obtain an isochronous plot.
		For creep testing at a single temperature, the minimum number of test specimens at each stress shall be two if four or more stress levels are used or three if fewer than four levels are used.
Test procedures	Unless the elongation is automatically and/or continuously measured, record the elongations at the following time schedule: 1, 3, 6, 12, and 30 min; 1, 2, 5, 10, 20, 50, 100, 200, 500, 1000 h.	Measure the extension of the specimens in accordance with the approximate time schedule: 1, 6, 12, and 30 min; 1, 2, 5, 50, 100, 200, 500, 700, and 1000 h.
Units	Tensile creep modulus at 1 h and at a strain <0.5% (MPa) Tensile creep modulus at 1000 h and at a strain <0.5% (MPa)	Tensile creep modulus (MPa) plotted versus time (h).
Flexural properties:		
Standard	ISO 178:93	D 790-98
Specimen	80 × 10 × 4 mm cut from the center of an ISO 3167 type A specimen. In any one specimen the thickness within the central one-third of length shall not deviate by more than 0.08 mm from its mean value, and the corresponding allowable deviation in the width is 0.3 mm from its mean value.	Specimens may be cut from sheets, plates, molded shapes or molded to the desired finished dimensions. The recommended specimen for molding materials is 127 × 12.7 × 3.2 mm.
Conditioning	Specimen conditioning, including any postmolding treatment, shall be carried out at 23°C ± 2°C and 50 ± 5% RH for a minimum length of time of 88 h, except where special conditioning is required as specified by the appropriate material standard.	At 23 ± 2°C and 50 ± 5% relative humidity for not less than 4 h prior to testing in accordance with D 618 Procedure A for those tests where conditioning is required. For hygroscopic materials, the material specification takes precedence over the preceding routine preconditioning requirements.
Apparatus	Support and loading nose radius, 5.0 ± 0.1 mm. (Figure 11.5)	Support and loading nose radius, 5.0 ± 0.1 mm. Figure 11.6 (unless otherwise specified or agreed upon by the interested parties, other chosen radii must be at least 3.2 mm with a maximum of 1.6 × specimen depth for loading nose).

TABLE 11.22 Comparison of Test Methods Between ISO 10350-1 and ASTM Approaches* (*Continued*)

Property	ISO/IEC methods as specified by ISO 10350-1:99	ASTM methods
Flexural properties	(*Continued*):	
	Parallel alignment of the support and loading nose must be less than or equal to 0.02 mm.	Parallel alignment of the support and loading noses may be checked by means of a jig with parallel grooves into which the loading nose and supports will fit if properly aligned.
Apparatus	Support span length, 60–68 mm. (Adjust the length of the span to within 0.5%, which is 0.3 mm for the span length specified previously.) Support span to specimen depth ratio, 16 ± 1; 1 mm/mm. (Specimens with a thickness exceeding the tolerance of ±0.5% of the mean thickness value shall be discarded and replaced by another one, sampled by chance.)	Support span length,* 49.5–50.5 mm. (Measure the span accurately to the nearest 0.1 mm for spans less than 63 mm. Use the measured span length for all calculations.) Support span to specimen depth ratio,* 16 (+4, −1); 1 mm/mm. *Recommended settings for molding materials, however, there exists 40 different span length and L/d combinations for test method I, and 80 for test method II.
Test procedures		Testing conditions indicated in material specifications take precedence; therefore, it is advisable to refer to the material specification before using the following procedures.
	Test speed, 2 mm/min.	Procedure A crosshead speed,* 1.3 mm/min Procedure B crosshead speed,* 13 mm/min *Procedure A must be used for modulus determinations, Procedure B may be used for flexural strength determination only.
	A minimum of five specimens shall be prepared in accordance with the relevant material standard. When none exists, or unless otherwise specified, specimens shall be directly compression or injection molded in accordance with ISO 293 or ISO 294-1. Test specimens that rupture outside the central one-third of the span length shall be discarded and new specimen shall be tested in their place.	A minimum of five test specimens are required. No specimen preparation conditions are given.
	Measure the width of the test specimen to the nearest 0.1 mm and the thickness to the nearest 0.01 mm in the center of the test specimen.	Measure the width and depth of the test specimen to the nearest 0.03 mm at the center of the support span.

	The reported flexural modulus is a chord modulus determined by drawing a straight line that connects the stress at 0.05% strain and the stress at 0.25% strain. There is no requirement for toe compensation in determining a corrected zero point, if necessary.	Tangent modulus is determined by drawing a tangent to the steepest initial straight line portion of the load-deflection curve and then dividing the difference in stress on any section of this line by the corresponding difference in strain.
		Secant modulus is the ratio of stress to corresponding strain at any given point on the stress-strain curve, or the slope of the straight line that joins the zero point and a selected point on the actual stress-strain curve. Toe compensation, if applicable as defined, is mandatory.
Values and units	Flexural modulus (MPA) Flexural strength, at rupture (MPa) Flexural strength, at maximum strain* (MPa) *At conventional deflection which is 1.5 × height: therefore 4 mm specimens would have a maximum strain at 3.5%.	Tangent modulus (MPa) Secant modulus (MPa) Flexural strength, (at rupture*) (MPa) Flexural yield strength (MPa) *Maximum allowable strain in the outer fibers is 0.05 mm/mm. **The point where the load does not increase with increased deflection, provided it occurs before the maximum strain rate.
Unnotched Charpy impact strength		
Standard	ISO 179-1:98 and ISO 179-2:97	No ASTM equivalent
Specimen	80 × 10 × 4 mm cut from the center of an ISO 3167 type A specimen; also referred to as an ISO 179/1eU specimen.	
Conditioning	Specimen conditioning, including any postmolding treatment, shall be carried out at 23°C ± 2°C and 50 ± 5% RH for a minimum length of time of 88 h, except where special conditioning is required as specified by the appropriate material standard.	
Apparatus	The machine shall be securely fixed to a foundation having a mass at least 20 times that of the heaviest pendulum in use and be capable of being leveled.	

TABLE 11.22 Comparison of Test Methods Between ISO 10350-1 and ASTM Approaches* (*Continued*)

Property	ISO/IEC methods as specified by ISO 10350-1:99	ASTM methods
Unnotched Charpy impact strength (*Continued*):		
	Striking edge of the hardened steel pendulums is to be tapered to an included angle of 30 ± 1° and rounded to a radius of 2.0 ± 0.5 mm.	
	The striking edge of the pendulum shall pass midway to within ± 0.2 mm between the specimen supports. The line of contact shall be within ± 2° of perpendicular to the longitudinal axis of the test specimen.	
	Pendulums with specified nominal energies shall be used: 0.5, 1.0, 2.0, 4.0, 5.0, 7.5, 15.0, 25.0, and 50.0 J.	
	Velocity at impact is 2.9 + 10% m/s for the 0.5 to 5.0 J pendulums and 3.8 ± 10% m/s for pendulums with energies from 7.5 to 50.0 J.	
	The support anvils line of contact with the specimen shall be 62.0 (+0.5, −0.0) mm.	
Test procedures	A minimum of 10 specimens shall be prepared in accordance with the relevant material standard. When none exists, or unless otherwise specified, specimens shall be directly compression or injection molded in accordance with ISO 293 or ISO 294-1.	
	Edgewise impact is specified.	
	Consumed energy is 10 to 80% of the pendulum energy, at the corresponding specified velocity of impact. If more than one pendulum satisfies these conditions, the pendulum having the highest energy is used. (It is not advisable to compare results obtained using different pendulums.)	
	Maximum permissible frictional loss without specimen: 0.02% for 0.5 to 5.0 J pendulum 0.04% for 7.5 J pendulum 0.05% for 15.0 J pendulum 0.10% for 25.0 J pendulum	

	0.20% for 50.0 J pendulum	
	Permissible error after correction with specimen: 0.01 J for 0.5, 1.0, and 2.0 J pendulums. No correction applicable for pendulums with energies > 2.0 J.	
	Four types of failure are defined: C—complete break; specimen separates into one or more pieces. H—hinge break; an incomplete break such that both parts of the specimen are only held together by a thin peripheral layer in the form of a hinge. P—partial break; an incomplete break which does meet the definition for a hinge break. NB—nonbreak; in the case of the nonbreak, the specimen is only bent and passed through, possibly combined with stress whitening.	
Values and units	The measured values of complete and hinged breaks can be used for a common mean value with remark. If in the case of partial breaks a value is required, it shall be assigned with the letter P. In case of nonbreaks, no figures are to be reported. (If within one sample the test specimens show different types of failures, the mean value for each failure type shall be reported.) Unnotched Charpy impact strength (kJ/m^2).	
Notched Charpy impact strength:		
Standard	ISO 179-1:98 and ISO 179-2:97	D 256-97
Specimen	80 × 10 × 4 mm cut from the center of an ISO 3167 type A specimen with a single notch A; also referred to as an ISO 179/1e A specimen. (See Figure 11.7)	124.5 to 127 × 12.7 mm × (*) mm specimen, *The width of the specimens shall be between 3.0 and 12.7 mm as specified in the material specification, or as agreed upon as representative of the cross section in which the particular material may be used. (Figure 11.8).

TABLE 11.22 Comparison of Test Methods Between ISO 10350-1 and ASTM Approaches* (*Continued*)

Property	ISO/IEC methods as specified by ISO 10350-1:99	ASTM methods
Notched Charpy impact strength (*Continued*):		
	Notch A has a 45° ± 1° included angle with a notch base radius of 0.25 ± 0.05 mm. The notch should be at a right angle to the principal axis of the specimen. The specimens shall have a remaining width of 8.0 ± 0.2 mm after notching. These machined notches shall be prepared in accordance with ISO 2818.	A single notch with 45° ± 1° included angle with a radius of curvature at the apex 0.25 ± 0.05 mm. The plane bisecting the notch angle shall be perpendicular to the face of the test specimen within 2°. The depth of the plastic material remaining in the bar under the notch shall be 10.16 ± 0.05 mm. The notches are to be machined.
Conditioning	Specimen conditioning, including any postmolding treatment, shall be carried out at 23°C ± 2°C and 50 ± 5% RH for a minimum length of time of 88 h, except where special conditioning is required as specified by the appropriate material standard.	At 23 ± 2°C and 50 ± 5% relative humidity for not less than 40 h prior to testing in accordance with D 618 Procedure A for those tests where conditioning is required. For hygroscopic materials, the material specification takes precedence over the above routine preconditioning requirements.
Apparatus	The machine shall be securely fixed to a foundation having a mass at least 20 times that of the heaviest pendulum in use and be capable of being leveled.	The machine shall consist of a massive base.
± 0.5 mm.	Striking edge of the hardened steel pendulums is to be tapered to an included angle of 30 ± 1° and rounded to a radius of 2.0 mm.	Striking edge of hardened steel pendulums is to be tapered to an included angle of 45 ± 2° and rounded to a radius of 3.17 ± 0.12 mm
	Pendulums with the specified nominal energies shall be used: 0.5, 1.0, 2.0, 4.0, 5.0, 7.5, 15.0, 25.0, and 50.0 J.	Pendulum with an energy of 2.710 ± 0.135 J is specified for all specimens that extract up to 85% of this energy. Heavier pendulums are to be used for specimens that require more energy; however, no specific levels of energy pendulums are specified.
	Velocity at impact is 2.9 ± 10% m/s for the 0.5 to 5.0 J pendulums and 3.8 ± 10% m/s for pendulums with energies from 7.5 to 50.0 J.	Velocity at impact is approximately 3.46 m/s, based on the vertical height of fall of the striking nose specified at 610 ± 2 mm.
	The support anvils line of contact with the specimen shall be 62.0 (+0.5, −0.0) mm.	The anvils line of contact with the specimen shall be 101.6 ± 0.5 mm.
Test procedures	A minimum of 10 specimens shall be prepared in accordance with the relevant material standard. When none exists, or unless otherwise specified, specimens shall be directly compression or injection molded in accordance with ISO 293 or ISO 294-1.	At least five, preferably 10 specimens shall be prepared from sheets, composites (not recommended), or molded specimen. Specific specimen preparations are not given or referenced.

	Edgewise impact is specified (Figure 11.9). Consumed energy is 10 to 80% of the pendulum energy, at the corresponding specified velocity of impact. If more than one pendulum satisfies these conditions, the pendulum having the highest energy is used. (It is not advisable to compare results obtained using different pendulum.)	Edgewise impact is specified (Figure 11.10).
Test procedures	Maximum permissible frictional loss without specimen: 0.02% for 0.5 to 5.0 J pendulum 0.04% for 7.5 J pendulum 0.05% for 15.0 J pendulum 0.10% for 25.0 J pendulum 0.20% for 50.0 J pendulum	Windage and friction correction are not mandatory, however, a method of determining these values is given.
	Permissible error after correction with specimen: 0.01 J for 0.5, 1.0, and 2.0 J pendulums No correction applicable for pendulums with energies >2.0 J	
	Four types of failure are defined: C—complete break; specimen separates into one or more pieces. H—hinge break; an incomplete break such that both parts of the specimen are only held together by a thin peripheral layer in the form of a hinge. P—partial break; an incomplete break which does not meet the definition for a hinge break. NB—nonbreak; in the case of the nonbreak, the specimen is only bent and passed through, possibly combined with stress whitening.	Four types of failure are specified: C—complete break; specimen separates into one or more pieces. H—hinge break; an incomplete break such that one part of the specimen cannot support itself above the horizontal when the other part is held vertically (less than 90° included angle). P—partial break; an incomplete break which does not meet the definition for a hinge break, but has fractured at least 90% of the distance between the vertex of the notch and the opposite side.

TABLE 11.22 Comparison of Test Methods Between ISO 10350-1 and ASTM Approaches* (*Continued*)

Property	ISO/IEC methods as specified by ISO 10350-1:99	ASTM methods
Notched Charpy impact strength (*Continued*):		
		NB—nonbreak; an incomplete break where the fracture extends less than 90% of the distance between the vertex of the notch and the opposite side.
Values and units	The measured values of complete and hinged breaks can be used for a common mean value with remark. If in the case of partial breaks a value is required, it shall be assigned with the letter P. In case of nonbreaks, no figures are to be reported. (If within one sample the tests specimen show different types of failures, the mean value for each failure type shall be reported.) Notched Charpy impact strength (kJ/m^2).	Only measured values for complete breaks can be reported. (If more than one type of failure is observed for a sample material, then report the average impact value for the complete breaks, followed by the number and percent of the specimen failing in that manner suffixed by the letter code.) Notched Charpy impact strength (J/m).
Tensile impact strength:		
Standard	ISO 8256:90	D 1822-93
Specimen	$80 \times 10 \times 4$ mm, cut from the center of an ISO 3167 type A specimen, with a double notch. Also referred to as an ISO 8256 type 1 specimen (Figure 11.11).	Type S or L specimen as specified by this standard (Figure 11.12). 63.50 mm length $\times$ 9.53 or 12.71 mm tab width $\times$ 3.2 mm (preferred thickness) type S has a nonlinear narrow portion width of 3.18 mm, whereas type L has a 9.53 mm length linear narrow portion width of 3.18 mm.
Conditioning	Specimen conditioning, including any postmolding treatment, shall be carried out at 23°C ± 2°C and 50 ± 5% RH for a minimum length of time of 88 h, except where special conditioning is required as specified by the appropriate material standard.	At 23 ± 2°C and 50 ± 5% relative humidity for not less than 40 h, prior to testing in accordance with Practice D618, procedure A. Material specification conditioning requirements take precedence.
Apparatus	The machine shall be securely fixed to a foundation having a mass at least 20 times that of the heaviest pendulum in use and be capable of being leveled.	The base and suspending frame shall be of sufficiently rigid and massive construction to prevent or minimize energy losses to or through the base and frame.
	Pendulums with the specified initial potential energies shall be used: 2.0, 4.0, 7.5, 15.0, 25.0, and 50.0 J.	No pendulums specified.

	Velocity at impact is 2.6 to 3.2 m/s for the 2.0 to 4.0 J pendulums and 3.4 to 4.1 m/s for pendulum with energies from 7.5 to 50.0 J	Velocity at impact is approximately 3.44 m/s, based on the vertical height of fall of the striking nose specified at 610 ± 2 mm.
	Free length between grips is 30 ± 2 mm.	Jaw separation is 25.4 mm.
	The edges of the serrated grips in close proximity to the test region shall have a radius such that they cut across the edges of the first serrations.	The edge of the serrated jaws in close proximity to the test region shall have a 0.40-mm radius to break the edge of the first serrations.
	Unless otherwise specified in the relevant material standard, a minimum of 10 specimens shall be prepared in accordance with that same material standard. When none exists, or unless otherwise specified, specimens shall be directly compression or injection molded in accordance with ISO 293 or ISO 294-1.	Material specification testing conditions take precedence; therefore, it is advisable to refer to the material specification before using the following procedures.
		At least five, preferably 10, sanded, machined, die cut, or molded in a mold with the dimensions specified for type S and L specimen.
Test procedures	Notches shall be machined in accordance with ISO 2818. The radius of the notch base shall be 1.0 ± 0.02 mm, with an angle of 45° ± 1°. The two notches shall be at right angles to its principal axis on opposite sides with a distance between the two notches of 6 ± 0.2 mm. The two lines drawn perpendicular to the length direction of the specimen through the apex of each notch shall be within 0.02 mm of each other.	Specimens are unnotched.
	The selected pendulum shall consume at least 20%, but not more than 80% of its stored energy in breaking the specimens. If more than one pendulum satisfies these conditions, the pendulum having highest energy is used.	Use the lowest capacity pendulum available, unless the impact values go beyond the 85% scale reading. If this occurs, use a higher capacity pendulum.
	Run three blank tests to calculate the mean frictional loss. The loss should not exceed 1% for a 2.0-J pendulum and 0.5% for those specified pendulums with a 4.0-J or greater energy pendulum.	A friction and windage correction may be applied. A nonmandatory appendix provides the necessary calculations to determine the amount of this type of correction.

TABLE 11.22 Comparison of Test Methods Between ISO 10350-1 and ASTM Approaches* (*Continued*)

Property	ISO/IEC methods as specified by ISO 10350-1:99	ASTM methods
Tensile impact strength (*Continued*):		
	Determine the energy correction, using method A or B, before one can determine the notched tensile impact strength, E_n: Method A—energy correction due to the plastic deformation and kinetic energy of the crosshead, E_q Method B—crosshead-bounce energy, E_b	The bounce correction factor may be applied. A nonmandatory appendix provides the necessary calculations to determine the amount of this correction factor. (A curve must be calculated for the cross head and pendulum used before applying in bounce correction factors.)
	Calculate the notched tensile impact strength, E_n by dividing the corrected energy (method A or B) by the cross-sectional area b]etween the two notches.	Calculate the corrected impact energy to break by subtracting the friction and windage correction and/or the bounce correction factor from the scale reading of energy to break.
Values and units	Notched tensile impact strength, E_n (kJ/m^2)	Tensile impact energy (J)
	Thermal Properties	
Melting temperature:		
Standard	ISO 11357-3:98	D 3418-97
Specimen	Molding compound	Powders, granules, pellets or molded part cut with a microtome, razor blade, hypodermic punch, paper punch or cork borer; slivers cut from films and sheets.
Apparatus	DSC	DSC or DTA
	Calibrate the temperature measuring system periodically over the temperature range used for the test.	Using the same heating rate to be used for specimen, calibrate the temperature scale with the appropriate reference materials covering the materials of interest.
Test procedures	Sample mass of up to 50 mg is recommended.	Sample weight of 5 mg is recommended. An appropriate sample will result in 25 to 95% of scale deflection.

	Perform and record a preliminary thermal cycle by heating the specimen at a rate of 10 K/min under inert gas from ambient to 30 K above the melting point to erase previous thermal history.	Perform and record a preliminary thermal cycle by heating the specimen at a rate of 10°C/min under nitrogen from ambient to 30°C above the melting point to erase previous thermal history.
	Hold for 10 min at temperature.	Hold for 10 min at temperature.
	Cool to 50°C below the peak crystallization temperature at a rate of 10 K/min.	Cool to 50°C below the peak crystallization temperature at a rate of 10°C/min.
	Immediately repeat heating under inert gas at a rate of 10 K/min.	Repeat heating as soon as possible under nitrogen at a rate of 10°C/min.
Values and units	T_p—peak melting point(s) from the second heat cycle (°C or K)	T_m—melting point(s) from the second heat cycle (°C)
Glass transition temperature:		
Standard	ISO 11357-2:98	D 3418-97
Specimen	Molding compound	Powders, granules, pellets, or molded part cut with a microtome, razor blade, hypodermic punch, paper punch or cork borer; slivers cut from films and sheets.
Apparatus	DSC	DSC or DTA
	Calibrate the temperature measuring system periodically over the temperature range used for the test.	Using the same heating rate to be used for specimen, calibrate the temperature scale with the appropriate reference materials covering the materials of interest.
Test procedures	Sample mass of 10–20 mg is satisfactory.	Sample weight of 10–20 mg is recommended.
	Perform and record an initial thermal cycle up to a temperature high enough to erase previous thermal history, by using a heating rate of 20° ± 1° K/min in 99.9% pure nitrogen or other inert gas.	Perform and record a preliminary thermal cycle by heating the specimen at a rate of 20°C/min in air or nitrogen from ambient to 30°C above the melting point to erase previous thermal history.
	Hold temperature until a steady state is achieved (usually 5–10 min).	Hold for 10 min at temperature.

TABLE 11.22 Comparison of Test Methods Between ISO 10350-1 and ASTM Approaches* (*Continued*)

Property	ISO/IEC methods as specified by ISO 10350-1:99	ASTM methods
Glass transition temperature (*Continued*):		
	Quench cool at a rate of at least (20° ± 1°) K/min to well below the T_g (usually 50 K below).	Quench cool to 50°C below the transition peak of interest.
	Hold temperature until a steady state is reached (usually 5–10 min)	
	Reheat at a rate (20° ± 1°) K/min and record heating curve until all desired transitions are recorded.	Repeat heating as soon as possible at a rate of 20°C/min until all desired transitions have been completed.
Values and units	T_{mg}, midpoint temperature (°C)	T_m (T_g), midpoint temperature (°C) T_f(T_g) extrapolated onset temperature (°C) For most applications T_f is more meaningful than T_m and may be designated as T_g in place of the midpoint of the T_g curve.
Temperature of deflection under load:		
Standard	ISO 75-1:93 and 75-2:93	D 648-98c
Specimen	Flatwise—80 × 10 × 4 mm, cut from the ISO 3167 type A specimen. Edgewise—110 × 10 × 4 mm.	Edgewise—120 ± 10 × 3 − 13 × 12.7 ± 0.3 mm
Conditioning	Specimen conditioning, including any postmolding treatment, shall be carried out at 23°C ± 2°C and 50 ± 5% RH for a minimum length of time of 88 h, except where special conditioning is required as specified by the appropriate material standard.	At 23 ± 2°C and 50 ± 5% relative humidity for not less than 40 h prior to testing in accordance with Procedure A of Method D618.
Apparatus	The contact edges of the supports and the loading nose radius are rounded to a radius of 3.0 ± 0.2 mm and shall be longer than the width of the test specimen. Specimen supports: 64 mm apart (flatwise specimens)	The contact edges of the supports and loading nose shall be rounded to a radius of 3.0 ± 0.2 mm. Specimen supports shall be 100 ± 2 mm apart.

Apparatus	Heating bath shall contain a suitable liquid (for example, liquid paraffin, glycerol, transformer oil, and silicone oils) which is stable at the temperature used and does not affect the material under the test (for example, swelling, softening, or cracking). An efficient stirrer shall be provided with a means of control so that the temperature can be raised at a uniform rate of 120 K/h ± 10 K/h. This heating rate shall be considered to be met if over every 6-min interval during the test, the temperature change is 12 K ± 1 K.	Immersion bath shall have a suitable heat-transfer medium (for example, mineral or silicone oils) which will not affect the specimen and which is safe at the temperatures used. It should be well stirred during the test and provided with means of raising the temperature at a uniform rate of 2 ± 0.2°C. This heating rate is met if over every 5-min interval the temperature of the bath shall rise 10 ± 1°C at each specimen location.
	A calibrated micrometer dial-gauge or other suitable measuring instrument capable of measuring to an accuracy of 0.01 mm deflection at the midpoint of the test specimen shall be used.	The deflection measuring device shall be capable of measuring specimen deflection to at least 0.25 mm and is readable to 0.01 mm or better.
Test procedures	At least two unannealed specimens.	At least two specimens shall be used to test each sample at each fiber stress of 0.455 MPa ± 2.5% or 1.820 MPa ± 2.5%.
	The temperature of the heating bath shall be 20 to 23°C at the start of each test, unless previous tests have shown that, for the particular materials under test, no error is introduced by starting at other temperatures.	The bath temperature shall be about room temperature at the start of the test unless previous tests have shown that, for a particular material, no error is introduced by starting at a higher temperature.
	Apply the calculated force to give the desired nominal surface stress. Allow the force to act for 5 min to compensate partially for the creep exhibited at room temperature when subjected to the specified nominal surface stress. Set the reading of the deflection measuring instrument to zero.	Apply the desired load to obtain the desired maximum fiber stress of 0.455 or 1.82 MPa to the specimen. Five minutes after applying load, adjust the deflection measuring device to zero/starting position.
	Heating rate (120 ± 10°C/h)	Heating rate (2.0 ± 0.2°C/min)
	Deflections 0.32 mm (edgewise) for 10.0 to 10.3 mm height 0.34 mm (flatwise) for height equal to 4 mm	The deflection when the specimen is positioned edgewise is 0.25 mm for a specimen with a depth of 12.7 mm

TABLE 11.22 Comparison of Test Methods Between ISO 10350-1 and ASTM Approaches* (*Continued*)

Property	ISO/IEC methods as specified by ISO 10350-1:99	ASTM methods
Temperature of deflection under load (*Continued*):		
	Note the temperature at which the test specimen reaches the deflection corresponding to height of the test specimen, as the temperature of deflection under load for the applied nominal surface stress.	Record the temperature at which the specimen has deflected the specific amount, as the deflection temperature at either 0.455 or 1.820 MPa.
Values and units	HDT at 1.8 MPa and 0.45 MPa or 8 MPa (°C)	HDT at 0.455 MPa or 1.820 MPa (°C)
Vicat softening temperature:		
Standard	ISO 306:94	D 1525-98
Specimen	10 × 10 × 4 mm, from middle region of the ISO 3167 multipurpose test specimen	The specimen shall be flat, between 3 and 6.5 mm thick, and at least 10 × 10 mm in area or 10 mm in diameter.
Conditioning	Specimen conditioning, including any postmolding treatment, shall be carried out at 23°C ± 2°C and 50 ± 5% RH for a minimum length of time of 88 h, except where special conditioning is required as specified by the appropriate material standard.	If conditioning of the test specimens is required, then condition at 23 ± 2°C and 50 ± 5% relative humidity for not less than 40 h prior to testing in accordance with test method D618.
Apparatus	The indenting tip shall preferably be of hardened steel 3 mm long, of circular cross section 1.000 ± 0.015 mm^2 fixed at the bottom of the rod. The lower surface of the indenting tip shall be plane and perpendicular to the axis of the rod and free from burrs.	A flat-tipped hardened steel needle with a cross-sectional area of 1.000 ± 0.015 mm^2 shall be used. The needle shall protrude at least 2 mm at the end of the loading rod.
	Heating bath containing a suitable liquid (for example, liquid paraffin, glycerol, transformer, and silicone oil) which is stable at the temperature used and does not affect the material under test (for example, swelling or cracking) in which the test specimen can be immersed to a depth of at least 35 mm is used. An efficient stirrer shall be provided.	Immersion bath containing the heat transfer medium (for example, silicone oil, glycerine, ethylene glycol, and mineral oil) that will allow the specimens to be submerged at least 35 mm below the surface.
Test procedures	At least two specimens to test each sample.	Use at least two specimens to test each sample. Molding conditions shall be in accordance with the applicable material

		specification or should be agreed upon by the cooperating laboratories. Specimens shall be annealed only if required in the material specification.
	Specimens tested flatwise.	Specimens tested flatwise.
	The temperature of the heating equipment should be 20 to 23°C at the start of each test, unless previous tests have shown that, for the material under test, no error is caused by starting at another temperature.	The bath temperature shall be 20 to 23°C at the start of the test unless previous tests have shown that, for a particular material, no error is introduced by starting at a higher temperature.
	Mount the test specimen horizontally under the indenting tip of the unloaded rod. The indenting tip shall at no point be nearer than 3 mm to the edge of the test specimen.	Place the specimen on the support so that it is approximately centered under the needle. The needle should not be nearer than 3 mm.
	Put the assembly in the heating equipment.	Lower the needle rod (without extra load) and then lower the assembly into the bath.
	After 5 min with the indenting tip still in position, add the weights to the load-carrying plate so that the total thrust on the test specimen is 50 ± 1 N.	Apply the extra mass required to increase the load on the specimen to 10 ± 0.2 N (loading 1) or 50 ± 1.0 N (loading 2).
	Set the micrometer dial gauge reading to zero.	After waiting 5 min, set the penetration indicator to zero.
	Increase the temperature of the heating equipment at a uniform rate heating rate (50 ± 5°C/h).	Start the temperature rise at one of these rates: 50 ± 5°C/h (rate A) or 120 ± 12°C/h (rate B). The rate selection shall be agreed upon by the interested parties.
	Note the temperature at which the indenting tip has penetrated in to the test specimen by 1 ± 0.01 mm beyond the starting position, and record it as the Vicat softening temperature of the test specimen.	Record the temperature at which the penetration depth is 1 mm. If the range of the temperatures recorded for each specimen exceeds 2°C, then record the individual temperatures and rerun the test.
Values and units	Vicat softening temperature (°C)	Vicat softening temperature (°C)

TABLE 11.22 Comparison of Test Methods Between ISO 10350-1 and ASTM Approaches* (*Continued*)

Property	ISO/IEC methods as specified by ISO 10350-1:99	ASTM methods
Coefficient of linear thermal expansion (CLTE):		
Standard	ISO 11359-2:99	E 831-93
Specimen	Prepared from ISO 3167 multipurpose test specimen cut from the middle parallel region.	Specimen shall be between 2 and 10 mm in length and have flat and parallel ends to within ±25 μm. Lateral dimensions shall not exceed 10 mm.
Conditioning	No conditioning requirements given. If the specimens are heated or mechanically treated before testing, then it should be noted in the report.	No conditioning requirements given. If the specimens are heated or mechanically treated before testing, then it should be noted in the report.
Apparatus	TMA	TMA
Test procedures	Three specimens are required.	Three specimens are required.
	Measure the initial specimen length in the direction of the expansion to ±25 μm at room temperature.	Measure the initial specimen length in the direction of the expansion to ±25 μm at room temperature.
	Place the specimen in the specimen holder in the furnace. If measurements at subambient temperatures are to be made, then cool the specimen to at least 20°C below the lowest temperature of interest.	Place the specimen in the specimen holder in the furnace. If measurements at subambient temperatures are to be made, then cool the specimen to at least 20°C below the lowest temperature of interest.
	Heat the specimen at a constant heating rate of 5°C/min over the desired temperature range and record changes in specimen length and temperature to all available decimal places.	Heat the specimen at a constant heating rate of 5°C/min over the desired temperature range and record changes in specimen length and temperature to all available decimal places.
	Determine the instrument baseline by repeating the two steps above without a specimen present.	Determine the instrument baseline by repeating the two steps above without a specimen present.
	The measured change in length of the specimen should be corrected for the instrument baseline.	The measured change in length of the specimen should be corrected for the instrument baseline.
	Record the secant value of the expansion vs. temperature over the temperature range of 23°C to 55°C.	Select a temperature range from a smooth portion of the thermal curves in the desired temperature range, then obtain the change in length over that temperature range.

Values and units	Coefficient of linear thermal expansion [μm/(m·°C)]	Coefficient of linear thermal expansion [μm/(m·°C)]
Flammability (linear burning rate of horizontal specimens):		
Standard	ISO 1210:92, Method A	D 635-98
Specimen	125 × 13 × 3 mm. (Additional specimen thickness <3 mm may be used.)	125 × 12.5 mm in thickness normally supplied (3–12 mm cut from sheet or molded).
Conditioning	Specimen conditioning, including any postmolding treatment, shall be carried out at 23°C ± 2°C and 50 ± 5% RH for a minimum length of time of 88 h, except where special conditioning is required as specified by the appropriate material standard.	As received, unless otherwise specified.
Apparatus	Laboratory burner in accordance with ISO 10093	Laboratory burner in accordance with D 5025-94.
Test procedures	Three specimens.	At least 10 specimens.
Values and units	Burning rate (mm/min) (If additional specimens with thicknesses <3 mm are tested, the specimen thickness must also be reported.)	Average time of burning (s) Average extent of burning (mm)
Flammability (after flame and after-glow times of vertical specimens):		
Standard	ISO 1210:92 Method B	D 3801-96
Specimen	125 × 13 × 3 mm. (Additional specimen thickness <13 mm may be used.)	125 × 13 × 3 mm. (Additional specimen thickness <13 mm may be used.)
Conditioning	Two sets of five specimens at 23 ± 2°C and 50 ± 5% relative humidity for at least 48 h; two sets of five specimens at 70 ± 1°C for 168 h ± 2 h.	One set of five specimens at 23 ± 2°C and 50 ± 5% relative humidity for at least 48 h; second set of five specimens at 70 ± 1°C for 168 h.
Apparatus	Laboratory burner in accordance with ISO 10093. Barrel length is 100 ± 10 mm and inside diameter 9.5 ± 0.3 mm.	Bunsen Tirrill type burner of tube length 95 ± 6 mm and inside diameter 9.5 (+1.6 mm, −0.0 mm).
Test procedures	Technical grade methane gas or natural gas having a heat content of approximately 37 MJ/m^3.	Technical grade methane gas or natural gas having energy density approximately 37 MJ/m^3.
Values and units	Flame classification.	Flame classification.

TABLE 11.22 Comparison of Test Methods Between ISO 10350-1 and ASTM Approaches* (*Continued*)

Property	ISO/IEC methods as specified by ISO 10350-1:99	ASTM methods
Ignitability:		
Standard	ISO 4589-2:96, Procedure A	D 2863-97
Specimen	80 × 10 × 4 mm cut from the center of the ISO 3167 multipurpose specimen (ISO 4589-2, Form 1).	70 − 150 × 6.5 × 3 mm.
Conditioning	Specimen conditioning, including any post molding treatment, shall be carried out at 23°C ± 2°C and 50 ± 5% RH for a minimum length of time of 88 h, except where special conditioning is required as specified by the appropriate material standard.	As received, unless otherwise agreed upon.
Apparatus	Test chimney dimensions of 450 mm in height × 75 mm minimum diameter cylindrical bore. The upper outlet shall be restricted as necessary to produce an exhaust velocity of at least 90 mm/s from a flow rate within the chimney of 30 mm/s. The base of the chimney will preferably have a layer of glass beads (3–5 mm in diameter) between 80 and 100 mm deep. The specimen shall be held by a small clamp which is at least 15 mm away from the nearest point at which the specimen may burn.	Test column of heat-resistant glass tube 450 mm in height × 75 mm minimum inside diameter. The base of the column contains a noncombustible material which can evenly distribute the gas mixture. A layer of glass beads (3–5 mm in diameter) between 80 and 100 mm deep has been found suitable. The specimen shall be held by a small clamp that will support the specimen at its base and hold it vertically in the center of the column.
	The moisture content of the gas entering the chimney shall be <0.1% (m/m) and the variation in oxygen concentration rising in the chimney, below the level of the test specimen, is <0.2% (V/V).	The flow control and measuring devices shall be such that the volumetric flow of each gas into the column is within 1% of the range being used.
	The flame ignitor is a tube with an outlet of 2 ± 1 mm diameter which projects a 16 ± 4 mm flame vertically downward from the outlet when the tube is vertical within the chimney. The flame fuel shall be propane without premixed air.	The flame ignitor is a tube with a small orifice 1–3 mm in diameter, which projects a flame 6 to 25 mm long. The flame fuel can be propane, hydrogen or other gas flame.
Test procedures	A minimum of 15 specimens shall be prepared in accordance with the relevant material standard. When none exists, or unless otherwise specified, specimens shall be directly compression or injection molded in accordance with ISO 293 or ISO 294-1.	A sufficient number of specimens (normally five to 10).
	Test specimens shall be marked 50 mm from the end to be ignited.	No marking indicated.

	Select the initial concentration of oxygen to be used based on experience with similar materials, or ignite the specimen in air and note the burning behavior. Select an initial concentration approx. 8%, approx. 21% or 25% (V/V) depending on the burning behavior.	Select the initial concentration of oxygen to be used based on experience with similar materials, or ignite the specimen in air and note the burning behavior. Select an initial concentration approx. 18%, or approx. 25% depending on the burning behavior.
	Specimen is mounted such that the top of the specimen is at least 100 mm below the top of the chimney, and the lowest exposed part of the specimen is 100 mm above the top of the gas distribution device.	Specimen is mounted vertically in approximate center of the column with the top of specimen at least 100 mm below the top of the column.
	Gas flow rate of 40 ± 10 mm/s, must flow at least 30 min prior to ignition.	Gas flow rate of 40 ± 10 mm/s, must flow at least 30 min prior to ignition.
	Apply the flame, with a sweeping motion to the top of the specimen for up to 30 s, removing it every 5 s to determine if the top is burning.	Ignite the entire top of the specimen so that the specimen is well lit, then remove the flame.
	Commence timing the period of burning. If the burning ceases but spontaneous combustion occurs in <1 s, continue timing. If period and extent of burning does not exceed 180 s and 50 mm, then the oxygen concentration would need to be incrementally increased. Adjust the oxygen concentration either up or down until there are two concentrations which differ by <1.0% and in which one specimen met the criteria and the other did not. Repeat the test on four more specimens.	Start timing. If the burning time and the extent of burning does not exceed 180 s and 50 mm, then the oxygen concentration would need to be incrementally increased. Adjust the oxygen either up or down until the critical concentration of oxygen is determined. This is the lowest level which meets the 180-s/50-mm criteria. At the next lower oxygen concentration that will give a difference in oxygen index of 0.2% or less, the specimen should not meet the 180-s/50-mm criteria. Repeat the test on three more specimens, but starting at a slightly different flow rate, yet within 30–50% (V/V).
Values and units	Average oxygen index [% (V/V)]	Average oxygen index [% (V/V)]
	Electrical Properties	
Relative permittivity:		
Standard	IEC 60250:69	D 150-95

TABLE 11.22 Comparison of Test Methods Between ISO 10350-1 and ASTM Approaches* (*Continued*)

Property	ISO/IEC methods as specified by ISO 10350-1:99	ASTM methods
Relative permittivity (*Continued*):		
Specimen geometry	>80 × >80 × 1 mm. (Greater thickness may be used for those materials that cannot be molded reliably at 1-mm thickness.)	Test specimens are of suitable shape and thickness determined by the material specification or by the accuracy of measurement required, and the frequency at which the measurements are to be made.
Conditioning	Specimen conditioning, including any postmolding treatment, shall be carried out at 23°C ± 2°C and 50 ± 5% RH for a minimum length of time of 88 h, except where special conditioning is required as specified by the appropriate material standard.	Clean the test specimen with a suitable solvent or as prescribed in the material specification. Use Recommended Practice D1371 as a guide to the choice of suitable cleaning procedures.
Test procedures	100 Hz and 1 MHz.	Frequency not specified but recorded.
	Null methods are used at frequencies up to 50 MHz and results are compensated for electrode edge effects.	Null method with resistive or inductive ratio arm capacitance bridge suggested for frequencies of <1 Hz to a few MHz.
Values and units	Relative permittivity (unitless)	Relative permittivity (unitless)
Dissipation factor:		
Standard	IEC 60250:69	D 150-95
Specimen geometry	>80 × >80 × 1 mm. (Greater thickness may be used for those materials that cannot be molded reliably at 1-mm thickness.)	Test specimens are of suitable shape and thickness determined by the material specification or by the accuracy of measurement required, and the frequency at which the measurements are to be made.
Conditioning	Specimen conditioning, including any postmolding treatment, shall be carried out at 23°C ± 2°C and 50 ± 5% RH for a minimum length of time of 88 h, except where special conditioning is required as specified by the appropriate material standard.	Clean the test specimen with a suitable solvent or as prescribed in the material specification. Use Recommended Practice D1371 as a guide to the choice of suitable cleaning procedures.
Test procedures	100 Hz and 1 MHz.	Frequency not specified but recorded.
	Null methods are used at frequencies up to 50 MHz and results are compensated for electrode edge effects.	Null method with resistive or inductive ratio arm capacitance bridge suggested for frequencies of <1 Hz to a few MHz.

Values and units	Dissipation factor (unitless)	Dissipation factor (unitless)
Volume/surface resistivity:		
Standard	IEC 60093:80	D 257-93
Specimen geometry	>80 × >80 × 1 mm. (Greater thickness may be used for those materials that cannot be molded reliably at 1-mm thickness.)	Minimum 50-mm diameter × up to 3-mm-thick specimens.
Conditioning	Specimen conditioning, including any postmolding treatment, shall be carried out at 23°C ± 2°C and 50 ± 5% RH for a minimum length of time of 88 h, except where special conditioning is required as specified by the appropriate material standard.	At 23 ± 2°C and 50 ± 5% relative humidity for not less than 40 h, according to D 618-96.
Test procedures	100 V DC applied for 1 min for surface resistivity and 100 min for volume resistivity.	500 ± 5 V DC for 1 min (surface and volume resistivity).
Values and units	Volume resistivity (Ω-m) Surface resistivity (Ω)	Volume resistivity (Ω-cm) Surface resistivity (Ω)
Electric strength:		
Standard	IEC 60243-1:88	D 149-95a, Method B
Specimen	>80 × >80 × 1 mm or 3 mm, sufficiently wide to prevent discharge along the surface.	Thickness not specified but measured. It shall be of sufficient size to prevent flashover under the conditions of the test.
Conditioning	Specimen conditioning, including any postmolding treatment, shall be carried out at 23°C ± 2°C and 50 ± 5% RH for a minimum length of time of 88 h, except where special conditioning is required as specified by the appropriate material standard.	If not specified in the applicable material specification, follow the procedures in Practice D 618–96.

TABLE 11.22 Comparison of Test Methods Between ISO 10350-1 and ASTM Approaches* (*Continued*)

Property	ISO/IEC methods as specified by ISO 10350-1:99	ASTM methods
Electric strength (*Continued*):		
Apparatus	Two coaxial cylinder electrodes (25 mm diameter × 25 mm and 75 mm diameter × 15 mm) with edges rounded to 3-mm radius.	Electrode type 6: Two coaxial cylinder electrodes (25 mm diameter × 25 mm and 75 mm diameter × 15 mm) with edges rounded to 3 mm radius.
Test procedures	Immersion in transformer oil in accordance with IEC60296.	Immersion in mineral oil, meeting D3487 type I or II requirements.
	Power frequencies between 48–62 Hz.	60 Hz, unless otherwise specified.
	20 s step-by-step.	60 ± 5 s step-by-step.
Values and units	Dielectric strength (kV/mm)	Dielectric strength (kV/mm)
Comparative tracking index:		
Standard	IEC 60112:79	D 3638-93
Specimen geometry	>15 × >15 × 4 mm from the shoulder of the ISO 3167 multipurpose test specimen.	Sample size is 50- or 100-mm disk with minimum thickness of 2.5 mm. Thin samples are to be clamped together to get minimum thickness.
Conditioning	Specimen conditioning, including any postmolding treatment, shall be carried out at 23°C ± 2°C and 50 ± 5% RH for a minimum length of time of 88 h, except where special conditioning is required as specified by the appropriate material standard.	In accordance with Procedure A of Practice D618–96
Apparatus	Two platinum electrodes of rectangular cross section 5 × 2 mm, with one end chisel edged with an angle of 30° and slightly rounded.	Two platinum electrodes of rectangular cross section 5 × 2 mm, with one end chisel edged with an angle of 30° and slightly rounded.
	Electrodes are symmetrically arranged in a vertical plane, the total angle between them being 60° and with opposing faces vertical and 4.0 ± 0.1 mm apart on the specimen surface. Force exerted on the surface by the electrode is 1.0 ± 0.05 N.	Position the electrodes so that the chisel edges contact the specimen at a 60° angle and the chisel faces are parallel in the vertical plane and are separated by 4 ± 0.2 mm.
Test procedures	0.1 ± 0.002% by mass ammonium chloride in distilled or deionized water (solution A) with a resistivity of 395 ± 5 Ω-cm at 23 ± 1°C.	0.1 ± 0.002% by mass ammonium chloride in distilled or deionized water (solution A) with a resistivity of 395 ± 5 Ω-cm at 23 ± 1°C.
	Voltage between 100 and 600 V at frequency between 46–60 Hz.	Voltage should be limited to 600 V at a frequency of 60 Hz.

	Determine maximum voltage at which no failure occurs at 50 drops on five sites. This is the CTI provided no failure occurs below 100 drops when the voltage is dropped by 25 V.	Plot the number of drops of electrolyte at breakdown versus voltage. The voltage which corresponds to 50 drops is the CTI.
	At least five test sites (can be on one specimen).	At least five specimen of each sample shall be tested.
Values and units	CTI (V)	CTI (V)

Other Properties

Water absorption:

Standard	ISO 62:99	D 570-98
Specimen and geometry	50 ± 1 mm square or diameter disks × 3 ± 0.2 mm thick for 24 h immersion and <1 mm thick for saturation values.	50.8-mm-diameter × 3.2-mm disk for molded plastics. The thickness shall be measured to the nearest 0.025 mm.
Conditioning	Dry specimens in an oven for 24 ± 1 h at 50 ± 2°C, allow to cool to ambient temperature in the desiccator and weigh to the nearest 1 mg.	Specimens of a material whose water absorption value is appreciably affected by temperatures close to 110°C, shall be dried in an oven for 24 h at 50 ± 3°C, cooled in a desiccator, and immediately weighed to the nearest 0.001 g. Specimens of a material whose water absorption value is not appreciably affected by temperatures up to 110°C, shall be dried in an oven for 1 h at 105 to 110°C. (No weighing requirement is given in the method; however the specimen should be weighed immediately to the nearest 0.001 g.) When data comparisons with other plastics are desired, the specimens shall be dried in oven for 24 h at 50 ± 3°C, cooled in a desiccator, and immediately weighed to the nearest 0.001 g.
Test procedures	Three specimens shall be prepared in accordance with the relevant material standard. When none exists, or unless otherwise specified, specimens shall be directly compression or injection molded in accordance with ISO 293 or ISO 294-1.	Three specimens shall be tested. No specimen preparation conditions are given.

TABLE 11.22 Comparison of Test Methods Between ISO 10350-1 and ASTM Approaches* (*Continued*)

Property	ISO/IEC methods as specified by ISO 10350-1:99	ASTM methods
Water absorption	(*Continued*):	
	The volume of water shall be at least 8 mL/cm^2 of the total surface area of the test specimen.	No specifics given on the volume of water required.
	Place the conditioned specimens in a container of distilled water, controlled at 23°C with a tolerance of ±0.5 or ±2.0°C according to the relevant material standard. In absence of such standard, the tolerance shall be ±0.5°C. After immersion for 24 ± 1 h, take the specimens from the water and remove all surface water with a clean, dry cloth or with filter paper. Reweigh the specimens to the nearest 1 mg within 1 min of taking them out of the water (method 1).	The conditioned specimens shall be placed in a container of distilled water maintained at 23 ± 1°C, and shall rest on edge and be entirely immersed. At 24 (+0.5, −0) h, the specimens shall be removed one at a time and wiped off with a dry cloth and weighed to the nearest 0.0001 g immediately.
	Saturation values in water or air at 50% relative humidity at 23°C.	Long-term immersion—To determine the saturation value, the specimens are tested according to the 24-h procedure, except after weighing the specimen are replaced in the water. The weighings shall be repeated at the end of the first week and every 2 weeks thereafter until the increase in weight per 2-week period, as shown by three consecutive weighings, averages less than 1% of the total increase in weight or 5 mg, whichever is greater. The difference between the saturated and dry weight shall be considered the water absorbed when substantially saturated.
	If it is desired to allow for the presence of water-soluble matter, dry the test specimens again for 24 ± 1 h in the oven controlled at 50 ± 2°C, after completion of method 1. Allow the specimen to cool to ambient temperature in the desiccator and reweigh to the nearest 1 mg (method 2). The percentage of water absorbed is a total of the % weight increase after immersion either by methos 1 or 2.	Materials with known or suspected to contain appreciable amounts of water-soluble ingredients shall be reconditioned for the same time and at the same temperature as used for conditioning the specimen originally. If the weight of the specimen is less than the original conditioned weight, then that difference in weight shall be considered as water-soluble matter lost during the immersion test. The percentage of water absorbed is a total of the % weight increase (to be noted whether it is 24 h or saturation) and the % soluble matter lost.

Values and units	Water absorption (24 h) (% wt)	Water absorption (24 h or saturation) (% wt)
Density:		
Standard	ISO 1183:87 Methods A, B, C, or D	D 792-98 Method A or B, or D1505-88 (Reapproved in 1990)
Specimen geometry	The specimen shall be of convenient size to give adequate clearance between the specimen and the beaker (a mass of 1–5 g is often convenient). This specimen shall be taken from the center portion of the ISO 3167 multipurpose test specimen.	Single piece of material of any size or shape that can be conveniently prepared, provided that its volume shall be not less than 1 cm^3 and its surface and edges made smooth, with a thickness of at least 1 mm for each 1 g weight (D 792 requirements).
		The test specimens should have dimensions that permit the most accurate position measurement of the center of volume of the suspended specimen. If it is suspected that interfacial tension affects the equilibrium position of the specimens in the thickness range from 0.025–0.051 mm, then films not less than 0.127 mm thick should be tested (D 1505 requirements).
Conditioning	Specimen conditioning, including any postmolding treatment, shall be carried out at 23°C ± 2°C and 50 ± 5% RH for a minimum length of time of 88 h, except where special conditioning is required as specified by the appropriate material standard.	At 23 ± 2°C and 50 ± 5% relative humidity for not less than 40 h according to D 618-96.
Test procedures	Any one of the four methods—immersion in liquid, pyknometer, titration, or density gradient column.	D 792—displacement in water (method A) or other liquids (method B). D1505—density gradient column.
	Three specimens are required for method D (density gradient column).	Several specimens are required for D792. Three specimens are required for D1505.
Values and units	Density (kg/m^3)	Density (kg/m^3)

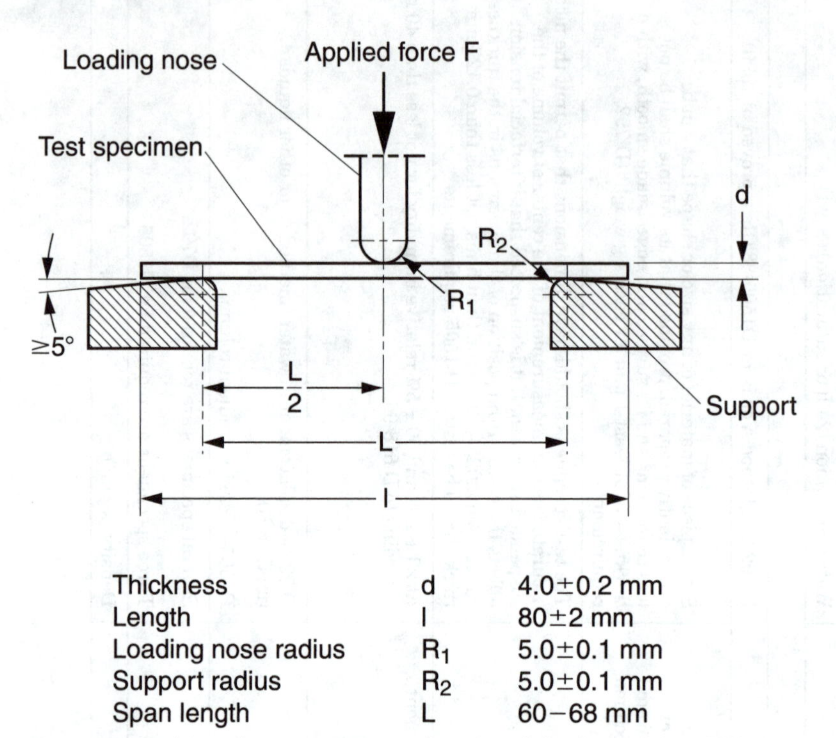

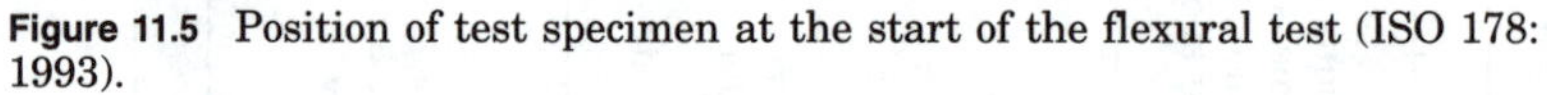

Figure 11.5 Position of test specimen at the start of the flexural test (ISO 178: 1993).

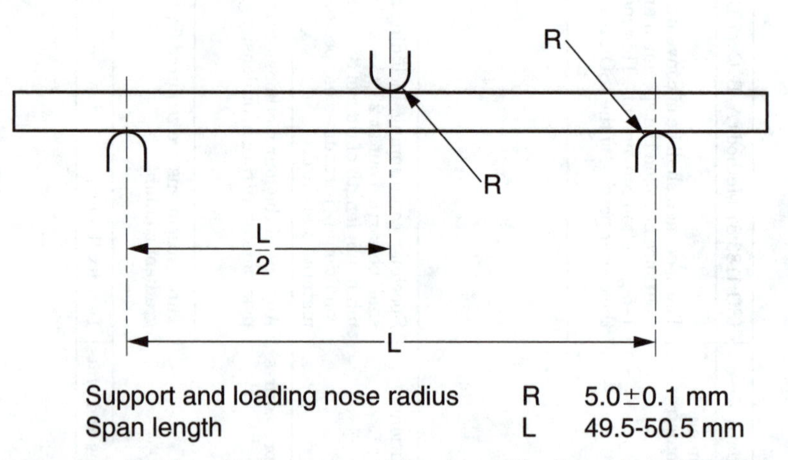

Figure 11.6 Position of test specimen at the start of the flexural test (ASTM D 790-98).

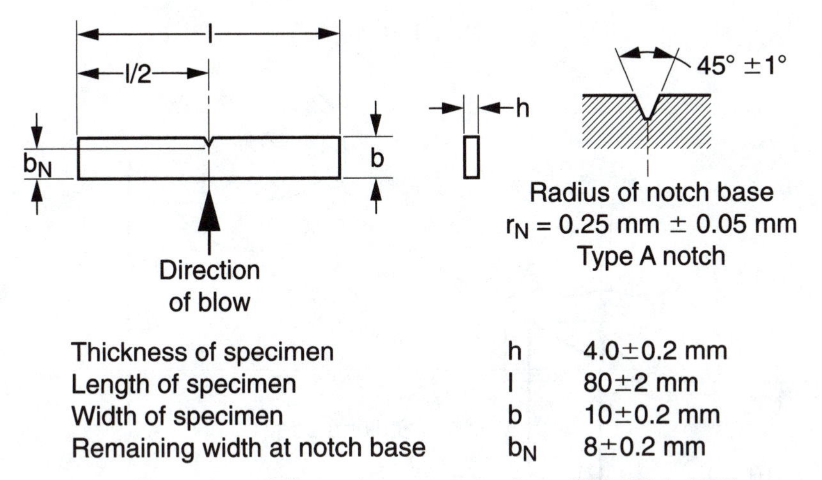

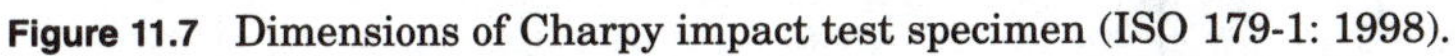
Figure 11.7 Dimensions of Charpy impact test specimen (ISO 179-1: 1998).

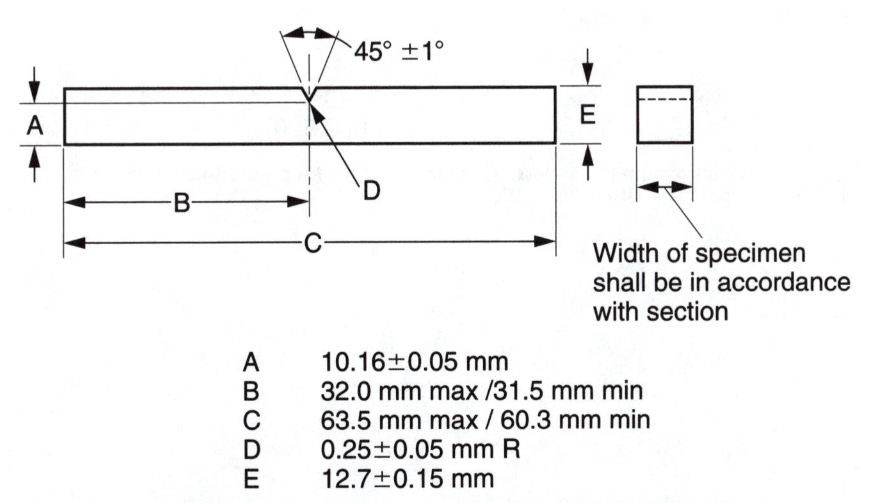

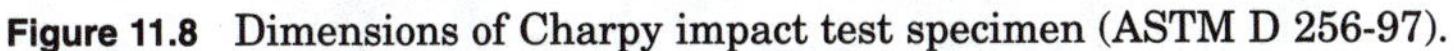
Figure 11.8 Dimensions of Charpy impact test specimen (ASTM D 256-97).

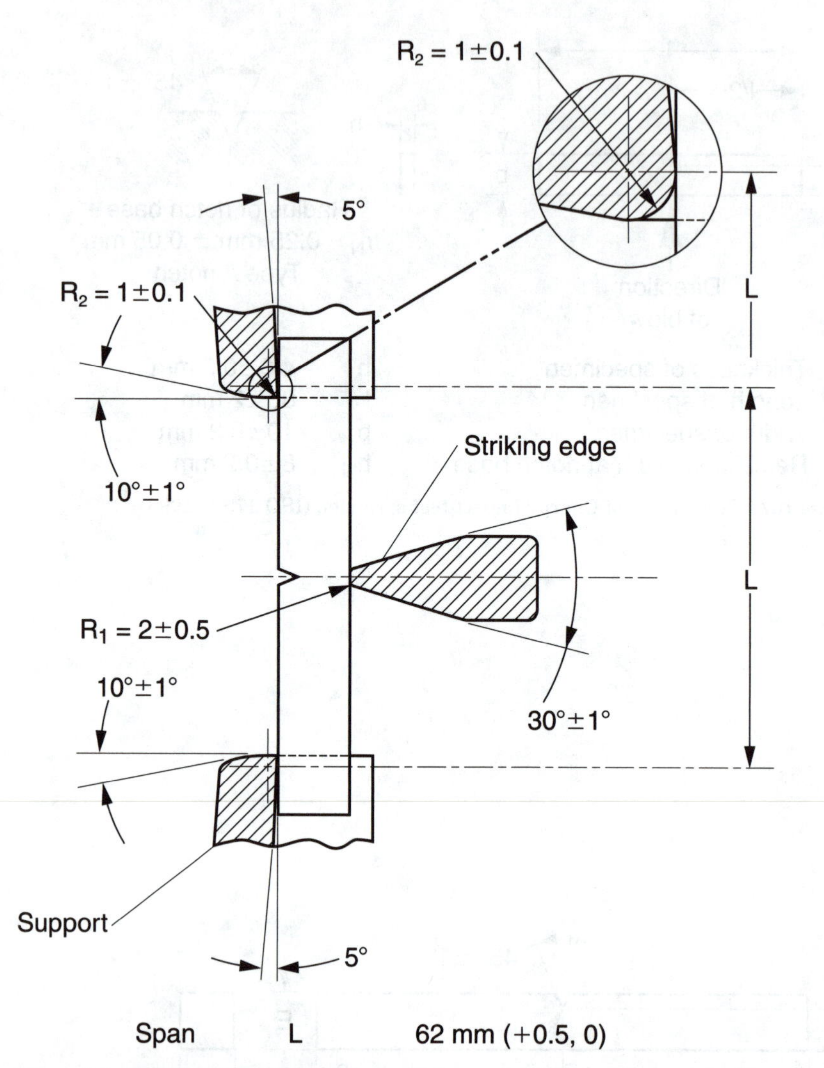

Figure 11.9 Relationship of support, specimen, and striking edge to each other in Charpy impact test (ISO 179-1: 1998).

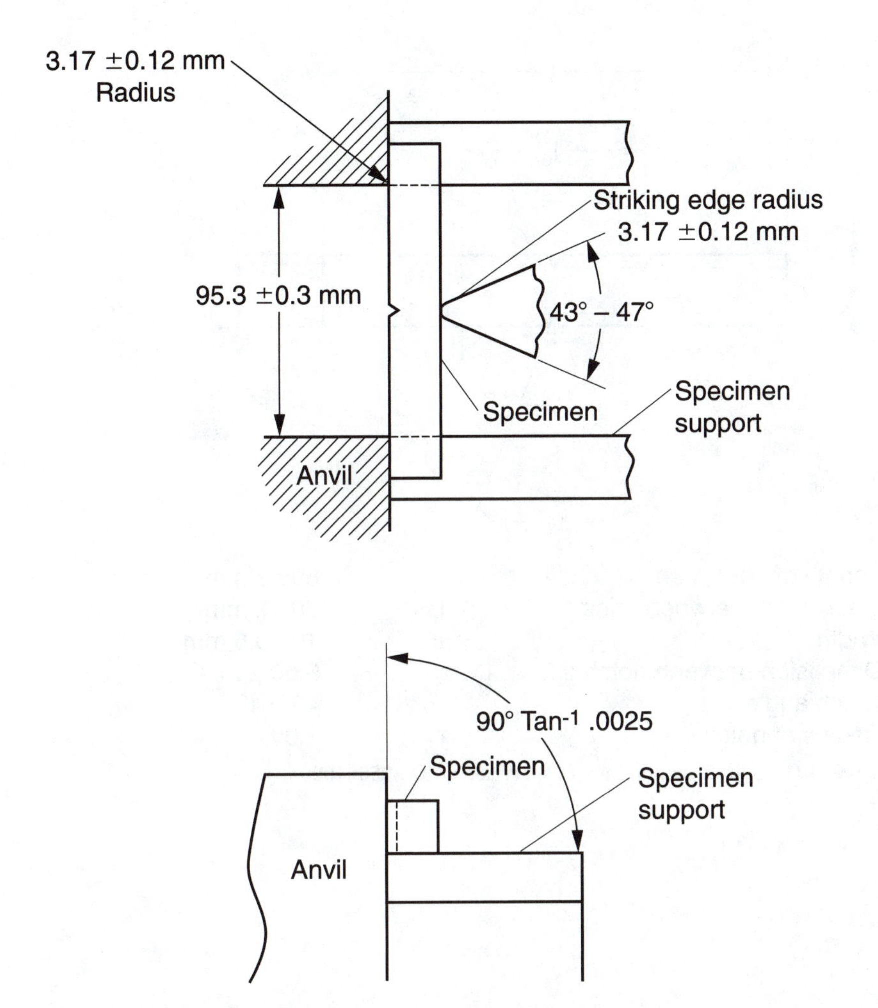

Figure 11.10 Relationship of anvil, specimen, and striking edge to each other in Charpy impact test (ASTM D 256-97).

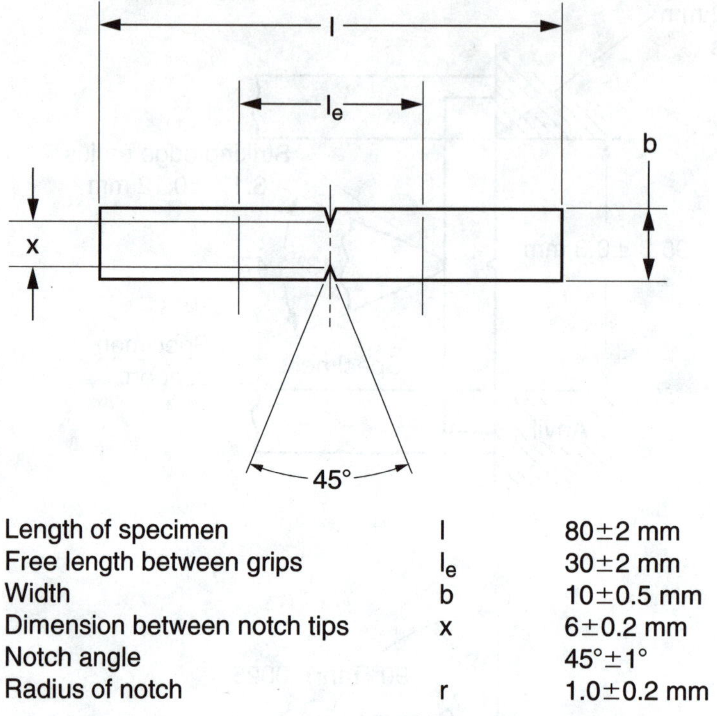

Length of specimen	l	80±2 mm
Free length between grips	l_e	30±2 mm
Width	b	10±0.5 mm
Dimension between notch tips	x	6±0.2 mm
Notch angle		45°±1°
Radius of notch	r	1.0±0.2 mm

Figure 11.11 Type 1 tensile impact specimen (ISO 8256: 1990).

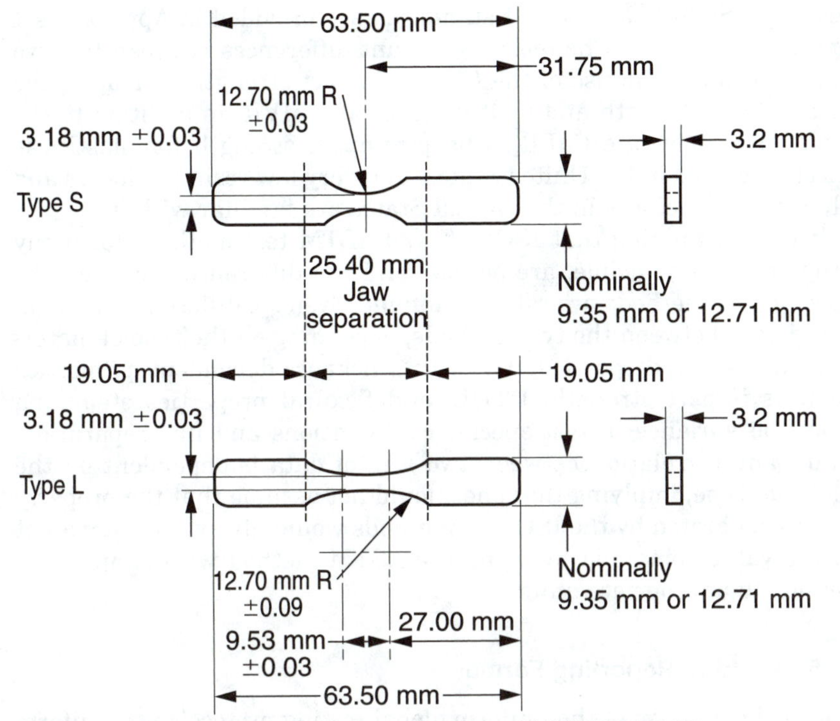

Figure 11.12 Type S and L tensile impact specimens (ASTM D 1822-93).

relevant ISO/IEC and ASTM standards are provided in Appendices 1 and 2, respectively. The most significant differences between the two approaches are the use of the Charpy impact strength instead of the Izod impact strength and melt volume rate (MVR) in addition to the melt mass flow rate (MFR). Charpy impact strength for plastics is rarely reported in the United States, and very few resin producers and almost no processors in the United States are familiar with it.

It is worth noting that ISO/IEC and ASTM test methods for many properties, in principle, are similar and the differences between the specific test methods are rather minimal. Even so, differences in measured data between the two methods, depending on the type of material, are expected, particularly with thickness-dependent properties such as impact strength, DTUL, and flexural properties stemming from the variance in test specimen dimensions and its preparation. Thus, any correlation between two sets of data is dependent on the material type, implying that one should not assume that the property values generated by the ISO test methods would always be equivalent to the values obtained by using the ASTM method with simple conversion to appropriate units.

11.5 Uniform Reporting Format

A logical extension of the uniform global testing protocols is a uniform data reporting format. Standardization of the format for reporting data is deemed essential to facilitate the much desired "apples-to-apples" comparison among thousands of materials available from different suppliers around the world. Part 2 of the ISO material standards for each material attempts to address this issue by employing a uniform data template selected from Table 2 in ISO 10350-1 standard for comparable data.

ISO 10350-1 is also the basis for acquisition and presentation of single-point data of the most successful and widely used materials database for plastics known as CAMPUS* (Computer Aided Material Preselection by Uniform Standards). Comparable multipoint data standards ISO 11403-1 and 2 are the basis for a comprehensive set of multipoint data in CAMPUS, such as:

- Isothermal stress-strain curves over wide range of temperatures
- Shear modulus–temperature curve
- Creep modulus versus time curves at 23°C and elevated temperatures at several stress levels

*Registered trademark of Chemie Wirtschaftförderungs-Gesellschaft mbH (CWFG), Frankfurt am Main, Germany.

- Isochronous creep curves at 23°C and elevated temperatures
- Viscosity–shear rate curves at three temperatures within the recommended processing temperatures range
- Specific enthalpy versus temperature curves
- Pressure–specific volume–temperature (pVT) curves

CAMPUS, led by BASF AG, Bayer AG, Degussa-Hüls AG, Ticona, The Dow Chemical Company, and DuPont Engineering Polymers, has the distinction of being the first and the only material data system that provides truly comparable data specifically developed through international cooperation among more than 50 resin suppliers worldwide and available *free* of charge directly from each participating resin producer.

Simultaneous display of all the relevant product data on a single screen is a convenient feature of the current Version 4.1 (Fig. 11.13). As one scrolls through the product listing in the top left window, data for the product corresponding to the cursor location is instantaneously displayed in the other three windows. The program has advanced search features to help find one or more material that can meet the criteria set by the user. Besides the ability to conduct standard queries by specify-

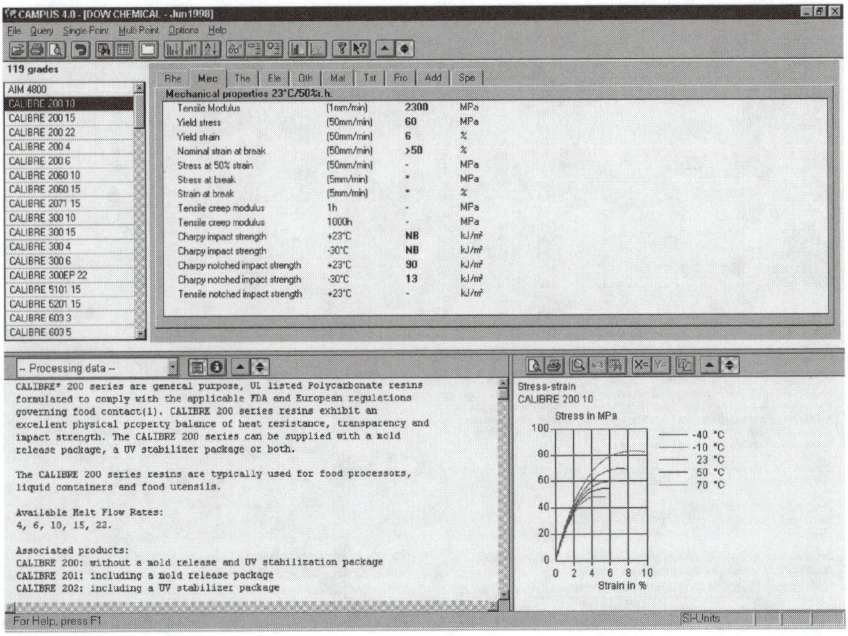

Figure 11.13 Layout of the screen display of product list, single-point data, multipoint data, and product description in separate window simultaneously.

ing minimum and maximum values for one or more properties or to find a grade which is within a specified bandwidth of the properties of another grade, the user has the ability to conduct queries graphically. Comparison of products in a tabular form on the basis of selected single-point properties is illustrated in Fig. 11.14. The graphical comparison feature of the program which allows overlaying selected multipoint data for up to 10 marked grades is also demonstrated in Fig. 11.14. CAMPUS also includes advanced graphical representation features like

- Two-dimensional scatter diagrams (two-dimensional property correlation maps) for displaying selected pairs of properties, as illustrated by Fig. 11.15
- Bubble charts
- Spider diagrams to quickly visualize and compare multiple products on the basis of selected set of properties (Fig. 11.16)

The content and presentation formats of CAMPUS have been continuously upgraded since its first introduction in 1988. Multipoint data was integrated in the program in Version 2 in 1989, and conversion from a DOS format to a Windows-based format was implemented with the introduction of Version 3. The thermal parameters essential

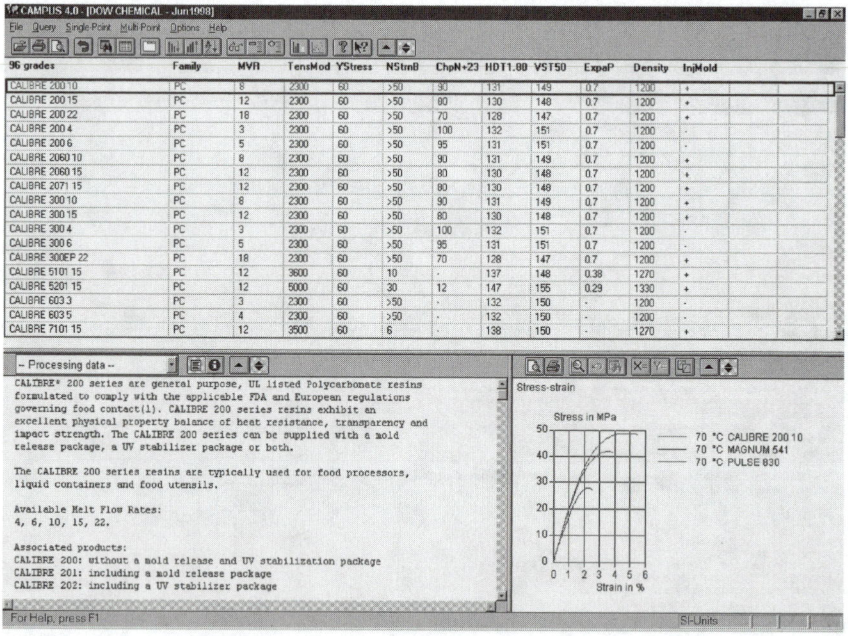
CAMPUS 4.0 - [DOW CHEMICAL - Jun1998]
File Query Single-Point Multi-Point Options Help
96 grades | Family | MVR | TensMod | YStress | NStrnB | ChpN+23 | HDT1.80 | VST50 | ExpaP | Density | InjMold
CALIBRE 200 10 | PC | 8 | 2300 | 60 | >50 | 90 | 131 | 149 | 0.7 | 1200 | +
CALIBRE 200 15 | PC | 12 | 2300 | 60 | >50 | 80 | 130 | 148 | 0.7 | 1200 | +
CALIBRE 200 22 | PC | 18 | 2300 | 60 | >50 | 70 | 128 | 147 | 0.7 | 1200 | +
CALIBRE 200 4 | PC | 3 | 2300 | 60 | >50 | 100 | 132 | 151 | 0.7 | 1200 | -
CALIBRE 200 6 | PC | 5 | 2300 | 60 | >50 | 95 | 131 | 151 | 0.7 | 1200 | -
CALIBRE 2060 10 | PC | 8 | 2300 | 60 | >50 | 90 | 131 | 149 | 0.7 | 1200 | +
CALIBRE 2060 15 | PC | 12 | 2300 | 60 | >50 | 80 | 130 | 148 | 0.7 | 1200 | +
CALIBRE 2071 15 | PC | 12 | 2300 | 60 | >50 | 80 | 130 | 148 | 0.7 | 1200 | +
CALIBRE 300 10 | PC | 8 | 2300 | 60 | >50 | 90 | 131 | 149 | 0.7 | 1200 | +
CALIBRE 300 15 | PC | 12 | 2300 | 60 | >50 | 80 | 130 | 148 | 0.7 | 1200 | +
CALIBRE 300 4 | PC | 3 | 2300 | 60 | >50 | 100 | 132 | 151 | 0.7 | 1200 | -
CALIBRE 300 6 | PC | 5 | 2300 | 60 | >50 | 95 | 131 | 151 | 0.7 | 1200 | -
CALIBRE 300EP 22 | PC | 18 | 2300 | 60 | >50 | 70 | 128 | 147 | 0.7 | 1200 | +
CALIBRE 5101 15 | PC | 12 | 3600 | 60 | 10 | - | 137 | 148 | 0.38 | 1270 | +
CALIBRE 5201 15 | PC | 12 | 5000 | 60 | 30 | 12 | 147 | 155 | 0.29 | 1330 | +
CALIBRE 603 3 | PC | 3 | 2300 | 60 | >50 | - | 132 | 150 | - | 1200 | -
CALIBRE 603 5 | PC | 4 | 2300 | 60 | >50 | - | 132 | 150 | - | 1200 | -
CALIBRE 7101 15 | PC | 12 | 3500 | 60 | 6 | - | 138 | 150 | - | 1270 | +
-- Processing data --
CALIBRE* 200 series are general purpose, UL listed Polycarbonate resins formulated to comply with the applicable FDA and European regulations governing food contact(1). CALIBRE 200 series resins exhibit an excellent physical property balance of heat resistance, transparency and impact strength. The CALIBRE 200 series can be supplied with a mold release package, a UV stabilizer package or both.
The CALIBRE 200 series resins are typically used for food processors, liquid containers and food utensils.
Available Melt Flow Rates:
4, 6, 10, 15, 22.
Associated products:
CALIBRE 200: without a mold release and UV stabilization package
CALIBRE 201: including a mold release package
CALIBRE 202: including a UV stabilizer package
Stress-strain
Stress in MPa
50
40
30
20
10
0
0 1 2 3 4 5 6
Strain in %
70 °C CALIBRE 200 10
70 °C MAGNUM 541
70 °C PULSE 830
For Help, press F1
SI-Units

Figure 11.14 Illustration of the ability to compare properties of products side by side in a tabular form (top half window) and in graphical form (bottom right window).

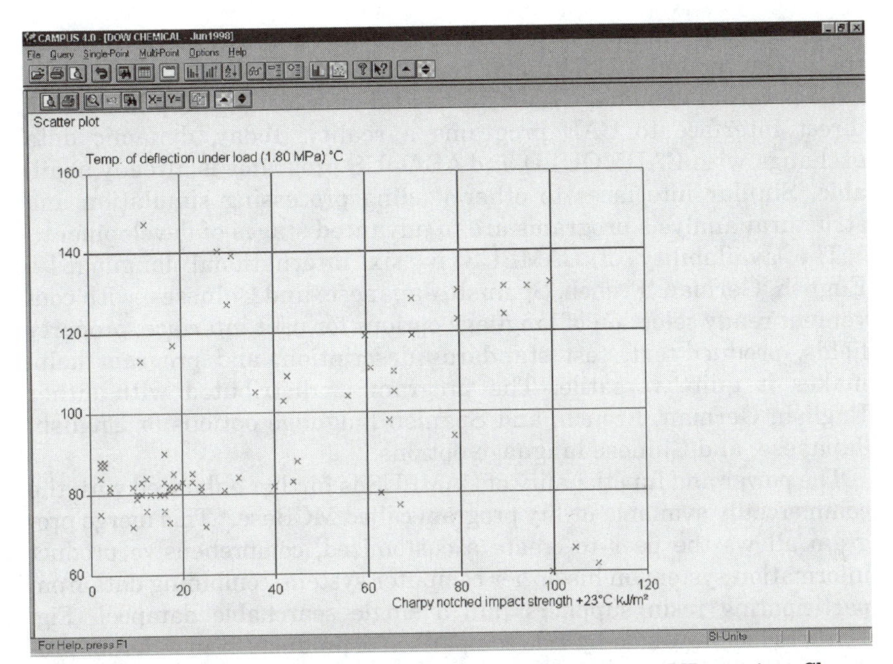

Figure 11.15 Illustration of the scatter diagram of DTUL at 1.8 MPa against Charpy notched impact strength for selected products.

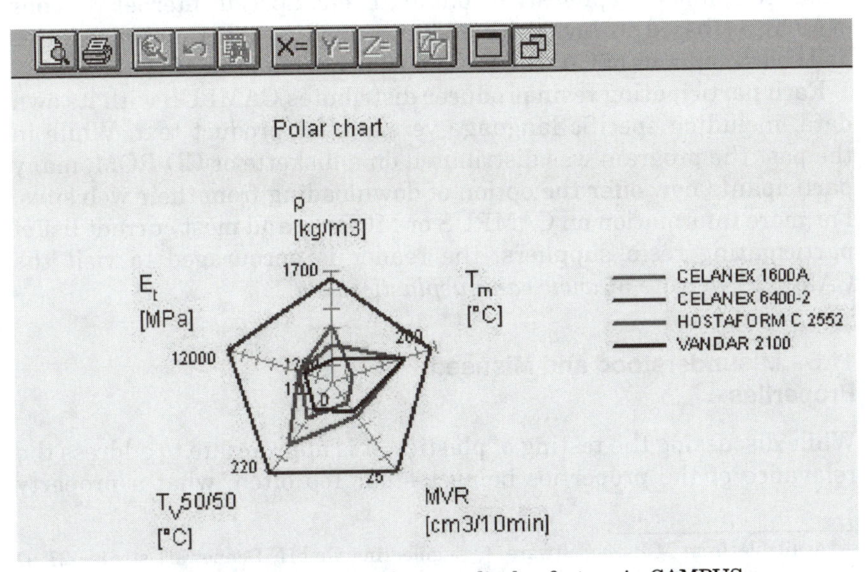

Figure 11.16 Illustration of the spider diagram display feature in CAMPUS.

for processing simulation, pvT data, and DSC curves were included in the latest version 4.1. Through close cooperation with designers and CAE program vendors, CAMPUS has taken a leading step in making direct interface to CAE programs a reality. Today, dynamic data exchange with CADMOULD and ABAQUS programs is already available. Similar interfaces to other leading processing simulation and structural analysis programs are in advanced stages of development.

The availability of CAMPUS in six international languages—English, German, French, Spanish, Japanese, and Chinese—with convenient ready selection of language options for user interface, property fields, product text, test standards description, and program help, makes it quite versatile. The program is distributed with either English, German, French, and Spanish language options or English, Japanese, and Chinese language options.

The power and functionality of CAMPUS is further enhanced with the commercially available utility program called MCBase.* This merge program allows the user to create a customized, comprehensive product information system on his or her computer system, combining data from participating resin suppliers into a single searchable datapool (Fig. 11.17) to facilitate side-by-side comparison of products from different producers. While offering the same query and display capabilities as CAMPUS, this utility offers additional search features by producer and polymer family (listed by uniform abbreviations in accordance with international guidelines). Additional features of the merge program include a data export function to CAE programs and other design software tools, links to applications database, etc. Special Internet versions of MCBase (based on Java) with full functionality—all graphical features and search options of CAMPUS—is now available from M-Base.

Each participating resin producer distributes CAMPUS with its own data, including specific language versions for product text. While in the past the program was distributed on a diskette or CD-ROM, many participants now offer the option of downloading from their web sites. For more information on CAMPUS or MCBase and most current list of participating resin suppliers, the reader is encouraged to visit the CAMPUS website at *www.campusplastics.com.*

11.6 Misunderstood and Misused Properties

While discussing the testing of plastics, it is appropriate to address the relevance of the properties being tested. Too often, what a property

*Available from M-Base Software + Engineering GmbH, Dennewart strasse 27, D 52068, Aachen, Germany, and its global distributor network.

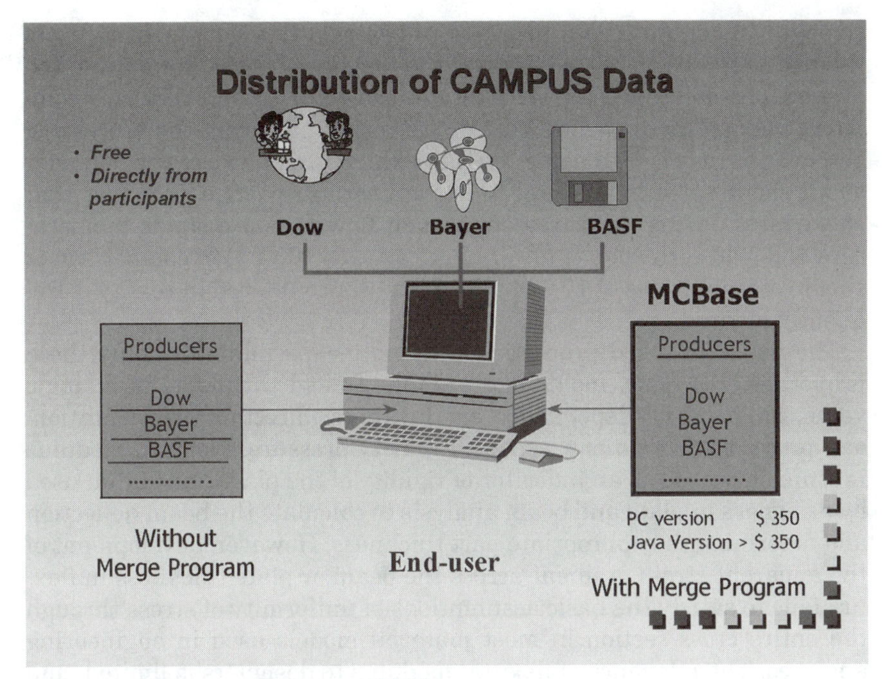

Figure 11.17 Data merge features of MCBase program.

really means and how it is applicable in practice is greatly misunderstood. Although it may provide little indication of the performance of the plastic under actual use conditions, there seems to be a reluctance in the plastics industry to discontinue relying heavily on these properties to determine the suitability of a resin for the intended application. Among the properties that are often misunderstood are those commonly reported in the product datasheets, such as

Property	As indicator of
Notched impact strength	Toughness
Hardness	Scratch and abrasion resistance
HDT @ 0.45 MPa and 1.8 MPa (66 and 264 psi) and Vicat softening temperature	Maximum use temperature or continuous use temperature
MFR	Processability

Although both DTUL at 0.45 and 1.8 MPa are arbitrarily defined reference temperatures, they are mistakenly used as an indicator of the maximum-use temperature or continuous-use temperature of the plastic material. Limitations of DTUL in describing the temperature dependence of modulus is discussed extensively by Sepe[34–36] and Nunnery.[37] Similarly, notched impact strength is often used in the industry as a differentiating criteria among materials for toughness. Notched impact strength, though a useful property to describe the notch sensitivity of the material, does not shed light on the toughness of the material in the absence of a notch or when stress states are not triaxial. Processors commonly rely on the melt flow rate of the resin and expect the processing behavior of resins with same MFR to exhibit similar viscosity behavior at all shear rates in the processing temperature range. The fact that shear rates during measurement of melt flow rate at a single temperature and load corresponds to very low range ($<100\ s^{-1}$) in comparison to shear rates in excess of $10{,}000\ s^{-1}$ typically encountered in reality is frequently ignored.

The list of misused properties is by no means limited to only these properties. Too many moldmakers think of mold shrinkage as a single value, ignoring its dependence on thickness, direction of orientation, and processing conditions, especially cavity pressure. Flexural modulus is commonly used as an indicator of rigidity of the plastic and often used by designers in plate and beam analysis to calculate the beam deflection and determine the appropriate part thickness. However, development of the apparent stress gradient across the beam or plate thickness in flexure fails to satisfy the basic assumptions of uniformity of stress through the entire cross section in most material models used in engineering analyses. The relevance of flexural modulus to designers is limited, and tensile modulus is more appropriate for these computations.[38] For the same reason, the most useful creep data to the designer is creep modulus under tension mode, although creep in bending is reported frequently.

In the case of filled or reinforced plastics, the effect of anisotropy is usually magnified, often with significant differences in properties between flow and the transverse directions and the properties measured from large molded parts generally differ from those obtained with standard test specimens. It is, therefore, essential to evaluate the properties in tension, creep, and coefficient of linear thermal expansion, etc., in both directions for the filled or reinforced materials.

Suitable alternatives to the misunderstood properties discussed previously are

Misused property	Suggested alternative
Flexural modulus	Tensile modulus
HDT @ 0.45 MPa and 1.8 MPa (66 and 264) and Vicat softening temperature	Modulus versus temperature curve obtained from dynamic mechanical analysis (DMA)
Molding shrinkage	Shrinkage as a function of thickness, cavity pressure, and injection velocity
Melt flowrate	Viscosity-shear rate data within the processing temperatures range

11.7 Costs of Data Generation

The resin suppliers are often criticized by processors, OEMs, and CAE program vendors alike that there is an unwillingness on their part to provide comprehensive data on the products they offer. At the outset, this criticism may appear to be a legitimate one. However, this misconception stems from

- A lack of understanding of the cost of data generation by the designers and especially CAE program vendors.
- A false sense of confidence among designers who feel that having more data will automatically translate to better predictions. Very often, possible contributions to any discrepancies between predicted versus measured values from various assumptions that are made in the chosen models are ignored.

The on-going proliferation of new models by CAE program vendors in their codes on a routine basis, leaves the designers confused with a large number of material models to choose from with very little guidance on which of those models are applicable to plastics for prediction of the intended response to various end-use conditions. To make matters worse, the program codes will not allow the analysis to proceed unless some value is entered against each input parameter. Consequently, the resin suppliers are faced with an unrealistic demand for significant amounts of data, without consideration of how much it really costs. Since not all designers employ the same CAE software tool, whether they are performing structural analysis or processing simulation, the resin suppliers have to generate the data required for all of the CAE software tools available in the marketplace. This means the resin suppliers have to bear the additional burden of cost in generating multiple sets of nonstandardized data.

The arguments for comprehensive sets of material data are very compelling. A comprehensive property profile for the particular grade chosen for a given application of course is expected to enhance the confidence of the design engineer. By allowing a more realistic assessment of the material under close to actual service environments, it may avoid premature failure of the designed component. Additionally, it could also eliminate use of larger "design safety factors," which result in "overengineering" or "overdesign." This is not only unwarranted but adds to the total part cost. However, the application-specific nature of performance requirements and the enormous time and cost involved in developing such exhaustive amount of data, including multiple sets of nonstandard data for each resin grade available today in the marketplace, is simply impractical. As resin suppliers are under mounting pressure to reduce their costs, it is important to put the direct costs of material data measurement in perspective.

11.7.1 How expensive is material data measurement?

Just how expensive is material data generation? The findings from the survey of 12 independent testing laboratories—eight testing facilities from the United States, three organizations from Europe, and one from Latin America—by Shastri and Baur[39] are reproduced in Table 11.23 to illustrate the individual costs of testing each property. The large variations in the cost estimates among the laboratories surveyed is readily evident.

For a single lot of material, just the cost of generating single-point data alone is reported to exceed $2600 (Table 11.24), while the cost of generating a multipoint data package comprising tensile stress-strain curves at five temperatures (23°C and four additional temperatures), dynamic modulus versus temperature at 1-Hz frequency, and tensile creep curves at four stress levels at each of the three temperatures (23°C and two elevated temperatures) amount to nearly $9000, as shown in Table 11.25.

The individual costs of generating material data needed for performing injection-molding simulation, including advanced shrinkage/warpage analysis (Table 11.26), comes to almost $8400 compared to a discounted price of $3000 to $4000 for group tests offered by some of the independent laboratories.

If the cost of specimen molding is added, the minimum cost of material data generation easily adds up to $20,000 per material (Table 11.27). Even if one takes advantage of group test discounts, it is reasonable to expect the cost of data generation would be at least $15,000 per material.

Testing a minimum of five production lots of the material (automotive certification packages require testing 30 production lots for some properties) to produce data that is truly representative, the total cost can easily exceed $75,000 per material.

There are various reasons why the cost of data generation is so expensive, the chief among them being:

- Large number of tests involved
- High costs of monitoring long-term tests like tensile creep, environmental influences, fatigue, etc.
- Risks associated with test failure either due to power failure or equipment damage, especially at the later stages of long-term tests
- High energy costs associated with testing at elevated temperatures
- Frequency with which the test is performed
- Whether applicable test standard exists

TABLE 11.23 Survey Results on Cost of Material Data Measurement

Test matrix	Range, $	Average cost, $
I. Single-Point Data		
Rheological properties:		
Melt volume rate	50–145	75
Mechanical properties:		
Tensile properties:		
Tensile modulus	40–340	140
Tensile stress at yield		
Yield strain	40–240	125
Elongation at break		
Tensile stress at break		
Impact properties:		
Unnotched Charpy impact strength	35–150	100
Notched Charpy impact strength @ 23°C	35–200	115
Notched Charpy impact strength @ −30°C	35–250	135
Puncture impact @ 23°C	90–195	145
Puncture impact @ −30°C	115–220	165
Tensile impact energy	90 90	
Thermal properties:		
Melt temperature	70–220	125
Glass transition temperature	70–220	135
DTUL @ 0.45 MPa	60–175	115
DTUL @ 1.8 MPa	60–175	115
Vicat softening temperature	60–145	100
CLTE	35–250	160
Electrical properties:		
Relative permittivity	35–160	100
Dissipation factor	35–160	100
Volume resistivity	25–155	100
Surface resistivity	25–155	100
Electrical strength	145–155	150
Comparative tracking index	35 35	
Other properties:		
Density	25–125	60
Water absorption	50–300	120
II. Multipoint Data		
Mechanical properties:		
Tensile stress-strain curves at @ 23°C	75–372.50	900
Tensile stress-strain curves @ additional temperature	75–372.50	900
Tensile creep data @ 23°C	800–3000	2000
Tensile creep data @ temperature > 23°C	2075–3000	2500
Dynamic modulus	90–350	220
Thermal properties:		
DSC data	90–255	170
Rheological properties:		
Viscosity-shear rate data @ three temperatures	255–855	400
III. Test Specimen Molding	75–260	155

TABLE 11.24 Cost of Generating Single-Point Data from Survey of 12 Independent Testing Facilities

Property	Cost range, $	Average cost per material, $
General properties	75–425	180
Rheological properties	50–145	75
Mechanical properties	480–1685	1015
Thermal properties	355–1185	750
Electrical properties	300–820	585
Total	1260–4260	2605

TABLE 11.25 Cost of Generating Multipoint Data Package from Survey of 12 Independent Testing Facilities

Property	Cost range, $	Average cost per material, $
Time- and rate-dependent mechanical properties	5,415–11,212.5	8357.5
Thermal properties	90–255	170
Viscosity–shear rate data	255–855	400
Total	5,760–12,322.5	8,927.5

TABLE 11.26 Individual Costs of Generating Data Needed for Injection-Molding Simulation

Property	Cost range, $	Average cost per material, $
Mold filling simulation:		
Viscosity-shear rate data @ three temperatures	750–900	825
Melt density	165–175	170
Thermal conductivity	175–220	200
Specific heat	100–220	160
No flow temperature	150–165	160
Ejection temperature	100–110	105
Shrinkage/warpage simulation:		
Shrinkage coefficients	3200–4000	3600
Isothermal pvT data	1500–2200	1850
CLTE in parallel and perpendicular directions	300–330	315
Anisotropic effect on tensile modulus	300–350	325
Poisson's ratio	385–550	470
In-plane shear modulus	180–250	215
Total	7305–9470	8395

TABLE 11.27 Cost of Generating Single-Point Data from Survey of 12 Independent Testing Facilities

Property	Cost range, $	Average cost per material, $
Specimen molding	75–260	155
Single-point data	1,260–4,260	2,605
Multipoint data package	5,760–12,322.5	8,927.5
Molding simulation data package	7,305–9,470	8,395
Total	14,400–26,312.5	20,082.5

11.7.2 A pragmatic solution

In view of the enormous cost associated with material data measurement, a pragmatic approach for resin suppliers is to measure only the data which is really necessary and useful to the design engineers. For example,

1. Make core sets of material data that essentially constitute the minimum data required for carrying out the engineering analyses available for "key" products in their portfolio, since not all commercial grades are intended for large-volume applications involving complex molded parts.
2. Provide additional data on other products on "as-needed basis."
3. Assist customers in identifying/providing data for an "equivalent" product as similarities and minor variations within a given product portfolio are rather common.

When customers do not find the data for a specific product, the resin suppliers encourage the customers to contact them directly and seek assistance in identifying and obtaining data for an equivalent product. If the application volume justifies generating additional material data needed, no resin supplier is unwilling to invest in the data measurement.

Appendix 11.1 Selected ISO/IEC Standards/Documents*

Standard	*Title*
ISO 62: 1999	Plastics—Determination of Water Absorption
ISO 75-1: 1993	Plastics—Determination of Temperature of Deflection under Load—Part 1: General Test Method

* Available from American National Standards Institute, 11 West 42nd Street, New York, NY 10035.

ISO 75-2: 1993	Plastics—Determination of Temperature of Deflection under Load—Part 2: Plastics and Ebonite
ISO 75-3: 1993	Plastics—Determination of Temperature of Deflection under Load—Part 3: High-Strength Thermosetting Laminates and Long-Fibre Reinforced Plastics
ISO 178: 1993	Plastics—Determination of Flexural Properties
ISO 179-1: 1998	Plastics—Determination of Charpy Impact Properties—Part 1: Noninstrumented Impact Test
ISO 179-2: 1997	Plastics—Determination of Charpy Impact Properties—Part 2: Instrumented Impact Test
ISO 180: 1998	Plastics—Determination of Izod Impact Strength of Rigid Materials
ISO 291: 1997	Plastics—Standard Atmospheres for Conditioning and Testing
ISO 293: 1999	Plastics—Compression Moulding Test Specimens of Thermoplastic Materials
ISO 294-1: 1996	Plastics—Injection Moulding of Test Specimens of Thermoplastic Materials—Part 1: General Principles, Multipurpose-Test Specimens (ISO Mould Type A) and Bars (ISO Mould Type B)
ISO 294-3: 1996	Plastics—Injection Moulding of Test Specimens of Thermoplastic Materials—Part 3: Plates (ISO Moulds Type D)
ISO 294-4: 1997	Plastics—Injection Moulding of Test Specimens of Thermoplastic Materials—Part 4: Determination of Moulding Shrinkage
ISO 306: 1994	Plastics—Determination of Vicat Softening Temperature
ISO 527-1: 1993	Plastics—Determination of Tensile Properties Part 1: General Principles
ISO 527-2: 1993	Plastics—Determination of Tensile Properties Part 2: Test Conditions for Moulding and Extrusion Plastics
ISO 527-4: 1997	Plastics—Determination of Tensile Properties Part 4: Test Conditions for Isotropic and Orthotropic Fiber Reinforced Plastics
ISO 604: 1993	Plastics—Determination of Compressive Properties
ISO 899-1: 1993	Plastics—Determination of Creep Behaviour—Part 1: Tensile Creep
ISO 899-2: 1993	Plastics—Determination of Creep Behaviour—Part 2: Flexural Creep by Three-Point Bending

ISO 1133: 1997	Plastics—Determination of the Melt Mass-Flow Rate and the Melt Volume-Flow Rate of Thermoplastics
ISO 1183: 1987	Plastics—Methods for Determining the Density and Relative Density of Non-Cellular Plastics
ISO 1210: 1992	Plastics—Determination of Flammability Characteristics of Plastics in the Form of Small Specimens in Contact with a Small Flame
ISO 1628-1: 1998	Guidelines for the Standardization of Methods for the Determination of Viscosity Number and Limiting Viscosity Number of Polymers in Dilute Solution—Part 1: General Conditions
ISO 1628-2: 1999	Guidelines for the Standardization of Methods for the Determination of Viscosity Number and Limiting Viscosity Number of Polymers in Dilute Solution—Part 2: Poly(vinyl Chloride) Resins
ISO 1628-3: 1991	Guidelines for the Standardization of Methods for the Determination of Viscosity Number and Limiting Viscosity Number of Polymers in Dilute Solution—Part 3: Polyethylenes and Polypropylenes
ISO 1628-4: 1999	Guidelines for the Standardization of Methods for the Determination of Viscosity Number and Limiting Viscosity Number of Polymers in Dilute Solution—Part 4: Polycarbonate (PC) Moulding and Extrusion Materials
ISO 1628-5: 1998	Guidelines for the Standardization of Methods for the Determination of Viscosity Number and Limiting Viscosity Number of Polymers in Dilute Solution—Part 5: Poly(alkylene Terephthalates)
ISO 1628-6: 1990	Guidelines for the Standardization of Methods for the Determination of Viscosity Number and Limiting Viscosity Number of Polymers in Dilute Solution—Part 6: Methyl Methacrylate Polymers
ISO 2577: 1984	Plastics—Thermosetting Moulding Materials—Determination of Shrinkage
ISO 2818: 1994	Plastics—Preparation of Test Specimens by Machining
ISO 3167: 1993	Plastics—Multipurpose-Test Specimens
ISO 4589-2: 1996	Plastics—Determination of Flammability by Oxygen Index
ISO 6603-2: 1998	Plastics—Determination of Puncture Impact Behaviour of Rigid Plastics—Part 2: Instrumented Puncture Test

ISO 6721-1: 1994	Plastics—Determination of Dynamic Mechanical Properties. Part 1: General Principles
ISO/ 6721-2: 1994	Plastics—Determination of Dynamic Mechanical Properties. Part 2: Torsion Pendulum
ISO 6721-5: 1995	Plastics—Determination of Dynamic Mechanical Properties. Part 5: Flexural Vibration—Non-Resonance Method
ISO 6721-7: 1997	Plastics—Determination of Dynamic Mechanical Properties. Part 7: Torsional Vibration—Non-Resonance Method
ISO 6721-10: 1998	Plastics—Determination of Dynamic Mechanical Properties. Part 10: Complex Shear Viscosity Using a Parallel Plate Oscillatory Rheometer
ISO 8256: 1990	Plastics—Determination of Tensile-Impact Strength
ISO 10350-1: 1999	Plastics—The Acquisition and Presentation of Comparable Single-Point Data
ISO 10351: 1999	Plastics—Determination of the Burning Behaviour of Specimens Using a 500 W Flame Source
ISO 11357-2: 1998	Plastics—Differential Scanning Calrimetry (DSC)—Part 2: Determination of Glass Transition Temperature
ISO 11357-3 :1998	Plastics—Differential Scanning Calrimetry (DSC)—Part 3: Determination of Temperature and Enthalpy of Melting and Crystallisation
ISO 11357-4: 1998	Plastics—Differential Scanning Calrimetry (DSC)—Part 4: Determination of Specific Heat Capacity
ISO 11357-5: 1999	Plastics—Differential Scanning Calrimetry (DSC)—Part 5: Determination of Reaction Temperatures, Reaction Times, Heat of Reaction and Degrees of Conversion
ISO/CD 11357-7	Plastics—Differential Scanning Calrimetry (DSC)—Part 7: Determination of Crystallisation Kinetics
ISO 11359-1: 1999	Plastics—Thermomechanical Analysis (TMA)—Part 1: General Principles
ISO 11359-2: 1999	Plastics—Thermomechanical Analysis (TMA)—Part 2: Determination of Coefficient of Linear Thermal Expansion and Glass Transition Temperature
ISO 11403-1: 1994	Plastics—The Acquisition and Presentation of Multipoint Data. Part 1. Mechanical Properties

ISO 11403-2: 1995	Plastics—The Acquisition and Presentation of Multipoint Data. Part 2. Thermal and Processing Properties
ISO 11403-3: 1999	Plastics—The Acquisition and Presentation of Multipoint Data. Part 3. Environmental Influences
ISO 11443: 1995	Plastics—Determination of the Fluidity of Plastics Using Capillary and Slit Die Rheometers
ISO 13586-1: 1999	Plastics—Determination of Fracture Toughness (G_C and K_C). Part 1: Linear Elastic Fracture Mechanics (LEFM) Approach
IEC 60093: 1980	Methods of Test for Volume Resistivity and Surface Resistivity of Solids Electrical Insulating Materials
IEC 60112: 1979	Method for Determining Comparative and the Proof Tracking Indices of Solid Insulating Materials under Moist Conditions
IEC 60243-1: 1988	Methods of Test for Electric Strength of Solid Insulating Materials—Part 1: Tests at Power Frequencies
IEC 60250: 1969	Recommended Methods for the Determination of the Permittivity and Dielectric Dissipation Factor of Electrical Insulating Materials at Power, Audio, and Radio Frequencies

Material standards

ISO 1163-1: 1995	Plastics—Unplasticized Polyvinyl Chloride (PVC-U) Moulding and Extrusion Materials—Part 1: Designation System and Basis for Specifications
ISO 1163-2: 1991	Plastics—Unplasticized Polyvinyl Chloride (PVC-U) Moulding and Extrusion Materials—Part 2: Preparation of Test Specimens and Determination of Properties
ISO 1622-1: 1994	Plastics—Polystyrene (PS) Moulding and Extrusion Materials—Part 1: Designation System and Basis for Specifications
ISO 1622-2: 1994	Plastics—Polystyrene (PS) Moulding and Extrusion Materials—Part 2: Preparation of Test Specimens and Determination of Properties
ISO 1872-1: 1993	Plastics—Polyethylene (PE) and Ethylene Copolymer Thermoplastics—Part 1: Designation System and Basis for Specifications

ISO 1872-2: 1997	Plastics—Polyethylene (PE) Moulding and Extrusion Materials—Part 2: Preparation of Test Specimens and Determination of Properties
ISO 1873-1: 1995	Plastics—Polypropylene (PP) Moulding and Extrusion Materials—Part 1: Designation System and Basis for Specifications
ISO 1873-2: 1997	Plastics—Polypropylene (PP) Moulding and Extrusion Materials—Part 2: Preparation of Test Specimens and Determination of Properties
ISO 1874-1: 1996	Plastics—Polyamides (PA) Moulding and Extrusion Materials—Part 1: Designation System and Basis for Specifications
ISO 1874-2: 1995	Plastics—Polyamides (PA) Moulding and Extrusion Materials—Part 2: Preparation of Test Specimens and Determination of Properties
ISO 2580-1: 1999	Plastics—Acrylonitrile/Butadiene/Styrene (ABS) Moulding and Extrusion Materials—Part 1: Designation System and Basis for Specifications
ISO 2580-2: 1994	Plastics—Acrylonitrile/Butadiene/Styrene (ABS) Moulding and Extrusion Materials—Part 2: Preparation of Test Specimens and Determination of Properties
ISO 2897-1: 1997	Plastics—Impact Resistant Polystyrene (PS-I) Moulding and Extrusion Materials—Part 1: Designation System and Basis for Specifications
ISO 2897-2: 1994	Plastics—Impact Resistant Polystyrene (PS-I) Moulding and Extrusion Materials—Part 2: Preparation of Test Specimens and Determination of Properties
ISO 2898-1: 1997	Plastics—Plasticized Compounds of Homopolymers and Copolymers of Vinyl Chloride (PVC-P)—Part 1: Designation System and Basis for Specifications
ISO 2898-2: 1997	Plastics—Plasticized Compounds of Homopolymers and Copolymers of Vinyl Chloride (PVC-P)—Part 2: Preparation of Test Specimens and Determination of Properties
ISO 4894-1: 1997	Plastics—Styrene/Acrylonitrile (SAN) Moulding and Extrusion Materials—Part 1: Designation System and Basis for Specifications

ISO 4894-2: 1994	Plastics—Styrene/Acrylonitrile (SAN) Moulding and Extrusion Materials—Part 2: Preparation of Test Specimens and Determination of Properties
ISO 7391-1: 1995	Plastics—Polycarbonate (PC) Moulding and Extrusion Materials—Part 1: Designation System and Basis for Specifications
ISO 7391-2: 1995	Plastics—Polycarbonate (PC) Moulding and Extrusion Materials—Part 2: Preparation of Test Specimens and Determination of Properties
ISO 7792-1: 1998	Plastics—Thermoplastic Polyester (TP) Moulding and Extrusion Materials—Part 1: Designation System and Basis for Specifications
ISO 7792-2: 1998	Plastics—Thermoplastic Polyester (TP) Moulding and Extrusion Materials—Part 2: Preparation of Test Specimens and Determination of Properties
ISO 8257-1: 1998	Plastics—Poly(Methyl Methacrylate) (PMMA) Moulding and Extrusion Materials—Part 1: Designation System and Basis for Specifications
ISO 8257-2: 1998	Plastics—Poly(Methyl Methacrylate) (PMMA) Moulding and Extrusion Materials—Part 2: Preparation of Test Specimens and Determination of Properties
FDIS 9988-1: 1999	Plastics—Polyoxymethylene (POM)—Part 1: Designation System and Basis for Specifications
ISO 9988-2: 1998	Plastics—Polyoxymethylene (POM)—Part 2: Preparation of Test Specimens and Determination of Properties
ISO 14910-1: 1997	Plastics—Thermoplastic Polyester/Ester and Polyether/Ester Elastomers for Moulding and Extrusion. Part 1: Designation System and Basis for Specifications
ISO 14910-2: 1998	Plastics—Thermoplastic Polyester/Ester and Polyether/Ester Elastomers for Moulding and Extrusion. Part 2: Preparation of Test Specimens and Determination of Properties
FDIS 15103-1: 1999	Plastics—Polyphenylene Ether (PPE) Moulding and Extrusion Materials. Part 1: Designation System and Basis for Specifications
FDIS 15103-2: 1999	Plastics—Polyphenylene Ether (PPE) Moulding and Extrusion Materials. Part 2: Preparation of Test Specimens and Determination of Properties

FDIS 15526-1: 1999	Plastics—Polyketone (PK) Moulding and Extrusion Materials. Part 1: Designation System and Basis for Specifications
FDIS 15526-2: 1999	Plastics—Polyketone (PK) Moulding and Extrusion Materials. Part 2: Preparation of Test Specimens and Determination of Properties

Appendix 11.2 Selected ASTM Standards*

Standard	*Title*
D149-97a	Standard Test Method for Dielectric Breakdown Voltage and Dielectric Strength of Solid Electrical Insulating Materials at Commercial Frequencies
D150-95	Standard Test Methods for A-C Loss Characteristics and Permittivity (Dielectric Constant) of Solid Electrical Insulation
D256-97	Standard Test Methods for Determining the Izod Pendulum Impact Resistance of Plastics
D257-93	Standard Test Methods for DC Resistance or Conductance of Insulating Materials
D570-98	Standard Test Method for Water Absorption of Plastics
D618-96	Standard Practice for Conditioning Plastics and Electrical Insulating Materials for Testing
D635-98	Standard Test Method for Rate of Burning and/or Extent and Time of Burning of Self Supporting Plastics in a Horizontal Direction
D638-98	Standard Test Method for Tensile Properties of Plastics
D648-98c	Standard Test Method for Deflection Temperature of Plastics Under Flexural Load in the Edgewise Position
D695-96	Standard Test Method for Compressive Properties of Rigid Plastics
D790-98	Standard Test Methods for Flexural Properties of Unreinforced and Reinforced Plastics and Electrical Insulating Materials
D792-98	Standard Test Methods for Density and Specific Gravity (Relative Density) of Plastics by Displacement
D955-89 (*Reapproved 1996*)	Standard Test Method for Measuring Shrinkage from Mold Dimensions of Molded Plastics
D1238-98	Standard Test Method for Flow Rates of Thermoplastics by Extrusion Plastometer

*Available from ASTM, 100 Barr Harbor Drive, West Conshohocken, PA 19428.

D1505-98	Standard Test Method for Density of Plastics by the Density-Gradient Technique
D1525-98	Standard Test Method for Vicat Softening Temperature of Plastics
D1822-93	Standard Test Method for Tensile-Impact Energy to Break Plastics and Electrical Insulating Materials
D1895-96	Standard Test Methods for Apparent Density, Bulk Factor, and Pourability of Plastics Materials
D2863-97	Standard Test Method for Measuring the Minimum Oxygen Concentration to Support Candle-like Combustion of Plastics (Oxygen Index)
D2990-95	Standard Test Methods for Tensile, Compressive, and Flexural Creep and Creep-Rupture of Plastics
D3028-95	Standard Test Method for Kinetic Coefficient of Friction of Plastic Solids
D3417-97	Standard Test Method for Enthalpies of Fusion and Crystallization of Polymers by Differential Scanning Calorimetry (DSC)
D3418-97	Standard Test Method for Transition Temperatures of Polymers by Thermal Analysis
D3638-93	Standard Test Method for Comparative Tracking Index of Electrical Insulating Materials
D3641-97	Standard Practice for Injection Molding Test Specimens of Thermoplastic Molding and Extrusion Materials
D3801-96	Standard Test Method for Measuring the Comparative Extinguishing Characteristics of Solid Plastics in a Vertical Position
D3835-96	Standard Test Method for Determination of Properties of Plastics by Means of a Capillary Rheometer
D4473-95a	Standard Practice for Measuring the Cure Behavior of Thermosetting Resins using Dynamic Mechanical Procedures
D5025-94	Standard Specification for a Laboratory Burner Used for Small-Scale Burning Tests on Plastics Materials
D5045-96	Standard Test Methods for Plane-Strain Fracture Toughness and Strain Energy Release Rate of Plastic Materials
D5279-95	Standard Test Method for Measuring the Dynamic Mechanical Properties of Plastics in Torsion
D5592-94	Standard Guide for Material Properties Needed in Engineering Design Using Plastics
D5930-97	Standard Test Method for Thermal Conductivity of Plastics by Means of a Transient Line-Source Method

E831-93	Standard Test Method for Linear Thermal Expansion of Solid Materials by Thermomechanical Analysis

Material standards

D788-96	Standard Classification System for Poly(Methyl Methacrylate) (PMMA) Molding and Extrusion Compounds
D1784-97	Standard Specification for Rigid Poly(Vinyl Chloride) (PVC) Compound and Chlorinated Poly(Vinyl Chloride) (CPVC) Compounds
D2287-96	Standard Specification for Non Rigid Vinyl Chloride Polymer and Copolymer Molding and Extrusion Compounds
D3935-94	Standard Specification for Polycarbonate (PC) Unfilled and Reinforced Material
D4066-98	Standard Specification for Nylon Injection and Extrusion Materials (PA)
D4101-98a	Standard Specification for Polypropylene Plastic Injection and Extrusion Materials
D4181-98	Standard Specification for Acetal (POM) Molding and Extrusion Materials
D4203-95	Standard Specification for Styrene-Acrylonitrile (SAN) Injection and Extrusion Materials
D4349-96	Classification System for Polyphenylene Ether (PPE) Materials
D4549-98	Standard Specification for Polystyrene Molding and Extrusion Materials (PS)
D4550-92	Standard Specification for Thermoplastic Elastomer-Ether-Ester (TEEE)
D4673-98	Standard Specification for Acrylonitrile-Butadiene-Styrene (ABS) Molding and Extrusion Materials
D4976-98	Standard Specification for Polyethylene Plastics Molding and Extrusion Materials
D5927-97	Standard Specification for Thermoplastic Polyester (TPES) Injection and Extrusion Materials Based on ISO Test Methods
D5990-96	Standard Classification System for Polyketone Injection and Extrusion Materials (PK)

Appendix 11.3 List of Resources

ISO standards

International Organization for Standardization
1, rue de Varembé
Case Postale 56
CH 1211 Genève 20
Switzerland
www.iso.ch
+41 22 749 01 11
+41 22 733 34 30 (fax)
central@iso.ch

American National Standards Institute
11 West 42nd Street
New York, NY 10036
www.ansi.org
+1 212 642 4900
+1 212 398 0023 (fax)

Global Engineering Documents
15 Inverness Way East
Englewood, CO 80112-5776
www.global.ihs.com
+1 303 397 7956
+1 303 397 2740 (fax)

ASTM standards

ASTM
100 Barr Harbor Drive
West Conshohocken, PA 19428-2959
www.astm.org
+1 610 832 9585
+1 610 832 9555 (fax)

Test specimen molds

Axxicon Molds Rochester Inc.
150 Park Center Drive
West Hanrietta, NY 14586
www.axxicon.com
+1 716 427 9410
+1 716 427 9438 (fax)
aim@axxicon.com

Master Unit Die Products, Inc.
P. O. Box 520
Greenville, MI 48838
+1 616 754 4601
+1 616 754 7478 (fax)
mud@iserv.net

Axxicon Moulds Helmond BV
Kanaaldijk z.w. 7b 5706 LD
P. O. Box 237
5700 AE Helmond
The Netherlands
www.axxicon.com
+31 492 598888
+31 492 533825 (fax)
aim@axxicon.com

Testing equipments

Ceast S.p.A.
Via Airauda 12
10044 Pianezza
Torino, Italy
www.ceast.com
+39 011 966 4038
+39 011 966 2902
dg@ceast.com

Ceast USA, Inc.
377 Carowinds Boulevard
Suite 207
Fort Mill, SC 29715
www.ceast.com
+1 803 548 6093
+1 803 548 1954 (fax)
SalesUSA@ceast.com

Haake
53 W. Century Boulevard
Paramus, NJ 07652-1482
www.haake-usa.com
+1 201 265 7865
+1 201 265 1977 (fax)

Tinius Olsen Testing Machine Co., Inc.
Easton Road
P. O. Box 429

Willow Grove, PA 19090-0429
www.TiniusOlsen.com
+1 215 675 7100
+1 215 441 0899 (fax)
info@TiniusOlsen.com

TMI
2910 Expressway Drive South
Islandia, NY 11722-1407
www.testingmachines.com
516/842-5400
516/842-5220 (fax)
tmi@testingmachines.com

Zwick GmbH & Co.
August Nagel strasse 11
D89079 Ulm
Germany
www.zwick.com
+49 7305 10 0
+49 7305 10 200 (fax)
info@zwick.de

Independent testing facilities

Plastics Technology Labs Inc.
50 Pearl Street
Pittsfield, MA 01201
www.ptli.com
+ 1 413 499 0983
+ 1 413 499 2339 (fax)

Datapoint Testing Services
95 Brown Road, #164
Ithaca, NY 14850
www.datapointlabs.com
+ 1 607 266 0405
+ 1 607 266 0168 (fax)
mail@datapointlabs.com

The Madison Group
505 South Rosa Road
Madison, WI 53719
www.madisongroup.com
+ 1 608 231 1907
+ 1 608 231 2694 (fax)
paul@prowler.madisongroup.com

RAPRA Technology Ltd.
Shawsbury, Shrewsbury
Shropshire SY4 4NR
U.K.
www.rapra.net
+44 1939 250 383
+44 1939 251 118 (fax)

Bodycote Broutman, Inc.
3424 South State Street
Chicago, IL 60616
www.broutman.com
+1 312 842 4100
+1 312 842 3583 (fax)
infor@broutman.com

ARDL - Plastics Testing
2887 Gilchrist Road
Akron, OH 44305
www.ardl.com/plastic.htm
+1 330 794 6600
+1 330 794 6610 (fax)
info@ardl.com

Polymer Diagnostics, Inc.
604 Moore Road
Avon Lake, OH 44012-9930
www.polymerdignostics.com
+1 440 930 1852
+1 440 930 1644 (fax)

Specialized Technology Resources, Inc.
10 Water Street
Enfield, CT 06082
www.strlab.com
+1 860 749 8371
+1 860 749 7533 (fax)

Polymer Testing of St. Louis, Inc.
16239 Westwoods Business Park
Ellisville, MO 63021
www.polymertesting.com
+1 314 394 1480
+1 314 394 1406 (fax)
PolymerTesting@msn.com

Detroit Testing Laboratory, Inc.
7111 E. Eleven Mile
Warren, MI 48092-2709
www.dettest.com
+1 810 754 9000
+1 810 754 9045 (fax)
testing_services@dettest.com

Polyhedron Laboratories, Inc.
10628 Kinghurst Street
Houston, TX 77099
www.polyhedron.com
+1 281 879 8600
+1 281 879 8666 (fax)
techsales@polyhedron.com

Applied Technical Services, Inc.
1190 Atlanta Industrial Drive
Marietta, GA 30066
www.atslab.com
+1 770 423 1400
+1 770 424 6415 (fax)

INDESCA
Investigación y Desarrollo, C.A.
Complejo Petroquimico Zulla
El Tablazo
VENEZUELA
+58 61 909403
+58 61 909481 (fax)

Impact Analytical
Michigan Molecular Institute
1910 W. St. Andrews Road
Midland, MI 48640
www.impactanalytical.org
+1 517 832 5555 ext. 563
+1 517 832 5560 (fax)
wood@impactanalytical.org

Georgia Tech Research Institute
EOEML/Materials Analysis Center
271 Baker Building
Atlanta, GA 30332-0827
+1 440 894 0590
+1 440 894 6199 (fax)

NTS—Sacramento
P.O. Box 857
North Highlands, CA 95660
+1 916 779 3100
+1 916 779 3105 (fax)
www.ntscorp.com

Martin Engineering Labratories
P.O. Box 2019
North Highlands, CA 95660
+1 916 563 1745
+1 916 563-1745 (fax)
www.martinengineering.net

Chemir/Polytech Laboratories
2672 Metro Blvd.
Maryland Heights, MO 63043
+1 314 291-6620
+1 314 291-6630 (fax)
www.chemir.com

Appendix 11.4 Some Unit Conversion Factors

To convert from U.S. units	To SI units	Multiply by
°F	°C	0.5556 × (F − 32)
ft-lb/in	J/m	53.38
ft-lb/in	kJ/m^2	5.235*
ft-lb/in^2	kJ/m^2	2.102
in	mm	25.4
in-lb	J	0.113
psi	MPa	0.006895
g/cm^3	kg/m^3	1000
Poise	Pa·s	0.1
V/mil	kV/mm	0.0394
To convert from SI units	**To U.S. units**	**Use conversion factor**
°C	°F	1.8 C + 32
J/m	ft-lb/in	0.0187
kJ/m^2	ft-lb/in^2	0.476
mm	in	0.0394
J	in-lb	8.8496
MPa	lb/in^2	145
kg/m^3	g/cm^3	0.001
Pa·s	Poise	10
kV/mm	V/mil	25.381

*Taking a depth under the notch of 10.2 mm.

References

1. S. B. Driscoll and C. M. Shaffer, "What Does the Property Datasheet Really Tell You?" ASTM Symposium on Limitations of Test Methods for Plastics, Norfolk, Virginia, November 1, 1998.
2. D. Rackowitz, "Beyond the Datasheet: A Designers Guide to the Interpretation of Datasheet Properties," *Challenging the Status Quo in Design, '94 Design RETEC Proceedings,* March 1994, pp. 28–31.
3. D. Rackowitz, "Looking Beyond the Material Datasheet; Understanding Plastic Material Properties," *SAE Tech. Paper 1999-01-275,* Society of Automotive Engineers International Congress and Exposition, Detroit, Michigan, March 1–4, 1999.
4. R. Shastri, "Material Properties Needed in Engineering Design Using Plastics," *International Plastics Engineering and Technology,* vol. 1, 1995, pp. 53–60.
5. R. Shastri, "Material Properties Needed in Engineering Design Using Plastics," ANTEC '94, 1994, pp. 3097–3101.
6. ASTM D 5592 - 94, "Standard Guide for Material Properties Needed in Engineering Design Using Plastics," 1994.
7. ISO/CD 17282: 1998, "Plastics—The Acquisition and Presentation of Design Data for Plastics."
8. SAE J1639, "Classification System for Automotive Polyamide (PA) Plastics."
9. SAE J1685, "Classification System for Automotive Acrylonitrile/Butadiene/Styrene (ABS) and ABS + Polycarbonate Blends (ABS + PC) Based Plastics."
10. SAE J1686, "Classification System for Automotive Polypropylene (PP) Plastics."
11. SAE J1687, "Classification System for Automotive Thermoplastic Elastomeric Olefins (TEO)."
12. SAE J2250, "Classification System for Automotive Polymethyl Methacrylate (PMMA) Plastics."
13. SAE J2273, "Classification System for Automotive Thermoplastic Polyester Plastics."
14. SAE J2274, "Classification System for Automotive Acetal (POM) Plastics."
15. SAE J2323, "Classification System for Automotive Polycarbonate Plastics."
16. SAE J2324, "Classification System for Automotive Polyethylene Plastics."
17. SAE J2325, "Classification System for Automotive Poly(Phenylene ether) (PPE) Plastics."
18. SAE J2326, "Classification System for Automotive S/MA (Styrene-Maleic Anhydride) Plastics."
19. SAE J2460, "Classification System for Automotive Thermoplastic Elastomeric Polyesters."
20. "The U.S. Plastics Industry: A Call for Uniform Global Testing Standards," The Society of the Plastics Industry, Inc., June 9, 1994.
21. "Uniform Global Testing Standards—A Technical Primer," The Society of the Plastics Industry, Inc., February 20, 1996.
22. "Uniform Global Materials Testing Standards—A Primer for Managers," The Society of the Plastics Industry, Inc., June 1996.
23. V. Wigotsky, "The Road to Standardization," *Plastics Engineering,* April 1995, pp. 22–28.
24. M. C. Gabriele, "Global Standards Could Resolve Inconsistencies," *Plastics Technology,* June 1993, pp. 48–55.
25. La Verne Leonard, "Comparable Data for Plastic Materials—Help Is on the Way," *Plastics Design Forum,* January/February 1993, pp. 37–40.
26. ISO 10350-1: 1999, "Plastics—The Acquisition and Presentation of Comparable Single-Point Data. Part 1: Moulding Materials."
27. ISO 11403-1: 1994, "Plastics—The Acquisition and Presentation of Comparable Multipoint Data Part 1. Mechanical Properties."
28. ISO11403-2: 1995, "Plastics—The acquisition and Presentation of Comparable Multipoint Data Part 2. Thermal and Processing Properties."
29. ISO 3167: 1993, "Plastics—Multipurpose Test Specimens."
30. ISO 294-1: 1996, "Plastics—Injection Moulding of Test Specimens of Thermoplastic Materials—Part 1: General Principles, Multipurpose-Test Specimens (ISO Mould Type A) and Bars (ISO Mould Type B)."

31. "ISO 2580-2: 1994, Plastics—Acrylonitrile/Butadiene/Styrene (ABS) Moulding and Extrusion Materials—Part 2: Preparation of Test Specimens and Determination of Properties."
32. ISO 527-2: 1993, "Plastics—Determination of Tensile Properties—Part 2: Test Conditions for Moulding and Extrusion Plastics."
33. ISO 11403-3: 1999, "Plastics—The Acquisition and Presentation of Comparable Multipoint Data—Part 3. Environmental Influences."
34. M. Sepe, "The Usefulness of HDT and a Better Alternative to Describe the Temperature Dependence of Modulus," *SAE Tech. Paper 1999M-221,* Society of Automotive Engineers International Congress and Exposition, Detroit, Michigan, March 1–4, 1999.
35. M. P. Sepe, "Dynamic Analysis Pinpoints Plastics Temperature Limits," *Advanced Materials & Processes,* April 1992, pp. 32–41.
36. M. P. Sepe, "Material Selection for Elevated Temperature Applications: An Alternative to DTUL," *Society of Petroleum Engineers ANTEC '91,* pp. 2257–2262.
37. L. E. Nunnery, "HDT and Izod—Past Their Prime?" *Plastics Design Forum,* May/June 1992, pp. 38–41.
38. H. Breuer, "Relevance of Flexural Modulus to the Design Engineer," *SAE Tech. Paper 1999-01-0276,* Society of Petroleum Engineers International Congress and Exposition, Detroit, Michigan, March 1–4, 1999.
39. R. Shastri and E. Baur, "Costs of Material Data Measurement," *SAE Tech. Paper 1999-01-0278,* Society of Automotive Engineers International Congress and Exposition, Detroit, Michigan, March 1–4, 1999.

Suggested Reading

Brown, R. P.: *Handbook of Polymer Testing—Physical Methods,* Marcel Dekker, New York, 1999.

Clements, L.: "Testing and Characterization," in *Engineered Materials Handbook, vol. 2. Engineering Plastics,* ASM International, 1988, pp. 515–609.

Hawley, S. W.: "Physical Testing of Thermoplastics," *RAPRA Review Report No. 60,* RAPRA, 1992.

Shah, Vishu: *Handbook of Plastics Testing,* 2d ed., John Wiley & Sons, New York, 1998.

Chapter 12

Plastics Recycling and Biodegradable Plastics

Susan E. Selke, Ph.D.
Michigan State University
East Lansing, Michigan

12.1 Introduction

The attention the public gives to environmental issues has long been recognized to wax and wane through time. For plastics, the first significant environmental pressure came during the mid- to late 1970s, when oil prices rose dramatically. At first, plastics were targeted as made from oil and, therefore, environmentally suspect. As time went on, however, the attributes brought by plastics in terms of energy efficiency became more widely recognized. In the recycling area, however, the rise in oil prices and, consequently, in the base price of plastic resins had a significant impact. The scrap from plastics manufacturing processes became too valuable to simply discard. Use of regrind in manufacturing of plastic products increased. The literature of this time has a variety of publications addressing the concerns which arose from this practice, and they look at the effects of degradation and, to a lesser degree, of contaminants in the feedstock. Slowly but surely, the use of regrind in plastics manufacture became routine, just as the use of in-house cullet in glass manufacture and edge trim in paper manufacture had become routine. Even when oil prices fell again, the economic benefits of using regrind were now recognized, and such use continued.

The next significant wave of environmental concern to impact the plastics industry began in the mid-1980s, and it had a considerably different focus—solid waste. There was a perception in the United States that, as a nation, we were running out of landfill space, and further that plastics were a particular problem because their nonbiodegradable nature meant they were occupying the limited landfill capacity available for seemingly an eternity. This also brought about an interest in plastics recycling, although this time the interest was primarily in recycling of products at the end of their useful lives, rather than of manufacturing scrap. Use of regrind, in fact, had become such a normal practice that it was no longer considered true recycling; rather it was just good business practice. At this same time, there was pressure for use of biodegradable plastics as a replacement for the synthetic nonbiodegradable polymers that were perceived as filling up valuable landfill space.

As time went on, landfill costs in the United States, which had risen dramatically, declined again. Fewer but bigger landfills relieved the capacity crunch. Studies about what really goes on in modern landfills demonstrated that even readily biodegradable materials, such as food, often degraded only very slowly. Further, some plastic products which had been marketed as biodegradable proved to have only very limited degradability. Interest in biodegradability decreased, while pressures and opportunities for recyclability increased. Nonetheless, technical progress toward the production of a greater variety of truly biodegradable plastics continued. Also, plastics recycling continued to grow. More and more U.S. households had access to curbside recycling programs which accepted a few or many plastics, primarily bottles.

By the mid-1990s, critics of plastic recycling began to get more attention. The high costs of adding plastic to curbside recycling programs were cited. Burgeoning production of virgin resin at different times caused falling prices for the two most widely recycled plastics, high-density polyethylene (HDPE) and polyethylene terephthalate (PET). The economic viability of plastics recycling was called into question. Nonetheless, the general public remained supportive of plastics recycling. Few communities dropped curbside recycling or dropped plastics from their recycling programs. In fact, interest in recycling plastic materials began a significant spread beyond packaging materials to the durable goods arena. The American public, by and large, has become convinced of the value of recycling. Many feel that it is one thing they, on a personal level, can do to help the environment, and they feel good about participating. Thus, though the solid waste "crisis" was over in the United States by the mid-1990s (and some argue it was never real in the first place), recycling seems to be here to stay.

During the 1990s, attention to recycling of postindustrial plastics also grew. While it got little public attention compared to recycling of postconsumer plastics, many producers of resins with recycled content relied heavily on industrial waste as feedstock. These waste streams were not the clean single-resin regrind, but rather typically consisted of multiresin materials, materials which combined plastics and non-plastics, or materials which were contaminated in some other way, and which therefore had been going to disposal rather than being reused in house. While some of these streams were heavily contaminated and difficult to use, many were relatively clean, uniform in content, and more economical to collect than postconsumer materials. Producers of such scrap found that they were able to avoid paying for disposal of these materials by arranging to feed them to a recycler, and often could make a little money on the exchange as well.

Along with the changes in public concern about plastics use and recycling, there were changes during this time period in legislative pressures. During the "solid waste crisis," the first wave of legislation often focused on bans of materials or products, particularly plastic packaging, which were seen as a particular problem. For example, nondegradable ring connectors for beverage cans were banned, first by a variety of states, and then throughout the United States. Mandatory recycling programs were instituted, sometimes at a statewide level and other times in individual communities. Taxes on plastic packaging were proposed and sometimes instituted. Grant and loan programs were instituted to help facilitate new recycling businesses as well as community education about recycling. Some states banned disposal of recyclable materials. Many of the legislative initiatives which were proposed never passed, but their sheer number was overwhelming. Most major plastics and packaging companies found it necessary to designate one or more people to devote all, or at least a substantial amount, of their time to environmental and recycling issues.

As time passed, the tenor of legislative initiatives became focused more on recycling. An interest in bans and taxes gave way to efforts to push for markets for recycled materials, including plastics. The pace of legislative activity decreased, but the issues did not go away.

In Europe, where in many countries the issue of landfill scarcity was much more real, a very different approach emerged than in the United States. First in Germany, and then throughout the European Union, the *producer responsibility principle* was adopted. This says, in essence, that the manufacturers of products are responsible for the disposal first, of the packaging for the products, and increasingly for the products themselves. Further, landfill disposal or incineration are not to be the main methods of disposal. Mandatory recycling quotas are imposed. This approach was first applied, after packaging, to the automotive industry, and it is now spreading to a variety of consumer products. In the United States, we have had up to now only a few isolated attempts to institute the producer responsibility philosophy, but it has spread, in modified form, from Europe to Canada and to some parts of Asia, and is making inroads in Latin America as well.

On the biodegradability side, along with the technical work to develop truly biodegradable plastics, the growth of composting as an accepted companion to recycling has opened at least limited opportunities to make use of biodegradability of materials as an asset in their ultimate disposal.

Along with all of these changes, the last 10 years have brought an increasing recognition of the complexity of environmental decision making. Most experts now agree, at least in principle, that decisions on what is best to do from an environmental perspective cannot be based on a single attribute, but must instead be based on an analysis of all the environmental impacts from the decision through the whole life cycle of the products or processes involved. This type of cradle-to-grave analysis is termed *life-cycle assessment.* While there is general agreement on the philosophy, turning that philosophy into a useful decision-making tool is not an easy task. Despite the current existence of several competing computer models which will produce an analysis on demand, there are a number of fundamental questions which have not yet been adequately answered. Simply put, how does one balance x amount of impact A against y amount of impact B, when A and B differ significantly. On a more concrete basis, how many grams (micrograms? picograms?) of dioxin emitted into the Mississippi River is equivalent to how many grams (kilograms?) of suspended particulates (and what kind of particulates?) emitted into the air (how high up?) over Salt Lake City? Thus life-cycle assessment must be regarded as a still-emerging tool for help in decision making.

In this chapter, we will attempt to portray the current status of environmental issues as they relate to plastics recycling and biodegradable plastics, the current status of legislative requirements which have an impact in these areas, how we are doing and where we are headed in recycling of plastics, and the current status and prospects for biodegradable plastics. Issues related to the technique and practice of life-cycle assessment, except in the general context of our look at environmental issues, are beyond the scope. Similarly, in our look at plastics recycling we will focus primarily on postconsumer plastic (plastic which has served its intended use and been discarded), with some attention to postindustrial plastics. Routine use of scrap in the form of regrind will not be addressed.

12.1.1 Solid waste issues

As already mentioned, in the mid-1980s solid waste disposal emerged as a "crisis" in the United States. Many major metropolitan areas, particularly on the east coast, were very close to being out of disposal capacity for municipal solid waste (MSW). Disposal costs were rising astronomically, reaching over $100/ton in New Jersey for tipping fees (the amount charged by the disposal facility for accepting the waste) alone. The public's attention was captured by the voyage of the garbage scow *Mobro,* which sailed from Long Island around a good part of the western hemisphere, searching for a home for its cargo, before finally ending up back on Long Island, with the garbage sent to an incineration facility.

Governments at various levels, from individual communities to whole states, were struggling to find ways to deal with ensuring continuation of the necessary public service of garbage disposal, while containing the costs that were threatening to ruin their budgets—and the chances of re-election for the responsible officials. Acronyms such as NIMBY (not in my backyard), NIMTO (not in my term of office), and PITBY (put it in their backyard) were coined.

Some communities and states solved at least their immediate problem by shipping their garbage to adjoining communities or states—or even farther. Garbage from Long Island, N.Y., reached landfills as far away as Michigan. Predictably, "host" communities were not always happy with their role. Many states tried to write laws to prohibit the import of "outsiders'" waste, only to have them struck down based on the free trade between states provisions in the U.S. Constitution. A number of incineration facilities for municipal solid waste were built, but public resistance to these facilities soon became even greater than resistance to landfills, and their costs were typically much higher than landfills as well. Recycling programs were started up around the country, first in the hundreds and then in the thousands. In contrast to incineration, recycling proved to be very popular politically.

At the same time, slowly but surely, new landfills were sited and built. Due to new regulations, these landfills were constructed much differently than the old landfills which were being shut down. They contained liners—often double liners—to protect against groundwater pollution, and caps to help prevent ingress of water. More care was given to locating them in geologically appropriate areas as well. The cost of these new landfills was also higher, but with increase in capacity and decrease in demand (as recycling increased), the average tipping fees in landfills actually declined in many areas from the record highs set in the early 1990s. For example, in New Jersey the average landfill tipping fee in 1997 was $61/ton.[1] While the absolute number of landfills in the United States continued to decline, to 2514 in 1997, capacity increased. In 1988, 14 states reported having less than 5 years of disposal capacity remaining. In 1997, only one state (Vermont) reported less than 5 years of capacity, and over half the states reported 10 years capacity or more.[1] The average landfill tipping fee in the United States was $31.75/ton in 1997.[1]

Incineration increased in the last half of the 1980s, but then leveled off in the face of growing public resistance. New York City, for example, at one time planned to build five large incineration facilities, but found its plans tied up for years because of public opposition, and eventually scrapped the idea. Incineration rates have been relatively steady at about 16 to 17% since 1990.[2]

Recycling rates have increased steadily in the United States since the mid-1980s, as many new recycling programs were begun. In 1997, the number of curbside recycling programs in the United States reached 8937.[1] The proportion of municipal solid waste which was recycled reached nearly 22%, with an additional 5% recovered by composting, for a total recovery rate of over 27%.[2]

During the mid- to late 1990s, another factor also began to reduce the amount of MSW destined for landfill. The overall generation rate for MSW began to fall. Initially, the decline could be seen on a per capita basis only, as the rising population made overall MSW generation go up, even when the amount generated per person declined slightly. By 1995, however, declines were seen in total tonnage as well. The U.S. Environmental Protection Agency (EPA) estimates that, in 1996, 209,660 thousand tons of municipal solid waste were generated in the United States, down from 214,170 thousand tons in 1994. Discards to landfill fell to 116,240 thousand tons, down from 139,730 thousand tons in 1990.[2] Historical trends in generation and disposal of MSW in the United States are shown in Fig. 12.1.

In much of Europe, as mentioned, the lack of landfill capacity was more real than in the United States. Many countries had been heavily dependent on incineration for a long time, since space for landfill was very hard to find. However, public resistance to incineration was increasing. These ongoing problems led to increased reliance on composting and recycling as alternatives to incineration and landfill.

Other parts of the world, too, have experienced problems with continuing to dispose of materials as had been done in the past. In much of the developing world, the usual method of waste disposal is open dumps. A considerable amount of unorganized recycling is common in these societies, with individuals scavenging reusable materials from the dump sites. As more modern forms of waste disposal are implemented in efforts to curtail the problems resulting from open dumping, recycling in a more organized fashion is becoming part of the solid waste management strategy.

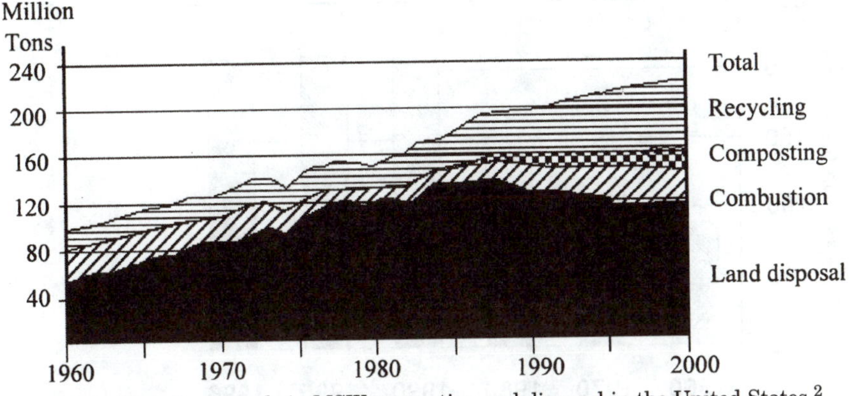

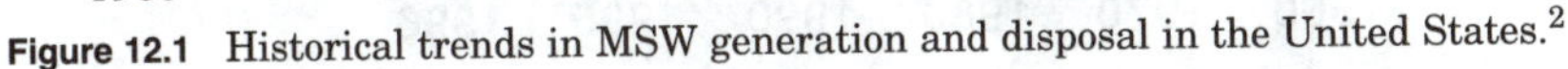

Figure 12.1 Historical trends in MSW generation and disposal in the United States.[2]

Thus, around the world there is increasing reliance on recycling, not as the only method for handling solid waste, but as an important part of what has become known as *integrated solid waste management*—the mix of strategies used to handle disposal of the wastes we generate.

An important consideration is how significant plastics are in contributing to problems with solid waste generation and disposal. There is no doubt that the amount of plastics entering the municipal solid waste stream has increased markedly in the last two decades, and continues to increase, as illustrated in Fig. 12.2 for the United States. It should be noted that the proportions shown are based on weight (see Fig. 12.3). When landfill is the disposal method most commonly used, the desired measurement is contribution by volume rather than weight. For a variety of reasons, such estimates are exceedingly difficult to determine accurately. The EPA has estimated the volume percent of plastics in materials going to disposal (that is, including landfill and incineration but excluding recycling and composting) as 25.1% in 1996 (Fig. 12.4).

12.1.2 Other environmental considerations

As was mentioned earlier, solid waste disposal is not the only environmental impact that should be considered when evaluating process or product alternatives. One environmental concern is resource depletion. Are we using up irreplaceable natural resources? During the flurry of environmental interest in the mid-1970s, when oil prices were rising, one of the key concerns was that we, as a planet, were running out of oil. Plastics were attacked as representing an unnecessary use of this valuable resource. As time went on, and the benefits of plastics in conserving energy were realized, this concern diminished. As new

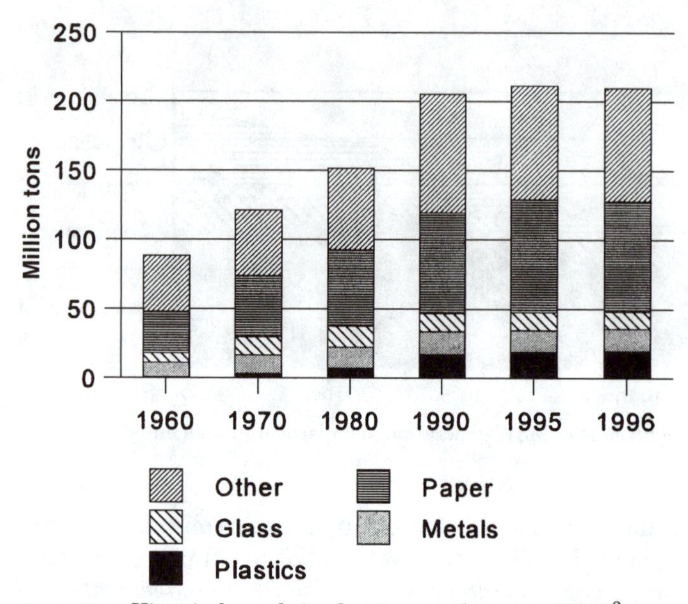

Figure 12.2 Historical trends in plastics contribution to MSW.[2]

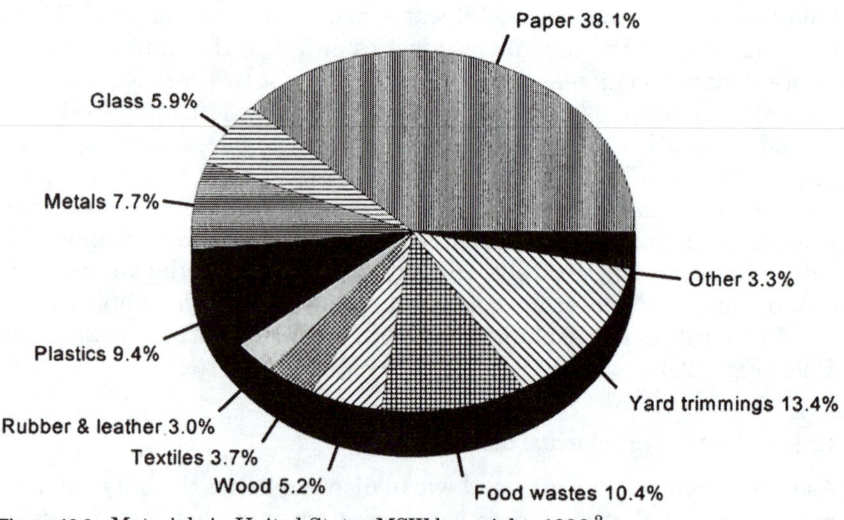

Figure 12.3 Materials in United States MSW by weight, 1996.[2]

reserves of oil were discovered, it died away still more. The fact remains, however, that the supply of petroleum on the planet is limited, and petroleum, along with natural gas, is the major feedstock in the manufacture of most plastics. Processes have been developed to substitute renewable resources (biomass) for petroleum as a plastics

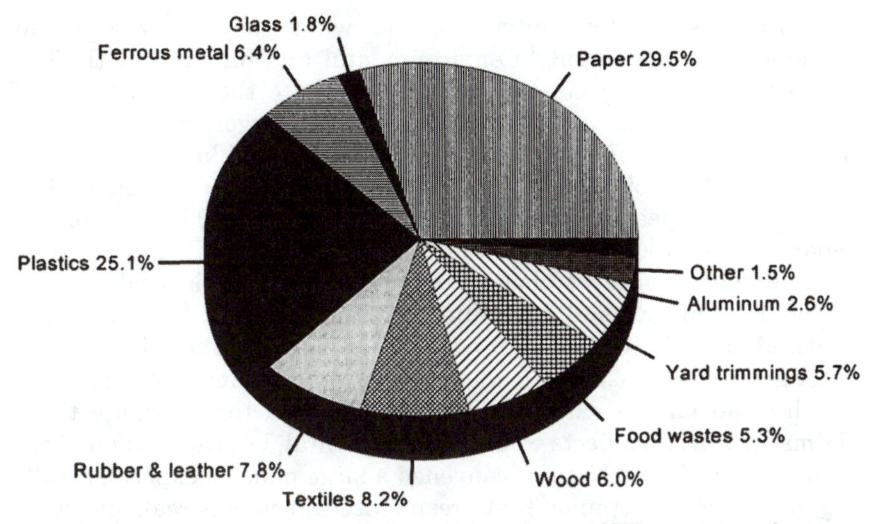

Figure 12.4 Volume percent of materials in United States MSW remaining after recycling, 1996.[2]

feedstock. At present, they are not economical, but they are available should petroleum supplies diminish and/or prices rise significantly.

Pollution is another environmental issue. Does the production, use, transportation, or disposal of the material result in damaging emissions? No human activity (including breathing) is totally free of such emissions. The important questions are what is the amount and type of the emissions, to what extent can they be controlled, and what are their effects on the environment, in comparison to the alternatives. As time passes, legislative controls on emissions of pollutants have tended to become more and more stringent. For example, in 1997, controversial new standards for particulates and ozone in the air were issued.[3] In 1998, the U.S. EPA modified methods for estimating styrene emissions in certain industries, and drew attention to overall increases in such emissions, with the message that new regulations could be coming if industry failed to reduce emissions of styrene.[4]

At the present time, there is considerable concern about the effects of two categories of pollutants which overlap considerably, but are not identical. One focus is emissions of organic chemicals containing chlorine. The other is on emissions of chemicals which are mimics or antagonists of natural hormones, particularly estrogen. Within the plastics industry, polyvinyl chloride (PVC) and polyvinylidene chloride (PVDC) are the polymers most affected on both fronts. Many of the suspected hormone-like chemicals are chlorinated, and some of those which are not chlorinated are used as plasticizers in PVC. The vinyl industry has been attacked by Greenpeace and some other organizations for its contribution to what they portray as an extremely serious threat to environmental and human health. The scientific evidence needed to reliably evaluate the risk, or lack of risk, of the thousands of chemicals in these categories is still being developed. Epidemiological evidence is limited and largely inconclusive. At the time of this writing, the issue has receded somewhat but has not disappeared. For example, in August 1998, Greenpeace reported that Nike Inc. plans to phase out its use of PVC because of environmental concerns about its manufacture and disposal.[5]

Another controversial environmental issue is related to CO_2 emissions. Many believe that man-made emissions of carbon dioxide and other greenhouse gases are leading to an overall warming of the planet, which could have a variety of adverse effects. Others dispute these claims, or that the effects would be detrimental, but this is a minority opinion. The United Nations convened a large panel of experts, resulting in a consensus opinion that greenhouse warming is real, and steps need to be taken to curb it.[6] A subsequent meeting in Kyoto resulted in an international agreement to reduce greenhouse gas emissions. This agreement has not yet been ratified by the U.S. Congress, although a number of other countries have agreed to abide by it, and some call for even more stringent measures. The potential impact on the plastics industry of energy conservation measures is hard to evaluate. Increased taxes on oil and natural gas would drive up costs. On the other hand, efforts to achieve more energy efficiency might lead to increased use of plastics, as their light weight and relatively energy efficient production could make them highly competitive.

12.1.3 Legislative requirements

In the United States, legislative requirements related to plastics recycling are in effect predominantly at the state, rather than federal, level. There are almost no regulations related to biodegradable plastics at any level of government. There are federal regulations, as well as laws in a number of states, which require that plastic ring connectors used for bundling of beverage cans be degradable, but the materials in commercial use to satisfy this requirement are photodegradable rather than biodegradable. Legislation was passed in a few states in the late 1980s to require plastic bags to be biodegradable, but nearly all has since been rescinded.

Legislation and regulations related to plastics recycling fall into several categories. A number of states have some kind of requirement that recycling opportunities be available to residents. Some require residents to participate in recycling, requiring that the target recyclable materials be kept out of the waste stream and instead diverted to recycling. Some do not mandate recycling per se, but prohibit the disposal, by landfill or incineration, of the target recyclables. Others require that communities incorporate recycling as part of their solid waste management strategy. Still others simply require that consideration be given to recycling as an option.[7,8]

Several states have considered the establishment of taxes or fees to promote recycling. Bottle deposit legislation can be put in this category. States with mandatory deposits on certain containers, typically carbonated beverages (see Table 12.1), achieve high rates of return, typically 90% or more, facilitating the recycling of the containers. In 1993, Florida instituted an advance disposal fee on containers which did not meet a minimum recycling rate. The fee had a 1995 sunset date, and was not extended. During the time this fee was in effect, major soft drink bottlers, including Coke and Pepsi, distributed their products in bottles containing 25% recycled PET within the state. When the fee ended, so did the use of recycled content in soft drink bottles.

TABLE 12.1 Bottle Deposit Legislation in the United States[7]

State	Containers covered	Characteristics
Connecticut	Beer, malt beverages, carbonated soft drinks, soda water, mineral water	5¢ deposit
Delaware	Nonalcoholic carbonated beverages, beer, and other malt beverages	5¢ deposit, aluminum cans exempt
Iowa	Beer, soda, wine, liquor	5¢ deposit
Maine	Beer, soft drinks, distilled spirits, wine, juice, water, and other noncarbonated beverages	5¢ deposit, 15¢ on wine and liquor, no deposit on milk
Massachusetts	Carbonated soft drinks, mineral water, beer, and other malt beverages	5¢ deposit; containers 2 gal or larger exempt
Michigan	Beer, soda, canned cocktails, carbonated water, mineral water	10¢ deposit, 5¢ on some refillable bottles
New York	Beer, soda, wine cooler, carbonated mineral water, soda water	5¢ deposit
Oregon	Beer, malt beverages, soft drinks, carbonated and mineral water	5¢ deposit, 3¢ on standard refillable bottles
Vermont	Beer and soft drinks, liquor	5¢ deposit, 15¢ on liquor bottles, all glass bottles must be refillable

Note: California has a refund system for beverage containers but it is not a true deposit system.

A number of states have instituted grant or loan programs to assist in the establishment of recycling. Funds from such programs have been used in a variety of ways, from developing educational materials for children to convince them of the value of recycling to buying equipment for processing recyclable materials or for manufacturing products from these materials.

Three states have passed laws requiring minimum recycled content in packaging materials, minimum recycling rates, or source reduction. Wisconsin requires plastic containers, except for food, beverages, drugs, and cosmetics, effective in 1995, to consist of at least 10% recycled or remanufactured material by weight.[8] Reportedly, there is little enforcement of this legislation. Oregon requires rigid plastic containers, except for food and medical packaging, to contain 25% recycled content, meet target 25% recycling rates (defined in a variety of ways), or be source-reduced by 10%.[8] Since this law went into effect in 1995, the aggregate recycling rate for all plastic containers in Oregon has been above the required 25%, so all plastic containers satisfy the law's requirements automatically. In 1996, the estimated recycling rate was 33.3% in the state.[9]

California has a law very similar to that in Oregon. In 1998, the California Waste Management Board determined that the aggregate recycling rate for rigid plastic containers in California fell below the required 25%, and began to ask a selected group of manufacturers to certify to the Board how they were meeting the requirements of the law.[10] Early results from that survey indicate many companies failed to respond by the deadline, and a significant number of respondents failed to demonstrate compliance. The previous year, the Board had, after considerable controversy, adopted a recycling rate range that spanned the required 25%, so no enforcement of the law was needed. Manufacturers were surprised at the outcome in 1998, and expressed disbelief that the amount of plastics recycling in the state had fallen, as the survey figures suggested. Therefore, it seems likely that any enforcement activity by the Waste Management Board will be challenged.

The majority of U.S. states require coding of plastic containers to identify the type of resin used (Table 12.2). The regulations specify use of the coding system developed by The Society of the Plastics Industry (SPI), consisting of a triangle formed from chasing arrows, with a number code inside the triangle and a letter code underneath. This system has been controversial since its inception, with complaints by many environmental groups that consumers misinterpret the identification symbol as an indication that the container is recyclable, or even that it contains recycled content. The problem is aggravated by the use of the symbols on a variety of objects other than rigid plastic containers. At the same time, the identification symbol has been criticized by recyclers as not providing enough information. For example, it does not differentiate between high- and low-melt flow grades of HDPE, even though the two are incompatible in a recycling system, and blending can result in a product which no end users find appropriate for their needs. There was a long series of meetings between representatives of environmental groups, recyclers, and plastics industry representatives to try to develop a solution to these problems, but the effort eventually failed.

As was mentioned, a different approach to MSW management and recycling was taken in Europe. Policies were put in place, first in Germany and then in the entire European Union, that made companies responsible for the proper disposal of the packages for their products, with requirements that certain percentages of such packaging be collected, and that a certain percentage of collected materials be recycled.[11] Incineration with energy recovery is counted as recycling in some countries but not in others. In most cases, industry responded by forming industry organizations to collectively handle the collection and recovery of the packaging, so that they did not have to do it individually. The first such organization, in Germany, was Duales System Deutschland (DSD), commonly known as the Green Dot system. At this writing, some European countries have well-organized systems for collecting and recycling packaging waste, while others are just getting started. Manufacturers' responsibility has also been implemented for automobiles, with a requirement that limits to 5% the amount of new cars that can be landfilled. Automobile manufacturers are responding by changing the design of their products to make them more recyclable, and are also using more recycled materials in their construction. The same philosophy is expected to be extended to household appliances.[12]

Canada has adopted a National Packaging Protocol, with a requirement to reduce the amount of packaging waste reaching disposal to 50% of 1988 levels by 2000.[13] To the surprise of many, this target was

TABLE 12.2 States Requiring the Society of the Plastics Industry (SPI) Code on Rigid Plastic Containers[7]

Alaska	Illinois	Minnesota	Rhode Island
Arizona	Indiana	Mississippi	South Carolina
Arkansas	Iowa	Nebraska	South Dakota
California	Kansas	Nevada	Tennessee
Colorado	Kentucky	New Jersey	Texas
Connecticut	Louisiana	North Carolina	Virginia
Delaware	Maine	North Dakota	Washington
Florida	Maryland	Ohio	Wisconsin
Georgia	Massachusetts	Oklahoma	
Hawaii	Michigan	Oregon	

reached by 1996, when packaging waste disposal was reported to be 51.2% less than in 1988.[14] Various Canadian provinces have their own regulations in support of this goal, including deposits and fees, landfill bans, and requirements for the use of refillable containers.[15]

Japan has had a deposit system for beer and sake bottles for many years. In 1995, a law was passed to require businesses to recycle designated packaging wastes, beginning in 1997. PET bottles and other containers are covered, and non-PET plastic containers will be included in 2000. Industry responded by creating the Japan Container and Package Recycling Association, a third-party organization, similar to the Green Dot system, which collects a fee in exchange for handling the recycling of the packaging waste.[8]

South Korea has a deposit system on most containers which is designed to encourage the use of reusable packaging and promote recycling of nonrefillable containers. It has also adopted guidelines intended to reduce the volume of polystyrene cushioning used in packaging.[8]

One municipal council in Malaysia began in July 1997 to restrict the use of plastic packaging because of concern about disposal of plastics and about adverse effects on wildlife from littered plastics, especially those which enter bodies of water.[16]

Israel passed a law in 1997 which requires local councils to recycle at least 15% of their solid waste by the year 2000 and to recycle 25% by 2008. The recycling rate in 1996 was slightly over 25% nationwide.[17]

A variety of other countries around the world have adopted, or are adopting, policies aimed at promoting the recycling of packaging materials and thus reducing their disposal burdens. In many cases, they are following the European producer responsibility approach.

12.2 Overview of Recycling

For plastics recycling (or recycling of other materials) to occur, three basic elements must be in place. First, there must be a system for collecting the targeted materials. Second, there must be a facility capable of processing the collected recyclables into a form which can be utilized by manufacturers to make a new product. Third, new products made in whole or part from the recycled material must be manufactured and sold. While the end uses differ substantially for different plastics, there are some similarities in collection and processing which can usefully be discussed in a generic fashion.

12.2.1 Collection of materials

The first step in recycling, obviously, is to gather together the materials to be recycled. Here there are three main approaches: (1) go out and get the material, (2) create conditions such that the material will be brought to you, or (3) use some combined approach.

Plastics recycling in the United States got its start with the recycling of PET beverage bottles in states with deposit legislation. The 5 or 10¢/container deposit proved to be a sufficient incentive to get consumers to bring in 90% or more of the covered containers to centralized collection points (retail stores). When there was a desire to increase recycling beyond PET soft drink bottles to other types of plastic containers, this was one model which could be followed. However, only Maine, to date, has expanded its deposit law much beyond carbonated beverages. In 1990 Maine extended the law to containers for most beverages, excluding milk, explicitly to facilitate recycling of those containers. Deposit redemption centers most often involve a person who counts the containers and issues the refund, but systems using reverse vending machines have also been developed and are in reasonably widespread use. The primary advantages of deposit systems are their high rates of return of the targeted containers, and relatively low levels of contamination, since each container is examined, at least superficially, by either a person or the scanning and verification functions built into the reverse vending machines. The primary disadvantages are the relatively high cost of such systems, and hygiene issues related to bringing in dirty containers to a retail establishment, often one which sells food. The latter disadvantage would increase markedly in importance if containers other than those for carbonated beverages were included.

The other primary way to get consumers to deliver their plastic objects for recycling is to establish drop-off facilities. In the 1980s, a number of multimaterial drop-off recycling centers were established, primarily by the beverage industry, as part of their efforts to prevent the passage of deposit legislation in additional states. These beverage industry recycling program (BIRP) centers typically accepted PET bottles along with glass bottles, newspaper, and sometimes other materials. They often provided a theme park atmosphere, in an effort to make a visit to the center fun and, therefore, likely to be repeated. This type of large attended drop-off center has mostly given way to a proliferation of smaller, simpler drop-off facilities, largely unattended, which attempt to encourage participation in recycling by being conveniently located and readily accessible. Some are multimaterial centers, usually consisting of a collection of bins or roll-off containers. Others, such as the barrels for collection of polyethylene (PE) bags found in the front of many retail stores, accept only one material. Such drop-off facilities have the advantage of being reasonably low in cost, especially if they are unattended. Their primary disadvantages are relatively low rates of participation and relatively high rates of contamination with undesired materials. Drop-off facilities are the primary means of collecting recyclables in much of Europe. In the United States, BioCycle counted 12,699 drop-off recycling programs in 1997.[1]

In the United States, most recycling of postconsumer materials is done through curbside collection. A BioCycle survey counted 8937 curbside recycling programs in 1997.[1] The U.S. EPA counted 8817 curbside recycling programs in the United States in 1996, serving 134.6 million people, 51% of the U.S. population.[2] In these systems, households set their recyclables out for collection in much the same way as they do their garbage, often at the same place, and on the same day. Many of these systems provide a bin (usually colored blue) to consumers as a collection container. In most systems, the consumer places a variety of recyclables in the bin, perhaps with others bundled alongside, and the materials are sorted in a material recovery facility (MRF). Sometimes the sorting is done at truckside instead. In other systems, the consumers must sort the materials into designated categories before they are picked up. Both the latter systems rely on the use of a compartmented recycling vehicle to keep the materials from intermingling. Some curbside systems use a bag (also usually blue) rather than a bin. In some of these, garbage and recyclables are collected at the same time, in the same vehicle, and the recyclables are sorted out after the load is dumped. One of the problems with curbside collection of plastics is the high volume occupied by the plastic, which is usually bottles, compared to its value. Many communities urge consumers to step on the bottles to flatten them before they are placed at the curb, though with only limited success. Even flattened bottles occupy a lot of space. Some collection programs have used on-truck compactors to densify the loads. Others have experimented with on-truck grinders, but this leads to difficulty in effectively sorting the plastics, as will be discussed in Sec. 12.2.4. Many systems focus on collection of recyclables from business or industrial generators, rather than from individual consumers.

Because collection systems enhance convenience for the generator of the waste materials, participation in recycling is typically higher in these systems than in drop-off systems, where the individuals must make more of an effort to feed the materials into the recycling system. Deposit systems are an exception; here the added incentive of the monetary reward, plus the fact that the redemption center is typically located in a retail establishment, where the consumers will be going anyway to buy their groceries, more than makes up for the little extra effort involved.

In some countries, recycling collection occurs primarily through the activities of scavengers. Estimates of plastic recycling in India, for example, are as high as 2.2 million lb/day, all due to the activities of rag pickers who scavenge waste dumps, collecting 7 to 11 lb of plastics per day and selling them to one of several plastic waste collection centers. In New Delhi, for example, about 5000 dealers trade in waste plastics and about 200 processors manufacture products from recycled plastics. Plastics may be recycled as many as 4 times.[18]

12.2.2 Processing of recyclables

Processing of recyclables is necessary to transform the collected materials into raw materials for the manufacture of new products. While the details of the processing are often specific to an individual plastic, or even to an individual product, three general categories of processing can be identified: (1) physical recycling, (2) chemical recycling, and (3) thermal recycling.

Physical recycling. *Physical recycling* involves changing the size and shape of the materials, removing contaminants, blending in additives if desired, and similar activities that change the appearance of the recycled material, but do not alter (at least not to a large extent) its basic chemical structure. Within this category, the usual processing methods for plastic containers, for example, include grinding, air classification to remove light contaminants, washing, a gravity-based system to separate components heavier than water from those lighter than water, screening, rinsing, drying, and often melting and pelletization, perhaps with the addition of colorants, heat stabilizers, or other ingredients. The vast majority of plastics recycling operations in existence today involve physical recycling.

Chemical recycling. *Chemical recycling* involves breaking down the molecular structure of the polymer, using chemical reactions. The products of the reaction then can be purified and used again to produce either the same or a related polymer. An example is the glycolysis process sometimes used to recycle PET, in which the PET is broken down into monomers, crystallized, and repolymerized. Condensation polymers, such as PET, nylon, and polyurethane, are typically much more amenable to chemical recycling than are addition polymers such as polyolefins, polystyrene, and PVC. Most commercial processes for depolymerization and repolymerization are restricted to a single polymer, which is usually PET, nylon 6, or polyurethane.

Thermal recycling. *Thermal recycling* also involves breaking down the chemical structure of the polymer. In this case, instead of relying on chemical reactions, the primary vehicle for reaction is heat. In pyrolysis, for example, the polymer (or mixture of polymers) is subjected to high heat in the absence of sufficient oxygen for combustion. At these elevated temperatures, the polymeric structure breaks down. Thermal recycling can be applied to all types of polymers. However, the typical yield is a complex mixture of products, even when the feedstock is a single polymer resin. If reasonably pure compounds can be recovered, products of thermal recycling can be used as feedstock for new materials. When the products are a complex mixture which is not easily separated, the products are most often used as fuel. There are relatively few commercial operations today which involve thermal recycling of plastics, though research continues. Germany has the largest number of such feedstock recycling facilities due to its requirements for recycling of plastics packaging.

A consortium of European plastic resin companies, the Plastics to Feedstock Recycling Consortium, has a pilot plant for thermal recycling in Grangemouth, Scotland, and hopes to use the technology in a full-scale commercial plant by late 2000. The system uses fluidized bed cracking to produce a waxlike material from mixed plastic waste. The product, when mixed with naptha, can be used as a raw material in a cracker or refinery to produce feedstocks such as ethylene and propylene.[19]

Incineration with energy recovery is a thermal process which is not generally regarded as recycling, although it is a way of recovering some value from the discarded materials. Incineration without energy recovery is rarely practiced in modern facilities. Those facilities which do operate in this manner are typically old and lack modern emission controls, so they are slowly but surely being shut down, at least in the industrialized countries. Incineration is relatively common around the world, but almost always operates on a mixed waste stream, not on plastics alone.

12.2.3 Separation and contamination issues

When plastics are collected for recycling, they are virtually never in a pure homogeneous form. The collected materials will contain product residues, dirt, labels, and other materials. Often the material will contain more than one base polymer, and resins with a variety of additives, including coloring agents. Usefulness of the material is enhanced if it can be cleaned and purified. Therefore, technologies for cleaning and separating the materials are an important part of most plastics recycling systems.

It is useful to differentiate between separation of plastic from nonplastic contaminants, and separation of plastics of one type from those of another type. Separation of plastics from nonplastics typically relies on a variety of fairly conventional processing techniques. Typically the plastic is granulated, sent through an air classifier to remove light materials, such as label fragments, washed with hot water and detergent to remove product residues and remove or soften adhesives, and screened to remove small, heavy contaminants such as dirt. If necessary, magnetic separation can be used to remove ferrous metals, and techniques such as eddy current separators or electrostatic separators can be used to remove other metals. Many of these techniques were developed in the minerals processing industries and have been adapted for use with plastics.

A particular concern for recycled plastics which are to be used in food-contact applications is the potential presence of materials which may be dissolved in the recycled plastic and later migrate out into a product. Special care is needed in the design of recycling processes to ensure that potentially hazardous substances do not migrate from recycled plastic into food products in amounts which might adversely impact human health. Companies desiring to produce recycled resins suitable for food contact generally challenge the process with known amounts of contaminant simulants, and then determine whether the processing is able to adequately remove the contaminants. The U.S. Food and Drug Administration (FDA), while it does not formally approve recycled resins for food contact, has issued "letters of nonobjection" to a few processes which have demonstrated, to the satisfaction of the U.S FDA, the ability to reduce contamination levels below the "threshold of regulation" of 0.5-ppb dietary concentration which the U.S. FDA regards as an acceptable level of protection for most potential contaminants.[20] Another approach which has been accepted by the U.S. FDA is to interpose a barrier layer of virgin polymer between the recycled polymer and the food product. The amount of barrier which is sufficient depends on the mass transfer characteristics of the polymer and the intended use of the resin, among other factors.

12.2.4 Sorting

Separation of different types of polymers from each other is often a required or a desired part of plastics recycling processes. Such separation procedures can be classified as macrosorting, microsorting, or molecular sorting. *Macrosorting* refers to the sorting of whole or nearly whole objects. *Microsorting* refers to sorting of chipped or granulated plastics. *Molecular sorting* refers to sorting of materials whose physical form has been wholly disrupted, such as by dissolving the plastics.

Macrosorting. Examples of macrosorting processes include separation of PVC bottles from PET bottles, separation of polyester carpet from nylon carpet, and sorting of automobile components by resin type. Much of this sorting is still done by hand, using people who pick materials off a conveyor belt and place them in the appropriate receptacle. However, a lot of effort has gone into development of more mechanized means of sorting, in order to make this process both more economical and more reliable, and the use of such mechanized systems is increasing.

Various devices are now commercially available to separate plastics by resin type. They typically rely on differences in the absorption or transmission of certain wavelengths of electromagnetic radiation. Many of these systems can be used to separate plastics by color as well as by resin type. For example, the process used at the plastics recovery facility in Salem, Oreg., which was developed by Magnetic Separation Systems (MSS) of Nashville, Tenn., sorts 2 to 3 bottles/s, using four sensors and seven computers to separate plastic bottles according to resin and color. X-ray transmission is used to detect PVC, an infrared-light high-density array to separate clear from translucent or opaque plastics, a machine vision color sensor to identify bottle color (ignoring the label), and a near-infrared spectrum detector to identify resin type.[21]

Frankel Industries of Edison, N.J., developed a system which combines manual sorting with differences in optical dispersion and refraction for separating PET and PVC from each other and from glycol-modified PET (PETG) and polystyrene. In this system, a special light shines on containers on a conveyor, and workers wear special goggles which gives the different resins a distinctive appearance.[22]

Particularly for recycling of appliances, carpet, and automobile plastics, several companies have developed equipment to scan the plastic, usually with infrared light, and compare its spectrum, using a computer, to known types of plastic, resulting in identification of the plastic resin. One such device is the Portasort, developed jointly by Ford and the University of Southampton, Highfield, Southampton, United Kingdom. It compares the spectrum of the unknown plastic against a library of 200 or more different polymers. A larger version, called the PolyAna system, can identify nearly 1000 different plastics, including blends and fillers. The same group developed the Tribo-pen, which uses triboelectric technology for plastics identification. This equipment, which has a sensing device about the size of a small flashlight, was developed for sorting automotive plastics. It comes in two basic types—the first identifies nylon, polypropylene, ABS, and polyacrylite, and the second designed for more limited sorting, such as differentiating between PE and PVC.[23-25]

Microsorting. The first step in microsorting is a size-reduction process, like chipping or grinding, to reduce the plastic articles to small pieces which will then be separated by resin type, and perhaps also by color. One of the oldest examples is separation of high-density polyethylene base cups from PET soft drink bottles using a sink-float tank. More modern separation processes, such as the use of hydrocyclones, also rely primarily on differences in the density of the materials for the separation.

A number of other attributes have also been used as the basis for microsorting systems, including differences in melting point and in triboelectric behavior.[26] In many of these systems, proper control over the size of the plastic flakes is important in being able to reliably separate the resins. Some systems rely on differences in the grinding behavior of the plastics combined with sieving or other size-based separation mechanisms for sorting. Sometimes cryogenic grinding is used to facilitate grinding and to generate size differences.

Systems which use electromagnetic radiation are under development and have had limited commercial application. SRC Vision, Inc., of Medford, Oreg., has an optical-based technology, originally developed for sorting of foods, which is used primarily by large processors for color sorting single resins, such as in separating green from clear PET. Union Carbide has used an SRC Vision system to separate colored HDPE flake into red, yellow, blue, green, black, and white product streams. The full SRC system uses x-rays, ultraviolet light, visible light, infrared light, reflectance, a monochromatic camera, and a color camera which is reported to be able to separate 16 million colors of red, green, and blue combinations.[26,27] ESM International, Inc. of Houston, Tex., also has developed an optical-based system.[26]

A novel European process being used for separation of plastics from durable goods separates the materials, including laminated structures, by blowing them apart at supersonic speeds. Various materials deform differently, permitting the use of sieving and classifying based on differences in size, geometry, specific gravity, and ballistic behavior, using fluid bed separators and other equipment.[28] The Multi-Products Recycling Facility operated by wTe Corporation is designed to recover engineering plastics (as well as metals) from durable goods. It uses air classifiers to remove light materials, including foam and fiber, and a series of sink/float classifiers operating with water solutions at different specific gravities to separate chipped plastics by density, as well as using infrared technology to identify plastics before grinding.[29] KHD Humboldt Wedag AG in Cologne, Germany, has designed a system for separation of plastics by density using centrifuges and water or salt solutions. The intense turbulence in the centrifuges also helps clean the flaked plastics as well as dewatering it.[30] Recovery Processes International, of Salt Lake City, Utah, has a froth flotation system designed to separate PET from PVC.[26]

Molecular sorting. Molecular sorting involves the complete destruction of the physical structure of the plastic article prior to separation of the resins. Such systems typically use dissolution in solvents and reprecipitation, using either a single solvent at multiple temperatures or combinations of solvents. Because of the use of organic solvents and, consequently, the need to control emissions and to recover the solvents, costs of such systems tend to be high. There is also a concern about residual solvent in the recovered polymer, and its tendency to leach into products. There are at present no commercial systems using this approach.

Chemical tracers. Some research effort has focused on facilitating plastics separation by incorporating chemical tracers into plastics, particularly packaging materials, so that they can be more easily identified and separated. One such effort, funded by the European Union, has resulted in a pilot plant for separating PVC, PET, and HDPE bottles, using fluorescent trace compounds which have been incorporated into the bottles.[31]

Commingled plastics. An alternative to relying on separation processes to generate reasonably pure streams of recycled plastic is the development of processes and applications which do not require pure feedstocks. A variety of such systems have been developed for plastics, with the majority falling in the general categories of plastic lumber and replacements for concrete. Recycling of commingled plastics will be addressed in Sec. 12.4.11.

12.2.5 Uses of recycled plastic

Recycled plastics are used in a variety of applications, including automobiles, housewares, packaging, and construction. More information about uses is found in the sections on recycling of individual types of plastics. Recycled materials, including plastics, also are an important segment of world trade activities. For example, in 1995, recycled plastic exports from the United States alone amounted to 652.8 million lb, for a value of about $205 million. Most of these exports went to Hong Kong, and much of that material probably went on to China.[32] The Far East is also an important market for other countries.

12.3 Design for Recycling

It has become obvious that many of the difficulties of recycling plastics are related to difficulties in separating plastics from other wastes and in sorting plastics by resin type. Design of products can do a lot to either exacerbate or minimize these difficulties. Therefore, increasing attention is being given to designing products with recycling in mind. The general philosophy is to simplify identification of plastics by resin type, and further, either make it relatively easy to separate the various plastic streams from each other, as well as from nonplastic components, or to use plastics which are compatible with each other and, therefore, do not need to be separated.

One of the relatively early efforts to produce guidelines for "design for recycling" was the City/Industry Plastic Bottle Redesign Project, established in early 1994 to reach a consensus on design changes for plastic bottles, aimed at improving the economics of recycling. The "city" representatives included Dallas, Jacksonville, Milwaukee, New York, San Diego, and Seattle. "Industry" participants included Avery Dennison, Enviro-Plastic, Johnson Controls, Owens Illinois, Procter & Gamble, SC Johnson Wax, and St. Jude Polymer. The study received funding from the U.S. EPA as well as the states of Wisconsin and New York. The focus of the project was to assist plastics recyclers as well as to maximize the return to cities collecting postconsumer resin. Therefore, balancing costs of making package design changes against the recycling benefits was part of the effort. Recommendations for plastic bottle design included making caps, closures, and spouts on high-density polyethylene (HDPE) bottles compatible with the bottles, ensuring that any aluminum seals used on plastic bottles pull off completely when the bottle is opened by the consumer, using unpigmented caps on natural HDPE bottles, phasing out the use of aluminum caps on plastic bottles and HDPE base cups on polyethylene terephthalate bottles, using water-dispersable adhesives for labels, not using metallized labels on plastic bottles with a specific gravity greater than 1.0, not printing directly on unpigmented containers, using PVC and PVDC labels only on PVC containers, making all layers in multilayer plastic bottles sufficiently compatible for use of the recyclate in high value end markets, and avoiding use of PVC bottles for products that are also packaged in other resins which look like PVC. The industry participants abstained from this final recommendation.[33]

The automotive industry has also directed efforts toward improving the recyclability of automotive plastics by change in automobile design. Efforts include designing components for ease of disassembly, as well as efforts to ensure that all resins in a component are compatible. During 1999, the Polyurethane Division of the Society of the Plastics Industry is holding a competition which will focus on ways to design automotive seats that allow reuse of polyurethane. The competition is also being supported by the Industrial Designers Society of America.[34] Honda Motor Company in November 1997 announced a process for manufacturing fully recyclable automobile instrument panels, which include a change from their current design using a combination of ABS and PVC resins to a single polyolefin for both the panel base and its outer skin.[35]

The American Plastics Council, along with other parties, has developed a guide for the design for recyclability for the information technology industry, including data processing equipment and communications equipment.[29]

Dell Computer Corporation announced in 1997 that it would make its personal computers marketed to business and government more recyclable by using plastic materials which do not contain fillers and coatings, which can inhibit the recyclability of the materials. Dell also changed the chassis design for their computers, making metal and plastic parts easily separable. Plastic components are marked with international ISO standard codes. Dell also has a take-back program for its large corporate customers, accepting computers of any brand and giving discounts for the purchase of new Dell computers. Much of the equipment, instead of being recycled, is upgraded and resold in other countries.[36]

IBM has been coding plastic components to promote their recyclability since 1992, when it became the first computer manufacturer to do so.[36] Computer and business equipment companies are also increasing their use of recycled-content resins in production.[37] For example, Kobe Steel Ltd., announced in 1998 that it had developed a practical sandwich technique for applying virgin resin to a core of recovered plastic, which allows increased use of recycled resin. They plan to market the technology to manufacturers of office machines, home electronics, and toiletries. The Japanese unit of IBM is manufacturing some personal computers with about 20% recycled plastic from older IBM PCs.[38]

12.4 Recycling of Major Polymers

The amount of plastics entering the municipal solid waste stream in the United States in 1996 is estimated at 19.7 million tons. The amount which was recycled was 1.1 million tons, or 5.4% of plastics generation (see Figs. 12.5 and 12.6). The remaining 18.7 million tons amounted to 12.3% by weight of total MSW discards.[2] Recycling rates are significantly higher for some plastic materials than for others, and for some types of plastic products, as will be discussed in more detail in the following sections (see Figs. 12.7 and 12.8). Many recycling programs for plastics focus on plastic containers, or even more narrowly on plastic bottles. The American Plastics Council (APC) calculated the 1996 recycling rate for rigid plastic containers as 21.2%. The rate for 1997 fell to 20.2%, although the tonnage of plastic collected for recycling increased from 1.321 billion to 1.375 billion lb. Use of virgin plastic increased at a higher rate, from 6.221 billion to 6.800 billion lb during the same period. The APC reported the recycling rate for plastic bottles as 23.7% in 1997, down from 24.5% in 1996. The 1996 recycling rate for flexible plastic packaging was only 2.8%.[39]

In Europe, recycling of plastics amounted to 1.6 million metric tons out of an overall consumption of 15.9 million tonnes, for an overall recycling rate of 10%, according to the Association of Plastics Manufacturers in Europe (APME).[40]

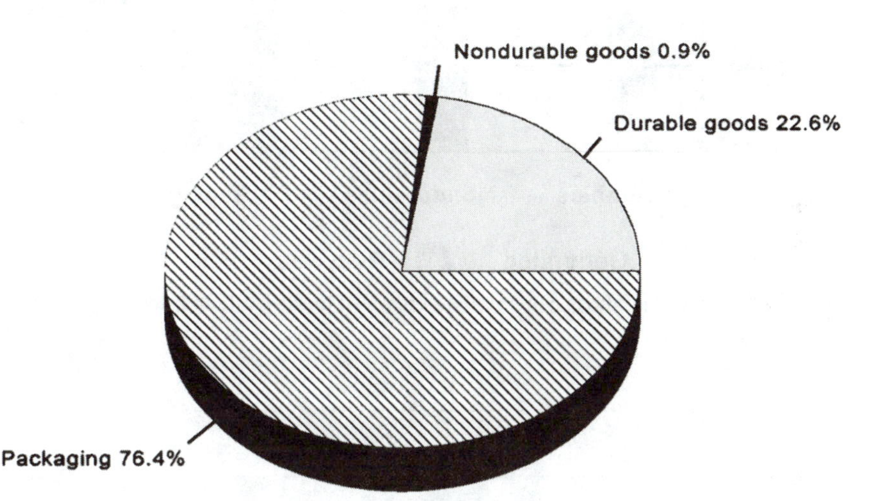

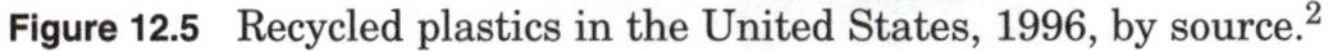

Figure 12.5 Recycled plastics in the United States, 1996, by source.[2]

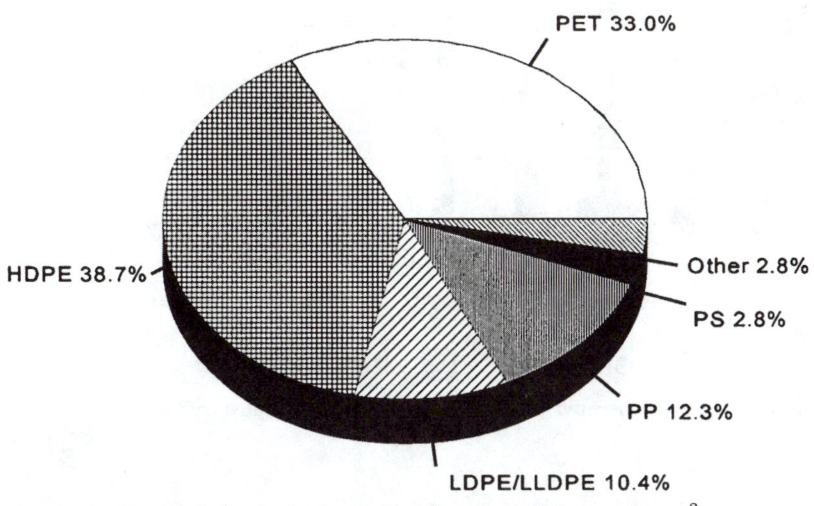

Figure 12.6 Recycled plastics in the United States, 1996, by resin type.[2]

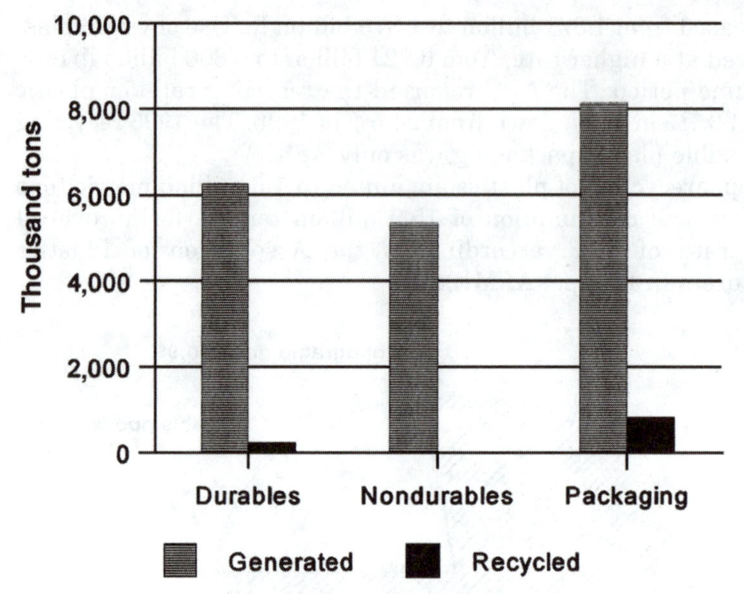

Figure 12.7 Plastics products generated and recycled in the United States, 1966.[2]

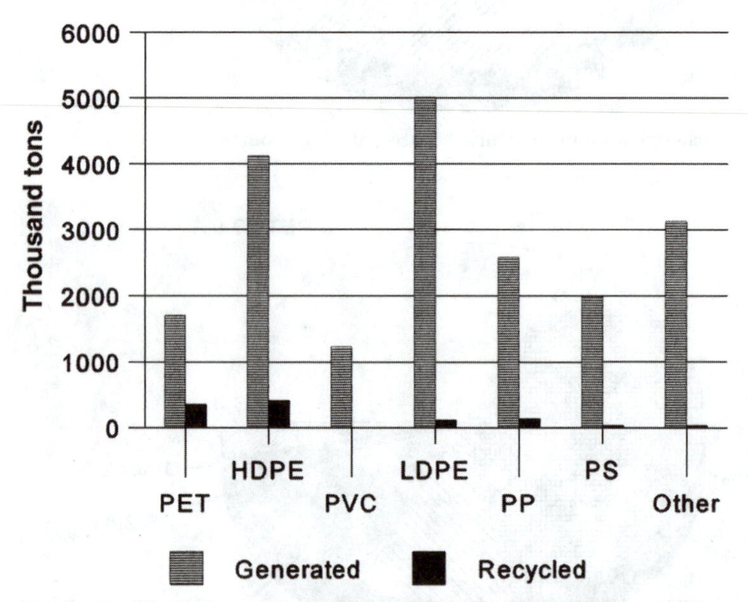

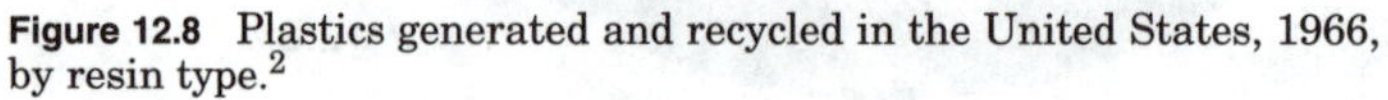

Figure 12.8 Plastics generated and recycled in the United States, 1966, by resin type.[2]

RECOUP reported that 7000 tonnes of plastic bottles were collected in the United Kingdom in 1997, a total of 140 million bottles. More than one in three local solid waste authorities ran a plastic bottle recycling scheme, including over 3000 plastic bottle banks and nearly 2 million homes served by curbside collection.[41]

In Germany, under the DSD green dot system, more than 5 million metric tons of postconsumer packaging were collected in 1995, 79% of the total amount of plastic packaging generated by households and small businesses. Almost 90% of that came from curbside recycling collection systems, with the remainder coming from drop-off systems at supermarkets, gas stations, and public buildings. Slightly less than 55% of the collected material was recycled, with the remainder used as an energy source.[42]

How recycling rates are calculated is itself a source of controversy. There have been charges in the past that surveys which ask recyclers for data receive inflated figures, and thus inflate recycling rates. Surveying organizations take various steps to avoid this problem, but cannot completely eliminate it. On the other hand, some organizations which do recycling may be missed in the survey, thus decreasing recycling rates. Even if the accuracy of the data can be guaranteed, a more fundamental problem remains. What should be counted as being recycled? The two fundamental options are measuring the amount of material collected for recycling or measuring the amount of material actually reused. Since 5 to 15% of collected material is lost during processing of the material, mostly because it is some type of contaminant, such as paper labels, product residues, undesired types of plastic, etc., there is a significant difference in recycling rates between the two approaches. The APC is the major source of information about plastics recycling rates in the United States. In 1997, APC switched from using the amount of cleaned material ready for use to the amount of material collected for processing. Their justification was that the latter method is more in keeping with the way recycling rates are calculated for other materials. Since this resulted in inflating recycling rates at a time when recycling rates, if calculated by comparable measures, were declining, this move brought considerable criticism. For instance, the PET bottle recycling rate in 1997 was 27.1% if based on material collected, but only 22.7% if based on clean material ready for reuse.[43] Criticism of APC was further heightened by their decision to restrict distribution of their annual plastic recycling report to APC members. The Environmental Defense Fund (EDF), in response, issued a report titled "Something to Hide: The Sorry State of Plastics Recycling," in which they used APCs numbers to highlight the decline in plastic recycling rate which was evident when 1996 data was compared to 1995 on the same basis. They also noted that polystyrene food service items were deleted from APCs definition of plastic packaging, beginning in 1995—a move which further shored up plastics recycling rates. EDF calculated that the recycling rate for plastic packaging in 1996 was only 9.5%, and would have been only 8.5% if polystyrene food service items had been included.[39]

How to count material generated by the industrial sector is also controversial. In general, there is agreement that scrap material which is routinely ground and reused as part of the normal manufacturing process, such as flash from extrusion blowmolding, should not be counted as recycled. On the other hand, there is a strong body of opinion that materials which used to be disposed of, but are now being collected and processed into usable materials, should be counted as recycled. As an example, defective in-mold–labeled bottles with paper labels were routinely disposed of until processes were developed which could remove the paper contaminant, leaving usable plastic. The problem that arises is that there is not a clear line between "routine" reuse and recycling. For example, if a thermoformer buys sheet, and sends the trim material back to the sheet supplier, where it is flaked and mixed with virgin resin in the manufacture of new sheet, is that recycling or routine reuse? No clear "rules" for these decisions have been promulgated. One answer is to consider only postconsumer material as being "real" recycling. This stance has the benefit of drawing clear lines, since then only material which has served its intended use is counted as being recycled. However, it has been criticized for ignoring the very real contributions that can be made to reducing burdens on disposal systems by recovery of various types of industrial waste.

To further complicate matters, some people distinguish between primary, secondary, and tertiary (sometimes also quaternary) recycling, but there is no consensus on how to define these terms. For example, the U.S. FDA defines primary recycling of plastics as the use of clean in-plant process scrap which is fed back into the same process in which it was generated, tertiary recycling as techniques involving depolymerization and repolymerization, and secondary recycling as everything else. In contrast, many recycling authorities define primary recycling as recycling back into the same or similar products; secondary recycling as recycling into a downgraded, less stringent use; and tertiary recycling as production of feedstock chemicals (and sometimes energy). Those who accept use as an energy source as a legitimate form of recycling sometimes classify it as quaternary recycling, rather than tertiary recycling. Because of these differing definitions, if these terms are used in communication, there needs to be agreement on what is meant by them.

12.4.1 Polyethylene terephthalate (PET)

PET soft drink bottles were the first postconsumer plastic containers to be recycled on a large scale. As was mentioned earlier, the existence of bottle deposit legislation caused large numbers of these containers to become available in reasonably centralized locations, creating an opportunity for the development of systems to take advantage of the value of this material. One of the first companies to successfully develop systems for recycling PET soft drink bottles was Wellman, which began processing clear PET bottles in 1979 and is still the largest PET recycler in the United States. St. Jude was another early entrant into PET recycling, beginning about the same time as Wellman but on a smaller scale, and concentrating on the green bottles, while Wellman concentrated on clear bottles.[44]

PET beverage bottles are the largest single source of PET in municipal solid waste, and packaging accounts for more PET in MSW than either durable or nondurable goods, as shown in Fig. 12.9.

The existence of bottle deposit legislation continues to be an important factor in PET recycling. It was estimated that, in 1997, 54% of the PET soft drink bottles recycled came from bottle-bill states, while these states accounted for only 29% of the population. Recycling rates for soft drink bottles in deposit states range from 76 to 90%.[47] It should be noted that the low end represents California, which has a bottle refund system rather than a true deposit. The monetary incentive in California is lower than in true deposit states, and the refund system is less convenient.

PET recycling rates. PET recycling grew rapidly from its beginnings in 1979, but was confined almost exclusively to deposit states, which

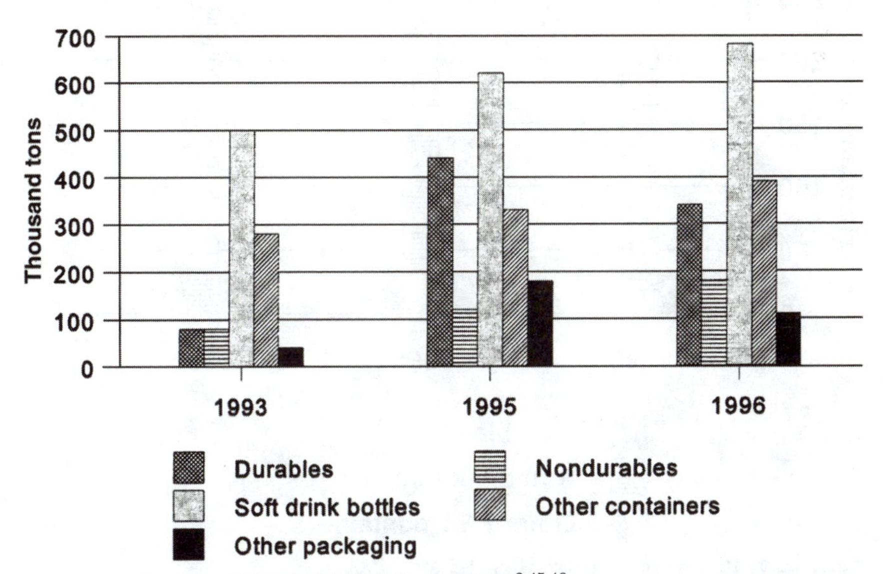

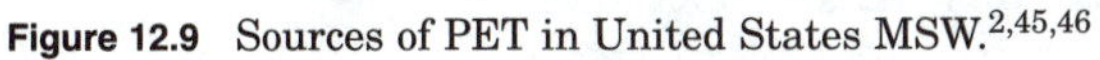
Figure 12.9 Sources of PET in United States MSW.[2,45,46]

typically attained a 90% or better recycling rate until the mid-1990s. When concerns about solid waste disposal led to the creation of a large number of new recycling programs, many of them providing curbside collection, many of these programs included PET soft drink bottles and HDPE milk bottles in the mix of materials they accepted for recycling. This significantly increased the available amount of PET. At the same time, uses of PET bottles began to expand significantly outside the soft drink bottle market. These "custom" bottles began to be included as accepted materials, along with the soft drink bottles. In the deposit states, where PET soft drink bottles were not included in curbside collection programs since they were collected through the deposit system, programs began to add PET to the collected materials. The result was a significant increase in both the amount of PET bottles potentially available for recycling and the amount which was actually collected and recycled. During the late 1980s and early 1990s, both the overall tonnage of PET recycled and the recycling rate continued to grow, with the soft drink bottle recycling rate higher than the rate for custom bottles, and the rates for bottles very much higher than the rates for other forms of PET (Figs. 12.10 and 12.11).

During the mid-1990s, the growth in use of PET, both in packaging applications and elsewhere, led a number of companies to invest in new facilities worldwide for production of virgin PET. As these facili-

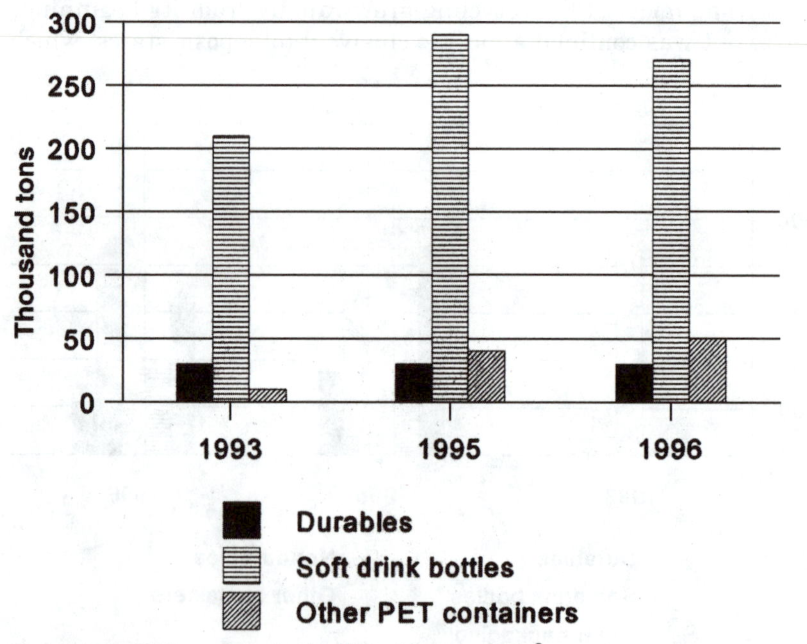

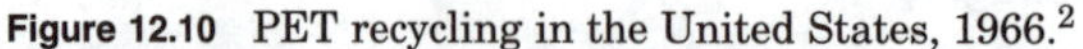
Figure 12.10 PET recycling in the United States, 1966.[2]

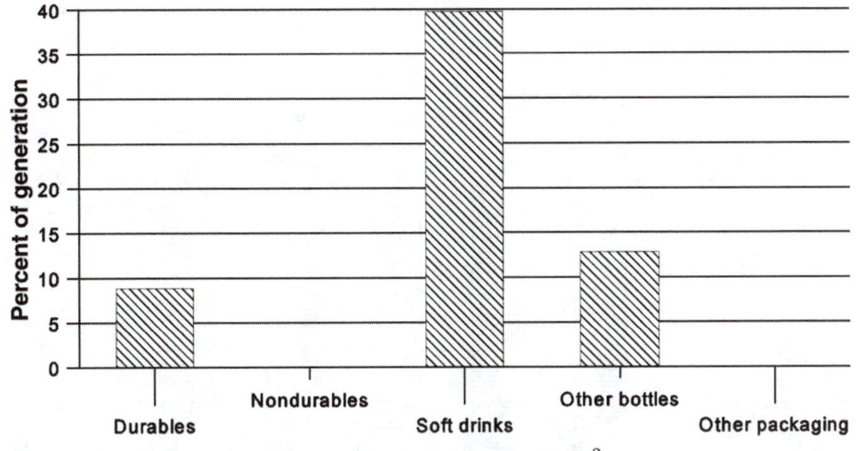

Figure 12.11 PET recycling rates in the United States, 1996.[2]

ties entered production, the supply of PET increased at a faster rate than the markets, and prices fell. Additionally, during startup, these facilities produced large amounts of off-spec resin, which was sold at very low prices. At the same time, there was a decrease in legislative pressures to use recycled plastic, particularly in Florida and California, and export markets decreased. The result, in mid-1996, was a drastic fall in the price at which recycled PET could be sold. Some PET recyclers shut down because their costs for processing the material were higher than the price they could obtain for it. A few recycling collection programs stopped accepting plastics. During this same time period, there was increasing use of PET in small single-serve beverage bottles, and it became evident that the willingness of consumers to divert these containers for recycling was less, on average, than with the larger size bottles. Much of this probably is because these bottles are more likely to be consumed away from home, where they may be tossed into the trash instead of taken home to the recycling bin. The combination of factors in the United States resulted in a decrease in both the total tonnage of PET recycled and, of course, in the recycling rate (Fig. 12.12). Late 1997 brought a small increase in the value of recycled PET and other signs of recovery, but the recycling rate remained below the highs reached earlier in the decade. Early reports for 1998 indicate little change from 1997. In Europe, where PET recycling is driven by government mandates, recycling rates and amounts continued to increase during this period despite the low prices.

In addition to recycling of PET bottles, there is some recycling of PET strapping. In nonpackaging applications, some PET photographic film, including x-ray film, is recycled. In that case, PET is obtained as

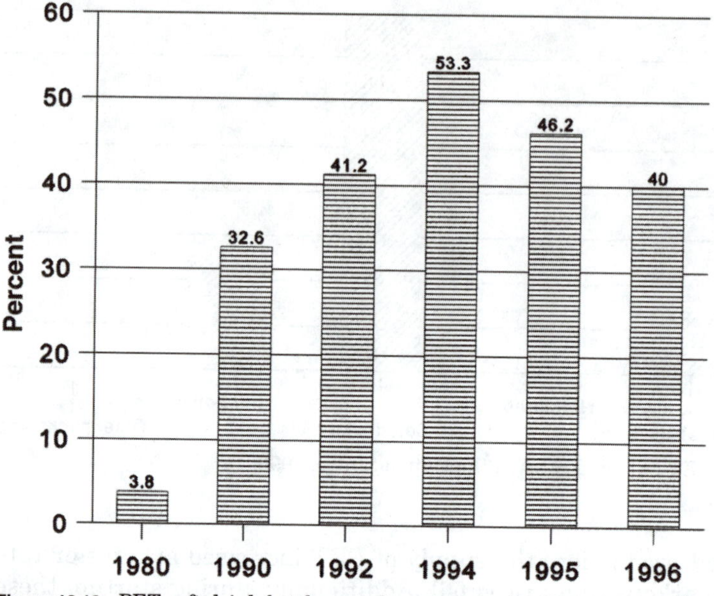

Figure 12.12 PET soft drink bottle recycling rates in the United States.[2]

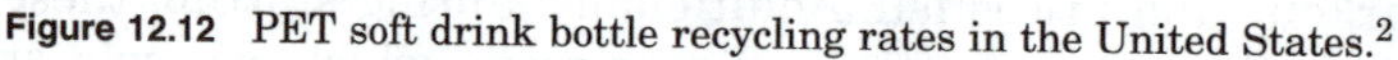

a byproduct of silver recovery. Recovery of PET from durable goods was estimated at 30,000 tons in 1996 or 8.8% of the amount discarded. Recovery of PET in nondurable goods was insignificant. Recovery of PET soft drink bottles was 270,000 tons, or 39.7%, and of other PET containers 50,000 tons, or 12.8%. Recovery of other PET packaging was negligible. Overall, 320,000 tons of PET packaging were recovered, or 27.1%; total recovery of PET from MSW was 350,000 tons, or 20.6%[2] (see Fig. 12.13). The APC reported a PET soft drink bottle recycling rate of 35.8% in 1997, down from 38.6% in 1996, a slightly lower rate than that reported by the U.S. EPA. The overall PET bottle recycling rate, according to the APC, was 25.4% in 1997, down from 27.8% in 1996.[47] The National Association for PET Container Resources (NAPCOR) calculated somewhat different rates, 31.7% in 1996 and 27.1% in 1998.[48]

In an effort to increase recycling of PET, NAPCOR announced it will provide a 2-year grant and technical assistance to a recycler to buy and reprocess unusually colored translucent and opaque PET bottles as well as glycol-modified and extrusion blow molded bottles.[49] NAPCOR is also sponsoring the placement of "big bin" collection containers in locations such as stadiums, convenience stores, and amusement parks in an effort to capture more of the single-serving PET bottles which are consumed away from home.[48]

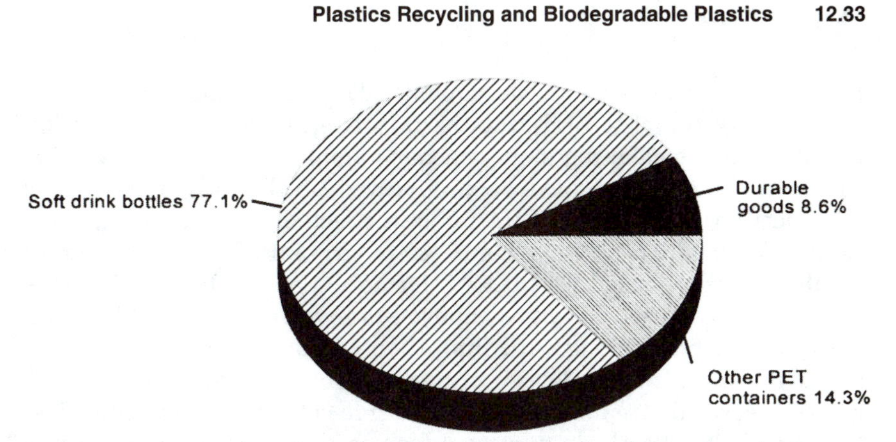

Figure 12.13 Sources of recycled PET in the United States, 1996.[2]

Physical recycling of PET. Most PET is processed by physical recycling. In these systems, the PET is typically first color-sorted to separate clear from green and other bottles, since the clear PET has a higher value. Next the PET is chipped, washed, and purified in various ways so that a pure resin can be obtained. One of the major issues is separation of PET from PVC. Because both are transparent plastics, they are difficult to separate reliably by manual sorting. Further, their densities overlap, so they cannot be separated by conventional float-sink methods. To complicate the matter, PVC may be present in the recycled stream in the form of labels, or as inner liners in caps, in addition to bottles. This presents major problems to recyclers, since very small amounts of PVC contamination, 4 to 10 ppm, can cause significant adverse effects on PET properties.[44] At PET melt temperatures, PVC decomposes, generating hydrogen chloride (HCl) which can catalyze PET decomposition, as well as leaving black specs in the recovered material. Thus, both performance and appearance can be significantly damaged.

Another contamination issue stems from the adhesives which may be used to attach labels or base cups. Often, not all of the adhesive residue can be removed by washing. These residues can cause color changes in the PET. Further, the ethylene vinyl acetate can decompose, releasing acetic acid, which along with the rosin acids in some adhesives can catalyze PET decomposition. Thus, these contaminants also can detract from both performance and appearance of the recycled material.

PET is also sensitive to degradation from the additional heat history and exposure to moisture during recycling. This commonly shows up as a decrease in intrinsic viscosity (IV). It is possible to subject the material to solid-stating, much as is done in resin manufacture, to increase the molecular weight (and, consequently, IV) back to the desired level.

Physically recycled PET from certain operations which add additional intensive cleaning steps, perhaps along with controls over the source of the material, have been approved (in the form of a letter of nonobjection) by the U.S. FDA for unrestricted food-contact applications, either alone or in a blend with virgin PET. The companies involved have released very little information about the details of the cleaning procedures. They are believed to involve intensive high-temperature washing, along with limitation of the incoming material to soft drink bottles from deposit states, which are known to provide a cleaner recycled stream than does curbside collection. Less intensively cleaned PET has been approved for use as a buried inner layer in food packaging, with virgin PET used as a barrier to prevent migration of contaminants from the recycled layer.

On the whole, recycled PET retains very good properties, and can be used for a variety of applications. Markets will be discussed further later in this section.

Chemical recycling of PET. Chemical recycling of PET depends on chemical reactions which break down the PET into small molecules, which can then be used as chemical feedstocks, either for repolymerizing PET or for manufacturing related polymers. Two procedures, glycolysis and methanolysis, are in commercial use. Both can be used to produce PET which is essentially chemically identical to virgin polymer, and which have been approved for food-contact use.[50,51]

The first of these processes to receive a "letter of nonobjection" from the U.S. FDA was Goodyear's glycolysis process in 1991 (later sold to Shell). Later that same year Eastman Chemical and Hoechst-Celanese received approval for their methanolysis processes. The glycolysis processes typically produce partial depolymerization, which is followed by purification and repolymerization. Methanolysis processes provide full depolymerization, followed by purification by crystallization and then repolymerization. Glycolysis cannot remove colorants and certain impurities which can be removed by methanolysis. DuPont also operated a methanolysis facility for recycling PET, but indicated recently that it is discontinuing the operation for economic reasons.

Markets for recycled PET. Historically, the first large market for recycled PET was in fiber applications, particularly polyester fiberfill for use in ski jackets, sleeping bags, pillows, and similar products. While there are now many additional markets for recycled PET bottles, fiber markets still dominate, as shown in Fig. 12.14. These fibers now have substantial use in carpet and even in clothing. Recently, a contest held by the Toronto-based Environment and Plastics Industry Council (EPIC) featured wedding dresses made from recycled plastic, with the average entry requiring 80 soft drink bottles to make. Half the polyester carpet manufactured in the United States now contains recycled PET.

Some PET is used in the manufacture of new bottles. For a time, PET soft drink bottles made from 25% repolymerized PET were being used in parts of the United States. However, the higher cost of the repolymerized (chemically recycled) PET which was being used caused such applications to disappear when legislative and consumer pressure to use packages with recycled content declined. Veryfine, headquartered in Westford, Mass., is one of a very few U.S. users of recycled PET in food or beverage bottles. Veryfine packages all its juice and juice drinks in bottles containing recycled PET in a middle layer, surrounded by ethylene vinyl alcohol for oxygen barrier and containing layers of virgin PET on the inner and outer surfaces. Recycled PET makes up about 35% of the container. Heinz USA uses essentially the same structure for ketchup bottles. Coca Cola uses recycled PET in soft drink bottles in some overseas markets, such as Australia, New Zealand, Saudi Arabia, and parts of Europe, but does not use any in the United States.[52]

As mentioned, the U.S. FDA has cleared the use of physically recycled PET from certain specific recycling systems for unrestricted use in food packaging, although little such use is presently occurring.[53] Recycled PET is also found in thermoformed trays for uses such as packaging eggs, fresh produce, or pastries. In these applications, purity standards are less stringent since there is less of a tendency for migration of contaminants to the food product. In fact, egg cartons were the earliest

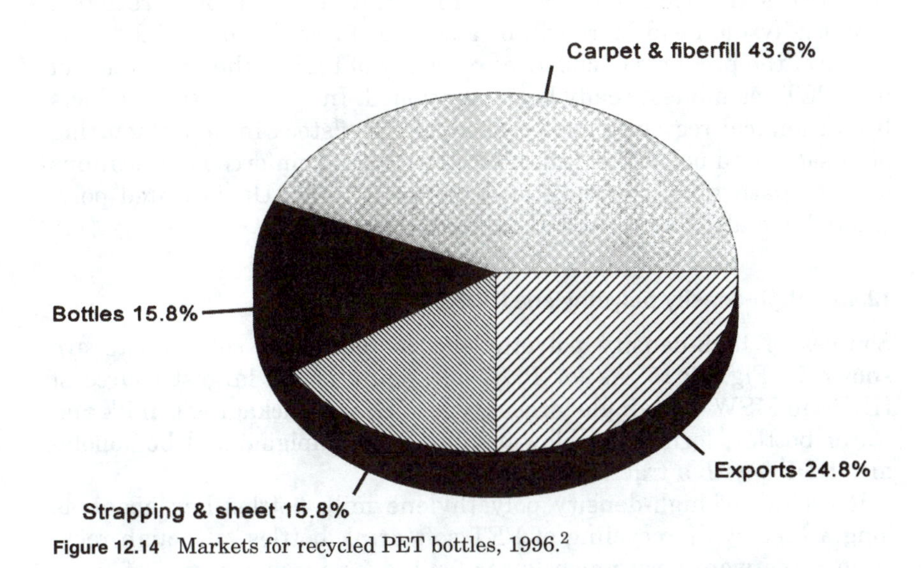

Figure 12.14 Markets for recycled PET bottles, 1996.[2]

food-contact application for recycled PET, using essentially the same grade of recyclate as nonfood packaging applications.

Use of physically reprocessed PET in nonfood containers is more common than for food packaging. Up to 100% recycled PET can be used, or the recycled material can be blended with virgin. For example, Clorox uses about 50% postconsumer recycled PET in bottles for its Pine-Sol cleaner, after having tried and abandoned use of 100% recycled content due to processing problems.[52] Most often, the recycled material in these containers is blended with virgin, since this is less costly than using multilayer technology.

Recycled PET is also used in sheet and strapping. For example, Precision Packaging Products, Inc., in Rochester, N.Y., supplies thermoformed blister packages made with recycled PET to pet products, electronics, cellular phones, health-care, and personal care markets.[54] Physically recycled PET is sometimes used in a buried inner layer, in either sheet or bottles, for food-contact applications. It is also used either alone, or blended with virgin PET, for a variety of nonfood applications. PET films with recycled content are also available. In 1998, the GrassRoots Recycling Network began a campaign asking consumers to mail empty PET bottles back to the Coca Cola Company, in an effort to convince them to use recycled resin in soft drink bottles, as well as urging a boycott of Coke products until the company begins using recycled PET.[55]

The automotive industry is increasing its use of recycled PET. Several companies, including Lear-Donnelly, Johnson Controls, and United Technologies, now manufacture headliners which incorporate recycled soft drink bottles, as an alternative to polyurethane. Eventually, old headliners will be a source of recycled material.[56]

Use of the products of chemical recycling of PET in the production of new PET resin has already been mentioned. In addition, the products from chemical recycling can be used as a feedstock in manufacturing of unsaturated polyesters, often for glass fiber–reinforced applications such as bath tubs, shower stalls, and boat hulls. Unsaturated polyesters have also found uses in polymer concrete.

12.4.2 High-density polyethylene (HDPE)

Sources of high-density polyethylene in municipal solid waste are shown in Fig. 12.15. As for PET, packaging is the largest source of HDPE in MSW. The single largest type of HDPE packaging is milk and water bottles, formed by blowmolding from unpigmented homopolymer HDPE with a fractional melt index.

Recycling of high-density polyethylene milk bottles has about as long a history as recycling of PET soft drink bottles, although recycling rates were very much lower for a considerable period of time,

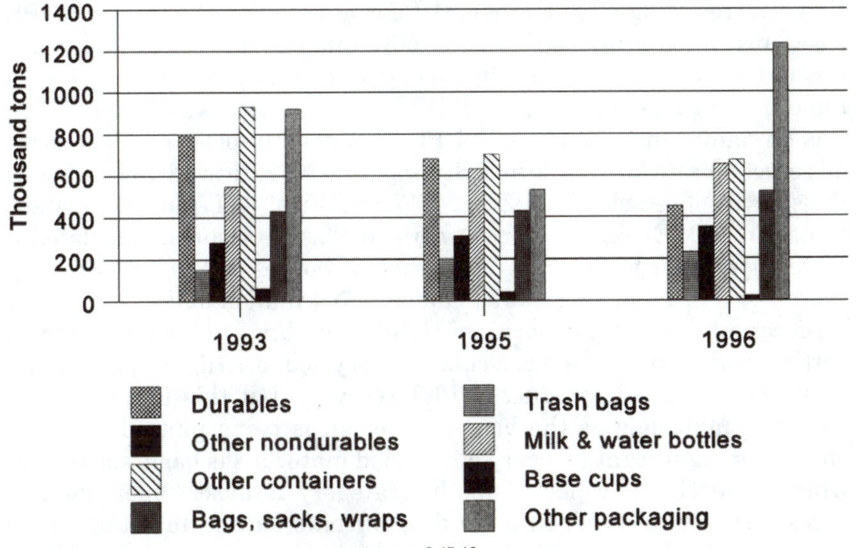

Figure 12.15 HDPE in United States MSW.[2,45,46]

and they remain lower today. While deposit programs provided the impetus for soft drink bottle recycling, no such programs existed for milk bottles. Therefore, milk bottle recycling got its start with drop-off programs, relying on the willingness of individuals to deliver the bottles for recycling. In the early years, the presence of paper labels on the bottles was a major problem, since many recyclers did not have technology which could successfully remove the paper. Many recycling programs requested that participants remove the labels from the bottles, some even suggesting placing a small amount of water in the bottles and heating them in a microwave to soften the adhesive so the labels could be peeled off. Such requests met with little success. In one of the early HDPE milk bottle recycling programs in Grand Rapids, Mich., an employee cut out the label-bearing part of each bottle with a utility knife and discarded it before feeding the rest of the bottle into the shredder. Such solutions obviously entailed high labor costs, as well as loss of potentially recyclable materials, and kept most HDPE milk bottle recycling programs on the borderline of profitability, at best.

As technology developed to better handle this and other contamination issues, and as pressure to recycle plastics mounted, HDPE milk bottle recycling expanded and many programs began to included nonmilk bottle HDPE containers. Now the majority of curbside and drop-off collection programs for recyclables include blow-molded HDPE bottles as one of the materials collected. The recycling rate for HDPE milk and water bottles in the United States in 1996 was 30.8% according to the U.S

EPA. The recycling rate for other HDPE containers was 20.9%, for HDPE base cups on soft drink bottles it was 50%, and for bags, sacks, and wraps it was 1.9%. Overall, the HDPE packaging recycling rate was 11.7%. In the durable goods category, the HDPE recycling rate was 11.1%. There was no significant recycling of HDPE from nondurable goods. The overall recycling rate for HDPE in MSW was 10.0%.[2] Figure 12.16 illustrates the sources of recycled HDPE in MSW, and Figs. 12.17 and 12.18 show trends in HDPE recovery. The American Plastics Council calculated a recycling rate for high-density polyethylene bottles of 24.7% in 1997, up from 24.4% in 1996, for a 1997 tonnage of 704 million lb.[47]

As can be seen, the majority of HDPE recycling, as for PET, is from bottles. Soft drink bottle base cups are recycled at a higher percentage than bottles, as a byproduct of PET recovery, but this is a declining source of materials as the base cup design is being phased out. The only other significant category of recycled materials is bags, sacks, and wraps. The HDPE collected in this category is mostly merchandise sacks, usually collected through drop-off bins located in retail stores which accept plastic bags of all types. There is also some recycling of HDPE envelopes.

Collected HDPE is, typically, first sorted to separate the natural (unpigmented) containers, which have higher value, from pigmented containers. Separation is typically a manual process, although automated systems have been developed. It is also possible to further separate the pigmented HDPE into various color categories, either automatically or manually, but this is still relatively uncommon. Sorting out of the natural HDPE is often done prior to baling the materials for delivery to a processor, although it can be done at a later stage.

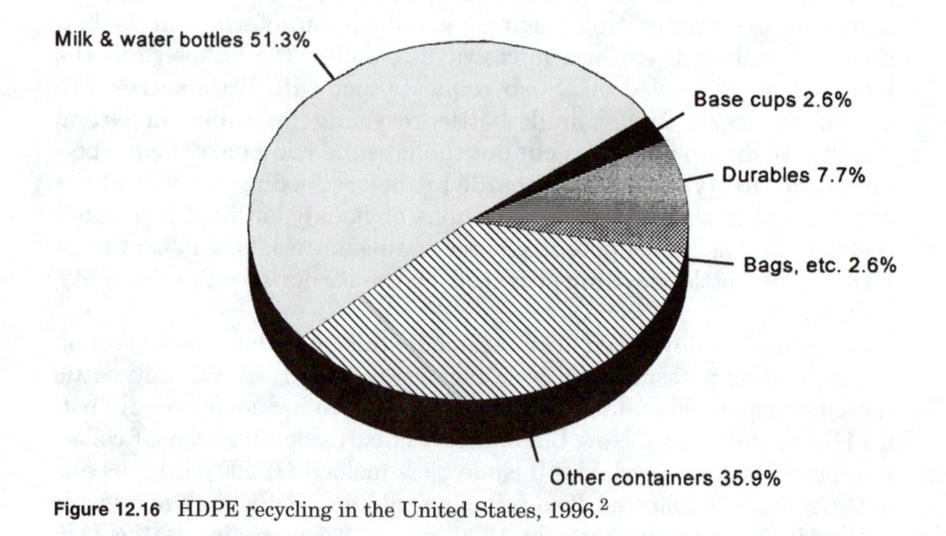

Figure 12.16 HDPE recycling in the United States, 1996.[2]

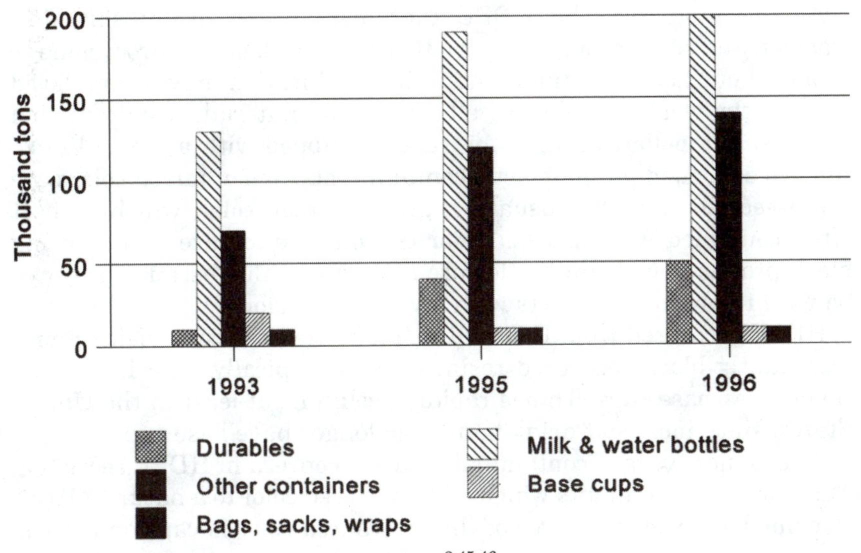

Figure 12.17 Trends in U.S. HDPE recycling.[2,45,46]

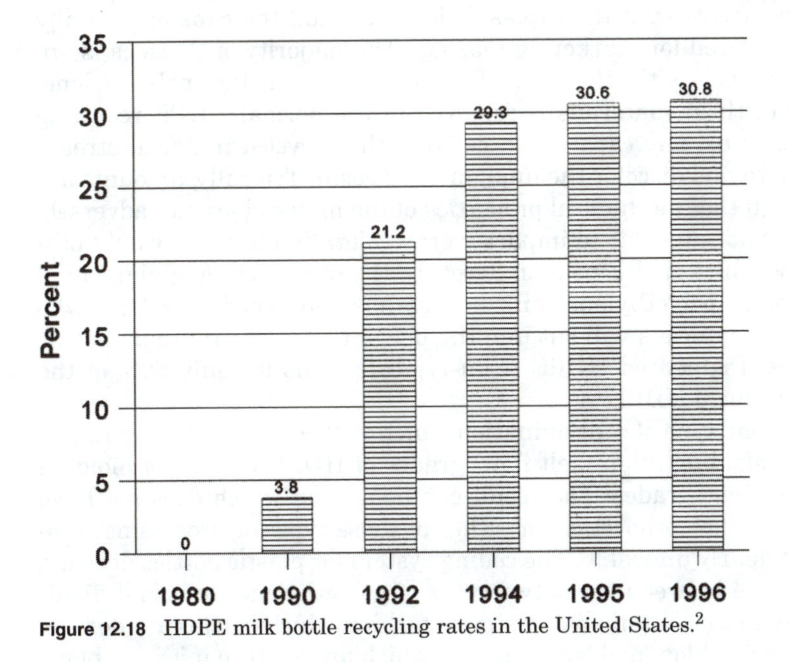

Figure 12.18 HDPE milk bottle recycling rates in the United States.[2]

At the processor, the HDPE containers are typically shredded, washed, and sent through either a float/sink tank or a hydrocyclone to separate out heavy contaminants. Air classification may be employed prior to the washing step as well. The clean materials are dried, and then usually pelletized in an extruder equipped with a melt filter to remove any residual nonplastic contaminants. When mixed colors are processed, the result is usually a grayish-green color, which is most often combined with a black color concentrate for use in producing black products. Natural bottles are of greater value because they can be used to produce products with a variety of colors.

HDPE recovered from PET soft drink bottle base cups yields a predominantly black recovered resin, which is typically used in production of new base cups. This is rapidly declining, at least in the United States, since most soft drink bottles no longer have base cups.

Four major types of contamination are a concern in HDPE recycling. The first is contaminants which add undesired color to a natural HDPE stream. The primary source of this unwanted color is caps on bottles. While nearly all recycling programs ask consumers to remove the caps before placing the bottles in the collection system, a significant number of bottles arrive with the caps still in place, and the caps are usually brightly colored for marketing reasons. The majority of these caps are polypropylene, with the next largest fraction being polyethylene. Neither of these materials is removed in the normal HDPE recycling systems. Thus, any caps which get into the recycled material stream will remain and discolor the unpigmented resin. Typically, amounts are low enough that mechanical properties of the material are not adversely affected, but sufficient to impart a gray coloration to what would otherwise be white HDPE pellets. Recently, the introduction of pigmented high-density polyethylene milk bottles has concerned recyclers, who fear these materials will cut into the use of the more profitable natural bottles. Pigmented HDPE resin typically sells for only 60% of the price of natural HDPE.[57]

The second type of contamination which is of concern is the mixing of injection molding (high melt flow) grades of HDPE with blow molding (low melt flow) grades. The result can be a resin which does not have the desired flow properties for either of these types of processing, rendering it nearly unusable. The coding system for plastic bottles does not differentiate between these two types of polyethylene, so it is difficult to convey to consumers in any simple fashion which bottles are desired (the extrusion blow-molded ones) and which are not (the injection blow-molded ones). Some collection programs attempt to instruct consumers to place for collection the bottles "with a seam" and not the ones which do not have this characteristic. Other programs ignore the issue and simply accept the resulting contamination and its adverse effects on properties. Fortunately, the vast majority of HDPE bottles, particularly in larger sizes, are extrusion blow molded. A few years ago, however, when extrusion blow-molded base cups were introduced as an alternative to injection-molded base cups, some recyclers found themselves with HDPE resins which they could not sell because the materials were not suitable for processing into new base cups or desired for other applications, due to the mixing of the different grades of resin.

A third significant contamination issue is the mixing of polypropylene into the HDPE stream. The polypropylene arises primarily from caps which, as discussed previously, are included in the recycling stream despite requests that consumers remove them. Some polypropylene (PP) also arises from fitments on detergent bottles and from inclusion of PP bottles with HDPE bottles when materials are collected. The density-based separation systems commonly employed in HDPE recycling do not separate PP from HDPE, since both are lighter than water. Fortunately, in most applications a certain level of PP contamination can be tolerated. However, particularly in the pigmented HDPE stream, levels of PP contamination are often sufficient to limit the amount of recycled HDPE which can be used, forcing manufacturers to blend the postconsumer materials with other scrap which is free of PP, or with virgin (often off-grade) HDPE. Commercially viable systems for separating PP from HDPE are, at least for the most part, not yet available.

The fourth type of contamination which is an issue is contamination of the HDPE with chemical substances which may later migrate from an HDPE container into the product. This is a more serious issue for HDPE than for PET for two reasons. First, the solubility of foreign substances of many types is greater in HDPE than in PET. Therefore, there is often more potential for migration. Second, the diffusivity of many substances is greater in HDPE than in PET. Consequently, the ability of substances to move through the HDPE and reach a contained product is greater. The strategies for dealing with this potential problem are essentially the same as for PET.

First, a combination of selection of materials and processing steps can be used to minimize the contamination levels in the HDPE. The U.S. FDA has issued letters of nonobjection for recycling systems for HDPE which permit those material to be used in some food contact applications. The first company to obtain a letter of nonobjection from the U.S. FDA for such purposes was Union Carbide. The technology was later sold to Ecoplast, which also received a letter of nonobjection.[58]

Second, the recycled HDPE can be used in a multilayer structure which provides a layer of virgin polymer as the product contact phase. This approach was first used for laundry products when problems were encountered with migration of odorous substances from recycled plastic to the products. The inner layer of virgin polymer provided a sufficient barrier to solve the problem. In these same applications, problems were also encountered with the appearance of the bottle. This was solved by incorporating a thin layer of virgin polymer on the outside of the polymer to carry the pigment. One added benefit was that this minimized the amount of (often expensive) pigment required to achieve the desired marketing image. The layer of virgin polymer on the inside of the container also provided an added benefit by reducing the tendency to environmental stress cracking in these containers. Since the recycled layer being incorporated was most often homopolymer HDPE from milk bottles, it did not have the stress-crack resistance of the copolymer HDPE typically used for detergents. Later, with the development of better washing technology, it was found to be possible to package such products in single-layer bottles formed from a blend of virgin and recycled HDPE. Nonetheless, such three-layer bottles, with the inner layer containing a combination of recycled milk bottles and regrind from bottle manufacture, remain standard for laundry detergents and similar products.

There are a variety of markets for recycled HDPE bottles. In the early days, the major market was agricultural drainage pipe. Today, this market accounts for only about 18% of recycled HDPE, with containers the largest market, followed by pallets and plastic lumber (see Fig. 12.19). Film, mostly merchandise sacks and trash bags, is also a significant market.

Procter & Gamble (P&G), which pioneered the use of three-layer bottles with an inner layer of recycled HDPE between outer layers of virgin material for its fabric softener and liquid detergent, is now the largest user of recycled plastic in the United States P&G packaging typically contains between 25 and 100% recycled HDPE, depending on product requirements. Clorox is another major user of recycled HDPE in bottles, as is DowBrands.[52]

DuPont uses 25% recycled HDPE in its Tyvek envelopes. The company also operates a program for recycling of used envelopes. For small users, the system involves selecting one envelope to be filled with other used envelopes, and mailing them back to the company. For large users, other systems can be put in place.[59]

12.4.3 Low-density polyethylene (LDPE)

Because of the similarity in properties and uses between low-density polyethylene (LDPE) and linear low-density polyethylene (LLDPE), and because they are often blended in a variety of applications, use and recycling of LDPE and LLDPE are often both reported and carried out together. Therefore, in the remainder of this discussion we will use

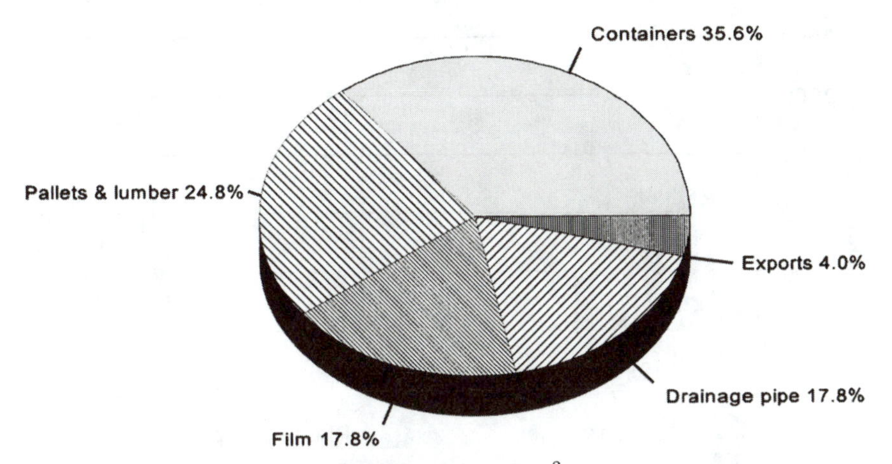

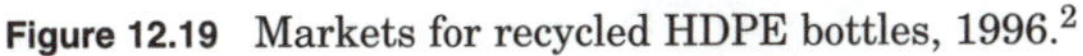
Figure 12.19 Markets for recycled HDPE bottles, 1996.[2]

the term low-density polyethylene, or LDPE, to refer to both LDPE and LLDPE. About half of the LDPE found in municipal solid waste comes from packaging. Another sizable fraction comes from nondurable goods, especially trash bags (see Figs. 12.20 and 12.21).

Recycled LDPE comes from two main sources—stretch wrap and merchandise bags. In contrast to PET and HDPE, curbside collection does not play a significant role in LDPE recycling systems in the United States. Stretch wrap is collected primarily from warehouses, retailers, and similar establishments where large quantities of goods arrive on pallets, with the loads stabilized by use of stretch wrap. These materials must be disposed of, so separating the stretch wrap and sending it for recycling avoids the disposal costs that would otherwise be incurred. In this way, recycling of stretch wrap is much like recycling of corrugated boxes.

Merchandise sacks are collected primarily through drop-off locations. Many retailers maintain a bin or barrel near the front of the store, where customers can bring plastic bags for recycling. The majority of these bags are LDPE, though a significant amount of HDPE is usually present as well. A few communities have experimented with adding plastic bags to curbside collection programs, but this remains very rare in the United States. Most multimaterial drop-off facilities do not include plastic bags in the materials they accept either. In recent years, there appears to have been some decrease in the availability of merchandise bag recycling. Some merchants have discontinued programs because of contamination of the stream with undesired materials, unfavorable economics, or for other reasons.

Another source of recycled LDPE is garment bags, collected from department stores in a similar manner to collection of stretch wrap.

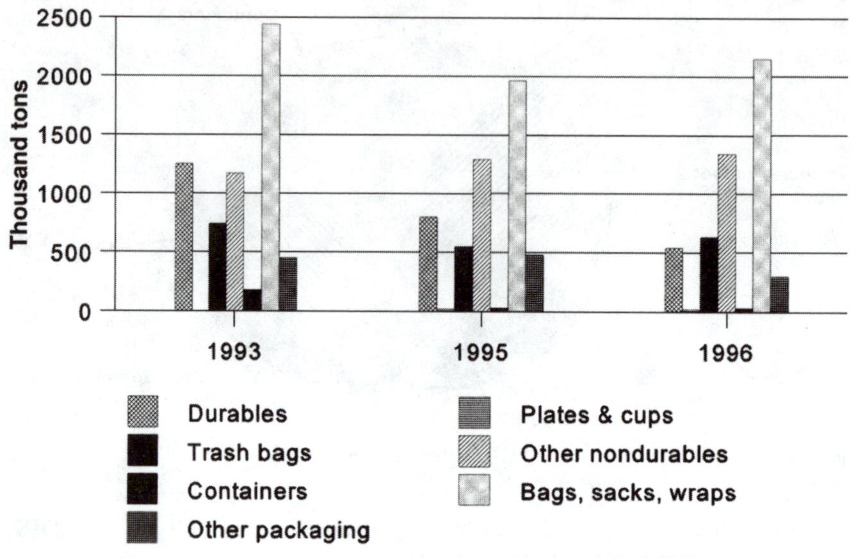

Figure 12.20 Sources of LDPE and LLDPE in United States MSW.[2,45,46]

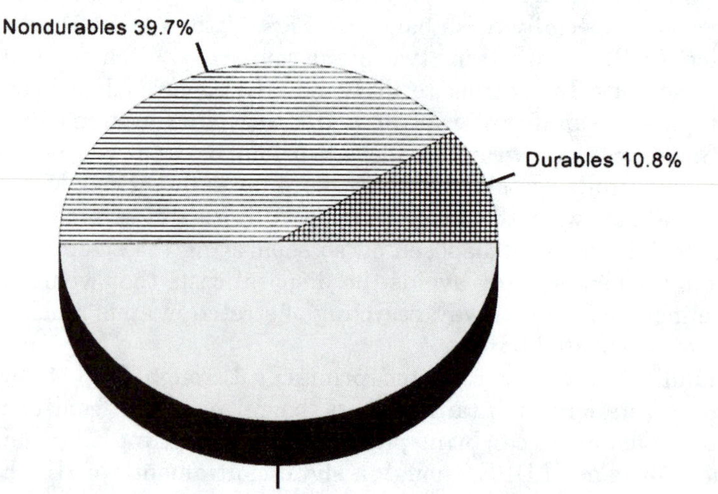

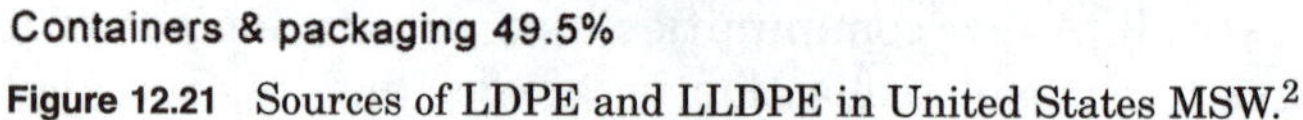

Figure 12.21 Sources of LDPE and LLDPE in United States MSW.[2]

One such program, in Riverside, Calif., relies on a program for people with severe developmental disabilities to do the sorting of collected material.[60]

Recovery of LDPE and LLDPE bags, sacks, and wraps in the United States in 1996 was 90,000 tons, or 4.4%, and HDPE bags, sacks, and wraps was 10,000 tons, or 2.0% of discards, according to the U.S. EPA.

Recovery of LDPE in other categories of packaging was negligible. About 20,000 tons of LDPE and LLDPE in durable goods were recovered, or 3.8% of discards. The U.S. EPA reported no information on the source of this material. There was no significant recovery in the nondurable goods category. The overall recovery rate for LDPE was 2.2% (see Figs. 12.22 and 12.23).[2]

In Canada, the Plastic Film Manufacturers Association of Canada and the Environment and Plastics Institute of Canada (EPIC) have

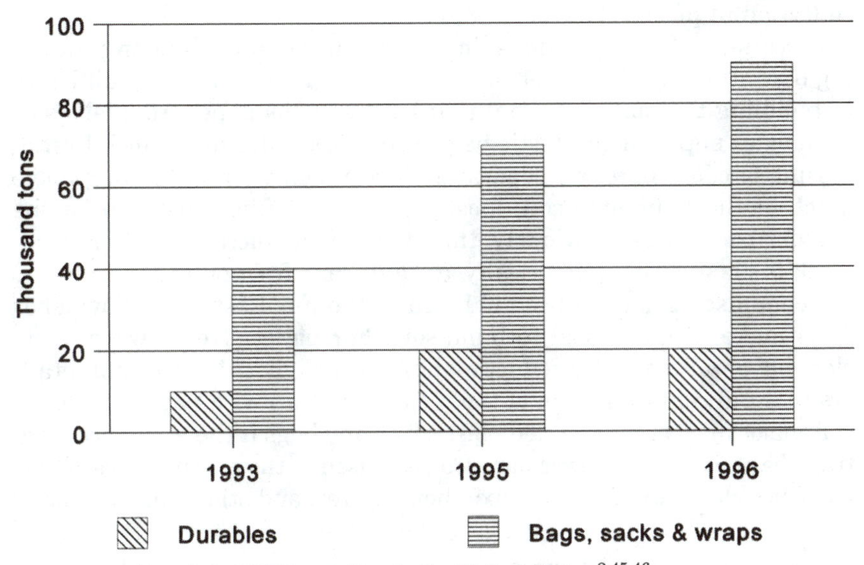

Figure 12.22 Trends in LDPE recycling in the United States.[2,45,46]

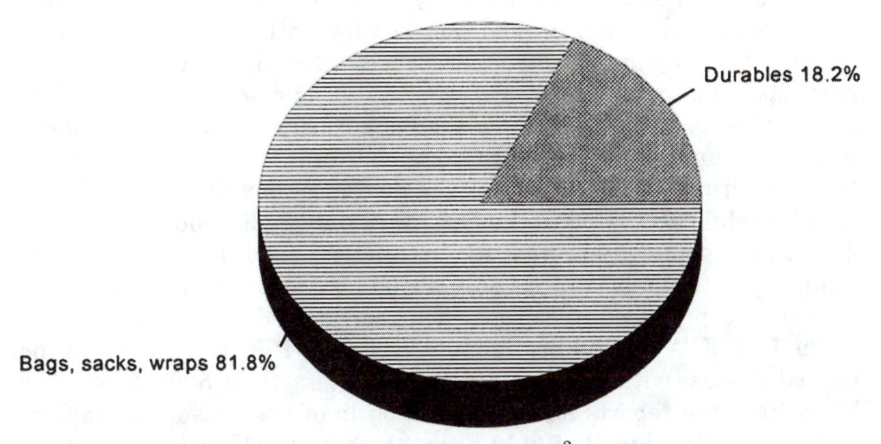

Figure 12.23 LDPE recycling in the United States, 1996.[2]

sponsored curbside recycling programs for plastic film of all types. In 1994, over 100 communities in Ontario and five in Quebec were participating.[61] By 1996, the program had grown to 19 municipalities in the Montreal area and 146 in Ontario.[62] In 1998, EPIC published "The Best Practices Guide for the Collection and Handling of Polyethylene Plastic Bags and Film in Municipal Curbside Recycling Programs." This 27-page guide, which offers step-by-step instructions for "best practices" for successful curbside recycling of plastic bags, is intended for use by Canadian municipalities which already have curbside film collection in place.[63]

Processing of film plastic is, in general, more difficult than processing of containers. The lower bulk density of the film leads to difficulty in handling the material. Contaminants, such as paper from labels or from sales slips left in plastic bags, are also harder to remove. In processing of containers, air separation is commonly employed to remove much of the light material, mostly paper and film plastic, from the heavier containers. Obviously, this will not be successful if the feedstock is plastic film. Historically, a significant fraction of the collected merchandise bags have been shipped to the Far East where low labor costs make it economical to hand-sort. For pallet stretch wrap, recyclers have worked with product manufacturers, distributors, and retailers to avoid contamination of the recovered wrap with paper labels.

A major market for recycled plastic film and bags is the manufacture of trash bags. Recycled plastic has also been used in the manufacture of new bags, bubble wrap, plastic lumber, housewares, and other applications.

12.4.4 Polyvinyl chloride (PVC)

Polyvinyl chloride in U.S. municipal solid waste originates most often in nondurable goods, followed closely by durable goods and packaging (Figs. 12.24 and 12.25). In addition, a substantial amount of PVC is found in building and construction debris, which is not categorized as municipal solid waste. Materials found in this category include vinyl siding, pipe, roofing, and floor tile, among others. A substantial majority of vinyl production goes into such long-term uses. According to the Vinyl Institute, 13.3 billion lb of vinyl were sold by the United States in 1996, while according to the U.S. EPA, only 1230 thousand tons, or 2.4 billion pounds, entered the municipal solid waste stream.[2,64] Similarly, estimates are that only about 18% of the 15 million metric tons of PVC produced in Europe each year is used in packaging.[65]

The U.S. EPA reports negligible recycling of PVC from MSW in the United States, with negligible defined as less than 5000 tons.[2] The Vinyl Institute reports about 9.5 million lb of postconsumer vinyl (of all types) recycled in 1995.[64] In Europe, where PVC has had wider use

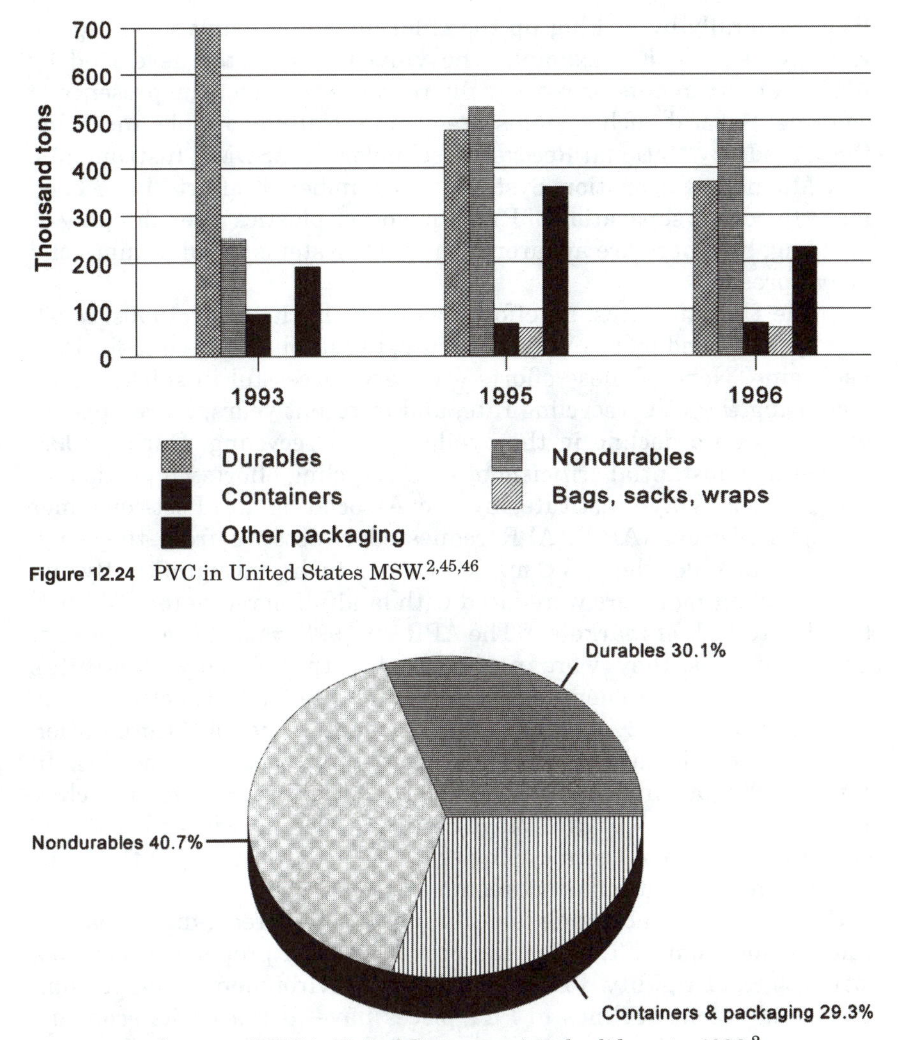

Figure 12.24 PVC in United States MSW.[2,45,46]

Figure 12.25 Sources of PVC in United States municipal solid waste, 1996.[2]

in packaging, particularly in water bottles, PVC recycling has a longer history, and has been more successful. About half of the 4 billion PVC mineral water bottles used in France each year are reported to be recycled, for example.[64]

One of the issues facing PVC recycling is separation of PVC from a stream of mixed plastics. This is particularly important in the United States for PVC packaging, since PVC, where it is collected at all, is generally collected in a program which targets all types of plastic bottles. Several systems have been developed to automatically identify

PVC, generally by picking up the chlorine signal using some type of radiative signal. For example, the vinyl bottle sorter developed by ASOMA Instruments uses x-ray fluorescence to detect the presence of chlorine. Several such systems are commercially available, including those made by National Recovery Technologies, ASOMA Instruments, and Magnetic Separation Systems.[64] A number of efforts have been made to permit separation of PVC from other plastics after chipping or grinding, but there are apparently no such systems in full commercial use at present.

In the United States, an effort was made in the early 1990s, much of it spearheaded by the Vinyl Institute, to establish recycling for PVC packaging. None of these efforts was very successful in substantially increasing the PVC recycling rate, and in recent years, there appears to have been a decline in the availability of recycling. During 1998, there was substantial criticism of PVC recycling efforts, in particular, those of the Vinyl Institute, by the Association for Post-consumer Plastic Recyclers (APR). APR requested assistance from the Vinyl Institute in 1996 when PVC markets began to dry up significantly and many of their members were faced with landfilling recovered PVC bottles due to lack of markets.[66] The APR, in 1998, raised the issue publicly again, as they were not pleased with the Vinyl Institute's response to their problems.[67] The recycling rate for PVC bottles fell to 0.1% in 1997, from 2% in 1996, after Occidental Chemical Corporation ended its subsidization of a program to buy back PVC bottles. In August 1998, a Vinyl Institute subsidized trial program for recycling PVC into floor tiles began. Vinyl Institute funding will end with the end of the trial, and the recycling effort will continue only if the firms involved feel that it makes economic sense.[68]

Use of PVC in packaging is declining in some areas, due in part to competition from PET, which can provide similar properties of clarity, strength, and rigidity. PET has a better environmental image than PVC, and recent declines in PET prices have increased its economic competitiveness. In some cases, some of the new very clear polypropylene bottles are being chosen as an alternative to PVC as well. In 1998, three Japanese firms announced that they will discontinue all use of PVC containers, switching to PET and PP. The firms are facing a government requirement for recycling their packaging beginning in April 2000.[69]

Recycling of vinyl siding and other preconsumer scrap, on the other hand, may have better success. A number of pilot projects have been carried out for recycling of vinyl siding, mostly focusing on scrap from building construction or remodeling, and often with financial support from the Vinyl Institute. As a result of a pilot project in Grand Rapids, Mich., recycling of vinyl siding waste is included in the "Residential

Construction Waste Management: A Builders' Field Guide," which was funded by the EPA.[70] Among the most high profile of these pilot projects are those involving Habitat for Humanity, which builds housing for low-income families. The Vinyl Institute and other PVC-related industry organizations have donated money and materials to some of these projects, in addition to supporting recycling efforts for the vinyl scrap generated during construction.

Polymer Reclaim and Exchange, in Burlington, N.C., recycles about 300,000 lb/month of vinyl siding from construction debris. Drop-off sites are located at landfills and near manufacturers of mobile and manufactured homes, and material is collected from as far as 500 miles away. The collected materials are cleaned and flaked, and then sold to molders, extruders, and compounders.[71]

In addition to vinyl siding, recycling efforts for building-related PVC wastes have focused on window profiles, carpet backing, pipe, and automotive scrap. The Vinyl Institute estimates that about 300 million lb of preconsumer vinyl scrap are recycled each year in the United States, far exceeding the 9.5 million lb of postconsumer vinyl. It further estimates that recycling of postindustrial vinyl of all types amounts to over 500 million lb annually in North America.[64] In France, the Autovinyle recycling program for PVC automotive scrap recycled 1740 metric tons of PVC in its first year of operation, 1997–1998. Its goal is 5000 metric tons by the end of 1999.[72]

Several firms recycle PVC (as well as PE) wire and cable insulation, primarily from the telecommunications industry. A primary focus is the recovery of the copper and aluminum wire and cable, with the plastic insulation being recovered as a byproduct. Since the recovered plastic typically contains small amounts of metal, it is suitable only for applications where high purity is not required. Uses include truck mud flaps, flower pots, traffic stops, and reflective bibs for construction workers. Most of the wire and cable originate in phone and business equipment wiring which is being replaced by fiber optic cable.[73]

Some recycling of PVC intravenous bags from hospitals is going on. One participating hospital is Beth Israel Medical Center in New York City, which is one of the pioneers in hospital recycling.[74]

A number of uses are possible for recycled PVC, depending on its source and purity. Most often, the recycled material is blended with virgin PVC. Applications include packaging, both bottles and blister packages, siding, pipe, floor tiles, and many more. Rhovyl, a French clothing manufacturer, is producing sweaters from postconsumer PVC mineral water bottles combined with wool in a 70/20% vinyl/wool blend. Collins & Aikman Floorcoverings uses discarded carpeting for parking stops and industrial flooring. Crane Plastics uses scrap from vinyl windows and siding to make retaining walls and bulkheads. In the United Kingdom, IBM has achieved closed-loop recycling of PVC monitor housings into 100% recycled content PVC computer keyboard backs. Philip Environmental in Hamilton, Ontario, Canada, is recycling about 125 million lb/year of wire and cable scrap into products such as sound-deadening panels for cars, truck mud flaps, and floor mats. Conigliaro Industries of Massachusetts recycles 500,000 lb/year of postconsumer PVC medical plastics, along with roofing membrane and other PVC scrap, into checkbook covers, plastic binders, and other products.[64] In India, PVC shoes are often recycled into new PVC shoes.[75]

12.4.5 Polystyrene (PS)

Slightly more than half of the polystyrene in municipal solid waste originates in packaging materials (Figs. 12.26 and 12.27). Nondurable goods, particularly plastic plates and cups, are the next largest category. As was true for PVC, a substantial amount of PS is used in the building and construction industry, in this case mostly for insulation materials.

According to the U.S. EPA, about 10,000 tons of PS were recovered in 1996 from durable goods, about 10,000 tons from plastic plates and cups, and about 10,000 tons from packaging materials, for an overall recycling rate for PS in MSW of 1.5% (see Figs. 12.28 and 12.29). The Polystyrene Packaging Council reported a similar value, a total of 54 million lb of polystyrene recycled in the United States in 1996, with 10 million lb of food service polystyrene, 23 million lb of transport and protective packaging, and 21 million lb of other nonpackaging polystyrene applications, including audio and video cassettes, CD jewel cases, insulation board, and other products.[76]

During the mid- to late 1980s, PS was under attack on a variety of fronts, including ozone depletion and litter as well as the perception that it contributed a great deal to solid waste problems. In response, eight polystyrene resin suppliers formed the National Polystyrene Recycling Company (NPRC) in 1989 to concentrate on recycling of food service polystyrene. Plans were to operate six plants around the United States to recycle these materials, with a goal of achieving a 25% recycling rate by 1995. High levels of contamination with food wastes and inability to operate profitably plagued the facilities. In 1990, a highly publicized decision by McDonalds, which had instituted pilot recycling programs in several of its facilities, to abandon the PS clamshells and discontinue PS recycling dealt a further blow to recycling of food service PS. By 1997, the NPRC was down to two facilities and five PS resin company owners. The remaining plants are located in Chicago and in Corona, Calif. While NPRC was still forecasting eventual profitability

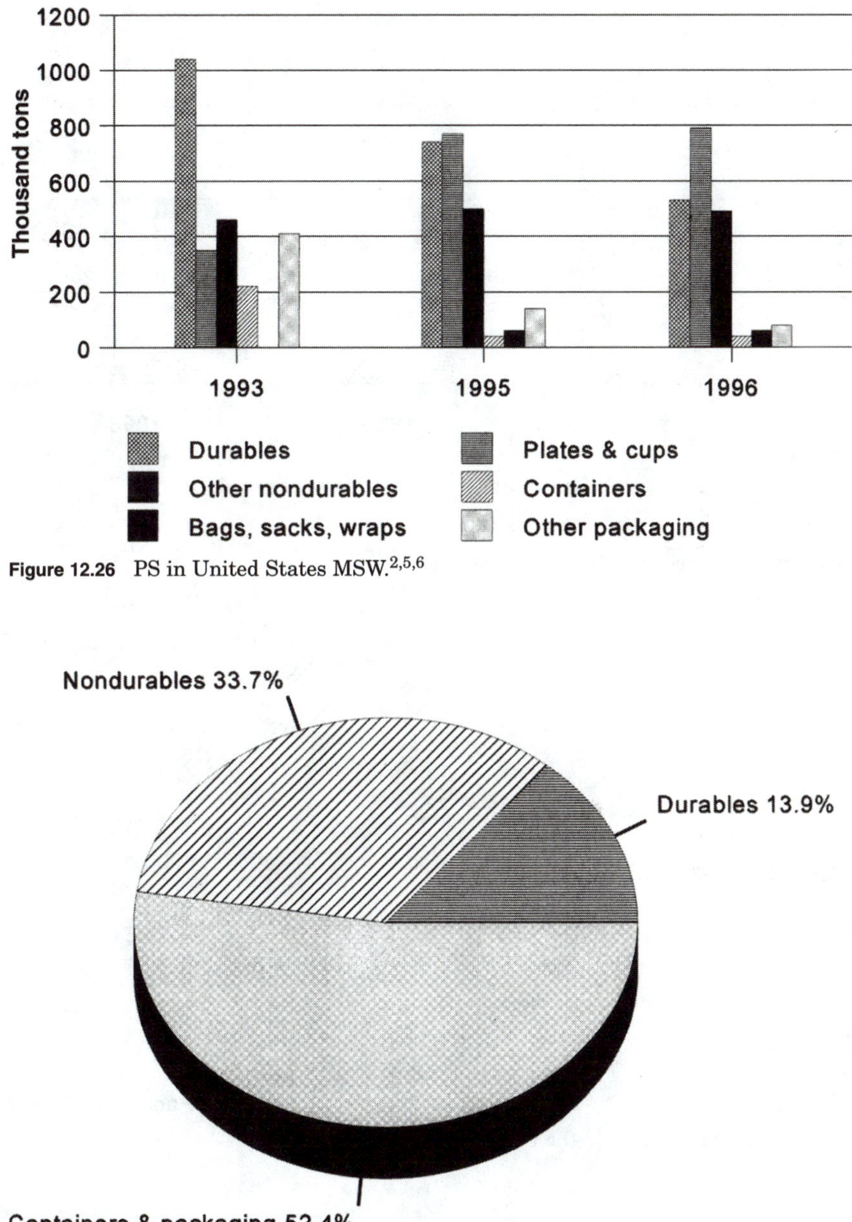

Figure 12.26 PS in United States MSW.[2,5,6]

Figure 12.27 Sources of PS in United States municipal solid waste, 1996.[2]

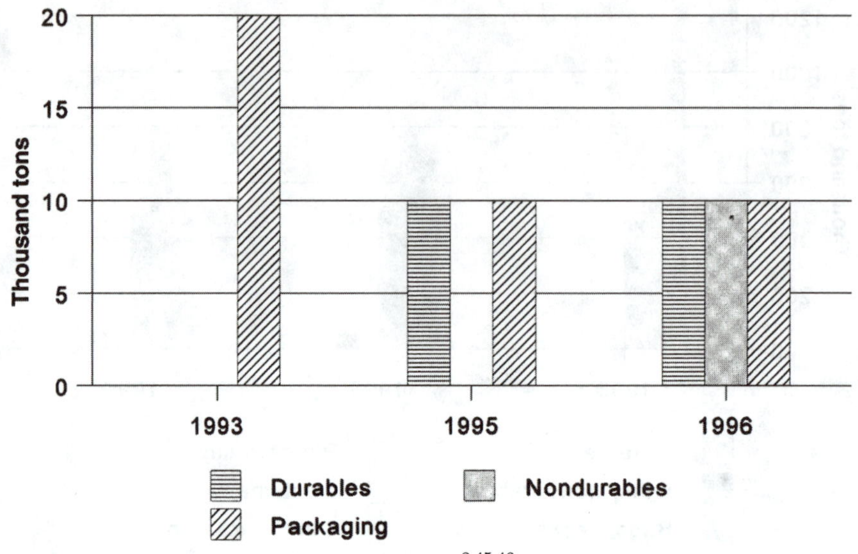

Figure 12.28 PS recycling in the United States.[2,45,46]

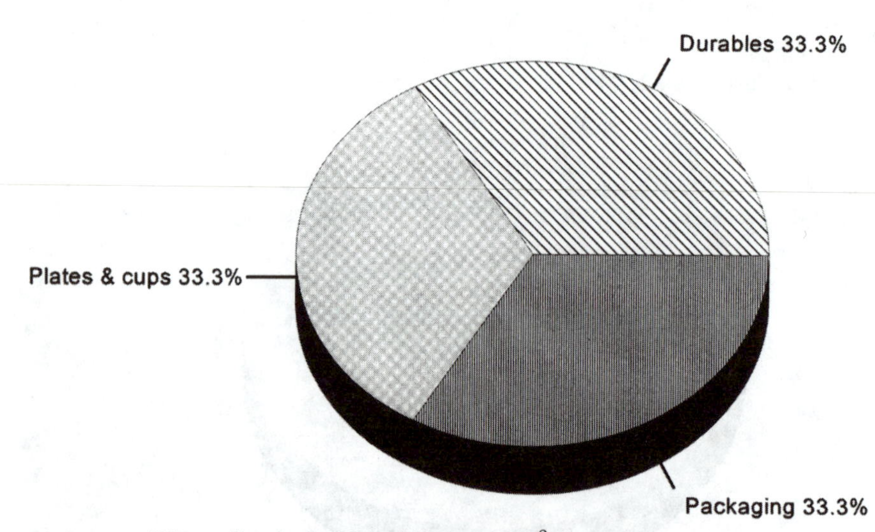

Figure 12.29 PS recycling in the United States, 1996.[2]

for these facilities, in part by discontinuing paying for PS in areas served by the California facility, and instituting a per-pound charge for PS in Chicago, the company has lost money through almost all of its history.[76, 77] The current economic woes are attributed, in part, to a worldwide oversupply of virgin general purpose polystyrene, which has driven down prices. With virgin resin selling at 40 to 50¢/lb, recycled resin sells for 38 to 45¢/lb. Since it costs 10 to 50¢/lb to sort, clean, and remanufacture the recycled polystyrene, depending on its quality and cleanliness, it is difficult for recycled PS to compete with virgin. Processing costs for food service PS tend to be at the high end of this range, so the Polystyrene Packaging Council states it is not economical to recycle at this time.[76] In the United States, collection of food service PS has focused on large generators of material such as school, office, and other institutional cafeterias. Little attempt has been made to collect PS at curbside. Drop-off facilities for PS have been initiated, often located at retailers where they were coupled with drop-offs for merchandise bags. In Ontario, Canada, however, PS has been included in about 20% of curbside collection programs.[78]

Recycling of PS from nonfood service applications has been more successful than that from food service. PS recycling from audio and video cassettes, CD jewel cases, and insulation board increased almost 70% between 1994 and 1996.[76] Another source of recycled polystyrene is clothing hangers used by department stores.[79]

Recycling of expanded polystyrene (EPS) protective foam packaging reportedly amounted to 23.3 million lb, or 10.9% of production in 1993.[80] The Association of Foam Packaging Recyclers was formed in 1991 to promote recycling of protective foam forms and peanuts. In 1994, they reported 185 collection locations for such materials around the United States.[81] AFPR has also worked with members to provide recycling opportunities for those without direct access to a PS recycling facility, sometimes by arranging for prepaid United Parcel Service (UPS) shipment of cushioning materials back to the manufacturer. Reuse is also part of the strategy for dealing with EPS protective foam. Some molded cushions are reused by manufacturers. Molded shapes are collected for reuse, among other places, at a network of Mail Boxes Etc. facilities. Largely due to the efforts of the membership of the Alliance of Foam Packaging Recyclers, the recycling rate for EPS transport packaging has held fairly steady at about 10% since 1993.[82] The rate climbed in 1995, reaching 12.7%, but fell in 1996 to 10.4%. The Alliance of Foam Packaging Recyclers report that over 62% of the EPS packaging recycled is recycled by molders of EPS protective packaging, and that 45% of all EPS recycled is made into new packaging. Further, they report that nearly 30% of all loose-fill EPS is reused. For mailing services, the reuse rate is as high as 50%. AFPR also points with pride to source reduction activities, stating that the amount of polystyrene source reduced in 1994 amounted to an energy savings equivalent to recycling 24% of the polystyrene packaging and disposables produced in that year.[83]

End uses for recycled PS vary. For cushioning materials, the most frequent use is back into cushioning materials. Studies of recycled cushioning showed that 25% recycled content EPS (more in some cases) can be used without adversely impacting cushion performance. In one study, foam containing 25% recycled content actually outperformed virgin EPS.[84] Insulation board containing recycled PS is also available. However, Amoco Foam Products, which had sold such material since 1991, discontinued use of recycled PS in insulation in 1997, stating that recycled PS was more expensive than virgin resin, and buyers were not willing to pay more for the recycled product.[85] Other applications for recycled PS include other types of packaging, housewares, and durable goods such as cameras and video cassette casings.

EPS recycling also occurs around the world. In many countries, however, recovery for combustion as fuel is included in the reported rates. In Japan, the Japan Expanded Polystyrene Recycling Association reported a recycling rate of 27.3% in 1995, or 107.8 million lb, which were recycled into pellets, soil improvers, and fuel.[86] In 1996, the recycling rate reached 28.7%.[87] The recycling rate in Korea was reported to be 21% in 1994. Taiwan has government regulations which require recycling of polystyrene, and reported a rate of over 56%, or 8.6 million lb, in 1995.[86] Australia has programs for recycling of EPS boxes for fruit and vegetables, with a reported collection rate of 2.2 million lb, or 10% of all boxes, in 1996.[88]

In the United Kingdom, recycling of polystyrene packaging has increased faster than most forecasts, from 11.8% in 1993 to 28% in 1996, for a total of 6000 tonnes. Forecasts for mechanical recycling of EPS by 2010 are now set at 35 to 40%, or 16,600 tonnes. The faster than expected growth was attributed to more rapid than anticipated growth in virgin EPS markets, along with advances in technology which enabled recyclers to handle contaminated EPS material such as fish boxes and horticultural trays. End uses include EPS foam cushioning materials, nonfoam applications such as CD and video cases, and extruded EPS applications such as hardwood and slate replacements. In addition to mechanical recycling, recovered EPS is used as a fuel for energy generation.[89]

One of the problems faced by recyclers of foamed PS, the kind most commonly used, is the very low bulk density of the materials. Compaction and baling are commonly employed to increase the bulk density, but the material is still expensive to ship for long distances. Recently, several companies have focused their efforts on transforming the PS foam into a gel, with a much higher density, to improve the transport economics. This method also can allow removal of contaminants by filtering the liquid material. One such company is International Foam Solutions, Inc., of Delray, Fla., which has developed STYRO SOLVE, a citrus-based biodegradable solvent for EPS foam and other PS products. The gel is diluted after it is received at the processing facility so that it can be filtered. Resource Recovery Technologies, located near Philadelphia, Pa., uses a nonflammable chemical solvent to dissolve the PS and separate it from contaminants. The solvent is recovered for reuse.[90]

The most common recycling of nonpackaging, nonfoam PS is recycling of disposable camera bodies. The recovered camera bodies are ground, mixed with virgin resin, and used in the production of new disposable cameras. The internal frame and chassis of the cameras, which are also polystyrene, are recovered intact and reused in new cameras.[91] Eastman Kodak reported in 1996 that 77% of its disposable cameras are recycled under a program in which photofinishers are reimbursed for returning the cameras.[92]

Polystyrene can also be recovered from appliances. Philips, in Hamburg, Germany, studied the recycling of nonflame-retardant PS from TV sets, using the recovered material to injection mold equipment housings, and concluded that there was no significant reduction in the properties of the recovered material.[93]

It is possible to use chemical methods to depolymerize PS into monomer, which could then be used for making new polystyrene. Some years ago, the Toyo Dynam company in Japan developed a prototype system in which polystyrene foam was ground and then sprayed with styrene monomer to dissolve the PS and separate it from contaminants, including other plastics and food scraps. The resultant solution was cracked and vaporized in a heated reflux vessel.[94]

12.4.6 Polypropylene (PP)

While packaging is the major source of polypropylene in municipal solid waste (see Figs. 12.30 and 12.31), durable goods are the primary source of recycled polypropylene (Figs. 12.32 and 12.33). Much of this material comes from recycling of polypropylene automotive battery cases. Due to concerns about the effect of lead emissions from disposal of such batteries, many states have prohibited their disposal in landfills or by incineration, and several states have instituted deposits to help ensure that batteries are collected for recycling rather than going to disposal. The recovery of the polypropylene is a side benefit of lead recycling, which is the driver for battery recycling. PP makes up about 7% of the battery by weight. The largest market for the recovered material blended with virgin PP is new battery cases. The U.S. EPA estimated recovery of lead from batteries as 93.3% in 1996.[2] The Battery Council International reported a 1996 recycling rate for lead-acid batteries of 96.5%.[95] The recovery of PP from battery cases was at about the same rate, since it is a routine part of battery reprocessing.

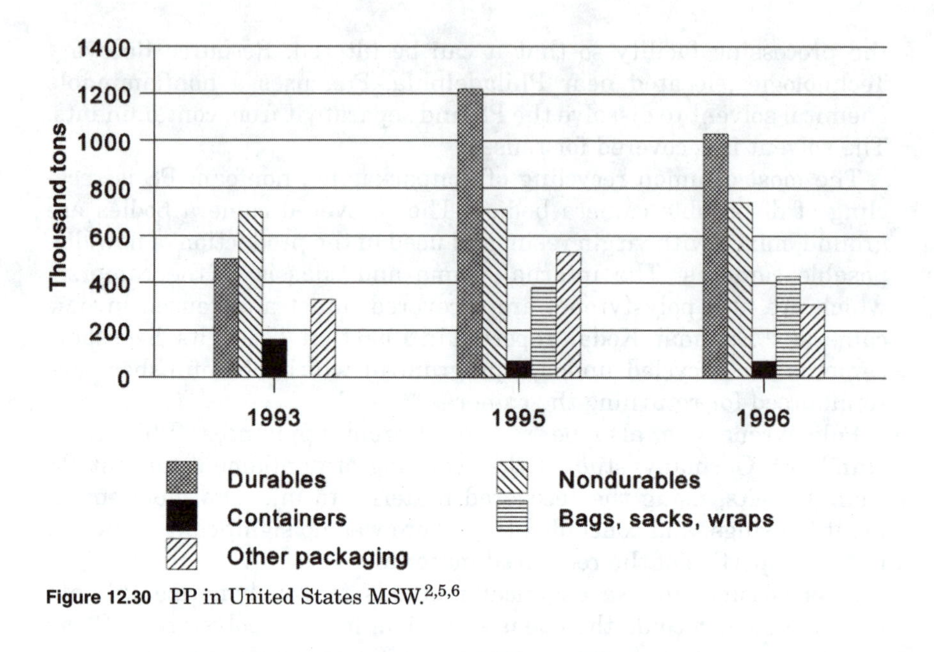

Figure 12.30 PP in United States MSW.[2,5,6]

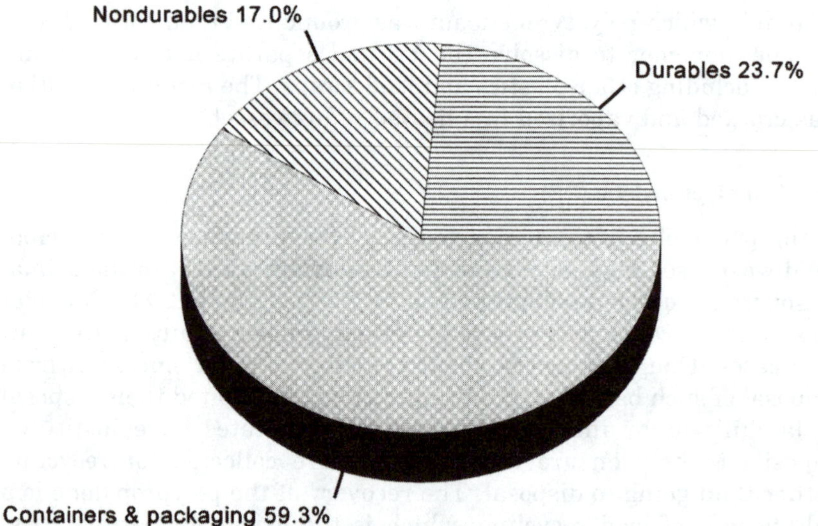

Figure 12.31 Sources of PP in United States municipal solid waste, 1996.[2]

Recycling of PP from durable goods, most of which were battery cases, was 100,000 tons in 1996. No recovery of PP from nondurable goods was reported. Recycling of PP from packaging was about 30,000 tons, all from the "other plastics packaging" category in the U.S. EPA report. Recovery of PP from containers was listed as negli-

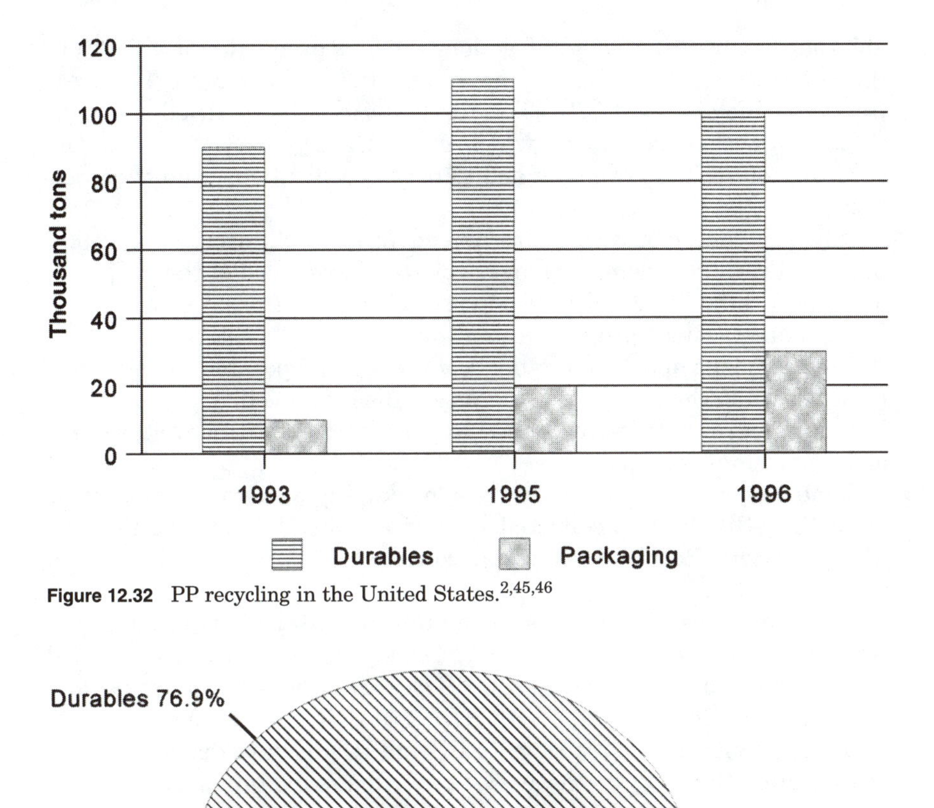

Figure 12.32 PP recycling in the United States.[2,45,46]

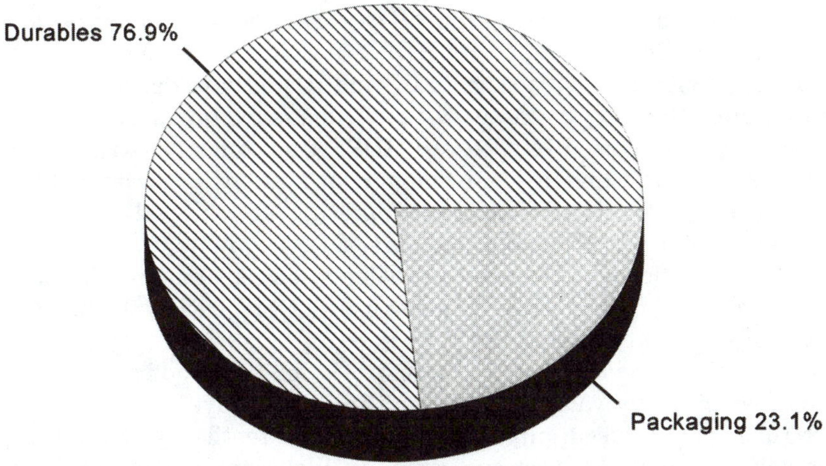

Durables 76.9%

Packaging 23.1%

Figure 12.33 PP recycling in the United States, 1996.[2]

gible (under 5000 tons), and no recycling of PP from the bags, sacks, and wraps category was reported. The total of 130,000 tons recycled gives an overall PP recycling rate of 5%.[2]

Recycling of PP battery cases also takes place outside the United States. Germany began such recycling in 1984, yielding about 3000 tonnes/year of polypropylene. The process involves crushing the batteries and then separating the light fraction, which is predominantly PP, from the heavy fraction, containing lead and other components. The light fraction, which is about 97% PP, is further size-reduced, dried, and sent through a cyclone separator, which increases its purity to 99.5%. The resulting material is compounded to user specifications, and then pelletized.[96]

Polypropylene can also be recovered from appliances. A recycling plant in Frankfurt reports recovery of about one-third of the weight of discarded coffee machines as polypropylene, with greater than 99% purity and excellent property retention.[93]

Polypropylene spools and wheel counters in disposable cameras are removed when the returned cameras are disassembled, and used again in new cameras.[91] Polypropylene hangers from department stores are also sometimes recycled.[60]

Some recycling of polypropylene packaging is carried out, most often targeting bottles collected as part of an "all plastic bottle" collection system. Recycled resin from collected containers has been used in soap bottles.[97]

Packaging uses of PP are approximately evenly divided between film, container, and closure applications. Little deliberate recycling of PP film or closures is done. However, PP closures and fitments are sometimes inadvertently recycled with HDPE containers, where they are an undesired contaminant. Since both PP and HDPE have densities lower than water, the gravity-based systems (float/sink tanks or hydrocyclones) commonly used to separate HDPE from other plastics do not separate out PP. Currently, no commercial systems are available for separating HDPE from PP once the materials are chipped or ground. While the presence of a small amount of PP contamination does not result in significant performance problems, larger amounts do make the recycled resin unsuitable for some uses, or require it to be blended with virgin resin to dilute the PP.

One use of polypropylene by industry is for the production of dye tubes, which carry yarn as it passes through the dye process. These dye tubes can be used only once, because the textile dye penetrates the polypropylene. At least one company, Wellmark Inc. of Asheboro, N.C., is recycling these tubes. First, residual scrap yarn is removed, and then the tubes are separated by type of polypropylene and granulated. Air separation is used to remove paper labels and other contaminants. The resulting PP is used in injection molding processes. Wellmark recycles 3500 tons/year of scrap tubes. Because of the high volume of the hollow tubes, the company estimates that the annual amount of tubes they process is enough to fill five football fields stacked 10 ft high.[98]

12.4.7 Nylon

Nylon recycling has increased substantially in the last several years. Most recycling efforts have focused on recovery of carpet. According to the U.S. Department of Energy, about 3.5 billion lb of waste carpet are discarded each year in the United States, with about 30% of them made from nylon 6. (For more on carpet recycling, see Sec. 12.4.15.) Recycling systems for condensation polymers, such as nylon and PET, can more effectively use chemical depolymerization techniques than can systems for addition polymers such as polyolefins and PVC. Most of the efforts directed at nylon recycling have taken this route.

The U.S. Department of Energy provided support to AlliedSignal, as well as carrying out research at the National Renewable Energy Laboratory, for the development of a process for chemical conversion of nylon 6 to caprolactam, which could then be repolymerized to nylon 6 for a variety of applications.[99,100] The economics of this process were particularly promising. A commercial-size plant recycling 100 million lb of waste carpet could produce high-grade caprolactam for about 15 to 20¢/lb, while caprolactam from virgin materials currently sells for 90¢ to $1/lb.[100]

Several companies are now engaged in commercial recycling of nylon 6 carpet. BASF Corporation collects only its own nylon 6 carpet, depolymerizes and purifies it, and then uses it in the manufacture of new carpet.[101] In June 1998, Evergreen Nylon Recycling LLC, a joint venture between AlliedSignal and DSM Chemicals North America, began construction of a new facility to recycle more than 200 million lb of nylon carpet each year. The facility is expected to be completed in fall, 1999. It will depolymerize the nylon to caprolactam, which will then be repolymerized and used for a variety of applications, including carpet and engineering plastics for automobiles. In addition to permitting closed-loop recycling, it will save 4.9 trillion Btus each year compared with conventional caprolactam production. Evergreen Recycling has also developed a laser device for resin identification to facilitate the sorting of nylon from other types of products. The company plans to adapt its process in the near future to recycle nylon automobile parts as well.[102]

DuPont is recycling mineral-reinforced nylon 66 carpet, including that generated by Ford Motor Company that contains 25% postconsumer recycled content to manufacture engine air-cleaner housings. A proprietary process is used to reclaim the face fiber, which is melt-recycled and compounded with virgin nylon.[103]

An exception to the general method of using depolymerization is the Lear Corporation's system for recycling scrap carpet pieces. Relatively clean scrap carpet made from nylon 6, nylon 66, and other resins is baled and then pelletized, producing a black-colored resin which they hope to sell to automobile manufacturers for nonvisible applications such as acoustical parts and the backs of floor mats. The plant is currently recycling more than 1 million lb of scrap per year.[56]

Recycling of nylon 6 is not limited to carpets and automobile parts. Toray, a Japanese company, in 1995 began recycling apparel made from nylon 6 by depolymerization and repolymerization.[104] Used U.S. Postal Service nylon mailbags which are not repairable, are being recycled into pellets used for, among other applications, automobile parts.[105]

An unusual recycling system is that of *Preserve* brand toothbrushes, manufactured by Recycline, Inc., of Somerville, Mass. Buyers of the toothbrushes, which have handles made from a blend of recycled and virgin polypropylene and nylon bristles, can send the brushes back to the company in a postage-paid return envelope. The brushes are then recycled into plastic lumber, with the claim that the blend of PP and nylon actually strengthens the lumber.[106]

12.4.8 Polyurethane (PU)

Polyurethane recycling, like that for nylon, has often taken a chemical approach. This is particularly important to polyurethane, since most of its applications are for cross-linked material which cannot, therefore, be melted and reformed. The first North American facility for chemical recycling of molded polyurethanes opened in September 1997 in Detroit, Mich. The 10-million-lb/year facility is owned and operated by BASF Corporation and Philip Services Corporation. While targeted first at automotive waste streams, it could eventually include appliances and construction materials, as well as recreational products such as bowling balls and pins. The facility uses BASF's patented glycolysis process to chemically reduce PU to polyol, which can be used like a virgin polyol, combined with isocyanate to produce polyurethane products. Various types of PU can be mixed together in the process, including polyurethane paint on polyurethane parts. BASF will use the recovered polyol for manufacture of new polyurethanes, for use in automotive, construction, and other applications.[107]

ICI Polyurethanes is working on development of a split-phase glycolysis chemical recycling technology for polyurethane, and plans to build a full-scale plant in Britain to recycle polyurethane foam from mattresses, furniture, and automotive seat cushioning.[108]

Other approaches to recycling polyurethane have focused on chopping up the flexible varieties and using them as filler in new polyurethane products. Rebonded seat foam materials have been studied for use in floor carpet underlayment in vehicles, and found to be comparable to currently used materials.[109]

Bayer Corporation of Pittsburgh, through its Hennecke Machinery division, together with the German company Greiner Schaumstofftechnik, has developed a process for producing molded parts from scrap polyurethane foam. The foam is shredded to flakes and then mixed with a prepolymer adhesive. The mixture is filled into molds, using a system which precisely and individually meters the amount fed into each mold, to allow all molds to fill completely. It is reported that at least one European company is using the system commercially. Another process developed by Hennecke grinds polyurethane foam cuttings into a fine powder, 2 μ in size, which is then added to the polyol in quantities up to 30% during polyurethane production.[110]

12.4.9 Polycarbonate (PC)

Some recycling of polycarbonate from products, such as automobile bumpers, compact disks, computer housings, and telephones, is being carried out. General Electric began buying back polycarbonate from 5-gal water bottles several years ago, and tried for a considerable amount of time to introduce its idea of a cascade of uses for polycarbonate, beginning with reusable packaging and ending in various product applications such as automobile parts. GE also initiated a pilot buyback recycling program in 1995 for polycarbonate scrap and other resins generated by its customers, including recovered PC parts from used automobiles if they were manufactured from GE resins.[111]

Bayer, in Leverkusen, Germany, in 1995 built Europe's first polycarbonate CD recycling facility. The PC is separated from aluminum coatings, protective layers, and imprinting, and the pulverized material is combined with virgin resin and sold for various uses.[112]

Polycarbonate winding levers and outer covers on disposable cameras are recovered intact during processing of the cameras, and reused in new cameras.[91]

12.4.10 Acrylonitrile butadiene styrene copolymers (ABS)

Recycling of ABS is primarily focused on appliances. British Telecommunications PLC recovers and recycles about 2.5 million telephones every year. About 300,000 are refurbished for reuse, and the remainder are recycled. The recovered ABS from the telephone housing is typically used in molded products such as printer ribbon cassettes and car wheel trims.[113] Similarly, AT&T Bell Laboratories recovers ABS phone housings, and recycles them primarily into mounting panels for business telephone systems.[114] GE Specialty Chemicals recycles ABS from refrigerator liners.[115]

Hewlett-Packard recycles computer equipment, using some of the recovered ABS in the manufacture of printers, blended with virgin ABS. The recycled content in the printer cases is reported to be at least 25%.[116]

HM Gesellschaft für Wertstoff-Recycling recovers ABS from vacuum cleaners, reporting recovery of more than 25% of the weight as ABS with better than 99% purity, and properties comparable to impact-resistant polystyrene.[93]

Blends of ABS and polycarbonate (PC) are also recycled, with the usual source again being used appliances. Siemens Nixdorf Informationsysteme (SNI) in Munich, Germany, recycles a variety of equipment made by the company such as computers and peripherals and automatic teller machines. Customers are charged a fee for the reprocessing. Much of the recovered material, which is manually disassembled and resin type identified by molded-in markings or by infrared (IR) spectroscopy, is PC/ABS blends, which are ground and shipped to Bayer AG in Leverkusen, Germany, where they are blended with virgin resin and used by SNI to mold equipment housings.[93]

12.4.11 Commingled plastics and plastic lumber

In some cases, separation of collected plastics by resin type is either not feasible with current technology or is economically unattractive. For such streams, use of the plastics in a commingled form may be preferable. A number of applications have been developed for commingled plastics. Many fall in the general category of plastic lumber, while others are replacements for concrete or other materials. Some of these processes are also able to handle substantial concentrations of nonplastic contaminants, reducing the need for cleaning the collected material. Some plastic lumber products use single-resin recycled plastics or single-resin plastics combined with fillers or reinforcers.

Most processes for recycling of commingled plastics depend on a predominance of polyolefin in the mix to serve as a matrix within which the other components are dispersed. One of the early processes was the Mitsubishi Reverzer, which was developed in Japan. This process used a short-time, high-temperature, high-shear machine which could handle up to 50% filler. The plastics were first softened in a hopper and then mixed in an extruder. Products were formed in a variety of shapes by flow molding, extrusion, or compression molding. However, the equipment was never successfully marketed.[117] The Klobbie process, developed in the Netherlands in the 1970s, consisted of an extruder coupled to several long linear molds mounted on a rotating turret. The molds were filled directly from the extruder, without any extrusion nozzle or filter pack, and then rotated into a tank of cooling water. After solidification, the products were removed using air pressure. This process is often called an *intrusion process,* since it is a cross between injection molding and extrusion. Klobbie also patented the use of foaming agents in the equipment.[117] The Klobbie process is the father of several commercial extrusion processes in use today. Like the Reverzer, these processes can generally incorporate substantial amounts of nonplastic contaminants, which tend to migrate toward the middle of the thick cross-sectional items being produced.

Another early process was the Recycloplast process, developed in Germany, in which materials are heated in a "plastificator," using kneading to produce a paste. The paste is then extruded into roll-shaped loaves, which are either immediately compression molded or granulated for further use. The British Regal Converter system is another compression molding process, in which commingled plastics are granulated, distributed on a steel belt, melted in an oven, and then compacted into a continuous board.[117]

The ET/1 system developed by Advanced Recycling Technology of Brakel, Belgium, is a more modern variant of the Klobbie process, and it has been used by a number of operations in several different countries. The feed must contain a minimum of 50 to 60% polyolefin. The short, high-speed adiabatic screw extruder melts the polyolefin and fills molds up to 12 ft long, mounted on a revolving turret which is cooled in a water bath. Products formed are in the plastic lumber category. Properties of lumber formed from mixed plastic bottles can be improved by adding 10 to 30% recycled polystyrene.[117]

The Superwood Process, developed in Ireland by Superwood International, is another Klobbie variant, in which fillers or blowing agents are commonly added to vary the product properties. Hammer's Plastic Recycling in Iowa developed another Klobbie variant, in which closed molds, a heated nozzle, and a screen pack are used to increase molding pressure.[117,118] Other processes include those of WormserKunststoffe Recycling GmbH in Germany, and C. A. Greiner and Sohne in Austria.[117]

Electrolux, in Stockholm, Sweden, is reportedly using 40 to 80% commingled recycled plastics from sources such as used telephones, car parts, and other appliances, in the manufacture of vacuum cleaner housings. The company is also moving toward closed-loop reuse of in-plant scrap in future product designs.[119]

Tipco Industries Ltd., in Bombay, India, produces Tipwood, a composite material from mixed plastic waste and 20% filler for use as a wood substitute for applications such as pallets, benches, fencing, and road markers.[18]

Hettinga Technologies of Des Moines, Iowa, has used commingled plastics to make injection-molded panels, using a controlled density molding process which accommodates a range of plastics. The panels are reported to have a thin, smooth outer skin around an integral core which is foamed in situ. Costs are reported to be about 60% of the cost of plywood for panels which handle like pine, hold nails and screws like hardwood, and resist rot and mildew. The process involves incorporating a dissolved blowing agent into the mold, which expands the part once it is removed from the mold.[120]

In recent years, the diversity of properties of "plastic lumber" produced from different mixes of materials, by different processes, has resulted in calls for development of standards. In late 1997, the American Society for Testing and Materials (ASTM) issued five new test methods for plastic lumber intended to help establish a benchmark for minimum plastic lumber performance. The standards include tests for compressive properties, flexural properties, bulk density and specific gravity, compressive and flexural creep and creep rupture, and tests for performance of mechanical fasteners.[121]

Plastic lumber is used for fencing, decks, benches, picnic tables, and other applications as an alternative to wood lumber. The plastic products generally are more expensive to purchase than wood but have a considerably longer life. Further, they do not need to be painted or varnished. Since wood products in these applications generally must use treated lumber, interest in them is expected to grow as concern about the environmental effects of the compounds used to treat the lumber increases. Some plastic lumber products are used in applications such as parking stops and machinery bases, where they replace concrete. While not all such applications use commingled recycled plastics, many do. Some use single-resin recycled plastics, usually HDPE or LDPE, while others use virgin plastic. In some cases, composites of plastic (virgin or recycled) are used with fibrous reinforcements, usually wood or glass fiber. The fibers may be recycled material as well.

Another application of plastics lumber is in pallets. A number of designs have emerged, most using high-density polyethylene, but some using fibrous reinforcements or steel inserts. Recently, researchers at Battelle developed a pallet made from recycled plastic, including both HDPE and commingled plastics, which can hold up to 10 tons. It is intended to replace wood and steel for storing 55-gal drums of hazardous and radioactive materials. In these applications, the plastic pallets are less likely to absorb contaminants and easier to clean than wood and metal alternatives.[122]

Most recently, plastic lumber has broken into two potentially large-scale applications—plastic railroad ties and bridges. Plastic rail ties have been used for some time in Japan, where wood is scarce, but these ties are made from virgin foamed polyurethane with a continuous glass fiber reinforcement. The ties now being tested in the United States are made from recycled plastics which are predimpled to sit better on the rocks used as ballast in the rail lines, and have the same size and appearance as traditional wood ties. Initial test results are positive.[123]

The U.S. Army Corps of Engineers has used 13,000 lb of commingled recycled plastics to build a bridge at Fort Leonard Wood in St. Robert, Mo. The bridge is primarily used for pedestrian traffic, but can hold up to 30 tons, allowing it to also support light trucks. The Army expects the bridge to last 50 years without maintenance, significantly longer than the 15 years for treated wood or the 5 year life of untreated wood. Even the joists are made of plastic, opening the door to structural applications which used to be considered unsuitable for plastic lumber. The material used for the joists is a polystyrene-modified lumber, formed from recycled postconsumer polystyrene foam along with recycled high-density polyethylene, which was produced by Polywood Inc. of South Plainfield, N.J. It has a modulus $2^1/_2$ times that of conventional plastic lumber and, therefore, can withstand higher loading. The deck planks were produced by Plastic Lumber Company of Akron, Ohio, and the railing by Hammer's Plastic Recycling Corporation of Iowa Falls, Iowa, and Renew Plastics Inc. of Luxemburg, Wisc. One unique property of the plastic lumber is that it melts around screws when they are driven into the boards. When the plastic cools, it hardens around the screw, locking it into place and ensuring a long-lasting bond.[124]

Research on composite materials made of wood fiber or flour or recovered paper fiber, together with recycled plastics, has been going on in several localities for a number of years. Investigators include CSIRO and the Cooperative Research Centre for Polymers in Australia, the University of Toronto and University of Quebec in Canada, and the Forest Products Research Laboratory, the Risk Reduction Engineering Laboratory in Cincinnati, and Michigan State University in the United States among others. Polymers used include low- and high-density polyethylene, polypropylene, polystyrene, and PVC, along with commingled plastics. For example, the U.S. Department of Agriculture (USDA) Forest Products Laboratory has studied the use of recovered paper fiber with a mix of PP, LDPE, HDPE, and PVC. This mix has been investigated by a plastic lumber manufacturer for use as a core layer in plastic lumber for pallets.[125] Boise Cascade is researching the use of plastic film, recovered from residents in the Seattle area, and wood waste to make a composite which they intend as a replacement for virgin wood material in building construction.[126]

On a commercial scale, Advanced Environmental Recycling Technologies Inc. (AERT) of Rogers, Ark., has for several years been producing window frames from a composite of recycled polyethylene film and wood fiber.[127] Atma Plastics Pvt. Ltd., of Chandigarh, India, produces a recycled PE/PP/wood composite for use as a wood substitute or as a filler in cast polyester for furniture.[18] Natural Fiber Composites Inc. of Baraboo, Wisc., has commercialized a pelletized wood fiber-filled plastic, using recovered paper or wood fibers along with PP, HDPE, or PS.[128,129] Comptrusion Corporation of Richmond Hill, Ontario, makes polyethylene wood flour composites, and is working on a PVC/wood composite. Formtech Enterprises Inc. of Stow, Ohio, manufactures a PVC wood-fiber composite.[130] Whether these materials use recycled plastic was not clear. Mikron Industries is manufacturing a wood-plastic composite for window frames which is based on an undisclosed mixture of plastics, and uses material which otherwise would be headed for landfill disposal.[131]

While many processes for using commingled plastics look at collection of bottles from curbside as the primary source of materials, they frequently combine these materials with industrial waste stream plastics, including coextruded scrap and other examples of multiresin, perhaps contaminated, materials. Other companies focus entirely on these types of waste streams. For example, Northern Telecom Ltd., based in Toronto, has a recycling facility for plastic reclaimed from wire and cable, along with materials from phone, fax, business machine, and pager equipment. The material, which contains small amounts of residual copper and aluminum, is used for truck mud flaps, flower pots, traffic stops, reflective bibs for construction workers, and other applications where high purity is not required.[132]

12.4.12 Compatibilization of commingled plastics

As discussed previously, one approach to the use of commingled plastics is to use them in wood-substitute and similar applications where the incompatibility between the various ingredients of the mix does not present severe problems. A different approach is to make the resins compatible in some way, hence improving the performance characteristics of the blend. The compatibilization can be done by the way the blend is processed, perhaps including some degree of control over the resins in the mix, or by additives.

Northwestern University has developed a process they refer to as *solid-state shear extrusion pulverization.* In this process, a twin-screw extruder is used to convert mixed plastics and scrap rubber into a uniform fine powder, which can then be used in a variety of products.[133]

Manas Laminations, of New Delhi, India, has developed compatibilizers that enable PET/PE waste streams to be processed into useful products, including flooring, office partitions, corrugated roofing, and slates for benches.[18]

A number of other companies are also marketing or developing compatibilizing additives. For example, Dexco Polymers, of Houston, Tex., sells a line of styrenic block copolymers which can enhance properties of mixed streams of polyolefins. BASF is developing compatibilizers for mixtures of polystyrene with polyethylene or polypropylene and for mixtures of ABS with polypropylene. DuPont sells compatibilizers based on maleic anhydride-grafted polyolefins for compatibilization of mixed polyolefins.[115]

12.4.13 Energy from mixed plastics

As mentioned earlier, another approach to recycling of mixed plastics is to use thermal processing to convert the plastics to a mixture of gases and liquids which can serve as a fuel or to use the plastics in unprocessed form as a fuel source. As noted, many do not consider these applications to be true recycling, while others argue that the replacement of fossil fuels with waste plastics does generate value from the plastic materials and so should be counted.

A variety of schemes for use of plastics, either with or without further processing, as an energy source have been developed. The simplest approach is to leave the plastics with other waste materials and process the mixed solid waste in a mass burn facility which recovers energy from the heat of combustion. Other processes use a separated stream of mixed plastics as an energy source. For example, the steelmaker Stahlwerke Bremen GmbH in Germany uses waste plastic as a reducing agent in place of heavy oil in its blast furnace.[134]

Processes designed to use thermal treatment to decompose the plastics into liquid and/or gaseous materials have often had mixed technical and economic success. For example, BASF planned to build a 661-million-lb/year facility for recycling mixed plastic wastes into hydrocarbon feedstocks, but eventually canceled the plans.[135]

12.4.14 Automotive plastics recycling

The automotive industry is under pressure, in various countries, to make automobiles more recyclable, and to use more recycled materials in automobile manufacture.[136] For example, European Union regulations require that only 5% of automobiles, by weight, be landfilled beginning in 2015. Many countries are incorporating, either by regulation or voluntarily, the philosophy of extended product responsibility on automobile manufacturers. With the increasing use of plastics in automobiles, this has forced attention to the problem of recycling these automotive plastics.

While some recycling of auto parts has been covered in the sections on recycling of individual plastic resins, it is also useful to look at automotive recycling in a more unified way. Recycling of automobiles has a very long history, but most of the effort in the past has focused on recovery of metals—steel in particular. The plastics in cars remained in the "auto shredder residue" or fluff, and have routinely been destined for disposal rather than recycling. Recycling of rubber, both from tires and from car parts, is an important part of automotive recycling efforts, but will not be addressed here.

Recovery of plastics from fluff, which contains a variety of plastic and nonplastic components, has not been achieved commercially. Much as recovery of plastics from household waste is more successful when the targeted materials are diverted from the garbage rather than trying to separate them out from a mixed waste stream, recycling of plastics from automobiles is easier when parts are disassembled than when the whole car is ground up. However, in the case of automobiles, even separating functional units such as bumpers or dashboards does not always result in a single type of plastic, since many of these components use a combination of materials. Further, the cost of disassembly is significantly higher than the cost of processing the auto body in the usual way. Automobile manufacturers are engaged in design changes to simplify the recovery of car parts for recycling. Efforts include improving the ease of removal of automobile subassemblies, reducing the number of different resins used in auto parts, as well as efforts to ensure that compatible resins are used in parts where multiple resins are necessary, so that they need not be separated for recycling. Design of hand-held systems to reliably identify resin types is an important adjunct to these efforts. While newer cars have plastic parts with molded-in resin identification, most cars being scrapped now were manufactured before those systems were in place.

Automakers have made substantial strides in the use of parts containing recycled plastic in the building of automobiles. As mentioned previously, recycled nylon is being used in air-cleaner housing, as well as in fan assemblies and other automotive parts.[137] Recycled ABS and polyester/polycarbonate alloys are being used for brackets to hold radio antennae, splash shields, and small under-the-hood parts.[138] Recycled polyurethane is being used in reaction injection-molded bumper fascias.[139] Recycled polycarbonate/polybutylene terephthalate bumpers are being incorporated into new bumpers.[140] Thermoplastic olefin (TPO) bumpers are being recycled into bumper fascias, splash shields, air dams, and claddings.[140] Recycled polypropylene is used in power-train applications, fender liners, air conditioning evaporative housings, vents, and other applications.[137,140,141] Recycled PET is being used in a variety of parts, including headliners and engine covers.[141,142] Recycled polycarbonate has found use in instrument panel covers.[140] Other examples can be found as well, and recycled plastics are also being used in packaging for parts used in manufacturing automobiles. Cost remains a key issue for the automobile industry, with most manufacturers insisting that parts made from recycled material deliver comparable performance at no higher cost than virgin materials.

12.4.15 Carpet recycling

In the past several years, interest in the recycling of carpeting has grown substantially. Most efforts have focused on commercial carpeting. It is estimated that about 1.7 million tons of waste carpet are landfilled in the United States each year, most of it during construction or remodeling of office space. Carpet is typically formed by bonding a face fiber onto a backing fiber, most commonly using a styrene-butadiene adhesive which incorporates fillers such as calcium carbonate. The face fibers are generally nylon, polypropylene, polyester, wool, or acrylic. The backing is usually polypropylene, PVC, nylon, or jute. In addition to the complexity of the structures themselves, used carpet is generally heavily contaminated with dirt, staples, food, and other materials.[143] Many efforts to recycle carpeting have focused on nylon carpet, which makes up about two-thirds of the face fiber market, particularly on nylon 6, because of the ease of chemical depolymerization of this material (see Sec. 12.4.7). The next most commonly used face fiber is polypropylene.[143]

One company, Interface Flooring Systems of LaGrange, Ga., has embraced the producer responsibility concept by leasing rather than selling its carpet, retaining responsibility for carpet care and ultimate disposal. The company works with Custom Cryogenic Grinding Corporation, of Simcoe, Ontario, to process the returned carpet using a cryogenic grinding process to make the carpet brittle, facilitating its separation into nylon face fibers and PVC backing.[143]

Monsanto will recycle all types of carpeting which its customers replace with Monsanto nylon carpet. Nylon carpet face fiber, latex, and backing are recycled into thermoplastic pellets, which are reused industrially. The company is also exploring opportunities for reuse of carpet in fuel recovery systems, as well as for respinning postconsumer nylon carpet face fiber.[144]

In Minnesota, over 60 businesses, along with government groups and the University of Minnesota participate in the Minnesota CA-RE (carpet recycling) Program. Some of the collected material is used by resin and fiber companies, and some is used by United Recycling, a subsidiary of Environmental Technologies USA, to make "grey felt" which is used for padding installed under commercial floor coverings, as well as for sound insulation in automobiles.[145]

Collins & Aikman Floorcoverings, of Dalton, Ga., uses vinyl-backed carpet to make solid commingled plastic products such as car stops and highway sound-wall barriers. They are now using up to 75% reclaimed carpet materials to make a nylon-reinforced backing for new carpet for modular tile products.[143]

DuPont is investigating the use of ammonolysis to depolymerize mixed nylon 6 and nylon 6/6 from used carpet. The company is also using reclaimed fiberized material from nylon carpet to make nylon building products for use in wet environments such as kitchens and bathrooms.[143]

Researchers in Georgia are investigating an unconventional use of carpet fibers, incorporating them into the surface of unpaved roads to improve road performance.[146]

To simplify the task of carpet material identification, the Carpet and Rug Institute has developed a seven-part universal coding system where a code on the carpet backing can be used to describe the components of the carpet, including facing, backing, adhesive, and fillers. As of 1997, it was estimated that 85% of the carpet now being made in the United States uses this code. However, the average 10-year life span of carpet means that for the next several years, most carpet entering the recycling stream will not be so labeled.[143] Thus, as for automobile parts, equipment for identifying carpet materials will continue to be needed.

12.4.16 Other plastics

While the major types of plastics recycling have been addressed, there are a variety of other types of plastics recycling going on, often on a small-scale or experimental basis. For example, Arco Chemical Company has a process for recycling glass-reinforced styrene maleic anhydride from industrial scrap.[147] The University of Nottingham has a project for developing recycling techniques for thermoset materials, including polyesters, vinyl esters, epoxies, phenolics, and amino resins along with glass and carbon-fiber reinforced resins.[148] The Fraunhofer Institute in Teltow is developing a process for recycling thermosets using an amine-based reagent in a one-step process which requires little added heat. The process is said to be applicable to almost all thermosets.[149] Imperial Chemical Industries plc and Mitsubishi Rayon Company Ltd. are developing technology for recycling of acrylics by chemical depolymerization and repolymerization.[150] The introduction of polyethylene naphthalate (PEN) in U.S. packaging markets was delayed by the perceived need to develop processes for automatic separation of PEN from PET as well as technology for recycling of PEN. Other examples could also be cited. The field of plastics recycling is constantly evolving, in response to changing demands and opportunities, as well as the emergence of new resins and new applications.

12.5 Overview of Plastics Degradation

Until the early 1970s, attention to plastics degradation, including biodegradation, was focused primarily on ensuring that the plastic materials being used were resistant to such degradation so that they could maintain their usability. Biodegradation and other types of degradation, such as photodegradation, were not always clearly differentiated. Further, the extent of degradation was frequently measured based on loss of useful properties, such as tensile strength, rather than on chemical changes in the polymer structure.

In the mid-1980s, when concerns about solid waste disposal were increasing rapidly, there was again a flurry of interest in biodegradation, stemming from a perception that disposal problems could be alleviated substantially if we stopped filling up our landfills with nonbiodegradable plastics and instead switched to biodegradable materials. As information increased about both the composition and the behavior of solid waste in landfills, it became clear that this was a misperception. First, the majority of material in landfills was, in fact, biodegradable, consisting of paper, food waste, and yard waste. Second, conditions in modern landfills, designed to keep materials dry to reduce problems with groundwater contamination, were not conducive to rapid biodegradation. Pictures of grass clippings, vegetables, and hot dogs, still recognizable after 10 to 20 years in a landfill, reinforced this reality, as did the statement by landfill researchers such as William Rathje of the University of Arizona that landfill waste was often dated simply by reading the dates on the still-legible newspapers contained in the garbage. Additionally, it became clear that the "out of sight, out of mind" approach to plastics degradation could not be justified. In other words, mechanical disintegration of a plastic product into plastic dust was not equivalent to chemical breakdown and return of the carbon and other elements to global cycles.

Nonetheless, during the time when biodegradability was perceived as a potent selling attribute for products such as merchandise sacks and garbage bags, a number of products were introduced which were composed of a mixture of starch, usually about 6% by weight, and low-density polyethylene. Manufacturers of these materials claimed they were biodegradable, based on the fact that the starch component would be consumed by microorganisms if the materials were buried in soil and the argument that the then-fragmented plastic would also be biodegraded. Some manufacturers added pro-oxidants to enhance the further degradation of the materials. The discovery that even materials such as food wastes often biodegraded very slowly under landfill conditions cast doubt on the claims, since even if the materials were biodegradable, it was not clear that this degradability would serve any useful purpose. Further, evidence suggested that the mere increase in exposed surface did not materially increase the biodegradability of the polyethylene remnants. The value of pro-oxidants in the anaerobic environment of a landfill was also questionable. As a result, some environmental organizations began calling for boycotts of these products. Finally, some manufacturers of these products were charged by some state attorneys general with misleading consumers, under statutes related to fair trade practices, and ordered to pay fines and cease making such claims. Ultimately, these products disappeared, having succeeded primarily in giving biodegradable plastics a bad name.

Since that time, two important changes have occurred. First, a variety of truly biodegradable plastics have been formulated, and their usefulness in niche markets, such as where plastics are likely to become a litter problem, particularly in bodies of water, has been recognized. Secondly, the use of composting as a waste management practice has grown dramatically. Composting is designed and managed to promote rapid biodegradation, so biodegradable products have assets in this scheme that they do not have where landfill or incineration are the usual approach to solid waste disposal. This, then, is the motivation for an examination of biodegradable plastics.

12.5.1 Definitions and tests

Biodegradability of a plastic means that living organisms can use the plastic as a food source, transforming its chemical structure within a reasonable period of time. In practice, the organisms which we rely on to accomplish this task are microorganisms, and the transformation of chemical structure results in conversion of most of the carbon in the polymer to carbon dioxide, methane, or other small molecules, along with some incorporation of the carbon into the cell mass of the microorganisms as they grow and reproduce. The time period involved is usually several weeks to several months.

Unfortunately, as indicated earlier, there has been abundant misuse of the terms "biodegradable" and "biodegradability," with considerable confusion resulting. One of the main points of confusion has been the misidentification of photodegradable polymers as biodegradable. Photodegradation refers to loss of physical properties induced by exposure to light. A significant part of polymer research in the decades since these materials were introduced into commerce has been in the determination of ways to minimize loss in strength in polymers which are exposed to sunlight. Virtually all polymers have some tendency to photodegrade, so outdoor use of many polymers depends on the inclusion of appropriate stabilizers to extend the plastic's lifetime. On the other hand, it is sometimes desirable to hasten this light-induced degradation. Appropriate modification of the polymer's chemical structure or the use of additives can accomplish this accelerated degradation. Such plastics are properly termed photodegradable, but have been misidentified in many instances as biodegradable. Photodegradable plastics are outside the scope of this chapter.

Another point of confusion has been the amount of chemical change needed to classify a plastic as biodegradable. Early work in this area followed the practice initiated by those who were seeking to conserve the performance attributes of plastics and who measured degradation by loss of physical properties such as tensile strength. If one is interested in using a plastic for some purpose, the end point is clearly the point at which the plastic no longer has useful properties. In this context, defining degradation in terms of loss of strength is perfectly reasonable. However, if one is interested in total decomposition of the polymer, it is not reasonable, since relatively few bond cleavages in a polymer backbone can destroy the polymer's strength, while still preserving most of its chemical structure. Many of the early discussions of biodegradable polymers failed to clearly make this crucial distinction.

Another point of confusion has been the "reasonable time" part of the definition. As is also true for photodegradation, the time required for biodegradation is a function of exposure conditions (as well as a function of the extent of degradation defined as the end point). Time to reach the defined end point after disposal can be markedly different if the object degrades in sewage sludge, in a compost pile, or in a landfill, even under the same climatic conditions. To this variation, then, must be added differences in ambient temperatures, rainfall, etc.

Faced with a proliferation of environmental claims about products and packaging, several state attorneys general produced guidelines for environmental claims, culminating in the issuing of the Green Report II, and took legal action against companies which they saw as making misleading claims. In 1992, the U.S. Federal Trade Commission (FTC) issued 16 CFR Part 260, "Guides for the Use of Environmental Marketing Claims," which was modified in 1996 and 1998. It includes guidelines for the use of degradability claims.[151]

The FTC guidance on the use of degradable, biodegradable, and photodegradable is that such claims can be made only if there is "competent and reliable scientific evidence that the entire product or package will completely break down and return to nature, i.e., decompose into elements found in nature within a reasonably short period of time after customary disposal." Unqualified claims about compostability can be made only if "all materials in the product or package will break down into, or otherwise become part of, usable compost (e.g., soil-conditioning material, mulch) in a safe and timely manner in an appropriate composting program or facility, or in a home compost pile or device."[151]

There has also been activity in the area of definitions, environmental claims, and testing of such materials by standards-setting organizations on both national and international levels. The American Society for Testing and Materials has issued several standards relating to degradability in a variety of environments. For example, ASTM D5338-92, "Standard Test Method for Determining Aerobic Biodegradation of Plastic Materials Under Controlled Composting Conditions," provides for measuring evolution of carbon dioxide after inoculation with composting microorganisms. The percent of biodegradation relative to a cellulose reference is reported.[152] In Germany, the Fraunhofer Institute for Process Engineering and Packaging (IVV) formulated a standard in 1998 for the compostability of biodegradable plastics, Standard DIN V 54900.[153] The European Committee for Standardization (CEN) has "Requirements for Packaging Recoverable through Composting and Biodegradation—Test Scheme and Evaluation Criteria for the Final Acceptance of Packaging" in draft form.[154] In Japan, the Biodegradable Plastics Society has developed several standards for testing biodegradability in different environments.[155]

The Degradable Polymers Council of SPI adopted definitions of biodegradable and compostable for plastic collection bags in 1998. Their standard is that "biodegradable" and "compostable" bags "should, at a minimum, satisfy ASTM D5338 and D6002 tests showing conversion to carbon dioxide at 60 percent for a single polymer and 90 percent for other materials in 180 days or less, and leave no more than 10 percent of the original weight on a 3/8″ screen after 12 weeks."[156]

Some early studies of biodegradation of packaging materials which used growth of microorganisms as a measure came to misleading conclusions. For example, PVC was incorrectly determined to be biodegradable since it supported growth of microorganisms. However, later studies showed that it was the plasticizers in PVC which were being metabolized, not the PVC itself. Care must be taken to avoid such misleading assessments. In particular, a limited amount of degradation in a short time cannot be extrapolated to a conclusion of substantial degradation at a later time. In addition to the problem of additives, some parts of the structure of a polymer may well be more resistant to degradation than other parts. For example, crystalline regions in a semicrystalline polymer will be more resistant to degradation than amorphous regions. Also, chain ends are typically less resistant than middle regions.

12.5.2 Effects of environment/exposure conditions

As was mentioned earlier, rates of biodegradation are very sensitive to environmental conditions. One result of this sensitivity is that degradation rates in modern municipal solid waste landfills tend to be quite slow. A key factor is the amount of water. When landfills were found to often be the source of pollutants entering groundwater systems, regulations were tightened to ensure that landfills would not generate substantial amounts of liquid effluent which could make their way into water systems. Liners at the bottom of landfills and caps on the top were added to the requirements, and it was no longer possible to build landfills in some high-moisture areas. The caps, in particular, are designed to create a barrier to the ingress of water. In addition, leachate (the liquid effluent from a landfill) must be pumped out and treated before it is discharged. Thus, relatively little water enters a modern landfill, and what does enter is routinely removed, so the landfill environment is relatively dry. Microorganisms do not grow as rapidly in such environments as they do when moisture is plentiful. In addition to the moisture factor, landfills within a few years become largely anaerobic. Trapped oxygen is used up, and cannot be replaced rapidly enough to maintain aerobic conditions. Therefore, the types of organisms predominating in an oxygen-rich environment will not be identical to those which predominate in an oxygen-poor environment, and growth rates in general will be lower. In addition to differences in the amounts and types of microorganisms, some change their metabolism in response to oxygen availability. The end result is slower rates of decay, and a change from generation of carbon dioxide to generation of methane. While the methane can be collected, concentrated, and used as an energy source, it is also a potentially explosive gas and an air pollutant. A modern landfill will continue to generate methane for a substantial length of time after it is closed and capped. Slow biodegradation occurring over many years results in sinking and settling of the landfill area as well as generation of methane. In addition, the leachate from a landfill can contain a variety of undesirable chemicals. Therefore, biodegradation in a landfill environment can have negative environmental consequences.

If it is desired to speed up decomposition in a landfill environment, leachate recirculation can be used. In such systems, instead of pumping out, treating, and discarding any liquid effluent which reaches the top of the landfill liner, the liquid which is pumped out is reintroduced into the top of the landfill, creating an environment much richer in water than is otherwise the case. In such circumstances, biodegradation will occur more rapidly, and the landfill will reach a stable condition in a much shorter time. The U.S. EPA has been reluctant to permit such schemes, however, because of the significantly greater potential for groundwater contamination which also results.

In contrast to landfills, compost operations, which will be addressed further in Sec. 12.5.3, are designed to ensure rapid biodegradation of susceptible materials. In composting, a water and oxygen-rich environment is maintained, along with a suitable inoculation of microorganisms to start the decay process. The time required for production of usable compost varies, but is most often less than a year. The composting operation can use open piles or windrows or closed containers. It can run on a very large scale as either a municipal or privately owned operation, or it can be a small compost pile in someone's backyard. It can be an open-air facility, contained inside a building, or operate in closed containers. It can contain yard waste only, source-separated organics, or mixed municipal solid waste. Composting of yard waste has grown rapidly in the United States over the last decade, significantly influenced by legislation in many states which prohibits the landfilling of yard waste. In Europe, the scarcity of landfill space led to adoption in many areas of systems for collecting source-separated organics which are then composted. In the United States, composting facilities which accept more than just yard waste are still relatively rare.

In addition to moisture levels and the amount of oxygen, temperature plays an important role in determining the rate of biodegradation. Increases or decreases in temperature affect the growth rate and activity levels of microorganisms, and hence the rate of biodegradation. Different types of organisms are at their best at different temperatures. Usually, the activity increases with increasing temperature, as long as the temperatures do not get too hot. Thus rates of degradation in landfills are usually higher in warm climates than in colder ones. In a compost operation, degradation results in the evolution of a substantial amount of heat, which raises the temperature significantly above ambient conditions. A well-designed system will maintain a sufficiently high temperature for long enough to kill pathogens which may be present, so the compost does not spread either disease or weeds.

Some biodegradable plastics may enter a different waste stream—liquid waste rather than solid waste. Sewage treatment facilities, like composting operations, are designed to hasten biodegradation of the collected wastes. For plastic products or packages which may end up in sewage systems, biodegradation is a significant asset. Rates of degradation in sewage are generally different than rates of degradation in compost. Holding times are also different; a product which degrades reasonably well in composting may not degrade fast enough in sewage treatment facilities to avoid causing problems.

We can conclude that in situations where landfill is the predominant disposal option, biodegradable plastics generally offer no significant advantage. Rates of degradation are slow, and the products of degradation, while they may include generation of methane to be used as fuel, are largely undesirable. Similarly, where incineration, with or without energy recovery, is practiced, biodegradable plastics are not advantageous. On the other hand, biodegradable plastics can play two types of roles in compost operations. First, if composting of a mixed organic stream is taking place, these materials will degrade along with the paper, food waste, and other biodegradable components. Second, even if yard waste is the only material composted, biodegradable plastics can be used as the bag in which compostable materials are collected. Biodegradable plastics can also be advantageous for products or packages which are likely to be disposed of in sewage.

All the previous discussion is based on products or packages entering a regulated waste stream. Products and packages can also be littered or illegally dumped. In such cases, if the item is not degradable, it may stay in the environment for a very long time. In some cases, this presents no problem. The steel may slowly rust away at the bottom of the lake, the plastic bag may first be buried in leaf litter, and then eventually be covered up with soil. Too often, however, the object is less innocuous. The beverage ring connector may entrap a duck. The floating plastic bag may be swallowed by a sea turtle who thinks it is a jellyfish. A skunk may get its head caught in the yogurt container. These encounters with wildlife can result in injury or death of the animal, either through entrapment in or ingestion of plastic items. The problem appears to be particularly acute when the plastics reach bodies of water, either by being thrown into the water, or being carried to the water by runoff or storm sewer overflow. Thus there is also a potentially valuable role for degradable plastics in items which are frequently littered.

An additional application for biodegradable plastics is in ships. Annex V of the International Convention for the Prevention of Pollution from Ships (MARPOL) prohibits ships from disposing of plastics into the ocean. While navies are exempt from the treaty, many nations, including the United States, have committed themselves to require their military forces to abide by the treaty. Ships of all sizes and types often have difficulty in appropriately storing or disposing of wastes. The problem is particularly acute for ships carrying large numbers of people, ships which spend long periods at sea, or ships operating in areas where ports are not equipped to off-load and dispose of the refuse. An aircraft carrier, for example, has about 6000 crewmembers and serves 18,000 meals a day, generating substantial amounts of trash. Nuclear submarines spend months continuously at sea. Development of biodegradable plastics, which in some areas could be disposed of in the ocean along with food wastes, could be a considerable asset.[157]

12.5.3 Composting

Composting of municipal solid waste in the United States is still in its infancy. According to BioCyle, 18 such facilities are currently in operation, and two are under construction.[158] Some facilities process a mixed waste stream and others process source-separated organics (that is, the generators of the waste separate the compostable organics from the noncompostable wastes). Generally, the source-separated organics stream contains paper and food waste, and does not include any plastics. Some institutional composting of source-separated organics, such as of waste from fast-food restaurants, has included biodegradable plastics used for food containers and cutlery. The mixed-waste composting facilities do include plastics among the materials collected. The EPA counted 14 such mixed waste composting facilities in 1996, handling a total of about 900 tons/day.[2] In such systems, the waste is generally processed to remove large items, ferrous metals, and sometimes other components before processing. The compost produced in these facilities is generally higher in levels of contaminants, including undesirables such as heavy metals, than compost originating from source-separated streams. At their current level of use combined with the current level of nonyard-waste composting, biodegradable plastics play only an insignificant role. For biodegradable plastics to have a significant impact on waste management, a higher level of use and a large growth in composting programs would be required. If source-separation is part of the system, significant consumer education efforts would also be needed to get people to divert only the right kinds of plastics to the compost stream.

Composting of yard waste is more prevalent in the United States than composting of other waste streams. In 1996, the U.S. EPA reported 3260 yard waste composting programs, handling approximately 25,500 tons/day.[2] Some facilities collect yard waste only in paper bags, which can be composted along with the yard waste, but which can cause problems if they become wet and consequently weaken. Some facilities collect leaves in loose form using a vacuum system. Some facilities permit plastic bags, but remove the yard waste from the bags before composting. This can add significantly to the cost of composting. A number of facilities have tried using biodegradable plastic bags, with mixed results. The variability in types of bags tested along with variability in composting conditions make it difficult to assess the suitability of biodegradable bags without field testing. For example, seven brands of biodegradable bags were field tested in Northville, Mich. Four bags were found to leave no visible residues after that facility's standard 8-week composting time, while three bags were readily visible and identifiable. Investigators pointed out that in differently designed facilities, results could have been quite different.[159] In Michigan, legislation was introduced in 1998 and passed by the state House to ban the use of most plastic bags for yard waste collection. The bill originally banned the use of any type of plastic bag, but was modified to exempt bags which are certified by the state as meeting compostability standards. The bill is now under consideration in the state Senate.[154] A number of yard waste composting operations, in Michigan and elsewhere, already prohibit the use of plastic bags, based on difficulties they have encountered in the past.

Minnesota is the state with the largest number of municipal composting facilities in operation. More than 400 composting facilities were started as a result of state laws passed in 1978 and 1980 that required development of alternatives to landfills. Most process only yard waste.[160] Eight of the top 10 solid waste management companies in North America had an involvement in composting programs by 1996. Many operated compost programs on the same site as landfills.[161]

In 1998, Portland, Oreg., became the first major U.S. city to consider mandatory food waste composting. The proposal recommended to the City Council would require participation by some grocery stores, restaurants, and food processors, but not by individual residents. The city says commercial food waste represents the largest single material in the current solid waste stream, and so it is important to target in attempts to meet the city's 60% recovery goal.[162]

Composting is much more prevalent in some countries than in the United States. In Canada, for example, increasing composting was an important part in successfully reducing waste disposal to 50% of the 1988 amount. Although more prevalent (proportionally) than in the United States, facilities handling more than yard waste are relatively new, and operators are still learning about the most effective ways to structure the systems.[163] Montreal collects compostable food residuals and yard trimmings from 17,000 households in 25-L brown plastic bins, with disposable plastic liners, and also provides a 7-L container for kitchen discards. Compostables are collected at curbside weekly.[164]

Composting has a significantly longer history in Europe, where the first composting plants for mixed municipal solid waste date to the 1970s, and it is estimated that more than 10 million tons of compost are produced per year.[165,166] In 1998, the European Union adopted a "Common Position," passed by the Environmental Council in March 1999, intended to reduce the amount of biodegradable material entering landfills. It requires reduction of landfilling of biodegradable municipal solid waste to 75% of 1995 levels by 2006, with further targets culminating in reduction to 35% of 1995 levels by 2016. European Union members have 2 years to include these requirements in national laws. One result is expected to be a significant growth in composting. Despite this history, the overall rate of composting of municipal waste in western Europe is estimated at only 4%. Many of the original compost facilities have closed due to lack of markets because of poor quality of the compost, including high contents of heavy metals and of noncompostables such as glass, plastic, and metal. One consequence was the development in several countries of organizations which certify the type and quality of compost generated.[165,166] Another is a shift from facilities treating mixed MSW to those treating source-separated organics. Germany is reported to have more than 400 such facilities, with France having about 200, Italy 60, and Spain 30. These countries, along with Denmark and The Netherlands, are regarded as leaders in composting of source-separated organics, and generally produce high-quality compost.[167] Composting rates for household organics are significantly higher than the overall compost rate in the European Union, although they vary considerably among countries. In the Netherlands, it is estimated that 90% of household organics are recovered through composting, Germany recovers about 45%, and Britain only 6%, despite a five-fold increase in composting between 1993 and 1997.[168]

The Netherlands achieves its high rate of composting by providing source-separated collection of residential organics, including yard, garden, and kitchen wastes, in virtually all municipalities, on at least a biweekly basis. As of January 1994, municipalities are mandated to provide this service, and residential organics are banned from landfill. It is estimated that such organics make up as much as 60% of residential waste in some communities, although the average is in the 40 to 50% range. Sixty to 75% is estimated to be garden waste, with the remainder food waste. Collection is typically provided in rolling carts collected by semi-automated compactors in single-family housing, with organics collection alternating on a weekly or biweekly schedule with collection of mixed waste. Residents in most municipalities are required to wheel the carts to the corner for collection.[169]

12.6 Natural Biodegradable Polymers

One way to obtain biodegradable plastics is to use natural polymers, that is, those formed by living organisms. There is, in nature, a general rule that what is formed by organisms can be decomposed by other organisms as part of the natural cycling of carbon in our environment. Thus, even very large polymer molecules such as lignin and cellulose can be decomposed by a variety of microorganisms.

While cellulose and cellophane are not plastics, there are plastics such as cellulose acetate, cellulose butyrate, cellulose acetate-butyrate, etc., which are derived from cellulose. Few studies of the degradability of these materials have been carried out, and they are not used in large quantities. A general rule is that the greater the extent of chemical substitution in the molecular structure, the slower the degradation rate of the material. Little attention has been given to these materials in efforts to develop biodegradable plastics. Slow degradation rates, high cost, and processes which generate some noxious discharges are likely the reason. There are other natural biodegradable polymers which offer more promise.

12.6.1 Bacterial polyesters

One of the early truly biodegradable polymers was polyhydroxybutyrate/valerate (PHBV). This is a member of the polyester family which is produced by certain types of bacteria when they have a diet which is carbon-rich but poor in some essential nutrient. Under these conditions, they produce polyhydroxybutyrate (PHB) as a food store to be called upon when carbon sources are less available. With manipulation of the diet, the bacteria can be induced to form a copolymer, PHBV, which has more useful properties than PHB.

While the polyesters grown in bacteria have been known for over 60 years, and their use had been investigated by some companies such as W. R. Grace & Company during the 1960s, modern development of bacteria-grown polyesters began with Imperial Chemical Industries (ICI) in Britain, around 1978.[170,171] ICI used the bacterium *Alcaligenes eutrophus* grown on glucose to produce poly(3-hydroxybutyrate) (PHB) in 80% yield, based on dry weight of the fermentation broth. While PHB was comparable to polypropylene homopolymer in melt point and tensile strength, and could be used to form small items such as golf tees by extrusion, it had a high glass-transition temperature and low extension at break. In 1991, ICI directed efforts toward purifying the polymer and adding plasticizers and fillers to alter its properties. Costs were already recognized as a problem, since the cheapest glucose source, corn, was almost 3 times as expensive as raw materials derived from oil or natural gas.[171] By 1987, ICI was manufacturing copolymers as well as PHB, and was marketing the resins on a limited scale in Europe, mostly in Germany, under the trade name Biopol. The resins could be processed into blown film, cast film, or calendered, and in trial runs, had been made into bottles by both injection and extrusion blow molding. Rates of degradation had been examined, with the finding that anaerobic sewage provided the most rapid degradation, followed by well-watered soil, seawater, sediments, aerobic sewage, cattle rumen, sea water, in vivo, and lastly with only a negligible degradation rate in moist air.[172] The polyester was described as a highly crystalline, stiff material with a melt point of about 175°C and a glass transition temperature of about 0°C.[173]

By 1987, ICI had focused on the random copolymer of butyrate and valerate, PHBV, and had patented a process for making the material. The feedstock for the copolymer was glucose plus propionate.[174] Copolymerization provided significantly improved properties, including reduced stiffness and a broader processing window. The homopolymer degrades at a temperature only 5 to 10° above its melting point, for example, while the copolymer has a much wider processing window. Marlborough Biopolymers, Ltd., an English subsidiary of ICI, produced the polymer in ton quantities for development of applications in surgical devices, personal hygiene products, and packaging. Properties were described as intermediate between PP and PVC. Degradation rates ranged from several days for films in anaerobic sewage to 9 months for a bottle in soil.[170] By 1991, PHBV was being produced at about 25 tons/year, and was being used by Wella, a German company, for shampoo bottles. The $15/lb price, when PP was selling for 50¢/lb, was a significant deterrent to widespread use.[174] The promise was sufficient, however, that ICI announced an expansion of capacity to 300 tons/year, with plans for a commercial-size plant with a 10,000-ton capacity in the next 3 to 5 years.[174]

The general process for manufacturing bacteria-grown polyesters, specifically polyhydroxyalkanoates (PHAs), is to deplete a growing culture of a nutrient, such as nitrogen, which it requires to grow, causing growth to stop. Then one or more carbon sources, such as glucose, are added to the fermenter. The bacteria accumulate pellets of polyester, about 2 μm in diameter and of irregular shape, within the cells. Under appropriate conditions, up to 80% of the total cell mass, or 80 g/L, of polyester can be accumulated. To harvest the polyester, the granules are extracted from the cells, purified, and formed into pellets. Crystallinity of PHBV is typically 70% or more, and molecular weights of about 600,000 are desired.[174]

By 1991, a large amount of research was going on at universities and other companies around the world in efforts to develop natural biodegradable polyester polymers and copolymers. One of the early entries was btF mbH, a state-owned subsidiary of Petrochemie Danubia, in Austria. In 1991, they were using *Alcaligenes latus* to produce 20 tons of PHB per year for controlled-release drug-delivery systems, and planned to increase production to 1 ton/week. One difference with this strain of bacteria is that it produces PHB throughout its growth phases, rather than only during nutrient deprivation.[174]

Japan at this time had two entries in the biodegradable polyester race. The Seibu Gas Company Ltd. used *A. eutrophus* in very pure carbon dioxide made from synthetic natural gas to produce PHB. The Research Laboratory of Resources Utilization at the Tokyo Institute of Technology used *A. eutrophus* to produce a copolymer P(3HB)-co(4HB), or 3-hydroxybutyrate with 4-hydroxybutyrate.[174]

In the United States, the applied microbiology lab at the Massachusetts Institute of Technology (MIT) was working on bioengineering of polyhydroxyalkanoate (PHA) biopolymers using recombinant DNA. Researchers at James Madison University successfully cloned the genes for making PHB in *A. eutrophus* and transferred them to *E. coli.* However, they failed to get PHBV from the modified *E. coli* when it was fed the same types of sugars and other nutrients as *A. eutrophus.* Researchers at the University of Massachusetts were working on developing novel bacteria-grown aliphatic polyesters, and produced a biodegradable elastomer using *P. oleovorans.* Researchers at MIT were working on bioengineering PHA polymers using recombinant DNA.[174] In 1992, researchers at Michigan State University and James Madison University reported the transfer of the three genes responsible for synthesis of PHB from *A. eutrophus* to a plant related to mustard, and succeeded in getting production of the plastic in the plants, although very little plastic was produced, and growth of the plants was significantly reduced.[175] Further genetic modification significantly increased production of plastic.[176] Researchers at MIT received a patent for transfer of PHBV genes to bacteria and crop plants.[177] Among their accomplishments was production of cotton fibers containing a core which, rather than being hollow, contains PHB, although in amounts less than 1%. Researchers at the Volcani Center of the Israel Ministry of Agriculture have also produced these materials.[178]

In 1992, ICI, with its production capacity now at 600,000 lb/year in the United Kingdom, found its first North American market in blow-molded bottles and injection-molded caps for hair-care products. The \$8 to \$10/pound price was expected to fall to \$4/lb as planned increased capacity came on-line.[179] In 1993, ICI got out of the biopolymer field, transferring Biopol to its spin-off business unit, Zeneca.[180] In 1996, Zeneca's Biopol unit was sold to Monsanto, which continued to market PHBV under the Biopol trade name. In addition to using bacteria, Monsanto investigated production of PHBV in plants, such as soybeans and cannola, using recombinant DNA.[181]

One high-profile use for Biopol is credit cards for Greenpeace, first available in 1997, in Europe.[182] Other uses include disposable cups, eating utensils, planters, fish nets, compost bags, and packaging for beauty products, with applications continuing to be limited by the high price of the resin compared to competitive materials.[183] A unique product, manufactured in Denmark, is a twin-blade razor with handle and replaceable heads both made with PHA. Interestingly, the packaging for the razor is a combination of polystyrene and paper.[184] In late 1998, Canada Trust chose Biopol as the material for some of their credit cards at about the same time that Monsanto announced that it would stop both production of Biopol and research and development on it by the end of the year.[185,186]

Metabolix, located in Cambridge, Mass., is producing sample quantities of PHB and of other PHA plastics. One member of this family, polyhydroxyoctanoate (PHO), is a rubbery elastomer, with an extension at break of 380%, compared to only 5% for PHB. It also has lower tensile strength than PHB.[187]

Researchers at the Universiti Sains Malaysia have been working on production of PHB from palm oil, and have found a number of local microorganisms which can be used to do so. Efforts include identifying the genes responsible for manufacture of the polymer.[16]

More than 100 PHA polymers are known to be produced by bacteria. Polymer properties are a function of chemical composition and of average molecular weight. Researchers at the Massachusetts Institute of Technology have been able to control the molecular weight of PHB by transferring genes from *A. eutrophus* to *E. coli,* and inserting additional genes to produce a larger amount of PHA synthase, which is the enzyme which links the individual polymer units together.[188]

Researchers at the Japanese Institute of Physical and Chemical Research, Riken, in 1997 announced the development of a 3-hydroxybutane acid/3-hydroxyhexanoic acid copolymer, which has a higher density and higher strength than other PHAs. Its melting point was listed as about 300°C.[189]

PHB and other PHAs can also be blended with other biodegradable polymers. In Japan, the Ministry of International Trade and Industry's Biological Industry Institute in 1990 announced the development of blends of PHB and polycaprolactone (PCL). The blends can be processed with conventional equipment, and the ratio of the two polymers determines the rate of degradation.[190]

Chemical synthesis of biodegradable PHAs is limited by the need to have stereo-specific polymers to obtain desired performance properties. The cost of single-isomer raw materials is quite high, rendering the polymers as expensive as those obtained by the biological route, so this approach has not been utilized commercially.[191] L-isomers of PHB, which are not made in bacteria, can cause spontaneous abortions, while D-isomers are nontoxic and can be safely used in the body.[174] Conventional chemical processes cannot, to date, produce pure D-isomers.

In contrast, biological systems produce polymers with all monomers in the D configuration. The general chemical structure of polyhydroxyalkanoates made by natural processes is -[O-CHR-CH_2-CO]$_n$- where the side-chain R is a straight chain, and contains between 1 and 11 carbons. The monomers from which natural PHAs are polymerized are all 3-hydroxyalkanes. A large range of bacteria types are known to synthesize PHAs. Low molecular weight PHB has also been found to naturally occur in plants and animals, including humans. In PHA plastics, molecular weights range between 2×10^5 and 3×10^5. Within the bacterial cells, PHB is in an amorphous form. However, it crystallizes during purification processes, treatment with solvents, heat, or if all water is removed. One possible explanation for this fact is that the rate of crystallization is very slow, unless nucleating agents are present, in which case it is very rapid. Another hypothesis is that water or lipids present in the cell act as plasticizer, preventing crystallization.[191]

Several types of bacteria are capable of forming PHB or of producing PHA/PHB copolymers from a mixed feedstock containing appropriate precursor molecules. The amount of hydroxyvalerate (HV) depends on the feed and also on the species of bacteria which is used. However, only the *Pseudomonas* species of bacteria can produce and store polymers from monomers containing 8 or more carbons.[191]

A. eutrophus, P. oleovorans, and several other organisms are able to form and store polyesters from monomers other than 3-hydroxyalkanes or alkanoic acids under appropriate conditions. *P. oleovorans* can use a much wider variety of substrates than microorganisms in the *Alcaligenes* family. In addition to homopolymers and copolymers, terpolymers have been created in this manner. For example *P. acidovorans* fed with 1-4-butanediol and pentanol creates a terpolymer of 3 hydroxybutyrate (3HB), 4HB, and 3HV. It is also possible to incorporate potentially reactive substituents in the R side chain, which might be used for later cross-linking reactions or some other type of derivatization to modify polymer properties.[191]

PHAs containing longer chain length monomers than PHBV tend to be elastomeric rubbery materials. Little is known about the properties of polymers formed from unusual monomers which yield aromatic, branched, or substituted side chains.[191]

Commercial production of PHB and PHBV is carried out in batch systems as large as 200,000 L in capacity. The medium used in the fermenter in the production stage contains glucose, a desired amount of phosphate to produce the required amount of biomass, and an excess of all other nutrients. The culture is then inoculated with microorganisms, which reproduce until phosphate is depleted, and then begin to store polymer inside the cells. Glucose is fed to maintain the polymer production, if PHB is being produced. If PHBV is desired, a mixture of glucose and propionic acid is fed to the fermenter. The proportion of HV units in the random copolymer is controlled by adjusting the ratio of glucose and propionic acid. An early problem with this scheme was that the bacteria converted much of the propionate to acetate or to carbon dioxide. Since propionate is more expensive, this increased costs. A mutant form of the bacteria was identified which cannot metabolize propionate to acetate, thus improving conversion efficiency and lowering costs.[191]

The first route used for harvesting the polymer from the bacterial cells involved the use of large quantities of solvent, about 20 times the amount of polymer recovered. This large excess was needed because of the high viscosity of the solution, even when very dilute. The most effective solvents were chlorinated alkanes. Such heavy use of these environmentally problematic solvents was undesirable, and eventually an aqueous system was developed to wash the polymer free from cell debris. The result is production of a white powder, which is subsequently melted, extruded, and pelletized. The aqueous effluent from harvesting is suitable for conventional activated sludge treatment before discharge.[191]

Some properties of PHB, PHBV copolyesters, and PHO are summarized in Table 12.3. PHB is a highly crystalline material, forming a right-handed helix when crystalline or in chloroform solution. PHBV copolymers with HV content up to 30% have HB units in the crystal lattice, and HV units are excluded. At HV content above 30%, HV occupies the crystal lattice, and HB is excluded. The copolymers have been shown to be random.[191]

A variety of microorganisms, including both bacteria and fungi, have been shown to degrade PHB and PHBV. Degradation rates are faster for PHBV than for PHB in all types of soils. In water, rates for the homopolymer and copolymer are identical. Degradation rates are much faster in salt water than in freshwater. In simulated compost and landfill conditions, PHBV degradation has been shown to be faster in anaerobic conditions than in aerobic ones.[191]

12.6.2 Starch-based plastics

As mentioned, some of the plastics which purported to be biodegradable in the late 1980s and early 1990s contained polyethylene blended with about 6% starch. While the starch was biodegradable, there was no convincing evidence presented that the polyethylene itself biodegraded to any significant degree. The term *biodeterioration* has been coined to reflect what happens in these materials: they lose strength and are fragmented, but the molecular structure is not much affected.

TABLE 12.3 Properties of PHB, PHBV, and PHO[174,187,191]

Polymer	Melting point, °C	Tensile strength, MPA	Flexural modulus, GPa	O_2 permeability, $cm^2 \cdot 25\ \mu m/m^2 \cdot atm \cdot d$
PHB	177	40, 100*	3.5	45
PHB/10%HV	150	25	1.2	
PHB/17%HV		54*		60
PHN/20%HV	135	20	1.2	
PHO	61	9	0.008†	

*Stress at break.
†Young's modulus.

This led some environmental groups, such as the Environmental Defense Fund (EDF) and the Environmental Action Foundation (EAF), to call for a consumer boycott of degradable plastics.[192] In contrast to these materials, starch-based plastics have been developed which are truly biodegradable. Some contain nearly 100% starch, and others are blends of starch with other biodegradable components. Most use carefully controlled amounts of water as a plasticizer to convert the starch into a thermoplastic, along with carefully controlled temperature and pressure.

An early entry into the all-starch biodegradable plastics was Warner-Lambert, which in 1990 announced the creation of the first biodegradable plastic from starch. According to the company, either potato, corn, rice, or wheat starch could be used. These materials were sold under the trade name Novon, and were water-soluble as well as biodegradable. Novon contained about 70% branched starches and 30% linear starch, along with a glyceride as an internal lubricant. Properties included tensile properties comparable to crystal polystyrene, optical properties like those of polyethylene, and elongation at break of about 20%.[193] Products included compostable bags and cutlery. In 1993, after Warner-Lambert was unable to sell the Novon business unit, it suspended operation. Then, in 1995, EcoStar International acquired the unit, and formed Novon International, which shortly thereafter was acquired as a wholly owned subsidiary of Churchill Technology, Inc.[194] Novon customers included Doane Products Company of Joplin, Mo., which used Novon for dog bones, and Eco-Turf Inc. in Chicago which used Novon for biodegradable turf tacks for golf courses, landscaping, and agriculture.[195] In late 1996, Churchill Technology filed for bankruptcy protection under Chapter 11, and announced the intent to reorganize with biodegradable plastic as their main product. The company had not reported a profit in the last 2 years, since its merger with Ecostar International.[196] No current information about its status was found.

StarchTech, Inc., of Golden Valley, Minn., sells biodegradable starch-based resins for injection molding and other manufacturing processes, with a claim that in volume applications they approach the cost of standard plastics. Target markets are landscape products, disposable food service items, golf tees, and some pet products. Polysheaf resins are based on wheat starch and intended for injection molding. The company also sells Re-NEW, Bio-Lapse, and Bio-Revert resins.[197]

The Natick Research Development and Engineering Center (NRDEC) has investigated the use of starch-based blown film, among other products, to help the U.S. Navy comply with MARPOL requirements (no discharge of plastics at sea).[157]

Researchers at the Central Tuber Crops Research Institute (CTCRI) in India have developed a starch-based plastic film for use in nursery bags for seedlings. The country has a massive forestation program underway, which requires about 65,000 tonnes of nursery bags per year. The raw materials for the film are tapioca or corn starch, urea, and water, along with a coupling agent. It can be formed on conventional blown film equipment.[198]

One of the most visible demonstrations of starch-based biodegradable plastics was at the 1998 Winter Olympics in Nagano, Japan, where the "Eco-Friendly Kentucky Fried Chicken" booth used biodegradable plastic bags, flatware, cups and coated paper products, all based on corn. The effort was sponsored by the U.S. Feed Grains Council.[199]

Research on starch-based plastics has taken place in many countries around the world. The Australian government funded a 1995 research project on development of starch-based plastics from corn and wheat, using water and glycerine as a plasticizer.[200] In Japan, the Biodegradable Plastics Society was formed in 1989, with 48 member companies located mainly in Japan. By 1990, the membership had expanded to 69 companies, and included a significant number of non-Japanese members.[201] In 1992, the U.S. Bio/Environmentally Degradable Polymer Society was formed, and had over 200 members by 1998.[202]

Several companies sell starch-based foam peanuts for loose fill packaging as a substitute for expanded polystyrene. One of the first was American Excelsior, which manufactured Eco-Foam loose fill, and then expanded their line to include sheet and laminated structures. Originally, the materials contained 95% starch and 5% polyvinyl alcohol.[203] Later, American Excelsior manufactured Eco-Foam materials containing over 99% corn starch from a special hybrid corn.[204] These products, in addition to being biodegradable and water-soluble, also had the advantage of dissipating static charges much more effectively than expanded polystyrene. The company also claims that, because of the purity and absence of odor of their materials, bugs and rodents are not attracted like they are to other natural cushioning materials.[200] Clean Green Packing, of Minneapolis, Minn., a subsidiary of StarchTech, Inc., began marketing starch-based loose fill cushioning materials in about 1992.[197,205] FP International sells Flo-Pak Bio 8 loosefill, made from corn, wheat, or potato starch, along with polystyrene loosefill, including several grades made with high percentages of recycled PS.[206]

Starch-based materials have also been designed for applications which conventionally use urethane-based foam-in-place packaging. In 1996, Environmental Packaging L.P. developed a device which propels a wheat starch-based material into a container holding an item to be shipped. The foam pieces are molded together using water, which is sprayed lightly on the foam pieces as they enter the container. Development of the material involved a collaboration of Enpac, which is a DuPont and ConAgra Company joint venture, and the Norel Paper Co.[207,208]

Biodegradable plastics which contain a mixture of starch and other biodegradable polymers, either natural or synthetic, have also been developed. For example, Yuking Company in Korea in 1997 announced the development of a biodegradable plastic containing polycaprolactone and more than 40% starch. The starch component was included primarily to lower the price.[209] Biotec, a German company, developed a biodegradable material made of starch combined with a biodegradable plastic for use in biodegradable garbage bags. Fardem Belgium, which manufactured the bags, received the first European "OK Compost" label for garbage bags. This label is certification that the product is completely compostable, leaving no plastic remnants after the compost process is completed.[210] In the United States, BioPlastics, headquartered in Michigan, manufactured biodegradable yard waste bags manufactured from starch and polycaprolactone for a trial program testing their use in a yard waste composting operation.[211] The Mellita Group in Europe sells biodegradable compost bags under the Swirl trade name. The bags are made from potato starch and polycaprolactam, and are reported to have properties similar to polyethylene. The bags degrade to carbon dioxide, water, and humus in about 40 days in compost piles. The bags sell for approximately 5¢ each, compared to 3¢ each for comparable LDPE bags. Melitta also markets resin, which it sold for about $2.40/lb in 1995.[212]

Novamont, a subsidiary of the Italian company Montedison, marketed Mater-Bi biodegradable film for composting and waste disposal. A typical resin contains 60% starch along with other materials. Mater-Bi resins containing starch and poly-ϵ-caprolactone can be handled on conventional film blowing and sealing equipment for LDPE, with minor modifications.[213] One U.S. manufacturer of trash bags made with this material is Biocorp USA, which sells them as

"Materbags."[214,215] The bags have been demonstrated as suitable for vermicomposting (using earthworms) in addition to standard composting. In vermicomposting, the biodegradation was essentially complete in 20 to 22 days, compared to 20 to 35 days in standard composting operations.[215] Resins are available in both molding and film grades.[216] A high-tech application for starch-based biodegradable materials is in composites formed from starch polymers and bonelike ceramics for medical applications. Mater-Bi plastics containing a cornstarch/ethylene vinyl alcohol blend reinforced with hydroxylapatite particles produced biodegradable materials with promise for some such applications.[217] In 1996, Montedison sold the Novamont unit, which creates and markets Mater-Bi, to a group of commercial banks headquartered in Italy. In 1997, production capacity was doubled, and Novamont GmbH in Germany was started to expand the European market. Biocorp, Inc., of Redondo, Calif., is the exclusive North American distributor for Mater-Bi, selling products, including bags and cutlery, under the reSource name.[154]

EarthShell Corporation, in Santa Barbara, Calif., markets packaging made by combining limestone, starch, and cellulose fiber, mixing the materials with water and then baking them in a mold. The evaporation of the water and the resulting steam thermoforms the product.[218] Containers reportedly dissolve completely in water after they are broken, and are compostable. The company also claims the containers use only half the energy of competing products, and generate fewer pollutants.[219]

12.6.3 Protein-based plastics

Rather than focusing on starch-based plastics, some investigators have worked on developing protein-based plastics from corn or other sources. Film made from zein, a protein found in corn gluten meal after production of ethanol, has been studied at the University of Illinois at Urbana-Champaign. They have produced a biodegradable, water-resistant shrink wrap. The first test application was to protect hay bales from wet weather, substituting for plastic covers that must be removed and disposed of. In addition to being biodegradable, the zein-based wrap has nutritional value, and can be eaten by livestock.[220] Another application studied was mixing zein with fatty acids and flax oil to produce water-resistant food containers, plates, sealing material, and trays. Untreated zein was found to be brittle and to absorb too much water. The addition of fatty acids and flax oil as plasticizers solved that problem, without interfering with the biodegradability of the material. The cost in 1997, however, was over $5/lb.[221]

Researchers at the National Food Research Institute in Japan and at Clemson have also studied zein from corn as a biodegradable packaging material, in film form and as a paper coating. Zein films and zein-coated paper were shown to be heat-sealable and compostable, as well as suitable for animal feed. The investigators also claimed that zein-coated paper is recyclable with ordinary paper. The material could also be modified with plasticizer.[222]

R. Narayan at Michigan State University has investigated the use of soy protein to manufacture biodegradable plastic films. While early soy-protein films were weak and brittle, he has extrusion-blended soy proteins with selected aliphatic polyesters to produce water-resistant biodegradable films with higher elongation and tensile strength. Potential applications range from plastic cutlery to containers for seedlings. Some blends have elongation at break of 500%, with tensile strengths around 2000 lb/in^2.[223]

Investigators at North Dakota State University have produced films from soy isolate plus a plasticizer, and are identifying chemical agents to produce cross-linking of the proteins through the lysine residues they contain in an effort to improve tensile strength and elongation.[224]

Iowa State University had a research project from 1991 to 1994 on using soy protein to make plastics similar to thermosets, using various aldehydes and acidic anhydrides as cross-linking agents. They also investigated the use of cellulose fibers as fillers in the plastic, using starch as an ingredient, and soy-protein-based foams. Closed-cell foams with densities of about 0.2 g/cm^3 were successfully produced.[225]

Researchers at Auburn University and the University of Alabama at Birmingham have produced a synthetic gene which induces the formation of protein-based polymers in microorganisms, and have successfully transferred the gene to tobacco cells. The plants showed polymerlike inclusions in the tobacco leaves.[226]

Showa Highpolymer Company of Tokyo, Japan, has developed a biodegradable thermosetting resin described as an amino protein resin. Its biodegradability is 31% weight loss after 31 weeks in standard soil, reportedly greater than that of natural wood. Mechanical properties are claimed to be on a par with or superior to those of conventional thermoset resins.[227]

12.6.4 Polysaccharides

A limited amount of research has been directed at polysaccharides. Cellulose and thermoplastics derived from cellulose are members of this family, of course, as are starch-based polymers. It is well known, as mentioned earlier, that biodegradation is influenced by the degree of modification of the basic structure of the cellulose as well as by its crystallinity. In 1990, Japanese researchers were investigating the use of selected natural polysaccharides to produce biodegradable plastics with controlled rates of degradation, as well as improved elongation, tear strength, and transparency.[228]

Hayashibara in Japan has produced Pullulan polysaccharide-based biodegradable films. Properties were reported to be similar to polystyrene. Aisero Chemical in Japan has produced biodegradable films and gels made by reacting chitin, found in the shells of crustaceans, with concentrated bases to form chitosan. Films manufactured from chitosan and cellulosic fiber have been reported to be hydrophilic, but not water-soluble, and to be good oxygen barriers.[222]

12.6.5 Wood-derived plastics

A few investigators have been able to modify wood to produce biodegradable thermoplastics. In 1992, for example, graft copolymers of lignin residue from paper manufacture and styrene were made. The materials are formable by conventional processes, and they are biodegradable. Degradation rates increase with lignin content, which varied from 10 to 50%.[229]

In 1996, Austrian researchers developed a thermoplastic in which waste wood chips, corn, and cellulose-derived resins were combined to produce a biodegradable and melt-processable wood.[230]

Researchers in Korea, in 1997, announced the development of a biodegradable plastic made from fibers in genetically engineered aspen trees. Their methodology included the use of chemical catalysts as well as modification of the genetic content of the tree.[231]

12.7 Synthetic Biodegradable Polymers

Most synthetic polymers are not biodegradable. However, there are a few synthetic polymers which are truly biodegradable. Some are water-soluble and become biodegradable once dissolved. Others are insoluble in water. The major families are lactic acid–based polymers, polycaprolactone, other synthetic polyesters, and polyvinyl alcohol.

12.7.1 Lactic acid–based polymers

Polymers based on lactic acid are inherently biodegradable. They have been available on a small scale for a number of years. Some absorbable sutures, for example, are made from lactic acid. The polymer is degraded by contact with moisture. In 1954, DuPont patented the ring-opening polymerization process for lactic acid polymers, following conversion of lactic acid to a cyclic dimer in the first reaction stage.[232] However, the cost of production of the monomers, especially if stereochemically pure enantiomers were desired, was a significant deterrent to widespread commercial development of the polymers until fairly recently. New technologies have lowered costs, fueling extensive research around the world. For example, researchers at Argonne National Laboratory, in Argonne, Ill., developed a process for forming lactic acid from potato waste which cut the greater than 100-h processing time down to less than 10 h.[233] One of the key developments was the ability to control the ratio and distribution of the *d*- and *l*-forms of lactic acid in the polymer backbone by modifying polymerization conditions.

Lactic acid can be formed by either chemical or biological processes. Fermentation processes provide more ability to control the enantiomers being produced. Bacteria have been identified which form both L and D enantiomers. Some preferentially form D, others preferentially form L, and still others form significant amounts of both enantiomers.[232] The ratio of the two forms in the polymer affects crystallization kinetics, melting temperature, and polymer rheology.[234] L enantiomers are also known to be present in mammalian systems and easily assimilated by humans.[232]

One of the earliest companies to extensively develop polylactic acid (PLA) polymers was Cargill, which began working on them before 1987, and began production of pilot plant quantities in 1992. Cargill marketed biodegradable lactic acid polymers under the EcoPLA trade name. In December 1997, Dow Chemical and Cargill, after 15 months of joint investigation, formed a joint venture, Cargill Dow Polymers, to further commercialize PLA polymers. Cargill already had a 4000-metric ton/year facility near Minneapolis, and was building a 35,000-metric ton/year plant in Blair, Neb. The joint venture plans to build a 125,000-metric ton/year plant by 2001.[235,236]

The starting material for PLA is starches or sugars from corn, sugar beets, or other sources. The starch is converted into sugar, and then fermented to yield lactic acid. Next, water is removed to yield a lactide. Solvent-free polymerization is then used to produce PLA. Its properties make PLA suitable for a wide range of applications. Its gloss, clarity, temperature stability, and processability are comparable to polystyrene, and its grease and oil resistance and odor and flavor barrier properties are comparable to PET. It is heat sealable at a lower temperature than polyethylene and polypropylene. PLA can be formulated to be flexible or rigid, and can be copolymerized with other materials to further modify its properties. It is not water-soluble. Other properties are listed in Table 12.4. Stereochemistry can also be varied to produce amorphous, semicrystalline, or crystalline polymers. It can be processed by most conventional plastics forming operations, including blown film, thermoforming, extrusion, and injection molding. The selling price for the resin in 1997 was above $1.00/lb,

TABLE 12.4 Properties of Pure Poly(L-Lactide) and Dainippon's CPLA[232,241]

Polymer	Melting point, °C	Tensile strength, MPA	Flexural modulus, GPa	O_2 permeability, $cm^2 \cdot 25\ \mu m/m^2 \cdot atm \cdot d$
Poly(L-lactide)	215	49	3.2*	
CPLA—hard	162	54	2.5	380
CPLA—soft	161	21	1.0	820

*Tensile modulus.

but Cargill and Dow expected it to fall to about 50¢/lb by 2001, with increases in production.[235,237] Polylactic acid has Generally Regarded As Safe (GRAS) status for food packaging.[234]

Potential applications for PLA include cast, blown, and oriented films; rigid containers; and coating for paperboard. A major application is compost bags. PLA resins were also being used for consumer packaging in Japan, and injection molding applications in both Japan and Europe. Other PLA manufacturers include Mitsui Toatsu and Shimatsu in Japan, Chronopol in Colorado, and Neste Oy in Finland.[235,237] Mitsui Toatsu in 1998 introduced an improved generation of its PLA called Lacea. Chronopol had a semicommercial resin trade-named Heplon, with a world scale plant planned within the next few years.[238]

Duro Bag Manufacturing Company EcoPLA bags consisted of three-layer blown film, with an interior layer of polylactide sandwiched between layers of a proprietary biodegradable aliphatic polyester. The bags were intended to replace paper bags for collection of compostables.[239]

Paper drinking cups made with an extrusion coating of PLA have greater stiffness than polyethylene-coated cups. However, unmodified PLA does not perform well in extrusion coating because of its low melt elasticity. Therefore, for such applications, it is desirable to modify the melt elasticity by modifying the molecular weight distribution and degree of branching. Investigators from Cargill Dow were able to increase branching by using epoxidized soybean oil as a multifunctional comonomer and also by reacting PLA with peroxide. They found the peroxide reaction to be the most favorable method for inducing the formation of a highly branched molecular structure which was suitable for extrusion coating of paper.[234]

CornCard International manufactures a degradable polylactic acid-based polymer called Mazin. It also contains a proprietary additive polymer. The lactic acid is derived from corn. The polymer has been tested at the University of Nebraska to verify its biodegradability.[240]

Dainippon Ink and Chemicals, Inc. (DIC), of Japan, is developing biodegradable plastics using lactic acid copolymerized with a biodegradable aliphatic polyester that is termed CPLA. The aliphatic polyester is formed from dicarboxylic acid, glycol, and other substances. CPLA is reported to break down into a low molecular weight polymer by hydrolysis after 3 to 6 months of exposure in soil, sludge, or seawater and to begin decomposing in 6 to 12 months by microbial action. If composted with food garbage, it begins to break down in about 2 weeks. The materials can be transparent and flexible or rigid, depending on the amount of the polyester. The soft flexible material contains a substantial amount of polyester, on the order of "a few tens percent." The hard rigid material has only a small percentage of polyester. Both materials have a density of about 1.25 g/cm^3, and a melting point at just over 160°C. The glass transition temperature for the hard material is 60°C, and 51°C for the soft material. The resins can be molded by extrusion or injection, including blow-molding, and can also be used to produce foam or spun to produce fiber. They are said to be stable at temperatures up to 200°C, and to be easily vacuum formed, heat sealed, and printed. Other properties are shown in Table 12.4.[241]

Many other copolymers can also be created from lactide or lactic acid and other appropriate monomers. Crystalline copolymers of L-lactide and glycolide are used as medical sutures. Copolyesters of lactide and glycolide are sold under the name Vicryl and have a melting point of 210°C and a glass transition temperature of 43°C. Amorphous copolymers of L-lactide and D,L-lactide with glycolide have been suggested for biodegradable drug delivery systems. Block copolymers of L-lactide and ϵ-caprolactone may also be suited for biomedical uses.

As mentioned, the enantiomeric purity of lactic acid polymers significantly affects their properties. Pure poly (D-lactide) and pure poly (L-lactide) are crystalline materials. Poly (L-lactide) typically has about 40% crystallinity. Poly (D,L-lactide) is totally amorphous. Copolymers of D-lactide or L-lactide with D,L-lactide may or may not be crystalline, depending on the amount of comonomer. Blends of poly (L-lactide) and poly (D-lactide) are reported to have better mechanical properties than either of the homopolymers alone.[232]

Unmodified lactic acid polymers tend to degrade rather slowly. One application which makes use of this fact is fishing line. Such lines will completely decompose after 1 y in seawater. Biodegradable shrink films have also been produced by melt-mixing of PLA with ethylene vinyl alcohol copolymers. Foams can also be manufactured.[232]

12.7.2 Polycaprolactone

Polycaprolactone is a member of the polyester family which has been used in relatively small quantities for a long time. The most well-known supplier is Union Carbide of Danbury, Conn., which sells polycaprolactone resins under the brand name Tone. These polymers are compostable but not water soluble.[216] A primary application is for compostable bags for yard waste or other organics collection. They can be utilized as homopolymers, copolymers, or in blends. For example, R. Narayan, of Michigan State University, has investigated Envar, formed from polycaprolactone and thermoplastic starch, as a substrate for biodegradable compost bags.[215] Other companies using starch/polycaprolactone blends are discussed in Sec. 12.5.2.

12.7.3 Other polyesters

DuPont has created a family of synthetic polyesters which are degraded by a combination of hydrolysis and microbial action. These materials, sold under the Biomax name, are similar to polyethylene terephthalate, but incorporate as many as three different proprietary aliphatic monomers into the structure. The monomers create weak spots which are susceptible to hydrolysis. The much smaller molecules which result are biodegradable. DuPont claims that in composting operations the materials are totally harmless and cannot be undetected by the unaided eye after about 8 weeks. The resins can be manufactured with existing equipment and existing bulk monomers, so they are only marginally more expensive than PET. Applications include household wipes, yard waste bags, components in disposable diapers, disposable eating utensils, geotextiles, agricultural films, plant pots, coated paper products, adhesives, and more. Biomax can be thermoformed, blow molded, and injection molded. Properties can be customized to meet the demands of the application. Melting points are around 200°C, which is significantly higher than in many biodegradable materials. Strength can be formulated to be as low as LDPE or as high as half the strength of polyester film. Elongation can range from 50 to 500%.[242]

Eastman, in 1998, announced the development of Eastar Bio copolyester 14766. The patented polymer is derived from conventional diacids and glycols but is completely biodegradable and compostable. Projected markets include lawn and garden bags, food packaging, and horticultural applications. Properties are similar to low-density polyethylene, and it can be processed on conventional equipment to make blown or cast film, extrusion coated, or woven into fiber and netting. It is water-resistant, imparts no odor or taste to food or beverages, and has no plasticizers or other migratory substances. The compost time is comparable to newspaper, that is, 60 to 90 days. The material meets standards for a variety of food-contact applications. It is being targeted to areas where composting is common. The polymer can be used alone, blended with starch or wood flour, or as a coating on paper. Cost is reported to be low, and the material is recyclable.[243]

Other companies manufacturing biodegradable synthetic aliphatic polyesters include Showa Highpolymer of Japan, which has manufactured Bionolle biodegradable synthetic polyester for several years. Bionolle can be processed on conventional polyolefin equipment at 160 to 260°C, and can be blown, extruded, or injection molded. It is heat-sealable, and has better resistance to water and organic solvents than many other biodegradable plastics. Bionolle has good environmental stress crack resistance, higher yield strength than polyethylene, and stiffness between HDPE and LDPE. Izod impact strength was reported to be adequate for most end uses. In compost conditions, bottles disappeared from view after 4 weeks. Films buried in moist soil nearly disappeared after 1 year. At least four types of Bionolle have been produced, with somewhat different properties, including degradation rates.[244] Some of the materials are extended with diisocyanate.[245]

Bayer, in 1995, produced a synthetic polyester amide material which is biodegradable. It is reported to have a melt temperature of 125°C, to have good resistance to solvents and light, and to be strong and extremely tough. Its biodegradability meets the German DIN standard. It can be extruded into films, injection molded, blow molded, or spun into fibers. Properties are similar to low-density polyethylene. The material is described as crystalline and translucent.[246] Bayer began marketing BAK 1095 in the United States, Europe, and Japan in September 1997, and was developing BAK 2195, which was available only in sample quantities.[154] In 1998, Bayer announced the development of a new biodegradable plastic, which is recyclable as well as biodegradable. Tests on 200-μm film showed biodegradability in under 70 days in compost conditions. Compostability reportedly is not affected by multiple recycling. The chemical makeup of the resin was not described.[247] In early 1999, the company lists biodegradable polyester amide only as a developmental product, with BAK as its provisional designation.[248]

In 1995, BASF announced the development of a biodegradable thermoplastic copolyester based on starch feedstock. A plasticizer reportedly is blended on-line with the copolyester to form a material which is antistatic, has good toughness, and good elongation at break.[249] In 1999, BASF listed Ecoflex as the trade name for their biodegradable polyesters.[250]

12.8 Water-Soluble Polymers

Several water-soluble polymers are known which are stable in the solid state but will biodegrade once they are dissolved. These include both synthetic and natural polymers, such as polyvinyl alcohol, cellulose esters and ethers, acrylic acid polymers, polyacrylamides, and polyethylene glycol, along with natural polymers derived from starch and some polylactides. Polyethylene oxide is biodegradable at molecular weights below 500.[251] For many of these materials, the degree of water solubility can be altered by changes in the polymer formulation. Several of these polymers have already been discussed. Others are discussed here.

12.8.1 Polyvinyl alcohol

Among the first manufacturers of polyvinyl alcohol (PVOH) were Air Products & Chemicals of Allentown, Pa., and ChrisCraft Industrial Products, Inc., of Gary, Ind. Currently, Air Products, which sells polyvinyl alcohol under the Airvol trade name, targets its resins primarily at adhesives and paper coatings. ChrisCraft sells PVOH film under the MonoSol trade name.[251-253] Nippon Gohsei claims to have the second largest polyvinyl alcohol capacity in the world at 6000 tons/month. Its resins are used in textiles, paper processing, emulsifiers, sizing, and adhesives as well as for biodegradable materials.[254] Italway, of Italy, sells polyvinyl alcohol polymers under the name Hydrolene.[255]

Polyvinyl alcohol is produced by hydrolysis of polyvinyl acetate, since the vinyl alcohol monomer is unstable. The degree of solubility can be controlled by modifying the extent of hydrolysis and the molecular weight of the polymer. Films can be produced which are readily soluble in water of any temperature or soluble in hot water only. Applications include hospital laundry bags and soluble pouches for chemicals and detergents. Most applications require the pouch to be contained in an outer barrier package to prevent moisture from reaching the material prematurely.

One difficulty in using this polymer is that it is generally not melt processable, since its decomposition temperature is lower than its melting point. Films are typically prepared by a solution casting process from a water solution. Plasticizers are incorporated in some resins. The films are readily biodegraded in wastewater streams or in compost. It is printable using either water-based or solvent-based inks. The materials are resistant to most organic liquids, including solvents, and to mineral oils. Film can be used to package products having low water contents but not high ones.[252]

12.8.2 Polyoxyethylene

Polyoxyethylene is a water-soluble polymer which has been known for a number of years but has seen little use. Reports about its biodegradability disagree. Some experts claim the polymer is inherently biodegradable, while others state it only biodegrades at molecular weights under

500. Polyoxyethylene was available in the United States from Planet Polymer Technologies through Mitsubishi in 1993.[256]

12.9 Summary

Recycling of plastics continues to grow around the world, although it has had some setbacks in the United States. Financial support from the plastics industry has declined in the United States, making it crucial that recycling programs be economically viable to survive. Developments in further mechanization of plastics sorting can lower operating costs, but sometimes require a financial investment that is out of reach. Regulatory pressure for recycling of all types has decreased in the United States in the last several years, but there is the potential for it to rapidly increase again if the public perceives a significant decline in the availability of recycling. In several other areas of the world, including the European Union, both recycling rates and regulatory pressure continue to increase.

The philosophy of producer responsibility seems to have gained a strong position in Europe, and is making substantial impact in Canada, Asia, and Latin America. The first rumblings of this movement have also been heard in the United States, primarily at the state level. There are serious consequences for industry in states continuing to take the lead in waste-related areas. In addition to having to make its case in 50 places rather than one, industry always runs the risk that one state may prohibit exactly the thing that another state requires.

Use of life-cycle assessment techniques to analyze material choices, processes, and waste disposal continues to increase. Some countries in Europe require life-cycle analysis before products are introduced. The U.S. EPA and the Department of Energy have jointly sponsored research to develop the tools and information needed for life-cycle analysis-based decisions about solid waste management strategies. The results of this project have already undergone peer review by experts, and are scheduled to be released in 2000. This study includes both economic and environmental aspects, and will have relevance internationally as well as in the United States.[257]

On a related note, concerns about greenhouse warming may lead to significantly higher prices for fossil fuel or even to limitations on energy use. To the extent that the plastics industry can take steps ahead of time to modify operations in ways that make them more energy efficient, the industry will be better prepared for whatever action results. Further, if it is determined that no action need be taken, the industry will still reap the economic benefits of lower fuel expenditures.

As composting of source-separated organics becomes more common, the potential markets for biodegradable plastics are likely to increase.

How much actual markets increase will be a function of cost. To the extent that biodegradable plastics become cost competitive with commodity plastics, their use will grow. If their prices do not decrease significantly, they are likely to continue to have a role only in minor niche markets. Of course, regulatory pressure could change this as well.

References

1. Glenn, J., "The State of Garbage in America, Part I," *BioCycle,* April 1998, pp. 32–43.
2. U.S. Environmental Protection Agency, "Characterization of Municipal Solid Waste in the United States: 1997 Update," EPA 530-R-98-007, May 1998.
3. U.S. Environmental Protection Agency, "Updated Air Quality Standards: Fact Sheet," Washington, D.C., 1997.
4. Toloken, Steve, "EPA to revise styrene emission standards," *Plastics News,* July 13, 1998, p. 23.
5. Toloken, Steve, "Nike Nixing its Vinyl Use," *Plastics News,* August 31, 1998, p. 5.
6. Intergovernmental Panel on Climate Change, *IPCC Second Assessment Synthesis of Scientific-Technical Information Relevant to Interpreting Article 2 of the UN Framework Convention on Climate Change,* 1995, United Nations, New York, 1995.
7. Thompson Publishing Group, *Environmental Packaging: U.S. Guide to Green Labeling, Packaging and Recycling,* Washington, D.C., 1998.
8. Raymond Communications, *State Recycling Laws Update,* College Park, Md., 1998.
9. "Oregon," *Resource Recycling,* February 1996, p. 10.
10. "Plastic Recycling Rate in California," *Reuse/Recycle,* April 1998, p. 30.
11. "German Green Dot System: Putting Responsibility on Industry," *BioCycle,* June 1993, pp. 60–63.
12. Myers, John, "Recyclers of Appliances, Durables Looking to Germany's Proposals," *Modern Plastics,* March 1995, pp. 14–15.
13. Canadian Council of Ministers of the Environment, *National Packaging Protocol,* CCME-TS/WM-FS 020, Toronto, Ontario, Canada, 1990.
14. "Packaging Waste in Canada Cut by Half," *Plastics News,* March 2, 1998, p. 13.
15. Gies, Glenda, "The State of Garbage in Canada," *BioCycle,* February 1996, pp. 46–52.
16. Majid, Mohn Isa Abd, "Restricting the Use of Plastic Packaging," *Professional Bulletin of the National Poison Centre,* Malaysia, June 1997, pp. 1–4, available online at *prn.usm.my/ bulletin/1997/prn13.html.*
17. "Mandatory Recycling Coming to Israel," *BioCycle,* February 1998, p. 71.
18. Moore, S., "Plastics Recycling Profit Soars in India," *Modern Plastics,* June 1995, pp. 19–21.
19. Higgs, R., "New Technology to Boost European Recycling," *Plastics News,* June 16, 1997, p. 56.
20. U.S. Food and Drug Administration, *Recommendations for Chemistry Data for Indirect Food Additive Petitions,* Center for Food Safety & Applied Nutrition, Washington, D.C., 1995.
21. Powell, J., "The PRFect Solution to Plastic Bottle Recycling," *Resource Recycling,* February 1995, pp. 25–27.
22. "Plastics Recyclers Stay on the Cutting Edge," *BioCycle,* May 1996, pp. 42–46.
23. Colvin, R., "Sorting Mixed Polymers Eased by Hand-Held Unit," *Modern Plastics,* April 1995, p. 34.
24. Stambler, I., "Plastic Identifiers Groomed to Cut Recycling Roadblocks," *R&D Magazine,* October 1996, pp. 29–30.
25. Smith, S., "PolyAna System Identifies Array of Plastics," *Plastics News,* December 15, 1997, p. 14.
26. Apotheker, S., "Flake and Shake, Then Separate," *Resource Recycling,* June 1996, pp. 20–27.

27. Smith, S., "Fry-Sorting Process Promising in Post-Consumer Plastics Use," *Plastics News,* November 11, 1996, p. 27.
28. Schut, J. H., "Process for Reclaiming Durables Takes Off in U.S.," *Modern Plastics,* March 1998, pp. 56–57.
29. Maten, A., "Recovering Plastics from Durable Goods: Improving the Technology," *Resource Recycling,* September 1995, pp. 38–43.
30. Schut, J. H., "Centrifugal Force Puts New Spin on Separation," *Plastics World,* January 1996, p. 15.
31. "Packaging Research Case Study," *Management & Technology,* June 1996, p. 23.
32. King, R., "U.S. Exports of Waste Plastics Climb," *Plastics News,* March 25, 1996, p. 1.
33. Anderson, P., S. Kelly, and T. Rattray, "Redesigning for Recycling," *BioCycle,* July 1995, pp. 64–65.
34. Renstrom, Roger, "Competition Fosters Design for Recycling," *Plastics News,* October 5, 1998, p. 19.
35. Hersch, Paul, "Honda's Technology Allows Fully Recyclable Instrument Panels," *Solid Waste Online,* December 9, 1997, available online at *www.solidwaste.com.*
36. Goldsberry, C., "OEMs Undertake Quest for Computer Afterlife," *Plastics News,* January 13, 1997, p. 9.
37. Grande, J. A., "Computer Manufacturers Make In-Roads in Use of Recyclate," *Modern Plastics,* November 1995, pp. 35–39.
38. "Japan Firms Reusing Plastics in Electronics," *Plastics News,* October 5, 1998, p. 49.
39. Denison, R. A., *Something to Hide: The Sorry State of Plastics Recycling,* Environmental Defense Fund, Washington, D.C., 1997.
40. Association of Plastics Manufacturers in Europe available online at *www.apme.org,* 1998.
41. RECOUP, "Plastic Bottle Recycling Statistics 1997," available online at *www.tecweb.com / recoup.*
42. Bilitewski, B. and C. Copeland, "Packaging Take-Back in Germany: The Plastics Recycling Picture," *Resource Recycling,* February 1997, pp. 46–52.
43. Toloken, Steve, "Contaminants Seep into Rates for Recycling," *Plastics News,* August 31, 1998, p. 1.
44. Babinchak, S., "Current Problems in PET Recycling Need to be Resolved," *Resource Recycling,* October 1997, pp. 29–31.
45. U.S. Environmental Protection Agency, "Characterization of Municipal Solid Waste in the United States: 1994 Update," EPA 530-R-94-042, November 1994.
46. U.S. Environmental Protection Agency, "Characterization of Municipal Solid Waste in the United States: 1996 Update," EPA 530-R-97-007, May 1997.
47. Toloken, Steve, "Plastic Bottle Recycling Rate Keeps Sliding," *Plastics News,* August 24, 1998, p. 1.
48. National Association for PET Container Resources, available online at *www.napcor.com.*
49. "Group Seeks Recycler of Uncommon Bottles," *Plastics News,* July 27, 1998, p. 15.
50. Barham, Vernon F., "Closing the Loop for P.E.T. Soft Drink Containers," *Journal of Packaging Technology,* January/February 1991, pp. 28–29.
51. Bakker, Marilyn, "Using Recycled Plastics in Food Bottles: The Technical Barriers," *Resource Recycling,* May 1994, pp. 59–64.
52. Newcorn, David, "Plastics' Broken Loop," *Packaging World,* June 1997, pp. 22–24.
53. "Process Yields Recycled PET in Food Packaging," *Modern Plastics,* November 1994, p. 13.
54. Smith, Sarah S., "Precision Using PET as Recycling Alternative," *Plastics News,* January 12, 1998, p. 10.
55. Toloken, Steve, "Campaign Asks Coke to Use Recycled PET," *Plastics News,* September 14, 1998, p. 6.
56. Pryweller, Joseph, "Lear Corp. Offers Automakers Recycled Plastic for Vehicles," *Plastics News,* September 21, 1998, p. 21.
57. Toloken, Steve, "Recyclers Worried Over Opaque Milk Bottles," *Plastics News,* October 27, 1997, p. 7.
58. "Ecoplast Gets FDA Nod," *Plastics News,* April 6, 1998.
59. DuPont, available online at *www.dupont.com / tyvek.*

60. Renstrom, R., "Disabled Employees Recycle Bags, Hangers," *Plastics News,* March 25, 1996, p. 21.
61. Apotheker, S., "Film at 11; A Picture of Curbside Recovery Efforts for Plastic Bags," *Resource Recycling,* May 1995, pp. 35–40.
62. "Montreal-Area Citizens Begin Film Program," *Plastics News,* February 12, 1996, p. 16.
63. Environment and Plastics Industry Council, "EPIC Moving Ahead on Several Initiatives," available online at *www.plastics.ca.*
64. Vinyl Institute, available online at *www.vinylinfo.org,* 1998.
65. The PVC Centre, European Vinyl Corporation, available online at *www. ramsay.co.uk/pvc/pvcenvir. htm,* 1998.
66. Smith, Sarah S., "APR Urges Vinyl Institute to Find Markets," *Plastics News,* July 14, 1997, p. 6.
67. Toloken, Steve, "PVC Bottle Recyclers Chastise Vinyl Institute," *Plastics News,* August 18, 1997, p. 36.
68. Toloken, Steve, "Recycling Program in the Works for PVC," *Plastics News,* August 31, 1998, p. 5.
69. "Japanese Packagers to Phase Out PVC Bottles," *Modern Plastics,* March, 1998, p. 20.
70. Wisner, D'Lane, "Recycling Post-Consumer Durable Vinyl Products," presented at The World Vinyl Forum, September 7–9, 1997, Akron, Ohio.
71. Alaisa, C, "Giving a Second Life to Plastic Scrap," *Resource Recycling,* February 1997, pp. 29–31.
72. Vinyl Institute, "In France, Vinyl Automotive Recycling Proving Successful," *Environmental Briefs,* August 1998, available online at *www.vinylinfo.org.*
73. Ford, T., "Nortel May Add to Its Recycling Process," *Plastics News,* December 18, 1995, p. 10.
74. Kiser, J., "Hospital Recycling Moves Ahead," *BioCycle,* November 1995, pp. 30–33.
75. Moore, Stephen, "Plastics Recycling Profit Soars in India," *Modern Plastics,* June 1995, pp. 19–21.
76. Ehrlich, Raymond J., "The Economic Realities of Recycling," *Polystyrene News,* Fall 1997, pp. 1, 3–4.
77. Toloken, Steve, "NPRC to Shut Failing PS Recycling Plant," *Plastics News,* August 4, 1997, pp. 1, 39.
78. Canadian Polystyrene Recycling Association, "CPRA Collection Program Update," *News from Canada's First Polystyrene Recycling Facility,* Mississaugua, Ontario, Spring 1994.
79. "A&E Starts Program to Rescue Hangers," *Plastics News,* December 2, 1996, p. 9.
80. "Post-Consumer EPS Foam Packaging Third Most Recycled Plastic Product," *Packaging Technology & Engineering,* November/December 1994, p. 16.
81. Alliance of Foam Packaging Recyclers, press release, Crofton, Md., December 12, 1994.
82. de Campos, Betsy, "More to the PS Story," *Resource Recycling,* January 1998, p. 10.
83. Alliance of Foam Packaging Recyclers, available online at *www.presstar.com/afpr.*
84. Hornberger, Lee and Timothy Hight, "How Recycled Content Affects Foam Cushioning Performance," *Packaging Technology & Engineering,* April, 1998, pp. 48–51.
85. "Recycled PS Insulation Is No More," *Resource Recycling,* April 1997, p. 59.
86. "Malaysia Establishes EPS Recycling Group," *Modern Plastics,* June, 1996, p. 25.
87. Japan Information Network, available online at *www.jinjapan.org/stat/stats/19ENV51.html,* 1998.
88. Tilley, K., "Aussies Upgrade EPS Program," *Plastics News,* November 11, 1996, p. 25.
89. UK EPS Information Service, available online at *www1.mailbox.co.uk/www.eps.co.uk,* 1998.
90. Alliance of Foam Packaging Recyclers, "EPS Recycling—What's Next," *Molding the Future,* October, 1998, p. 1.
91. "Eastman Kodak Recycles 50 Million Cameras," *Plastics News,* August 14, 1995, p. 10.
92. "Snapshot of Recycling," *BioCycle,* June 1996, p. 88.
93. Myers, S., "Recyclers of Appliances, Durables Looking to Germany's Proposals," *Modern Plastics,* March 1995, pp. 14–15.

94. "A Noncatalytic Process Reverts Polystyrene to Its Monomer," *Chemical Engineering,* February 1993, p. 19.
95. Battery Council International, "Lead-Acid Batteries Head List of Recycled Products," press release, Chicago, Ill., November 12, 1998.
96. Heil, K., and R. Pfaff, "Quality Assurance in Plastics Recycling by the Example of Polypropylene; Report on the Experience Gathered with a Scrap Battery Recycling Plant," in F. LaMantia, ed., *Recycling of Plastic Materials,* ChemTec Publications, Ontario, 1993, pp. 171–185.
97. "Murphy Scrubs Virgin PP Bottle in Favor of PCR," *Packaging Digest,* March 1995.
98. "Recovering Dye Tubes," *BioCycle,* May 1996, p. 25.
99. U.S. Department of Energy, "Waste Carpet Recycling," available online at *www.oit.doe.gov/chemicals/citar96/CITAR96p31.htm,* 1998.
100. Texas Society of Professional Engineers, "Research into Chemical Recycling Could Open New Opportunities," available online at *www.tspe.org/recycle-6.htm.*
101. BASF, "6ix Again®: Technology That Sustains the Earth," available online at *www.basf.com/businesses/fibers/sixagain/index2.html,* 1998.
102. AlliedSignal, Inc., press release, Morristown, N.J., June 25, 1998.
103. "Ford Parts Incorporate Reclaimed Nylon Carpet," *Modern Plastics,* March 1997, p. 14.
104. Toray, available online at *www.toray.co.jp/e/kankyou/risai.html,* 1998.
105. Federal Prison Industries, Inc., available online at *www.unicor.gov/_vti_bin/shtml.exe/unicor/environtextiles.html/map,* 1998.
106. Recycline, available online at *www.recycline.com/recinfo.html,* 1998.
107. BASF Corporation Philip Services Corporation, press release, Mount Olive, N.J., September 16, 1997.
108. Higgs, R., "ICI to Recycle PU Foam Waste in U.K.," *Plastics News,* October 20, 1997, p. 24.
109. Duranceau, C. M., G. R. Winslow, and P. Saha, "Recycling of Automotive Seat Foam: Acoustics of Post Consumer Rebond Seat Foam for Carpet Underlayment Application," Society of Automotive Engineers, 1998, available online at *www.polyurethan.org/PURRC/REPORT1/ index.html.*
110. "Bayer Introduces Recycling Processes," *Plastics News,* January 13, 1997, p. 11.
111. White, K., "GE Plastics Begins Buyback Program Aimed at Plastic Auto Scrap," *Waste Age's Recycling Times,* March 7, 1995, p. 8.
112. "CD Recycling Plant is Europe's First," *Modern Plastics,* September 1995, p. 13.
113. British Telecommunications plc, available online at *www.bt.co.uk/corpinfo/enviro/phones.htm,* 1998.
114. Texas Society of Professional Engineers, available online at *www.tspe.org/recycle-4.HTM,* 1998.
115. Graff, G., "Additive Makers Vie for Reclaimed Resin Markets," *Modern Plastics,* February 1996, pp. 51–53.
116. Ford, T., "Hewlett-Packard Printers Use Recycled ABS," *Plastics News,* August 7, 1995, p. 40.
117. Bisio, A. L., and M. Xanthos, eds., *How to Manage Plastics Waste: Technology and Market Opportunities,* Hanser Publications, Munich, 1994.
118. Van Ness, K. E., and T. J. Nosker, "Commingled plastics," in R. Ehrig, ed., *Plastics Recycling: Products and Processes,* Hanser Publications, Munich, 1992, pp. 187–229.
119. "Electrolux Is Putting Recyclate in Vacuums," *Modern Plastics,* January 1997, p. 22.
120. Mapleston, P., "Housing May Be Built from Scrap in Low-Pressure Process," *Modern Plastics,* September 1995, p. 21.
121. "Five Standard Test Methods on Plastic Lumber Approved," *ASTM Standardization News,* November 1997, p. 11.
122. "Plastic Pallets Rival Wood, Steel Models," *Plastics News,* November 17, 1997, p. 30.
123. Bregar, B., "Plastic Rail Ties Gaining Favor," *Plastics News,* May 11, 1998.
124. Urey, C., "Uncle Sam Recruits Recycled Plastic Lumber," *Plastics News,* July 13, 1998, p. 4.
125. Solomon-Hess, J., "Process for Plastic-Paper Mix Tested," *Plastics News,* December 4, 1995, p. 13.
126. "Boise Cascade Eyes Film-Wood Composite." *Plastics News,* October 5, 1998, p. 11.
127. Bregar, B., "AERT Wins Major Contract for Its Recycled Material," *Plastics News,*

May 20, 1991, p. 1.
128. "Wood-Fiber Composite Is Now Commercial," *Modern Plastics,* May 1997, p. 16.
129. Lavendel, Brian, "Recycled Wood and Plastic Composites Find Markets," *BioCycle,* December 1996, pp. 39–43.
130. Urey, C., "Wood Composites Make Show at Meeting," *Plastics News,* September 29, 1997, p. 9.
131. Urey, Craig, "Mikron Invests in Wood-Plastic Composite," *Plastics News,* August 17, 1998, p. 8.
132. Ford, Tom, "Nortel May Add to Its Recycling Process," *Plastics News,* December 19, 1995, p. 10.
133. White, K., "Research Center to Demonstrate New Commingled Plastic Processing System," *Waste Age's Recycling Times,* October 4, 1994, p. 9.
134. "Waste Plastic Replaces Heavy Oil," *The National Environmental Journal,* July/August 1995, p. 10.
135. "BASF Cancels Plans for Feedstock Recycling Plant," *C&EN,* August 14, 1995, p. 10.
136. Wilt, C., and L. Kincaid, "There Auto Be a Law: End-of-Life Vehicle Recycling Policies in 21 Countries," *Resource Recycling,* March 1997, pp. 42–50.
137. Pryweller, J., "Ford Driving Recycled Nylon Applications," *Plastics News,* February 24, 1997, pp. 7, 9.
138. Pryweller, J., "Cost Is King in Auto-Related Recycling," *Plastics News,* September 8, 1997, pp. 1, 8.
139. Pryweller, J., "Recycling Center Pitching PU to Carmakers," *Plastics News,* September 22, 1997, p. 35.
140. Sherman, L. M., "Compounders Take the Lead in Post-Use Bumper Recycling," *Plastics Technology,* March 1996, pp. 27–29.
141. Grande, J. A., "Ford Is Targeting 50% Use of Recycle-Content Resin by 2002," *Modern Plastics,* July 1996, pp. 32–33.
142. Pryweller, J., "Projects Could Turn Plastics into a Recycling `Headliner,'" *Plastics News,* February 23, 1998, pp. 2, 9.
143. Powell, J., "Magic Carpet Ride: The Coming of a New Recyclable," *Resource Recycling,* April 1997, pp. 42–46.
144. Monsanto, available online at *www.floorspecs.com/reference_Library/Monsanto_Recycling/index.htm,* 1998.
145. United Recycling, available online at *www.cais.net/publish/stories/0596haz9.htm,* 1998.
146. "Rolling Out the Red Carpet for Recycling," *BioCycle,* February 1998, pp. 12–13.
147. "Arco Chemical Devises SMA Recycling Process," *Plastics News,* March 11, 1996, p. 6.
148. "British Project for Recycling Polymer Composites Launched," *C&EN,* June 19, 1995, p. 11.
149. "Thermoset Reclaim Bets on Chemical Process," *Modern Plastics,* April 1995, p. 13.
150. Higgs, R., "ICI, Mitsubishi Research Acrylic Recycling," *Plastics News,* May 26, 1997, p. 58.
151. U.S. Federal Trade Commission, "Guides for the Use of Environmental Marketing Claims," 16 CFR Part 260.
152. American Society for Testing and Materials, "Standard Test Method for Determining Aerobic Biodegradation of Plastic Materials Under Controlled Composting Conditions," ASTM D5338-92, West Conshohocken, Pa., 1992.
153. "New Standard for Biodegradable Plastic," available online at *www.british-dgtip.de/technology/98i5052.htm.*
154. Riggle, D., "Moving Towards Consensus on Degradable Plastics," *BioCycle,* March 1998, pp. 64–70.
155. *Bio/Environmentally Degradable Polymer Society News,* Winter 1998, available online at *www.bedps.org/newslet.html.*
156. Degradable Polymers Council, "SPI's Degradable Polymers Council Recognizes Standards for Biodegradable/Compostable Plastic Bags," press release, Washington, D.C., February 13, 1998.
157. Natick Research Development and Engineering Center, "Performing a Balancing Act: NRDEC Creates Biodegradable Materials That Perform," available online at *www-sscom.army.mil/warrior/97/jan/bio.htm.*

158. Glenn, Jim, "Solid Waste Composting Trending Upward," *BioCycle,* November 1998, pp. 65–72.
159. Tyler, Rod, "Breaking Down Biodegradable Bags," *BioCycle,* December 1997, pp. 54–55.
160. Aquino, John T., "Composting: The Next Step?" *Waste Age,* March 1996, pp. 47–58.
161. Ballister-Howells, Pegi, "Major Waste Management Firms Become Composters," *BioCycle,* June 1996, pp. 35–38.
162. "Mandatory Food Composting Proposed," *Resource Recycling,* March 1998, p. 13.
163. Walker, M., "So—You're Thinking of Composting," *Resource Recycling,* June 1998, pp. 37–40.
164. "Montreal, Canada: Collecting Food Residuals and Yard Trimmings," *BioCycle,* November 1997, p. 20.
165. Barth, J., and Kroeger, B., "Marketing Compost in Europe," *BioCycle,* October 1998, pp. 77–78.
166. de Bertoldi, Marco, "Composting in the European Union," *BioCycle,* June 1998, pp. 74–75.
167. Evans, G., "Keeping Organics Out of Landfills," *BioCycle,* October 1998, pp. 72–74.
168. Holland, Biona, and Alec Proffitt, "Overview of Composting in the U.K.," *BioCycle,* February 1998, pp. 69–71.
169. Scheinberg, Anne, "Going Dutch: Collecting Residential Organics in the Netherlands," *Resource Recycling,* January 1996, pp. 33–40.
170. Johnson, Regina, "An SPI Overview of Degradable Plastics," *Proceedings of Symposium on Degradable Plastics,* The Society of the Plastics Industry, Inc., Washington, D.C., June 10, 1987, pp. 6–12.
171. Mannon, J. H., "British Route to Polymer Hinges on Bacteria," *Chemical Engineering,* May 4, 1981, p. 41.
172. "A New Biodegradable Plastic," *European Packaging Newsletter,* vol. 20, no. 11, November 1987, pp. 3–4.
173. Leaversuch, Robert, "Industry Weighs Need to Make Polymer Degradable," *Modern Plastics,* August 1987, pp. 52–55.
174. Keeler, Robert, "Plastics Grown in Bacteria Inch Toward the Market," *R&D Magazine,* January 1991, pp. 46–52.
175. Poirier, Y., D. Dennis, K. Klomparens, and C. Somerville, "Polyhydroxybutyrate, A Biodegradable Thermoplastic, Produced in Transgenic Plants, *Science,* vol. 256, 1992, pp. 520–522.
176. "Making Plastics (and Other Fine Things) from Plants," National Science Foundation, available online at *www.nsf.gov/bio/pubs/arabid/arab-pl.htm.*
177. "Biotechnology Patent Could Lead to Polyester Harvests," *C&EN,* July 29, 1996, p. 40.
178. "Can Genetically Engineered Plants Produce Polymers?" *Modern Plastics,* November 1997, p. 16.
179. "ICI Begins North American Sales of Biodegradable Plastic," *C&EN,* June 15, 1992, p. 8.
180. Graff, G., "Firm Launches Degradables While Others Are Bailing Out," *Modern Plastics,* April 1994, p. 28.
181. Smith, S. S., "Biopol Deal Puts Monsanto in Plants-to-Plastic Venture," *Plastics News,* May 6, 1996, p. 30.
182. "English Bank Opts for 'Green' Credit Card," *Modern Plastics,* July 1997, p. 14.
183. Urey, C., "Greenpeace Supports Plastic Visa Card: Monsanto's Biopol Resin Replacing PVC in Credit Cards," *Plastics News,* June 2, 1997, p. 4.
184. Technical University of Denmark, "Razor, Biodegradable," available online at *www.ipt.dtu.dk/*~tl/inspsite/htmsider/k008.htm.
185. Faulkner & Gray, "Canada Trust Goes Green" available online at *ccm.faulknergray.com,* December 10, 1998.
186. "Monsanto Disposes Biodegradable Plastics," *Chemical Week Executive Edition,* December 4, 1998, available online at *www.chemweek.com.*
187. Metabolix, available online at *www.metabolix.com.*
188. Wu, C., "Weight Control for Bacterial Plastic," *Science News,* vol. 151, no. 23, 1997.
189. "High-Strength Biodegradable Plastic Developed," *Alexander's Gas & Oil Connections,* available online at *www.gasandoil.com/goc/features/fex74412.htm,* September 18, 1997.

190. "Cheap Biodegradable Polymer for Packaging," *New Materials Japan,* March 1990, p. 95.
191. Byrom, David, "Polyhydroxyalkanoates," in David Mobley, ed., *Plastics from Microbes: Microbial Synthesis of Polymers and Polymer Precursors,* Hanser Publications, Munich, 1994, pp. 5–33.
192. "Boycott 'Degradable' Plastics," *Environmental Action,* January/February 1990, p. 32.
193. Wilder, R. V., "Degradable Resin is 100% Starch-Based," *Modern Plastics,* March 1990, pp. 22–24.
194. "Degradable Additives for Plastic Compost Bags," *BioCycle,* March 1995, pp. 77–78.
195. Lauzon, Michael, "Novon Degradable Polymer Pleasing to Dogs' Palates," *Plastics News,* October 14, 1996, p. 26.
196. "Churchill Seeks Protection from Its Creditors," *Plastics News,* January 6, 1997, p. 8.
197. StarchTech, Inc., available online at *www.usda.gov/agency/aarc/srbk/3200.html.*
198. "Starch-Based Biodegradable Plastic Film Developed," available online *www.hindubusinessline.com/bline/1996/05/27/BLFP15.html.*
199. "Council Demonstrates U.S. Corn-Based Biodegradables at Olympics," U.S. Feed Grains Council, press release, Washington, D.C., December 8, 1997.
200. Tilley, Kate, "Researchers Seeking Degradable Plastics," *Plastics News,* June 26, 1995, p. 8.
201. BPS, Biodegradable Plastics Society, fact sheet, undated.
202. *Bio/Environmentally Degradable Polymer Society News,* Winter 1998, available online at *www. bedps.org/newslet.html.*
203. "Loose-Fill an Environmentalist Can Love," *Packaging Digest,* April 1991, pp. 44–46.
204. American Excelsior Company, "Eco-Foam, The Original All Natural Starch Based Loose Fill," Arlington, Tex., undated.
205. "Clean Green Packing Product Summary," Clean Green Packing, Minneapolis, Minn., undated.
206. FP International, "Recycled-Content Polystyrene Loosefill and Starch-Based Loosefill," available online at *fpintl.com/loosefill%20+.htm.*
207. "Dispensing System Material Biodegradable," *Plastics News,* December 23, 1996, p. 7.
208. King, Robert, "Enpac Combines Loosefill, Foamed-in Methods," *Plastics News,* December 23, 1996, p. 20.
209. Kim, Y. H., "Monthly Situation Analysis Report," U.S. Feed Grains Council, available online at *www.grains.org/veg/kmar1997.htm.*
210. "OK Compost Label Given to Biodegradable Bag," *BioCycle,* July 1996, p. 18.
211. "Biodegradable Bag Pilot," *BioCycle,* July 1996, p. 21.
212. Colvin, Robert, "Biodegradable Polymers Make Small-Scale Return," *Modern Plastics,* April 1995, p. 17.
213. "Italy's Montedison Begins Marketing Biodegradable Film," *C&EN,* January 22, 1996, p. 11.
214. Biocorp USA, "Biocorp Trash Bags: A Living Chemistry," available online at *www.biocorpusa.com/html/living_resource.html.*
215. Garnham, Peter, "Slow Progress for Biodegradable Plastics," *BioCycle,* February 1997, pp. 64–67.
216. Lindsay, Karen, " 'Truly Degradable' Resins Are Now Truly Commercial," *Modern Plastics,* February 1992, pp. 62–64.
217. Reis, R. L., A. M. Cunha, and M. J. Bevis, "Using Nonconventional Processing to Develop Anisotropic and Biodegradable Composites of Starch-Based Thermoplastics Reinforced with Bone-like Ceramics," *Medical Plastics and Biomaterials,* November 1997, available online at *www. devicelink.com.*
218. " 'All Natural' Packaging Material Debuts," *Packaging Strategies,* October 31, 1996, p. 3.
219. "Composite of Natural Materials Developed for Fast-Food Containers," *C&EN,* October 28, 1996, p. 29.

220. Miles, Irene, "Researchers Develop Corn-Based Plastic," *University of Illinois ACES News,* August 7, 1997, available online at *spectre.ag.uiuc.edu/news/articles/871669331.html.*
221. Esposito, Frank, "Scientists Turn Corn into Plastic Packaging," *Plastics News,* June 16, 1997, p. 73.
222. Selke, S., *Biodegradation and Packaging,* 2d ed., Pira International Reviews of Packaging, Surrey, United Kingdom, 1996.
223. United Soybean Board, "Plastic Film Made Out of Soybeans," available online at *www.ag.uiuc.edu/~usb/newsletters/fsv313g.html.*
224. Chang, Sam, "Chemical Modifications of Soybeans for Making Biodegradable Films: A Preliminary Study," North Dakota Soybean Council, available online at *www.ag.uiuc.edu/~nd-qssb/chang2_97.html.*
225. Jane, Jay-lin, "Soy Protein Plastics: A Non-Petrochemical Alternative," *Stratsoy,* available online at *stratsoy.ag.uiuc.edu.*
226. Daniell, H., S. Zhang, B. Guda, and D. Urry, "Plastics from Plants: AAES Researchers Implant Tobacco Plants with Synthetic Biodegradable Polymer Genes," available online at *www.ag.auburn.edu/aaes/information/highlights,* fall 1995.
227. "Thermoset Amino Protein Resin Is Said to Degrade Naturally," *Modern Plastics,* May 1996, p. 36.
228. "New in Biodegradable Plastics," *European Packaging Newsletter and World Report,* July 1990, p. 8.
229. Meister, J. J., M. J. Chen, and F. F. Chang, "Make Polymers from Biomass," *Chemtech,* July 192, pp. 430–435.
230. Colvin, Robert, "For Molders Seeking a Niche, There's Melt-Processable Wood," *Modern Plastics,* November 1996, pp. 38–40.
231. "New Biodegradable Plastic from Trees," *Plastics Distributor & Fabricator Magazine,* available online at *www.plasticsmag.com,* 1997.
232. Kharas, G. B., F. Sanchez-Rivera, and D. K. Severson, "Polymers of Lactic Acids," in David Mobley, ed., *Plastics from Microbes: Microbial Synthesis of Polymers and Polymer Precursors,* Hanser Publications, Munich, 1994, pp. 93–137.
233. "Transforming Those Wastes into Plastic," *Food Engineering,* December 1990, p. 58.
234. Ryan, C. M., M. H. Hartmann, and J. F. Nangeroni, "Poly(lactic Acid): Increasing Its Melt Strength for Extrusion Coating," *Packaging Technology & Engineering,* March 1998, pp. 39–47.
235. Thayer, Ann, "Polylactic Acid Is Basis of Dow, Cargill Venture," *C&EN,* December 8, 1997, pp. 14–16.
236. Cargill, "Cargill Developing Degradable Polymers Made from Corn," press release, Minneapolis, Minn., October 15, 1991.
237. Esposito, Frank, "Cargill, Dow Team Up to Make Biopolymers," *Plastics News,* December 1, 1997, p. 5.
238. Leaversuch, Robert, "Polylactic Acid Venture Set to Ferment Biopolymer Use," *Modern Plastics,* February 1998, pp. 38–40.
239. Lauzon, Michael, "Duro Adding Compostable Bags," *Plastics News,* October 14, 1996, p. 14.
240. CornCard International, "Mazin," available online at *ianrwww.unl.edu/ianr/iapc/mazin.htm.*
241. Dainippon Ink and Chemicals, Inc., "Transparent and Flexible Biodegradable Plastics (CPLA)," available online at *www.dic.co.jp/green/index-e.html.*
242. DuPont, "Raised on a Diet of Plastic Cups, Snack Bags and Gum Wrappers," available online at *www.dupont.com/corp/biomax.html.*
243. Eastman, "Biodegradable Thermoplastics," available online at *www.eastman.com.*
244. Showa Highpolymer Co., Ltd., "Bionolle Biodegradable Aliphatic Polyester," Technical data sheet, Tokyo, Japan, undated.
245. WTEC, "Site Report, Showa High Polymer Co, Ltd," available online at *144.126.176.216/ biopoly/showa.htm.*
246. Bayer, "Polyester Amide—A New Biodegradable Plastic," press release, Leverkusen, Germany, July 1995.
247. Bayer, "Bayer Markets New Biodegradable Plastic," Leverkusen, Germany, press release, January 12, 1998.

248. Bayer, available online at *www.bayer.com.*
249. "Biodegradable Plastics Offer Tailored Rates of Degradation," *Modern Plastics,* January 1996, p. 95.
250. BASF, available online at *www.basf.com.*
251. Lo, F., J. Petchonka, and J. Hanly, J., "Water-Soluble Polymers: Trend Setters for the 21st Century?" *Chemical Engineering Progress,* July 1993, pp. 55–58.
252. ChrisCraft Industrial Products, Inc., available online at *www.monosol.com.*
253. Air Products, available online at *www.airproducts.com.*
254. WTEC, "Site Report, Nippon Gohsei," available online at *144.126.176.213/biopoly/gohsei.htm.*
255. Italway, "Hydrolene: The Only Plastic That Mother Nature Likes?" available online at *www.italway.it.*
256. McCarthy-Bates, L., "Biodegradables Blossom into Field of Dreams for Packagers," *Plastics World,* March 1993, pp. 22–27.
257. Thorneloe, Susan and Keith Weitz, "Tradeoffs of Integrated Waste Management Strategies," *Resource Recycling,* January 1999, pp. 30–33.

Appendix

A

Glossary of Terms and Definitions*

A-stage Stage in which thermosetting reactants are mixed, but at which the polymerization reaction has not yet begun.

adhesive Film or coating, such as a mold release, applied to a surface to prevent adhesion or sticking.

ablative plastics Plastics or resins, the surface layers of which decompose when surface is heated, leaving a heat-resistant layer of charred material. Successive layers break away, exposing a new surface. These plastics are especially useful in applications such as outer skins or spacecraft, which heat up to high temperatures on reentry into the earth's atmosphere.

abrasion resistance Capability of a material to withstand mechanical forces such as scraping, rubbing, or erosion, that remove material from the surface.

ABS plastics Abbreviated phrase referring to acrylonitrile-butadiene-styrene copolymers; elastomer-modified styrene.

absorption Penetration of one substance into the mass of another, such as moisture or water absorption of plastics.

accelerated test Test in which conditions are intensified to obtain critical data in shorter time periods, such as accelerated life testing.

accelerator Chemical used to speed up a reaction or cure. Term is often used interchangeably with promoter. For example, cobalt naphthanate is used to accelerate the reaction of certain polyester resins. An accelerator is often used along with a catalyst, hardener, or curing agent.

activation Process, usually chemical, of modifying a surface so that coatings will more readily bond to that surface.

activator Chemical material used in the activation process. (*See* **activation.**)

*Reprinted with permission from *Handbook of Plastics, Elastomers, and Composites*, 3d ed., Charles A. Harper (ed.), © McGraw-Hill, New York, 1996.

addition reaction of polymerization Chemical reaction in which simple molecules (monomers) are added to each other to form long-chain molecules without forming by-products.

additive Substance added to materials, usually to improve their properties. Prime examples are plasticizers, flame retardants, or fillers added to plastic resins.

adhere To cause two surfaces to be held together by adhesion.

adherend Body held to another body by an adhesive.

adhesion State in which two surfaces are held together by interfacial forces, which may consist of valence forces or interlocking action, or both.

adhesive Substance capable of holding materials together by surface attachment.

adhesive, anaerobic Adhesive that sets only in the absence of air; for instance, one that is confined between plates or sheets.

adhesive, contact Adhesive that is apparently dry to the touch and will adhere to itself instantaneously upon contact; also called contact-bond adhesive or dry-bond adhesive.

adhesive, heat-activated Dry adhesive film that is rendered tacky or fluid by application of heat or heat and pressure to the assembly.

adhesive, pressure-sensitive Viscoelastic material that in solvent-free form remains permanently tacky. Such material will adhere instantaneously to most solid surfaces with the application of very slight pressure.

adhesive, room-temperature-setting Adhesive that sets in the temperature range of 20 to 30°C (68 to 86°F), in accordance with the limits for standard room temperature as specified in ASTM Methods D 618, Conditioning Plastics and Electrical Insulating Materials for Testing.

adhesive, solvent Adhesive having a volatile organic liquid as a vehicle.

adhesive, solvent-activated Dry adhesive film that is rendered tacky just prior to use by application of a solvent.

aging Change in properties of a material with time under specific conditions.

air vent Small gap in a mold to avoid gases being entrapped in the plastic part during the molding process.

airless spraying High-pressure spraying process in which pressure is sufficiently high to atomize liquid coating particles without air.

alcohols Characterized by the fact that they contain the hydroxyl (—OH) group, they are valuable starting points for the manufacture of synthetic resins, synthetic rubbers, and plasticizers.

aldehydes In general, volatile liquids with sharp, penetrating odors that are slightly less soluble in water than the corresponding alcohols. The group (—CHO), which characterizes all aldehydes, contains the most active form of the carbonyl radical and makes the aldehydes important as organic synthetic agents. They are widely used in industry as chemical building blocks in organic synthesis.

aliphatic hydrocarbon *See* **hydrocarbon.**

allowables Statically derived estimate of a mechanical property based on repeated tests. Value above which at least 99 percent of the population of values is expected to fall with a confidence of 95 percent.

alloy Blend of polymers, copolymers, or elastomers under controlled conditions.

alpha particle Heavy particle emitted during radioactive decay consisting of two protons and two neutrons bound together. It has the lowest penetration of the various emitted particles and will be stopped after traversing through only a few centimeters of air or a very thin solid film.

ambient temperature Temperature of the surrounding cooling medium, such as gas or liquid, that comes into contact with the heated parts of the apparatus.

amine adduct Products of the reaction of an amine with a deficiency of a substance containing epoxy groups.

amorphous plastic Plastic that is not crystalline, has no sharp melting point, and exhibits no known order or pattern of molecule distribution.

anhydride Organic compound from which water has been removed. Epoxy resins cured with anhydride-curing agents are generally characterized by long pot life, low exotherm during cure, good heat stability, and good electrical properties.

antioxidant Chemical used in the formulation of plastics to prevent or slow down the oxidation of material exposed to the air.

antistatic agents Agents that, when added to the molding material or applied onto the surface of the molded object, make it less conducting (thus hindering the fixation of dust).

aramid Generic name for highly oriented organic material derived from a polyamide but incorporating an aromatic ring structure.

arc resistance Time required for an arc to establish a conductive path on the surface of an organic material.

areal weight Weight of fabric or prepreg per unit width.

aromatic amine Synthetic amine derived from the reaction of urea, thiourea, melamine, or allied compounds with aldehydes, that contains a significant amount of aromatic subgroups.

aromatic hydrocarbon *See* **hydrocarbon.**

aspect ratio Ratio of length to width for a flat form, or of length to diameter for a round form such as a fiber.

assembly Group of materials or parts, including adhesive, that has been placed together for bonding or that has been bonded together.

atactic State when the radical groups are arranged heterogeneously around the carbon chain. (*See also* **isotactic.**)

atomic oxygen resistance Ability of a material to withstand atomic oxygen exposure. This is related to the use of plastic and elastomers in space applications.

autoclave Closed vessel for conducting a chemical reaction or other operation under pressure and heat.

autoclave molding After lay-up, the entire assembly is placed in a steam autoclave at 50 to 100 lb/in^2. Additional pressure achieves higher reinforcement loadings and improved removal of air.

average molecular weight Molecular weight of most typical chain in a given plastic. There will always be a distribution of chain sizes and, hence, molecular weights in any polymer.

B-allowable Statistically derived estimate of a mechanical property based on numerous tests; value above which at least 90 percent of the population of values is expected to fall with a confidence of 95 percent.

B-stage Intermediate stage in the curing of a thermosetting resin. In this stage, resins can be heated and caused to flow, thereby allowing final curing in the desired shape. The term *A-stage* is used to describe an earlier stage in the curing reaction, and the term *C-stage* is sometimes used to describe the cured resin. Most molding materials are in the B-stage when supplied for compression or transfer molding.

bag molding Method of applying pressure during bonding or molding in which a flexible cover, usually in connection with a rigid die or mold, exerts pressure on the material being molded, through the application of air pressure or drawing of a vacuum.

bagging Appling an impermeable layer of film over an uncured part and sealing the edges so that a vacuum can be drawn.

Barcol hardness Hardness value obtained using a Barcol hardness tester, which gages hardness of soft materials by indentation of a sharp steel point under a spring load.

basket weave Weave where two or more warp threads cross alternately with two or more filling threads. The basket weave is less stable than the plain weave but produces a flatter and stronger fabric. It is also a more pliable fabric than the plain weave, and a certain degree of porosity is maintained without too much sleaziness, but not as much as with the plain weave.

benzene ring Basic structure of benzene, which is a hexagonal, six-carbon-atom structure with three double bonds; also basic aromatic structure in organic chemistry. Aromatic structures usually yield more thermally stable plastics than do aliphatic structures. (*See also* **hydrocarbon.**)

binder Resin or plastic constituent of a composite material, especially a fabric-reinforced composite.

blister Raised area on the incompletely hardened surface of a molding caused by the pressure of gases inside it.

bismaleimide Type of polyimide that cures by addition rather than a condensation reaction; generally of higher temperature resistance than epoxy.

blind fastener Fastener designed for holding two rigid materials, with access limited to one side.

blow molding Method of fabrication of thermoplastic materials in which a parison (hollow tube) is forced into the shape of the mold cavity by internal air pressure.

blowing agent Chemical that can be added to plastics and generates inert gases upon heating. This blowing, or expansion, causes the plastic to expand, thus forming a foam; also known as foaming agent.

bond Union of materials by adhesives; to unite materials by means of an adhesive.

bond strength Unit load, applied in tension, compression, flexure, peel, impact, cleavage, or shear, required to break an adhesive assembly, with failure occurring in or near the plane of the bond.

boron fibers High-modulus fibers produced by vapor deposition of elemental boron onto tungsten or carbon cores. Supplied as single strands or tapes.

boss Projection on a plastic part designed to add strength, to facilitate alignment during assembly, to provide for fastenings, and so on.

bridging Suspension of tensioned fiber between high points on a surface, resulting in uncompacted laminate.

bulk density Density of a molding material in loose form (granular, nodular, and the like), expressed as a ratio of weight to volume (for instance, g/cm^3 or lb/ft^3).

bulk rope molding compound Molding compound made with thickened polyester resin and fibers less than $^1/_2$ in. Supplied as rope, it molds with excellent flow and surface appearance.

capacitance (capacity) That property of a system of conductors and dielectrics that permits the storage of electricity when potential difference exists between the conductors. Its value is expressed as the ratio of the quantity of electricity to a potential difference. A capacitance value is always positive.

carbon, fibers Fiber produced by the pyrolysis of organic precursor fibers such as rayon, polyacrylonitrile (PAN), or pitch, in an inert atmosphere.

cast To embed a component or assembly in a liquid resin, using molds that separate from the part for reuse after the resin is cured. Curing or polymerization takes place without external pressure. (*See* **embed, pot.**)

catalyst Chemical that causes or speeds up the cure of a resin, but does not become a chemical part of the final product. Catalysts are normally added in small quantities. The peroxides used with polyesters are typical catalysts.

catalytic curing Curing by an agent that changes the rate of the chemical reaction without entering into the reaction.

caul plate Rigid plate contained within vacuum bag to impart a surface texture or configuration to the laminate during cure.

cavity Depression in a mold that usually forms the outer surface of the molded part; depending on the number of such depressions, molds are designated as single-cavity or multicavity.

centrifugal casting Fabrication process in which the catalyzed resin is introduced into a rapidly rotating mold where it forms a layer on the mold surfaces and hardens.

charge Amount of material used to load a mold for one cycle.

chlorinated hydrocarbon Organic compound having hydrogen atoms and, more important, chlorine atoms in its chemical structure. Trichloroethylene, methyl chloroform, and methylene chloride are chlorinated hydrocarbons.

circuit board Sheet of copper-clad laminate material on which copper has been etched to form a circuit pattern. The board may have copper on one (single-sided) or both (double-sided) surfaces. Also called printed-circuit board or printed-wiring board.

cleavage Imposition of transverse or "opening" forces at the edge of adhesive bond.

coat To cover with a finishing, protecting, or enclosing layer of any compound (such as varnish).

coefficient of thermal expansion (CTE) Change in unit length or volume resulting from a unit change in temperature. Commonly used units are 10^{-6} cm/cm/°C.

cohesion State in which the particles of a single substance are held together by primary or secondary valence forces. As used in the adhesive field, the state in which the particles of the adhesive (or the adherend) are held together.

cold flow (creep) Continuing dimensional change that follows initial instantaneous deformation in a nonrigid material under static load.

cold-press molding Molding process where inexpensive plastic male and female molds are used with room-temperature-curing resins to produce accurate parts. Limited runs are possible.

cold pressing Bonding operation in which an assembly is subjected to pressure without the application of heat.

compaction In reinforced plastics and composites, application of a temporary press bump cycle, vacuum, or tensioned layer to remove trapped air and compact the lay-up.

composite Homogeneous material created by the synthetic assembly of two or more materials (a selected filler or reinforcing elements and compatible matrix binder) to obtain specific characteristics and properties.

compound Some combination of elements in a stable molecular arrangement.

compression molding Technique of thermoset molding in which the molding compound (generally preheated) is placed in the heated open mold cavity and the mold is closed under pressure (usually in a hydraulic press), causing the material to flow and completely fill the cavity, with pressure being held until the material has cured.

compressive strength Maximum compressive stress a material is capable of sustaining. For materials that do not fail by a shattering fracture, the value is arbitrary, depending on the distortion allowed.

condensation polymerization Chemical reaction in which two or more molecules combine with the separation of water or other simple substance.

condensation resins Any of the alkyd, phenol-aldehyde, and urea-formaldehyde resins.

conductivity Reciprocal of volume resistivity.

conformal coating Insulating coating applied to printed-circuit-board wiring assemblies that covers all of the components and provides protection against moisture, dust, and dirt.

copolymer *See* **polymer.**

corona resistance Resistance of insulating materials, especially plastics, to failure under the high-voltage state known as partial discharge. Failure can be erosion of the plastic material, decomposition of the polymer, or thermal degradation, or a combination of these three failure mechanisms.

coupling agent Chemical or material that promotes improved adhesion between fiber and matrix resin in a reinforced composite, such as an epoxy-glass laminate or other resin-fiber laminate.

crazing Fine cracks that may extend in a network on or under the surface or through a layer of a plastic material.

creep Dimensional change with time of a material under load following the initial instantaneous elastic deformation; time-dependent part of strain resulting from force. Creep at room temperature is sometimes called cold flow. See ASTM D 674, Recommended Practices for Testing Long-Time Creep and Stress-Relaxation of Plastics under Tension or Compression Loads at Various Temperatures.

cross linking Process where chemical links set up between molecular chains of a plastic. In thermosets, cross linking makes one infusible supermolecule of all the chains, contributing to strength, rigidity, and high-temperature resistance. Thermoplastics (like polyethylene) can also be cross-linked (by irradiation or chemically through formulation) to produce three-dimensional structures that are thermoset in nature and offer improved tensile strength and stress-crack resistance.

crowfoot satin Type of weave having a 3-by-1 interlacing; that is, a filling thread floats over the three warp threads and then under one. This type of fabric looks different on one side than on the other. Such fabrics are more pliable than either the plain or the basket weave and, consequently, are easier to form around curves. (*See* **four-harness satin.**)

crystalline melting point Temperature at which the crystalline structure in a material is broken down.

crystallinity State of molecular structure referring to uniformity and compactness of the molecular chains forming the polymer and resulting from the formation of solid crystals with a definite geometric pattern. In some resins, such as polyethylene, the degree of crystallinity indicates the degree of stiffness, hardness, environmental stress-crack resistance, and heat resistance.

cull Material remaining in a transfer chamber after the mold has been filled. Unless there is a slight excess in the charge, the operator cannot be sure the cavity is filled. The charge is generally regulated to control the thickness of the cull.

cure To change the physical properties of a material (usually from a liquid to a solid) by chemical reaction, by the action of heat and catalysts, alone or in combination, with or without pressure.

curing agent *See* **hardener.**

curing temperature Temperature at which a material is subjected to curing.

curing time In the molding of thermosetting plastics, the time it takes for the material to be properly cured.

cycle One complete operation of a molding press from closing time to closing time.

damping In a material, the ability to absorb energy to reduce vibration.

decorative laminates High-pressure laminates consisting of a phenolic-paper core and a melamine-paper top sheet with a decorative pattern.

deflashing Any finishing technique used to remove the flash (excess unwanted material) from a plastic part; examples are filing, sanding, milling, tumbling, and wheelabrating. (*See* **flash.**)

deflection temperature Formerly called heat-distortion temperature (HDT).

degas To remove gases, usually air, from liquid resin mixture, usually achieved by placing mixture in a vacuum. Entrapped gases or voids in a cured plastic can lead to premature failure, either electrically or mechanically.

delamination Separation of the layers of material in a laminate, either locally or in a large area. Can occur during cure or later during operational life.

denier Numbering system for fibers or filaments equal to the weight in grams of a 9000-meter-long fiber or filament. The lower the denier, the finer the yarn.

design allowables Tested, statistically defined material properties used for design. Usually refers to stress or strain.

dessicant Substance that will remove moisture from materials, usually due to absorption of the moisture onto the surface of the substance; also known as a drying agent.

diallyl phythalate Ester polymer resulting from reaction of allyl alcohol and phthalic anhydride.

dielectric constant (permittivity, specific inductive capacity) That property of a dielectric that determines the electrostatic energy stored per unit volume for unit potential gradient.

dielectric loss Time rate at which electric energy is transformed into heat in a dielectric when it is subjected to a changing electric field.

dielectric-loss angle (dielectric-phase difference) Difference between 90° and the dielectric-phase angle.

dielectric-loss factor (dielectric-loss index) Product of the dielectric constant and the tangent of the dielectric-loss angle for a material.

dielectric-phase angle Angular difference in phase between the sinusoidal alternating potential difference applied to a dielectric and the component of the resulting alternating current having the same period as the potential difference.

dielectric-power factor Cosine of the dielectric-phase angle (or sine of the dielectric-loss angle).

dielectric sensors Sensors that use electrical techniques to measure the change in loss factor (dissipation) and in capacitance during cure of the resin in a laminate. This is an accurate measure of the degree of resin cure or polymerization.

dielectric strength Voltage that an insulating material can withstand before breakdown occurs, usually expressed as a voltage gradient (such as volts per mil).

differential scanning Calorimetry measurement of the energy absorbed (endotherm) or produced (exotherm) as a resin system is cured.

diluent Ingredient usually added to a formulation to reduce the concentration of the resin.

diphenyl oxide resins Thermoplastic resins based on diphenyl oxide and possessing excellent handling properties and heat resistance.

dissipation factor (loss tangent, loss angle, tan δ, approximate power factor) Tangent of the loss angle of the insulating material.

doctor blade Straight piece of material used to spread and control the amount of resin applied to roving, tow, tape, or fabric.

domes In a cylindrical container, that portion that forms the integral ends of the container.

dry To change the physical state of an adhesive on an adherend through the loss of solvent constituents by evaporation or absorption, or both.

drying agent *See* **dessicant.**

ductility Ability of a material to deform plastically before fracturing.

E-glass Family of glasses with low alkali content, usually under 2.0 percent, most suitable for use in electrical-grade laminates and glasses. Electrical properties remain more stable with these glasses due to the low alkali content. Also called electrical-grade glasses.

eight-harness satin Type of weave having a 7-by-1 interlacing; that is, a filling thread floats over seven warp threads and then under one. Like the crowfoot weave, it looks different on one side than on the other. This weave is more pliable than any of the others and is especially adaptable where it is necessary to form around compound curves, such as on radomes.

elastic limit Greatest stress a material is capable of sustaining without any permanent strain remaining when the stress is released.

elasticity Property of a material by virtue of which it tends to recover its original size and shape after deformation. If the strain is proportional to the applied stress, the material is said to exhibit Hookean or ideal elasticity.

elastomer Material that at room temperature can be stretched repeatedly to at least twice its original length and, upon release of the stress, will return with force to its approximate original length. Plastics with such or similar characteristics are known as elastomeric plastics. The expression is also used when referring to a rubber (natural or synthetic) reinforced plastic, as in elastomer-modified resins.

electric strength (dielectric strength, disruptive gradient) Maximum potential gradient that a material can withstand without rupture. The value obtained for the electric strength will depend on the thickness of the material and on the method and conditions of test.

electrode Conductor of the metallic class through which a current enters or leaves an electrolytic cell, at which there is a change from conduction by electrons to conduction by charged particles of matter, or vice versa.

elongation Increase in gage length of a tension specimen, usually expressed as a percentage of the original gage length. (*See also* **gage length.**)

embed To encase completely a component or assembly in some material—a plastic, for instance. (*See* **cast, pot.**)

encapsulate To coat a component or assembly in a conformal or thixotropic coating by dipping, brushing, or spraying.

engineering plastics Plastics, the properties of which are suitable for engineered products. These plastics are usually suitable for application up to 125°C, well above the thermal stability of many commercial plastics. The next higher grade of plastics, called high-performance plastics, is usually suitable for product designs requiring stability of plastics above 175°C.

environmental-stress cracking Susceptibility of a thermoplastic article to crack or craze under the influence of certain chemicals, aging, weather, or other stress. Standard ASTM test methods that include requirements for environmental stress cracking are indexed in Index of ASTM Standards.

epoxy Thermosetting polymers containing the oxirane group; mostly made by reacting epichlorohydrin with a polyol such as bisphenol A. Resins may be either liquid or solid.

eutectic Mixture, the melting point of which is lower than that of any other mixture of the same ingredients.

exotherm Characteristic curve of resin during its cure that shows heat of reaction (temperature) versus time. Peak exotherm is the maximum temperature on this curve.

exothermic Chemical reaction in which heat is given off.

extender Inert ingredient added to a resin formulation chiefly to increase its volume.

extrusion Compacting of a plastic material and forcing of it through an orifice.

fabric Planar structure produced by interlacing yarns, fibers, or filaments.

failure, adhesive Rupture of an adhesive bond such that the separation appears to be at the adhesive-adherend interface.

fiber washout Movement of fiber during cure because of large hydrostatic forces generated in low-viscosity resin systems.

fiberglass Individual filament made by attenuating molten glass. A continuous filament is a glass fiber of great or indefinite length; a staple fiber is a glass fiber of relatively short length (generally less than 17 in).

filament winding Process for fabricating a composite structure in which continuous reinforcements (filament, wire, yarn, tape, or other), either previously impregnated with a matrix material or impregnated during the winding, are placed over a rotating and removable form or mandrel in a previously prescribed way to meet certain stress conditions. Generally the shape is a surface of revolution and may or may not include end closures. When the right number of layers is applied, the wound form is cured and the mandrel removed.

fill *See* **weft.**

filler Material, usually inert, that is added to plastics to reduce cost or modify physical properties.

film adhesive Thin layer of dried adhesive; also class of adhesives provided in dry-film form with or without reinforcing fabric, that are cured by heat and pressure.

finish, fiber Mixture of materials for treating glass or other fibers to reduce damage during processing or to promote adhesion to matrix resins.

fish paper Electrical-insulation grade of vulcanized fiber in thin cross section.

flame retardants Materials added to plastics to improve their resistance to fire.

flash Extra plastic attached to a molding along the parting line. Under most conditions it would be objectionable and must be removed before the parts are acceptable.

flexibilizer Material that is added to rigid plastics to make them resilient or flexible. It can be either inert or a reactive part of the chemical reaction. Also called a plasticizer in some cases.

flexural modulus Ratio, within the elastic limit, of stress to the corresponding strain. It is calculated by drawing a tangent to the steepest initial straight-line portion of the load-deformation curve and calculating by the following equation:

$$E_B = \frac{L^3 m}{4bd^3}$$

where E_B = modulus
b = width of beam tested
d = depth of beam
m = slope of tangent
L = span, inches

flexural strength Strength of a material in bending expressed as the tensile stress of the outermost fibers of a bent test sample at the instant of failure.

fluorocarbon Organic compound having fluorine atoms in its chemical structure. This structure usually lends chemical and thermal stability to plastics.

four-harness satin Fabric, also named crowfoot satin because the weaving pattern when laid out on cloth design paper resembles the imprint of a crow's foot. In this type of weave there is a 3-by-1 interlacing; that is, a filling thread floats over the three warp threads and then under one. This type of fabric looks different on one side than on the other. Fabrics with this weave are more pliable than either the plain or basket weave and, consequently, are easier to form around curves. (*See* **crowfoot satin.**)

fracture toughness Measure of the damage tolerance of a matrix containing initial flaws or cracks G_{1c} and G_{2c} are the critical strain energy release rates in the 1 and 2 directions.

gage length Original of that portion of the specimen over which strain is measured.

gate Orifice through which liquid resin enters mold in plastic molding processes.

gel Soft rubbery mass that is formed as a thermosetting resin goes from a fluid to an infusible solid. This is an intermediate state in a curing reaction, and a stage in which the resin is mechanically very weak.

gel coat Resin applied to the surface of a mold and gelled prior to lay-up. The gel coat becomes an integral part of the finished laminate and is usually used to improve surface appearance and so on.

gel point The point in a curing reaction at which gelatin begins. (*See* **gel.**)

gelation Point in resin cure when the viscosity has increased to a point where the resin barely moves when probed with a sharp point.

glass-transition point Temperature at which a material loses its glasslike properties and becomes a semiliquid. (*See also* **glass-transition temperature.**)

glass-transition temperature Temperature at which a plastic changes from a rigid state to a softened state. Both mechanical and electrical properties degrade significantly at this point, which is usually a narrow temperature range, rather than a sharp point, as in freezing or boiling.

glue line (bond line) Layer of adhesive that attaches two adherends.

glue-line thickness Thickness of the fully dried adhesive layer.

glycol Alcohol containing two hydroxyl (—OH) groups.

graphite fibers High-strength, high-modulus fibers made by controlled carbonization and graphitization of organic fibers, usually rayon, acrylonitrile, or pitch.

hand lay-up Process of placing in position (and working) successive plies of reinforcing material or resin-impregnated reinforcement on a mold by hand.

hardener Chemical added to a thermosetting resin for the purpose of causing curing or hardening. Amines and acid anhydrides are hardeners for epoxy resins. Such hardeners are a part of the chemical reaction and a part of the chemical composition of the cured resin. The terms hardener and curing agent are used interchangeably. Note that these can differ from catalysts, promoters, and accelerators. (*See* **catalyst, promoter, accelerator.**)

hardness *See* **indentation hardness.**

heat-deflection temperature *See* **heat-distortion point.**

heat-distortion point Temperature at which a standard test bar deflects 0.010 in under a stated load of either 66 or 264 lb/in^2. See ASTM D 648, Standard Method of Test for Deflection Temperature of Plastics under Load.

heat sealing Method of joining plastic films by simultaneous application of heat and pressure to areas in contact. Heat may be applied conductively or dielectrically.

helical pattern Pattern generated when a filament band advances along a helical path, not necessarily at a constant angle except in the case of a cylinder, in a filament-wound object.

high-frequency preheating Plastic to be heated forms the dielectric of a condenser to which a high-frequency (20 to 80 MHz) voltage is applied. Dielectric loss in the material is the basis. The process is used for sealing vinyl films and preheating thermoset molding compounds.

high-performance plastics In general, plastics that are suitable for use above 175°C. (*See also* **engineering plastics.**)

homopolymer Polymer resulting from polymerization of a single monomer. (*See also* **monomer.**)

honeycomb Manufactured product of resin-impregnated sheet material or metal foil, formed into hexagonal cells. Skins are bonded to top and bottom surfaces to achieve strength.

hot-melt adhesive Thermoplastic adhesive compound, usually solid at room temperature, that is heated to a fluid state for application.

hydrocarbon Organic compound having hydrogen and carbon atoms in its chemical structure. Most organic compounds are hydrocarbons. Aliphatic hydrocarbons are straight-chained hydrocarbons, and aromatic hydrocarbons are ringed structures based on the benzene ring. Methyl alcohol, trichloroethylene, and the like are aliphatic; benzene, xylene, toluene, and the like are aromatic.

hydrolysis Chemical decomposition of a substance involving the addition of water.

hydrophilic Materials having a tendency to absorb water or to be wetted by water.

hydrophobic Materials having a tendency to repel water; usually materials exhibiting a low surface energy, measured by wetting angle.

hydroxyl group Chemical group consisting of one hydrogen atom plus one oxygen atom.

hygroscopic Tending to absorb moisture.

immiscible Two fluids that will not mix to form a homogeneous mixture, or that are mutually insoluble.

impact strength Strength of a material when subjected to impact forces or loads.

impregnate To force resin into every interstice of a part. Cloths are impregnated for laminating, and tightly wound coils are impregnated in liquid resin using air pressure or vacuum as the impregnating force.

in-situ joint Joint between a composite and another surface that is formed during cure of the composite.

indentation hardness Hardness evaluated from measurements of area or indentation depth caused by pressing a specified indentation into the material surface with a specified force.

inhibitor Chemical added to resins to slow down the curing reaction. Inhibitors are normally added to prolong the storage life of thermosetting resins.

injection molding Molding procedure whereby a heat-softened plastic material is forced from a cylinder into a cavity that gives the article the desired shape. Used with all thermoplastic and some thermosetting materials.

inorganic chemicals Chemicals, the structure of which is based on atoms other than the carbon atom.

insulation, electrical Protection against electrical failure in an electrical product.

insulation resistance Ratio of applied voltage to total current between two electrodes in contact with a specified insulator.

insulator Material that provides electrical insulation in an electrical product.

interpenetrating network (IPN) Two or more polymers that have been formed together so that they penetrate each other in the final polymer form.

isotactic Molecules that are polymerized in parallel arrangements of radicals on one side of the carbon chain. (*See also* **atactic.**)

Izod impact strength Measure of the toughness of a material under impact as measured by the Izod impact test.

Izod impact test One of the most common ASTM tests for testing the impact strength of plastic materials.

joint Location at which two adherends are held together with a layer of adhesive.

joint, lap Joint made by placing one adherend partly over another and bonding together the overlapped portions.

joint, scarf Joint made by cutting away similar angular segments of two adherends and bonding the adherends with the cut areas fitted together.

joint, starved Joint that has an insufficient amount of adhesive to produce a satisfactory bond.

Kevlar Trademark for a group of DuPont aromatic polyimides that are frequently used as fibers in reinforced plastics and composites. Major characteristics are low thermal expansion, light weight, and good electrical properties, coupled with stiffness in laminated form. One special application area is in high-performance circuit boards requiring low *x-y* axis thermal expansion.

laminae Set of single plies or layers of a laminate (plural of *lamina*).

laminate To unite sheets of material by a bonding material, usually with pressure and heat (normally used with reference to flat sheets); product made by so bonding.

latent curing agent Curing agent that produces long-time stability at room temperature but rapid cure at elevated temperature.

lay-up As used in reinforced plastics, the reinforcing material placed in position in the mold; resin-impregnated reinforcement; process of placing the reinforcing material in position in the mold.

leno weave Locking-type weave in which two or more warp threads cross over each other and interlace with one or more filling threads. It is used primarily to prevent shifting of fibers in open fabrics.

limited-coordination specification (or standard) Specification (or standard) that has not been fully coordinated and accepted by all the interested activities. Limited-coordination specifications and standards are issued to cover the need for requirements unique to one particular department. This applies primarily to military agency documents.

liquid-crystal polymer (LCP) Polymers that spontaneously order themselves in the melt, allowing relatively easy processing at relatively high temperatures. They are characterized as rigid rods. Kevlar and Nomex are examples, as is Xydar thermoplastic.

liquid injection molding Fabrication process in which catalyzed resin is metered into closed molds.

loss angle *See* **dissipation factor.**

loss tangent *See* **dissipation factor.**

macerate To chop or shred fabric for use as a filler for a molding resin; molding compound obtained when so filled.

mandrel Form around which resin-impregnated fiber is wound to make pipes, tubes, or vessels by the filament-winding process.

mat Reinforcing material composed of randomly oriented short, chopped fibers. Manufactured in sheet or blanket form, and commonly used as alternative to woven fabric (especially glass) in fabrication of laminated plastic forms.

matched metal molding Method of molding reinforced plastics between two close-fitting metal molds mounted in a hydraulic press.

matrix Essentially homogeneous material in which the fiber system of a composite resides.

matrix manipulations Mathematical method of relating stresses and strains.

mechanical properties Material properties associated with elastic and inelastic reactions to an applied force.

melamines Thermosetting resins made from melamine and formaldehyde and possessing excellent hardness, clarity, and electrical properties.

melt Molten plastic, in the melted phase of material during a molding cycle.

microcracks Cracks formed in composites when thermal stresses locally exceed strength of matrix. These cracks generally do not penetrate or cross fibers.

micrometer (micron) Unit of length equal to 10,000 Å, 0.0001 cm, or approximately 0.000039 in.

mock leno weave Open-type weave that resembles a leno and is accomplished by a system of interlacings that draws a group of threads together and leaves a space between the next group. The warp threads do not actually cross each other as in a real leno and, therefore, no special attachments are required for the loom. This type of weave is generally used when a high thread count is required for strength and, at the same time, the fabric must remain porous.

modifier Chemically inert ingredient added to a resin formulation that changes its properties.

modulus of elasticity Ratio of unidirectional stress to the corresponding strain (slope of the line) in the linear stress-strain region below the proportional limit. For materials with no linear range, a secant line from the origin to a specified point on the stress-strain curve or a line tangent to the curve at a specified point may be used.

moisture absorption Amount of water pickup by a material when that material is exposed to water vapor. Expressed as percent of original weight of dry material.

moisture resistance Ability of a material to resist absorbing moisture, either from the air or when immersed in water.

moisture vapor transmission Rate at which moisture vapor passes through a material at specified temperature and humidity levels. Expressed as grams per mil of material thickness per 24 h per 100 in^2.

mold Medium or tool designed to form desired shapes and sizes; to process a plastics material using a mold.

mold release Lubricant used to coat a mold cavity to prevent the molded piece from sticking to it, and thus to facilitate its removal from the mold. Also called release agent.

mold shrinkage Difference in dimensions, expressed in inches per inch, between a molding and the mold cavity in which it was molded, both the mold and the molding being at room temperature when measured.

molecular weight Sum of the atomic masses of the elements forming the molecule.

monofilament Single fiber or filament.

monomer Small molecule that is capable of reacting with similar or other molecules to form large chainlike molecules called polymers.

multilayer printed circuits Electric circuits made on thin copper-clad laminates, stacked together with intermediate prepreg sheets and bonded together with heat and pressure. Subsequent drilling and electroplating through the layers result in a three-dimensional circuit.

necking Localized reduction of the cross-sectional area of a tensile specimen that may occur during loading.

NEMA standards Property values adopted as standard by the National Electrical Manufacturers Association.

notch sensitivity Extent to which the sensitivity of a material to fracture is increased by the presence of a disrupted surface such as a notch, a sudden change in section, a crack, or a scratch. Low notch sensitivity is usually associated with ductile materials and high notch sensitivity with brittle materials.

nuclear radiation resistance Ability of a material to withstand nuclear radiation and still perform its designated function.

nylon Generic name for all synthetic polyamides. These are thermoplastic polymers with a wide range of properties.

olefin Family of unsaturated hydrocarbons with the formula C_nH_n, named after the corresponding paraffins by adding "ene" or "ylene" to the stem; for instance, ethylene. Paraffins are aliphatic hydrocarbons. (*See* **hydrocarbon.**)

oligomer Polymer containing only a few monomer units, such as a dimer or trimer.

orange peel Undesirably uneven or rough surface on a molded part, resembling the surface on an orange.

organic Composed of matter originating in plant or animal life, or composed of chemicals of hydrocarbon origin, either natural or synthetic. Used in referring to chemical structures based on the carbon atom.

orthotropic Having three mutually perpendicular planes of elastic symmetry.

parting agent *See* **mold release.**

paste Adhesive composition having a characteristic plastic-type consistency; that is, a high order or yield value such as that of a paste prepared by heating a mixture of starch and water and subsequently cooling the hydrolyzed product.

peel Imposition of a tensile stress in a direction perpendicular to the adhesive bond line, to a flexible adherend.

peel strength Strength of an adhesive in peel; expressed in pounds per inch of width.

penetration Entering of one part or material into another.

permanence Resistance of a given property to deteriorating influences.

permeability Ability of a material to allow liquid or gaseous molecules to pass through a film.

permittivity *See* **dielectric constant.**

phenolic Synthetic resin produced by the condensation of an aromatic alcohol with an aldehyde, particularly of phenol with formaldehyde.

phenylsilane Thermosetting copolymer of silicone and phenolic resins; furnished in solution form.

pitch fibers Fibers made from high-molecular-weight residue from the destructive distillation of coal or petroleum products.

plain weave The most simple and commonly used weave, in which the warp and filling threads cross alternately. Plain woven fabrics are generally the least pliable, but they are also the most stable. This stability permits the fabrics to be woven with a fair degree of porosity without too much sleaziness.

plastic Material containing an organic substance of large molecular weight that is solid in its final condition and that, at some earlier time, was shaped by flow. (*See* **resin, polymer.**)

plastic deformation Change in dimensions of an object under load that is not recovered when the load is removed; opposed to elastic deformation.

plasticity Property of plains that allows the material to be deformed continuously and permanently without rupture upon the application of a force that exceeds the yield value of the material.

plasticize To soften a material and make it plastic or moldable by means of either a plasticizer or the application of heat.

plasticizer Material incorporated in a resin formulation to increase its flexibility, workability, or distensibility. The addition of a plasticizer may cause a reduction in melt viscosity, lower the temperature of second-order transition, or lower the elastic modulus of the solidified resin.

plastisols Mixtures of vinyl resins and plasticizers that can be molded, cast, or converted to continuous films by the application of heat. If the mixtures contain volatile thinners, they are also known as organosols.

Poisson's ratio Absolute value of the ratio of transverse strain to axial strain resulting from a uniformly applied axial stress below the proportional limit of the material.

polyacrylonitrile (PAN) Synthetic fiber used as base material or precursor in manufacture of certain carbon fibers.

polyesters Thermosetting resins produced by reacting unsaturated, generally linear, alkyd resins with a vinyl-type active monomer such as styrene, methyl styrene, or diallyl phthalate. Cure is effected through vinyl polymerization using peroxide catalysts and promoters or heat to accelerate the reaction. The resins are usually furnished in liquid form.

polyimide High-temperature resins made by reacting aromatic dianhydrides with aromatic diamines.

polymer Compound formed by the reaction of simple molecules having functional groups that permit their combination to proceed to high molecular weights under suitable conditions. Polymers may be formed by polymerization (addition polymer) or polycondensation (condensation polymer). When two or more monomers are involved, the product is called a copolymer. Also, any high-molecular-weight organic compound, the structure of which consists of a repeating small unit. Polymers can be plastics, elastomers, liquids, or gums and are formed by chemical addition or condensation of monomers. (*See also* **addition reaction or polymerization, condensation polymerization.**)

polymerize To unite chemically two or more monomers or polymers of the same kind to form a molecule with higher molecular weight.

pot To embed a component or assembly in a liquid resin, using a shell, can, or case that remains as an integral part of the product after the resin is cured. (*See* **embed, cast.**)

pot life Time during which a liquid resin remains workable as a liquid after catalysts, curing agents, promoters, and the like, are added; roughly equivalent to gel time; sometimes also called working life.

power factor Cosine of the angle between the voltage applied and the current resulting.

precursor PAN or pitch fibers from which carbon graphite fiber is made.

preform Pill, tablet, or biscuit used in thermoset molding. Material is measured by volume, and the bulk factor of powder is reduced by pressure to achieve efficiency and accuracy.

preheating Heating of a compound prior to molding or casting in order to facilitate operation, reduce cycle, and improve product.

premix Molding compound prepared prior to and apart from the molding operations and containing all components required for molding: resin, reinforcement fillers, catalysts, release agents, and other compounds.

prepolymer Polymer in some stage between that of the monomers and the final polymer. The molecular weight is, therefore, also intermediate. As used in polyurethane production, reaction product of a polyol with excess of an isocyanate.

prepreg Ready-to-mold sheet that may be cloth, mat, or paper-impregnated with resin and stored for use. The resin is partially cured to a B-stage and supplied to the fabricator, who lays up the finished shape and completes and cure with heat and pressure. (*See also* **B-stage.**)

pressure-bag molding Process for molding reinforced plastics in which a tailored flexible bag is placed over the contact lay-up on the mold, sealed, and clamped in place. Fluid pressure, usually compressed air, is placed against the bag, and the part is cured.

primer Coating applied to a surface prior to the application of an adhesive to improve the performance of the bond.

printed-circuit board *See* **circuit board.**

printed-circuit laminates Laminates, either fabric- or paper-based, covered with a thin layer of copper foil and used in the photofabrication process to make circuit boards.

printed-wiring board *See* **circuit board.**

promoter Chemical, itself a weak catalyst, that greatly increases the activity of a given catalyst.

proportional limit Greatest stress a material can sustain without deviating from the linear proportionality of stress to strain (Hooke's law).

pultrusion Reversed "extrusion" of resin-impregnated roving in the manufacture of rods, tubes, and structural shapes of a permanent cross section. The roving, after passing through the resin dip tank, is drawn through a die to form the desired cross section.

qualified products list (QPL) List of commercial products that have been pretested and found to meet the requirements of a specification, especially government specifications.

reactive diluent As used in epoxy formulations, a compound containing one or more epoxy groups that functions mainly to reduce the viscosity of the mixture.

refractive index Ratio of the velocity of light in a vacuum to its velocity in a substance; also ratio of the sine of the angle of incidence to the sine of the angle of refraction.

regrind Excess of waste material in a thermoplastic molding process that can be reground and mixed with virgin raw material, within limits, for molding future parts.

reinforced molding compound Plastic to which fibrous materials such as glass or cotton have been added to improve certain physical properties such as flexural strength.

reinforced plastic Plastic with strength properties greatly superior to those of the base resin, resulting from the presence of reinforcements in the composition.

reinforced thermoplastics Reinforced molding compounds in which the plastic is thermoplastic.

relative humidity Ratio of the quantity of water vapor present in the air to the quantity which would saturate it at any given temperature.

release agent *See* **mold release.**

release paper Impermeable paper film or sheet that is coated with a material to prevent adhering to prepreg.

resin High-molecular-weight organic material with no sharp melting point. For general purposes, the terms *resin, polymer,* and *plastic* can be used interchangeably. (*See* **polymer.**)

resistivity Ability of a material to resist passage of electric current, either through its bulk or on a surface. The unit of volume resistivity is the ohm-centimeter, and the unit of surface resistivity is ohms per square.

rheology Study of the change in form and flow of matter, embracing elasticity, viscosity, and plasticity.

rigidsol Plastisol having a high elastic modulus, usually produced with a cross-linking plasticizer.

Rockwell hardness number Number derived from the net increase in depth of impression as the load on a penetrator is increased from a fixed minimum load to a higher load and then returned to minimum load. Penetrators include steel balls of several specified diameters and a diamond-cone penetrator.

rotational casting (or molding) Method used to make hollow articles from thermoplastic materials. Material is charged into a hollow mold capable of being rotated in one or two planes. The hot mold fuses the material into a gel after the rotation has caused it to cover all surfaces. The mold is then chilled and the product stripped out.

roving Collection of bundles of continuous filaments, either as untwisted strands or as twisted yarns. Rovings may be lightly twisted, but for filament winding they are generally wound as bands or tapes with as little twist as possible.

rubber Elastomer capable of rapid elastic recovery; usually natural rubber, Hevea. (*See* **elastomer.**)

runners All channels in the mold through which molten or liquid plastic raw materials flow into mold.

S glass Glass fabric made with very high tensile strength fibers for high-performance-strength requirements.

sandwich construction Panel consisting of some lightweight core material bonded to strong, stiff skins on both faces. (*See also* **honeycomb.**)

separator ply *See* **shear ply.**

set, mechanical Strain remaining after complete release of the load producing the deformation.

set, polymerization To convert an adhesive into a fixed or hardened state by chemical or physical action, such as condensation, polymerization, oxidation, vulcanization, gelation, hydration, or evaporation of volatile constituents.

shape factor For an elastomeric slab loaded in compression, ratio of loaded area to force-free area.

shear Action or stress resulting from applied forces that causes two contiguous parts of a body or two bodies to slide relative to each other in a direction parallel to their plane of contact.

shear ply Low-modulus layer, rubber, or adhesive interposed between metal and composite to reduce differential shear stresses.

shear strength Maximum shear stress a material is capable of sustaining. In testing, the shear stress is caused by a shear or torsion load and is based on the original specimen dimensions.

sheet molding compound Compression-molding material consisting of glass fibers longer than $^1/_2$ in and thickened polyester resin. Possessing excellent flow, it results in parts with good surfaces.

shelf life Time a molding compound can be stored without losing any of its original physical or molding properties.

Shore hardness Procedure for determining the indentation hardness of a material by means of a durometer. Shore designation is given to tests made with a specified durometer instrument.

silicones Resinous materials derived from organosiloxane polymers, furnished in different molecular weights, including liquids and solid resins.

sink mark Depression or dimple on the surface of an injection-molded part due to collapsing of the surface following local internal shrinkage after the gate seals. May also be an incipient short shot.

sizing agent Chemical treatment containing starches, waxes, and the like, that is applied to fibers, making them more resistant to breakage during the weaving process. The sizing agent must be removed after weaving, as its presence would cause delamination and moisture pickup problems if it remained in the final laminate made from the woven fiber.

slipping Lateral movement of tensioned fiber on a surface to a new unanticipated fiber angle.

slush molding Method for casting thermoplastics in which the resin in liquid form is poured into a hot mold where a viscous skin forms. The excess slush is drained off, the mold is cooled, and the molding is stripped out.

solvent Any substance, usually a liquid, that dissolves other substances.

specific heat Ratio of a material's thermal capacity to that of water at 15°C.

specific modulus Young's modulus divided by material density.

specific strength Ultimate tensile strength divided by material density.

spiral flow test Test method for measuring flow properties of a resin wherein the resin flows along a spiral path in a mold in the molding press. The variation in flow for different resins or different molding conditions can be compared. Flow is expressed in inches of flow in a standard spiral flow mold.

spray-up Process in which fiber reinforcement is wetted with resin applied from a spray gun. The fiber is fed through a chopper and into a stream of resin that is sprayed onto a form or into a mold.

stabilizers Chemicals used in plastics formulation to assist in maintaining physical and chemical properties during processing and service life. A specific type of stabilizer, known as an ultraviolet stabilizer, is designed to absorb ultraviolet rays and prevent them from attacking the plastic.

stacking sequence Sequence of laying plies into mold. Different stacking sequences have a great effect on off-axis mechanical properties.

starved area Part of a laminate or reinforced plastic structure in which resin has not completely wetted the fabric.

storage life *See* **shelf life.**

strain Deformation resulting from a stress, measured by the ratio of the change to the total value of the dimension in which the change occurred; unit change, due to force, in the size or shape of a body referred to its original size or shape. Strain is nondimensional but frequently expressed in inches per inch or centimeters per centimeter.

strength, dry Strength of an adhesive joint determined immediately after drying under specified conditions or after a period of conditioning in the standard laboratory atmosphere.

strength, wet Strength of an adhesive joint determined immediately after removal from a liquid in which it has been immersed under specified conditions of time, temperature, and pressure.

stress Unit force or component of force at a point in a body acting on a plane through the point. Stress is usually expressed in pounds per square inch.

stress relaxation Time-dependent decrease in stress for a specimen constrained in a constant strain condition.

substrate Material upon the surface of which an adhesive-containing substance is spread for any purpose, such as bonding or coating; broader term than adherend. Material upon the surface of which a circuit is formed.

surface preparation Physical and/or chemical preparation of an adherend to render it suitable for adhesive joining.

surface resistivity Resistance of a material between two opposite sides of a unit square of its surface. Surface resistivity may vary widely with the conditions of measurement.

syntactic foams Lightweight systems obtained by the incorporation of prefoamed or low-density fillers in the systems.

tack Property of an adhesive that enables it to form a bond of measurable strength immediately after adhesive and adherend are brought into contact under low pressure.

tan δ *See* **dissipation factor.**

tensile strength Maximum tensile stress a material is capable of sustaining. Tensile strength is calculated from the maximum load during a tension test carried to rupture and the original cross-sectional area of the specimen.

thermal conductivity Ability of a material to conduct heat; physical constant for the quantity of heat that passes through a unit cube of a material in a unit of time when the difference in temperatures of two faces is 1°C.

thermal stress cracking Crazing and cracking of some thermoplastic resins resulting from overexposure to elevated temperatures.

thermoforming Process of creating a form from a flat sheet by combinations of heat and pressure, which first soften the sheet and then form the sheet into some three-dimensional shape. This is one of the simplest, most economical plastic forming processes. There are numerous variations of this process.

thermoplastic Plastics capable of being repeatedly softened or melted by increases in temperature and hardened by decreases in temperature. These changes are physical rather than chemical.

thermoset Material that will undergo, or has undergone, a chemical reaction by the action of heat, catalysts, ultraviolet light, and the like, leading to a relatively infusible state that will not remelt after setting.

thermosetting Classification of resin that cures by chemical reaction when heated and, when cured, cannot be remelted by heating.

thinner Volatile liquid added to an adhesive to modify the consistency or other properties.

thixotropic Material that is gel-like at rest but fluid when agitated.

time, assembly Time interval between spreading of the adhesive on the adherend and application of pressure or heat, or both, to the assembly.

time, curing Period of time during which an assembly is subjected to heat or pressure, or both, to cure the adhesive.

tow Untwisted bundle of continuous fibers. Commonly used in reference to synthetic fibers, particularly carbon and graphite, but also glass and aramid. A tow designated as 12K has 12,000 filaments.

tracking Conductive carbon path formed on surface of a plastic during electrical arcing. (*See also* **arc resistance.**)

transfer molding Method of molding thermosetting materials in which the plastic is first softened by heat and pressure in a transfer chamber and then forced or transferred by high pressure through suitable sprues, runners, and gates into a closed mold for final curing.

transformation Mathematical (tensor analysis) method of obtaining stress from strain values or vice versa, or to find angular properties of a laminate.

transverse isotropy Having essentially identical mechanical properties in two directions but not the third.

transverse, properties Properties perpendicular to the axial (x,1 or 1,1) direction. May be designated as Y or Z, 2 or 3 directions.

twill weave Basic weave characterized by a diagonal rib or twill line. Each end floats over at least two consecutive picks, permitting a greater number of yarns per unit area than a plain weave while not losing a great deal of fabric stability.

twist Spiral turns about its axis per unit length for a textile strand; expressed as turns per inch.

ultrasonic bonding Bonding of plastics by vibratory mechanical pressure at ultrasonic frequencies due to melting of plastics being joined, heat being generated as frictional heat.

ultraviolet Shorter wavelengths of invisible radiation that are more damaging than visible light to most plastics.

ultraviolet stabilizers Additives mixed into plastic formulations for the purpose of improving resistance of plastic to ultraviolet radiation.

undercured State of a molded article that has not been adequately polymerized or hardened in molding process, usually due to inadequate temperature-time-pressure control in molding process.

vacuum bag Impermeable film applied to outside of lay-up to facilitate conformability to mold form and air removal during cure.

vacuum-bag molding Process for molding reinforced plastics in which a sheet of flexible transparent material is placed over the lay-up on the mold and sealed. A vacuum is applied between sheet and lay-up. The entrapped air is mechanically worked out of the lay-up and removed by the vacuum, and the part is cured.

vacuum-injection molding Molding process where, using a male and a female mold, reinforcements are placed in the mold, a vacuum is applied, and a room-temperature-curing liquid resin is introduced which saturates the reinforcement.

vent Small opening placed in a mold for allowing air to exit mold as molding material enters. This eliminates air holes, voids, or bubbles in finally molded part.

Vicat softening point One standard test for measuring temperature at which a thermoplastic will soften, involving the penetration of a flat-ended needle into the plastic under controlled conditions.

viscoelastic Characteristic mechanical behavior of some materials that is a combination of viscous and elastic behaviors.

viscosity Measure of the resistance of a fluid to flow (usually through a specific orifice).

void Air bubble that has been entrapped in a plastic part during molding process. (*See also* **vent.**)

volume resistivity (specific insulation resistance) Electrical resistance between opposite faces of a 1-cm cube of insulating material, commonly expressed in ohm-centimeters. The recommended test is ASTM D 257-54T.

vulcanization Chemical reaction in which the physical properties of an elastomer are changed by causing it to react with sulfur or other cross-linking agents.

vulcanized fiber Cellulosic material that has been partially gelatinized by action of a chemical (usually zinc chloride) and then heavily compressed or rolled to required thickness, leached free from the gelatinizing agent, and dried.

warp Fibers that run lengthwise in a woven fabric. (*See also* **weft.**)

warpage Dimensional distortion in a plastic object after molding.

water absorption Ratio of the weight of water absorbed by a material to the weight of the dry material.

water-extended polyester Casting formulation in which water is suspended in the polyester resin.

weave Pattern in which a fabric is woven. There are standard patterns, usually designated by a style number.

weft Fibers that run perpendicular to warp fibers; sometimes also called fill or woof. (*See also* **warp.**)

wet lay-up Reinforced plastic structure made by applying a liquid resin to an in-place woven or mat fabric.

wetting Ability to adhere to a surface immediately upon contact.

woof *See* **weft.**

working life *See* **pot life.**

woven fabric Flat sheet formed by interwinding yarns, fibers, or filaments. Some standard fabric patterns are plain, satin, and leno.

woven roving Heavy glass-fiber fabric made by the weaving of roving.

***x-y* axis** Directions parallel to fibers in a woven-fiber-reinforced laminate. Thermal expansion is much lower in the *x-y* axis, since this expansion is more controlled by the fabric in the laminate. (*See also* ***z* axis.**)

yield value (yield strength) Lowest stress at which a material undergoes plastic deformation. Below this stress, the material is elastic; above it, the material is viscous. Also, stress at which a material exhibits a specified limiting deviation from the proportionality of stress to strain.

Young's modulus Ratio of normal stress to corresponding strain for tensile or compressive stresses at less than the proportional limit of the material.

***z* axis** Direction perpendicular to fibers in a woven-fiber-reinforced laminate; that is, through the thickness of the laminate. Thermal expansion is much higher in the *z* axis, since this expansion is more controlled by the resin in the laminate. (*See also* ***x-y* axis.**)

Appendix

B

Some Common Abbreviations Used in the Plastics Industry*

AAGR	average annual growth rate
AS	atomic absorption spectroscopy
ABA	acrylonitrile-butadiene-acrylate
ABS	acrylonitrile-butadiene-styrene copolymer
ACM	acrylic acid ester rubber
ACS	acrylonitrile-chlorinated PE-styrene
AES	acrylonitrile-ethylene-propylene-styrene
AMMA	acrylonitrile-methyl methacrylate
AN	acrylonitrile
AO	antioxidant
APET	amorphous polyethylene terephthalate
APP	atactic polypropylene
ASA	acrylic-styrene-acrylonitrile
ASTM	American Society for Testing and Materials
ATH	aluminum trihydrate
AZ(O)	azodicarbonamide
BATF	Bureau of Alcohol, Tobacco, and Firearms
BM	blow molding
BMC	bulk molding compounds
BMI	bismaleimide
BO	biaxially-oriented (film)
BOPP	biaxially-oriented polypropylene
BR	butadiene rubber
BS	butadiene styrene rubber
CA	cellulose acetate
CAB	cellulose acetate butyrate
CAD	computer aided design
CAE	computer aided engineering
CAM	computer aided manufacturing
CAP	cellulose acetate propionate
CAP	controlled atmosphere packaging
CBA	chemical blowing agent
CF	cresol formaldehyde
CFA	chemical foaming agent
CFC	chlorofluorocarbons
CFR	Code of Federal Regulations
CHDM	cyclohexanedimethanol
CIM	computer integrated manufacturing
CN	cellulose nitrate
COP	copolyester
COPA	copolyamide
COPE	copolyester
CP	cellulose propionate
CPE	chlorinated polyethylene
CPET	crystalline polyethylene terephthalate
CPP	cast polypropylene
CPVC	chlorinated polyvinyl chloride
CR	chloroprene rubber
CS	casein
CSD	carbonated soft drink
CTA	cellulose triacetate
CVD	chemical vapor deposition
DABCO	diazobicyclooctane
DAM	days after manufacture
DAM	diallyl maleate
DAP	diallyl phthalate
DCPD	dicyclopentadiene
DE	diotamaceous earth
DEA	dielectric analysis
DETDA	diethyltoluenediamine
DMA	dynamic mechanical analysis
DSC	differential scanning analysis
DMT	dimethyl ester of terephthalate
DWV	drain, waste, vent (pipe grade)
EAA	ethylene acrylic acid
EB	electron beam
EBA	ethylene butyl acrylate
EC	ethyl cellulose
ECTFE	ethylene-chlorotrifluoroethylene copolymer
EEA	ethylene-ethyl acrylate
EG	ethylene glycol
EMA	ethylene-methyl acrylate
EMAA	ethylene methacrylic acid
EMAC	ethylene-methyl acrylate copolymer
EMC	electromagnetic compatibility
EMI	electromagnetic interference

*Reprinted with permission from *Modern Plastics Encyclopedia '99,* William A. Kaplan (Ed.), © McGraw-Hill, New York, 1998.

EMPP	elastomer modified polypropylene
EnBA	ethylene normal butyl acrylate
EP	epoxy resin, also ethylene-propylene
EPA	Environmental Protection Agency
EPDM	ethylene-propylene terpolymer rubber
EPM	ethylene-propylene rubber
EPS	expandable polystyrene
ESCR	environmental stress crack resistance
ESI	ethylene-styrene copolymers
ETE	engineering thermoplastic elastomers
ETFE	ethylene-tetrafluoroethylene copolymer
ETP	engineering thermoplastics
EVA(C)	polyethylene-vinyl acetate
EVOH	polyethylene-vinyl alcohol copolymers
FDA	Food and Drug Administration
FEP	fluorinated ethylene propylene copolymer
FPVC	flexible polyvinyl chloride
FR	flame retardant
FRP	fiber reinforced plastic
GIM	gas injection molding
GIT	gas injection technique
GMT(P)	glass mat reinforced thermoplastics
GPC	gel permeation chromotography
GPPS	general purpose polystyrene
GRP	glass fiber reinforced plastics
GTP	group transfer polymerization
HALS	hindered amine light stabilizer
HAS	hindered amine stabilizers
HB	Brinell hardness number
HCFC	hydrochlorofluorocarbons
HCR	heat-cured rubber
HDI	hexamethylene diisocyanate
HDPE	high-density polyethylene
HDT	heat deflection temperature
HFC	hydrofluorocarbons
HIPS	high-impact polystyrene
HMDI	diisocyanato dicyclohexylmethane
HMW	high molecular weight
HNP	high nitrile polymer
IM	injection molding
IMC	in-mold coating
IMD	in-mold decoration
IPI	isophorone diisocyanate
IV	intrinsic viscosity
LCP	liquid crystal polymers
LIM	liquid injection molding
LDPE	low-density polyethylene
LLDPE	linear low-density polyethylene
LP	low-profile resin
MAP	modified atmosphere packaging
MbOCA	3,3'-dichloro-4,4-diamino-diphenylmethane
MBS	methacrylate-butadiene-styrene
MC	methyl cellulose
MDI	methylene diphenylene diisocyanate
MEKP	methyl ethyl ketone peroxide
MF	melamine formaldehyde
MFI	melt flow index
MIS	management information systems
MMA	methyl methacrylate
MPE	metallocene polyethylenes
MPF	melamine-phenol-formaldehyde
MPR	melt-processable rubber
MRP	manufacturing requirement planning
MWD	molecular weight distribution
NBR	nitrile rubber
NDI	naphthalene diisocyanate
NDT	nondestructive testing
NR	natural rubber
ODP	ozone depleting potential
OFS	organofunctional silanes
OPET	oriented polyethylene terephthalate
OPP	oriented polypropylene
O-TPV	olefinic thermoplastic vulcanizate
OEM	original equipment manufacturer
OSA	olefin-modified styrene-acrylonitrile
PA	polyamide
PAEK	polyaryletherketone
PAI	polyamide imide
PAN	polyacrylonitrile
PB	polybutylene
PBA	physical blowing agent
PBAN	polybutadiene-acrylonitrile
PBI	polybenzimidazole
PBN	polybutylene naphthalate
PBS	polybutadiene styrene
PBT	polybutylene terephthalate
PC	polycarbonate
PCC	precipitated calcium carbonate
PCD	polycarbodiimide
PCR	post-consumer recyclate
PCT	polycyclohexylenedimethylene terephthalate
PCTA	copolyester of CHDM and PTA
PCTFE	polychlorotrifluoroethylene
PCTG	glycol-modified PCT copolymer
PE	polyethylene
PEBA	polyether block polyamide
PEC	chlorinated polyethylene
PEDT	3,4 polyethylene dioxithiophene
PEEK	polyetheretherketone
PEI	polyether imide
PEK	polyetherketone
PEL	permissible exposure level
PEKEKK	polyetherketoneetherketoneketone
PEN	polyethylene naphthalate
PES	polyether sulfone
PET	polyethylene terephthalate
PETG	PET modified with CHDM
PF	phenol formaldehyde
PFA	perfluoroalkoxy resin
PI	polyimide
PID	proportional, integral, derivative
PIBI	butyl rubber
PIM	powder injection molding
PLC	programmable logic controller
PMDI	polymeric methylene diphenylene diisocyanate
PMMA	polymethyl methacrylate
PMP	polymethylpentene
PO	polyolefins
POM	polyacetal
PP	polypropylene
PPA	polyphthalamide
PPC	chlorinated polypropylene
PPE	polyphenylene ether, modified
ppm	parts per million
PPO	polyphenylene oxide
PPS	polyphenylene sulfide
PPSU	polyphenylene sulfone
PS	polystyrene
PSU	polysulfone
PTA	purified terephthalic acid
PTFE	polytetrafluoroethylene
PU	polyurethane
PUR	polyurethane
PVC	polyvinyl chloride
PVCA	polyvinyl chloride acetate
PVDA	polyvinylidene acetate

PVDC	polyvinylidene chloride
PVDF	polyvinylidene fluoride
PVF	polyvinyl fluoride
PVOH	polyvinyl alcohol
QMC	quick mold change
RFI	radio frequency interference
RHDPE	recycled high density polyethylene
RIM	reaction injection molding
RPET	recycled polyethylene terephthalate
RTD	resistance temperature detector
RTM	resin transfer molding
RTV	room temperature vulcanizing
SI	silicone plastic
SAN	styrene acrylonitrile copolymer
SB	styrene butadiene copolymer
SBC	styrene block copolymer
SBR	styrene butadiene rubber
SMA	styrene maleic anhydride
SMC	sheet molding compound
SMC-C	SMC-continuous fibers
SMC-D	SMC-directionally oriented
SMC-R	SMC-randomly oriented
SPC	statistical process control
SQC	statistical quality control
SRIM	structural reaction injection molding
TA	terephthalic acid
TDI	toluene diisocyanate
TEO	thermoplastic elastomeric olefin
TGA	thermogravimetric analysis
TLCP	thermoplastic liquid crystal polymer
TMA	thermomechanical analysis
TMC	thick molding compound
T/N	terephthalate/naphthalate
TPA	terephthalic acid
TP	thermoplastic
TPE	thermoplastic elastomer
TPO	thermoplastic olefins
TPU	thermoplastic polyurethane
TPV	thermoplastic vulcanizate
TS	thermoset
TWA	time-weighted average
UF	urea formaldehyde
UHMW	ultrahigh molecular weight
ULDPE	ultralow-density polyethylene
UP	unsaturated polyester resin
UR	urethane
UV	ultraviolet
VA(C)	vinyl acetate
VC	vinyl chloride
VDC	vinylidene chloride
VLDPE	very low-density polyethylene
VOC	volatile organic compounds
ZNC	Ziegler-Natta catalyst

Important Properties of Plastics and Listing of Plastic Suppliers*

*Reprinted with permission from *Modern Plastics Encyclopedia '99,* William A. Kaplan (Ed.), © McGraw-Hill, New York, 1998.

Materials	Properties	ASTM test method	ABS						
				Flame-retarded grades, molding and extrusion					Injection molding grades
			Extrusion grade	ABS	ABS/PVC	ABS/PC	ABS/Nylon	ABS/PC injection molding and extrusion	Heat-resistant
Processing	1a. Melt flow (gm./10 min.)	D1238	0.4-6.86	1.2-1.7; 6	1.9				1.1-1.8
	1. Melting temperature, °C. T_m (crystalline)								
	T_g (amorphous)		88-120	110-125					110-125
	2. Processing temperature range, °F. (C = compression; T = transfer; I = injection; E = extrusion)		E: 350-500	C: 350-500 I: 380-500	370-410	I: 425-520	I: 460-520	I: 460-540 E: 450-500	C: 325-500 I: 475-550
	3. Molding pressure range, 10^3 p.s.i.			8-25		10-20	8-25	10-20	8-25
	4. Compression ratio		2.5-2.7	1.1-2.0	2.0-2.5	1.1-2.5	1.1-2.0	1.1-2.5	1.1-2.0
	5. Mold (linear) shrinkage, in./in.	D955	0.004-0.007	0.004-0.008	0.003-0.006	0.004-0.007	0.003-0.010	0.005-0.008	0.004-0.009
Mechanical	6. Tensile strength at break, p.s.i.	D638[b]	2500-8000	3300-8000	5800-6500	5800-9300	4000-6000	5800-7400	4800-7500
	7. Elongation at break, %	D638[b]	2.9-100	1.5-80		20-70	40-300	50-125	3-45
	8. Tensile yield strength, p.s.i.	D638[b]	4300-7250	4000-7400	4300-6600	7700-9000	4300-6300	3500-8500	4300-7000
	9. Compressive strength (rupture or yield), p.s.i.	D695	5200-10,000	6500-7500		11,000-11,300			7200-10,000
	10. Flexural strength (rupture or yield), p.s.i.	D790	4000-14,000	6200-14,000	7900-10,000	12,000-14,500	8800-10,900	8700-13,000	9000-13,000
	11. Tensile modulus, 10^3 p.s.i.	D638[b]	130-420	270-400	325-380	350-455	260-320	350-380	285-360
	12. Compressive modulus, 10^3 p.s.i.	D695	150-390	130-310		230			190-440
	13. Flexural modulus, 10^3 p.s.i. 73° F.	D790	130-440	300-600	320-400	350-400	250-310	290-375	300-400
	200° F.	D790							
	250° F.	D790							
	300° F.	D790							
	14. Izod impact, ft.-lb./in. of notch (1/8-in. thick specimen)	D256A	1.5-12	1.4-12	3.0-18.0	4.1-14.0	15-20	6.4-12.0	2.0-6.5
	15. Hardness Rockwell	D785	R75-115	R100-120	R100-106	R115-119	R93-105	R95-120	R100-115
	Shore/Barcol	D2240/ D2583			Shore D-73				
Thermal	16. Coef. of linear thermal expansion, 10^{-6} in./in./°C.	D696	60-130	65-95	46-84	67	90-110	62-72	60-93
	17. Deflection temperature under flexural load, °F. 264 p.s.i.	D648	170-220	158[g]; 181; 190-225 annealed	169-200 annealed	180-220	130-150	210-240	220-240 annealed 181-193[g]
	66 p.s.i.	D648	170-235	210-245 annealed		195-244	180-195	220-265	230-245 annealed
	18. Thermal conductivity, 10^{-4} cal.-cm./sec.-cm.^{2}-°C.	C177							4.5-8.0
Physical	19. Specific gravity	D792	1.02-1.08	1.16-1.21	1.13-1.25	1.17-1.23	1.06-1.07	1.07-1.15	1.05-1.08
	20. Water absorption (1/8-in. thick specimen), % 24 hr.	D570	0.20-0.45	0.2-0.6		0.24		0.15-0.24	0.20-0.45
	Saturation	D570							
	21. Dielectric strength (1/8-in. thick specimen), short time, v./mil	D149	350-500	350-500	500	450-760		430	350-500
		SUPPLIERS[a]	Albis; American Polymers; Ashley Polymers; Bamberger Polymers; BASF; Bayer Corp; Diamond Polymers; Dow Plastics; Federal Plastics; GE Plastics; LG Chemical; RSG Polymers; A. Schulman; Shuman	Albis; Ashley Polymers; BASF; Bayer Corp.; ComAlloy; Diamond Polymers; DSM; Federal Plastics; GE Plastics; LG Chemical; M.A. Polymers; Polymer Resources; RSG Polymers; RTP; A. Schulman; Shuman	ComAlloy; CONDEA Vista; Novatec; A. Schulman; Shuman	American Polymers; Ashley Polymers; Bayer Corp.; ComAlloy; Diamond Polymers; Dow Plastics; GE Plastics; LG Chemical; M.A. Polymers; Polymer Resources; RTP	Bayer Corp.	Ashley Polymers; Bayer Corp.; ComAlloy; Diamond Polymers; Dow Plastics; GE Plastics; LG Chemical; M.A. Polymers; Polymer Resources; RTP	American Polymers; Ashley Polymers; BASF; Bayer Corp.; Diamond Polymers; Dow Plastics; Federal Plastics; GE Plastics; LG Chemical; Polymer Resources; RSG Polymers; RTP; A. Schulman; Shuman

	ABS (Cont'd)										
	Injection molding grades (Cont'd)								EMI shielding (conductive)		
	Medium-impact	High-impact	Platable grade	Transparent	20% glass fiber-reinforced	30% glass fiber-reinforced	20% long glass fiber-reinforced	40% long glass fiber-reinforced	6% stainless steel fiber	7% stainless steel fiber	10% stainless steel fiber
1a.	1.1-34.3	1.1-18	1.1								
1.											
	102-115	91-110	100-110	120	100-110	100-110	100-110	100-110	100-110	100-110	100-110
2.	C: 325-350 I: 390-525	C: 325-350 I: 380-525	C: 325-400 I: 350-500	455-500	C: 350-500 I: 350-500	I: 400-460	I: 400-460	I: 400-460	I: 400-460	I: 400-460	I: 400-460
3.	8-25	8-25	8-25		15-30						
4.	1.1-2.0	1.1-2.0	1.1-2.0								
5.	0.004-0.009	0.004-0.009	0.005-0.008	0.009-0.067	0.001-0.002	0.002-0.003	0.001-0.002	0.001	0.004-0.006	0.004	0.004-0.006
6.	5500-7500	4400-6300	5200-6400	5000	10,500-13,000	13,000-16,000	13,000	16,000	6300-9200	6000	7100
7.	2.2-60	3.5-75		20	2-3	1.5-1.8	2.0	1.5	3.8	3.8	2.5
8.	5000-7200	2600-6300	6700	7000							
9.	1800-12,500	4500-8000			13,000-14,000	15,000-17,000	14,000	17,000			
10.	7100-13,000	5400-11,000	10,500-11,500	10,000	14,000-17,500	17,000-19,000	20,000	25,000	8700-11,000	10,000-12,000	12,100
11.	300-400	150-350	320-380	290	740-880	1000-1200	900	1000	300-400		400
12.	200-450	140-300			800						
13.	310-400	179-375	340-390		650-800	1000	850	1100	280-410	430	500
14.	3.0-9.6	6.0-10.5	4.0-8.3	1.5-2.0	1.1-1.4	1.2-1.3	2.0	2.5	1.2-1.4	1.0-1.1	1.4
15.	R102-115	R85-106	R103-109	R94	M85-98, R107	M75-85	M85-95	M90-100			
16.	80-100	95-110	47-53	60-130	20-21						
17.	174-220 annealed	205-215 annealed; 192 unannealed	190-222 annealed	194	210-220	215-230	210	215	190	185-200	190
	192-225 annealed	210-225 annealed	215-222 annealed	207	220-230	230-240	225	225			
18.					4.8						
19.	1.03-1.06	1.01-1.05	1.04-1.07	1.08	1.18-1.22	1.29	1.23	1.36	1.10-1.28	1.12	1.14
20.	0.20-0.45	0.20-0.45		0.35	0.18-0.20	0.3	0.2	0.2	0.4	0.4	
21.	350-500	350-500	420-550		450-460						
	Albis; American Polymers; Ashley Polymers; Bamberger Polymers; BASF; Bayer Corp.; Diamond Polymers; Dow Plastics; Federal Plastics; GE Plastics; LG Chemical; Polymer Resources; RSG Polymers; RTP; A. Schulman; Shuman	Albis; American Polymers; Ashley Polymers; Bamberger Polymers; BASF; Bayer Corp.; Diamond Polymers; Dow Plastics; Federal Plastics; GE Plastics; LG Chemical; Polymer Resources; RSG Polymers; RTP; A. Schulman; Shuman	American Polymers; Ashley Polymers; Bamberger Polymers; Bayer Corp.; Diamond Polymers; Dow Plastics; Federal Plastics; GE Plastics; LG Chemical; RSG Polymers; A. Schulman	BASF; Bayer Corp.; Diamond Polymers; GE Plastics; LG Chemical; A. Schulman	Albis; American Polymers; Ashley Polymers; ComAlloy; Diamond Polymers; DSM; Ferro; LG Chemical; LNP; M.A. Polymers; RTP; A. Schulman; Thermofil	Albis; American Polymers; Ashley Polymers; ComAlloy; Diamond Polymers; DSM; Ferro; LG Chemical; LNP; M.A. Polymers; RTP	DSM	DSM	Federal Plastics; Ferro; LNP; RTP; Ticona	DSM; Ferro; LNP; RTP; Ticona	RTP; Ticona

Materials	Properties	ASTM test method	ABS (Cont'd)				Acetal		
			EMI shielding (conductive) (Cont'd)			Rubber-modified			
			20% PAN carbon fiber	20% graphite fiber	40% aluminum flake	Injection molding and extrusion grades	Homo-polymer	Copolymer	Impact-modified homo-polymer
Processing	1a. Melt flow (gm./10 min.)	D1238					1-20	1-90	0.5-7.0
	1. Melting temperature, °C. T_m (crystalline)						172-184	160-175	175
	T_g (amorphous)		100-110						
	2. Processing temperature range, °F. (C = compression; T = transfer; I = injection; E = extrusion)		I: 415-500	I: 420-530	I: 400-550		I: 380-470	C: 340-400 I: 360-450	I: 380-420
	3. Molding pressure range, 10^3 p.s.i.		15-30				10-20	8-20	6-12
	4. Compression ratio						2.0-4.5	3.0-4.5	
	5. Mold (linear) shrinkage, in./in.	D955	0.0005-0.004	0.001	0.001		0.018-0.025	0.020 (Avg.)	0.012-0.019
Mechanical	6. Tensile strength at break, p.s.i.	D638[b]	15,000-16,000	15,200-15,800	3300-4200	5400-7100	9700-10,000		6500-8400
	7. Elongation at break, %	D638[b]	1.0-2.0	2.0-2.2	1.9-5	20-30	10-75	15-75	60-200
	8. Tensile yield strength, p.s.i.	D638[b]				6100	9500-12,000	8300-10,400	5500-7900
	9. Compressive strength (rupture or yield), p.s.i.	D695	17,000	16,000-17,000	6500		15,600-18,000 @ 10%	16,000 @ 10%	7600-11,900 @ 10%
	10. Flexural strength (rupture or yield), p.s.i.	D790	23,000-25,000	23,000	6200	8900-13,000	13,600-16,000	13,000	5800-10,000
	11. Tensile modulus, 10^3 p.s.i.	D638[b]	1800-2000	1660	370	310	400-520	377-464	190-350
	12. Compressive modulus, 10^3 p.s.i.	D695					670	450	
	13. Flexural modulus, 10^3 p.s.i. 73° F.	D790	890-1800	1560	400-600	260-380	380-490	370-450	150-350
	200° F.	D790					120-135		50-100
	250° F.	D790					75-90		33-60
	300° F.	D790							
	14. Izod impact, ft.-lb./in. of notch (1/8-in. thick specimen)	D256A	1.0	1.3	1.4-2.0	3.12-7.34	1.1-2.3	0.8-1.5	2.0-17
	15. Hardness Rockwell	D785	R108		R107	90-105	M92-94, R120	M75-90	M58-79
	Shore/Barcol	D2240/ D2583							
Thermal	16. Coef. of linear thermal expansion, 10^{-6} in./in./°C.	D696	18	20	40		50-112	61-110	92-117
	17. Deflection temperature under flexural load, °F. 264 p.s.i.	D648	215-225	216	190-212	181-212	253-277	185-250	148-185
	66 p.s.i.	D648	225-230	240	220		324-342	311-330	293-336
	18. Thermal conductivity, 10^{-4} cal.-cm./sec.-cm.^{2}-°C.	C177	9.6				5.5	5.5	
Physical	19. Specific gravity	D792	1.13-1.14	1.17	1.54-1.61	103-119	1.42	1.40	1.32-1.39
	20. Water absorption (1/8-in. thick specimen), % 24 hr.	D570	0.17	0.15	0.23		0.25-1	0.20-0.22	0.30-0.44
	Saturation	D570					0.90-1	0.65-0.80	0.75-0.85
	21. Dielectric strength (1/8-in. thick specimen), short time, v./mil	D149					400-500 (90 mil)	500 (90 mil)	400-480 (90 mil)
		SUPPLIERS[a]	DSM; Ferro; LNP; RTP	Albis; LNP; RTP; Thermofil	ComAlloy; Thermofil	Diamond Polymers; RSG Polymers	Ashley Polymers; DuPont; RTP; Shuman	American Polymers; Ashley Polymers; BASF; LG Chemical; M.A. Hanna Eng.; Network Polymers; RTP; Ticona	Ashley Polymers; DuPont

Acetal (Cont'd)

	Impact-modified copolymer	Mineral-filled copolymer	Extrusion and blow molding grade (terpolymer)	Copolymer with 2% silicone, low wear	UV stabilized copolymer	20% glass-reinforced homo-polymer	25% glass-coupled copolymer	40% long glass fiber-reinforced	21% PTFE-filled homo-polymer	2-20% PTFE-filled copolymer	1.5% PTFE-filled homo-polymer
1a.	5-26		1.0	9	2.5-27.0	6.0			1.0-7.0		6
1.	160-170	160-175	160-170	160-170	170	175-181	160-180	190	175-181	160-175	175
2.	I: 360-425	I: 360-450	E: 360-400	I: 360-450	I: 360-450	I: 350-480	I: 365-480	380-430	I: 370-410	I: 350-445 I: 325-500 E: 360-500	I: 400-440
3.	8-15	10-20		10-20	10-20	10-20	8-20	8-12	10-20	8-20	
4.	3.0-4.5	3.0-4.0	3.0-4.0	3.0-4.0	3.0-4.0		3.0-4.5	3.0-4.0		3.0-4.5	
5.	0.018-0.020	0.015-0.019	0.02	0.022	0.022	0.009-0.012	0.004 (flow) 0.018 (trans.)	0.003-0.010	0.020-0.025	0.018-0.029	
6.		6400-11,500				8500-9000	16,000-18,500	17,400	6900-7600	8300	10,000
7.	60-300	5-55	67	60	30-75	6-12	2-3	1.3	10-22	30	13
8.	3000-8000	6400-9800	8700	7400	8800-9280	7500-8250	16,000		6900-7600	8300	10,000
9.			16,000		16,000	18,000 @ 10%	17,000 @ 10%	20,400	13,000 @ 10%	11,000-12,600	
10.	7100	12,500-13,000	12,800	12,000	13,000	10,700-16,000	18,000-28,000	27,000	11,000	11,500	13,900
11.	187-319	522-780			410-910	900-1000	1250-1400	1700	410-420	250-280	450
12.					450						
13.	120-300	430-715	350	350	375-380	600-730	1100	1430	340-380	310-360	430
						300-360			110-120		150
						250-270			80-85		95
14.	1.7-4.7	0.9-1.2	1.7	1.1-1.4	1.12-1.5	0.5-1.0	1.0-1.8	6.9	0.7-1.2	0.5-1.0	1.0
15.	M35-70; R110	M83-90	M84	M75	M80-85	M90	M79-90, R110		M78, M110	M79	M93
16.	130-150	80-90				33-81	17-44		75-113	52-68	
17.	132-200	200-279	205	230	221-230	315	320-325	320	210-244	198-225	277
	308-318	302-325	318	316	316-320	345	327-331		300-334	280-325	342
18.										4.7	
19.	1.29-1.39	1.48-1.64	1.41	1.4	1.41	1.54-1.56	1.58-1.61	1.72	0.15-1.54	1.40	1.42
20.	0.31-0.41	0.20	0.22	0.21	0.22	0.25	0.22-0.29		0.20	0.15-0.26	0.19
	1.0-1.3	0.8-0.9	0.8	0.8	0.8	1.0	0.8-1.0		0.72	0.5	0.90
21.						490 (125 mil)	480-580		400-460 (125 mil)	400-410	450 (90 mils)
	Ashley Polymers; BASF; LG Chemical; M.A. Hanna Eng.; Network Polymers; Ticona	BASF; LG Chemical; M.A. Hanna Eng.; Network Polymers; Ticona	Network Polymers; Ticona	LG Chemical; M.A. Hanna Eng.; Network Polymers; RTP; Ticona	BASF; LG Chemical; M.A. Hanna Eng.; Network Polymers; Ticona	ComAlloy; DSM; DuPont; Ferro; LNP; RTP	BASF; ComAlloy; DSM; Ferro; LNP; M.A. Hanna Eng.; Network Polymers; RTP; Ticona	Ticona	Adell; DSM; DuPont; Ferro; LNP; RTP	ComAlloy; DSM; Ferro; LG Chemical; LNP; M.A. Hanna Eng.; RTP; Ticona	DuPont

Materials	Properties	ASTM test method	Acetal (Cont'd)						Acrylic
							EMI shielding (conductive)		Sheet
			Chemically lubricated homo-polymer	UV stabilized homo-polymer	UV stabilized, 20% glass-filled homo-polymer	Extrusion grade homo-polymer	30% carbon fiber	10% PAN carbon fiber, 10% PTFE-filled copolymer	Cast
Processing	1a. Melt flow (gm./10 min.)	D1238	6	1-6	6	1-6			
	1. Melting temperature, °C. T_m (crystalline)		175	175	175	175	166	163-175	
	T_g (amorphous)								90-105
	2. Processing temperature range, °F. (C = compression; T = transfer; I = injection; E = extrusion)		I: 400-440	I: 400-440	I: 400-440	E: 380-420	I: 350-400	I: 350-410	
	3. Molding pressure range, 10^3 p.s.i.						10-20		
	4. Compression ratio						3-4		
	5. Mold (linear) shrinkage, in./in.	D955					0.003-0.005	0.002	1.7
Mechanical	6. Tensile strength at break, p.s.i.	D638[b]	9500	10,000	8500	10,000	7500-11,500	12,000	66-11,000
	7. Elongation at break, %	D638[b]	40	40-75	12	40-75	1.5-2	1.3	2-7
	8. Tensile yield strength, p.s.i.	D638[b]	9500	10,000	8500				
	9. Compressive strength (rupture or yield), p.s.i.	D695							11,000-19,000
	10. Flexural strength (rupture or yield), p.s.i.	D790	13,000	14,100-14,300	10,700	14,100-14,300	12,500-16,500	14,000	12,000-17,000
	11. Tensile modulus, 10^3 p.s.i.	D638[b]	450	400-450	900	400-450	1350	1300	310-3100
	12. Compressive modulus, 10^3 p.s.i.	D695							390-475
	13. Flexural modulus, 10^3 p.s.i. 73° F.	D790	400	420-430	730	420-430	1100-1200	1000	320-3210
	200° F.	D790	130	130-135	360	130-135			
	250° F.	D790	80	90	270	90			
	300° F.	D790							
	14. Izod impact, ft.-lb./in. of notch (1/8-in. thick specimen)	D256A	1.4	1.5-2.3	0.8	1.5-2.3	0.7-0.8	0.7	0.3-0.4
	15. Hardness Rockwell	D785	M90	M94	M90	M94			M80-102
	Shore/Barcol	D2240/ D2583							
Thermal	16. Coef. of linear thermal expansion, 10^{-6} in./in./°C.	D696							50-90
	17. Deflection temperature under flexural load, °F. 264 p.s.i.	D648	257	257-264	316	257-264	320	320	98-215
	66 p.s.i.	D648	329	334-336	345	334-336	325	325	165-235
	18. Thermal conductivity, 10^{-4} cal.-cm./sec.-cm.^{2}-°C.	C177							4.0-6.0
Physical	19. Specific gravity	D792	1.42	1.42	1.56	1.42	1.43-1.53	1.49	1.17-1.20
	20. Water absorption (1/8-in. thick specimen), % 24 hr.	D570	0.27	0.25	0.25	0.25	0.22-0.26	0.25	0.2-0.4
	Saturation	D570	1.00	0.90	1.00	0.90			
	21. Dielectric strength (1/8-in. thick specimen), short time, v./mil	D149	400 (125 mils)	500 (90 mils)	490 (125 mils)	500 (90 mils)			450-550
		SUPPLIERS[a]	DuPont	DuPont	DuPont	DuPont	DSM; LNP; RTP; Ticona	DSM; Ferro; LNP; M.A. Hanna Eng.; RTP	Aristech; AtoHaas; Cyro; DuPont; ICI Acrylics

	Acrylic (Cont'd)							Acrylonitrile			Allyl
	Sheet (Cont'd)	Molding and extrusion compounds									
	Coated	Acrylic/PC alloy	PMMA	MMA-styrene copolymer	Impact-modified	Heat-resistant	Acrylic multi-polymer	Extrusion	High-impact extrusion	Injection	Allyl diglycol carbonate cast sheet
1a.		3-4	1.4-27	1.1-24	1-11	1.6-8.0	2-14	3	3	12	
1.										135	Thermoset
	90-110	140	85-105	100-105	80-103	100-165	80-105			95	
2.		I: 450-510 E: 430-480	C: 300-425 I: 325-500 E: 360-500	C: 300-400 I: 300-500	C: 300-400 I: 400-500 E: 380-480	C: 350-500 I: 400-625 E: 360-550	I: 400-500 E: 380-470	380-420	380-410	C: 320-345 I: 410 E: 380-410	
3.		5-20	5-20	10-30	5-20	5-30	5-20	25	25	20	
4.			1.6-3.0			1.2-2.0		2-2.5	2-2.5	2-2.5	
5.		0.004-0.008	0.001-0.004(flow) 0.002-0.008(trans.)	0.002-0.006	0.002-0.008	0.002-0.008	0.004-0.008	0.002-0.005	0.002-0.005	0.002-0.005	
6.	10,500	8000-9000	7000-10,500	8100-10,100	5000-9000	9300-11,500	5500-8200			9000	5000-6000
7.	3	58	2-5.5	2-5	4.6-70	2-10	5-28	3-4	3-4	3-4	
8.			7800-10,600		5500-8470	10,000		9500	7500	7500-9500	
9.	18,000		10,500-18,000	11,000-15,000	4000-14,000	15,000-17,000	7500-11,500	12,000	11,500	12,000	21,000-23,000
10.	16,000	11,300-12,500	10,500-19,000	14,100-16,000	7000-14,000	12,000-18,000	9000-13,000	14,000	13,700	14,000	6000-13,000
11.	450	320-350	325-470	430-520	200-500	350-650	300-430	500-550	450-500	500-580	300
12.	450		370-460	240-370	240-370	450					300
13.	450	320-350	325-460	450-460	200-430	450-620	290-400	490	390	500-590	250-330
						150-440					
						350-420					
14.	0.3-0.4	26-30	0.2-0.4	0.3-0.4	0.40-2.5	0.2-0.4	1.0-2.5	5.0	9.0	2.5-6.5	0.2-0.4
15.	M105	46-49	M68-105	M80-85	M35-78	M94-100	22-56	M60	M45	M60-M78	M95-100
16.	40	52	50-90	60-80	48-80	40-71	44-50	66	66	66	81-143
17.	205	214	155-212	208-211	165-209	190-310	180-194	156	151	151-164	140-190
	225	253	165-225		180-205	200-315		170	160	166-172	
18.	5.0		4.0-6.0	4.0-5.0	4.0-5.0	2.0-4.5	5.3	6.1	6.1	6.1-6.2	4.8-5.0
19.		1.15	1.17-1.20	1.06-1.13	1.11-1.18	1.16-1.22	1.11-1.12	1.15	1.11	1.11-1.15	1.3-1.4
20.	<0.4	0.3	0.1-0.4	0.11-0.17	0.19-0.8	0.2-0.3	0.3			0.28	0.2
21.	500		400-500	450	380-500	400-500		220-240	220-240	220-240	380
	DuPont	Cyro	American Polymers; AtoHaas; Continental Acrylics; Cyro; DuPont; ICI Acrylics; LG Chemical; Network Polymers; Plaskolite; RTP	Network Polymers; NOVA Chemicals	AtoHaas; Continental Acrylics; Cyro; DuPont; ICI Acrylics; Network Polymers; RTP	AtoHaas; Cyro; ICI Acrylics; Network Polymers; Plaskolite; RTP	Cyro	BP Chemicals	BP Chemicals	BP Chemicals	PPG

Materials			Allyl (Cont'd)		Cellulosic				
			DAP molding compounds			Cellulose acetate		Cellulose acetate butyrate	Cellulose acetate propionate
	Properties	ASTM test method	Glass-filled	Mineral-filled	Ethyl cellulose molding compound and sheet	Sheet	Molding and extrusion compound	Molding and extrusion compound	Molding and extrusion compound
Processing	1a. Melt flow (gm./10 min.)	D1238							
	1. Melting temperature, °C. T_m (crystalline)		Thermoset	Thermoset	135	230	230	140	190
	T_g (amorphous)								
	2. Processing temperature range, °F. (C = compression; T = transfer; I = injection; E = extrusion)		C: 290-360 I: 300-350	C: 270-360	C: 250-390 I: 350-500		C: 260-420 I: 335-490	C: 265-390 I: 335-480	C: 265-400 I: 335-515
	3. Molding pressure range, 10^3 p.s.i.		2000-6000	2500-5000	8-32		8-32	8-32	8-32
	4. Compression ratio		1.9-10.0	1.2-2.3	1.8-2.4		1.8-2.6	1.8-2.4	1.8-3.4
	5. Mold (linear) shrinkage, in./in.	D955	0.0005-0.005	0.002-0.007	0.005-0.009		0.003-0.010	0.003-0.009	0.003-0.009
Mechanical	6. Tensile strength at break, p.s.i.	D638[b]	6000-11,000	5000-8000	2000-8000	4500-8000	1900-9000	2600-8100	2000-7800
	7. Elongation at break, %	D638[b]	3-5	3-5	5-40	20-50	6-70	40-88	29-100
	8. Tensile yield strength, p.s.i.	D638[b]					2500-7600	1600-7200	
	9. Compressive strength (rupture or yield), p.s.i.	D695	25,000-35,000	20,000-32,000			3000-8000	2100-7500	2400-7000
	10. Flexural strength (rupture or yield), p.s.i.	D790	9000-20,000	8500-11,000	4000-12,000	6000-10,000	2000-16,000	1800-10,100	2900-11,400
	11. Tensile modulus, 10^3 p.s.i.	D638[b]	1400-2200	1200-2200				50-200	60-215
	12. Compressive modulus, 10^3 p.s.i.	D695							
	13. Flexural modulus, 10^3 p.s.i. 73° F.	D790	1200-1500	1000-1400			1000-4000	90-300	120-350
	200° F.	D790							
	250° F.	D790							
	300° F.	D790							
	14. Izod impact, ft.-lb./in. of notch (1/8-in. thick specimen)	D256A	0.4-15.0	0.3-0.8	0.4	2.0-8.5	1.0-7.8	1.0-10.9	0.5-No break
	15. Hardness Rockwell	D785	E80-87	E60-80E	R50-115	R85-120	R17-125	R11-116	R10-122
	Shore/Barcol	D2240/ D2583							
Thermal	16. Coef. of linear thermal expansion, 10^{-6} in./in./°C.	D696	10-36	10-42	100-200	100-150	80-180	110-170	110-170
	17. Deflection temperature under flexural load, °F. 264 p.s.i.	D648	330-550+	320-550	115-190		111-195	109-202	111-228
	66 p.s.i.	D648					120-209	130-227	147-250
	18. Thermal conductivity, 10^{-4} cal.-cm./sec.-cm.^{2}-°C.	C177	5.0-15.0	7.0-25	3.8-7.0	4-8	4-8	4-8	4-8
Physical	19. Specific gravity	D792	1.70-1.98	1.65-1.85	1.09-1.17	1.28-1.32	1.22-1.34	1.15-1.22	1.17-1.24
	20. Water absorption (1/8-in. thick specimen), % 24 hr.	D570	0.12-0.35	0.2-0.5	0.8-1.8	2.0-7.0	1.7-6.5	0.9-2.2	1.2-2.8
	Saturation	D570							
	21. Dielectric strength (1/8-in. thick specimen), short time, v./mil	D149	400-450	400-450	350-500	250-600	250-600	250-475	300-475
		SUPPLIERS[a]	Cosmic Plastics; Rogers Corp.	Cosmic Plastics; Rogers Corp.	Dow Chem.; Federal Plastics; Rotuba	Rotuba	Albis; Eastman; Rotuba	Albis; Eastman; Rotuba	Albis; Eastman; Rotuba

	Cellulosic (Cont'd)	Chlorinated PE	Epoxy								
			Bisphenol molding compounds			Sheet molding compound (SMC)		Novolak molding compounds		Casting resins and compounds	
	Cellulose nitrate	30-42% Cl extrusion and molding grades	Glass fiber-reinforced	Mineral-filled	Low density glass sphere-filled	Glass fiber-reinforced	Carbon fiber-reinforced	Mineral- and glass-filled, encapsulation	Mineral- and glass-filled, high temperature	Unfilled	Silica-filled
1a.											
1.		125	Thermoset	Thermoset	Thermoset	Thermoset	Thermoset	Thermoset	Thermoset	Thermoset	Thermoset
								145-155	155-195		
2.	C: 185-250	E: 300-400	C: 300-330 T: 280-380	C: 250-330 T: 250-380	C: 250-300 I: 250-300	C: 250-330 T: 270-330	C: 250-330 T: 270-330	C: 280-360 I: 290-350 T: 250-380	T: 340-400		
3.	2-5		1-5	0.1-3	0.1-2	0.5-2.0	0.5-2.0	0.25-3.0	0.5-2.5		
4.			3.0-7.0	2.0-3.0	3.0-7.0	2.0	2.0		1.5-2.5		
5.			0.001-0.008	0.002-0.010	0.006-0.010	0.001	.001	0.004-0.008	0.004-0.007	0.001-0.010	0.0005-0.008
6.	7000-8000	1400-3000	5000-20,000	4000-10,800	2500-4000	20,000-35,000	40,000-50,000	5000-12,500	6000-15,500	4000-13,000	7000-13,000
7.	40-45	300-900	4			0.5-2.0	0.5-2.0			3-6	1-3
8.											
9.	2100-8000		18,000-40,000	18,000-40,000	10,000-15,000	20,000-30,000	30,000-40,000	24,000-48,000	30,000-48,000	15,000-25,000	15,000-35,000
10.	9000-11,000		8000-30,000	6000-18,000	5000-7000	50,000-70,000	75,000-95,000	10,000-21,800	10,000-21,800	13,000-21,000	8000-14,000
11.	190-220		3000	350		2000-4000	10,000	2100	2300-2400	350	
12.				650					660		
13.			2000-4500	1400-2000	500-750	2000-3000	5000	1400-2400	2300-2400		
						1500-2500					
						1200-1800					
14.	5-7		0.3-10.0	0.3-0.5	0.15-0.25	30-40	15-20	0.3-0.5	0.4-0.45	0.2-1.0	0.3-0.45
15.	R95-115		M100-112	M100-M112				M115		M80-110	M85-120
		Shore A60-76				B: 55-65	B: 55-65	Barcol 70-75	Barcol 78		
16.	80-120		11-50	20-60		12	3	18-43	35	45-65	20-40
17.	140-160		225-500	225-500	200-250	550	550	300-500	500	115-550	160-550
18.	5.5		4.0-10.0	4-35	4.0-6.0	1.7-1.9	1.4-1.5	10-31	17-24	4.5	10-20
19.	1.35-1.40	1.13-1.26	1.6-2.0	1.6-2.1	0.75-1.0	0.10	0.10	1.6-2.05	1.85-1.94	1.11-1.40	1.6-2.0
20.	1.0-2.0		0.04-0.20	0.03-0.20	0.2-1.0	1.4	1.6	0.04-0.29	0.15-0.17	0.08-0.15	0.04-0.1
								0.15-0.3			
21.	300-600		250-400	250-420	380-420			325-450	440-450	300-500	300-550
	P.D. George	Dow Plastics	Amoco Electronic; Cytec Fiberite	Amoco Electronic; Cytec Fiberite	Cytec Fiberite	Quantum Composites	Quantum Composites	Amoco Electronic; Cosmic Plastics; Cytec Fiberite; Rogers	Amoco Electronic; Cosmic Plastics; Cytec Fiberite; Rogers	Ciba Specialty Chemicals; Conap; Dow Plastics; Emerson & Cuming; Epic Resins; ITW Devcon; Shell; United Mineral	Conap; Emerson & Cuming; Epic Resins; ITW Devcon

Materials	Properties	ASTM test method	Epoxy (Cont'd) — Casting resins and compounds (Cont'd): Aluminum-filled	Epoxy (Cont'd) — Casting resins and compounds (Cont'd): Flexibilized	Epoxy (Cont'd) — Casting resins and compounds (Cont'd): Cyclo-aliphatic	Ethylene vinyl alcohol	Fluoroplastics: Polychloro-trifluoro-ethylene	Fluoroplastics — Polytetrafluoroethylene: Granular	Fluoroplastics — Polytetrafluoroethylene: 25% glass fiber-reinforced
Processing	1a. Melt flow (gm./10 min.)	D1238				0.8-14.0			
	1. Melting temperature, °C. T_m (crystalline)		Thermoset	Thermoset	Thermoset	142-191		327	327
	T_g (amorphous)					49-72	220		
	2. Processing temperature range, °F. (C = compression; T = transfer; I = injection; E = extrusion)					I: 365-480 E: 365-480	C: 460-580 I: 500-600 E: 360-590		
	3. Molding pressure range, 10^3 p.s.i.						1-6	2-5	3-8
	4. Compression ratio					3-4	2.6	2.5-4.5	
	5. Mold (linear) shrinkage, in./in.	D955	0.001-0.005	0.001-0.010			0.010-0.015	0.030-0.060	0.018-0.020
Mechanical	6. Tensile strength at break, p.s.i.	D638[b]	7000-12,000	2000-10,000	8000-12,000	5405-13,655	4500-6000	3000-5000	2000-2700
	7. Elongation at break, %	D638[b]	0.5-3	20-85	2-10	180-330	80-250	200-400	200-300
	8. Tensile yield strength, p.s.i.	D638[b]				7385-10,365	5300		
	9. Compressive strength (rupture or yield), p.s.i.	D695	15,000-33,000	1000-14,000	15,000-20,000		4600-7400	1700	1000-1400 @ 1% strain
	10. Flexural strength (rupture or yield), p.s.i.	D790	8500-24,000	1000-13,000	10,000-13,000	230-285	7400-11,000		2000
	11. Tensile modulus, 10^3 p.s.i.	D638[b]						58-80	200-240
	12. Compressive modulus, 10^3 p.s.i.	D695		1-350	495	300-385	150-300	60	
	13. Flexural modulus, 10^3 p.s.i. 73° F.	D790					170-200	80	190-235
	200° F.	D790					180-260		
	250° F.	D790							
	300° F.	D790							
	14. Izod impact, ft.-lb./in. of notch (1/8-in. thick specimen)	D256A	0.4-1.6	2.3-5.0		1.0-1.7	2.5-5	3	2.7
	15. Hardness Rockwell	D785	M55-85				R75-112		
	Shore/Barcol	D2240/ D2583		Shore D65-89			Shore D75-80	Shore D50-65	Shore D60-70
Thermal	16. Coef. of linear thermal expansion, 10^{-6} in./in./°C.	D696	5.5	20-100			36-70	70-120	77-100
	17. Deflection temperature under flexural load, °F. 264 p.s.i.	D648	190-600	73-250	200-450			115	
	66 p.s.i.	D648					258	160-250	
	18. Thermal conductivity, 10^{-4} cal.-cm./sec.-cm.^{2}-°C.	C177	15-25				4.7-5.3	6.0	8-10
Physical	19. Specific gravity	D792	1.4-1.8	0.96-1.35	1.16-1.21	1.12-1.20	2.08-2.2	2.14-2.20	2.2-2.3
	20. Water absorption (1/8-in. thick specimen), % 24 hr.	D570	0.1-4.0	0.27-0.5		6.7-8.6	0	<0.01	
	Saturation	D570							
	21. Dielectric strength (1/8-in. thick specimen), short time, v./mil	D149		235-400			500-600	480	320
		SUPPLIERS[a]	Conap; Emerson & Cuming; Epic Resins; ITW Devcon	Conap; Dow Plastics; Emerson & Cuming; Epic Resins; ITW Devcon	Ciba Specialty Chemicals; Union Carbide	Eval Co. of America	Ciba Specialty Chemicals; Elf Atochem N.A.	Ausimont; DuPont; Dyneon; ICI Americas	Ausimont; DuPont; Dyneon; ICI Americas; RTP

Fluoroplastics (Cont'd)

		Fluorinated ethylene propylene		Polyvinyl fluoride film		Polyvinylidene fluoride			Modified PE-TFE		
	PFA fluoro-plastic	Unfilled	20% milled glass fiber	Unfilled	Filled	Molding and extrusion	Wire and cable jacketing	EMI shielding (conductive); 30% PAN carbon fiber	Unfilled	25% glass fiber-reinforced	THV-200
1a.											
1.	300-310	275	262	192	192	141-178	168-170		270	270	120
						–60 to –20	–30 to –20				
2.	C: 625-700 I: 680-750	C: 600-750 I: 625-760	I: 600-700			C: 360-550 I: 375-550 E: 375-550	E: 420-525	I: 430-500	C: 575-625 I: 570-650	C: 575-625 I: 570-650	E: 450
3.	3-20	5-20	10-20			2-5	1.10		2-20	2-20	
4.	2.0	2.0				3	3				
5.	0.040	0.030-0.060	0.006-0.010			0.020-0.035	0.020-0.030	0.001	0.030-0.040	0.002-0.030	
6.	4000-4300	2700-3100	2400	6000-16,000	6000-16,000	3500-7250	7100	14,000	6500	12,000	3500
7.	300	250-330	5	100-250	75-250	12-600	300-500	0.8	100-400	8	600
8.	2100					2900-8250	4460				
9.	3500	2200				8000-16,000	6600-7100		7100	10,000	
10.			4000			9700-13,650	7000-8600	19,800	5500	10,700	
11.	70	50		300-380	300-380	200-80,000	145-190	2800	120	1200	
12.						304-420	180				
13.	95-120	80-95	250			170-120,000	145-260	2100	200	950	12
									80	450	
									60	310	
									20	200	
14.	No break	No break	3.2			2.5-80	7	1.5	No break	9.0	
15.						R79-83, 85	R77		R50	R74	
	Shore D64	Shore D60-65				Shore D80, 82 65-70	Shore D75		Shore D75		44
16.	140-210		22			70-142	121-140		59	10-32	
17.			150			183-244	129-165	318	160	410	
	166	158				280-284			220	510	
18.	6.0	6.0		0.0014-0.0017	0.0014-0.0017	2.4-3.1	2.4-3.1		5.7		
19.	2.12-2.17	2.12-2.17		1.38-1.40	1.38-1.72	1.77-1.78	1.76-1.77	1.74	1.7	1.8	1.95
20.	0.03	<0.01	0.01			0.03-0.06	0.03-0.06	0.12	0.03	0.02	
21.	500	500-600		2000-3300 *(D150-81)	2000-3300 *(D150-81)	260-280	260-280		400	425	
	Ausimont; DuPont; Dyneon	DuPont	RTP	DuPont	DuPont	Ausimont; Elf Atochem N.A.; Solvay Polymers	Ausimont; Elf Atochem N.A.; Solvay Polymers	RTP	DuPont	Ausimont; DuPont; RTP	Dyneon

Materials	Properties	ASTM test method	Fluoroplastics (Cont'd)			Ionomer		Ketones	
								Polyaryletherketone	
			THV-400	THV-500	PE-CTFE	Molding and extrusion	Glass- and rubber-modified; molding and extrusion	Unfilled	30% glass fiber-reinforced
Processing	1a. Melt flow (gm./10 min.)	D1238						4-7	15-25
	1. Melting temperature, °C. T_m (crystalline)		150	180	220-245	81-96	81-220	323-381	329-381
	T_g (amorphous)								
	2. Processing temperature range, °F. (C = compression; T = transfer; I = injection; E = extrusion)		E: 470	E: 480	C: 500 I: 525-575 E: 500-550	C: 280-350 E: 300-450	C: 300-400 I: 300-550 E: 350-525	I: 715-805	I: 715-805
	3. Molding pressure range, 10^3 p.s.i.				5-20	2-20	2-20	10-20	10-20
	4. Compression ratio					3	3	2	2
	5. Mold (linear) shrinkage, in./in.	D955			0.020-0.025	0.003-0.010	0.002-0.008	0.008-0.012	0.001-0.009
Mechanical	6. Tensile strength at break, p.s.i.	D638[b]	3400	3300	6000-7000	2500-5400	3500-7900	13,500	23,700-27,550
	7. Elongation at break, %	D638[b]	500	500	200-300	300-700	5-200	50	2.2-3.4
	8. Tensile yield strength, p.s.i.	D638[b]			4500-4900	1200-2300	1200-4500	15,000	
	9. Compressive strength (rupture or yield), p.s.i.	D695						20,000	30,000
	10. Flexural strength (rupture or yield), p.s.i.	D790			7000			18,850-24,500	34,100-36,250
	11. Tensile modulus, 10^3 p.s.i.	D638[b]			240	to 60		520-580	1410-1754
	12. Compressive modulus, 10^3 p.s.i.	D695							
	13. Flexural modulus, 10^3 p.s.i. 73° F.	D790		30	240	3-55	8-700	530	1600
	200° F.	D790						530	1520
	250° F.	D790						520	1460
	300° F.	D790						500	1390
	14. Izod impact, ft.-lb./in. of notch (1/8-in. thick specimen)	D256A			No break	7-No break	2.5-1.8 No break	1.6-2.7	1.8-1.9
	15. Hardness Rockwell	D785			R93-95	R53		M98	M102
	Shore/Barcol	D2240/ D2583	53	54	Shore D75	Shore D25-66	Shore D43-70	Shore D86	Shore D90
Thermal	16. Coef. of linear thermal expansion, 10^{-6} in./in./°C.	D696			80	100-170	50-100	41-44.2	18.5-20
	17. Deflection temperature under flexural load, °F. 264 p.s.i.	D648			170	93-100		323-338	619-662
	66 p.s.i.	D648			240	113-125	131-400	482-582	644-662
	18. Thermal conductivity, 10^{-4} cal.-cm./sec.-cm.^{2}-°C.	C177			3.8	5.7-6.6		7.1	
Physical	19. Specific gravity	D792	1.97	1.78	1.68-1.69	0.93-0.96	0.95-1.2	1.3	1.47-1.53
	20. Water absorption (1/8-in. thick specimen), % 24 hr.	D570			0.01	0.1-0.5	0.1-0.5	0.1	0.07
	Saturation	D570						0.8	0.5
	21. Dielectric strength (1/8-in. thick specimen), short time, v./mil	D149			490-520	400-450		355	370
		SUPPLIERS[a]	Dyneon	Dyneon	Ausimont	DuPont; Exxon; Network Polymers	DuPont; Network Polymers; A. Schulman	Amoco Polymers	RTP

	Ketones (Cont'd)						Liquid Crystal Polymer				
	Polyaryletherketone (Cont'd)			Polyetheretherketone							
	40% glass fiber-reinforced	Modified, 40% glass	30% carbon fiber	Unfilled	30% glass fiber-reinforced	30% carbon fiber-reinforced	45% glass fiber-reinforced, HDT	40% glass fiber-reinforced	30% mineral-filled	Glass fiber-reinforced for SMT	Unfilled medium melting point
1a.	15-25	15-25	15-25								
1.	329	324	329	334	334	334					280-421
2.	I: 715-805	I: 715-765	I: 715-805	I: 660-750 E: 660-720	I: 660-750	I: 660-800	I: 660-730	I: 590-645	I: 660-690	I: 610-680	I: 540-770
3.	10-20	10-20	10-20	10-20	10-20	10-20			4-8		1-16
4.	2	2	2	3	2-3	2	2.5-3	2.5-3	2.5-3	2.5-3	2.5-4
5.	0.001-0.009	0.001-0.009	0.002-0.008	0.011	0.002-0.014	0.0005-0.014					0.001-0.006
6.	25,000	22,500	30,000	10,200-15,000	22,500-28,500	29,800-33,000	18,200	21,240	15,950	19,600	15,900-27,000
7.	2	1.5	1.5	30-150	2-3	1-4	2.1	1.5	4.0	1.6	1.3-4.5
8.				13,200							
9.	32,500	33,000	33,800	18,000	21,300-22,400	25,000-34,400	10,900				6200-19,000
10.	40,000	34,000	40,000	16,000	33,000-42,000	40,000-48,000	22,400	27,000	18,415	25,000	19,000-35,500
11.	1900	2250	2700		1250-1600	1860-3500	2170	2890		2700	1400-2800
12.							491		590		400-900
13.	2100	2100	2850	560	1260-1600	1860-2600	22.4	2280	1400	1950	1770-2700
	2010	1960	2790	435	1400	1820					1500-1700
	1910	1750	2480		1350	1750					1300-1500
	1790	1500	2100	290	1100	1400					1200-1450
14.	2	1.2	1.5	1.6	2.1-2.7	1.5-2.1	2.0	8.3	3.0	1.8	1.7-10
15.	M102	M103									M76; R60-66
16.	18	18.5	7.9	<150°C: 40-47 <150°C: 108	<150°C: 12-22 >150°C: 44	<150°C: 15-22 >150°C: 5-44	12.5-80.2		8-22		5-7
17.	619	586	634	320	550-599	550-610	572	466	455	520	356-671
	644	643	652			615					
18.	10.5	8.8			4.9	4.9	2.02				2
19.	1.55	1.6	1.45	1.30-1.32	1.49-1.54	1.42-1.44	1.75	1.70	1.63	1.6	1.35-1.84
20.	0.05	0.04	<0.2	0.1-0.14	0.06-0.12	0.06-0.12	<.1			<.1	0-<0.1
				0.5	0.11-0.12	0.06					<0.1
21.	420	385					900				800-980
	RTP	Amoco Polymers; RTP	Amoco Polymers; RTP	Victrex USA	DSM; LNP; RTP; Victrex USA	DSM; LNP; RTP; Victrex USA	RTP	Amoco Polymers; RTP	DuPont; RTP	Amoco Polymers; RTP	Ticona

Materials		ASTM test method	Liquid Crystal Polymer (Cont'd)						
	Properties		30% carbon fiber-reinforced	50% mineral-filled	30% glass fiber-reinforced	30% glass fiber-reinforced, high HDT	Unfilled platable grade	PTFE-filled	15% glass fiber-reinforced
Processing	1a. Melt flow (gm./10 min.)	D1238							
	1. Melting temperature, °C. T_m (crystalline)		280	327	280-680	355		281	280
	T_g (amorphous)								
	2. Processing temperature range, °F. (C = compression; T = transfer; I = injection; E = extrusion)		555-600	I: 605-770	I: 555-770	I: 625-730	I: 600-620		
	3. Molding pressure range, 10^3 p.s.i.		1-14	1-14	1-14	4-8			
	4. Compression ratio		2.5-4	2.5-4	2.5-4	2.5-4	3-4	3:1	3:1
	5. Mold (linear) shrinkage, in./in.	D955	0-0.002	0.003	0.001-0.09	-0.01		0-3	0-3
Mechanical	6. Tensile strength at break, p.s.i.	D638[b]	35,000	10,400-16,500	16,900-30,000	18,000-21,025	13,500	24,500-25,000	28,000
	7. Elongation at break, %	D638[b]	1.0	1.1-2.6	1.7-2.7	1.7-2.2	2.9	3.0-5.2	3
	8. Tensile yield strength, p.s.i.	D638[b]			19,600				
	9. Compressive strength (rupture or yield), p.s.i.	D695	34,500	6800-7500	9900-21,000	9600-12,500			
	10. Flexural strength (rupture or yield), p.s.i.	D790	46,000	14,200-23,500	21,700-25,000	24,000-25,230		18,500-29,000	30,000
	11. Tensile modulus, 10^3 p.s.i.	D638[b]	5400	1500-2700	700-3000	2330-2600	1500	1100-1600	1100-2100
	12. Compressive modulus, 10^3 p.s.i.	D695	4800	490-2016	470-1000	447-770			
	13. Flexural modulus, 10^3 p.s.i. 73° F.	D790	4800	1250-2500	1660-2100	1800-2050	1500	1000-1400	1600
	200° F.	D790							
	250° F.	D790			900	1100			
	300° F.	D790			800	1100			
	14. Izod impact, ft.-lb./in. of notch (1/8-in. thick specimen)	D256A	1.4	0.8-1.5	2.0-3.0	2.0-4.2	0.6	2.1-3.8	5.5
	15. Hardness Rockwell	D785	M99	82	77-87, M61	M63			
	Shore/Barcol	D2240/ D2583							
Thermal	16. Coef. of linear thermal expansion, 10^{-6} in./in./°C.	D696	-2.65	9-65	4.9-77.7	14-36			
	17. Deflection temperature under flexural load, °F. 264 p.s.i.	D648	440	429-554	485-655, 271°C	518-568	410	352-435	430
	66 p.s.i.	D648			400-530				
	18. Thermal conductivity, 10^{-4} cal.-cm./sec.-cm.^{2}-°C.	C177		2.57	1.73	1.52			
Physical	19. Specific gravity	D792	1.49	1.84-1.89	1.60-1.67	1.6-1.66		1.50-1.62	1.5
	20. Water absorption (1/8-in. thick specimen), % 24 hr.	D570	<0.1	<0.1	<0.1%, 0.002	<0.1	0.03		
	Saturation	D570			0.05				0.2
	21. Dielectric strength (1/8-in. thick specimen), short time, v./mil	D149		900-955	640-1000	900-1050	600		
		SUPPLIERS[a]	RTP; Ticona	RTP; Ticona	Amoco Polymers; Dupont; RTP; Ticona	Amoco Polymers; Dupont; RTP; Ticona	Ticona	RTP; Ticona	RTP; Ticona

	Liquid Crystal Polymer (Cont'd)	Melamine formaldehyde		Phenolic						
				Molding compounds, phenol-formaldehyde						
									Impact-modified	
	Glass/mineral-filled	Cellulose-filled	Glass fiber-reinforced	Glass	PAN carbon	Woodflour-filled	Woodflour- and mineral-filled	High-strength glass fiber-reinforced	Cotton-filled	Cellulose-filled
1a.										
1.	280	Thermoset	Thermoset	Thermoset	Thermoset	Thermoset	Thermoset	Thermoset	Thermoset	Thermoset
2.		C: 280-370 I: 200-340 T: 300	C: 280-350			C: 290-380 I: 330-400	C: 290-380 I: 330-390 T: 290-350	C: 300-380 I: 330-390 T: 300-350	C: 290-380 I: 330-400	C: 290-380 I: 330-400
3.		8-20	2-8			2-20	2-20	1-20	2-20	2-20
4.	3:1	2.1-3.1	5-10			1.0-1.5		2.0-10.0	1.0-1.5	1.0-1.5
5.	0-3	0.005-0.015	0.001-0.006	0.0012	0.0016	0.004-0.009	0.003-0.008	0.001-0.004	0.004-0.009	0.004-0.009
6.	21,000-25,000	5000-13,000	5000-10,500	13,500	29,750	5000-9000	6500-7500	7000-18,000	6000-10,000	3500-6500
7.	1.4-2.3	0.6-1	0.6	1.6	1.15	0.4-0.8		0.2	1-2	1-2
8.										
9.	18,000-21,000	33,000-45,000	20,000-35,000	55,600	62,700	25,000-31,000	25,000-30,000	16,000-70,000	23,000-31,000	22,000-31,000
10.	31,000-34,000	9000-16,000	14,000-23,000	31,700	49,000	7000-14,000	9000-12,000	12,000-60,000	9000-13,000	5500-11,000
11.	2400-3100	1100-1400	1600-2400	1280	2490	800-1700	1000-1800	1900-3300	1100-1400	
12.	2000-2800							2740-3500		
13.	2200-2900	1100		1570	2800	1000-1200	1200-1300	1150-3300	800-1300	900-1300
14.	1.6-3.8	0.2-0.4	0.6-18	0.3	0.4	0.2-0.6	0.29-0.35	0.5-18.0	0.3-1.9	0.4-1.1
15.	76-79	M115-125	M115			M100-115	M90-110	E54-101	M105-120	M95-115
								Barcol 72		
16.	6-8	40-45	15-28	10-51	5-50	30-45	30-40	8-34	15-22	20-31
17.	437	350-390	375-400			300-370	360-380	350-600	300-400	300-350
18.		6.5-10	10-11.5			4-8	6-10	8-14	8-10	6-9
19.	1.68-1.89	1.47-1.52	1.5-2.0	1.50-1.52	1.37-1.40	1.37-1.46	1.44-1.56	1.69-2.0	1.38-1.42	1.38-1.42
20.		0.1-0.8	0.09-1.3			0.3-1.2	0.2-0.35	0.03-1.2	0.6-0.9	0.5-0.9
	0.2							0.12-1.5		
21.	840	270-400 175-215 @100°C	130-370			260-400	330-375	140-400	200-360	300-380
	RTP; Ticona	Cytec Fiberite; Patent Plastics; Perstorp; Plastics Mfg.	Cytec Fiberite	Cytec Fiberite	Cytec Fiberite	Amoco Electronic; OxyChem; Plaslok; Plastics Eng.; Rogers	Amoco Electronic; OxyChem; Plaslok; Plastics Eng.; Rogers	Cytec Fiberite; OxyChem; Plastics Eng.; Quantum Composites; Resinoid; Rogers	Bayer Corp.; Cytec Fiberite; OxyChem; Plaslok; Plastics Eng.; Resinoid; Rogers	Amoco Electronic; Cytec Fiberite; OxyChem; Plaslok; Plastics Eng.; Resinoid; Rogers

Materials			Phenolic (Cont'd)				Polyamide	
			Molding compounds, phenol-formaldehyde (Cont'd)		Casting resins		Nylon alloys	Nylon, Type 6
			Impact-modified (Cont'd)	Heat-resistant				
	Properties	ASTM test method	Fabric and rag-filled	Mineral- or mineral- and glass-filled	Unfilled	Mineral-filled	Ceramic and glass fiber-reinforced	Molding and extrusion compound
Processing	1a. Melt flow (gm./10 min.)	D1238	0.5-10					0.5-10
	1. Melting temperature, °C. T_m (crystalline)		Thermoset	Thermoset	Thermoset	Thermoset		210-220
	T_g (amorphous)							
	2. Processing temperature range, °F. (C = compression; T = transfer; I = injection; E = extrusion)		C: 290-380 I: 330-400 T: 300-350	C: 270-350 I: 330-380 T: 300-350				I: 440-550 E: 440-525
	3. Molding pressure range, 10^3 p.s.i.		2-20	2-20				1-20
	4. Compression ratio		1.0-1.5	2.1-2.7				3.0-4.0
	5. Mold (linear) shrinkage, in./in.	D955	0.003-0.009	0.002-0.006			0.002	0.003-0.015
Mechanical	6. Tensile strength at break, p.s.i.	D638[b]	6000-8000	6000-10,000	5000-9000	4000-9000	25,000-30,000	6000-24,000
	7. Elongation at break, %	D638[b]	1-4	0.1-0.5	1.5-2.0		3-4	30-100[c]; 300[d]
	8. Tensile yield strength, p.s.i.	D638[b]						13,100[c]; 7400[d]
	9. Compressive strength (rupture or yield), p.s.i.	D695	20,000-28,000	22,500-36,000	12,000-15,000	29,000-34,000		13,000-16,000[c]
	10. Flexural strength (rupture or yield), p.s.i.	D790	10,000-14,000	11,000-14,000	11,000-17,000	9000-12,000	39,000-45,000	15,700[c]; 5800[d]
	11. Tensile modulus, 10^3 p.s.i.	D638[b]	900-1100	2400	400-700			380-464[c]; 100-247[d]
	12. Compressive modulus, 10^3 p.s.i.	D695						250[d]
	13. Flexural modulus, 10^3 p.s.i. 73° F.	D790	700-1300	1000-2000			1800-2150	390-410[c]; 140[d]
	200° F.	D790						
	250° F.	D790						
	300° F.	D790						
	14. Izod impact, ft.-lb./in. of notch (1/8-in. thick specimen)	D256A	0.8-3.5	0.26-0.6	0.24-0.4	0.35-0.5	1.6-2.0	0.6-2.2[c]; 3.0[d]
	15. Hardness Rockwell	D785	M105-115	E88	M93-120	M85-120	R-120	R119[c]; M100-105[c]
	Shore/Barcol	D2240/ D2583		Barcol 70				
Thermal	16. Coef. of linear thermal expansion, 10^{-6} in./in./°C.	D696	18-24	19-38	68	75		80-83
	17. Deflection temperature under flexural load, °F. 264 p.s.i.	D648	325-400	275-475	165-175	150-175	410-495	155-185[c]
	66 p.s.i.	D648					420-515	347-375[c]
	18. Thermal conductivity, 10^{-4} cal.-cm./sec.-cm.^{2}-°C.	C177	9-12	10-24	3.5			5.8
Physical	19. Specific gravity	D792	1.37-1.45	1.42-1.84	1.24-1.32	1.68-1.70	1.59-1.81	1.12-1.14
	20. Water absorption (1/8-in. thick specimen), % 24 hr.	D570	0.6-0.8	0.02-0.3	0.1-0.36		0.35-0.50	1.3-1.9
	Saturation	D570		0.06-0.5				8.5-10.0
	21. Dielectric strength (1/8-in. thick specimen), short time, v./mil	D149	200-370	200-350	250-400	100-250		400[c]
		SUPPLIERS[a]	Cytec Fiberite; OxyChem; Resinoid; Rogers	Amoco Electronic; Cytec Fiberite; OxyChem; Plaslok; Plastics Eng.; Resinoid; Rogers	Ametek; Schenectady; Solutia; Union Carbide	Schenectady; Solutia	Network Polymers; RTP; Thermofil	Adell; Albis; AlliedSignal; ALM; Ashley Polymers; BASF; Bamberger Polymers; Bayer Corp.; ComAlloy; Custom Resins; DuPont; EMS; Federal Plastics; M.A. Hanna Eng.; Network Polymers; Nyltech; Polymer Resources; Polymers Intl.; A. Schulman; Thermofil; Ticona; Wellman

Polyamide (Cont'd)

Nylon, Type 6 (Cont'd)

								Semiaromatic semi-crystalline copolymer	
	15% glass fiber-reinforced	25% glass fiber-reinforced	30-35% glass fiber-reinforced	50% glass fiber-reinforced	30% long glass fiber-reinforced	40% long glass fiber-reinforced	50% long glass fiber-reinforced	35% glass fiber-reinforced	45% glass fiber-reinforced
1a.									
1.	220	220	210-220	220	210-220	210-271	220	300	300
2.	520-555	520-555	I: 460-550	535-575	I: 460-550	I: 460-550	I: 480-540	610-625	610-625
3.			2-20		10-20	10-20	10-20	5-20	5-20
4.			3.0-4.0		3.0-4.0	3.0-4.0	3-4	3	3
5.	$2\text{-}3 \times 10^{-3}$	2×10^{-3}	0.001-0.005	1×10^{-3}	0.003-0.009	0.002-0.010	0.002-0.008	0.002-0.003	0.001-0.002
6.	18,900^c; 10,200^d	23,200^c; 14,500^d	24-27,600^c; 18,900^d	33,400^c; 23,200^d	25,200-26,000^c	30,400-31,300	35,400-36,200^c	31,000-30,500	35,500-33,500
7.	3.5^c; 6^d	3.5^c; 5^d	2.2-3.6^c	3.0^c; 3.5^d	2.3-2.5^c	2.2-2.3	2.0-2.1^c	2.4-2.2	2.2-2.2
8.									
9.			19,000-24,000^c		24,000-32,200^c	33,800-37,400	39,700-39,900^c	48,600	48,600
10.			34-36,000^c; 21,000^d		38,800-40,000^c	45,700	53,900^c	44,100	47,500
11.	798^c; 508^d	1160^c; 798^d	1250-1600^c; 1090^d	2320^c; 1740^d	1300	1800	2200-2270^c	1,750,000	2,230,000
12.								550,000	560,000
13.	700^c; 420^d	910^c; 650^d	1250-1400^c; 800-950^d	1700^c; 1570^d	1200^c	1600	1930-2000^c	1,500,000	2,000,000
								1,450,000	1,770,000
								740,000	850,000
14.	1.1	2.0	2.1-3.4^c; 3.7-5.5^d		4.2^c	6.2-6.4	8.4-8.6^c	2.1	2.2
15.	M92^c; M74^d	M95^c; M83^d	M93-96^c; M78^d	M104^c; M93^d	M93-96	M93		124	124
16.	52	40	16-80	30	22^c			15-48	15-48
17.	374	410	392-420^c	419	420^c	405	415	500	502
	419	428	420-430^c	428	425^c				
18.			5.8-11.4						
19.	1.23	1.32	1.35-1.42	1.55	1.4	1.45	1.56	1.47	1.58
20.	2.6	2.3	0.90-1.2	1.5	1.3			0.4	0.27
	8.0	7.1	6.4-7.0	4.8				3.5	2.8
21.			400-450^c		400				
	Ashley Polymers; BASF; Bayer Corp.; ComAlloy; EMS; M.A. Hanna Eng.; M.A. Polymers; Network Polymers; Nyltech; Polymers Intl.; RTP; A. Schulman; Wellman	Ashley Polymers; BASF; Bayer Corp.; EMS; M.A. Hanna Eng.; Network Polymers; Nyltech; Polymers Intl.; RTP; A. Schulman	Adell; Albis; AlliedSignal; ALM; Ashley Polymers; BASF; Bamberger Polymers; Bayer Corp.; ComAlloy; DSM; EMS; Ferro; LNP; M.A. Hanna Eng.; M.A. Polymers; Network Polymers; Nyltech; Polymers Intl.; RTP; A. Schulman; Thermofil; Ticona; Wellman	Ashley Polymers; BASF; Bayer Corp.; ComAlloy; EMS; M.A. Hanna Eng.; Network Polymers; Nyltech; Polymers Intl.; RTP	Adell; ALM; DSM; Ferro; LNP; RTP; Ticona	Adell; ALM; DSM; Ferro; LNP; RTP; Ticona	RTP; Ticona	DuPont	DuPont

Materials	Properties	ASTM test method	Polyamide (Cont'd) — Nylon, Type 6 (Cont'd) — Toughened — Unreinforced	Toughened — 33% glass fiber-reinforced	Flame-retarded grade — 30% glass fiber-reinforced	40% mineral- and glass fiber-reinforced	40% mineral-reinforced
Processing	1a. Melt flow (gm./10 min.)	D1238					
	1. Melting temperature, °C. T_m (crystalline)		210-220	210-220	210-220	210-220	210-220
	T_g (amorphous)						
	2. Processing temperature range, °F. (C = compression; T = transfer; I = injection; E = extrusion)		I: 520-550	I: 520-550	I: 520-560	I: 450-550	I: 450-550
	3. Molding pressure range, 10^3 p.s.i.				12-25	2-20	2-20
	4. Compression ratio				3.0-4.0	3.0-4.0	3.0-4.0
	5. Mold (linear) shrinkage, in./in.	D955	0.006-0.02	0.001-0.003	0.001	0.003-0.006	0.003-0.006
Mechanical	6. Tensile strength at break, p.s.i.	D638[b]	6500-7900[c]; 5400[d]	17,800[c]	18,800-22,000[c]	17,400[c]; 19,000[d]	11,000-11,300
	7. Elongation at break, %	D638[b]	65.0-150[c]	4.0[c]	1.7-3.0[c]	3[c]; 2-6[d]	3.0
	8. Tensile yield strength, p.s.i.	D638[b]			22,000[c]	19,000-20,000	
	9. Compressive strength (rupture or yield), p.s.i.	D695			23,000[c]	14,000-18,000[c]	
	10. Flexural strength (rupture or yield), p.s.i.	D790	9100[c]	25,800[c]	28,300-31,000[c]	23,000-30,000[c]	
	11. Tensile modulus, 10^3 p.s.i.	D638[b]	290[c]; 102[d]		1200[c]-1700	1160-1400[c]; 725[d]	
	12. Compressive modulus, 10^3 p.s.i.	D695					
	13. Flexural modulus, 10^3 p.s.i. 73° F.	D790	250[c]	1110[c]	1160-1400[c]	900-1300[c]; 650-996[d]	650-700
	200° F.	D790					
	250° F.	D790					
	300° F.	D790					
	14. Izod impact, ft.-lb./in. of notch (1/8-in. thick specimen)	D256A	16.4[c]	3.5[c]	1.5[c]-2.2	0.6-4.2[c]; 5.0[d]	1.8-2.0
	15. Hardness Rockwell	D785				R118-121[c]	
	Shore/Barcol	D2240/ D2583					
Thermal	16. Coef. of linear thermal expansion, 10^{-6} in./in./°C.	D696				11-41	
	17. Deflection temperature under flexural load, °F. 264 p.s.i.	D648	135[c]; 122	400[c]	380-400[c]	390-405[c]	270-285
	66 p.s.i.	D648	158	430[c]	420	410-425[c]	
	18. Thermal conductivity, 10^{-4} cal.-cm./sec.-cm.^{2}-°C.	C177					
Physical	19. Specific gravity	D792	1.07; 1.08	1.33	1.62-1.7	1.45-1.50	1.45
	20. Water absorption (1/8-in. thick specimen), % 24 hr.	D570		0.86	0.5-0.6	0.6-0.9	
	Saturation	D570				4.0-6.0	
	21. Dielectric strength (1/8-in. thick specimen), short time, v./mil	D149				490-550[c]	
		SUPPLIERS[a]	Adell; Albis; AlliedSignal; Ashley Polymers; BASF; Bamberger Polymers; Bayer Corp.; Custom Resins; DSM; EMS; Ferro; M.A. Hanna Eng.; Network Polymers; Nyltech; Polymers Intl.; A. Schulman; Wellman	Adell; AlliedSignal; Ashley Polymers; BASF; Bamberger Polymers; Bayer Corp.; ComAlloy; DSM; EMS; Ferro; LNP; M.A. Hanna Eng.; Network Polymers; Polymers Intl.; RTP; A. Schulman; Wellman	AlliedSignal; ComAlloy; DSM; Ferro; LNP; M.A. Hanna Eng.; Network Polymers; Nyltech; Polymers Intl.; RTP; Ticona	Adell; AlliedSignal; Ashley Polymers; BASF; Bayer Corp.; ComAlloy; DSM; EMS; Ferro; LNP; M.A. Hanna Eng.; M.A. Polymers; Network Polymers; Nyltech; Polymers Intl.; RTP; A. Schulman; Thermofil; Wellman	Albis; AlliedSignal; Ashley Polymers; Bayer Corp.; M.A. Hanna Eng.; Network Polymers; Polymers Intl.; RTP; A. Schulman; Wellman

	Polyamide (Cont'd)									
	Nylon, Type 6 (Cont'd)									Nylon, Type 66
	High-impact copolymers and rubber-modified compounds	Unfilled with molybdenum disulfide	Impact-modified; 30% glass fiber-reinforced	EMI shielding (conductive); 30% PAN carbon fiber	Cast	Cast, heat-stabilized	Cast, oil-filled	Cast, plasticized	Cast, Type 612 blend	Molding compound
1a.	1.5-5.0									
1.	210-220	215	220	210-220	227-238	227-238	227-238	227-238	227-238	255-265
2.	I: 450-580 E: 450-550	I: 460-500	I: 480-550	I: 520-575						I: 500-620
3.	1-20	5-20	3-20							1-25
4.	3.0-4.0	3.0-4.0	3.0-4.0							3.0-4.0
5.	0.008-0.026	1.1	0.003-0.005	0.001-0.003						0.007-0.018
6.	6300-11,000[c]	11,500[c]	21,000[c]; 14,500[d]	30,000-36,000[c]	12,000-13,500	12,000-13,500	9,500-11,000	9,700-10,800	8,700-11,500	13,700[c]; 11,000[d]
7.	150-270[c]	60-80	5[c]-8[d]	2-3[c]	20-45	20-30	45-55	25-35	25-80	15-80[c]; 150-300[d]
8.	9000[c]-9500	11,500-12,300								8000-12,000[c]; 6500-8500[d]
9.	3900[c]			29,000[c]	16,000-18,000	16,000-18,000	12,000-14,000	22,000-25,000		12,500-15,000[c] (yld.)
10.	5000-12,000[c]	13,000		46,000-51,000[c]	15,500-17,500	15,000-16,000	14,000-16,000	12,000-13,000	15,000-20,000	17,900-1700[c]; 6100[d]
11.		440	1220[c]-754[d]	2800-3000[c]	485-550	485-550	375-475	375-440	240-330	230-550[c]; 230-500[d]
12.					300-350	300-310	275-375	250-275		
13.	110-320[c]; 130[d]	400	1160[c]-600[d]	2500-2700[c]	420-440	420-440	275-375	330-350	285-385	410-470[c]; 185[d]
	60-130[c]									
14.	1.8-No break[c] 1.8-No break[d]	0.9-1.0	2.2[c]-6[d]	1.5-2.8[c]	0.7-0.9	0.7-0.9	1.4-1.8	0.82-0.91	0.9-1.4	0.55-1.0[c]; 0.85-2.1[d]
15.	R81-113[c]; M50	R119-120		E70[c]	115-125	110-115	110-115	110-115	100-115	R120[c]; M83[c]; M95-M105[d]
					78-83	76-78	74-78	76-78	75-81	
16.	72-120		20-25	14.0[c]	50	45	35	45	40-45	80
17.	113-140[c]	200	410[c]	415-490[c]	330-400	330-400	330-400	330-400	330-400	158-212[c]
	260-367[c]	230	428[c]	425-505[c]	400-430	400-430	400-430	400-430	400-430	425-474[c]
18.										5.8
19.	1.07-1.17	1.17-1.18	1.33	1.28	1.15-1.17	1.15-1.17	1.14-1.15	1.14-1.16	1.10-1.13	1.13-1.15
20.	1.3-1.7	1.1-1.4	2.0	0.7-1.0	0.5-0.6	0.5-0.6	0.5-0.6	0.5-0.6	0.5-0.6	1.0-2.8
	8.5		6.2		5-6	5-6	2-2.5	5-6	4-5	8.5
21.	450-470[c]				500-600	500-600	500-600	500-600	500-600	600[c]
	Adell; AlliedSignal; Ashley Polymers; BASF; Bamberger Polymers; Bayer Corp.; Custom Resins; EMS; M.A. Hanna Eng.; M.A. Polymers; Network Polymers; Nyltech; Polymers Intl.; RTP; A. Schulman; Wellman	Ashley Polymers; DSM; LNP; M.A. Hanna Eng.; Nyltech; RTP; Ticona	Adell; Albis; AlliedSignal; Ashley Polymers; BASF; Bamberger Polymers; Bayer Corp.; ComAlloy; EMS; Ferro; LNP; M.A. Hanna Eng.; M.A. Polymers; Network Polymers; Polymers Intl.; RTP	ComAlloy; DSM; Ferro; LNP; Nyltech; RTP; Thermofil	Cast Nylons Ltd.	Cast Nylons Ltd.	Cast Nylons Ltd.	Cast Nylons Ltd.	Cast Nylons Ltd.	Adell; Albis; ALM; Ashley Polymers; BASF; Bamberger Polymers; Bayer Corp.; ComAlloy; DSM; DuPont; EMS; M.A. Hanna Eng.; MRC; Network Polymers; Nyltech; Polymer Resources; Polymers Intl.; A. Schulman; Solutia; Thermofil; Ticona; Wellman

Materials			Polyamide (Cont'd)				
			Nylon, Type 66 (Cont'd)				
	Properties	ASTM test method	13% glass fiber-reinforced, heat-stabilized	15% glass fiber-reinforced	30-33% glass fiber-reinforced	50% glass fiber-reinforced	30% long glass fiber-reinforced
Processing	1a. Melt flow (gm./10 min.)	D1238					
	1. Melting temperature, °C. T_m (crystalline)		257	260	260-265	260	260-265
	T_g (amorphous)						
	2. Processing temperature range, °F. (C = compression; T = transfer; I = injection; E = extrusion)		I: 520-570	535-575	I: 510-580	555-590	I: 530-570
	3. Molding pressure range, 10^3 p.s.i.		7-20		5-20		10-20
	4. Compression ratio		3.0-4.0		3.0-4.0		
	5. Mold (linear) shrinkage, in./in.	D955	0.005-0.009	$4x10^{-3}$	0.002-0.006	$1x10^{-3}$	0.003
Mechanical	6. Tensile strength at break, p.s.i.	D638[b]	15,000-17,000	18,900[c]; 11,600[d]	27,600[c]; 20,300[d]	33,400[c]; 26,100[d]	24,000-28,000[c]
	7. Elongation at break, %	D638[b]	3.0-5	3[c]; 6[d]	2.0-3.4[c]; 3-7[d]	2[c]; 3[d]	2.1-2.5[c]
	8. Tensile yield strength, p.s.i.	D638[b]			25,000[c]		
	9. Compressive strength (rupture or yield), p.s.i.	D695			24,000-40,000[c]		28,000-34,200[c]
	10. Flexural strength (rupture or yield), p.s.i.	D790	27,500[c]; 15,000[d]	25,600[c]; 17,800[d]	40,000[c]; 29,000[d]	46,500[c]; 37,500[d]	40,000-40,300[c]
	11. Tensile modulus, 10^3 p.s.i.	D638[b]		870[c]; 653[d]	1380[c]; 1090[d]	2320[c]; 1890[d]	
	12. Compressive modulus, 10^3 p.s.i.	D695					
	13. Flexural modulus, 10^3 p.s.i. 73° F.	D790	700-750	720[c]; 480[d]	1200-1450[c]; 800[d]; 900	1700[c]; 1460[d]	1200-1300[c]
	200° F.	D790					
	250° F.	D790					
	300° F.	D790					
	14. Izod impact, ft.-lb./in. of notch (1/8-in. thick specimen)	D256A	0.95-1.1	1.1	1.6-4.5[c]; 2.6-3.0[d]	2.5	4.0-5.1[c]
	15. Hardness Rockwell	D785	95M/R120	M97[c]; M87[d]	R101-119[c]; M101-102[c]; M96[d]	M102[c]; M98[d]	E60[c]
	Shore/Barcol	D2240/ D2583					
Thermal	16. Coef. of linear thermal expansion, 10^{-6} in./in./°C.	D696		52	15-54	0.33	23.4
	17. Deflection temperature under flexural load, °F. 264 p.s.i.	D648	450-470	482	252-490[c]	482	485-495
	66 p.s.i.	D648	494	482	260-500[c]	482	505
	18. Thermal conductivity, 10^{-4} cal.-cm./sec.-cm.^{2}-°C.	C177			5.1-11.7		
Physical	19. Specific gravity	D792	1.21-1.23	1.23	1.15-1.40	1.55	1.36-1.4
	20. Water absorption (1/8-in. thick specimen), % 24 hr.	D570	1.1		0.7-1.1		0.9
	Saturation	D570	7.1	7	5.5-6.5	4	
	21. Dielectric strength (1/8-in. thick specimen), short time, v./mil	D149			360-500		500
		SUPPLIERS[a]	Adell; Albis; ALM; Ashley Polymers; ComAlloy; DSM; LNP; M.A. Hanna Eng.; Network Polymers; Polymers Intl.; RTP; A. Schulman; Ticona; Wellman	Ashley Polymers; BASF; EMS; M.A. Hanna Eng.; Network Polymers; Nyltech; Polymers Intl.; RTP; Wellman	Adell; Albis; ALM; Ashley Polymers; BASF; Bamberger Polymers; Bayer Corp.; ComAlloy; DSM; DuPont; EMS; Ferro; LNP; M.A. Hanna Eng.; MRC; Network Polymers; Nyltech; Polymer Resources; Polymers Intl.; RTP; A. Schulman; Solutia; Thermofil; Ticona; Wellman	Ashley Polymers; BASF; ComAlloy; EMS; M.A. Hanna Eng.; Network Polymers; Nyltech; Polymers Intl.; RTP	Ashley Polymers; DSM; Ferro; LNP; RTP; Ticona

Polyamide (Cont'd)

	Nylon, Type 66 (Cont'd)							
				Toughened			Flame-retarded grade	
	40% long glass fiber-reinforced	50% long glass fiber-reinforced	60% long glass fiber-reinforced	Unreinforced	15-33% glass fiber-reinforced	Modified high-impact, 25-30% mineral-filled	Unreinforced	20-25% glass fiber-reinforced
1a.								
1.	257-265	260	260	240-265	256-265	250-260	249-265	200-265
2.	I: 520-570	I: 550-580	I: 560-600	I: 520-580	I: 530-575	I: 510-570	I: 500-560	I: 500-560
3.	10-20	10-20	10-20	1-20				
4.	3.0-4.0	3-4				10-20		
5.	0.002-0.10	0.002-0.007	0.002-0.006	0.012-0.018	0.0025-0.0045[c]	0.01-0.018	0.01-0.016	0.004-0.005
6.	32,800-32,900	37,200-38,000[c]	40,500-41,800[c]	7000[c]-11,000; 5800[d]	10,900-20,300[c]; 14,500[d]	9000-18,900[c]	8500[c]-10,600	20,300[c]-14,500[d]
7.	2.1-2.5	2.0-2.1[c]	2.0-2.1[c]	4-200[c]; 150-300[d]	4.7[c]; 8[d]	5-16[c]	4-10.0[c]	2.3-3[c]
8.				7250-7500[c]; 5500[d]			10,600	
9.	37,700-42,700	42,900-44,900[c]	47,000-47,900[c]		15,000[c]-20,000		25,000[c]	
10.	49,100	54,500-57,000[c]	65,000[c]	8500-14,500[c]; 4000[d]	17,400-29,900[c]	20,000[c]	14,000-15,000[c]	23,000[c]
11.	1700-1790	2270-2300[c]	2900-3170[c]	290[c]-123[d]	1230[c]; 943[d]		420[c]	1230[c]; 870[d]
12.								
13.	1560-1600	1900-1920[c]	2580-2600[c]	225-380[c]; 125-150[d]	479-1100[c]	600[c]-653	400-420[c]	1102
	810	990	1440					
14.	6.2; 6.6[c]; 6.9	8.0-9.2[c]	10.0-11.0[c]	12.0[c]-19.0; 3-N.B.[c]; 1.4-N.B.[d]	>3.2-5.0	1.0-3.0[c]	0.5-1.5[c]	1.1[c]
15.				M60[c]; R100[c]; R107[c]; R113; R114-115[c]; M50[d]	R107[c]; R115; R116; M86[c]; M70[d]	R120; M86[c]	M82[c]; R119	M98[c]; M90[d]
16.				80	43	30		50
17.	490	500	505	140-175[c]	446-470	300-470	170-200	482
	490			385-442[c]	480-495	399-460	410-415	482
18.								
19.	1.45	1.56	1.69	1.06-1.11	1.2-1.34	1.28-1.4	1.25-1.42	1.3-1.51
20.				0.8-2.3	0.7-1.5	0.9-1.1	0.9-1.1	0.7
				7.2	5			
21.							520	430
	DSM; Ferro; LNP; RTP; Ticona	Ashley Polymers; RTP; Ticona	RTP; Ticona	Adell; ALM; Ashley Polymers; BASF; Bamberger Polymers; ComAlloy; DSM; DuPont; Ferro; LNP; M.A. Hanna Eng.; MRC; Network Polymers; Nyltech; Polymers Intl.; RTP; A. Schulman; Solutia; Ticona; Wellman	Adell; ALM; Ashley Polymers; BASF; Bamberger Polymers; ComAlloy; DSM; DuPont; EMS; Ferro; LNP; M.A. Hanna Eng.; Network Polymers; Nyltech; Polymers Intl.; RTP; Solutia; Ticona; Wellman	ALM; Ashley Polymers; ComAlloy; DSM; Ferro; LNP; M.A. Hanna Eng.; Network Polymers; RTP; Wellman	Ashley Polymers; BASF; ComAlloy; DSM; DuPont; EMS; M.A. Hanna Eng.; Nyltech; Polymers Intl.; RTP; Wellman	BASF; ComAlloy; DSM; DuPont; Ferro; LNP; M.A. Hanna Eng.; Nyltech; Polymers Intl.; RTP; Wellman

Materials	Properties	ASTM test method	Polyamide (Cont'd)					
			Nylon, Type 66 (Cont'd)					
					EMI shielding (conductive)			
			40% glass- and mineral-reinforced	40-45% mineral-filled	30% graphite or PAN carbon fiber	40% aluminum flake	5% stainless steel, long fiber	6% stainless steel, long fiber
Processing	1a. Melt flow (gm./10 min.)	D1238						
	1. Melting temperature, °C. T_m (crystalline)		250-260	250-265	258-265	265	260-265	260-265
	T_g (amorphous)							
	2. Processing temperature range, °F. (C = compression; T = transfer; I = injection; E = extrusion)		I: 510-590	I: 520-580	I: 500-590	I: 525-600	I: 530-570	I: 530-570
	3. Molding pressure range, 10^3 p.s.i.		9-20	5-20	10-20	10-20		5-18
	4. Compression ratio		3-4	3.0-4.0				
	5. Mold (linear) shrinkage, in./in.	D955	0.001-0.005	0.012-0.022	0.001-0.003	0.005	0.004	0.004-0.006
Mechanical	6. Tensile strength at break, p.s.i.	D638[b]	15,500-31,000[c]; 13,100[d]	14,000[c]; 11,000[d]	27,600-35,000[c]	6000[c]	10,000[c]	10,000-11,300[c]
	7. Elongation at break, %	D638[b]	2-7[c]	5-10[c]; 16[d]	1-4[c]	4[c]	5.0[c]	2.9; 5.0[c]
	8. Tensile yield strength, p.s.i.	D638[b]		13,900				
	9. Compressive strength (rupture or yield), p.s.i.	D695	18,000-37,000[c]	15,500-22,000[c]	24,000-29,000[c]	7500[c]		
	10. Flexural strength (rupture or yield), p.s.i.	D790	24,000-48,000[c]	22,000[c]; 9000[d]	45,000-51,000[c]	11,700[c]	16,000[c]	16,000-38,700[c]
	11. Tensile modulus, 10^3 p.s.i.	D638[b]		900[c]; 500[d]	3200-3400[c]	720[c]	450[c]	450-1500[c]
	12. Compressive modulus, 10^3 p.s.i.	D695		370[c]				
	13. Flexural modulus, 10^3 p.s.i. 73° F.	D790	985-1750[c]; 600[d]	900-1050[c]; 400[d]	1500-2900[c]	690[c]	450[c]	500-1400
	200° F.	D790	600[c]					
	250° F.	D790						
	300° F.	D790						
	14. Izod impact, ft.-lb./in. of notch (1/8-in. thick specimen)	D256A	0.6-3.8[c]	0.9-1.4[c]; 3.9[d]	1.3-2.5[c]	2.5[c]	1.3[c]	0.7; 2.2
	15. Hardness Rockwell	D785	M95-98[c]	R106-121[c]	R120[c]; M106[c]	R114[c]		
	Shore/Barcol	D2240/ D2583						
Thermal	16. Coef. of linear thermal expansion, 10^{-6} in./in./°C.	D696	20-54	27	11-16	22		
	17. Deflection temperature under flexural load, °F. 264 p.s.i.	D648	432-485[c]	300-438	470-500	380	285	175-480
	66 p.s.i.	D648	480-496[c]	320-480[c]	500-510	400	295	
	18. Thermal conductivity, 10^{-4} cal.-cm./sec.-cm.^{2}-°C.	C177	11	9.6	24.1			
Physical	19. Specific gravity	D792	1.42-1.55	1.39-1.5	1.28-1.43	1.48	1.27	1.19-1.45
	20. Water absorption (1/8-in. thick specimen), % 24 hr.	D570	0.4-0.9	0.6-0.55	0.5-0.8	1.1	0.12	
	Saturation	D570	5.1	6.0-6.5				
	21. Dielectric strength (1/8-in. thick specimen), short time, v./mil	D149	300-525	450[c]				
		SUPPLIERS[a]	Adell; ALM; Ashley Polymers; BASF; ComAlloy; DSM; DuPont; EMS; Ferro; LNP; M.A. Hanna Eng.; MRC; Network Polymers; Nyltech; RTP; Solutia; Thermofil; Ticona; Wellman	Adell; Albis; ALM; Ashley Polymers; ComAlloy; DSM; DuPont; Ferro;LNP; M.A. Hanna Eng.; MRC; Network Polymers; Nyltech; RTP; Solutia; Thermofil; Ticona; Wellman	ComAlloy; DSM; Ferro; LNP; RTP; Thermofil	Thermofil	DSM; RTP	RTP; Ticona

Polyamide (Cont'd)										
	Nylon, Type 66 (Cont'd)									
	EMI shielding (conductive) (Cont'd)							Lubricated		
	10% stainless steel, long fiber	50% PAN carbon fiber	30% pitch carbon fiber	40% pitch carbon fiber	15% nickel-coated carbon fiber	40% nickel-coated carbon fiber	Antifriction molybdenum disulfide-filled	5% silicone	10% PTFE	30% PTFE
1a.										
1.	257-265	260-265	260-265	260-265	260-265	260-265	249-265	260-265	260-265	260-265
2.	I: 530-570	I: 530-570	I: 530-570	I: 530-570	I: 530-570	I: 530-570	I: 500-600	I: 530-570	I: 530-570	I: 530-570
3.	5-20						5-25			
4.	3.0-4.0									
5.	0.003-0.004	0.0005	0.003	0.002	0.005	0.001	0.007-0.018	0.015	0.01	0.007
6.	11,500	38,000^c	15,500^c-15,600	17,500^c; 19,250	14,000^c	20,000^c	10,500-13,700^c	8500^c	9500^c	5500^c
7.	2.6	1.2^c	2.0^c	1.5^c	1.6^c	2.5^c	4.4-40^c			
8.										
9.			20,500	24,000			12,000-12,500^c			
10.	18,100	54,000^c	24,800-26,000^c	28,000^c-29,000	21,000^c	27,000^c	15,000-20,300^c	15,000^c	13,000^c	8000^c
11.	600	5000^c	1400-2000^c	2250-2600^c	1100^c		350-550^c			
12.										
13.	500	4200^c	1200-1500^c	1800-2000^c	1000^c	2000^c	420-495^c	300^c	420^c	460^c
14.	0.7; 1.9^c	2.0^c	0.6-0.7^c	0.7-0.8^c	0.7^c	1.0^c	0.9-4.5^c	1.0^c	0.8^c	0.5^c
15.							R119^c			
16.			16.0-19.0	9.0-14.0			54	63.0	35.0	45.0
17.	175-480	495	465-490	475-490	460	470	190-260^c	170	190	180
		505	490-500	498-500			395-430			
18.										
19.	1.24	1.38	1.30-1.31	1.36-1.38	1.20	1.46	1.15-1.18	1.16	1.20	1.34
20.		0.5	0.6	0.5	1.0	0.8	0.8-1.1	1.0	0.7	0.55
							8.0			
21.							360^c			
	RTP; Ticona	DSM; LNP; RTP	ComAlloy; LNP; Polymers Intl.; RTP	ComAlloy; LNP; RTP	DSM; RTP	DSM; RTP	Adell; ALM; Ashley Polymers; ComAlloy; DSM; LNP; M.A. Hanna Eng.; Nyltech; RTP; Thermofil	ComAlloy; DSM; Ferro; LNP; M.A. Hanna Eng.; RTP	ComAlloy; DSM; Ferro; LNP; M.A. Hanna Eng.; RTP	ComAlloy; DSM; Ferro; LNP; M.A. Hanna Eng.; RTP

Materials	Properties	ASTM test method	Polyamide (Cont'd)					
			Nylon, Type 66 (Cont'd)		Nylon, Type 610			
			Lubricated (Cont'd)					Flame-retarded grade
			5% Molybdenum disulfide and 30% PTFE	Copolymer	Molding and extrusion compound	30-40% glass fiber-reinforced	30-40% long glass fiber-reinforced	30% glass fiber-reinforced
Processing	1a. Melt flow (gm./10 min.)	D1238						
	1. Melting temperature, °C. T_m (crystalline)		260-265	200-255	220	220	220	220
	T_g (amorphous)							
	2. Processing temperature range, °F. (C = compression; T = transfer; I = injection; E = extrusion)		I: 530-570	I: 430-500	I: 445-485 E: 480-500	I: 510-550	I: 510-550	I: 500-560
	3. Molding pressure range, 10^3 p.s.i.			1-15	1-19			
	4. Compression ratio				3-4			
	5. Mold (linear) shrinkage, in./in.	D955	0.01	0.006-0.015	0.005-0.015	0.015-0.04	0.013-0.03	0.002
Mechanical	6. Tensile strength at break, p.s.i.	D638[b]	7500[c]	7400-12,400[c]	10,150[c]; 7250[d]	22,000-26,700[c]	22,400-25,400[c]	19,000[c]
	7. Elongation at break, %	D638[b]		40-150[c]; 300[d]	70[c]; 150[d]	4.3-4.7[c]	4.0-4.1[c]	3.5[c]
	8. Tensile yield strength, p.s.i.	D638[b]						
	9. Compressive strength (rupture or yield), p.s.i.	D695				20,400-21,000[c]	20,000[c]	23,000[c]
	10. Flexural strength (rupture or yield), p.s.i.	D790	1200[c]	12,000	350[c]; 217[d]	32,700-38,000[c]	34,000-37,400[c]	28,000[c]
	11. Tensile modulus, 10^3 p.s.i.	D638[b]		150-410[c]		800[c]	950-1600[c]	
	12. Compressive modulus, 10^3 p.s.i.	D695						
	13. Flexural modulus, 10^3 p.s.i. 73° F.	D790	400[c]	150-410[c]		1150-1500[c]	1200-1430[c]	1230[c]
	200° F.	D790						
	250° F.	D790						
	300° F.	D790						
	14. Izod impact, ft.-lb./in. of notch (1/8-in. thick specimen)	D256A	0.6[c]	0.7[c]; No break[d]		1.6-2.4	3.2-4.2[c]	1.5[c]
	15. Hardness Rockwell	D785		R114-119; R83[d]; M75[c]		E43-48[c]	E42-56[c]	M89[c]
	Shore/Barcol	D2240/ D2583						
Thermal	16. Coef. of linear thermal expansion, 10^{-6} in./in./°C.	D696						
	17. Deflection temperature under flexural load, °F. 264 p.s.i.	D648	185	135-170[c]		410-415	425	390
	66 p.s.i.	D648		430[c]-440		430	430	
	18. Thermal conductivity, 10^{-4} cal.-cm./sec.-cm.^{2}-°C.	C177						
Physical	19. Specific gravity	D792	1.37	1.08-1.14		1.3-1.4	1.33-1.39	1.55
	20. Water absorption (1/8-in. thick specimen), % 24 hr.	D570	0.55	1.5-2.0	1.4	0.17-0.19	0.17-0.21	0.16
	Saturation	D570		9.0-10.0	3.3			
	21. Dielectric strength (1/8-in. thick specimen), short time, v./mil	D149		400[c]			500	
		SUPPLIERS[a]	ComAlloy; DSM; M.A. Hanna Eng.; RTP	AlliedSignal; Ashley Polymers; BASF; DuPont; EMS; M.A. Hanna Eng.; Nyltech; Polymers Intl.; Solutia	EMS; M.A. Hanna Eng.	Ashley Polymers; DSM; Ferro; LNP; M.A. Hanna Eng.; RTP	DSM; LNP	DSM; LNP; RTP

Polyamide (Cont'd)

	Nylon, Type 612									Nylon, Type 46	
				Toughened		Flame-retarded grade	Lubricated				
	Molding compound	30-35% glass fiber-reinforced	35-45% long glass fiber-reinforced	Unreinforced	33% glass fiber-reinforced	30% glass fiber-reinforced	10% PTFE	15% PTFE, 30% glass fiber-reinforced	10% PTFE, 30% PAN carbon fiber	Extrusion	Unreinforced
1a.											
1.	195-219	213-217	195-217	195-217	195-217	195-217	195-217	195-217	195-217	295	295
2.	I: 450-550 E: 464-469	I: 450-550	I: 510-550	I: 510-550	I: 510-550	I: 500-560	I: 510-550	I: 510-550	I: 510-550	E: 560-590	I: 570-600
3.	1-15	4-20									5-15
4.										3-4	3-4
5.	0.011	0.002-0.005	0.001-0.002				0.012	0.002-0.003	0.0013		0.018-0.020
6.	6500-8800[c]	22,000[c]; 20,000[d]	26,000-29,000[c]	5500[c]	18,000[c]	18,000-19,000[c]	7000[c]	19,500-20,000[c]	28,000[c]	8,500	14,400
7.	150[c]; 300[d]	4[c]; 5[d]	2.9-3.2[c]	40[c]	5[c]	2.0-3.5[c]		2.5		60	25
8.	5800-8400[c]; 3100[d]										
9.		22,000[c]	23,000[c]			15,000-21,000[c]		19,000			13,000
10.	11,000[c]; 4300[d]	32,000-35,000[c]	39,000-44,000[c]	6500[c]	27,000[c]	28,000[c]		30,500-31,000[c]	42,000[c]	11,500	21,700
11.	218-290[c]; 123-180[d]	1200[c]; 900[d]				1000[c]-1400		1200		250	435
12.											319
13.	240-334[c]; 74-100[d]	1100-1200[c]; 900[d]	1200-1500[c]	195[c]	1050[c]	1200-1250[c]	100[c]	1100-1200[c]	2600[c]	270	460
14.	1.0-1.9[c]; 1.4-No break[d]	1.8-2.6[c]	4.2-6.3[c]	12.5[c]	4.5[c]	1.0-1.5[c]	1.0[c]	2.5-3.0[c]	2.4[c]	17	1.8
15.	M78[c]; M34[d]; R115	M93[c]; E40-50[d]; R116	E40[c]			M89[c]					R113
	D72-80[c]; D63[d]										D85
16.			21.6-25.2					18.0			
17.	136-180[c]	390-425[c]	410-415	135	385	385-390	202[c]	385	390	194	320
	311-330[c]	400-430[c]	420-425			400					
18.	5.2	10.2									
19.	1.06-1.10	1.30-1.38	1.34-1.45	1.03	1.28	1.55-1.60	1.13	1.42-1.45	1.30	1.10	1.18
20.	0.37-1.0	0.2	0.2	0.3	0.2	0.16	0.2	0.13	0.15	1.84	2.3
	2.5-3.0	1.85									
21.	400[c]	520[c]				450					673
	ALM; Ashley Polymers; Creanova; DuPont; EMS; M.A. Hanna Eng.; A. Schulman	ALM; Ashley Polymers; ComAlloy; DSM; DuPont; Ferro; LNP; M.A. Hanna Eng.; RTP	DSM; RTP	DSM; DuPont; M.A. Hanna Eng.; RTP	DSM; DuPont; LNP; M.A. Hanna Eng.; RTP	ComAlloy; DSM; LNP; M.A. Hanna Eng.; RTP	ComAlloy; DSM; Ferro; LNP; M.A. Hanna Eng.; RTP	ComAlloy; DSM; Ferro; LNP; M.A. Hanna Eng.; RTP	ComAlloy; DSM; LNP; RTP	DSM	DSM

Materials	Properties	ASTM test method	Polyamide (Cont'd) Nylon, Type 46 (Cont'd) Super-tough	15% glass-reinforced	15% glass-reinforced, V-0	30% glass-reinforced	30% glass-reinforced, V-0	50% glass-reinforced	50% glass and mineral-filled
Processing	1a. Melt flow (gm./10 min.)	D1238							
	1. Melting temperature, °C. T_m (crystalline)		295	295	295	295	295	295	295
	T_g (amorphous)								
	2. Processing temperature range, °F. (C = compression; T = transfer; I = injection; E = extrusion)		I: 570-600	I: 570-600	I: 570-600	I: 570-600	I: 570-600	I: 570-600	I: 570-600
	3. Molding pressure range, 10^3 p.s.i.		5-15	5-15	5-15	5-15	5-15	5-15	5-15
	4. Compression ratio		3-4	3-4	3-4	3-4	3-4	3-4	3-4
	5. Mold (linear) shrinkage, in./in.	D955	0.018-0.020	0.005-0.009	0.006-0.009	0.004-0.006	0.004-0.006	0.002-0.004	0.003-0.006
Mechanical	6. Tensile strength at break, p.s.i.	D638[b]	8,500	21,500	16,500	30,000	23,000	34,000	20,000
	7. Elongation at break, %	D638[b]	60	3	8	4	3	3	2
	8. Tensile yield strength, p.s.i.	D638[b]							
	9. Compressive strength (rupture or yield), p.s.i.	D695				33,000	34,000		
	10. Flexural strength (rupture or yield), p.s.i.	D790	11,500	31,900	27,000	43,000	34,000	50,750	34,000
	11. Tensile modulus, 10^3 p.s.i.	D638[b]	250	841	1,000	1,300	1,500	2,320	2,100
	12. Compressive modulus, 10^3 p.s.i.	D695				507	688		
	13. Flexural modulus, 10^3 p.s.i. 73° F.	D790	270	798	1125	1,200	1,300	2,030	1700
	200° F.	D790							
	250° F.	D790							
	300° F.	D790							
	14. Izod impact, ft.-lb./in. of notch (1/8-in. thick specimen)	D256A	17	1.6	.5	2.0	1.3	2.2	1.1
	15. Hardness Rockwell	D785				R120	R120		
	Shore/Barcol	D2240/ D2583				D89	D88		
Thermal	16. Coef. of linear thermal expansion, 10^{-6} in./in./°C.	D696							
	17. Deflection temperature under flexural load, °F. 264 p.s.i.	D648	194	480	480	545	545	545	545
	66 p.s.i.	D648							
	18. Thermal conductivity, 10^{-4} cal.-cm./sec.-cm.^{2}-°C.	C177							
Physical	19. Specific gravity	D792	1.10	1.3	1.47	1.41	1.68	1.62	1.6
	20. Water absorption (1/8-in. thick specimen), % 24 hr.	D570	1.84			1.5	0.9	1.15	
	Saturation	D570							
	21. Dielectric strength (1/8-in. thick specimen), short time, v./mil	D149				863	838		
		SUPPLIERS[a]	DSM	DSM	DSM	DSM	DSM	DSM	DSM

	Polyamide (Cont'd)								Polyamide-imide	
	Nylon, Type 11	Nylon, Type 12	Aromatic polyamide		Partially aromatic					
	Molding and extrusion compound	Molding and extrusion compound	Amorphous transparent copolymer	Aramid molded parts, unfilled	35% glass fiber-reinforced impact modified	40% glass fiber-reinforced	50% glass fiber-reinforced	60% glass fiber-reinforced	Unfilled compression and injection molding compound	30% glass fiber-reinforced
1a.										
1.	180-190	160-209		275	260	260	260	260		
		125-155							275	275
2.	I: 390-520 E: 390-475	I: 356-525 E: 350-500	I: 480-610 E: 520-595		270-310	270-310	270-310	270-310	C: 600-650 I: 610-700	C: 630-650 I: 610-700
3.	1-15	1-15	5-20						2-40	15-40
4.	2.7-3.3	2.5-4							1.0-1.5	1.0-1.5
5.	0.012	0.003-0.015	0.004-0.007		0.15/0.95	0.15/0.95	0.15/0.90	0.10/0.80	0.006-0.0085	0.001-0.0025
6.	8000[c]-9,500	5100-10,000[c]; 8000[d]	7600-14,000[c]; 13,000[d]	17,500[c]	26,000	31,500	34,700	36,200	22,000	
7.	300[c]-400	250-390[c]	40-150[c]; 260[d]	5[c]	5	3	3	3	15	7
8.		3000-6100[c]	11,000-14,861[c]; 11,000[d]						27,800	29,700
9.	7300-7800		17,500[c]; 14,000[d]	30,000[c]			28,800		32,100	38,300
10.		1400-8100[c]	10,000-16,400[c]; 14,000[d]	25,800[c]	39,400	48,900	54,400	57,100	34,900	48,300
11.	185[c]	36-180[c]	275-410[c]; 270[d]						700	1560
12.	180[c]		340[c]	290[c]						1150
13.	44-180[c]	27-190[c]	306-400[c]; 350[d]	640[c]	1,270	1,634	2,106	2,680	730	1700
	20									
14.	1.8[c]-N.B.	1.0-N.B.	0.8-3.5[c]; 1.8-2.7[d]	1.4[c]	2.9	2.3	2.3	2.3	2.7	1.5
15.	R108[c]; R80	R70-109[c]; 105[d]	M77-93[c]	E90[c]	86	89	89	91	E86	E94
		D58-75[d]	D83[c]; D85[d]							
16.	100	61-100	28-70	40	18	14	14	11	30.6	16.2
17.	104-126	95-135[c]	170-268	500[c]	440	455	460	460	532	539
	300[c]	158-302[c]	261-330[c]							
18.	8	5.2-7.3	5	5.2					6.2	8.8
19.	1.03-1.05	1.01-1.02	1.0-1.19	1.30	1.39	1.47	1.58	1.72	1.42	1.61
20.	0.4	0.25-0.30	0.4-1.36	0.6	0.59	0.56	0.45	0.36		
	1.9	0.75-1.6	1.3-4.2		5.0	4.6	4.0	3.80	0.33	0.24
21.	650-750	450[c]	350[c]	800[c]					580	840
	Elf Atochem N.A.	ALM; Ashley Polymers; Creanova; Elf Atochem N.A.; EMS	AlliedSignal; Ashley Polymers; Bayer Corp.; Creanova; DuPont; EMS; M.A. Hanna Eng.; Nyltech	DuPont	EMS	EMS	EMS	EMS	Amoco Polymers	Amoco Polymers

Materials	Properties	ASTM test method	Polyamide-imide (Cont'd)						Poly-buta-diene
			Graphite fiber-reinforced	Bearing grade	High compressive strength	Wear resistant for speeds	Stiffness and lubricity	Cost/performance ratio	Casting resin
Processing	1a. Melt flow (gm./10 min.)	D1238							
	1. Melting temperature, °C. T_m (crystalline)								Thermoset
	T_g (amorphous)		275	275					
	2. Processing temperature range, °F. (C = compression; T = transfer; I = injection; E = extrusion)		C: 630-650 I: 610-700	I: 580-700					
	3. Molding pressure range, 10^3 p.s.i.		15-40	15-40					
	4. Compression ratio		1.0-1.5	1.0-1.5					
	5. Mold (linear) shrinkage, in./in.	D955	0.000-0.0015	0.0025-0.0045	3.5-6.0	3.5-6.0	0.0-1.5		
Mechanical	6. Tensile strength at break, p.s.i.	D638[b]							
	7. Elongation at break, %	D638[b]	6	7	7	9	6	7	
	8. Tensile yield strength, p.s.i.	D638[b]	26,000	22,000	23,700	17,800	26,000	31,800	
	9. Compressive strength (rupture or yield), p.s.i.	D695	36,900		24,100	18,300	30,300	46,700	
	10. Flexural strength (rupture or yield), p.s.i.	D790	50,700	27,100-31,200	31,200	27,000	40,100	52,000	8000-14,000
	11. Tensile modulus, 10^3 p.s.i.	D638[b]	3220	870-1130	950	870		2020	560
	12. Compressive modulus, 10^3 p.s.i.	D695	1430						
	13. Flexural modulus, 10^3 p.s.i. 73° F.	D790	2880	910-1060	1000	910	2440	2100	
	200° F.	D790							
	250° F.	D790							
	300° F.	D790							
	14. Izod impact, ft.-lb./in. of notch (1/8-in. thick specimen)	D256A	0.9	1.2-1.6	1.2	1.3	1.0	1.5	
	15. Hardness Rockwell	D785	E94	E66-E72	72	66		107	R40
	Shore/Barcol	D2240/ D2583							
Thermal	16. Coef. of linear thermal expansion, 10^{-6} in./in./°C.	D696	9.0	25.2-27.0	14	15	7	7	
	17. Deflection temperature under flexural load, °F. 264 p.s.i.	D648	540	532-536	534	532	534	536	
	66 p.s.i.	D648							
	18. Thermal conductivity, 10^{-4} cal.-cm./sec.-cm.^{2}-°C.	C177	12.7		3.7				
Physical	19. Specific gravity	D792	1.48	1.46-1.51	1.46	1.50	1.50		0.97
	20. Water absorption (1/8-in. thick specimen), % 24 hr.	D570		0.17-0.33					0.03
	Saturation	D570	0.26	0.33	0.28	0.17		0.21	
	21. Dielectric strength (1/8-in. thick specimen), short time, v./mil	D149					Conductive	490	630
		SUPPLIERS[a]	Amoco Polymers	Amoco Polymers	Amoco Polymers	Amoco Polymers	Amoco Polymers	Amoco Polymers	OxyChem

	Polybutylene			Polycarbonate							
				Unfilled molding and extrusion resins		Glass fiber-reinforced					
	Extrusion compound	Film grades	Adhesive resin	High viscosity	Low viscosity	10% glass	30% glass	20-30% long glass fiber-reinforced	40% long glass fiber-reinforced	50% long glass fiber-reinforced	35% random glass mat
1a.				3-10	10-30	7.0					
1.	126	118	90								
				150	150	150	150	150	150	150	
2.	C: 300-350 I: 290-380 E: 290-380	E: 380-420	300-350	I: 560	I: 520	I: 520-650	I: 550-650	I: 590-650	I: 575-620	570-620	C: 560-600
3.	10-30			10-20	8-15	10-20	10-30	10-30	10-20	10-20	2-3.5
4.	2.5	2.5		1.74-5.5	1.74-5.5					3	
5.	0.003 (unaged) 0.026 (aged)			0.005-0.007	0.005-0.007	0.002-0.005	0.001-0.002	0.001-0.003	0.001-0.003	0.001-0.003	0.002-0.003
6.	3800-4400	4000	3500	9100-10,500	9100-10,500	7000-10,000	19,000-20,000	18,000-23,000	23,100	25,100-26,000	18,000
7.	300-380	350	500	110-120	110-150	4-10	2-5	1.9-3.0	1.4-1.7	1.3-1.5	2.5
8.	1700-2500	1700	600	9000	9000	8500-11,600					18,000
9.				10,000-12,500	10,000-12,500	12,000-14,000	18,000-20,000	18,000-29,800	31,600	32,700	14,800
10.	2000-2300			12,500-13,500	12,000-14,000	13,700-16,000	23,000-25,000	22,000-36,900	36,400	40,800	30,000
11.	30-40	30	10-15	345	345	450-600	1250-1400	1200-1500	1700	2200	1100
12.	31			350	350	520	1300				
13.	45-50		13	330-340	330-340	460-580	1100	800-1500	1500-1700	2100	1000
				275	275	440	960				
				245	245	420	900				
14.	No break	No break	No break	12-18 @ 1/8 in. 2.3 @ 1/4 in.	12-16 @1/8 in. 2.0 @1/4 in.	2-4	1.7-3.0	3.5-4.7	5.0	6.6	11.8
15.		D45	A90	M70-M75	M70-M75	M62-75; R118-122	M92, R119	M85-95			
	Shore D55-65		D25								
16.	128-150			68	68	32-38	22-23				
17.	130-140			250-270	250-270	280-288	295-300	290-300	305	310	290
	215-235			280-287	273-280	295	300-305	305			
18.	5.2			4.7	4.7	4.6-5.2	5.2-7.6				
19.	0.91-0.925	0.909	0.895	1.2	1.2	1.27-1.28	1.4-1.43	1.34-1.43	1.52	1.63	1.39
20.	0.01-0.02			0.15	0.15	0.12-0.15	0.08-0.14	0.09-0.11			
				0.32-0.35	0.32-0.35	0.25-0.32					
21.	>450			380->400	380->400	470-530	470-475				528
	Shell	Shell	Shell	Albis; American Polymers; Ashley Polymers; Bamberger Polymers; Bayer Corp.; Dow Plastics; Federal Plastics; GE Plastics; MRC; Network Polymers; Polymers Intl.; Polymer Resources; RTP; Shuman	Albis; American Polymers; Ashley Polymers; Bayer Corp.; Dow Plastics; Federal Plastics; GE Plastics; MRC; Network Polymers; Polymers Intl.; Polymer Resources; RTP; Shuman	Albis; American Polymers; Ashley Polymers; Bayer Corp.; ComAlloy; DSM; Dow Plastics; Federal Plastics; Ferro; GE Plastics; LNP; M.A. Polymers; MRC; Network Polymers; Polymers Intl.; Polymer Resources; RTP	Albis; American Polymers; Ashley Polymers; ComAlloy; DSM; Ferro; GE Plastics; LNP; M.A. Polymers; MRC; Network Polymers; Polifil; Polymers Intl.; Polymer Resources; RTP	DSM; RTP; Ticona	RTP; Ticona	RTP; Ticona	Azdel

Materials	Properties		ASTM test method	Polycarbonate (Cont'd)						
				Flame-retarded grade	High-heat			Conductive polycarbonate	EMI shielding (conductive)	
				20-30% glass fiber-reinforced	Polyester copolymer	Polycarbonate copolymer	Impact-modified polycarbonate/ polyester blends	6% stainless steel fiber	10% stainless steel fiber	20% PAN carbon fiber
Processing	1a. Melt flow (gm./10 min.)		D1238							
	1. Melting temperature, °C.	T_m (crystalline)								
		T_g (amorphous)		149	160-195	160-205		150	150	150
	2. Processing temperature range, °F. (C = compression; T = transfer; I = injection; E = extrusion)			I: 530-590	I: 575-710	I: 580-660	I: 475-560		I: 590-650	I: 590-650
	3. Molding pressure range, 10^3 p.s.i.				8-20	8-20	15-20	10-20		
	4. Compression ratio				1.5-3	2-3	2-2.5	3		
	5. Mold (linear) shrinkage, in./in.		D955	0.002-0.004	0.007-0.010	0.007-0.009	0.006-0.009	0.004-0.006	0.003-0.006	0.001
Mechanical	6. Tensile strength at break, p.s.i.		D638[b]	14,000-20,000	9500-11,300	8300-10,000	7600-8500	9800	10,110-11,000	18,000-20,000
	7. Elongation at break, %		D638[b]	2.0-3.0	50-122	70-90	120-165	4.7	4.0	2.0
	8. Tensile yield strength, p.s.i.		D638[b]		8500-9800	9300-10,500	7400-8300			
	9. Compressive strength (rupture or yield), p.s.i.		D695	18,000-21,000	11,500		7000			18,500
	10. Flexural strength (rupture or yield), p.s.i.		D790	21,000-30,000	10,000-13,800	12,000-14,000	10,900-12,500	14,000	16,300-17,000	27,000-28,000
	11. Tensile modulus, 10^3 p.s.i.		D638[b]	1000-1100	320-340	320-340		410	500	2000
	12. Compressive modulus, 10^3 p.s.i.		D695							
	13. Flexural modulus, 10^3 p.s.i.	73° F.	D790	900-1200	294-340	320-340	280-325	400	500	1500-1800
		200° F.	D790							
		250° F.	D790							
		300° F.	D790							
	14. Izod impact, ft.-lb./in. of notch (1/8-in. thick specimen)		D256A	1.8-2.0	1.5-10	1.5-12	2-18	0.8-1.7	1.1-1.7	1.4-2.0
	15. Hardness	Rockwell	D785	M77-85	M74-92	M75-91	R114-122			
		Shore/Barcol	D2240/ D2583							
Thermal	16. Coef. of linear thermal expansion, 10^{-6} in./in./°C.		D696		70-92	70-76	80-95		1410	180
	17. Deflection temperature under flexural load, °F.	264 p.s.i.	D648	288-305	285-335	284-354	190-250	270	270-295	290-295
		66 p.s.i.	D648	295	305-365	306-383	223-265			300
	18. Thermal conductivity, 10^{-4} cal.-cm./sec.-cm.^{2}-°C.		C177		4.7-5.0	4.7-4.8	4.3			
Physical	19. Specific gravity		D792	1.36-1.45	1.15-1.2	1.14-1.18	1.20-1.22	1.28	1.26-1.35	1.28
	20. Water absorption (1/8-in. thick specimen), %	24 hr.	D570	0.15-0.17	0.15-0.2	0.15-0.2	0.12-0.16		0.12	0.2
		Saturation	D570				0.35-0.60			
	21. Dielectric strength (1/8-in. thick specimen), short time, v./mil		D149		509-520	>400	440-500			
			SUPPLIERS[a]	ComAlloy; DSM; Ferro; GE Plastics; LNP; Polymer Resources; RTP	Ferro; GE Plastics	Bayer Corp.	Bayer Corp.; Eastman; GE Plastics; MRC; Polymer Resources	RTP; Ticona	DSM; LNP; RTP; Ticona	ComAlloy; DSM; Ferro; LNP; RTP

	Polycarbonate (Cont'd)						Polydicyclopentadiene	Polyester, thermoplastic		
	EMI shielding (conductive) (Cont'd)		Lubricated					Polybutylene terephthalate		
	30% graphite fiber	40% PAN carbon fiber	10-15% PTFE	30% PTFE	10-15% PTFE, 20% glass fiber-reinforced	2% silicone, 30% glass fiber-reinforced	RIM solid; unfilled	Unfilled	30% glass fiber-reinforced	30% long glass fiber-reinforced
1a.										
1.						Thermoset	Thermoset	220-267	220-267	235
	149-150		150	150	150	150	90-165			
2.	I: 540-650	I: 580-620	I: 590-650	I: 590-650	I: 590-650	I: 590-650	T: <95-<100	I: 435-525	I: 440-530	I: 480-540
							<0.050	4-10	5-15	10-20
3.	15-20	15-20								3-4
4.										
5.	0.001-0.002	0.0005-0.001	0.007	0.009	0.002	0.002	0.008-0.012	0.009-0.022	0.002-0.008	0.001-0.003
6.	20,000-24,000	23,000-24,000	7500-10,000	6000	12,000-15,000	16,000	5300-6000	8200-8700	14,000-19,500	20,000
7.	1-5	1-2	8-10		2		5-70	50-300	2-4	2.2
8.							5000-6700	8200-8700		
9.	19,000-26,000	22,000	110,000		11,000		8500-9000	8600-14,500	18,000-23,500	26,000
10.	30,000-36,000	34,000-35,000	11,000-11,500	7600	18,000-23,000	22,000	10,000-11,000	12,000-16,700	22,000-29,000	35,000
11.	250-2150	3000-3100	340		1200		240-280	280-435	1300-1450	1400
12.								375	700	
13.	240-1900	2800-2900	250-300	460	850-900	900	260-280	330-400	850-1200	1300
14.	1.8	1.5-2.0	2.5-3.0	1.3	1.8-3.5	3.5	5.0-9.0	0.7-1.0	0.9-2.0	5.7
								M68-78	M90	
15.	R118, R119	R119					D72-D84			
16.	9	11.0-14.4	58.0		21.6-23.4	12.0	46-49 in./in./°F.	60-95	15-25	
17.	280-300	295-300	270-275	260	280-290	290	217-240	122-185	385-437	405
	295	300			290		239	240-375	421-500	
								4.2-6.9	7.0	
18.	16.9	17.3								
19.	1.32-1.33	1.36-1.38	1.26-1.29	1.39	1.43-1.5	1.46	1.03-1.04	1.30-1.38	1.48-1.54	1.56
20.	0.04-0.08	0.08-0.13	0.13	0.06	0.11	0.12	0.09	0.08-0.09	0.06-0.08	
								0.4-0.5	0.35	
								420-550	460-560	
21.										
	ComAlloy; DSM; LNP; RTP; Thermofil	ComAlloy; DSM; Ferro; LNP; RTP	ComAlloy; DSM; GE Plastics; LNP; Polymer Resources; RTP	DSM; Polymer Resources; RTP	ComAlloy; DSM; Ferro; GE Plastics; LNP; Polymer Resources; RTP	ComAlloy; DSM; Ferro; LNP; RTP	BFGoodrich; Hercules	Albis; Ashley Polymers; BASF; ComAlloy; Creanova; DuPont; GE Plastics; M.A. Hanna Eng.; RTP; Ticona	Albis; Adell; Ashley Polymers; BASF; ComAlloy; Creanova; DSM; DuPont; Ferro; GE Plastics; LNP; M.A. Hanna Eng.; Polymer Resources; RTP; Ticona	Ticona

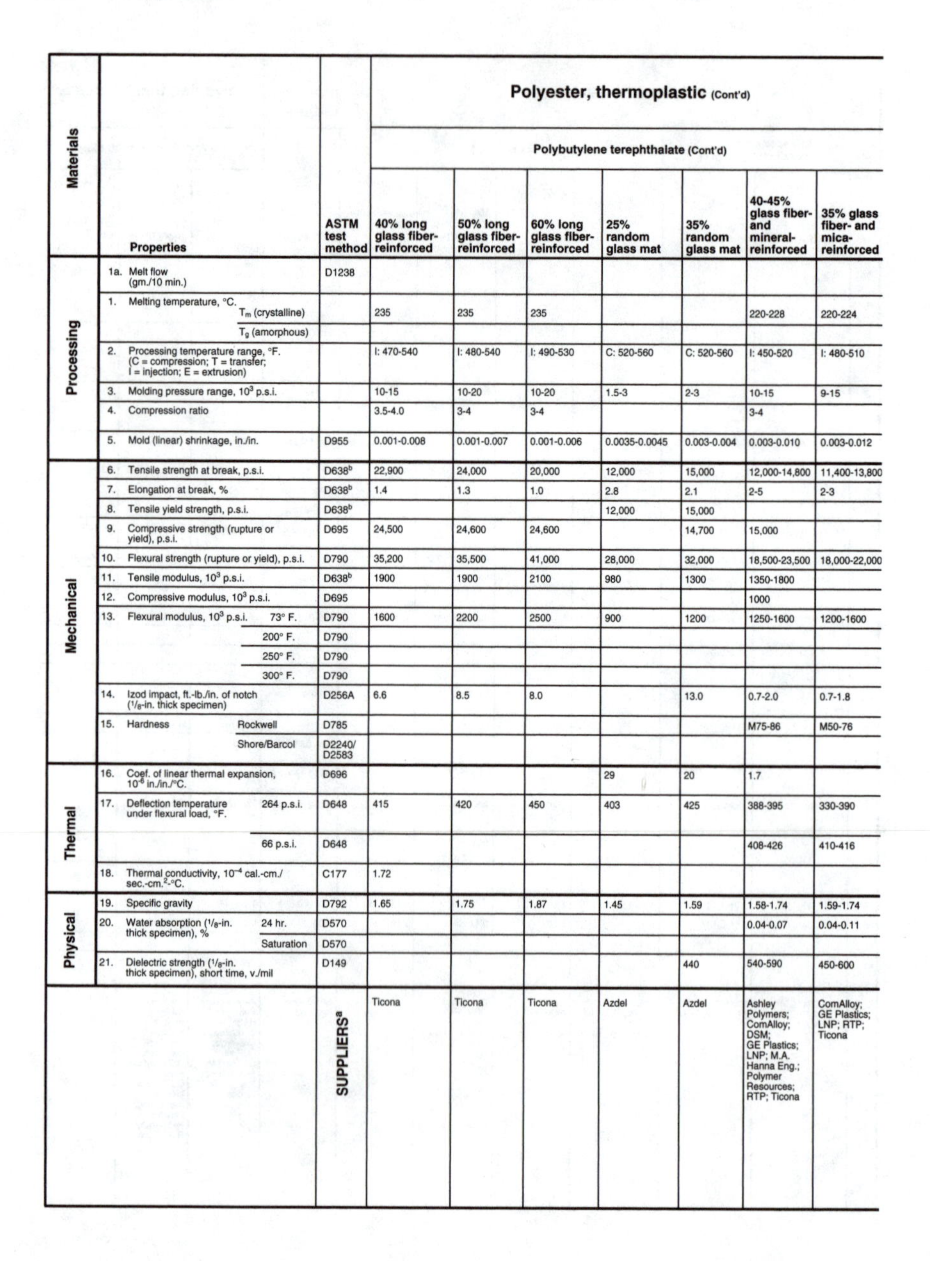

Materials								
		Polyester, thermoplastic (Cont'd)						
		Polybutylene terephthalate (Cont'd)						
Properties	ASTM test method	40% long glass fiber-reinforced	50% long glass fiber-reinforced	60% long glass fiber-reinforced	25% random glass mat	35% random glass mat	40-45% glass fiber- and mineral-reinforced	35% glass fiber- and mica-reinforced
Processing								
1a. Melt flow (gm./10 min.)	D1238							
1. Melting temperature, °C. T_m (crystalline)		235	235	235			220-228	220-224
T_g (amorphous)								
2. Processing temperature range, °F. (C = compression; T = transfer; I = injection; E = extrusion)		I: 470-540	I: 480-540	I: 490-530	C: 520-560	C: 520-560	I: 450-520	I: 480-510
3. Molding pressure range, 10^3 p.s.i.		10-15	10-20	10-20	1.5-3	2-3	10-15	9-15
4. Compression ratio		3.5-4.0	3-4	3-4			3-4	
5. Mold (linear) shrinkage, in./in.	D955	0.001-0.008	0.001-0.007	0.001-0.006	0.0035-0.0045	0.003-0.004	0.003-0.010	0.003-0.012
Mechanical								
6. Tensile strength at break, p.s.i.	D638[b]	22,900	24,000	20,000	12,000	15,000	12,000-14,800	11,400-13,800
7. Elongation at break, %	D638[b]	1.4	1.3	1.0	2.8	2.1	2-5	2-3
8. Tensile yield strength, p.s.i.	D638[b]				12,000	15,000		
9. Compressive strength (rupture or yield), p.s.i.	D695	24,500	24,600	24,600		14,700	15,000	
10. Flexural strength (rupture or yield), p.s.i.	D790	35,200	35,500	41,000	28,000	32,000	18,500-23,500	18,000-22,000
11. Tensile modulus, 10^3 p.s.i.	D638[b]	1900	1900	2100	980	1300	1350-1800	
12. Compressive modulus, 10^3 p.s.i.	D695						1000	
13. Flexural modulus, 10^3 p.s.i. 73° F.	D790	1600	2200	2500	900	1200	1250-1600	1200-1600
200° F.	D790							
250° F.	D790							
300° F.	D790							
14. Izod impact, ft.-lb./in. of notch (1/8-in. thick specimen)	D256A	6.6	8.5	8.0		13.0	0.7-2.0	0.7-1.8
15. Hardness Rockwell	D785						M75-86	M50-76
Shore/Barcol	D2240/ D2583							
Thermal								
16. Coef. of linear thermal expansion, 10^{-6} in./in./°C.	D696				29	20	1.7	
17. Deflection temperature under flexural load, °F. 264 p.s.i.	D648	415	420	450	403	425	388-395	330-390
66 p.s.i.	D648						408-426	410-416
18. Thermal conductivity, 10^{-4} cal.-cm./sec.-cm.^{2}-°C.	C177	1.72						
Physical								
19. Specific gravity	D792	1.65	1.75	1.87	1.45	1.59	1.58-1.74	1.59-1.74
20. Water absorption (1/8-in. thick specimen), % 24 hr.	D570						0.04-0.07	0.04-0.11
Saturation	D570							
21. Dielectric strength (1/8-in. thick specimen), short time, v./mil	D149					440	540-590	450-600
	SUPPLIERS[a]	Ticona	Ticona	Ticona	Azdel	Azdel	Ashley Polymers; ComAlloy; DSM; GE Plastics; LNP; M.A. Hanna Eng.; Polymer Resources; RTP; Ticona	ComAlloy; GE Plastics; LNP; RTP; Ticona

	Polyester, thermoplastic (Cont'd)									
	Polybutylene terephthalate (Cont'd)					Polyester alloy			PCT	
			Flame-retarded grade							
	Impact-modified	50% glass fiber-reinforced	7-15% glass fiber-reinforced	30% glass fiber-reinforced	EMI shielding (conductive); 30% carbon fiber	Unfilled	7.0-30% glass fiber-reinforced	Unreinforced flame retardant	15% glass fiber-reinforced	30% glass fiber-reinforced
1a.										
1.	225	225	220-260	220-260	222					
2.	I: 482-527	I: 482-527	I: 490-560	I: 490-560	I: 430-550	I: 475-540	I: 475-540	I: 470-510	I: 565-590	I: 555-595
3.					5-20	5-17	5-17	5-17		
4.						3.0-4.0	3.0-4.0	3.0-4.0		
5.	$20x10^{-3}$	$4x10^{-3}$	0.005-0.01	0.002-0.006	0.001-0.004	0.016-0.018	0.003-0.014		0.001-0.004	0.001-0.004
6.		21,800	11,500-15,000	17,400-20,000	22,000-23,000	4900-6300	7200-12,000	5000	13,700	18,000-19,500
7.		2.5	4.0	2.0-3.0	1-3	150-300	4.5-45	50	2.0	1.9-2.3
8.	6530									
9.			17,000	18,000						
10.			18,000-23,500	30,000	29,000-34,000	7100-8700	12,400-19,000	8500	23,900	24,000-29,800
11.	276	2470	800	1490; 1700	3500					
12.										
13.			580-830	1300-1500	2300-2700	210-280	395-925	260	812	1200-1450
14.	3.3	1.4	0.6-1.1	1.3-1.6	1.2-1.5	NB	2.8-4.1	NB	0.76	1.3-1.8
15.			M79-88	M88; M90	R120	R101-R109; >R115	R104-R111; >R115	R105	M88	>R115
16.	135	25	5	1.5		20	18			20
17.	122	419	300-450	400-450	420-430	105-125	170-374	130	475	500
	248	428	400-490	425-490		180-260	379-417	240	518	>500
18.					15.8	6.9	8.3			6.9
19.	1.20	1.73	1.48-1.53	1.63	1.41-1.42	1.23-1.25	1.30-1.47	1.31	1.33	1.45
20.			0.06	0.06-0.07	0.04-0.45	0.1	0.11	0.10		0.04-0.05
	0.3	0.3								
21.			460	490		460	435		462	440-460
	Ashley Polymers; BASF; M.A. Hanna Eng.	Ashley Polymers; BASF	Albis; Ashley Polymers; ComAlloy; DSM; DuPont; GE Plastics; LNP; M.A. Hanna Eng.; M.A. Polymers; Polymer Resources; RTP; Ticona	Albis; Ashley Polymers; BASF; ComAlloy; DSM; DuPont; Ferro; LNP; M.A. Hanna Eng.; M.A. Polymers; Polymer Resources; RTP; Ticona	ComAlloy; DSM; LNP; RTP	Eastman; GE Plastics; Ticona	Albis; GE Plastics; Ticona	Ticona	Eastman	Eastman; GE Plastics

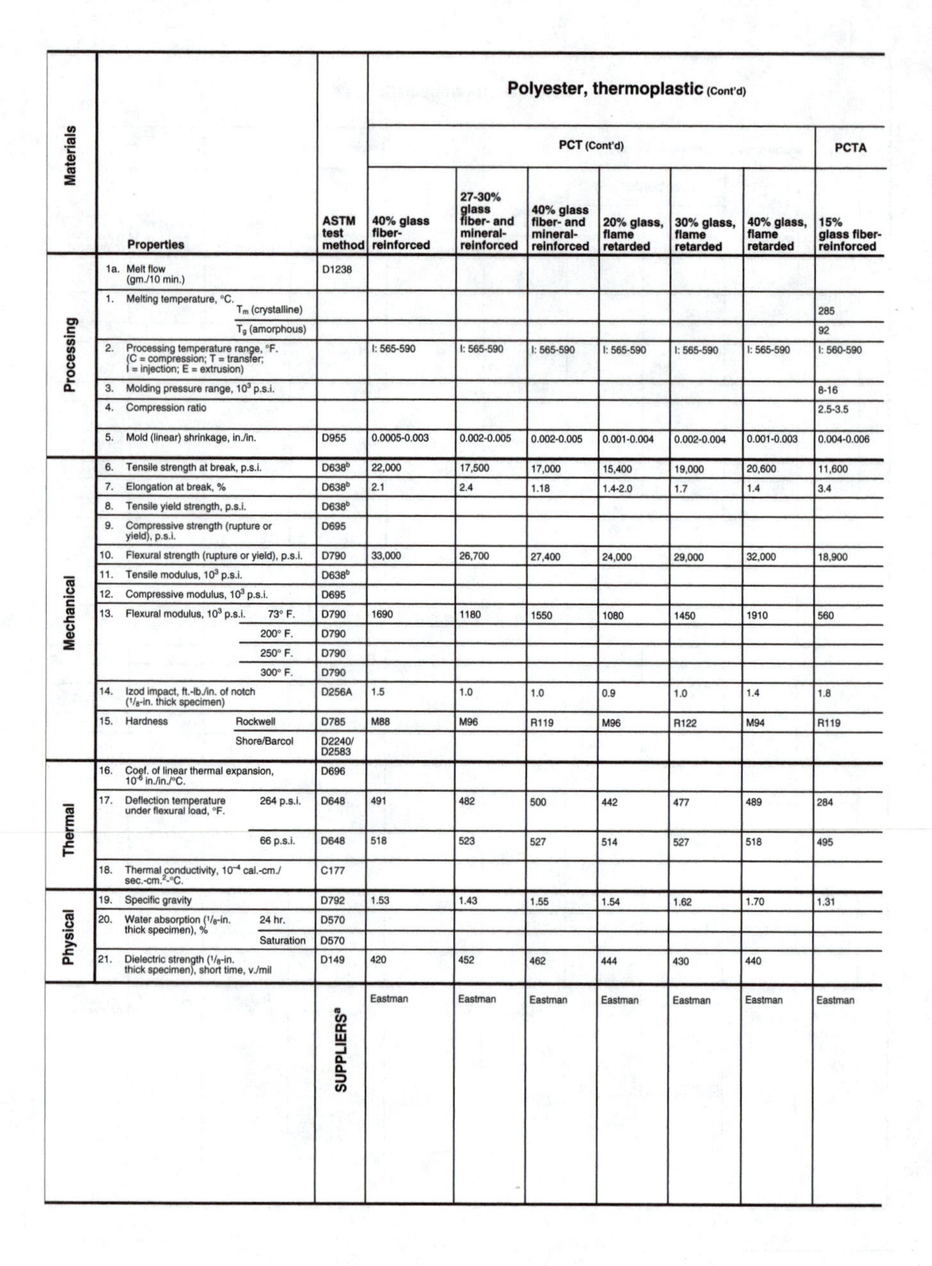

Materials	Properties	ASTM test method	Polyester, thermoplastic (Cont'd)						
			PCT (Cont'd)						PCTA
			40% glass fiber-reinforced	27-30% glass fiber- and mineral-reinforced	40% glass fiber- and mineral-reinforced	20% glass, flame retarded	30% glass, flame retarded	40% glass, flame retarded	15% glass fiber-reinforced
Processing	1a. Melt flow (gm./10 min.)	D1238							
	1. Melting temperature, °C. T_m (crystalline)								285
	T_g (amorphous)								92
	2. Processing temperature range, °F. (C = compression; T = transfer; I = injection; E = extrusion)		I: 565-590	I: 565-590	I: 565-590	I: 565-590	I: 565-590	I: 565-590	I: 560-590
	3. Molding pressure range, 10^3 p.s.i.								8-16
	4. Compression ratio								2.5-3.5
	5. Mold (linear) shrinkage, in./in.	D955	0.0005-0.003	0.002-0.005	0.002-0.005	0.001-0.004	0.002-0.004	0.001-0.003	0.004-0.006
Mechanical	6. Tensile strength at break, p.s.i.	D638[b]	22,000	17,500	17,000	15,400	19,000	20,600	11,600
	7. Elongation at break, %	D638[b]	2.1	2.4	1.18	1.4-2.0	1.7	1.4	3.4
	8. Tensile yield strength, p.s.i.	D638[b]							
	9. Compressive strength (rupture or yield), p.s.i.	D695							
	10. Flexural strength (rupture or yield), p.s.i.	D790	33,000	26,700	27,400	24,000	29,000	32,000	18,900
	11. Tensile modulus, 10^3 p.s.i.	D638[b]							
	12. Compressive modulus, 10^3 p.s.i.	D695							
	13. Flexural modulus, 10^3 p.s.i. 73° F.	D790	1690	1180	1550	1080	1450	1910	560
	200° F.	D790							
	250° F.	D790							
	300° F.	D790							
	14. Izod impact, ft.-lb./in. of notch (1/8-in. thick specimen)	D256A	1.5	1.0	1.0	0.9	1.0	1.4	1.8
	15. Hardness Rockwell	D785	M88	M96	R119	M96	R122	M94	R119
	Shore/Barcol	D2240/ D2583							
Thermal	16. Coef. of linear thermal expansion, 10^{-6} in./in./°C.	D696							
	17. Deflection temperature under flexural load, °F. 264 p.s.i.	D648	491	482	500	442	477	489	284
	66 p.s.i.	D648	518	523	527	514	527	518	495
	18. Thermal conductivity, 10^{-4} cal.-cm./sec.-cm.^{2}-°C.	C177							
Physical	19. Specific gravity	D792	1.53	1.43	1.55	1.54	1.62	1.70	1.31
	20. Water absorption (1/8-in. thick specimen), % 24 hr.	D570							
	Saturation	D570							
	21. Dielectric strength (1/8-in. thick specimen), short time, v./mil	D149	420	452	462	444	430	440	
		SUPPLIERS[a]	Eastman	Eastman	Eastman	Eastman	Eastman	Eastman	Eastman

Polyester, thermoplastic (Cont'd)

	PCTA (Cont'd)			Polyethylene terephthalate								
	20% glass fiber-reinforced	30% glass fiber-reinforced	Unfilled	Unfilled	15% glass fiber-reinforced	30% glass fiber-reinforced	40-45% glass fiber-reinforced	35% glass, super toughened	15% glass, easy processing	30% glass, flame retarded, V-0 1/32	15% glass, flame retarded, V-0 1/32
1a.											
1.	285	285	285	212-265		245-265	252-255	245-255	245-255	245-255	245-255
	92	92	92	68-80							
2.	I: 560-590	I: 560-590	I: 299-316 E: 299-302	I: 440-660 E: 520-580	I: 540-570	I: 510-590	I: 500-590	I: 525-555	I: 525-555	I: 525-555	I: 525-555
3.	8-16	8-16		2-7		4-20	8-12	8-18	8-18	8-18	8-18
4.	2.5-3.5	2.5-3.5		3.1	2-3	2-3	4	3:1	3:1	3:1	3:1
5.	0.004-0.006	0.003-0.005	0.004	0.002-0.030	0.001-0.004	0.002-0.009	0.002-0.009	0.002-0.009	0.003-0.010	0.002-0.009	0.003-0.010
6.	12,600	14,100		7000-10,500	14,600	20,000-24,000	14,000-27,500	15,000	11,500	22,000	15,500
7.	3.3	3.1	25-250	30-300	2.0	2-7	1.5-3	6.0	6.0	2.3	2.6
8.			5900-9000	8600		23,000					
9.				11,000-15,000		25,000	20,500-24,000	11,700	13,500	25,000	25,000
10.	20,600	22,700		12,000-18,000	20,000	30,000-36,000	21,000-42,400	21,000	13,500	32,000	23,000
11.				400-600		1300-1440	1800-1950				
12.											
13.	690	980	240-285	350-450	830	1200-1590	1400-2190	1000	525	1500	850
						520	489	360	185	620	350
						390	320	275	155	420	220
14.	2.2	2.8	1.5-NB	0.25-0.7	1.9	1.5-2.2	0.9-2.4	4.4	2.2	1.6	1.2
15.	R118	R113	105-122	M94-101; R111	R121	M90; M100	R118; R119	M62; R107	M58, R111	M100, R120	M88, R120
16.			5.8×10^{-5}	65×10^{-6}		18-30	18-21	1.5	1.0	1.1	1.0
17.	381	430	69-95	70-150	400	410-440	412-448	428	405	435	410
	507	514	83-142	167	464	470-480	420-480	475	454	475	471
18.			5×10^{-4} or 5	3.3-3.6		6.0-7.6	10.0				
19.	1.33	1.41	1.195-1.215	1.29-1.40	1.33	1.55-1.70	1.58-1.74	1.51	1.39	1.67	1.53
20.				0.1-0.2		0.05	0.04-0.05	0.25	0.24	0.05	0.07
				0.2-0.3							
21.			422-441	420-550	475	405-650	415-600	530	450	430	490
	Eastman	Eastman	Eastman	DuPont; Eastman; M.A. Polymers; A. Schulman; Ticona; Wellman	Eastman	Albis; AlliedSignal; ComAlloy; DSM; DuPont; EMS; Eastman; Ferro; GE Plastics; M.A. Polymers; MRC; RTP; Ticona; Wellman	Albis; AlliedSignal; ComAlloy; DSM; DuPont; EMS; Eastman; Ferro; GE Plastics; RTP; Ticona	DuPont	DuPont	Albis; DuPont	DuPont

Materials		Polyester, thermoplastic (Cont'd)						
		Polyethylene terephthalate (Cont'd)						
Properties	ASTM test method	15-20% glass, flame retarded	30% glass, flame retarded	40% glass, flame retarded	35-45% glass fiber- and mica-reinforced	30% long glass fiber-reinforced	40% long glass fiber-reinforced	50% long glass fiber-reinforced
Processing								
1a. Melt flow (gm./10 min.)	D1238							
1. Melting temperature, °C. T_m (crystalline)					252-255	275	275	275
T_g (amorphous)								
2. Processing temperature range, °F. (C = compression; T = transfer; I = injection; E = extrusion)		I: 540-560	I: 540-560	I: 540-560	I: 500-590	I: 470-530	480-540	I: 470-530
3. Molding pressure range, 10^3 p.s.i.					5-20	10-20	10-20	10-20
4. Compression ratio		2-3	2-3	2-3	4	3-4		3-4
5. Mold (linear) shrinkage, in./in.	D955	0.0015-0.004	0.001-0.004	0.001-0.004	0.002-0.007	0.001-0.008	0.001-0.005	0.001-0.008
Mechanical								
6. Tensile strength at break, p.s.i.	D638[b]	13,700-15,700	18,600	19,000	14,000-26,000	20,200	23,200	23,500
7. Elongation at break, %	D638[b]	2.2	1.8	1.5	1.5-3	1.4	1.4	1.0
8. Tensile yield strength, p.s.i.	D638[b]							
9. Compressive strength (rupture or yield), p.s.i.	D695				20,500-24,000	31,000	34,200	35,100
10. Flexural strength (rupture or yield), p.s.i.	D790	20,000-23,200	27,500	29,300	21,000-40,000	29,300	35,400	36,500
11. Tensile modulus, 10^3 p.s.i.	D638[b]				1800-1950	1700	2100	2400
12. Compressive modulus, 10^3 p.s.i.	D695							
13. Flexural modulus, 10^3 p.s.i. 73° F.	D790	850-1090	1540	2020	1400-2000	1500	1900	2100
200° F.	D790				489			
250° F.	D790							
300° F.	D790				320			
14. Izod impact, ft.-lb./in. of notch (1/8-in. thick specimen)	D256A	1-1.2	1.4	1.6	0.9-2.4	4.0	5.0	6.2
15. Hardness Rockwell	D785	M83	M84	M79	R118, R119			
Shore/Barcol	D2240/ D2583							
Thermal								
16. Coef. of linear thermal expansion, 10^{-6} in./in./°C.	D696				18-21			
17. Deflection temperature under flexural load, °F. 264 p.s.i.	D648	383-409	425	429	396-440	470	475	480
66 p.s.i.	D648	455-462	459	466	420-480			
18. Thermal conductivity, 10^{-4} cal.-cm./sec.-cm.^{2}-°C.	C177				10.0			
Physical								
19. Specific gravity	D792	1.60-1.63	1.71	1.78	1.58-1.74	1.61	1.70	1.85
20. Water absorption (1/8-in. thick specimen), % 24 hr.	D570				0.04-0.05			
Saturation	D570							
21. Dielectric strength (1/8-in. thick specimen), short time, v./mil	D149	437-460	427	399	550-687			
	SUPPLIERS[a]	Eastman	Eastman	Eastman	Albis; AlliedSignal; Bayer Corp.; ComAlloy; DSM; DuPont; M.A. Polymers; MRC; RTP; Ticona; Wellman	Ticona	Ticona	Ticona

Polyester, thermoplastic (Cont'd)												
	Polyethylene terephthalate (Cont'd)					PETG	PCTG	Polyester/polycarbonate blends			Wholly aromatic (liquid crystal)	
	60% long glass fiber-reinforced	EMI shielding (conductive); 30% PAN carbon fiber	Recycled content, 30% glass fiber	Recycled content, 45% glass fiber	Recycled content, 35% glass/ mineral-reinforced	Unfilled	Unfilled	High-impact	30% glass fiber-reinforced	Flame retarded	Unfilled medium melting point	Unfilled high melting point
1a.												
1.	270-280										280-421	400
						81-91						
2.	I: 500-530	I: 550-590	I: 530-550	I: 530-550	I: 530-550	I: 480-520 E: 490-550	I: 530-560	I: 460-630	I: 470-560	I: 480-550	I: 540-770	I: 700-850
3.	10-20					1-20	1-20	8-18	10-18	10-20	1-16	5-18
4.	3-4		2-3	2-3	2-3	2.4-3	2.4-3				2.5-4	2.5-3
5.	0.001-0.007	0.001-0.002				0.002-0.005	0.002-0.005	0.0005-0.019	0.003-0.009	0.005-0.007	0.001-0.008	0-0.002
6.	23,500	25,000	24,000	28,500	14,000-15,000	4100	7600	4500-9000	12,000-13,300	8900	15,900-27,000	12,500
7.	0.9	1.4	2.0	2.0	2.1-2.2	110	330	100-175		130	1.3-4.5	2
8.						7300	6500	5000-8100		7400		
9.	35,100							8600-10,000	10,860-11,600		6200-19,000	10,000
10.	40,100	38,000	35,500	45,000	21,500-22,000	10,200	9600	8500-12,500	20,000	14,100	19,000-35,500	19,000
11.	3000	3600						240-325			1400-2800	1750
12.											400-900	309
13.	2600	2700	1400	2100	1400	300	260	310	780-850	380,000	1770-2700[e]	1860
											1500-1700[e]	
											1300-1500[e]	450F: 870
											1200-1450[e]	575F: 450
14.	8.0	1.5	1.5	2.0	1.1	1.9	NB	12-19-No break	3.1-3.2	13	1.7-10	1.2
15.						R106	R105	R112-116	R109-110	R122	M76; R60-66	R97
16.								58-150	25	0.64	5-7	8.9
17.	480	430	435	445	395-420	147	149	140-250	300-330	212	356-671	606
		470				158	165	210-265	400-415	239		
18.								5.2			2	
19.	1.91	1.42	1.58	1.70	1.60	1.27	1.23	1.20-1.26	1.44-1.51	1.3	1.35-1.84	1.79
20.		0.05				0.13	0.13	0.13-0.80	0.09-0.10	0.07	0-<0.1	0
								0.30-0.62		0.22	<0.1	0.02
21.			565	540	450-550			396-500		660	600-980	470
	Ticona	ComAlloy; DSM; Ferro; RTP	Ticona; Wellman	Ticona	Ticona; Wellman	Eastman	Eastman	Bayer Corp.; ComAlloy; Eastman; Ferro; GE Plastics; M.A. Polymers; MRC	ComAlloy; Ferro; GE Plastics; M.A. Polymers; MRC; RTP	Bayer Corp.	Amoco Polymers; Ticona	Amoco Polymers

Materials	Properties	ASTM test method	Polyester, thermoplastic (Cont'd)						
			Wholly aromatic (liquid crystal) (Cont'd)						
			30% carbon fiber-reinforced	40% glass fiber-filled	40% glass plus 10% mineral-filled	30-50% mineral-filled	30% glass fiber-reinforced	30% glass-reinforced, high HDT	Unfilled platable grade
Processing	1a. Melt flow (gm./10 min.)	D1238							
	1. Melting temperature, °C. T_m (crystalline)		280			327	280	355	
	T_g (amorphous)								
	2. Processing temperature range, °F. (C = compression; T = transfer; I = injection; E = extrusion)		555-600	I: 660-770	I: 660-770	I: 605-770	I: 555-770	I: 625-730	I: 600-620
	3. Molding pressure range, 10^3 p.s.i.		1-14	5-14	5-14	1-14	1-14	4-8	
	4. Compression ratio		2.5-4	2.5-3	2.5-3	2.5-4	2.5-4	2.5-4	3-4
	5. Mold (linear) shrinkage, in./in.	D955	0-0.002			0.003	0.001-0.09		
Mechanical	6. Tensile strength at break, p.s.i.	D638[b]	35,000	13,600	14,200	10,400-16,500	16,900-30,000	18,000-21,000	13,500
	7. Elongation at break, %	D638[b]	1.0	1.8	2.3	1.1-4.0	1.7-2.7	1.7-2.2	2.9
	8. Tensile yield strength, p.s.i.	D638[b]							
	9. Compressive strength (rupture or yield), p.s.i.	D695	34,500	10,400	9700	6900-7500	9900-21,000	9800-12,500	
	10. Flexural strength (rupture or yield), p.s.i.	D790	46,000	20,500	19,700	14,200-23,500	21,700-24,600	24,000-26,000	
	11. Tensile modulus, 10^3 p.s.i.	D638[b]	5400	1870	1870	1500-2700	700-3000	2330-2600	1500
	12. Compressive modulus, 10^3 p.s.i.	D695	4800	420	473	490-2016	470-1000	447-700	
	13. Flexural modulus, 10^3 p.s.i. 73° F.	D790	4600	1320	1730	1250-2500	1680-2100	1800-2050	1500
	200° F.	D790							
	250° F.	D790					900	1100	
	300° F.	D790					800	1100	
	14. Izod impact, ft.-lb./in. of notch (1/8-in. thick specimen)	D256A	1.4	1.6	1.9	0.8-3.0	2.0-3.0	2.0-2.5	0.6
	15. Hardness Rockwell	D785	M99	79		82	61-87	63	
	Shore/Barcol	D2240/ D2583							
Thermal	16. Coef. of linear thermal expansion, 10^{-6} in./in./°C.	D696	–2.65	14.9	12.9-52.8	9-65	4.9-77.7	14-36	
	17. Deflection temperature under flexural load, °F. 264 p.s.i.	D648	440	606	493	429-554	485-655	518-568	410
	66 p.s.i.	D648					489-530		
	18. Thermal conductivity, 10^{-4} cal.-cm./sec.-cm.^{2}-°C.	C177			2.5	2.57	1.73	1.52	
Physical	19. Specific gravity	D792	1.49	1.70	1.78	1.63-1.89	1.60-1.67	1.6-1.66	
	20. Water absorption (1/8-in. thick specimen), % 24 hr.	D570	<0.1	<0.1	<0.1	<0.1	<0.1	<0.1	0.03
	Saturation	D570							
	21. Dielectric strength (1/8-in. thick specimen), short time, v./mil	D149		510	1145	900-955	640-900	900-1050	600
		SUPPLIERS[a]	RTP; Ticona	Amoco Polymers; RTP	Amoco Polymers; RTP	Amoco Polymers; DuPont; RTP; Ticona	Amoco Polymers; DuPont; Eastman; RTP; Ticona	Amoco Polymers; DuPont; RTP; Ticona	Ticona

	Polyester, thermoset and alkyd									
	Cast		Glass fiber-reinforced							
	Rigid	Flexible	High	Unidirectional Glass	Preformed, chopped roving	Premix, chopped glass	Woven cloth	SMC	SMC, BMC low-density	SMC low-pressure
1a.										
1.	Thermoset	Thermoset			Thermoset	Thermoset	Thermoset		Thermoset	Thermoset
2.			C: 260-320	C: 260-320	C: 170-320	C: 280-350	C: 73-250	C: 270-380 I: 280-310 T: 280-310	C: 270-330 I: 270-350 T: 270-350	C: 270-330
3.			0.2-2.0	0.4-2.0	0.25-2	0.5-2	0.3	0.3-2	0.5-2	0.25-0.8
4.			1.0	1.0	1.0	1.0		1.0	1.0	1.0
5.			0.001	0.000	0.0002-0.002	0.001-0.012	0.0002-0.002	0.0005-0.004	0.0002-0.001	0.0002-0.001
6.	600-13,000	500-3000	50,000	90,000	15,000-30,000	3000-10,000	30,000-50,000	7000-25,000	4000-20,000	7000-25,000
7.	<2.6	40-310	1-2	1-2	1-5	<1	1-2	3	2-5	3
8.										
9.	13,000-30,000		42,000		15,000-30,000	20,000-30,000	25,000-50,000	15,000-30,000	15,000-30,000	15,000-30,000
10.	8500-23,000		90,000	160,000	10,000-40,000	7000-20,000	40,000-80,000	10,000-36,000	10,000-35,000	10,000-36,000
11.	300-640		3500		800-2000	1000-2500	1500-4500	1400-2500	1400-2500	1400-2500
12.			2700							
13.	490-610		3000	6000	1000-3000	1000-2000	1000-3000	1000-2200	1000-2500	1000-22,000
							660			
							430			
							270			
14.	0.2-0.4	>7	35	70	2-20	1.5-16	5-30	7-22	2.5-18	7-24
15.										
	Barcol 35-75	Shore D84-94	60		Barcol 50-80	Barcol 50-80	Barcol 60-80	Barcol 50-70		Barcol 40-70
16.	55-100		15.5	6	20-50	20-33	15-30	13.5-20	6-30	6-30
17.	140-400		>500	>500	>400	>400	>400	375-500	>375	375-500
18.										
19.	1.04-1.46	1.01-1.20	1.9	1.95	1.35-2.30	1.65-2.30	1.50-2.10	1.65-2.6	1.0-1.5	1.65-2.30
20.	0.15-0.6	0.5-2.5	0.08	0.10	0.01-1.0	0.06-0.28	0.05-0.5	0.1-0.25	0.4-0.25	0.1-0.25
21.	380-500	250-400	310		350-500	345-420	350-500	380-500	300-400	380-500
	AOC; Aristech Chem; ICI Americas; Interplastic; Reichhold	AOC; Aristech Chem.; ICI Americas; Interplastic; Reichhold	Quantum Composites	Quantum Composites	Glastic; Haysite; Jet Moulding; Premix; Reichhold; Rostone	Bulk Molding Compounds; Cytec Fiberite; Glastic; Haysite; Jet Moulding; Premix; Reichhold; Rostone	Glastic; Haysite; Jet Moulding; Premix; Reichhold; Rostone	Budd; Haysite; Interplastic; Jet Moulding; Plastics Mfg.; Polyply; Premix; Rostone	Cytec Fiberite; Interplastic; Jet Moulding; Rostone	Interplastic; Jet Moulding; Rostone

Materials	Properties	ASTM test method	Polyester, thermoset and alkyd (Cont'd)						
			Glass fiber-reinforced (Cont'd)		EMI shielding (conductive)		High-strength SMC	Alkyd molding compounds	
			SMC low-shrink	BMC, TMC	SMC, TMC	BMC	50% glass fiber-reinforced	Granular and putty, mineral-filled	Glass fiber-reinforced
Processing	1a. Melt flow (gm./10 min.)	D1238							
	1. Melting temperature, °C. T_m (crystalline)		Thermoset	Thermoset	Thermoset	Thermoset	Thermoset	Thermoset	Thermoset
	T_g (amorphous)								
	2. Processing temperature range, °F. (C = compression; T = transfer; I = injection; E = extrusion)		C: 270-330 I: 270-380	C: 280-380 I: 280-370 T: 280-320	C: 270-380 I: 270-370 T: 280-320	C: 310-380 I: 300-370 T: 280-320	C: 270-330	C: 270-350 I: 280-390 T: 320-360	C: 290-350 I: 280-380
	3. Molding pressure range, 10^3 p.s.i.		0.5-2	0.4-1.1	0.5-2		0.5-2	2-20	2-25
	4. Compression ratio		1.0	1.0	1.0		1.0	1.8-2.5	1-11
	5. Mold (linear) shrinkage, in./in.	D955	0.0002-0.001	0.0003-0.007	0.0002-0.001	0.0005-0.004	0.002-0.003	0.003-0.010	0.001-0.010
Mechanical	6. Tensile strength at break, p.s.i.	D638[b]	4500-20,000	3000-13,000	7000-8000	4000-4500		3000-9000	4000-9500
	7. Elongation at break, %	D638[b]	3-5	1-2			3-5		
	8. Tensile yield strength, p.s.i.	D638[b]					27,000-30,000		
	9. Compressive strength (rupture or yield), p.s.i.	D695	15,000-30,000	14,000-30,000	20,000-24,000	18,000		12,000-38,000	15,000-36,000
	10. Flexural strength (rupture or yield), p.s.i.	D790	9000-35,000	11,000-33,000	18,000-20,000	12,000	45,000-50,000	6000-17,000	8500-26,000
	11. Tensile modulus, 10^3 p.s.i.	D638[b]	1000-2500	1500-2500			1200-1600	500-3000	2000-2800
	12. Compressive modulus, 10^3 p.s.i.	D695						2000-3000	
	13. Flexural modulus, 10^3 p.s.i. 73° F.	D790	1000-2500	1500-1800	1400-1500	1400-1500	900-1400	2000	2000
	200° F.	D790							
	250° F.	D790							
	300° F.	D790							
	14. Izod impact, ft.-lb./in. of notch (1/8-in. thick specimen)	D256A	2.5-15	2-18.5	10-12	5-7	22-28	0.3-0.5	0.5-16
	15. Hardness Rockwell	D785						E98	E95
	Shore/Barcol	D2240/ D2583	Barcol 40-70	Barcol 50-65	Barcol 45-50	Barcol 50	50-60		
Thermal	16. Coef. of linear thermal expansion, 10^{-6} in./in./°C.	D696	6-30	20			6-30	20-50	15-33
	17. Deflection temperature under flexural load, °F. 264 p.s.i.	D648	375-500	320-536	395-400+	400+	>425	350-500	400-500
	66 p.s.i.	D648							
	18. Thermal conductivity, 10^{-4} cal.-cm./sec.-cm.^{2}-°C.	C177		18-22				12-25	15-25
Physical	19. Specific gravity	D792	1.6-2.4	1.72-2.10	1.75-1.80	1.80-1.85	1.77-1.83	1.6-2.3	2.0-2.3
	20. Water absorption (1/8-in. thick specimen), % 24 hr.	D570	0.01-0.25	0.1-0.45			0.19-0.25	0.05-0.5	0.03-0.5
	Saturation	D570			0.5	0.5			
	21. Dielectric strength (1/8-in. thick specimen), short time, v./mil	D149	380-450	300-390				350-450	259-530
		SUPPLIERS[a]	Budd; Haysite; Interplastic; Jet Moulding; Polyply; Premix; Rostone	BP Chemicals; Budd; Bulk Molding Compounds; Cytec Fiberite; Epic Resins; Glastic; Haysite; Jet Moulding; Plaslok; Polyply; Premix; Rostone	Jet Moulding; Premix; Rostone	Jet Moulding; Premix; Rostone	Jet Moulding	Cytec Fiberite; Plastics Eng.; Premix	Cosmic Plastics; Cytec Fiberite; Plastics Eng.; Premix; Rogers

	Polyester, thermoset and alkyd (Cont'd)	Polyetherimide			Polyethersulfone					Polyethylene and ethylene copolymers	
	Vinyl Ester BMC									Low and medium density	
	Glass fiber-reinforced	Unfilled	30% glass fiber-reinforced	EMI shielding (conductive); 30% carbon fiber	10% glass fiber-reinforced	20-30% glass fiber-reinforced	20% mineral-filled	30% carbon-filled	Polyether-sulfone unreinforced	Branched homopolymer	Linear copolymer
1a.									12.5-30	0.25-27.0	
1.	Thermoset									98-115	122-124
		215-217	215	215						—25	
2.	C: 290-350 I: 290-350	I: 640-800	I: 620-800	I: 600-780	I: 680-715	I: 660-765	I: 662-716	I: 680-734		I: 300-450 E: 250-450	I: 350-500 E: 450-600
3.	0.5-2	10-20	10-20	10-30						5-15	5-15
4.	1.0	1.5-3	1.5-3	1.5-3						1.8-3.6	3
5.	0-2	0.005-0.007	0.001-0.004	0.0005-0.002	0.005-0.006	0.004-0.006	0.007-0.008		0.007	0.015-0.050	0.020-0.022
6.	9-12	14,000	23,200-28,500	29,000-34,000	16,500	20,000-22,000	7975	15,800	12,000	1200-4550	1900-4000
7.	1-2	60	2-5	1-3	4.3	2.1-2.8	5	1.4		100-650	100-965
8.		15,200	24,500							1300-2100	1400-2800
9.	23,000-28,000	21,900	23,500-30,700	32,000							
10.	20,000-31,000	22,000	33,000	37,000-45,000	24,500	26,500-28,500			16,100		
11.		430	1300-1600	2600-3300	740	1150-1550	522	1740	385	25-41	38-75
12.		480	550-938								
13.		480	1200-1300	2500-2600	650	980-1300			420	35-48	40-105
		370	1100								
		360	1060								
		350	1040								
14.	5-15	1.0-1.2	1.7-2.0	1.2-1.6	1.3	1.4-1.7	1.5		1.6	No break	1.0-No break
15.		M109-110	M114, M125, R123	M127	M94	M96-M97					
	45-65									Shore D44-50	Shore D55-56
16.	20	47-56	20-21		19	12-14	47	12	49	100-220	
17.	500+	387-392	408-420	405-420	414	419	399	433	400		
		405-410	412-415	410-425	419	430	414	440		104-112	
18.	15-25	1.6	6.0-9.3	17.6						8	
19.	1.7-1.9	1.27	1.49-1.51	1.39-1.42	1.45	1.53-1.60	1.52	1.53	1.37	0.917-0.932	0.918-0.940
20.	0.1-0.2	0.25	0.16-0.20	0.18-0.2					1.85	<0.01	
		1.25	0.9		1.9	1.5-1.7	1.7				
21.	300-400	500	495-630						380	450-1000	
	Cytec Fiberite	GE Plastics	ComAlloy; DSM; Ferro; GE Plastics; LNP; Polymer Resources; RTP	ComAlloy; DSM; Ferro; LNP; RTP	Amoco Polymers; BASF; RTP	BASF; RTP	BASF; RTP	BASF; RTP	Amoco Polymers	American Polymers; Bamberger Polymers; Chevron; Dow Plastics; DuPont; Eastman; Equistar; Exxon; Huntsman; Mobil; Network Polymers; NOVA Chemicals; RSG Polymers; A. Schulman; Union Carbide; Wash. Penn; Westlake	Bamberger Polymers; Chevron; Dow Plastics; DuPont; Eastman; Equistar; Exxon; Exxon Chemical Canada; Mobil; Montell NA; Network Polymers; NOVA Chemicals; RSG Polymers; A. Schulman; Solvay Polymers; Union Carbide

Materials			Polyethylene and ethylene copolymers (Cont'd)						
			Low and medium density (Cont'd)			High density			
			LDPE copolymers					Copolymers	
	Properties	ASTM test method	Ethylene-vinyl acetate	Ethylene-ethyl acrylate	Ethylene-methyl acrylate	Polyethylene homopolymer	Rubber-modified	Low and medium molecular weight	High molecular weight
Processing	1a. Melt flow (gm./10 min.)	D1238	1.4-2.0			5-18			5.4-6.8
	1. Melting temperature, °C. T_m (crystalline)		103-110		83	130-137	122-127	125-132	125-135
	T_g (amorphous)								
	2. Processing temperature range, °F. (C = compression; T = transfer; I = injection; E = extrusion)		C: 200-300 I: 350-430 E: 300-380	C: 200-300 I: 250-500	E: 200-620	I: 350-500 E: 350-525	E: 360-450	I: 375-500 E: 300-500	I: 375-500 E: 375-475
	3. Molding pressure range, 10^3 p.s.i.		1-20	1-20		12-15		5-20	
	4. Compression ratio					2		2	
	5. Mold (linear) shrinkage, in./in.	D955	0.007-0.035	0.015-0.035		0.015-0.040		0.012-0.040	0.015-0.040
Mechanical	6. Tensile strength at break, p.s.i.	D638[b]	2200-4000	1600-2100	1650	3200-4500	2300-2900	3000-6500	2500-4300
	7. Elongation at break, %	D638[b]	200-750	700-750	740	10-1200	600-700	10-1300	170-800
	8. Tensile yield strength, p.s.i.	D638[b]	1200-6000		1650	3800-4800	1400-2600	2600-4200	2800-3900
	9. Compressive strength (rupture or yield), p.s.i.	D695		3000-3600		2700-3600		2700-3600	
	10. Flexural strength (rupture or yield), p.s.i.	D790							
	11. Tensile modulus, 10^3 p.s.i.	D638[b]	7-29	4-7.5	12	155-158		90-130	136
	12. Compressive modulus, 10^3 p.s.i.	D695							
	13. Flexural modulus, 10^3 p.s.i. 73° F.	D790	7.7			145-225		120-180	125-175
	200° F.	D790							
	250° F.	D790							
	300° F.	D790							
	14. Izod impact, ft.-lb./in. of notch (1/8-in. thick specimen)	D256A	No break	No break		0.4-4.0		0.35-6.0	3.2-4.5
	15. Hardness Rockwell	D785							
	Shore/Barcol	D2240/ D2583	Shore D17-45	Shore D27-38		Shore D66-73	Shore D55-60	Shore D58-70	Shore D63-65
Thermal	16. Coef. of linear thermal expansion, 10^{-6} in./in./°C.	D696	160-200	160-250		59-110		70-110	70-110
	17. Deflection temperature under flexural load, °F. 264 p.s.i.	D648							
	66 p.s.i.	D648				175-196		149-176	154-158
	18. Thermal conductivity, 10^{-4} cal.-cm./sec.-cm.^{2}-°C.	C177				11-12		10	
Physical	19. Specific gravity	D792	0.922-0.943	0.93	0.942-0.945	0.952-0.965	0.932-0.939	0.939-0.960	0.947-0.955
	20. Water absorption (1/8-in. thick specimen), % 24 hr.	D570	0.005-0.13	0.04	0.0	<0.01		<0.01	
	Saturation	D570							
	21. Dielectric strength (1/8-in. thick specimen), short time, v./mil	D149	620-760	450-550		450-500		450-500	
		SUPPLIERS[a]	AT Plastics; Chevron; DuPont; Equistar; Exxon; Federal Plastics; Huntsman; Mobil; Network Polymers; A. Schulman; Union Carbide; Westlake	Network Polymers; Union Carbide	Chevron; Exxon; Network Polymers; A. Schulman	American Polymers; Bamberger Polymers; Chevron; Dow Plastics; Eastman; Equistar; Exxon; Exxon Chemical Canada; Federal Plastics; Fina; M.A. Polymers; Mobil; Network Polymers; NOVA Chemicals; Phillips; RSG Polymers; A. Schulman; Shuman; Solvay Polymers; Ticona; Union Carbide	Exxon; M.A. Polymers; Network Polymers	AlphaGary; American Polymers; Bamberger Polymers; Chevron; Dow Plastics; Eastman; Equistar; Exxon; Exxon Chemical Canada; Fina; Mobil; Network Polymers; NOVA Chemicals; Phillips; RSG Polymers; A. Schulman; Shuman; Ticona; Union Carbide	AlphaGary; Amoco Chemical; Bamberger Polymers; BASF; Chevron; Dow Plastics; Equistar; Exxon; Fina; Mobil; Network Polymers; NOVA Chemicals; Phillips; Solvay Polymers; Ticona; Union Carbide

	Polyethylene and ethylene copolymers (Cont'd)				Polyimide					
	High density (Cont'd)		Crosslinked		Thermoplastic					
	Ultra high molecular weight	30% glass fiber-reinforced	Molding grade	Wire and cable grade	Unfilled	30% glass fiber-reinforced	30% carbon fiber-reinforced	30% carbon fiber-reinforced, crystallized	15% graphite-filled	40% graphite-filled
1a.					4.5-7.5					
1.	125-138	120-140			388	388	388		388	
					250-365	250	250		250	365
2.	C: 400-500	I: 350-600	C: 240-450 I: 250-300	E: 250-400	C: 625-690 I: 734-740 E: 734-740	I: 734-788	I: 734-788	I: 734-788	I: 734-788	C: 690
3.	1-2	10-20			3-20	10-30	10-30	10-30	10-30	3-5
4.					1.7-4	1.7-2.3	1.7-2.3	1.7-2.3	1.7-2.3	
5.	0.040	0.002-0.006	0.007-0.090	0.020-0.050	0.0083	0.0044	0.0021		0.006	
6.	5600-7000	7500-9000	1600-4600	1500-3100	10,500-17,100	24,000	33,400	31,700	8000-8400	7600
7.	350-525	1.5-2.5	10-440	180-600	7.5-90	3	2	1	3.5	3
8.	3100-4000			1200-2000	12,500-13,000					
9.		6000-7000	2000-5500		17,500-40,000	27,500	30,200		25,000	
10.		11,000-12,000	2000-6500		10,000-28,800	35,200	46,700	43,600	11,000-14,100	18,000
11.		700-900	50-500		300-400	1720	3000			14,000
12.			50-150		315-350	458	573		330	
13.	130-140	700-800	70-350	8-14	360-500	1390	2780	3210	460-500	
										700
					210	1175	2450		260	
14.	No break	1.1-1.5	1-20		1.5-1.7	2.2	2.0	2.4	1.1	0.7
15.	R50	R75-90			E52-99,R129,M95	R128, M104	R128, M105			E27
	Shore D61-63		Shore D55-80	Shore D30-65						
16.	130-200	48	100	100	45-56	17-53	6-47		41	38
17.	110-120	250	105-145	100-173	460-680	469	478	>572	680	680
	155-180	260-265	130-225							
18.		8.6-11			2.3-4.2	8.9	11.7			41.4
19.	0.94	1.18-1,28	0.95-1.45	0.91-1.40	1.33-1.43	1.56	1.43	1.47	1.41	1.65
20.	<0.01	0.02-0.06	0.01-0.06	0.01-0.06	0.24-0.34	0.23	0.23		0.19	0.14
					1.2					0.6
21.	710	500-550	230-550	620-760	415-560	528			250	
	Montell NA; Network Polymers; Ticona	ComAlloy; DSM; Ferro; LNP; M.A. Polymers; RTP; A. Schulman; Thermofil	Equistar; Mobil; Phillips; A. Schulman	AlphaGary; AT Plastics; Equistar; A. Schulman; Union Carbide	Ciba Specialty Chemicals; DuPont; Mitsui Chemicals America; Solutia	Mitsui Chemicals America; RTP	Mitsui Chemicals America; RTP	Mitsui Chemicals America; RTP	DuPont; Mitsui Chemicals America	DuPont

Materials		Poly-imide (Cont'd)	Polyketone			Poly-methylpentene		Poly-phenylene oxide,
		Thermoset						Alloy with polystyrene
Properties	ASTM test method	Unfilled	Unfilled	30% carbon	Polyketone	Unfilled	Filled	Low glass transition
Processing								
1a. Melt flow (gm./10 min.)	D1238				6.0	26	30	
1. Melting temperature, °C. T_m (crystalline)		Thermoset			428°F	230-240	240	
T_g (amorphous)								100-112
2. Processing temperature range, °F. (C = compression; T = transfer; I = injection; E = extrusion)		460-485	715-805	715-805	I: 480-510	I: 510-610 E: 510-650	I: 510-610	I: 400-600 E: 420-500
3. Molding pressure range, 10^3 p.s.i.		7-29				1-10	1-10	12-20
4. Compression ratio		1-1.2	gp screw	gp screw	2.5-3.0:1	2.0-3.5	2.0-3.5	1.3-3
5. Mold (linear) shrinkage, in./in.	D955	0.001-0.01	0.008-0.012	0.002-0.008	0.028	0.016-0.021	0.014-0.017	0.005-0.008
Mechanical								
6. Tensile strength at break, p.s.i.	D638[b]	4300-22,900	13,500	30,000	8000	2300-2500	2400	6800-7800
7. Elongation at break, %	D638[b]	1	50	1.5	7300	20-120	25	48-50
8. Tensile yield strength, p.s.i.	D638[b]	4300-22,900			8700	2200-3400	3400	6500-7800
9. Compressive strength (rupture or yield), p.s.i.	D695	19,300-32,900	17,200	33,800				12,000-16,400
10. Flexural strength (rupture or yield), p.s.i.	D790	6500-50,000	24,500	40,000	8000	6300-8300		8300-12,800
11. Tensile modulus, 10^3 p.s.i.	D638[b]	460-4650	520	2700	230	160-280		310-380
12. Compressive modulus, 10^3 p.s.i.	D695	421				114-171		
13. Flexural modulus, 10^3 p.s.i. 73° F.	D790	422-3000	2000	10,000	230	70-190	270	325-400
200° F.	D790					36		260
250° F.	D790					26		
300° F.	D790	1030-2690				17		
14. Izod impact, ft.-lb./in. of notch (1/8-in. thick specimen)	D256A	0.65-15	1.6	1.5		2-3		3-6
15. Hardness Rockwell	D785	110M-120M			105	R35-85	90	R115-116
Shore/Barcol	D2240/ D2583				75D			
Thermal								
16. Coef. of linear thermal expansion, 10^{-6} in./in./°C.	D696	15-50			1X10-4	65		38-70
17. Deflection temperature under flexural load, °F. 264 p.s.i.	D648	572->575	323	634	221	120-130		176-215
66 p.s.i.	D648		582	652	410	180-190	230	230
18. Thermal conductivity, 10^{-4} cal.-cm./sec.-cm.^{2}-°C.	C177	5.5-12				4.0		3.8
Physical								
19. Specific gravity	D792	1.41-1.9	1.30	1.45	1.24	0.833-0.835	1.08	1.04-1.10
20. Water absorption (1/8-in. thick specimen), % 24 hr.	D570	0.45-1.25	0.10	0.20	0.40	0.01	0.11	0.06-0.1
Saturation	D570				2.1			
21. Dielectric strength (1/8-in. thick specimen), short time, v./mil	D149	480-508	355	Conductive	320	1096-1098		400-665
	SUPPLIERS[a]	Ciba Specialty Chemicals	Amoco Polymers	Amoco Polymers	Shell	Mitsui Petrochemical	Mitsui Petrochemical	Ashley Polymers; GE Plastics; Polymer Resources; Shuman

	Polyphenylene oxide, modified (Cont'd)									Polyphenylene sulfide	
	Alloy with polystyrene (Cont'd)								Alloy with nylon		
							EMI shielding (conductive)				
	High glass transition	Impact-modified	15% glass fiber-reinforced	20% glass fiber-reinforced	30% glass fiber-reinforced	Mineral-filled	30% graphite fiber	40% aluminum flake	Un-reinforced	Unfilled	10-20% glass fiber-reinforced
1a.											
1.										285-290	275-285
	117-190	135	100-125	100-125	100-125	110-135				88	
2.	I: 425-670 E: 460-525	I: 425-550	I: 400-630 E: 460-525	I: 400-630 E: 460-525	I: 400-630 E: 460-525	I: 540-590 E: 470-530	I: 500-600	I: 500-600		I: 590-640	I: 600-675
3.	12-20	10-15	10-40	10-40	10-40	12-20	10-20	10-20		5-15	
4.	1.3-3					2-3				2-3	
5.	0.006-0.008	0.006	0.002-0.004	0.002-0.004	0.001-0.004	0.005-0.007	0.001	0.001	0.013-0.016	0.006-0.014	0.002-0.005
6.	9600	7000-8000	10,000-12,000	13,000-15,000	15,000-18,500		18,700	6500	8500	7000-12,500	7500-14,000
7.	60	35	5-8	8.0	2-5	25	2.5	3.0	60	1-6	1.0-1.5
8.	7000-9000				14,500	9500-11,000					
9.	16,400	10,000			17,900		20,000	6000		16,000	17,000-20,000
10.	9500-14,000	8200-11,000			20,000-23,000		24,000	9500	315	14,000-21,000	9500-20,000
11.	355-380	345-360			1000-1300		1150	750		480	850-1200
12.											
13.	330-400	325-345	450-500	760	1100-1150	425-500	1100	850	290-315	550-600	900-1200
	305				1000						
								35			
14.	5	6.8	1.1-1.3	1.5	1.7-2.3	3-4	1.3	0.6	3.8	<0.5	0.7-1.2
15.	R118-120	L108, M93	R106-110	R115	R115-116	R121	R111	R110		R123-125	R121
16.	33-77				14-25		11	11		27-49	16-20 Trans. dir.: 15.36-45
17.	225-300	190-275	252-260	262-275	275-317	190-230	240	230	250	212-275	440-480
	279	205-245	273-280	280-290	285-320		265	250	350	390	500-520
18.	5.2				3.8-4.1					2.0-6.9	
19.	1.04-1.09	1.27-1.36			1.27-1.36	1.24-1.25	1.25	1.45	1.10	1.35	1.39-1.47
20.	0.06-0.12	0.1-0.07			0.06	0.07	0.04	0.03	0.3	0.01-0.07	0.05
									1.0		
21.	500-700	530			550-630	490				380-450	
	Creanova; GE Plastics; Polymer Resources	GE Plastics; Polymer Resources	Ashley Polymers; Polymer Resources	Ashley Polymers; Polymer Resources	Ashley Polymers; ComAlloy; Ferro; GE Plastics; LNP; Polymer Resources; RTP	ComAlloy; GE Plastics; Polymer Resources; RTP	ComAlloy; LNP; RTP	ComAlloy	Ashley Polymers; GE Plastics	Phillips; Ticona	Akzo; LNP; RTP

Materials	Properties	ASTM test method	Polyphenylene sulfide (Cont'd) 30% glass fiber-reinforced	40% glass fiber-reinforced	53% glass/mineral-reinforced, high elong. and impact	30% glass fiber, 15% PTFE	30% long glass fiber-reinforced	40% long glass fiber-reinforced	50% long glass fiber-reinforced
Processing	1a. Melt flow (gm./10 min.)	D1238		30-41	50	30			
	1. Melting temperature, °C. T_m (crystalline)		275-285	275-290	354	275-290	310	299	315
	T_g (amorphous)		90	88-90					
	2. Processing temperature range, °F. (C = compression; T = transfer; I = injection; E = extrusion)		I: 590-640	I: 600-675	I: 590-620	640-660	580-620	590-620	I: 580-620
	3. Molding pressure range, 10^3 p.s.i.		8-12	5-20	7-15		8-10	7-20	8-10
	4. Compression ratio		3	3	3-4		3	3-4	3
	5. Mold (linear) shrinkage, in./in.	D955	0.003-0.005	0.002-0.005	0.001-0.003	0.002-0.004	0.001-0.007	0.001-0.003	0.001-0.005
Mechanical	6. Tensile strength at break, p.s.i.	D638[b]	22,000	17,500-29,100	24,000	21,000	21,000	23,000	23,200
	7. Elongation at break, %	D638[b]	1.5	0.9-4	1.5	1.8	1.2	1.1	1.0
	8. Tensile yield strength, p.s.i.	D638[b]							
	9. Compressive strength (rupture or yield), p.s.i.	D695		21,000-31,200	30,000		32,400	32,000-34,000	34,200-35,000
	10. Flexural strength (rupture or yield), p.s.i.	D790	28,000	22,700-43,600	35,000	29,000	32,900	35,000-36,400	37,300-38,000
	11. Tensile modulus, 10^3 p.s.i.	D638[b]		1100-2100		1340	1800	2200	2600
	12. Compressive modulus, 10^3 p.s.i.	D695							
	13. Flexural modulus, 10^3 p.s.i. 73° F.	D790	1700	1700-2160	2300	1360	1700	2300	2400
	200° F.	D790							
	250° F.	D790		1000					
	300° F.	D790		730					
	14. Izod impact, ft.-lb./in. of notch (1/8-in. thick specimen)	D256A	1.3	1.1-2.0	1.5	1.6	4.6	4.8-5.3	5.0-5.5
	15. Hardness Rockwell	D785	M102.7-M103	R123, M100-104	100M	R116			
	Shore/Barcol	D2240/ D2583							
Thermal	16. Coef. of linear thermal expansion, 10^{-6} in./in./°C.	D696		Flow dir: 12.1-22 Trans. dir.: 14.4-45	Flow: 19-32 Trans.: 32-80				
	17. Deflection temperature under flexural load, °F. 264 p.s.i.	D648	507	485-515	510	480	490	500	505
	66 p.s.i.	D648	534	536					
	18. Thermal conductivity, 10^{-4} cal.-cm./sec.-cm.^{2}-°C.	C177		6.9-10.7					
Physical	19. Specific gravity	D792	1.38-1.58	1.60-1.67	1.80	1.60	1.52-1.62	1.62	1.72
	20. Water absorption (1/8-in. thick specimen), % 24 hr.	D570	<0.03	0.004-0.05	0.02	0.005			
	Saturation	D570							
	21. Dielectric strength (1/8-in. thick specimen), short time, v./mil	D149	10^6; 380	347-450	300				
		SUPPLIERS[a]	Ferro; GE Plastics; LNP; Phillips; RTP; Ticona	Albis; DSM; Ferro; GE Plastics; LNP; Phillips; RTP; Ticona	Ticona	Ferro; GE Plastics; LNP; RTP	RTP; Ticona	RTP; Ticona	RTP; Ticona

	Polyphenylene sulfide (Cont'd)										Polyphthalamide
							EMI shielding (conductive)				
	60% long glass fiber-reinforced	65% mineral- and glass-filled	50% glass and mineral-reinforced	55% glass and mineral-reinforced	60% glass and mineral-reinforced	Encapsulation grades	60% stainless steel	30% carbon fiber	40% PAN carbon fiber	30% PAN carbon fiber, 15% PTFE	Extra-tough
1a.		20-35	97	85	63	287					
1.	315	285-290	275-285	275-285	275-285	275-285	299	275-285	275-285	275-285	310
		88	90								
2.	I: 580-620	I: 600-675	I: 600-675	I: 600-675	I: 600-675	I: 600-675	I: 590-620	I: 500-675	I: 600-675	I: 600-675	I: 610-660
3.	8-10	5-20	4-12				7-20	5-20			5-15
4.	3	3	3				3.0-4.0				2.5-3
5.	0.001-0.005	0.001-0.004	0.002-0.004	0.003-0.005	0.007	0.006-0.007	0.005-0.007	0.005-0.003	0.0005	0.0008	0.015-0.020
6.	22,500	13,000-23,100	18,000-22,000	12,700	8100	9800	8400	20,000-27,000	26,000-29,000	25,500	11,000
7.	1.0	1.3; <1.4	1.0-3.5	1.6	2.0	1.0		0.5-3	1.0-1.5	2.6	30
8.		11,000									10,500
9.	30,800	11,000-32,300						26,000	27,000		
10.	38,300	17,500-33,900	28,000-33,700	20,300	11,200	15,100	17,800	26,000-36,000	40,000	34,000	17,000
11.	3000							2500-3700	4400-4800	3600	350
12.											
13.	3000	1800-2400	2240-2400	1750	1441	1574	590	2450-3300	3900-4100	3000	368
		2000									
		1200									
14.	5.5	0.5-1.37	0.8-1.3	0.7	0.5	0.4	0.2	0.8-1.2	1.2-1.5	1.2	18
15.		R121, M102	R120	M66	M85	M98		R123	R123		120
16.		12.9-20 14.3	12.2-14.6		14.3-14.8	18.0-18.2		6-16	8		
17.	510	500-510	500-510	491	340	328	450	500-505	505	500	243
		534						>505	505		
18.								8.6-17.9			
19.	1.84	1.78-2.03	1.78-1.8	1.82	1.90	1.86	1.37	1.42-1.47	1.46-1.49	1.47	1.15
20.		0.02-0.07	0.02-0.07	0.07	0.08	0.03		0.01-0.02	0.02	0.06	0.68
21.		280-450	280-343			338					
	Ticona	Ferro; LNP; Phillips; RTP; Ticona	Albis; LNP; RTP; Ticona	LNP; Phillips; RTP	Albis; LNP; Phillips; RTP	Phillips	Ticona	DSM; Ferro; LNP; RTP	DSM; Ferro; LNP; RTP	DSM; Ferro; LNP; RTP	Amoco Polymers

Materials	Properties	ASTM test method	Polyphthalamide (Cont'd) 45% glass-reinforced	33% glass-reinforced, V-0	40% mineral-reinforced	40% glass/mineral-reinforced, heat-stabilized	50% glass/mineral reinforced, V0	51% glass/mineral reinforced	15% glass-reinforced
Processing	1a. Melt flow (gm./10 min.)	D1238							
	1. Melting temperature, °C. T_m (crystalline)		310	310	310	310	310	310	
	T_g (amorphous)								
	2. Processing temperature range, °F. (C = compression; T = transfer; I = injection; E = extrusion)		I: 610-660	I: 610-660	I: 610-660	I: 620-650	I: 610-640	I: 610-640	I: 610-640
	3. Molding pressure range, 10^3 p.s.i.		5-15	5-15	5-15				
	4. Compression ratio		2.5-3	2.5-3	2.5-3				
	5. Mold (linear) shrinkage, in./in.	D955	0.002-0.006	0.002-0.004	0.008-0.010	0.004	0.003-0.3		0.006-0.007
Mechanical	6. Tensile strength at break, p.s.i.	D638[b]	38,000-39,400	26,000-27,000	15,800-17,000	23,700	22,200	17,200	
	7. Elongation at break, %	D638[b]	2.0-2.6	1.3	1.6	2.2-2.5	1.3	1.9-2	2.0
	8. Tensile yield strength, p.s.i.	D638[b]				23,500	21,000	16,200	19,000
	9. Compressive strength (rupture or yield), p.s.i.	D695	45,500		26,000	33,000			30,000
	10. Flexural strength (rupture or yield), p.s.i.	D790	54,000-55,100	35,900-37,300	28,600-30,000	31,500-34,300	31,000	24,000-28,400	23,700
	11. Tensile modulus, 10^3 p.s.i.	D638[b]	2500	2000	1300	1600	2400	1530	
	12. Compressive modulus, 10^3 p.s.i.	D695							
	13. Flexural modulus, 10^3 p.s.i. 73° F.	D790	2100-2160	1900-1930	1080-1300	1200-1410	2040	1300	1090
	200° F.	D790							
	250° F.	D790							
	300° F.	D790							
	14. Izod impact, ft.-lb./in. of notch (1/8-in. thick specimen)	D256A	2.1-2.5	1.1-1.5	0.8	0.8-0.9	1.2	1.3	1.1
	15. Hardness Rockwell	D785	125	125	125	125			127
	Shore/Barcol	D2240/ D2583							
Thermal	16. Coef. of linear thermal expansion, 10^{-6} in./in./°C.	D696	8		19	2.9-4.6			3.5
	17. Deflection temperature under flexural load, °F. 264 p.s.i.	D648	527-549	511-523	316-361	478-527	490-505	482	531
	66 p.s.i.	D648							
	18. Thermal conductivity, 10^{-4} cal.-cm./sec.-cm.^{2}-°C.	C177	2.6		2.6				
Physical	19. Specific gravity	D792	1.56-1.60	1.71	1.53-1.54	1.54	1.82	1.66-1.68	1.26
	20. Water absorption (1/8-in. thick specimen), % 24 hr.	D570	0.12	0.18	0.14	0.16	0.12		0.30
	Saturation	D570							
	21. Dielectric strength (1/8-in. thick specimen), short time, v./mil	D149	560	458	>560	505	635		480
		SUPPLIERS[a]	Amoco Polymers; LNP; RTP	Amoco Polymers; RTP	Amoco Polymers; RTP	Amoco Polymers; RTP	Amoco Polymers	Amoco Polymers; RTP	RTP

	Polyphthalamide (Cont'd)				Polypropylene					
					Homopolymer					
	15% glass-reinforced, V0	45% glass-reinforced, V0	40% glass/mineral reinforced	40% mineral-reinforced, heat-stabilized	Unfilled	10-40% talc-filled	10-40% calcium carbonate-filled	10-30% glass fiber-reinforced	10-50% mica-filled	40% glass fiber-reinforced
1a.					0.4-100	0.1-30.0	0.1-30.0	1-20	4-10	1-20
1.	310	310		310	160-175	158-168	168	168	168	168
					—20					
2.	I: 610-640	I: 610-640	I: 610-650	I: 610-650	I: 375-550 E: 400-500	I: 350-550	I: 375-525	I: 425-475		I: 450-550
3.			5-15		10-20	10-20	8-20			10-25
4.			2.5-3		2.0-2.4					
5.	0.005	0.002	0.004-0.007		0.010-0.025	0.008-0.022	0.007-0.018	0.002-0.008	0.002-0.015	0.003-0.005
6.	18,600	28,100		16,500	4500-6000	3545-5000	3400-4500	6500-13,000		8400-15,000
7.	2.0		2.5	1.6	100-600	3-60	10-245	1.8-7	3-10	1.5-4
8.	18,400	29,500	23,500	17,000	4500-5400	3500-5000	3000-4600	7000-10,000	4700-6500	
9.			33,000	24,000	5500-8000	7500	3000-7200	6500-8400		8900-9800
10.	24,400-26,600	39,000-41,100	31,500	28,000-29,800	6000-8000	7000-9200	5500-7000	7000-20,000		10,500-22,000
11.	1400			1300	165-225	450-575	375-500	700-1000		1100-1500
12.					150-300					
13.	1000-1700	2500-2600	1200	1150-1200	170-250	210-670	230-450	310-780	420-1150	950-1000
					50	400	320			
					35					
14.	0.6-0.8	1.24-1.7	0.9	0.7-0.9	0.4-1.4	0.4-1.4	0.5-1.0	1.0-2.2	0.50-0.85	1.4-2.0
15.			125	125	R80-102	R85-110	R78-99	R92-115	R82-100	R102-111
									066-78	
16.			2.9-4.6	3.8-4.1	81-100	42-80	28-50	21-62		27-32
17.	466-504	522-527	527	315-325	120-140	132-180	135-170	253-288		300-330
					225-250	210-290	200-270	290-320		330
18.					2.8	7.6	6.9	5.5-6.2		8.4-8.8
19.	1.58	1.78-1.79	1.54	1.54-1.57	0.900-0.910	0.97-1.27	0.97-1.25	0.97-1.14	0.99-1.35	1.22-1.23
20.	0.28	0.17	0.16	0.14	0.01-0.03	0.01-0.03	0.02-0.05	0.01-0.05	0.01-0.06	0.05-0.06
										0.09-0.10
21.			505	455	600	500	410-500			500-510
	Amoco Polymers; RTP	Amoco Polymers; RTP	Amoco Polymers; RTP	Amoco Polymers; RTP	American Polymers; Amoco Chemical; Aristech Chem.; Bamberger Polymers; ComAlloy; Dow Plastics; Epsilon; Equistar; Exxon; Federal Plastics; Ferro; Fina; Huntsman; M.A. Polymers; Montell NA; Network Polymers; Phillips; RSG Polymers; A. Schulman; Shuman; Solvay Polymers; Union Carbide; Wash. Penn	Adell; Albis; Amoco; Bamberger Polymers; ComAlloy; DSM; Exxon; Federal Plastics; Ferro; M.A. Polymers; Montell NA; MRC; Network Polymers; Polifil; RSG Polymers; RTP; A. Schulman; Spartech Polycom; Thermofil; Wash. Penn	Adell; Albis; Bamberger Polymers; ComAlloy; DSM; Federal Plastics; Ferro; M.A. Polymers; Montell NA; Network Polymers; Polifil; RSG Polymers; RTP; A. Schulman; Spartech Polycom; Thermofil; Wash. Penn	Adell; Albis; Amoco Polymers; Bamberger Polymers; ComAlloy; DSM; Federal Plastics; Ferro; LNP; M.A. Polymers; Montell NA; MRC; Network Polymers; Polifil; RSG Polymers; RTP; A. Schulman; Spartech Polycom; Thermofil; Wash. Penn	Network Polymers; Polifil; RTP; A. Schulman; Spartech Polycom; Washington Penn	Adell; Albis; Amoco Polymers; ComAlloy; DSM; Ferro; LNP; M.A. Polymers; Montell NA; MRC; Network Polymers; Polifil; RTP; A. Schulman; Thermofil

Materials			Polypropylene (Cont'd)					
			Homopolymer (Cont'd)					Copolymer
	Properties	ASTM test method	20-30% long glass fiber-reinforced	40% long glass fiber-reinforced	30% random glass mat	Impact-modified, 40% mica-filled	EMI shielding (conductive); 30% PAN carbon fiber	Unfilled
Processing	1a. Melt flow (gm./10 min.)	D1238						0.6-100
	1. Melting temperature, °C. T_m (crystalline)		168	163	168	168	168	150-175
	T_g (amorphous)							–20
	2. Processing temperature range, °F. (C = compression; T = transfer; I = injection; E = extrusion)		I: 360-440	I: 370-410	C: 420-440	I: 350-470	I: 360-470	I: 375-550 E: 400-500
	3. Molding pressure range, 10^3 p.s.i.			6-12	1-2			10-20
	4. Compression ratio			3-4				2-2.4
	5. Mold (linear) shrinkage, in./in.	D955	0.0025-0.004	0.001-0.003	0.002-0.003	0.007-0.008	0.001-0.003	0.010-0.025
Mechanical	6. Tensile strength at break, p.s.i.	D638[b]	7500-14,100	10,500-15,600	12,000	4500	6800	4000-5500
	7. Elongation at break, %	D638[b]	2.1-2.2	1.7	3	4	0.5	200-500
	8. Tensile yield strength, p.s.i.	D638[b]			12,000			3000-4300
	9. Compressive strength (rupture or yield), p.s.i.	D695	6500-14,400	10,400-15,900				3500-8000
	10. Flexural strength (rupture or yield), p.s.i.	D790	10,000-22,800	20,800-25,200	20,000	7000	9000	5000-7000
	11. Tensile modulus, 10^3 p.s.i.	D638[b]	750-900	970-1120	670	700	1750	130-180
	12. Compressive modulus, 10^3 p.s.i.	D695						
	13. Flexural modulus, 10^3 p.s.i. 73° F.	D790	550-800	920-1000	620	600	1650	130-200
	200° F.	D790						40
	250° F.	D790						30
	300° F.	D790						
	14. Izod impact, ft.-lb./in. of notch (1/8-in. thick specimen)	D256A	3.5-7.8	8.0-10.04	12.2	0.7	1.1	1.1-14.0
	15. Hardness Rockwell	D785	R105-117					R65-96
	Shore/Barcol	D2240/ D2583						Shore D70-73
Thermal	16. Coef. of linear thermal expansion, 10^{-6} in./in./°C.	D696			15			68-95
	17. Deflection temperature under flexural load, °F. 264 p.s.i.	D648	250-295	300	310	205	245	130-140
	66 p.s.i.	D648	305					185-220
	18. Thermal conductivity, 10^{-4} cal.-cm./sec.-cm.^{2}-°C.	C177	2.35					3.5-4.0
Physical	19. Specific gravity	D792	1.04-1.17	1.21	1.1	1.23	1.04	0.890-0.905
	20. Water absorption (1/8-in. thick specimen), % 24 hr.	D570	0.05				0.12	0.03
	Saturation	D570						
	21. Dielectric strength (1/8-in. thick specimen), short time, v./mil	D149						600
		SUPPLIERS[a]	DSM; LNP; Montell NA; RTP; Ticona	LNP; Montell NA; RTP; Ticona	Azdel	Albis; ComAlloy; DSM; Federal Plastics; Ferro; M.A. Polymers; Polifil; A. Schulman; Spartech Polycom	ComAlloy; DSM; LNP; RTP	American Polymers; Amoco Chemical; Aristech Chem.; Bamberger Polymers; ComAlloy; Dow Plastics; Epsilon; Equistar; Exxon; Federal Plastics; Ferro; Fina; Huntsman; M.A. Polymers; Montell NA; Network Polymers; NOVA Chemicals; Phillips; RSG Polymers; A. Schulman; Shuman; Solvay Polymers; Spartech Polycom; Union Carbide; Wash. Penn

	Polypropylene (Cont'd)			Polystyrene and styrene copolymers						
	Copolymer (Cont'd)			Polystyrene homopolymers			Rubber-modified		Styrene copolymers	
									Styrene-acrylonitrile (SAN)	
	Unfilled, impact-modified	10-20% glass fiber-reinforced	10-40% talc-filled	High and medium flow	Heat-resistant	20% long and short glass fiber-reinforced	Flame-retarded, UL-V0	High-impact	Molding and extrusion	Olefin rubber-modified
1a.		0.1-20	0.1-30					5.8	1.4	
1.	150-168	160-168								
	−20			74-105	100-110	115		9.3-105	100-200	−55/110
2.	I: 390-500 E: 400-500	I: 350-480	I: 350-470 E: 425-475	C: 300-400 I: 350-500 E: 350-500	C: 300-400 I: 350-500 E: 350-500	I: 400-550	I: 400-450 E: 375-425	I: 350-525 E: 375-500	C: 300-400 I: 360-550 E: 360-450	I: 480-530 E: 435-460
3.	10-20		15-20	5-20	5-20	10-20	6-15	10-20	5-20	1-2
4.	2-2.4		2-2.5	3	3-5		3	4	3	2.7-3.2
5.	0.010-0.025	0.003-0.01	0.009-0.017	0.004-0.007	0.004-0.007	0.001-0.003	0.003-0.006	0.004-0.007	0.003-0.005	0.005-0.007
6.	3500-5000	5000-8000	3000-3775	5200-7500	6440-8200	10,000-12,000	2650-4100	1900-6200	10,000-11,900	5100
7.	200-700	3.0-4.0	20-50	1.2-2.5	2.0-3.6	1.0-1.3	30-50	20-65	2-3	15-30
8.	1600-4000		2800-4100		6440-8150		3100-4400	2100-6000	9920-12,000	5000-6000
9.	3500-6000	5500-5600		12,000-13,000	13,000-14,000	16,000-17,000			14,000-15,000	
10.	4000-6000	7000-11,000	4500-5100	10,000-14,600	13,000-14,000	14,000-18,000	4500-7500	3300-10,000	11,000-19,000	7700-8900
11.	50-150			330-475	450-485	900-1200	240-300	160-370	475-560	300
12.				480-490	495-500				530-580	
13.	60-160	355-510	160-400	380-490	450-500	950-1100	280-330	160-390	500-610	280-300
14.	2.2-No break	0.95-2.7	0.6-4.0	0.35-0.45	0.4-0.45	0.9-2.5	1.9-3.3	0.95-7.0	0.4-0.63	13-15
15.	R50-60	R100-103	R83-88	M60-75	M75-84	M80-95, R119	R38-65	R50-82; L-60	M80, R83, 75	R100-102
	Shore D45-55									
16.	68-95			50-83	68-85	39.6-40	45	44.2	65-68	80
17.	115-135	260-280	100-165	169-202	194-217	200-220	180-205	170-205	203-220	197-200
	167-192	305	195-260	155-204	200-224	220-230	176-181	165-200	220-224	
18.	3.5-4.0			3.0	3.0	5.9			3.0	
19.	0.880-0.905	0.98-1.04	0.97-1.24	1.04-1.05	1.04-1.05	1.20	1.15-1.17	1.03-1.06	1.06-1.08	1.02
20.	0.03	0.01	0.02	0.01-0.03	0.01	0.07-0.01	0.0	0.05-0.07	0.15-0.25	0.09
				0.01-0.03	0.01	0.3			0.5	
21.	500			500-575	500-525	425	550		425	420
	American Polymers; Amoco Polymers; ComAlloy; Dow Plastics; Epsilon; Equistar; Exxon; Federal Plastics; Huntsman; LG Chemical; M.A. Polymers; Montell NA; Network Polymers; Phillips; A. Schulman; Solvay Engineered Polymers; Solvay Polymers; Spartech Polycom; Wash. Penn	Adell; Albis; ComAlloy; DSM; Federal Plastics; Ferro; LG Chemical; LNP; M.A. Polymers; Montell NA; Polifil; RTP; A. Schulman; Solvay Engineered Polymers; Spartech Polycom; Thermofil; Wash. Penn	Adell; Albis; Bamberger Polymers; ComAlloy; DSM; Federal Plastics; Ferro; LG Chemical; M.A. Polymers; Montell NA; Polifil; RTP; A. Schulman; Solvay Engineered Polymers; Spartech Polycom; Wash. Penn	American Polymers; Bamberger Polymers; BASF; Chevron; Deltech Polymers; Dow Plastics; Federal Plastics; Fina; Huntsman; LG Chemical; Mobil; Network Polymers; RSG Polymers; A. Schulman	American Polymers; BASF; Chevron; Deltech Polymers; Dow Plastics; Federal Plastics; Fina; Huntsman; LG Chemical; Mobil; Network Polymers; A. Schulman	DSM; Ferro; LG Chemical; LNP; M.A. Polymers; Mobil; RTP	BASF; Dow Plastics; Huntsman; LG Chemical; Mobil; Network Polymers; RTP; A. Schulman	American Polymers; Bamberger Polymers; BASF; Bayer Corp.; Chevron; Dow Plastics; Federal Plastics; Fina; Huntsman; LG Chemical; Mobil; NOVA Chemicals; RSG Polymers; RTP; A. Schulman; Shuman	Albis; American Polymers; BASF; Bamberger Polymers; Bayer Corp.; Dow Plastics; Federal Plastics; Huntsman; LG Chemical; Network Polymers; RSG Polymers; A. Schulman	Dow Plastics; Huntsman; Network Polymers

Materials			Polystyrene and styrene copolymers (Cont'd)					
			Styrene copolymers (Cont'd)					
			Styrene-acrylonitrile (SAN) (Cont'd)	Clear styrene-butadiene copolymers (SBC)	Acrylate-styrene-acrylonitrile (ASA)		Styrene-maleic anhydride (SMA)	
	Properties	ASTM test method	20% glass fiber-reinforced		ASA extrusion, blow molding, injection molding grades	Clear styrene-butadiene copolymers	Molding and extrusion	Impact-modified
Processing	1a. Melt flow (gm./10 min.)	D1238		6-18		7-15		
	1. Melting temperature, °C. T_m (crystalline)							
	T_g (amorphous)		120			108	114	
	2. Processing temperature range, °F. (C = compression; T = transfer; I = injection; E = extrusion)		I: 400-550	I: 375-425 E: 375-427	E: 380-450 I: 400-470	I: 380-450 E: 380-440	I: 430-510 E: 400-500	I: 450-550 E: 425-525
	3. Molding pressure range, 10^3 p.s.i.		10-20		9-15		12-17	12-17
	4. Compression ratio				2.8-3.0			2.5-2.7
	5. Mold (linear) shrinkage, in./in.	D955	0.001-0.003	0.002-0.008	0.004-0.006	0.004-0.010	0.004-0.006	0.004-0.006
Mechanical	6. Tensile strength at break, p.s.i.	D638[b]	15,500-18,000		4000-7500		8100	4500-5500
	7. Elongation at break, %	D638[b]	1.2-1.8	20-180	25-40	20-180	1.8-30	10-35
	8. Tensile yield strength, p.s.i.	D638[b]		1900-4500	5200-5600	1900-4400	5200-8100	4500-6400
	9. Compressive strength (rupture or yield), p.s.i.	D695	17,000-21,000					
	10. Flexural strength (rupture or yield), p.s.i.	D790	20,000-22,700	2700-6400	6000-8000	2700-6400	8000-14,200	7600-13,000
	11. Tensile modulus, 10^3 p.s.i.	D638[b]	1200-1710				340-390	270-360
	12. Compressive modulus, 10^3 p.s.i.	D695						
	13. Flexural modulus, 10^3 p.s.i. 73° F.	D790	1000-1280	150-240	220-341	153-215	320-470	280-490
	200° F.	D790						
	250° F.	D790						
	300° F.	D790						
	14. Izod impact, ft.-lb./in. of notch (1/8-in. thick specimen)	D256A	1.0-3.0	0.3-1.5 NB	9-11	0.4-1.4 NB	0.4-2.0	2.5-6
	15. Hardness Rockwell	D785	M89-100, R122		R85-90		R106-109	R75-109
	Shore/Barcol	D2240/ D2583		Shore D 60-65				
Thermal	16. Coef. of linear thermal expansion, 10^{-6} in./in./°C.	D696	23.4-41.4	~65	59		80	47-88
	17. Deflection temperature under flexural load, °F. 264 p.s.i.	D648	210-230	140-170	185-190	143-170	~214-245	198-235
	66 p.s.i.	D648	220		200-210			
	18. Thermal conductivity, 10^{-4} cal.-cm./sec.-cm.^{2}-°C.	C177	6.6					
Physical	19. Specific gravity	D792	1.22-1.40	1.00-1.01	1.05-1.06	1.01	1.05-1.08	1.05-1.09
	20. Water absorption (1/8-in. thick specimen), % 24 hr.	D570	0.1-0.2	0.08	0.2-0.3	0.08	0.1	0.1-0.5
	Saturation	D570	0.7					
	21. Dielectric strength (1/8-in. thick specimen), short time, v./mil	D149	500		490	300		415-480
		SUPPLIERS[a]	ComAlloy; DSM; Ferro; LG Chemical; LNP; Network Polymers; RTP; Thermofil	Phillips	BASF; Bayer Corp.; Diamond Polymers; GE Plastics; LG Chemical; Network Polymers	Network Polymers	Bayer Corp.; DSM; NOVA Chemicals	Bayer Corp.; NOVA Chemicals

	Polystyrene and styrene copolymers (Cont'd)						Polyurethane			
	Styrene copolymers (Cont'd)						Thermoset		Thermoplastic	
	SMA (Cont'd)	High heat-resistant copolymers					Casting resins			
	20% glass fiber-reinforced	Injection molding	Impact-modified	20% glass fiber-reinforced	Styrene methyl methacrylate	EMI shielding (conductive); 20% PAN carbon fiber	Liquid	Unsaturated	Unreinforced molding	30% long glass fiber-reinforced
1a.			1.0-1.8							
1.							Thermoset	Thermoset	75-137	240
					100-105					
2.	I: 400-550	I: 450-550	I: 425-540 E: 400-500	I: 425-550	I: 375-475	I: 430-500	C: 43-250		I: 370-500 E: 370-510	I: 440-500
3.	10-20				5-20		0.1-5		6-15	10-20
4.					2.5-3.5				3	3
5.	0.002-0.003	0.005	0.003-0.006	0.003-0.004	0.002-0.006	0.0005-0.003	0.020		0.004-0.010	0.002-0.006
6.		7100-8100	4600-5800	10,000-14,000	8100-10,100	14,000	175-10,000	10,000-11,000	4500-9000	24,200
7.	2-3	1.7-1.9	10-20	1.4-3.5	2.1-5.0	1	100-1000	3-6	60-550	3.0
8.	8100-11,000								7,800-11,000	
9.		11,400-14,200					20,000			24,600-25,700
10.	16,300-17,000		8500-10,500	16,300-22,000	14,100-16,000	20,700	700-4500	19,000	10,200-15,000	35,300
11.	750-880	440-490	280-330	850-900	440-520	2000	10-100		190-300	1100
12.					440-480		10-100			
13.	720-800	450-490	320-370	800-1050	450-460	1900	10-100	610	4-310	1100
14.	2.1-2.7	0.4-0.6	1.5-4.0	2.1-2.6	0.3-0.4	0.7	25 to flexible	0.4	1.5-1.8-No break	6.2-7.5
15.	R73		R75-95		M80-85				R >100; M48	
							Shore A10-13,D90	Barcol 30-35	Shore 75A-70D	
16.		65-67	67-79	20	40-72		100-200		0.5-0.8	
17.	229-245	226-249	230-260	231-247	205-210	220	Varies over wide range	190-200	158-260	185
						230			115-275	
18.							5			
19.	1.21-1.22	1.07-1.10	1.05-1.08	1.20-1.22	1.08-1.13	1.14	1.03-1.5	1.05	1.12-1.24	1.43
20.	0.1	0.1	0.1	0.1	0.11-0.17	0.1	0.2-1.5	0.1-0.2	0.15-0.19	
									0.5-0.6	
21.							300-500		400	
	Bayer Corp.; DSM; NOVA Chemicals; RTP	NOVA Chemicals	Bayer Corp.; NOVA Chemicals	ComAlloy; DSM; LNP; M.A. Polymers; NOVA Chemicals; RTP	Network Polymers; NOVA Chemicals	DSM; LNP; RTP	Bayer Corp.; Cabot; Conap; Dow Plastics; Emerson & Cuming; ITW Devcon; Polyurethane Specialties; Union Carbide	Dow Plastics; Emerson & Cuming; Polyurethane Specialties	Bayer Corp.; BF Goodrich; Dow Plastics	RTP; Ticona

Materials		Polyurethane (Cont'd)		Polyvinylidene chloride copolymers			Silicone	
		Thermoplastic (Cont'd)			Barrier film resins		Casting resins	Liquid injection molding
Properties	ASTM test method	Long glass-reinforced molding compound	EMI shielding (conductive); 30% PAN carbon fiber	Injection molding	Unplasticized	Plasticized	Flexible (including RTV)	Liquid silicone rubber
Processing								
1a. Melt flow (gm./10 min.)	D1238							
1. Melting temperature, °C. T_m (crystalline)		75		172	160	172	Thermoset	Thermoset
T_g (amorphous)				−15	0.2	-15		
2. Processing temperature range, °F. (C = compression; T = transfer; I = injection; E = extrusion)		I: 450-500	I: 360-450	C: 260-350 I: 300-400 E: 300-400	E: 320-390	E: 340-400		I: 360-420
3. Molding pressure range, 10^3 p.s.i.				5-30	3-30	5-30		1-2
4. Compression ratio				2.5	2-2.5	2-2.5		
5. Mold (linear) shrinkage, in./in.	D955	0.001	0.001-0.002	0.005-0.025	0.005-0.025	0.005-0.025	0.0-0.006	0.0-0.005
Mechanical								
6. Tensile strength at break, p.s.i.	D638[b]	27,000-33,000	13,000	3500-5000	2800	3500	350-1000	725-1305
7. Elongation at break, %	D638[b]	2	20	160-240	350-400	250-300	20-700	300-1000
8. Tensile yield strength, p.s.i.	D638[b]	27,000-33,000		2800-3800		4900		
9. Compressive strength (rupture or yield), p.s.i.	D695			2000-2700				
10. Flexural strength (rupture or yield), p.s.i.	D790	45,000-57,000	9000	4200-6200				
11. Tensile modulus, 10^3 p.s.i.	D638[b]	1700-2500	500	50-80	50-80	50-80		
12. Compressive modulus, 10^3 p.s.i.	D695			55-95				
13. Flexural modulus, 10^3 p.s.i. 73° F.	D790	1500-2200	500	55-95				
200° F.	D790							
250° F.	D790							
300° F.	D790							
14. Izod impact, ft.-lb./in. of notch (1/8-in. thick specimen)	D256A	8-16	10	0.4-1.0	0.3-1.0	0.3-1.0		
15. Hardness Rockwell	D785			M60-65	R98-106	R98-106		
Shore/Barcol	D2240/ D2583						Shore A10-70	Shore A20-70
Thermal								
16. Coef. of linear thermal expansion, 10^{-6} in./in./°C.	D696	0.6-0.8		190	190	190	10-19	10-20
17. Deflection temperature under flexural load, °F. 264 p.s.i.	D648	200-260	180	130-150	130-150	130-150		
66 p.s.i.	D648	230-370						
18. Thermal conductivity, 10^{-4} cal.-cm./sec.-cm.^{2}-°C.	C177			3	3	3	3.5-7.5	
Physical								
19. Specific gravity	D792		1.33	1.65-1.72	1.65-1.70	1.68-1.72	0.97-2.5	1.08-1.14
20. Water absorption (1/8-in. thick specimen), % 24 hr.	D570			0.1	0.1	0.1	0.1	
Saturation	D570							
21. Dielectric strength (1/8-in. thick specimen), short time, v./mil	D149			400-600		400-600	400-550	
	SUPPLIERS[a]	Dow Plastics; RTP	LNP; RTP	Dow Plastics	Dow Plastics	Dow Plastics	Bayer Corp.; Dow Corning; Emerson & Cuming; GE Silicones	Bayer Corp.

	Silicone (Cont'd)			Sulfone polymers						
	Molding and encapsulating compounds	Silicone/poly-amide pseudo-interpenetrating networks[f]		Polysulfone						
	Mineral- and/or glass-filled	Silicone/ nylon 66	Silicone/ nylon 12	Injection molding, flame-retarded, extrusion	Mineral-filled	Extrusion/ injection molding grade	20% glass fiber-reinforced	EMI shielding (conductive); 30% carbon fiber	Injection molding platable grade	Polyaryl-sulfone
1a.				3.5-9	7-8.5	3-20			12-18	10-30
1.	Thermoset	Interpenetrating network	Interpenetrating network							
				187-190	190	190	190	190		220
2.	C: 280-360 I: 330-370 T: 330-370	I: 460-525	I: 360-410	I: 625-750 E: 600-700	I: 675-775	I: 610-735 E: 600-680	I: 660-715	I: 550-700	I: 690-750	I: 630-800 E: 620-750
3.	0.3-6	8-15	7-12	5-20	10-20			10-20	5-100	5-20
4.	2.0-8.0			2.5-3.5	2.5-3.5			2.5-3.5	2	
5.	0.0-0.005	0.004-0.007	0.005-0.007	0.0058-0.007	0.004-0.005	0.005-0.007	2.8×10^{-3}	0.0005-0.001	0.0025	0.007-0.008
6.	500-1500	10,100-12,500	5200-7400		9500-9800		16,700	23,000-23,500	1100	9000
7.	80-800	5-20	10	50-100	2-5	40-100	2.4	1.5-2	2.3	40-60
8.				10,200-11,600		11,500				10,400-12,000
9.				13,900-40,000				25,000		
10.		14,000-15,900	8000-8300	15,400-17,500	14,300-15,400	17,500	21,000	31,000-35,000	7700	12,400-16,100
11.				360-390	550-650	390	1020	2150-2800		310-385
12.				374						
13.		360-410	200-216	390	600-750	370	850	1900-2300	881	330-420
				370	570-710					
				350	550-690					
				310	510-650					
14.		0.8-0.9	0.6-0.7	1.0-1.3	0.65-1.0	1.0-1.2	1.5	1.2-1.8		1.6-12
15.				M69	M70-74	M69	M83	M80		
	Shore A10-80									
16.	20-50			56	34-39	31	25	6		31-49
17.	>500			345	345-354	340	363	360-365	410	400
				358		360	369	365-380		
18.	7.18			6.2						
19.	1.80-2.05	1.12-1.13	1.01-1.02	1.24-1.25	1.48-1.61	1.24	1.40	1.36-1.7	1.68	1.29-1.37
20.	0.15	0.6-0.8	0.12-0.15	0.3		0.3		0.15-0.25		
	0.15-0.40			0.8	0.5-0.6	0.8	0.6			1.1-1.85
21.	200-550			425	450				380	370-380
	Cytec Fiberite; Dow Corning; GE Silicones	LNP	LNP	Amoco Polymers; BASF	Amoco Polymers; RTP	Amoco Polymers; BASF	Amoco Polymers; BASF; ComAlloy; RTP	DSM; Ferro; LNP; RTP	Amoco Polymers	LNP; RTP

Materials			Sulfone polymers						
			Polyethersulfone				Modified polysulfone		
	Properties	ASTM test method	Unfilled	10% glass fiber-reinforced	20% glass fiber-reinforced	EMI shielding (conductive); 30% carbon fiber	Poly-phenyl sulfone	Poly-phenyl sulfone, unrein-forced	Mineral-filled
Processing	1a. Melt flow (gm./10 min.)	D1238		12	10		14-20	11.5-17.0	
	1. Melting temperature, °C. T_m (crystalline)								
	T_g (amorphous)		220-230	225	220-225	225	225	220	
	2. Processing temperature range, °F. (C = compression; T = transfer; I = injection; E = extrusion)		C: 645-715 I: 590-750 E: 625-720	I: 660-715	C: 610-750 I: 630-735 E: 570-650	I: 600-750	I: 680-735	680-735	I: 575-650
	3. Molding pressure range, 10^3 p.s.i.		6-20	50-100	6-100	10-20	10-20		5-20
	4. Compression ratio		2-2.5	2:1	2.0-3.5	2.5-3.5	2.2		2.5-3.5
	5. Mold (linear) shrinkage, in./in.	D955	0.006-0.007	0.5	0.002-0.005	0.0005-0.002	0.007	0.007	0.006-0.007
Mechanical	6. Tensile strength at break, p.s.i.	D638[b]	9800-13,800	16,000	15,200-20,000	26,000-30,000	10,100	10,100	
	7. Elongation at break, %	D638[b]	6-80	4.1	2-3.5	1.3-2.5	60-120	60-120	50-100
	8. Tensile yield strength, p.s.i.	D638[b]	12,200-13,000		18,000-18,800				10,500
	9. Compressive strength (rupture or yield), p.s.i.	D695	11,800-15,600		19,500-24,000	22,000			
	10. Flexural strength (rupture or yield), p.s.i.	D790	17,000-18,700	21,000; 24,500	23,500-27,600	36,000-38,000	13,200	13,200	16,500
	11. Tensile modulus, 10^3 p.s.i.	D638[b]	350-410	555; 740	825-1130	2120-2900	340	340	400
	12. Compressive modulus, 10^3 p.s.i.	D695							
	13. Flexural modulus, 10^3 p.s.i. 73° F.	D790	348-380	590; 650	750-980	2000-2600	350	350	480
	200° F.	D790			812-850				370
	250° F.	D790	330		840				340
	300° F.	D790	280		580-842				320
	14. Izod impact, ft.-lb./in. of notch (1/8-in. thick specimen)	D256A	1.4-No break	0.9; 1.3	1.1-1.7	1.2-1.6	13	13.0	1.1
	15. Hardness Rockwell	D785	M85-88	M94	M96-99	R123			M74
	Shore/Barcol	D2240/ D2583							
Thermal	16. Coef. of linear thermal expansion, 10^{-6} in./in./°C.	D696	55	34	23-32	10	17	5.6	53
	17. Deflection temperature under flexural load, °F. 264 p.s.i.	D648	383-397	437	408-426; 437	415-420	405	405	325
	66 p.s.i.	D648	410	423	410-430	420-430			
	18. Thermal conductivity, 10^{-4} cal.-cm./sec.-cm.^{2}-°C.	C177	3.2-4.4						
Physical	19. Specific gravity	D792	1.37-1.46	1.43; 1.45	1.51-1.53	1.47-1.48	1.29-1.3		1.30
	20. Water absorption (1/8-in. thick specimen), % 24 hr.	D570	0.12-1.7		0.15-0.4	0.29-0.35	0.37	0.37	
	Saturation	D570	1.8-2.5	1.9	1.65-2.1; 1.7		1.1		0.8
	21. Dielectric strength (1/8-in. thick specimen), short time, v./mil	D149	400	440	375-500		360	380	460
		SUPPLIERS[a]	Amoco Polymers; BASF	Amoco Polymers; BASF; RTP	Amoco Polymers; BASF; DSM; LNP; RTP	DSM; LNP; RTP	Amoco Polymers	Amoco Polymers	RTP

	Thermoplastic elastomers								
	Polyolefin							Silicone-based, pseudo-interpenetrating networks[f]	
	Low and medium hardness	High hardness	Copolyester ether	Polyester	Polyether/ amide block copolymers	Block copolymers of styrene and butadiene or styrene and isoprene	Block copolymers of styrene and ethylene and/or butylene	Silicone/ polyamide	Silicone/ polyester
1a.	0.4-20.0	0.4-10.0	0.5-20 @190-240	3-12 @190-240		0.5-20			
1.			150-223	148-230	148-209			Interpenetrating network	Interpenetrating network
	165	163-165	–3						
2.	C: 350-450 I: 360-475 E: 380-450	C: 370-450 I: 350-480 E: 380-475	E: 400-500 I: 435-500	I: 340-500 E: 340-500	C: 365-480 I: 340-482 E: 340-460	C: 250-325 I: 300-425 E: 370-400	C: 300-380 I: 350-480 E: 330-380	I: 360-410 I: 355-580 E: 320-365	I: 360-450 E: 380-450
3.	4-19	6-10	1-15	1-15	8-12	0.3-3	1.5-20	5-25	6-14
4.	2-3.5	2-3.5	2-4	3-3.5		2.0-4.0	2.5-5.0	3-5	2.5-4
5.	0.015-0.020	0.007-0.021	0.004-0.02	0.003-0.018		0.001-0.022	0.003-0.022	0.004-0.007	0.004-0.015
6.	650-2500	950-4000	1000-10,000	1000-6800	2000-7000	100-4350	600-3000	5200-12,500	5000
7.	150-780	20-600	200-1000	170-900	350-680	20-1350	600-940	5-275	950-1050
8.		1600-4000	1370-1750	1350-3900	3000-3500	3700-4400			
9.		22-39							
10.					1100-1900	5100-6400	0.1-100	2200-15,900	1700
11.	1.1-16.4	0.34-34.1	16-18	1.1-130	2-60	0.8-235			18
12.				7.5-48		3.6-120			
13.	1..5-30	2.7-300	18	5-175	2.9-66	4-215	0.1-100	35-410	25
14.	No break	5.0-16.0 No break	NB (–30C)	2.5-No break	4.3-No break	No break	No break	No break; 0.6-0.9	No break
15.				104					
	Shore A64-92	Shore D40-70	55/95 (D/A); 30-85	Shore D35-72	Shore A75-D72	Shore A20-95 D63-75	Shore A5-95	Shore D60	Shore D49-51
16.		36-110	150	85-190	210-230	67-140			
17.					45-135	<0-170			
				111-284	158-257	<0-190			
18.	4.5-5.0		5-6	3.6-4.5		3.6			
19.	0.88-0.98	0.90-1.15	1-1.3	1.10-1.28	1.0-1.03	0.90-1.2	0.9-1.2	1.01-1.21	1.21-1.16
20.	0.01	0-0.28	0.4-50	0.2-3.6	1.01-1.36	0.009-0.39	0.1-0.42	0.5-0.15	0.3-0.8
21.	410-445	390-465	350	350-600		300-520	450-800		
	AlphaGary; DuPont; Exxon; M.A. Polymers; Montell NA; Network Polymers; A. Schulman; Solutia; Solvay Engineered Polymers; Teknor Apex; Union Carbide; Vi-Chem	AlphaGary; Equistar; M.A. Polymers; Montell NA; Network Polymers; A. Schulman; Solutia; Solvay Engineered Polymers; Teknor Apex; Vi-Chem; Wash. Penn	DuPont; Eastman	DuPont; Ticona	Creanova; Elf Atochem N.A.; EMS	Fina; Firestone Synthetic Rubber; Network Polymers; A. Schulman; Shell	AlphaGary; Dow Plastics; Network Polymers; Shell; Teknor Apex	Creanova; LNP	Creanova; LNP

Materials	Properties	ASTM test method	Thermoplastic elastomers (Cont'd)				
			Silicone-based, pseudo-interpenetrating networks[f] (Cont'd)				Polyurethane
					Silicone/polyurethane		Molding and extrusion compounds
							Polyester
			Silicone/ polyolefin	Silicone/ polystyrene ethylene butadiene-styrene	Aromatic and aliphatic polyether and polyester urethane, unfilled	Aromatic poly-ether urethane, 15% carbon fiber-reinforced	Low and medium hardness
Processing	1a. Melt flow (gm./10 min.)	D1238					
	1. Melting temperature, °C. T_m (crystalline)		Interpenetrating network	Interpenetrating network	Interpenetrating network	Interpenetrating network	
	T_g (amorphous)				150-225		120-160
	2. Processing temperature range, °F. (C = compression; T = transfer; I = injection; E = extrusion)		I: 340-425	I: 390-490 E: 325-460	I: 325-450 E: 325-475	I: 410-475	I: 340-435 E: 340-410
	3. Molding pressure range, 10^3 p.s.i.		6-11	6-20	5-15	6-15	0.8-1.4
	4. Compression ratio		2.5-3.5	2-3.5	2.5-3.5	2.5-3.5	
	5. Mold (linear) shrinkage, in./in.	D955	0.010-0.020	0.005-0.040	0.010-0.020	0.011-0.014	0.008-0.015
Mechanical	6. Tensile strength at break, p.s.i.	D638[b]	475-1000	1000-3300	2100-6500	18,500	3300-8400
	7. Elongation at break, %	D638[b]	120-1000	500-1000	400-1300	5	410-620
	8. Tensile yield strength, p.s.i.	D638[b]					
	9. Compressive strength (rupture or yield), p.s.i.	D695					
	10. Flexural strength (rupture or yield), p.s.i.	D790			7000-8000	18,000	
	11. Tensile modulus, 10^3 p.s.i.	D638[b]			9-19		
	12. Compressive modulus, 10^3 p.s.i.	D695					
	13. Flexural modulus, 10^3 p.s.i. 73° F.	D790			2-27	800	4000-6990
	200° F.	D790					
	250° F.	D790					
	300° F.	D790					
	14. Izod impact, ft.-lb./in. of notch (1/8-in. thick specimen)	D256A	No break	No break	No break	1.6	
	15. Hardness Rockwell	D785					
	Shore/Barcol	D2240/ D2583	Shore A57-65	Shore A50-84	Shore A55-87; D55-60	Shore D70	Shore A55-94
Thermal	16. Coef. of linear thermal expansion, 10^{-6} in./in./°C.	D696					
	17. Deflection temperature under flexural load, °F. 264 p.s.i.	D648					
	66 p.s.i.	D648					
	18. Thermal conductivity, 10^{-4} cal.-cm./sec.-cm.^{2}-°C.	C177					
Physical	19. Specific gravity	D792	0.96-0.97	0.90-0.97	1.04-1.19	1.24	1.17-1.25
	20. Water absorption (1/8-in. thick specimen), % 24 hr.	D570	0.01	0.05-0.3	0.3-0.6	0.4	
	Saturation	D570					
	21. Dielectric strength (1/8-in. thick specimen), short time, v./mil	D149					
		SUPPLIERS[a]	LNP	Creanova; LNP	Creanova	LNP	BASF; Bayer Corp.; Dow Plastics; BFGoodrich; A. Schulman

	Thermoplastic elastomers (Cont'd)					Vinyl polymers and copolymers			
	Polyurethane (Cont'd)			Elastomeric alloys		PVC and PVC-acetate MC, sheets, rods, and tubes			
	Molding and extrusion compounds (Cont'd)								
	Polyester (Cont'd)	Polyether							
	High hardness	Low and medium hardness	High hardness	Low and medium hardness	High hardness	Rigid	Rigid, tin-stabilized	Flexible, unfilled	Flexible, filled
1a.		Thermoset							
1.				165	165				
	120-160	120-160	120-160			75-105		75-105	75-105
2.	I: 400-440 E: 370-410	I: 350-430 E: 340-410 C: 72-120	I: 400-435 E: 380-440	I: 350-450	I: 350-450	C: 285-400 I: 300-415	I: 380-400	C: 285-350 I: 320-385	C: 285-350 I: 320-385
3.	0.8-1.4	0.6-1.2	1-1.4	6-10	6-10	10-40		8-25	1-2
4.				2.5-3.5	2.5-3.5	2.0-2.3	2.0	2.0-2.3	2.0-2.3
5.	0.005-0.015	0.008-0.015	0.008-0.012	1.5-2.5	1.5-2.5	0.002-0.006		0.010-0.050	0.008-0.035 0.002-0.008
6.	4000-11,000	158-6750	6000-7240	650-3200	2300-4000	5900-7500		1500-3500	1000-3500
7.	110-550	475-1000	340-600	300-750	250-600	40-80		200-450	200-400
8.					1600	5900-6500	5140-7560		
9.						8000-13,000		900-1700	1000-1800
10.						10,000-16,000	6300-13,600		
11.				0.7-5.0	16-100	350-600	310-435		
12.									
13.	9150-64,000	2480-6190	4000-44,900	1.5-6.6	15-50	300-500	305-420		
14.				No fracture	No fracture	0.4-22	2.1-20.0	Varies over wide range	Varies over wide range
15.									
	Shore D46-78	Shore A13-92	Shore D55-75	Shore 55-95A	Shore 32-72D	Shore D65-85	Shore D69-78	Shore A50-100	Shore A50-100
16.				82	82	50-100	1.0-18.1	70-250	
17.						140-170	69-163		
						135-180	161-166		
18.						3.5-5.0		3-4	3-4
19.	1.15-1.28	1.02	1.14-1.21	0.90-1.39	0.90-1.31	1.30-1.58	1.32-1.45	1.16-1.35	1.3-1.7
20.	0.3					0.04-0.4		0.15-0.75	0.5-1.0
21.		470	470	400-600	400-600	350-500		300-400	250-300
	BASF; Bayer Corp.; Dow Plastics; BFGoodrich; A. Schulman	BASF; Bayer Corp.; Cabot; Dow Plastics; BFGoodrich; A. Schulman	BASF; Bayer Corp.; Dow Plastics; BFGoodrich; A. Schulman	AlphaGary; Goodyear; M.A. Hanna; Montell NA; Network Polymers; Novatec; A. Schulman; Solutia; Vi-Chem	AlphaGary; M.A. Hanna; Montell NA; Network Polymers; A. Schulman; Solutia	AlphaGary; Colorite; CONDEA Vista; Creanova; Formosa; Georgia Gulf; Keysor-Century; LG Chemical; Novatec; OxyChem; Rimtec; RSG Polymers; Shintech; Synergistic; Union Carbide; Vi-Chem	AlphaGary; Colorite; CONDEA Vista; Georgia Gulf; Keysor-Century; Oxychem; Rimtec; Shintech; Synergistic	AlphaGary; Colorite; CONDEA Vista; Creanova; Keysor-Century; LG Chemical; Novatec; Rimtec; A. Schulman; Shintech; Shuman; Synergistic; Teknor Apex; Union Carbide; Vi-Chem	AlphaGary; Colorite; CONDEA Vista; Creanova; Keysor-Century; Novatec; Rimtec; Shintech; Synergistic; Teknor Apex; Union Carbide; Vi-Chem

Materials	Properties	ASTM test method	Vinyl polymers and copolymers (Cont'd) — Molding and extrusion compounds: Chlorinated polyvinyl chloride	Vinyl butyral, flexible	PVC/acrylic blends
Processing	1a. Melt flow (gm./10 min.)	D1238			
	1. Melting temperature, °C. T_m (crystalline)				
	T_g (amorphous)		110-127	49	
	2. Processing temperature range, °F. (C = compression; T = transfer; I = injection; E = extrusion)		C: 350-400 I: 395-440 E: 360-430	C: 280-320 I: 250-340	I: 360-390 E: 390-410
	3. Molding pressure range, 10^3 p.s.i.		15-40	0.5-3	2-3
	4. Compression ratio		1.5-2.8		2-2.5
	5. Mold (linear) shrinkage, in./in.	D955	0.003-0.007		0.003
Mechanical	6. Tensile strength at break, p.s.i.	D638[b]	6500-9000	500-3000	6400-7000
	7. Elongation at break, %	D638[b]	4-100	150-450	35-100
	8. Tensile yield strength, p.s.i.	D638[b]	6000-8200		
	9. Compressive strength (rupture or yield), p.s.i.	D695	9000-22,000		6800-8500
	10. Flexural strength (rupture or yield), p.s.i.	D790	11,500-17,000		10,300-11,000
	11. Tensile modulus, 10^3 p.s.i.	D638[b]	326-475		340-370
	12. Compressive modulus, 10^3 p.s.i.	D695	212-600		
	13. Flexural modulus, 10^3 p.s.i. 73° F.	D790	334-450		350-380
	200° F.	D790			
	250° F.	D790			
	300° F.	D790			
	14. Izod impact, ft.-lb./in. of notch (1/8-in. thick specimen)	D256A	1.0-10.0	Varies over wide range	1-12
	15. Hardness Rockwell	D785	R109-117	A10-100	R106-110
	Shore/Barcol	D2240/ D2583	82-88		
Thermal	16. Coef. of linear thermal expansion, 10^{-6} in./in./°C.	D696	52-78		44-79
	17. Deflection temperature under flexural load, °F. 264 p.s.i.	D648	194-234		167-185
	66 p.s.i.	D648	215-247		172-189
	18. Thermal conductivity, 10^{-4} cal.-cm./sec.-cm.^{2}-°C.	C177	3.3		
Physical	19. Specific gravity	D792	1.39-1.58	1.05	1.26-1.35
	20. Water absorption (1/8-in. thick specimen), % 24 hr.	D570	0.02-0.16	1.0-2.0	0.09-.016
	Saturation	D570			
	21. Dielectric strength (1/8-in. thick specimen), short time, v./mil	D149	600-625	350	480
		SUPPLIERS[a]	BF Goodrich; Elf Atochem N.A.; Georgia Gulf	Solutia; Union Carbide	AlphaGary

Notes for Appendix C

a. See list below for addresses of suppliers.
b. Tensile test method varies with material; D638 is standard for thermoplastics; D651 for rigid thermosetting plastics; D412 for elastomeric plastics; D882 for thin plastics sheeting.
c. Dry, as molded (approximately 0.2 percent moisture content).
d. As conditioned to equilibrium with 50 percent relative humidity.
e. Test method in ASTM D4092.
f. *Pseudo* indicates that the thermosetting and thermoplastic components were in the form of pellets or powder prior to fabrication.
g. Dow Plastics samples are unannealed.

Names and Addresses of Suppliers Listed in Appendix C

Adell Plastics, Inc.
4530 Annapolis Rd.
Baltimore, MD 21227
800-638-5218, 410-789-7780
Fax: 410-789-2804

Ain Plastics, Inc.
249 E. Sandford Blvd.
P.O. Box 151-M
Mt. Vernon, NY 10550
800-431-2451, 914-668-6800
Fax: 914-668-8820

Albis Corp.
445 Hwy. 36 N.
P.O. Box 711
Rosenberg, TX 77471
800-231-5911, 713-342-3311
Fax: 713-342-3058
Telex: 166-181

Alliedsignal Inc.
Alliedsignal Engineered Plastics
P.O. Box 2332, 101 Columbia Rd.
Morristown, NJ 07962-2332
201-455-5010
Fax: 201-455-3507

A L M Corp.
55 Haul Rd.
Wayne, NJ 07470
201-694-4141
Fax: 201-831-8327

Alpha/Owens-Corning
P.O. Box 610
Collierville, TN 38027-0610
901-854-2800
Fax: 901-854-1183

AlphaGary Corp.
170 Pioneer Dr.
P.O. Box 808
Leominster, MA 01453
800-232-9741, 508-537-8071
Fax: 508-534-3021

American Polymers
P.O. Box 366
53 Milbrook St.
Worcester, MA 01606
508-756-1010
Fax: 508-756-3611

Ametek, Inc.
Haveg Div.
900 Greenbank Rd.
Wilmington, DE 19808
302-995-0400
Fax: 302-995-0491

Amoco Chemical Co.
200 E. Randolph Dr.
Mail Code 7802
Chicago, IL 60601-7125
800-621-4590, 312-856-3200
Fax: 312-856-4151

Applied Composites Corp.
333 N. Sixth St.
St. Charles, IL 60174
708-584-3130
Fax: 708-584-0659

Applied Polymer Systems, Inc.
P.O. Box 56404
Flushing, NY 11356-4040
718-539-4425
Fax: 718-460-4159

Arco Chemical Co.
3801 West Chester Pike
Newtown Square, PA 19073
800-345-0252 (PA Only),
215-359-2000

Aristech Chemical Corp.
Acrylic Sheet Unit
7350 Empire Dr.
Florence, KY 41042
800-354-9858
Fax: 606-283-6492

Ashley Polymers
5114 Ft. Hamilton Pkwy.
Brooklyn, NY 11219
718-851-8111
Fax: 718-972-3256
Telex: 42-7884

AtoHaas North America Inc.
100 Independence Mall W.
Philadelphia, PA 19106
215-592-3000
Fax: 215-592-2445

Ausimont USA, Inc.
Crown Point Rd. & Leonards Lane
P.O. Box 26
Thorofare, NJ 08086
800-323-2874, 609-853-8119
Fax: 609-853-6405

Bamberger, Claude P., Molding Compounds Corp.
111 Paterson Plank Rd.
P.O. Box 67
Carlstadt, NJ 07072
201-933-6262
Fax: 201-933-8129

Bamberger Polymers, Inc.
1983 Marcus Ave.
Lake Success, NY 11042
800-888-8959, 516-328-2772
Fax: 516-326-1005
Telex: 6711357

BASF Corp., Plastic Materials
3000 Continental Dr. N.
Mount Olive, NJ 07828-1234
201-426-2600

Bayer Corp.
100 Bayer Rd.
Pittsburgh, PA 15205-9741
800-662-2927, 412-777-2000

BFGoodrich Adhesive Systems Div.
123 W. Bartges St.
Akron, OH 44311-1081
216-374-2900
Fax: 216-374-2860

BFGoodrich Specialty Chemicals
9911 Breckville Rd.
Cleveland, OH 44141-3247
800-331-1144, 216-447-5000
Fax: 216-447-5750

Boonton Plastic Molding Co.
30 Plain St.
Boonton, NJ 07005-0030
201-334-4400
Fax: 201-335-0620

Borden Packaging, Div.
Borden Inc.
One Clark St.
North Andover, MA 01845
508-686-9591

BP Chemicals (Hitco) Fibers and Materials
700 E. Dyer Rd.
Santa Ana, CA 92705
714-549-1101

BP Chemicals, Inc.
4440 Warrensville Center Rd.
Cleveland, OH 44128
800-272-4367, 216-586-5847
Fax: 216-586-5839

BP Performance Polymers Inc.
Phenolic Business
60 Walnut Ave.
Suite 100
Clark, NJ 07066
908-815-7843
Fax: 908-815-7844

Budd Chemical Co.
Pennsville-Auburn Rd.
Carneys Point, NJ 08069
609-299-1708
Fax: 609-299-2998

Budd Co.
Plastics Div.
32055 Edward Ave.
Madison Heights, MI 48071
810-588-3200
Fax: 810-588-0798

Bulk Molding Compounds Inc.
3N497 N. 17th St.
St. Charles, IL 60174
708-377-1065
Fax: 708-377-7395

Cadillac Plastic & Chemical Co.
143 Indusco Ct.
P.O. Box 7035
Troy, MI 48007-7035
800-488-1200, 810-583-1200
Fax: 810-583-4715

Cast Nylons Ltd.
4300 Hamann Pkwy.
Willoughby, OH 44092
800-543-3619, 216-269-2300
Fax: 216-269-2323

Chevron Chemical Co.
Olefin & Derivatives
P.O. Box 3766
Houston, TX 77253
800-231-3828, 713-754-2000

Ciba-Geigy Corp., Ciba Additives
540 White Plains Rd.
P.O. Box 20005
Tarrytown, NY 10591-9005
800-431-2360, 914-785-2000
Fax: 914-785-4244

Color-Art Plastics, Inc.
317 Cortlandt St.
Belleville, NJ 07109-3293
201-759-2400

ComAlloy International Co.
481 Allied Dr.
Nashville, TN 37211
615-333-3453
Fax: 615-834-9941

Conap, Inc.
1405 Buffalo St.
Olean, NY 14760
716-372-9650
Fax: 716-372-1594

Consolidated Polymer Technologies, Inc.
11811 62nd St. N.
Largo, FL 34643
800-541-6880, 813-531-4191
Fax: 813-530-5603

Cook Composites & Polymers
P.O. Box 419389
Kansas City, MO 64141-6389
800-821-3590, 816-391-6000
Fax: 816-391-6215

Cosmic Plastics, Inc.
27939 Beale Ct.
Valencia, CA 91355
800-423-5613, 805-257-3274
Fax: 805-257-3345

Custom Manufacturers
858 S. M-18
Gladwin, MI 48624
800-860-4594, 517-426-4591
Fax: 517-426-4049

Custom Molders Corp.
2470 Plainfield Ave.
Scotch Plains, NJ 07076
908-233-5880
Fax: 908-233-5949

Custom Plastic Injection Molding Co. Inc.
3 Spielman Rd.
Fairfield, NJ 07004
201-227-1155

Cyro Industries
100 Enterprise Dr., 7th fl.
Rockaway, NJ 07866
201-442-6000

CYTEC Industries Inc.
5 Garret Mt. Plaza
West Paterson, NJ 07424
800-438-5615, 201-357-3100
Fax: 201-357-3065

CYTEC Industries Inc.
12600 Eckel Rd.
P.O. Box 148
Perrysburg, OH 43551
800-537-3360, 419-874-7941
Fax: 419-874-0951

Diamond Polymers
1353 Exeter Rd.
Akron, OH 44306
216-773-2700
Fax: 216-773-2799

Dow Chemical Co.
Polyurethanes
2040 Willard H. Dow Center
Midland, MI 48674
800-441-4369

Dow Corning Corp.
P.O. Box 0994
Midland, MI 48686-0994
517-496-4000
Fax: 517-496-4586

Dow Corning STI
47799 Halyard Dr.
Suite 99
Plymouth, MI 48170
313-459-7792
Fax: 313-459-0204

Dow Plastics
P.O. Box 1206
Midland, MI 48641-1206
800-441-4369

DSM Copolymer, Inc.
P.O. Box 2591
Baton Rouge, LA 70821
800-535-9960, 504-355-5655
Fax: 504-357-9574

DSM Engineering Plastics
2267 W. Mill Rd.
P.O. Box 3333
Evansville, IN 47732
800-333-4237, 812-435-7500
Fax: 812-435-7702

DSM Thermoplastic Elastomers, Inc.
29 Fuller St.
Leominster, MA 01453-4451
800-524-0120, 508-534-1010
Fax: 508-534-1005

DuPont Engineering Polymers
1007 Market St.
Wilmington, DE 19898
800-441-7515, 302-999-4592
Fax: 302-999-4358

Eagle-Picher Industries, Inc.
C & Porter Sts.
Joplin, MO 64802
417-623-8000
Fax: 417-782-1923

Eastman Chemical Co.
P.O. Box 511
Kingsport, TN 37662
800-327-8626

Elf Atochem North America, Inc.
Fluoropolymers
2000 Market St.
Philadelphia, PA 19103
800-225-7788, 215-419-7000
Fax: 215-419-7497

Elf Atochem North America, Inc.,
Organic Peroxides
2000 Market St.
Philadelphia, PA 19103
800-558-5575, 215-419-7000
Fax: 215-419-7591

Emerson & Cuming, Inc./Grace Specialty Polymers
77 Dragon Ct.
Woburn, MA 01888
800-832-4929, 617-938-8630
Fax: 617-935-0125

Epic, Inc.
654 Madison Ave.
New York, NY 10021
212-308-7039
Fax: 212-308-7266

Epsilon Products Co.
Post Rd. and Blueball Ave.
P.O. Box 432
Marcus Hook, PA 19061
610-497-8850
Fax: 610-497-4694

EVAL Co. of America
1001 Warrenville Rd.
Suite 201
Lisle, IL 60532-1359
800-423-9762, 708-719-4610
Fax: 708-719-4622

Exxon Chemical Co.
13501 Katy Freeway
Houston, TX 77079-1398
800-231-6633, 713-870-6000
Fax: 713-870-6970

Federal Plastics Corp.
715 South Ave. E.
Cranford, NJ 07016
800-541-4424, 908-272-5800
Fax: 908-272-9021

Ferro Corp.
Filled & Reinforced Plastics Div.
5001 O'Hara Dr.
Evansville, IN 47711
812-423-5218
Fax: 812-435-2113

Ferro Corp., World Headquarters
1000 Lakeside Ave.
P.O. Box 147000
Cleveland, OH 44114-7000
216-641-8580
Fax: 216-696-6958
Telex: 98-0165

Fina Oil & Chemical Co.
Chemical Div.
P.O. Box 2159
Dallas, TX 75221
800-344-3462, 214-750-2806
Fax: 214-821-1433

Firestone Canada
P.O. Box 486
Woodstock, Ontario, Canada N4S 7Y9
800-999-6231, 519-421-5649
Fax: 519-537-6235

Firestone Synthetic Rubber & Latex Co.
P.O. Box 26611
Akron, OH 44319-0006
800-282-0222
Fax: 216-379-7875
Telex: 67-16415

Formosa Plastics Corp. USA
9 Peach Tree Hill Rd.
Livingston, NJ 07039
201-992-2090
Fax: 201-716-7208

Franklin Polymers, Inc.
P.O. Box 481
521 Yale Ave.
Pitman, NJ 08071-0481
800-238-7659, 609-582-6115
Fax: 609-582-0525

FRP Supply Div.
Ashland Chemical Co.
P.O. Box 2219
Columbus, OH 43216
614-790-4272
Fax: 614-790-4012
Telex: 24-5385 ASHCHEM

GE Plastics
One Plastics Ave.
Pittsfield, MA 01201
800-845-0600, 413-448-7110

GE Silicones
260 Hudson River Rd.
Waterford, NY 12188
800-255-8886
Fax: 518-233-3931

General Polymers Div.
Ashland Chemical Co.
P.O. Box 2219
Columbus, OH 43216
800-828-7659
Fax: 614-889-3195

George, P. D., Co.
5200 N. Second St.
St. Louis, MO 63147
314-621-5700

Georgia Gulf Corp.
PVC Div.
P.O. Box 629
Plaquemine, LA 70765-0629
504-685-1200

Glastic Corp.
4321 Glenridge Rd.
Cleveland, OH 44121
216-486-0100
Fax: 216-486-1091

GLS Corp.
Thermoplastic Elastomers Div.
740B Industrial Dr.
Cary, IL 60013
800-457-8777 (not IL), 708-516-8300
Fax: 708-516-8361

Goodyear Tire & Rubber Co.
Chemical Div.
1485 E. Archwood Ave.
Akron, OH 44316-0001
800-522-7659, 216-796-6253
Fax: 216-796-2617

Grace Specialty Polymers
77 Dragon Ct.
Woburn, MA 01888
617-938-8630

Haysite Reinforced Plastics
5599 New Perry Hwy.
Erie, PA 16509
814-868-3691
Fax: 814-864-7803

Hercules Inc.
Hercules Plaza
Wilmington, DE 19894
800-235-0543, 302-594-5000
Fax: 412-384-4291

Hercules Moulded Products
R.R. 3
Maidstone, Ontario, Canada N0R 1K0
519-737-6693
Fax: 519-737-1747

Hoechst Celanese Corp., Advanced Materials Group
90 Morris Ave.
Summit, NJ 07901
800-526-4960, 908-598-4000
Fax: 908-598-4330
Telex: 13-6346

Hoechst Celanese Corp.
Hostalen GUR Business Unit
2520 S. Shore Blvd.
Suite 110
League City, TX 77573
713-334-8500

Huls America Inc.
80 Centennial Ave.
Piscataway, NJ 08855-0456
908-980-6800
Fax: 908-980-6970

Huntsman Chemical Corp.
2000 Eagle Gate Tower
Salt Lake City, UT 84111
800-421-2411, 801-536-1500
Fax: 801-536-1581

Huntsman Corp.
P.O. Box 27707
Houston, TX 77227-7707
713-961-3711
Fax: 713-235-6437
Telex: 227031 TEX UR

Hysol Engineering Adhesives
Dexter Distrib. Programs
One Dexter Dr.
Seabrook, NH 03874-4018
800-767-8786, 603-474-5541
Fax: 603-474-5545

ICI Acrylics Canada Inc.
7521 Tranmere Dr.
Mississauga, Ontario, Canada L5S 1L4
800-387-4880, 905-673-3345
Fax: 905-673-1459

ICI Acrylics Inc.
10091 Manchester Rd.
St. Louis, MO 63122
800-325-9577, 314-966-3111
Fax: 314-966-3117

ICI Americas Inc.
Tatnall Bldg.
P.O. Box 15391
3411 Silverside Rd.
Wilmington, DE 19850-5391
800-822-8215, 302-887-5536
Fax: 302-887-2089

ICI Fiberite
Molding Materials
501 W. Third St.
Winona, MN 55987-5468
507-454-3611
Fax: 507-452-8195
Telex: 507-454-3646

ICI Polyurethanes Group
286 Mantua Grove Rd.
West Deptford, NJ 08066-1732
800-257-5547, 609-423-8300
Fax: 609-423-8580

ITW Adhesives
37722 Enterprise Ct.
Farmington Hills, MI 48331
800-323-0451, 313-489-9344
Fax: 313-489-1545

Jet Composites Inc.
405 Fairall St.
Ajax, Ontario, Canada L1S 1R8
416-686-1707
Fax: 416-427-9403

Jet Plastics
941 N. Eastern Ave.
Los Angeles, CA 90063
213-268-6706
Fax: 213-268-8262

Keysor-Century Corp.
26000 Springbrook Rd.
P.O. Box 924
Santa Clarita, CA 91380-9024
805-259-2360
Fax: 805-259-7937

Kleerdex Co.
100 Gaither Dr.
Suite B
Mt. Laurel, NJ 08054
800-541-7232, 609-866-1700
Fax: 609-866-9728

Laird Plastics
1400 Centrepark
Suite 500
West Palm Beach, FL 33401
800-610-1016, 407-684-7000
Fax: 407-684-7088

LNP Engineering Plastics Inc.
475 Creamery Way
Exton, PA 19341
800-854-8774, 610-363-4500
Fax: 610-363-4749

Lockport Thermosets Inc.
157 Front St.
Lockport, LA 70374
800-259-8662, 504-532-2541
Fax: 504-532-6806

M.A. Industries Inc.
Polymer Div.
303 Dividend Dr.
P.O. Box 2322
Peachtree City, GA 30269
800-241-8250, 404-487-7761

Mitsui Petrochemical Industries Ltd.
3-2-5 Kasumigaseki, Chiyoda-ku
Tokyo, Japan 100
03-3593-1630
Fax: 03-3593-0979

Mitsui Plastics, Inc.
11 Martine Ave.
White Plains, NY 10606
914-287-6800
Fax: 914-287-6850

Mitsui Toatsu Chemicals, Inc.
2500 Westchester Ave.
Suite 110
Purchase, NY 10577
914-253-0777
Fax: 914-253-0790

Mobil Chemical Co.
1150 Pittsford-Victor Rd.
Pittsford, NY 14534
716-248-1193
Fax: 716-248-1075

Mobil Polymers
P.O. Box 5445
800 Connecticut Ave.
Norwalk, CT 06856
203-854-3808
Fax: 203-854-3840

Monmouth Plastics Co.
800 W. Main St.
Freehold, NJ 07728
800-526-2820, 908-866-0200
Fax: 908-866-0274

Monsanto Co.
800 N. Lindbergh Blvd.
St. Louis, MO 63167
314-694-1000
Fax: 314-694-7625
Telex: 650-397-7820

Montell Polyolefins
Three Little Falls Centre
2801 Centerville Rd.
Wilmington, DE 19850-5439
800-666-8355, 302-996-6000
Fax: 302-996-5587

Morton International Inc.
Morton Plastics Additives
150 Andover St.
Danvers, MA 01923
508-774-3100
Fax: 508-750-9511

Network Polymers, Inc.
1353 Exeter Rd.
Akron, OH 44306
216-773-2700
Fax: 216-773-2799

Nova Polymers, Inc.
P.O. Box 8466
Evansville, IN 47716-8466
812-476-0339
Fax: 812-476-0592

Novacor Chemicals Inc.
690 Mechanic St.
Leominster, MA 01453
800-225-8063, 508-537-1111
Fax: 508-537-5685

Novacor Chemicals Inc.
Clear Performance Plastics
690 Mechanic St.
Leominster, MA 01453
800-243-4750, 508-537-1111
Fax: 508-537-6410

Novatec Plastics & Chemicals Co. Inc.
P.O. Box 597
275 Industrial Way W.
Eatontown, NJ 07724
800-782-6682, 908-542-6600

Nylon Engineering
12800 University Dr.
Suite 275
Ft. Myers, FL 33907
813-482-1100
Fax: 813-482-4202

Occidental Chemical Corp.
5005 LBJ Freeway
Dallas, TX 75244
214-404-3800

Patent Plastics Inc.
638 Maryville Pike S.W.
P.O. Box 9246
Knoxville, TN 37920
800-340-7523, 615-573-5411

Paxon Polymer Co.
P.O. Box 53006
Baton Rouge, LA 70892
504-775-4330

Performance Polymers Inc.
803 Lancaster St.
Leominster, MA 01453
800-874-2992, 508-534-8000
Fax: 508-534-8590

Perstorp Compounds Inc.
238 Nonotuck St.
Florence, MA 01060
413-584-2472
Fax: 413-586-4089

Perstorp Xytec, Inc.
9350 47th Ave. S.W.
P.O. Box 99057
Tacoma, WA 98499
206-582-0644
Fax: 206-588-5539

Phillips Chemical Co.
101 ARB Plastics Technical Center
Bartlesville, OK 74004
918-661-9845
Fax: 918-662-2929

Plaskolite, Inc.
P.O. Box 1497
Columbus, OH 43216
800-848-9124, 614-294-3281
Fax: 614-297-7287

Plaskon Electronic Materials, Inc.
100 Independence Mall West
Philadelphia, PA 19106
800-537-3350, 215-592-2081
Fax: 215-592-2295
Telex: 845-247

Plaslok Corp.
3155 Broadway
Buffalo, NY 14227
800-828-7913, 716-681-7755
Fax: 716-681-9142

Plastic Engineering & Technical Services, Inc.
2961 Bond
Rochester Hills, MI 48309
313-299-8200
Fax: 313-299-8206

Plastics Mfg. Co.
2700 S. Westmoreland St.
Dallas, TX 75223
214-330-8671
Fax: 214-337-7428

Polymer Resources Ltd.
656 New Britain Ave.
Farmington, CT 06032
800-423-5176, 203-678-9088
Fax: 203-678-9299

Polymers International Inc.
P.O. Box 18367
Spartanburg, SC 29318
803-579-2729
Fax: 803-579-4476

Polyply Composites Inc.
1540 Marion
Grand Haven, MI 49417
616-842-6330
Fax: 616-842-5320

PPG Industries, Inc.
Chemicals Group
One PPG Place
Pittsburgh, PA 15272
412-434-3131
Fax: 412-434-2891

Premix, Inc.
Rte. 20 & Harmon Rd.
P.O. Box 281
North Kingsville, OH 44068
216-224-2181
Fax: 216-224-2766

Prime Alliance, Inc.
1803 Hull Ave.
Des Moines, IA 50302
800-247-8038, 515-264-4110
Fax: 515-264-4100

Prime Plastics, Inc.
2950 S. First St.
Clinton, OH 44216
216-825-3451

Progressive Polymers, Inc.
P.O. Box 280
4545 N. Jackson
Jacksonville, TX 75766
800-426-4009, 903-586-0583
Fax: 903-586-4063

Quantum Composites, Inc.
4702 James Savage Rd.
Midland, MI 48642
800-462-9318, 517-496-2884
Fax: 517-496-2333

Reichhold Chemicals, Inc.
P.O. Box 13582
Research Triangle Park, NC 27709
800-448-3482, 919-990-7500
Fax: 919-990-7711

Reichhold Chemicals, Inc.
Emulsion Polymer Div.
2400 Ellis Rd.
Durham, NC 27703-5543
919-990-7500
Fax: 919-990-7711

Resinoid Engineering Corp.
P.O. Box 2264
Newark, OH 43055
614-928-6115
Fax: 614-929-3165

Resinoid Engineering Corp.
Materials Div.
7557 N. St. Louis Ave.
Skokie, IL 60076
708-673-1050
Fax: 708-673-2160

Rhone-Poulenc
Rte. 8
Rouseville Rd.
P.O. Box 98
Oil City, PA 16301-0098
814-677-2028
Fax: 814-677-2936

Rimtec Corp.
1702 Beverly Rd.
Burlington, NJ 08016
800-272-0069, 609-387-0011
Fax: 609-387-0282

Rogers Corp.
One Technology Dr.
Rogers, CT 06263
203-774-9605
Fax: 203-774-9630

Rogers Corp.
Molding Materials Div.
Mill and Oakland Sts.
P.O. Box 550
Manchester, CT 06045
800-243-7158, 203-646-5500
Fax: 203-646-5503

Ronald Mark Associates, Inc.
P.O. Box 776
Hillside, NJ 07205
908-558-0011
Fax: 908-558-9366

Rostone Corp.
P.O. Box 7497
Lafayette, IN 47903
317-474-2421
Fax: 317-474-5870

Rotuba Extruders, Inc.
1401 Park Ave. S.
Linden, NJ 07036
908-486-1000
Fax: 908-486-0874

R.S.G. Polymers Corp.
P.O. Box 1677
Valrico, FL 33594
813-689-7558
Fax: 813-685-6685

RTP Co.
580 E. Front St.
P.O. Box 5439
Winona, MN 55987-0439
800-433-4787, 507-454-6900
Fax: 507-454-2041

Schulman, A., Inc.
3550 W. Market St.
Akron, OH 44333
800-547-3746, 216-668-3751
Fax: 216-668-7204
Telex: SCHN 6874 22

Shell Chemical Co.
One Shell Plaza
Rm. 1671
Houston, TX 77002
713-241-6161

Shell Chemical Co.
Polyester Div.
4040 Embassy Pkwy.
Suite 220
Akron, OH 44333
216-798-6400
Fax: 216-798-6400

Shintech Inc.
Weslayan Tower
24 Greenway Plaza
Suite 811
Houston, TX 77046
713-965-0713
Fax: 713-965-0629

Shuman Co.
3232 South Blvd.
Charlotte, NC 28209
704-525-9980
Fax: 704-525-0622

Solvay Polymers, Inc.
P.O. Box 27328
Houston, TX 77227-7328
800-231-6313, 713-525-4000
Fax: 713-522-2435

Sumitomo Plastics America, Inc.
900 Lafayette St.
Suite 510
Santa Clara, CA 95050-4967
408-243-8402
Fax: 408-243-8405

Synergistics Industries (NJ) Inc.
10 Ruckle Ave.
Farmingdale, NJ 07727
908-938-5980
Fax: 908-938-6933

Syracuse Plastics, Inc.
400 Clinton St.
Fayetteville, NY 13066
315-637-9881
Fax: 315-637-9260

Teknor Apex Co.
505 Central Ave.
Pawtucket, RI 02861
800-554-9892, 401-725-8000
Fax: 401-724-6250

Tetrafluor, Inc.
2051 E. Maple Ave.
El Segundo, CA 90245
310-322-8030
Fax: 310-640-0312

Texapol Corp.
177 Mikron Rd.
Lower Nazareth Comm. Park
Bethlehem, PA 18017
800-523-9242, 610-759-8222
Fax: 610-759-9460

Thermofil, Inc.
6150 Whitmore Lake Rd.
Brighton, MI 48116-1990
800-444-4408, 810-227-3500
Fax: 810-227-3824

Union Carbide Corp.
39 Old Ridgebury Rd.
Danbury, CT 06817-0001
800-335-8550, 203-794-5300

Vi-Chem Corp.
55 Cottage Grove St. S.W.
Grand Rapids, MI 49507
800-477-8501, 616-247-8501
Fax: 616-247-8703

Vista Chemical Co.
900 Threadneedle
P.O. Box 19029
Houston, TX 77079
713-588-3000

Washington Penn Plastic Co.
2080 N. Main St.
Washington, PA 15301
412-228-1260
Fax: 412-228-0962

Wellman Extrusion
P.O. Box 130
Ripon, WI 54971-0130
800-398-7876, 414-748-7421
Fax: 414-748-6093

Westlake Polymers Corp.
2801 Post Oak Blvd.
Suite 600
Houston, TX 77056
800-545-9477, 713-960-9111
Fax: 713-960-8761

Appendix

D

Sources of Specifications and Standards for Plastics and Composites*

As in other material categories, the reliable use of plastics and composites requires specifications to support procurement of these materials and standards to establish engineering and technical requirements for processes, procedures, practices, and methods for testing and using these materials. These specifications and standards may be general industry-wide, or they may be specific to one industry, such as the spaceborne industry, wherein plastics and composites must exhibit special stability in the harsh environments of temperature and vacuum. Specifications and standards for this industry, for instance, would be controlled by documents from the National Aeronautics and Space Administration (NASA). Increasingly, the development and use of standards is becoming international. Since the total sum of these specifications and standards is voluminous, no attempt is made herein to itemize all of them. However, the names and addresses of organizations that are sources of these documents, and information concerning them, are listed below. Also, one comprehensive reference book that can be recommended to readers with interest in this area is given as Ref. 1 at the end of this listing. Another excellent reference book dealing with the major issues of flammability standards of plastics and composites is listed as Ref. 2.

*Reprinted with permission from *Handbook of Plastics, Elastomers, and Composites*, 3d ed., Charles A. Harper (ed.), © McGraw-Hill, New York, 1996.

Names and Addresses of Organizational Sources of Specifications and Standards for Plastics and Composites

1. Industry standards

American National Standards
Institute, Inc. (ANSI)
1430 Broadway
New York, NY 10018

or
Global Engineering Documents
1990 M Street N.W., Suite 400
Washington, DC 20036

Global Engineering Documents
2805 McGaw Avenue
P.O. Box 19539
Irvine, CA 92714

National Standards Association
1200 Quince Orchard Boulevard
Gaithersburg, MD 20878

2. Federal standards and specifications

General Services Administration
(GSA)
Seventh and D Streets, S.W.
Washington, DC 20407
National Standards Association
(See 1. Industry Standards, above)

Global Engineering Documents
(See 1. Industry Standards, above)
Document Engineering
Company, Inc.
15210 Stagg Street
Van Nuys, CA 91405

3. Military specifications and standards

DODSSP—Customer Service
Standardization Document
Order Desk
Building 4D
700 Robbins Avenue
Philadelphia, PA 19111-5094

Global Engineering Documents
(See 1. Industry Standards, above)

National Standards
Association (See 1.
Industry Standards, above)

Document Engineering
Company, Inc.
(See 2. Federal Standards
and Specifications, above)

4. Spaceborne plastics and composites specifications and standards (NASA)

For plastics:
Individual NASA Space Centers
For composites:
NASA
ACEE Composites Project Office
Langley Research Center
Hampton, VA 23665-5225

5. Foreign or international standards

American National Standards
Institute, Inc. (See 1.
Industry Standards, above)

Global Engineering Documents
(See 1. Industry Standards, above)

International Standards and
Law Information (ISLI)
P.O. Box 230
Accord, MA 02018

International Organization for
Standardization
1 Rue de Varembe
Case Postale 56
CH-1211 Geneva 20,
Switzerland

6. National and international electrical standards

National Electrical Manufacturers
Association (NEMA)
2101 L Street, N.W.
Washington, DC 20037

National Standards Association
(See 1. Industry Standards
and 22. Specifications, Standards,
and Industry Approvals for
Plastics and Testing of Plastics)

International Electrotechnical
Commission (IEC)
3 Rue de Varembe
Case Postale 131
CH-1211 Geneva 20,
Switzerland

7. International standards

International Standards
Organization (ISO)
(Located in Geneva, Switzerland
but documents and information
available in the United States from
American National Standards
Institute [see 1. Industry Standards, above])

8. Japanese standards and specifications

Japanese Standards Association
1-24, Akasaka 4-chome, Minatoku-ku
Tokyo 107, Japan

9. German (DIN) standards

Beuth Verlag GmBH
Burggrafenstrasse 6
D-1000 Berlin 30, Germany

Deutsches Institut für Normung e.V. (DIN)
Postfach 1107
D-1000 Berlin 30, Germany

10. European aerospace specifications and standards

European Association of Aerospace Manufacturers (AECMA)
88, Boulevard Malesherbes
F-75008 Paris, France

Generally available in U.S. from National Standards Association—see 1. Industry Standards, above

11. European plastics manufacturing standards

Association of Plastics Manufacturers in Europe (APME)
Avenue Louise 250
Box 73
B-1050 Brussels, Belgium

12. Plastics standards in United Kingdom

British Standards Institute
2 Park Street
London W1A 2BS, U.K.

British Plastics Federation
5 Belgrave Square
London SWIX 8 PH, U.K.

13. European plastics machinery standards

European Committee of Machinery Manufacturers for the Plastics and Rubber Industry (EUROMAP)
Kirchenweg 4
CH-8032 Zurich, Switzerland

14. Rubber and plastic standards in United Kingdom

RAPRA Technology, Ltd.
Shawbury, Shrewsbury
Shropshire SY4 4NR, U.K.

15. Plastics standards in France

French Association for Standardization (AFNOR)
Tour Europe, Cedex 7
92080 Paril La Défense
Paris, France

16. Specifications and standards for aerospace products

Aerospace Industries Association of America, Inc. (AIA)
1250 I Street, N.W.
Washington, DC 20005

17. National standards for U.S. standards developing bodies

American National Standards Institute, Inc.
(ANSI) (See 1. Industry Standards, above)

18. Standards for testing

American Society for Testing and
Materials (ASTM)
100 Barr Harbor Drive
West Conshohocken, PA 19428-2959

19. Specifications and standards for the plastics industry

Society of the Plastics Industry (SPI)
355 Lexington Avenue
New York, NY 10017
or
1275 K Street, N.W., Suite 400
Washington, DC 20005

20. Specifications and standards for composites materials characterization

Composite Materials Characterization, Inc.
Attn: Cecil Schneider
Lockheed Martin Aeronautical Systems Company
Advanced Structures and Materials Division
Marietta, GA 30063

21. Standards in measurements

National Institute of Standards and Technology (NIST)
Office of Standard Reference Materials
Room B311, Chemistry Building
Gaithersburg, MD 20899
(Note: Previously known as National Bureau of Standards [NBS])

22. Specifications, standards, and industry approvals for plastics and testing of plastics

Underwriters Laboratories, Inc. (UL)
333 Pfingston Road
Northbrook, IL 60062
(Offices also in Melville, Long Island, NY; Santa Clara, CA; Tampa, FL)

23. Standardization of composite materials test methods

Suppliers of Advanced Composite Materials Association (SACMA)
1600 Wilson Boulevard, Suite 1008
Arlington, VA 22209

References

1. Traceski, F. T., *Specifications and Standards for Plastics and Composites,* ASM International, Materials Park, OH, 1990.
2. Hilado, C. J., *Flammability Handbook for Plastics,* 4th ed., Technomic Publishing Co., Inc., Lancaster, PA, 1990.

Appendix

Plastics Associations*

*Reprinted with permission from *Handbook of Plastics, Elastomers, and Composites*, 3d ed., Charles A. Harper (ed.), © McGraw-Hill, New York, 1996.

Algeria

National Enterprise of Plastics and Rubber
BP 452-453 Zone Industrielle
Setif 19000
Phone: (213) 590 8157
Fax: (213) 590 0665

Argentina

Argentine Chamber of the Plastics Industry
Jeronimo Salguero 1939/41
1425 Buenos Aires
Phone: (54) 1 821 9603
Fax: (54) 1 826 5480

Australia

Plastics and Chemicals Industries Association, Inc.
P.O. Box 1610M
Melbourne, Victoria 3001
Phone: (61) 03 9699 6299
Fax: (61) 03 9699 6717

Austria

Austrian Chemical Industries Federation
Wiedner Hauptstrasse 63
P.O. Box 325
1045 Vienna 4
Phone: (43) 1 50105 33
Fax: (43) 1 50206 280

National Guild of Plastics Processors
Wiedner Hauptstrasse 63
1045 Vienna 4
Phone: (43) 1 50105 32
Fax: (43) 1 50206 291

Belgium

Association of Belgian Plastics Converters
Rue des Drapiers 21
B-1050 Brussels
Phone: (32) 2 510 2506
Fax: (32) 2 510 2562

Association of Plastics Manufacturers in Europe
Av. E. Van Nieuwenhuyse 4
P.O. Box 3
1160 Brussels
Phone: (32) 2 675 3297
Fax: (32) 2 675 3935

Belgian Association of Manufacturers of Extruded Thermoplastic Pipes
Rue des Drapiers 21
B-1050 Brussels
Phone: (32) 2 510 2507
Fax: (32) 2 510 2562

Belgian Association of Manufacturers of Technical Plastic Components
Rue des Drapiers 21
B-1050 Brussels
Phone: (32) 2 510 2506
Fax: (32) 2 510 2562

Belgian Plastic Joinery Manufacturers' Association
Rue des Drapiers 21
B-1050 Brussels
Phone: (32) 2 510 2507
Fax: (32) 2 510 2562

Belgian Reinforced Plastics Association
Rue des Drapiers 21
B-1050 Brussels
Phone: (32) 2 510 2506
Fax: (32) 2 510 2562

European Association of Flexible Polyurethane Foam Blocs Manufacturers
c/o FIC Square
Marie-Louise 49
1040 Brussels
Phone: (32) 2 238 9711
Fax: (32) 2 238 9998

European Organization of Reinforced Plastics/Composite Materials
Rue des Drapiers 21
B-1050 Brussels
Phone: (32) 2 510 2506
Fax: (32) 2 510 2562

European Plastics Converters
Av. de Cortenbergh 66
1040 Brussels
Phone: (32) 2 732 4124
Fax: (32) 2 732 4218

Plastics Manufacturers and Plastics Processors
49 Square Marie-Louise
1040 Brussels
Phone: (32) 2 238 9711
Fax: (32) 2 238 9998

Bolivia

Aniplast c/o Plasmar
Casilla 942
La Paz
Fax: (591) 2 850 581

Brazil

National Institute of Plastics
Avenida Paulista 1313 cj. 702
01311-923 Sao Paulo SP
Phone: (55) 11 289 6287
Fax: (55) 11 289 6287

Transforming Plastics Material Association
Avenida Paulista 2439 8° andar
01311 São Paulo SP
Phone: (55) 11 282 8288
Fax: (55) 11 282 8042

Canada

Canadian Association of Moldmakers
424 Tecumseh Rd. E.
Windsor, Ontario N8X 2R6
Phone: (1) 519 255 7863
Fax: (1) 519 255 9446
Internet: www.cdnmolds.com

Canadian Plastics Industry Association (CPIA)
5925 Airport Road, Suite 500
Mississauga, Ontario, L4V 1W1
Phone: (1) 905 678 7748
Fax: (1) 905 678 0774
Internet: www.plastics.ca

Environment & Plastics Institute of Canada
(see Canadian Plastics Industry Association)

Chile

Chilean Plastics Association
Av. Andrés Bello 2777
5o Piso, Oficina 507
Las Condes, Santiago
Phone: (56) 2 203 3342
Fax: (56) 2 203 3343

China

Chiu Chau Plastics Manufacturers Association Co. Ltd.
Rm 603, Fu Hing Bldg.
9-11 Jubilee Street
Hong Kong
Phone: (852) 545 0384
Fax: (852) 545 0539

Federation of Hong Kong Industries
Hankow Centre, 4th fl.
5-15 Hankow Rd, Tsim Sha Tsui
Kowloon, Hong Kong
Phone: (852) 2732 3188
Fax: (852) 2721 3494
Internet: www.fhki.org.hk

Hong Kong & Kowloo Plastic Products Merchants United Association Ltd.
13/F, 491 Nathan Road
Kowloon, Hong Kong
Phone: (852) 384 0171
Fax: (852) 781 0107

Hong Kong Plastics Manufacturers Association Ltd.
1/F, Flat B, Fo Yuen,
39-49 Wanchai Road
Wanchai, Hong Kong
Phone: (852) 574 2230
Fax: (852) 574 2843

Hong Kong Plastic Material Suppliers Association Ltd.
1/F, 11 Lai Yip Street,
Kwun Tong,
Kowloon, Hong Kong
Phone: (852) 757 9331
Fax: (852) 796 8885

Hong Kong Plastics Technology Centre Co. Ltd.
U 509, Hong Kong Polytechnic,
Hom, Kowloon, Hong Kong
Phone: (852) 2766 5577
Fax: (852) 2766 0131
Internet: www.plastics-ctr.org.hk

Colombia

Colombian Association of Plastics Industries
Calle 69 No 5-33
Santafe de Bogota
Phone: (57) 1 346 0655
Fax: (57) 1 249 6997

Costa Rica

Costa Rican Association of the Plastic Industry
Apartado 8247
1000 San Jose
Phone: (506) 55 0961
Fax: (506) 55 0961

Cyprus

The Cyprus Plastics Processors Association
P.O. Box 1455
Nicosia
Phone: (357) 2 449 500
Fax: (357) 2 467 593

Czech Republic

Confederation of Plastics Industry of the Czech Republic
Mikulandská 7
CZ-113 61 Praha (Prague) 1
Phone: (42) 2 2491 5678
Fax: (42) 2 297 896

Czech Packaging Association
Jecná 11
CZ-120 00 Praha (Prague) 2
Phone/Fax: (42) 2 292 370

Denmark

Plastics Industries of Denmark
Radhuspladsen 55
1550 Copenhagen
Phone: (45) 331 33 022
Fax: (45) 339 10 898

Ecuador

Ecuadorian Association of Plastics
c/o Polimalla SA,
Via a Daule, km. 9.5,
C.P. 10830, Guayaquil
Phone: (593) 4 253 850
Fax: (593) 4 253 833

Finland

Finnish Plastics Industries Federation
Eteläranta 10, 7th floor,
P.O. Box 4
00130 Helsinki
Phone: (358) 0 172 841
Fax: (358) 0 171 164

France

European Adhesive Tapes Manufacturers Association
60, rue Auber
94408 Vitry sur Seine
Phone: (33) 49 60 5757
Fax: (33) 45 21 0350

International Committee for Plastics in Agriculture
65, rue de Prony
75854 Paris Cedex 17
Phone: (33) 1 44 01 1648
Fax: (33) 1 44 01 1655

National Syndicate of Rubber, Plastics and Associated Industries
60, rue Auber
94408 Vitry sur Seine
Phone: (33) 49 60 5757
Fax: (33) 45 21 0350

Professional Syndicate of Plastic Materials Producers
14, rue de la Republique
92800 Puteaux
Phone: (33) 46 53 1053
Fax: (33) 46 53 1073

Symacap
Phone: (33) 1 47 17 6358
Fax: (33) 1 47 17 6360

Germany

Association of the German Electroplaters
Horionplatz 6
40213 Dusseldorf
Phone: (49) 211 132 381
Fax: (49) 211 327 199

Association for PVC and the Environment
Pleimesstr. 3
5321 Bonn
Phone: (49) 228 917 830
Fax: (49) 228 538 95 94

Association for Engineering Plastics
Am Hauptbahnhoff 10
60329 Frankfurt am Main
Phone: (49) 69 250 920
Fax: (49) 69 250 919

Euromap
P.O. Box 71 08 64
60498 Frankfurt
Phone: (49) 69 6603 1831
Fax: (49) 69 6603 1840
Internet: www.euromap.org

European Group of Manufacturers of Plastics Technical Parts
Am Hauptbahnhof 12
60329 Frankfurt
Phone: (49) 69 27 105
Fax: (49) 69 23 98 36

Institute for Design Theory and Plastic Machinery (KKM)
Schützenbahn 70
D-45127 Essen
Phone: (49) 201 183 3975
Fax: (49) 201 183 2877

Institute for Plastics Processing (IKV)
Pontstr. 49
D-52062 Aachen
Phone: (49) 241 8038 06
Fax: (49) 241 8888 262

Machinery Manufacturers for the Plastics and Rubber Industries (VDMA)
P.O. Box 71 08 64
D-60498 Frankfurt/M
Phone: (49) 69 6603 1831
Fax: (49) 69 6603 1840
Internet: www.guk.vdma.org

Plastics Packaging Association (IK)
Kaiser-Friedrich-Promenade 43
D-61348 Bad Homburg,
Phone.: (49) 6172 926601
Fax: (49) 6172 926670

Society of Independent Plastics Engineers and Consultants
Am Hauptbahnhof 12
60329 Frankfurt
Phone: (49) 69 271 050
Fax: (49) 69 232 799

Greece

Hellenic Plastics Industries Association
64 Michalakopoulou
115 28 Athens
Phone: (30) 1 77 94 519
Fax: (30) 1 77 94 518

Hungary

Association of the Hungarian Plastics Industry
Magyar Münyagipari Szövetség
H-1406 Budapest 76 POB.40
Phone: (36) 1 343 0759
Fax: (36) 1 343 0759

India

All India Plastics Manufacturers' Association
A-52, Street No. 1, M.I.D.C., Marol, Andheri (East)
Bombay 400 093
Phone: (91) 22 821 7324
Fax: (91) 22 821 6390

Organization of Plastics Processors of India
Moderna House, Ground Floor,
88-C, Old Prabhadevi Rd
Bombay 400 025
Phone: (91) 22 430 5106
Fax: (91) 22 430 6022

Plastindia Foundation
2, Leela Apts., 355 S.V. Road,
Vile Parle (West)
Bombay 400 056
Phone: (91) 22 671 2500
Fax: (91) 22 671 1906

Indonesia

Indonesia Plastics Recycling Industries Association
IV/114A, Brigjen Katamso St.
Waru-Sidoarjo 61256
Phone: (62) 31 853 6415
Fax: (62) 31 853 6356

Industrial Plastics Federation (FIPLASIN)
Jalan Cempaka Putih Tengah 20B, No. 8
10510 Jakarta Pusat
Phone: (62) 21 420 9126
Fax: (62) 21 420 9126

Iraq

Iraqi Federation of Industries
Khulani Square, P.O. Box 5665
Baghdad
Phone: (964) 1 888 009
Fax: (964) 1 888 2305

Ireland

Plastics Industries Association, Irish Business and Employers Confederation
Confederation House
84-86 Lower Baggot St.
Phone: (353) 1 660 1011
Fax: (353) 1 660 1717

Israel

Society of Israel Plastics & Rubber Industry
29 Hamered Street,
P.O. Box 50022
61500 Tel Aviv
Phone: (972) 3 519 8846
Fax: (972) 3 519 8717

Italy

Assogomma
Via S. Vittore 36
20123 Milano
Phone: (39) 2 466 020
Fax: (39) 2 435 432

Assoplast
c/o Federchimica,
Via Accademia 33
20131 Milano
Phone: (39) 2 268 101
Fax: (39) 2 268 10311
Internet: www.plastica.it

Italian Plastics and Rubber Processing Machinery and Mold Manufacturers Association
P.O. Box 24
20090 Assago (MI)
Phone: (39) 2 575 12700
Fax: (39) 2 575 12490
Internet: www.assocomaplast.com

National Union of the Plastics Processing Industry
Via Petitti 16
20149 Milano
Phone: (39) 2 392 10425
Fax: (39) 2 392 66548

Japan

Japan Expanded Polystyrene Recycling Association
6F FAX Bldg.,
2-20 Kanda Sakuma-cho,
Chiyioda-ku, Tokyo 101
Phone: (81) 3 3861 9046
Fax: (81) 3 3861 0096

Japan Pet Bottle Association
Ishikawa CO Bldg., 1-9-11, Kajicho
Chiyoda-ku, Tokyo 101
Phone: (81) 3 5294 7591
Fax: (81) 3 5294 2823

Japan Plastics Industry Federation
5-18-17 Roppongi
Minato-ku, Tokyo 106
Phone: (81) 3 3586 9761
Fax: (81) 3 3586 9760

Japan Polystyrene Foamed Sheet Industry Association
Tokon Bldg.
26 Higashikonya,
Chiyoda-ku
Tokyo 101
Fax: (81) 3 3257 3339

Plastic Waste Management Institute
Fukide Bldg, 1-13, 4 chome
Toranomon, Minato-ku
Tokyo 105
Phone: (81) 3 3437 225
Fax: (81) 3 3437 5270

Korea

Korea Plastic Industry Corp.
146-2, Sangnim-dong
Chung-ku, Seoul 100-400
Phone: (82) 2 275 7991
Fax: (82) 2 277 5150

Korea Reclaimed Plastic Industry Corp.
94-121, Youngdeungpo-dong,
Youngdeungpo-ku,
Seoul 150-020
Phone: (82) 2 677 0331
Fax: (82) 2 671 6136

Lebanon

General Syndicate of Plastics Materials Processors
P.O. Box 11
6635 Beirut
Phone: (961) 1 86 06 9
Fax: (961) 1 2124 781 532

Malaysia

Malaysian Plastics Manufacturers Association
37, Ground Floor, Jalan
20/14, Paramount Garden
46300 Petaling Jaya
Phone: (60) 3 776 3027
Fax: (60) 3 776 8352

Malta

Malta Federation of Industry
Development House,
St. Anne Street
Floriana, VLT 01
Phone: (356) 222 074
Fax: (356) 240 702

Mexico

Asociacion Nacional de Industrias del Plastico A.C.
Av. Parque Chapultepec 66
Desp. 301-3 Piso,
Col. El Parque
Naucalpan de Juarez 53390
Phone: (52) 5 576 5547
Fax: (52) 5 576 5548

Camara Nacional de la Industria de la Transformacion
Ave. San Antonio 256
Col. Empl. Napoles
03849 Mexico D.F.
Phone: (52) 5 563 3400 ext 247
Fax: (52) 5 598 8020

Instituto Mexicano del Plastico Industrial, S.C.
Insurgentes No. 954
Primer Piso
Col. Del Valle,
03100 Mexico, D.F.
Phone: (52) 5 669 3325
Fax: (52) 5 687 4960

Netherlands, The

PET Container Recycling Europe
Strawinskylaan 3051
1077 ZX Amsterdam
Phone: (31) 20 301 2332
Fax: (31) 20 301 2343

Union of the Metal and
Electrotechnical Industry
Boerhaavelaan 40
P.O. Box 190
2700 AD Zoetermeer
Phone: (31) 79 353 1100
Fax: (31) 79 353 1365

New Zealand

Composites Association of
New Zealand, Inc.
1/7 Musick Point Road
PO Box 54-160
Bucklands Beach,
Auckland, New Zealand
Phone: (64) 9 535 6494
Fax: (64) 9 535 6494

The Plastics Institute of
New Zealand
PO Box 76-378,
Manukau City
Auckland, New Zealand
Phone: (64) 9 262 3773
Fax: (64) 9 262 3850

Norway

Norwegian Plastics
Federation
Stensberg gt. 27
0170 Oslo
Phone: (47) 22 96 10 0
Fax: (47) 22 96 10 99

Pakistan

Pakistan Plastic
Manufacturers Association
H-16, Textile Avenue,
S.I.T.E.
Karachi 75700
Phone: (92) 21 257 2884
Fax: (92) 21 257 6928

Paraguay

Paraguayan Chamber of the
Plastic Industry
Herrera 564, entre Mexico
y Paraguari
Asuncion
Phone: (595) 21 448 03

Paraguayan Industrial
Union
Cerro Cora 1038
Asuncion
Phone: (595) 21 21 2556
Fax: (595) 21 21 3360

Peru

National Society of
Industries Committee of
Plastics Producers
365 Los Claveles, San Isidro
Lima 27
Phone: (51) 14 21 8830
Fax: (51) 14 40 2001

Philippines

Philippine Plastics Industrial
Association Inc.
317 Rizal Ave. Extension,
Rearblock,
Solid Bank Bldg.
Caloocan City
Metro Manila
Phone: (63) 2 361 1160
Fax: (63) 2 361 1168

Poland

Polish Plastics Converters
Association
UL Berbeckiego 6
PL-44 100 Gliwice, Poland
Phone: (483) 231 3031
Fax: (483) 231 3719

Portugal

Portugese Association for
the Mold Industry
Av. Victor Gallo, 21-3° Dt°
2430 Marinha Grande
Phone: (351) 44 56 7955
Fax: (351) 44 56 9359

Portuguese Association of
Plastic Industries
Rua D. Estefania 32-2 Esq.
1000 Lisbon
Phone: (351) 1 315 0633
Fax: (351) 1 314 7760

Rep. of South Africa

Plastics Federation of South
Africa
(includes the following:
Assn. of Plastics Processors,
Plastics Institute, Plastics
Manufacturers Assn.,
Machinery Suppliers Assn.
for the Plastics and Allied
Industries, Polymer
Importers Assn., Plastics
Mold Makers Assn., Assn. of
Rotational Molders, and Expanded PolystyreneAssn.)

18 Gazelle Ave, Corporate
Park
Old Pretoria Rd, Midrand.
Private Bag X68, Halfway
House, 1685
Phone: (27) 11 314 4021
Fax: (27) 11 314 376

Romania

Aspaplast
Str. Ziduri Mosi No 23,
Sector 2
73342 Bucharest
Phone: (40) 0 635 160
Fax: (40) 0 635 0615

Saudi Arabia

Council of Saudi Chambers of Commerce and Industry
P.O. Box 16683
Riyadh 11474
Phone: (966) 1 405 3200
Fax: (966) 1 402 4747

Slovenia

Association of Chemical and Rubber Industries
G25/Kemija Demiceva 9
SLO-61000 Ljubljana, Slovenia
Phone: (38) 61 21 6122
Fax: (38) 61 21 8380

Spain

Spanish Association of Machinery Manufacturers for Plastics and Rubber
Riera Sant Miquel 3
08006 Barcelona
Phone: (34) 3 415 0422
Fax: (34) 3 416 0980
e-mail: asoc@amec.es
Internet: www.amec.es

Spanish Confederation of Plastics Enterprises
Av. Brasil 17-13A
28020 Madrid
Phone: (34) 1 556 7575
Fax: (34) 1 556 499

Sweden

Plastics and Chemicals Federation
Box 105
101 22 Stockholm
Phone: (46) 8 402 1360
Fax: (46) 8 411 4526

Swedish Plastics Industry Association
Barnhusgatan 3,
P.O. Box 1133
111 81 Stockholm
Phone: (46) 8 440 1170
Fax: (46) 8 249 530

Switzerland

Association of Swiss Rubber and Thermoplastic Industries
Schachenallee 29
5000 Arrau
Phone: (41) 62 823 0970
Fax: (41) 62 823 0762

Swiss Association of Machinery Manufacturers
Kirchenweg 4
8032 Zurich
Phone: (41) 1 384 4844
Fax: (41) 1 384 4848

Swiss Plastics Association
Schachenallee 29
5000 Aarau
Phone: (41) 62 823 0863
Fax: (41) 62 823 0762

Taiwan

Taiwan Plastics Industry Association
8F, 162 Chang-An E. Rd, Sec. 2
Taipei
Phone: (886) 2 771 911
Fax: (886) 2 731 5020

Taiwan Regional Association of Synthetic Leather Industries
5F, 30 Nanking W. Rd
Taipei
Phone: (886) 2 559 020
Fax: (886) 2 559 8823

Taiwan Synthetic Resins Manufacturers Association
4F, 82 Chienkuo N. Rd, Sec. 2
Taipei
Phone: (886) 2 504 0879
Fax: (886) 2 502 9406

Thailand

Plastic Industry Club, Federation of Thai Industries
394/14 Samsen Rd., Dusit
Bangkok 10300
Phone: (66) 2 2800 0951
Fax: (66) 2 2800 0959

Thai Plastics Industries Association
127/2 Phaya Mai Rd., Somdejchaophaya, Klongsan
Bangkok 10600
Phone: (66) 2 438 9457-8
Fax: (66) 2 437 2850

Tunisia

Syndicate Chamber of Plastic, The
17, rue Abderrahman El Zaziri
1002 Tunis
Phone: (216) 1 791 882
Fax: (216) 1 790 526

Turkey

Research, Development and Training Foundation of Turkish Plastics Industrialists
9-10, Kisim, S1-C Blok, Daire: 63
Ataköy 34750 Istanbul
Phone: (90) 0212 560 4397
Fax: (90) 0212 560 6749

U.K.

British Independent Plastic Extruders Association
89 Cornwall St.
GB-Birmingham B3 3BY
Phone: (44) 121 236 1866
Fax: (44) 121 200 1389

British Laminated Plastic Fabricators Association
6, Bath Place, Rivington St.
London EC2A 3JE
Phone: (44) 171 457 5000
Fax: (44) 171 457 5045

British Plastics Federation
6, Bath Place, Rivington St.
London EC2A 3JE
Phone: (44) 171-457 5000
Fax: (44) 171-457 5045

European Resin Manufacturers' Association
Queensway House, 2 Queensway
Redhill, Surrey RH1 1QS
Phone: (44) 737 76 8611
Fax: (44) 737 76 1685

Institute of Materials
1 Carlton House Terrace
London SW1Y 5DB
Phone: (44) 71-976 133
Fax: (44) 71-839 2078

Plastics Machinery Distributors Association Ltd.
P.O. Box 1414
Dorchester, Dorset DT2 8YH
Phone: (44) 1305 250 002
Fax: (44) 1305 250 996

The Welding Institute (TWI)
Abington Hall
Cambridge CB1 6AL
Phone: (44) 1223 891 162
Fax: (44) 1223 892 588

USA

Alliance of Foam Packaging Recyclers
1801 K Street, N.W., Suite 600
Washington DC 20006
Phone: 202-974-5205
Fax: 202-296-7354

American Architectural Manufacturers Association
1827 Walden Office Square, Suite 104
Schaumburg, IL 60173
Phone: 847-303-5664
Fax: 847-303-5774

American Mold Builders Association
P.O. Box 404
Medinah, IL 60157
Phone: 847-303-5664
Fax: 847-303-5774
Internet: www.amba.org

American Plastics Council
1801 K St., N.W., Suite 600
Washington, DC 20006
Phone: 202-974-5400
Fax: 202-296-7119
Internet: www.plasticsresource.com

American Polyolefin Association Inc.
2312 E. Mall
Ardentown, DE 19810
Phone: 302-475-1450
Fax: 302-475-2357

American Society for Plasticulture
P.O. Box 860238
St. Augustine, FL 32086
Phone: 904-794-2870
Fax: 904-794-2870

American Society of Electroplated Plastics, Inc.
1767 Business Center Dr., Suite 302
Reston, VA 22090
Phone: 703-438-8292
Fax: 703-438-3113

Association of Home Appliance Manufacturers
20 North Wacker Dr.
Chicago, IL 60606
Phone: 312-984-5800
Fax: 312-984-5823

Association of the Nonwoven Fabrics Industry
1001 Windstead Dr., Suite 460
Cary, NC 27513
Phone: 919-677-0060
Fax: 919-677-0211
Internet: www.inda.org

Association of Rotational Molders
2000 Spring Rd. Suite 511
Oak Brook, IL 60523
Phone: 630-571-0611
Fax: 630-571-0616
Internet: www.rotomolding.org

Center of Excellence for Composites Manufacturing Technology
8401 Lake View Pkwy., Suite 200
Kenosha, WI 53142-7403
Phone: 414-947-8900
Fax: 414-947-8919

Chemical Fabrics & Film Association, Inc.
1300 Sumner Ave.
Cleveland, OH 44115
Phone: 216-241-7333
Fax: 216-241-0105
Internet: www.taol.com/cffa

Composites Fabricators Association
1655 N. Ft. Meyer Dr., Suite 510
Arlington, VA 22209
Phone: 703-525-0511
Fax: 703-525-0743

Composites Institute, The, Div. of SPI
355 Lexington Ave.
New York, NY 10017
Phone: 212-351-5410
Fax: 212-370-1731
Internet: www.socplas.org

Flexible Packaging Association (FPA)
1090 Vermont Ave., N.W.
Washington, D.C. 20005
Phone: 202-842-3880
Fax: 202-842-3841

GLCC, Inc.
103 Trade Zone Dr., Suite 26C
West Columbia, SC 29170
Phone: 803-822-3700
Fax: 803-822-3710

Industrial Fabrics Association International
1801 County Road B West
Roseville, MN 55113
Phone: 800-225-4324; 651-222-2508
Fax: 651-631-9334
E-mail: generalinfo@ifai.com
Internet: www.ifai.com

Institute of Scrap Recycling Industries, Inc. (ISRI)
325 G St., N.W., Suite 1000
Washington, D.C. 20005-3104
Phone: 202-737-1770
Fax: 202-626-0900
Internet: www.isri.org

International Association of Plastics Distributors
4707 College Blvd., Suite 105
Leawood, KS 66211-1611
Phone: 913-345-1005
Fax: 913-345-1006
Internet: www.iapd.org

International Cast Polymer Association
8201 Greensboro Dr., Suite 300,
Alexandria, VA 22102
Phone: 703-610-9034
Fax: 703-610-9005
Internet: www.icpa-hq.co

National Association for Plastic Container Recovery
3770 NationsBank Corporate Ctr.,
100 N. Tryon St.,
Charlotte, NC 28202
Phone: 704-358-8882
Fax: 704-358-8769
Internet: www.napcor.com

National Plastics Center and Museum
P.O. Box 639
Leominster, MA 01453
Phone: 508-537-9529
Fax: 508-537-3220
Internet: npcm.plastics.com

National Tooling & Machining Association
9300 Livingston Rd.
Fort Washington, MD 20744
Phone: 301-248-6200
Fax: 301-248-7104
Internet: www.ntma.org

National Recycling Coalition
1727 King Street, Suite 105
Alexandria, VA 22314-2720
Phone: 703-683-9025
Fax: 703-683-9026

Plastic Bag Association
355 Lexington Ave.
New York, NY 10017
Phone: 212-661-4261
Fax: 212-370-9047
Internet: www.plasticbag.com

Plastic Pipe Institute
(see Society of the Plastics Industry)

Plastics Institute of America, Inc.
333 Aiken St.
Lowell, MA 01854
Phone: 508-934-3130
Fax: 508-459-9420
Internet:
www.eng.umi.edu/dept/pia

Polymer Processing Institute, The
Stevens Institute of Technology,
Castle Point on the Hudson
Hoboken, NJ 07030
Phone: 201-216-5539
Fax: 201-216-8243
Internet:
www.ppi.stevens-tech.edu

Polystyrene Packaging Council, The
1275 K St., N.W., Suite 400
Washington, D.C. 20005
Phone: 202-371-5269
Fax: 202-371-1284

Polyurethane Foam Association
P.O. Box 1459
Wayne, NJ 07474-1459
Phone: 201-633-9044
Fax: 201-628-8986
Internet: www.pfa.org

Polyurethanes Division, SPI
355 Lexington Ave.
New York, NY 10017
Phone: 212-351-5425
Fax: 212-697-0409
Internet: www.socplas.org

Society for the Advancement of Material and Process Engineering (SAMPE)
P.O. Box 2459
Covina, CA 91722
Phone: 626-331-0616
Fax: 626-332-8929
Internet: www.sampe.org

Society of Plastics Engineers (SPE)
14 Fairfield Dr.
Brookfield, CT 06804-0403
Phone: 203-775-0471
Fax: 203-775-8490
Internet: www.4spe.org

Society of the Plastics Industry, Inc., The (SPI)
1801 K. St., N.W., Suite 600K
Washington DC 20006
Phone: 202-974-5200
Fax: 202-296-7005
Internet: www.socplas.org

Styrene Information and Resource Center
1801 K. St., N.W., Suite 600K
Washington DC 20006
Phone: 202-974-5314
Fax: 202-296-7159

Underwriters Laboratories Inc.
333 Pfingsten Rd.
Northbrook, IL 60062
Phone: 847-272-8800
Fax: 847-272-8189
Internet: www.ul.com

Vinyl Institute, The, Div. of SPI
65 MadisonAvenue
Morristown, NJ 07960
Phone: 973-898-6699
Fax: 2973-898-6633

Uruguay

Uruguayan Association of Plastic Industries
Av. General Rondeau 1665
11100 Montevideo
Phone: (598) 2 923 405
Fax: (598) 2 922 567

Venezuela

Venezuelan Association of Plastics Industries
Edif.Multicentro
Macaracuay, 7th Fl
Av. Principal de
Macaracuay
Caracas 1070
Phone: (58) 2-256 3345
Fax: (58) 2-256 2867

Venezuelan Bureau of Small and Medium Plastic, Rubber, and Related Industries
Av. Principal de
La Cooperativa,
Qta. Maria Elisa, Detras
de Malariologia
Maracay 2101,
Estado Aragua
Phone: (58) 43-41 542
Fax: (58) 43-41 70 63

Yugoslavia

Plastic Industry Business Association (Juplas)
Sv. Save 1 - Hotel "Slavija"
YU-11000 Belgrad
Phone: (381) 11 444 6144
Fax: (381) 11 444 6144

Index

ABOUT THE EDITOR

Charles A. Harper is a leading authority on plastics, plastic fabrication, and plastic applications. President of Technology Seminars, Inc., an organization committed to advancing technology through expert presentations, he is Series Editor for both the McGraw-Hill Materials Science and Engineering Series and the McGraw-Hill Electronic Packaging and Interconnection Series. He serves on the advisory board of several professional and business organizations. A former Johns Hopkins University engineering professor, with an esteemed career in industry, Harper has given numerous technical and product application presentations and seminars before professional societies on plastics, elastomers, and composites. He has been a leader in the Society of Plastics Engineers, Society for the Advancement of Material and Process Engineers (SAMPE), and the IEEE.